*North Sea Oil and*
*Gas Reservoirs—II*

# *ORGANIZATION*

## *Host and Organizers*

University of Trondheim
The Norwegian Institute of Technology (NTH)

## *Scientific Organizing and Editorial Committee*

| | |
|---|---|
| **Antony T. Buller (Chairman)** | Research Adviser, Statoil, Trondheim<br>Adjunct Professor of Petroleum Geology, NTH |
| **Eivind Berg** | Senior Research Geophysicist, Statoil, Trondheim |
| **Odd Hjelmeland** | Managing Director, Reservoir Laboratories AS (ResLab), Trondheim |
| **Jon Kleppe** | Professor of Reservoir Engineering, NTH |
| **Ole Torsæter** | Associate Professor of Reservoir Engineering, NTH |
| **Jan Ole Aasen** | Research Manager, The Continental Shelf and Petroleum Technology Research Institute A/S (IKU), Trondheim<br>Adjunct Professor of Numerical Mathematics, NTH |

## *Secretariat*

Vibeke Halaas
Ann Kristin Engan
Madelein Wold
The Norwegian Institute of Technology

## Sponsors

Norwegian Research Council for Science and Humanities (NAVF), Norway
Phillips Petroleum Company, Norway
Saga Petroleum a.s., Norway
Statoil, Norway
The Royal Norwegian Council for Scientific and Industrial Research (NTNF), Norway

# *North Sea Oil and Gas Reservoirs—II*

Proceedings of the 2nd *North Sea Oil and Gas Reservoirs* Conference
organized and hosted by the Norwegian Institute of Technology (NTH),
Trondheim, Norway, May 8–11, 1989

*Edited by*

**A.T. Buller, E. Berg, O. Hjelmeland,
J. Kleppe, O. Torsæter and J.O. Aasen**

**Graham & Trotman**
A member of the Kluwer Academic Publishers Group
LONDON/DORDRECHT/BOSTON

Graham & Trotman Limited
Sterling House
66 Wilton Road
London SW1V 1DE
UK

Kluwer Academic Publishers Group
101 Philip Drive
Assinippi Park
Norwell, MA 02061
USA

ISBN-13: 978-94-010-6843-7
DOI: 10.1007/978-94-009-0791-1

e-ISBN-13: 978-94-009-0791-1


First published in 1990

Softcover reprint of the hardcover 1st edition 1990

British Library Cataloguing in Publication Data
North Sea Oil and Gas Reservoirs, *International Conference*,
2nd (1989), Trondheim, Norway
North Sea oil and gas reservoirs II.
1. North Sea. Natural gas deposits and petroleum deposits
I. Title II. Buller, A.T. (Antony T) III. Norges
tekniske høgskole
553.280916336

Library of Congress Cataloging-in-Publication Data
International Conference North Sea Oil and Gas Reservoirs
(2nd: 1989: Trondheim, Norway)
North sea oil and gas reservoirs II: proceedings of the 2nd International
Conference North Sea Oil and Gas Reservoirs organized by the Norwegian Institute
of Technology and held in Trondheim, Norway, 8–11 May, 1989/
edited by A.T. Buller... [*et al.*].
p. cm.
Includes bibliographical references.

1. Petroleum—Geology—North Sea—Congresses. 2. Gas, Natural—Geology—
North Sea—Congresses. 3. Oil fields—North Sea—Congresses. 4. Gas reservoirs—
North Sea—Congresses. I. Buller, A.T. (Antony Treverton) II. Norges tekniske høgskole.
III. Title.
TN870.5.I49 1989
553.2'8'0916336—dc20 90–2870
CIP



Typeset in Baskerville and printed and bound in Malta by Interprint Ltd.

# Contents

# Foreword

The first *North Sea Oil and Gas Reservoirs Conference* was held in Trondheim in 1985 as part of the Norwegian Institute of Technology's 75th anniversary celebrations. Favourable reactions from the delegates prompted the Committee to re-run the event some three and a half years later, and it is now intended that the Conference be held on a regular basis as long as there is a demand for this type of gathering.

The objectives of the 1989 Conference, which were broadly similar to those of the previous one, were: (a) to bring together those engaged in various geoscientific and reservoir engineering aspects of North Sea Oil and gas reservoirs in one forum; (b) to demonstrate wherever possible the interdependence of the various disciplines and specializations; (c) to promote innovative, synergistic approaches to research and development programmes aimed at North Sea conditions; and (d) to reflect current trends in the reservoir sciences. Naturally there was no place for specialist parallel sessions in a Conference aimed at encouraging interdisciplinary integration and awareness.

The Conference format was essentially the same as before. It contained a blend of North Sea field reviews (by B.P., Shell, Saga, Statoil and Maersk), keynote addresses (by Professor Larry Lake, Professor Norman Wardlaw and Dr. Wafik Beydoun) and 31 research papers. There was, however, one significant difference. All of the presentations at the 1985 Conference were by invitation only. This time the research papers were selected from an overwhelming response to a Call for Papers in accordance with the aims of the Conference and their technical contents.

The programme, which attracted 230 scientists from 11 countries, embraced a wide variety of topics within improved seismic methods for detailed reservoir description, characterization of the near-well environment, reservoir mechanisms and processes, and reservoir heterogeneities, modelling and simulation. These themes have been maintained in the structuring of this refereed Proceedings Volume.

Any successful conference requires the generous assistance of a great many people. We would like to extend a special "Thank you" to the devoted ladies of our Secretariat (Vibeke Halaas, Ann Kristin Engan and Madelein Wold) who shouldered the main burden of the administration in a highly professional and willing manner. We would also like to thank: our speakers, co-authors and those companies and institutions who sanctioned the release of proprietary information; our honoured guests, Professor Dag Kavlie (Rector of the Norwegian Institute of Technology) who made the Welcoming Address and Professor Jan Øverli (Director of Statoil's Research and Development Centre in Trondheim) who made the main speech at the Conference Dinner; and our chairmen and referees—I. Akervoll, J. Alvestad, L. Amundsen, B. Arntsen, H.O. Augedal, K.A. Berteussen, K. Bjørlykke, R.K. Bratli, J.D. Collinson, V. Dalen, S. Ehrenberg, Å. Espedal, L. Fält, J. Feder, G. Fladmark, R. Gabrielsen, C.D. Griffiths, S. Hagen, H.H. Haldorsen, J. Hauge, A. Henríquez, P. Horsrud, O.K. Jensen, R. Kanestrøm, A.G. Koestler, C. Kossack, A. Krill, O. Langeland, A. MacDonald, D. Malnes, J.H. Martin, T. Meyer, J.P. Nystuen, H. Omre, L. Rasmussen, R. Risnes, T. Rudshaug, H.G. Rueslåtten, N. Saigal, R. Sinding–Larsen, P. Skalle, A. Skauge, S.M. Skjæveland, S. Snow, O. Steinert, L. Sønneland, S.B. Torp, R.H. Utseth, O. Vikane and C.H. Whitson.

It is also important at a conference to generate a pleasant social atmosphere amongst the delegates. This was helped to a considerable degree by the introduction of a little local culture into the social events. In this respect, we would like to thank the Baroque Ensemble of the Ringve Museum of Music History in Trondheim, the NTH student glee club—Pirum—and the young people of the Byåsen School Band, dressed in uniforms based on the old Trondheim Guard. The speakers, chairmen and special guests also received a small token of the Committee's appreciation in the form of a glass ornament. This comprised a model of a production platform encased in a glass "oil droplet" and set on a base of Berea sandstone. The glasswork was most skilfully carried out by Mr. Kjell Ramsøskar—NTH's own glassblower—and the bases were prepared by Lars Øyno, a NTH research student.

Finally we would like to acknowledge financial support from the following government institutions and companies: the Norwegian Research Council for Science and Humanities (NAVF), Phillips Petroleum Company (Norway), Saga Petroleum a.s., Statoil, and The Royal Norwegian Council for Scientific and Industrial Research (NTNF). Their generosity enabled us to defray our costs significantly, thereby keeping the Conference registration fee at a reasonable level.

*Antony T. Buller*
Conference Chairman

# Field Review 1

# The Forties Field and the Evolution of a Reservoir Management Strategy

*J.M. Wills and D.K. Peattie*

**BP Petroleum Development Ltd, Farburn Industrial Estate, Aberdeen, UK**

**The Forties Field is located 180 km ENE of Aberdeen, predominantly in UK Licence Block 21/10. It was discovered in 1970, when an exploration well encountered hydrocarbons in Palaeocene sandstone within an anticlinal structure. Four appraisal wells confirmed the existence of a major oilfield, with an area of approximately 90 km². Initial estimates of the oil-in-place were 4600 million stb, with a recovery factor of 40%.**

**The initial reservoir development plan involved producing the oil from four platforms with a plateau rate of 400 000 stbd. Because of uncertainties concerning the performance of the underlying aquifer, water-injection facilities were installed on the platform to provide additional pressure support via peripheral water injectors. Information gained early during the development phase led to the oil-in-place estimates being revised downwards to approximately 4000 million stb, but to the recovery factor being increased upwards to 45% and the plateau rate revised to 500 000 stbd.**

**As more geological and engineering data has become available it has been possible gradually to increase the estimate of the recovery factor, although estimates of oil-in-place have remained essentially unchanged. The current estimate is approximately 57% with artificial lift, with the predominant oil displacement mechanism being bottom water drive. At the end of 1987, 75% of the total reserves had been produced and as a result of prudent reservoir management the annual average offtake rate had only declined to some 330 000 stbd. However, there is increasing evidence available which indicates that the Forties reservoir is more heterogeneous than previously thought.**

**The evolution of a coherent strategy aimed at securing remaining.reserves and recognising, quantifying and recovering any additional incremental oil, represents a considerable challenge in mature field reservoir management. This requires an integrated approach by geophysicists, geologists, reservoir engineers and production engineers and a much more detailed description of the reservoir than is commonly required during the early plateau period of the life of a North Sea field.**

## INTRODUCTION

Discovered in 1970, the Forties Field has been producing oil since September 1975. Stock Tank Oil Initially in Place (STOIIP) has been calculated as 4343 mmstb*, and original reserves estimated as 2470 mmstb, representing an overall recovery of 57%. The field came off plateau production of 500 mstbpd in 1981, and by late 1988 production had declined to about 280 mstbpd. After 13 years production, the field has produced 1907 mmstb, and remaining reserves are 563 mmstb, which includes recovery via artificial lift. Field life has been projected to extend beyond the year 2000.

Water injection into the aquifer commenced in 1976, and continues at the present day at an average rate of 390 mbwpd. There is also a significant contribution to pressure support from the underlying aquifer. Water production started in 1979 and has steadily increased to a current rate of 100 mbpd. An important reservoir management issue is to determine the origins of this water and effectively control its production so as to maximize oil recovery.

At present there are 103 available wells in the field, 81 producers and 22 injectors. At any one time, a number of wells may be temporarily shut-in, suspended or dead. Proper reservoir management has a significant impact on the economic worth of the field. This paper describes significant geological, geophysical and reservoir engineering aspects of the field, brought up to date since previous publications, and also provides a view of the future management of the reservoir. The emphasis is on a multi-disciplinary team approach involving geologists, geophysicists, and reservoir and production engineers.

The original licencees of the field were BP (100%) in Block 21/10, and Shell and Esso (50% each) in Block 22/6. In 1983 and 1984 BP assigned certain fixed interests to minor partners, numbering 22 companies. Subsequent company and interest changes have resulted in 23 present licencees in the field. Unit Participation Interests have been agreed as 94.78% in Block 21/10 (BP 83.13%, 20 minor partners 11.65%) and 5.22% in Block 22/6a (Shell 2.61%, Esso 2.61%).

*Prefixes: m = $10^3$; mm = $10^6$.

*North Sea Oil and Gas Reservoirs—II*

# RESERVOIR GEOLOGY

## Location of field

The Forties Field is located in the central North Sea approximately 180 km ENE of Aberdeen, Scotland (Fig. 1). Most of the field underlies UKCS Block 21/10, licenced to BP, with a southeastern extension into Shell/Esso's Block 22/6a. Water depth ranges from 104 to 128 m.

## Regional geological setting

The field is situated close to the crest of the Forties-Montrose Ridge, at the northern end of the Central Graben (Fig. 1).

In the North Sea, the early Tertiary marked the transition from Late Cretaceous and Danian pelagic carbonate sedimentation with low terrigenous input to the accumulation of vast thicknesses of clastic sediment deposited in deltaic, shelf and submarine-fan environments (Stewart, 1987). The distribution of early Tertiary sediments was initially controlled by the pre-Tertiary structural configuration of the basin, so that early fan sediments accumulated in the axial regions of the Mesozoic grabens, with no or only thin sedimentation over the previously emergent highs such as the Forties-Montrose Ridge. By late Palaeocene times, however, clastic sedimentation had smothered all topographic features so that the regional structure was expressed as a series of monoclinal ramps. This resulted in the development of a thick, extensive sand-rich submarine fan system (the Forties Fan) which both dips and thins in a southeasterly direction along the Central Graben, and also thins westward towards the Scottish coast. The Forties reservoir is part of this major fan system. At the end of the Palaeocene, a relative rise in sea level resulted in the deposition of the regional basinal mudstones of the Sele Formation. These fine-grained sediments form the seal to the Forties reservoir.

The pattern of sedimentation is believed to have been influenced by the relative position of sea-level in the basin (Stewart, 1987). Low sea-level stands corresponded to periods of high coarse clastic input, whereas high sea-level stands corresponded to quiescent intervals of mudstone deposition (see chronostratigraphic diagram, Fig. 2). Further fluctuations between highstand mudstones and lowstand submarine fan clastic deposition occurred throughout the Eocene in the central North Sea. In the Forties area, extensive basin-slope erosional channels were developed during the Middle Eocene, some of which are sand-filled and oil-bearing.

From the Late Eocene to the Holocene, continuing mudstone deposition onlapped and overlapped the underlying shelf-delta-fan system in the Forties area and infilled the entire basin.

## Exploration and development history

BP was awarded Block 21/10 in the UK Second Licencing Round, in November, 1965. The field was discovered in October, 1970, by well 21/10-1, drilled on the axis of a large structural nose recognized on seismic at Base Tertiary level (Thomas *et al.*, 1974). The well encountered 119 m of oil-bearing Palaeocene sands at a depth of 2131 m. Following the acquisition of a 1.5 × 1.5 km seismic survey over the structure, a further four appraisal wells were drilled during 1971–72. Shell/Esso also drilled a successful well in Block 22/6 that demonstrated the southeastern extension of the field.

The development plan evolved during 1972–75, and at the same time reservoir studies were undertaken to assess likely field performance (Hillier *et al.*, 1978). Four fixed steel drilling/production platforms were installed on the Main Field in Block 21/10 with the facility for 108 wells, a producing system capable of around 500 000 bpd, and a seawater injection system with a capacity of 600 000 bwpd. Development drilling began in June 1975 and production started in September 1975, with the plateau being reached during 1978. A fifth platform was added over the southeastern extension (South East Forties) in Block 22/6a during 1986, and production commenced through this facility in March, 1987.

Further major seismic data were acquired over the field in 1980–81 and a regional grid overlapping the periphery of the field in 1986. A full 3D seismic survey was acquired over the entire field during the summer of 1988.

Early geological descriptions of the field were published by Thomas *et al.* (1974), Hill and Wood (1980), and Carman and Young (1981).

## Structural development/trap configuration

The field has a broad, four-way dip-closed, anticlinal structure, about 90 km$^2$ in area, with a vertical relief of about 180 m. Depth to crest is 2030 m subsea. The structure at top reservoir level has two culminations (Fig. 3). The main crest is located west of the FA platform, and a secondary crest lies to the east of the FC platform. There is an approximate coincidence between the structural spill point and the initial oil-water contact at 2217 m subsea. The structure is very gentle and nowhere in the field are dips in excess of 10° recorded (Fig. 4). The structural style is similar at all underlying levels within the reservoir and aquifer.

The structure results from a combination of sedimentary drape, synsedimentary faulting, and compaction over the Forties-Montrose Ridge.

Minor faults are believed to penetrate the reservoir, particularly on the southern flank where evidence of syndepositional slumping exists. However, these features are difficult to resolve with the current seismic database and well spacing, and are not believed significantly to influence reservoir continuity or production.

The reservoir seal is provided by the overlying and conformable Eocene-age mudstones of the Sele Formation.

## Field stratigraphy

The stratigraphic column for the field discovery well, 21/10-1, is shown in its local context in Fig. 4.

The oil accumulation occurs in thick sandstones of Late Palaeocene age, deposited in two major sand-rich submarine fan sequences of the Forties and Andrew Formations. They lie beneath a thick, monotonous section of mudstones, ranging in age from Late Palaeocene to Holocene. Thin, and occasionally oil-bearing, sandstones also occur in the overlying middle Eocene. The fan sequences overlie the interbedded mudstones, sandstones and thin limestones of the Maureen Formation, which in turn overlie the basal Palaeocene (Danian) and Maastrichtian micritic limestones of the Ekofisk and Tor Formations respectively.

Figure 5 summarizes the lithostratigraphy currently in use over the Forties area, which is consistent with the regional nomenclature established by Deegan and Scull (1977).

## Seismic-stratigraphic framework

Reservoir subdivision in the field has been achieved by the use of seismic-stratigraphic techniques.

Prior to the 1988 3D survey, the most recent seismic survey across the field was shot in 1980/81 in a triangular

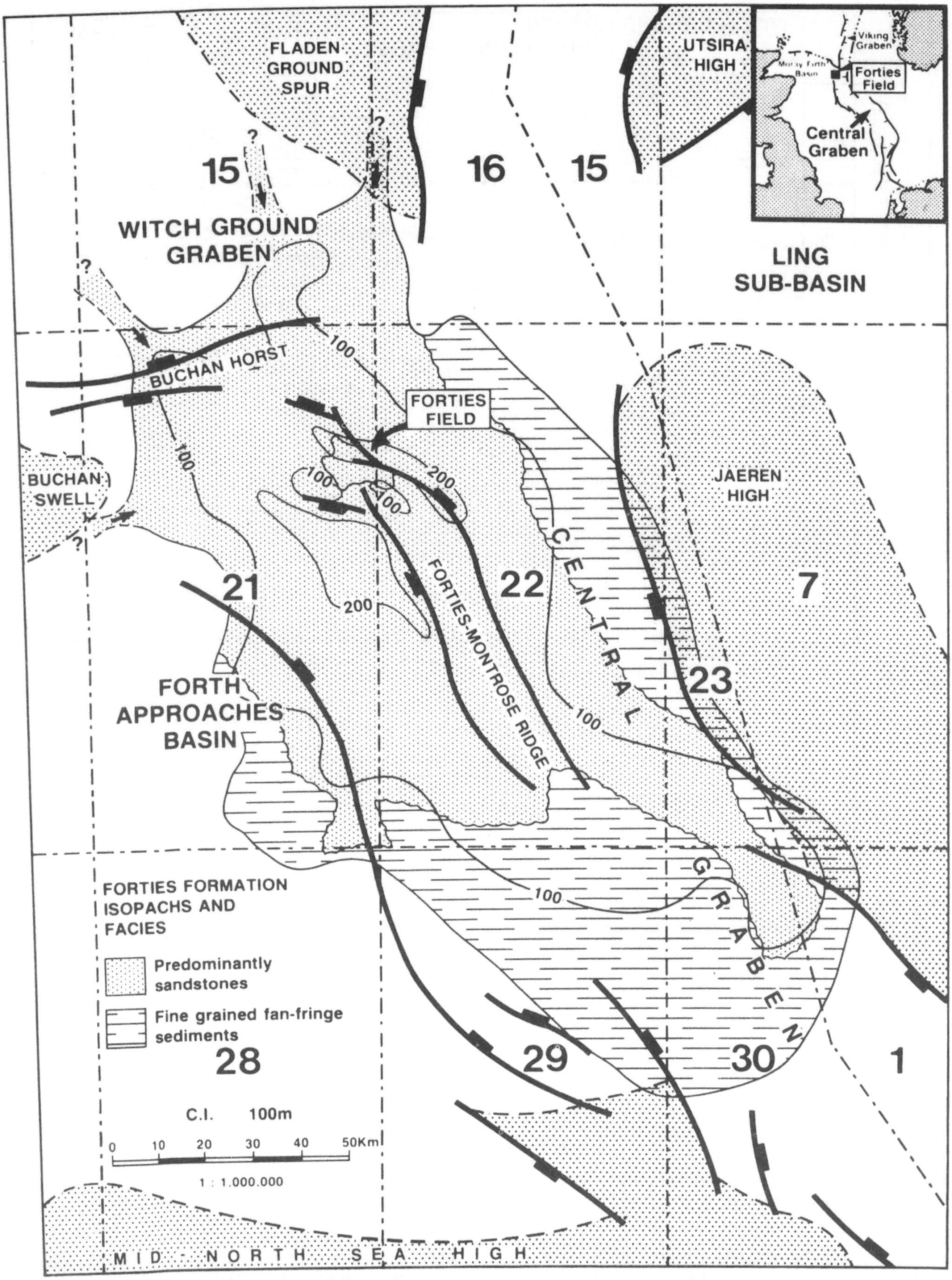


**Fig. 1.** Structural setting of the Forties Field at the northern end of the Forties-Montrose Ridge, a basement high in the Central Graben.

grid of approximately 0.5 to 1 km spacing. The data were reprocessed in 1985. A later regional survey in 1986 provided data over peripheral areas of the field. Total seismic coverage is approximately 750 line km, and the seismic data have been a fundamental part of a number of seismic-stratigraphic reservoir models developed since 1982, culminating with the current reservoir model.

The 3D seismic programme, consisting of 3865 sail line

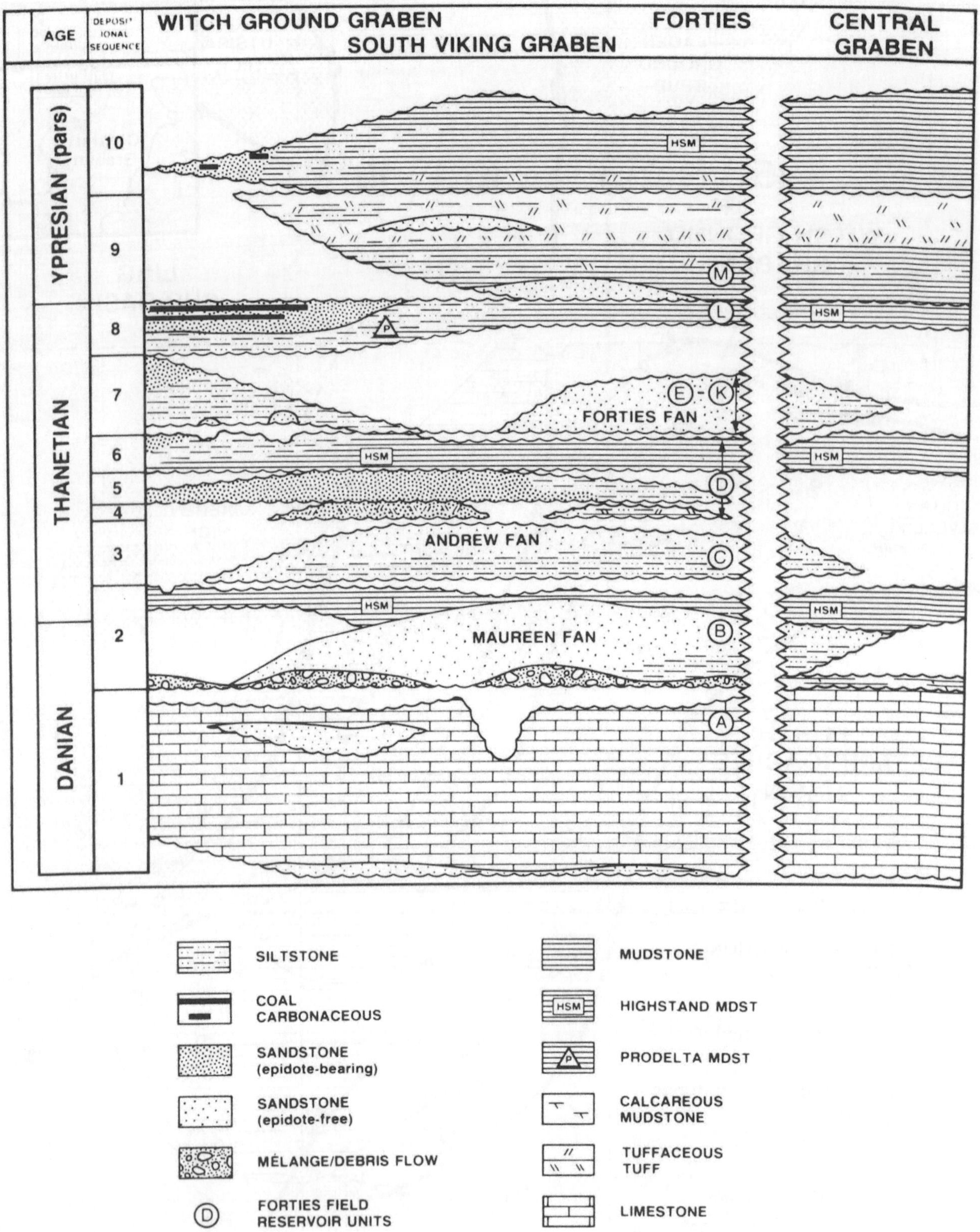


**Fig. 2.** Composite chronostratigraphic diagram for the Early Palaeogene of the central North Sea (from Stewart, 1987). Letters A–M indicate geological units—see text.

kilometres of dual-streamer data acquired in the summer of 1988, will significantly enhance the seismic-stratigraphic definition of the reservoir.

The most reliable seismic event is that corresponding to the Top Sele Formation, and a depth map on this horizon, closely controlled by the abundant well data, has been produced. All other reservoir maps have been generated by isopaching down from the Top Sele using a combination of well data, seismic sequence and time isopach mapping.

Eight intrareservoir seismic horizons have been interpreted and tied, using synthetic seismograms and well logs, to a comprehensive micropalaeontological/palynological biozonation (utilizing foraminifera, diatoms, dinocysts and pollen) in key cored wells.

Recognizable sequence and subsequence boundaries were combined with observed sedimentological sequence breaks in cored wells and production barriers recognized from reservoir monitoring data in order to establish

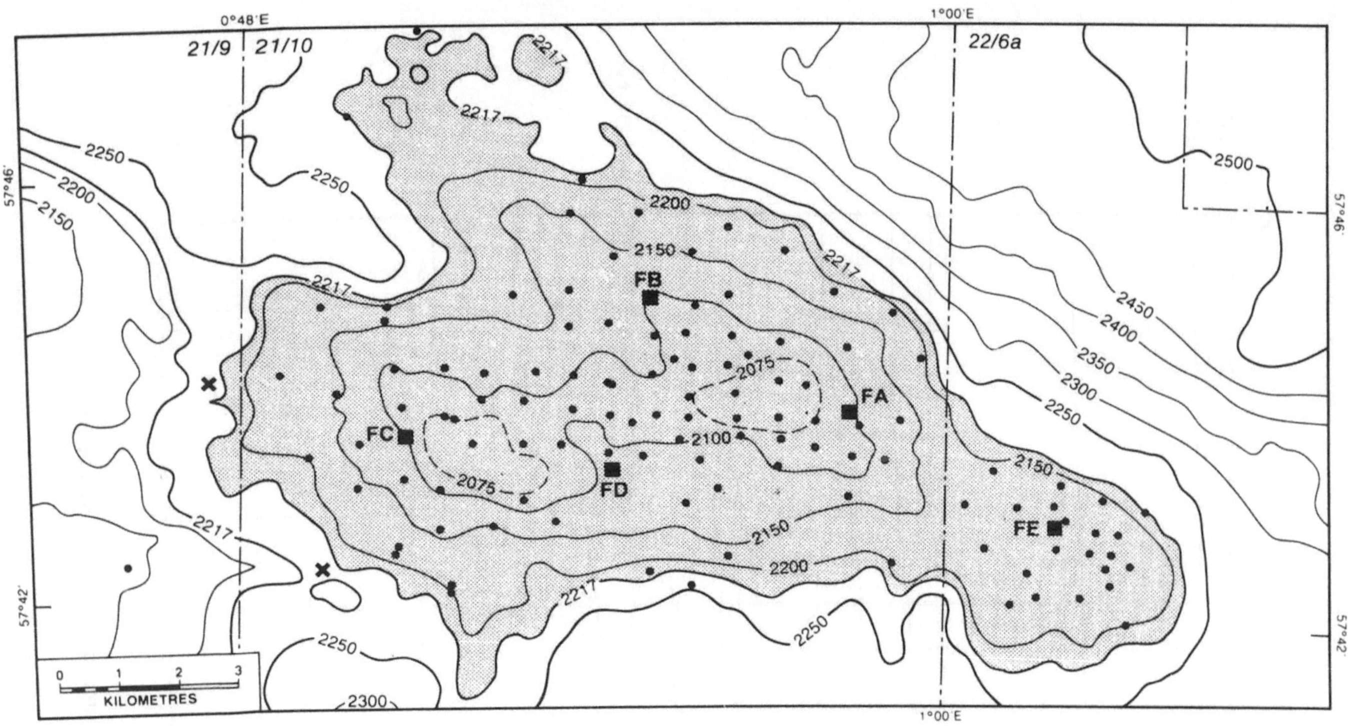


**Fig. 3.** Reservoir structure map showing platforms, wells (●) and spill points (×).

a field-wide stratigraphy. Seismic sequence correlation and log correlation were finally used to extend the subdivision within the overall stratigraphic framework to all wells in the field. Figure 6 shows the stratigraphic framework used for reservoir subdivision in the field, and Fig. 7 illustrates the relationship between seismic, lithostratigraphic and biostratigraphic picks over the reservoir interval. Figure 8 is an example of a geoseismic section through the reservoir which clearly shows typical thickness variations of the reservoir within the field.

## Facies analysis and the depositional model

The Forties Field occupies a minor part of a large submarine fan system (Stewart, 1987; Fig. 1). This fan system consists mostly of stacked sandstone bodies with fine-grained sediments restricted to a narrow distal fringe. It comprises a sand-dominated ramp depositional system (Heller and Dickinson, 1985).

Eight lithofacies are recognized from a study of the available data. These are summarized in Fig. 9. The fine-to medium-grained sandstones of lithofacies B constitute the major lithology within the field (approximately 80%).

Sands were deposited in elongate, topographically confined belts or fairways which generally trend NW–SE or N–S (Fig. 10). The fairways are composed of a complex anastomosing network of braided "channels", with the massive sandstones being deposited from high-density turbidity currents. These sandstones are characterized by a topography-controlled, elongate sheet-like geometry, with high net-to-gross ratios, and good lateral and vertical connectivity. Classical turbidites of interbedded mudstones and sandstones were deposited away from the fairways, often on topographic highs.

The overall sedimentological model is one of topographic lows being filled with massive sandstones. The controlling influences on the thickness and direction of the fairways are the boundary faults of the Forties-Montrose Ridge and differential compaction over previously-deposited sequences. Lobes and offshoots from the main fairways occur as a result of overspill from one filled low to an adjacent low; these would still be essentially topographically controlled. In contrast, depositionally-controlled lobate sandbodies were formed in a waning-flow period at the end of a sand depositional cycle, or at the distal end of a sand fairway where individual flows became topographically unconfined.

## Petrography and diagenesis

The petrography of the reservoir units is uniform throughout the field. The sandstones comprise very fine-, fine- and medium-grained sandstones, moderately sorted, with a low detrital clay content. Locally they grade into pebbly sandstones and conglomerates. Argillaceous debris flows, however, are much more poorly sorted, with a high detrital clay content. The sandstones are mostly sublithic to lithic, and locally subarkosic in composition.

The major detrital component is monocrystalline quartz, with subsidiary feldspar, polycrystalline quartz and lithic fragments. Rare mica, heavy minerals, plant fragments and glauconite also occur. Detrital grains are generally subrounded to angular.

The sandstones are generally poorly lithified, displaying only minor cementation. SEM, XRD and thin section analysis have identified the following cement phases:

(1) *Clay minerals*, including kaolinite, illite, mixed layer illite-smectite and chlorite. Thicker aggradational sand bodies deposited by high-density turbidites generally contain less than 10% clay matrix.

(2) *Quartz overgrowths*. This cement infills a minor part of the primary porosity.

(3) *Carbonates*. Coarsely crystalline calcite occurs as a poikilotopic cement in concretionary bands up to 3 m thick. Very finely crystalline ferroan calcite or dolomite aggregates are locally present as a minor pore-filling cement.

(4) *Feldspar and pyrite* are usually present only as trace amounts.

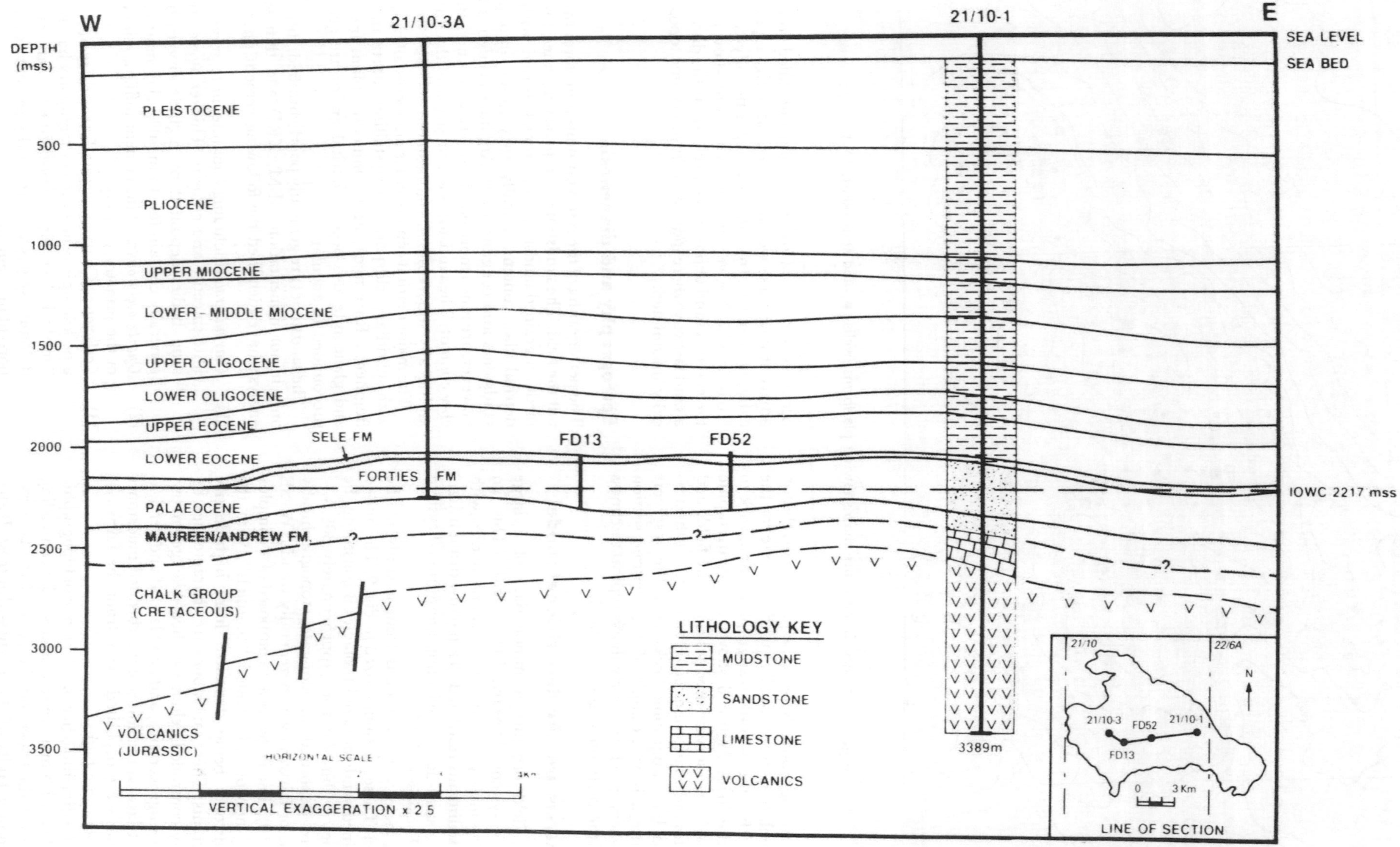


**Fig. 4.** East-west cross-section across the Forties Field. General stratigraphy illustrated by discovery well 21/10-1.

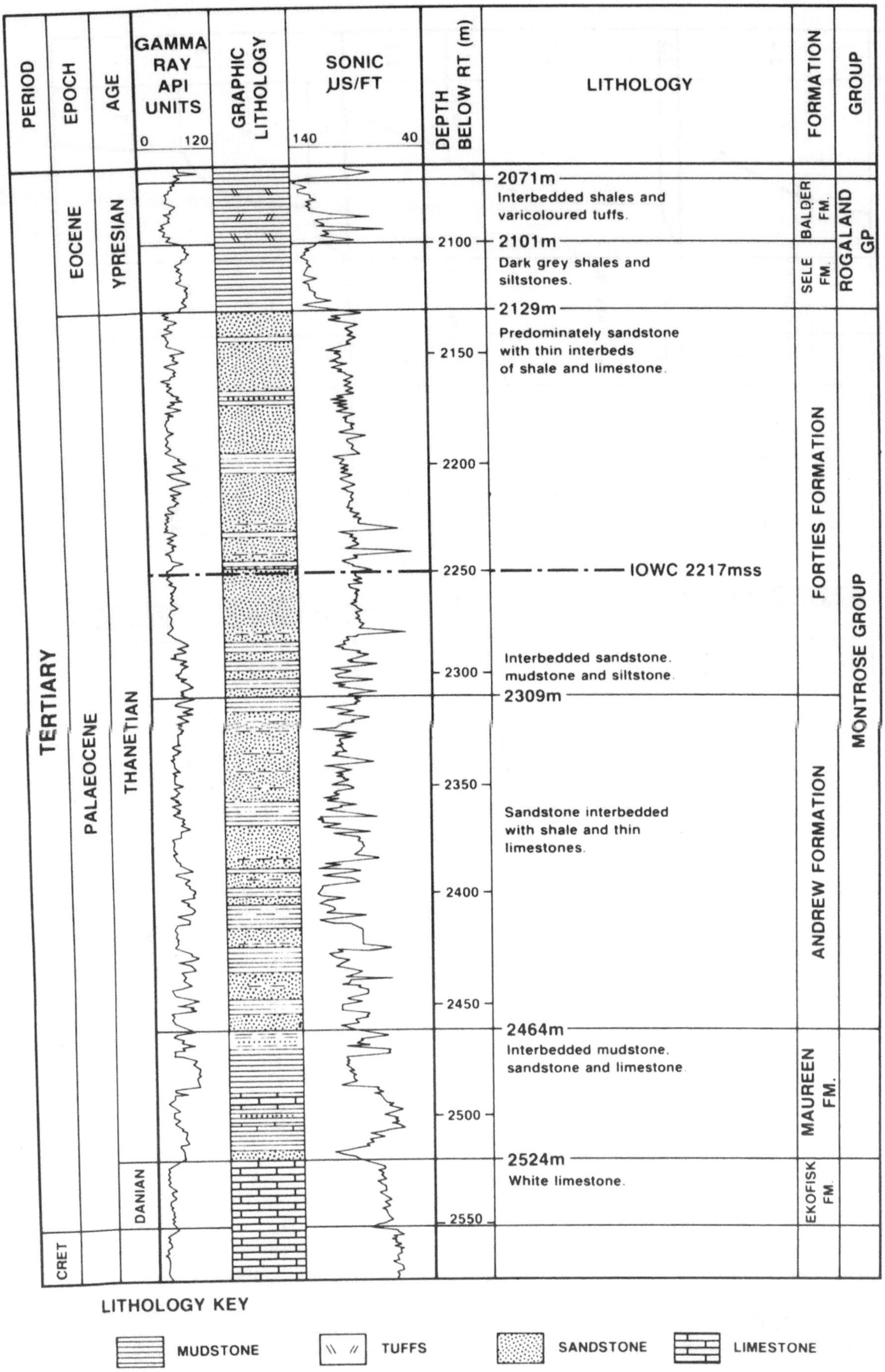


**Fig. 5.** Palaeocene lithostratigraphy of the Forties area.

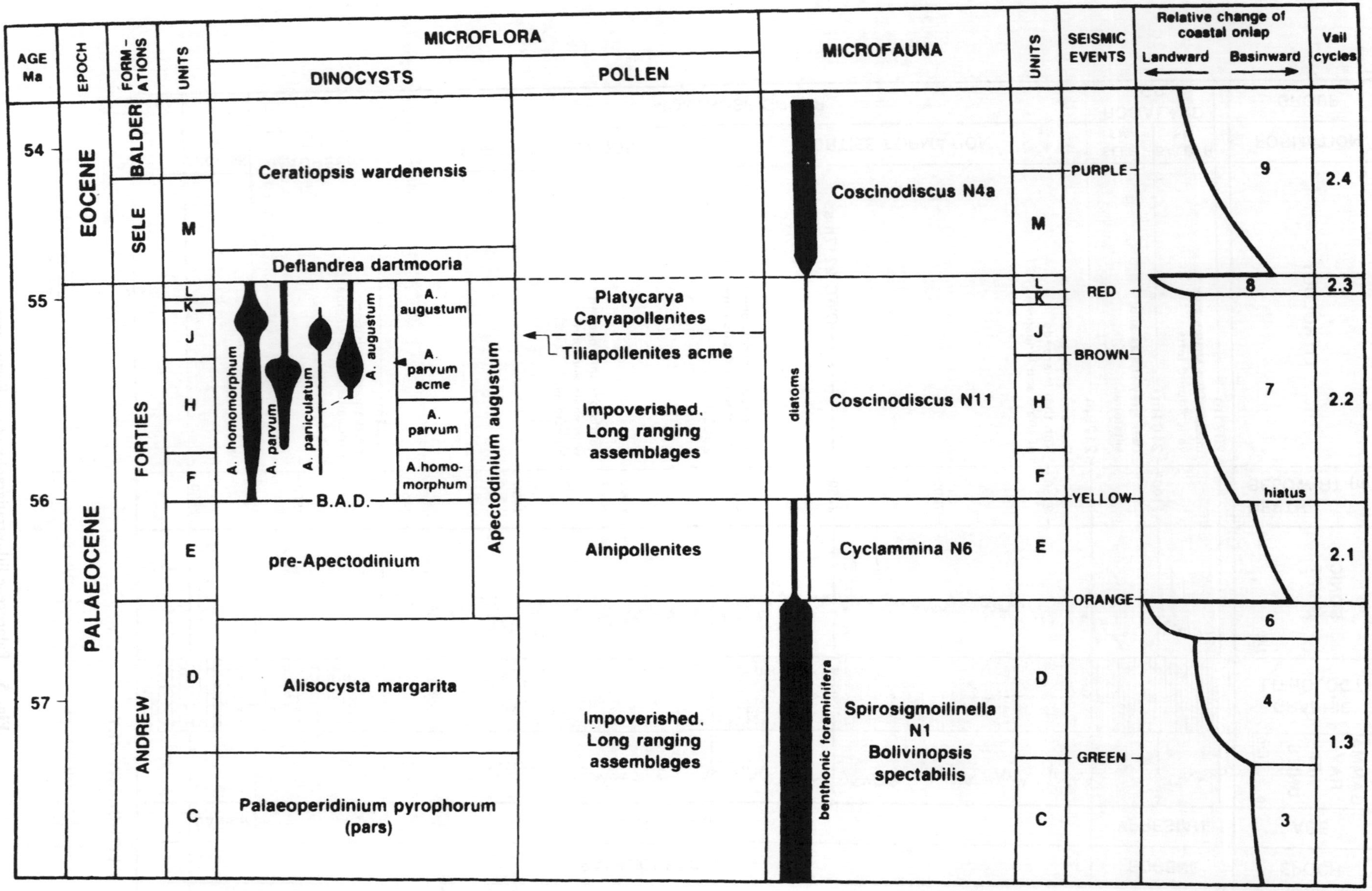


**Fig. 6.** Stratigraphic framework for reservoir subdivision in the Forties Field. Relative change of coastal onlap curve is taken from Stewart (1987), and illustrates the relationships between the Forties Field reservoir stratigraphy and the regional Central North Sea seismic stratigraphy. The "Vail cycles" are taken from Haq *et al.* (1987). B.A.D. = Base Apectodinium Datum

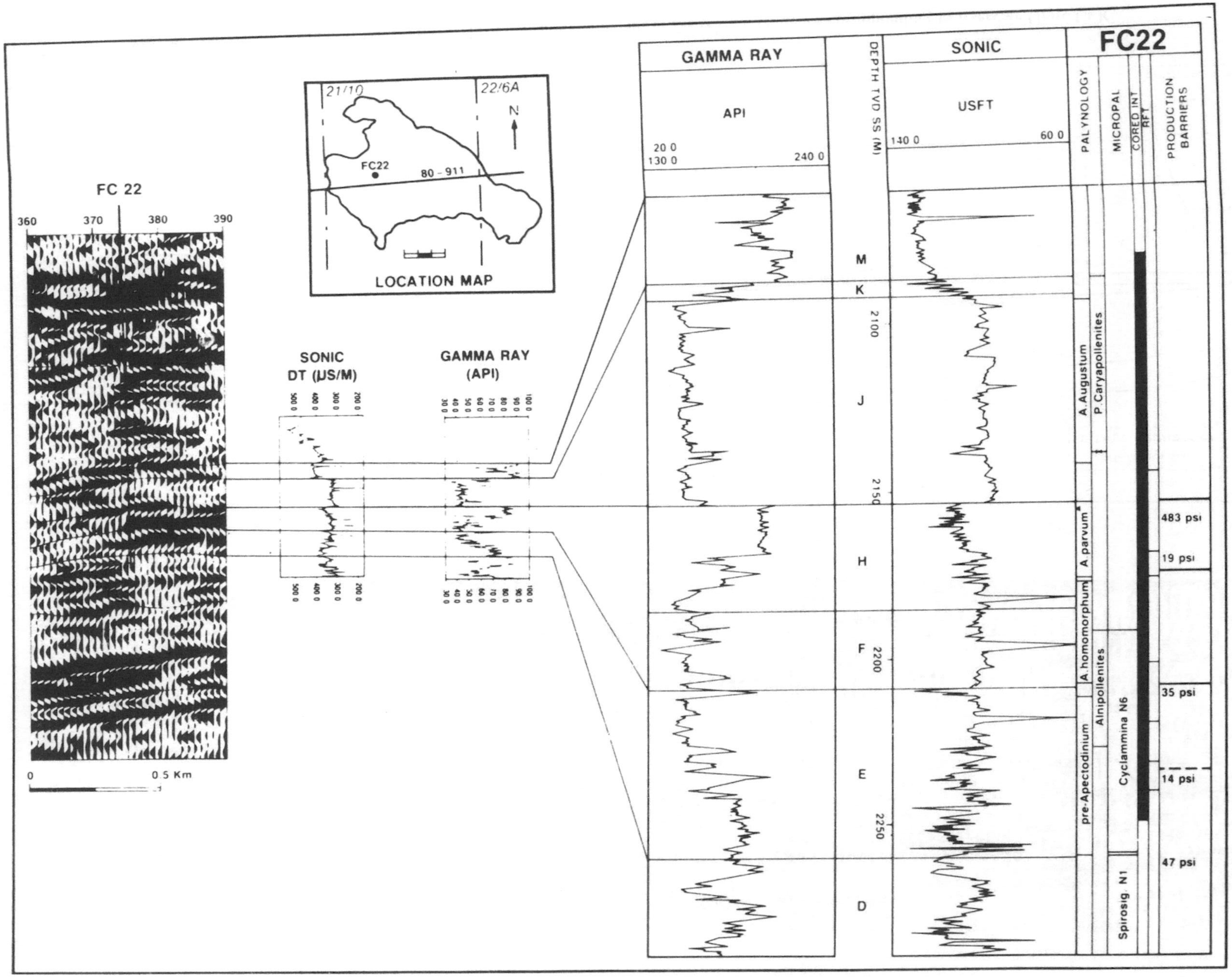


**Fig. 7.** Relationships between the palaeontologically defined units and seismic.

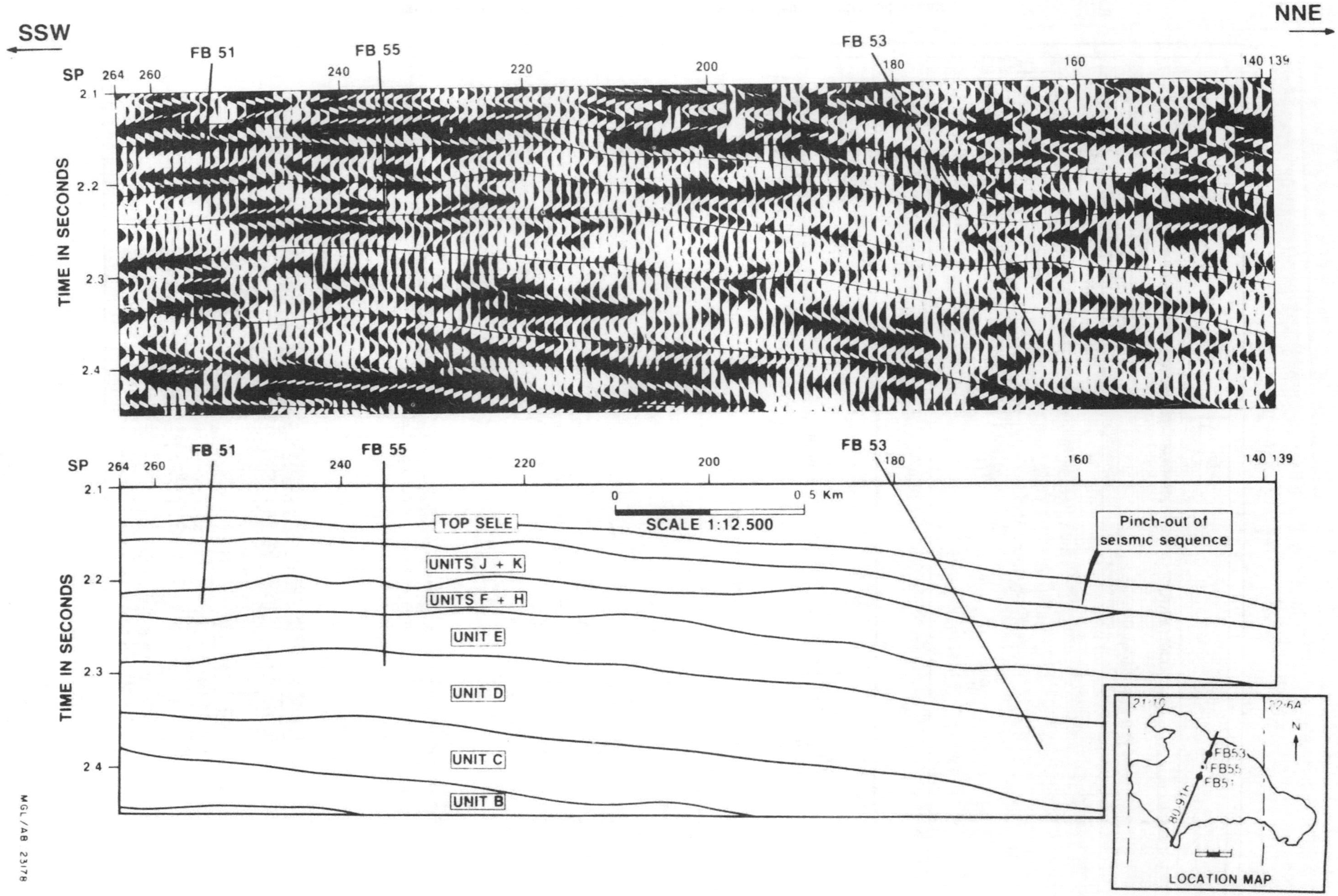


**Fig. 8.** Geoseismic section in the Forties Field. Line shows pinchout of the seismic sequence corresponding to reservoir Units J+K.

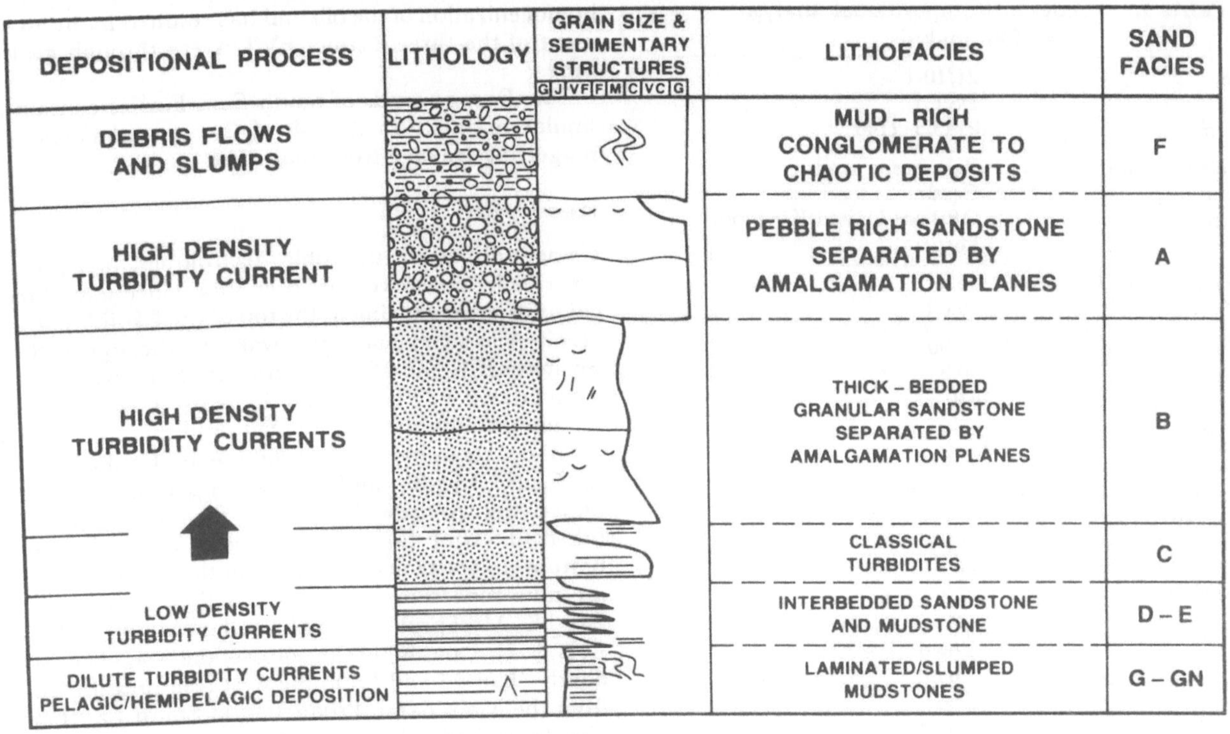


A. Coarse grained pebbly sandstones to pebble grade, clast supported conglomerates containing intra- and extra- formational clasts in a sandstone matrix. Clasts generally 1-10cm. Beds up to 8m thick, are non-graded or exhibit fining or coarsening upwards. Sharp, erosive bases.
B. Mostly fine to medium grained sandstone. Beds usually 1-3m thick but may be up to 7m. Occur as amalgamated sequences but are locally separated by thin mudstones or debris flows. Sedimentary structures include indistinct horizontal laminae, dish structures and other irregular dewatering structures. Sharp, erosive or loaded bases.
C. Interbedded mudstone, siltstone and very fine to fine grained sandstone. Ratio of sandstone to mudstone 80:20. Individual sandstone beds approx. 20-30cm thick. May exhibit parallel or ripple cross laminations, ideally in classical Bouma Tb-Te sequences.
D. Interbedded mudstone, siltstone and very fine to fine grained sandstone. Ratio of sandstone to mudstone about 50:50. Individual sandstone beds approx. 5-30cm thick. Exhibits ripple cross laminations and local bioturbation. (Bouma Tc-Te)
E. Interbedded mudstone, siltstone and very fine to fine grained sandstone. Ratio of sandstone to mudstone about 25:75. Individual sandstone beds approx. 0.5-10cm thick. Exhibits horizontal lamination, ripple cross-lamination and bioturbation. (Bouma Tc-Te)
F. Highly deformed and contorted mixtures of the other facies, exhibiting slump-folding, soft sediment faulting, ball and pillow structures. Also very poorly sorted argillaceous sandstone/sandy mudstone (debris flows).
G. Laminated grey mudstone with thin siltstone and sandstone beds (< 10cm thick).
GN. Grey green 'waxy' mudstone with pervasive 'Zoophycos' and 'Chondrites' burrows.

**Fig. 9.** Forties Field lithofacies.

## Hydrocarbon composition, source and migration

Forties crude is a low-sulphur, medium-wax, light oil of 37° API gravity. Table I presents the geochemical analysis of oil recovered from the discovery well 21/10-1. Gas chromatogram and GC–MS analyses indicate a moderately mature crude derived from a clastic marine source.

All evidence suggests that the oil was sourced from the Jurassic Kimmeridge Clay Formation, which was deposited to the north and south of the Forties-Montrose Ridge. The present depth of the Kimmeridge defines two source "kitchens": a southern area which drains towards South East Forties, and a northeastern area which drains into the northern area of the field (Fig. 11). These kitchens generated oil which migrated vertically and then laterally through the Palaeocene sandstones into the trap. Thermal modelling suggests that the main phase of oil expulsion was from Middle Eocene to Miocene times (50–10 Ma) in the southern kitchen, and from Late Eocene/Oligocene times to the present day (40–0 Ma) for the northern kitchen.

Although analysis of several oils from across the field reveals no major compositional differences, there are variations in the oil properties (see section on Fluid Description). Three distinct oil types correspond to three areal regions of the field, which correlate to the source kitchen locations and maturities. The more mature, more gaseous oil in the South East Forties area contrasts with the less gaseous, less mature oil in the western area of the field. Between these areas the oil has constant but intermediate properties. The occurrence of two major permeability barriers in the reservoir has prevented the complete

**Table I.** Forties oil compositional analysis
(a) Oil analysis

| | |
|---|---|
| Sample | 21/10-1 DST 1 |
| Depth (m) | 2143.5–2148 |
| Reservoir/formation | Palaeocene sands |
| Sample type | Crude |
| Location | Offshore United Kingdom |
| Field | Forties |
| API gravity @ 15°C | 36.5 |
| Wax (% wt) | 5.00 |
| Asphaltenes (% wt) | 1.90 |
| Sulphur (% wt) | 0.3 |
| Nitrogen (ppm) | 830 |
| Nickel (ppm) | <2 |
| Vanadium (ppm) | 3 |
| Type analysis by HPLC on de-asphaltened residue >200°C | |
| Saturates (% wt) | 57.9 |
| Aromatics (% wt) | 33.6 |
| Residue (% wt) | 8.5 |
| Carbon isotope ratios per mil | |
| Total oil | −28.9 |
| Asphaltenes | −27.6 |
| Saturates | −29.3 |
| Aromatics | −28.6 |
| Residue | −28.2 |
| Standard NBS 22–29.8 | |
| n-Alkane CPI | 1.05 |
| Pristane/phytane | 1.30 |
| PR/nC17 | 0.76 |
| PH/nC18 | 0.65 |

**Table I.** Forties oil compositional analysis
(b) Reservoir fluid compositions for Forties "type well"

| *Component* | *Main field mol%* | *Forties Echo platform mol%* |
|---|---|---|
| Nitrogen | 0.41 | 0.39 |
| $CO_2$ | 0.32 | 0.67 |
| Methane | 17.09 | 21.91 |
| nC2 | 5.49 | 6.62 |
| nC3 | 8.75 | 10.03 |
| iC4 | 1.73 | 1.65 |
| nC4 | 4.96 | 4.23 |
| iC5 | 1.91 | 1.77 |
| nC5 | 2.88 | 2.91 |
| nC6 | 4.12 | 3.76 |
| nC7 | 6.26 | 4.83 |
| nC8 | 7.32 | 4.97 |
| nC9 | 5.29 | 3.69 |
| BP 160 | 2.99 | 2.91 |
| BP 180 | 2.89 | 2.82 |
| BP 200 | 2.60 | 2.53 |
| BP 240 | 6.85 | 6.66 |
| BP 300 | 5.86 | 5.69 |
| BP 360 | 4.17 | 4.06 |
| BP 420 | 2.89 | 2.81 |
| BP 480 | 2.06 | 2.01 |
| BP 540 | 1.19 | 1.16 |
| BP 600 | 0.81 | 0.79 |
| BP 660 | 0.49 | 0.47 |
| BP 720 | 0.30 | 0.29 |
| BP 780 | 0.19 | 0.18 |
| BP 840 | 0.12 | 0.12 |
| BP 885 | 0.05 | 0.05 |
| $H_2S$ | 0.00 | 0.00 |

homogenization of the oils and has resulted in the establishment of the three distinct PVT areas through geological time.

The Eocene sands of South East Forties contain oil of similar maturity to the oils of the central Forties Field, suggesting leakage from that area.

## Reservoir units

Using seismic-stratigraphic techniques, as previously described, the reservoir has been subdivided into 11 geological units, commencing at the top of the Ekofisk Formation (Unit A) and extending upward to the top of the Sele Formation (Unit M). The principal reservoir units are summarized in Figs 12 and 13, with a schematic cross-section through the field and net sand maps for each unit.

Oil occurs in five units of the Forties Formation (Units E, F, H*, J, K) and in the uppermost unit of the underlying Andrew Formation in the crestal area of the field (Unit D). After 13 years of production and bottom-up water flooding, most of the remaining reserves are now concentrated in the crestal regions of Units H, J and K. Most production is from the "Main Sand" (Units D, E, F, H, J and K) in the centre of the field. The "Charlie Sand" (Units J and K) in the west is in pressure isolation from the Main Sand. Production in South East Forties is also from Units J and K, which are also in pressure isolation at reservoir level. Each unit is briefly described, starting with the uppermost Sele Formation.

*Unit M.* A prominent laminated basinal mudstone/siltstone horizon (lithofacies G) forms the caprock to the reservoir.

*Unit L.* The top of the Forties Formation, comprising mudstone within thinly interbedded sandstone (lithofacies E). This unit represents the abandonment phase of the Forties submarine fan system.

*Unit K* (Fig. 12). This unit comprises a thin sandstone/mudstone sequence lying immediately above the thick Unit J sandstones. Two main elongate sandstone bodies (lithofacies B) are flanked by areas of thin-bedded sandstone and mudstone (lithofacies C, D, E) with thin debris flows (lithofacies F). This unit represents the last major phase of coarse clastic sedimentation in the fan system.

*Unit J* (Fig. 12). The reservoir unit for the bulk of the oil in the field. It comprises two major sandstone bodies (lithofacies B)—the "Main Sand" and the "Charlie Sand", separated by a thin area of low net-to-gross, thin-bedded sandstone and mudstone (lithofacies C, D, E, G).

*Unit H* (Fig. 12). Unit H comprises fining-upward sandstone (lithofacies B) or sandstone/mudstone sequences (lithofacies. E, G). The top of the unit is marked by a prominent mudstone (the "Charlie Shale") in the west and centre of the field which has a significant pressure discontinuity measured across it. In the east, particularly in South East Forties, the mudstone is much thinner and discontinuous, partly due to erosion by overlying units. In South East Forties, the unit thickens abruptly and comprises large-scale slump deposits (lithofacies F).

*Unit F* (Fig. 13). Thick sandstones (lithofacies B) and sandstone/shale sequences forming two channel-like bodies across the field, with two areas of thin, mudstone-dominated sequences in the southwest and southeast (lithofacies D, E).

* A Unit G was originally recognized, being the net pay interval below the "Charlie Shale" (Unit H), but was subsequently incorporated with the overlying mudstone as part of Unit H.

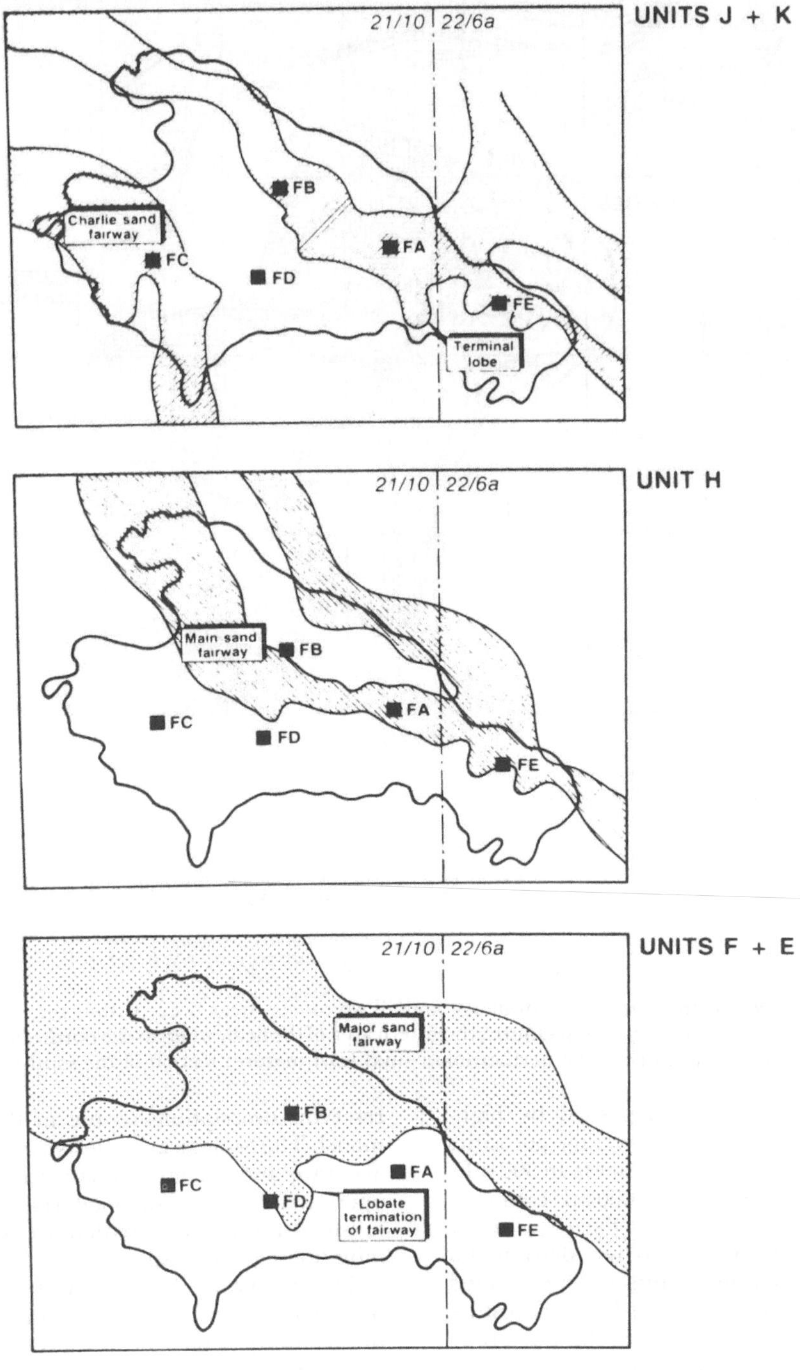


**Fig. 10.** Distribution of major sand bodies in the Forties Field.

*Unit E* (Fig. 13). This unit has a similar distribution of lithofacies and thicknesses as the overlying Unit F, with a major area of thick-bedded sandstone (lithofacies B) occupying the northern part of the field, and interbedded sandstones and mudstones (lithofacies C, D, E), thick sandstones (lithofacies B), and interbedded debris flows (lithofacies F) occupying the southwest and southeast.

*Unit D* (Fig. 13). The top of the unit is identified by a thick mudstone sequence (lithofacies G–GN) with a prominent pressure discontinuity across it. Underlying the mudstone are thin, bioturbated sandstones (lithofacies C, D), structureless sandstones (lithofacies B), and deformed slump and debris flow deposits (lithofacies F). The top of Unit D corresponds to the top of the Andrew Formation.

*Unit C.* This unit corresponds to the lower part of the Andrew Formation and consists predominantly of sandstone with thin mudstone interbeds and limestone stringers, capped by a prominent shale horizon.

*Unit B.* A mixture of sandstones, mudstones, siltstones and re-worked limestones characterize this unit, which corresponds to the Maureen Formation.

*Unit A.* The massive white limestones of the Ekofisk Formation.

### *Controls on porosity and permeability*

Primary depositional characteristics are the most important parameters that control the poroperm distribution in the reservoir sandstones. Essentially, porosity and permeability decrease as the grain size decreases and the detrital illite/illite-smectite content increases. Diagenetic kaolinite, illite and chlorite are not abundant and have only a minor detrimental effect on reservoir quality.

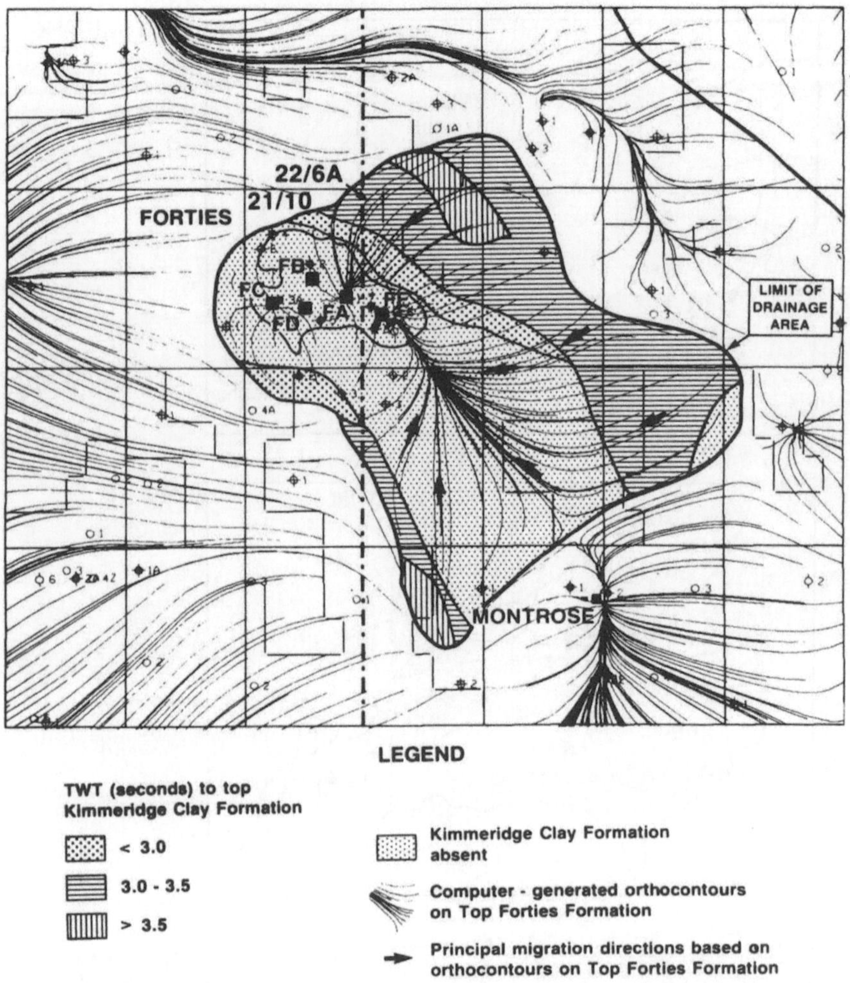


**Fig. 11.** Location of source areas within the Forties drainage area. Two source areas occur within the drainage area. Oil from the southern area (which expelled oil earlier) only accesses the south-eastern area of the field, whereas the northern area oil can access all parts of the Field. The FC area is only sourced from the northernmost (and least mature) section of the southern area.

The coarsest-grained sandstones, deposited by high-density turbidity currents, tend to have average porosities around 27% and permeabilities often greater than 300 mD, with high permeability zones in excess of 1000 mD common. The pebble content of pebbly sandstones (Facies A) causes an overall decrease in porosity, but is less detrimental to permeability. The finer-grained sandstones, deposited by low-density turbidity currents, have lower porosities (23–25%) and significantly lower permeabilities (100–200 mD). Siltstones and mudstones deposited by waning low-density turbidity currents are considered as non-pay.

Mudstone and shale intervals can be laterally extensive, resulting in permeability barriers to vertical fluid flow with significant pressure differentials across them. Several important barriers to lateral fluid movement also exist in the reservoir where sandstones thin and pinch-out into areas of low net-to-gross siltstones and mudstones. The most significant lateral barriers are those between the "Main Sand" in the centre of the field and the "Charlie Sand" to the west, and between the Main Field in Block 21/10 and South East Forties (Fig. 12).

Evidence suggests that carbonate-cemented concretionary bands have only limited extent, and therefore are unlikely to form significant barriers to vertical fluid flow.

## Application of reservoir geological model to reservoir engineering

The principal objective of creating the geological model was to provide an accurate, up-to-date and detailed reservoir description for use in the reservoir simulation model. The geological model was required to be sufficiently accurate and technically correct to be used both as an operational model for good reservoir management and full-field production predictions, and for Final Redetermination equity studies.

The reservoir description has incorporated all the pertinent available technical data to establish and describe the geometry of the reservoir units. Prominent restrictions to vertical flow have been identified and used as unit boundaries, and sand body connectivity maps generated to model fluid flow between adjacent units. Recognizable restrictions to horizontal flow have been used to control net-to-gross, porosity and permeability maps, honouring geological fairways where these are believed to exist. Unit mapping has been extended to model the extensive aquifer surrounding and underlying the reservoir.

For each geological unit, depth, isopach and connectivity (vertical permeability) maps were constructed. These, together with petrophysical maps of net-to-gross, porosity and permeability, were input to an ECLIPSE reservoir simulator. All maps were constructed using the Radian CPS 1 computer mapping system. The reservoir units modelled in the simulator were D, E, F, H, J, K + L. The reservoir simulation is discussed later in more detail.

There will be a continual requirement for updating and improvement of the geological model to match the demand for continued reservoir simulation studies. The acquisition of approximately 7500 km of 3D seismic data over the entire field in the summer of 1988 will lead to a major

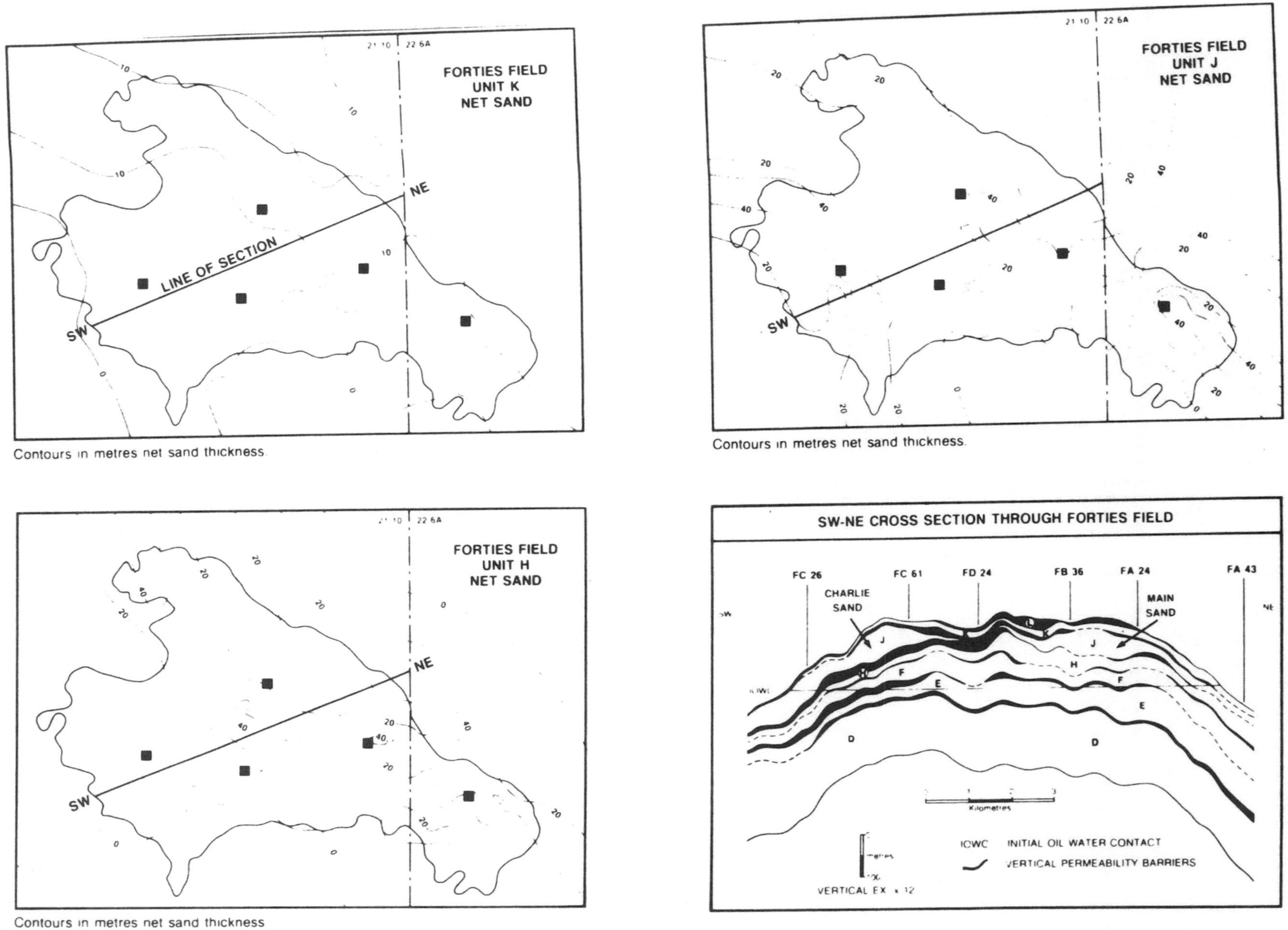


**Fig. 12.** Reservoir subdivision and net sand maps, Units K, J and H. Contours in metres net sand thickness.

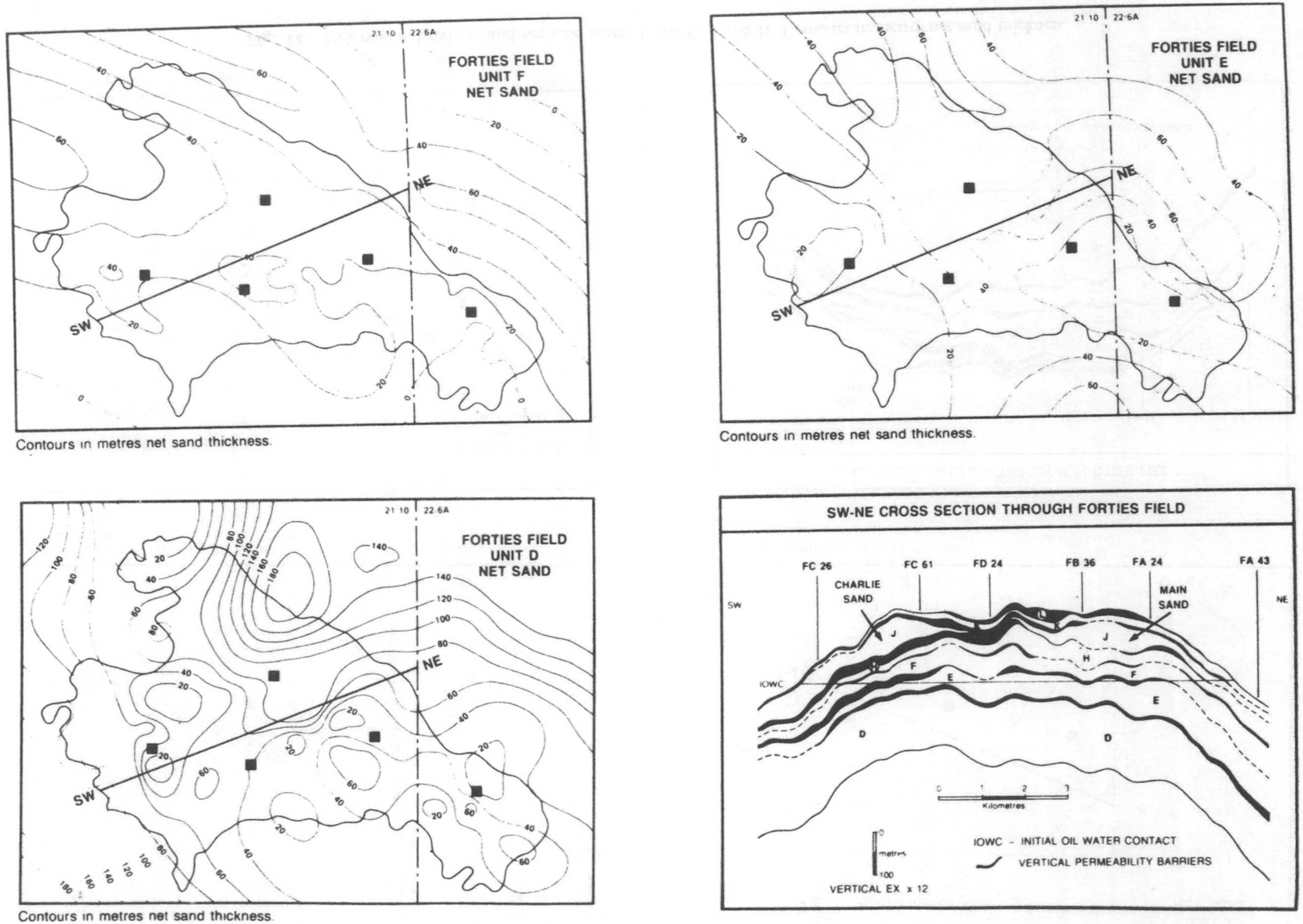


**Fig. 13.** Reservoir subdivision and net sand maps, Units F, E and D. Contours in metres net sand thickness.

improvement in the reservoir description. The large increase in data provided by this survey will be instrumental to the better understanding of reservoir performance, and will enable the detailed mapping required to define the small accumulations of bypassed and incremental oil that will be left in the final years of field life. Innovative processing techniques, e.g. amplitude-versus-offset, and associated detailed interpretation, can provide further control on the lateral extent of sandstones and their fluid content. The work to defne and access these accumulations will be a fundamental part of the evolving reservoir management strategy.

## RESERVOIR ENGINEERING

The following two sections present the rock and fluid properties of the field. Overall, the field is characterized by very clean, high-permeability sands containing high quality, undersaturated crude oil. As a result, it has been sufficient to study the rock and fluid properties by routine methods. These properties are presented with descriptions of the methods employed.

### Rock properties

The properties of the rock have been studied extensively through core analysis, well testing and log interpretation. The following sections present the rock properties in detail together with the methods used to derive them.

*Porosity*

Routine core porosity data were corrected to in-situ reservoir conditions based upon regression analysis of special core analysis data. These core porosities were then regressed against the bulk density log (RHOB).

Porosity was thus defined by the equation:

$$\phi = (2.654 - \text{RHOB})/1.694 \qquad (1)$$

where $\phi$ is porosity and the matrix density of 2.654 g/cm$^3$ is the mean value of core grain density measurements.

Typical values of porosity are in the range 25–27%.

*Net reservoir estimation*

The estimation of net reservoir rock was made by identifying non-reservoir rock due either to shale intervals or calcite cemented low-porosity sandstones.

Non-reservoir rock due to shale was identified through the normalization of gamma ray data relative to a range of 30 to 95 API units. A cut-off of 62.5 API units was chosen, corresponding to the 50% level between the clean sand and shale values.

Low-porosity non-reservoir rock was defined primarily by a density log reading of 2.45 g/cm$^3$. Where density logs were unavailable or unreliable, other logs were used, notably the neutron, sonic and thermal decay time (TDT) logs.

The ratio of net sand to gross rock volume is generally about 0.6.

*Water saturation*

Different water-saturation equations have been investigated and the most appropriate relationship has been found to be the Archie equation:

$$S_w^n = F\frac{R_w}{R_t} \qquad (2)$$

where

| | Units |
|---|---|
| $S_w$ = water saturation | decimal fraction |
| $R_w$ = formation water resistivity | ohm m |
| $R_t$ = true formation resistivity | ohm m |
| $n$ = saturation exponent | — |
| $F$ = formation resistivity factor $\phi^{-m}$ | — |
| $m$ = cementation exponent = 1.81 | — |

The above equation has been found to be most appropriate because special core analysis indicates Forties to be an essentially clean reservoir. The cation exchange capacity data from wet chemistry measurements were insufficiently reliable to enable their use in fieldwide predictions. Also, laboratory measurements of Resistivity Index versus water saturation consistently showed no measurable clay effect.

Formation water resistivity is 0.034 ohm m at 204.6°F (96°C) based on water samples, corrected for temperature using the Arps formula with a temperature gradient of 6.56°F/100m (3.64°C/100m).

True formation resistivity values were obtained from the application of borehole and invasion corrections to laterolog data.

The saturation exponent $n$ was defined by:

$$n = -0.24 * \log_{10}(K_{kl}) + 2.5334 \qquad (3)$$

where $K_{kl}$ = Klinkenberg corrected air permeability (mD). This equation resulted from laboratory measurements and was applied by means of a fieldwide predictor of permeability from log data.

*Water saturation versus height*

A non-linear regression of net sand data points was used to derive a relationship between $S_w$ and height of the form:

$$S_w = \frac{1.0}{1.0 + \exp(-Y)} \qquad (4)$$

where

$S_w$ = predicted water saturation
$Y = A + BZ + (A_1 + B_1 Z)X + A_2 X^2 + A_3 X^3$
$Z = \exp(-C.h)$
$X = 10(\phi - 0.19)$
$\phi$ = calculated porosity
$h$ = vertical height above 2217 mtvd ss (the initial oil–water contact)
$A, B, A_1, B_1, A_2, A_3, C$ are constants

For example, at a height of 80 m above the oil-water contact (OWC) for a porosity of 26%, $S_w = 14\%$.

*Permeability*

Permeability values were obtained by a regression of air permeability from core against log measurements. These values were corrected to reservoir conditions through a comparison between log and test-derived permeability thickness ($K_h$). The range of values is from 200 mD to 2000 mD.

*Compressibility*

Rock compressibility has been estimated from measurements and standard correlations as

$$C_f = 3.2 \times 10^{-6}\ \text{psi}^{-1}$$

## Fluid description

### *Oil*

The Forties crude is a light oil of 37° API gravity. The oil is undersaturated at reservoir conditions, resulting in all the gas being in solution. Accordingly, there is no gas cap.

The oil is underlain by a massive aquifer that has proved to be very mobile.

A summary of the fluid properties is given in Table II. The fluid composition is given in Table I.

**Table II.** Summary of fluid properties

| | *Western PVT region (Charlie Sand)* | *Central PVT region* | *Southeast PVT region* |
|---|---|---|---|
| Saturation pressure psia at 96°C | 1079 | 1145 | 1376 |
| Density of stock tank oil $lb\,ft^{-3}$ at 14.7 psia and 15.6°C | 52.8 | 52.6 | 52.2 |
| Density of produced gas $lb\,ft^{-3}$ at 14.7 psia and 15.6°C | 0.078 | 0.078 | 0.079 |
| Gas–oil ratio scf/stb | 300 | 303 | 392 |
| Formation volume factor (rb/stb at saturation pressure and 96°C) | 1.228 | 1.242 | 1.289 |
| Oil viscosity (cP at 3230 psia and 96°C) | 0.78 | 0.76 | 0.66 |
| Oil compressibility ($psi^{-1}$ at 3230 psi and 96°C) | $9.5 \times 10^{-6}$ | $9.6 \times 10^{-6}$ | $10 \times 10^{-6}$ |

### *Formation water*

The formation water properties are summarized below:

| | |
|---|---|
| Density ($lbft^{-3}$ at 3230 psia and 96°C) | 63.9 |
| Formation volume factor (rb/stb at 3230 psia and 96°C) | 1.029 |
| Viscosity (cP at 3230 psia and 96°C) | 0.36 |
| Compressibility ($psi^{-1}$ at 3230 psia and 96°C) | $3 \times 10^{-6}$ |
| Total dissolved solids (mg/l) | 90 000 |
| Resistivity at 20°C (ohm m) | 0.13 |

## Initial fluid distribution

The initial field oil water contact was at 2217 mtvdss. The fluid distribution per reservoir layer (Fig. 12) is given below.

| *Reservoir unit* | *Oil-in-place as percentage of total* |
|---|---|
| K | 12 |
| J | 37 |
| H | 17 |
| F | 22 |
| E | 10 |
| D | 2 |
| | 100 |

The initial connate water saturation is in the range 12–21%.

## Reservoir performance

### *General*

The Forties reservoir performance has been characterized by excellent vertical sweep through the clean sands by the underlying aquifer supported by peripheral water injection. Development of the reservoir was initially planned with approximately 80 production and 20 injection wells from four main field platforms. The well locations are shown in Fig. 14. Figure 15(a) shows the pressure differences between the different regions of the field, and Fig. 15(b) shows the history of reservoir pressure for the Main and Charlie Sands. The production/injection history for the field is shown in Fig. 16.

### *Charlie Sand*

The Charlie Sand is defined as the low-pressured K and J Units in the west of the field. There is a sufficient permeability restriction between this region and the Main Sand to allow for a significant pressure difference between the two regions. The management of the reservoir has been based on treating the two regions separately. In broad terms there is more significant edge water movement in the Charlie Sand than in the Main Sand.

### *Reservoir heterogeneities*

Whilst the reservoir performance is essentially that of a clean sand being displaced by good bottom-water drive, there are a number of areas where anomalies occur. These are described below.

(a) *Faulting* There is some evidence for boundaries within the reservoir from pressure build-up testing. In addition, there is evidence from the large difference in OWC encountered in two neighbouring wells (FA 2-1 and FA 2-3), and from PBU analysis, that a fault may exist in this part of the field.

(b) *Water-overrun* Over the last 2 years there has been an increasing number of wells in the north east of the field which exhibit water overrun in the upper reservoir units. There are currently 15 wells in this category. This has been identified on TDT/PLT logs that are run for monitoring purposes. In addition, log data from two of the South East Forties wells suggest that this phenomenon is also present in this region.

(C) *Irregular seawater breakthrough* The pattern of seawater breakthrough from injectors to producers has also shown some anomalies. In the FD platform region, for instance, there have been wells that have exhibited seawater breakthrough before wells that are located closer to injectors and would have expected to have shown seawater breakthrough earlier.

### *Production of sand and scaling products*

The Forties reservoir sand is friable in most areas of the field. This has resulted in significant sand production causing erosion to surface facilities and the requirement periodically to wash out the sand from the separators. Since the onset of seawater breakthrough, there has been an increase in the deposition of barium sulphate scale in downhole tubulars and some surface pipework.

## Reservoir monitoring

Reservoir monitoring comprises four main elements: wireline logs, pressure surveys, well tests and produced fluid analyses. These are considered in more detail in the following sections.

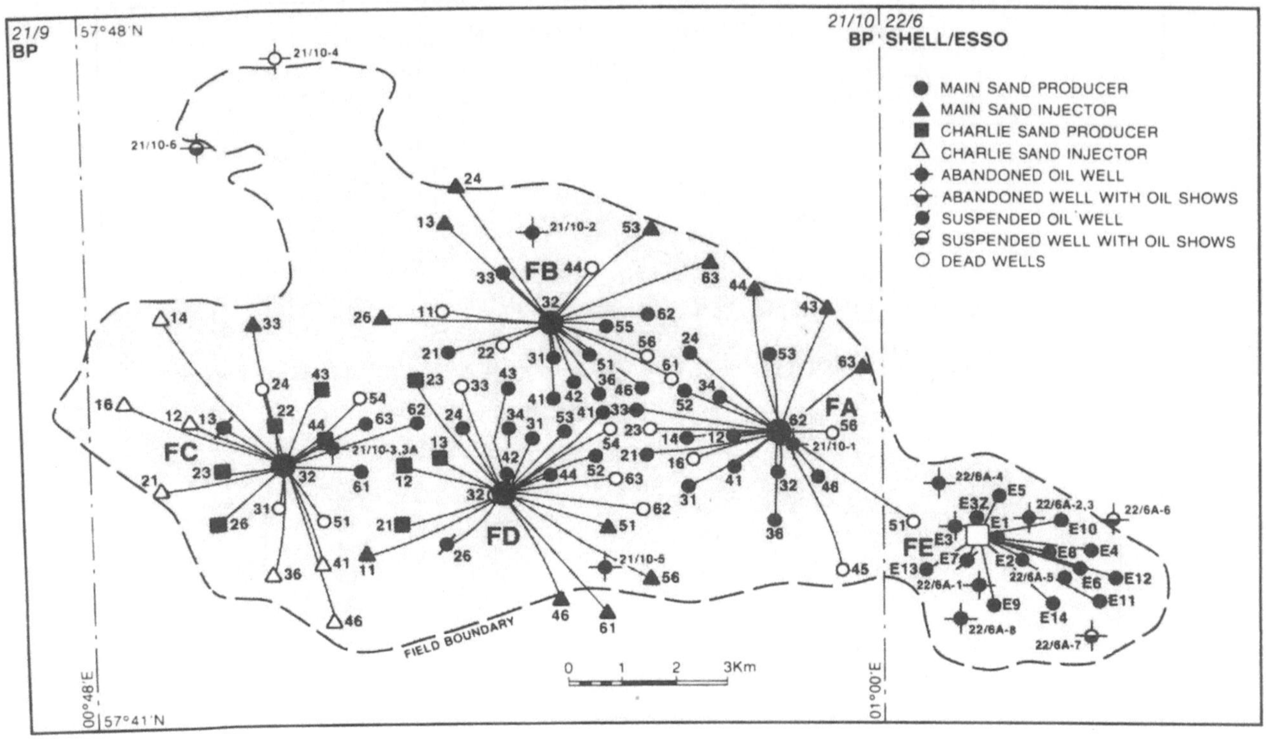


**Fig. 14.** Forties Field well locations.

*Production logging*

Routine monitoring of the OWC movement is carried out through the use of production logs (PLT) and thermal decay time (TDT) logs. One of each log is run on all producing wells approximately once every 2 years. These logs are also used to aid workover and recompletion planning.

*Pressure surveys*

Pressure surveys are conducted on each well on a regular basis, approximately twice per year for each well. The surveys consist of pressure build-up tests (PBU), pressure fall-off tests (PFO) and bottom hole closed in pressure surveys (BHCIP). The tests are conducted on producers, injectors and dead wells respectively.

*Well tests*

Individual well flowrates and water-cuts are assessed by production testing to test separators on the platforms. The Echo wells are tested on the Alpha platform through a dedicated test line between the two platforms. Each well is tested approximately once per month.

*Produced fluid analysis*

Water samples are taken from water-producing wells for laboratory analysis. Each well is sampled two or three times a month. These results help to define patterns of seawater breakthrough, to track changes in water cut and to identify wells producing water.

## Reservoir studies

*Computer simulation*

Field management has been supported by a series of simulation models. The early studies included tank models, cross-sections and 3D models based on the BETA II simulator.

Recent studies consist of a full-field model supported by several cross-section studies and single well models. These have been carried out using the ECLIPSE simulator. The current model has been history-matched to the end of 1987. This, and previous full-field models, have been used to support field-development decisions in several areas: artificial lift selection, future production profiles for government consent, future water-handling requirements, optimization of water injection, pipeline requirements and equity studies.

The full-field model consists of a relatively coarse areal grid (300 m cells) and 6 layers which represent the main geological units.

Recent acquisition by BP of a 4-megaword Cray supercomputer allowed the previous full-field model to be substantially upgraded. As a result of the extra memory, it has been possible to increase the model size to around 13 000 cells (from 7700), and also to increase considerably the flexibility in the specification of cell and fluid flow parameters. This, in turn, has allowed an improved representation of several important aspects of the physical processes involved.

The dominant displacement mechanism, vertical equilibrium (VE), was not able to be fully modelled using a commercial package because of significant variations in displacement mechanism in some areas of the field. In the present model, fluid flow is controlled by pseudorelative permeability curves which are specified, on a cell-by-cell basis, for each of six cell faces to allow for all possible directions of flow (for example, up and down dip).

Also, owing to the complex deposition of the main sands, the description of the local aquifer outside the OWC is particularly important. With the additional Cray memory available, a more rigorous description of the aquifer was achieved by modelling the aquifer explicitly, rather than relying on the previously used numerical or analytical functions, which had proved difficult to tune.

History matching the full-field model proved to be a major exercise despite the relatively simple concept of VE. One of the main geological uncertainties was the description of the vertical transmissibility of interlayer shales, and this was a dominant factor in the history-

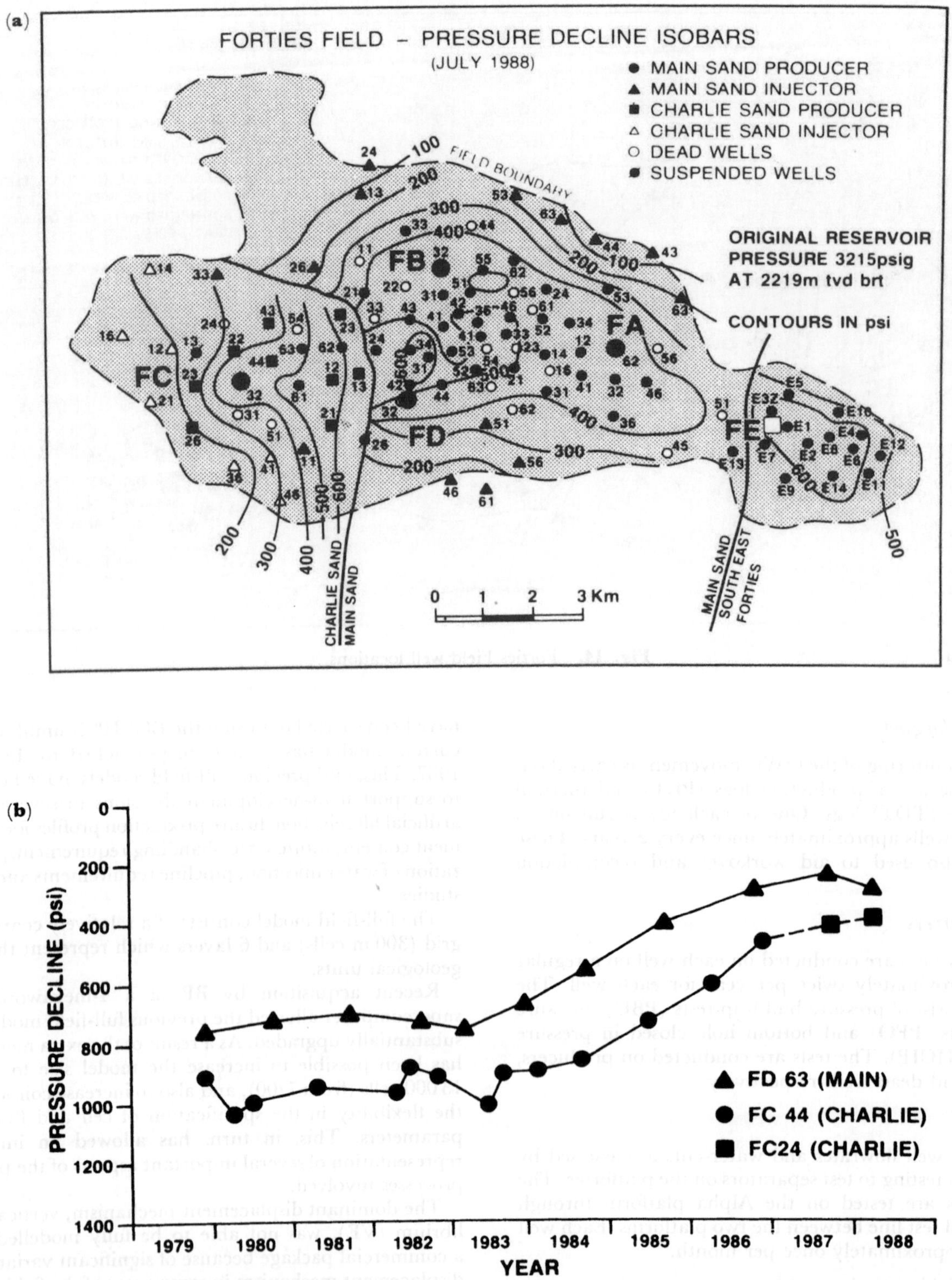


**Fig. 15.** (a) Pressure differences between the Charlie Sand and the central PVT regions of the Forties Field. (b) Comparison of the Main Sand and the Charlie Sand pressure decline.

matching process. The fact that an acceptable match was finally achieved, however, gives some confidence that the displacement mechanisms and certain key parameters have been satisfactorily understood.

History matching has also improved our understanding of injection well performance, and thus allowed prediction studies of water-injection optimization schemes to be carried out, with increased confidence in the results.

*Combined reservoir engineering/geological studies*

Over the last few years the development of a reservoir-management strategy has involved the use of an integrated team approach to understanding the reservoir behaviour. This team, comprising engineers, geologists and geophysicists has been working closely together to generate the basis for continued reservoir management and to aid in the planning of future development. The emphasis is on the multidisciplinary team being in daily contact to discuss problems and to suggest solutions that are consistent with all the constraints as identified by the different disciplines. This has proved an effective way of ensuring that the maximum benefit is obtained from the experience of the various specialists.

Future studies will concentrate on smaller localized areas of the field, in order to be able to assist with

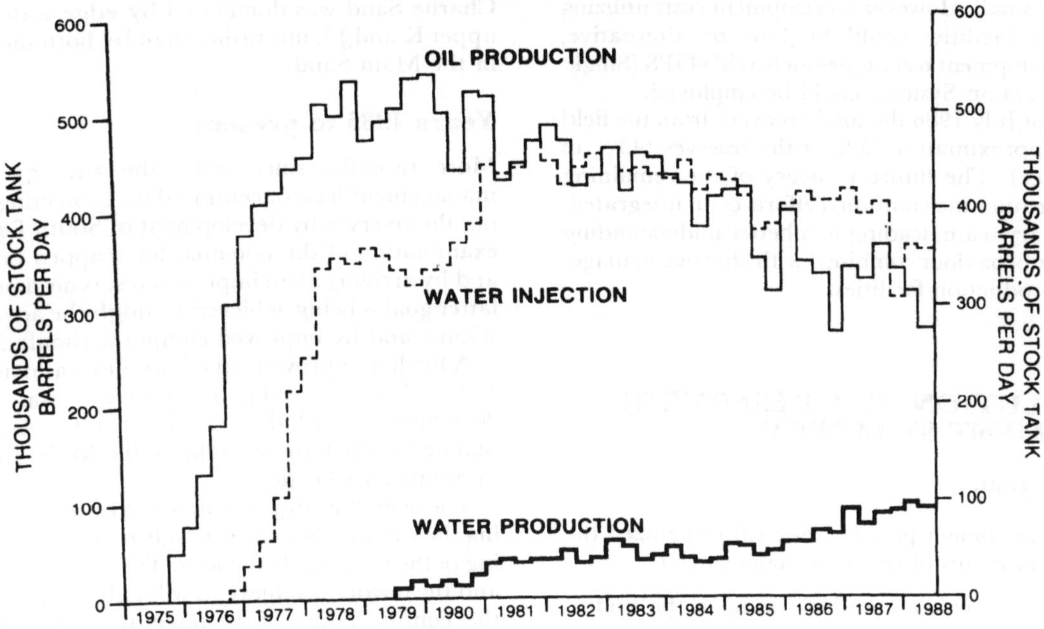


**Fig. 16.** Forties Field production and injection profiles, as at 31 July 1988.

reservoir-management decisions as the upper units are flooded out. This future work will involve cross-sections and detailed sector models. The acquisition of 3D seismic data should be a great assistance with these models.

## Injection experience

Water injection was planned for Forties from the outset. A total of 22 water injectors have been drilled on the periphery of the main field. Water injection began at the end of 1976, approximately 1 year after production started. Each platform was designed to inject 150 mbd of treated seawater. At the end of July 1988 the field injection rate was 390 mrbd, compared to a fluid offtake rate of 480 mrbd. The production and injection profiles are shown in Fig. 16. There are no plans for water injection in the South East region of the field as there is adequate aquifer support to sustain the optimum production rate.

There has been a requirement since early 1987 to work over injection wells to repair completions that have suffered from corrosion and erosion of the tubulars.

The availability of some dead production wells in regions of the field that have been fully swept has permitted conversion of these wells to injectors. This has enabled water injection to be further optimized by improving areal sweep and by adding additional pressure support.

In the future, the optimization of water injection will be an essential part of the reservoir development during the artificial-lift phase of the field life.

## Future development and improved recovery potential

The future development of the field will concentrate on maximizing ultimate oil recovery within the economic life of the field.

### *Artificial lift*

The implementation of an artificial lift programme was begun in 1988 with the installation of electrical submersible pumps (ESPs) in the South East Forties wells. Plans to install gas-lift completions in all suitable production wells in the main field are being implemented. Gas lift is preferred to ESPs because of greater reliability, more flexibility, and the more favourable economics of gas lift compared to ESPs.

### *Enhanced recovery*

Various enhanced oil-recovery techniques have been considered for the Forties reservoir. Surfactant flooding appeared to be the most suitable, as miscible floods with hydrocarbon gas, nitrogen or carbon dioxide would not be possible in the Forties reservoir owing to very poor miscibility.

An experimental programme of surfactant flooding was conducted between 1982 and 1985 which culminated in a field trial. Considerable uncertainties remained regarding the application of a surfactant flood on a full-field basis. The main areas of concern were in defining the baseline waterflood performance, and the errors in field measurements of water cut leading to an inability to measure the effect of the surfactant. Finally, the economic analysis yielded unfavourable results at an oil price below 70 US $ per barrel.

The results of the 3D seismic survey combined with the analysis of well data are expected to identify areas of trapped or bypassed oil within the reservoir. Oil may be left behind in small-scale intrareservoir traps that are not effectively swept by the advancing flood front, or may be bypassed owing to areal reservoir heterogeneities and the stream lines set up by the production/injection well pattern. TDT log evidence and detailed geological mapping indicate that four-way dip-closed structures below intrareservoir shales constitute the most important trapping mechanism. Other traps could occur in intrareservoir stratigraphic pinchouts, isolated sand lenses in areas of low net-to-gross, faulted and slumped areas, hydrodynamic traps, and attic structures located updip from producing wells. This oil will only be recovered either through workovers and sidetracks of existing wells or by drilling new wells into intrareservoir traps.

Further potential for increased recovery from within the field probably exists in field extensions, where downdip areas have not been produced and may not have been effectively swept, and in the Eocene. A small accumulation has already been established overlying South East Forties, testing oil rates of 1620 bopd. The reservoir sands are thin and tenuous, and all Eocene accumulations within Forties are likely to be small. However, development costs utilizing available field facilities could be low, or alternative, innovative development techniques such as SWOPS (Single Well Oil Production System) could be employed.

By the end of July 1988 the total recovery from the field represented approximately 75% of the reserves (43% of the oil in place). The future recovery of the remaining reserves will require the continued efforts of an integrated, multidisciplinary team, leading to a better understanding of the reservoir behaviour combined with effective management of the production facilities.

## THE EVOLUTION OF A RESERVOIR-MANAGEMENT STRATEGY

### Years 1975–1985

The initial development plan for the field recognized the following key elements of reservoir management

- the need for water injection owing to the undersaturated oil having no gas cap;
- the likelihood of good aquifer support;
- well spacing such that an even distribution of wells is balanced with achieving target production rates.

The reservoir was managed for the first 3 years to bring offtake rates up to a plateau of 500 mstbd. This involved continued pattern drilling of production and injection wells.

By the end of 1983 the field production rate was 445 mstbd of oil with water cuts of around 12%. Average well rates were 11 000 stbd with a range of 3–24 000 stbd. Average injection rates were 25 000 bd with a range between 5000 and 40 000 bd. Reservoir pressure declined rapidly in the first 5–7 years to 800–1000 psi below original (3210 psig at 2175 m tvdss). Since 1984, reservoir pressure has been rising at rates of 5–10 psi per month.

Two to three infill wells were drilled in each of the years to 1985, by which time the oil production rate had declined to 375 mstbd at 16% water cut. The reservoir continued to outperform expectations and upward revisions to reserves were made. The most recent estimate is for a recovery of 2470 mmstb (including artificial lift). This represents a recovery factor of 57%, close to recovering all of the mobile oil.

The reservoir management strategy developed in the early to mid 1980s was directed at minimizing water production from the wells. Such remedies were employed as cement squeezes, casing patches and wireline bridge plugs to isolate water-producing intervals. When the water cut exceeds 30–40% in those wells with 7-inch completion tubing, continuous production becomes more difficult owing to slugging and the wells' inability to return to production after shutdowns without a coiled-tubing nitrogen gas-lift workover.

Recompletion of the 7-inch wells to 4.5-inch tubing combined with squeezing and patching of water-producing intervals has enabled production to continue and allowed the continued drainage of all areas of the reservoir.

### Charlie Sand

Within a year of production start up in 1975, it became apparent from pressure data that the Charlie Sand and Main Sand were behaving quite differently. The relatively poor pressure support to the Charlie Sand from the aquifer necessitated additional water injection.

As more data became available, it was clear that the Charlie Sand was dominated by edge-water drive in the upper K and J Units rather than by bottom-water drive as in the Main Sand.

### Years 1985 to present

More recently, since 1985, the strategy for reservoir management has concentrated on protecting and improving the reserves by development of South East Forties, by examination of the potential for trapped or bypassed oil and by arriving at an improved reservoir description. This latter goal is being achieved through the acquisition of 3D seismic and by improved computer simulation.

A further improvement of reserves was made possible in 1987 when the estimated recovery was increased to 2470 mmstb (including artificial lift). Artificial lift is planned in the form of gas lift in the Main Field and ESPs in South East Forties.

The overall strategy continues to be directed at maximizing the value of the asset through an improved understanding of the reservoir behaviour. The success of this strategy, and the means to achieve it, will only be obtained through the continued use of an integrated team of geologists, geophysicists, reservoir engineers and production engineers.

## CONCLUSIONS

(1) The Forties Field is a giant amongst North Sea oil fields. After 13 years of production, the field still has 563 mmstb remaining reserves. Current field production is about 280 mstbpd and field life is expected to extend beyond the year 2000.

(2) The field is defined by a four-way dip closure at Top Palaeocene level, and comprises stacked Palaeocene turbidite sandstones deposited as part of a large submarine fan system. The trap was created as a result of sedimentary drape, synsedimentary faulting, and sediment compaction over the Forties-Montrose Ridge.

The current reservoir model has evolved from previous models and has incorporated all of the available technical data. The reservoir has been subdivided into 11 geological units within a seismic-stratigraphic framework, using a combination of well log data, micropalaeontology, palynology, sedimentology, reprocessed seismic data, petrophysics, reservoir behaviour.

(3) Field performance and available well and geological data suggest that potential incremental reserves probably exist in areas of intrareservoir trapped and bypassed oil, possible field extensions, and in the overlying Eocene sandstones. Further work is required to define and quantify these potential targets.

(4) The development of a reservoir-management strategy is an evolving process throughout the life of a field, combining reservoir monitoring, simulation and the day-to-day management of the wells.

(5) As more field data become available, an improved understanding of reservoir behaviour allows better operational decisions to be made and leads to more confident predictions of future performance.

(6) Routine monitoring of all wells, combined with operational flexibility to unexpected behaviour, are vital for efficient reservoir management.

(7) An integrated team in daily contact is the most effective approach to improving the understanding of the reservoir behaviour.

## ACKNOWLEDGEMENTS

This paper was developed from many unpublished studies conducted by the staff of BP as part of the development of the Forties Field. The authors are indebted to many colleagues and coworkers, too numerous to name individually, who have contributed directly or indirectly to this paper. Special acknowledgements, however, are due to D.J. Webster, N.A. Wegerhoff and J. Wreford, who have played an intrinsic part in the development of the current reservoir model.

The authors wish to thank the Directors and Managements of BP Exploration Co., Ltd., Shell UK Exploration and Production, and Esso Petroleum Company Limited, for permission to publish this paper.

## REFERENCES

Carman G.J. and Young, R. (1981). Reservoir geology of the Forties oilfield. In: L.V. Illing and G.D. Hobson (eds.), *Petroleum Geology of the Continental Shelf of North-West Europe*, pp. 371–391. Heyden, London.

Deegan, C.E. and Scull, B.J. (1977). A standard lithostratigraphic nomenclature for the Central and Northern North Sea. *Institute of Geological Sciences Report 77/25.* HMSO, London.

Haq, B., Hardenbol, J. and Vail, P.R. (1987). Chronology of fluctuating sea levels since the Triassic. *Science*, **235**, 1156–1167.

Heller, P. and Dickinson, W.R. (1985). Submarine-ramp-facies-model for delta-fed, sand-rich turbidite systems. *Bull. Am. Assoc. Petroleum Geol.* **69**, 960–976.

Hill, P.J. and Wood, G.V. (1980). Geology of the Forties Field, UK Continental Shelf, North Sea. *Bull. Am. Assoc. Petroleum Geol.* **64**, 81–93.

Hillier, G.R.K., Cobb, R.M. and Dimmock, P.A. (1978). Reservoir Development planning for the Forties Field. *Proceedings of the European Offshore Petroleum Conference*, Vol. II, pp. 325–335. Society of Petroleum Engineers, Dallas.

Stewart, I.J. (1987). A revised stratigraphic interpretation of the Early Palaeogene of the Central North Sea. In: J. Brooks and K.W. Glennie (eds.), *Petroleum Geology of North West Europe* Vol. 1, pp. 557–576. Graham and Trotman, London.

Thomas, A.N., Walmsley, P.J. and Jenkins, D.A.L. (1974). Forties Field, North Sea. *Bull. Am. Assoc. Petroleum Geol.*, **58**, 396–405.

# Field Review 2

# Fulmar: The Development of a Large North Sea Field

*A.A. van der Helm, D.I. Gray, M.A. Cook and A.M. Schulte*

Shell UK Expro, UEAC/4, Aberdeen, UK

The Fulmar field, discovered in 1975 by well 30/16–6 and appraised by 30/16–7 in 1977, is located in the Central North Sea, UK sector. It is situated mainly in block 30/16, licensed to Shell/Esso, and extends northwards into block 30/17b, at 01.01.1989 licensed to Amoco/Texas Eastern/Amerada Hess. The field is unitized and Shell operates the field on behalf of the Shell/Esso North Sea Venture and the Fulmar Unit.

The Fulmar field is a pronounced detached anticlinal closure. It lies on a fault-bounded terrace within the western margin of the south-west Central Graben, adjacent to the Auk Horst. The highly permeable reservoir sands are encountered between 9900 ft (3017m) tvss and 11 500 (3505m) ft tvss. They comprise an Upper Jurassic Kimmeridgian sequence of shallow marine sandstones, the Fulmar Formation, and overlying sandstone turbidites contained within the Kimmeridge Clay Formation. The Fulmar sands are composed of regressive cycles interpreted as a complex series of offshore sand ridges, separated by thin transgressive units of poorer reservoir quality. Sandstone turbidites were deposited following a widespread marine transgression, with re-activation of the adjacent Auk Horst providing a local sediment source. Fulmar is of relatively small areal extent, only 11.3 $km^2$ and has an undersaturated oil column in excess of 900 ft (274m). The API gravity of the oil is 40°. The reservoir was initially overpressured by some 1100 psi above hydrostatic.

Development takes place from an eight-leg steel platform containing 36 slots and a six-slot wellhead jacket placed over a template. Development drilling commenced in 1978 from the template to facilitate early production start-up. In 1980 the Fulmar platform was installed, from which 18 wells were drilled between 1982 and 1984. Drilling was resumed in 1987 and is presently ongoing. Production started on 11 February 1982. Oil evacuation takes place via the Fulmar Floating Storage Unit, a converted VLCC, and shuttle tankers. Gas export takes place via the Fulmar Gas Line, commissioned in mid 1986. Until that time, produced gas was re-injected for conservation, thereby creating a secondary gas cap. At 01.01.1989, a total of 32 wells were drilled on the structure, comprising 10 water injectors, one gas injector and 17 oil producers.

Total hydrocarbons initially in place are estimated at 815 mmstb oil and 500 Bscf gas. The main drive is water drive. As there is a limited aquifer, the primary drive is supplemented by injecting water in a flank water drive. The reservoir exhibits a stable displacement of oil by water. Favourable permeabilities and high net-over-gross ratios allow a high sweep efficiency in the better sands. The ultimate recovery of the field is expected to reach 462 mmstb oil and 264 Bscf wet gas. These estimates are fully supported by simulation studies.

At 01.01.1989, 319 mmstb oil, 121 Bscf gas and 20 mmbbl water had been produced. Cumulative water injection amounted to 360 mmbbl, cumulative gas injection stood at 80 Bscf.

## PART ONE: RESERVOIR GEOLOGY

### Geographic location and field dimensions

The Fulmar field is located in the Central North Sea (UK Sector) in water depths of around 270 ft (83m). The field is some 170 miles south-east of Aberdeen and 7 miles north-east of the Auk field (Fig. 1), occurring mostly within block 30/16 (Shell/Esso), but extending northwards into block 30/11b (Amoco Group). The field is unitized and is operated by Shell UK Ltd. on behalf of the Shell/Esso North Sea Venture and the Fulmar Unit.

Fulmar is of relatively small areal extent (11.3 $km^2$) with a hydrocarbon column in excess of 900 ft (274m). Total hydrocarbons initially in place are estimated at 815 mmstb oil and 500 Bscf wet gas. The ultimate recovery of hydrocarbons from Fulmar is currently estimated at 462 million bbl oil and 264 Bscf of associated wet gas.

*North Sea Oil and Gas Reservoirs—II*


### Discovery and subsequent drilling history

The Fulmar field was the first commercial discovery of oil in Jurassic reservoirs in the North Sea Central Graben. It established a new play, which has continued to prove successful in the Clyde, Kittiwake and Guillemot oil fields and in a number of other less well-defined discoveries (Fig. 1). The Fulmar field lies on a fault-bounded terrace within the western margin of the south-west Central Graben adjacent to the Auk Horst, off the northern flank of the Mid North Sea High (Ziegler, 1982).

*Exploration and appraisal*

The discovery resulted from a review of an area that had already passed through an initial phase of exploration for Permian targets. The proposal to drill on the structure was based on a closure mapped at base Cretaceous level, which

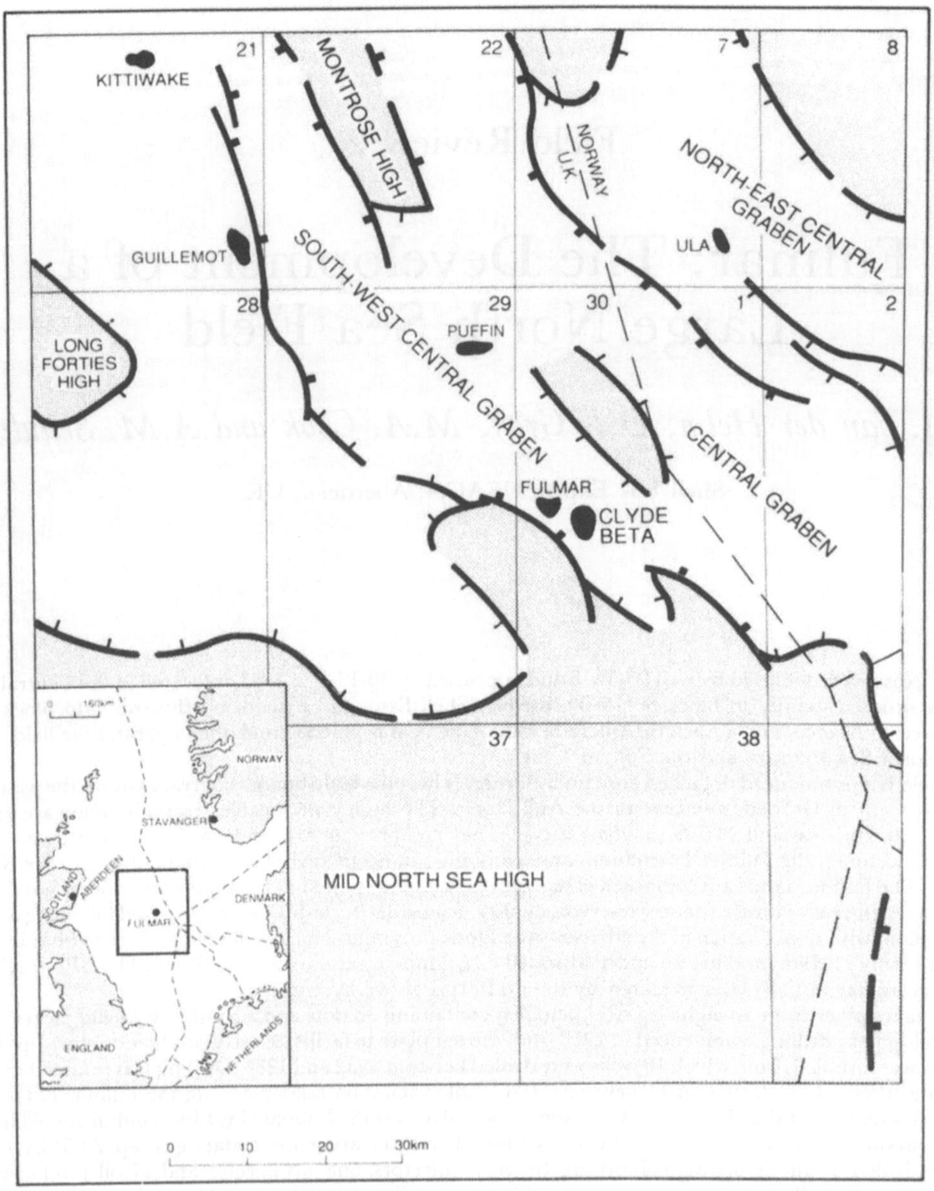


**Fig. 1.** Location map showing regional structural elements and the Jurassic oil fields Fulmar, Clyde, Ula, Puffin, Kittiwake and Guillemot.

was thought to be salt induced, and an acknowledged potential in Upper Jurassic sandstones in the area. By 1974, Upper Jurassic sands had been encountered in wells 30/17a-1 (Gulf/Phillips), 29/20-1 (Shell/Esso), and 30/24-6 (Hamilton Brothers). In the latter, a 23 ft interval flowed 1200 b/d (Pennington, 1975).

The Fulmar discovery well, 30/16–6, was spudded during August 1975 by the semisubmersible Staflo. Some 990 ft of gross sand (607 ft net oil sand) were encountered below 10 162 ft tvss. At this time the sequence was ascribed to the "Auk Sands" and approximatey 500 ft of this was cored. The uppermost 89 ft were thinly bedded with intercalated shales; the remaining 901 ft being fine-grained well-sorted clean sands. A clearly defined oil–water contact (OWC) was observed at 10 830 ft tvss. The well was not tested owing to hole problems.

The first appraisal well, 30/16–7, was drilled in 1977 some 700 ft south-west of the discovery well. The Ribble Sand Member (upper heterolithic unit), here some 139 ft thick, was separated from a water bearing 'Main Reservoir' by 94 ft of Kimmeridge Shale (Johnson *et al.*, 1986). The 'water up to' level at 10 860 ft was 30 ft deeper than the OWC seen in 30/16–6. The well tested the Ribble sands at 8.5 Mb/d of 40° API oil. At this stage STOIIP figures for the field were estimated at 800 to 1000 mmstb. Cores from these two wells caused considerable debate as to the shallow or deep water origin of the predominantly massive Upper Jurassic sand sequence. In hindsight, this debate resulted from the differences between the Fulmar Formation and the Ribble Sand Member.

*Pre-production development*

A first phase of development drilling ensued during 1978–9 with the pre-drilling of four oil producers from a six-slot subsea template. A well head jacket was installed over the template to facilitate tieback of the wells. Data from these wells, several of which were cored, formed the basis of a reservoir geological model subsequently used in development drilling and for unitization. A consistent model was established. The "Main Reservoir" Sands were interpreted as a stacked series of shallow marine deposits, overlain in the west by deeper water mass-flow sands encased within the Kimmeridge Shale (Johnson *et al.*, 1986). At this time, a member status nomenclature was adopted based on the "FULMAR" mnemonic using British Rivers: **F**orth, **U**sk, **L**ydell, **M**ersey, **A**von and **R**ibble.

*Further development*

By 1980, a 36-slot platform was installed on Fulmar, from which 18 wells were drilled between 1982 and 1984. The first round of development drilling was suspended after completing 12 oil producers (five on the template), 10 water injectors and one gas injector, since their combined production potential exceeded the capacity of the surface facilities. In April 1987, as daily oil production started to fall below processing capacity, a second phase of development drilling and workovers was initiated. By the end of 1988, a further five wells had been drilled including the first set targeted at the thinner and more complex reservoirs of the Ribble and Lower Usk. Development continues on this primary and secondary front, with interest focusing mainly on the secondary reservoir targets.

The Fulmar field has recently been reviewed by Mehenni and Roodenburg (in preparation). This paper builds on their contribution. Differences in common material between their contribution and this paper reflect the changes to the field development and interpretation during 1988 and early 1989.

## Seismic and structural interpretation

The Fulmar field is dome-shaped, subtriangular in outline, with flanks dipping at 18—25° to the south-west and at 10° to the south and east (Fig. 2). The northern flank is defined by the Base Cretaceous Unconformity, the 'X' Unconformity, dipping northwards at ca 12°. The lateral and upper seals are formed by the Kimmeridge Clay to the south and west and by the Chalk Group to the north and east.

*Seismic framework*

The original 3D seismic survey in the Fulmar field area was shot and processed in 1977. The areal coverage of the survey (Plate 1) is approximately 62 km$^2$. The seismic was provisionally interpreted in 1978 and re-interpreted in 1981, 1984 and 1986 to tie in the results from development drilling. In 1987, the seismic data were reprocessed and this data set is currently being assimilated into the field modelling. The significant seismic reflectors that can be mapped within the Fulmar field are: Top Chalk, Base Cretaceous Unconformity, Top Ribble, Top Main Sand and Triassic (Plate 1). These reflectors have been identified from the available well velocity survey data via synthetic seismograms produced for all the wells. The Top Ribble and Top Main Sand are characteristic and continuous seismic horizons. Towards the eastern part of the field they are obscured by the overlying strong reflector of the Base Cretaceous Unconformity. The top Triassic reflector is ill-defined in the west and south-western area, where well-ties are used for mapping.

*Structural history*

The field has been shaped by three major tectonic events: (1) the continued and large vertical displacement along the Auk Horst Boundary Fault and development of related synthetic and antithetic faults along the western flank; (2) Zechstein salt halokinesis; and (3) strike-slip faulting along Caledonide trends (ENE).

*Pre-late Cimmerian* Rifting of the North Sea and formation of the Central Graben commenced in the early Permian and continued into the Cretaceous. The Auk Horst

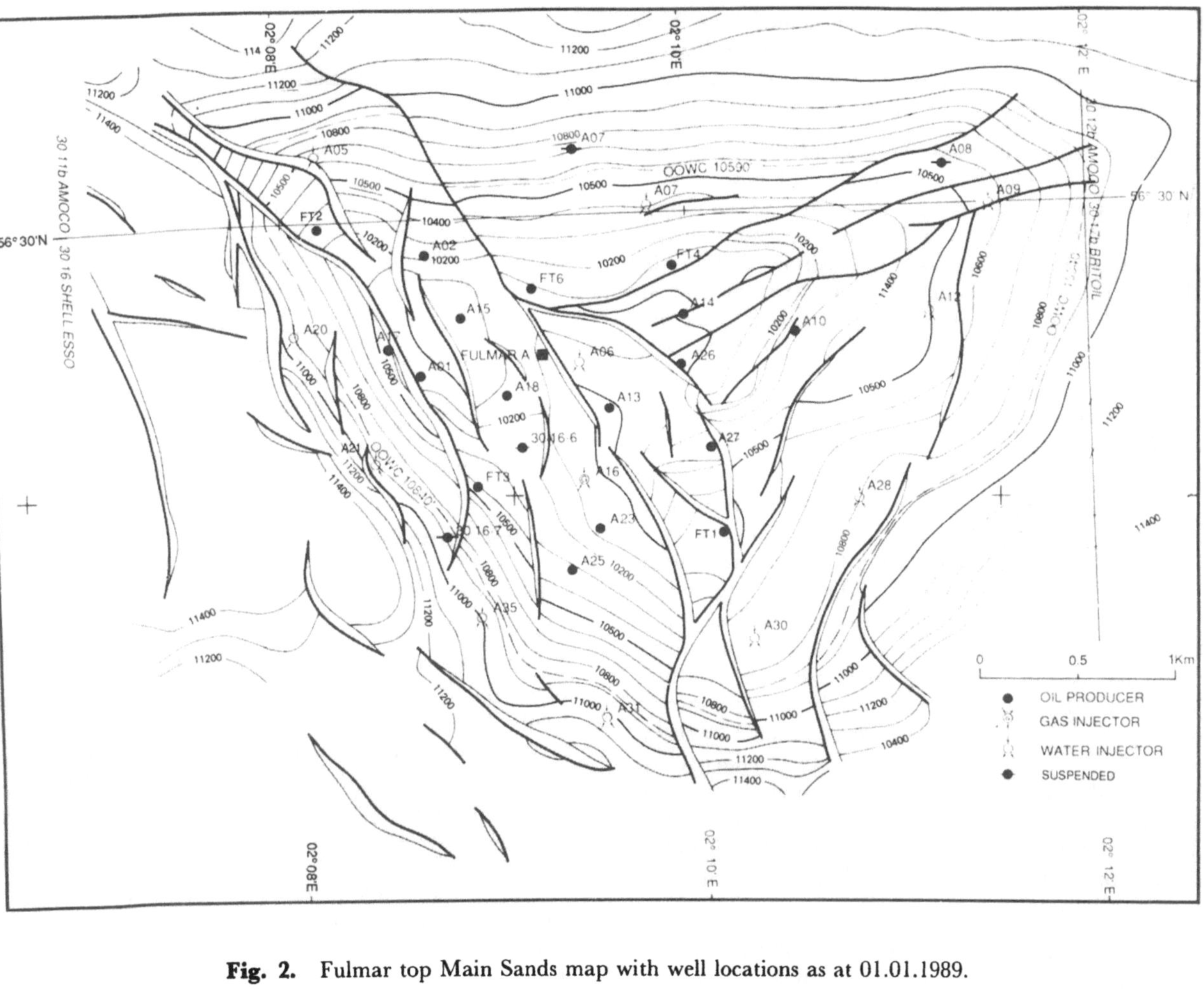


**Fig. 2.** Fulmar top Main Sands map with well locations as at 01.01.1989.

Boundary fault, lying between the Auk and Fulmar fields, forms the western boundary to the Central Graben. The large vertical displacement (ca. 3000 ft) has induced a strong trend of NW–SE synthetic and antithetic faults, with throws ranging up to 300 ft. The antithetic faults have a compressional component, which has been more clearly highlighted with the introduction of seismic work stations to aid 3D interpretation. The synthetic faults probably initiated the salt movement and provided at an early stage a division into a north-eastern "platform" and south-western "depocentre". Coupled with continued salt removal and activation of the Auk Horst Boundary Fault, there was a gradual westward migration of the depocentre during "Main Sand" deposition, enhancing the "hinge-line" effects.

*Late Cimmerian* Following the Kimmeridge transgression, either strong movement along the Auk Horst Boundary Fault or a change in base level allowed sand transport off the Auk Horst and introduction of Ribble sands into the developing Kimmeridge basin in the south-west of Fulmar. This phase was complemented by sinistral strike-slip movement developed along an ENE trend. This movement is interpreted to be linked to the "transfer faults" (Gibbs, 1984) north of Auk, defining the northern boundary to the Auk Horst (Fig. 1). This fault suite is considered responsible for sealing and separating a shallower OWC in the north of the field. After Ribble deposition further rotation occurred, tilting the Fulmar block to the south and providing relief for erosion in the late Cimmerian. No significant displacement is evident on the seismic map at Base Cretaceous level, implying that strike-slip movement was active before Cimmerian erosion.

## Stratigraphic setting of the Fulmar Field

The Fulmar area is located adjacent to the northern flank of the Mid North Sea High, towards the southern margin of the northern Permian Basin. The general lithostratigraphy is shown in Fig. 3.

*Permo-Triassic* Lower Permian Rotliegendes aeolian sands are known to be developed in the adjacent Clyde cluster. Deposition of a thick evaporite sequence during the Upper Permian was followed by non-marine Triassic. Fluvio-lacustrine conditions led to salt dissolution and to the initiation of sedimentary depocentres in rim synclines in the Fulmar/Clyde area. These gradually grounded, became topographically expressed and provided further material for re-working.

*Jurassic* Widespread erosion of the Jurassic and Triassic during the Early Cimmerian was followed by an extensive late Oxfordian marine transgression. This resulted in the deposition of a storm dominated shallow marine sand complex: the Fulmar Formation. The Upper Jurassic sands of the Fulmar Formation are inferred to have been derived from the nearby continental Triassic sequence of the Auk Horst. Strong differential subsidence, caused in part by the salt removal, led to the accumulation of a thick sand sequence. Rapid basin deepening followed, resulting in anoxic conditions and deposition of the organic-rich Kimmeridge Clay Formation. Activation of the Auk Horst Boundary Fault and the development of a palaeorelief, shortly after the initial phase of deepening, led to the deposition of a turbiditic sand wedge (the Ribble Sand Member) within the Kimmeridge Clay Formation. Discontinuities are present in the claystone dominated sequence overlying the Ribble.

*Cretaceous and post-Cretaceous* Regionally developed over the Central and Northern North Sea, the Base Cretaceous Unconformity marks a significant fall in sea level. In the Fulmar area this truncates the Jurassic sequence, cutting northwards into the Triassic. Differential compaction, clay draping and salt removal, coupled with strong antithetic and synthetic faulting related to the Auk Horst Boundary Fault, accentuated the inversion of the Fulmar structure. Dips at top Main Sand level on the west flank were in excess of 20° prior to Chalk deposition. Chalk sedimentation began in the Cenomanian and continued into the Danian and was influenced by regional subsidence patterns. This depositional pattern continued into the Tertiary.

## Lithostratigraphy of the Fulmar reservoir sequence

### Reservoir unit nomenclature

A dual nomenclature is maintained for Fulmar linking Member status with reservoir units (Fig. 4). This followed a review of reservoir geology initiated in 1987 to refine the existing reservoir model and to provide the geological framework for the construction of a full field 3D numerical reservoir-simulation model. The study was based on all available core data (ca. 6500 ft), well log and production data. Fourteen reservoir units were distinguished based on lithofacies and reservoir quality (see tabulation below) and can be identified from a combination of density/neutron, gamma-ray/sonic and induction logs (type log, Fig. 5). They also have well-defined porosity–permeability relationships (Fig. 6a). A palynological study provided valuable support in establishing a stratigraphic framework for correlation. Further field mapping has allowed creation of a compatible suite of unit maps. Figure 7a exemplifies these in cross section.

The summarized reservoir division is tabulated below.

| | | *Reservoir Quality* |
|---|---|---|
| **Kimmeridge Formation** | | |
| Ribble Member | Unit 1 | Good |
| Avon Member | | Non-reservoir |
| **Fulmar Formation** | | |
| Clyde Member | Unit 2 | Poor |
| Mersey Member | Unit 3.1 | Good |
| | Unit 3.2 | Good |
| | Unit 3.3 | Variable |
| | Unir 3.4 | Good |
| | Unit 3.5 | Poor to non-reservoir |
| Lydell Member | Unit 4 | Good |
| Usk Member U | Unit 5.1 | Moderate to poor |
| M | Unit 5.2 | Variable to poor |
| L | Unit 5.3 | Good |
| Forth Member | Unit 6.1 | Non-reservoir |
| | Unit 6.2 | Non-reservoir |
| **Triassic Smith Bank Formation** | | |

### General variations in lithofacies

The areal variations in Fulmar sand quality, facies and thickness are related primarily to a NE–SW trending hinge line (Fig. 7b), effectively dividing the field into two equal areas. The platform area to the north-east is a readily correlateable sand sequence, whereas the depocentre to the south-west contains a thick succession of poorly differentiated massive sandstones. Three transgressive/regressive sequences have been recognized within the "Main Reservoir Sands". The Ribble Sands (Unit 1) were deposited from mass-flows, which onlap the structure from the south-west, thinning to zero over the crest. The cross-section in Fig. 7a shows the geographic relationships of the facies.

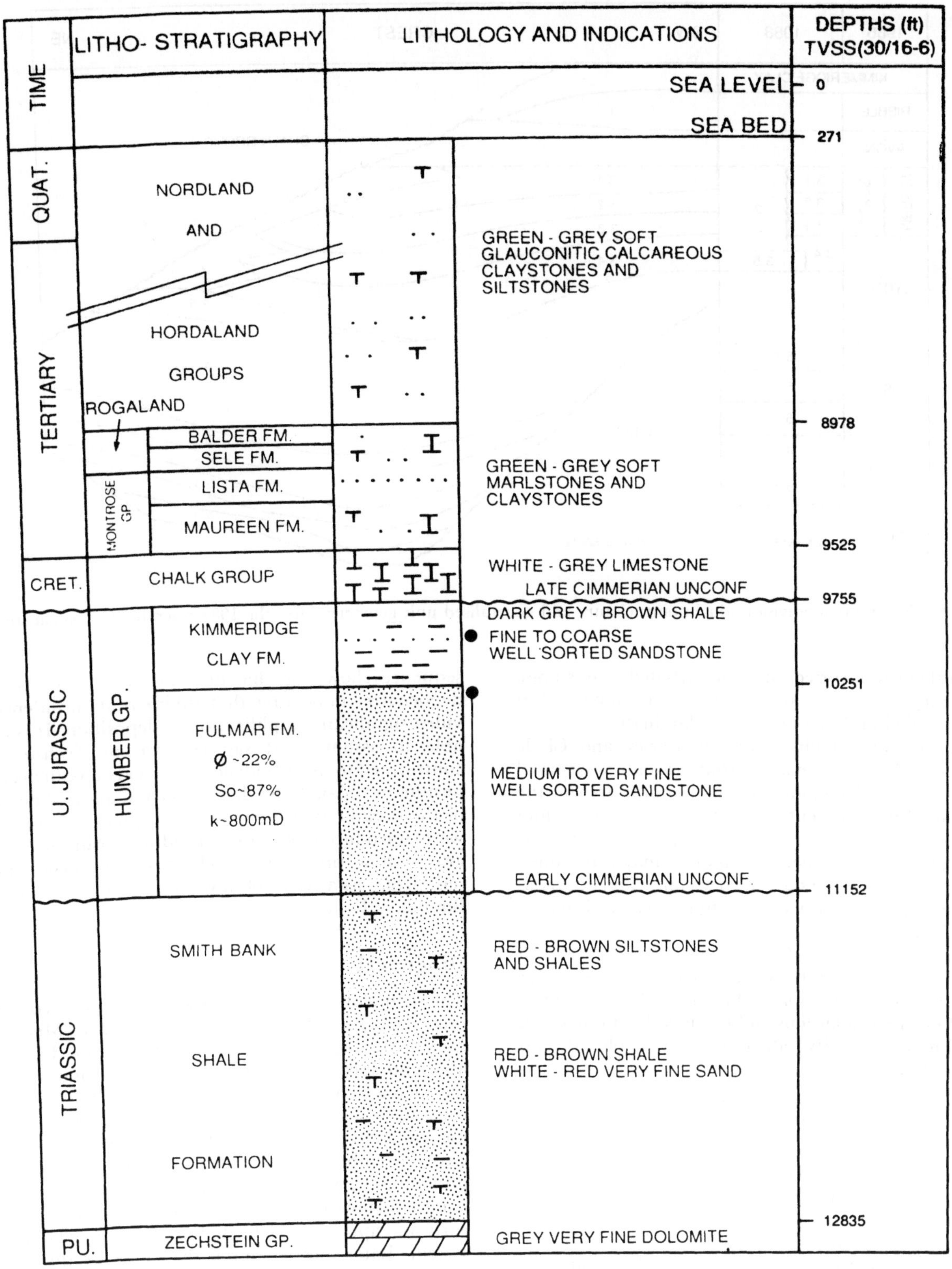


**Fig. 3.** General lithostratigraphy of the Fulmar area: the prospective sequence is contained in the Upper Jurassic Humber Group.

## *Description of the reservoir sequence*

*Main Sands* The Forth (Unit 6) forms a transgressive sequence of fine-grained argillaceous and glauconitic sandstones onlapping the Triassic to the south-west. They lie mostly below the original OWC and carry no reserves. Unit 5.3 (Lower Usk) is a clean well-sorted fine-grained sand, interpreted as a regressive phase of shallow marine sand ridge development. It can be correlated throughout the north-eastern area and is a significant secondary reservoir target in terms of STOIIP. Small faults, however, have an impact on reservoir continuity and its production potential. The top of the Unit is represented by a sharp break into transgressive phase deposition of the Middle Usk (Unit 5.2).

The second cycle commences with a Middle Usk (Unit 5.2) sequence of interbedded sands and shales that form a localized restriction to fluid flow. The major regressive element of this cycle is represented by Units 5.1 and 4 (Upper Usk and Lydell). The latter comprises an "amplified sequence", greater than 350 ft thick in the depocentre, of fine- to medium-grained, generally massive

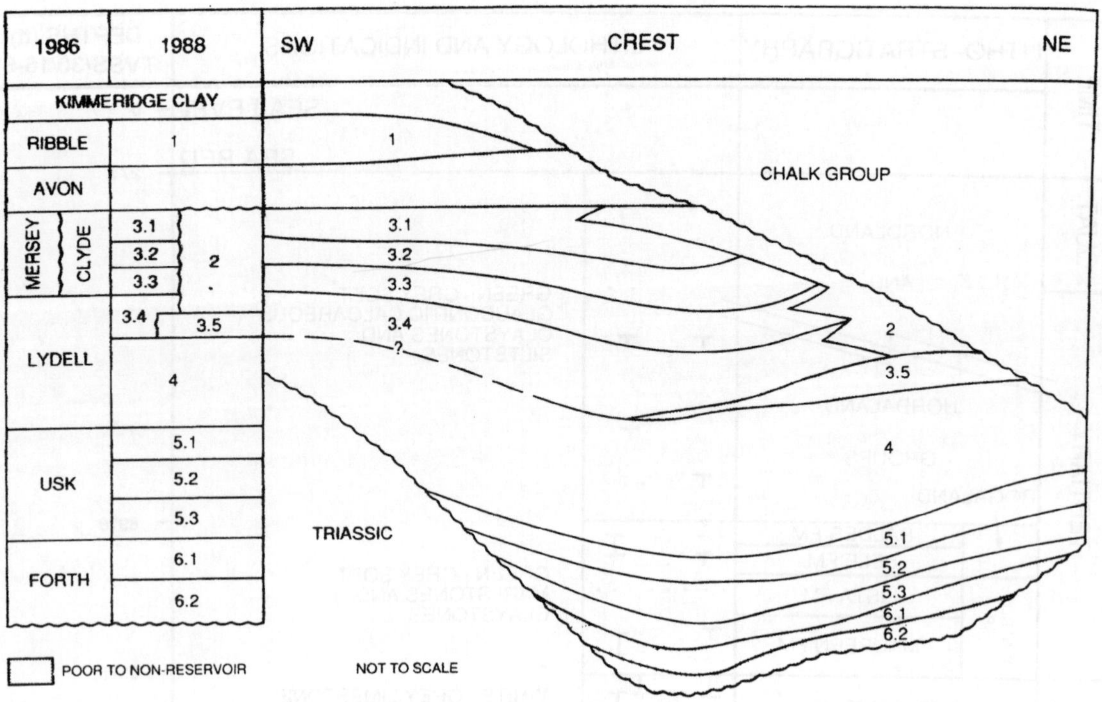


**Fig. 4.** Reservoir subdivision, dual nomenclature and generalised unit geometry after the 1987 reservoir geological study.

and mottled, well-sorted and bioturbated sandstones. Sedimentary structures are relatively uncommon, but include parallel and low-angle cross-lamination.

The third cycle (Units 3 and 2; Mersey and Clyde Members) includes a thin and variable argillaceous basal sandstone Unit 3.5 confined to the platform area in the north-east. Where present, it provides a barrier to fluid flow. The upper remaining Mersey sequence (Units 3.4, 3.3, 3.2 and 3.1) comprises once more a major amplified regressive sequence, with stacked coarsening (and occasionally fining) upwards sequences, at times clearly visible on porosity logs.

To the north and east, Clyde Member sandstones and siltstones interfinger with Mersey facies. They are finer grained and more poorly sorted than their Mersey counterparts. They are argillaceous and extensively bioturbated, with a history of porosity reduction through dolomite and silica cementation.

*Avon and Ribble sequence* A major transgression caused an abrupt end to Mersey sedimentation. The Avon shale is the lowest member of the Kimmeridge Clay Formation, separating the Main Sand from the Ribble sands (Unit 1). It is a finely laminated non-bioturbated shale, thickening south-west towards the Auk Horst.

Unit 1 comprises fine- to medium-grained sandstones, quite without structure, interbedded with thin laminated shales and siltstones. The sands are variably sorted and contain abundant clay clasts. They are interpreted as a sequence of mass-flow sands built out from the Auk Horst and thinning towards the north-east. Much of the current field development is targeted at this heterolithic reservoir.

## Palaeo-environment modelling

The Fulmar Formation is interpreted as a major regressive sequence developed in a moderate energy offshore to near-shore shallow marine environment. No shoreline has been identified. Coarsening-upward sequences represent progradation of shallow marine ridges or sand sheets, formed by periodic, storm-induced currents. Individual sequences show large thickness variations from platform to depocentre, indicating that differential subsidence provided significant amplification to deposition. It is reasonable to assume that salt removal at Zechstein level contributed to the subsidence. A period of basin deepening terminated Main Sand deposition and allowed deposition of the organic-rich Kimmeridge sequence. Unit 1 sands were introduced into the subsiding basin by sediment gravity-flow processes, which may have been initiated seismically as a result of movement along the Auk Horst Boundary Fault.

## Diagenesis

The sediments are generally arkosic. Uranium concentrations occur, notably within Units 3 and 4 (Mersey and Lydell Members), providing in places a poorly correlatable and spiky gamma-ray response. Dolomite provides the prime diagenetic control to permeability. Dolomite cementation is extensive in some areas, significantly reducing effective permeability. Units 1 and 4 (Ribble and Lydell Members) contain siliceous sponge spicules, locally providing significant secondary mouldic porosity and a source of mobile silica. Most diagenetic changes occurred early and appear to be due to burial in a closed diagenetic system. Later, deeper burial diagenesis produced grain dissolution and minor illitization of feldspars. Unit 2 (Clyde Member) is particularly prone to diagenetic deterioration of reservoir properties.

## Hydrocarbon setting

### *Composition and origin of hydrocarbons*

The Kimmeridge Clay Formation is the dominant source rock for Central Graben generated hydrocarbons. It is a mature, organic-rich shale. The Triassic has no source potential. No Silesian sources are present in this area.

The Kimmeridge has an average total organic carbon content of 5%, type II kerogen, vitrinite reflectance 0.75 and an average yield of 0.032 v/v oil.

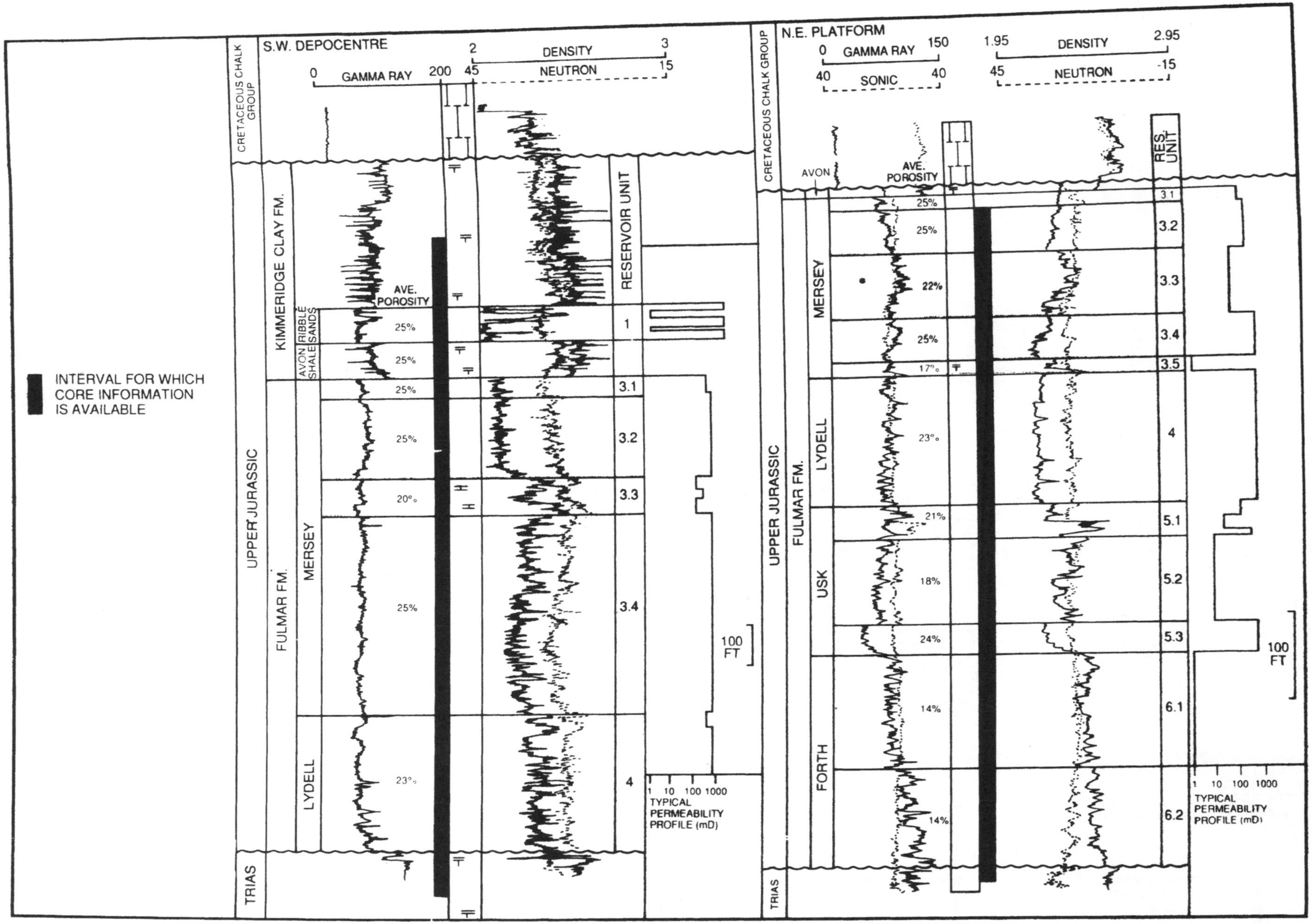


**Fig. 5.** Type logs in Fulmar showing the difference in log character between SW depocentre and NE platform wells. Note the sequence amplification in the depocentre and the loss of character wit iin the well logs over tl:e Mersey/Lydell Units 3 and 4 intervals between the platform and the depocentre.

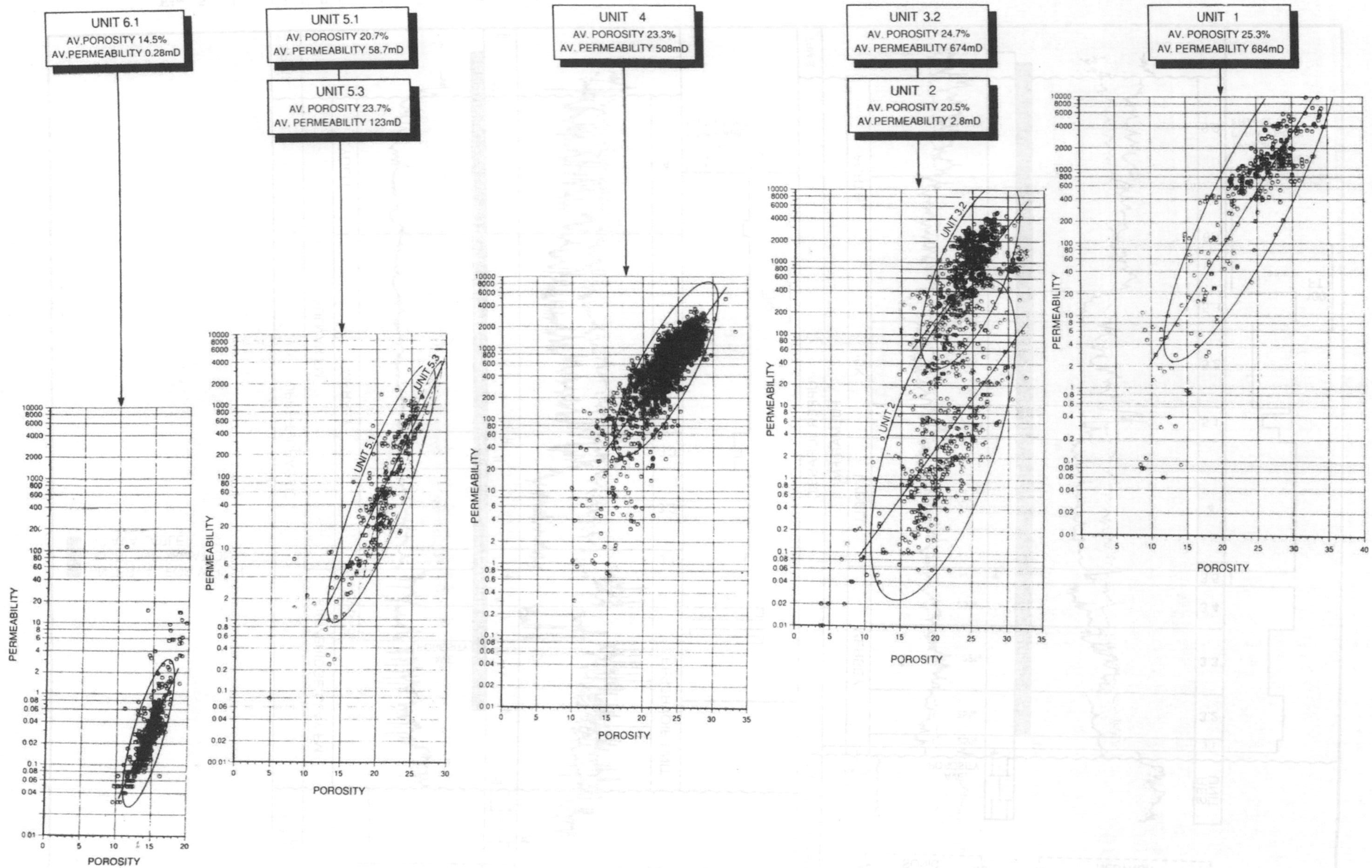


**Fig. 6a.** Representative porosity permeability plots for the Fulmar Units. The ellipses represent the 90% confidence level. The scatter in the data points is significant in most good quality sands and generally less in the poorer reservoir units.

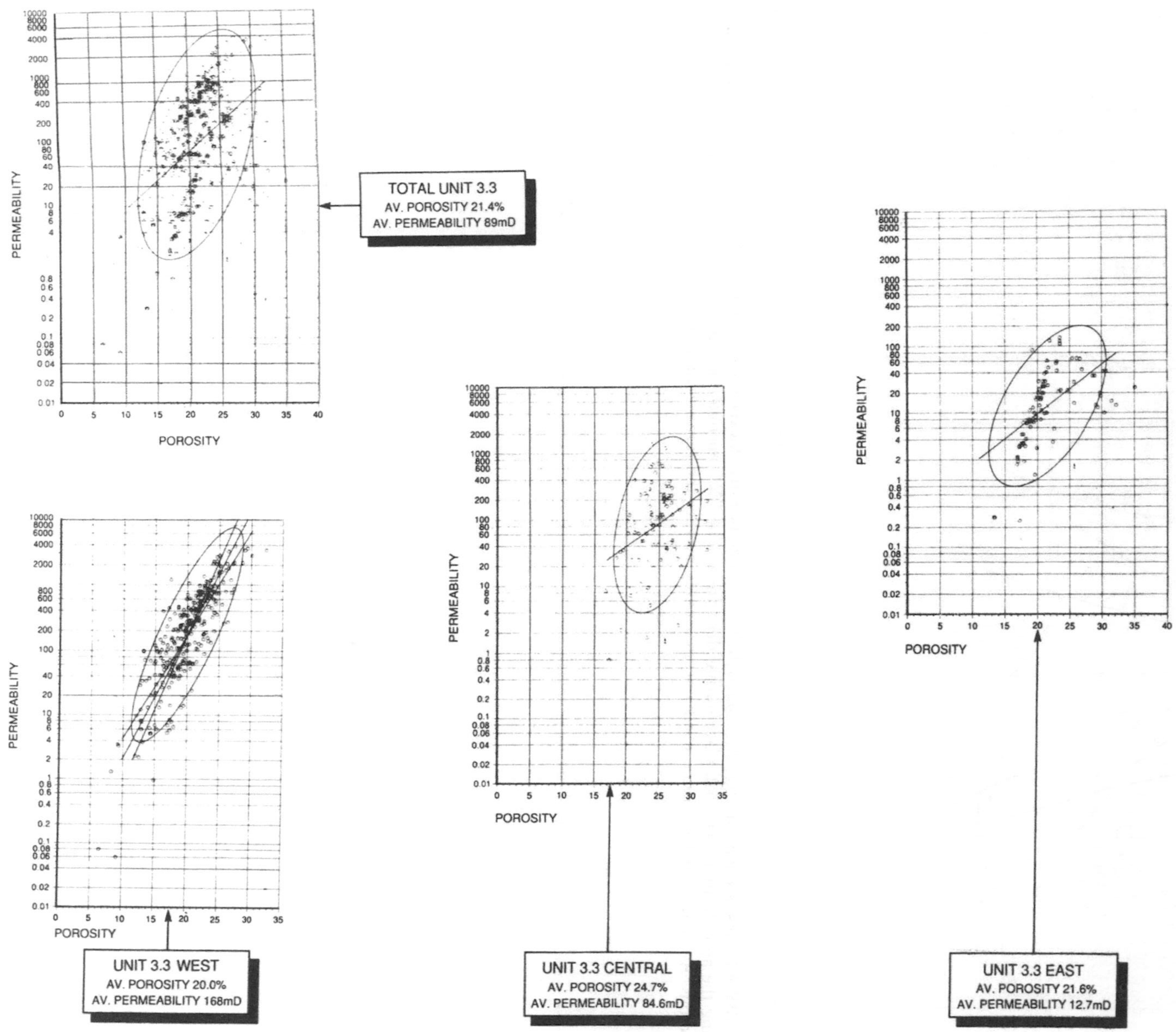


**Fig. 6b.** Geographic variation in rock properties of Unit 3.3. In this Unit the recognition of areal variation led to a significantly improved description of the reservoir properties and improved the history match of the reservoir performance.

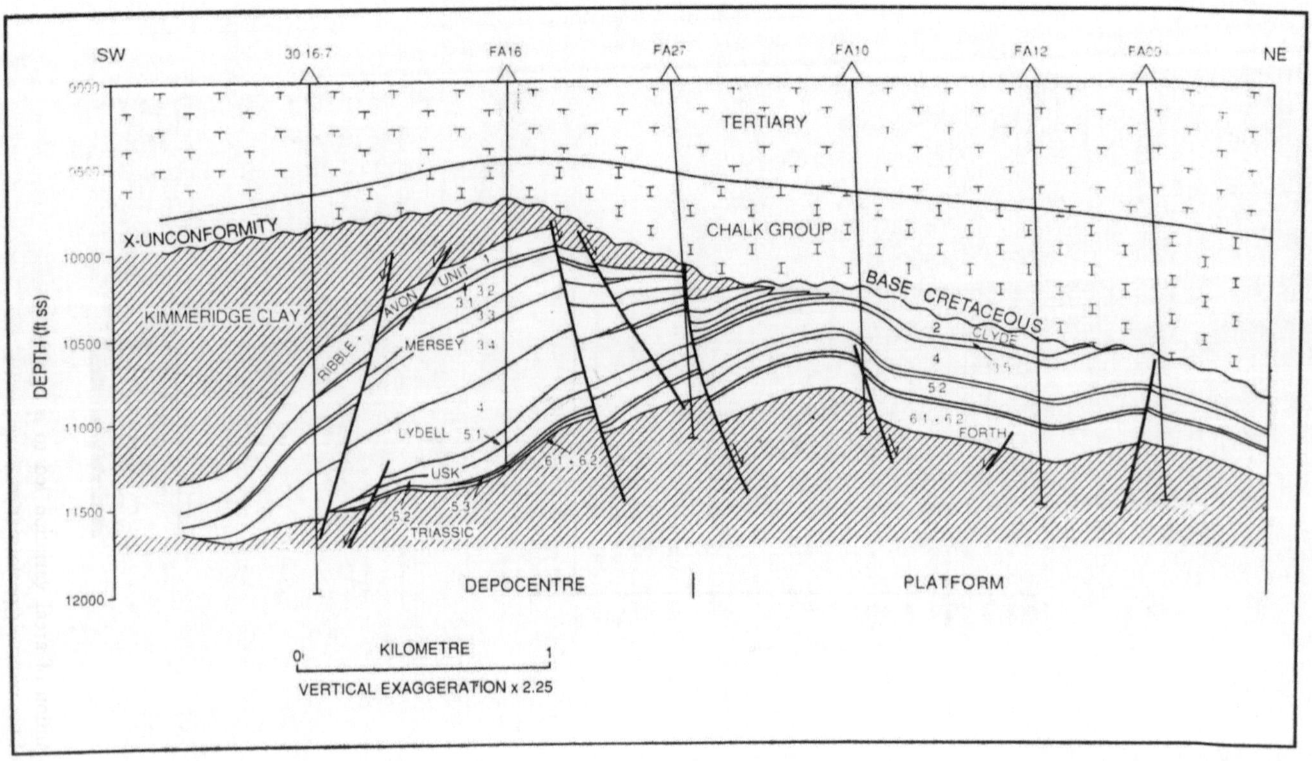


**Fig. 7a.** SW/NE cross-section through the Fulmar field. The significantly thicker reservoir section SW of FA-27 is clearly shown, emphasizing the amplification that took place during deposition. The "X" unconformity is an erosion surface marking the base of the Cretaceous. As shown, the Cretaceous chalk unconformably overlies the Jurassic sands and clays.

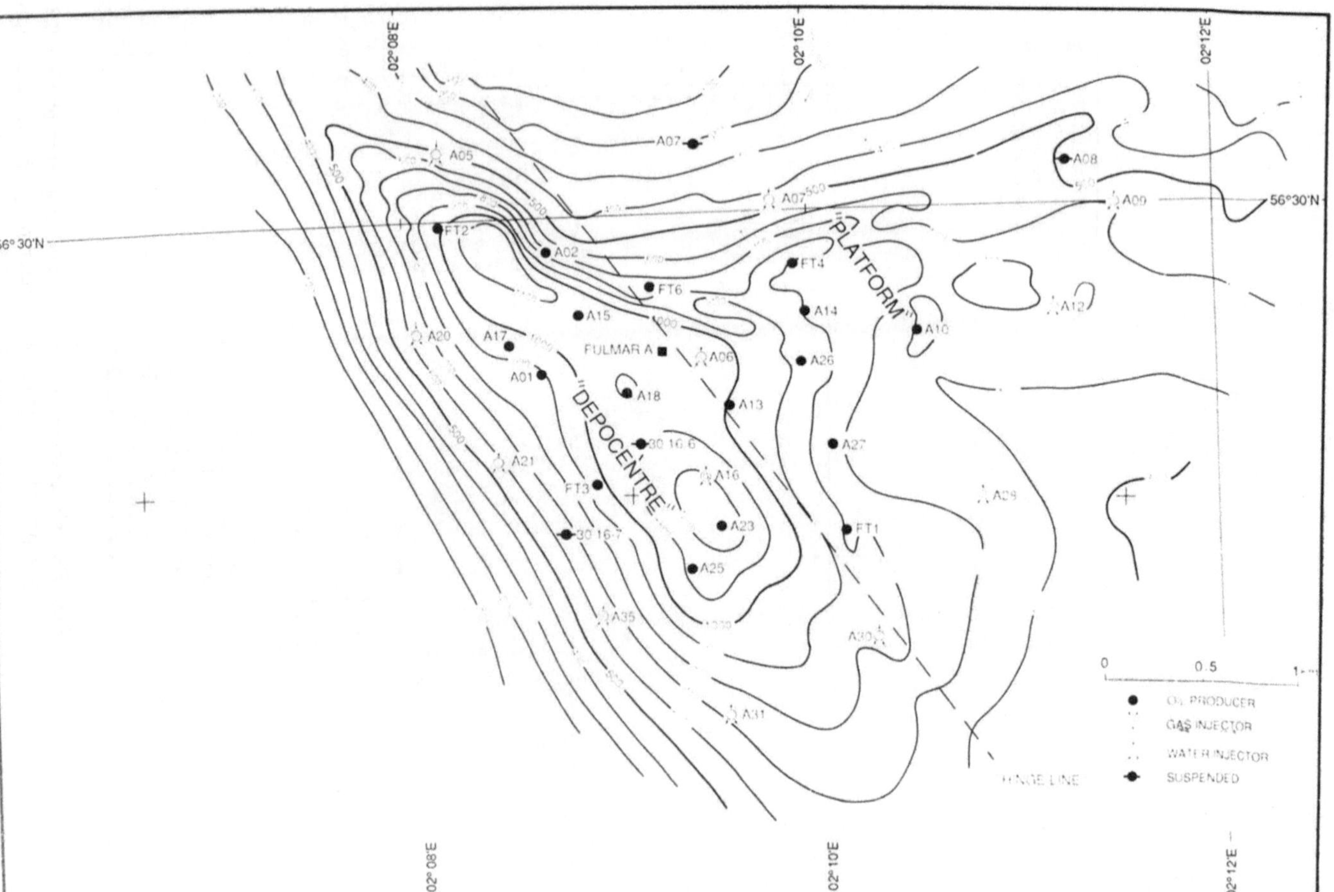


**Fig. 7b.** Gross isochore map of the Main Sands showing the depocentre, platform and hinge line.

*Migration history of hydrocarbons*

Fulmar is one of a group of structures that show progressive westwards and upwards stepping in spill point from Clyde through Fulmar to Auk (Fig. 8). They are believed to have been filled to spill point via regional migration from the east and north-east during the early Cenozoic. Entrapment of the hydrocarbons took place in early to mid Tertiary, postdating the deposition of the Chalk Group. The cap rock is formed by the Kimmeridge shales in the south and west and by the Chalk in the north and east.

The Fulmar field has a limited aquifer. The rapid burial, together with the good basal seal formed by the Triassic Smith Bank Formation, led to the retention of abnormally high pore pressures. As a consequence, the field was originally overpressured by some 1100 psi over hydro-static.

## PART TWO: RESERVOIR ENGINEERING

### Rock properties

Rock properties are determined from some 6500 ft of core from 17 wells and from wireline log data acquired from wells drilled on the structure to the end of 1987. Permeabilities in the good-quality sand units (1, 3.1–3.4 and 4) are typically in the order of 1 D, with the exception of the poorer-quality Unit 3.3 in the east, while the lesser Units 2, 5.1 and 5.2 exhibit typical permeabilities of 50 and 100 mD respectively. As discussed, Units 6.1 and 6.2 are considered to be non-reservoir.

Porosity-permeability relationships have been generated for all units using core data. The revised subdivision of the Fulmar sands proposed in 1987 reduced the scatter on the porosity–permeability cross-plots, but the correlation is still poor in the good-quality units. A typical cross-plot for Unit 4 is shown in Fig. 6a. In the poorer-quality units, the correlation is generally better. In Unit 3.3 a major reduction of the scatter in the data was achieved by dividing the reservoir into three discrete areas—west, central and east—and deriving separate porosity–permeability relationships for each area (Fig. 6b). The scatter in the full data set is believed to originate from poorer sorting of the sands towards the east, resulting in lower permeability values for the same porosities.

On an individual core plug basis, the measured vertical and horizontal permeabilities are equal, owing to the heavily bioturbated nature of the Fulmar sands. However, within the reservoir units, heavily cemented bands reduce the overall $K_v/K_h$ ratio. The magnitude of this reduction is the subject of current considerations in history matching the full-field model.

Air/mercury capillary pressure measurements performed on 174 samples have been averaged using the Heseldin technique (Heseldin, 1974). They are in reasonable agreement with averaged log-derived saturations. An example of averaged capillary pressure curves for Unit 4 is shown in Fig. 9. However, because of the small transition zone in the better-quality reservoirs and the accuracy of saturation measurements from logs at irreducible water saturation, log-derived saturation curves are considered to be more representative. A function relating saturation to porosity and height above the oil–water contact is derived from the log data using a non-linear fitting routine, an example of which is shown in Fig. 10. Log data are used for the determination of STOIIP where quantity and quality are sufficient. Where log data are insufficient, as for Units 2 and 5.3, the averaged capillary pressure curves are used. The major feature of the saturation profiles is that the transition zone in the Fulmar Field is usually less than 40 ft. Irreducible water saturation above the transition zone is a function of porosity, but for the better sands lies in the range 8–15%. Residual oil saturation has been measured by a variety of methods including core flooding experiments, sponge core analysis using Dean–Stark extraction, centrifuge experiments and TDT logging. These methods show a variation in the value of residual oil saturation, with average values in the range of 22–30%. Reservoir model studies assume a value of 25%.

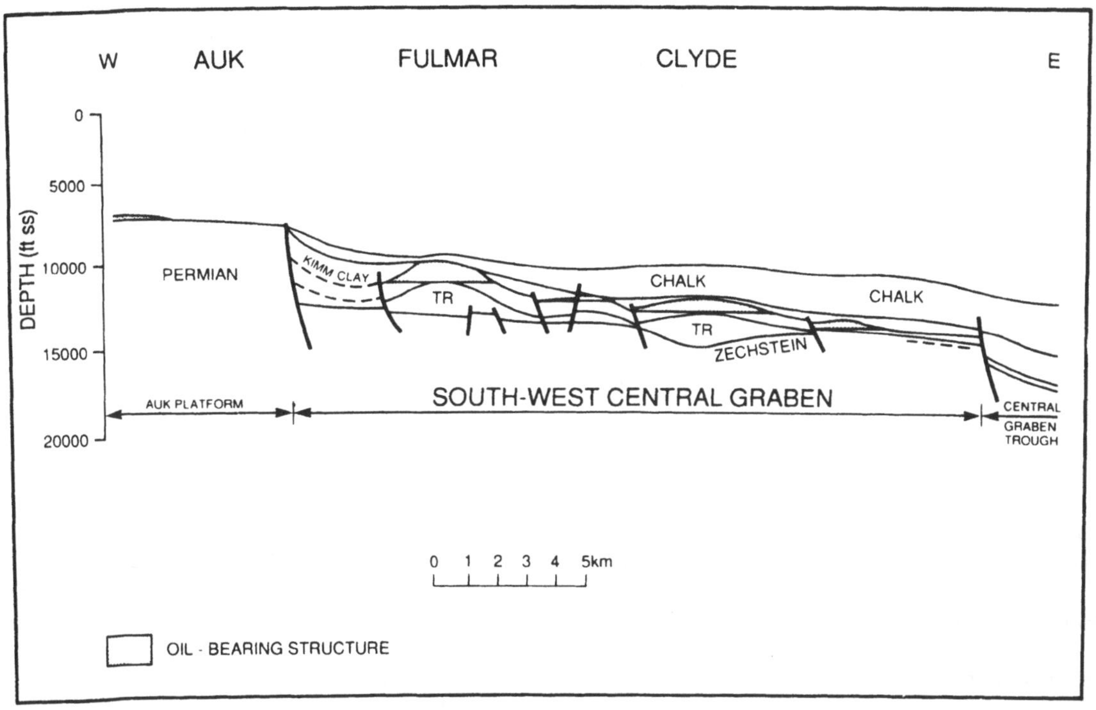


**Fig. 8.** Regional cross-section W–E through Auk, Fulmar and Clyde terraces illustrating the geometry of hydrocarbon accumulations and possible migration path from the Central Graben.

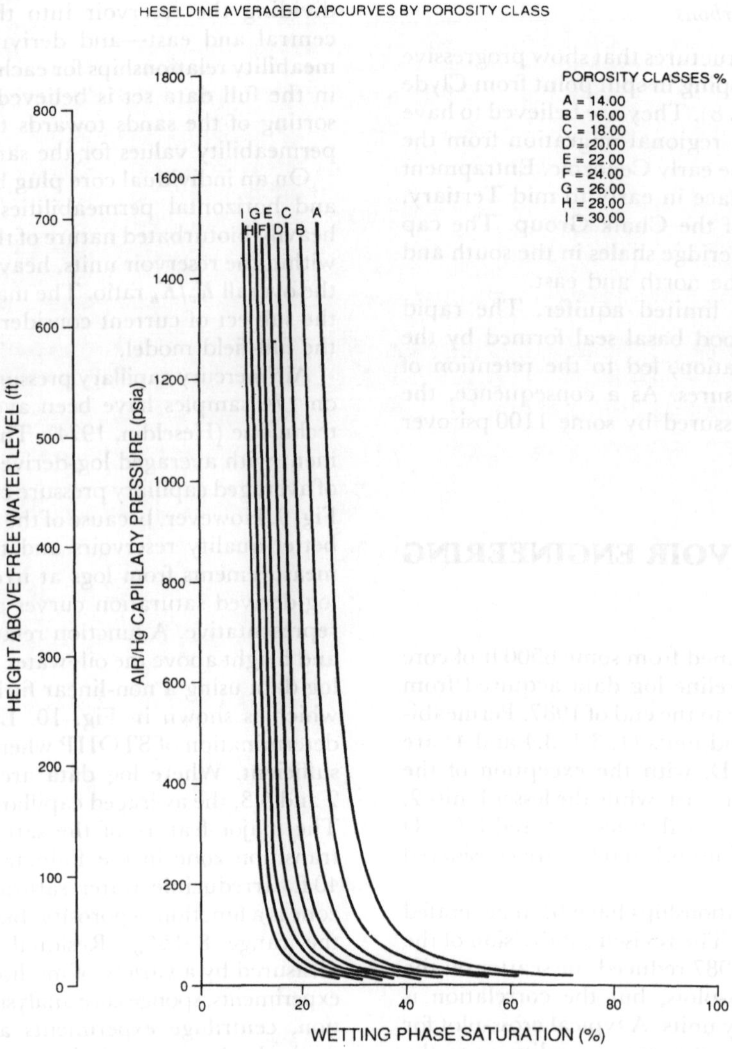


**Fig. 9.** Edited air/Hg capillary pressure curves for Unit 4. Note that the curves do not approach an asymptotic value at high elevation. For this reason air/Hg capillary pressure curves were generally rejected in favour of log-derived curves.

Using a varity of techniques, relative permeability measurements were performed on samples from 13 cored wells. Correlations for the relative permeability end points have been deduced. The observed end points result in a favourable mobility ratio for the oil–water displacement process of 0.5.

## Fluid description

Fulmar oil is highly undersaturated with a bubble point around 1800 psia, substantially below the initial reservoir pressure of 5700 psia at datum (10 500 ft tvss). The reservoir temperature is 285°F (140°C). The viscosity of the oil is around 0.44 cP at reservoir conditions. The average GOR derived from PVT samples is 614 scf/stb and the formation volume factor is 1.43 v/v. The reservoir fluid contains 0.7 mol% carbon dioxide and ca. 2 ppm hydrogen sulphide.

Miscibility experiments demonstrate that the separator gas is fully miscible with the reservoir oil at pressures above 4000 psia.

The formation water salinity is high at 150 g/l NaCl eq.

## Fluid distribution

The total field STOIIP is 815 mmstb, of which some 65% lies in Units 3.1–3.4 and 4. Unit 1, which for development purposes is considered to be a separate reservoir, contains some 74 mmstb, 9% of the total. Table I details the STOIIP contained in each of the reservoir units.

Original oil–water contacts vary across the field. These contacts average to yield 10 875 ft tvss in Unit 1, 10 840 ft tvss in the Main Sands (Units 3, 4 and 5) over the majority of the field and a "perched" contact at ca. 10 590 ft tvss in the Main Sands of the northern fault block.

**Table I.** Tabulation of STOIIP per reservoir in Fulmar

| | *STOIIP* | |
|---|---|---|
| *Unit* | *(mmstb)* | $10^6 m^3$ |
| Unit 1 | 74.0 | 11.77 |
| Unit 2 | 17.0 | 2.70 |
| Unit 3.1 | 23.9 | 3.80 |
| Unit 3.2 | 128.1 | 20.3 |
| Unit 3.3 | 65.1 | 10.35 |
| Unit 3.4 | 147.4 | 23.44 |
| Unit 3.5 | 8.1 | 1.29 |
| Unit 4 | 259.9 | 41.32 |
| Unit 5.1 | 42.0 | 6.68 |
| Unit 5.2 | 36.2 | 5.76 |
| Unit 5.3 | 13.3 | 2.11 |
| Total | 815.0 | 129.59 |

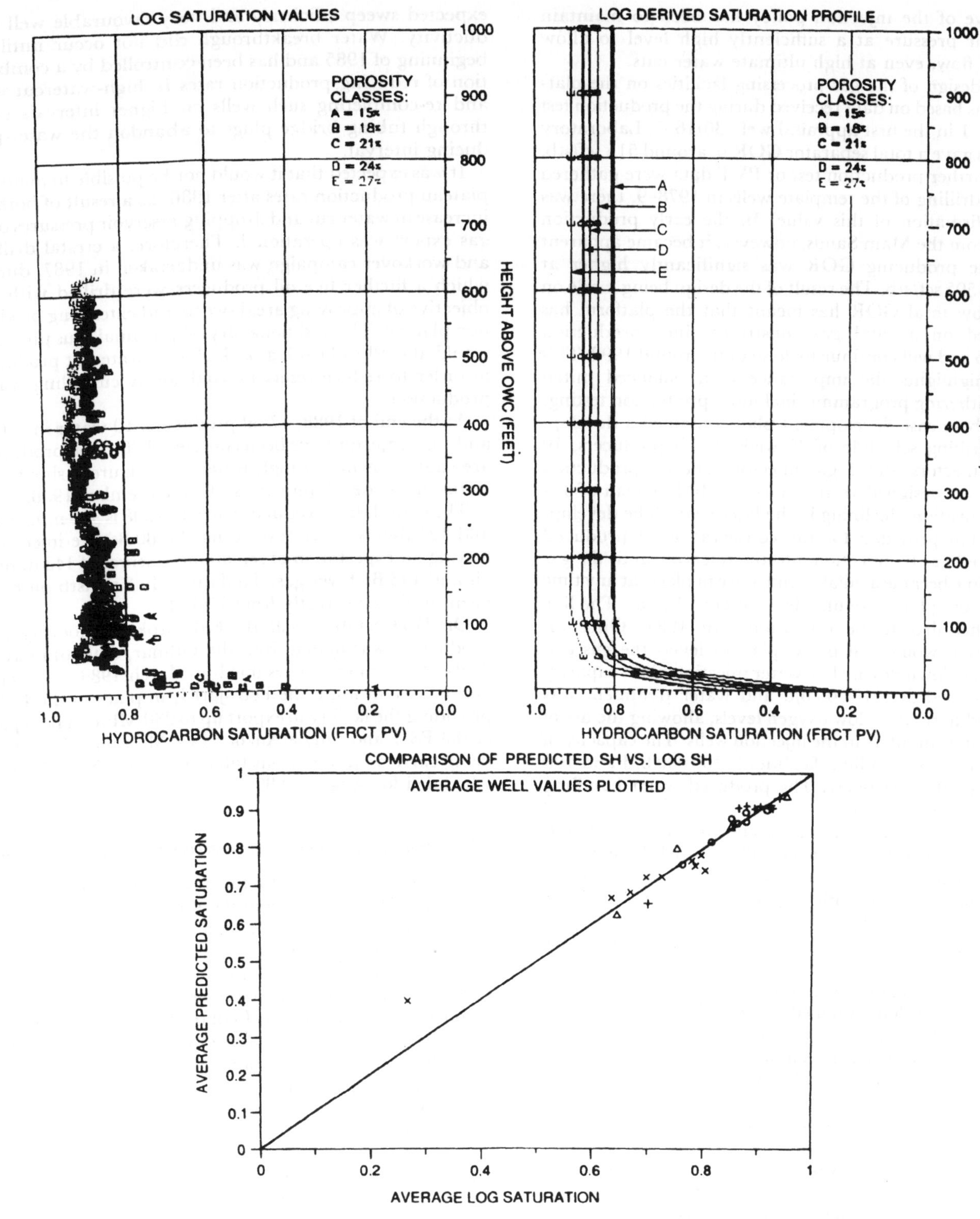


**Fig. 10.** Saturation versus height relationships derived from curve-fitted log data. In the good quality reservoirs these data were used for volumetric estimates and for reservoir model initialization. Noteworthy is the excellent agreement between calculated and observed saturation data in most sands.

## Reservoir development concepts and development scheme

For development purposes the Fulmar sands are subdivided into three intervals: Unit 1, the Main Sands (Units 3.1–5.2) and Unit 5.3, as the shape of the Fulmar structure does not allow maximizing ultimate recovery by commingling production. Primary development, which started in 1978, focused on the Main Sands. The development of Unit 1 commenced in 1988. A pilot development of Unit 5.3 was carried out to establish reservoir continuity. Further development of this Unit is subject to study and is planned in the 1989/90 drilling campaign.

From regional geological considerations, little aquifer support was anticipated. With the sharply dipping flanks and favourable mobility ratio, the field was considered suitable for development by flank water injection and crestal oil production from natural lift. In view of the anticipated high productivity index of the producers, no provision for artificial lift has been made. Instead, the objective of the injection policy has been to maintain reservoir pressure at a sufficiently high level to allow natural flow, even at high ultimate water cuts.

The design of the gas processing facilities on the platform was based on data received during the production test on Unit 1 in the first appraisal well, 30/16–7. Laboratory analysis gave a total separator GOR of around 515 scf/stb. As no further production test or PVT data were gathered during drilling of the template wells in 1978–9, there was no confirmation of this value. In the early production phase from the Main Sands, however, it became apparent that the producing GOR was significantly higher at around 595 scf/stb. The result of the design being made on a too low total GOR has meant that the platform has operated on a total gas constraint since production start-up and will continue to do so until around 1992. This point highlights the importance of a balanced initial data-gathering programme, including production testing.

The base case development plan contained a development drilling schedule of 35 wells: 24 oil producers, 10 water injectors and 1 gas injector. The oil production profile was designed with a 160 000 mb/d plateau of 3–4 years' duration, declining in the later years. The development plan provided for the conservation of produced associated gas by injection into the reservoir until the gas export line became available and, once in place, at any time the onshore plants are unable to receive the gas. The line was scheduled to be operational in 1987. Once oil production would fall below plateau levels the gas cap would be blown down. The water injection plant capacity was designed for 210 000 b/d and water treatment was expected to result in low oxygen levels, allowing the use of carbon steel tubulars in the injection wells. The capacity of the oil and gas handling facilities is 180 000 b/d and 100 million scf/d respectively. The produced water handling capacity is 80 000 b/d.

Oil export is via a floating storage unit (FSU), which is serviced by two shuttle tankers. The 20-inch Fulmar Gas Pipeline exports the wet gas to St Fergus in Scotland. Additionally these facilities are used by the neighbouring Clyde and Auk fields.

Early production data showed that aquifer support was even less than that assumed in the development stage. The reservoir pressure declined rapidly in 1982. Thereafter the pressure was well maintained in the early producing years as a result of combined water and gas injection. The very high uptime of the gas injection compressors resulted not only in low flare levels, but also increased injected volumes well above expectations. In order to limit the growth of the secondary gas cap and safeguard future oil production levels by minimizing the risk of coning and cusping, the gas export project was accelerated to allow export to commence in 1986. It was clear that, as soon as gas export were to commence, the capacity of the water-injection plant would be insufficient to maintain reservoir pressure. In order to be able to maintain the provision of full voidage replacement, the water injection plant needed upgrading. This work was carried out in 1985 to provide a maximum injection capacity of 272 000 b/d.

## Reservoir performance

### *General production history*

The production performance for the field is presented in Fig. 11. First oil production was in February 1982. The plateau production rate of 160 Mb/d was first achieved in 1984 and has been maintained to the end of 1988. The prolonged plateau period compared with the original forecast was caused by stable displacement, higher than expected sweep efficiency and very favourable well productivity. Water breakthrough did not occur until the beginning of 1985 and has been controlled by a combination of reducing production rates in high-watercut wells and re-completing such wells on higher intervals using through tubing bridge plugs to abandon the water-producing interval.

It was expected that it would not be possible to maintain plateau production rates after 1986, as a result of both an increase in water cut and dropping reservoir pressures once gas export was operational. Therefore, a crestal drilling and workover campaign was undertaken in 1987, during which a further five oil producers were drilled with the objective of improving areal sweep and extending production plateau. The additional dry oil potential thus provided would also allow beaning back of high-watercut producers in order to reduce reservoir voidage by curtailing water production.

At the end of 1988, 17 oil producers, 10 water injectors and 1 gas injector have been completed. Two oil producers are shut in owing to high water cut. Figure 2 shows the status of the development as at 31 December 1988.

The cumulative production to end 1988 is 319 mmstb oil and 121 Bscf wet gas (excluding the 80 Bscf re-injected), leaving an expectation of remaining reserves of 143 mmstb oil and 143 Bscf wet gas. To date, only 3 mmstb oil have been produced directly from Unit 1.

On 24 December 1988, the FSU broke its mooring and production was halted from the Fulmar, Auk and Clyde fields. Production was resumed in March 1989 via a temporary system using the Petrojarl I, in place of the FSU, providing the facility to export up to 230 Mb/d. The repair of the FSU and Single Anchor Leg Mooring (SALM) is being carried out and re-installation of the FSU/SALM is scheduled for August 1989.

### *Water injection, voidage and reservoir pressure performance*

First water injection into the reservoir commenced in November 1982, some 8 months after first oil production. Direct voidage replacement by injection was achieved in mid 1983. In the interim, average reservoir pressure at datum declined from an initial 5700 psia to around 4500 psia. Thereafter, the reservoir pressure has increased to some 5250 psia. The benefit of the water injection plant upgrade during the 1985 platform shutdown was partially offset by the end of gas injection in May 1986. Poor water-injection plant uptimes during 1987 and 1988, due to replacement of corroded injection flow lines and workover of water injection wells, meant that the target of full voidage replacement could not be met. Full injection support was restored in November 1988, at which time the Main Sands reservoir pressure had dropped to around 4650 psia. Cumulative reservoir voidage at the end of 1988 was 41 mmrb (Fig. 12). Reservoir pressure data have been used as input for both classical material balance calculations and for history matching of simulation models. Pressure data indicate good pressure communication in the oil leg throughout the Main Sands. Material balance studies have been performed to estimate the size and activity of the connected aquifer. These have been endorsed by full-field simulation models. The models are pressure-matched to the performance history using an aquifer size comparable to the reservoir volume, with a permeability of 500 mD, acting on the Main Sands in the south-east flank of the field. Cumulative aquifer influx to end 1988 is estimated to be some 22 mmrb, which is considered to be only a modest contribution to pressure support in the field, when compared to the injection of 80 Bscf gas,

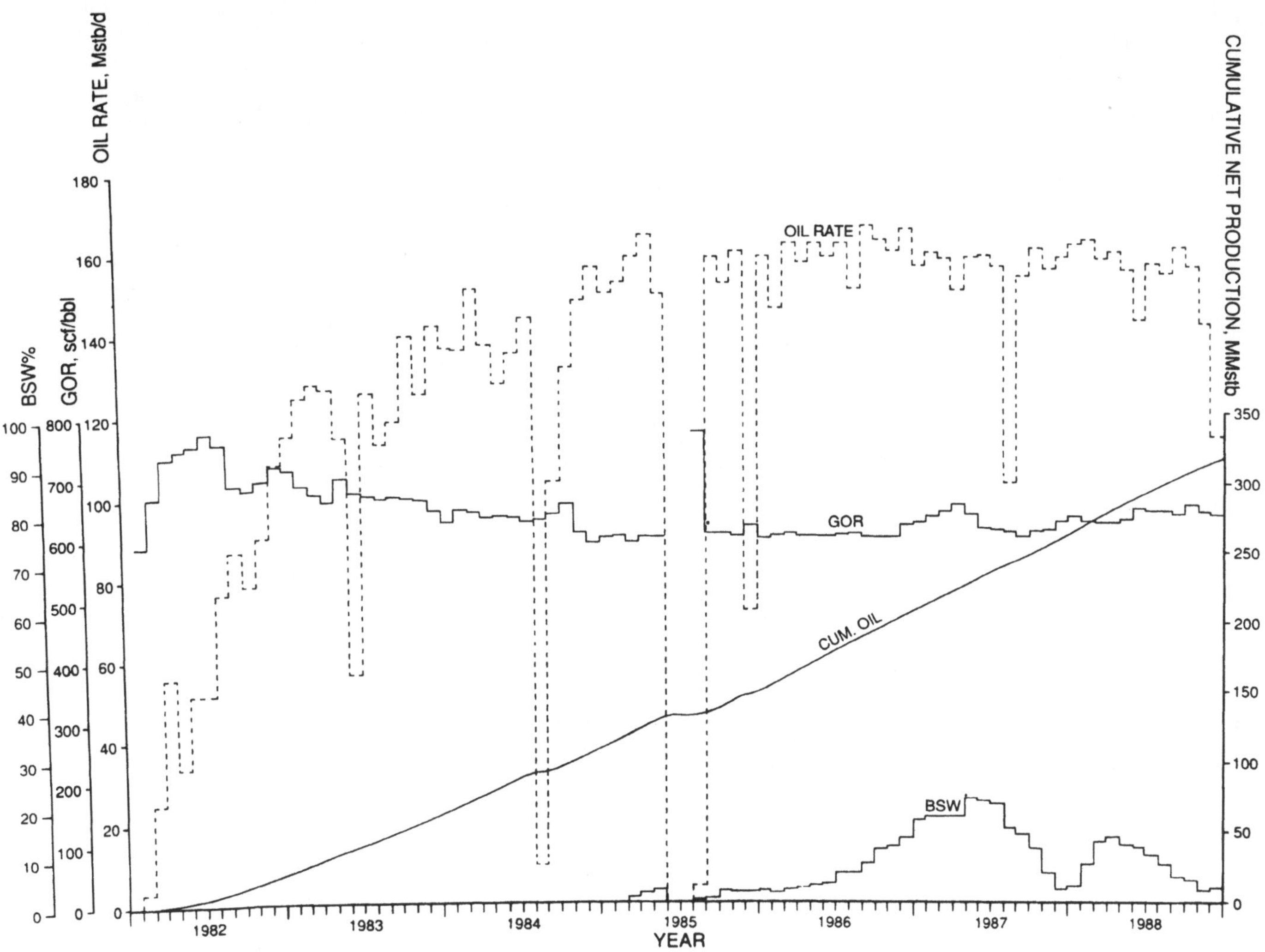


**Fig. 11.** Production performance plot for the total field showing the extended plateau period to end 1988. Major production interruptions took place in every year, with the exception of 1986 and 1988, for maintenance and modifications to the platform facilities. Elevated GOR figures in 1982 resulted from low gas process uptimes leading to higher flare levels and lower liquid recovery. The increased GOR in 1987 coincided with the second phase of development drilling. GOR levels and water cuts declined as dry, low-GOR production potential became available. Towards the end of 1988 a slow increase in both GOR and water cut was observed, resulting from the encroachment of gas and water into the producing intervals.

equivalent to 55 mmrb and some 360 mmbbls (382 mmrb) water.

Re-pressuring the reservoir to mantain natural flow becomes more important as water cut increases. Material balance calculations indicate that, with a realistic water injection and production forecast, original reservoir pressure could be restored during 1991.

### *Unit 1 performance*

First direct production form Unit 1 commenced in April 1988, when well FA-25, a watered-out Main Sands producer, was re-completed. Re-completion of template well FT-03 was performed later in the year. At the end of 1988, a further dedicated producer was completed.

Reservoir pressure monitoring indicates limited pressure communication between Unit 1 and the Main Sands, most likely through juxtaposition of sands at the crest of the reservoirs. Figure 13 shows how the pressure response follows the Main Sands pressure trend prior to any direct production from this Unit, but with a considerable time lag. The early PVT sample from well 30/16–7 and recent measurements have indicated that there is possibly a lower GOR in Unit 1 compared with the reservoir fluids from the Main Sands. This is further indication that this Unit should be treated as a separate reservoir.

### *Reservoir monitoring activity*

Individual well production/injection performance is routinely measured on the platform by passing producers through the test separator and by continuous metering on an individual well basis for water injectors. The aim is to have every well tested at least once per month. Reservoir pressure has been routinely monitored by static bottom hole pressure surveys, pressure build-up surveys, surface fall-off surveys and pressure measurements in new wells. A major surveillance campaign has been directed at monitoring fluid contact movement using time-lapse thermal decay time logging (TDT) techniques. Base TDT measurements were made in all oil producers at or soon after being drilled and in selected injectors where movement of the OWC in Unit 1 could be monitored. The TDT technique is very effective in the Fulmar field owing to the high formation water salinity (150 g/l NaCl eq), which allows a clear distinction of the OWC. It is also an important tool in tracking the gas–oil contact (GOC). Production logging (PLT) surveys are run to allocate production and injection to individual reservoir intervals.

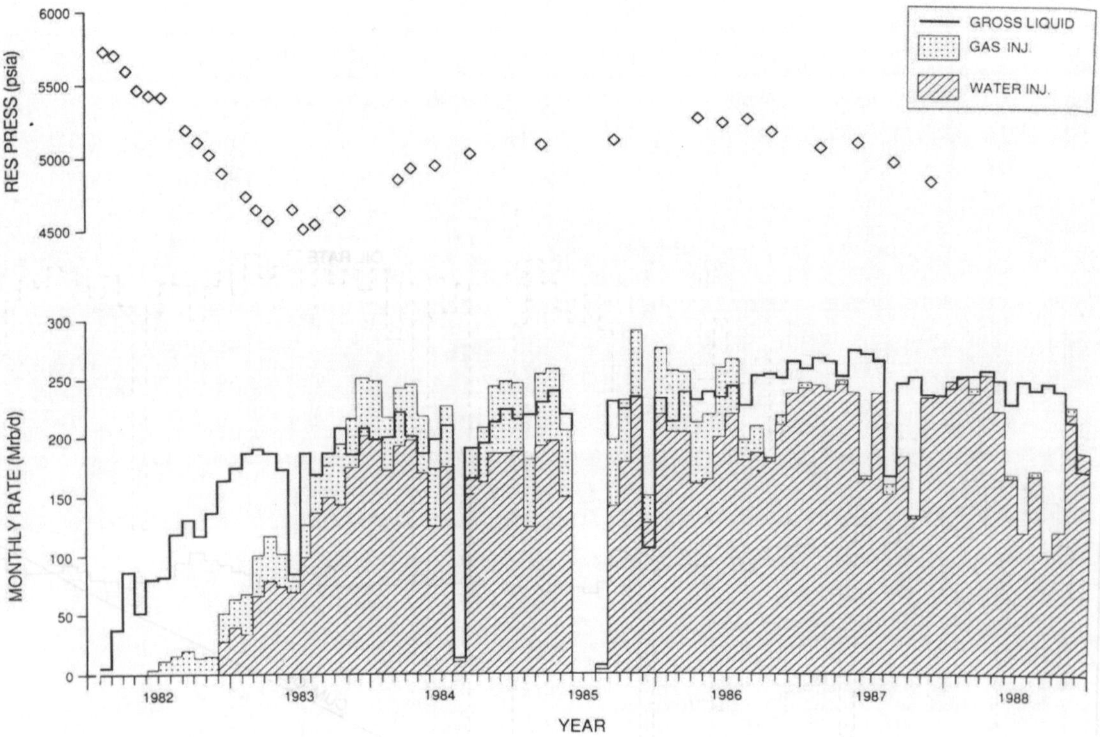


**Fig. 12.** Reservoir voidage and pressure data from the field, showing that full voidage replacement by water and gas injection was first achieved in 1983. Gas export started in 1986 and is reflected in insufficient voidage replacement and dropping reservoir pressures. Full voidage replacement by water only was achieved in late 1987, but varying uptimes of the water-injection facilities resulted in a significant drop in reservoir pressure later in 1988.

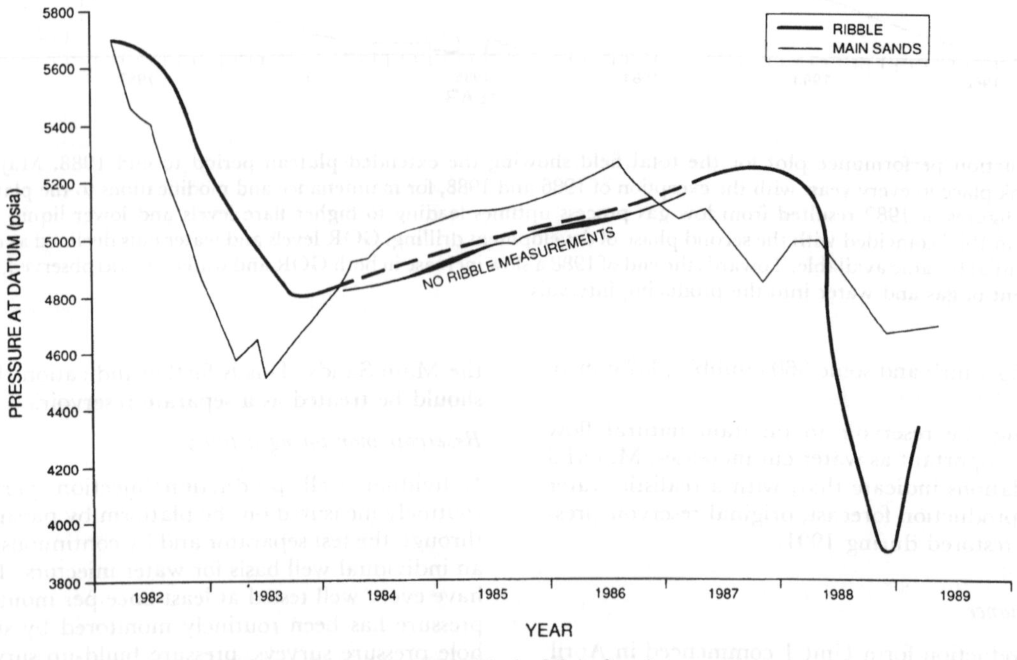


**Fig. 13.** The Ribble (Unit 1) shows limited communication with the Main Sands, most likely through juxtaposition of reservoir sands at the crest of the structure. Without direct offtake from Unit 1, the pressure follows the Main Sands pressure, albeit with significant time lag.

### *Displacement*

Monitoring of the fluid contact movement with time-lapse TDT runs has provided a detailed picture of the fluid displacement efficiency in the Main Sands. In general, a very stable displacement of oil by water has been observed, such that the field-wide OWC has risen quite uniformly through the homogeneous Units 3.1–3.4 and Unit 4 at a rate of approximately 70 ft/yr (Fig. 14). No movement of the OWC has been observed in the northern area.

Unit 2 is of poor reservoir quality and is not expected to contribute to field ultimate recovery. At present there are no plans to attempt dedicated development of this Unit.

Unit 5.3 contains relatively thin beds of reservoir-quality rock, easily disrupted by faulting. Some difficulty in providing pressure support has been experienced in the area between FA-26 and FA-13. Also the production

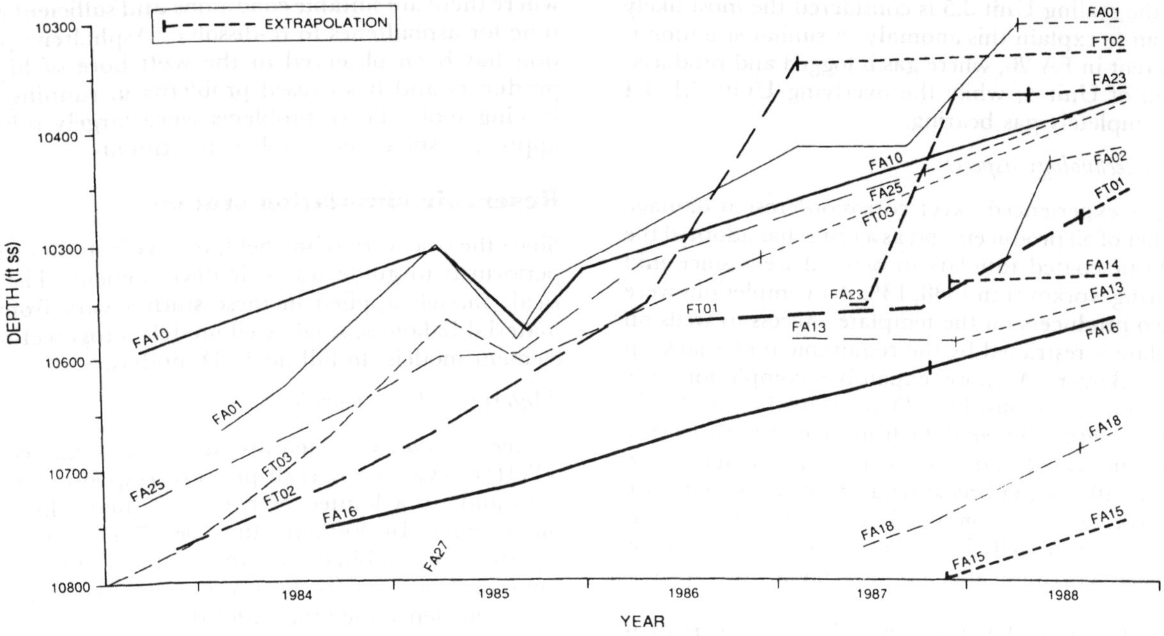


**Fig. 14.** The rise in the OWC as a result of water injection shows clear areal trends across the field, with the fastest rise observed on the west flank. The time-lapse TDT measurements also show significant coning effects, especially in wells FT-03, FA-01 and FA-10 in surveys performed just prior to and immediately after the extended summer shutdown in 1985. Coning is also significant in FT-02, FA-23 and FA-02.

performance of FA-26, which died after only a month's production, indicated that the reservoir may be broken up in relativey small blocks. Nevertheless, this Unit is seen as a secondary development target in view of the significant STOIIP (13.3 mmstb) it contains. Some overrunning of oil by injection water has been observed in Unit 5.3 on the eastern flank producers. The east flank injectors are perforated on Unit 5.3, but no significant quantities of water are actually injected into this unit due to its permeability contrast with the better quality upper Main Sands units perforated in the same wells.

The sweep efficiency of the Main Sands during production history was estimated by monitoring the movement of the fluid contacts and the cumulative produced volume. The historical fluid contact data are of sufficient quality to map the contact level changes over time. By applying the appropriate discounts to the resulting maps for the non-productive and poorer sands, the (high) sweep efficiency within the best quality Main Sands could be estimated. Further discounting for incomplete sweep at the crest of the structure the present expectation of ultimate recovery is 462 mmstb. This estimate has been backed up by full field reservoir simulation studies. The sweep by the secondary gas cap is generally uniform. No residual oil is expected in the gas zone owing to the fully miscible system. Continued monitoring of the GOC is important to assist in deciding on re-completion intervals to optimize low water cut, low GOR production. In line with the original development plan, the gas cap will be blown down in order to maintain gas-export plateau levels after the oil production starts to decline. The potential loss in recovery as a result of re-saturation of the gas cap is more than offset by the more favourable abandonment position of the reduced oil column.

### *GOR performance*

Producing GOR remains essentially constant as the reservoir pressure is well above the bubble point and owing to the fact that up to now little gas from the gas cap has been produced. Fluctuations in the measured and reported GOR reflect the uptime of the gas-processing plant. With compression in operation, either for export or injection, heavier components from the gas stream are recycled into the separator liquids stream. Without compression, those heavier components remain in the gas stream and are flared. As uptimes of the NGL recycle have improved, the overall process GOR has decreased. Current GOR on the platform is around 630 scf/stb, indicating direct production of gas from the secondary gas cap. Blowing down the gas cap will temporarily increase the producing GOR. Gas cap blow down is expected to be completed during 1993, after which time only associated gas will be produced. The average field life GOR is estimated to be 614 scf/stb.

### *Coning and cusping*

The combination of the domal structure of the Fulmar field, the flank water drive mechanism from injection and the presence of a secondary gas cap create a situation in which cusping and coning of both gas and water must be managed. This was recognized in the initial development and formed the basis for the policy of perforating oil producers in the middle of the oil column. As production progresses, the thickness of the oil column, located between encroaching water and the existing gas cap, diminishes to the point where the challenge becomes one of optimizing oil production in the presence of water and gas coning. The effects of water coning have been experienced directly, with water cones in excess of 100 ft being observed in FA-01, −10, −23 and FT−02 (Fig. 14).

A re-completion policy using through tubing bridge plugs to combat high water cuts has been effective to date, as water cut for the field is currently less than 5%. This achievement cannot of course be sustained and, when water cut starts to increase, a rapid rise is anticipated, heralding a sharp decline from what is still effectively plateau production rate. The onset of this decline is predicted to occur at the end of 1989. Excess gas production as a result of gas coning has not been observed. Although gas is not logged in northern block producer FT-04, the well produces at high GOR. Gas cusping beneath the sealing Unit 3.5 is considered the most likely mechanism to explain this anomaly. A similar situation is also apparent in FA-26, where gas is logged and produced at the top of Unit 4, while the overlying Units 3.1–3.4 are not completely gas bearing.

*Production technology aspects*

Fulmar has experienced sweet corrosion/erosion damage in a number of oil producers and as a result has adopted the use of plastic-coated tubulars in new oil wells since mid 1987. During workover in 1988, 13%-Cr completions were run in two producers on the template. Access to wells on the template is restricted by the requirement of a jack-up rig for workover. A more expensive completion was therefore considered justified. Data from these wells, in conjunction with calliper data from a number of other wells, have indicated that severe corrosion damage (up to 12 mm/year after water breakthrough) occurs in high-rate wet wells owing to the combined effects of corrosion and erosion. In low-rate wells, however, where only corrosion is taking place, corrosion damage and metal loss are less (1–2 mm/year).

During 1987 and 1988 it became evident that corrosion had occurred in the tubing, Xmas trees and surface pipework of 10 water-injection wells. Although the integrity of the water-injection facilities was not immediately affected, a major workover campaign was undertaken to repair these wells, resulting in a prolonged reduction in water-injection capacity during the third quarter of 1988.

The corrosion experienced in the Xmas trees affected the sealing surfaces behind the valve seats of the flowing wing valves and upper master gate valves. An in-situ machining technique was developed and successfully used to repair the Xmas trees on 7 wells during late 1988. This was done without the need for a well kill. A programme of flowline replacements is ongoing through 1988–89, in tandem with the water-injector workover activities.

Scale build-up in producers and hydrogen sulphide are further production technology aspects of the field that cause concern and are associated with injection water breakthrough. Evidence has accumulated in several wells that sea-water breakthrough in a producing well results in deposition of a barium/strontium sulphate scale in the well bore. The high concentration of strontium ions in the formation water (Sr 660 mg/l, Ba 6 mg/l) means that strontium sulphate scale will predominate. This can occur very rapidly as evidenced by FA-02, which over a 2–3 month period in early 1988 went from dry production to 27% BS&W, with 42% sea-water breakthrough. This resulted in a reduction in liner ID, due to scale, of up to 1 inch. Scale dissolver soaks have proved to be unsuccessful and it has been necessary to re-instate the more expensive preventive scale-inhibition treatments.

Fulmar export gas contains 10–15 ppm hydrogen sulphide. Its concentration in the export gas is expected to rise as the field water cut increases. Predictions based on the current production forecast suggest that hydrogen sulphide concentrations will not exceed 20 ppm and overall landed quantities are expected to remain within the capacity of the pipeline and the onshore sweetening plant.

Fulmar crude is virtually saturated with asphaltenes at reservoir conditions. With reduction in pressure and, to a lesser extent, temperature, asphaltenes come out of solution with solubility reaching a minimum at bubble point. Prior to significant water production, problems were experienced with asphaltene deposits building up in the dead area at the bottom of the production separators. With increased water production this problem has disappeared and the asphaltenes are carried over to the FSU where there are suitable conditions and sufficient residence time for asphaltenes to re-dissolve. Asphaltene precipitation has been observed in the well bore of high-GOR producers and has caused problems in running wireline logging tools. These problems were largely solved after applying asphaltene-dissolver treatments.

## Reservoir simulation studies

Since the discovery of the field, reservoir studies have been performed to guide reservoir development. The (numerical) models applied in these studies vary from simple material balances, single well models, cross-sectional and element models, to full field 3D models.

*Material balance models*

Material balance models are used to estimate connected STOIIP, evaluate overall pressure response and aquifer sizes and, to a limited extent, to estimate fluid contact movements. In Fulmar, the use of material balance equations in combination with the observed fluid contacts from TDT surveys has been very successful in estimating sweep efficiencies for the water drive.

*Single well models*

In the first phase of drilling, a single well numerical simulation model was used to select initial perforation intervals that would not induce water or gas coning during the first 5 years of the field life. At present, after 69% of the total initial oil reserves have been produced, the gas cap is still largely in place. The water cut is around 5% as a result of an active isolation and re-perforation policy. A detailed performance simulation of each subject well with a single well model forms the basis for each re-perforation. The boundary conditions are supplied by field observations and the 3D full-field models. Single well models will play an even greater role in the coming years when optimization of oil production will be strongly dependent on detailed knowledge of near-wellbore effects.

*Cross-sectional/element models*

Cross-sectional numerical models have been and are being used to improve understanding of the displacement processes in various parts of the field and to support full field 3D simulations. Cross-sectional models are used when a high resolution for geological and structural detail is more important than a third flow direction. Early studies highlighted the problems to be expected in the development of oil from the tight Unit 2 and the need for dedicated injection and drainage points for Unit 5.3. More recently, 2D models have been used to study various ways of modelling faults and to study the dynamics of the gravity-dominated (stable) water/oil displacement, in particular the tilt angles and saturations of the displacement front. These studies were carried out to support the development of a new full-field model.

Since 3D element studies concentrate on a section of the total field only, the resolution that can be obtained is higher than that of a full-field model. In addition, running times are often much shorter. Therefore, element simulation studies are complementing full-field simulations and are used in Fulmar for the short- to medium-term operational planning.

*Full-field numerical models*

Full field 3D reservoir simulation studies have always played an important role in the development planning of Fulmar. Their size, refinement and predictive capabilities have increased with the increasing modelling and computing capabilities within Shell Expro and, more importantly, increased production and geological information. Before the most recent study was initiated in 1987, earlier studies took place in 1978, 1981 and 1983. These studies formed the basis for the initial development plan and subsequent refinements. However, none of these models could be matched on a production history containing appreciable water cut or high GOR development. The 1983 model was matched on pressure data only.

The study initiated in 1987 ran concurrently with the review of reservoir geology and a petrophysical review incorporating the newly proposed subdivisions. The overall objectives of the study were to develop an accurate tool that can be used for medium- and long-term field development planning, reservoir management and production forecasting. It integrates the present knowledge of the field and helps to highlight the major deficiencies in that knowledge.

In order to meet the objectives, a coarse grid model (CGM) and a fine grid model (FGM) have been set up. The CGM ($15 \times 16 \times 5$ grid blocks) has been built to provide a fast simulation facility to undertake certain simple predictions and, more importantly, to be used as a "test bed" for the construction of the FGM. In particular, field observations and analytical and cross-sectional studies showed that the displacement in the Main Sands is characterized by vertical equilibrium and can be modelled as a stable, fully segregated flow. The CGM was used to ascertain that a satisfactory history match can be obtained using pseudorelative permeability curves based on these assumptions. In addition, the CGM was used to test the modelling of surface facilities. The latter is particularly important because Fulmar production is and will continue to be constrained by the gas-export facilities.

The FGM ($52 \times 55 \times 8$ grid blocks) has an areal resolution of 75 m by 75 m (246 ft by 246 ft). Table II gives the reservoir subdivision and permeabilities of the eight layers. The vertical resolution of the model is significantly better than suggested by the eight layers, owing to the application of the segregated flow concept in the high-permeability layers. A satisfactory history match of the FGM (Figs. 15 and 16) was obtained after a process in which the interaction between geology, petrophysics and reservoir engineering was extremely important. An excellent gas match was obtained, both in (high) GOR production and in areal extent of the gas cap as mapped by TDT measurements (Plate 2). The match of the reservoir pressure and the field water cut is good. The FGM has been used to investigate various development scenarios and to produce long-term production forecasts. A full-field model can only play an important role in reservoir management if it is updated on a regular basis to reflect the ever-increasing knowledge of the field. A revision of the 3D model is ongoing in order to incorporate the results of the Main Sands wells drilled in 1987 and 1988, the geological re-mapping of the field and the associated petrophysical study. In addition, the results of the new Unit 1 wells drilled in the last quarter of 1988 and early 1989 will be incorporated to refine the modelling of the field.

## Future development and improved recovery

### *Short- to medium-term plans*

The overall objective for the field development is to maximize economic ultimate recovery. Since the development to date has concentrated on the Main Sands, it is now important to develop the minor sands in both an economic and timely manner in order to ensure that their development is completed within the Main Sands producing life.

The largely undeveloped minor sands comprise Unit 1 and Unit 5.3. Of these, Unit 1 is the most significant, containing ca. 20% of remaining field reserves. The drilling programme for 1989 features its continued development with the completion of dedicated oil producers and water injectors. The long-term development plan includes a further water injector and oil producer on this Unit, in addition to re-completions of existing producers on the Main Sands.

The feasibility of developing Unit 5.3 with dedicated wells remains subject to further study. A pilot development was carried out during 1988 between a Unit 5.3 dedicated water injector (FA-06) and a proximal oil producer additionally perforated on Unit 5.3 (FA-13, some 1000 ft SE of FA-06). A multi-rate PLT demonstrated very limited communication, if any, between the two wells at Unit 5.3 level, as evidenced by the absence of a noticeable increase in production from this Unit in FA-13 and a significant pressuring up FA-06. This is generally discouraging in terms of further development.

Future development of the Main Sands will be directed towards optimizing the use of existing producers by re-completion of these wells to minimize water production and manage the secondary gas cap and possibly drilling a small number of infill wells to recover oil from unswept areas of the field.

### *Enhanced oil recovery studies*

The potential benefits of enhanced oil recovery (EOR) in Shell-operated fields are well recognized and a continuous research programme is in progress to investigate systematically the feasibility of the various EOR processes. In order to assess the scope for these processes, special studies have

**Table II.** Fulmar fine grid model reservoir subdivision and base case absolute permeabilities

| *Geological model (unit)* | | | *Numerical model (layer)* | *Absolute permeability (mD)* | |
|---|---|---|---|---|---|
| Ribble | 1 | | 1 | 600 | |
| Mersey | 3.1/3.2 | Clyde 2 | 2 | 800 | 1 |
| | 3.3 | | 3 | West: 150 | |
| | | | | Central: 75 | |
| | | | | East: 10 | |
| | 3.4 | | 4 | | |
| Lydell | 4 | | 5 | 800 | |
| Usk | 5.1 | | 6 | 100 | |
| | 5.2 | | 7 | 5 | |
| | 5.3 | | 8 | 200 | |

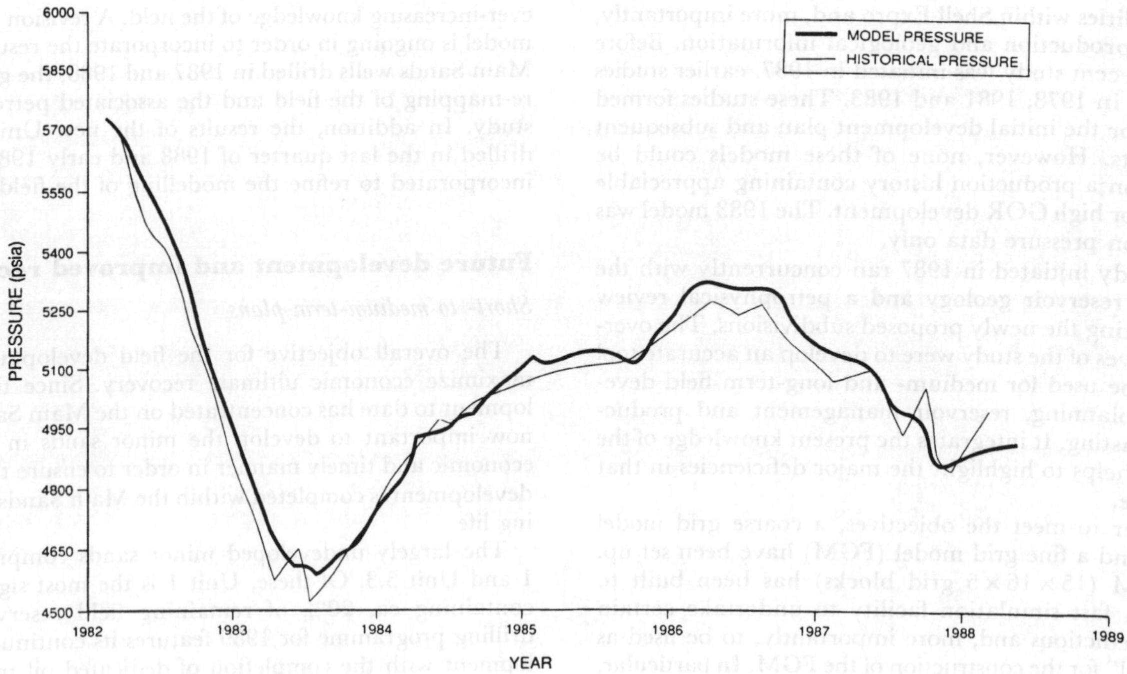


**Fig. 15.** **The pressure data from the full-field fine grid reservoir simulation are in excellent agreement with observed pressures.**

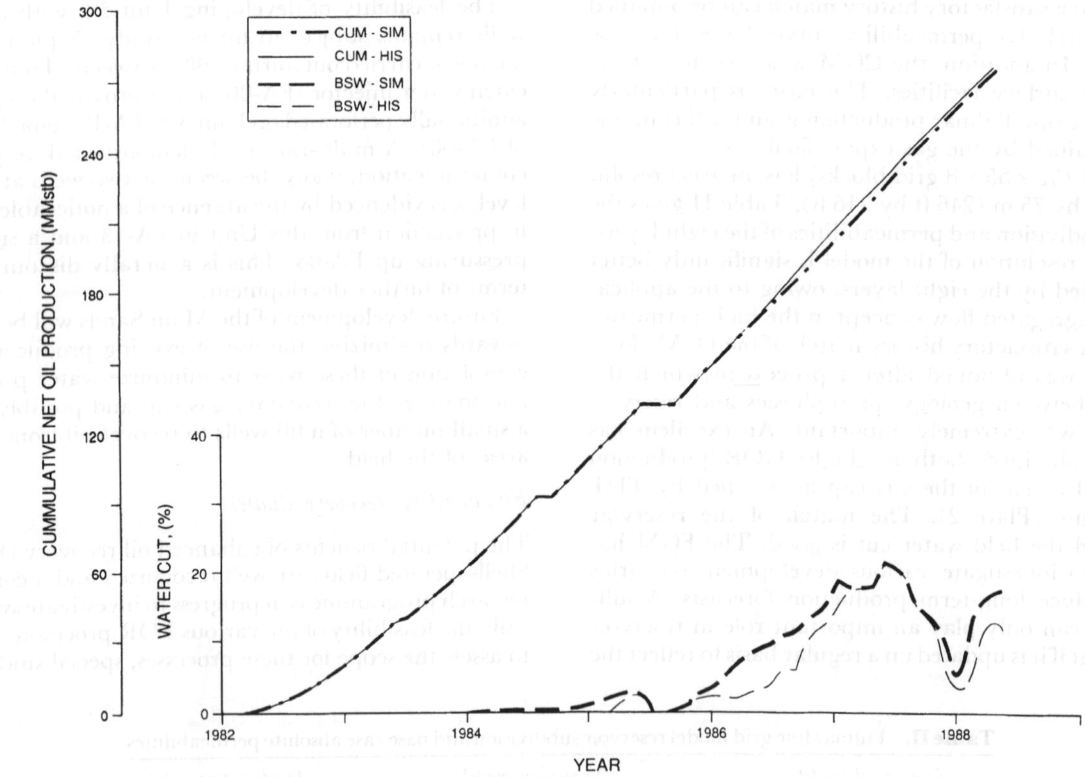


**Fig. 16.** **With the reservoir model constrained on maximum gross liquid offtake during production history an excellent match is achieved, both in net offtake levels and water cut development.**

been devoted to an accurate determination of the water flood residual oil saturation.

One of the more promising EOR processes identified for high-pressure, watered-out light oil reservoirs is tertiary nitrogen injection (Niko *et al.*, 1987). The process is based on a combination of gravity drainage of oil and water on the gas-invaded zone and phase-behaviour effects caused by multiple contacts between the nitrogen and the crude. Oil and water are accumulated in a bank that is displaced downwards. Although tertiary nitrogen injection would appear technically feasible as a recovery process, preliminary studies indicated that the project would not be economic under present assumptions regarding future oil price.

## CONCLUSIONS

The Fulmar reservoir is characterized by high-quality sands and a relatively simple geology. The shape of the reservoir with the steep flanks, coupled with the favourable mobility ratio, allows optimal use of water drive as a recovery mechanism. High recovery figures are expected. In the near future the presence of the secondary gas cap and the diminishing oil column will complicate production planning and forecasting. There is a requirement to engineer re-completions of existing producers carefully in order to keep GOR and BS&W under control.

The development of the Fulmar field has shown the benefits of a strongly integrated approach to petroleum and field engineering and has highlighted the requirement for pro-active and flexible planning. The development can be considered a successful business venture.

## ACKNOWLEDGEMENTS

The data and interpretations presented in this paper are the result of several years of integrated petroleum engineering work by numerous Shell staff and do not represent the views of all Fulmar Unit Owners. The authors wish to express their thanks to Shell Internationale Petroleum Maatschappij, Shell U.K. Exploration and Production, Esso Exploration and Production UK, Amoco (U.K.) Exploration Company, Texas Eastern North Sea Inc. and Amerada Hess Ltd for their permission to present and publish this paper.

## APPENDIX

### Abbreviations and symbols used in the text and on figures

| | |
|---|---|
| m, mm | thousands, millions |
| mmstb | millions of barrels at stock tank conditions |
| mmrb | millions of barrels at reservoir conditions |
| Bscf | billions ($10^9$) cubic feet at standard conditions |
| tvss | true vertical depth subsea |
| STOIIP | stock tank oil initially in place |
| $K_v$, $K_h$ | vertical, horizontal permeability |
| GOR | gas–oil ratio (scf/stb) |
| $S_h$ | hydrocarbon saturation (fraction on %) |
| TDT log | thermal decay time log, a neutron log through casing |
| OWC | oil–water contact |
| GOC | gas–oil contact |
| FA | Fulmar Alpha (platform) |
| FT | Fulmar template (wellhead jacket) |
| FSU | floating storage unit |
| NGL | natural gas liquids |
| BS&W | bottom sediment and water (water cut) |
| ppm | parts per million |
| ID | internal diameter |
| CGM | coarse grid model |
| FGM | fine grid model |
| EOR | enhanced oil recovery |
| $\phi$ | porosity (fraction or %) |

## REFERENCES

Gibbs, A.D. (1984). Structural evolution of extensional basin margins. *J. Geol. Soc. London* **141**, 609–620.

Heseldin, G.M. (1974). A method of averaging capillary pressure curves. *Trans. SPWLA*, Paper E.

Johnson, H.D., Mackay, T.A. and Stewart, D.J. (1986). The Fulmar oil field (Central North Sea): Geological aspects of its discovery, appraisal and development. *Marine Petroleum Geol.* **3**, 99–124.

Mehenni, M. and Roodenburg, W.Y. (in preparation). Fulmar Field Central North Sea U.K. Sector. Contribution to the *AAPG Atlas of Hydrocarbon Reservoirs.*

Niko, H., Schulte, A.M., Drohm, J.K. and Cottrell, C.W. (1987). The feasibility of tertiary nitrogen injection in waterflooded volatile oil reservoirs. *4th European Symposium on Enhanced Oil Recovery, Hamburg.*

Pennington, J.J. (1975). The geology of the Argyll Field. In A.W. Woodland (ed.), *Petroleum and the Continental Shelf of North West Europe*, pp. 285–294, Institute of Petroleum, London.

Ziegler P.A. (1982). *Geological Atlas of Western and Central Europe.* Elsevier, Amsterdam.

Field Review 3

# The Tyra Field

*M.C. Doyle and J.M. Conlin*

Mærsk Olie og Gas A/S, Copenhagen, Denmark

The Tyra field, located in the Danish sector of the North Sea, contains free wet gas with an initial condensate gas ratio of 50 stb/mmscf and solution gas in an underlying oil rim. The gas accumulation has a maximum column height of 225 ft while the oil rim, although containing a large amount of STOIIP, is present in a long transition zone of up to 200 ft thickness. The Tyra structure is fairly flat with structural dips ranging from 1.5 to 3.5°. The reservoir rocks are Danian and Maastrichtian chalks with the Danian chalk contaning nearly 60% of the GIIP. Discovered in 1968, a further three appraisal wells were drilled, the latest in 1976.

Following the signing of a gas supply contract in 1979, development of the field started in 1980 and first contractual gas deliveries were made in October 1984. A gas recycling project was initiated in mid 1987 to take advantage of excess gas availability relative to sales volumes. Current gas development planning and reservoir management is based on a full-field three-dimensional field simulation study carried out following development drilling (36 wells). Although field performance to date is broadly in line with predictions, it has reinforced the importance of reservoir characterization in the assessment of recovery efficiency and chalk compaction. Only limited development of the oil rim has taken place to date, although dedicated appraisal drilling is planned in the short term to clarify the nature of the oil rim and assess its real potential.

## INTRODUCTION

The Tyra gas field is located in the Danish sector of the North Sea, 230 km west of the port of Esbjerg (Fig. 1). Companies participating in the Tyra field are A.P. Møller, Shell and Texaco, forming the Dansk Undergrunds Consortium (DUC). Chevron was a member of the consortium until 1986. A.P. Møller is the sole concession holder and the A.P. Møller company, Mærsk Olie og Gas A/S is operator for the group.

In March, 1979 contracts were signed between the DUC partners and the Danish state energy company Dansk Olie og Naturgas A/S (DONG) for the sale and purchase of 2 Tcf (55 bcm) of natural gas over a contract period of 25 years. This agreement, which triggered the Tyra field development, covers gas reserves from four dedicated fields; Tyra, Roar, Dan and Gorm. Tyra and Roar are gas/condensate fields, while Dan and Gorm are oil fields, whose associated gas production is also a dedicated reserve under the Gas Sales Agreement. The Tyra field came on stream in 1984 while the Roar field is as yet undeveloped.

This chapter is based upon data gathered and experience gained during the development of Tyra and four contract years of production (1984–1987).

## FIELD HISTORY

The first exploration well in the Tyra field, E-1X, was drilled in 1968 with a primary objective in Tertiary/Upper Cretaceous chalks. The basis for the test was a mappable, broad, low-relief structure at base Tertiary level. Secondary objectives were Lower Cretaceous sandstones and limestones and Jurassic sandstones. The well was drilled on the eastern flank of the structure (Fig. 2) and encountered only traces of hydrocarbons in the chalk sequence. In the Lower Cretaceous and Jurassic intervals, no significant reservoir rock was present.

A second well, E-2X, located to the west of the crestal area, was drilled in the same year and encountered 161 ft of wet gas-bearing chalk. The two subsequent appraisal wells, E-3X and E-4X, drilled in 1974 and 1976 respectively, confirmed the presence of a wet gas accumulation overlying a thin movable black oil rim with a long (150–200 ft) oil/water transition zone.

Development drilling began in 1982 and was completed in 1984. The Tyra field was put on stream in October 1984 and dry gas recycling was initiated in June 1987. The production and evacuation facilities at Tyra and their position relative to other DUC installations in the Danish sector of the North Sea are diagramatically illustrated in Fig. 3.

## RESERVOIR GEOLOGY

### Regional setting

Denmark's oil and gas fields are located within the North Sea Central Graben. Rifting of the Graben was initiated in the Permian or Triassic and was followed by a long and complex tectonic history. The major elements of the Danish Central Graben are illustrated in Fig. 4.

The Tyra field is located at the southern end of the Tail End Graben, where it merges with the Southern Salt Dome Province. The extensional tectonic regime asso-

*North Sea Oil and Gas Reservoirs—II*

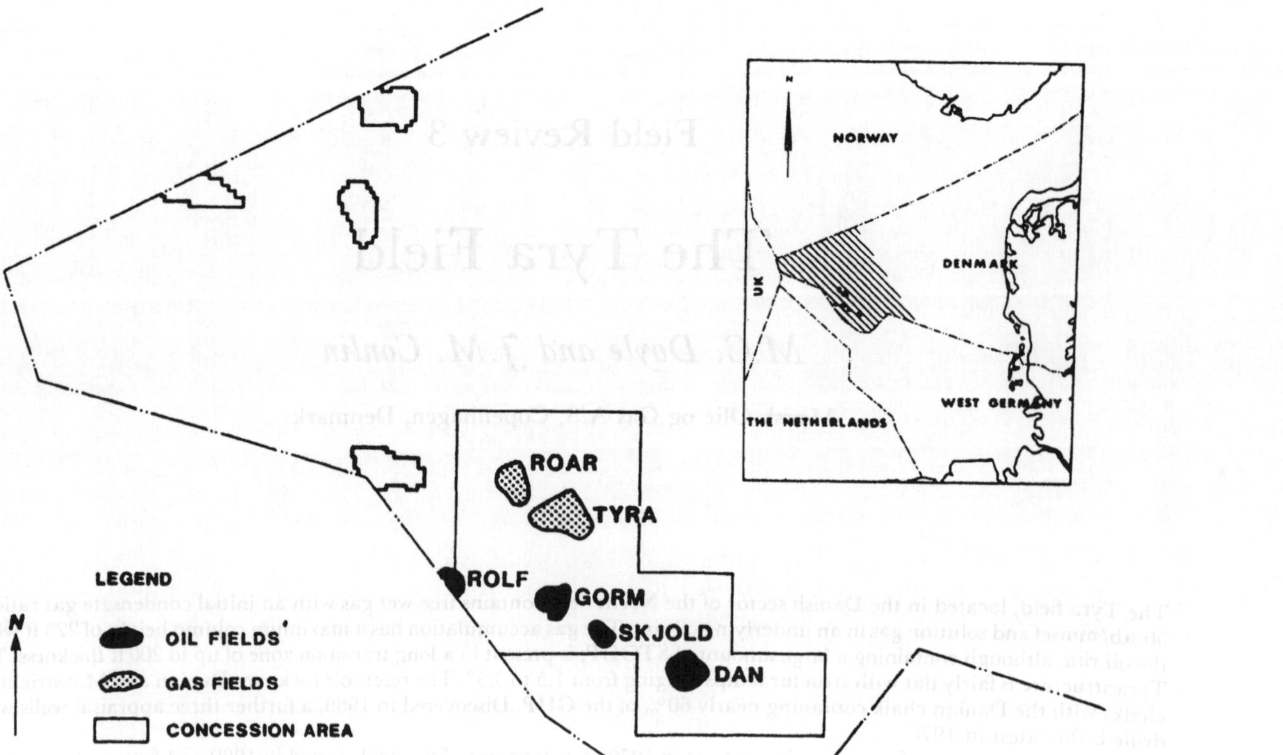


**Fig. 1.** Field location map.

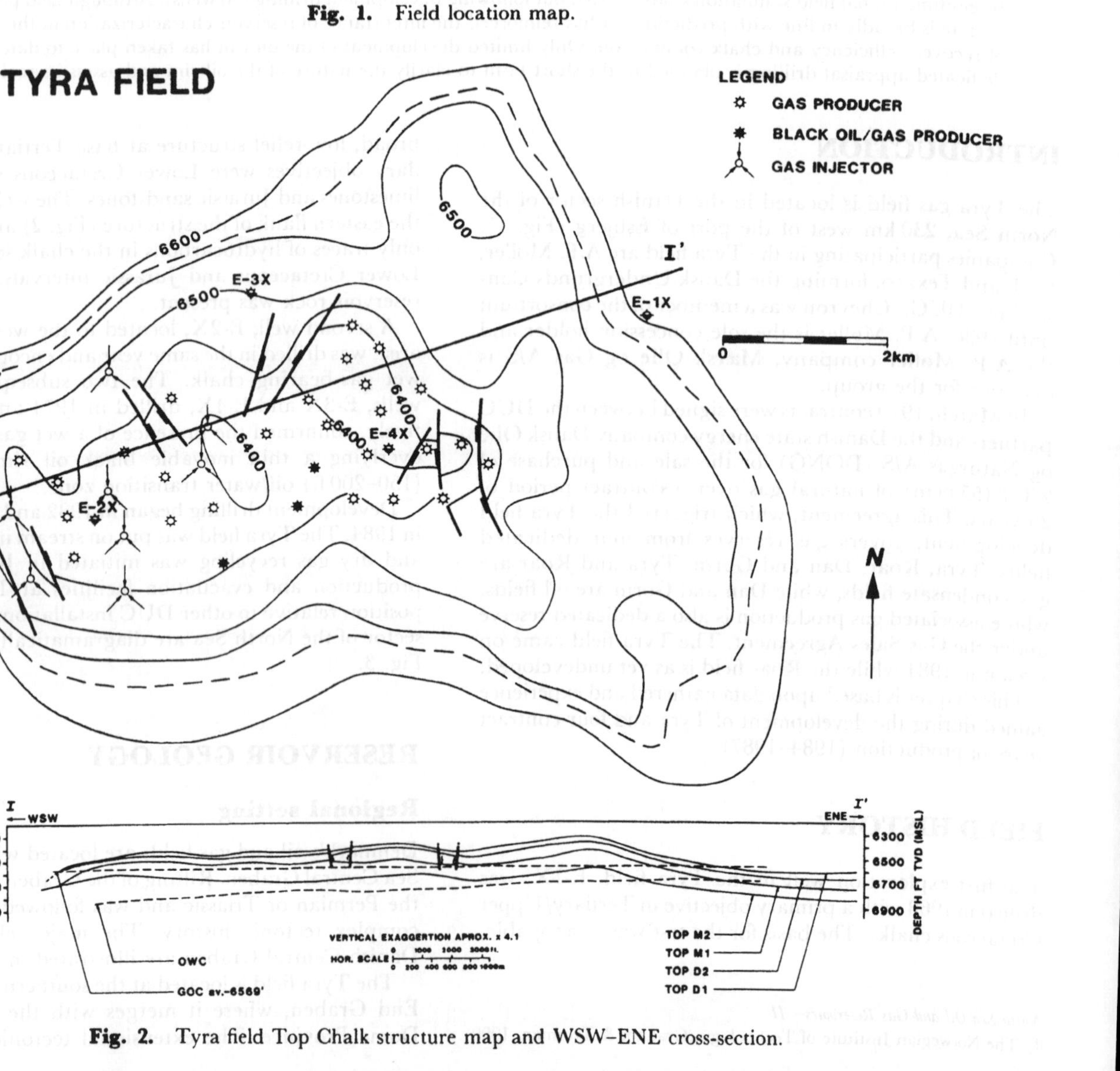


**Fig. 2.** Tyra field Top Chalk structure map and WSW–ENE cross-section.

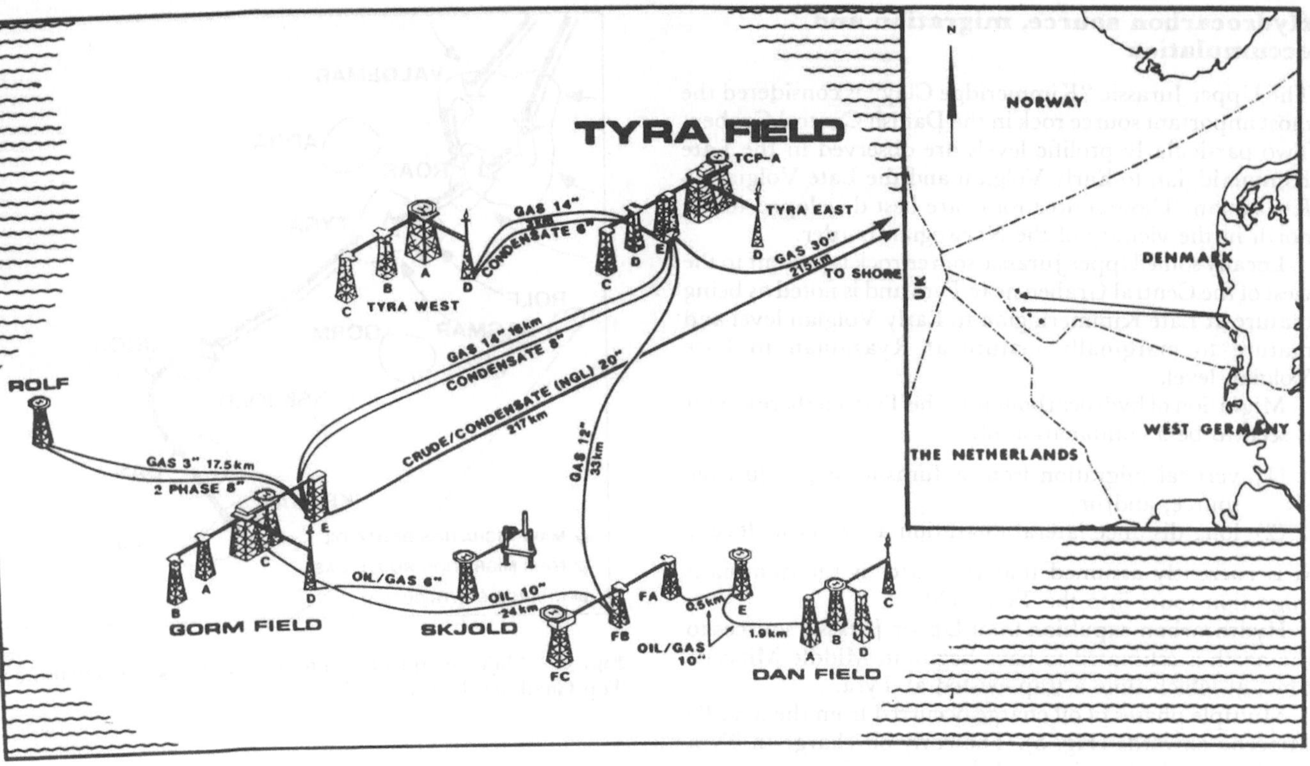


**Fig. 3.** Tyra facilities and pipelines.

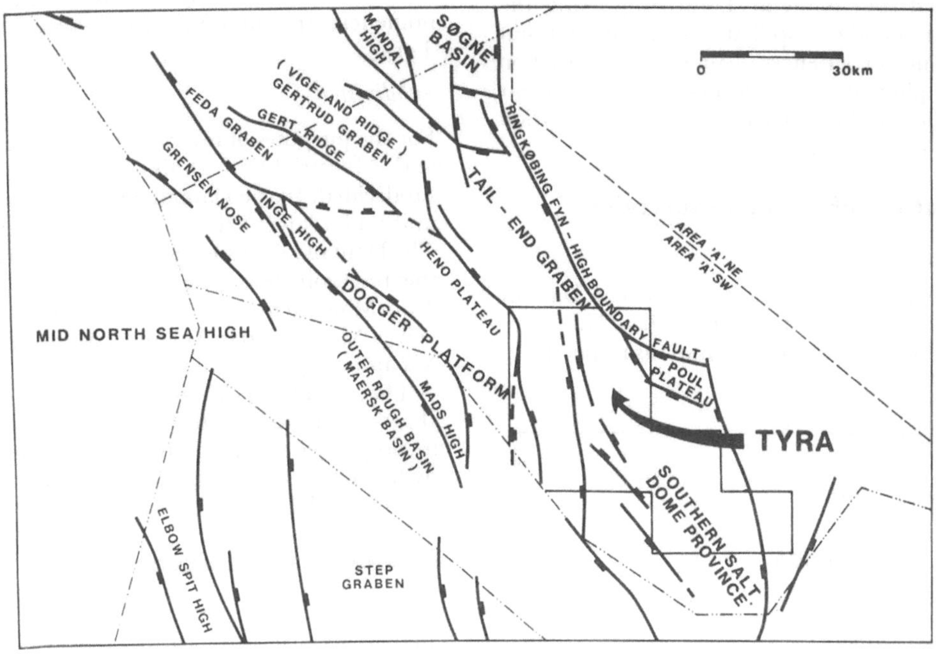


**Fig. 4.** Major tectonic features—Danish Central Graben.

ciated with rifting continued into the Cretaceous and sediment deposition was characterized by general regional subsidence. The Tail End Graben was one of the major depocentres in the Central Graben, with up to 11 km of post Rotliegendes sediment deposited (Michelsen, 1982).

During the Cretaceous and Early Tertiary the area was subject to an increasingly convergent strike-slip regime leading to basin inversion in multiple phases. Evolution of the Tyra structure is thought to be related to the compressional strike-slip regime and inversion tectonics. A major fault zone exists to the west of the Tail End Graben (Fig. 4). Strike-slip movement along this zone may have caused compression and flexuring of strata, sufficient to cause the structural expression seen at Tyra, which is known to have existed from the Middle Miocene.

## Structural configuration

The Tyra structure at top Danian chalk level is a broad low-relief domal anticline (Fig. 2). Maximum vertical relief is 250 ft and at the lowest closing contour of 6600 ft ss the structure covers an area of approximately 9 km by 7 km. The flanks dip between 1.5° and 3.5°. The crest of the structure is fairly flat and has the greatest vertical relief to the east. Faulting is of minor importance in Tyra.

## Hydrocarbon source, migration and accumulation

The Upper Jurassic "Kimmeridge Clay" is considered the most important source rock in the Danish Central Graben. Two particularly prolific levels are observed in the Late Kimmeridgian to Early Volgian and the Late Volgian to Ryazanian. Those source rocks are best developed to the north in the vicinity of the Norwegian border.

Locally some Upper Jurassic source rock is present to the west of the Central Graben near Tyra and is noted as being mature at Late Kimmeridgian to Early Volgian level and mature to marginally mature at Ryazanian to Late Volgian level.

Migration of hydrocarbons into the Tyra chalk reservoir is seen to be a combination of:

(1) vertical migration from a Jurassic or pre-Jurassic source, and/or
(2) long distance lateral migration at reservoir levels.

It is currently assumed that the latter is the dominant migration route into the Tyra field.

Hydrocarbon expulsion from Upper Jurassic sources to the north is estimated to have begun in Middle Miocene time, at which time a trap existed at Tyra.

Multiple phases of oil charge occurred from the Middle Miocene onwards (Fig. 5). The early oil charge in Tyra was later flushed to the south by gas/condensate charge, the oil spilling into the Dan and Kraka fields (Fig. 6).

Residual oil is encountered down to 6800 ft ss, 200 ft below the mapped spill point at Tyra, supporting the concept of migration and overspill and the likelihood that the reservoir has not experienced a unique hydrocarbon charge. A similar phenomenon is reported from the Roar field, which also lies on the major hydrocarbon migration routes illustrated in Figs. 5 and 6.

## Reservoir stratigraphy and subdivision

*Stratigraphy*

The reservoir rock in Tyra consists of Palaeocene Danian and Upper Cretaceous Maastrichtian chalks. These are equivalent to the Ekofisk and Tor Formations respectively

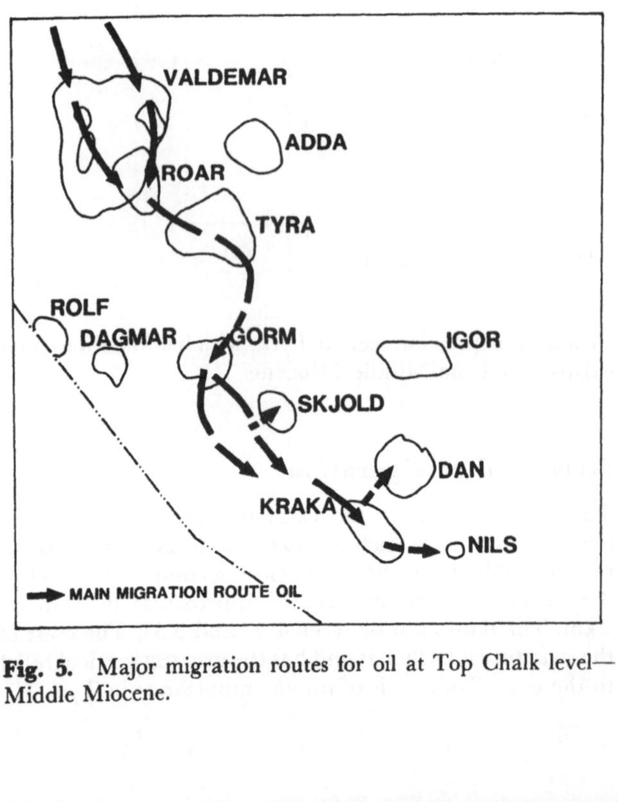


**Fig. 5.** Major migration routes for oil at Top Chalk level—Middle Miocene.

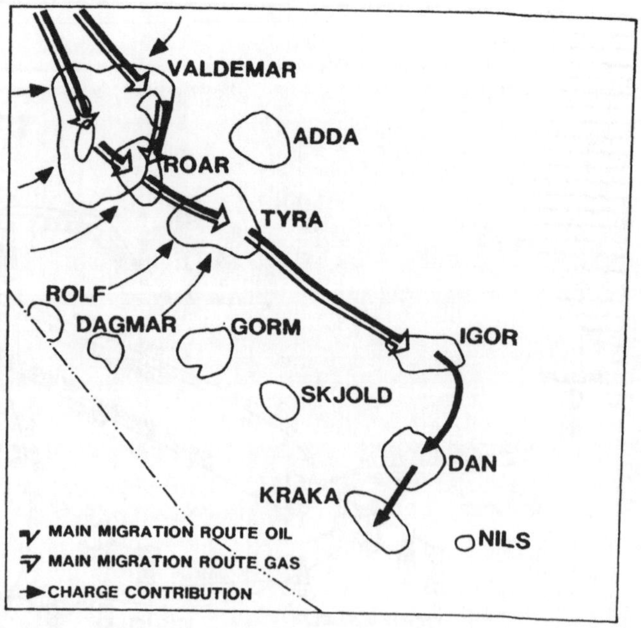


**Fig. 6.** Major migration routes for oil and gas/condensate at Top Chalk level—present day.

(Deegan and Scull, 1977) and Chalk Units 6 and 5 as defined by the Geological Survey of Denmark (DGU) (Michelsen, 1982). Figure 7 illustrates the stratigraphic nomenclature and designated reservoir units pertinent to Tyra.

*Subdivision*

For the purposes of reservoir characterization, the Danian and Maastrichtian chalks in Tyra have been further subdivided. In the Danian, an upper "porous" Danian D1 unit and a lower "tight" Danian D2 unit are recognized. The Danian D2 possesses four correlatable subunits but for this reservoir description they will be treated as one unit. Two Maastrichtian units are identified within the reservoir, a Maastrichtian M1 and a Maastrichtian/Campanian M2 unit (Fig. 7).

The definition of these units is based on reservoir characteristics as identified on logs and these can be related to properties of the chalk as seen in core material. The hydrocarbon accumulation at Tyra is mainly contained within the DI, D2 and M1 units.

## Lithology and diagenesis

*Lithology*

The chalk in Tyra is a fine-grained carbonate sediment consisting largely of the skeletal debris of pelagic unicellular algae. The skeletal material comprises spherical coccospheres of up to 30 $\mu$m that are made up of smaller elements or coccoliths. Most coccospheres are disaggregated and coccoliths are broken up either by depositional or compaction processes. The chalk is generally clean, although more clay-rich intervals occur.

The coccolith material forms a matrix in which coarse particles and grains are enclosed. Particulate matter includes foraminifera, calcispheres and sponge spicules.

*Diagenesis*

The Danian and Maastrichtian chalks in Tyra have undergone several processes of diagenetic alteration including compaction, cementation, silicification and stylolitization. These processes have generally been pore destructive.

| *Tyra field* | | | |
|---|---|---|---|
| *Chronostratigraphic age* | *Formal chalk formations (Deegan and Scull, 1977)* | *DGU central graben study* | *Maersk oil reservoir units* |
| | | | D1 |
| Danian | Ekofisk formation | Chalk-6 | D2 |
| Maastrichtian | Tor formation | Chalk-5 | M1 |
| | | | M2 |

**Fig. 7.** Tyra field stratigraphic nomenclature and reservoir units.

Three major types of cementation have been identified:

- Spot welding of grain-to-grain contacts that suggests a very early diagenetic origin (Mapstone 1975, Hardman 1982).
- Calcite or silica pore-fill cement, probably of late diagenetic origin.
- Selective calcite overgrowths on particles such as coccolith plates and microfossils, also of late diagenetic origin.

The lithological composition and subsequent diagenetic history of the Tyra chalk units are seen to have a fundamental influence on reservoir properties.

### *Chalk units*

*Danian D1* The D1 is a foraminiferal mudstone and contains 89% matrix components. Chert forms less than 1% of the bulk rock volume and is finely disseminated in the chalk.

The unit is characterized by loosely packed matrix components and cementation is restricted to grain-to-grain spot welding, with only minor pore-space cementation and selective calcite overgrowth.

The majority of the porosity is interparticle. Foraminifera and shell fragments are largely recrystallized. Some minor intraparticle porosity is seen within foraminifera chambers. Stylolites are occasionally present in the unit.

*Danian D2* The D2 is layered, contains argillaceous laminations, chert and pyrite and is of poor reservoir quality relative to the overlying Danian D1 and underlying Maastrichtian chalks. In places, the chalk is heavily bioturbated. Pore-fill cementation is advanced and is both calcitic and siliceous. A silica content of up to 20% has been recorded in places.

Chert is common in the D2 and frequently occurs in the form of massive nodules or discrete bands and layers.

*Maastrichtian M1* The M1 chalk is a foraminifera-rich and calcisphere-rich mudstone to wackestone. Interbedded in the unit are thin layers with angular floatstones that are most probably intraclasts.

The chalk is relatively soft, has undergone very little diagenetic alteration and has been mainly affected by mechanical compaction. This has caused fragmentation of the majority of coccospheres and coccoliths. Only minor amounts of pore-fill cementation and calcite overgrowth have been observed. Stylolites are abundant and have a much greater amplitude than the few observed in the D1 and contain dark argillaceous material, pyrite and hydrocarbon staining.

The upper boundary of the unit is defined by a hardground that is strongly cemented and can be traced as a regional event.

*Maastrichtian M2* The M2 is a dense, calcisphere-rich mudstone to wackestone. The matrix content is considerably less than that of the other Tyra reservoir chalks owing to rather high content of calcispheres. Sponge spicules and pyrite are rare and the unit is finely laminated.

The M2 is better cemented than the M1 unit and is therefore more competent. Both pore space and overgrowth cementation are seen. Intraparticle porosity, however, is significant.

There are limited core data available from the M2 in Tyra since it typically occurs towards the base of or below the hydrocarbon column.

Scanning electron microscope (SEM) photographs of the four reservoir units are shown in Fig. 8. The photographs illustrate variations in cementation and argillaceous content that affect reservoir quality of the chalk units.

## Reservoir properties

### *Porosity and permeability*

An outstanding feature of the Tyra chalk is the extremely high porosity, frequently in excess of 50%. The preservation of porosity in Tyra is probably a combination of early overpressuring, reducing the effect of mechanical compaction and pore-space reduction and the inhibition of later diagenetic processes owing to early hydrocarbon migration into and through the structure. Air permeabilities show a fairly wide range from 0.3 mD to 25 mD. All permeability values discussed are non-Klinkenberg-corrected.

The type log in Fig. 9 shows a typical reservoir profile for Tyra. The D1 is a fairly homogeneous unit, whereas the D2 shows considerable vertical variation, which had led to a more detailed reservoir subdivision. The Maastrichtian units are defined by subtle porosity changes. Between the D2 and M1 the Maastrichtian hardground is identified by a marked decrease in porosity.

Horizontal air permeability versus core porosity correlations have been derived for the Danian D1, D2 and

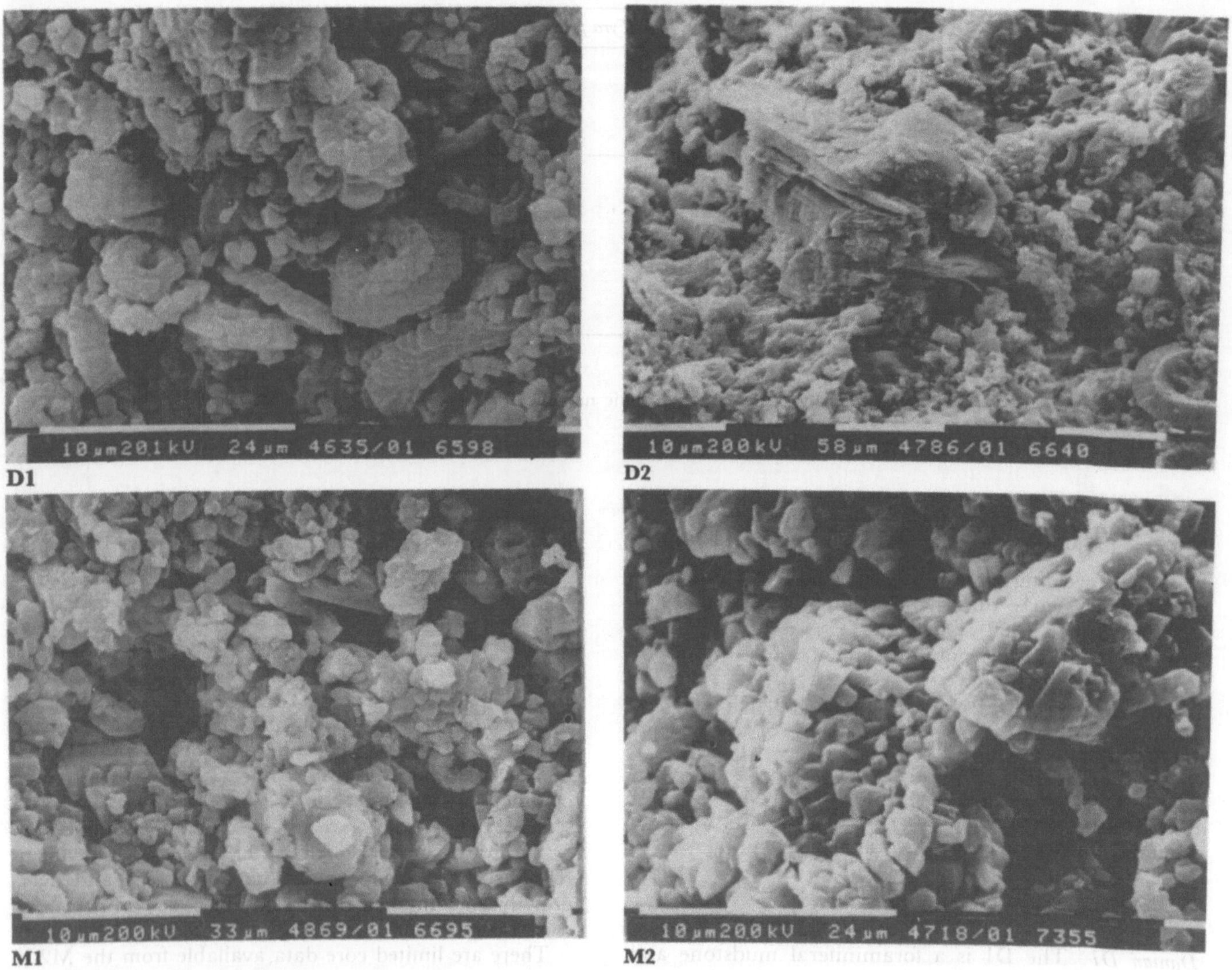


**Fig. 8.** SEM photographs of the Danian and Maastrichtian units, illustrating variations in reservoir quality.

Maastrichtian M1, M2 units. A composite plot of these correlations is shown in Fig. 10. The length of the line for each unit indicates the spread of porosity and permeability values measured. For given porosity values, Danian chalks possess lower average permeabilities than Maastrichtian chalks.

Within the Danian interval, the homogeneous D1 possesses the highest porosities, i.e. between 35% and 50% with air permeabilities generally ranging from 1 mD to 12 mD. The loose packing of the matrix components accounts for the high porosities in the D1. Because the D2 unit is more stratified and variable in lithology, a greater range in porosity values occurs, i.e. from 20%–45%. Air permeabilities are seen to range from 0.3 mD to 10 mD. The lower porosities seen in the D2 are due to the greater clay and silica content plus more advanced cementation as compared with other units. The greater degree of diagenesis in this unit is probably promoted by the higher clay content. Figure 10 indicates, however, that higher permeabilities can occur in the D2 as compared with the D1. This may be attributed to a higher content of sponge spicules, forams and other particulate matter increasing interconnected pore space. Nevertheless, the D2 as a whole is of poorer reservoir quality and the well-cemented intervals and interbedded argillaceous laminations can inhibit vertical fluid flow.

The M1, M2 units possess the best reservoir properties in Tyra. As with the Danian D1, the M1 interval has porosities between 25% and 50%, but permeabilities are higher, ranging from 3 mD to 25 mD. In the M2 interval porosities range from 25% to 40% and permeabilities from 2 mD to 15 mD. The high porosities in the M1 can be attributed to the fact that the unit has undergone predominantly mechanical compaction, but otherwise little destructive diagenesis. The lower porosity in the M2 unit is due to a greater degree of pore space cementation or crystal overgrowth. Surprisingly, for a given porosity, M2 chalk possesses a higher air permeability than the M1. As was the case for the D2 unit, the higher permeability reflects a greater percentage of particulate matter.

*Vertical and areal trends*

For each of the reservoir units, porosity and permeability are best preserved where the unit is in a structurally high position and deteriorate towards the flank. A porosity reduction of up to 8% from crest to flank is observed in the Danian and Maastrichtian within well control. This trend is seen in other Danish chalk fields and is attributed to the progressive filling of the reservoir with hydrocarbons. A minimum amount of porosity reducing diagenesis has thus occurred at the crest but this increases downflank towards the aquifer.

For the same reason, porosity and permeability also decrease vertically where a particular unit is located within the transition zone. The exploration well E-1X, drilled just outside the accumulation provides important information

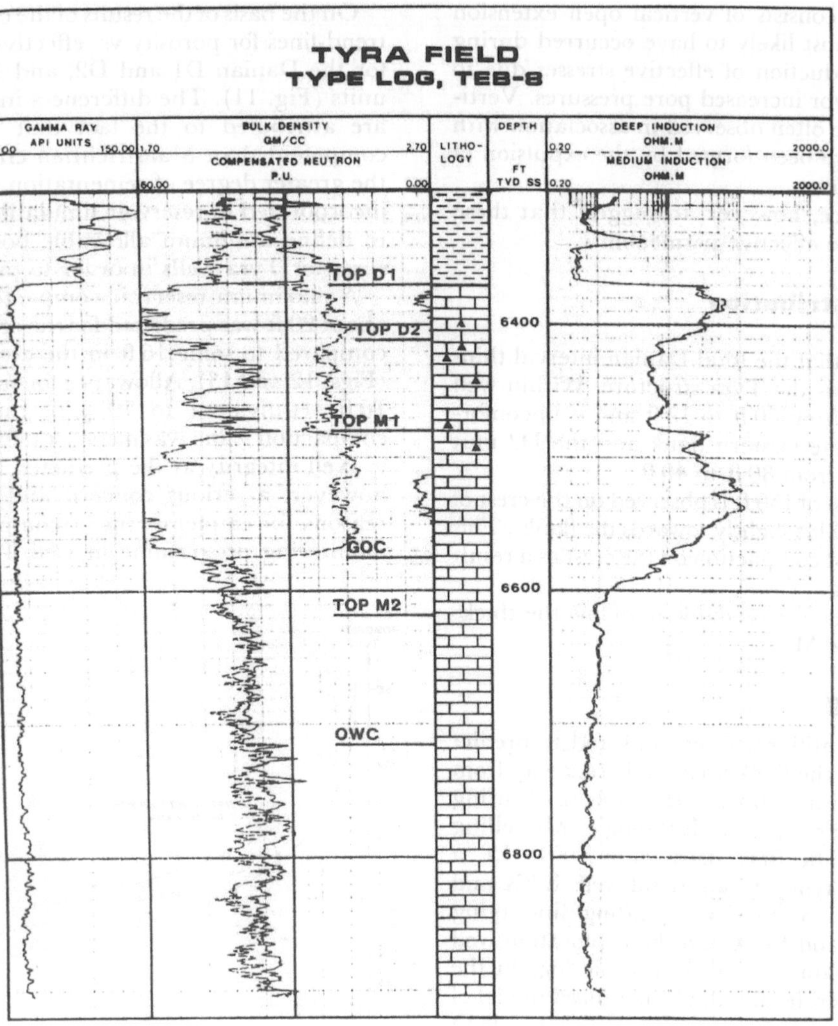


**Fig. 9.** Tyra field type log.

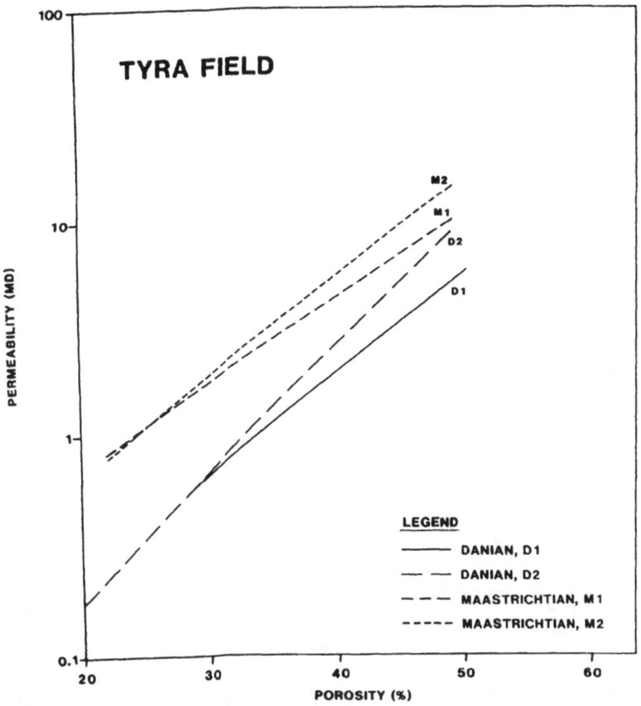


**Fig. 10.** Horizontal air permeability (non-Klinkenberg-corrected) vs. core porosity trends.

on the effect of continued diagenesis within the aquifer and confirms the trends observed.

Vertical permeability variations are of particular interest in relation to Danian/Maastrichtian drainage efficiency and are the key to oil rim and flank gas development.

There is little observed layering and vertical permeability variation in the major reservoir units D1 and M1. The D2, however, is discretely stratified, argillaceous and contains numerous chert beds. Although measured horizontal permeability in places is superior to that of D1, vertical permeability is reduced and may provide a vertical barrier to flow.

The Maastrichtian hardground at the top of the M1 unit is known to reduce drastically vertical flow and may form an effective local seal.

### *Fracturing*

Two systems of fractures have been observed in core material from Tyra. The first system comprises two types of natural hydraulic fractures, namely vertical extension and conjugate dilational shear fractures. These hydraulic fractures were formed relatively early after deposition, probably as a result of water escape during early compaction when only slight lithification of the chalk had taken place.

Both sets of fractures are completely healed with microcrystalline calcite, and therefore have no effective porosity and may reduce permeability locally.

The second system consists of vertical open extension fractures. These are most likely to have occurred during uplift as a result of reduction of effective stresses due to relative extension and/or increased pore pressures. Vertical tensile fractures are often observed in association with stylolites and may have been formed by the expulsion of fluids from the stylolites.

There is no evidence, however, to suggest that these open fractures improve effective permeability.

## Unit thickness distribution

Seismic data indicate that the total Danian interval thins from east to west across the Tyra structure. Within well control, the D1 thins from 60 ft to 12 ft and is unconformably truncated on the western flank and the D2 unit thins correspondingly from 80 ft to 40 ft.

A thick M1, in excess of 150 ft is observed on the crest of the structure, while it thins slightly towards the flanks. This could be due to reduced compaction on the crest as a result of early overpressuring.

Insufficient well data are available to define the thickness distribution of the M2.

## Seismic modelling

In an attempt to indentify reservoir units and to predict reservoir properties in the flank areas of Tyra, away from well control, two seismic lines were processed using Western Geophysical's Seismic Lithologic Modelling (SLIM) technique. The lines used were TY81–15, a NW–SE trending line tying the appraisal wells E-3X and E-4X, and TY81-17, a NE–SW trending line tying appraisal wells E-2X and E-4X and the exploration well E-1X. The velocity data derived from sonic logs in the wells were used to generate an initial generalized model of acoustic impedance for each line. The model was then manipulated by SLIM processing to best match the seismic lines being modelled. Well data were then used to define stratigraphic units. The impedance character of the units could then be compared and variation within units noted.

Plate 3 illustrates the results of processing the top Danian to base Upper Cretaceous interval on line TY81-15. Here the stratigraphic units identified from well data are colour coded. Acoustic impedances values calculated from SLIM are represented by colours ranging from dark blue for low, to bright red for high values. The large area of low impedance between shotpoints 250 and 450 occurs within the Maastrichtian and corresponds to the crestal area where, according to well control, porosities in the Maastrichtian are highest. The higher acoustic impedance on the flanks, indicates a reduction in porosity.

## Compaction

In view of the high porosity, the poor cementation and relatively soft nature of the chalk, the impact of reservoir compaction with regard to both reservoir (and well) performance and jacket/well integrity was addressed at the pre-development stage.

A series of laboratory experiments was initiated with the aim of developing predictive correlations for porosity/permeability as a function of declining reservoir pressure. Triaxial compaction tests indicated chalk compaction to be a function only of effective stress (overburden pressure minus pore pressure). No significant differences in compaction levels were seen when the pore space contained air, gas-oil, artificial formation water or sea water.

On the basis of the results of the compaction experiments, trend-lines for porosity vs. effective stress were established for the Danian D1 and D2, and Maastrichtian reservoir units (Fig. 11). The differences in compaction behaviour are attributed to the fact that Danian chalk is more competent than Maastrichtian chalk, probably owing to the greater degree of cementation. The relationships were incorporated in reservoir simulation studies and also used to define minimum allowable bottomhole flowing pressures for Tyra wells in order to prevent pore collapse.

A maximum reservoir compaction at abandonment of about 20 ft was predicted for the crestal area of Tyra East, compared to some 16 ft in the crestal area of Tyra West (Figs. 12 and 13). Allowance for subsidence levels of some 16 ft—equivalent to 70% of the maximum expected compaction value was incorporated into the jacket designs.

Well integrity at the predicted levels of compaction is, however, a serious concern and was one of the main reasons for implementng a comprehensive compaction monitoring programme in well TEB-9. Some 40 radio-

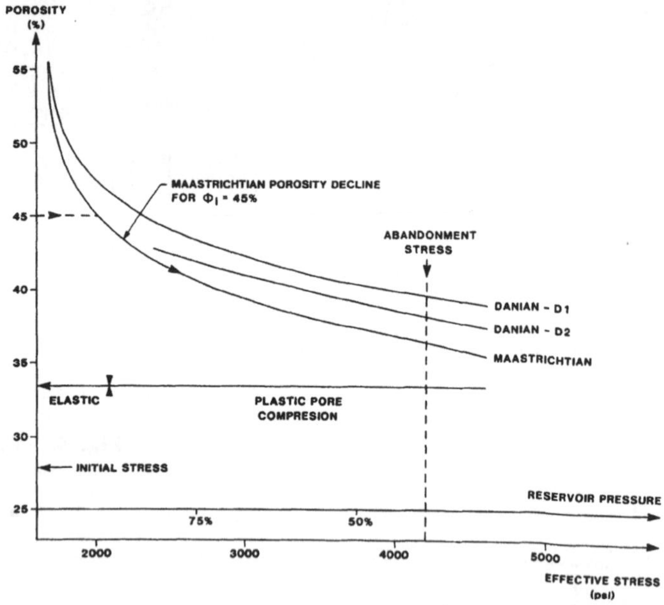


**Fig. 11.** Porosity reduction as a function of effective stress.

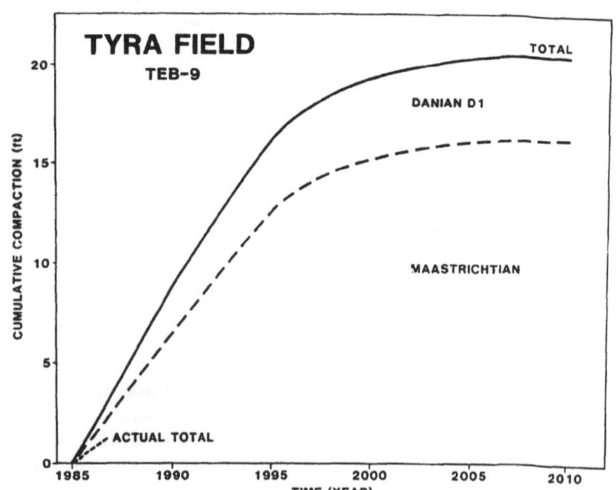


**Fig. 12.** Tyra field compaction predictions (TEB-9). The D2 formation does not contribute to the cumulative compaction. Actual compaction (1987 data) is significantly less than expected.

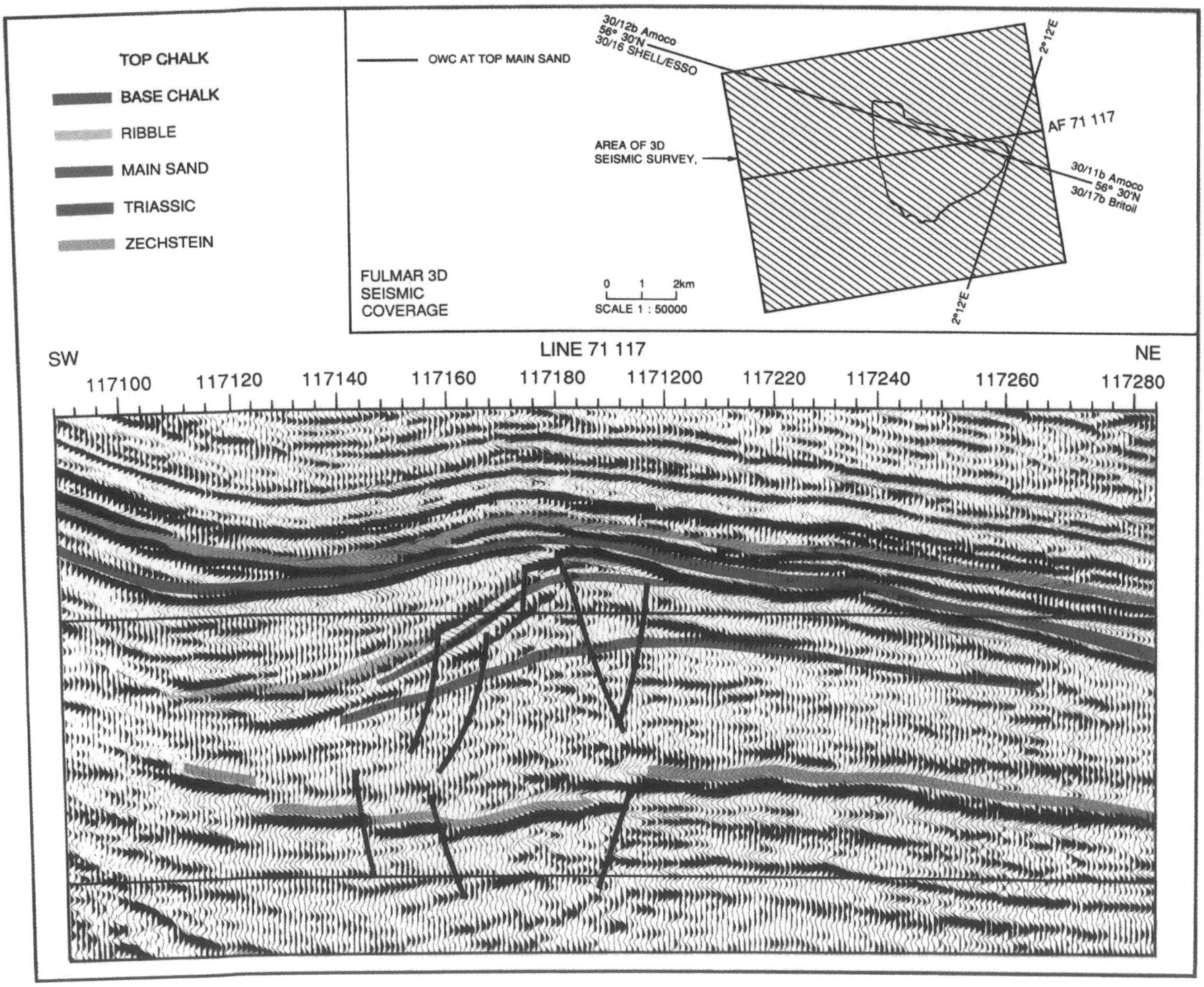


**Plate 1.** Fulmar seismic coverage and representative line interpretation highlighting the significant seismic reflectors.

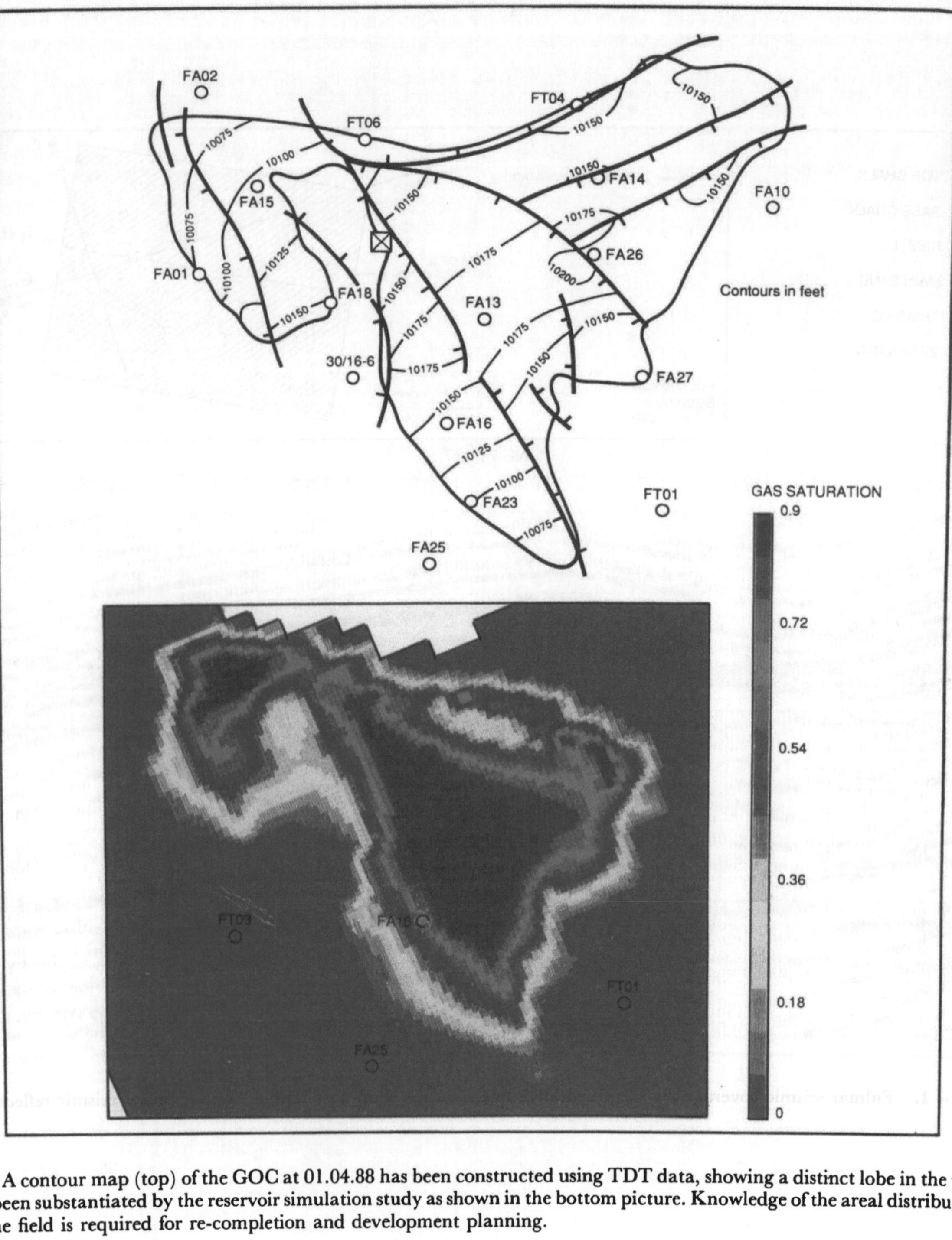


**Plate 2.** A contour map (top) of the GOC at 01.04.88 has been constructed using TDT data, showing a distinct lobe in the west. This shape has been substantiated by the reservoir simulation study as shown in the bottom picture. Knowledge of the areal distribution of the gas over the field is required for re-completion and development planning.

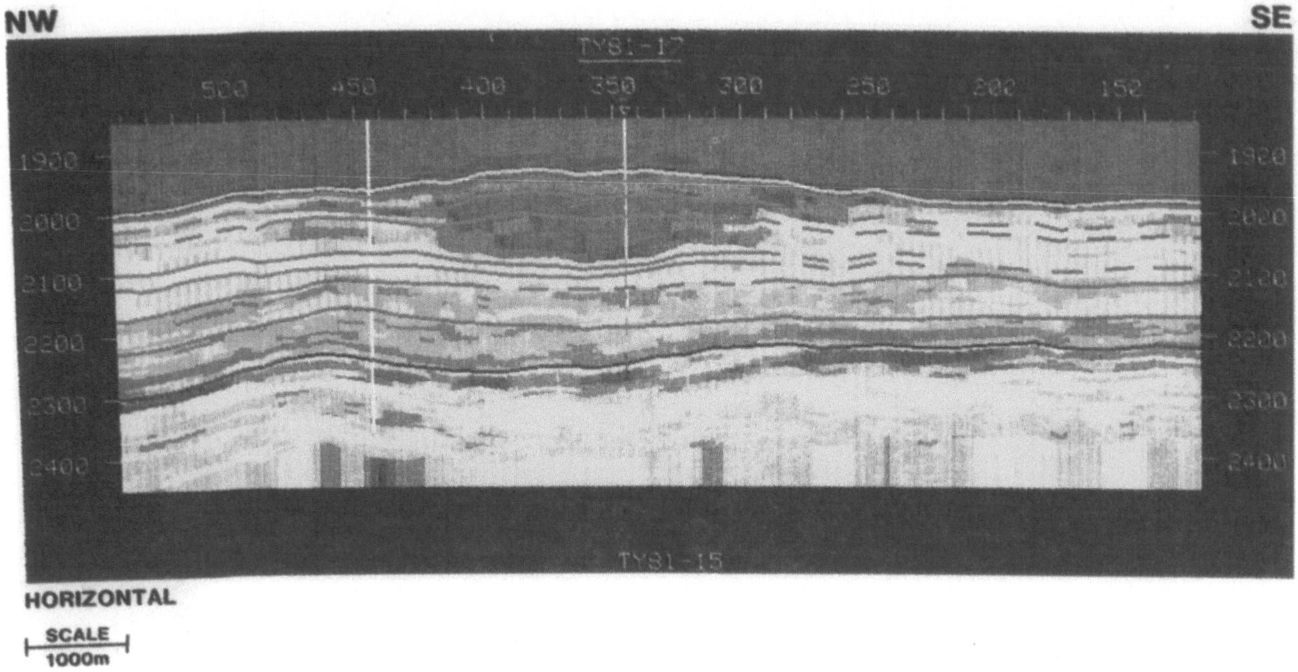


**Plate 3.** Tyra Field, seismic lithologic modelling on line TY81–15. In order from top to bottom, the interpreted horizons displayed are:

(1) Top chalk, Top Danian D1 (orange)
(2) Top Danian D2 (pink)
(3) Top Maastrichtian M1 (light blue)
(4) Top Maastrichtian M2 (light blue)
(5) Top Campanian (light green)
(6) Top Santonian (light green)
(7) Top Coniacian (light green)
(8) Top Turonian (light green)
(9) Base chalk, Base Upper Cretaceous (dark blue)

Acoustic impedance range: low—dark blue; high—bright red.

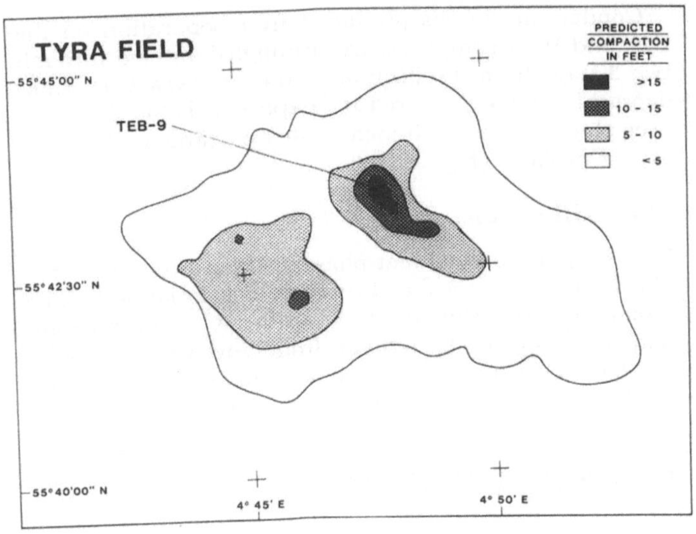


**Fig. 13.** Predicted compaction at field abandonment. Actual compaction is monitored in TEB-9 across the chalk section and the overlying shales.

active bullets were shot in the well, the majority being placed across the gas-bearing chalk from 50 ft above top chalk to 50 ft below the GOC. The rest of the bullets were placed across two intervals in the overlying shales.

A Formation Subsidence Monitoring Log (FSMT) was run in October 1984, and the distance between the bullets was determined prior to the start of production from Tyra.

TEB-9 was re-logged with the FSMT in November 1987. Total compaction of the gas-bearing chalk column was only 0.5 ft compared to approximately 2.5 ft predicted for the prevailing reservoir pressure. This low compaction in the early decline period is encouraging and reinforces our reservations regarding the simplistic assumptions on reservoir layering used in prediction models.

In addition to FSMT surveys in TEB-9 the exact height of the Tyra platforms is monitored using the Navstar/Global satellite positioning system. These surveys will thus monitor the seabed subsidence around the platforms.

## INITIAL FLUIDS IN PLACE

### Initial fluid distribution

The gas column which is contained almost exclusively in the D1, D2 and M1 reservoirs has a height of some 225 ft in Tyra East and 175 ft in Tyra West.

The GOC is assumed to be more or less horizontal throughout the field at a level of around 6570 ft ss. Away from the crest, where the GOC is located in the D2, precise definition of the contact level is hampered by the clay content of the D2 unit.

The oil zone thickness from the GOC to the 100% SW level is 100–200 ft. However, only oil at saturations above 50% is considered as movable. The 50% saturation level varies between 6570 ft ss and 6600 ft ss, indicating a maximum thickness of the movable oil rim of some 20–30 ft. The fieldwide distribution of the movable oil rim is poorly defined but in general thickens from north to south.

The Maastrichtian movable oil rim which contains virtually all the assumed mobile black oil is absent towards the northeast, and thickens to a maximum 30 ft towards the southwest.

### PVT properties

Accurate PVT data are essential for evaluation of gas/condensate reservoirs (especially when operated under a dry-gas recycling scheme). Difficulties in obtaining representative PVT samples arise in Tyra because the saturation pressure is close to initial reservoir pressure.

Sampling was carried out on three wells (TWB-8, TEB-8 and TWB-6) and compared with analyses of PVT samples from the exploration wells E-2x and E-4X. An average dew-point pressure close to initial reservoir pressure (4254 psia) is indicated with a maximum liquid dropout saturation (at 1250 psia) of ca. 5%. No retrograde behaviour is expected under depletion since the abandonment pressure will be substantially higher than 1250 psia.

Produced CGR values range from 30 to 43 bbl/mmscf across the Tyra field. No conclusive PVT evidence is available to diagnose whether areal (E–W) or vertical (Danian–Maastrichtian) differences in reservoir fluid composition exist.

Eastern wells, which are predominantly completed in the Danian, generally produce with lower CGR's than those to the west, where several wells have been perforated in the Maastrichtian. This lends support to the concept of a richer condensate in the Maastrichtian.

Alternatively, the key variable may be the flowing bottom hole pressure. Wells producing at a high drawdown (low FBHP) will experience a higher condensate dropout around the well, and consequently produce at a lower CGR. Generally, Maastrichtian completions produce at higher FBHPs than Danian completions, owing to generally higher PI. This issue is under review.

Tyra fluid properties can be summarized as follows:

| | |
|---|---|
| Reference elevation | 6500 ft ss |
| Initial reservoir pressure: | 4254 psia at 6500 ft ss |
| Initial reservoir temperature: | 160°F |
| Saturation pressure/dew point: | 4254 psia (assumed) |
| Gas gravity: | 0.64 (air = 1) |
| Gas FVF (Bgi): | 0.685 rb/mscf* |
| CGR (initial): | 50 stb/mmscf |
| Condensate gravity: | 56° API (initial) |
| Black oil FVF (Boi): | 1.4 rb/stb |
| Solution GOR (Rsi): | 960 scf/stb |
| Black oil gravity: | 33° API |
| Formation water salinity: | 26 000 ppm equivalent NaCl |
| Formation water resistivity: | 0.264 ohm. m. at 77°F (TWB-7) |

### Fluids in place

The distribution of hydrocarbons in place in the Tyra field, which has been calculated volumetrically, is summarized below.

#### *Gas in Place*

Of the gas in place, 73% consists of free gas, whereas 27% is solution gas. The distribution of the free GIIP, which has an important influence on recovery potential owing to

* Prefixes: m = $10^3$; mm = $10^6$

the thin gas column in the flank area, is as follows:

| | Crestal area | Flank area | Total |
|---|---|---|---|
| Danian D1 | 27 | 25 | 52 |
| Danain D2 | 27 | 2 | 29 |
| Maastrichtian | 19 | — | 19 |
| Total | 73 | 27 | 100 |

(All figures are per cent).

The flank area in this context is defined as being 1000 m or more away from current crestal production wells.

*Liquids in place*

The liquid in place consists of condensate (11%) and black oil (89%). The approximate distribution of the STOIIP between the movable oil rim ($S_0 > 50\%$) and the transition zone shown below confirms that the Maastrichtian rim is the key target area:

| | Movable rim | Transition zone | Total |
|---|---|---|---|
| Danian D1 | — | 9 | 9 |
| Danian D2 | — | 29 | 29 |
| Maastrichtian | 17 | 45 | 62 |
| Total | 17 | 83 | 100 |

(All figures are per cent).

## PRODUCTION AND PROCESSING FACILITIES

### Processing and evacuation

The Tyra field facilities that were completed in 1984 consist of two platform centres located to the east and west of the Tyra crestal area to allow optimum well spacing. The two centres are located more than 3 km apart. Each centre contains two 12-slot wellhead platforms bridge-connected to gas processing/accommodation platforms.

On each of the two main processing platforms, two 3-phase separation (gas, condensate and water) and dehydration trains have been installed. The Tyra East main processing platform has additional facilities for condensate stabilization and hydrocarbon dew-point control.

The gas and condensate processing capacities for the Tyra East and Tyra West gas process platforms are:

| | Gas (mmscf/d) | Condensate (stb/d) | Water (b/d) |
|---|---|---|---|
| Tyra East | 2 × 155 | 23 000 | 12 000 |
| Tyra West | 2 × 155 | — | 12 000 |

On Tyra West a re-injection compressor with a capacity of 180 mmscf/d has been in operation since mid 1987.

After separation and dehydration, the produced gas is dewpoint controlled in two chilling trains, prior to compression, metering and export via the 30-inch DONG-owned pipeline to shore.

The capacity of the five gas-export compressors on Tyra East depends upon the suction/discharge pressure levels. At 800 psi suction and 1600 psi discharge, the total capacity is some 600 mmscf/d.

The gas export system that is owned by DONG comprises a pig launcher and auxiliary equipment located on the riser platform Tyra East "E" and a 214 km 30-inch gas line to the Nybro treatment plant onshore. The pipeline has an ultimate capacity of 920 mmscf/d at 2000 psi pipeline pressure and an arrival pressure onshore of 1200 psi.

Condensate liquids produced from separation on the East and West complexes are comingled and fed through two 3-stage flash-stabilization trains on Tyra East. Condensate liquids are metered and exported via the Tyra East riser platform and an 8-inch subsea pipeline to Gorm for export to shore (Fig. 3).

### Slot utilization

In the initial development phase, 36 wells were drilled—nine from each of the four 12-slot wellhead platforms. The remaining three slots on each wellhead platform represented contingency for infill drilling. However, one slot on TE-B has been abandoned and one slot on TE-C has been suspended, leaving 10 spare slots in total.

## GAS DEVELOPMENT

### Development studies

In the lead-up to the sales contract negotiations in 1978/79, reservoir studies were carried out based on data from the four exploration/appraisal wells. A free gas recovery factor of some 60% was predicted and a field-development plan with 36 wells at approximately 140-acre spacing was defined.

Following development drilling but prior to significant production, a major reservoir simulation study was undertaken. The study included both single well and full-field 3D modelling.

*Single well modelling*

The objectives of the single well modelling were:

(1) to investigate oil and water coning behaviour;
(2) to evaluate the effects of pressure-dependent permeability and porosity on deliverability and gas recovery.

Single well models were set up for both Danian and Maastrichtian prototype completions, from which it was concluded that the Danian completions would achieve higher recoveries than the Maastrichtian ones. For both cases there was a considerable reduction in recovered gas when pore collapse was included. However, it was recognized that the simulation results were likely to be rather pessimistic in these single well (compaction) runs, since the relationships describing the impairment process were not strictly applicable for near well-bore conditions.

*Full-field modelling*

The objective of the 3D full-field model was to predict future gas availability and recovery. Specific topics to be investigated were:

- solution gas recovery from the oil rim;
- the effects (fieldwide) of pressure-dependent permeability and porosity.

A full-field model was constructed with the features listed below:

(1) a petrophysical and geological description based on the 36 development wells, which were incorporated with their original completion configuration;
(2) a compaction process modelled using an irreversible pressure-dependent permeability;
(3) a maximum drawdown constraint of 1500 psi;
(4) a modest level of aquifer support.

With field production having just commenced, no attempt was made to history-match.

Two important categories of undeveloped reserves were identified.

(1) The Maastrichtian completions were predicted to water out early, presenting an opportunity to recover additional gas through upward recompletion in the Danian.

(2) Considerable quantities of gas remained in the crestal areas of the field, to be picked up by a simple infill drilling campaign using the 10 spare slots (Fig. 14).

### *Ultimate gas recovery*

The expected ultimate gas recovery for Tyra, which reflects these undeveloped gas reserves, can thus be summarized as follows:

| | *UR* |
|---|---|
| Developed (existing 36 development wells) | 50 |
| Undeveloped | |
| Danian recompletions | 4.5 |
| Crestal infill wells | 4.5 |
| | 59 |

Ultimate recoveries are given as percentages of the free gas in place, thereby neglecting the small contribution from the solution gas.

The areal breakdown of recovery is illustated below:

| | *GIIP* | *RF* | *Total UR* |
|---|---|---|---|
| Crest | 73 | 72 | 53 |
| Flank | 27 | 22 | 6 |
| Free gas total | | | 59 |

The somewhat low overall recovery factor of 59% for free gas is explained by the low gas recovery from the eastern flank area (Fig. 15). Of the total free gas in place, some 27% is located in the so-called flank area, from where recovery is by migration into the crest leading to a recovery factor of only 22%. From this the potential of a flank development campaign is evident. The predicted recovery factor for free gas within the drainage area of the existing wells is some 72%.

## Reservoir performance and evaluation

### *Production performance*

First contractual gas deliveries were made from the Tyra field in October 1984—the start of the first "contract year". Gas sales, as nominated by DONG, follow a simple seasonal pattern as illustrated in Figs. 16 and 17. Tyra in effect plays the role of swing producer, supplementing the Dan and Gorm associated gas streams, which are also dedicated to the contract.

Gas potential (in late 1988) from 35 active wells is 706 mmscf/d, comprising 412 mmscf/d from Tyra West (average per well: 24 mmscf/d) and 294 mmscf/d from Tyra East (average per well: 16 mmscf/d). Excluding the present eight injectors, the potential of Tyra West is 215 mmscf/d. Total injectivity is 229 mmscf/d, compared to the injection compressor capacity of some 180 mmscf/d.

Actual reinjection on any one day is determined by what remains after the buyer's daily nomination. This nomination shows a strikingly similar cumulative distribution over each of the first four contract years in relation to the two key contract volumes: the MDR (maximum daily rate which the seller is obliged to be able to deliver) and the DCQ (annual contract volume on a daily basis). This almost linear relationship is used for forecasting the volume of recycling gas available as a function of the declining Tyra availability (Figs. 18 and 19).

### *Reservoir management and evaluation*

The main recovery mechanism is considered to be simple depletion supported by a very moderate water drive. The key issues as regards well performance are bottom water coning and well productivity deterioration due to pore collapse or condensate dropout.

After 4 years of production, some 8% of the free GIIP has been produced in Tyra. Hence, it is too early to make definitive statements about field behaviour. Nevertheless, some trends are evident from data acquired during pressure survey campaigns carried out routinely in the low-demand summer period. In Fig. 20, average well pressures are compared to a predicted pressure decline derived by assuming that only crestal GIIP volumes are active in the early producing years, mitigated later by

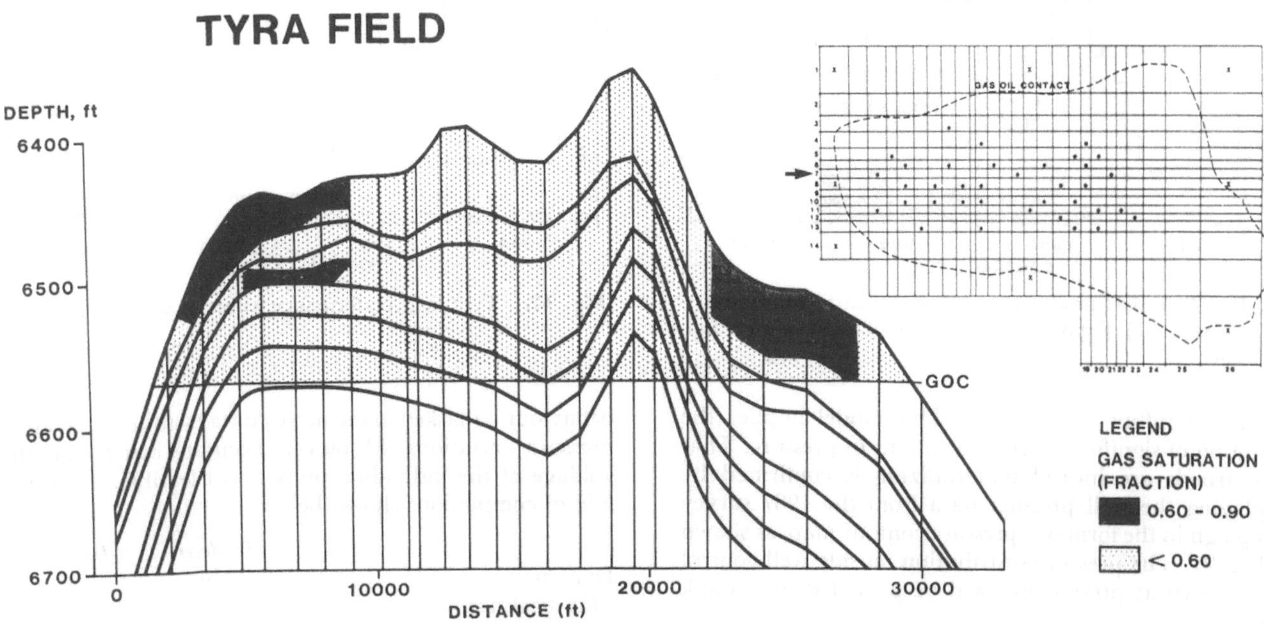


**Fig. 14.** Undepleted areas at field abandonment (36 well development).

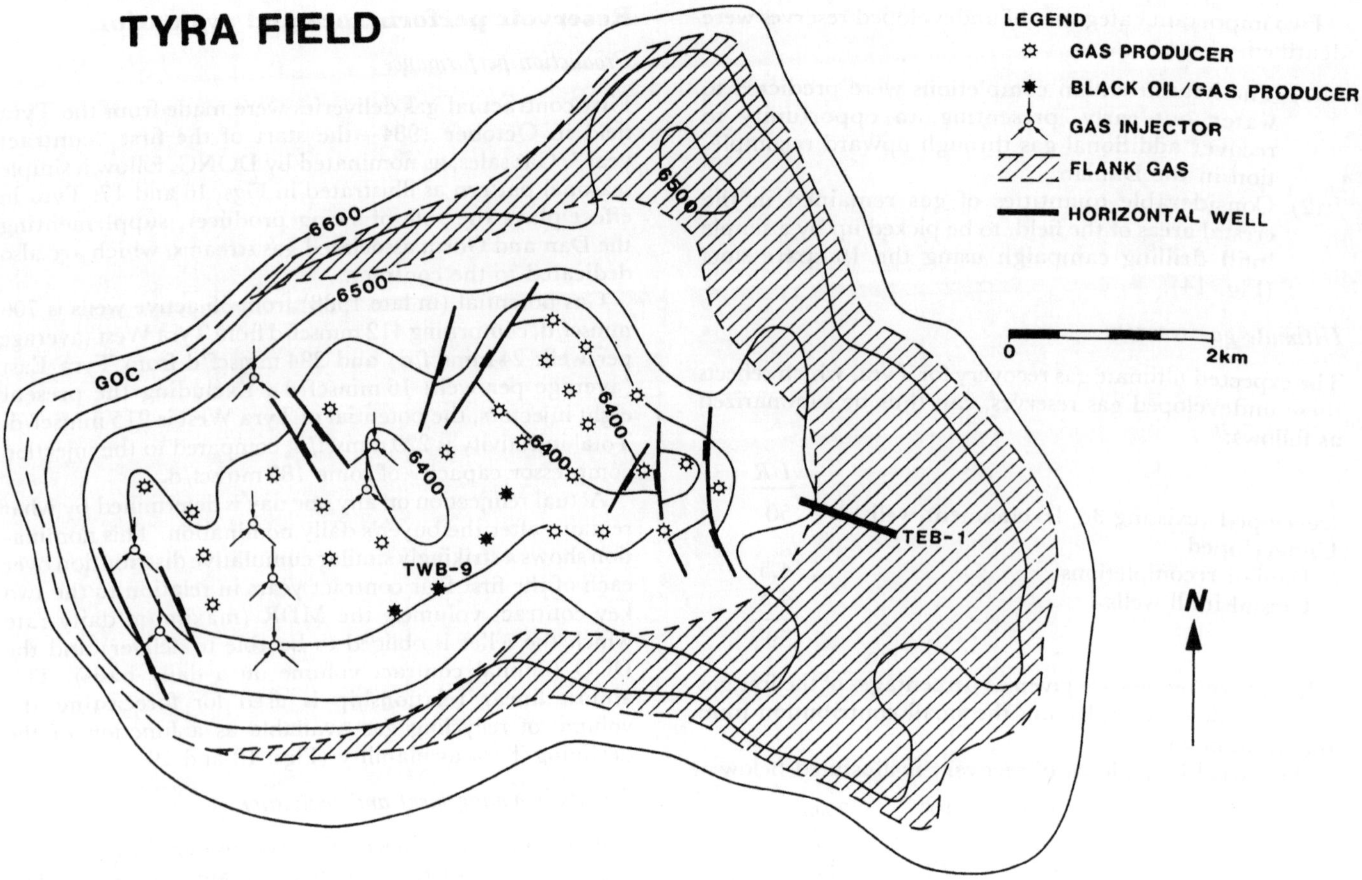


**Fig. 15.** Top Chalk structure map indicating crestal and flank areas. Flank area is defined as being 1000 m beyond well control. Location of two new wells: TWB-9 (black oil producer) and TEB-1 (horizontal flank gas producer) are shown.

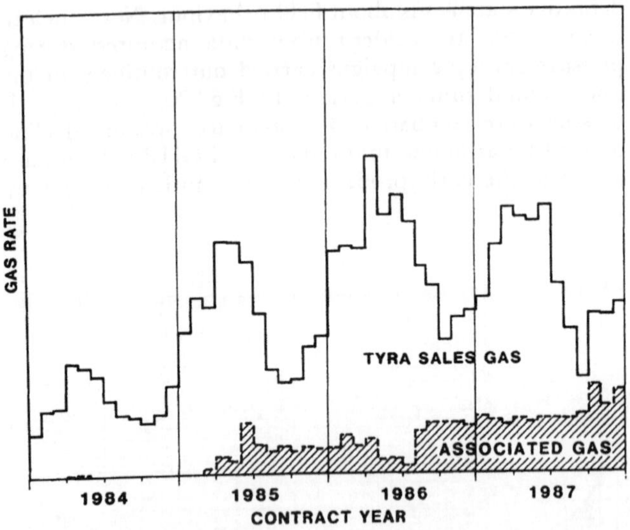


**Fig. 16.** Sales gas production during the first four contract years (1984–1987). Export gas is composed of associated gas and Tyra sales gas.

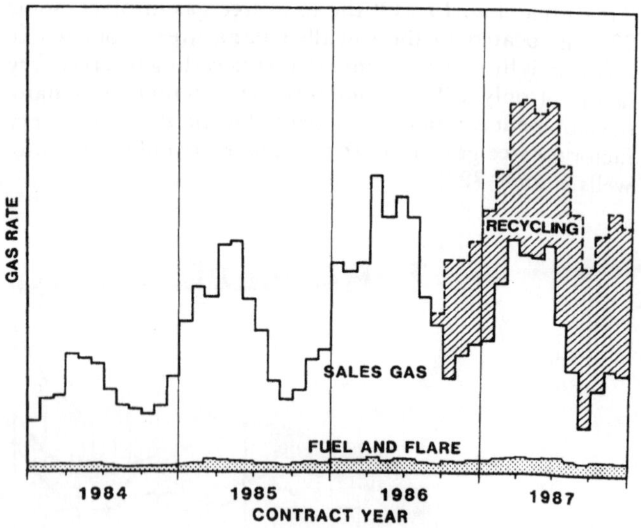


**Fig. 17.** Tyra field performance, contract years 1984–1987.

a delayed influx of flank gas. This model implies the evolution of significant areal variations in pressure. That such trends are indeed materializing is confirmed by displaying the well pressure data from the 1987 survey campaign in the form of a pressure contour map, as shown in Fig. 21. The pressure distribution outside well control is unknown at present but will be provided by a flank well.

The issue of whether the Danian/Maastrichtian permeability characterization is reflected in layered system behaviour is masked in the survey data by the evident areal pressure variations. Moreover, accurate control on the voidage of the individual units is not feasible, given the mix of completions shown below:

| | *Producers* | *Injectors* |
|---|---|---|
| Danian | 16 | 3 |
| Maastrichtian | 5 | 4 |
| Danian and Maastrichtian | 6 | 1 |
| | 27 | 8 |

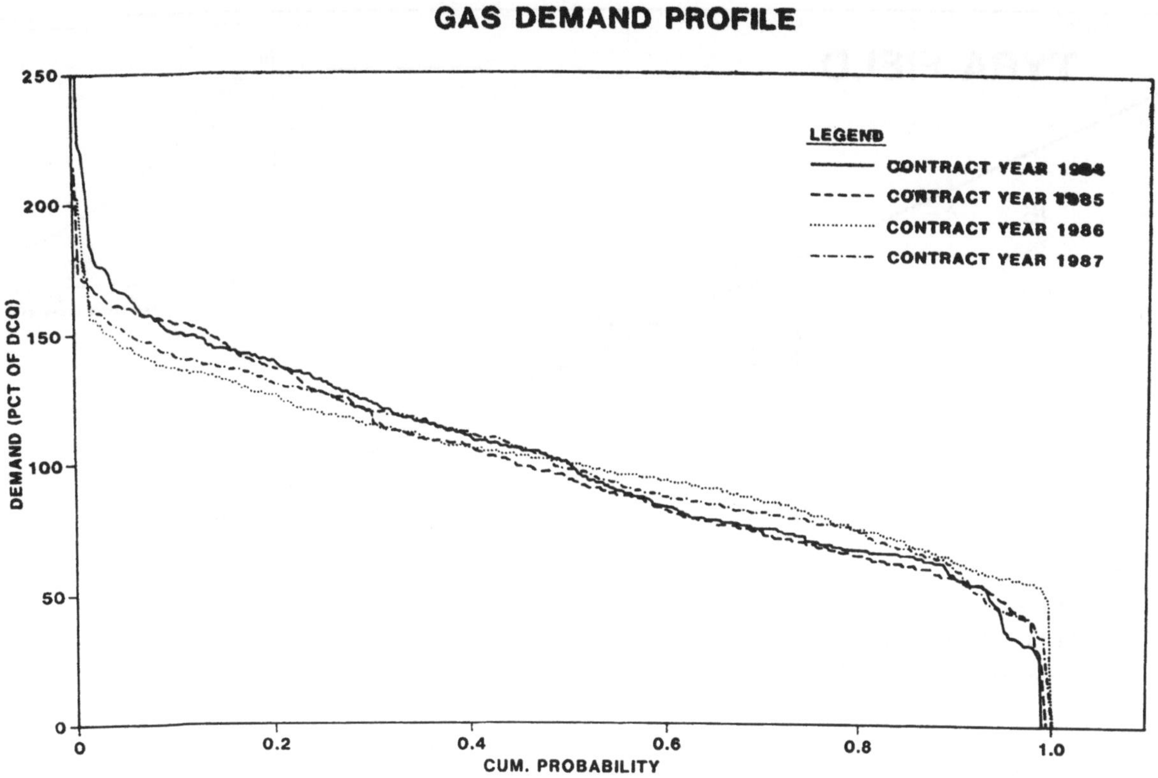


**Fig. 18.** Sales gas demand profile for the first four contract years.

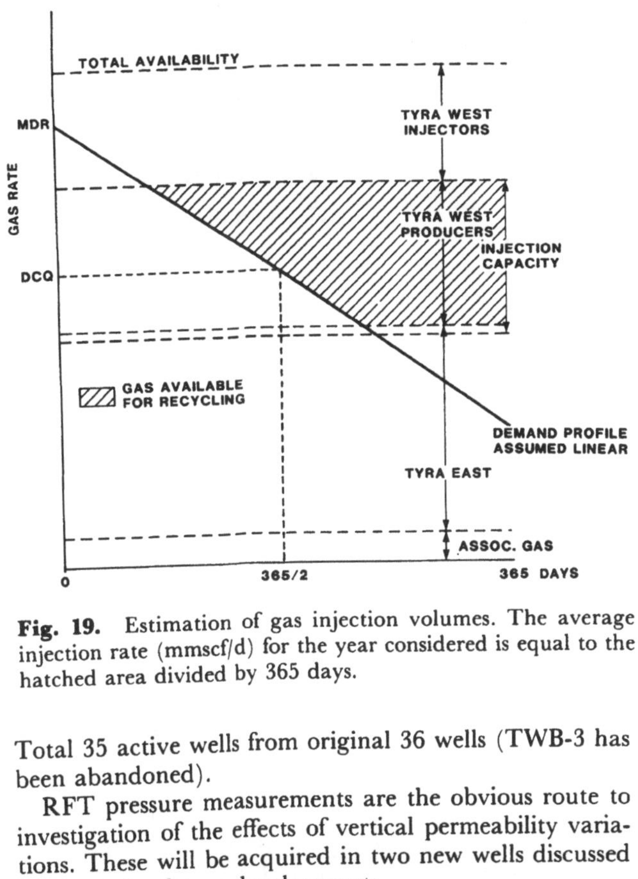


**Fig. 19.** Estimation of gas injection volumes. The average injection rate (mmscf/d) for the year considered is equal to the hatched area divided by 365 days.

Total 35 active wells from original 36 wells (TWB-3 has been abandoned).

RFT pressure measurements are the obvious route to investigation of the effects of vertical permeability variations. These will be acquired in two new wells discussed below under future developments.

The geological model, which predicts permeability decay into the aquifer, would appear to rule out substantial aquifer movement. On the other hand, significant water production has been observed in a small number of wells completed in the Maastrichtian, notably black oil producer TWB-7 and combined Danian/Maastrichtian gas producer TEC-12, and hence some limited aquifer influx could materialize. Again, RFT data will provide clues in this area.

Recognizing the damaging impact of pore collapse on well productivity, a relatively conservative policy of constrained minimum bottomhole flowing pressures has been implemented in Tyra since the start of production. This may well have pre-empted early problems but it is no guarantee in the longer term.

## Gas recycling

### *Project evaluation*

Owing to the low sales gas volumes required during the first few contract years, a dry-gas recycling scheme was proposed in Tyra West in order to utilize the excess gas production potential. Both accelerated and increased condensate recovery were anticipated—the latter since condensate is recovered prior to pressure depletion and consequent loss due to dropout in the reservoir.

To evaluate the sweep efficiency of gas recycling, analytical studies using a simple model for stratified layers were made. Thereafter the results from the analytical calculations were validated by numerical simulation of which the principal objective was to obtain a "feel" for the sweep efficiency of dry gas displacing wet gas in a layered reservoir. Condensation and evaporation of condensate were thus neglected. A prototype symmetry element was chosen, containing one quarter and two half producers together with one quarter and one half injector. Rates were based on average yearly rates as planned for the Tyra field.

The conclusions from the dry gas recycling study were that a good areal sweep could be achieved in both the Danian and Maastrichtian layers, even though there is a preference for gas to flow in the Maastrichtian owing to higher permeabilities. Sweep efficiency of the order of 75% was predicted, while the duration of the recycling project

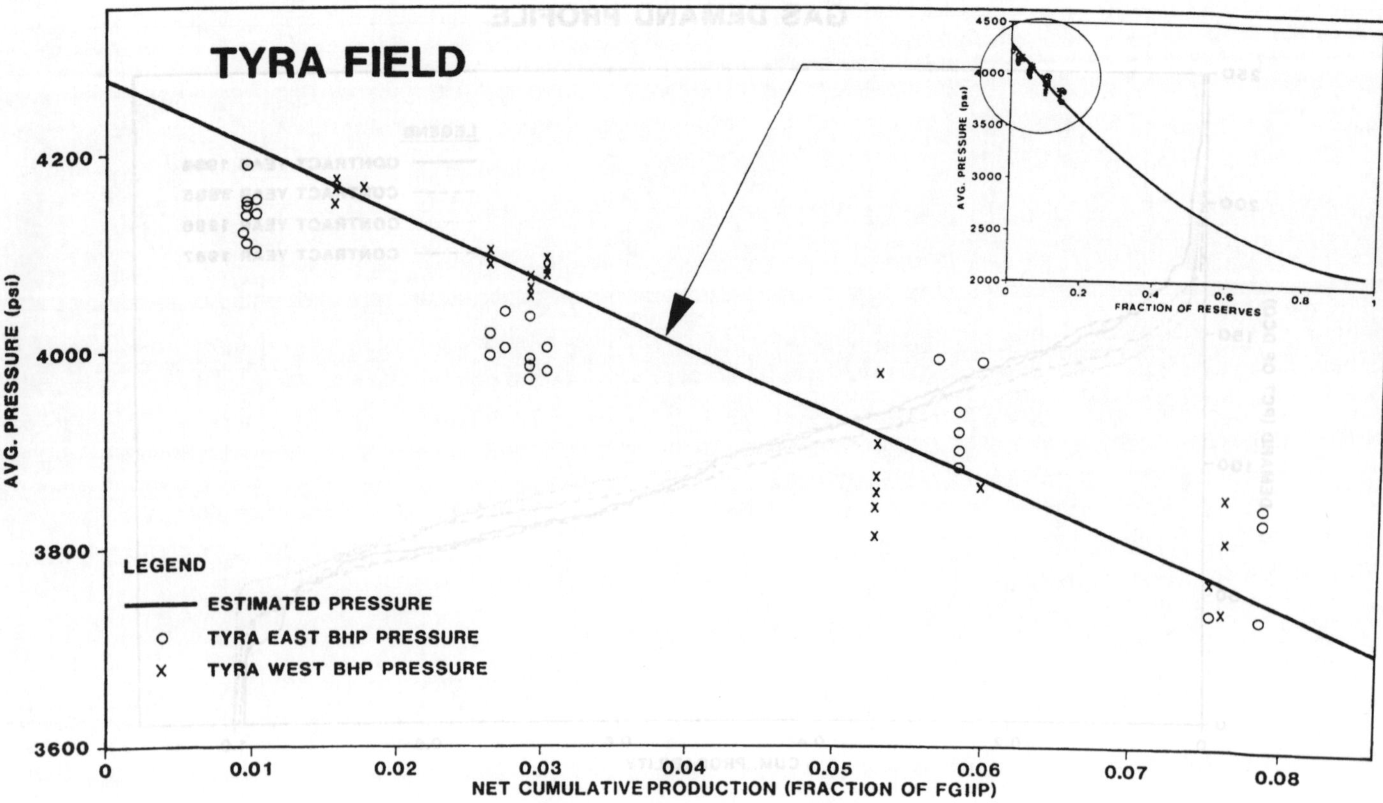


**Fig. 20.** Tyra well pressure vs. net cumulative production. Estimated pressure decline for existing development plus nine infill wells (expectation reserves). Actual measurements from BHP tests are indicated for comparison.

**Fig. 21.** Pressure contour map (psig) at 6500 ft ss (based on pressure data from 1987 BHP campaign) indicating a significant areal pressure variation in the field.

was anticipated to be some 5 years prior to significant dry-gas breakthrough.

*Implementation*

The initial plan was to inject 135 mmscf/d of dry produced gas into seven wells TWB-1, -3, -4, -7, TWC-1, -5 and -11. The original pattern was essentially a peripheral sweep plus one central injector. Late in 1986 the project was modified to an injection level of 180 mmscf/d into 10 possible wells—the original seven plus TWB-2, TWC-4 and TWC-7 (Fig. 22).

The project modification was driven by two events: firstly, a full-load compression test confirmed a compressor capacity of 180 mmscf/d; secondly, three out of the initial seven designated injectors continued to produce significant quantities of black oil (TWB-3, TWB-7 and TWC-11).

In June 1987 the Tyra West Gas Recycling Project was initiated, utilizing 8 out of the 10 nominated wells, the black oil wells TWB-7 and TWC-11 remaining on production (TWB-3 had by then watered out and had been converted into an injector). Shortly after start-up it was apparent that the total injection potential of the 8 wells was only some 135 mmscf/d, significantly lower than expected. A successful acid restimulation campaign was conducted, which increased the total injectivity into the 8 wells to 230 mmscf/d. By 1 October 1988 a cumulative total of 69 bcf had been injected into Tyra West. The average injection rate for contract year 1987 was 160 mmscf/d. The injectivity of the wells varies from approximately 19 to 42 mmscf/d (with an average of 28 mmscf/d). The present injection/production balance in the target area is illustrated in Fig. 22.

## Future gas development

Further gas development activities in Tyra can be considered in three areas:

- undeveloped reserves;
- scope for direct flank drainage;
- optimization of dry-gas recycling in Tyra West.

*Undeveloped reserves*

As discussed above, undeveloped reserves are carried for two categories of field activity:

(1) A campaign to recomplete Maastrichtian drainage points upwards into the Danian. This will be driven by the watercut behaviour of the existing Maastrichtian producers.

(2) A block of nine infill gas producers in the crestal area of the field. These will in principle be implemented to provide gas availability as required.

*Scope for direct flank drainage*

The Tyra flank contains 23% of the total free GIIP. Only some 22% of this is predicted to be recovered by migration to the crestal area and production from within the existing well pattern. A successful dedicated flank development could increase the total flank gas recovery factor to 50%.

Over 90% of the flank gas is contained in the D1 reservoir with a gas column height of 50–60 ft. The scope for improved flank gas recovery by direct offtake is critically determined by the coning characteristics of flank wells and the aquifer performance. This has naturally led to consideration of the potential benefit of horizontal wells

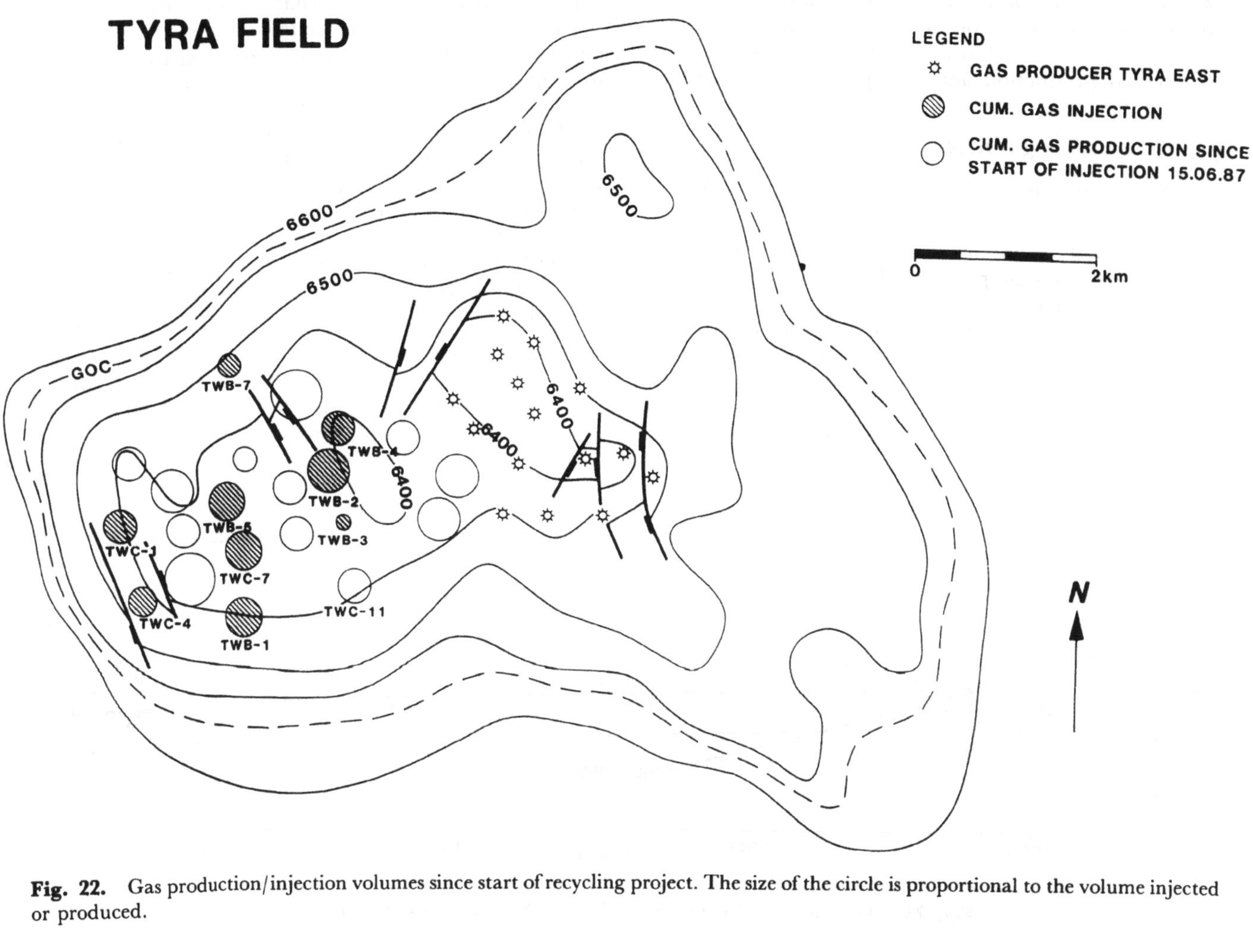


**Fig. 22.** Gas production/injection volumes since start of recycling project. The size of the circle is proportional to the volume injected or produced.

for flank drainage. In this area, Mrask Oil can draw on its considerable experience with horizontal wells in the Dan field.

As a first stage in the evaluation of the technical and economic feasibility of direct flank drainage, it is planned to drill a horizontal well on the eastern Tyra flank (TEB-1) (Fig. 15 and 23) where the D1 is most thickly developed. The well has both appraisal and production performance objectives, which are described in some detail to illustrate both the concept and the uncertainties.

The key objective of the well is to obtain long-term production performance data. The design criteria for this test of the viability of horizontal well development for the flank are aimed at minimizing the risk of coning in water. Hence, over the entire length (estimated 1500 ft) of the horizontal productive hole the GOC will be below the base D1. This should allow the relatively low-permeability D2 reservoir unit to act as a barrier to water coning. At the end of the productive hole the GOC in D1 is planned to be some 30 ft below the drainhole (Fig. 24).

To assist in positioning the horizontal trajectory, a high-angle pilot hole will be drilled to penetrate the full hydrocarbon column (Fig. 23). The well will then be sidetracked horizontally in the Danian D1. It is planned to drill approximately 3000 ft of horizontal hole for appraisal purposes, prior to plugging back to achieve a horizontal drainhole of some 1500 ft meeting the criteria outlined above.

The appraisal objectives of the well include:

- Formation tops for calibration of the seismic model and reservoir mapping away from existing well control;
- detailed fluid distribution data in the flank area;
- reservoir rock properties in the flank area;
- vertical and lateral pressure profiles that will clarify the depletion level in the flank area resulting from crestal production and also provide insight into aquifer performance.

*Optimization of dry-gas recycling*

Two main issues will be addressed in the short/medium term.

(1) Infill well drilling in order to maintain maximum recycling throughput as field potential falls and/or gas sales demand rises (Fig. 19).

(2) A related effort to improve the injection pattern. As mentioned above, the initially proposed injection pattern was modified to accommodate the black oil producers. More recently, injector TWB-3 has been abandoned for mechanical reasons while black oil producer TWB-7 watered out and was converted to an injector. Further scope clearly exists for improvement of the injection/production pattern and this will be actively pursued.

# OIL RIM DEVELOPMENT

## Development studies

Single well simulation studies on the Tyra oil rim predicted an extremely low recovery factor of less than 1%. These were considered no more than indicative of the complex challenge of developing the oil rim, not least because of the ill-defined three-phase relative permeability characteristics of the oil rim reservoir.

Thus, the strategy in the early development stages was directed towards the acquisition of oil rim production performance data.

## Reservoir performance

*Well results*

The results of early production tests on the oil zone in exploration well E-2X and development wells TWB-7 and TWB-6 were not encouraging—low oil rates at high water cuts.

The first sign of commercial black oil production was the performance of TEC-7, which initially produced only gas and condensate. Subsequently the well started to produce black oil in significant quantities from a completion zone in the Maastrichtian reservoir via perforations some 20 ft above the initially interpreted GOC. Petrophysical re-evaluation, however, indicates that the GOC is in fact somewhat higher than initially evaluated, i.e. the well is completed across the oil rim. The black oil production in TEC-7 has slowly declined from 1750 b/d to some 1400 b/d since March 1986.

The second enigmatic clue to successful black oil production was provided by TWB-7. Following the unsuccessful test in the oil–water transition zone mentioned above, the well was completed in the Danian reservoir some 56 ft above the GOC. Amazingly, the well produced black oil at an initial rate of 800 bopd.

The TEC-7 and TWB-7 experience prompted a number of activities to promote oil rim production. These can be grouped into three categories.

(1) *High-rate drawdown tests:* In reservoir and completion configuration, TEC-6 is considered a twin to well TWB-7. From May to December 1986 the well was produced with a high drawdown in an attempt

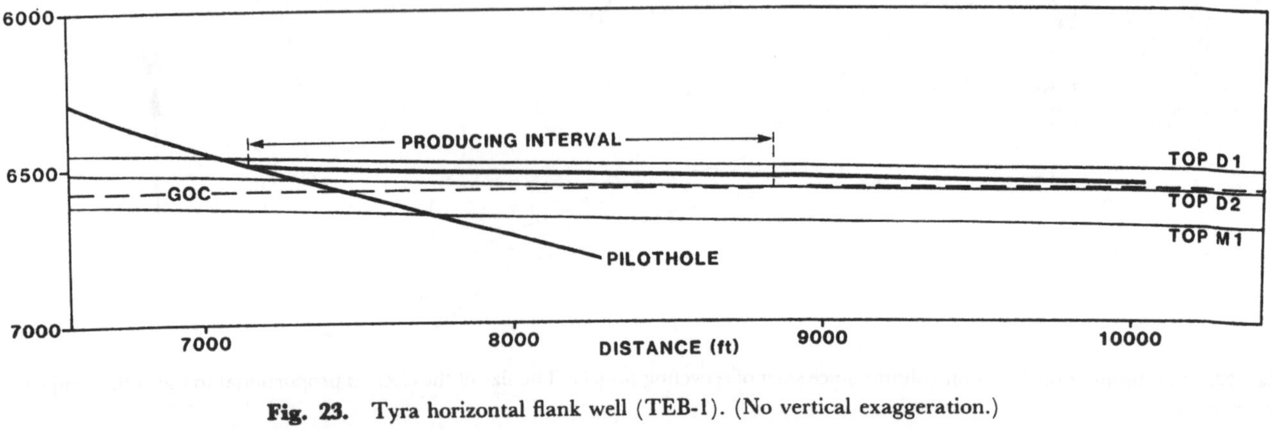


**Fig. 23.** Tyra horizontal flank well (TEB-1). (No vertical exaggeration.)

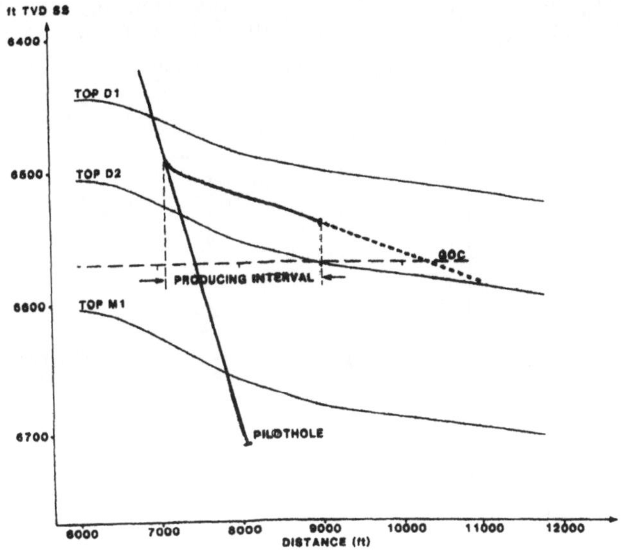


**Fig. 24.** Tyra horizontal flank well concept (Vertical exaggeration = 16.)

to initiate an oil cone up to the completion interval. No black oil or formation water was produced from this well, and the test was abandoned.

(2) *Additional perforations.* Four Danian gas producers were additionally perforated in the oil rim.

*TEC-4B* in a 15-ft interval in the Maastrichtian oil rim. After a period of almost 1 year without reaction, black oil production has started at a rate of some 400 b/d.

*TWC-11* across the GOC at the top of the Maastrichtian. A promising black oil production rate of 800 b/d has since materialized, but only after a significant time lag, as shown in Fig. 25. Dry-gas producer TWC-8 is shown for comparison in Fig. 26.

*TWC-4* just above the GOC. No black oil nor water was produced.

*TEC-12A* over an interval with $S_w$ ranging from 40–50%. Only traces of produced black oil were observed.

(3) *Selective completion on the oil rim.* Three wells that had to be worked over for mechanical reasons were selectively completed on the oil rim and the Upper Danian in 1986. For evaluation purposes wells TWB-3, -5 and -12 were perforated across the GOC, above the GOC and across the oil rim respectively.

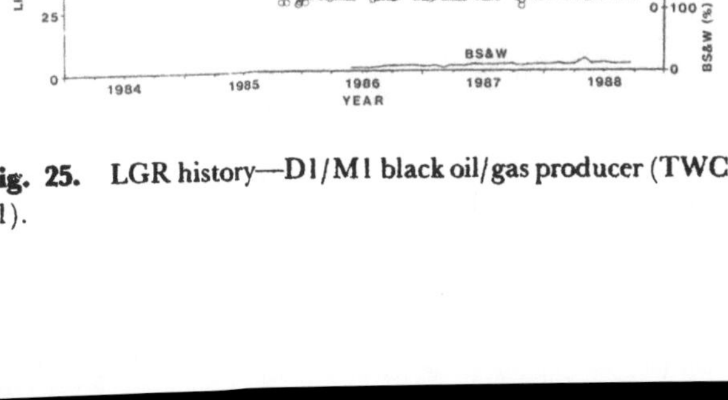


**Fig. 25.** LGR history—D1/M1 black oil/gas producer (TWC-11).

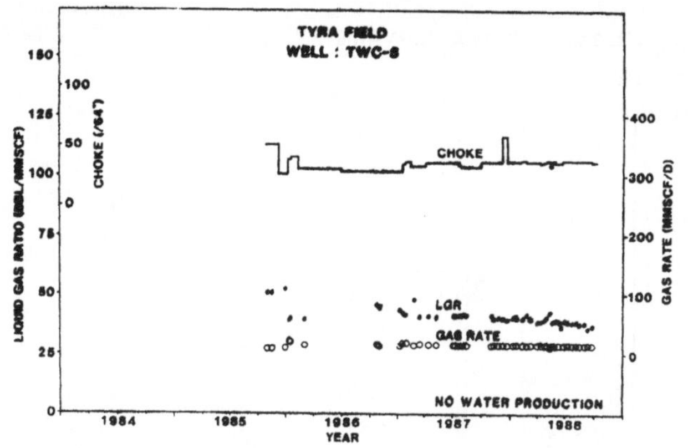


**Fig. 26.** LGR history—D1 gas producer(TWC-8).

Initially, a black oil production of 100 b/d was recorded in *TWB-3*, but ceased very quickly.

*TWB-5* was produced at maximum drawdown (FTHP = 1100 psi) in an attempt to establish an oil cone. Although the initial test indicated a black oil production of about 1000 b/d, subsequent tests showed no black oil production.

In *TWB-12* the production of black oil decreased from 300 b/d to a stable 100 b/d.

*Conclusions on oil rim performance*

From the above rather mixed set of performance data, which are summarized in Fig. 27, few clear conclusions can be drawn.

The most successful completions are in the following wells (current well status is shown):

| | *Oil Rate* (*b/d*) | *Cumulative Production* (*mmstb*) |
|---|---|---|
| TEC-7 | 1400 | 1.7 |
| TWC-11 | 800 | 0.7 |
| TEC-4B | 400 | 0.1 |
| TWB-7 | abandoned | 0.2 |

Of these only TEC-7 is a dedicated well perforated across the oil rim in the Masstrichtian.

TWC-11, TEC-7 and TEC-4B are all located in the southern part of the field and in a position where there is little to no Maastrichtian gas cap. It is believed that the Danian D2 and the Danian/Maastrichtian boundary inhibit flow of gas into the wellbore. Thus, the high drawdown rates necessary to promote reasonable black oil production can be achieved.

## Future oil rim development

It is planned to drill a dedicated oil rim appraisal well to upgrade our data set on the oil rim. Cores, RFT-pressures and dedicated production tests are envisaged. The area chosen for this test is between wells TWC-11 and TEC-7 (Fig. 28). One obvious concept to be considered is horizontal well drainage but it is too early to evaluate the potential of this.

It is, however, worthwhile to put the oil prospects in Tyra in their proper perspective. Although the STOIIP is large, it is clear that only 17% thereof (movable oil in the Maastrichtian) represents the real target. Recovery to date is some 3 mmstb, which at least exceeds the original estimates.

Recovery via a successful horizontal well campaign could realistically be in the 20–40 mmstb range. This is the justification for the appraisal programme, which, combined with a number of detailed studies, will be

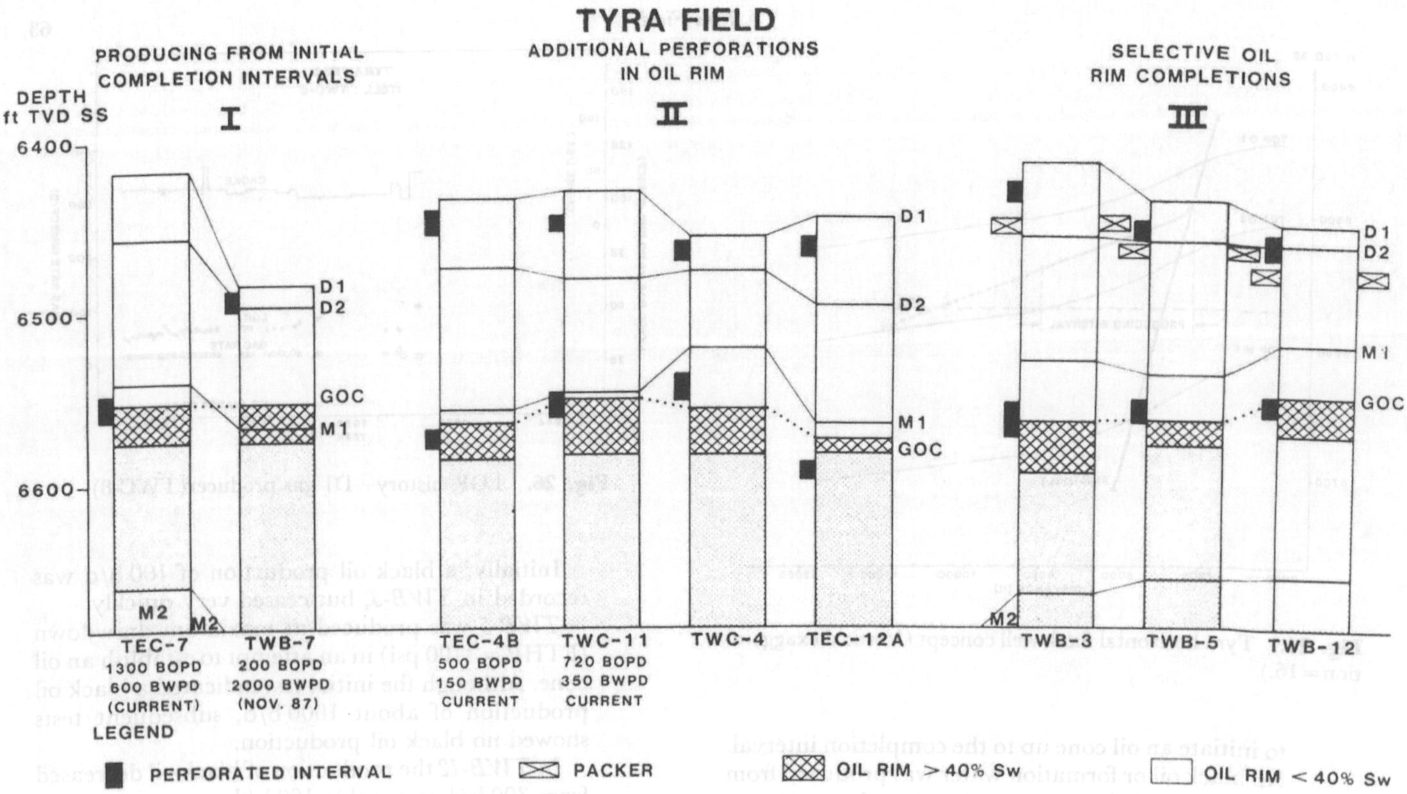


**Fig. 27.** Black oil production campaign.

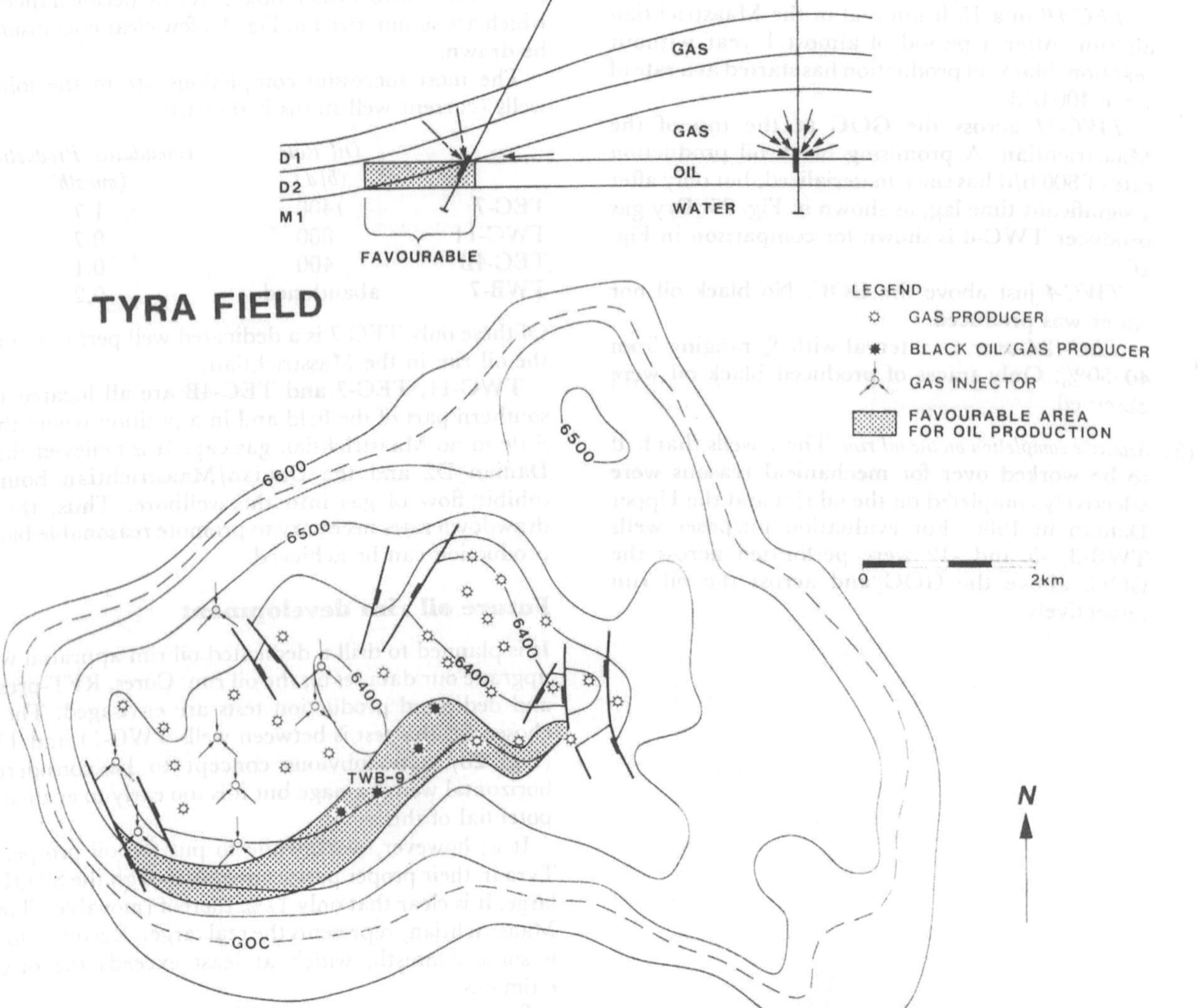


**Fig. 28.** Potential areas for black oil production. The Maastrichtian oil zone with no (or very limited) gas cap in the M1 is considered prospective for black oil production as the D2 and the Maastrichtian hardground will impede production of free gas. The concept is supported by the location of the black oil producers.

undertaken with the objective of developing a better understanding of both the nature of the oil rim and its production characteristics.

Figure 28 illustrates in a conceptual fashion those areas of the field considered favourable for black oil production, i.e. where the oil column is located just below the Danian/Maastrichtian boundary and there is little to no Maastrichtian gas. Since the oil column thickens towards the south, this area is considered as having the most potential.

## ACKNOWLEDGEMENTS

The authors thank members of the Dansk Undergrunds Consortium, Mærsk Olie og Gas A/S, Shell and Texaco for permission to publish this account of the Tyra field.

## REFERENCES

Deegan, C.E. and Scull, B.H. (1977). A proposed standard lithostratigraphic nomenclature for the central and northern North Sea. *Rep. Inst. Geol. Sci., No. 77/25.* Norwegian Petroleum Directorate. Bull. No. 1.

Hardman, R.F.P. (1982). Chalk Reservoirs of the North Sea. *Bulletin of the Geological Society of Denmark*, **30**, 119–137.

Mapstone, N.B. (1975). Diagenetic history of a North Sea chalk. *Sedimentology* **22**, 601–614.

Michelsen, O. (1982). Geology of the Danish Central Graben. *The Geological Survey of Denmark, Series B.*, Number 8, 133 pages.

Field Review 4

# The Gullfaks Field: Geology and Reservoir Development

*O. Petterson[a], A. Storli[a], E. Ljosland[a] and I. Massie[b]*

[a]Statoil, Bergen, Norway [b]Conoco, Stavanger, Norway

The Gullfaks oilfield is located in the north-eastern part of block 34/10 in the Norwegian sector of the North Sea. It represents the shallowest structural element of the Tampen Spur, and was formed during the Upper Jurassic to Lower Cretaceous as a sloping high, with a westerly structural dip gradually decreasing towards the east. The major north to south striking faults with easterly dipping fault planes divided the field into several rotated fault blocks. Central and eastern parts have been eroded by the early Cretaceous transgression.

The reservoir sands are comprised of the Middle Jurassic delta-deposited Brent Group, the Lower Jurassic shallow marine sandstones of the Cook Formation, and the Lower Jurassic fluvial channel and delta-plain deposits of the Statfjord Formation.

The complexity of the Gullfaks Field, the presence of gas in the post-Jurassic section and a variable water depth have complicated seismic interpretation. Nevertheless, the improved quality of the 1985 3D seismic survey and deliberate deepening of the development wells has resulted in a more accurate and complete structural interpretation.

The Brent reserves in the western part of the field are currently being developed by the Gullfaks A and B plaforms: the eastern part of the field will be developed by Gullfaks C. Water injection is the main drive mechanism maintaining reservoir pressure above the bubble point.

One of the most important factors in the reservoir development has been the effect of fault transmissibilities on lateral and vertical pressure distributions. A comprehensive data acquisition programme, including permanent downhole pressure gauges, has enabled accurate monitoring of the strategic injection and production plan. Radioactive tracer is injected into selected water injection wells in order to characterize fluid movement.

Long-term field development planning using a full-field simulation model has been supplemented successfully through the use of smaller, detailed fault block models. Resultant well locations and drilling schedules are continuously updated, incorporating information from new wells and the latest production data.

## INTRODUCTION

The Gullfaks Field is located in the central part of the East Shetland Basin on the western flank of the Viking Graben (Fig. 1). The field lies within the Norwegian offshore licence PL050 in Block 34/10, which was awarded in 1978 to a group comprising Statoil (85%), Norsk Hydro (9%) and Saga Petroleum (6%) with Statoil designated as operator. The block contains four separate hydrocarbon bearing structures (Fig. 2). The Gullfaks Field in the north-eastern corner is the most important and was declared commercial in 1980.

The areal extent of the Gullfaks reservoir is 50 km², which made it evident at an early stage that three platforms were needed to exploit the reserves. Gullfaks A and C are fully independent processing platforms, while Gullfaks B has only first-stage oil separation. Partially stabilized oil is transferred from Gullfaks B to Gullfaks A or C for complete processing. The platform locations and pipeline infrastructure for the field are shown in Fig. 3.

Crude oil is stored in the concrete base of Gullfaks A and C and loaded into tankers via two loading buoys. Produced gas is exported via the Statpipe system to Emden, Germany. Gas injection requirements are catered for by combined oil-producing/gas-injection wells in order to handle modulation of gas deliveries.

The A and B platforms have 42 drilling slots each, while Gullfaks C has 52 slots. In addition, a total of 21 subsea wells can be connected to the platforms. The current development plan (1988) includes 115 development wells for the whole field.

## DISCOVERY AND SUBSEQUENT DRILLING HISTORY

Initial exploration of block 34/10 was based on the interpretation of a 1 × 1 km 2D seismic survey undertaken in 1974 after the discovery of the Statfjord Field. The first interpretation recognized the presence of a large pre-

*North Sea Oil and Gas Reservoirs—II*

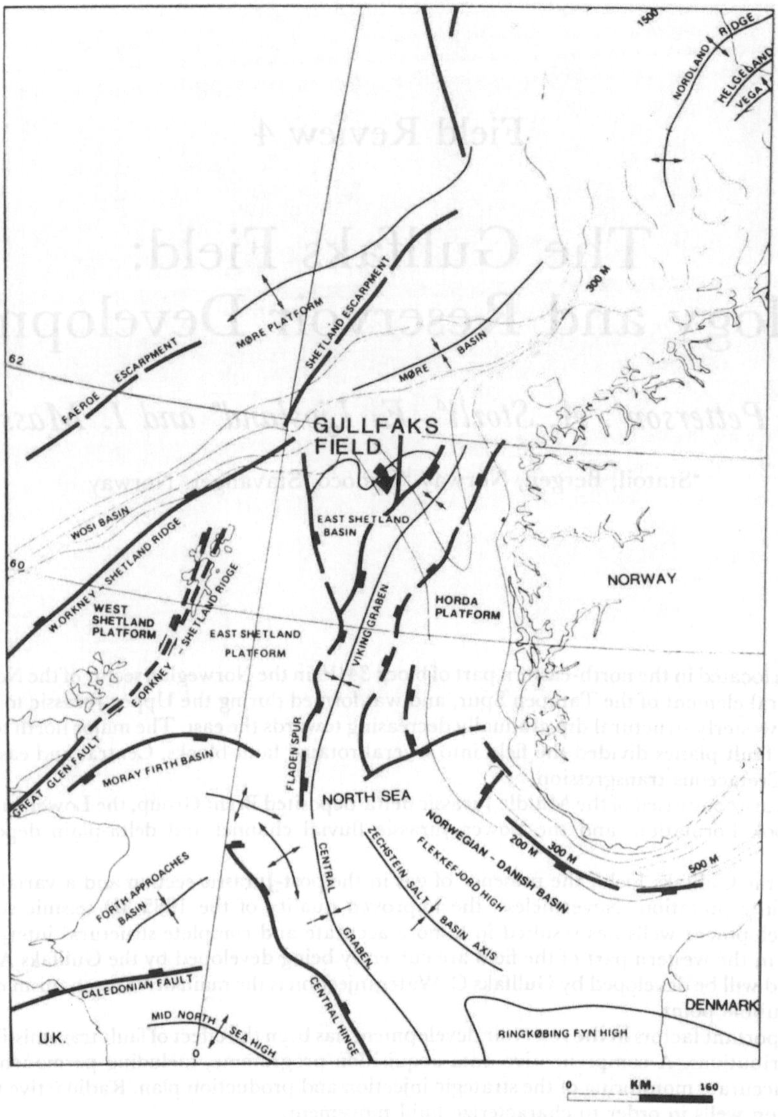


**Fig. 1.** Major tectonic elements of the northern North Sea.

Cretaceous structural high in the north-eastern quadrant of the block. Hopes based solely on the size of the structure and its proximity to the giant fields of Brent and Statfjord (Fig. 4) led to Block 34/10 being called the "Gold Block".

## Exploration and appraisal

The first well 34/10-1, on the Gullfaks structure was spudded on 20 June 1978, just after the licence was awarded. It penetrated the Jurassic section and encountered 162 m of oil filled Brent sandstone with oil down to base Brent. The reservoir properties were found to be excellent, with porosities ranging up to 34% and permeabilities up to 4000 mD. A 60 m thick regressive sand in the Cook Formation of the Dunlin Group as well as the Statfjord Formation were dry. The next four exploration wells (34/10-3, 4, 5 and 6) were drilled in the western part of the field. All except one (34/10-6) proved oil filled Brent sandstone. An oil–water contact (OWC) was established at 1947 mMSL. At this point two questions had to be resolved:

- What was the extent of the Brent sandstone in the eastern direction?
- If the Brent sand was eroded or not present, could oil be preserved in the Cook Formation or in the deeper Statfjord Formation?

The need for more and better seismic data was therefore essential before further exploration drilling in the easternmost part of the structure could start. Hence, a 3D seismic survey was shot over the structure in 1979. Improvement in basic data quality together with the large increase in quantity of data meant that unmapped areas in the eastern part of the structure could be mapped. Six wells (34/10-7, 8, 9, 11, 13 and 14) were then drilled in the eastern area and were all successful oil wells. Two appraisal wells (34/10-15 and 19) were now drilled in the south-eastern area in an effort to delineate the field in this direction. The Cook sandstone which was the main objective was found to be dry in the 34/10-15 and eroded in 34/10-19.

At the end of 1983 the exploration phase was brought to an end after a total of 14 wells were drilled on the structure. The status was 10 successful oil wells, 3 dry wells and 1 abandoned owing to problems with shallow gas. Three of the wells are key wells in understanding the structural position of the main reservoirs in the Gullfaks Field (Fig. 5):

- Well 34/10-3 in the western area discovered the OWC in the Brent reservoir at 1947 m MSL.
- Well 34/10-9 in the central area found oil bearing Cook Formation below the oil bearing Brent sandstone. An OWC in the Cook Formation was measured at 2090 m MSL.

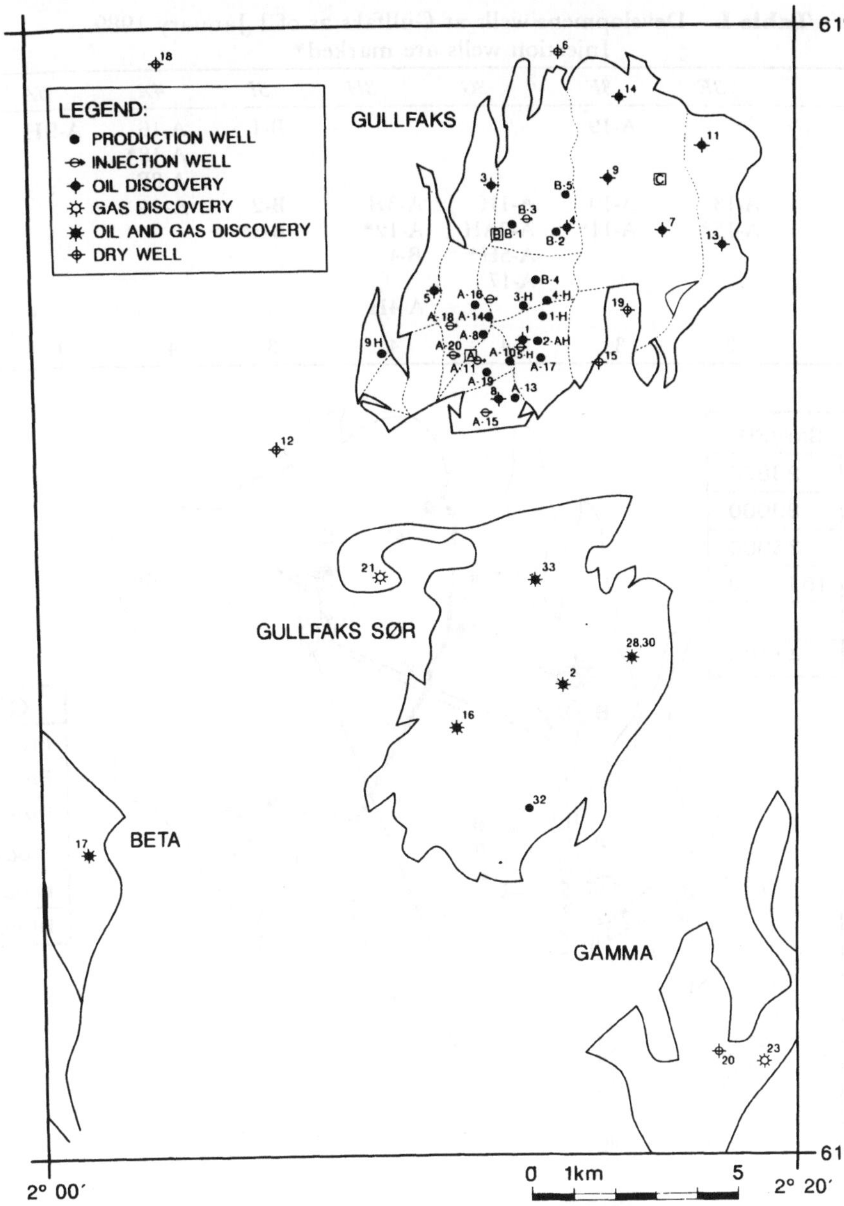


**Fig. 2.** Block 34/10.

- Well 34/10-11 was drilled into a horst block in the eastern area. It discovered oil bearing Statfjord Formation structurally 300 m higher than the adjacent western fault block.

## Development wells

Development drilling commenced in August 1985 with the first subsea well, A-1H. Five subsea wells were drilled in fault blocks 3G and 3H, north-east of Gullfaks A during 1985 and 1986. Of these, three are production wells in Lower Brent (Etive and Rannoch) and one is a Lower Brent injection well. These wells were subsequently completed and tied back to Gullfaks A in the second half of 1986. Well A-4H was completed as a production well in the Cook Formation, and later hooked up as a convertible gas injector. On 22 December 1986, Gullfaks A was brought on stream with production from subsea wells A-1H, A-2AH and A-4H. Well A-3H was put on production in February 1987. In order to provide pressure support to wells A-1H, A-2AH and A-3H, subsea water injection well A-5H was put on injection in April 1987.

In delineation well A-9H in fault block 6A oil was encountered in Upper Brent (Ness and Tarbert). The well was tied back to Gullfaks A as a subsea oil production well, and put on stream in May 1988.

Prior to installation of Gullfaks A, shallow wells were drilled in the vicinity of the proposed platform site to investigate the occurrence of shallow gas. The first two wells drilled from Gullfaks A were completed in the shallow gas formation, to deplete the pressure and minimize drilling problems in the zone. One of these wells is still on production, while the other has been shut in.

The first development well drilled from Gullfaks A was completed in June 1987. The first Gullfaks B well was completed and the platform came on stream in February 1988. By the end of 1988, 11 development wells have been drilled from Gullfaks A and five from Gullfaks B. In total, 15 oil production wells and 6 water injectors have been completed by 1 January 1989 as shown in Table I and Fig. 6.

**Table I.** Development wells at Gullfaks as of 1 January 1989. Injection wells are marked*

| *Fault block* | *3E* | *3F* | *3G* | *3H* | *3I* | *4A* | *6A* | *Total* |
|---|---|---|---|---|---|---|---|---|
| Upper Brent | | A-19 | | | B-1 | A-16<br>A-18*<br>A-8R | A-9H | 6 |
| Lower Brent | A-13<br>A-15* | A-10<br>A-11* | A-1H<br>A-2AH<br>A-5H*<br>A-17 | A-3H<br>A-12*<br>B-4 | B-2<br>B-3* | A-14 | | 14 |
| Cook | | | | A-4H | | | | 1 |
| Total | 2 | 3 | 4 | 4 | 3 | 4 | 1 | 21 |

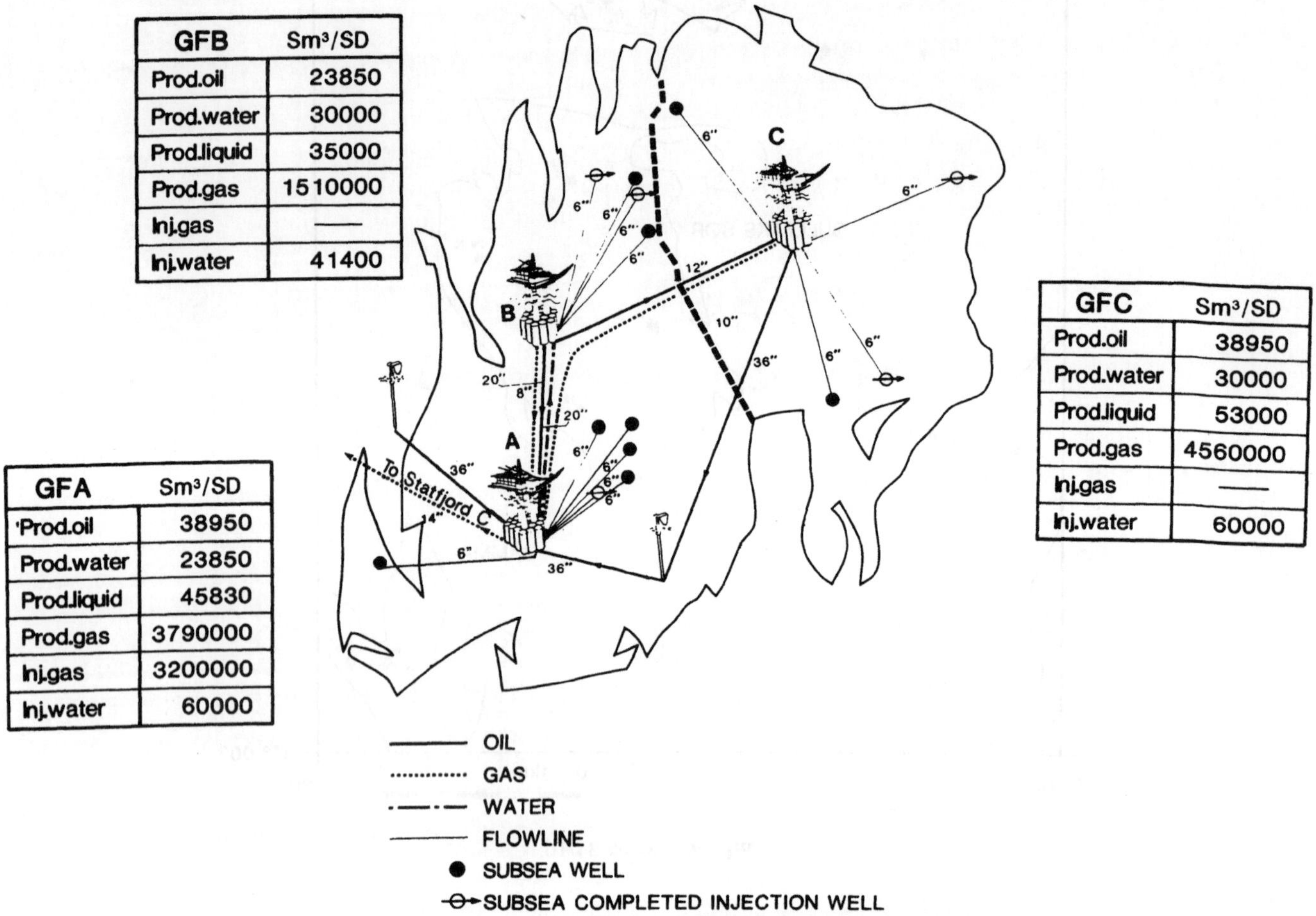


**Fig. 3.** Field facilities, design capacities and infrastructure. Subsea well locations for Gullfaks B and C are preliminary.

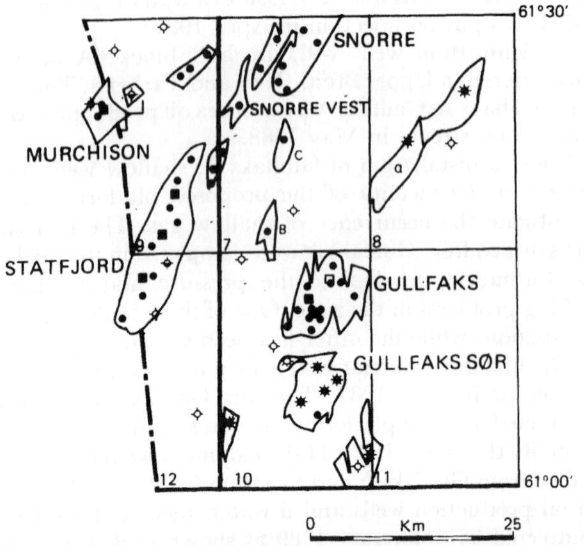


**Fig. 4.** The Gullfaks and Statfjord area.

## GEOPHYSICS

The Gullfaks Field has been described as the most complex structure so far developed in Norwegian waters. There are several reasons for this, but the main one is the complex fault pattern that intersects and divides the field into many small fault blocks. A total of 32 fault blocks have been mapped (Fig. 6) This large number of fault blocks has required a rather high number of exploration and appraisal wells in addition to high-quality seismic to give some degree of confidence in the interpretation and structural mapping. Several sets of seismic data have been acquired over the Gullfaks Field. Since the initial interpretation based on a 1 × 1 km conventional 2D seismic from 1974, two 3D seismic surveys have been collected (1979 and 1985).

In order to improve the data quality and the reliability of the structural maps prior to development drilling, the acquisition of a new 3D data set was carried out in 1985. The survey consisted of 8700 km of seismic data divided

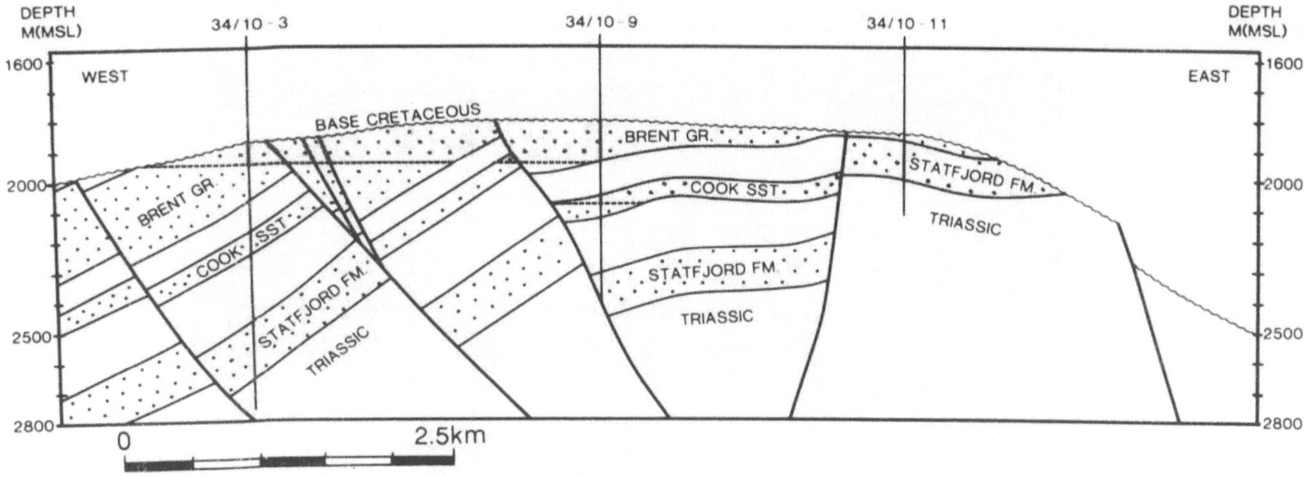


**Fig. 5.** Cross-section through key wells.

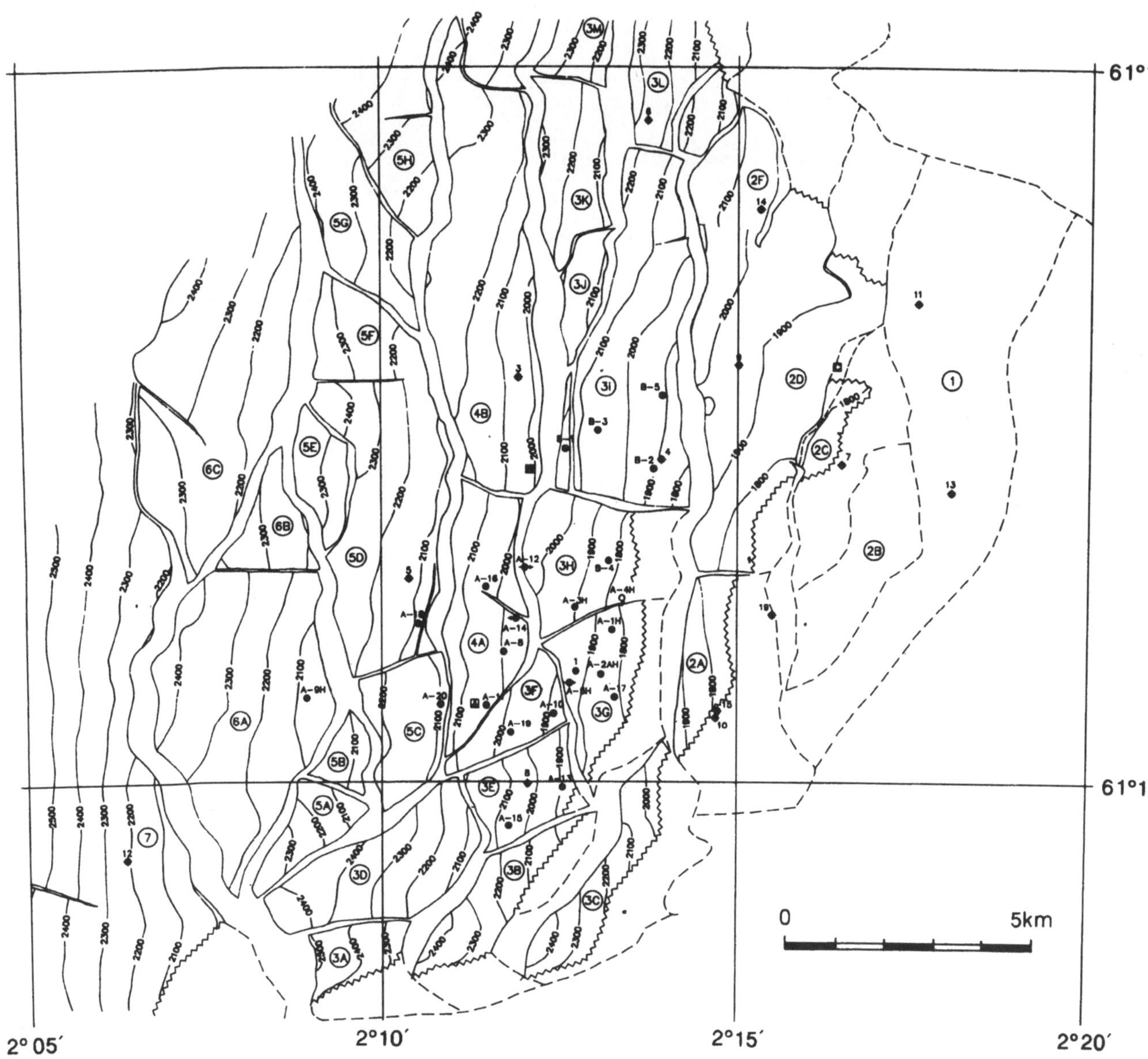


**Fig. 6.** Structural map of base Brent with well locations.

into 550 E–W lines with a spacing of 25 m. Processed data from this survey have proven to be of higher quality than the data collected in 1979 (Fig. 7). The line spacing of 25 m has in particular resulted in improvements in the cross-lines. This has provided better control of fault patterns and a much better definition of the reflectors. Despite the observed improvement in the 1985 survey, it must be emphasized that the seismic data are still of relatively low quality in comparison with many other fields in the region. The most important reasons for this are:

(a)

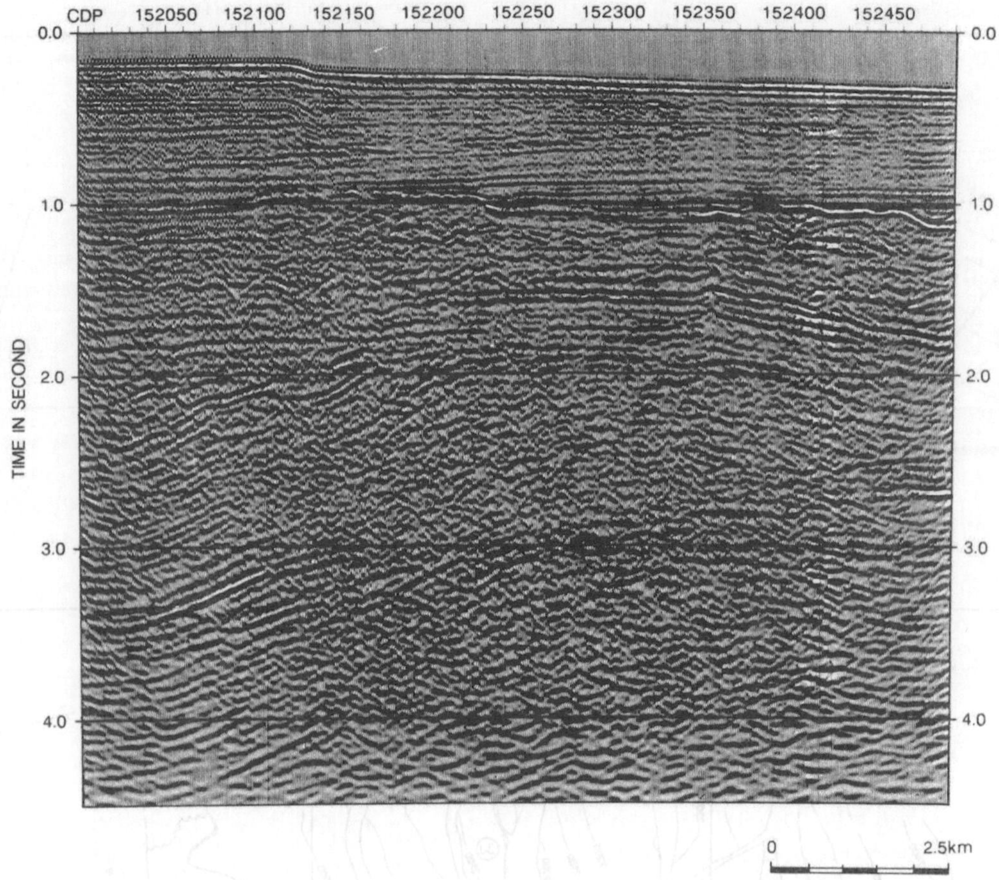


(b)

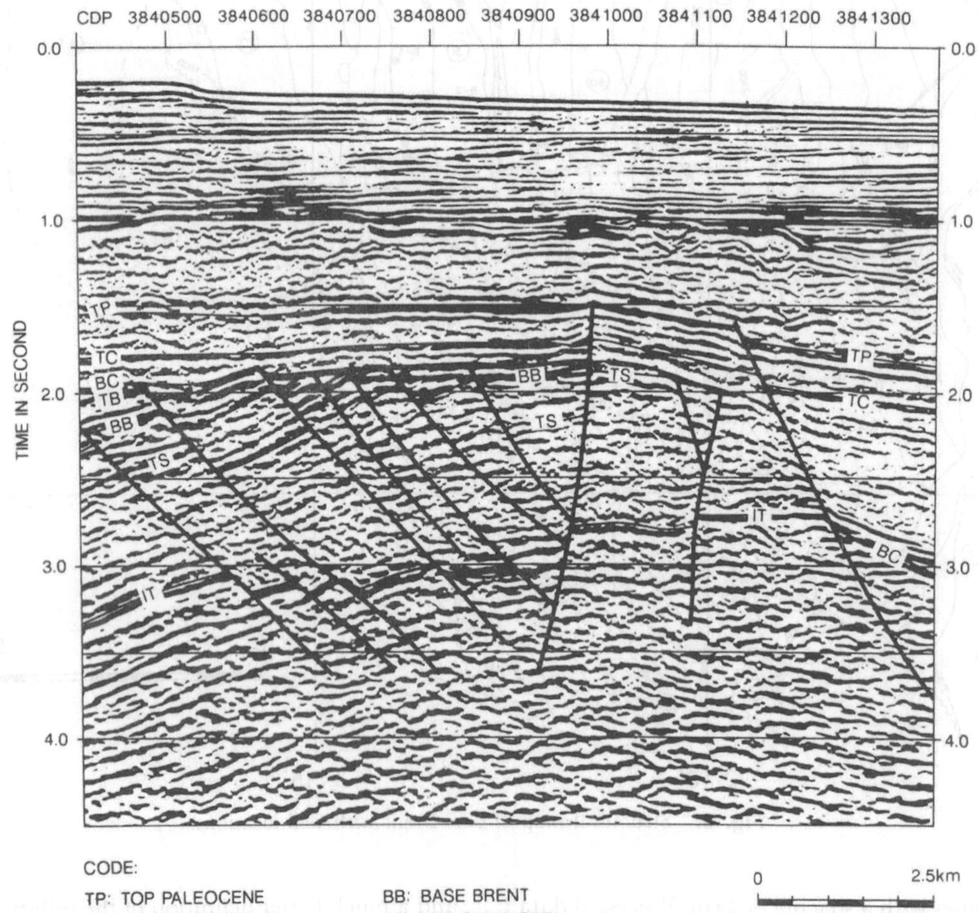


**Fig. 7.** (a) Seismic profile, 1979 3D survey. (b) Seismic profile, 1985 3D survey, with interpretation.

- Variation in water depth from 130 m in south-west to 230 m in north-east with a steep seabottom slope crossing the field.
- Presence of shallow gas down to approximately 450 m MSL.
- Presence of gas throughout a large interval in the post-Jurassic sections.
- Very strong reflectors at top Palaeocene and top Cretaceous reduce the amount of the original seismic energy penetrating the pre-Cretaceous section. These two reflectors also generate strong multiples, which interfere with the pre-Cretaceous reflections.
- Deterioration of the seismic data due to the large number of fault blocks.

In spite of these problems it can be concluded that the new 3D seismic data have provided important insights to the tectonic evolution of the Gullfaks field. Borehole information from 13 exploration and 23 development wells has in addition resulted in better control with respect to identification of the different reflectors and faults. To improve the accuracy of the mapping, most of the development wells have deliberately been deepened in order to penetrate potential faults and minimize uncertainties. However, areas in the central and eastern parts of the structure remain very difficult to interpret owing to discontinuous reflectors, noise and strong multiples within the seismic sections. This is particulary evident in the detection, orientation and extension of faults.

## STRUCTURAL DEVELOPMENT

The structural genesis of the Gullfaks Field took place throughout Jurassic times, culminating during Late Jurassic to Early Cretaceous times in connection with regional tectonic movements.

The extensional tectonics that seem to have controlled the development of the Gullfaks structure resulted in sagging with normal faulting and block rotation in the western part of the structure. The central area underwent "keystone" faulting that resulted in a graben feature, while the eastern part remained as an elevated horst structure. The structural development took place within the framework of a half graben with sliding and rotation of the fault blocks on a deep detachment zone (Fig. 8). The intense faulting throughout the period formed a very marked N–S oriented fault system. The width of the fault blocks vary from 0.5 to 1 km, while the major fault throws vary from 50 to 150 m.

Most faults terminate against the Kimmerian unconformity, which seals off the Gullfaks reservoirs at approximately 1700 m MSL in the crestal area. To the north, east and south, the structure is limited by a major boundary fault separating it from the Tampen Spur rim (Block 34/8), the Viking Graben and the Gullfaks Sør structure, respectively.

The Gullfaks Field can be divided into three different structural regimes from west to east (Fig. 8):

(1) *The western area* comprises a series of rotated fault blocks. The normal faults, which strike in a N–S direction, are developed with fault planes dipping at a rather unusually low angle of approximately 30–40° to the east, while the formations are dipping at 10–15° towards the west. There is normally good agreement between dips detected from seismic lines and dipmeter logs. Usually, normal faults will develop with fault plane dip of around 60° (Fig. 9A). In the case of block rotation yielding a fault plane dip of 30–40°, bedding dip should be expected to be in the order of 20–30° (Fig. 9B), and not as on the Gullfaks Field, where the observed dip is 10–15° unless other deformations exists.

This unusual feature of low fault dip in combination with low bedding dip has been explained by the presence of two different deformation types (Koestler and Milnes, 1988, GEO-RECON A.S., confidential contract report for Statoil):

(a) the "domino-type" deformation of several fault blocks, with rotational deformation of the blocks situated above a detachment (Fig. 9B);

(b) an internal shearing within the separate blocks to maintain the low bedding dip observed (Fig. 9C).

An analogue to this deformation type has been produced in a sandbox experiment by McClay and Ellis (1987).

Space problems close to the detachment are expected to be solved by ductile deformation. Ductile deformation in

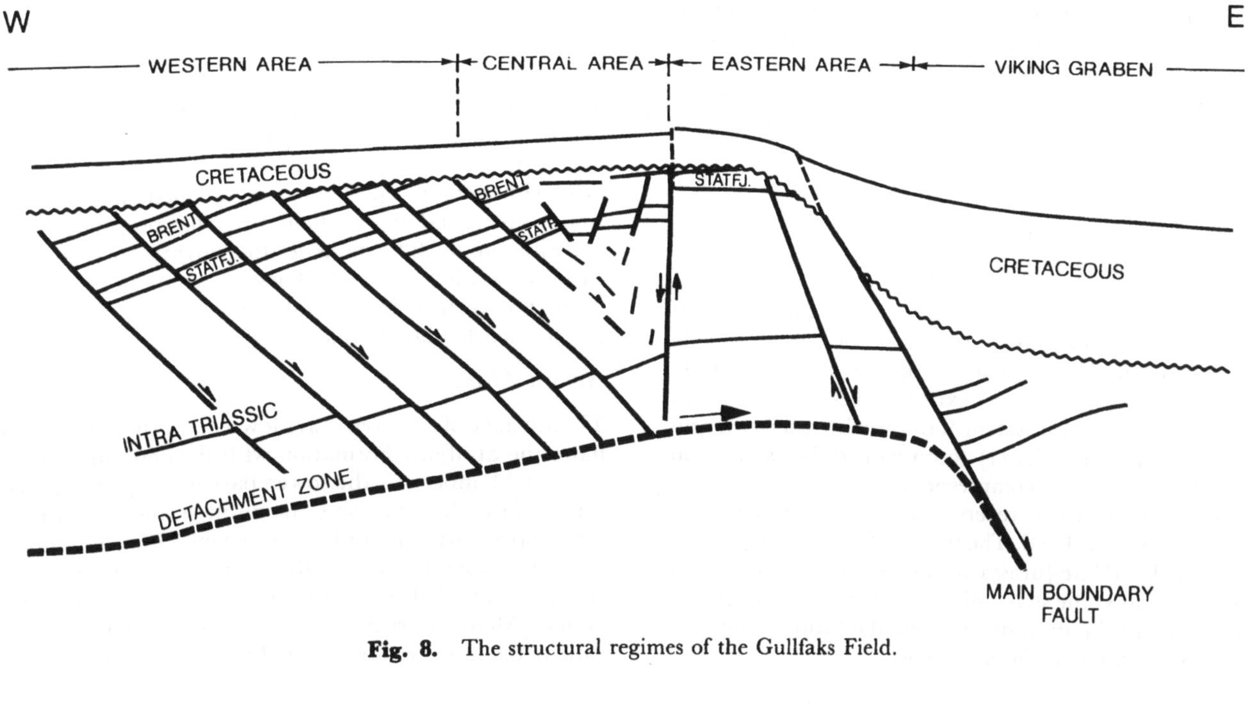


**Fig. 8.** The structural regimes of the Gullfaks Field.

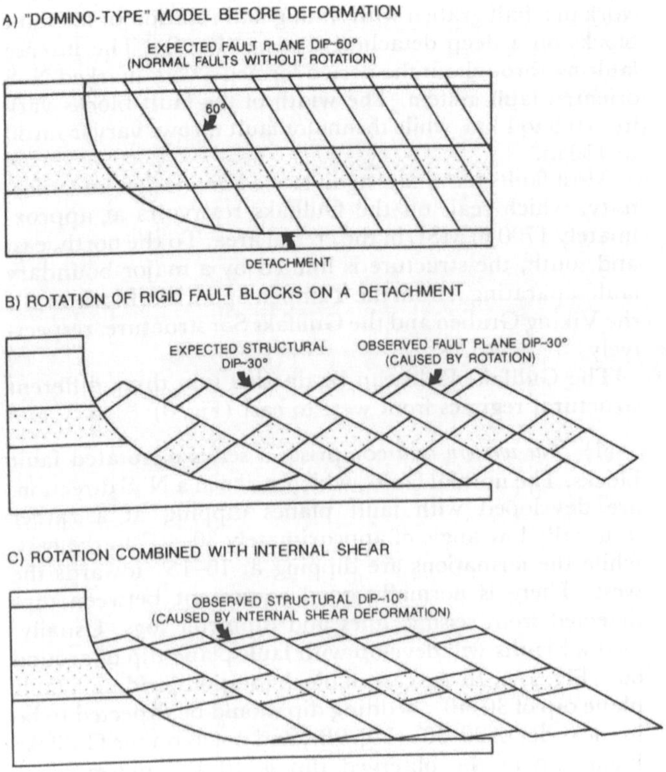


**Fig. 9.** Model of structural development in the western area.

this context means change of shape without observable discontinuities in the scale of the seismic resolution.

(2) *The central area* could structurally be described as a graben feature. The graben is interpreted as the result of a "keystone" collapse (Sæland and Simpson, 1982). Seismic sections normally show a poorly defined reflector pattern which is believed to be the expression of a more complexly deformed structure.

Reflectors show dip variations from approximately 10° to more subhorizontal in the eastward direction.

(3) *The eastern area* forms a horst structure, where the Statfjord Formation is lifted approximately 300 m compared to the central area. The elevated position of the horst has resulted in a complete erosion of the Brent and the Cook reservoirs. At the northern and southern flanks of the horst, the erosion can be followed down into the Hegre Group of the Upper Triassic (Fig. 10, cross-section B-B′), resulting in the erosion of more than 700 m of Jurassic strata.

The horst structure is bounded by an almost vertical fault from the central area (Fig. 5) and by a major boundary fault with a throw in the order of 1000–1500 m down to the Viking Graben in the east.

Most authors seem to favour the assumption that extensional stress prevailed throughout the late Jurassic to early Cretaceous. However, a new seismic/structural study (Fossen, 1989) argues that dextral strike-slip movement along the NE–SW fault system at the north-eastern margin of the Tampen Spur (Fig. 11) accompanied the extensional phases. This resulted in compressional structures, involving the formation of thrust, reverse faults and compressional folds at the Gullfaks Field. The timing of the transpressional tectonics is dated late Jurassic to early Cretaceous times. In a regional context the strike-slip movement may be related to dextral strike-slip movements along the Møre-Trøndelag Fault Zone further to the north-east.

## STRATIGRAPHY AND ENVIRONMENTAL MODEL

The recoverable oil in the Gullfaks Field is found in three different reservoir sandstones:

- Statfjord Formation (Type log presentation, Fig. 12);
- Cook Formation (Type log presentation, Fig. 13);
- Brent Group (Type log presentation, Fig. 14);

Minor reserves are also present in the Upper Triassic Hegre Group. The lithostratigraphic nomenclature used for the Gullfaks Field (Fig. 15) follows the subdivisions of Vollset and Doré (1984).

### Hegre Group

The oldest reservoir sandstones are the medium-grained Triassic fluvial sandstones of the Lunde Formation. Exploration well 34/10-13 has proven and tested reserves in this formation. However, the reservoir quality is considered moderate, and the potential of the reservoir low.

### Statfjord Formation

In the Gullfaks area, the Statfjord Formation is subdivided into the Raude, Eiriksson and Nansen Members (Fig. 12). The Statfjord Formation represents an alluvial environment that changed its character from a well-drained semi-arid setting with episodic sheet flood deposition to a more humid alluvial plain, with poorly drained, locally swampy overbank areas and channelized flows (Collinson, J.D. 1985, confidential contract report for Statoil).

*Raude Member*

The Raude Member is divided into the three reservoir units Raude 1, 2 and 3. The lower part, Raude 1, comprises a coarsening upwards shale/silt unit with poor reservoir quality and a net-to-gross ratio of only 0.1. Raude 2 and 3 comprise several sheet flood sands and rapidly deposited alluvial sands during short periods of flood in a generally dry climate. The sands and shales have been influenced by pedoturbation, and soil profiles as well as caliche are common. Raude 2 and 3 have medium reservoir quality owing to high content of kaolinite in the pores. Thickness varies from 100 m in the west to 80 m in the east.

*Eiriksson Member*

The Eiriksson 1, 2 and 3 reservoir units were deposited in an environment with meandering channels and swampy overbank areas in a gradually more humid climate. Each unit is defined by one, or several, stacked channels at the base with a shaly top. The coarse-grained channel sandstones give good to excellent reservoir properties, with porosities up to 28% and a net-to-gross ratio of 0.60 in the eastern oil-filled area. Thicknesses vary between 60 and 75 m. The vertical and lateral continuity is good owing to the stacked fluvial channels.

*Nansen Member*

The advance of a transgressive low-energy shoreline terminates the Statfjord Formation with the deposition of the Nansen Member. In the lower part it comprises a well-sorted, laterally extensive, medium-grained sandstone. The upper part is more heterogeneous with a mixture of different shoreline and embayment sediments of more restricted areal distribution. Reservoir qualities of the Nansen Member are good to moderate with a net-to-gross ratio of 0.6 in the eastern area. This 15–25 m thick sand is

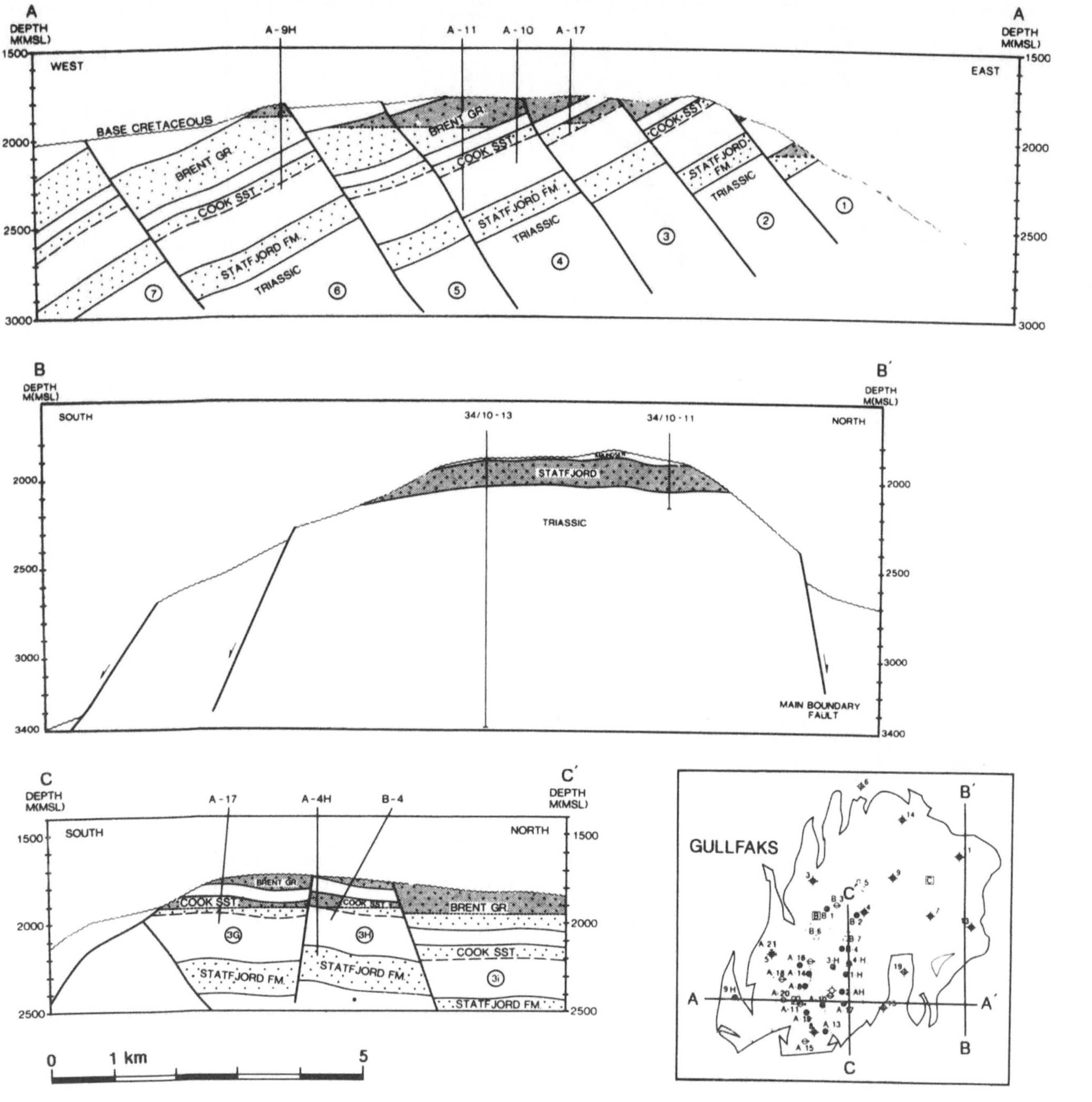


**Fig. 10.** Cross-sections through the Gullfaks Field.

likely to be laterally extensive, but may be wholly or partly broken by shale developments.

## Dunlin Group

The Dunlin Group is subdivided into the Amundsen, Burton, Cook and Drake Formations. The marine transgression that terminated the deposition of the Statfjord Formation progressed southwards and the 170–180 m thick claystones of the Amundsen and Burton Formations were deposited.

### *Cook Formation*

An Intra Dunlin regression together with increased tectonic activity caused silt and sand to be transported farther into the basin, depositing the Cook Formation. On the Gullfaks Field the Cook Formation is subdivided into Cook-1, Cook-2 and Cook-3 units (Fig. 13), where the lowermost open marine silty claystone of Cook-1 has no potential as a reservoir rock. The bioturbated muddy sandstones of Cook-2 were deposited on an inner marine shelf and generally have a thickness of 50 m over the field. The uppermost unit, Cook-3, consists of interbedded sand and shale with a thickness of 40 m in the southern area decreasing to 15 m in the north-east of the field. This unit shows cleaner sands upwards, forming an excellent reservoir at the top. It was deposited in a restricted marine setting, probably with anoxic bottom conditions. This is indicated by the absence of trace fossils, calcareous microfossils and a high hydrogen index in the shales.

Tightly calcite-cemented sandstones, which in general are 0.5–2 m thick, occur rather frequently throughout the Cook Formation. They are relatively non-porous (2–8% porosity) with permeabilities of less than 1 mD. Both nodular and layered zones are observed. Log examination suggests that the calcite-cemented zones, particularly in the Cook-2 unit, occasionally form continuous, essentially

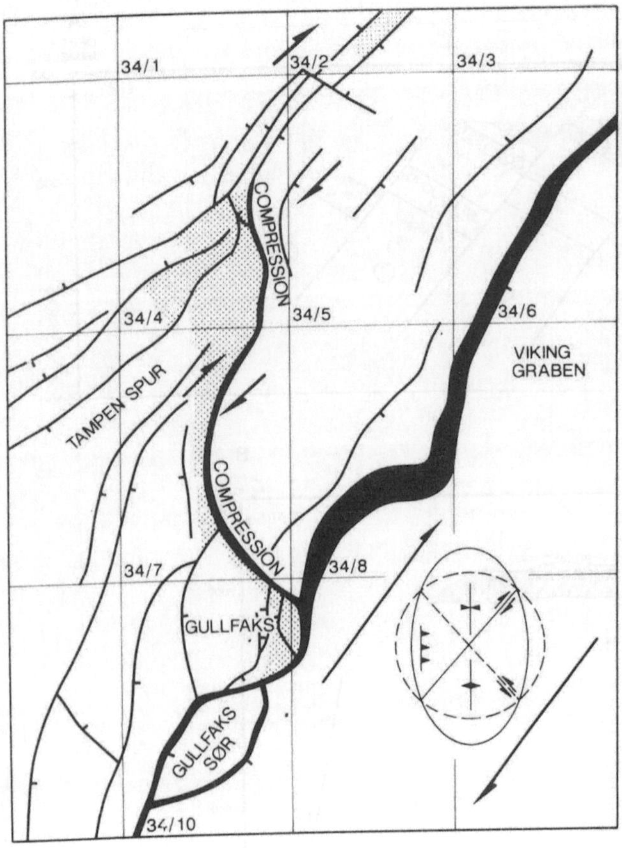


**Fig. 11.** Structural elements of the Tampen area. (Fossen (1989) *Marine and Petroleum Geology*, vol. 6, No. 1, Fig. 7, p. 26. By permission of the publishers, Butterworth and Co. (Publishers) Ltd. ©)

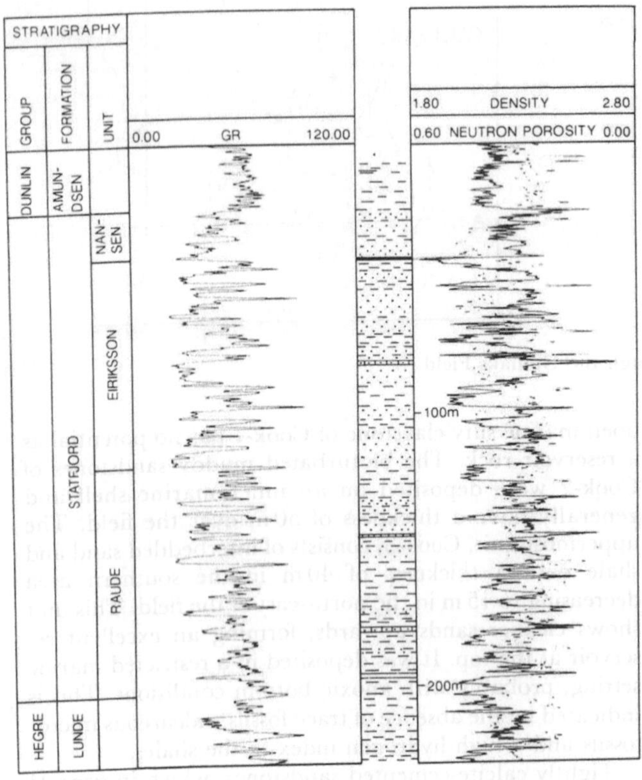


**Fig. 12.** Type log Statfjord Formation.

stratigraphic layers which can be correlated between wells over distances of more than 4 km. These layers may form laterally continuous permeability barriers that will be important in the development of the Cook Formation.

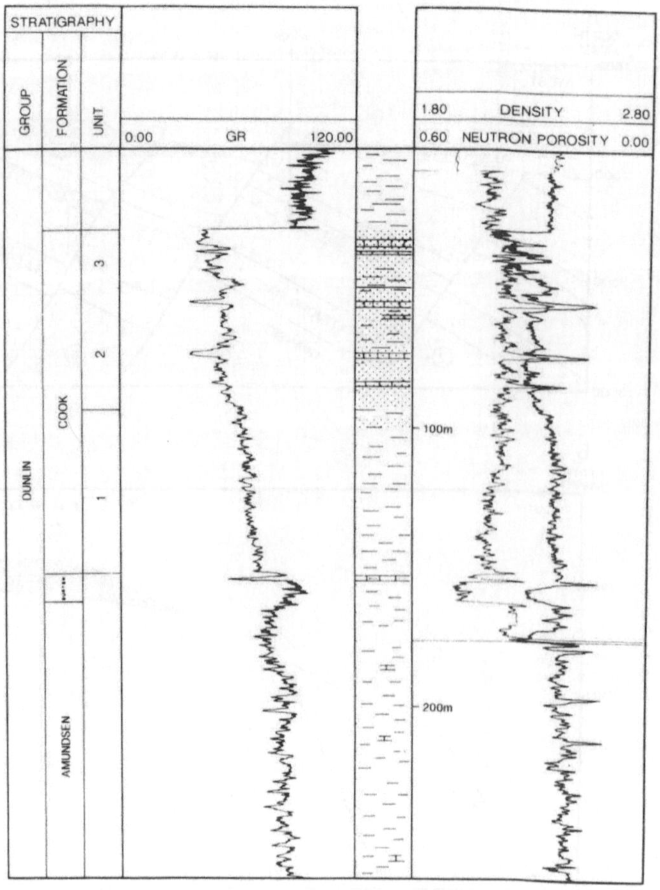


**Fig. 13.** Type log Cook Formation.

The general transgression in the area continued after the deposition of the Cook Formation, and the Drake Formation was deposited as a marine shale with varying amounts of silt. The thickness in the Gullfaks area varies from 70 m in the south to 120 m in the north.

## Brent Group

The Brent Group is subdivided into the Broom, Rannoch, Etive, Ness and Tarbert Formations (Fig. 14). All the formations in the Brent Group have been classified as deposits associated with a river delta system.

The vertical succession of the Brent Group is divided into two separate delta systems. A wave/fluvial-dominated delta interpretation is preferred for the Lower Brent Delta comprising the Broom, Rannoch, Etive and Ness, or the "Bajocian" delta lobe. The Upper Brent Delta is interpreted as a fluvial/tide dominated delta, and comprises the Tarbert Formation or the "Bathonian" delta lobe.

An idealized core description of the Brent Group is shown in Fig. 16, indicating that in the Gullfaks Field, all the "classical" Brent Group formations are present.

### *Broom Formation*

The Broom Formation is a 5–15 m thick mudstone/shale with thin layers of coarse sandstone and gravel beds, and is interpreted as prodelta deposits in an open marine environment. The thin sandstone beds are usually carbonate-cemented and in restricted communication with the overlying sandstones of the Rannoch Formation.

### *Rannoch Formation*

This formation, generally regarded as delta front shoreface micaceous sandstone, is subdivided into the three reservoir

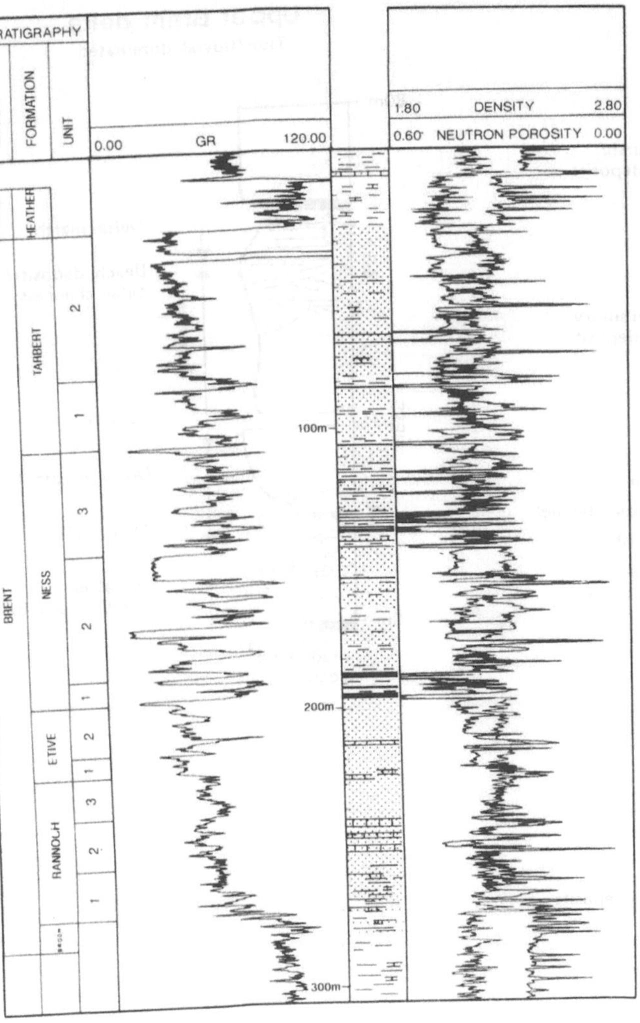


**Fig. 14.** Type log Brent Group.

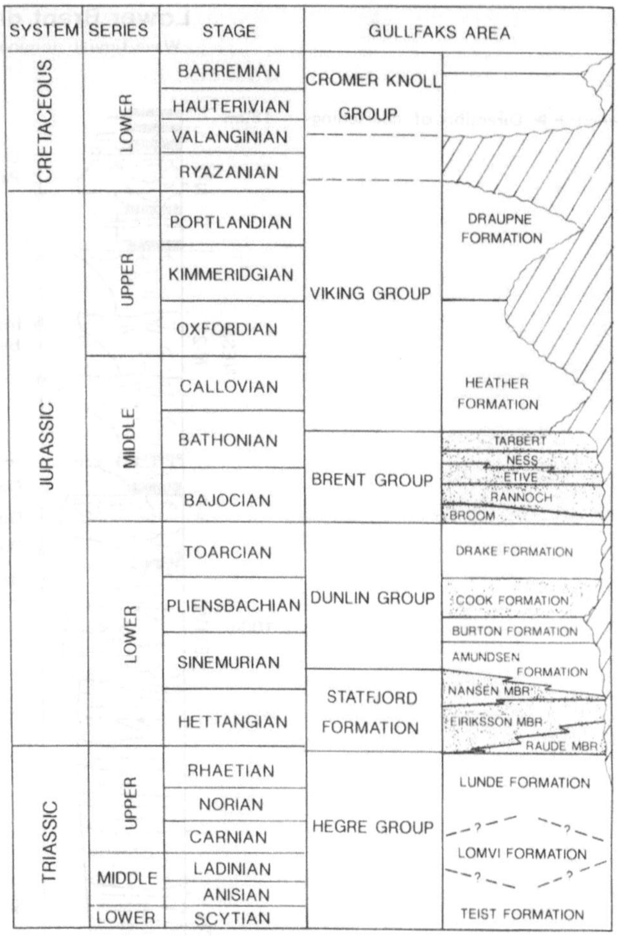


**Fig. 15.** Lithostratigraphic nomenclature of the Gullfaks Field. (Modified after Deegan and Scull, 1977; Vollset and Doré, 1984.)

units, R-1, R-2 and R-3. The Rannoch Formation comprises an overall upwards-coarsening sequence in a sheet-formed sandstone body thinning to the south, with a total thickness of 50–90 m. Abundant hummocky cross-stratification shows that wave processes were important during the deposition of this formation.

The lowest unit R-1 is characterized by a decreasing content of mud in a well-defined upwards-coarsening trend. Bioturbation, small-scale ripples, microhummocky cross-bedding and lenticular bedding are characteristic. Minor carbonate-cemented beds occur, and may be laterally extensive.

Units R-2 and R-3 have good to excellent reservoir properties in the fine- to medium-grained sandstones. In cores, hummocky cross-stratification, minor burrows and small-scale ripples are seen. Carbonate-cemented beds are rather common, but are not correlatable between wells, and are therefore probably laterally restricted. In the topmost few metres of the Rannoch Formation the content of mica may be locally high, particularly in the southern part of the field. This reduces the vertical permeability in this part of the reservoir considerably.

*Etive Formation*

Barrier bar complexes are the depositional environment for the Etive Formation. On the basis of low-angle large-scale cross-stratification, grain size, heavy mineral concentrations and parallel lamination, a high-energy beach environment is suggested. The Etive Formation consists of medium- to coarse-grained sandstones with a thickness of 15–40 m. The lower part commonly reflects upwards-fining sequences that may represent channel-fill deposits. These sequences are interpreted as tidal inlets in the barrier bar systems. The reservoir quality is excellent owing to the clean, laterally persistent sandstones with minor mica and clay content and little calcite cementation. High porosities (30–36%) and permeabilities (1000-4500 mD) reflect the superb reservoir qualities.

*Ness Formation*

The first occurrence of a coal bed above the clean sands of the Etive Formation is taken as the base of the Ness Formation. On the Gullfaks Field, Ness is subdivided into three units, N-1, N-2 and N-3 with a total thickness of 85–110 m. A further subdivision into 11 subunits has also been carried through in a detailed correlation of this formation. A delta plain environment is suggested for the three units, predominantly in a lower delta plain setting.

The lower part, Ness-1, consists of interbedded coals, mudstones and sandstones, occasionally with thick sandstone channel deposits. These channel deposits are interpreted as low-sinuosity distributary channels that locally are expected to be in communication with the underlying Etive Formation. Ness-2 is characterized by several upwards-coarsening sandstone sequences with good reservoir quality. The depositional style suggests crevasse splay, crevasse channel, overbank flooding and minor mouth bar sequences, representing interdistributary bay fill deposits. Ness-3 marks the end of the Lower Brent Delta progradation in the Gullfaks area. Typical of this unit are

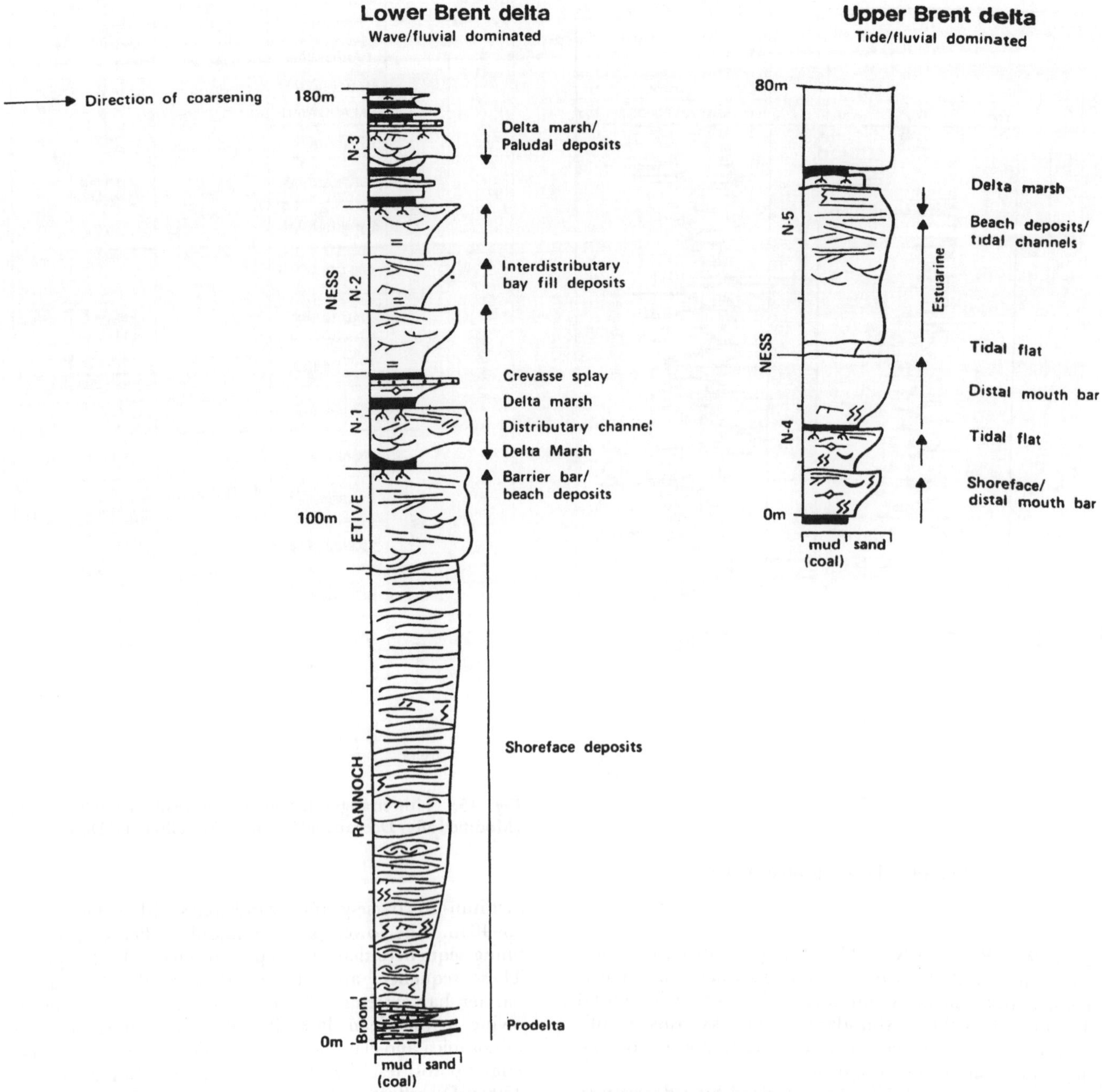


**Fig. 16.** Idealized core description of the Brent Group.

the common occurrences of soil profiles with siderite. The unit is dominated by siltstone/claystone and coal deposits, but also fluvial channel sands and lacustrine deposits are recorded. Where uneroded, N-3 contains laterally persistent permeability barriers across the Gullfaks Field. Only very locally may the fluviatile channel sands be in communication with the underlying Ness-2 sands.

### *Tarbert Formation*

The boundary between the Ness and Tarbert Formations is represented by a marine transgressive event that separates the two above-mentioned phases of delta building. The Tarbert is subdivided into two units, and for reservoir purposes a further subdivision into seven subunits has been carried through. Owing to the Kimmerian erosion, core and log data are limited, and restricted to the western part of the field only.

Sedimentary structures associated with tidal flat and shoreface deposits are found in the lower part. Lithology varies from shales, siltstones and coal beds to medium/coarse-grained sands, where calcite cementation is quite common. This lower part represents a permeability barrier that is likely to cover most of the field. Estuarine and bar complex deposits are suggested as the depositional environments for the upper part of the Tarbert Formation. Also, for this unit, the lithology is variable, but it is generally more homogeneous than the lower unit. Heavy minerals and carbonate cementation occur. Thick, clean sands of excellent reservoir properties are found locally.

Where uneroded by the Kimmerian unconformity, the Tarbert Formation is followed by the shales of the Heather Formation.

## PETROPHYSICS

A calibrated model for multimineral open-hole log evaluation provides a detailed representation of variations in sand types and characteristics on a fieldwide basis. The structural complexity and large areal variation in reservoir subzone characteristics render the assignment of zone parameters for field wide use complicated. To overcome this, an evaluation method for the Brent reservoir was developed to operate on a common set of parameters for all zones in all wells. Results of conventional and X-ray diffraction analysis of Gullfaks cores in addition to produced formation fluids analysis led to the reservoir being modelled as:

- minerals—quartz, feldspar, mica, carbonates, shale, coal;
- formation fluids—formation water, hydrocarbons, water-based mud filtrate.

The evaluations provide an estimation of porosity, saturation and permeability used in perforation interval selection, well performance prediction and, eventually, the reservoir-simulation model. In addition, the mineralogical/lithological description of the well is used to improve the correction and computation of porosity and saturation, enable easier interwell correlation, give qualitative indications of sand type and integrity (for the selection of perforation intervals) and generally improve the understanding of well characteristics.

This approach has allowed a degree of standardization between well evaluations to be maintained. The log evaluation of well B-1 is shown in Fig. 17 as an example. The diagram comprises the raw logs on the right side and the evaluated logs on the left side. A preliminary evaluation is processed offshore within 24 hours after the logging operation is terminated, and forms the basis for selection of perforation intervals.

## RESERVOIR GEOLOGY

The complexity of the Gullfaks Field has been confirmed by development drilling. These wells have contributed important knowledge about the reservoir geology.

- Well control has proved to be very important, since a large number of faults (both minor and major) have been penetrated and mapped. Many of the faults are smaller than the resolution of the seismic data.
- The existence of several E–W oriented faults has been established.
- The majority of the faults that have been mapped with sand-to-sand contact are non-sealing, thus providing limited restriction to fluid flow between fault blocks.
- The existence of sealing faults where sand to sand contact is mapped has been verified.
- The majority of the Brent reserves in the western area of the field are now proved to be located in the Ness and Tarbert Formations. Earlier, they were mapped in the Etive ad Rannoch Formations. This is due to a combination of remapped fault locations and finding the Brent reservoir deeper than earlier interpretation indicated.
- Oil-bearing Cook sandstone in fault blocks 3G and 3H has been established.
- In the western fault blocks, the Ness Formation is uneroded or only slightly eroded. Well control has shown that the Ness Formation can be correlated over wide areas. For use in reservoir simulations models, a detailed correlation containing 12 sand beds has been carried out.

Increased fault control and improved mapping of the reservoir sands across the faults have been extremely important in the production phase. An example of how development drilling has improved the interpretation of the geology is shown in Fig. 18 Wells B-1 and B-3 established the presence of three minor faults in addition to the main fault. This stepwise downthrowing of the sediments results in sand against sand contact between the fault blocks which was not originally mapped. Pressure information and production data have also confirmed that the faults between the main fault blocks 3 and 4 are not sealing, thus resulting in fluid movement over the faults. The understanding and mapping of sand against sand across the fault planes is of utmost importance in the placement of production and injection wells. A fieldwide mapping of the fault planes has therefore been carried out.

In contrast to the non-sealing faults between fault blocks 3 and 4, the results of subsea well A-9H in fault block 6A has confirmed the presence of a sealing fault. Faultplane mapping shows sand-against-sand contact eastwards against oil-bearing Brent sandstone in fault block 5B, but fault block 6A has different oil properties and a different oil–water contact (Fig. 10, cross-section A-A′).

Another important contribution from the development wells has been the increase of reserves by the establishment of oil in the Cook sandstone in the main fault block 3. Early mapping indicated potential hydrocarbons updip in blocks 3G and 3H. Production wells A-4H, B-4 and A-17 have confirmed the presence of oil in the Cook Formation (Fig. 10, cross-section C-C′). The oil–water contact in the Cook Formation is identical in both blocks, but shallower than in the overlying Brent. Fault-plane mapping shows oil-bearing Cook sandstone in block 3H in contact with oil-bearing Brent sandstone in block 3I (Fig. 19). Accordingly, the fault between these two blocks must be sealing at this level. Southwards the fault between blocks 3G and 3H shows Cook sandstone in contact with Cook sandstone (Fig. 20). The identical oil–water contact and pressure regime in these two blocks verify that this fault is non-sealing.

## HYDROCARBONS

### Source

The Upper Jurassic Draupne Formation (Fig. 15) is the most important source rock for generation of hydrocarbons in the East Shetland Basin. Source-rock analysis from wells in the northern North Sea indicates a total organic content in the order of 5–10%.

The Draupne Formation is often eroded on the structural highs, as seen at the Gullfaks structure, but reaches a thickness in the order of 200–400 m in the Viking Graben. Supplementary source rocks of less importance are assumed to be the shales of the Heather Formation as well as the marine shales of the Drake Formation. Other potential source rocks are the shales and coals within the Ness Formation of the Brent Group, but their volumes are restricted.

There are three important "kitchens" for maturation of hydrocarbons situated around the Tampen Spur area (Fig. 21). The "Oseberg kitchen" to the south, the "Troll kitchen" in the east and the "Møre kitchen" to the north of Snorre.

### Migration

Hydrocarbon generation and migration most likely started in Palaeocene to Eocene times. At that time the Gullfaks structure was sealed off by the Upper Cretaceous shales

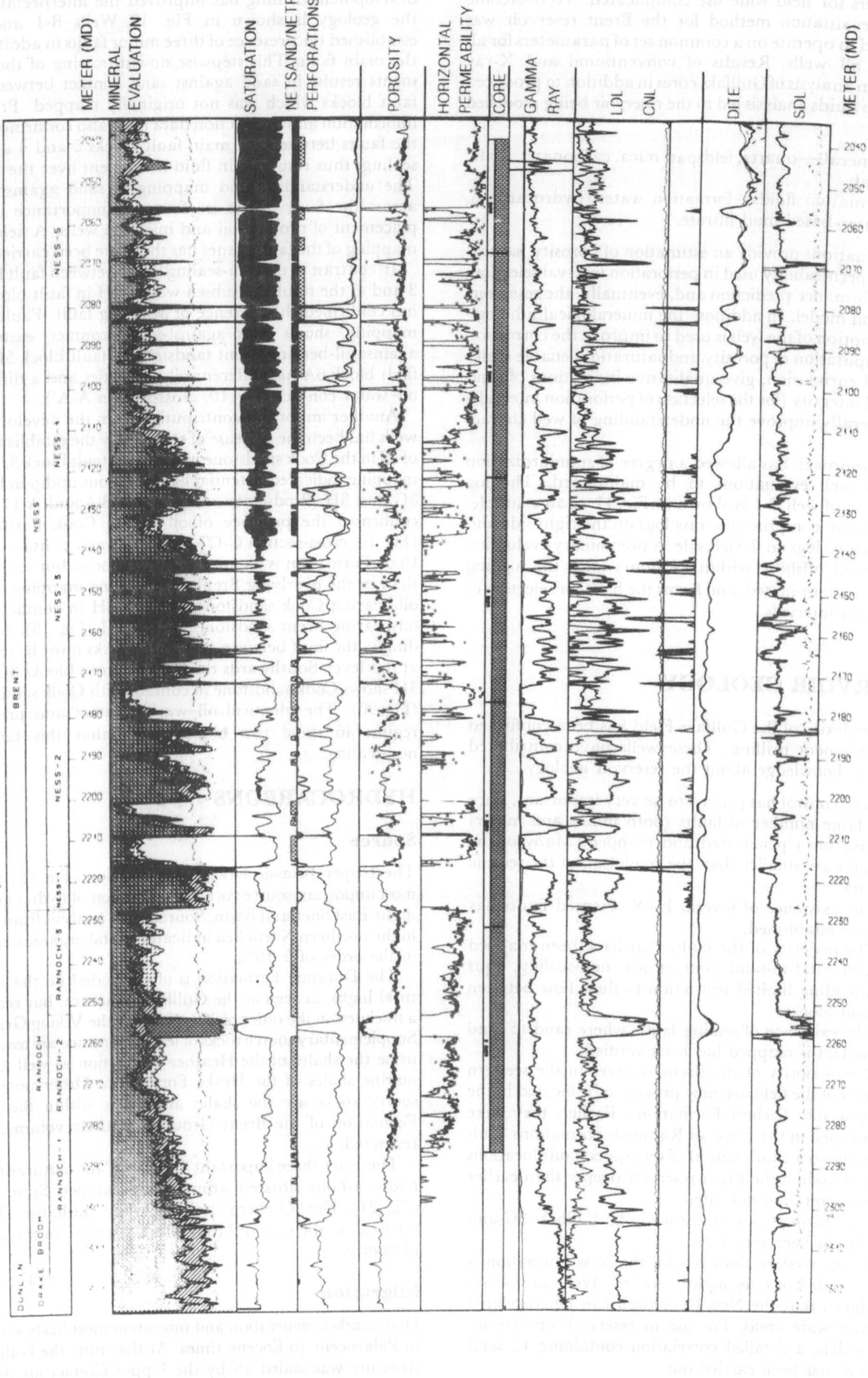


**Fig. 17.** Log evaluation for Brent, Well B-1.

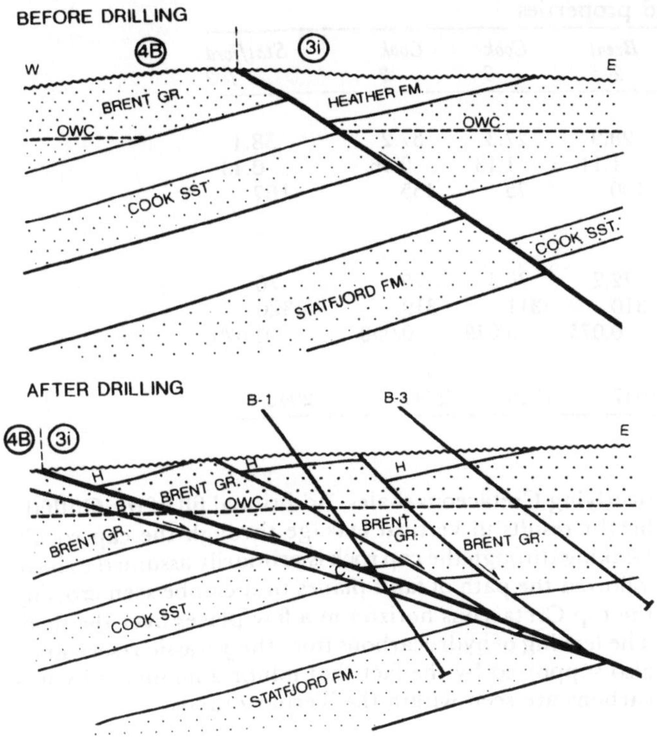


**Fig. 18.** Schematic cross-section through wells B-1 and B-3, prediction and drilling results.

and marls, which from other fields in the area were known to form an efficient seal (Skarpnes *et al.*, 1982).

Direct migration of hydrocarbons from the thick source rocks in the Viking Graben is assumed to have only partly contributed to the oil filling of the structure. This is due to the major boundary fault, which is believed to have restricted potential migration from the "Troll kitchen". However, the deepest oil is seen within the Cook, Statfjord and Hegre Formations in the eastern area of the Gullfaks Field in wells 34/10-7, -9, -11 and -13. This, together with the fact that two different oil types are observed (Table II), indicate that migration from the north-east (block 34/8), east and south (Gullfaks Sør) might all be regarded as potential important substitutes for the more general view of a "filling by spill only" migration from the filled neighbouring structures in the north-west.

Fault block 6A contains oil with a much higher GOR (195 $Sm^3/Sm^3$) than measured elsewhere in the Brent Group. The OWC in this fault block is found at 1891 mMSL which is 56 m shallower than in the main Brent reservoir. This anomaly may be attributed to migration from Gullfaks Sør.

One of the migration routes responsible for the filling of the Gullfaks structure is assumed to have been the spilling of oil along the strike of the North Alwyn–Brent–Statfjord–Snorre–Gullfaks trend as explained by Thomas *et al.* (1984) and Karlsson (1986). Data from these fields show that migration from structure to structure is possible. The amount of gas decreases updip from the Brent Field and

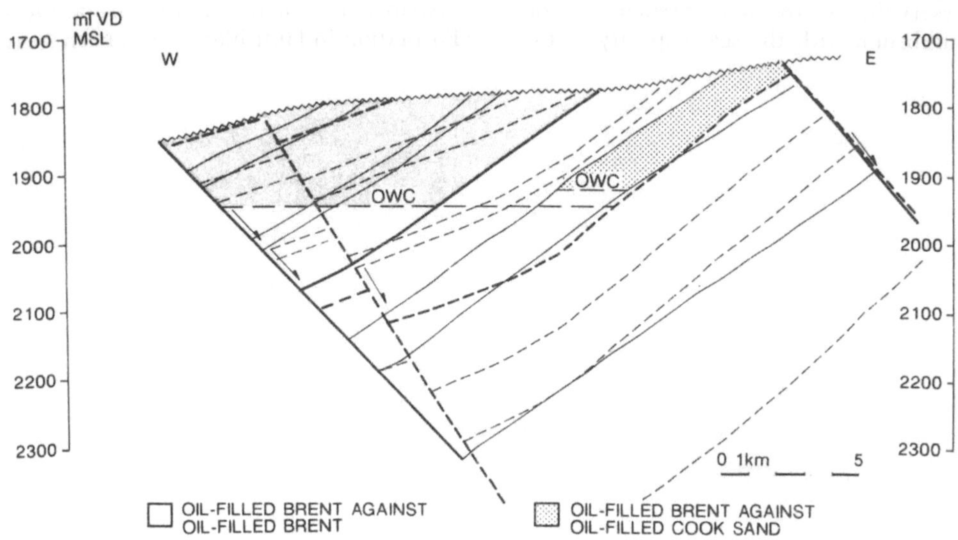


**Fig. 19.** Juxtaposition diagram along fault plane between fault blocks 3H and 3I.

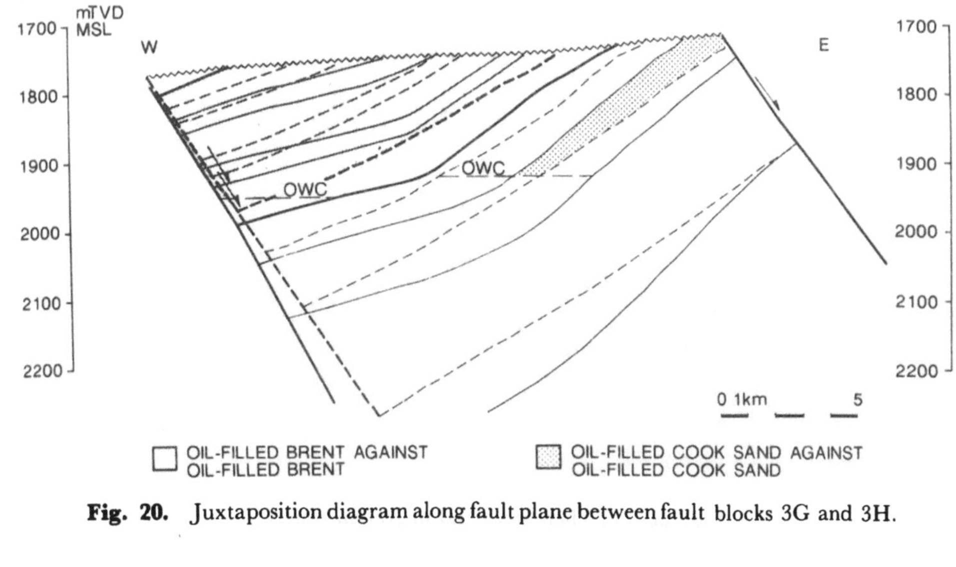


**Fig. 20.** Juxtaposition diagram along fault plane between fault blocks 3G and 3H.

**Table II.** Fluid properties

| *Formation:* | *Brent* | *Brent* | *Cook* | *Cook* | *Statfjord* |
|---|---|---|---|---|---|
| *Fault block:* | *6* | *2–5* | *3* | *2* | *1–2* |
| **Hydrocarbons** | | | | | |
| Oil gravity (°API) | 36 | 28.9 | 27.3 | 37.2 | 38.4 |
| Viscosity at bubble point (cP) | 0.38 | 1.11 | 1.13 | 0.46 | 0.43 |
| Gas–oil ratio ($Sm^3/Sm^3$) | 199 | 100 | 75 | 155 | 162 |
| **Reservoir conditions at 1850 m MSL** | | | | | |
| Temperature (°C) | 71.5 | 72.2 | 76.5 | 76 | 76 |
| Initial pressure (bar) | 309 | 310 | 311 | 319 | 320 |
| Pressure gradient (bar/m) | 0.066 | 0.075 | 0.079 | 0.058 | 0.079 |
| **Fluid contacts** | | | | | |
| Oil–water contact (m MSL) | 1891 | 1947 | 1920 | 2090 | 2043[a] |

[a]Encountered in fault block 2.

the hydrocarbon–water contacts become shallower. Oil densities increase from the Brent Field towards the Gullfaks Field. Migration beyond the Gullfaks structure, being the shallowest part of the Tampen Spur area, through pre-Cretaceous strata is impossible. At the time of initial exploration of the Gullfaks Field, the seal capacity was expected to control the hydrocarbon column of the structure. Throughout the exploration phase of the field, it soon became clear that the structure was not filled to the spill point. The reason for this has been very well explained by Karlsson (1986) who stated that "the formation integrity of the overlying Cretaceous shales is close to the pore pressure of the reservoir, i.e. the pore pressure of the reservoir is in equilibrium with the seal capacity of the overlying Cretaceous shales. Additional fill of the Gullfaks height results in vertical leakage through the cap rock". Leaking through the cap rock is primarily assumed to have followed the path of fault planes that can be seen crossing the top Cretaceous horizon in a few places over the field. The leaking of hydrocarbons from the Jurassic reservoirs is also supported by the fact that minor amounts of hydrocarbons are seen within the Tertiary sands.

## Fluid properties

Two hydrocarbon systems are present in Gullfaks. One system found in the Brent Formation and in the Cook Formation in fault block 3 has been heavily biodegraded

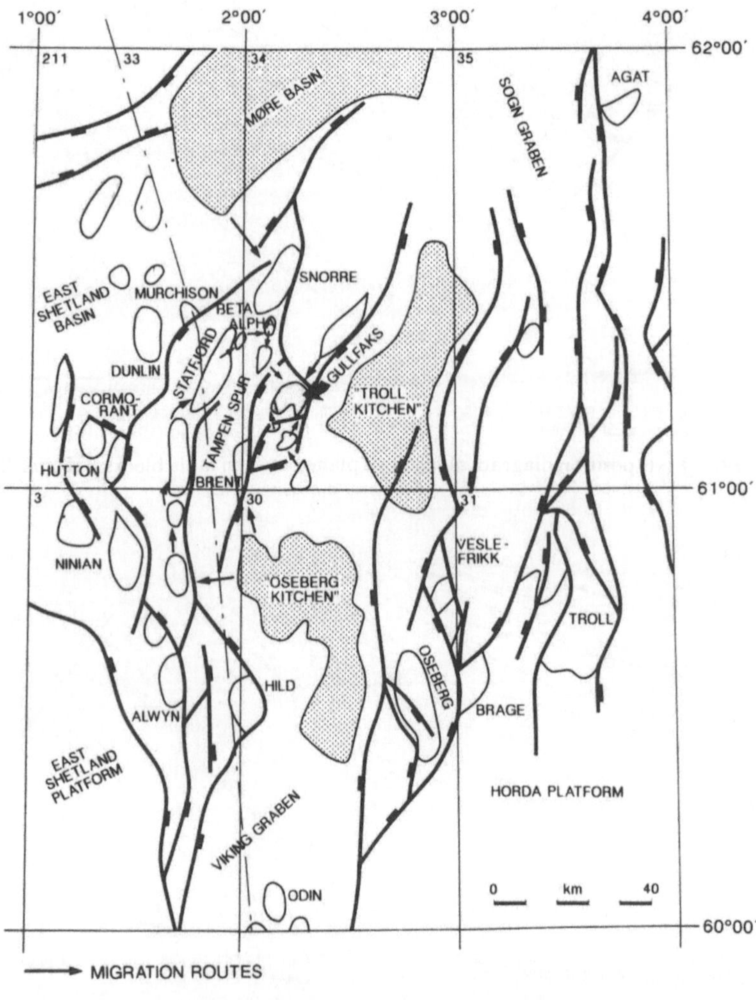


**Fig. 21.** Kitchen areas and migration routes.

and is characterized by a low concentration of normal paraffins. The hydrocarbons in the Cook and Statfjord Formations in fault blocks 1 and 2 have not been subjected to biodegradation, and are characterized by high concentrations of normal paraffins. Oil gravity and oil PVT properties vary throughout the field and are presented in Table II.

### Reserves

Gullfaks stock tank oil initially in place is estimated to be 480 million $Sm^3$ (3020 million bbl) broken down as follows:

| | |
|---|---|
| Brent Group: | 375 million $Sm^3$ (2365 million bbl) |
| Cook Formation: | 45 million $Sm^3$ (280 million bbl) |
| Statfjord Formation: | 60 million $Sm^3$ (375 million bbl) |

The areal distribution of the three reservoirs is shown in Fig. 22.

## ORIGINAL RESERVOIR DEVELOPMENT PLAN

In the commerciality report, dated October 1980, the western area of the field was planned to be developed first from Gullfaks A and B, while the eastern part was to be developed at a later stage from Gullfaks C. The two areas were referred to as Phase I in the west, comprising fault blocks 3–6, and Phase II in the east comprising fault blocks 1–2. The principles for Gullfaks reservoir development described below are taken from the development plan for Phase I, dated January 1986, and the December 1984 Phase II development plan. An updated fieldwide development plan was completed in November 1988.

Early studies showed pressure support by waterflooding to be the most suitable recovery mechanism for the hydrocarbon types and reservoir conditions present in all the Gullfaks reservoirs. Of the miscible flooding options investigated, only the injection of rich gas in the Cook Formation showed any potential but was rendered less favourable than water flooding.

In the field-development plans, each of the following were considered as separate reservoirs to be developed individually until field results prove communication with another reservoir:

- Upper Brent (Ness and Tarbert);
- Lower Brent (Rannoch and Etive);
- Cook;
- Statfjord.

Other important features of the early development plan are as follows.

(1) Reservoir pressure will be maintained above the bubble point in each of the reservoirs. Each reservoir will be developed by the necessary number of dedicated wells to ensure maximum recovery within the economic constraints of the Gullfaks project. For wells that penetrate two reservoirs, recompletion into the second reservoir will be utilized when this is favourable. Upper Brent sands will be developed after the Lower Brent wells have been drilled in each region. Using this development scenario, areal continuity and maximum information pertaining to the Upper Brent sands can be achieved prior to development of these sands.

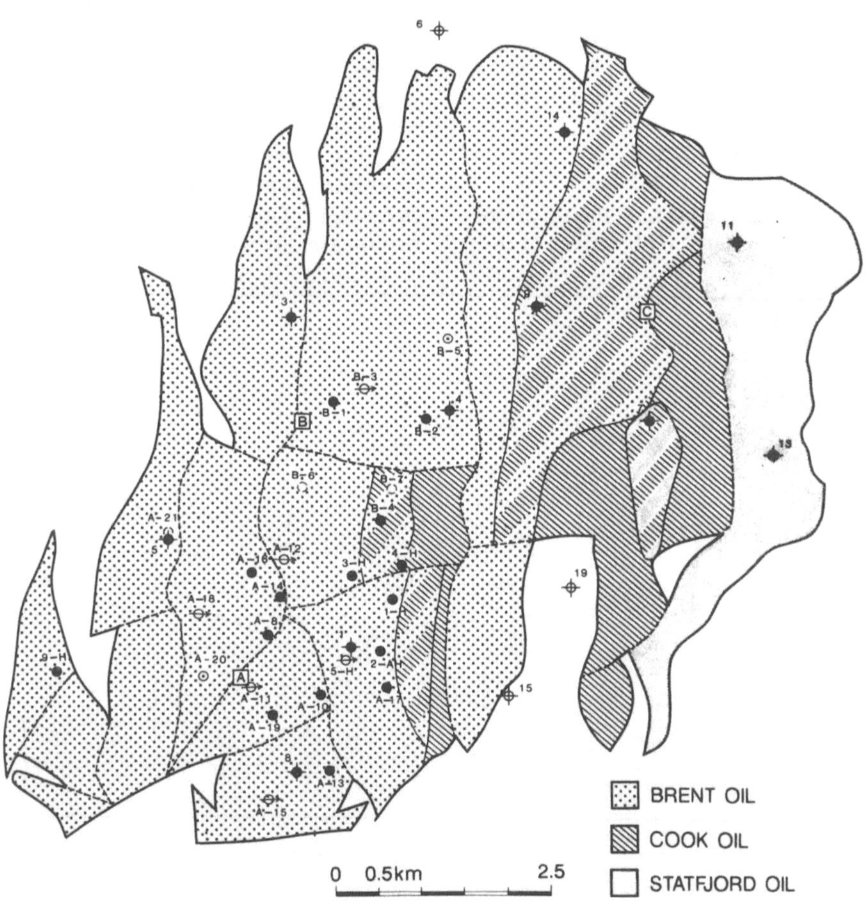


**Fig. 22.** Areal distribution of reserves.

(2) Until communication is observed in the field, each fault block will be developed as an isolated unit with the necessary number of production and water-injection wells.
(3) Lower Brent will be developed initially by production and injection wells perforated in the Rannoch sands only, in order to reduce water override in the more permeable Etive sands. Dedicated production and injection wells are to be completed in the Tarbert sands or the Ness sands in the Upper Brent. Where possible, injection wells will be located below the OWC, but it is estimated that 60% of the Ness and Tarbert injection wells must be perforated above the OWC (Fig. 23)
(4) Selective perforation will be the preferential choice of completion method. In sands with insufficient sand strength, gravel packing will be considered.

In the planning phase it was realized that the development of Gullfaks would not be straightforward. The extent of the faulting already mapped for the field, together with the knowledge of UK sector fields that were proving more complex than expected, reinforced the belief that the reservoir complexity should not be underestimated. In the field-development studies carried out between 1980 and 1985, it became apparent that certain reservoir properties influencing production performance were determinable only through production experience. Examples of such properties are:

- fault transmissibility;
- vertical transmissibility within each reservoir;
- sand strength;
- effect and flow pattern of injected water.

The development plan was deliberately based upon a pessimistic interpretation of these properties. Accordingly, production forecasts developed in the planning stage were conservative, taking into account the possibility of unpleasant surprises in the development phase.

## RESERVOIR MANAGEMENT

Correct formulation of a reservoir-management programme was important to ensure that the objectives of the field-development plan were achieved while securing the acquisition of strategic data as soon as the field came on production. Predevelopment planning studies highlighted activities that would, if implemented at an early stage, render maximum information on communication routes within the field.

### Early drilling schedule

The decision was made to accelerate production through the early start-up of the subsea wells in the Gullfaks A area. In addition to improving the economics of the field by reducing the time to plateau as specified in the field development plan, it also provided an earlier opportunity to reduce the degree of uncertainty in the field data.

Priority was given to early drilling of high-rate production wells in locations with minimum uncertainty. These wells were drilled systematically into previously undeveloped fault blocks and consequently the effects of production from adjacent fault blocks could be seen on the measured pressures. Water-injection wells were drilled when the bottomhole flowing pressures in production wells were at or just above bubble point. The exact timing of pressure support was dependent upon the depletion trends and producing drawdowns, which varied from one fault block to another.

### Fault location and communication

Whenever possible, well paths and bottomhole locations were chosen such that the well would intersect a fault within the reservoir section or could be deepened to intersect the fault. Pressure build-up surveys have been taken in all new wells and analysed for boundary effects.

The Repeat Formation Tester (RFT) pressures from the first well drilled in each fault block are very important in measuring depletion due to production from adjacent fault blocks. RFT surveys measure the pressures in the fault block at only one particular point in time. This limitation was recognized in the planning stage and the decision was taken to install a permanent downhole gauge with surface read-out in one key well in each fault block. The dynamic behaviour of individual fault blocks can be seen within the confines of the whole field, thus making the interaction of

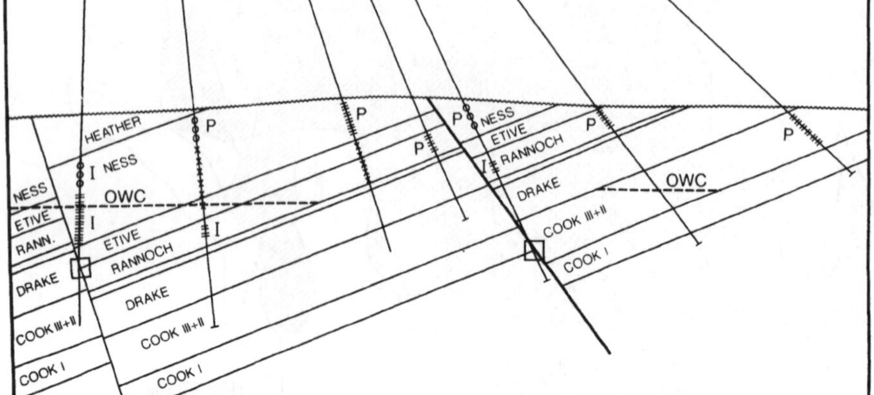


**Fig. 23.** Typical well placements in the Gullfaks A area.

these fault blocks easier to monitor and eventually history-match in the reservoir simulator. The most important benefits associated with the use of permanent downhole gauges versus the use of static pressure surveys have been the following.

- Bottomhole pressure can be monitored continuously.
- The effect of changes in production and injection rates from surrounding wells or fault blocks can be seen immediately.
- Planned and unplanned well shutdowns can be utilized for pressure build-up surveys and interference effects.
- The cost of installing such gauges has been justified by the saving in static pressure surveys.

Currently, there are permanent gauges installed in five wells. To supplement the permanent downhole gauges, long-term memory gauges have been run in selected production wells. These have also been used to measure the effect of injection and production from adjacent fault blocks or nearby wells.

Radioactive tracers have been injected into selected water-injection wells early in their injecting life. These Lower Brent wells are situated such that injected water could move into three different fault blocks and Upper Brent sands. Analysis of produced water in neighbouring wells in these adjacent fault blocks is used to identify water movement across faults.

### Vertical communication

RFT pressures have been used extensively to establish vertical communication within each reservoir and fault block. Through a carefully planned drilling and coring programme, good areal and vertical core coverage has been achieved. Integration of RFT pressures, core inspection and log evaluation results is used to map the areal and vertical continuity of sands and barriers.

### Water-injection programme

By monitoring reservoir pressure in all fault blocks simultaneously, the effects of the planned pressure-maintenance program has been evaluated. Additionally, introduction of radioactive tracers into early water-injection wells has been important in gaining an understanding of the communication routes between the fault blocks early in the field life. Tracer content in produced water is used as input parameter for reservoir simulation models to predict the water flow pattern. In wells where the sands to be investigated have not been perforated, the gamma spectroscopy tool (GST) is used to detect water movement behind casing.

### Sand strength

The knowledge of sand strength of each sand unit has been increased prior to production from each sand through the combined use of a numerical sand-production prediction simulator, the multimineral log evaluation model and core inspection. Through selective perforation, the balance of eliminating weak sands and achieving high well productivity has been maintained. To monitor the success of this approach, sonic sand detectors on each platform well flowline together with downhole detectors run as part of logging tools are used.

### Cook, fault block 3

The decision was made to develop the Cook Formation in fault block 3 through well A-4H by pressure depletion. The well was later hooked up as a convertible gas injector and tested as such in May and August 1988. The gas injection is a means to modulate the export profile, and repressurize the reservoir.

## FIELD PERFORMANCE

Produced and injected volumes from 22 December 1986 up to the end of 1988 are given in Table III. The largest proportion of the production has been from Lower Brent, where the majority of wells are perforated. In February 1988 the first Upper Brent production well B-1 was put on stream. Subsea well A-4H has to date produced nearly 1 million $Sm^3$ of oil from the Cook Formation.

The monthly average daily production and injection rates for oil and water are shown in Fig. 24. Reduced offtakes in accordance with production constraints imposed by the Norwegian government in late 1987 and mid 1988 are evident.

Compared to the January 1986 field-development plan forecast, cumulative production from the field has been accelerated as a result of the early start-up of the Gullfaks A subsea wells (Fig. 25). The produced oil volumes are in accordance with the prediction prepared in June 86 as part of the application for December 1986 start-up. This is attributable to a combination of higher production potential than predicted, and production restrictions imposed by the Norwegian government.

**Table III.** Produced and injected volumes. 22 December 1986 to 1 January 1989

| | *Produced oil* ($Sm^3$) | *Injected water* ($m^3$) | *Produced water* ($m^3$) |
|---|---|---|---|
| Upper Brent | 1 762 534 | 164 919 | |
| Lower Brent | 10 019 900 | 9 629 501 | 1 079 967 |
| Cook | 917 095 | | |
| Total | 12 699 529 | 9 794 420 | 1 079 967 |

Since the summer of 1987, field water production has been considerably higher than predicted. This is mainly due to well A-8, which was perforated in the upper Rannoch sand, 20 m above the OWC. Water production commenced the month after the well was put on stream. When the water cut reached 85% in November 1988, the well was plugged back in the Lower Brent and reperforated in the Upper Brent. While the actual field water cut has been higher than predicted, subtraction of well A-8 water results in a watercut development versus percent recovery comparable to the prediction (Fig. 26). The first well to experience injection water breakthrough was A-2AH in April 1988. Radioactive tracer that was injected into A-5H in April 1987 has also been detected in this well.

Current water-injection capacity does not allow for complete voidage replacement in every fault block or for the field overall. The pressure development in each fault block has been characterized by an initial depletion due to production from adjacent blocks, followed by a steep pressure decline after the first production well is put on stream. As injection starts, the pressure is stabilized. The data in Fig. 27 show clearly the effect of pressure support from fault block 3G to the adjacent fault block 3H, prior to the start of water injection in fault block 3H in 1988.

Gas export commenced in July 1987 after commissioning of the export gas compressors. All gas produced prior to July 1987 was flared.

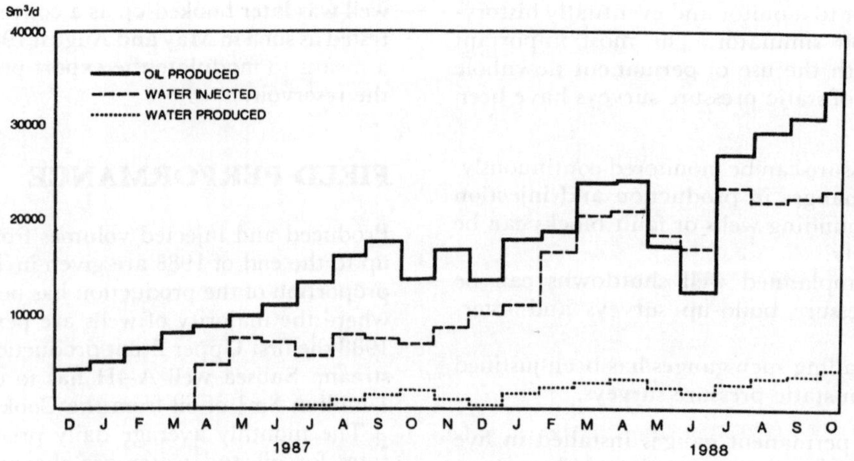


**Fig. 24.** Production and injection rates.

## EXPERIENCE FROM EARLY DEVELOPMENT

With 2 years of production and drilling results from 23 development wells, it is now possible to assess the results from the reservoir-management programme and the future applicability of the current field-development plan. It is important to highlight new field results that represent the most important contributions to the improved understanding of the reservoir.

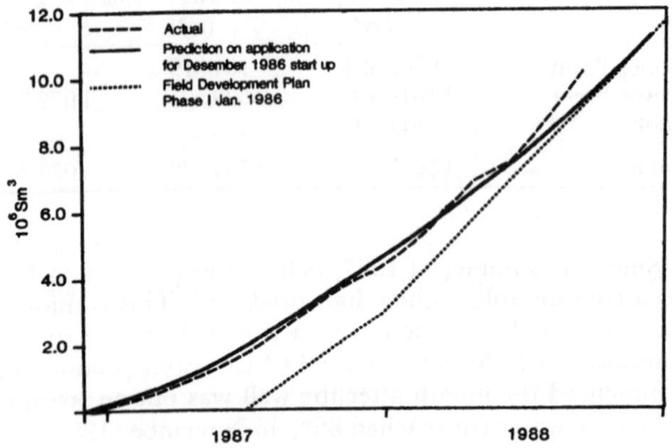


**Fig. 25.** Cumulative production at Gullfaks.

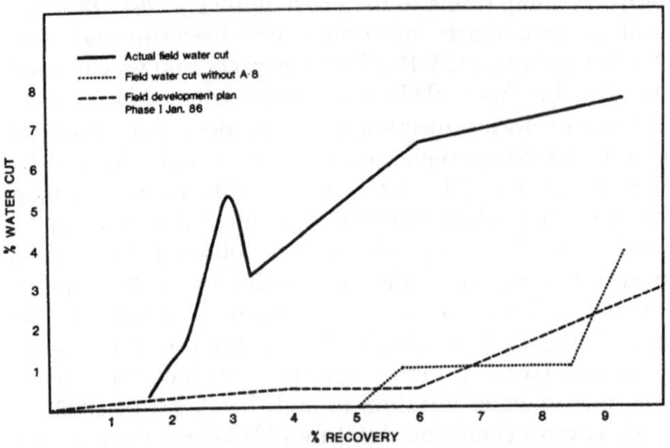


**Fig. 26.** Field watercut development.

### Well productivity

Uncertainty in much of the strategic data in the planning stage resulted in the deliberate creation of conservative field-performance predictions. This has been evident in the well production rates, where capacities of up to 5500 Sm³/d are typical compared to 4000Sm³/d used in the planning stage. Factors responsible for this are higher well productivity indices in certain parts of the field and pressure support across faults allowing larger well drawdowns without going below the bubble point.

### Field productivity

The design platform processing capacity of Gullfaks A (38950 Sm³/d) was predicted to be reached at the end of 1989, but already in August 1988 the production potential exceeded this limit. This is due to the following.

- Higher well productivity as explained above.
- Communication between fault blocks has supplemented pressure support by injection in the early stages of production.
- Higher water-injection well capacities have enabled more production wells to be drilled and sustain higher production rates.
- Average drilling time per well is now 60 days compared to the 80 days used in the planning stage.
- Operating efficiency of the production facilities has consistently been over 97%, compared to 85% used in the planning stage.

Since the production potential exceeded the design processing capacity 15 months ahead of schedule and was predicted to increase for at least another year, a process capacity test was carried out in October 1988. During this test a total rate of 51 115 Sm³/d was for one day produced through the existing facilities.

### Well injectivity

The initial philosophy for the operation of the water-injection wells included keeping the bottomhole injecting pressure below the formation fracture pressure. Under this constraint, planned injection rates were seldom above 5500 Sm³/d. To support the increased field production rates, water injection wells have been hydraulically fractured in the lower Rannoch sands resulting in well rates up to the erosional limit of 7700 Sm³/d. The decision to fracture the wells was taken in preference to drilling

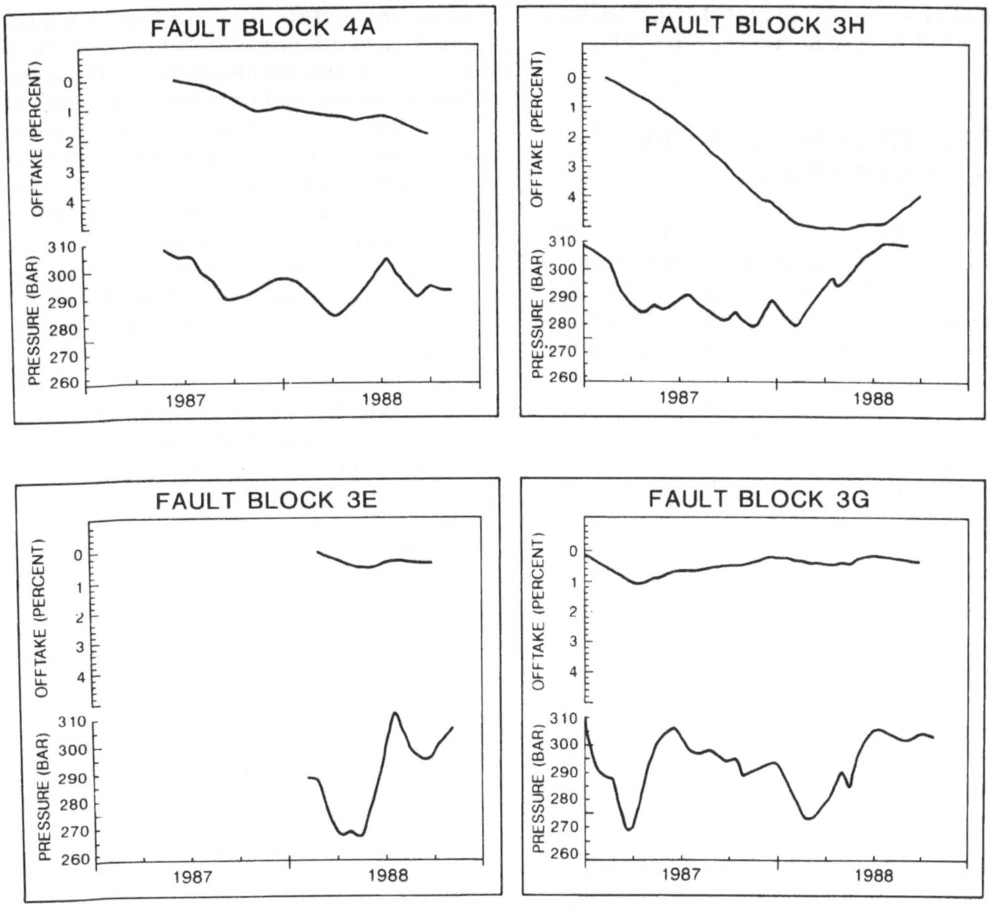


**Fig. 27.** Pressure development and voidage replacement in central Lower Brent fault blocks. The offtake from each block is taken as the net cumulative oil and water production in reservoir volume (gross production – water injection) as a percentage of fluid volume in Lower Brent in the respective fault block. Reservoir pressures are from RFT tests, permanent gauges, memory gauges and estimated from separator tests.

additional injection wells or alternatively perforating the Etive sand in existing wells. With this injection philosophy, water override in the Etive sand is minimized and increased sweep of the lower Rannoch sands can be achieved.

## Pressure communication

The strategy of developing the Lower Brent reserves by perforating the Rannoch sands only is successful to date. Pressure equilibrium exists between the Rannoch and Etive sands throughout the field. The vertical transmissibility between these two sands on a fieldwide basis appears to be higher than anticipated in the planning stage. Exceptions are calcite-cemented beds present in the upper Rannoch sands of some wells that prove uncorrelatable across the field. These beds act as vertical barriers local to the well with unknown lateral extent. Pressure build-up data show the characteristic rollover due to Etive sand contribution occuring at different times from well to well.

With limited production and injection experience from the Ness sands, it is difficult to infer any conclusions about the success of the proposed development strategy. From RFT data, maximum depletion has been seen in the Lower Brent reservoir and Lower Ness sands, owing to production from Lower Brent wells only. Tarbert sands have also shown some depletion (5 bar) prior to production from these or Ness sands.

From Fig. 27 it is evident that the central fault blocks of the Gullfaks A area receive pressure support from sources other than water injection. It is possible to quantify the magnitude of the additional support by a simple material balance, but how this support is apportioned to each sand in the adjacent fault blocks is difficult to resolve with the unknowns that exist at this stage of the development. In fault blocks 5B, 5C and 4B, only exploration wells have been drilled. Current pressures within these Brent sands are therefore unknown.

In summary, it is the current belief that pressure communication exists throughout the major proportion of the Brent reservoir where sand-to-sand contact exists across faults.

## Cook, fault block 3

Cumulative production from A-4H has exceeded early predictions. When the reservoir pressure approached the bubble-point pressure, a gas-injection test was carried out. In addition to other results, this test re-established production rates.

## Sand strength

Natural completion with selective perforation where necessary has proved to be the correct choice of completion method for Rannoch sand wells. Wells have been produced down to bubble-point pressure without experiencing sand production. After water breakthrough, sand production has been seen and wells have been choked back to approximately 75% of their previous production rate.

The limited production from the Ness sands to date

indicates that it may be necessary to gravel-pack Tarbert production wells drilled in specific areas of the field.

## USE OF SIMULATION MODELS IN DEVELOPMENT STRATEGY

Reservoir simulation has been a very important tool, both in the predevelopment studies and in the current development phase of Gullfaks. The types of studies that are currently being carried out reflect the progressive development of the reservoirs and fault blocks from the three platforms. In addition to the ongoing creation of production forecasts on a field basis, there has been continued necessity to evaluate development options of the field on a fault block and reservoir basis.

With the benefit of data from more wells, justification and confidence in the areal mapping of sands, shales and coals has increased. The balance between the inclusion of more geological detail and the size of the simulation model has been resolved through the use of three different model types.

- Planning models with fine layering are used to study detailed development scenarios for the individual reservoirs found in each fault block. These models are particulary beneficial for the Ness development, since significant differences exist in the total as well as the distribution of the Ness reserves mapped in each fault block.
- Interaction of the reservoirs across the faults is taken care of by reservoir models that are used to study the development of the individual reservoir group as a whole.
- Integration of all resevoirs in a full-field model provide the basis for long-term field planning and full-field production profiles. Using this model, the design capacities and interfield pipeline facilities are incorporated and limitations or restrictions on a fieldwide basis are evaluated.

Accurate representation of the sloping faults was a feature of the simulation work carried out up to 1987. Thereafter, all faults have been modelled as vertical, making updating and modifying much easier and quicker. Comparison of results from simulation studies using sloping and vertical faults show no difference within the uncertaintly inherent with the exact throw and geological interpretation of the fault.

Good matches for both Lower and Upper Brent pressures in all fault blocks under development have been achieved. Lack of production experience from the Upper Brent renders the quality of this match below that of the Lower Brent. The lack of data points from surrounding fault blocks not drilled into renders the history-matching process for the individual fault blocks non-unique. Major history-matching parameters have been the transmissibility and degree of sand-to-sand contact across the faults. Expansion of the data acquisition programme to include more local pressure monitoring by memory gauges to supplement the permanent gauges will continually improve the history match.

## FUTURE DEVELOPMENT

The performance of the field to date shows that there is no reason to deviate from the current reservoir-management philosophy and the basic strategy of the field-development plan. In accordance with these, future development will be to continue the development of the Lower Brent in the Gullfaks B area and simultaneously gain knowledge of the Upper Brent. In the Gullfaks A area, emphasis will be put on development of the Upper Brent. Upper Brent development wells will be used to monitor the success of the waterflood in the Lower Brent. In 1990, development of Brent from Gullfaks C will commence, followed by development of the Cook and Statfjord in the eastern area.

The Cook Formation in fault block 3 is currently the subject for detailed studies to determine the future production strategy. Water injection as well as gas injection is being investigated to determine the optimum production mechanism. The data from development wells drilled into this formation show that the previous reservoir description was too simple. A complete re-evaluation of reservoir data for the whole Cook Formation is being conducted.

An important goal in the futue development of the field is to reduce the number of wells without reducing recovery. The current plans include several approaches to reach this goal.

- The potential of each well to be reperforated in other formations is considered already at the planning stage. In some parts of the field this will lead to a reduction in the number of wells in each reservoir. In other parts it will lead to increased recovery.
- The conclusion that the majority of the faults with sand-to-sand contact are unsealing gives the possibility to pressure-support production wells in one fault block by injecting in an adjacent fault block in the reservoir that is in sand contact with the producing formation. This type of scenario is planned to give pressure support to the Upper Brent production well B-1 (Fig. 18) and will be considered as a possible strategy in several parts of the field.
- A third approach is to increase the capacity of each water-injection well. The perforation strategy practised to date for Lower Brent injectors will be continued. Lower Brent water injectors will be perforated in a relatively small interval in the lower part of Rannoch and fractured to obtain rates of the same magnitude as for current wells. Increasing the tubing dimension of the water injectors is considered in order to obtain higher rates.

Work is in progress to identify restrictions in the surface facilities connected to the processing and injection capacities. Later in the life of the field, well maintenance work will be emphasized as well as infill and kickoff drilling, to reach poorly swept and unflooded parts of the reservoir.

In the early development phase, the activities that are given highest preference to increase the oil recovery are reservoir description, building of detailed reservoir-simulation planning models, detailed well planning, and perforation strategy. To obtain a good reservoir description, a well-planned data acquisition programme is necessary. The data are used to continuously update the existing reservoir description. New information and interpretations are transferred to the different simulation models to optimize well targets and perforation strategy.

The possibility of chemical flooding with polymer or surfactant is under continuous evaluation. Laboratory experiments and reservoir simulations are being conducted to develop a suitable surfactant system. The results are technically promising, but induced costs are currently too high to make a surfactant flood economically feasible.

Extensive work has been conducted to evaluate the potential for polymer flooding of the Lower Brent sands.

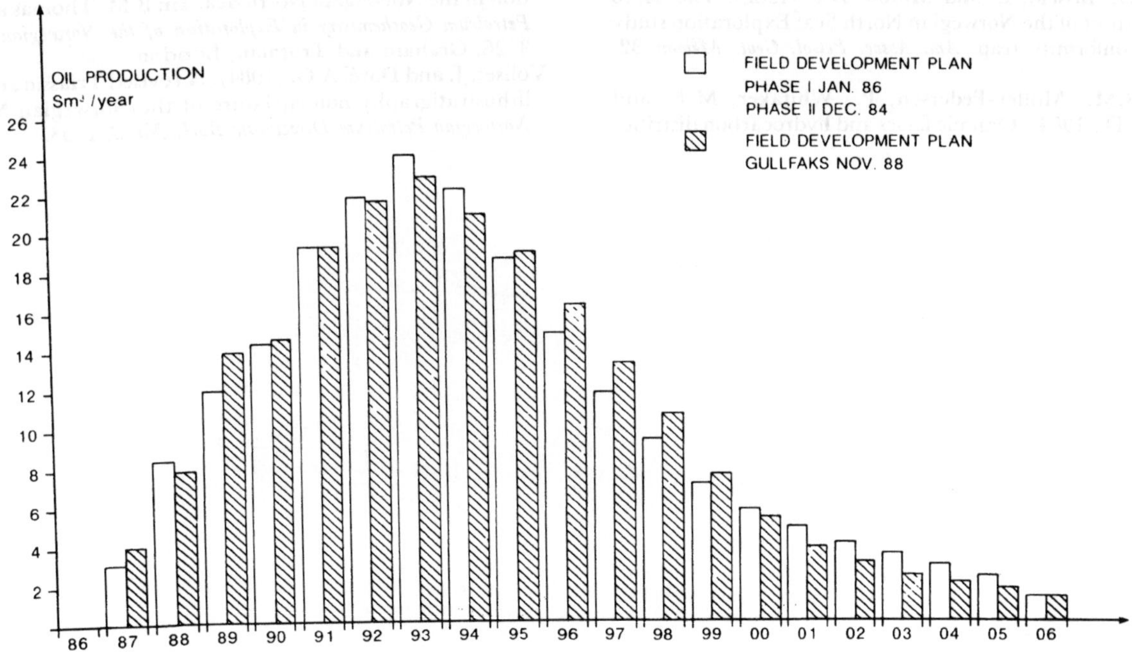


**Fig. 28.** Production forecast, Gullfaks Field.

The permeability ratio between Etive and Rannoch is in the order of 10. The possible water override in Etive is the background for the choice to perforate Lower Brent wells in the Rannoch sand only, and injection of polymer could be a way to increase the permeability control. The biopolymer Xanthan is rated as the most promising polymer for Gullfaks with respect to flow behavior. The studies show that it is technically feasible to inject the polymer in one or several wells, and that logistics can be catered for. Reservoir simulations, however, indicate that the extra recovery that could be attained is too small to cover the costs incurred with current oil prices.

### Production forecast

The production forecast for the life of the field is given in Fig. 28. The comparison between the updated November 1988 field-development plan forecast and the preceding development-plan forecasts for phases I and II shows a slight increase in the produced volumes in the first years of production and in the years 1995–1999. The low production in 1988 is due to a 7.5% cut induced by the Norwegian government and an agreement with the Statfjord licence group to take the Statfjord 7.5% cut on Gullfaks. The total predicted volume produced is unchanged.

## CONCLUSIONS

Development of the Gullfaks Field has been difficult owing to the complex fault pattern. New seismic data and results from development wells have led to improved knowledge and refinement in the fault mapping.

Carefully planned data acquisition and flexible drilling programmes have resulted in a sound understanding of the communication routes within the reservoir, early in the development.

Reservoir simulation has been used extensively to continuously evaluate production and development options for the field on a fault-block and reservoir basis.

Through this work, Gullfaks has been able to produce in accordance with or above its early production forecasts.

## ACKNOWLEDGEMENTS

The authors wish to express their thanks to Den norske stats oljeselskap a.s., Norsk Hydro a.s. and Saga Petroleum a.s. for their permission to publish this article. The views and opinions in this article are those of the authors, and are not necessarily reflective of the operator and partners in the Gullfaks Field. We also wish to express our thanks to everyone who has contributed, and in any way helped making the article a reality. This includes colleague geologists and reservoir engineers, drafting personnel and other technical staff.

## REFERENCES

Deegan, C.E. and Scull, B.J. (compilers)(1977). A proposed standard lithostratigraphic nomenclature for the Central and Northern North Sea. *Rep. Inst. Geol. Sci., No. 77/25*; Norwegian Petroleum Directorate Bull. No. 1.

Fossen, H. (1989). Indication of transpressional tectonics in the Gullfaks oilfield, northern North Sea. *Marine Petroleum Geol.* **6**, No. 1, 22–30.

Karlsson, W. (1986). The Snorre, Statfjord and Gullfaks oilfields and the habitat of hydrocarbons on the Tampen Spur, offshore Norway. In: A.M. Spencer *et al.* (eds.) *Habitat of Hydrocarbons on the Norwegian Continental Shelf*, pp. 181–197. Graham and Trotman, London.

McClay, K.R. and Ellis, P.G. (1987). Geometries of extensional fault systems developed in model experiments. *Geology* **15**, 341–344.

Sæland, G.T. and Simpson, G.S. (1982). Interpretation of 3D data in delineating a subconformity trap in block 34/10, Norwegian North Sea. *Am. Assoc. Petrol. Geol. Memoir* **32**, 217–235.

Skarpnes, O., Briseid, E. and Milton, D.I. (1982). The 34/10 Delta Prospect of the Norwegian North Sea: Exploration study of an unconformity trap. *Am. Assoc. Petrol. Geol. Memoir* **32**, 207–216.

Thomas, B.M., Møller-Pedersen, P., Whitaker, M.F. and Shawn, N.D. (1984). Organic facies and hydrocarbon distribution in the Norwegian North Sea. In: B.M. Thomas *et al.* (eds.), *Petroleum Geochemistry in Exploration of the Norwegian Shelf*, pp. 3–26. Graham and Trotman, London.

Vollset, J. and Doré, A.G. (1984). A revised Triassic and Jurassic lithostratigraphy nomenclature of the Norwegian North Sea. *Norwegian Petroleum Directorate Bull. No. 3*, 1–53.

Field Review 5

# I. Stochastic Modelling of Reservoir Sand Bodies for Input to Reservoir Simulation, Snorre Field, Northern North Sea, Norway

*K.O. Stanley*[a], *K. Jorde*[b], *N. Ræstad*[c] *and C.P. Stockbridge*[d]

[a]Exxon International, Houston, Texas, USA [b]Saga Petroleum a.s., Høvik, Norway
[c]Saga Petroleum a.s., Harstad, Norway [d]Esso Expro UK Ltd., London, UK

The Snorre Field of the northern North Sea contains oil in reservoirs of fluvial origin; the upper part of the Upper Triassic Lunde Formation and the Upper Triassic to Lower Jurassic Statfjord Formation. The formations were deposited by streams that ranged from braided to low–moderate sinuosity, and have resulted in reservoirs made up of a complex pattern of sand bodies intercalated in a silty mudstone matrix. A stochastic procedure was developed to construct three-dimensional computer models of selected parts of the field. This was done in order to make an estimate of average reservoir and well performance at an early stage in the delineation and evaluation of the field, while development planning was still underway.

Parameters used to define the geometry and spatial distribution of reservoir sand bodies in the computer models were established from log and core data, supplemented by data from modern analogues. For each model, several (10–20) three-dimensional computer representations were constructed by stochastically distributing sand bodies in a non-reservoir matrix, using the established parameters to constrain the distribution. Synthetic production and injection wells were introduced into the representations, and the volumes of sand bodies interconnected between the wells were calculated. The representation with sand-body interconnectedness closest to the median value was selected for reservoir simulation.

Agreement of the models with field conditions was verified by visual inspection of cross-sections through the representations, and by a close similarity between the result of an extended production test and a prediction obtained from the computer model.

## INTRODUCTION

The bulk of hydrocarbons discovered in the Norwegian sector of the northern North Sea (Fig. 1) is encountered in Jurassic Brent and Statfjord sandstones, with only a few oil shows in the red beds of the underlying Triassic Lunde Formation (Fig. 2). Snorre is the first giant oil field in the northern North Sea with significant reserves in these red beds. Early in the exploration and delineation of the Snorre Field in Norwegian Blocks 34/4 and 34/7 (Fig. 3), Snorre team geologists recognized the complexity of the Triassic reservoir and its potential impact on estimates of recoverable reserves, field development planning and field economics. To address these issues, Saga Petroleum developed 3D reservoir models that mimicked the Triassic and Statfjord Formation reservoirs, and that could be used as an input to reservoir simulation.

The Snorre Field is made up of a series of fault-bounded blocks, 1–2 km wide and 4–8 km long, that subdivide the Triassic and Lower Jurassic rocks into several reservoir compartments (Fig. 3). Each of these compartments is considered to be a relatively independent reservoir system requiring a staggered line drive pattern of water-injector and producer wells to maintain efficient production. The reservoir modelling of Snorre sought to model the reservoir within individual compartments, and this is the focus of this chapter. The results of the modelling were later integrated into a fieldwide model with the faults being incorporated.

Modelling the reservoir rocks in the Snorre Field is a geometric problem related to sand-body abundance, size, shape, orientation, communication and location in three-dimensional space. These reservoir properties are a product of the depositional setting, albeit modified by post-depositional burial and deformation. The inferred depositional environment for the reservoir and non-reservoir rocks can therefore be used to constrain the geometric variables. The problem is to define adequately the most likely values for the geometric variables from the available information, as well as the possible range around these values.

*North Sea Oil and Gas Reservoirs—II*

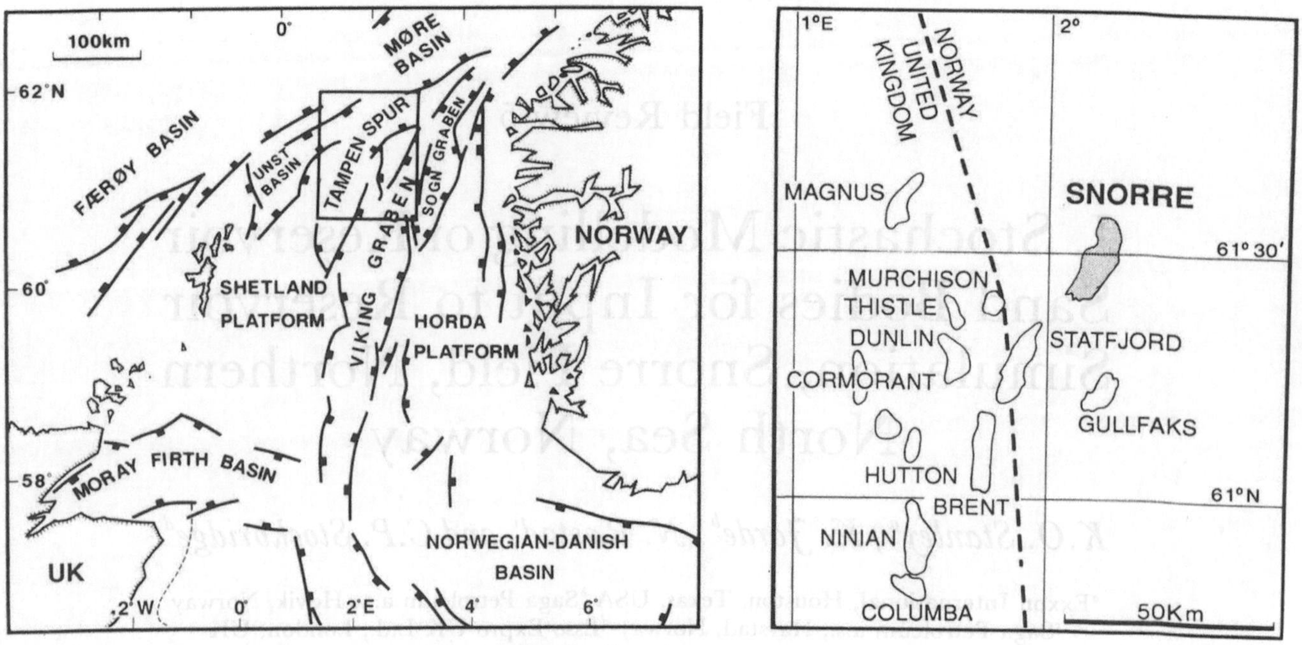


**Fig. 1.** Location map.

| CHRONO-STRATIGRAPHY | | | LITHOSTRATIGRAPHY | | |
|---|---|---|---|---|---|
| SYSTEM | SERIES | STAGE | GROUP | FORMATION | MEMBER |
| JURASSIC | LOWER | PLIENSBACHIAN | DUNLIN | AMUNDSEN | CALCAREOUS |
| | | SINEMURIAN | | STATFJORD | UPPER |
| | | HETTANGIAN | | | |
| TRIASSIC | UPPER | RHAETIAN | | | LOWER |
| | | | | | UPPER |
| | | NORIAN | | | MIDDLE |
| | | CARNIAN | HEGRE GROUP | LUNDE | LOWER |
| | MIDDLE | LADINIAN | | | |
| | | ANISIAN | | LOMVI | |
| | LOWER | SCYTHIAN | | TEIST | UPPER |
| | | | | | LOWER |

**Fig. 2.** Triassic–Lower Jurassic stratigraphy of the northern North Sea.

The purpose of this paper is to characterize the 3D models of the reservoir sand network for Snorre and to discuss how the models were constructed. Sensitivities of the models to geological parameters are outlined and discussed in relation to their impact on representations selected for reservoir simulation. The use of the models in reservoir simulation is described by Nybråten *et al.* (this volume).

Modelling of the Snorre Field began at the field delineation stage of the appraisal programme. Clearly, an exact model of the reservoir was impossible when so little was known about the field. However, simplified stochastic representations of the reservoir could be generated that would approximate average reservoir behaviour and recovery. For this procedure, a computer is used to distribute sand bodies within a volume with non-reservoir matrix until a specified sand content and variation of sand content are obtained. The location of each sand body is randomly chosen within the reservoir volume, while the size and orientation of the sand body is randomly chosen from a prescribed probability distribution.

The input parameters for this procedure are sand content and sand body geometry and orientation. The validity of the representations is dependent upon how well the parameters in the model mimic the actual reservoir. Because the results are only approximations, iteration of the model as new data become available will improve its accuracy and help to fine-tune the input parameters. As the field is developed, the stochastic model should be replaced by more deterministic models.

Stochastic modelling of reservoirs during field delineation is helpful for initial evaluations of economic viability, particularly in complex reservoirs in offshore settings where development costs are high.

## CONSTRAINTS ON MODELS

The Snorre Field is situated on the western side of the Viking Graben, north of the Gullfaks Field and northeast of the Statfjord Field (Fig. 1). It covers about 100 km², and is located in water depths of 300–350 m. The reservoir units dip a few degrees to the west, and are truncated by the Kimmerian unconformity with overlying Cretaceous shales sealing the reservoir (Fig. 3).

The large area of the field relative to the total field STOIIP of $490 \times 10^6$ Sm³ places constraints on the total number of wells that can be used. The 3D computer models of the reservoir sand network that were supplied to the engineers for simulation were reservoir element models, i.e. simulation models of restricted areal and stratigraphic extent (Nybråten *et al.*, this volume). They were designed to model seven separate segments of the field, two fault

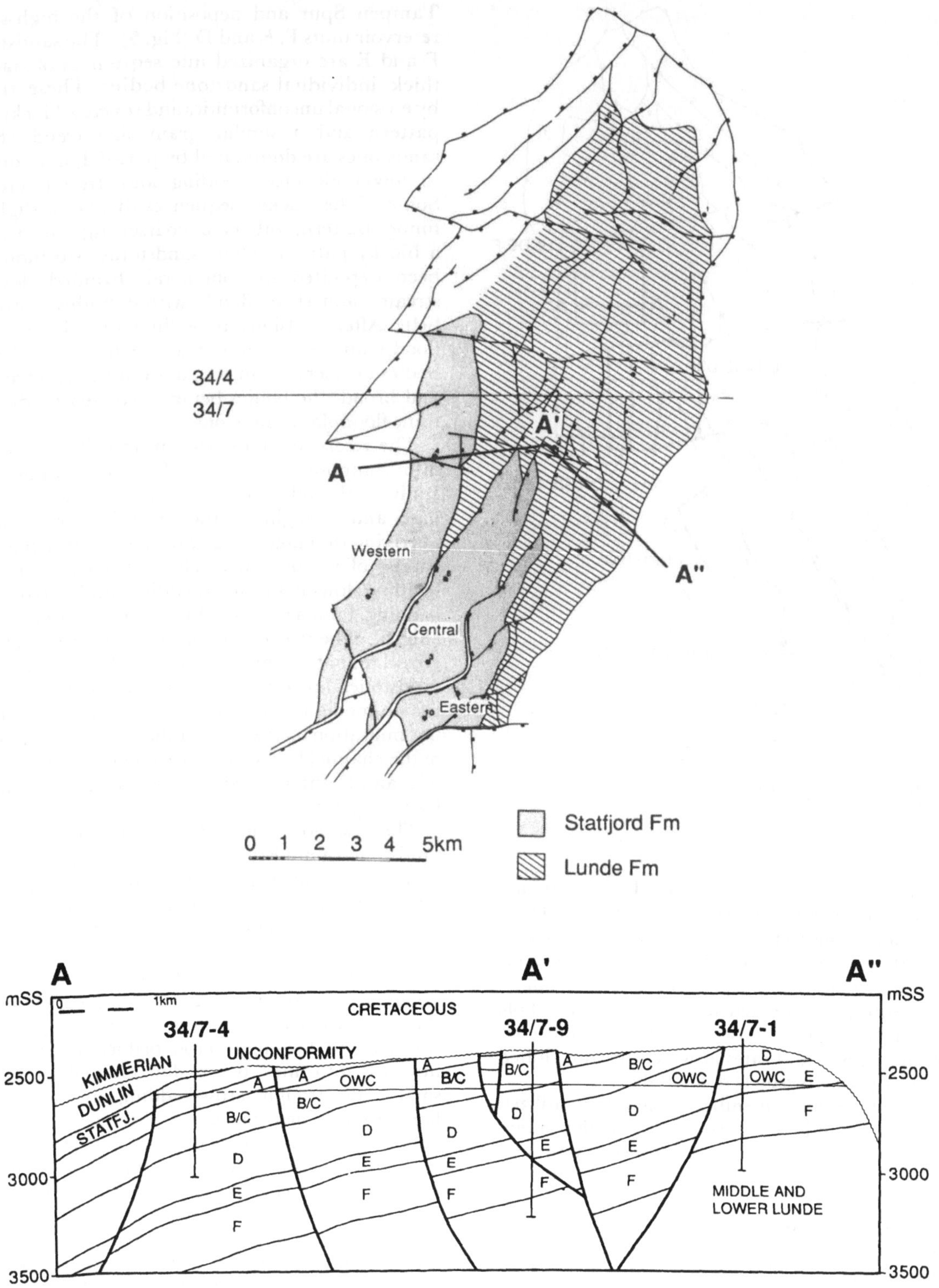


**Fig. 3.** The Snorre Field.

compartments that contained oil in the Statfjord Formation and five segments with oil in different parts of the Lunde Formation (Fig. 4). Data from the exploration and appraisal wells penetrating the fault compartments were used as input to the modelling.

## GEOLOGICAL MODEL

### Stratigraphy

The Snorre Field reservoirs are found within the Upper Triassic (Rhaetian) upper member of the Lunde Formation and the Upper Triassic (latest Rhaetian) – Lower Jurassic Statfjord Formation (Fig. 2). The Lunde red beds are the principal reservoirs in the eastern and northern parts of the field; the Statfjord is the main reservoir to the southwest (Hollander, 1987).

The upper member of the Lunde Formation consists of approximately 800 m of alternating red, sandy, silty mudstone and red to drab sandstone. Sand layers in the Lunde are about 4 m thick and few are greater than 10 m thick. They are interpreted as fluvial channel, channel belt and floodplain sands intercalated in floodplain mudstones.

The upper member of the Lunde is divisible into five reservoir zones, units A, B/C, D, E and F, based on the

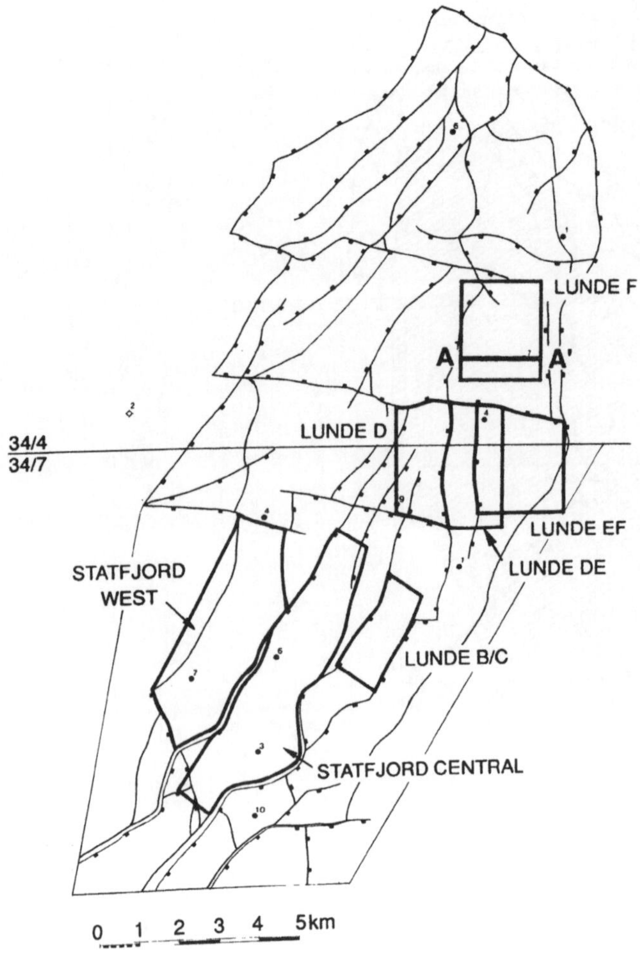


**Fig. 4.** Element model locations.

abundance of sandstone layers (Fig. 5). These units are recognizable over the entire Snorre Field. Their similarity in thickness and consistent variation of sand content suggests that they can be considered as chronostratigraphic units related to regional time-dependent sediment supply, and tectonic subsidence related to sagging of the whole Tampen Spur area, and/or base-level changes that controlled stream regimes and rates at which streams shifted across the floodplain (Nystuen *et al.*, 1989).

The models of the upper member of the Lunde utilized these thickness and sand-content trends as time-dependent parameters of sediment supply to define reservoir units and project stratigraphic architecture away from wells in each fault compartment.

The Statfjord Formation is divisible into three reservoir units in the Snorre Field: lower, middle and upper. The formation is about 100 m thick in the central fault block (Fig. 3) and about 80 m thick in the western block, both of which were modelled. The lower unit of the Statfjord is a sand-poor unit, while the upper reservoir unit is sand-rich. The sand-rich middle unit, which is observed in the central fault block only, contains a mudstone that appears to be locally continuous, at least over the area covered by the element model. These stratigraphic considerations were all incorporated into the Statfjord models by building the models with two and four layers respectively, the layer which represented the continuous mudstone being modelled as continuous.

## Depositional model

Regional lowering of base level at the end of Norian time resulted in a regression of the Triassic Sea from the Tampen Spur and deposition of the high-sand-content reservoir units F, E and D (Fig. 5). The sandstones of units F and E are organized into sequences of stacked, 2–6 m thick, individual sandstone bodies. These are underlain by erosional unconformities and reveal a blocky gamma-ray pattern and a similar grain size trend (Fig. 6). The sandstones are dominated by parallel, horizontal bedding to low-angle cross-bedding and trough cross-bedding. Some of the stacked sequences display a slight upwards-fining pattern, others a coarsening upwards trend or a blocky pattern. These sandstones are thought to have been deposited in ephemeral, braided, low-sinuousity streams and sheet floods within highly mobile channel belts. After avulsions, these fluvial sands were covered by floodplain mud. The overall geometry of the sandstone bodies in reservoir units F and E is interpreted as elongate and broad, the bodies being separated by discontinuous, thin, floodplain mudstones.

The reservoir sandstones in unit D are also organized into composite sequences of stacked sandstone bodies. Individual sandstone units, being defined by an erosional base and a slightly upwards-fining trend towards an overlying thin mudstone or the next succeeding sandstone, consist of various facies characterized by trough cross-bedding, low-angle cross-bedding, and horizontal parallel bedding, tabular cross-bedding and ripple cross-lamination (Fig. 6). The vertical order of these facies varies, but the cross-laminated interval is usually at the top. These sandstones are interpreted as having been deposited in low-to-moderate braided sinuous channels in which lateral bar migration has been a significant process. The geometry of the channel belt sandstones bodies of unit D is, like that of F and E, interpreted as elongate and broad, separated by floodplain mudstones.

The reservoir unit B/C is characterized by a lower sandstone-to-mudstone ratio than units F, E and D. The sandstones generally occur in markedly upwards-fining sequences that can be identified as single channel fills or multistorey bodies formed in mobile channel belts. A characteristic upward fining B/C sandstone commences at the base with an erosional boundary towards underlying floodplain mudstones (Fig. 6). A lag conglomerate with shale and carbonate clasts is succeeded by trough cross-bedded and low-angle cross-bedded intervals that fine upwards into fine- to very fine-grained sandstones with small-scale trough cross-bedding and current ripple lamination. These facies are succeeded by interbedded very fine-grained sandstones and mudstones that are stained brownish-red and may have root structures and palaeosols with calcrete nodules. The vertical profiles of the B/C sandstone bodies are very similar to the "classical" point-bar sequence. The sandstones of this unit are thus interpreted as having formed by lateral bar and channel migration in slightly meandering channels of low to moderate sinuousity. An overall ribbon-shaped geometry was modelled for the B/C sandstone bodies.

Unit A is dominated by floodplain mudstones and minor thin sandstones that can represent single sheet flood deposits, or small intrabasinal drainage channels. Unit A was modelled as a mudstone sequence.

The increased frequency of upwards-fining channel-fill sequences containing a lateral accretionary component, combined with thicker, intervening distal floodplain units through unit B/C and into unit A, suggests a decrease in depositional slope and a higher subsidence rate during the depositional time of these units (cf. the process–response models of Allen, 1978; Leeder, 1978; Bridge and Leeder, 1979). A decrease in slope was probably the result of a relative rise in base level, as indicated by brackish water,

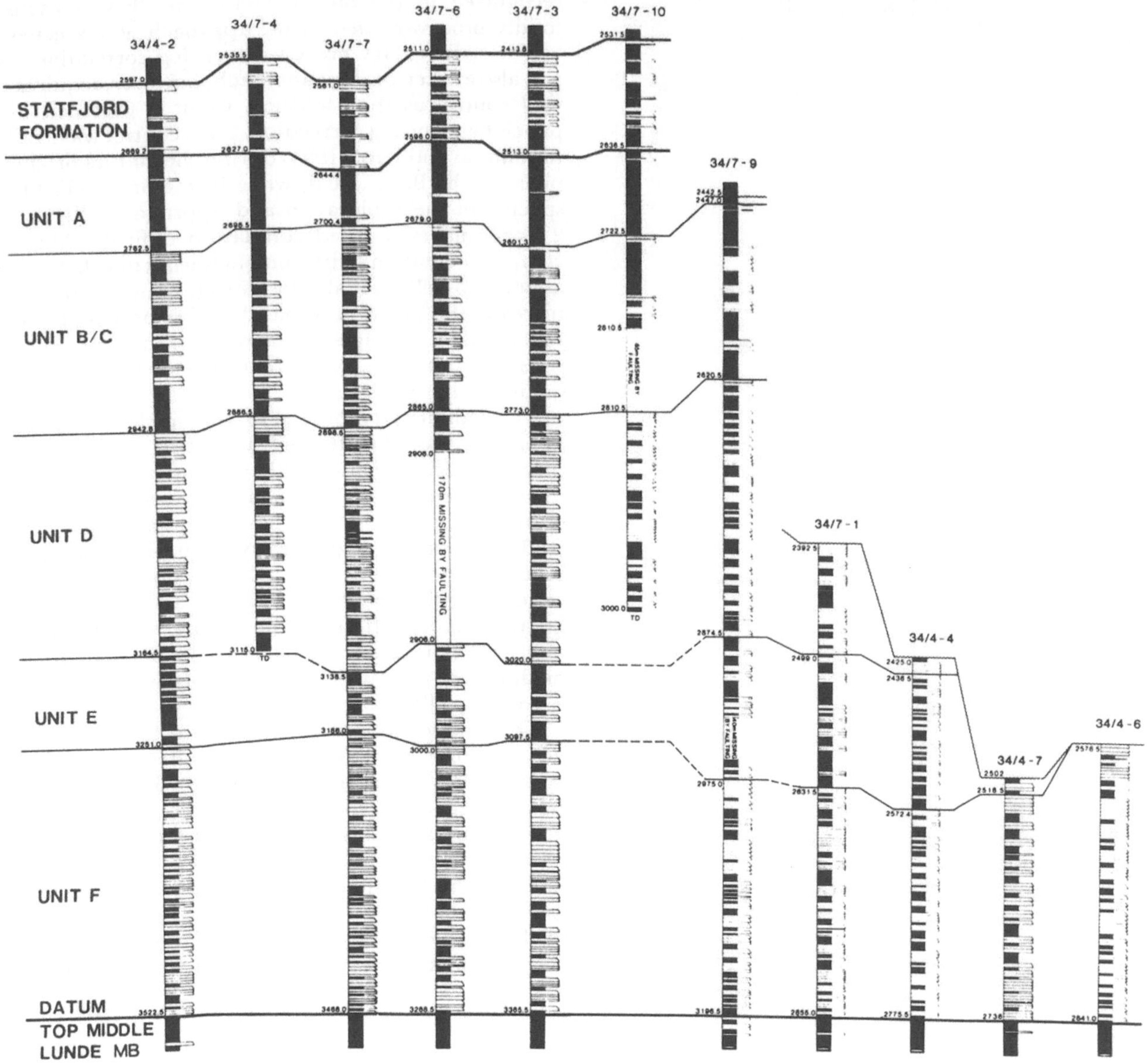


**Fig. 5.** Main reservoir subdivision.

marginal-marine limestone in unit A north of the Snorre Field in Block 34/4. This was an event of maximum marine flooding in the Tampen Spur area, followed by progradation of the fluvial deposits of the Statfjord Formation.

The sandstone bodies of the Statfjord Formation represent single-channel fills and channel belt sequences of several stacked individual channel deposits. The sandstone bodies are separated by mudstone of a floodplain facies; a more persistent shale unit in the middle part of the formation may represent a brackish water deposit formed during an intermittent transgressive event. The channel-type sandstones are typically multistorey and consist of several structures: medium- and small-scale trough cross-bedding, tabular cross-bedding, horizontal and low-angle cross-bedding, massive bedding and current ripple lamination. These bedsets are usually organized in upwards-fining sequences, but the order of succession of the facies varies considerably (Fig. 6).

There has been much debate concerning the depositional environment of the Statfjord Formation, which has a large regional extent within the Tampen Spur area. For the Brent and Statfjord Fields, Deegan and Scull (1977) suggested that the main part of the Statfjord Formation was formed of braided stream deposits on an alluvial plain (the Raude and Eiriksson Members), whereas the upper part (Nansen Member) had been deposited in a coastal to shallow marine environment. Moiola (in Chauvin and Valachi, 1980, and in Kirk, 1980) also suggested a fluvial origin, mainly in braided streams, for the Raude and Eiriksson Members. Johnson and Krol (1984) interpreted the sandstones of the Statfjord Formation as mainly representing coarse-grained, meandering channels of low to moderate sinuousity, and also a braided-stream origin for some sand-rich, multistorey intervals. From the same area, Røe and Steel (1985) considered the Raude Member as representing distal fan and coastal flood basin and lagoon-bay deposits, and the Eiriksson Member as representing alluvial fan bodies, coastal flood basin and lagoon-bay facies.

The data from the Snorre Field favour the interpretation of the sandstones in the Statfjord Formation as mainly of braided-stream origin. The uppermost formal member of the Statfjord Formation, the marine reworked sand of the Nansen Member, has not been identified in the Snorre Field. The braided-stream depositional model was therefore used to build the Statfjord models.

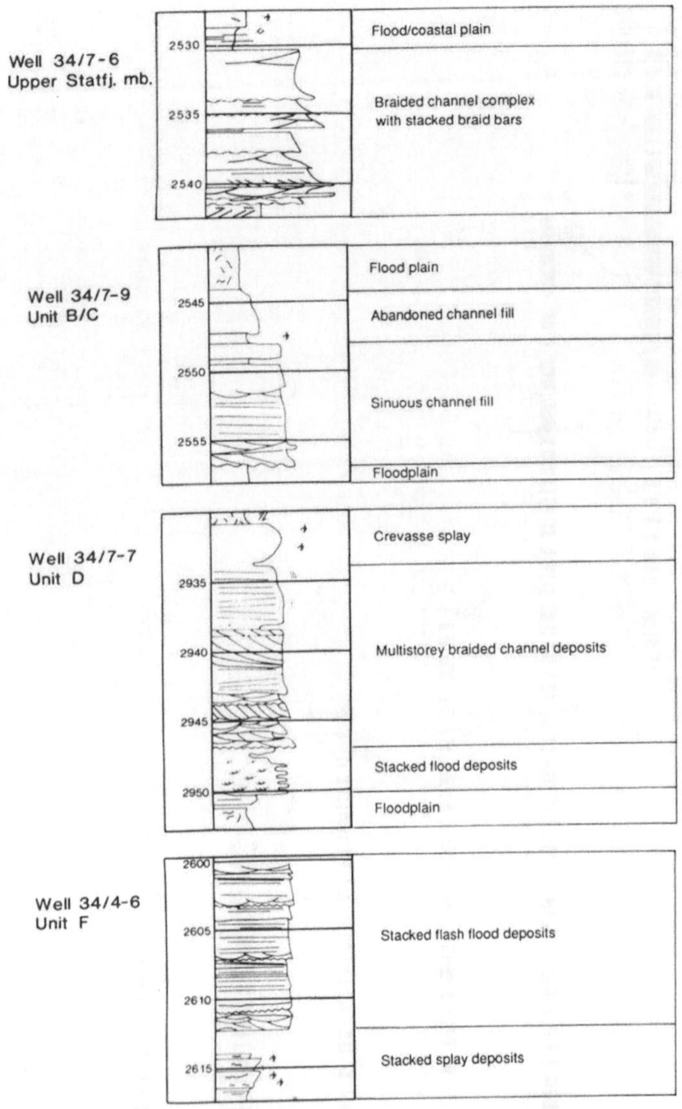


**Fig. 6.** Typical core descriptions of the main reservoir units.

## COMPUTER MODEL

Three-dimensional synthetic models of the Snorre reservoir sand network were constructed utilizing the stratigraphic and depositional models derived from core and log data from the exploration and appraisal wells on Blocks 34/4 and 34/7. Three approaches were explored: the representation of geological processes by mathematical and statistical models as developed by Bridge and Leeder (1979), a sedimentary process-driven log correlation model, such as that described by Johnson and Krol (1984) for the Brent Field, and a sedimentary process-driven stochastic model.

In the process-related procedure described by Bridge and Leeder (1979), sedimentary and tectonic processes involved in the deposition of fluvial sediments are described mathematically and simulated by computer. The resulting representations thus reflect the geological processes, thereby making it an attractive method. However, there is presently insufficient information available for the Snorre area to undertake this, and such information is unlikely to be available on many prospects under evaluation during an appraisal phase. Also, the present sedimentary process-related models are most suitable for meandering rivers, while greater uncertainties exist with regard to input parameters for the less predictable low-sinuosity rivers recognized in the Snorre Field. For these reasons the totally process-related model approach was rejected.

The sedimentary process-driven log correlation model was also attractive. This approach, however, requires good well control as the geological input is provided by 2D panels based on log correlation. Such control was available to Johnson and Krol (1984) during the mature production phase of the Brent Field, while in the Snorre Field wide spacing of the exploration and appraisal wells (average 2.7 km) makes detailed correlation difficult. Also, fault compartmentation, with compartments that have one or at best two wells available for stratigraphic control, further increases this problem. It has therefore only been possible to correlate packages of sand, broken down into the reservoir units A–F, rather than correlate individual sand bodies within these units.

The sedimentary process-driven stochastic modelling approach was also deemed attractive at the appraisal phase of field evaluation. It has the advantage of approximating average conditions, and can therefore provide an estimate of average well behaviour, possible problems in field development, and an early estimate of field producibility and recoverable reserves. Since stochastic modelling cannot predict the exact location and extent of sand bodies, it is less attractive for mature fields where more information is available, and results derived from this method need to be updated continually as new information becomes available.

The sedimentary process-driven stochastic models developed for Snorre utilized a computer program called "SISABOSA" (Augedal *et al.*, 1986) to make synthetic representations of the reservoir. This program creates a reservoir sand network by placing three-dimensional parallelepipeds of permeable sand into a volume of impermeable mudstone until a prescribed sand content and sand-content variation is achieved (Fig. 7). The length of the parallelepipeds is considered to be infinite, while their thicknesses are determined by a probability distribu-

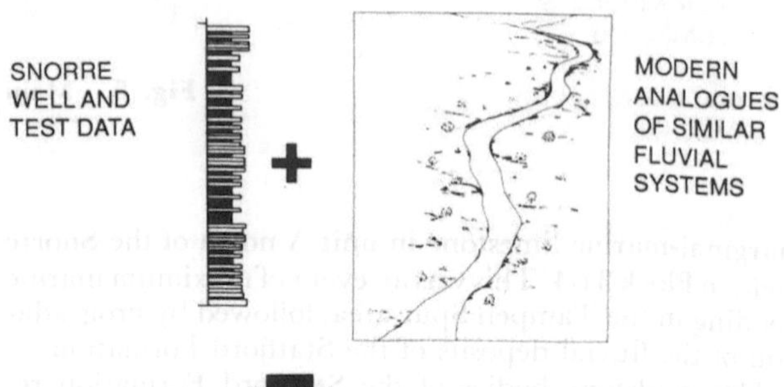


**Fig. 7.** Generation of a reservoir model.

tion. Widths are determined by a conditioned probability distribution that may be tied to thickness, while orientation varies according to a (simple) probability distribution. Snorre exploration and appraisal wells were assigned coordinates in the models and the sand parallelepipeds were emplaced according to the sand body depth and thickness observed in the wells.

The parallelepipeds represent an approximation of sand body shape. This simplification is considered reasonable for the relatively small areas (1–2 km × 4–8 km) covered by the models, particularly for the braided and low-sinuosity river systems.

To help constrain the geometric variables, each model was constructed using layers: 60–80 m thick for reservoir unit F (Fig. 8), 110–145 m for reservoir unit D, 30–60 m for unit B/C, 31–52 m for the Statfjord Formation in the western fault block, and 8–35 m for the Statfjord Formation in the central fault block. An average sand content was prescribed for each layer, and channel parallelepipeds were distributed randomly to give single and stacked sand bodies until the desired sand content and sand content variation was obtained.

Several representations were generated for each layer, and cross-sections of these representations were generated to evaluate stacking arrangements and the distribution of single-storey and multistorey sand bodies. The key elements in these models that determine how well they mimic the reservoir are sand content and sand body size and orientation.

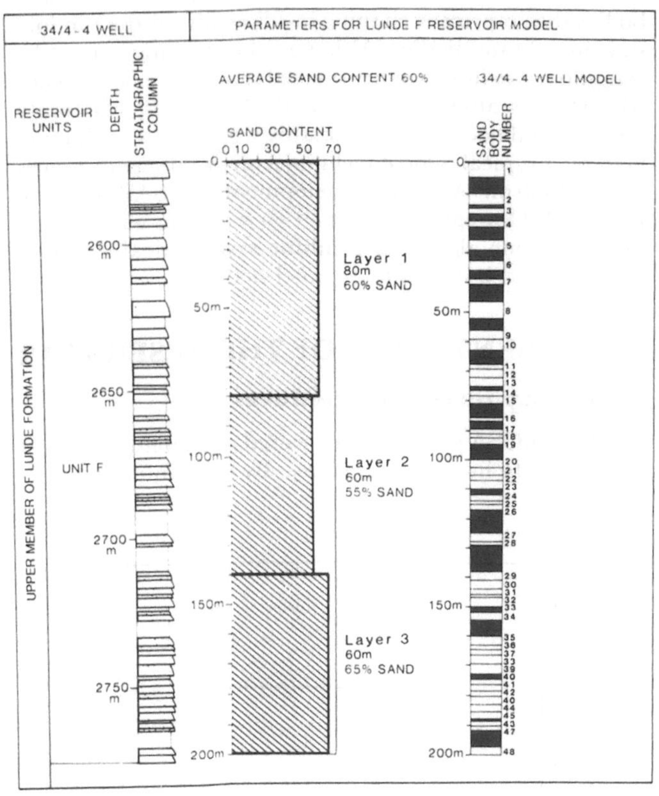


**Fig. 8.** Geological input for the upper Lunde unit F.

## Sand body abundance

Sand body abundance is the most easily established parameter, but it is complicated by the need to model all porous sands in the system whether they were formed in fluvial channels or on floodplains. For the Snorre modelling, sand content included (1) channel-fill sands, (2) non-channellized sands stacked with channel-fill units, and (3) some of the thicker, porous, non-channel sandstones within floodplain mudstones. For the sake of simplicity, all of these types of sandstones were modelled as parallelepipeds of different thicknesses.

Curves of sand content were generated for all exploration and appraisal wells as running averages of sand percentage (Fig. 9). For each reservoir unit in the Lunde and Statfjord Formations these curves were averaged to obtain a single curve of running averages. The average curves were then blocked into layers of constant sand content (Fig. 9), and these sand contents were used to build the layers in the 3D models.

Figure 10 shows an example of how the sand content was distributed in a representation of layer 2 of unit F (Fig. 8).

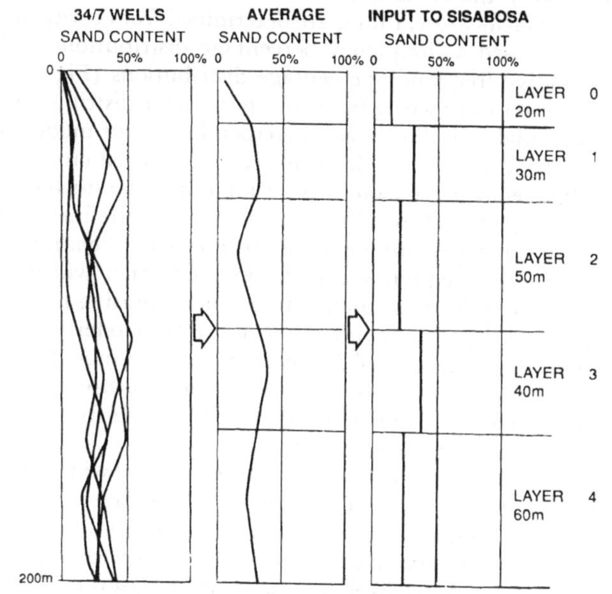


**Fig. 9.** Graphs of running average of sand content for the upper Lunde unit B/C in 34/7 wells showing how curves for all wells were used to construct an average curve that was blocked into layers with specific sand contents.

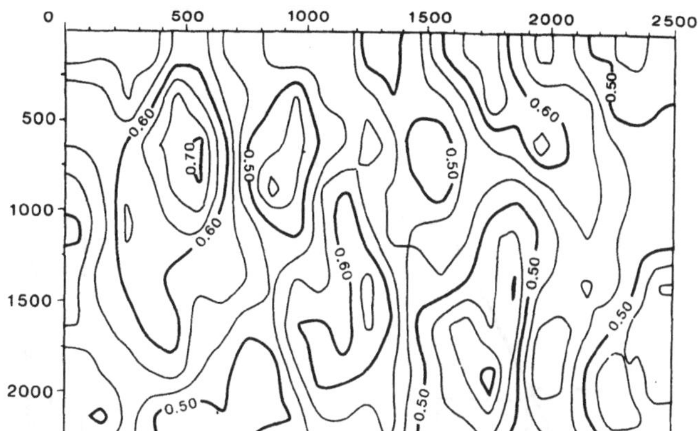


**Fig. 10.** Sand content map of layer 2 of upper Lunde unit F representation constructed by the SISABOSA computer model. Sand content is expressed as a fraction of the total reservoir volume. Contour trends reflect a stream direction of north ±25° input into the modelling.

The map illustrates how the sand content varies about an average value of 55% prescribed for this layer.

## Sand body size

Sand body thickness and width are critical parameters in the stochastic modelling of the Snorre Field reservoir units. Thicknesses can be derived directly from well data, while sand body widths must be deduced from indirect evidence, such as the widths of comparable modern and ancient depositional systems. Also, reservoir well-test results and seismic information from the field may provide additional data. Clearly, the widths of the sand bodies are a significant source of uncertainty. This uncertainty can be risked by varying widths over reasonable ranges of values for the most likely and alternative interpretations of the depositional environments. Reduction of this uncertainty is difficult, however, because of the nature of the available data.

The thicknesses and widths of mudstone-bounded sand bodies in the Lunde and Statfjord Formations represent the dimensions of the mobile channel belt deposits. Channel sand body thicknesses, i.e. the thicknesses of uneroded channel facies associations, are determined from cores and are expressed as frequency percentage distributions and cumulative frequency percentage distributions (Fig. 11). In reservoir units where frequency percentage distributions are bimodal, two thicknesses corresponding to the modes of the distribution are used. In this case, a blocked cumulative frequency percentage curve is input into the stochastic modelling procedure (Fig. 11). This curve defines the thickness and percentage abundance of each "channel" mode in the frequency distribution. Channel width is derived from the depositional model incorporating ranges in channel depth-to-width ratios of similar modern river systems (e.g. Leeder, 1973; Stear, 1983). The most likely channel widths are estimated from the depth-to-width ratio for modern analogues, supported by data from well tests and simulations of well tests.

Our studies indicate that estimates of channel width and thickness have the greatest impact on representations where the sand content is low, but do not significantly influence the reservoir simulation of models with a high sand content. Most likely depth-to-width ratios of 1 : 60 were used to model low-sinuosity streams, but values of 1 : 20 and 1 : 90 were also input to evaluate how these parameters influenced the models and simulations. For more sinuous streams, values of 1 : 25 and 1 : 50 were used.

## Orientation

The orientation of fluvial channels is input as a cumulative frequency percentage curve of channel trends (Fig. 8). Both the average direction and dispersion in channel trends are important, and were determined from dipmeter studies over cored intervals and from oriented cores (Fig. 12). These data were supplemented by data from regional structural analysis and the distributions of sandstone in stratigraphic sections.

All of these data indicate that the Lunde river systems flowed toward the north, down the axis of the basin in the area of the Snorre Field (Fig. 12). Only minor contributions from the basin margin are suggested by some dipmeter data that indicate transport from the east. For the Statfjord Formation the trend appears to change, with stream directions largely to the east or north-east. An analysis of oriented cores from the Lunde Formation in well 34/7-7 indicated a relatively low directional dispersion (±25°) which was used to model variations in channel trend over distances of a few thousand metres, as well as differences in trend between channels of different rivers or the shifting course of the same channel.

In the modelling, the orientation assigned to each sand body is determined by random sampling from the cumulative frequency curve for orientation, which reflects the average direction and the dispersion. The sand content map of Fig. 10 illustrates the northerly average trend of the sand bodies in a unit F model, as well as the dispersion of ±25° about this direction.

The random distribution of location of single sand bodies that represent channel in-fill results in the random stacking of sand bodies. Although the stacking cannot be conditioned to link together stacked sand bodies for extended distances along the sand body long axes, randomly stacked sand bodies can reasonably well represent channel belts over typical Snorre Field injector-producer distances of 800–1500 m. (A further development of SISABOSA, the FLUREMO program of Clemetsen *et al.*, this volume, has improved our ability to combine systematically single-channel sand bodies into channel belts).

# REPRESENTATION OF THE RESERVOIR

## Sand-body interconnectedness

The SISABOSA program can be used to establish estimates of sand-body interconnectedness. By superimposing pro-

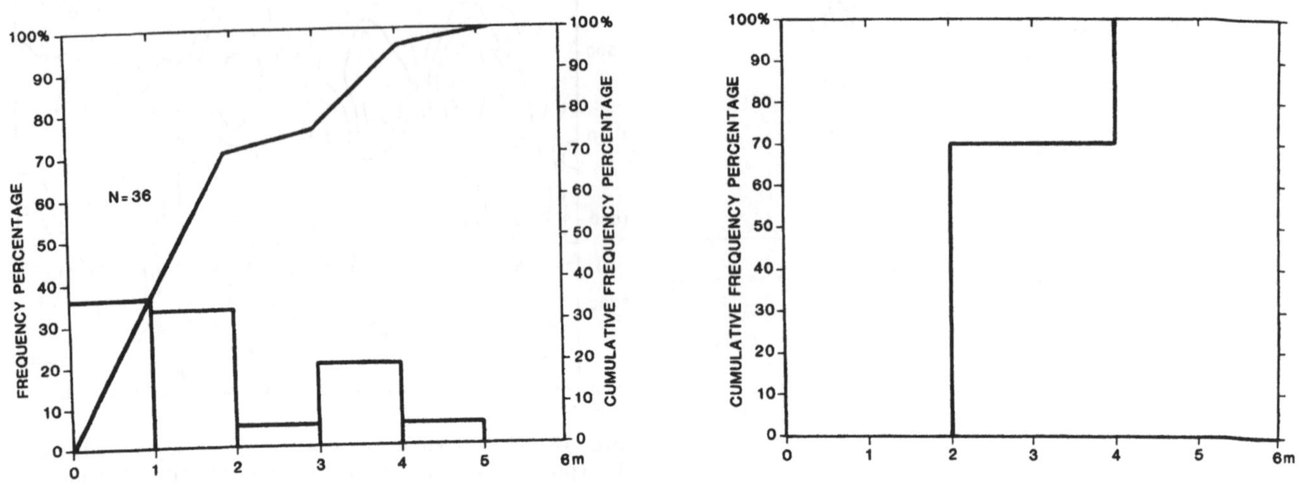


**Fig. 11.** Histogram and cumulative frequency curve of thickness of the upper Lunde unit B/C, and the bimodal distribution used to build reservoir models of the unit.

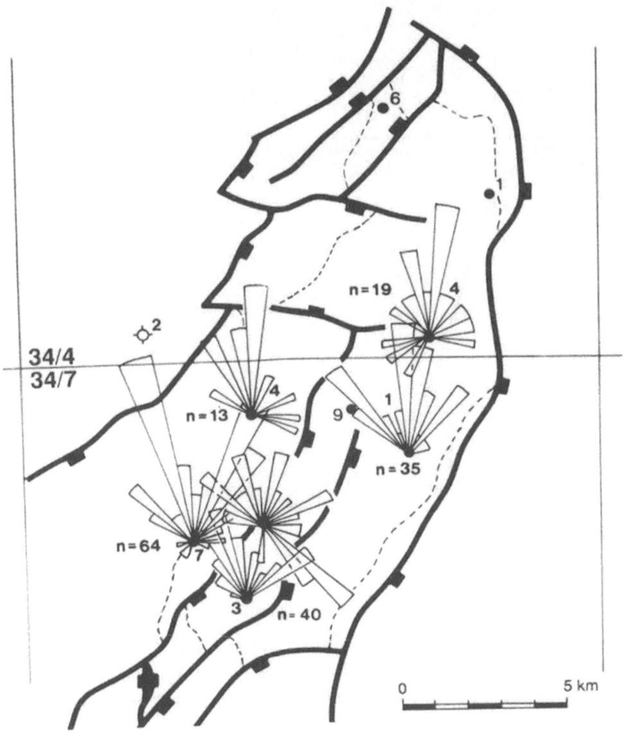


**Fig. 12.** The Snorre Field showing major faults and paleocurrent directions in the Lunde Formation based on dipmeter data from cored intervals and oriented cores.

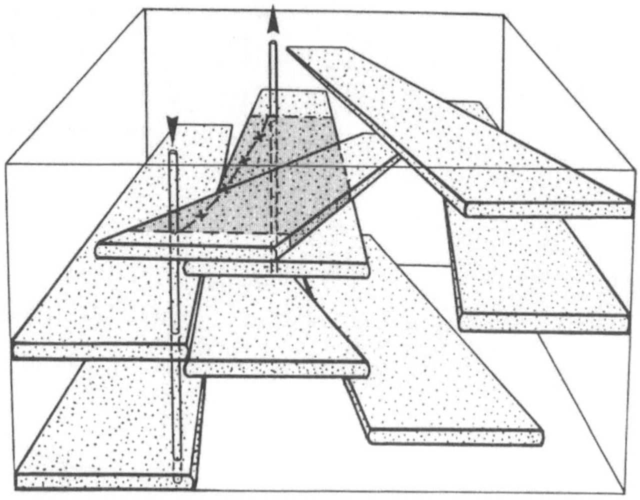


**Fig. 13.** Cartoon of a three-dimensional SISABOSA representation showing the relationship between sand bodies and wells used in the definition of continuity ratio and path length.

ducer–injector well pair locations on a synthetic geological framework, a well-pair-dependent interconnectedness can be calculated as the fraction of the total sand volume that is interconnected between the wells. Interconnectedness, in this restricted sense, is directly calculated by the SISABOSA program, and is termed the "continuity ratio" (Fig. 13). The continuity ratio is an indication of the maximum percentage of the reservoir volume that may be swept by a waterflood, and is hence the best criterion for the selection of representations to be used in reservoir simulation.

In order to restrict computer time, practical limitations were made as to the amount of sand that could be registered as being interconnected between wells. A maximum path length of 1375 m was allowed (Fig. 13), corresponding to the distance from an injector to its second-nearest producer with the 800 m well spacing used in the modelling. Also, any path containing more than four sand bodies was excluded from the continuity calculations.

From an examination of the synthetic cross-sections and continuity calculations, the tie between continuity ratio and stacking of the channel sandstones becomes evident, particularly for models with low sand content. The median value of continuity derived from multiple runs of the same model (Fig. 14) reflects a moderate number of stacked channels and a moderate number of single-channel sandstones. These match closely what is observed from cores in terms of channel sandstone stacking. In contrast, continuity ratios for representations with values smaller than the first standard deviation from the mean exhibit a minimum of stacked channels, as indicated on the left-hand side of Fig. 14. Correspondingly, in low-sand-content models, few if any stacked channels are observed. For such representations, well placement is a critical factor. As might be expected, continuities for representations with values larger than the first standard deviation from the mean exhibit a very large percentage of stacked channel sandstones, as indicated on the right-hand side of Fig. 14.

### Selection criteria

Ten synthetic representations were generated for each element model using the most likely set of input parameters. Five representations for high-case and five for low-case parameters were also run in order to evaluate the variability in the models. From the ten runs of the most likely case for each model, a normal distribution of continuity measurements was generated. The median continuity value was used to select the representation that was supplied to the engineers.

## VERIFICATION

The models and their representations supplied to the reservoir engineers are designed to mimic the reservoir units F, E, D and B/C of the upper member of the Lunde, and the lower, middle and upper reservoir units of the Statfjord Formation. We naturally do not consider them to be exact replications of the reservoir, but they provide a more realistic representation of reservoir heterogeneity than those that which have been created using conventional techniques.

A verification of the model involves a comparison between vertical profiles taken from the computer-generated stratigraphy with the actual wells drilled in the Snorre Field. An example is given in Fig. 15, which shows an E–W cross-section through a representation of sand bodies containing well 34/4-7. The similarity between the synthetic well profiles obtained by randomly selecting vertical lines through the cross-section and the actual well profile at well 34/4–7 indicates that the 3D computer models of the stratigraphy are consistent with the Snorre Field sedimentology as currently known.

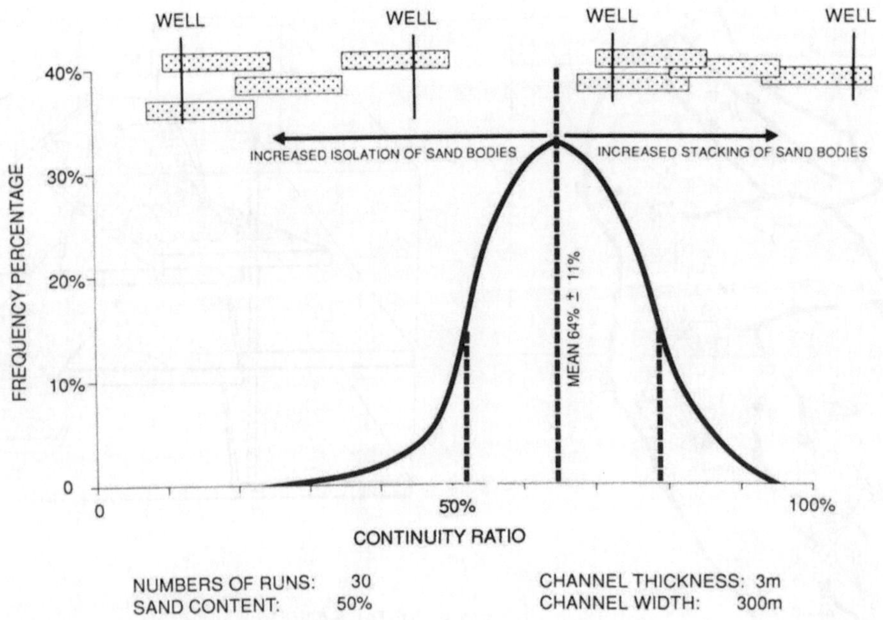


**Fig. 14.** Frequency distribution curve for 30 representations of a Snorre reservoir layer, with 50% sand content and channel dimensions of 3 × 300 m, showing the mean continuity ratio for the representations (64%) and the first standard deviation (±11%). Differences observed between representations with continuity ratios less and greater that the first standard deviation from the mean value are shown graphically at the top of the figure.

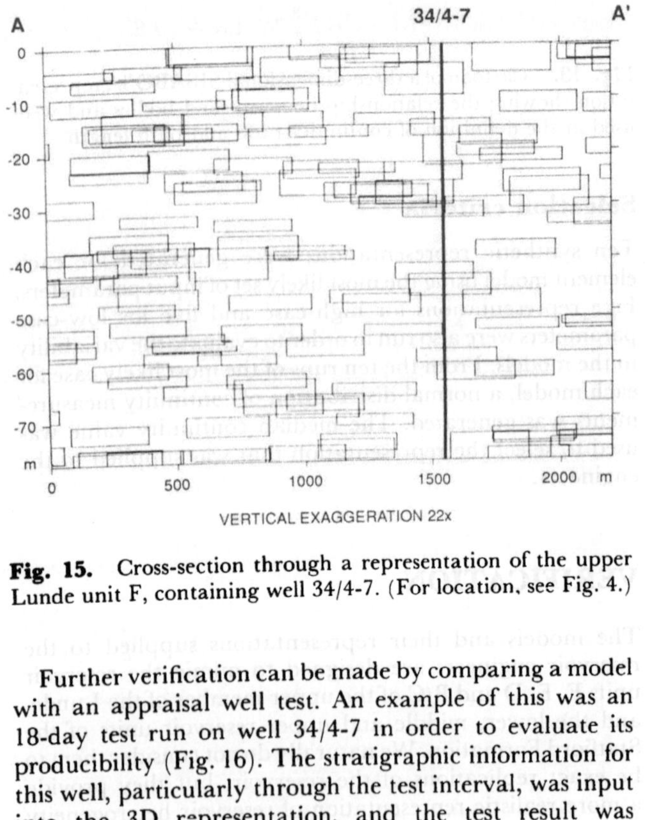


**Fig. 15.** Cross-section through a representation of the upper Lunde unit F, containing well 34/4-7. (For location, see Fig. 4.)

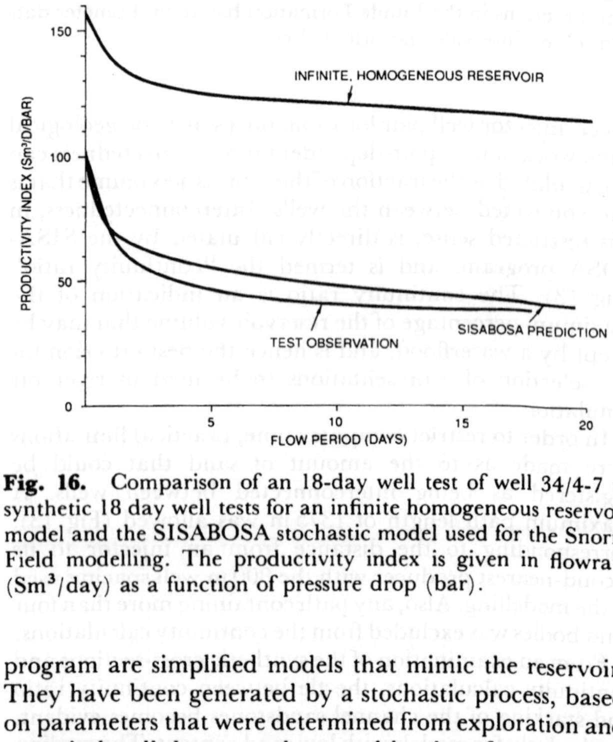


**Fig. 16.** Comparison of an 18-day well test of well 34/4-7 to synthetic 18 day well tests for an infinite homogeneous reservoir model and the SISABOSA stochastic model used for the Snorre Field modelling. The productivity index is given in flowrate ($Sm^3$/day) as a function of pressure drop (bar).

Further verification can be made by comparing a model with an appraisal well test. An example of this was an 18-day test run on well 34/4-7 in order to evaluate its producibility (Fig. 16). The stratigraphic information for this well, particularly through the test interval, was input into the 3D representation, and the test result was predicted with the reservoir simulator. The prediction shows a reasonable match with the actual curve shape and general productivity level (Fig. 16). Thus, the 3D model gave a realistic prediction, at least over this particular period. For comparison, a prediction based on an assumption of radial flow in an infinite, homogeneous sand body yields a significantly higher productivity level (Fig. 16).

## CONCLUSIONS

The 3D representations of the Snorre Field reservoir units that have been constructed by the SISABOSA computer program are simplified models that mimic the reservoir. They have been generated by a stochastic process, based on parameters that were determined from exploration and appraisal well data, supplemented by data from modern river analogues. The representations can be directly input to numerical simulators and thereby serve as an effective and convenient way of transferring reservoir geological data into element model simulations.

By using "most likely" input parameters, average reservoir and well performance can be simulated, whereas the effects of more adverse or more beneficial conditions can be illustrated by using low-case or high-case parameters respectively.

In spite of the simplicity of the constructed models, verification indicates that they reflect the complexity of the fluvial sand body network sufficiently well to allow reasonable estimates of reservoir and well performance. As this approach can be used at the delineation stage of field development, it permits the field operator to include this complexity in the evaluation of field commerciality and in initial field planning. As the field is developed, however, the additional information gathered should be used to replace the stochastic models with more deterministic ones.

## ACKNOWLEDGEMENTS

The authors would like to acknowledge the Norwegian Computing Center and, more specifically, Henning Omre who wrote the SISABOSA program. Nils B. Hollander and Bryan G. Pryce are acknowledged for their supervision and Geir W. Diesen, Ragnar Knarud, Johan Petter Nystuen and the reservoir engineers on the project for valuable input. Finally, we would like to thank the Snorre Field partners, Amerada Hess Norge A/S, Deminex (NORGE) A/S, Det Norske Oljeselskap A/S., Elf Aquitaine Norge A/S, Enterprise Oil Norge Ltd., Esso Norge A.S., Norsk Hydro Produksjon a.s., Den Norske Stats Oljeselskap a.s. (Statoil), for permission to publish this paper.

## REFERENCES

Allen, J.R.L. (1978). Studies in fluviatile sedimentation: an exploratory quantitative model for the architecture of avulsion-controlled alluvial suites. *Sedimen. Geol.* **21**, 129–147.

Augedal, H.O., Stanley, K.O. and Omre, H. (1986). SISABOSA, a programme for stochastic modelling and evaluation of reservoir geology. *Proceedings of conference on Reservoir Description and Simulation with Emphasis on EOR Oslo, September 1986*, Nitteberg, J. (ed.). Institute for Energy Technology.

Bridge, J.S. and Leeder, M.R. (1979). A simulation model of alluvial stratigraphy. *Sedimentology* **26**, 617–644.

Chauvin, A.L. and Valachi, L.Z. (1980). Sedimentology of the Brent and Statfjord Formation of Statfjord field. In. *The Sedimentation of the North Sea Reservoir Rocks*. Norwegian Petroleum Society, Geilo XVII, pp. 1–17.

Clemetsen, R., Hurst, A., Knarud, R. and Omre, H. A computer program for evaluation of fluvial reservoirs. *This volume.*

Deegan, C.E. and Scull, B.J. (1977). Lithostratigraphic nomenclature for the Central and Northen North Sea. *Rep. Inst. Geol. Sci. No. 77/25;* Norwegian Petroleum Directorate, Bull. No. 1.

Hollander, N.B, (1987). Snorre. In: Spencer, A.M. *et al.* (eds.), *Geology of Norwegian Oil and Gas Fields*, pp. 307–318. Graham & Trotman, London.

Johnson, H.D. and Krol, D.E. (1984). Geological modelling of a heterogeneous sandstone reservoir: Lower Jurassic Statfjord Formation, Brent Field. *Society of Petroleum Engineers,* **13050**, 1–13.

Kirk, R.H. (1980). Statfjord Field—a North Sea giant. In: Halbouty, M.T. (ed.), *Giant Oil- and Gasfields of the Decade 1968–1978*, pp. 95–117. American Association of Petroleum Geologists, Memoir 30.

Leeder, M.R. (1973). Fluviatile fining-upward cycles and the magnitude of palaeochannels. *Geol. Mag.* **110**, 265–276.

Leeder, M.R. (1978). A quantitative stratigraphic model for alluvium, with special reference to channel deposit density and interconnectedness. In: Miall, A.D. (ed.), *Fluvial Sedimentology*, pp. 587–596. Canadian Society of Petroleum Geologists, Memoir 5.

Nybråten, G., Skolem, E. and Østby, K. Reservoir simulation of the Snorre Field. *This volume.*

Nystuen, J.P., Knarud, R., Jorde, K. and Stanley, K.O. (1989). Correlation of Triassic to Lower Jurassic sequences, Snorre field and adjacent areas, northern North Sea. In: Collinson, J.D. (ed.), *Correlation in Hydrocarbon Exploration*, pp. 273–289. Graham and Trotman, London.

Røe, S.L. and Steel, R. (1985). Sedimentation, sea-level rise and tectonics at the Triassic-Jurassic boundary (Statfjord Formation), Tampen Spur, northern North Sea. *J. Petroleum Geol.* **8**(2), 163–186.

Stear, W.M., (1983). Morphological characteristics of ephemeral stream channel and overbank splay sandstone bodies in the Permian Lower Beaufort Group, Karoo Basin, South Africa. In: Collinson, J.D. and Lewin J. (eds.), *Modern and Ancient Fluvial Systems*, pp. 405–420. Special Publication No. 6 of the International Association of Sedimentologists.



[illegible]



Sedimentology of the Norwegian North Sea. Norwegian Petroleum Society, Graham & Trotman, [illegible]



[illegible] (1989) [illegible] of the [illegible] Formation, northern North Sea. [illegible]



# Field Review 5

# II. Reservoir Simulation of the Snorre Field

*G. Nybråten, E. Skolem and K. Østby*

Saga Petroleum a.s., Høvik, Norway

Reservoir simulation studies of the Snorre Field were initiated in 1985 for the purpose of formulating an optimum reservoir development strategy and predicting reservoir recovery and performance. The results formed the basis for the commerciality evaluation and the Field Development and Operating Plan.

The Snorre Field reservoirs are interpreted as fluvially deposited sandstones interbedded with shales. The reservoir simulation modelling of these complex reservoirs progressed from two-dimensional channel (single sand body) models and "panel" models (cross-sections with randomly distributed sand/shale sequences) to three-dimensional (3D) element models.

The 3D element models were the primary prediction tools. They were based on stochastic, spatial sand/shale distributions which were gridded and input as grid block multipliers to porosities and transmissibilities. Relatively fine grid blocks were required to model adequately the reservoir discontinuities present in the geological model. The 3D element models were therefore quite large (up to 18 000 active grid blocks).

The recovery and rate predictions from the element models were adjusted to account for intra-sandbody heterogeneities (based on channel model results) and other field heterogeneities that could not be represented directly owing to practical limitations of the model size. The element models covered approximately 44% of the field. The results were extrapolated to full-field production profiles by use of typical well performances and a well-summation routine. Pseudorelative permeabilities and well functions developed from the element models were used to build a full-field simulation model that confirmed the production profiles.

## INTRODUCTION

This chapter contains a description of the reservoir simulation work performed on the Snorre Field in the period 1985–1987. This work formed the basis for the well location and reservoir management plans, as well as production-profile estimates used both in the commerciality evaluation and as a basis for the Field Development and Operating Plan.

The objectives of the reservoir simulation work were:

- to establish an optimum reservoir production strategy, including drive mechanism, number, location and sequence of wells, well completion strategy, injection and production rates; and
- to estimate recoverable reserves and production profiles for a number of different development scenarios for use in development planning and commerciality evaluation work.

The objectives were met through a simulation programme using both two-dimensional (2D) and three-dimensional (3D) simulation models. The simulation work programme was tailored to provide a timely input to several milestone reports that required the preparation of some interim models.

The basic strategy in the simulation programme was to incorporate as many details and inhomogeneities as possible in the simulation models (each of which covered a limited area, or *element*, of the field and contained relatively small grid blocks), rather than attempting to construct a full-field model where a high degree of data averaging would be necessary. The development of the models progressed from relatively simple 2D models describing a single sand body with a total thickness of 5–10 m to larger 2- and 3D models involving 2–12 wells.

The final results were based on the 3D element models, which were 3–15 km$^2$ in areal extent and consisted of up to 18 000 active grid blocks. Correction factors were applied to the element model results to reflect the heterogeneities that could not be adequately incorporated in them. The results from the element models were extrapolated to unmodelled areas of the field, on the basis of reservoir quality and considerations of structural geometry.

A key part of the input data was the transfer of the geological model of a complex, fluvially deposited reservoir into a 3D Cartesian simulation grid. The geological modelling is described in a companion paper by Stanley *et al.* (this volume). As discussed by Augedal *et al.* (1986), stochastic representations of the geological model were generated using the SISABOSA (SImulation of SAnd BOdies for SAga) programme. These representations were discretized, and input into the reservoir simulation models through the use of multipliers to porosity and transmissibilities.

Figure 1 shows an overview of the simulation programme in relation to the major development planning milestone reports.

*North Sea Oil and Gas Reservoirs—II*

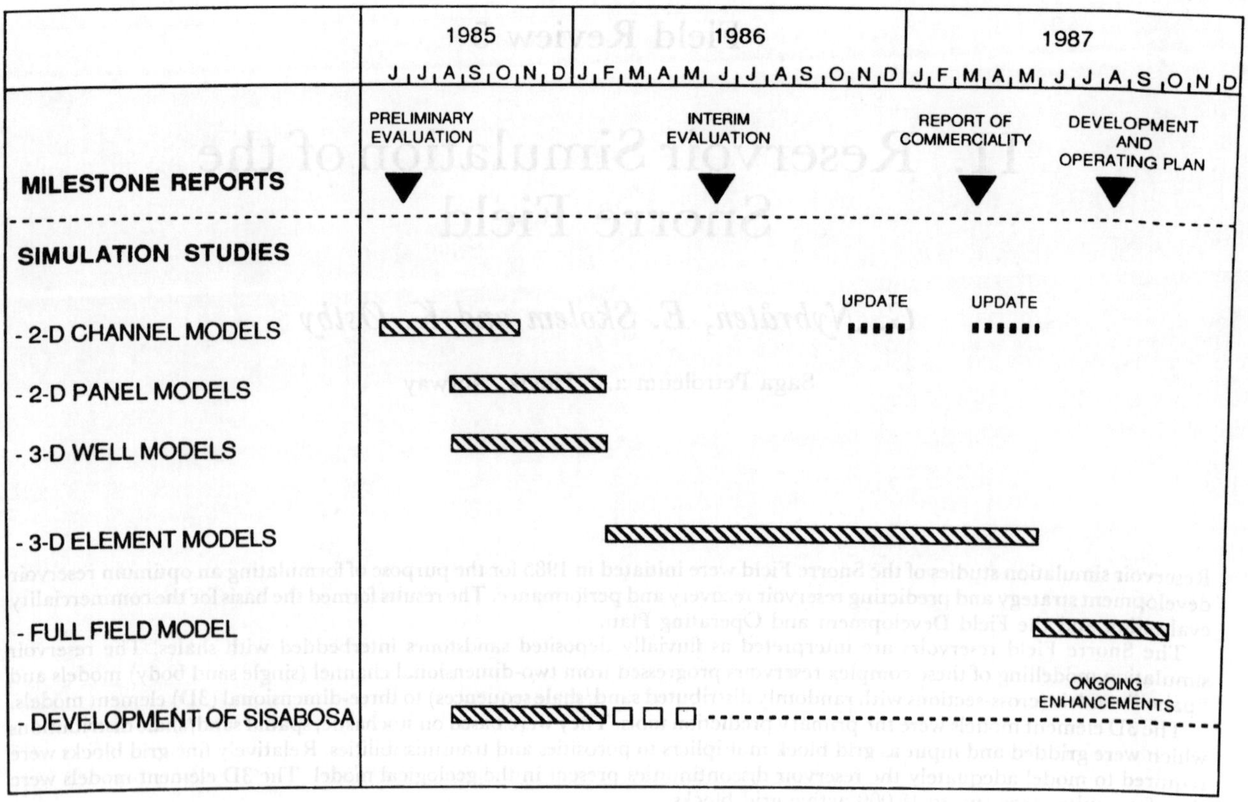


**Fig. 1.** Overview of the Snorre Field Reservoir Simulation Programme, with major development planning milestones.

## ROCK AND FLUID DATA

The Snorre Field reservoir is interpreted as fluvially deposited sandstones interbedded with shales. A total of 11 exploration/appraisal wells has been drilled on the field. An overview of the data collected is given in Fig. 2. All of the wells have been tested, and fluid samples of the undersaturated low-viscosity Snorre oil have been collected. A total of 1635 m of cores has been cut, and porosity and permeability measurements have been made. Well tests and permeability measurements made on the cores have indicated average oil permeabilities at initial water saturations ranging from 200 to 2000 mD, while the porosity ranges from 21 to 26%.

Special core analysis has been performed on a large number of core plugs as described by Lien and Nysetvold (1988). It was concluded from an evaluation of wettability that the reservoir rocks are of an intermediate to oil-wet nature. Hysteresis scanning curves for use in the simulation models were generated according to a history-dependent model for saturation functions.

## INTERFACING AND VERIFICATION OF GEOLOGICAL AND RESERVOIR SIMULATION MODELS

Stochastic sand/shale distributions generated by the SISABOSA modelling tool were transformed into discrete distributions suitable for direct input to the reservoir-simulation model. This was done by superimposing 3D grids onto the sand/shale distributions, and assigning representative values for sand content to each grid block. The grid block sand content values do not supply the model with information concerning where within the grid blocks the sand body "edges" exist. Two neighbouring grid blocks might both have non-zero sand content and still be intersected by a shale that would act as a barrier to flow. It was therefore necessary to supply the model with information about grid block connectedness as well. This was achieved by assigning transmissibility multipliers in all three directions to all grid blocks.

An example of how a vertical transmissibility multiplier is calculated is shown in Fig. 3. This figure shows a 3 × 1 × 2 grid block system superimposed on two sandbodies, and how a vertical transmissibility multiplier between the two central grid blocks is calculated. Nine horizontal planes intersect the grid blocks between the grid block centres, and the sand fraction found in each of these planes is recorded. Because the flow between the two grid blocks will be dominated by the smallest sand area, the transmissibility multiplier is taken as the harmonic average of the sand fractions from the planes. In the example, one plane with zero sand content was found, and hence a transmissibility multiplier of zero was calculated. The horizontal transmissibility multipliers are calculated in the same manner.

The sand content in a grid block is calculated as the arithmetic average of the sand content values of 10 vertical planes through the block, and is entered into the simulation model as a multiplier to porosity.

The treatment of wells penetrating grid blocks with a discontinuous sand/shale distribution has a significant impact on the reservoir-simulation results for some models. It is important to ensure that the model wells communicate with the sand bodies in accordance with the model's actu location of the sand bodies so that:

- The simulation model should complete the well only in sand bodies that are intersected by the well. Sand bodies that go through the well grid block, but are not actually intersected by the well, should not be in direct communication with the well.

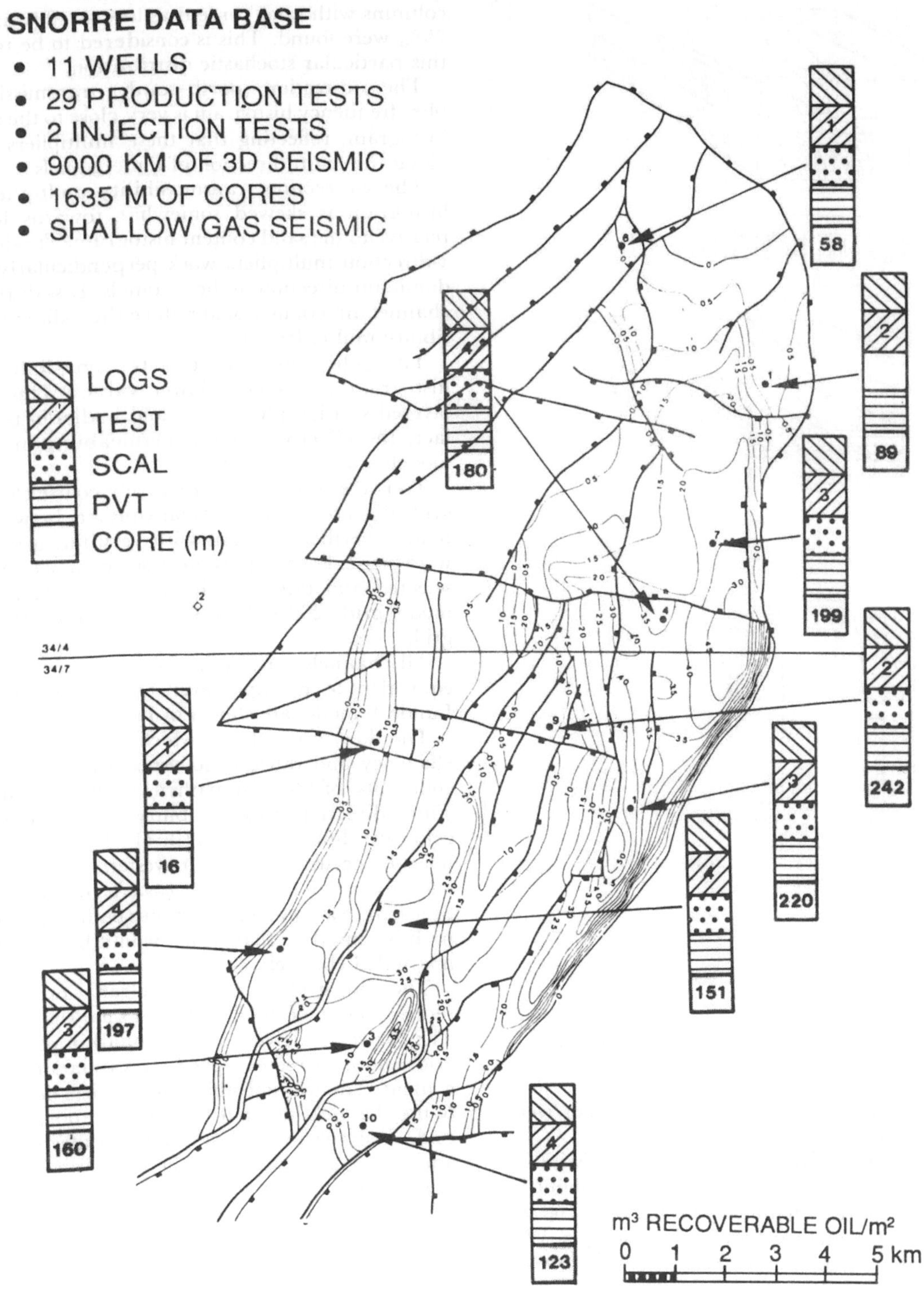


**Fig. 2.** Map of the Snorre Field showing the amounts and types of data collected in the exploration/appraisal wells.

- The simulation model should complete the well with a permeability thickness that represents only the sand bodies actually intersected.

The necessary information needed to ensure correct completion data in the simulation model is generated by combining a vertical line at the well location with the model-generated sand bodies. The resulting intersections are then transferred to simulation well-completion data. The procedure is illustrated in Fig. 4.

## Simulation model quality checks

Before using the SISABOSA generated sand/shale distributions in a simulation model designed to generate production forecasts, both the SISABOSA distribution itself, and the simulation model's capability of adequately representing it, were checked. The three main verification procedures were:

- visual inspections of plots of the SISABOSA distributions to check that they represented the geological understanding of the system;
- statistical checks of simulation model grid sand content and transmissibility values;
- simulations of parts of the reservoir with coarse and fine grids to determine the grid size that would adequately represent the SISABOSA distribution.

The Lunde F element model (see Reservoir Simulation

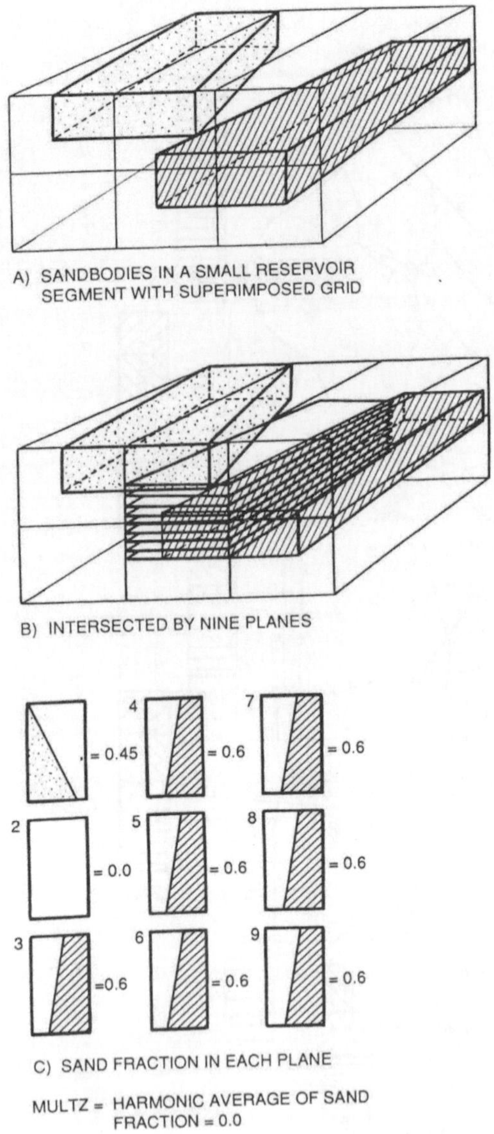


**Fig. 3.** Principle of transmissibility multiplier calculations.

Models) is now used as an example. The methods were similar for all of the models.

Average values for sand content and transmissibility multipliers were calculated for each vertical grid column in the models and plotted as frequency histograms (Fig. 5). The average sand content values in this example were concentrated around the input value of 0.6. However, columns with sand content as low as 40% and as high as 75% were found. This is considered to be reasonable for this particular stochastic distribution.

The *y*-direction (north–south) transmissibility multiplier frequency histogram is very close to the sand-content histogram, reflecting that these multipliers are working along the main direction of the channels.

The *x*-direction transmissibility multiplier frequency histogram is skewed somewhat towards lower values relative to the sand content histogram. This is because the *x*-direction multipliers work perpendicularly to the predominant direction of the channels. Less dispersion of the channel directions would reduce the *x*-direction transmissibility multipliers further.

The *z*-direction (vertical) transmissibility multiplier histogram shows even lower values. This reflects the layered stratigraphy of the sand bodies in the model. In fact, the effective vertical permeability in most of the models was close to zero.

To make sure that the relatively coarse element model grids did not create artificial contacts between the sand bodies, verification runs with very fine-gridded models were made. For this purpose, a segment of each model was selected and gridded with both the standard element model grid (200 × 100 × 3 m) and a fine (50 × 50 × 3 m) grid.

All channels existing in the uppermost 3 m (which is equivalent to the upper layer) of a selected segment of the Lunde F model are shown in Fig. 6(A).

The flow of the fluids, which will determine recovery efficiency and rates, is governed by the tortuosity and continuity of the flow paths (i.e. the transmissibilities). The *x*-direction transmissibility multipliers for the two grids are shown in Fig. 6 (B, C). It can be seen that if injection occurs in the north–western corner of the model, and production in the south–eastern corner, the fluids will be forced to follow a very tortuous path. Furthermore, this path is similar to that which would be expected for the ungridded channels.

The fine grid (Fig. 6(B)) clearly provides a good representation of the SISABOSA sand/shale distribution. The coarse grid gives the same fluids in place as the fine grid, and is therefore a good representation of total sand content. The transmissibility multipliers define the edges of the channels in about the same locations as in the fine grid. The coarse grid is therefore expected to give an adequate representation of the flow paths in the system.

As a final check of the adequacy of the coarse (element model) grid, both the coarse and the fine grids were used to simulate flow performance. The reservoir segment was simulated using an idealized geometry and idealized well

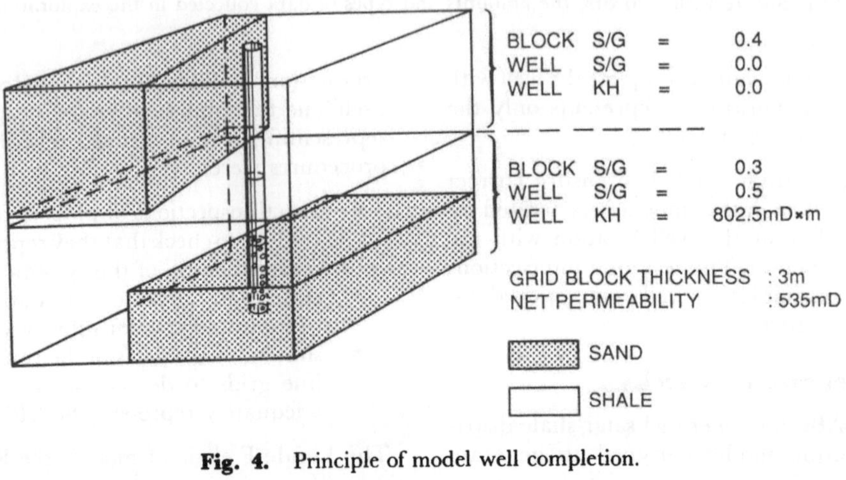


**Fig. 4.** Principle of model well completion.

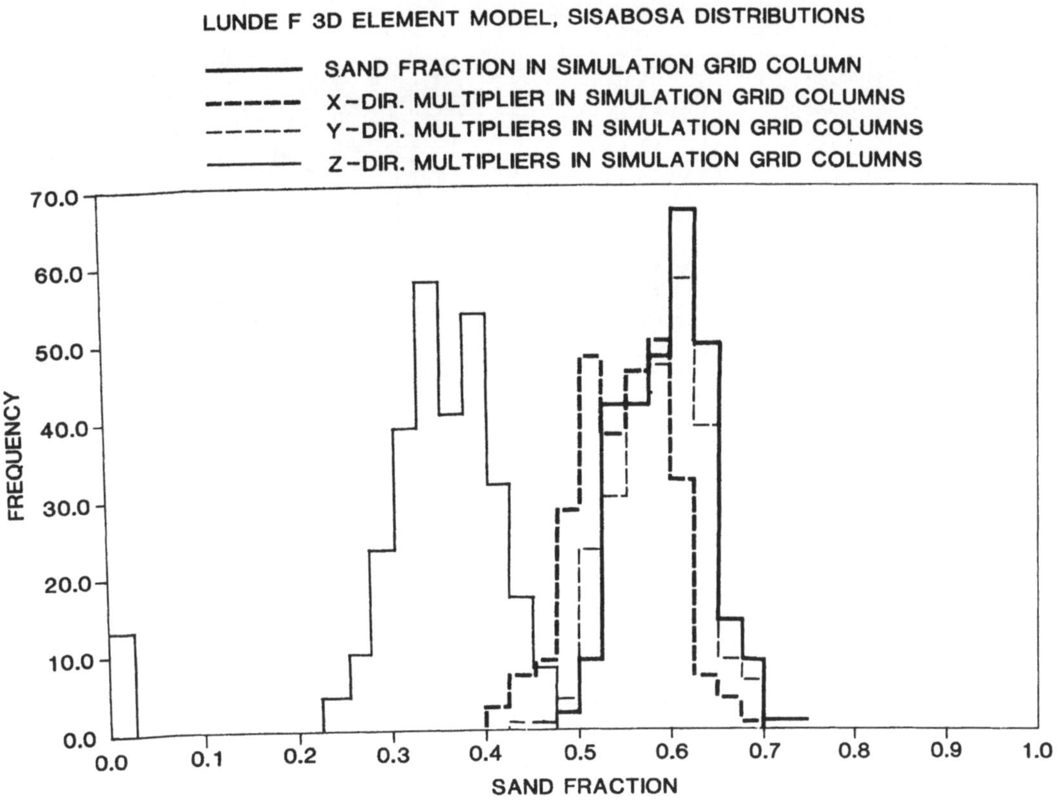


**Fig. 5.** Frequency histograms showing distribution of sand fraction and transmissibility multipliers.

constraints to minimize the effect of parameters other than the sand/shale distribution. The model was placed high above the OWC, with a constant dip angle of 8°.

The results of these simulations are summarized in Fig. 7. On the basis of the close agreement in both rate and recovery between the two models, it was concluded that the coarse grid adequately describes the sand/shale system for this particular reservoir unit. The same conclusion was reached for most of the models, but for reservoir units with low sand content, narrow channels and a large dispersion in channel directions, areal grid blocks as small as 75 × 75 m were required. Grid orientation sensitivities also showed that the main grid axis should be oriented parallel to the predominant channel direction to achieve the best possible definition of the sand body edges.

## RESERVOIR SIMULATION MODELS

The different types of simulation models used in the Snorre Field reservoir evaluation are illustrated in Fig. 8. The 3D element models were the main tool for predicting recovery efficiencies, well and field performance and completion strategy for the Field Development Plan. The 2D channel models were used to generate some of the correction factors applied to the 3D element model results. The 2D panel models and 3D well models were used to generate information for the interim evaluation reports (Fig. 1).

### Two-dimensional channel models

The 2D single sand body (channel) models represented either one or a few sand-filled channels. To simulate a typical sand body between an injector and a producer, the model dimensions were 800–1500 m in length, 100–300 m in width and 3–12 m in thickness, with horizontal grid block lengths of 8–15 m and vertical grid block thicknesses of 0.2–0.5 m. The results were used to adjust the recovery efficiencies in the 3D element models, where the vertical gridding was too coarse to include any vertical permeability variations within the sand bodies.

### Element models

Selected areas of the field, representing a total of 44% of the oil in place, were simulated using seven different element models (Fig. 9). These models were based on the 3D geological representations generated by the SISABOSA programme. The geological representations were then transformed into reservoir simulation model data as previously described. The gridding in the element models was kept sufficiently fine to represent adequately the complex sand/shale distribution. This required a large number of grid blocks, particularly in the vertical direction, where layer thicknesses of 3 m were used. The areal grid block size was typically around 200 × 100 m, but in some models a finer areal grid of 75 × 75 m was required. The resulting models were relatively large, having 7000–18 000 active grid blocks, 5–12 wells, and containing 8–80 million $Sm^3$ of oil in place for each of them.

The models were run using realistic well constraints consistent with the planned processing system and well design. Furthermore, in grid blocks with injection well completions, the water viscosity was assumed to be 1.0 cp, simulating the lower (surface) temperature of injected water forming a bank around the injectors.

## FIELD DEVELOPMENT CONSIDERATIONS

The element models were used interactively in the development planning process to form the basis for recommendations regarding drive mechanism, well requirements,

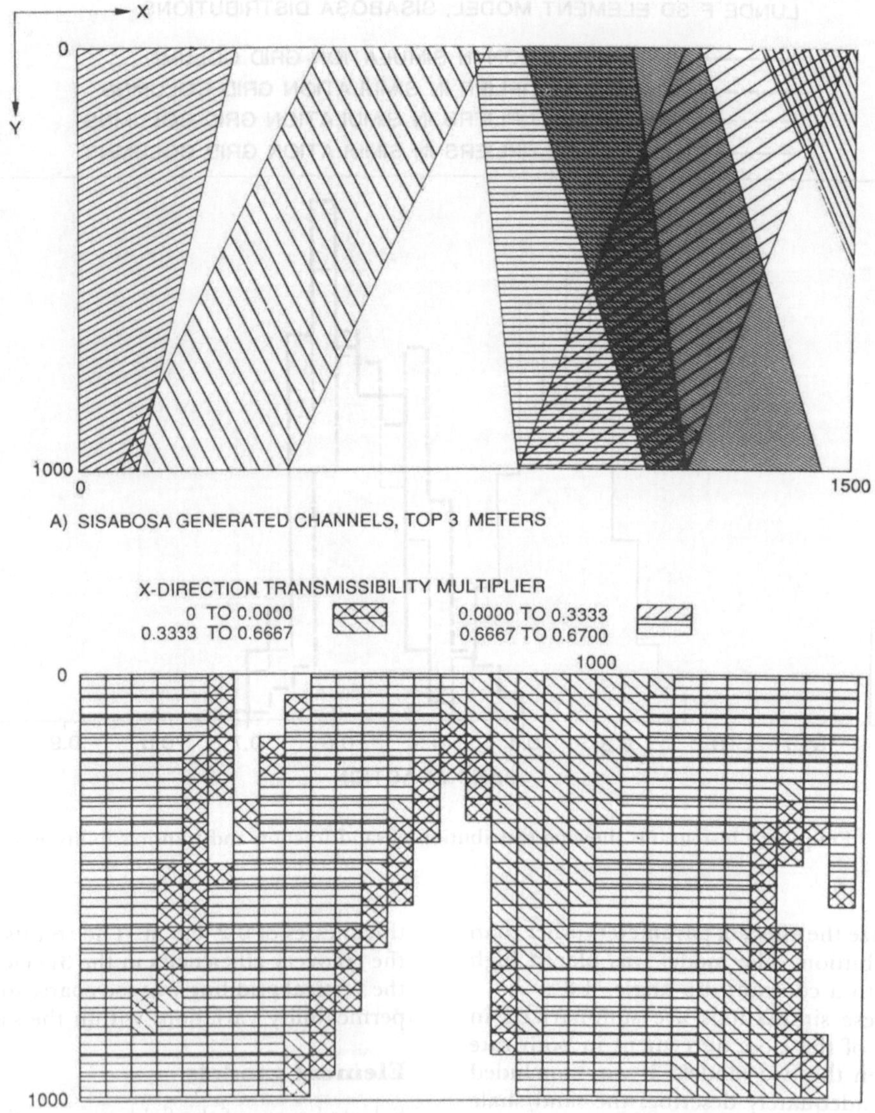


**Fig. 6.** Areal view of a reservoir segment, showing the sand bodies present and two simulation model representations: (a) SISABOSA generated channels; (b) fine-grid representation; (c) coarse-grid representation.

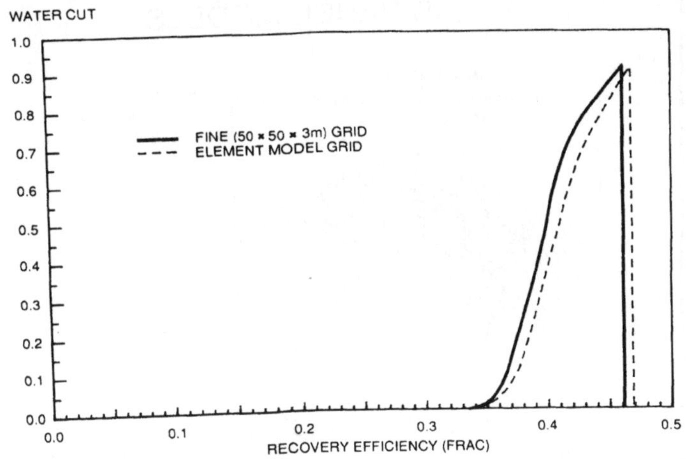


WATER CUT vs. RECOVERY EFFICIENCY

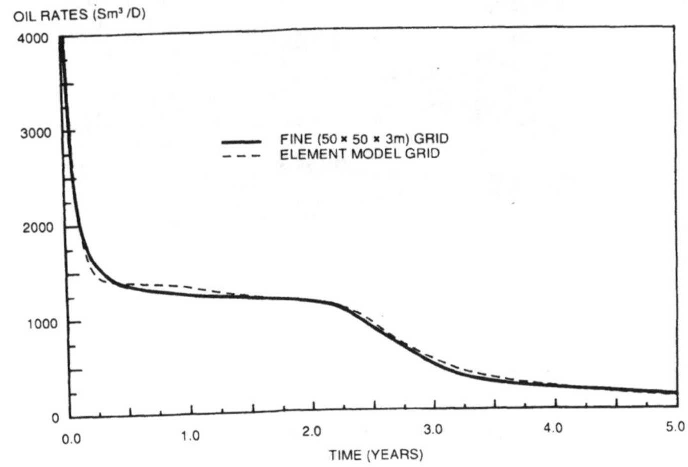


**Fig. 7.** Predicted reservoir performance for the reservoir segment using fine and coarse gridding: (a) water cut vs. recovery efficiency; (b) oil rate vs. time.

locations, sequence and completion as well as development strategy.

Full pressure maintenance by water injection was established as the most effective drive mechanism. In this reservoir, expansion drive would give a rapid fall in rates and low total recovery. Pressure maintenance by gas injection was studied for parts of the reservoir, but it was not found to be as attractive as water injection.

Element model sensitivities included a wide variety of well-location combinations to determine the well requirements and optimum well pattern. The injectors were located in the aquifer wherever possible. In the models, this created a uniform pressure support from the aquifer. Also, the rate reduction effect from residual oil was avoided around the injector wellbores.

In most areas, the updip/downdip spacing between injectors and producers was determined by the fault pattern and the reservoir geometry. The sensitivities indicated that the best sweep efficiency would be achieved through a staggered line drive pattern, with an along-strike well spacing of between 800 and 1400 metres. A total requirement of 93 wells was found when extrapolating the element model results to all economically producible areas of the field.

The timing of production and injection wells in an area was found to have some effect on the sweep efficiency. Based on sensitivities, a sequential development of the different fault bounded areas was recommended in order to get all of the wells in one area drilled within a relatively short space of time, and thereby maintain a stable flow situation and an optimum sweep efficiency.

Within reasonable limits, oil recovery was found to be virtually insensitive to varying well rates, provided that the volume produced was replaced by water injection. The field production profile could therefore be generated for various sensitivities to process capacity by use of a simple summation routine, as described in the following section.

The results formed the basis for the Report of Commerciality (March 1987) and the Field Development and Operating Plan (August 1987). The Field Development and Operating Plan was approved by all licencees in September 1987, and by the Norwegian authorities in May 1988.

## FIELD PRODUCTION PROFILES

Typical well performances were extracted from each of the element models by the use of dimensionless well profiles, expressed as dimensionless oil rate and water cut plotted versus recovery efficiency. Dimensionless oil rate was defined as a well's oil rate at any time divided by its maximum oil rate. The dimensionless well profile concept is illustrated in Fig. 10. The model well performances were averaged as shown in the figure to yield one or a few typical dimensionless well profiles for an area.

The following procedure was followed in order to convert the dimensionless well profiles into a full-field production profile.

- Maximum sustainable well rates from the models were identified and correlated to permeability-thickness. To ensure that none of the areas was produced under unrealistically high rates, it was defined that none of the wells should be predicted to produce at a rate higher than the rate corresponding to 50% per annum of the reserves within that well's drainage radius.
- To convert this information into a full-field production profile, each planned well was assigned a dimensionless well profile, a recoverable reserve number, and a rate based on its specific location. By doing this, well performance from the models was in effect extrapolated to all areas of the field.
- Using a well-summation routine (PRODGEN), the individual well profiles were coupled to a drilling schedule and the planned process capacities, and then summed to obtain oil and water rates versus time for the full-field profile.

As a quality check, the above procedure was also applied to the areas included in each of the simulation models. A comparison of a production profile obtained by simulation and by using this method is shown in Fig. 11 as an example. As can be seen, the results compare well, and this method was found to be adequate for generating production profiles under a number of different development scenarios for Snorre.

To verify the total field performance generated through extrapolation of the element model results, and to estimate the impact of the assumption of sealing faults implicit in the use of separate element models for different areas, a full-field model was constructed based on pseudo relative permeability and capillary pressure functions.

A comparison of the resulting field production profiles from the full-field model and the well-summation routine for a selected development scenario is shown in Fig. 12.

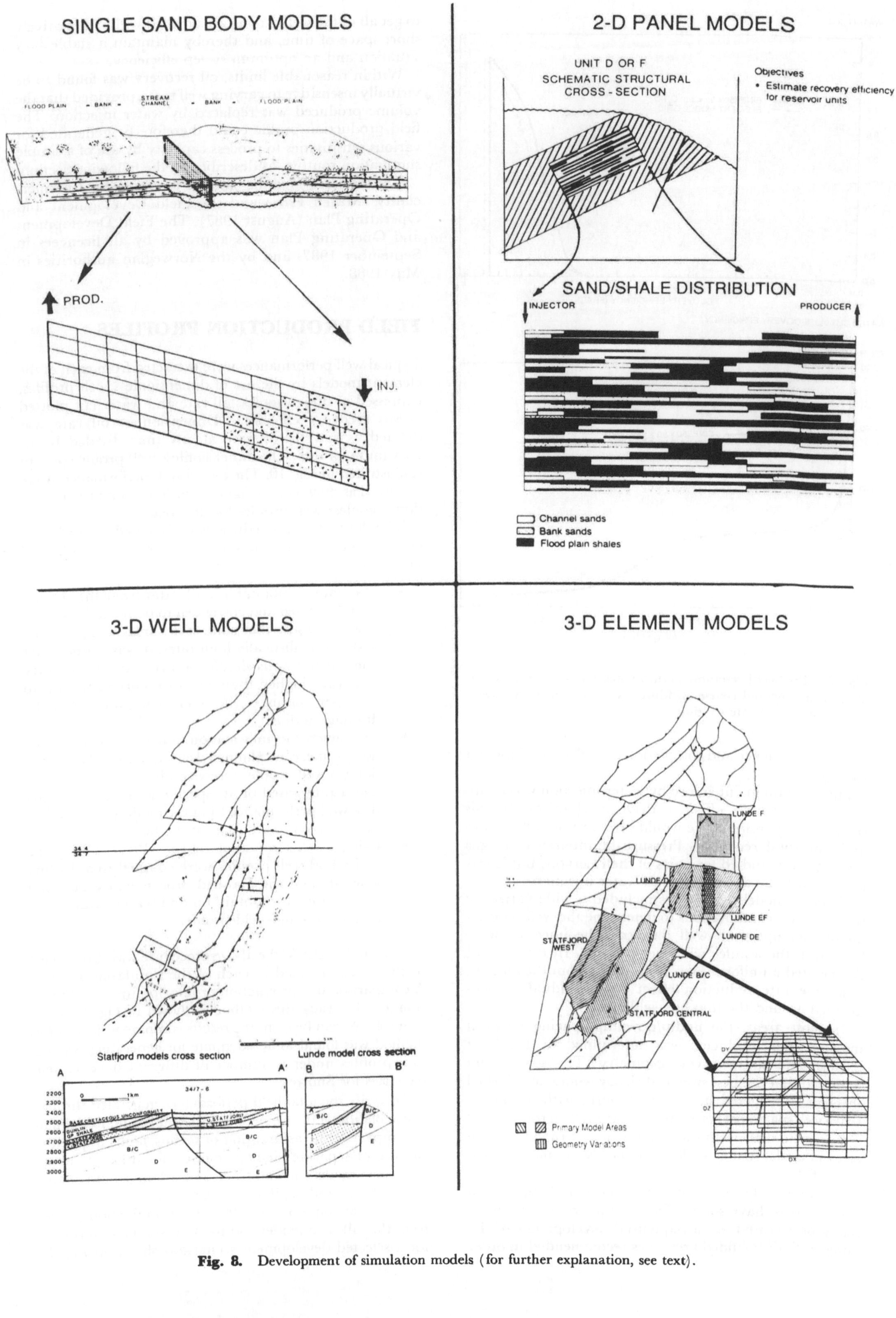


**Fig. 8.** Development of simulation models (for further explanation, see text).

**PERCENTAGE OF STOOIP MODELLED:**

| | PHASE 1 | | TOTAL | |
|---|---|---|---|---|
| | | % | | % |
| | STOOIP | MODELLED | STOOIP | MODELLED |
| STATFJORD | 120 | 79 | 123 | 77 |
| LUNDE DEF | 62 | 30 | 243 | 46 |
| LUNDE B/C | - | - | 105 | 9 |
| M/L LUNDE | - | - | 9 | - |
| UNIT A | - | - | 10 | - |
| | 182 | 66 | 490 | 44 |

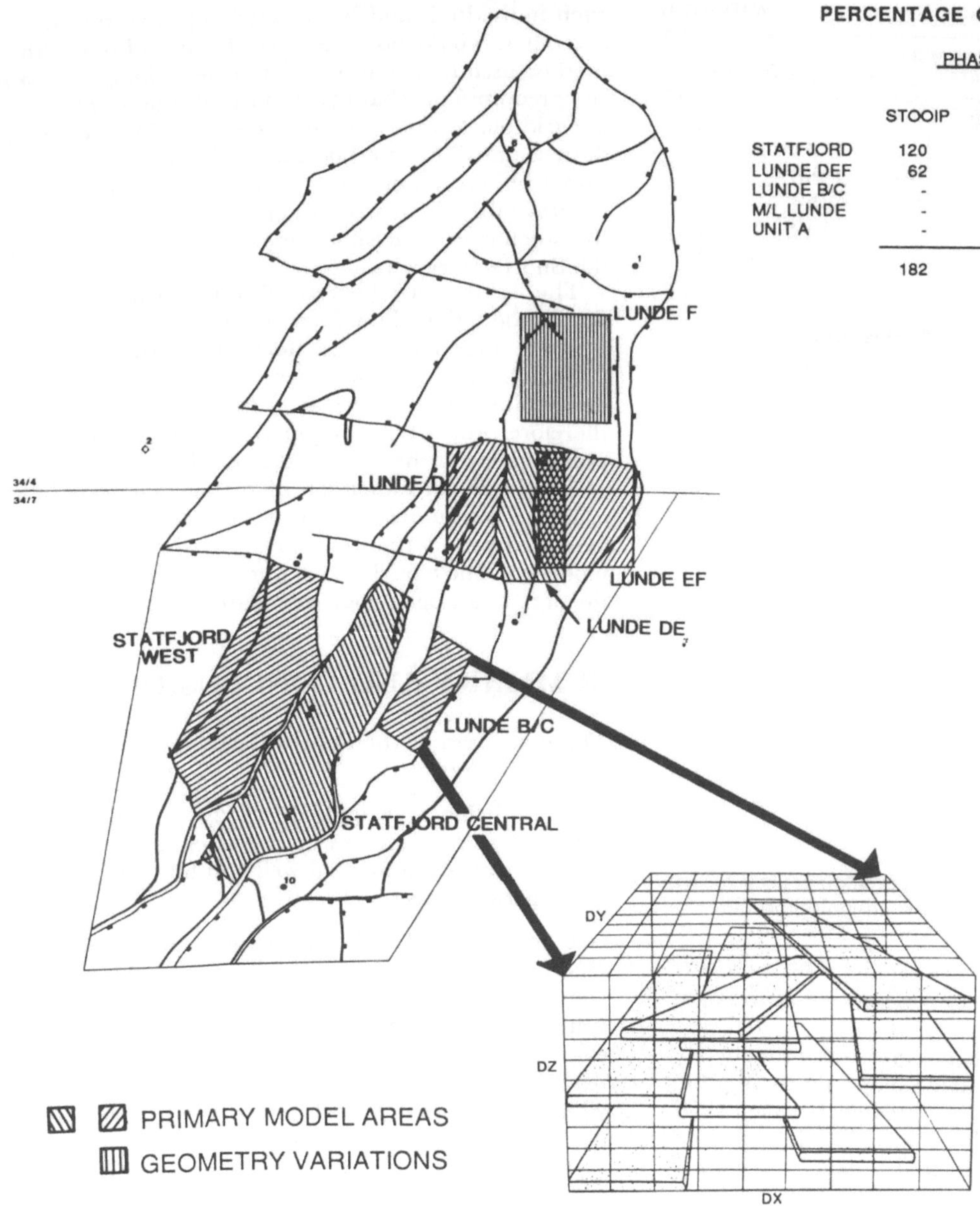


**Fig. 9.** Areal coverage of the element models.

The correspondence between the full-field model results and the aggregated element model results is excellent.

Although the assumption of sealing faults influenced a few individual wells, it was not found to be critical for the field production profile (Fig. 13). This is due to the repetitive well patterns and the chosen development strategy using full pressure maintenance by water injection.

## DISCUSSION

A large amount of high-quality data has been collected. These data indicate a complex, fluvially deposited reservoir with rapid variations in reservoir quality. However, the well data give a consistent picture that supports the geological model, thereby indicating that the general nature of the reservoir is reasonably well defined, and that the available data may be used to predict reservoir performance.

Even though the production data necessary to history-match the models are not yet available for the Snorre Field, the stochastic reservoir description is considered to give a more realistic understanding of the reservoir than conventional layered models. The generated reservoir model must, however, be in accordance with the geological model. The reduced rates and recovery efficiencies predicted by models based on stochastic sand/shale distributions (as compared to traditional layered models), result from the tortuous flow paths created by these models. This is considered to be consistent with the geological model.

It is important, however, to note that a stochastically generated reservoir model is only one of many possible realizations for the same reservoir. The model should not, therefore, be used to optimize parameters that are realization-dependent (such as locating a well where the model shows a local maximum in net/gross). Furthermore, in order to arrive at reasonable average reservoir models, it is necessary to generate a number of different reservoir realizations and to pick one average. For this purpose, sand body interconnectedness (see Stanley *et al.*, this volume) was calculated for the different realizations of the reservoir, and the realization with sand body intercon-

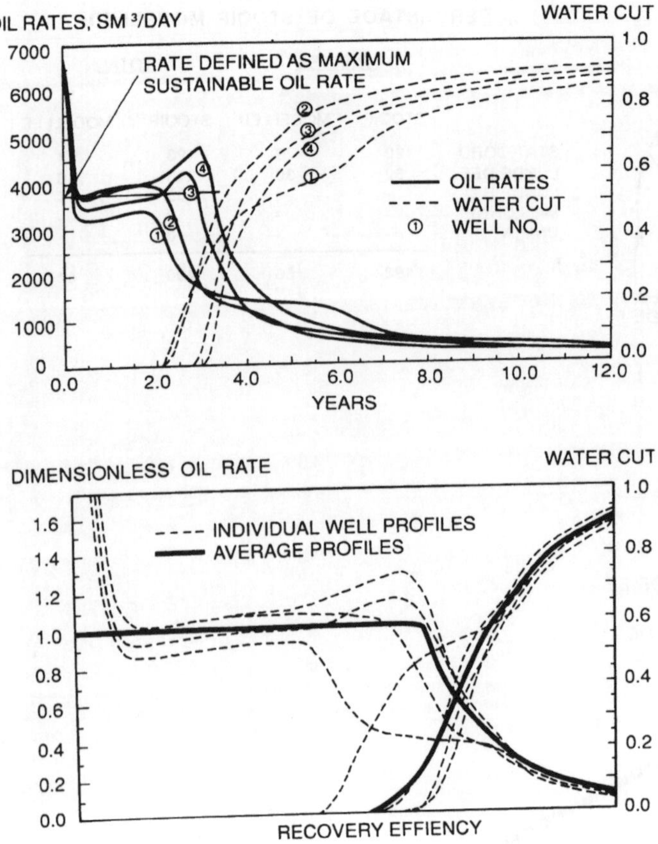


**Fig. 10.** Principle of dimensionless well profiles: upper—model predicted well performances; lower—dimensionless profiles.

nectedness closest to the median was selected as the most representative.

The reservoir simulation model grid must adequately reflect the stochastically generated sand/shale distribution so that no artificial contacts are created between sand bodies that are not physically in contact. The most obvious method would be to use a grid that is exactly overlying each individual sand body. The implementation of such a method would not, however, be straightforward. The method used in this study was to orient the grid parallel to the predominant channel (sand body) orientation and to use grid blocks that were small compared to the sand body dimensions. This method allowed the use of a conventional reservoir simulator with a Cartesian grid system. As previously discussed, this approach adequately models the stochastic sand/shale representations generated for the Snorre reservoirs.

The accuracy of this type of model will not be proved until production data from the field become available. Because these models are stochastic (rather than deterministic), they give average, rather than site-specific descriptions of the reservoir. A large number of wells will therefore probably be required to evaluate properly the performance of the described models.

To obtain an indication of the models' capability to estimate adequately reservoir performance, an 18-day extended production test performed in well 34/4–7 was modelled. The model results were in good agreement with the actual test behaviour (see Stanley *et al.*, this volume).

## SUMMARY AND CONCLUSIONS

The objectives of the Snorre Field reservoir simulation activities were:

- to define development strategy guidelines;
- to estimate recovery efficiencies; and
- to predict individual well and full-field production performances.

Geological models of stochastically distributed fluvial sand bodies have been successfully transferred to 3D reservoir simulation models with conventional, Cartesian grid systems using porosity and transmissibility multipliers calculated for each individual grid block. The porosity multipliers represent the sand fraction in each grid block, while transmissibility multipliers in all three directions describe the degree of communication between the sand bodies.

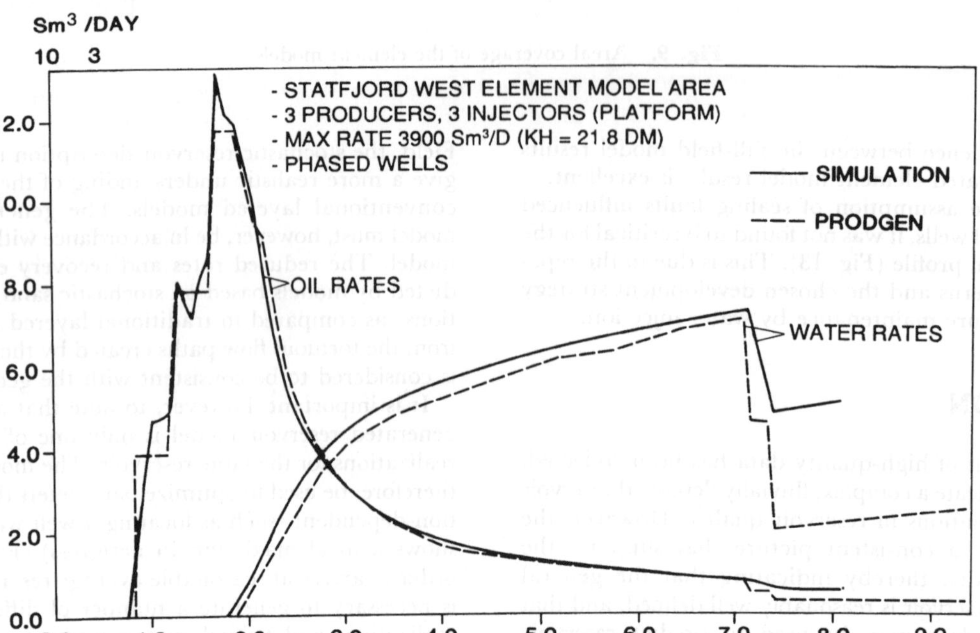


**Fig. 11.** Comparison between production profiles generated by a 3D element model and the well-summation programme.

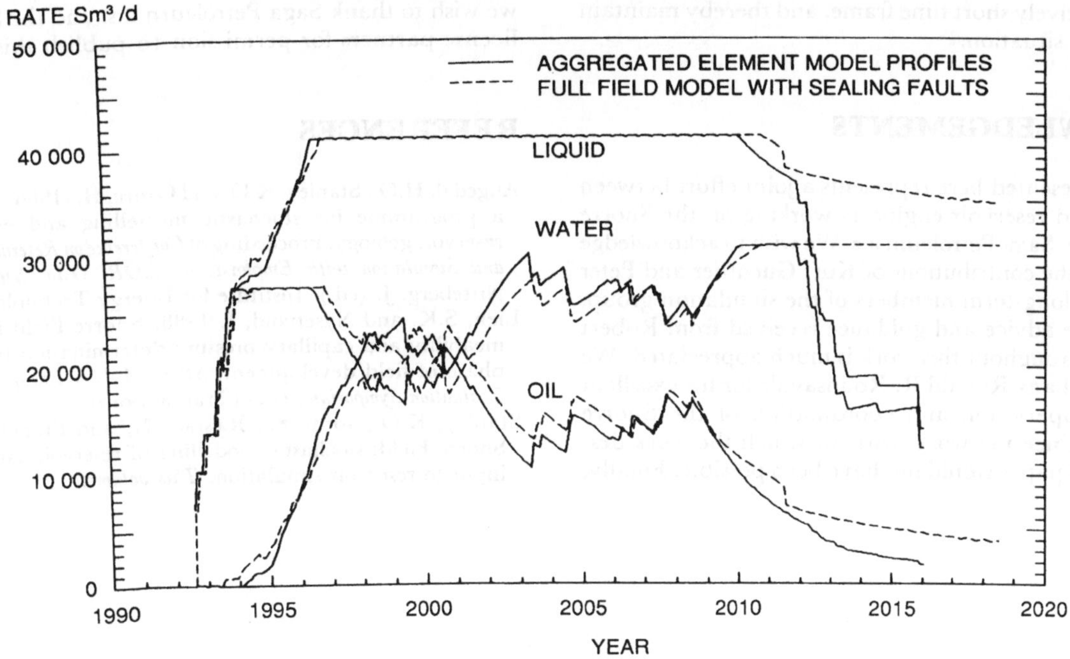


**Fig. 12.** Comparison between full-field production profiles based on the 3D element models and generated by the full-field model.

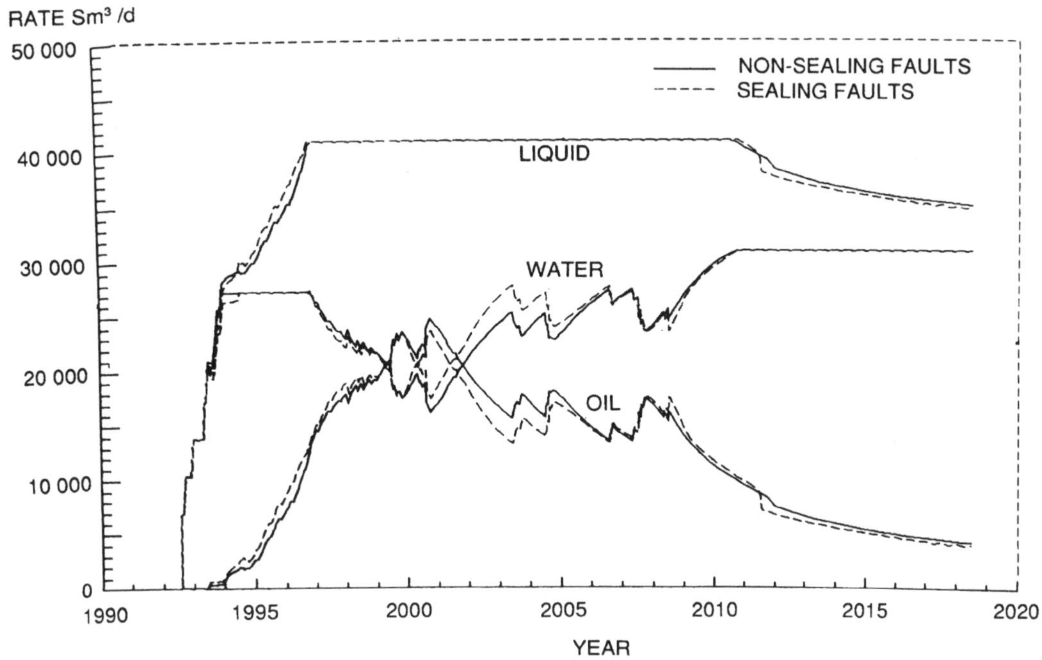


**Fig. 13.** Comparison between full-field model predictions with sealing and non-sealing faults.

Recovery efficiencies and production profiles for the Snorre Field have been based on 3D (stochastic) element models covering selected areas of the field. The model results were adjusted to account for heterogeneities that were not included in the models. For example, small 2D channel models were used to estimate the effects of intra-channel permeability variations.

Recovery efficiencies and well performances were extrapolated to non-modelled areas of the field, and well performances for all planned development wells were finally aggregated into fieldwide production profiles using a well-summation routine. This allowed a large number of production profiles for different development scenarios to be generated in an efficient manner.

A full-field model verified that the approach of simulating representative portions of the field and extrapolating the results to fieldwide performance was valid. It also showed that different assumptions concerning the sealing nature of the faults had only a minor impact on the predicted field performance.

It was found that a relatively regular staggered line well pattern is optimal for the development of the Snorre Field, with full pressure support from water injectors located downdip and in the water zone where possible. A total of 93 wells was recommended for the development of the economically producible areas of the field. A sequential development of the different fault-bounded areas is recommended in order to get all of the wells in one area drilled within a relatively short time frame, and thereby maintain a stable flow situation.

## ACKNOWLEDGEMENTS

The work presented here represents a joint effort between geologists and reservoir engineers working on the Snorre Project within Saga Petroleum a.s. We wish to acknowledge particularly the contributions of Kurt Guenther and Peter Dempsey as long-term members of the simulation group. The valuable advice and guidance received from Robert H. Rossen throughout this work is much appreciated. We also wish to thank Ronald R. Rounsaville for his excellent planning, supervision and coordination of the Snorre reservoir engineering work, without which the work described in this paper would not have been possible. Finally, we wish to thank Saga Petroleum a.s. and the Snorre Field license partners for permission to publish this paper.

## REFERENCES

Augedal, H.O., Stanley, K.O. and Omre, H. (1986). SISABOSA, a programme for stochastic modelling and evaluation of reservoir geology. Proceeding of *Conference on Reservoir Description and Simulation with Emphasis on EOR, Oslo, September 1986*, Nitteberg, J. (ed.). Institute for Energy Technology.

Lien, S.K. and Nysetvold, I. 1988. Snorre Field relative permeability and capillary pressure determination in evaluation phase of field development. *SPWLA 11th European Formation Evaluation Symposium, Oslo* (Transactions).

Stanley, K.O., Jorde K., Ræstad, N. and Stockbridge, C.P. Snorre Field: stochastic modelling of reservoir sand bodies as input to reservoir simulation. *This volume.*

Keynote Address 1

# Reservoir Characterization from Crosshole Seismic: New Directions

*W.B. Beydoun*

ARCO Oil and Gas Company, Plano, Texas, USA

The synergism that can be obtained by integrating engineering, geological and geophysical data requires a substantial effort but can significantly improve reservoir characterization. Three dimensional (3D) surface seismic surveys with standard data processing, combined with well logs, have been the conventional geophysical approach for describing relatively simple reservoirs. Seismic acquisition technology (e.g. multicomponent vertical seismic profiling (VSP), and crosshole experiments) and processing techniques (e.g. amplitude versus offset (AVO), elastic migration/inversion) have been, and are being, developed to enhance reservoir information and resolution prior to petrophysical evaluation.

The potential of crosshole seismic for interwell imaging was addressed nearly two decades ago. Since then, the technique of travel-time tomography has found some success in monitoring enhanced oil recovery processes. Tomograms have the capability of representing the smooth character of the velocity field in regions where the ray coverage is satisfactory. With the progress of supercomputers, 3D and anisotropy effects can now be incorporated but the technique persists in producing low-resolution images of the subsurface. To improve these images, current research has focused on using more of the full-waveform character of the crosshole seismic data. With increasing efforts in controlled-amplitude downhole source development and in preprocessing techniques, the scattered wavefield in crosshole data can now be used. These waves carry valuable lithological information, and when used in elastic migration/inversion or diffraction tomography techniques, yield high-resolution interwell images of rapid variations in elastic parameters (e.g., P and S wave impedances and density maps). Preliminary results with crosshole data generated by a weightdrop downhole source indicate some of the precautions and requirements necessary for obtaining reliable, detailed (of the order of 1 metre), interwell elastic information.

## INTRODUCTION

Characterization of porosity and permeability within reservoir traps constitutes a most ambitious geophysical initiative. This effort is necessary and more appropriate than ever with current progress in exploration and supercomputer technology, the increased understanding of complex wave phenomena, and the potential impact on exploration and production economics. To achieve this goal, i.e. efficient reservoir characterization and development, the integration of geophysical, geological and engineering data is required. A uniform terminology and adaptive human and technical networks are necessary to amalgamate the input and output of these three fields (Fig. 1). An initial strategy for this integration is shown in Fig. 2, where the preliminary common ground is reservoir geology.

The objective of this contribution is twofold. First, a rapid general overview of existing and forthcoming seismic methods for reservoir description is presented. This covers surface seismic, VSP and crosshole imaging techniques, followed by a brief analysis of why travel-time tomography techniques for crosshole data are limited to producing low-resolution images of the subsurface. The capability of using scattered waves in crosshole data to complement transmission tomograms with high-resolution information is subsequently addressed. The second part of the paper deals with an example of an existing elastic diffraction tomography technique that uses scattered waves as input. This technique is applied to synthetic and field data collected in the Paris Basin. The intention is to address practically some of the issues introduced earlier. Requirements, precautions and advantages of using a diffraction tomography technique for crosshole data to obtain high-resolution interwell depth images are developed.

## OVERVIEW

The current geophysical approach to studying reservoirs is 3D surface seismic surveys with conventional data processing (e.g. acoustic post-stack time migration), combined with existing well logs. Petrophysical parameters are inferred from the integration of velocity and amplitude analysis of seismic time sections with well log calibration, empirical or experimental relations, and/or simple rock/fluid physics models, assuming that all approximations

*North Sea Oil and Gas Reservoirs—II*

| | GEOPHYSICS | GEOLOGY | ENGINEERING |
|---|---|---|---|
| STUDY | PHYSICAL DESCRIPTION | STRATIGRAPHY AND FRAMEWORK | ROCKS AND FLUIDS DISTRIBUTION |
| TOOLS | SEISMIC, ELECTRICAL, MAGNETIC, GRAVITY, RADIATION, ... | CORES, OUTCROPS, WELL LOGS, SEISMIC | CORES, WELL TESTS, WELL LOGS, PRODUCTION LOGS |
| OUTPUT | Vp, Vs, $\rho$, RESISTIVITY, ... | DEPOSITIONAL ENVIRONMENT, DIAGENESIS, $\phi$, K, ... | $\phi$, K, S, $K_R$, $P_C$, ... |
| INFO | TIME VARYING, INTERWELL IMAGE | FLOW UNITS, PAY AND NON-PAY FORMATIONS | FLUIDS RATE, CUMULATIVE RECOVERY, SWEPT REGION DET. |
| NATURE OF INFO | AVERAGED 3D+t INDIRECT | DETAILED 1D+t INTERPOLATED 3D | AVERAGED 1D+t INTERPOLATED 3D |

**Fig. 1.** Definition of reservoir geophysics, geology and engineering within the context of reservoir characterization

involved hold. Some cases using this approach have recently been made public (e.g. Greaves and Fulp, 1987; Robertson, 1988; Abriel, 1988).

The ambition of exploiting the complete seismic wavefield appears to be beyond our current understanding. The propagation of seismic waves may be very complicated in a reservoir system that includes a distribution of fluids within the rock matrix. Generally, the wavefield can be viewed as a measure of the medium's elasticity, heterogeneity, viscosity, anisotropy, porosity, permeability, etc. However, seismic waves are not equally sensitive to all of these effects, and this sensitivity depends on the acquisition geometry (surface, VSPs, crosshole, full-waveform sonic, etc.) and the data processing scheme. As a consequence, each effect might not always be simple to detect and describe, particularly in the presence of noise. Seismic wave modelling techniques are also rapidly growing in complexity; e.g. 3D elastic heterogeneous anisotropic media (Mora, 1988), 2D viscoelastic media (Carcione *et al.*, 1988), and porous saturated media (Schmitt *et al.*, 1988). Amplitude versus offset (AVO), migration and inversion (M/I) techniques follow the same trend, but with some delay because pre-stack elastic processing is still not commonly used owing to high computation costs. AVO analysis is geared towards more realistic consideration of the overburden, in particular by performing amplitude-controlled migration operations to compensate for a heterogeneous background (Resnick *et al.*, 1987). Multiparameter migration/inversion techniques are becoming useful tools for detailed stratigraphic interpretation (e.g. Stolt and Weglein, 1985; Tarantola, 1987; Parsons, 1986; Beydoun and co-workers 1988, 1988a,b, 1989a,b). Examples of what would affect elastic image artefacts with depth migration/inversion of surface seismic data include:

(1) low-frequency content and poor S/N ratio data (poor reconstruction/stack, low-resolution, no distinguishable coherency in events); (2) multiples present in the input data (false reflectors with poor vertical resolution); (3) no amplitude control in the data (loss of multiparameter estimation; at best, images may represent an acoustic type migration); (4) the elastic background model is unsatisfactory (poor stack and vertical resolution, coupling between multiparameter images); and (5) the target zone suffers from limited aperture problems (poor horizontal resolution, distorted egdes, coupling between multiparameter images).

In parallel to this effort in processing, acquisition technology and preprocessing research is moving towards more accurate amplitude and polarization control.

Borehole seismology, and in particular crosshole seismic, is an area where substantial efforts in technology (e.g. downhole sources: SEG Borehole Source session, 1988) and processing techniques are necessary. Downhole seismic source technology still requires some years of development before reaching the level of maturity of surface sources. Although no single tool is expected to perform all tasks, the ideal source would be powerful but non-destructive, repetitive, reliable and depth-independent. This source should generate large frequency bandwidth P waves and S waves with a rapid cycle time (less than 30 seconds), operate on a wireline in high temperature conditions (up to 500°F), and in cased and open holes with inner diameters ranging from 5 to 12 inches. Although seismic downhole sources currently propagate signals up to an interwell range of about 300 m, within the next few years this limit is expected to increase to interwell ranges typical of North Sea oil and gas fields, namely in the order of 1000 m.

First arrival travel-time tomography, used to analyse crosshole data since 1971 (Bois *et al.*), has found some success in assisting enhanced oil recovery projects (Peterson *et al.*, 1985; Macrides *et al.*, 1988; Justice *et al.*, 1989). Geophysical and non-geophysical contributions to tomographic algorithm development and associated technical issues have been, and are still being made by literally hundreds of individuals. It is not the purpose here to review the important intricate details of these algorithms, but rather to outline, simply, the ultimate capability of travel-time transmission algorithms. From a mathematical point of view, within some regularity conditions, the exact unique reconstruction of an object profile (e.g. a velocity field) from an infinite set of its projections (e.g. travel-times) was proved in 1917 by Radon. Technically, this inverse problem is ill-posed with a formal rigorous solution. Any algorithm solving this problem may be regarded as approximate (Deans, 1983); existence of solution, uniqueness, stability, use of a-priori information, and optimality must be analysed.

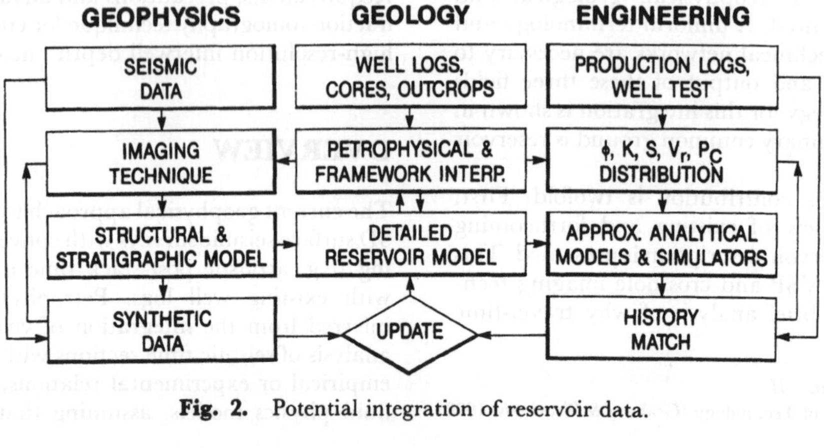


**Fig. 2.** Potential integration of reservoir data.

Adapting this framework for geophysical applications requires additional considerations and approximations. Some of these are:

(1) The back-projection is now the direct (curved) ray connecting a source–receiver pair. Ray approximations are assumed to be valid; this constrains the signal wavelength to be much smaller than any medium scale length (e.g. no diffractions or scattering are allowed). These arrivals contain essentially integrated or global medium information (along the ray); they are pure transmissions, and are not very sensitive to local changes in the velocity field.
(2) The object we wish to image might not be of finite extent nor be completely contained within the acquisition geometry (e.g. incomplete coverage of the target zone).
(3) The projections are discretized and finite in number (e.g. source and receiver have finite spatial sampling).
(4) A background medium is necessary for tracing rays and computing travel times that will be compared to the data (if an optimization iterative technique is used, e.g. ART, SIRT).

Consequently, first arrival travel-time tomograms represent, at best, the smooth aspect of the velocity field. Small scatterers or diffracting elements will not be imaged, even with complete coverage (Devaney, 1982a). Three-dimensional considerations with anisotropy can be incorporated (Guiziou and Haas, 1988), thereby rendering the technique more applicable, but its limitation to low-resolution images persists.

In contrast, scattered waves (e.g. reflection or converted transmission waveforms, diffractions) provide high-resolution images. The information gathered about the medium is complementary to first arrival travel times, in the sense that these waves are sensitive to the local properties from which they originated. The target zone is required to have a satisfactory scattered wave coverage, which is a somewhat less constraining requirement than for travel-time tomograms (i.e. good direct ray coverage). For example, imaging of events below the acquisition geometry is possible. However, these techniques generally require an input background medium adequately representing the large-scale variations of the subsurface and good-quality data. The background medium can be obtained either by travel-time tomography techniques and/or from prior information such as well logs or velocity analyses of surface reflection data (e.g. Yilmaz, 1987). VSP-CDP transform images from reflected crosshole data have been obtained (Iverson, 1986; Delvaux *et al.*, 1987; Beydoun *et al.*, 1989) and indicate the potential for high-resolution images.

Crosshole seismic data generally contain different wave types, which can interfere with each other physically (e.g. conversions of body wave to guided waves or vice versa) or overlap in time sections (e.g. tube waves and reflections). Figure 3 lists some of the wave types that can be of use for reservoir characterization. The task of identifying and reliably extracting only the primary elastic scattered wave part of the data is of crucial importance. This operation can be very difficult and delicate, particularly in complex media. Sophisticated preprocessing with amplitude control is required, but it is not necessarily achievable. For instance, the problem of filtering interbed multiples is still unsolved. The acquisition geometry also allows primary reflections to propagate up (from bottom reflectors) and down (from upper reflectors), indicating that standard one-way wave separation methods are inadequate. Wave equation-based preprocessing schemes, that have the advantage of not only correctly handling a particular phase or wave-direction in the data, but also the entire

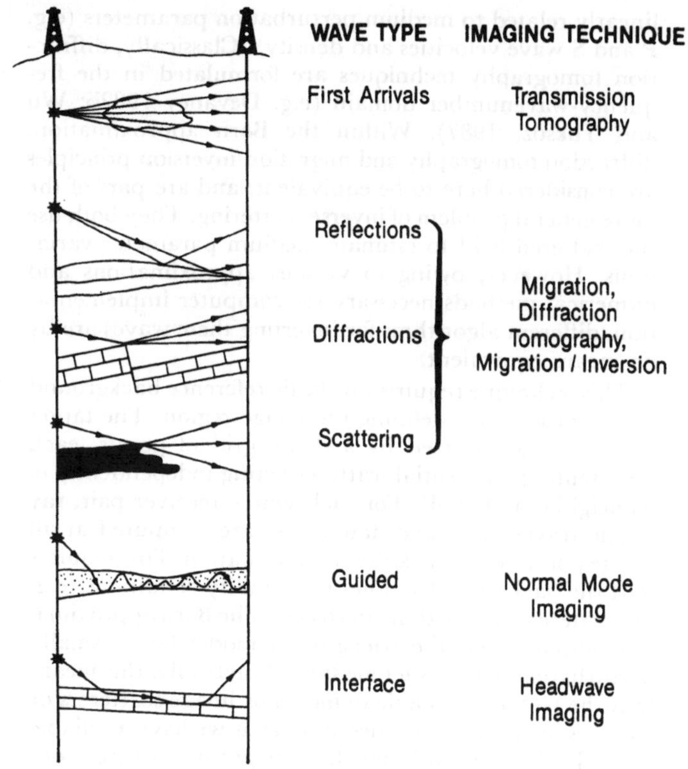


**Fig. 3.** Wave types encountered in crosshole seismics that are of interest for reservoir characterization.

waveform, are starting to be addressed (e.g. Foster and Mosher, 1988.) This should help in optimally isolating scattered waves while conserving their original amplitude variations.

## EXAMPLE OF DIFFRACTION TOMOGRAPHY

This section points out, with examples, some of the requirements, precautions and advantages of an elastic diffraction tomography technique as applied to crosshole data for reservoir characterization. Elastic, multicomponent, single-source, synthetic and real data examples in a heterogeneous background are presented. Other cases, for different acquisition geometries, have also been addressed in previous works (see Beydoun and co-workers 1988, 1988a,b, 1989a,b). Preliminary tests and results indicate that valuable high-resolution lithological information can be obtained from elastic depth images if certain (non-trivial) conditions, outlined in the previous section, are met. A synthetic example displays the effect on inter-well images of two different preprocessing schemes. Also analysed and discussed is a diffraction tomography experiment carried out by Elf Aquitaine with the IFP weight-drop downhole source at a test site in the Paris Basin.

### Technique

The imaging technique used for back-propagating the elastic scattered wavefield is derived in detail by Beydoun and Mendes (1989) and will be reviewed very briefly. This technique consists of a formulation of the inversion as a least-squares optimization problem in the space–time domain within the first Born approximation. The primary scattered wavefield (e.g. P–P, P–S, S–S, S–P) is then linearly related to medium perturbation parameters (e.g. P and S wave velocities and density). Classically, diffraction tomography techniques are formulated in the frequency-wavenumber domain (e.g. Devaney 1982b; Wu and Toksoz, 1987). Within the Born approximation, diffraction tomography and migration/inversion principles are considered here to be equivalent, and are part of the more general problem of inverse scattering. They both use the scattered field to estimate medium parameter variations. However, owing to various approximations and numerical methods necessary for computer implementation, different algorithms for inverting these waves are by no means equivalent.

This technique requires an elastic reference background with a target zone defining the image region. The target zone is parameterized by a dense grid of points, each representing a potential scatterer acting independently of its neighbour (Fig. 4). For each source-receiver pair, ray elastic travel times and amplitudes are computed at all image points using background information. This requires that the background satisfies ray validity conditions (e.g. no scatterers or very thin interfaces). The Born approximation imposes that the background model be a "small" perturbation of the exact medium. Practically, this means that the technique is able to model or image scatterers or changes in elastic properties, as long as we have a reliable ray-valid background and that images or perturbations are not too different from it (i.e. ideally less than half a seismic wavelength "away"); geological faults and some typical reservoirs could serve as examples. We will assume that the operation of filtering out everything except primary elastic scattered waves is feasible, bearing in mind that it is far from perfect and that induced errors will be difficult to quantify. The output of this preprocessing is the input for the diffraction tomography technique.

The technique consists first of depth-migrating the scattered data in the image region followed by an inversion

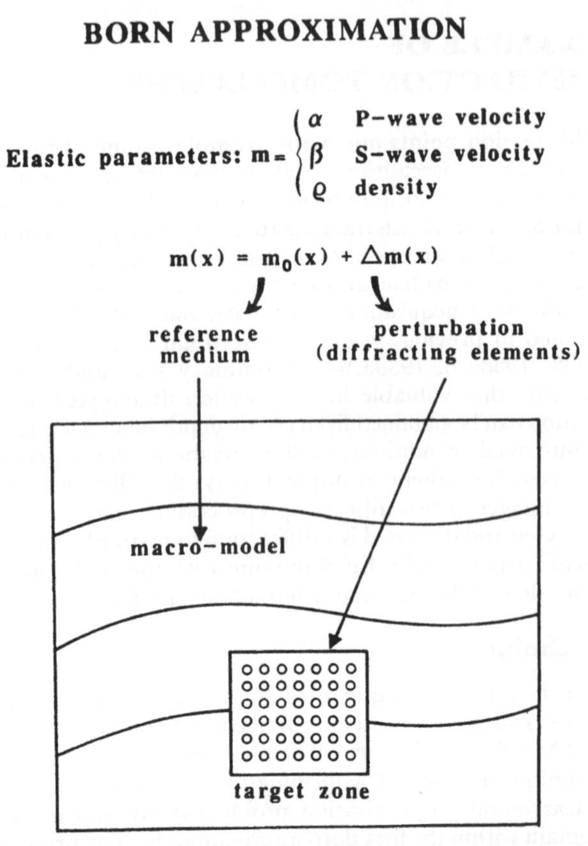


**Fig. 4.** Reference background with target (image) zone.

that acts as a spatial deconvolution filter (least-squares inversion with approximate elastic Hessian correction) compensating for wave spreading and multiparameter coupling. This is a one-step procedure (no iterations). If the acquisition geometry adequately illuminates the target zone, and the background model is accurate, final elastic images represent relative high-frequency variations of elastic parameters (e.g. P and S wave velocities and density maps) of the "exact" medium within the signal band.

These images, indicating relative changes in elastic properties, will lead to an estimation of relative changes in petrophysical parameters via empirical or experimental relations (e.g. Domenico, 1984) or simple rock/fluid physics models (e.g. Biot, 1956). This technique can also provide a detailed elastic depth section useful for assisting reservoir development and monitoring. In particular, if scattered events are present in the data, these images would be valuable for differential seismology experiments during enhanced oil recovery processes. Note, however, that even though the inversion will generally provide three elastic images, their confidences are not equal and depend, among other things, on the acquisition geometry, the reference model and the data quality. There are ways to measure these effects and to assess the degree of image confidence (Beydoun and Mendes, 1989a; Beydoun *et al.*, 1988b, 1989).

## Interwell imaging

An elastic model was derived using realistic petrophysical values from well logs and geological information in a Paris Basin test site. A crosshole geometry with a vertical force as the downhole source was defined and is shown in Fig. 5. A reservoir area is simulated at 600 m depth. Two-component ($x$, $z$) synthetic seismograms (Fig. 6) were generated with a 2D elastic finite-difference program, and body waves were identified. Note that the data are not necessarily ray-Born valid since they were generated by a numerical modelling method containing the full-wave field (including multiples). This is an interesting test since the diffraction tomography technique will assume that its input is ray-Born valid. Two different preprocessing methods were used for filtering downgoing P and S wave direct arrivals. Each uses a modified version of the wave-field separation technique of Seeman and Horowicz (1983), hereafter referred to as MSH. The preprocessing also filters out primary reflections from interfaces above the source, but since the target zone is below it, removal of these waves is necessary. The first preprocessing technique consists of filtering simultaneously downgoing P and S wave arrivals using the MSH technique, followed by some aesthetic clean-up, leading to an apparently continuous section of upgoing events (Fig. 7). The second preprocessing technique consists of filtering only the downgoing P wave with the MSH technique; the S wave arrival was muted in the time domain with a tapered window (Fig. 8).

The diffraction tomography technique was then applied to these two sets of data (preprocessing 1 and 2, for Figs. 7 and 8, respectively). We also computed S-wave VSP–CDP transforms for comparison. All images, along with reference and exact models, and an exact image, are shown in Fig. 9 (after preprocessing 1) and Fig. 10 (after preprocessing 2). Note that although preprocessing 1 appears "better" than preprocessing 2, it distorted relative amplitude information and added events, thus producing poorer results, particularly in the reservoir area. Preprocessing 2, although simpler, did a better job of preserving relative amplitudes, and consequently produced better

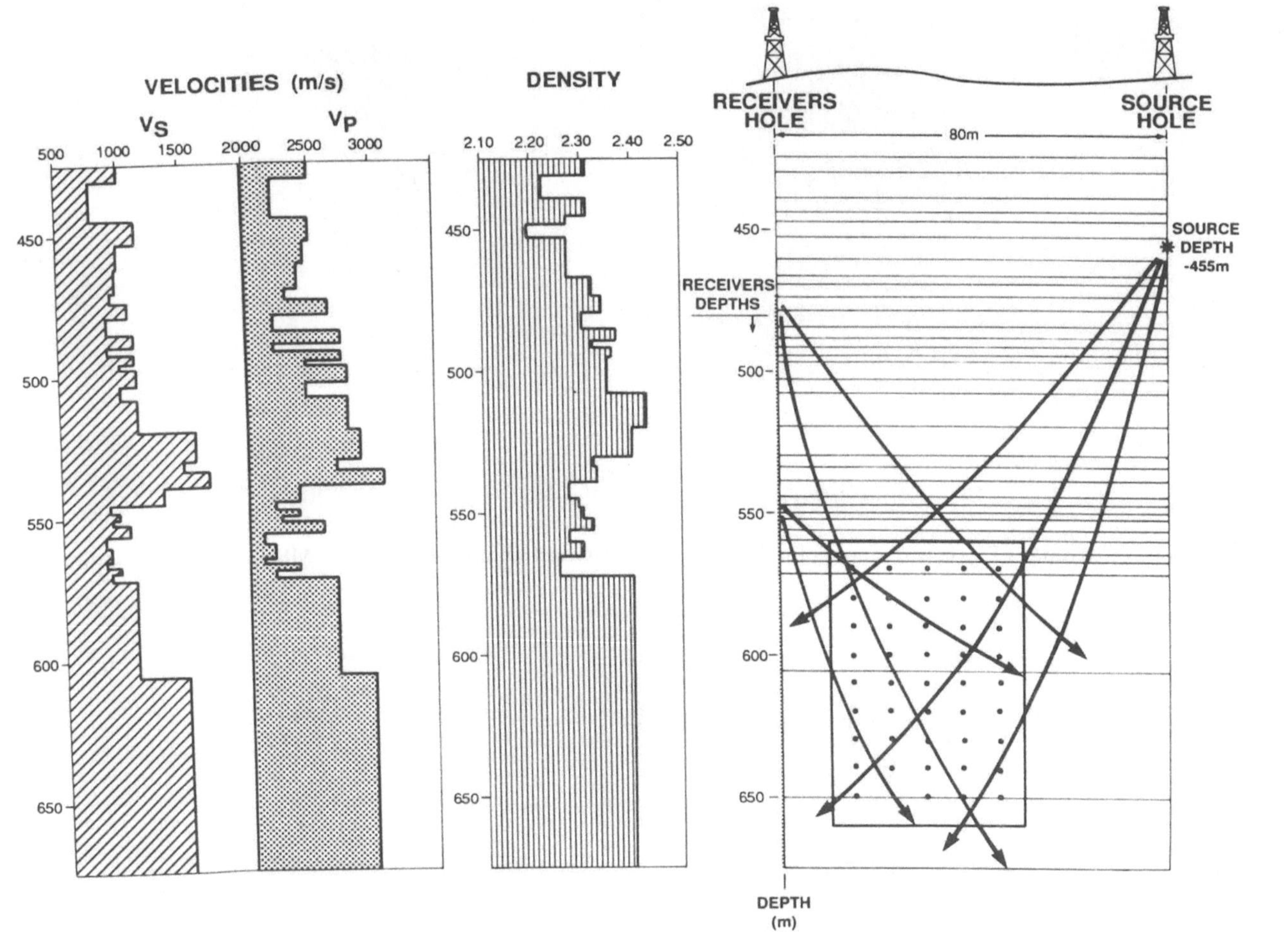


**Fig. 5.** Exact crosshole model for the synthetic example.

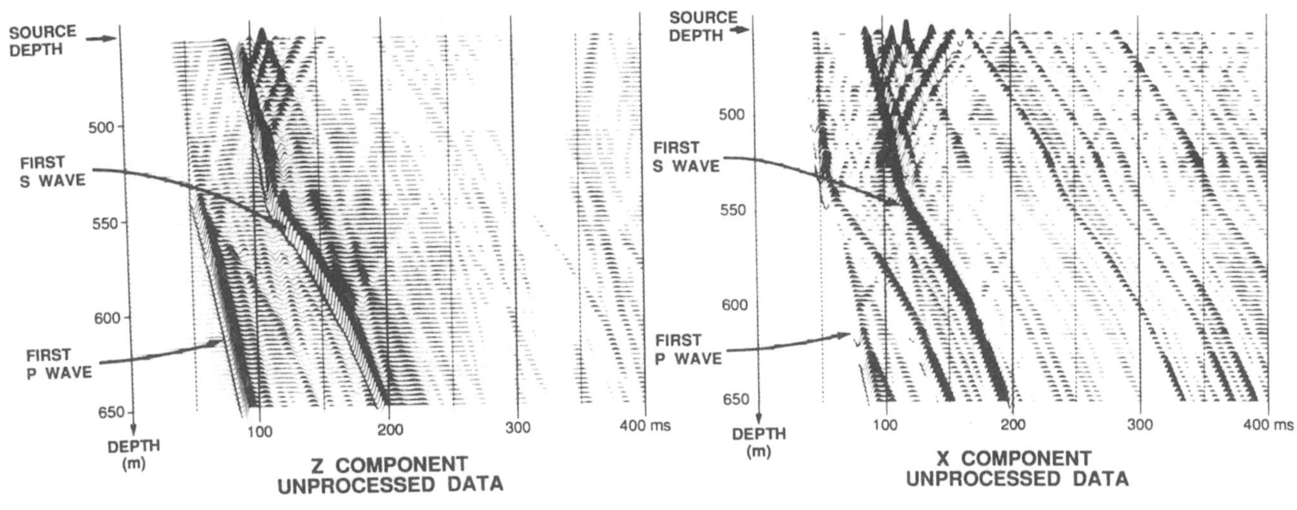


**Fig. 6.** Synthetic seismograms generated by an elastic finite-difference method; particle displacement components are $z$–vertical, $x$–horizontal.

results. Since multiples are present, we can see some of their effect on the reconstruction, in particular between 570 and 600 m. The strong artefacts at the top left of all of the images are due to poor coverage of reflected events from the acquisition geometry. Note that these result were obtained using only a one-shot gather, which limits the confidence region to a narrow strip about 25–30 metres wide, extending somewhat diagonally from the top right to the bottom left of the image (next to the borehole). VSP–CDP images are satisfactory; however, unrealistic structural undulations exist in the confidence region. There is a slight depth shift between the VSP-CDP and tomography images because the latter uses a zero-phase-like signal (maximum at the reflector), whereas the former uses the causal data (maximum right below the reflector). Multiple source levels must be considered to extend spatially the confidence region. The purpose here is to show that it is best to preprocess the data with the

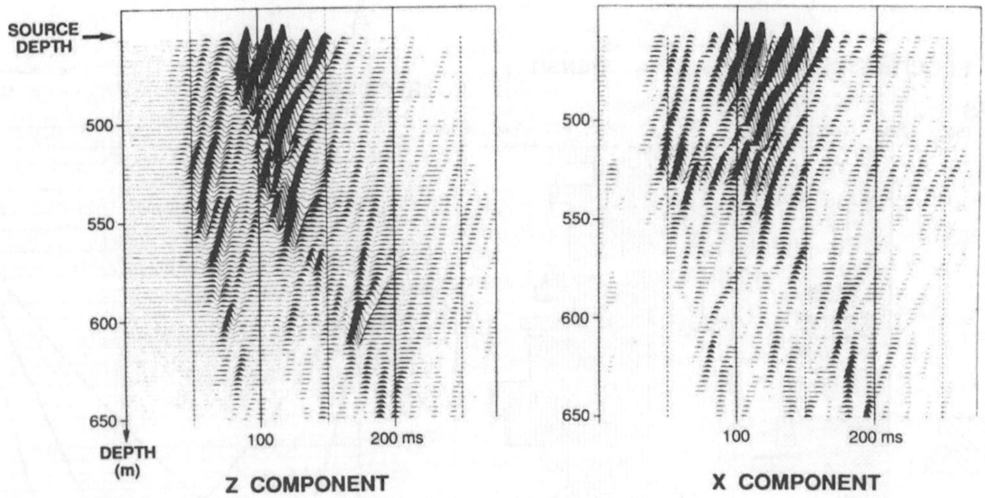


**Fig. 7.** Preprocessed data of Fig. 6, with downgoing P and S waves removed simultaneously using MSH technique (preprocessing 1)

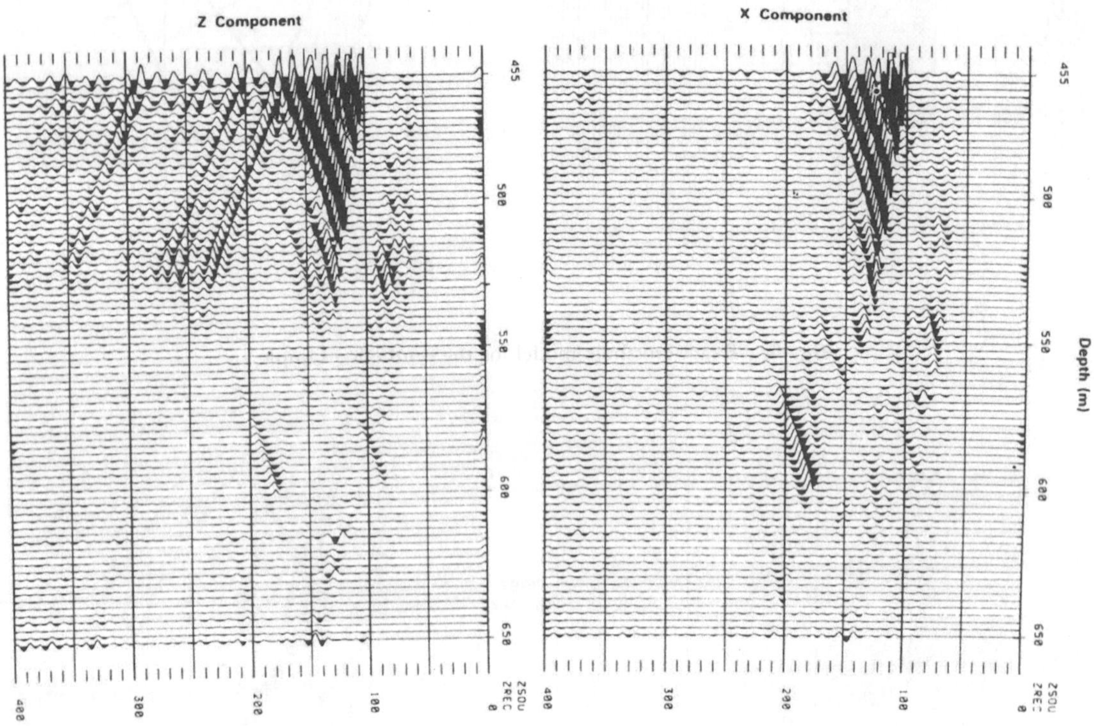


**Fig. 8.** Preprocessed data of Fig. 6, with downgoing P wave removed using MSH technique and downgoing S wave muted with a tapered time window (preprocessing 2).

fewest number of operations, particularly with regard to amplitudes changes. The best result is not necessarily obtained from preprocessed data that look continuous.

The real data example is a subset of a crosshole experiment performed in the Paris Basin (see Delvaux *et al.*, 1987; Beydoun *et al.*, 1989). The previous synthetic example was a model of this experiment. Figure 11 shows a plan view of the site, and Fig. 12, a geological section. Three Neocomian sand reservoirs designated as R1, R2 and R3 occur at depths 575, 583 and 600 m, respectively. The effective thickness of each reservoir does not exceed 5 m. A weightdrop downhole source (IFP) was fixed at 455 m in well CHU 162, and a multicomponent recording tool, Geolock (CGG), was lowered in well CHU 161, 79 m away from the source well. Owing to the tool mechanical resonance, the data were band limited to 150 Hz even though the source signals contained up to 350 Hz. Three-component data were recorded and reoriented. The two ($x$, $z$) in-plane components showed primarily strong S-wave arrivals. First S-wave arrival travel-time tomograms indicated that the velocity model was laterally fairly homogeneous. A velocity model was then defined using this and other information (Fig. 12)

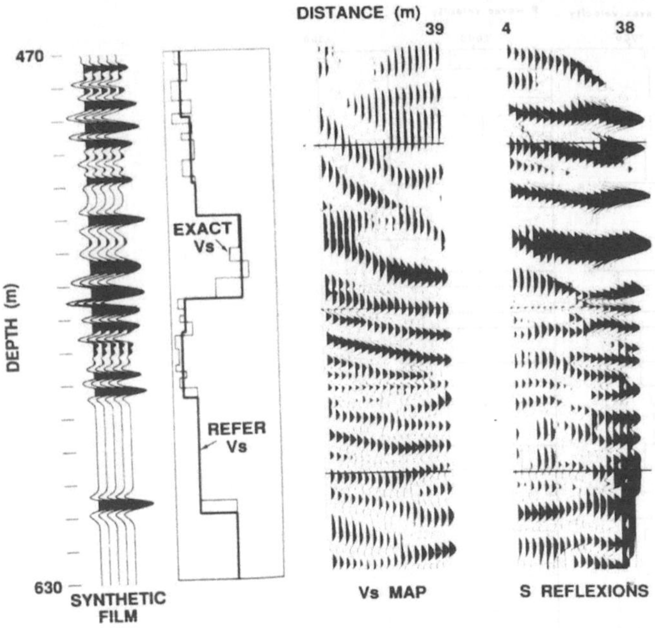


**Fig. 9.** Imaging results of the target zone after preprocessing 1. From left to right: exact S wave image obtained from convolution of a spatial source with the exact S wave log; exact and reference S wave log; S wave map from the diffraction tomography (left side closest to well); VSP–CDP image of S reflections.

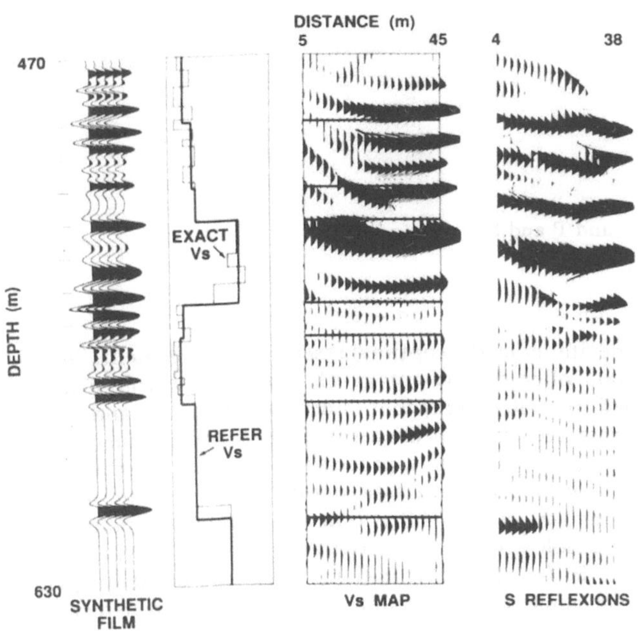


**Fig. 10.** Imaging results of the target zone after preprocessing 2. From left to right: Exact S wave image obtained from convolution of a spatial source with the exact S wave log; exact and reference S wave log; S wave map from the diffraction tomography (left side closest to well); VSP–CDP image of S reflections.

Removing the downgoing S wave using the MSH technique yielded upgoing events (Fig. 13), with largest amplitude S–S and S–P scattered waves. Consequently, only the S wave and density images can be obtained from the diffraction tomography (Fig. 14) since the P-wave image is sensitive to P–P scattering which does not seem to be present. The right side of the image is closest to the well, and therefore the confidence region is the reverse of the synthetic example (i.e. extending from the top left to the bottom right). At about 600 m, and next to the borehole, the three reservoir regions have been imaged. The width of the peak (shaded) at the 600 m level is about 5 m, indicating that the vertical resolution is at least if not better than this. This value is by no means a limit of the method, since other types of receivers (or sources) can certainly receive (or emit) higher-frequency signals. The interpretation of the elastic images within the confidence region must proceed with caution. Multilevel sources must be considered for better characterization. Conflicting dipping events at the centre of the images illustrate artefacts due to poor illumination by the source and receiver geometry of this portion of the images, and indicate that we are at the lower limit of multiparameter imaging requirements with only one source record. Nevertheless, this example demonstrates that current crosshole data and preprocessing techniques can already be used for diffraction tomography imaging. Calibration and integration of these images with other data (e.g. well logs) is essential to validate their interpretation.

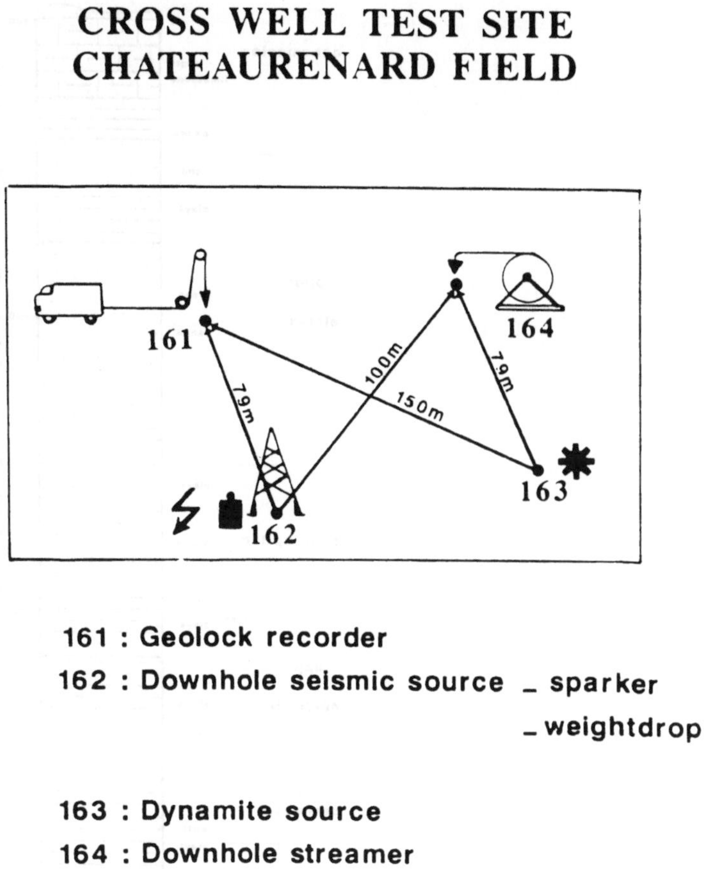


**Fig. 11.** Plan view of the test site in the Paris Basin. Wells and tools used are identified.

## CONCLUSIONS

Reservoir science and engineering research is heading towards a more uniform and integrated approach that will

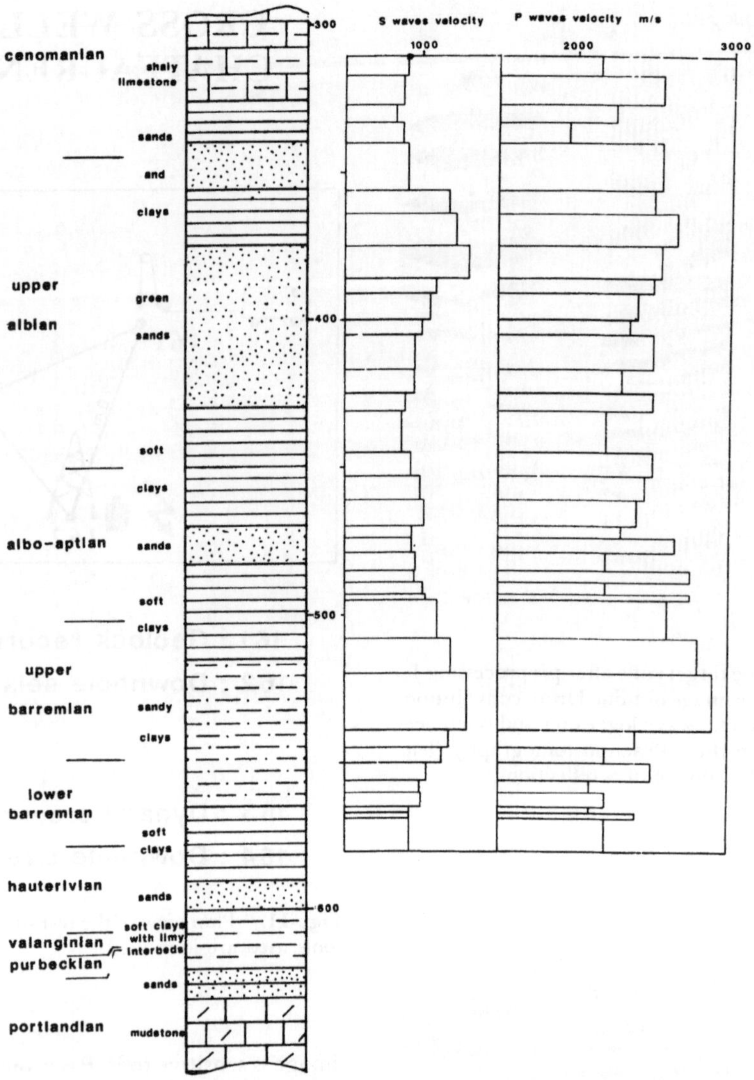


**Fig. 12.** Regional stratigraphy and P and S wave logs.

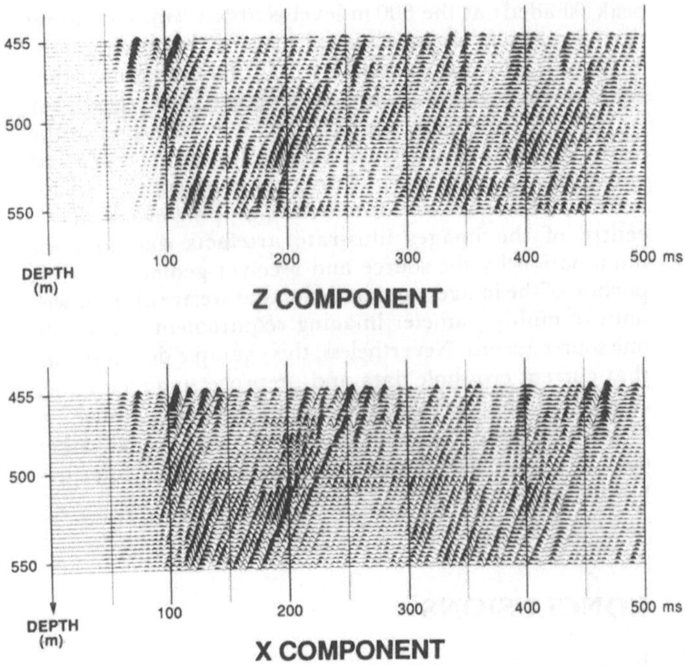


**Fig. 13.** Preprocessed two-component data after separation of downgoing S-wave with the MSH technique.

inevitably lead to optimal reservoir management. Seismic data offer the advantage of imaging reservoir interwell information in up to four dimensions (i.e. $x, y, z$ and $t$), with crosshole seismic being an area where substantial efforts in technology and data processing are being sought.

High-resolution elastic depth images can be obtained using an elastic diffraction tomography technique on crosshole seismic scattered data. Changes in elastic properties (e.g. lateral variations, faults) can be identified and characterized. The value of this description depends on several factors, the most important being the data and preprocessing quality, the coverage of the target zone by the acquisition geometry, and the background medium. Image artefacts (e.g. smearing, multiparameter coupling, unexplained events) strongly depend on how accurate these operations are.

A test of two preprocessing techniques indicated that some care must be taken for separating the scattered wavefield from the data. Preliminary (synthetic and field) results from single-shot data are encouraging and indicate that diffraction tomography techniques are now of practical use. With the advent of new downhole sources and the development of wave-based preprocessing schemes, elastic imaging techniques will be useful in assisting detailed (in the order of 1 metre) reservoir characterization and monitoring.

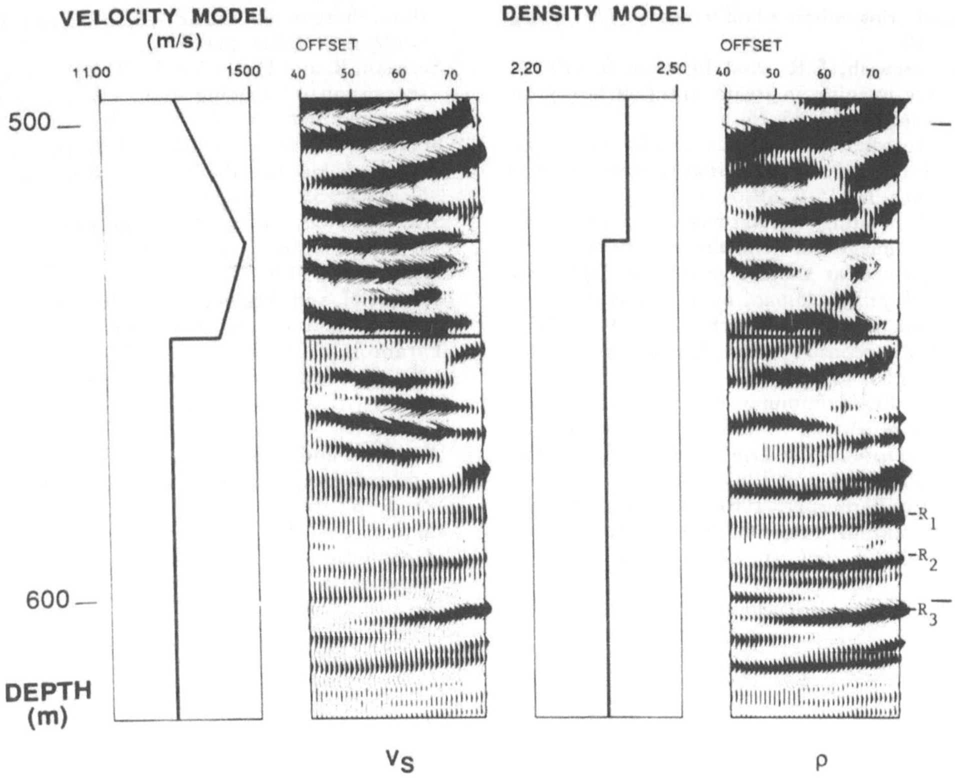


**Fig. 14.** S-wave and density images of the target zone (right sides closest to well) with correspondng background models. The offset is given in metres away from the source well.

## ACKNOWLEDGEMENTS

I wish to thank Elf Aquitaine for permission to publish the field data example. I would also like to acknowledge the assistance in data preprocessing, VSP–CDP calculations and Fig. 6–14 of J. Delvaux, G. Noual and E. Maocec from Elf Aquitaine. R. Greaves and T. Keho from ARCO provided useful comments.

## REFERENCES


Abriel, W.L. (1988). A case of reservoir discovery and description from seismic data, *Society of Exploration Geophysicists/European Association of Exploration Geophysicists Research Workshop on Reservoir Geophysics, Dallas* (abstract).

Beydoun, W.B. and Delvaux, J. (1988). Use of high resolution elastic images from crosshole data for reservoir characterization. Presented at the *Society of Exploration Geophysicists/European Association of Exploration Geophysicists Research Workshop on Reservoir Geophysics, Dallas* (abstract).

Beydoun, W.B. and Mendes, M. (1989). Elastic ray-Born $l_2$ – migration/inversion. *Geophys. J.* **97**, 151–160.

Beydoun, W.B., Blanco, J., Mendes, M. and Noual, G. (1988a). Elastic subsurface images from well seismic data. Presented at the *58th Society of Exploration Geophysicists, Anaheim Expanded Abstract*, paper No. S 6.1.

Beydoun, W.B., Mendes, M., Blanco, J. and Tarantola, A. (1988b). North Sea reservoir description: benefits of an elastic migration/inversion applied to multicomponent VSP data, submitted to *Geophysics*.

Beydoun, W.B., Delvaux, J., Mendes, M., Noual, G. and Tarantola, A. (1989). Practical aspects of an elastic migration/inversion of crosshole data for reservoir characterization: A Paris Basin example. *Geophysics*, December.

Biot, M.A. (1956). Theory of propagation of elastic waves in a fluid-saturated porous solid. *J. Acoustical Soc. Am. 28*, I: Low-frequency range, pages 168–178; II: Higher frequency range, pages 179–191.

Bois, P., Laporte, M., Lavergne, M. and Thomas, G. (1971). Essai de determination automatiques des vitesses sismiques par mesures entre puits. *Geophys. Prospect.* **19**, 42–83.

Carcione, J.M., Kosloff, D. and Kosloff, R. (1988). Full forward modelling in anisotropic and anelastic inhomogeneous media. Presented at the *Workshop Meeting on Seismic Waves in Laterally Inhomogeneous Media III, Prague* (abstract). Also in *Geophysics* **53**, 769–777.

Deans, S.R. (1983). *The Radon Transform and Some of Its Applications.* Wiley-Interscience, New York.

Delvaux, J., Nicoletis, L., Noual, G. and Dutzer, J.F. (1987). Acquisition techniques in cross-hole surveys. Presented at the *62nd Society of Petroleum Engineers Transactions, Dallas*, paper No. 16785.

Devaney, A.J. (1982a). Inverse scattering as a form of computed tomography. *Soc. Photo-Optical Instrum. Eng.: Appl. Math. Modern Optics*, **358**, 10–16.

Devaney, A.J. (1982b). A filtered backpropagation algorithm for diffraction tomography. *Ultrasonic Imaging*, **4**, 336–350.

Domenico, S.N. (1984). Rock lithology and porosity determination from shear and compressional wave velocity. *Geophysics* **49**, 1188–1195.

Foster, D.J. and Mosher, C.C. (1988) Up and downgoing wave field separation for multioffset single level VSP data. Presented at the *58th Society of Exploration Geophysicists, Anaheim, Expanded Abstract*, paper No. S 6.5.

Greaves, R.J. and Fulp, T.J. (1987). Three-dimensional seismic monitoring of an enhanced oil recovery process. *Geophysics* **52**, 1175–1187.

Guiziou, J.L. and Haas, A. (1988). Three-dimensional traveltime inversion in anisotropic media. Presented at the *58th Society of Exploration Geophysicts, Anaheim, Expanded Abstract*, pp. 1089–1092.

Iverson, W.P. (1986). Crosswell logging for acoustic impedances. Presented at the *61st Society of Petroleum Engineers Transactions, New-Orleans*, paper No. 15543.

Justice, J.H., Vassiliou, A.A., Singh, S., Logel, J.D., Hansen, P.A., Hall, B.R., Hutt, P.R., and Solanki, J.J. (1989). Acoustic tomography for monitoring enhanced oil recovery. *The Leading Edge*, February, 12–19.

Macrides, C.G., Kanasewich, E.R., and Bharata, S. (1988). Multiborehole seismic imaging in steam injection heavy oil recovery projects. *Geophysics* **53**, 65–75.

Mora, P. (1988) Wave equation modeling of 3D anisotropic waves in the earth. Presented at the *American Geophysical Union Transactions, Fall meeting, Dec. 5–9* (abstract).

Parsons, R.K. (1986). Estimating reservoir mechanical properties using constant offset images of reflection coefficients and incident angles. Presented at the *56th Society of Exploration Geophysicists, Houston, Expanded Abstract*, paper No. S 16.7

Peterson, J.E., Paulsson, B.N.P. and McEvilly, T.V. (1985). Applications of algebraic reconstruction techniques to crosshole seismic data. *Geophysics* **50**, 1566–1580.

Radon, J. (1917). Uber die Bestimmung von Funktionen durch ihre Integralwerte langs gewisser Mannigfaltigkeiten. *Berichte Sachsische Akademie der Wissenschagten, Leipzig, Math.-Phys. Kl.* **69**, 262–267 (also in English in Deans, 1983).

Resnick, J.R., Ng, P. and Larner, K. (1987). Amplitude Versus offset in the presence of dips. Presented at the *57th Society of Exploration Geophysicists New Orleans, Expanded Abstract*, No. S 10.2.

Robertson, J.D. (1988). Reservoir development using 3D seismic data. Presented at the *SEG/EAEG Research Workshop on Reservoir Geophysics, Dallas* (abstract).

Seeman, B. and Horowicz, L. (1983). Vertical seismic profiling: separation of upgoing and downgoing acoustic waves in a stratified medium. *Geophysics* **48**, 555–568.

Schmitt, D., Bouchon, M. and Bonnet, G. (1988). Full-wave synthetic logs in radially semi-infinite saturated porous media. *Geophysics* **53**, 807–823.

SEG Borehole Source (1988). Borehole Seismic Sources session DEV 2. *58th Society of Exploration Geophysicists, Anaheim, Expanded Abstract*, pp. 139–164.

Stolt, R.H. and Weglein, A.B. (1985). Migration and inversion of seismic data. *Geophysics* **50**, 2458–2472.

Tarantola, A. (1987). *Inverse Problem Theory: Methods for Data Fitting and Model Parameter Estimation*. Elsevier, Amsterdam.

White, J.E. (1965). *Seismic Waves Radiation, Transmission and Attenuation*. McGraw-Hill, New York.

Wu, R-S. and Toksoz, M.N. (1987). Diffraction tomography and multisource holography applied to seismic imaging. *Geophysics* **52**, 11–25.

Yilmaz, O. (1987). *Seismic Data Processing*. Investigation in Geophysics, Vol. 2, (*Doherty, S.M., ed.*) *Society of Exploration Geophysicists*, Tulsa.

# 1

# Use of Seismic Attributes in Reservoir Characterization

*L. Sønneland and O. Barkved*

GECO A.S., Stavanger, Norway

Since the mid-1970s seismic attributes have been used as hydrocarbon indicators. For certain geological provinces these indicators have given encouraging results. Attributes are typically derived from normal-incidence seismic data on a trace-by-trace basis and are presented in a cross-section. It is generally accepted that minor character changes observed in these displays can be related to real geological changes. However, it is not at all obvious how the geological significance of subtle lateral variations observed on a conventional 2D seismic cross-section can be deduced.

A 3D seismic data set provides an ideal basis for relating minor areal variation of seismic amplitudes to a field's stratigraphy in terms of depositional units and rock qualities. Furthermore, by relating seismic attributes to a particular structural interpretation or horizon, the appropriate attribute can be presented as a map to support the correlation of minor areal character or wavefield response changes to geological features.

In this contribution we review the use of conventional seismic attributes in reservoir characterization. Furthermore, we introduce the concept of model-driven attributes as a tool for refining structural interpretation and for stratigraphic applications. The model-driven attributes are derived from the 3D seismic wavefield in a horizon-consistent manner, and are designed to reveal areal variations in reflectivity.

The preconditioning required for the 3D seismic dataset to facilitate the application of the method is addressed, and the application of the model-driven attributes in the interpretation of a 3D seismic dataset is demonstrated. Finally, we point out the relationship between the model-driven attributes approach and image interpretation applied to other disciplines.

## INTRODUCTION

Several case studies have demonstrated the potential of using seismic attributes in stratigraphic interpretation. Since the mid-1970s a number of interpretation procedures have been developed, utilizing the wealth of information available in a reflection seismic data set. The approach presented by Vail *et al.* (1977a,b) in *AAPG Memoir* 26 is recognized as the state of the art. From maps constructed using these procedures the explorationist is able to make interpretations of gross depositional environments and depositional frameworks, recognize the distribution of reservoir quality rocks and develop models of the burial history of subsurface rock units (Hubbard, *et al.*, 1985 and Mitchum and Vail, 1977).

For stratigraphic applications the interpreted horizon can be segmented according to its reflector characteristics, which are not easily observed on the seismic section. Furthermore, the areal variations in reflector character might expose shapes directly relatable to palaeogeographic elements (Brown, 1985).

Lateral variations in amplitude maps from seismic horizons have revealed features directly relatable to stratigraphic events and have been used for the extrapolation of reservoir properties from well control under favourable conditions (Thadani *et al.*, 1987).

Having established the three-dimensional structural subsurface model from the 3D seismic image, further refinement can be gained by evaluating pre-stack data. The amplitude versus offset analysis (AVO) has been used in direct hydrocarbon detection for more than a decade. These methods take advantage of analysing the energy exchanges between elastic wave modes, as described by the Zoeppritz equations, at the elastic interfaces.

## COMPLEX TRACE ATTRIBUTES

Complex trace attributes were first used in signal analysis procedures developed in radio and optical applications for studying instantaneous signal characteristics. These features cannot be analysed by conventional Fourier transforms

*North Sea Oil and Gas Reservoirs—II*

with the signal represented by its amplitude and phase spectrum.

According to Bracewell (1978), the complex trace can be estimated from a signal and its Hilbert transform. Having a real signal function $g(t)$, the corresponding analytical signal is defined as

$$s(t) = g(t) - iG_H(t) \quad (1)$$

where $G_H(t)$ is the Hilbert transform of $g(t)$. Mathematically this represents an analytic continuation of the real function $g(t)$. It is well known that an equivalent polar representation of the complex signal function is:

$$s(t) = A(t)e^{i\theta(t)} \quad (2)$$

where

$$A(t) = (g(t)^2 + G_H(t)^2)^{1/2} = \text{reflection strength} \quad (3)$$

and

$$\theta(t) = \arctan\left(\frac{G_H(t)}{g(t)}\right) = \text{instantaneous phase} \quad (4)$$

The analytic signal can be described as a phasor of length $A(t)$ rotating along the time axis. The angular rate of rotation is

$$fi(t) = \frac{\partial\theta(t)}{\partial t} = \text{instantaneous frequency} \quad (5)$$

The polar representation of the seismic signal allows us to evaluate the seismic trace amplitude and phase characteristics in a sample-by-sample manner, often referred to as instantaneous attributes. Taner *et al.* introduced these attributes to geophysical applications in 1979.

One important consideration is that an exact estimation of the instantaneous attributes requires an infinitely broad frequency band, as the perfect spectral estimation requires an infinite time window (Gabor, 1946; Rihaczek, 1968). Based on this, Bodine (1984) defined an extension of complex trace attributes: *response phase* and *response frequency*. These parameters define the dominant properties of the waveform using the instantaneous attributes in the vicinities of the local peak amplitudes.

An additional attribute is apparent polarity, defined as the sign of the seismic trace controlled by the intensity of the reflection strength at the nearest local minimum.

In addition to the complex trace attributes, quantities like polarity, continuity, acoustic impedance and P and S velocities are commonly referred to as seismic attributes. Geophysicists with experience from a area are able to recognize specific characters in the seismic section and make use of them in the interpretation. To capture these characters in a kind of expert system, and to make use of them in computerized interpretation, has been difficult, but during the last decade case studies and systematic analyses have, to a certain degree, succeeded in establishing some relations between quantities derived from the seismic section and their geological significance. Even so, most of these relations are only indicative, and calibration with observations from wells is needed to draw any firm conclusion.

Table I lists the attributes commonly referred to in the literature on exploration and development work.

**Table I.** Geological significance of seismic attributes used in seismic stratigraphy

| *Attributes* | *Geological significance* |
|---|---|
| Areal association | Geological setting<br>Sediment source<br>Genesis of depositional sequence<br>Gross depositional environment |
| Reflection geometry<br>External form | Geological settings<br>Sediment source<br>Genesis of depositional sequence<br>Gross depositional environment |
| Internal configuration | Bedding pattern<br>Depositional process<br>Erosion patterns and paleotopography<br>Direction of sedimentary transport |
| Continuity | Bedding continuity<br>Depositional processes |
| Amplitude | Lithological contrasts<br>Bedding continuity<br>Bed spacing<br>Gross porosity |
| Inst. frequency | Bed thickness<br>Lithogic contrast<br>Fluid content |
| Interval velocity | Gross lithology<br>Gross porosity |
| Reflection strength | Lithologic contrast<br>Bedding continuity<br>Bed spacing<br>Gross porosity<br>Continuous or massive reflectors vs. composites |
| Inst. phase | Bedding continuity |
| Polarity | Sense of (lithologic) contrast<br>Polarity |
| Waveform | Lateral and vertical lithological variation<br>Porosity<br>Bed spacing and geometry |
| Amplitude vs. Offset | Fluid content<br>Lithology |

## ATTRIBUTE INTERPRETATION

The interpretation problem can in general be formulated as to describe physical objects in a scene from images of the scene. The backscattered seismic wavefield can be regarded as a 3D image of a scene of physical subsurface objects. These physical objects *do not* in general produce images that possess one unique characteristic. To help in the interpretation process—extracting facts about the objects—we need additional discriminating properties. One approach is to introduce attributes or classifiers to aid the interpretation. This is well known from other disciplines like remote sensing and medical diagnostic applications.

The challenge is to design a set of attributes with discriminatory properties relevant for the interpretation of a particular scene. The optimal set is considered to be the set requiring the minimal number of different attributes to identify that particular scene. This indicates another desirable feature with attributes: compaction of information. For the geoscientist, attribute-based interpretation should offer a more efficient representation of the data and thereby provide a greater potential for taking advantage of the wealth of information in the data. Our approach to attribute interpretation is outlined in Fig. 1. It is important to notice that attribute interpretation supplements conventional structural interpretation. In fact, as we will point out, we will take advantage of the existence of the structural interpretation.

The learning step is the first step in the attribute interpretation procedure. Here the interpreter should

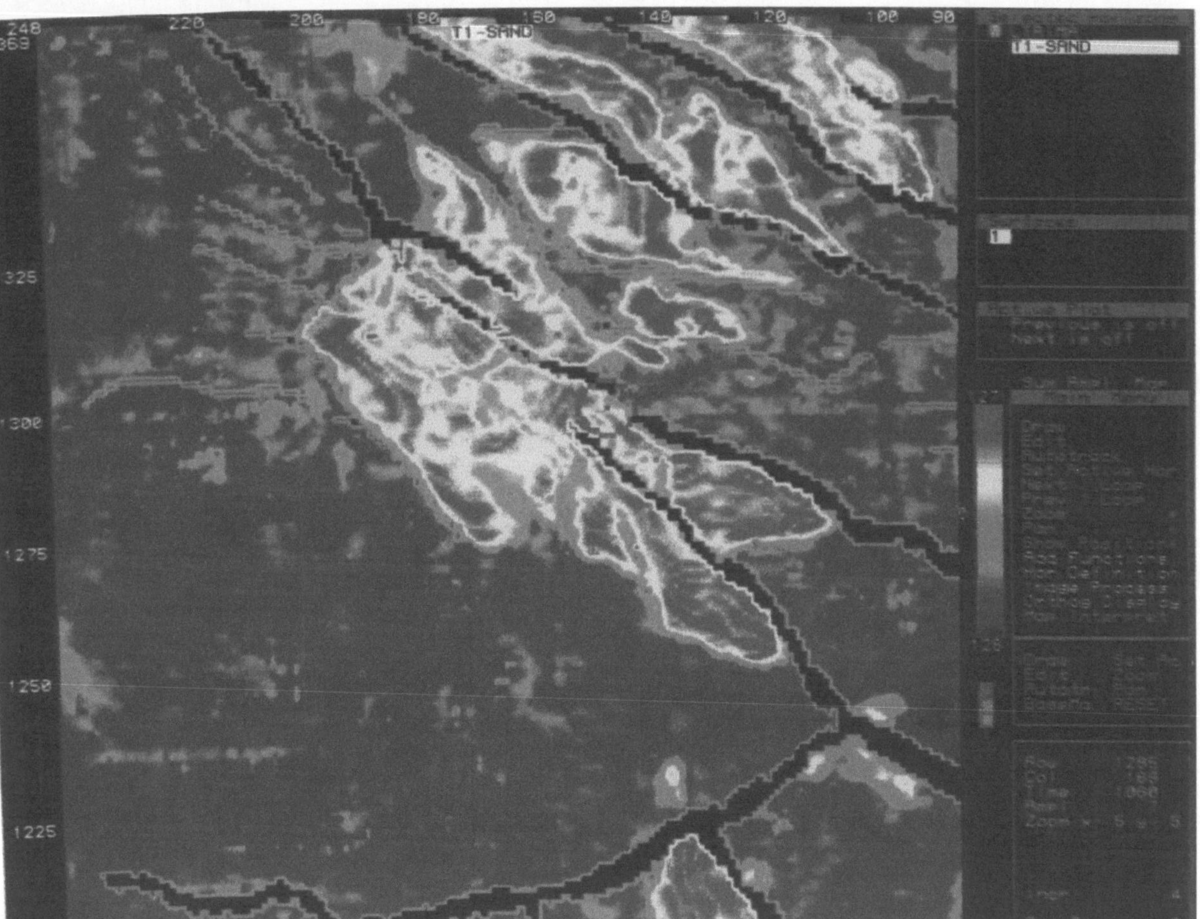


**Plate 4.** The intensity map.

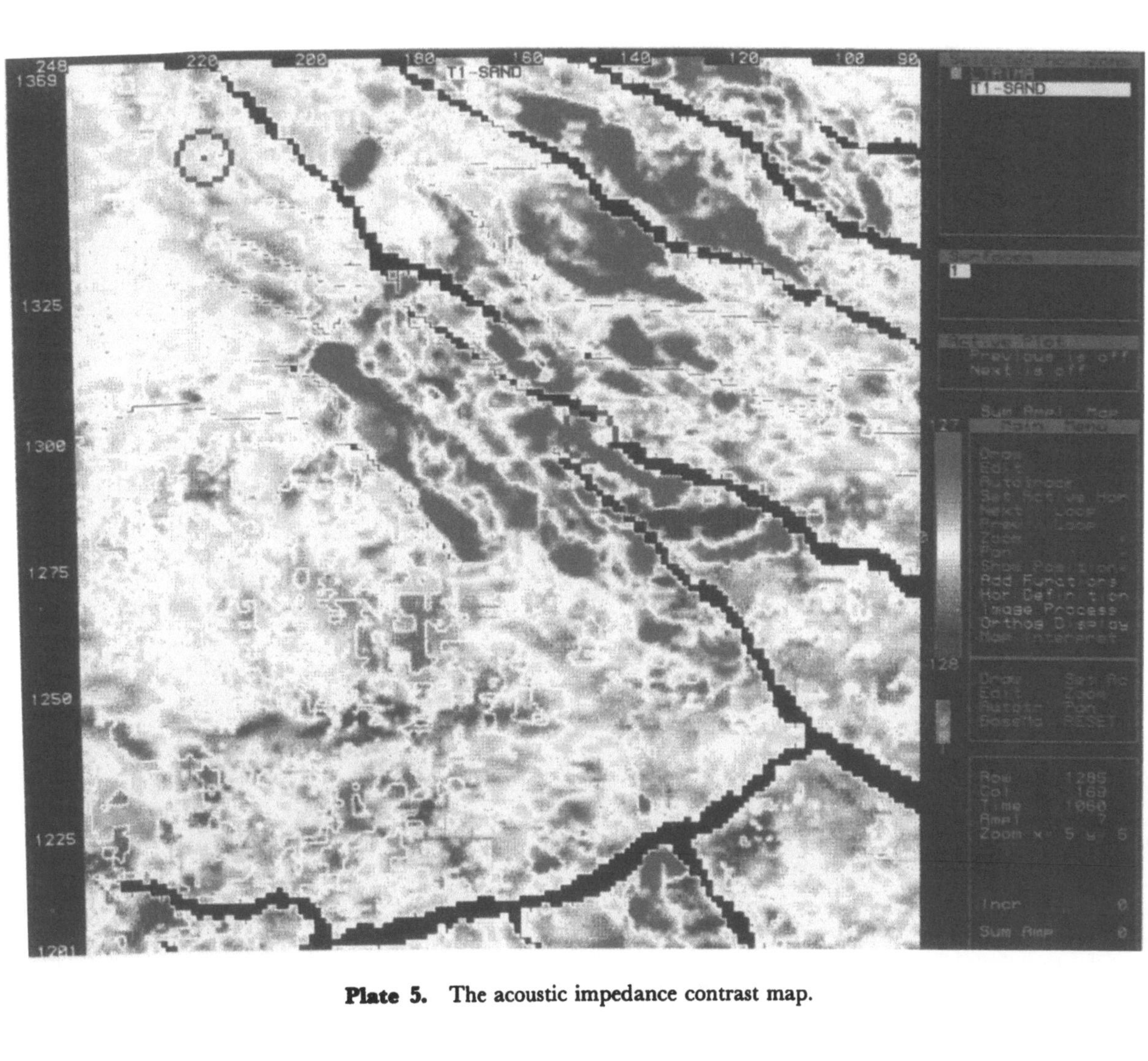


**Plate 5.** The acoustic impedance contrast map.

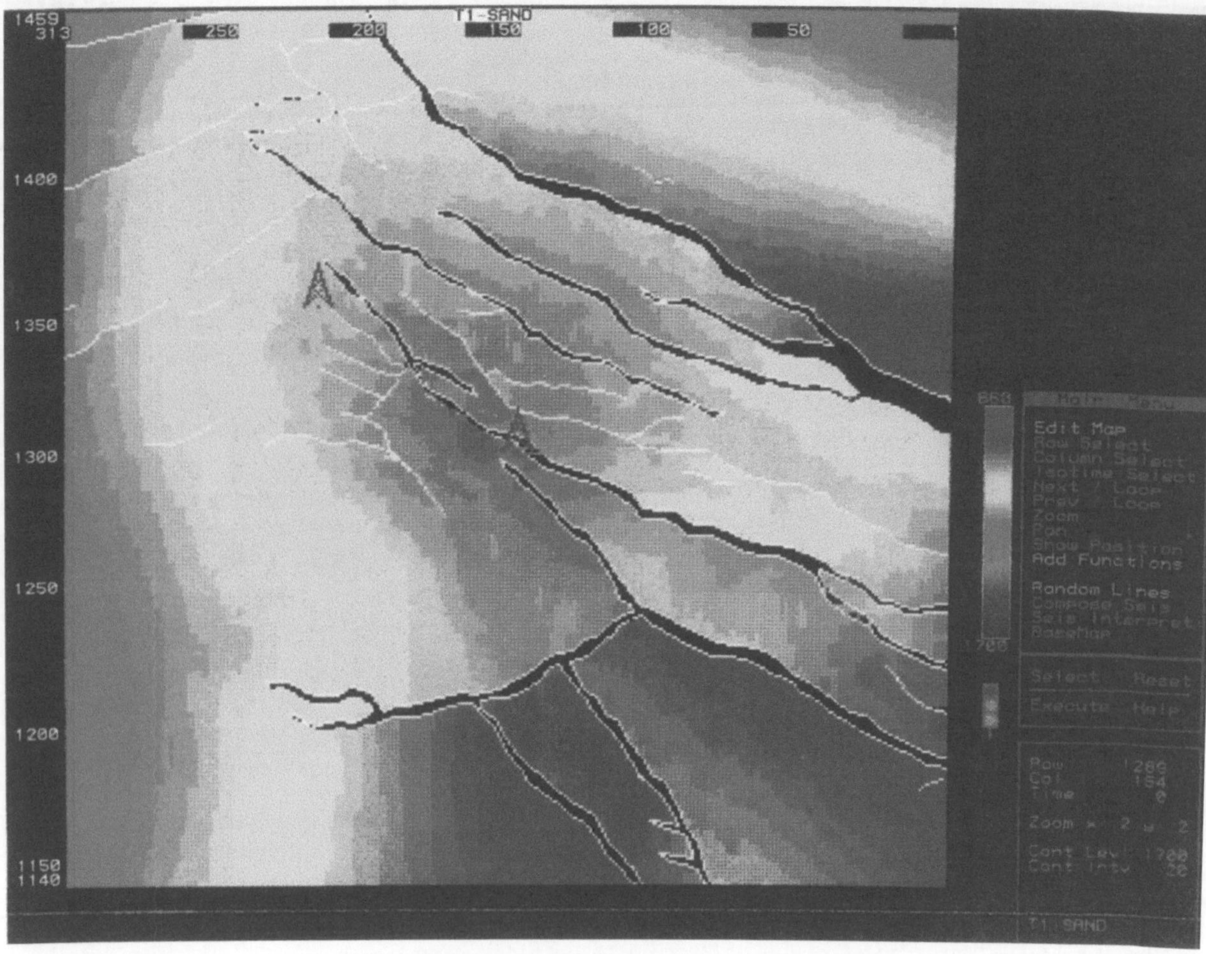


**Plate 6.** The structural map.

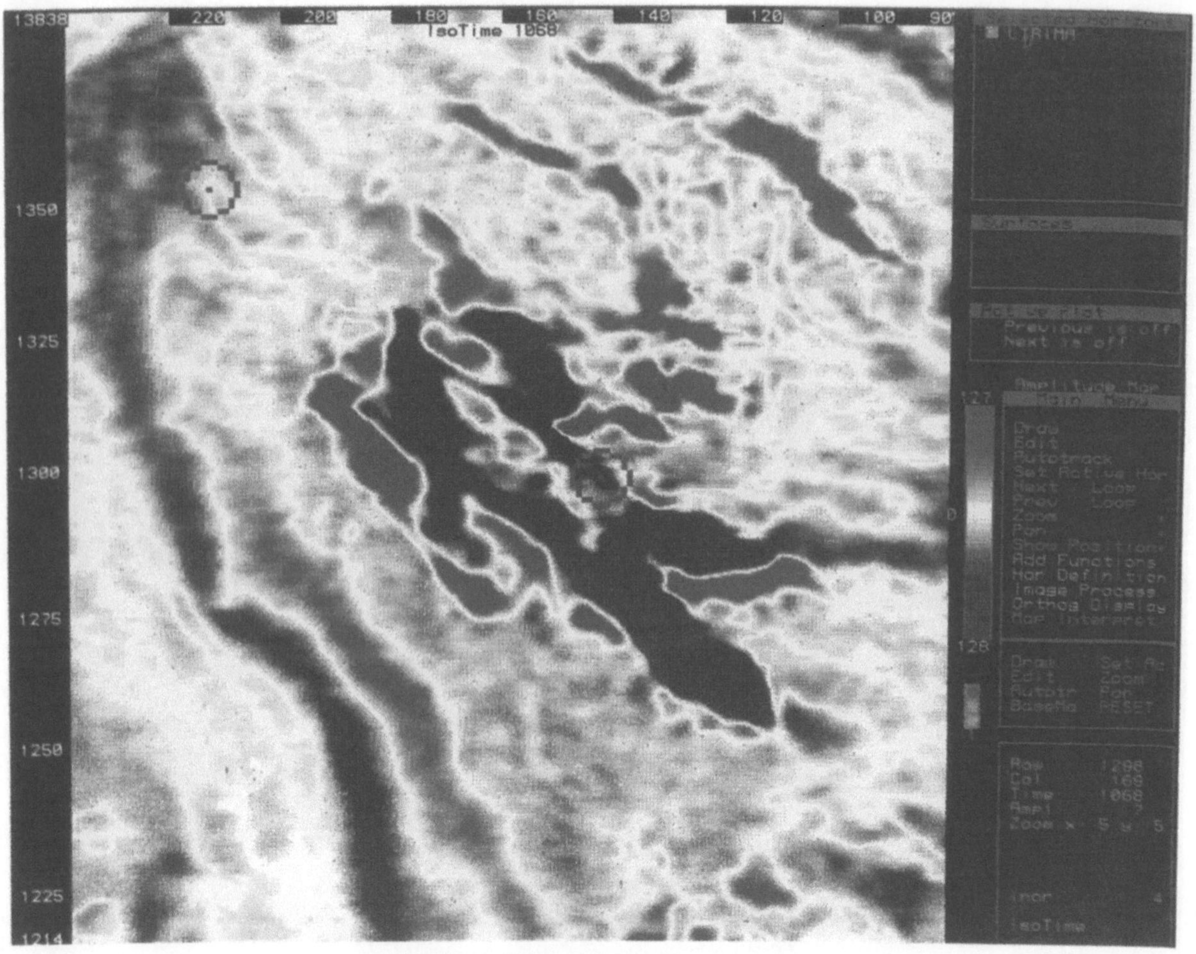


**Plate 7.** The isotime seismic section.

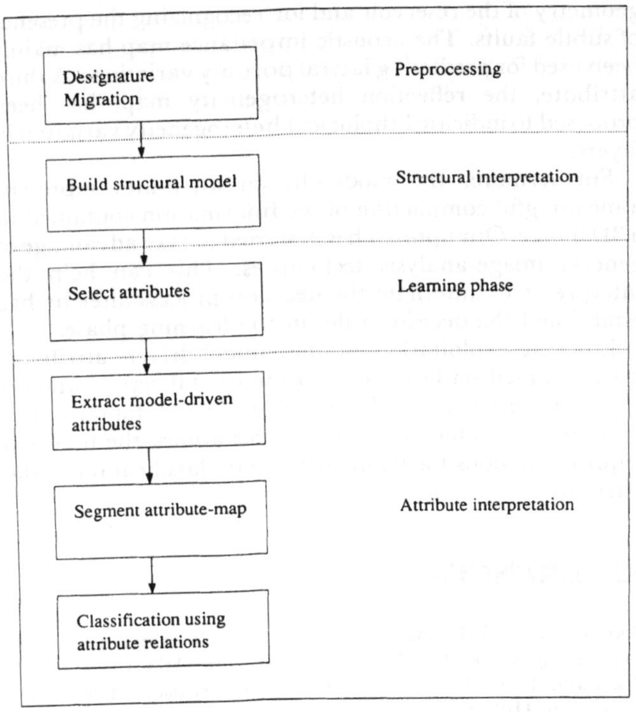


**Fig. 1.** Attribute interpretation procedure.

select and design the attributes most relevant for his particular problem. Simulated seismic response or modelling can be used to check the discriminating properties of the attribute set. Experience from interpretations of similar scenes is another learning option. Ultimately, well control and borehole seismic can provide further insight into the discriminating "power" of a set of attributes.

We will place emphasis on a new class of attributes—structural model-driven attributes.

If ideal designature processing and migration processing is done, we will assume that the seismic data $S$ represent the 3D reflectivity function $R$ of the subsurface given by

$$S = \mathbf{U}_{(x,y,z)\in V} R(x, y, z) \qquad (6)$$

where $x, y, z$ indicate voxels in the 3D sampling grid, $V$, and the union, $\mathbf{U}$.

Let the structural interpretation map of a horizon $i$ in $S$ be defined as

$$z_i = f_i(x, y) \qquad (7)$$

Then $A_{\mathbf{T}}$ is defined as a model-driven attribute map related to horizon $i$ if

$$A_{\mathbf{T}}(x, y, z_i) = \mathbf{T}[R(x, y, z_i)] \qquad (8)$$

where $\mathbf{T}$ represents an arbitrary transform from image-space (data-space) to attribute-grid, i.e. a transform from 3D to 2D. Obviously this is a wide class of attribute-transform taking advantage of the interpreter's first iteration in identifying the objects in the subsurface scene—*their range* from the reference datum represented by the structural map. Since our acoustic (or elastic) probe has a finite bandwidth, our interpretation is performed on a finite bandwidth version of the reflectivity function $R$.

The objects we try to identify are partly transparent with respect to the acoustic probe. We therefore have the option of taking advantage of "internal" reflections—i.e. reflections between two successive interpreted horizons or ranges—to add discriminating "power" to our analysis.

## MODEL-DRIVEN ATTRIBUTE MAPS

The intensity distribution over the interpreted horizon, $z_i = f(x, y)$, has proven to be a valuable attribute in our analysis. Our definition of the reflection-intensity map is:

$$A_E(x, y, z_i) = \int_{z_i}^{z_i + \Delta z} |R(x, y, z)|\, dz \qquad (9)$$

A special case of this attribute, $\Delta z = 0$, gives the amplitude map. But by working with a finite-bandwidth version of the reflectivity function, such amplitude maps might create spurious attribute properties.

In our application, the intensity maps have proved useful in the detection of subtle faults and buried channels. Furthermore, bright-spot analysis and texture analysis benefit from these maps.

Acoustic impedance contrast maps are helpful in analysing trends in impedance variations.

$$A_I(x, y, z_i) = e^{\int_{z_i}^{z_{i+1}} R(x,y,z)\, dz} \sim \frac{\rho v(x, y, z_{i+1})}{\rho v(x, y, z_i)} \qquad (10)$$

where $\rho v$ is the acoustic impedance. Since acoustic impedance variations can be related to porosity variations, these attribute maps are important in reservoir characterization.

The reflection heterogeneity map is designed to reveal the heterogeneous nature of any internal reflector pattern in a given volume segment. Indication of shale layers in sand bodies is an important application of this attribute. The definition of the reflection heterogeneity map is

$$A_H(x, y, z_i) = \int_{z_i}^{z_{i+1}} \sqrt{\left[1 + \left(\frac{dR(x, y, z)}{dz}\right)^2\right]}\, dz \qquad (11)$$

Examples of some attribute maps generated for a particular survey are displayed in Plates 4 and 5 which represent the intensity and acoustic impedance contrast maps, respectively. The structural map driving the definitions of these attributes is shown in Plate 6. The attribute maps are expanded in the area of interest around the wells.

On the intensity map, the bright spots are easily identified. Furthermore, this map helps to determine the delineation of the reservoir where this is not controlled by the fault system. The southernmost fault, running north-west into the major bright spot area on the intensity map, might be extended through the reservoir as indicated by the weak intensity pattern connecting the fault with the fault system on the north-west side of the reservoir. On the impedance contrast map, the trend is an increase in impedance in a north-westerly direction from the well indicating a decrease in porosity in this direction. For comparison we have included an isotime section (Plate 7) through the bright-spot area to illustrate the difference in information content relative to the attribute maps.

## SYSTEM ENVIRONMENT

We have demonstrated the additional value that attribute maps can bring to our interpretation. The discrimination of the subsurface objects' attribute features is reflected in the segmentation phase in our approach (Fig. 1). Segmentation means dividing the attribute map into regions of common characteristics. Examples of this process are the delineation of the reservoir using the increase in intensity (intensity threshold) as a criterion for the extension of the reservoir, or modification of the fault pattern in the reservoir based on the weak snake-like intensity behaviour (Plate 4). The segmentation might also involve colour-code calibrations as indicated on the acoustic impedance contrast map. By choosing the colour-code at the well, the user can relate the impedance contrast trend there relative to the rest of the reservoir using the colour table.

It is an important requirement of the system environment that the interpreter is allowed to segment interactively the attribute maps into regions, as indicated here.

As can be appreciated, attribute interpretation will result in an increased number of maps to keep track of. The inventory of such maps or grids and their internal relationship is a key requirement in the system environment. Our solution is a relational database structure as outlined in Fig. 2. The attribute maps are related in a hierarchical fashion to the structural map "driving" the extraction of the attributes.

## CONCLUSIONS

The use of model-driven attributes has improved our ability to discriminate relevant features of a subsurface object. The intensity map has been used for delineating the

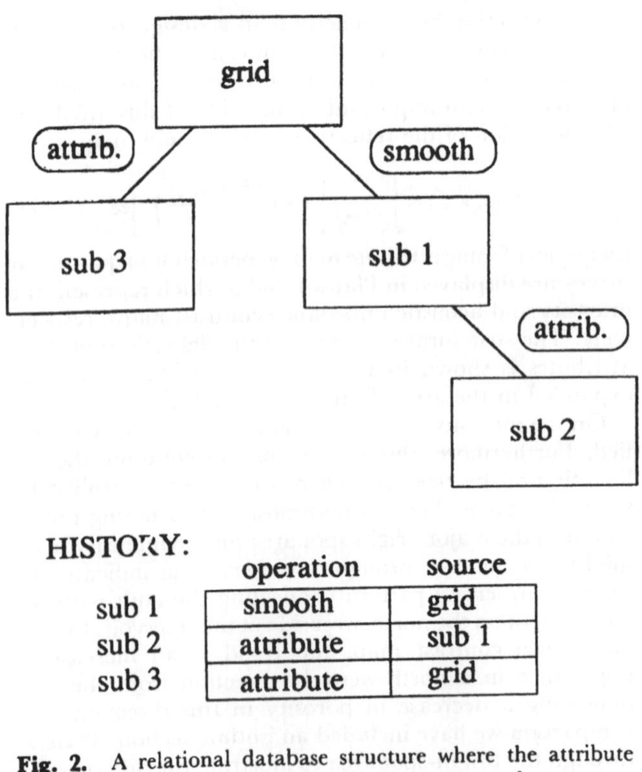


**Fig. 2.** A relational database structure where the attribute maps are related in a hierarchical grid pattern to the structure maps.

geometry of the reservoir and for recognizing the presence of subtle faults. The acoustic impedance map has mainly been used for predicting lateral porosity variations. A third attribute, the reflection heterogeneity map, has been proposed to indicate lithological heterogeneity variation in layers.

Furthermore, the model-driven attributes represent a meaningful compaction of the information contained in a 3D image. Our approach is designed to take advantage of general image-analysis techniques. This can help the interpreter to automate the decision process after he has established the decision rules in the learning phase.

Efficient exploitation of the model-driven attributes maps is based on fast access to the digital representation of the measurements. In addition to the functionality generally available on seismic workstations, the user will require functions for segmention and classification of the attributes.

## REFERENCES

Bodine, J.H. (1984). Waveform analysis with seismic attributes. *Proceedings of the 54th. Annual SEG* Meeting, Atlanta.

Bracewell, R.N. (1978). *The Fourier Transform and its Applications.* McGraw-Hill, New York.

Brown, A.R. (1985). The role of horizontal seismic sections in stratigraphic interpretation. In: Berg, O.R. and Woolverton, D.G. (eds.), *Seismic Stratigraphy II – an Integrated Approach to Hydrocarbon Exploration* pp. 37–47 AAPG Memoir 39.

Gabor, D. (1946). Theory of communications. *JIEE* **95,** 429–457.

Hubbard, J.R., Pape, J. and Roberts, D.G. (1985). Depositional sequence mapping as a technique to establish tectonic and stratigraphic framework and evaluate hydrocarbon potential on a passive continental margin. In: Berg, O.R. and Woolverton D.G. (eds), *Seismic Stratigraphy II – an Integrated Approach to Hydrocarbon Evaluation*, pp. 79–91. AAPG Memoir 39.

Mitchum, R.M. Jr. and Vail, P.R. (1977). Seismic stratigraphy and global changes of sea level, part 7: Seismic stratigraphic interpretation procedure. In: Payton, C.E. (ed), *Seismic Stratigraphy – Applications to Hydrocarbon Exploration*, pp. 135–143. AAPG Memoir 26.

Rihaczek, A.W. (1968). Signal energy distribution in time and frequency. *Trans. IEEE Information Theory* **IT–14**, 369.

Thadani, S.G., Alabert, F. and Journel, A.G. (1987). An integrated geostatistical/pattern recognition technique for characterization of reservoir spatial variability. *Proceedings of the 57th Annual SEG Meeting, New Orleans.*

Taner, M.T., Koehler, F. and Sheriff, R.E. (1979). Complex trace analysis. *Geophysics* **44**, 1041–1063.

Vail, P.R., Mitchum, R.M. Jr., Todd, R.G., Widmier, J.M., Thompson, S.III, Sangree, J.B., Bubb, J.N. and Hatleid, W.G. (1977a). Seismic stratigraphy and global changes of sea level. In: Payton, C.E. (ed.), *Seismic Stratigraphy–Applications to Hydrocarbon Exploration*, pp. 49–212. AAPG Memoir 26.

Vail, P.R., Todd, R.G. and Sangree, J.B. (1977b). Seismic stratigraphy and global changes of sea level, part 5: Chronostratigraphic significance of seismic reflections. In: Payton C.E. (ed), *Seismic Stratigraphy–Applications to Hydrocarbon Exploration*, pp. 99–116. AAPG Memoir 26.

# 2

# Experiences Gained using a Seismic Inversion Method for Detailed Reservoir Studies

*E. Berg, I. Brevik and A.T. Buller*

Statoil's Research & Development Centre, Trondheim, Norway

This chapter contains a summary of experiences gained while testing a high-resolution inversion algorithm designed for the estimation of reflection coefficient series from migrated seismic data. The algorithm is based on a one-dimensional seismic response model. When coupled to a closely and regularly spaced series of reflection coefficients obtained from well logs, the algorithm permits the stable estimation of reflection coefficient series away from a well with a potential resolution of about 5 m at depths of up to 2500 m.

The problem of non-uniqueness, which is typically associated with such high-resolution estimates using standard inversion algorithms, is largely overcome by the incorporation of the well log data and the addition of bandlimited noise. The validity of the basic assumptions emodied by the algorithm are dependent, however, on the (geological) structural complexity contained in the original datasets. Furthermore, the reliability of the estimates will generally decrease as a function of increasing distance away from the well location.

Tests have been carried out in areas close to wells for three North Sea fields that display varying degrees of stratigraphic and structural complexity. Where the degree of structuration is relatively modest and the seismic data are good, features displaying smooth and consistent variations from trace to trace away from a well can be interpreted with confidence. The stratigraphic/sedimentological significance of these features can then be determined by comparison with the well data. Sudden vertical shifts in these features may also be indicative of small-scale faults that are beyond the resolution of conventional seismic. Where the quality of the seismic data is poor, owing to structural complexity (fault zones; steeply dipping fault-bounded compartments) and/or poor processing, the estimates are rendered unstable and reliable interpretations cannot be made.

Finally it should be pointed out that the degree of achievable resolution under otherwise optimum conditions is also dependent upon the presence of sufficient acoustic impedance contrasts throughout the reservoir interval of interest.

## INTRODUCTION

Optimum field development and the maximization of hydrocarbon recovery are dependent upon the accuracy of the geological/petrophysical model and its effective translation into a reservoir engineering model. The geological modelling of deeply buried and complex offshore reservoirs is, however, an extremely difficult business. This is because many of the geological heterogeneities that should be incorporated in the model (e.g. intricate sand-body architectures, discontinuous shales and cemented horizons, small-scale faults) are not only below seismic resolution, but also are inadequately sampled by wells that may be very widely spaced, especially during phases of early development.

An increasing awareness of this problem and its potential impact on recovery has led to the emergence of two major research themes: the prediction of heterogeneity distributions in interwell volumes using probabilistic methods (e.g. Haldorsen *et al.*, 1987; Gundesø and Egeland, this volume); improvements in the acoustic imaging of reservoirs using seismic inversion techniques (e.g. Gelfand *et al.*, 1987; Poggiagliolmi, 1980; Van Riel *et al.*, 1987) and advancements in borehole geophysics (e.g. Balch and Lee, 1982; Krug *et al.*, 1981; Landgren and Grubb, 1983).

The subject of this paper is related to the second theme. It contains a summary of experiences gained while testing a high-resolution inversion algorithm designed for the estimation of reflection coefficient series from migrated seismic sections (the so-called HIGHRES method). The algorithm is based on a one-dimensional (1D) seismic response model that is a modification (by Brevik and Berg, 1986) of the one originally developed by Ursin and Holberg (1985). When coupled to a closely and regularly spaced series of reflection coefficients derived from well logs, the algorithm permits the stable estimation of

*North Sea Oil and Gas Reservoirs—II*

reflection coefficient series away from a well. The underlying tenet is that a seismic trace can be regarded and modelled as a convolution between a pulse and a reflection coefficient series, together with the addition of bandlimited noise. All of the estimates are carried out using a least-squares method.

## PROCEDURE

The first step in the HIGHRES procedure is the selection of an appropriate time window (usually 0.2 s two-way travel time (TWT)) from a migrated seismic section which passes through a well (Fig. 1). A pulse can then be estimated from a series of closely spaced reflection coefficients derived from well logs (density and sonic) and the seismic trace that corresponds to the well location. The interval between the reflection coefficients is kept at a constant and is usually taken as $2 \times 10^{-3}$ s (2 ms). Assuming that the estimated pulse does not vary significantly within the time window, the next step is to use this pulse to estimate reflection coefficient series for all of the traces making up the section.

In general, the reflection coefficient amplitudes when calculated over such short intervals are thought to be stable. This is because a set of estimated filter coefficients when convolved with the error series (the objective function in the least-squares sense) produces bandlimited (coloured) noise. As the method is operating beyond the bandwidth of the seismic signal, reliable estimations can only be made when the reflection coefficient series derived from the well logs is used as the initial values for the next trace away from a well.

The estimated series for this trace is then used as a basis for calculation for the next trace, and so on. In other words the procedure is performed trace by trace, starting at the well and then working outwards. The essential idea behind this is that small perturbations in the seismic trace are explained by small perturbations in the reflection coefficient series.

The product, a section consisting of estimated reflection coefficient series for each trace, serves as the foundation for the final step—the computation of a section composed of estimated, relative acoustic impedance profiles (Fig. 1). This is used for the interpretation, which involves the determination and delimitation of features displaying smooth and consistent variations from trace to trace away from the well. The gross stratigraphic/sedimentological significance of these features can be unravelled by comparison with the well data, preferably through a core–well log calibration exercise. Abrupt vertical shifts in the position of these features may be indicative of small-scale faults.

## FIELD TEST CASES

The HIGHRES method has been tested in areas close to wells for three North Sea fields. These fields were carefully selected so that they represented a variety of commonly occurring geological situations of varying degrees of complexity. The fields in question are referred to below as Fields A, B and C.

### Field A

The objective of this test was to see whether the method could be used for the detection of lithostratigraphic units and small-scale faults away from a single well (well $\alpha$). The data set consisted of 7 parallel lines taken from a 3D migrated seismic survey and covering a rectangular area of 875 m × 450 m. Two time windows were selected; one within the reservoir interval (2370–2500 ms TWT) and one above (2200–2320 ms TWT). The reservoir interval in well $\alpha$ had excellent core and log coverage, whereas the upper interval had only been logged.

*The upper window* (Plates 8 and 9)

Four lithostratigraphic/relative acoustic impedance units (hereafter referred to as "lithoimpedance" units) have

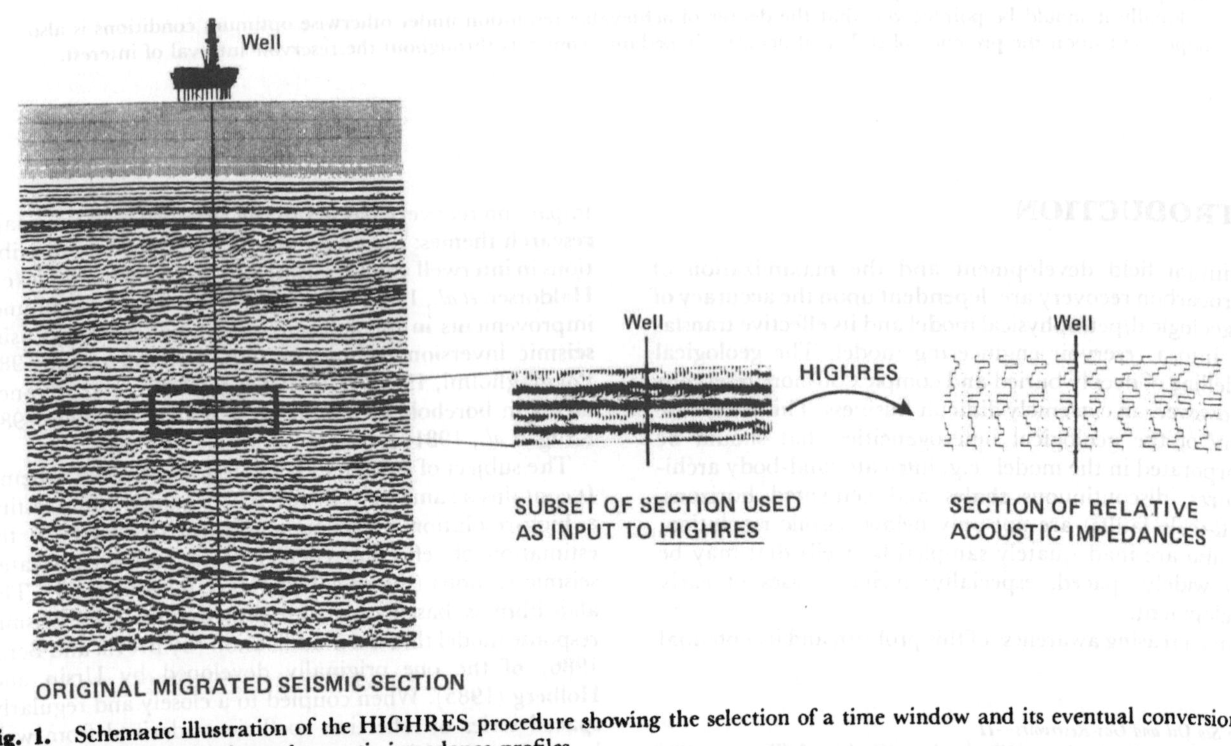


**Fig. 1.** Schematic illustration of the HIGHRES procedure showing the selection of a time window and its eventual conversion to a section composed of estimated acoustic impedance profiles.

been defined from an impedance profile derived from the density and sonic logs in the well, supplemented by cuttings descriptions (Plate 8). The upper unit (which extends above $S_1$ to the top of the time window) is a shaly sequence with two prominent limestone horizons towards the base (dark blue). This is underlain by a high impedance marly/limestone unit (light blue) between $S_1$ and $S_2$, and a low impedance shaly unit between $S_2$ and $S_3$. The lowest unit, which extends from $S_3$ downwards through to the base of the time window (at 2320 ms TWT), consists of sandy shales represented by intermediate impedances. The interpreted distribution of these units away from the well is shown in the upper HIGHRES relative acoustic impedance section in Plate 9. (These sections will hereafter be referred to as "HIGHRES" sections).

All four lithoimpedance units are traceable right across the section, with the marly unit being the most pronounced. The lower three are downfaulted and dip slightly towards the east. The positions and offsetting (between 2 to 12 ms TWT) of the small-scale faults are based on abrupt vertical and lateral changes in the impedance profiles. This distribution, which is consistent with the larger-scale structure of the field, is less evident from the original seismic section.

The upper unit, which is unfaulted, appears to onlap the $S_1$ (unconformable) boundary. This is particularly apparent from the attitude of the discontinuous, high-impedance limestone horizons. Up to five limestones are decipherable with thicknesses ranging from 2 to 6 ms TWT.

### *The lower window* (Plates 8 to 10)

Two relative acoustic impedance profiles are shown in Plate 8 for the lower time window, which covers the reservoir interval ($S_4$ to $S_7$) and portions of the over- and underlying shales. The profile on the left (which continues unbroken through to the upper window) was based on the density and sonic logs provided by the contractor. This, however, could not be used because the logs, although corrected to a certain extent, were still badly affected by caving in the $S_4$ to $S_5$ interval. It was therefore necessary to undertake some further (manual) corrections while taking guidance from core–log correlations in uncaved intervals. The revised acoustic impedance profile (shown to the right) was then used as an input for the HIGHRES test.

The reservoir interval has been subdivided into three major lithoimpedance units which are easily correlated to the revised profile. The lowest unit ($S_6$ to $S_7$), shown in yellow, has well-defined upper and lower boundaries and consists of a thick sequence of low-impedance, porous sandstones. Overlying this is a much thinner unit shown in red ($S_5$ to $S_6$). This comprises fine-grained micaceous sandstones and elicits a highly distinctive, blocky impedance response. The uppermost unit ($S_4$ to $S_5$ ) is lithologically complex. Nevertheless, it has been possible to subdivide it further into two lithoimpedance subunits. The "green" subunit represents rapidly alternating sequences of mudstones, coals and thin sandstones. These compositely yield higher impedances than the "orange" which are dominated by thicker, cleaner sandstone bodies.

The HIGHRES test for the lower window was initially carried out for the section passing through well α (Plate 9, lower), but was then extended to a number of parallel lines to the north and south (see Fig. 2 for locations). As can be seen (Plate 9, lower), the reservoir interval has been interpreted right across the HIGHRES section, apart from an area of instability between CDPs 622 and 632 and the time interval 2370 to 2400 ms TWT. This has been possible because the *general* trends of increasing or decreasing impedance are maintained from trace to trace and can be re-established across minor faults as the major unit boundaries are fairly conspicuous. The green and orange subunits within the uppermost unit display a more complex pattern that is believed to reflect the anticipated style of lithoimpedance architecture. The positioning of the faults, with offsets of a few milliseconds, is again based on vertical shifts in the profiles and the juxtaposition of contrasting impedances. The two on the left are somewhat uncertain because they impinge upon the area of instability.

The next HIGHRES section some 800 m to the south (line 3A) essentially displays the same picture. There are, however, slight differences in the lateral positions of the faults and as expected the configuration of the orange and green subunits. The original seismic data set for this section is shown for comparison (Plate 10, upper). The coarse pattern provided by the trend lines gives an imprecise impression that small-scale faults may be present as well as suggesting a dip to the east.

The consistent appearance of the small-scale normal faults along all of the seismic lines 1A to 7A allowed them to be mapped for both time windows. The isochron map for the upper window (Fig. 2, upper) is based on the $S_2$ boundary with a contour interval of 4 ms. The slightly sinuous faults display a N–S trend with the fault-bounded blocks clearly dipping to the east. Only the three more easterly faults were mapped for the lower time window, as the two westerly faults are masked by the area of instability. Again, the fault distribution on the isochron map of the $S_5$ boundary (Fig. 2, lower) shows a N–S trend; the fault

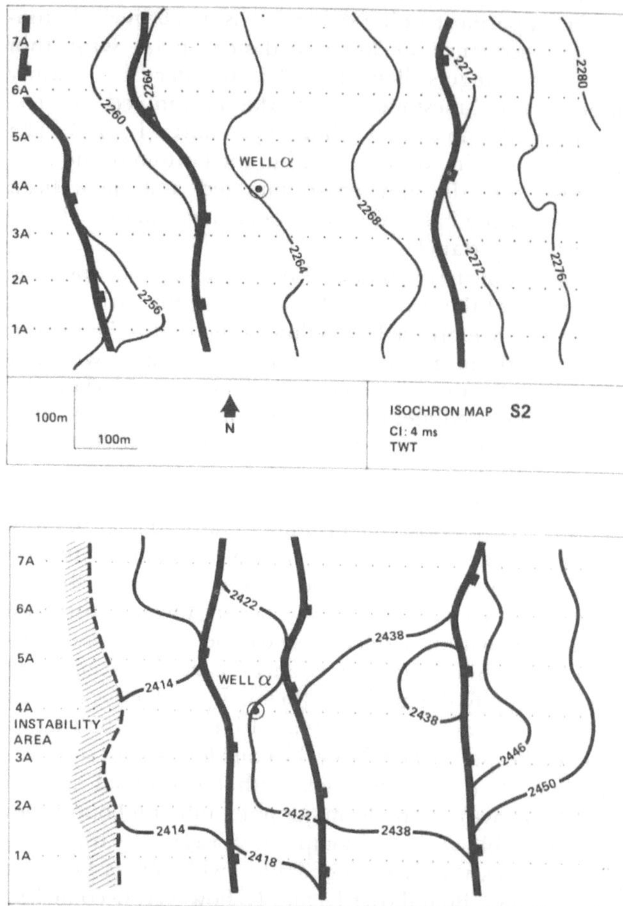


**Fig. 2.** Structural isochron maps based on the unit boundaries $S_2$ and $S_5$ (Plate 9). The maps have been made from HIGHRES interpretations for the seven parallel seismic lines 1A–7A.

blocks, however, are much flatter. It is also tempting to suggest that two of the faults extend from the lower to the upper time window. This is indicated on Plate 9 by the broken lines.

## Field B

The data set for the second test contained a portion of a seismic line passing through two wells ($\beta$ and $\gamma$) located towards either end (Plate 12). The distance between the wells is approximately 1.5 km.

The objectives of this test were twofold: (a) to produce a geoseismic reconstruction of the lithoimpedance stratigraphy between the two wells for a HIGHRES section based on data from well $\beta$, thereby giving us the opportunity of predicting the impedance stratigraphy for well $\gamma$ which, at this stage, had not been revealed; and (b) to use the data from the originally blind well $\gamma$ to produce another HIGHRES estimate back over the *same* seismic section in order to verify the sensitivity and stability of the method and the reproducibility of the first geoseismic interpretation. The time window extends from 1546 to 1692 ms TWT.

### *Test based on well* $\beta$ (Plates 11 and 12)

Five major lithoimpedance units (Plate 11) have been determined from a comparison between the well impedance profile and core and cuttings descriptions. Most of the units are composite (i.e. they incorporate more than one lithofacies) and have well-defined boundaries that correspond to strong impedance contrasts.

The uppermost unit (above $L_1$) is a relatively homogeneous sequence of low-impedance shales with thin limestone stringers. Beneath this is an interval of mainly impermeable limestones (hence the high impedance responses) with subordinate shales and marls ($L_1$–$L_2$). The three lower units, which extend to the bottom of the time window, are all sandstone-dominated and have been selected on the basis of their progressively higher (background) impedances.

The first of these sandstone units ($L_2$–$L_3$) occupies the upper half of the reservoir. Relatively slight variations in the impedance profile have allowed the unit to be broken down into three sandstone subunits (shc vn in brown, light yellow and orange), each of which is underlain by finer-grained sediments. The lower half of the resevoir is covered by unit $L_3$–$L_4$ and contains poorer-quality sandstones with carbonate-cemented horizons less than 1m thick. Although the gas–oil contact is not shown on the figure, it coincides with a prominent limy zone at the top. This zone is actually made up of two carbonate-cemented horizons that are recognizable on the impedance profile. The oil–water contact (OWC) occurs towards the base of the unit within a thicker sandstone association. Conspicuous, thin, high-impedance carbonate-cemented horizons typify the final unit (below $L_4$), which continues right to the base of the interval of interest at 1650 m.

The interpreted HIGHRES section for this test is shown in Plate 12 (lower) together with the original seismic section (above). All of the major units and unit boundaries have been followed with comparative ease away from well $\beta$ and across the two minor faults (with offsets between 5 and 10 ms). Boundaries $L_3$ and $L_4$ however, become less certain in the vicinity of well $\gamma$ (CDP 942), and the interpretation of $L_3$ was particularly difficult.

The four limestone horizons within the uppermost unit (above $L_1$) appear to be laterally discontinuous and are replaced by others coming in at different levels. Below this the $L_1$–$L_2$ interval, with its characteristic twin-peaked impedance amplitudes, thins slightly towards well $\gamma$. It also truncates some of the underlying $L_2$–$L_3$ unit towards the right of the section. Nevertheless, the three sandstone subunits with their intervening silty horizons seem to be traceable throughout, even though they have a tendency to pinch and swell.

The carbonate-cemented horizons within the two lower sandstone units ($L_3$–$L_4$ and below $L_4$) are also discontinuous. Those occurring within $L_3$–$L_4$ seem to intensify in concentration towards well $\gamma$, whereas those below $L_4$ seem to decrease. The intensification in $L_3$–$L_4$ may be an artefact resulting from some localized instability in the estimations. The OWC can only be traced for a limited distance away from the well.

The acoustic impedance profile derived from well logs run in the originally blind well $\gamma$ is shown to the left of the HIGHRES section (Plate 12 lower). Its vertical position has been adjusted so that it ties in with the *interpreted* top reservoir boundary $L_2$. The TWT mismatch of some 16 ms is due to a velocity variation above the time window that has not been accounted for in the processing. Once this match had been achieved there was very good agreement between the interpreted boundaries with their well $\gamma$ equivalents. The largest deviation occurs opposite the rather uncertain boundary $L_3$, which comes a few milliseconds too high.

Plate 13 shows an extension of the previous HIGHRES section to the right of well $\beta$, between CDPs 1024 and 1049. The purpose of including this illustration is to demonstrate the undesirable influence of a substantial fault zone on the stability of the HIGHRES estimations. In this case, the presence of the fault zone can be seen clearly on the original seismic data shown in the upper half of the figure.

The estimations towards the left-hand side of the fault zone are stable, but they break down into a chaotic pattern through the zone before re-stabilizing once the zone has been passed. The familiar lithoimpedance stratigraphy (to the left) cannot be re-established across the fault zone, even though the HIGHRES profiles can be crudely segmented. This may be due in part to the geological conditions being different on either side of the zone, but the main reason is that the reflection coefficient series estimated trace-by-trace away from the well ($\beta$) are no longer appropriate on the far side.

### *Test based on well* $\gamma$ *(Plates 14 and 15)*

The second test (as mentioned above) was based on well $\gamma$ (back towards well $\beta$). The impedance profile for the well (Plate 14) is a revised version because of severe caving problems that affected the logs over a 40 metres interval above the $L_2$ boundary. The logs were manually corrected through the offending interval.

Although the gross lithoimpedance stratigraphy is the same in both wells, there are significant differences with respect to the positioning of one of the boundaries and the internal composition of some of the units (cf. Plates 11 and 14). The $L_3$ boundary, which came in a little too high on the first test, has been re-defined somewhat lower down in well $\gamma$ and is designated $L_3'$. Unit $L_2$–$L_3'$ differs markedly from its well $\beta$ equivalent ($L_2$–$L_3$) as it contains only one impedance-defined sandstone association (rather than three subunits) and is characterized by numerous carbonate-cemented horizons. Furthermore, the units below $L_3'$ are relatively carbonate-free when compared to their $\beta$ counterparts.

The $\gamma$-based HIGHRES section (Plate 15) was produced without reference to either the previous one (the $\beta$-based section of Plate 12) or to the lithoimpedance stratigraphy in well $\beta$ (Plate 11). This was done in order to remain as objective as possible during the interpretation. The similarities and dissimilarities between the two interpretations are best appreciated by comparing the two sections (Plates 12 and 15).

The $L_1$, $L_2$ and $L_4$ boundaries are the most stable and tie into well $\beta$ with remarkable precision (Plate 15). Unit $L_1$–$L_2$ also reproduces well, and the two minor faults reappear at the same location and exhibit similar displacements. Some of the carbonate-cemented horizons below $L_4$ (whose estimated impedances intensify strongly towards $\beta$) also come right in on target. Above $L_1$, however, the limestones can only be followed for a short distance out from well $\gamma$ before a disturbing positive-impedance trend (caused by a modelling error between 1565 and 1590 ms TWT) precludes further interpretation.

Major discrepancies also occur throughout the reservoir zone ($L_2$–$L_4$) with respect to the lithoimpedance architecture. At first glance, the two interpretations seem to be almost irreconcilable, but the situation is not as serious as it seems when the HIGHRES methodology, the bases for the interpretations and the geological conditions within the reservoir are taken into account (see Discussion and Summary).

## Field C (Figs. 3, 4 and 5, Plate 16)

The objectives of the third and final test were (a) to investigate whether the HIGHRES method could be used for similar geological purposes in a strongly dipping portion of a structurally complex field, and (b) to ascertain the sensitivity of the method to noise and poorer seismic data quality caused in part by multiples. The data set consisted of two seismic lines taken from a 3D survey with a single well ($\delta$) located at their intersection. One of the lines was oriented parallel or subparallel to the overall dip direction (the inline); the other was oriented parallel or subparallel to the overall strike direction (the cross-line). The time window extends from 1780 to 2000 ms TWT.

A crude lithoimpedance stratigraphy for the well is shown in Fig. 3, and a highly simplified description of the units is given below. The field is overlain (above $H_1$) and underlain (below $H_7$) by shales.

- $H_1$–$H_2$ An interval containing two sandstone associations with an intervening sequence of coals and shales
- $H_2$–$H_3$ A sequence of rapidly alternating coals, shales and sandstones
- $H_3$–$H_4$ A relatively homogeneous sandstone succession
- $H_5$–$H_6$ A fieldwide shale with a characteristic high-impedance signature (a key horizon)
- $H_6$–$H_7$ A large-scale coarsening-upwards sequence of sandstones with occasional carbonate-cemented horizons

As will become apparent, a more sophisticated subdivision of the interval of interest is unnecessary.

A comparison between the log-derived and the estimated impedance profiles for the well (Fig. 4, arrowed) immediately indicates that there are much larger modelling errors involved in this test than in the previous cases. Yet even if the finer-scale impedance structure is lost almost straight away, the broader trends are just about strong enough to permit an extrapolation of the unit boundaries right across the left-hand side of the HIGHRES section away from the well. The most convincing boundaries are $H_1$, $H_2$, $H_5$ and $H_6$. Unit $H_4$–$H_5$ is, as expected, particularly prominent. This has proved most useful in guiding the rest of the interpretation and particularly the definition of the small faults. Even so, the fault on the extreme left of

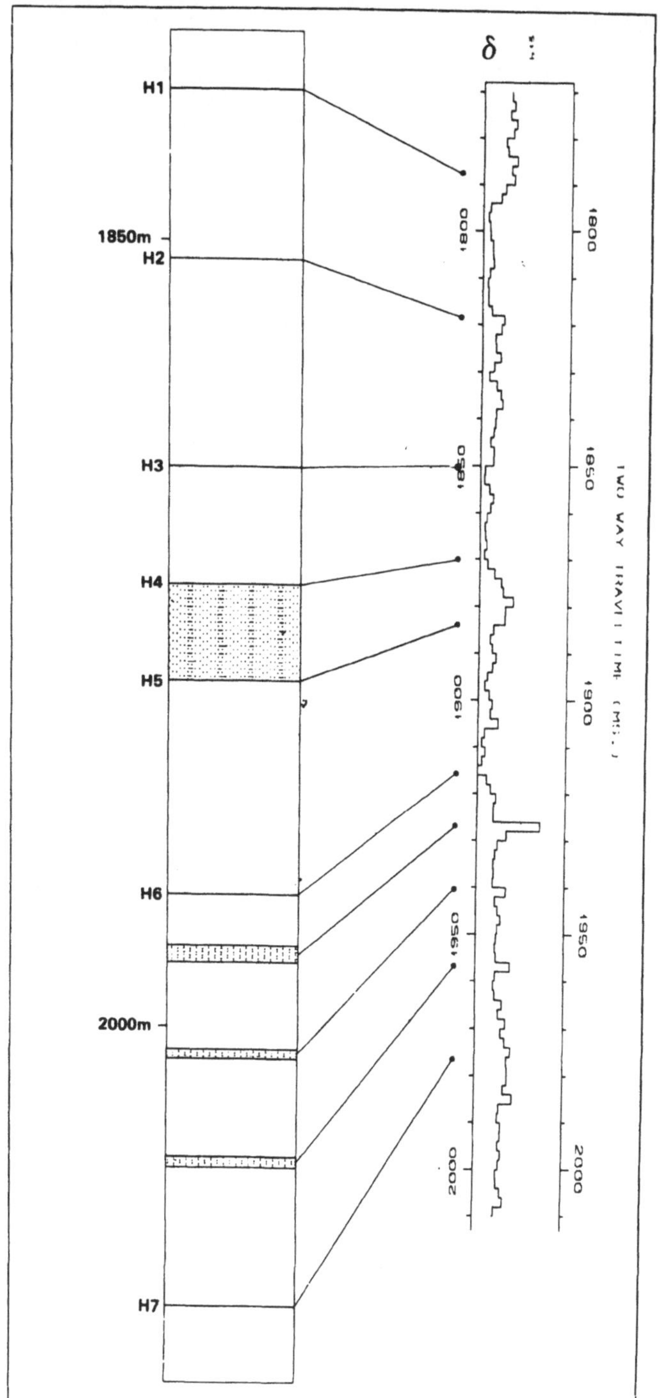


**Fig. 3.** Well log-derived acoustic impedance profile through the time window for well $\delta$(right), and the corresponding lithoimpedance stratigraphy used as the basis for the interpretations in Fig. 4 and Fig. 5.

the section was troublesome, as the impedance profiles on the downthrown side differ radically from those on the upthrown side.

A similar phenomenon occurs across the fault bounding the right-hand side of the small "horst" through which the well passes. All of the boundaries beyond this fault are uncertain (with the possible exception of $H_1$), and even the picking of $H_4$–$H_5$ was dubious. Incidentally, the pronounced positive impedance contrast at 1872 ms TWT cuts horizontally across the entire section. This, and others like it, are artefacts of the estimations.

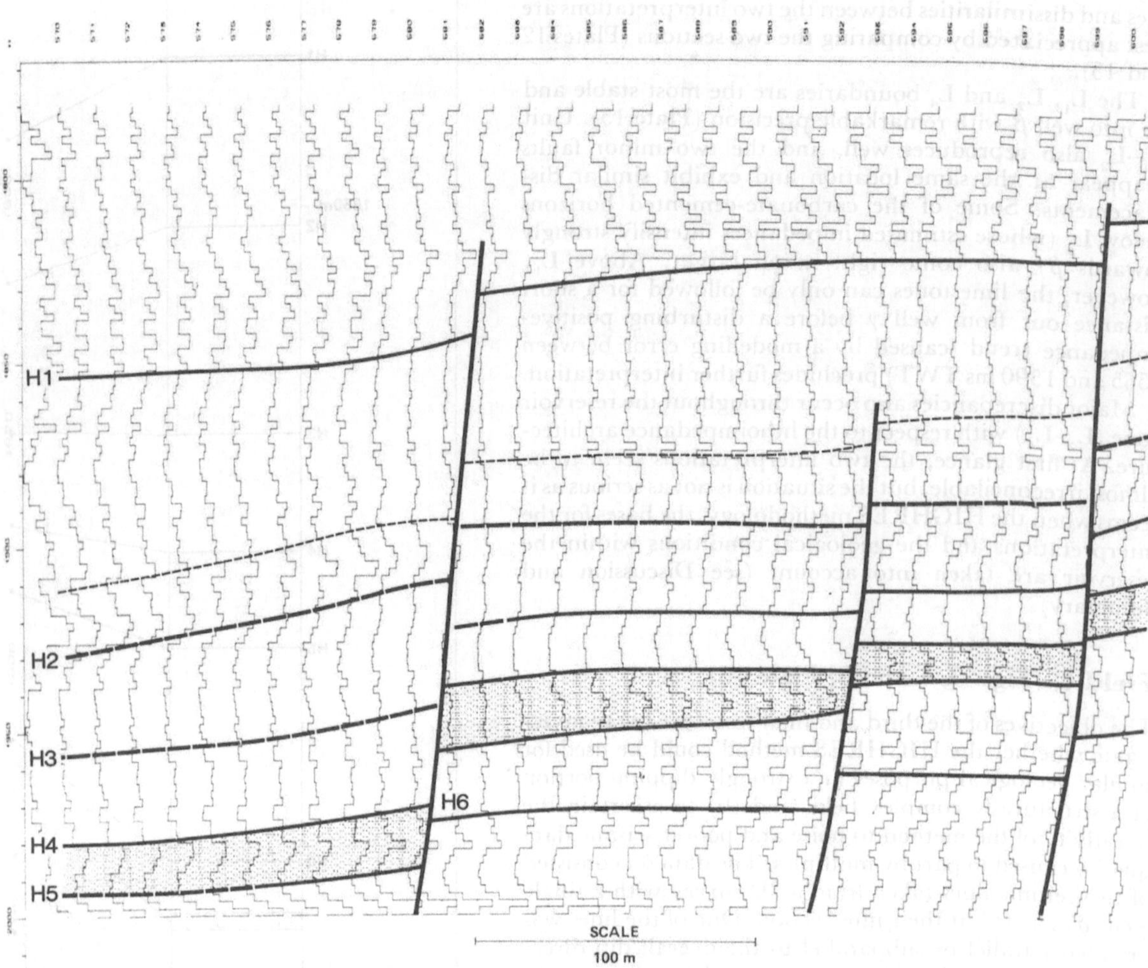


**Fig. 4.** Interpreted HIGHRES section for the "inline". The basis for the interpretation is given in Fig. 3. The position of well $\delta$ and the

Another way of appreciating the HIGHRES estimates and the original seismic traces is to view them as false-colour amplitude intensity panels (Plate 16). The upper panel shows the amplitude intensities for the original seismic traces within the time window, while the lower panel shows the amplitude intensities of the estimated relative acoustic impedance profiles. The main difference between the two is that the upper panel provides a clearer impression of the dip of the reservoir and the flattening out of the structure both within and to the right of the horst. The top of the reservoir, $H_1$, is very clear as it coincides with the top of the low-intensity blue belt (which diminishes slightly to the immediate left of the well location). The $H_4$–$H_5$ interval is also well represented as an almost continuous sigmoidal belt of high-intensity greens that appears at 1880 ms TWT at the well. The HIGHRES panel, however, gives a far better impression of the faulting, even if the picture is somewhat marred by the horizontal texture resulting from the method.

For the crossline (Fig. 5), the quality of the HIGHRES section is just about adequate to permit a tentative interpretation in the near-well region I. The quality then deteriorates dramatically in region II, and is so poor in region III that an interpretation has not been attempted. The thick vertical lines simply delimit the various "regions" and do not necessarily denote faults.

## DISCUSSION AND SUMMARY

From a analysis of the three test cases, the HIGHRES method seems to be most stable for those portions of reservoirs that are essentially flat-lying (or only dipping slightly) and are not dissected by major fault zones. The reliability of the estimated reflection coefficient series and acoustic impedance profiles is destroyed across fault zones and severely weakened by multiples and diffractions generated by steeply dipping surfaces (e.g. Field C).

Small-scale faults, which are not resolvable from ordinary seismic, seem to be identifiable in all but the most severe areas of instability. The basis for these is the wholesale shifting of gross impedance patterns (which are fairly conspicuous even where the finer impedance structure is unreliable) as well as the sudden juxtaposition of contrasting impedances across them. Field A provides a good example in which faults have been consistently followed from one seismic line to another. Offsets down to 2 ms TWT have been recognized.

The usefulness of the HIGHRES method for reconstructing the geoseismic (lithoimpedance) stratigraphy and architecture of a reservoir interval is directly dependent upon (a) the stratigraphic/sedimentological complexity of the reservoir and (b) the availability of sufficient

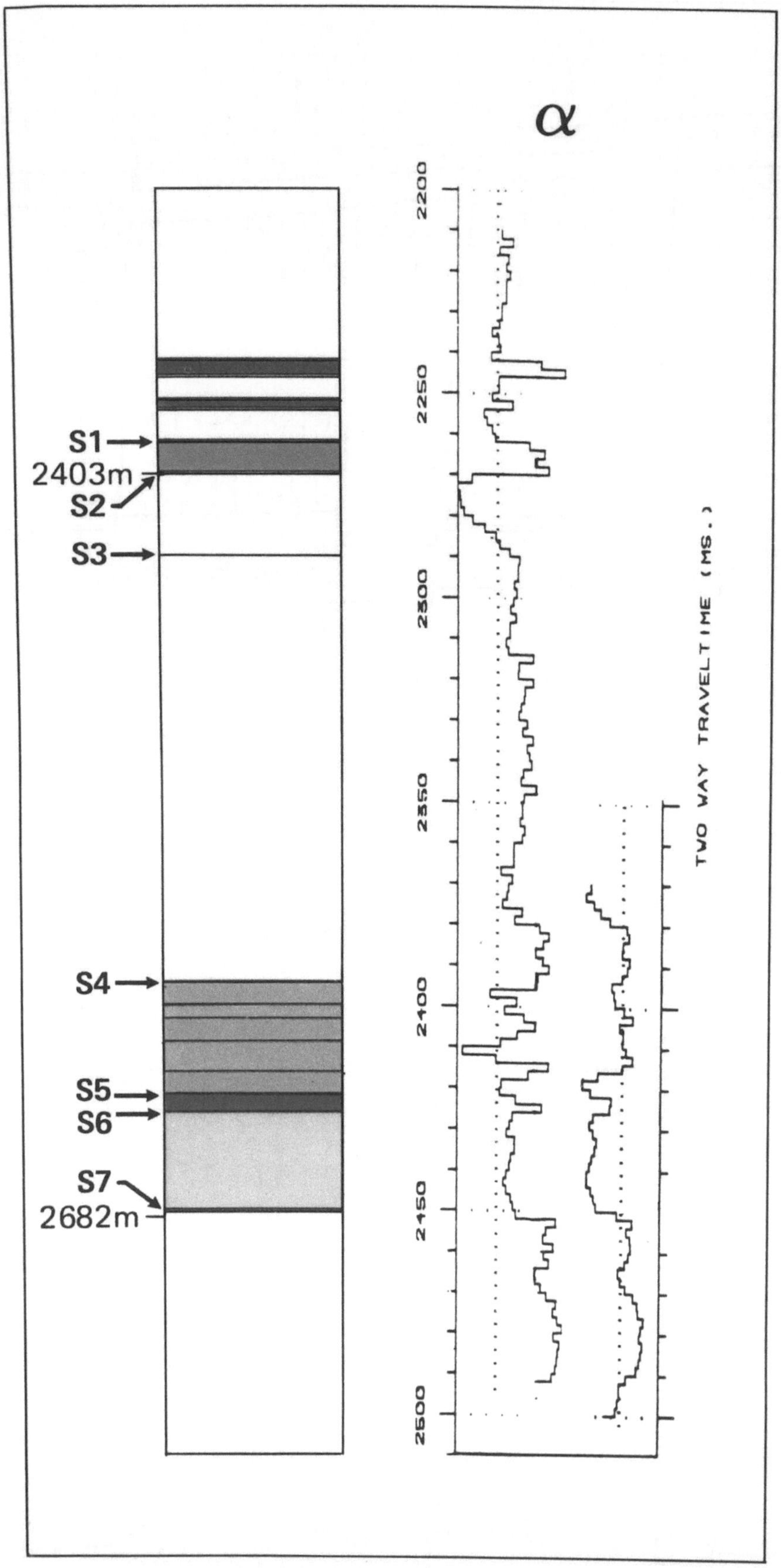


**Plate 8.** Well log-derived acoustic impedance profile for well α extending over two time windows (right), and the corresponding lithoimpedance stratigraphy (left) used as the basis for interpretation of the HIGHRES section in Plate 9. The repeated segment of the impedance profile is a revised version (see text for explanation).

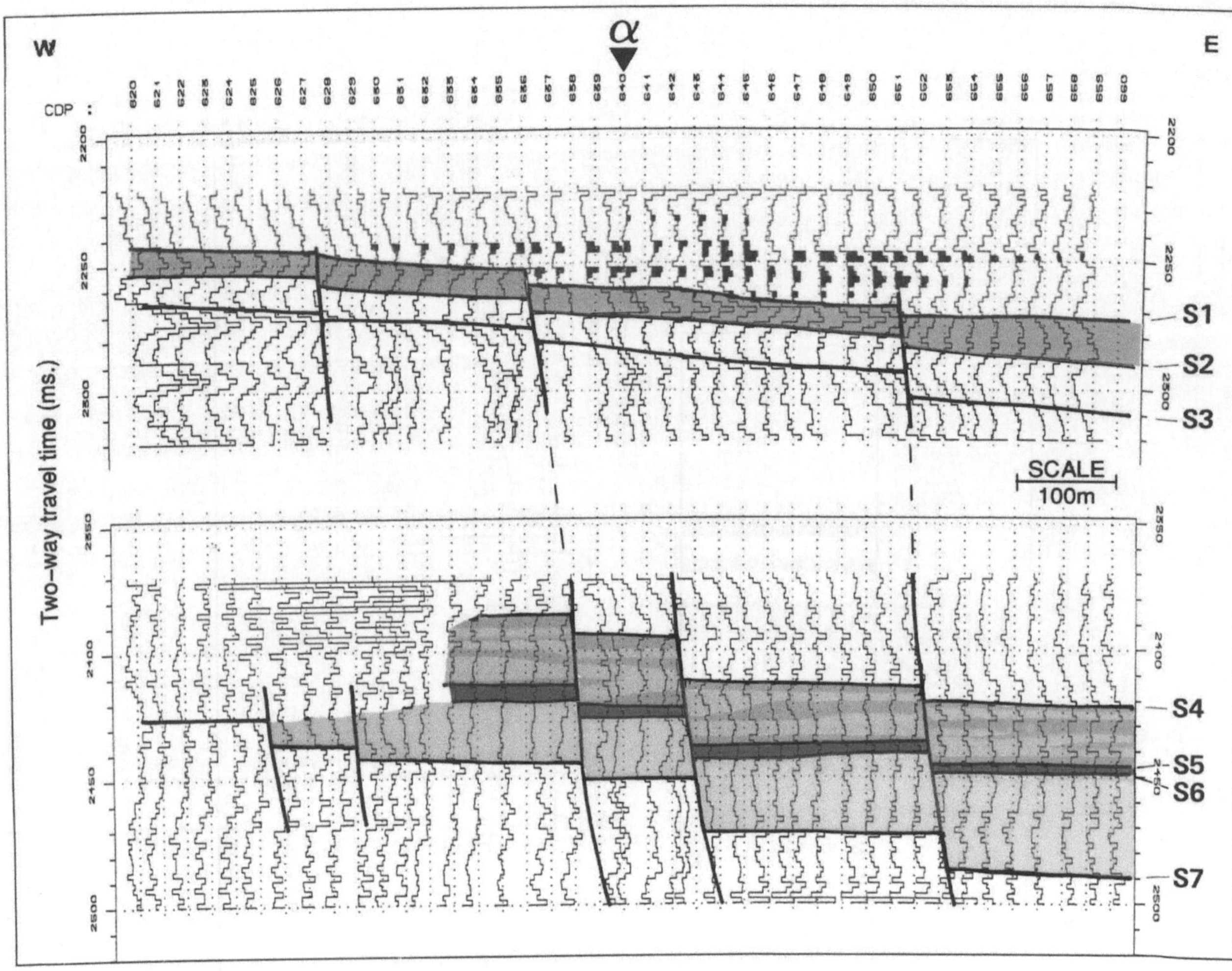

W
α
E
CDP :
Two-way travel time (ms.)
S1
S2
S3
SCALE
100m
S4
S5
S6
S7


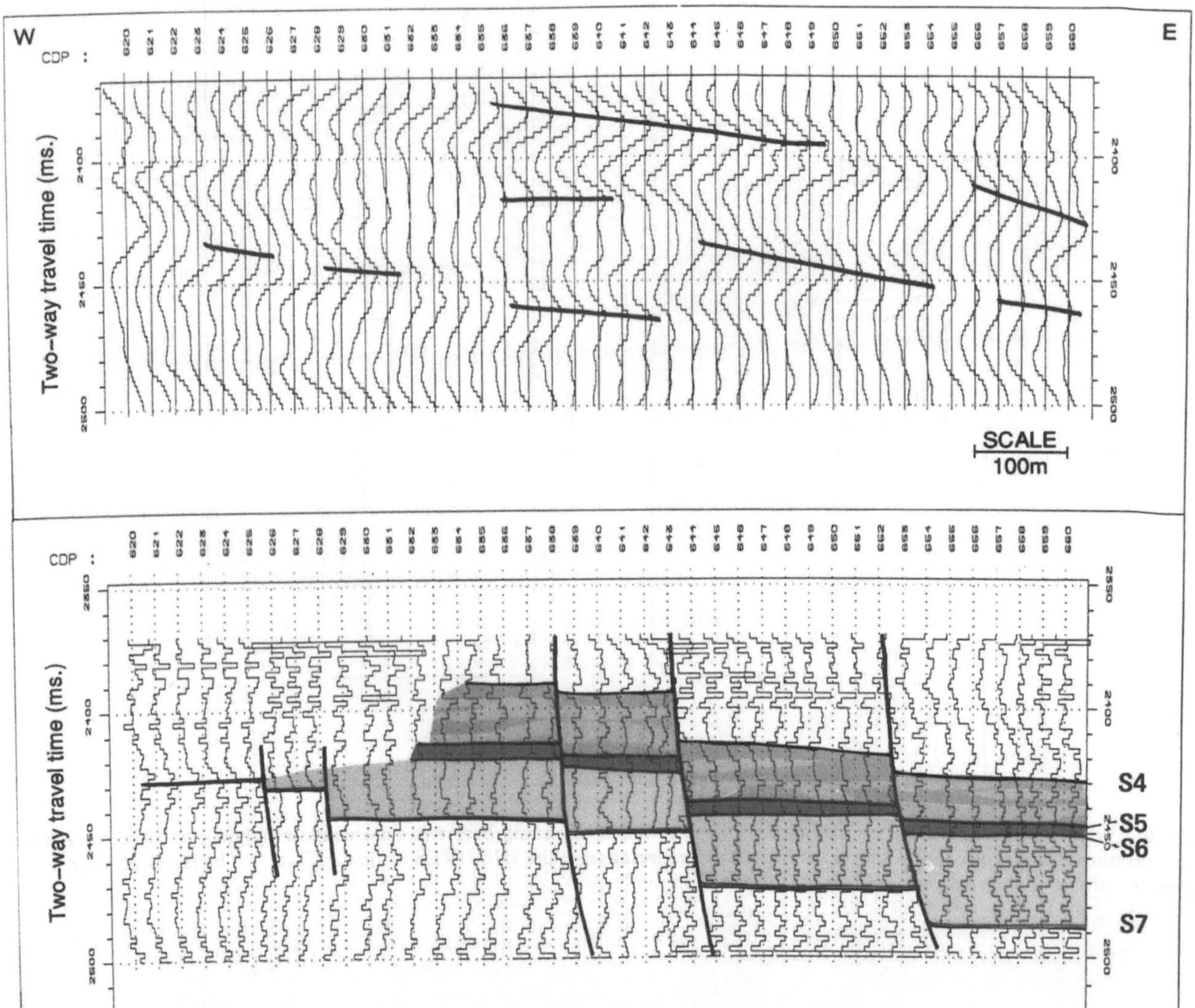

W
E
CDP :
Two-way travel time (ms.)
SCALE
100m
Two-way travel time (ms.)
S4
S5
S6
S7

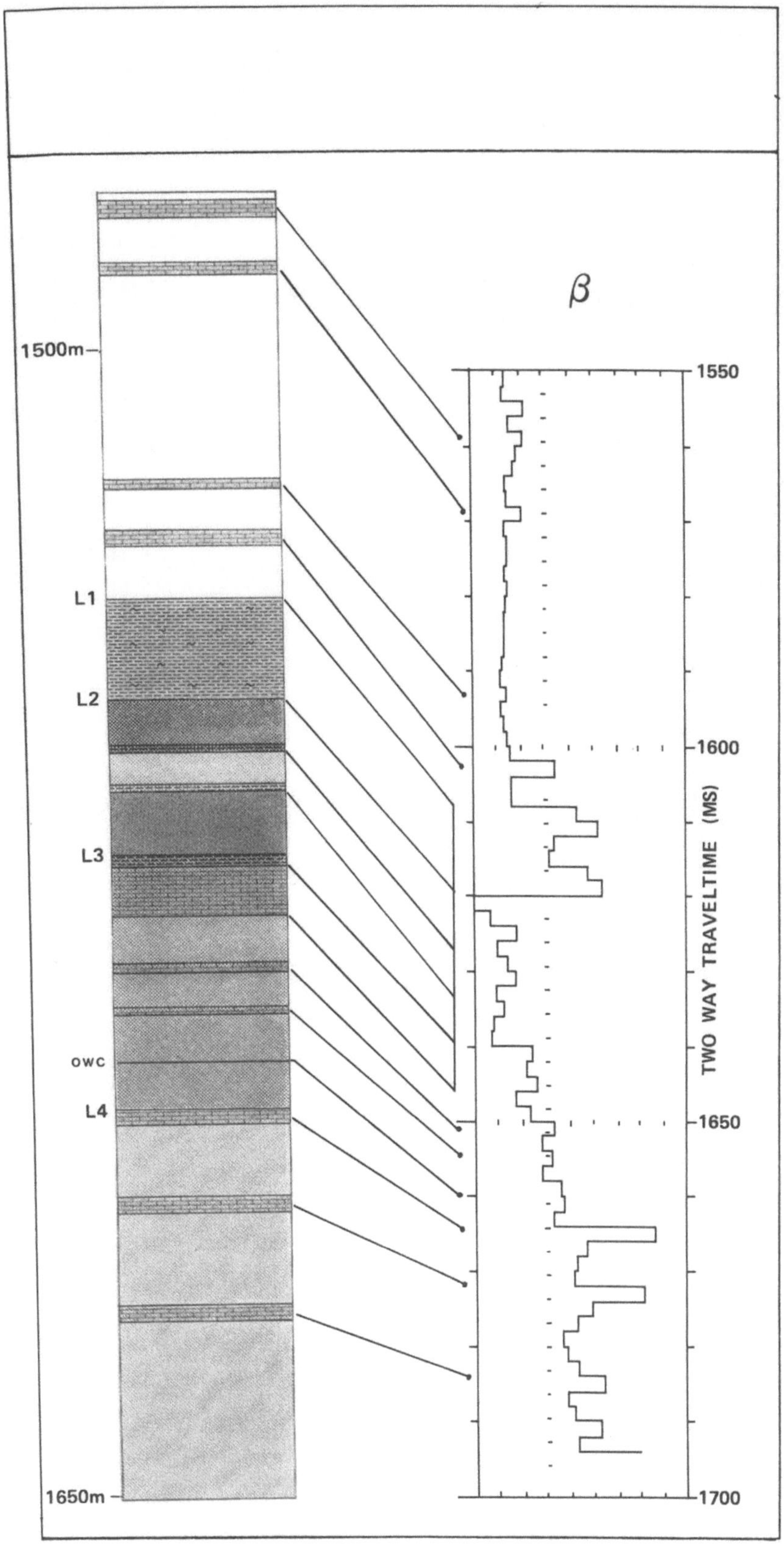


**Plate 11.** Well log-derived acoustic impedance profile through the time window for well $\beta$ (right), and the corresponding lithoimpedance stratigraphy (left) used as a basis for the interpretation of the HIGHRES section shown in Plate 12 (lower).

**Plate 9.** Interpreted HIGHRES sections for the upper and lower time windows for seismic line 4A (see Fig. 2 for location). Well $\alpha$ is located at CDP 640, and the basis for the interpretation is given in Plate 8.

**Plate 10.** *Upper*: original seismic data for line 3A (see Fig. 2 for location) for the lower time window. Trends are indicated by the thick black lines. *Lower*: interpreted HIGHRES section for the lower time window. The basis for the interpretation is given in Plate 8.

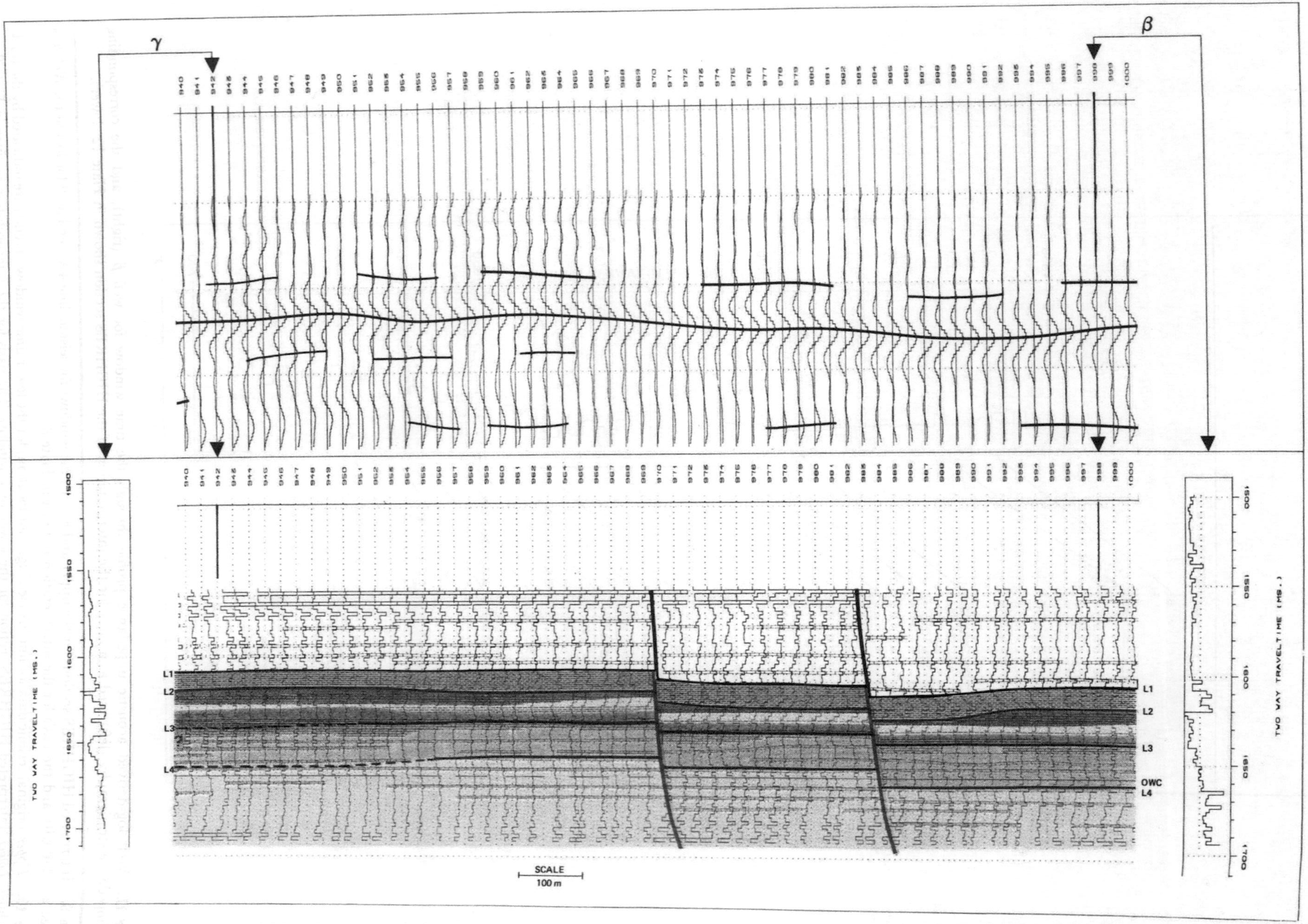


**Plate 12.** *Upper*: original seismic data for the time window. Trends are indicated by the thick black lines. *Lower*: interpreted HIGHRES section. The basis for the interpretation is given in Plate 11. The well log-derived acoustic impedance profile for wells $\beta$ and $\gamma$ are shown at either end of the section, and the well locations are indicated by the arrows.

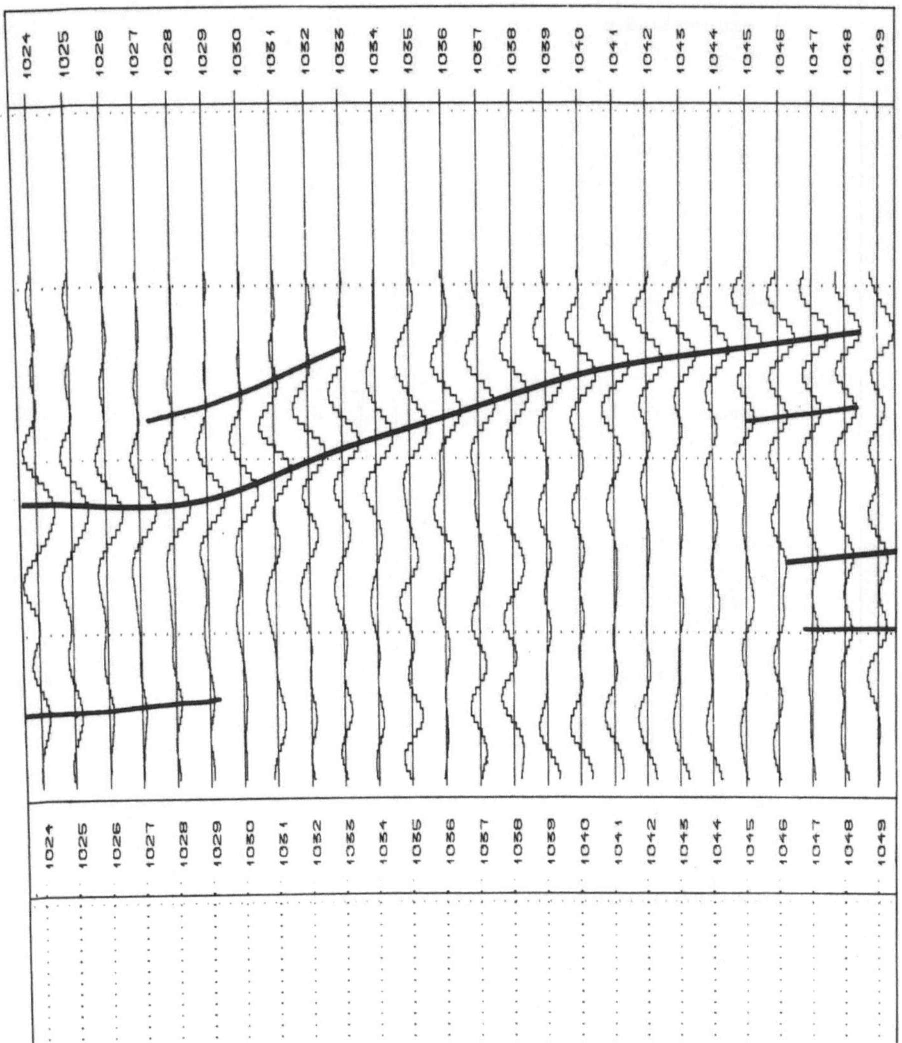


**Plate 13.** An extension of the previous section (Plate 12) to the right of well $\beta$. The wiggly lines delimit a broad area of instability in the estimates through a large fault zone.

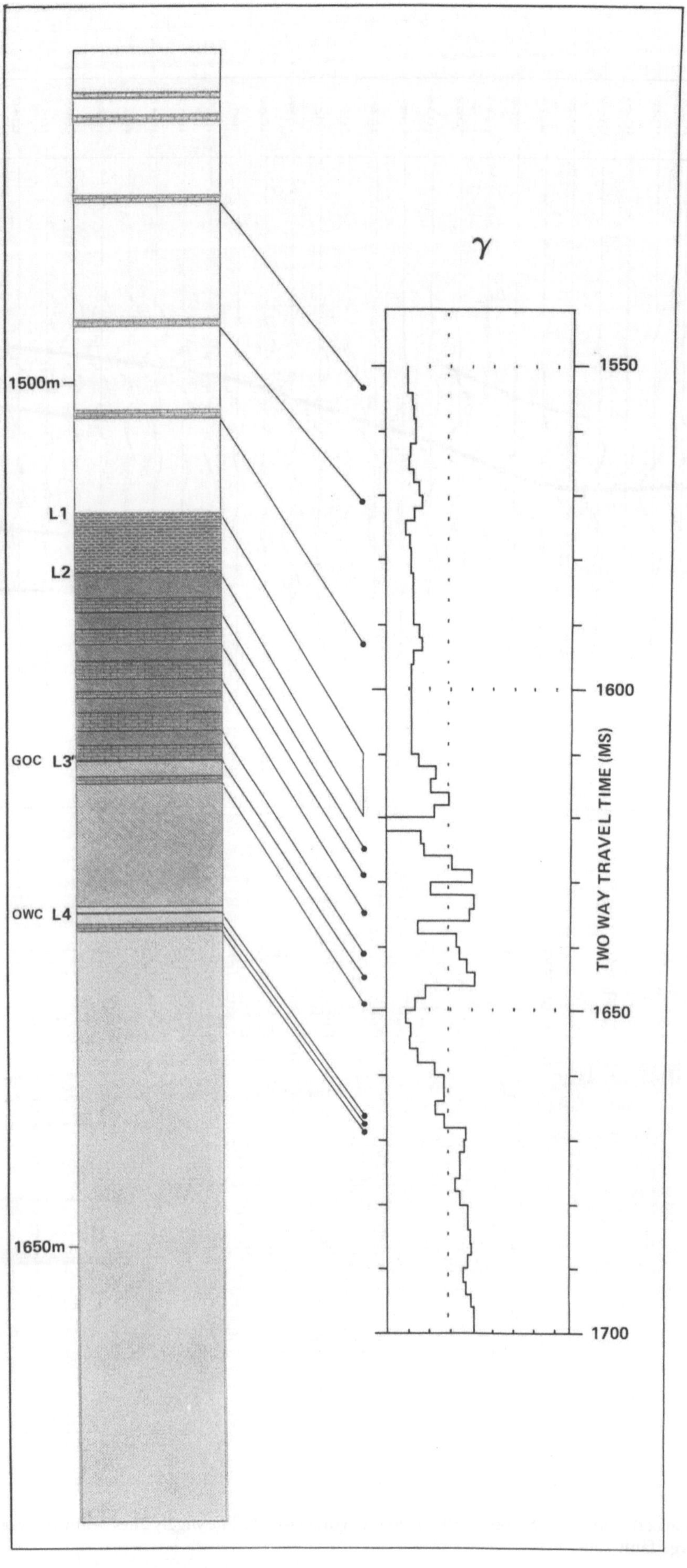


**Plate 14.** Well log-derived acoustic impedance profile through the time window for well γ (right), and the corresponding lithoimpedance stratigraphy used as a basis for the interpretation of the HIGHRES section shown in Plate 15.

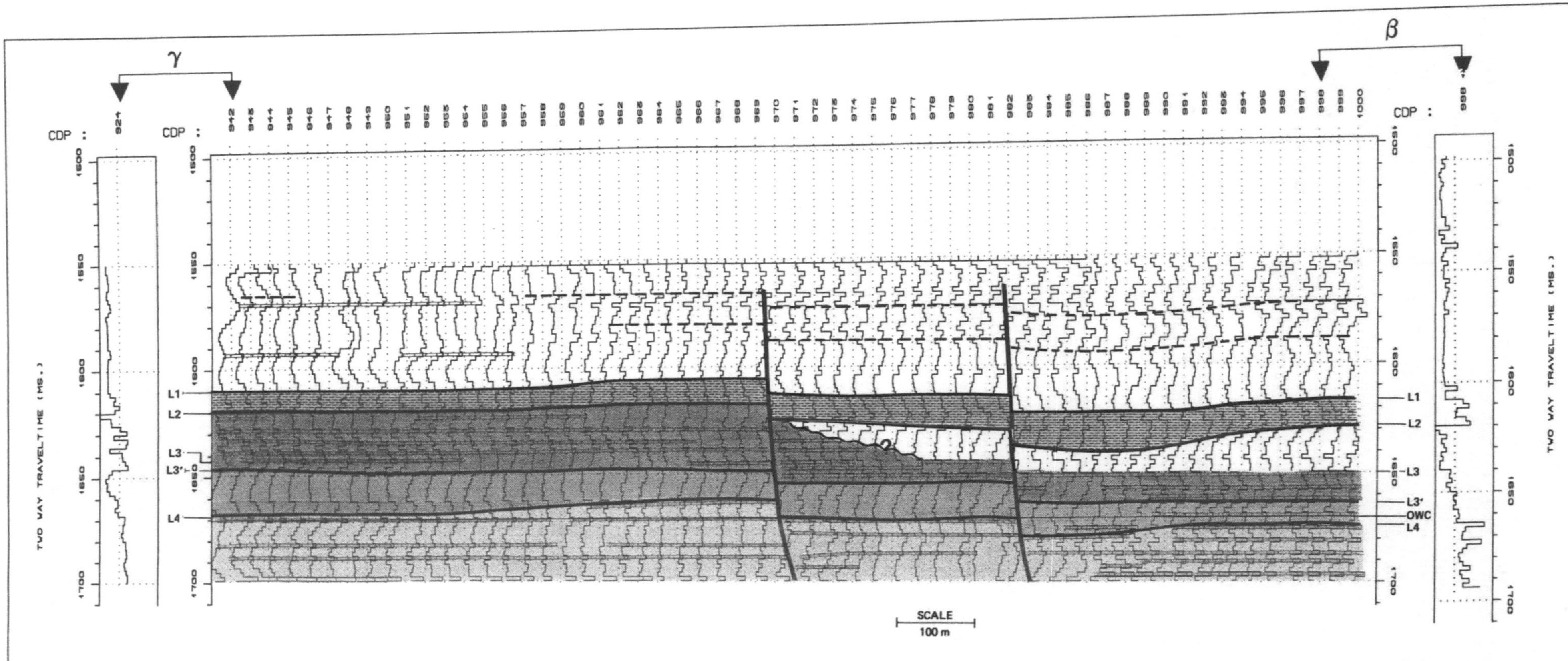


**Plate 15.** Interpreted HIGHRES section. The basis for the interpretation is given in Plate 14. The well log-derived acoustic impedance profiles for wells $\beta$ and $\gamma$ are shown on either side of the section. The well locations are indicated by the arrows.

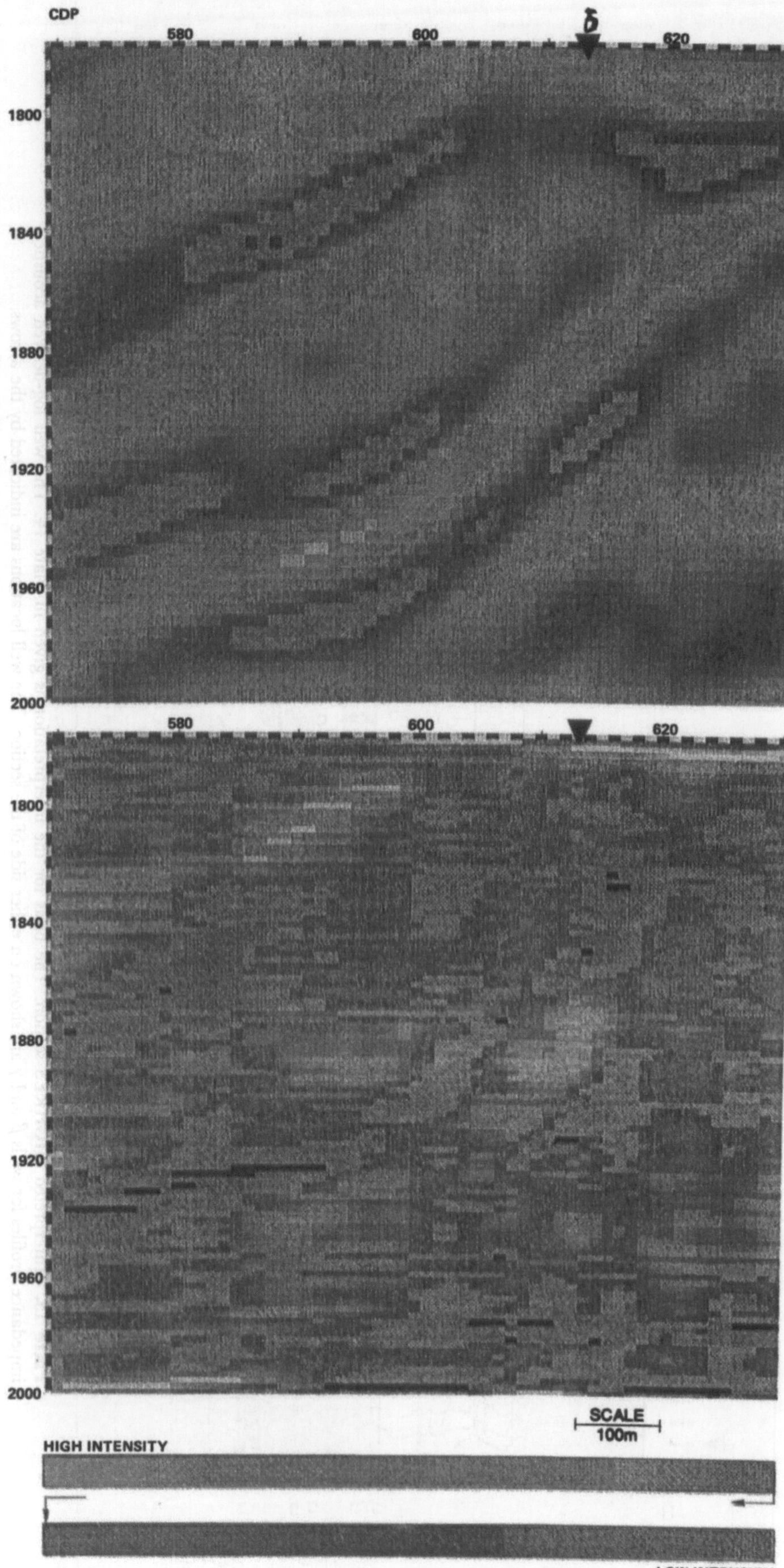


**Plate 16.** False-colour amplitude intensity panels (cf. Fig. 4). *Upper*: amplitude intensities for the original "inline" seismic traces for the time window. *Lower*: amplitude intensities of the HIGHRES estimated relative acoustic impedance profiles for the same section.

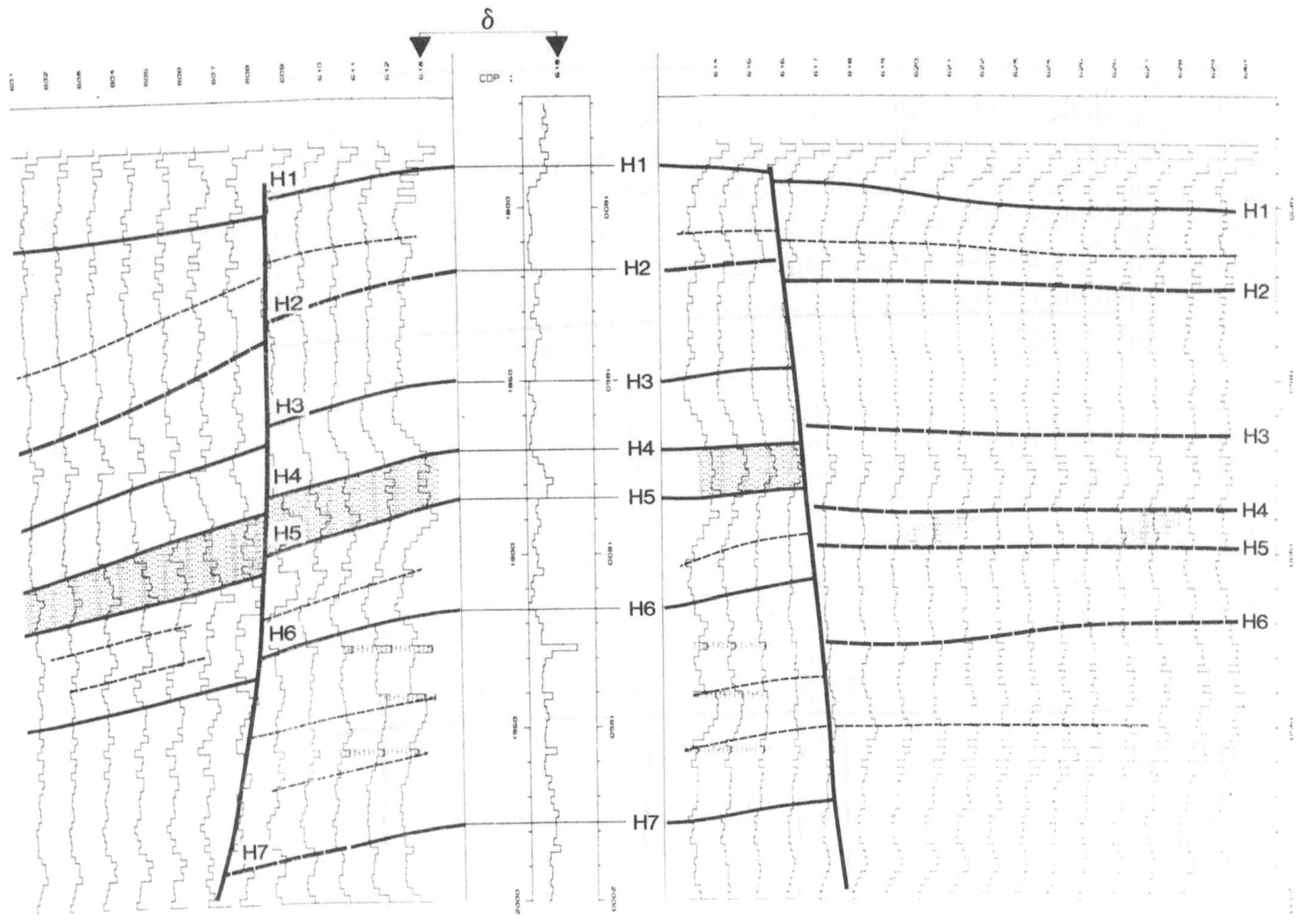


inset well log-derived acoustic impedance profile are marked by the arrows.

impedance contrasts between the various beds, facies associations or formations. If these conditions are met, both major "unit" boundaries (e.g. Fields A and B) and the internal lithoimpedance structure within some of the units (e.g. Field A) can be interpreted with confidence.

Difficulties of interpretation, however, arose within the reservoir interval for Field B (Plate 12 and Fig. 3). The reasons for this are believed to be as follows.

(1) The bases for the two interpretations (i.e. the well log-derived lithoimpedance stratigraphies for the two wells $\beta$ and $\gamma$) differ somewhat, and the interpretations of the HIGHRES sections were naturally biased towards whichever one was being used.
(2) The geological variations between the two wells are now thought to be rather too subtle for this type of exercise.
(3) The reliability of the HIGHRES estimates usually decreases as a function of distance away from a basis well.
(4) The interpreter may experience considerable difficulty when faced with lateral changes in the lithostratigraphy without a-priori knowledge of their possible existence.

In retrospect it might have been wiser to have undertaken the interpretations of the HIGHRES sections simultaneously, and with frequent cross-referencing. This has not yet been tried.

With reference to carbonate-cemented horizons, the results have been variable. One of the problem is that where the HIGHRES estimates break down, the impedance profiles are often dogged by sharp, rapidly alternating, horizontally aligned impedance contrasts that may look deceptively like hard layers (e.g. within unit $L_3$–$L_4$ close to well $\gamma$ in Plate 12). These clearly pose a considerable challenge to the interpreter who has to try to discriminate between the real events and artefacts of the methodology. Nevertheless, the tracing of carbonate horizons close to the wells is thought to be fairly reliable, and this is an important consideration in areas close to producers (e.g. for the avoidance of coning: Haldorsen *et al.*, 1987).

Apart from this specific case, the fact that artefacts and zones of instability are often so clearly and unequivocally recognizable that they can be delimited and mapped (e.g. Fig. 2) is indeed one of the strengths of the HIGHRES method. Naturally, difficulties will arise in transition zones between real and false estimates, but these can largely be overcome by remaining faithful to the interpretational process. This process, which in a sense serves as a control mechanism, naturally has to obey geological rules that force the interpreter to seek out plausible relations from

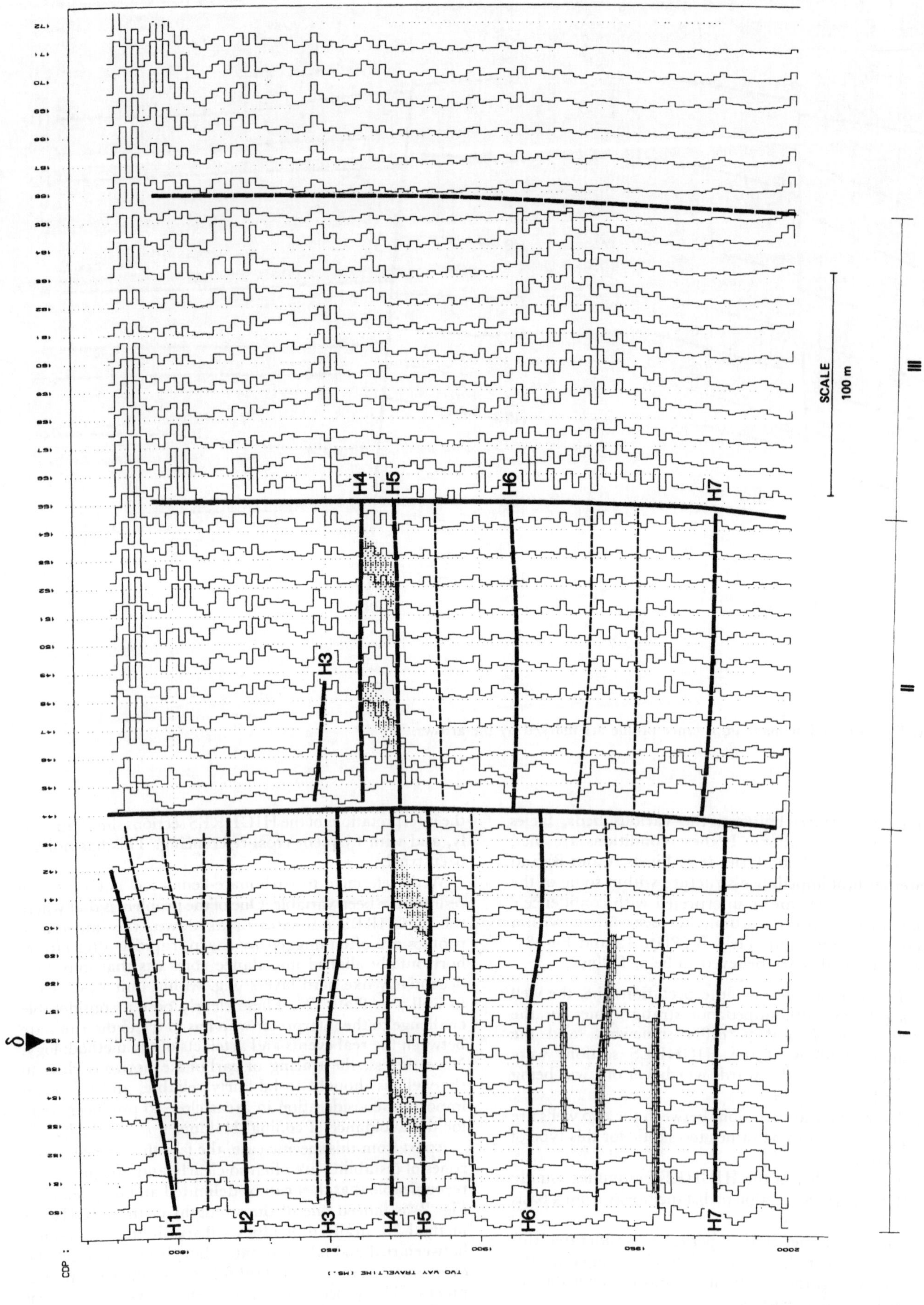


**Fig. 5.** Interpreted HIGHRES "crossline" section. The basis for the interpretation is given in Fig. 3, and the location of well δ is shown by the arrow.

one trace to another. Where these relations weaken, the interpreter should suspect that the data are unreliable and should put a stop to the interpretation at that juncture.

## CONCLUSION

The HIGHRES method shows considerable promise as a near-well and as an interwell reservoir description tool under optimum conditions of relative structural simplicity and with sufficient impedance contrasts through and within the interval of interest. Some small-scale faults and lithoimpedance features have been interpreted with confidence at depths of up to 2500 m and with a potential vertical resolution of some 5–10 m. Its reliability, however, is severely reduced where the original seismic data are poor owing to bad processing and/or fault zones and steeply dipping fault-bounded compartments. Lithoimpedance reconstructions will also be difficult in areas of complex stratigraphy characterized by subtle transitional variations and (perhaps) where contrasted sequences are juxtaposed across faults with considerable displacements.

## ACKNOWLEDGEMENT

We would like to thank the Norwegian State Oil Company, Statoil, for granting permission to publish this paper.

## REFERENCES

Balch, A.H. and Lee, M.W. (1982). Some considerations on the use of downhole sources in vertical seismic profiles. Paper presented at the *35th Annual SEG Midwestern Exploration Meeting*.

Brevik, I. and Berg, E. (1986). Maximum likelihood estimation of reflection coefficients from seismic data. A case study. *Proceedings of SEG*, pp. 273–277.

Gelfand, V., Ng, P., Nguyen, H. and Larner, K. (1987). Seismic lithologic modelling of amplitude vs. offset data. Presented at the *49th Annual Meeting of EAEG, Belgrade, June 9–12*, abstract.

Gundesø, R. and Egeland, O. SESIMIRA—a new geological tool for 3D modelling of heterogeneous reservoirs. *This volume*.

Haldorsen, H.H., Change, D.M. and Begg, S.H. (1987). Discontinuous vertical permeability barriers: a challenge to engineers and geologists. In: Kleppe *et al.*, (eds.), *North Sea Oil and Gas Reservoirs*, pp. 127–151. Graham & Trotman, London.

Krug, V., Baum, G. and Windgassen, W. (1981). Special borehole-seismic measurements used to study petrophysical parameters in the environment of deep boring. *Z. Geol. Wiss.* **9**(8), 851–861.

Landgren, K.M. and Grubb, K. (1983). Acquisition and processing of Gulf Coast VSP data. Paper OTC 4539, *Offshore Technology Conference, Houston, Texas*, pp. 229–236.

Poggiagliolmi, E. (1980). Detailed reservoir delineation by interactive seismic stratigraphic exploration. *European Offshore Petroleum Conference and Exhibition, EUR 198, London, October 21–24*, pp. 513–514.

Van Riel, P., Samilo, J.P., Mesdag, P.R. and Eckhardt, W. (1987). A new strategy for detailed reservoir delineation. Presented at the *49th Annual Meeting of EAEG, Belgrade, June 9–12*, abstract.

Ursin, B. and Holberg, O. (1985). Maximum likelihood estimation of seismic impulse responses. *Geophys. Prospect.* **33**, 233–251.

# 3

# Some Requirements from Seismic Methods for Use in Reservoir Simulation Models

*R.H. Seymour and J.S. Archer*

Department of Mineral Resources Engineering, Imperial College, London, UK

Geological controls in reservoir simulation models are limited to information gathered at wells and would be improved by a greater knowledge of sandbody continuity and stratigraphy in the reservoir. Any advance in interwell correlation and in the definition of boundaries will strengthen the validity of these models.

The concept in this work is to provide a link between seismic stratigraphic interpretation and reservoir simulation modelling. We have been able to set some bounds on resolution for current practice and indicate directions for future improvement. At this stage we have not yet addressed the use of downhole seismic methods or amplitude versus offset studies, but have concentrated on conventional surface practices and their influence on the thickness resolution of reservoir simulation cell blocks.

The problem was approached parametrically in this study by setting target reservoir simulation cell block thicknesses for three-dimensional modelling as 10 m, 20 m, and 40 m at a typical reservoir depth of 3000 m (MSL). In effect, the reservoir model cell character then defined the quality of the geological and seismic information required and allowed the generation of synthetic seismic sections. The synthetic sections were compared with sections at successively tighter frequency bandwidths in order to define thickness resolution limits.

Sensitivities at a given depth to lithology, porosity, shaliness, fluid saturation and pressure regime were investigated in terms of their effect on the resolution of grid cell thickness.

The results of our investigations show that seismic resolution to thicknesses in common North Sea usage in three-dimensional reservoir simulation is achievable in many cases, and that finer resolution is mainly controlled by absorptivity of the overburden. The scope for internal characterization of reservoir properties by seismic interpretation is indicated at reservoir model scale.

## INTRODUCTION

Surface seismic methods have been in use for over half a century to help delineate deeply buried geological structures in the exploration for oil and gas. Over this time, technological development has improved the results in terms of clarity, accuracy, resolution, and density of coverage to the point where it is now also realistic to expect that data from seismic surveys should contribute to the efficient development and management of discovered hydrocarbon reserves.

As with the "invasion" of any other technology into a new area of application, the extension of seismic methods into reservoir engineering needs to be treated with a balanced mix of optimism and scepticism. In particular, it is certainly the case that seismic methods can provide valuable additional information for reservoir models for use in reservoir simulation, but such use must be accompanied by engineering judgement based on the limitations of the methods.

The contribution that can be made to the structural picture of a reservoir is well known, as is the delineation of fluid contacts in those cases where conditions are favourable. Such an example is the Troll field (Birtles, 1986). What is less clear is the degree to which a seismic survey can be relied upon to help characterize the formation parameters in undrilled areas of a reservoir and between wells.

In a North Sea reservoir context, this work reviews the expected sensitivities of conventional seismic surveys to formation parameters, and the expected degree of resolution achievable. The intention is to help show what can be done now, and in what directions research could be aimed to improve the value of seismically derived information in the future.

### Current state of the art

A well-established method (Angeleri and Carpi, 1982) of using surface seismic data to help with reservoir character-

*North Sea Oil and Gas Reservoirs—II*

ization involves two distinct inversion steps after the completion of conventional seismic data processing. In the first step—seismic inversion—algorithms are applied to the seismic traces to synthesize estimates of acoustic impedance logs, or of seismic velocity logs if velocity/density relationships have been supplied. The second step, lithological inversion, predicts formation parameters such as porosity and shaliness from the synthesized acoustic or velocity logs by incorporating seismic/well-log relationships and other geological information.

In its simplest form, the seismic inversion step consists of the numerical integration of the seismic amplitude versus arrival time traces (which are assumed to be noise-free reflectivity logs). This yields logs of the logarithm of acoustic impedance, but these are lacking in low and high frequencies that are absent from seismic data, and have various other deficiencies.

Numerous much more sophisticated seismic inversion techniques exist, accounting for the more complex character of true seismic traces. Cornish and King (1988) described a particularly powerful technique that incorporates externally supplied information, such as sonic logs, to give broad-bandwidth synthesized acoustic logs. However, the qualitative concept of a seismic trace being like the derivative of an acoustic impedance log is a valuable one in estimating the likely utility and reliability of a result in a given geological interval.

The lithological inversion step exploits the controls provided by borehole information, and the results of seismic/geological interpretation. Probably the simplest implementation involves first tying (reconciling) the logs with the seismic traces, then cross-plotting porosity, for example, against the acoustic impedance within each formation or other geological unit. Finally, estimates of porosity, etc., are derived from the synthesized acoustic impedance traces throughout the survey area using regression lines from cross-plots for the appropriate (interpreted) formations. More sophisticated methods involve error minimization, complex multivariate mapping functions, and automatic harmonization with the seismic interpretation results.

A few years ago the state of the art of estimating reservoir parameters was summarized by Fillipone (1982). It was discussed in a more detailed manner for porosity estimation by Angeleri and Carpi (1982), and for pressure by Bilgeri and Ademeno (1982). Martinez (1985) showed how these techniques could be applied. Development in this area has advanced since then at a rapid pace, exploiting the revolution in microcomputing and graphics, and major improvements in the power, ease of application and controllability of lithological inversion are being developed.

In current practice, the product of the processing is a suite of reservoir maps and sections, usually in colour, portraying any or all of the following: predicted porosity, shaliness, net sand thickness, lithology, and overpressure. A further interpretation step is then required to incorporate this information into quantitative grid-block reservoir models. In the absence of well-founded averaging laws, this must allow for the application of judgement before re-gridding to reservoir simulation model cell dimensions.

### Problems and objectives

Seismic data can be interpreted in terms of reservoir parameter estimates, but it is not always obvious how reliable they are. Moreover, the thickness resolution of seismic data is frequently poorer than the zonation needed by reservoir engineers.

To help clarify these points, we estimated the probable sensitivities of conventional seismic results to formation parameters in a northern North Sea setting, with a view to evaluating which aspects of the reservoirs can be best characterized by seismic data.

The practical effects of the limited bandwidth of seismic reflection profiles on thickness resolution were also investigated. One purpose in this was to indicate what might be the finest layer resolution in a model that could be influenced primarily by seismic data. The interaction between the sensitivity and resolution of seismic data was then examined. For example, this showed the extent to which the porosity of a thin bed would be less well determined than that of a thicker bed.

## SEISMIC SENSITIVITIES AND RESOLUTION

There has been a very great deal of theoretical and laboratory work, and some field experimentation (principally with borehole seismic), relating rock seismic transmission responses with other rock properties and with the reservoir environment. Jones (1986) and Stainsby and Worthington (1985), for example, looked at seismic attenuation, while Han *et al.* (1986), Domenico (1984), Toksoz *et al.*, (1976) and others have made valuable contributions to seismic velocity dependency relationships.

Here we have collated and simplified the work so as to bring it into a common framework related to the conditions found in typical Jurassic and Cretaceous reservoirs of the North Sea.

### Compression-wave velocity

A sonic log is often considered to be a porosity log, so one should not be surprised to find a strong sensitivity of velocity to porosity. Wyllie *et al.* (1956) and Wyllie *et al.* (1958) are attributed with the "time-average equation", in which the transit time of seismic energy through a rock varies linearly with porosity between the limits of pure matrix transit time (0% porosity) and a value related to the fluid content of the pores ("100% porosity"). This relationship has been refined by various workers, extending its validity into the porosity range of 2–30% (Han, *et al.*, 1986). They proposed a linear relationship with empirical (but perhaps generally applicable) constants between seismic velocity and porosity in sandstones. Their simple linear relationship also contains a shale term, which was shown to give good agreement with measurements when clay fractions were in the range of 1–50%.

$$V_P = u_{mat} \cdot (1 - \phi_{is} - v_{sh}) + f_{phi} \cdot \phi_{is} + m_{sh} \cdot v_{sh} \qquad (1)$$

where

$V_P$ is the pressure-wave velocity at reservoir datum pressure

$u_{mat}$ is an empirical constant for the matrix mineral
$f_{phi}$ is an empirical constant for the pore fluid
$m_{sh}$ is an empirical constant for the shaliness
$\phi_{is}$ is the effective in-situ porosity (2–30%)
$v_{sh}$ is the clay content (1–50%)

We have extended this to carbonate rocks and to hydrocarbon pore fluids by choosing additional empirical constants based on matching Schlumberger (1986) sonic-porosity charts and other sources as detailed below.

$u_{mat}$, $f_{phi}$ and $m_{sh}$ for the case of water-saturated sandstone were derived from Han *et al*, (1986) as 5590, −1340, and 3410 m/s, respectively. (Note that the negative fluid-related constant reflects the fact that fluid-filled porosity has a disproportionately large effect on rock velocity.) The best fit of $u_{mat}$ for limestone beds was chosen around 16% greater than that for sandstone, as 6494 m/s.

In order to choose $f_{phi}$ values for oil-and gas-saturated rocks, a velocity was calculated using Eq. (1) with $\phi_{is}=20\%$, $v_{sh}=1\%$, and $f_{phi}$ (water) as $-1340$ m/s. According to the work of Toksoz *et al.*, (1976), and King (1966), if the differential pressure is around 350 bar (normal for 3000 m depth), then replacing the water with oil saturation would reduce the velocity by 1.0%. Gas saturation would give a further 1.5% velocity reduction. Back calculation of these changes through Eq. (1) gave a value for $f_{phi}$ for oil of $-1552$ m/s, and for gas of $-1863$ m/s.

### *Velocity variation with pressure*

Wyllie, *et al.*, (1958) and other workers found that the pressure-wave (P-wave) velocity is a function of the simple difference between overburden pressure and pore pressure ("differential" pressure). The gradient of this is quite large at low differential pressures, but at the reservoir depths considered, where differential pressures are in the range of 275–450 bar, the gradient is much smaller. King (1966) showed a total change of around 0.9% in the velocity over this range in a rock of 25% porosity. For the purposes of this study, this was taken as being linear with fractional differential pressure and with porosity:

$$V_{Ph}=V_P\cdot[1+N_k\cdot\phi_{is}\cdot(P_d-P_{dd})/P_{dd}] \qquad (2)$$

where

$V_{Ph}$ is the P-wave velocity expected at some point
$N_k$ is a dimensionless empirical constant that describes the velocity response to pressure
$P_d$ is the local differential pressure
$P_{dd}$ is the reservoir datum differential pressure

Back-calculation from King's data gave an $N_k$ value of 0.07. It is appreciated that these adjustments for fluid content and pressure are simplistic compared with the sophisticated approaches of Toksoz, Biot, and others. However, under the reservoir conditions of depth and pressure being considered, the sensitivities of velocity to these factors are so low as to make first-order approximations acceptable.

## Bulk density

The acoustic impedances of rock layers, which of course control the seismic reflectivities of the interfaces between layers, are proportional to the layers' bulk densities. For the purpose of this study, the simplest possible mixing law for bulk density was assumed to apply, namely that the bulk density of a mixture is the volume-weighted sum of the densities of the constituents. The constituents' densities (particularly those of hydrocarbons) are, in turn, sensitive to pressure through their compressibilities.

## Shales

Fillipone (1982) suggested that the seismic velocity is approximately linearly related to the ratio of pore pressure to overburden pressure in shales, and that this can be calibrated empirically using local well data. The underlying assumption is that, locally, the pore pressures of shales are directly related to their fluid volume fractions.

Gardner *et al.* (1974) found that for many formations, shales included, their densities were proportional to the fourth roots of their velocities. Thus, it would appear that the main sensitivities of both P-wave velocity and acoustic impedance to shale parameters are to the local degree of overpressuring.

Wyllie's time-average equation, in which the transit time through the whole is equal to the sum of the transit times through the volume fractions of constituents, although insufficiently accurate to deal with porosity, is still considered appropriate for normal transmission through layer-piles of different materials. Hence the seismic velocity response to various net/gross ratios of sand and shale may be estimated. Estimating the reflectivity response is more complex, but if the sand/shale layering is much finer than the seismic wavelength, then average properties over successive quarter-wavelengths may be used to approximate the acoustic impedance and thus reflectivity.

## Absorption

The main effects of seismic absorption are a progressive loss of high-frequency energy with depth, and an associated change in wavelet phase. In most circumstances, there is also a minor contribution to interface reflectivity, while dispersion linked with absorption sometimes causes poor ties between surface seismic and sonic logs. In the context of surface seismic recording, however, the loss of high-frequency energy is by far the most important effect remaining after processing.

Hamilton (1972) showed a non-linear relationship between absorptivity and grain size, with losses varying from 0.1 to 1.1 dB/wavelength (that is, the "quality factor" $Q$, the number of radians of phase taken for wave energy to decay to $1/e$ times its initial value, varied from 300 to 25, respectively). Others have found values for consolidated rocks to be within the same general range. Clean, well-cemented, low-porosity rocks seem to have high $Q$ values. So do similar rocks but of high porosity, if they are either dry or 100% liquid-saturated. Low $Q$ values seem associated with partial gas saturation, poorly consolidated rocks, and silty sands. An overall value of around $Q=100$ (losses of about 0.27 dB/wavelength) is thought to be fairly typical of the post-Cretaceous rocks of the northern North Sea.

A loss of 0.27 dB/wavelength gives an exponential, frequency-dependent, decay of amplitude during travel, to about 0.97 raised to the power of the frequency–travel time product (i.e. the distance travelled in wavelengths). Given the depth to a reservoir and the seismic velocity, plus knowledge of the limitations of spectral recovery by seismic deconvolution, an estimate can be made of what is the highest recoverable frequency in reservoir reflections. For example, reflections from a reservoir beneath 3000 m of 4000 m/sec overburden, when subjected to a deconvolution process capable of 20 dB (10:1 amplitude ratio) of spectral rebalancing could in principle yield a bandwidth of about 50 Hz. In practice, the presence of multiple reflections and scattered noise, and the filtering action of successions of closely spaced reflectors, would result in a lower figure, perhaps 30 Hz.

If deconvolution could be made twice as effective by substantially improving the seismic signal-to-noise ratio, so that 20:1 spectral amplitude imbalance could be compensated for, the theoretically recoverable bandwidth would be improved only from 50 Hz to 64 Hz. Thus, bandwidth expectations in the range of 30–60 Hz are realistic for surface seismic data from North Sea reservoirs.

A well-known corollary of the sampling theorem in information theory (Shannon, 1949) is that a bandlimited signal contains a total of precisely two independent data per cycle of the bandwidth frequency. This can be established from Fourier series analysis. If one then assumes a typical consolidated sand velocity of 4000 m/sec, one finds that a bandwidth of 60 Hz gives an average (one-way) thickness resolution of about 16 m, although finer detail than this may be resolved under some circumstances. For example, in the situation where there is a single thin layer within a homogeneous background (Farr, 1978), or where external information can be incorporated (Cornish and King, 1988), such finer detail is possible.

## HYPOTHETICAL FIELD MODEL

For demonstration purposes, a simple model of a hypothetical North Sea oilfield was created, and corresponding synthetic seismic data were computed and displayed. The overall approach was to design reservoirs with conditions and attributes roughly parallelling those found in the Cretaceous and Jurassic oil reservoirs in the northern North Sea, although all the features modelled would not be found in any one oilfield.

The model was designed as a 5000 m (E–W) by 3500 m (N–S) elliptical anticline, down-faulted just east of the crest. It was described using a grid of rectangular cells each about 80 m (E–W) by 95 m (N–S) by 10 m thick. The maximum dip was 6°W, and the crestal depth 2591 m (MSL).

Three separate reservoirs were described (as shown in Table I), each with a uniform 80 m gross thickness, and enclosed in homogeneous shales. The lower two were sandstone, and the top one was carbonate. Each reservoir comprised eight 10 m zones, in a cyclical sequence of pay, pay, and shale. It contained fairly light oil with a gas cap, and was surrounded by an aquifer. The bottom reservoir alone was overpressured progressively through successive (isolated) zones.

Successive 10 m pay zones were alternatively given totally uniform propeties, or had porosity, shaliness or net/gross ratio varying linearly only in the N–S direction. The best reservoir quality was always to the north.

Porosities varied from 1% to 19% in the limestone, and from 11% to 29% in the sandstones. Shale fractions varied from 2% to 38%, and the net-to-gross ratio from 4% to 100%. An E–W fluid saturation cross-section, and a N–S porosity section are shown in Figs 1 and 2.

For each cell, the P-wave velocity and the acoustic impedance (product of density and velocity) were estimated in a manner consistent with the seismic sensitivity relationships described in Table II. Figs 3 and 4 show E–W and N–S acoustic impedance cross-sections for the model.

Using the estimated velocities, the depth scales were converted to seismic reflection times, and the seismic reflectivity logs were calculated. These are shown unfiltered in Fig. 5, and after filtering to a set of successively narrower

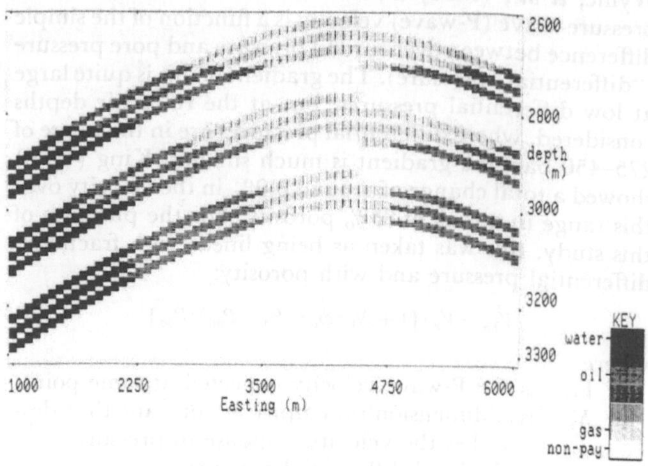


**Fig. 1.** E–W section through a hypothetical field model, showing gas and oil regions. Each of the three reservoirs comprises six layers of 10 m-thick cells and two 10 m shale zones. The top reservoir is carbonate, the lower two are sandstone.

**Table I.** Formation parameters

| Zone No. | Thickness (m) | Lithology | G.O.C.[a] | O.W.C.[b] | Porosity (%) | Shaliness (%) | Over pressure | Net-to-gross (%) |
|---|---|---|---|---|---|---|---|---|
| 1 | — | shale | | | | 100 | 1 | 0 |
| 2 | 10 | LST | 2702 | 2824 | 1–19[c] | 20 | 1 | 100 |
| 3 | 10 | LST | 2702 | 2824 | 10 | 20 | 1 | 100 |
| 4 | 10 | shale | | | | 100 | 1 | 0 |
| 5 | 10 | LST | 2702 | 2824 | 10 | 20 | 1 | 100 |
| 6 | 10 | LST | 2702 | 2824 | 1–19[c] | 20 | 1 | 100 |
| 7 | 10 | shale | | | | 100 | 1 | 0 |
| 8 | 10 | LST | 2702 | 2824 | 10 | 20 | 1 | 100 |
| 9 | 10 | LST | 2702 | 2824 | 10 | 39–1[c] | 1 | 100 |
| 10 | 80 | shale | | | | 100 | 1 | 0 |
| 11 | 10 | SST | 2862 | 2984 | 11–29[c] | 20 | 1 | 100 |
| 12 | 10 | SST | 2862 | 2984 | 20 | 20 | 1 | 100 |
| 13 | 10 | shale | | | | 100 | 1 | 0 |
| 14 | 10 | SST | 2862 | 2984 | 20 | 20 | 1 | 100 |
| 15 | 10 | SST | 2862 | 2984 | 11–29[c] | 20 | 1 | 100 |
| 16 | 10 | shale | | | | 100 | 1 | 0 |
| 17 | 10 | SST | 2862 | 2984 | 20 | 20 | 1 | 100 |
| 18 | 10 | SST | 2862 | 2984 | 20 | 39–1[c] | 1 | 100 |
| 19 | 80 | shale | | | | 100 | 1 | 0 |
| 20 | 10 | SST | 2962 | 3084 | 20 | 20 | 1 | 100 |
| 21 | 10 | SST | 2962 | 3084 | 20 | 20 | 1 | 100 |
| 22 | 10 | shale | | | | 100 | 1.25 | 0 |
| 23 | 10 | SST | 2992 | 3114 | 20 | 20 | 1.5 | 100 |
| 24 | 10 | SST/shale | 2992 | 3114 | 20 | 20/100 | 1.5 | 4–96[c] |
| 25 | 10 | shale | | | | 100 | 1.75 | 0 |
| 26 | 10 | SST | 3022 | 3144 | 20 | 20 | 2 | 100 |
| 27 | 10 | SST/shale | 3022 | 3144 | 11–29[c] | 20/100 | 2 | 4–96[c] |
| 28 | – | shale | | | | 100 | 2 | |

[a] Gas–oil contact.
[b] Oil-water contact.
[c] Varied spatially.

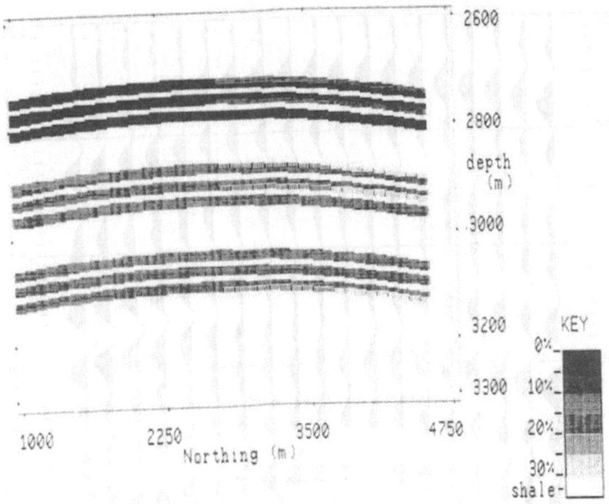


**Fig. 2.** N–S section through the oil region of the hypothetical field model, showing cell average porosities varying linearly with distance in some zones. Some other zones had linear variations in their shale content.

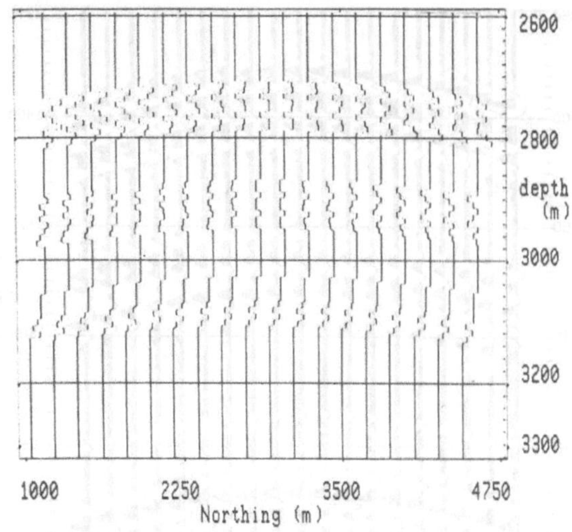


**Fig. 4.** Synthetic acoustic impedance logs for the N–S section.

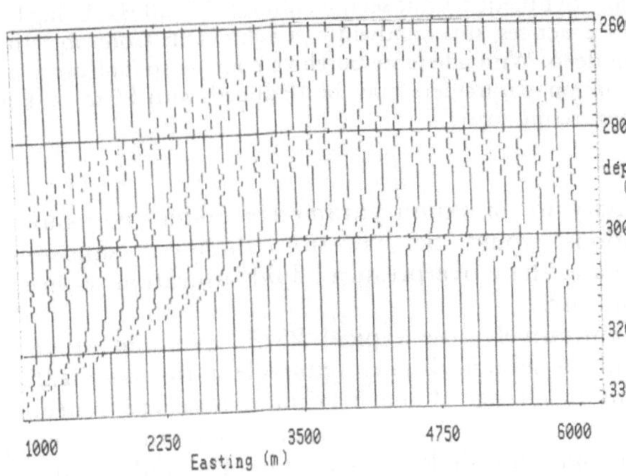


**Fig. 3.** Synthetic acoustic impedance logs for the E–W section through the model.

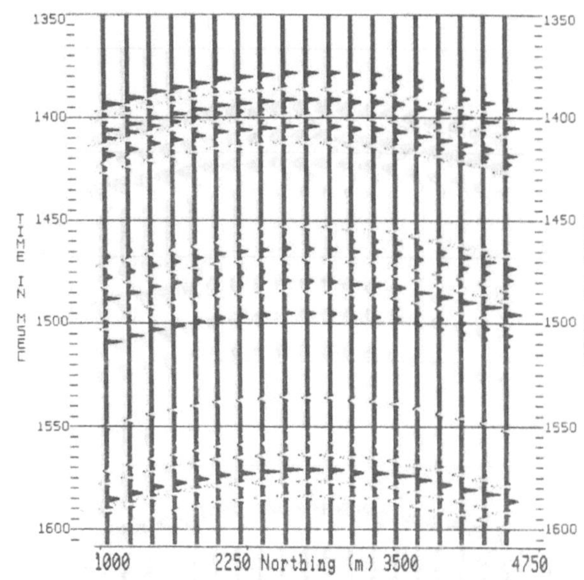


**Fig. 5.** Synthetic seismic reflectivity section (400 Hz bandwidth) computed from the acoustic impedance logs of Fig. 4. The separate responses to individual interfaces can be seen.

bandwidths (Figs 6, 7, 8). As previously mentioned, the achievable limit for bandwidth under the modelled conditions is around 30–60 Hz, so that Fig. 8 (7.5–50 Hz) is representative of a practical result. The wider bandwidths of Figs 6 and 7 were generated to simulate the results that might have been achieved under more favourable conditions, or with more effective *Q*-compensation processing.

**Table II.** Seismic sensitivity summary for field model

| *Parameter* | *Velocity* (*m/s*) | *Bulk density* (*kg/m*$^3$) | *Acoustic impedance* ($\times 10^6$) |
|---|---|---|---|
| Porosity = 29% | 3070 | 2050.9 | 6.295 |
| Porosity = 20% | 3730 | 2227.7 | 8.308 |
| Porosity = 11% | 4390 | 2404.7 | 10.555 |
| Shaliness = 39% | 3327 | 2191.1 | 7.291 |
| Shaliness = 20% | 3731 | 2228.0 | 8.312 |
| Shaliness = 2% | 4135 | 2265.1 | 9.366 |
| 100% Sandstone | 3783 | 2242.0 | 8.481 |
| 100% Limestone | 4329 | 2277.9 | 9.860 |
| 90% Gas saturation | 3665 | 2144.7 | 7.860 |
| 90% Oil saturation | 3730 | 2227.7 | 8.308 |
| 100% Water saturation | 3775 | 2290.8 | 8.648 |
| 50% overpressured | 3711 | 2204.4 | 8.181 |
| Normal-pressured | 3733 | 2228.4 | 8.318 |
| Shale 50% overpressured | 3319 | 2349.3 | 7.797 |
| Shale normally pressured | 3929 | 2450.0 | 9.626 |

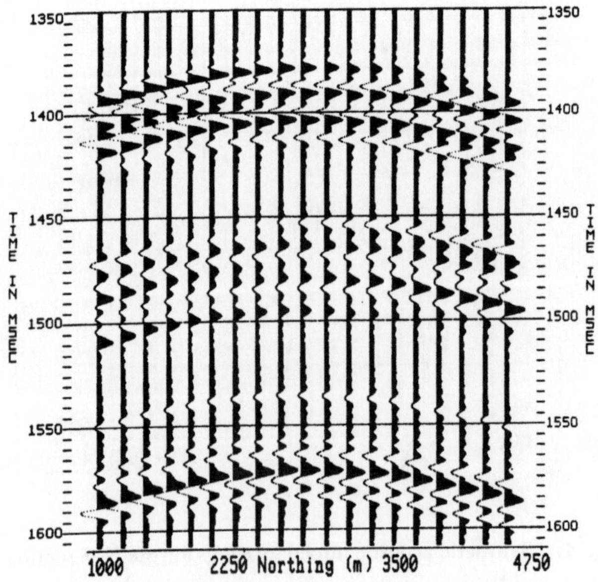


**Fig. 6.** Synthetic seismic section, filtered 30–200 Hz.

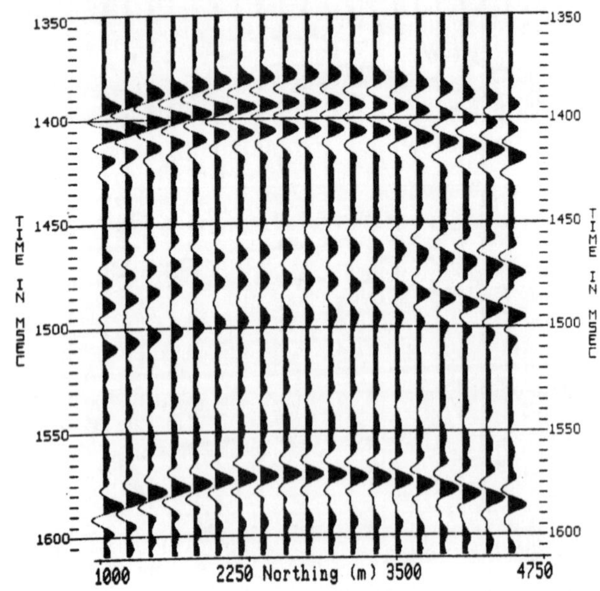


**Fig. 7.** Synthetic seismic section, filtered 15–100 Hz.

## RESULTS

### Sensitivity

The expected ranges of variation of velocity and density, in response to individually varying formation parameters and pressure, were plotted relative to a base case. This "norm" was a sandstone reservoir of 20% porosity, 20% shaliness, with 90% oil saturation, and normally pressured at about 3000 m (MSL). This was compared with cases in which each parameter in turn attained its maximum and minimum likely values. The "norm" for the shale overpressure sensitivity was a normally pressured massive shale.

In Fig. 9, the "norm" or base case acoustic impedance is placed on the vertical "0%" scale line. The percentage changes in velocity and density in response to changes of porosity, shaliness, etc., from their extreme values are represented by the lengths of the bars to the left or right.

From Fig. 9, it can be inferred that the rank order for

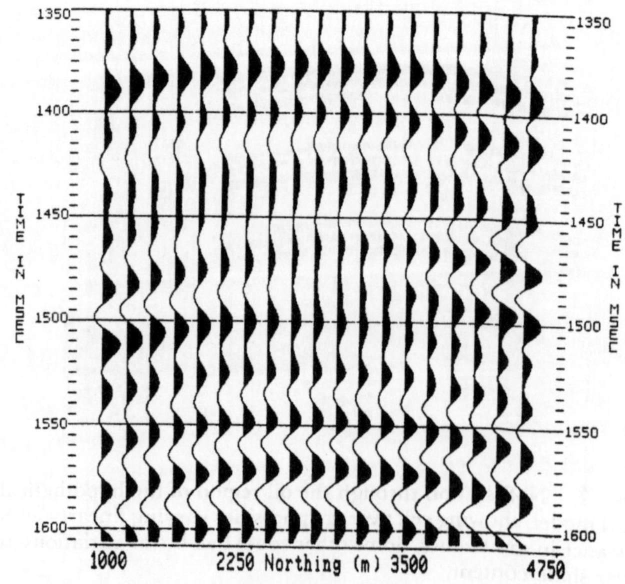


**Fig. 8.** Synthetic seismic section, filtered 7.5–50 Hz. Using best current practices, this bandwidth is achievable from objectives 3000 m deep. With adequate borehole control, spatial variations of formation parameters may be deduced with 20 m or finer thickness resolution.

sensitivity to formation parameters under the assumed conditions would be: porosity; shaliness; sandstone/carbonate; shale overpressure; fluid saturation; reservoir overpressure.

Quantitatively, the typical 25 dB dynamic range of a standard seismic display (equivalent to about 2.25% reflectivity change if the maximum observed reflection coefficients were ±20%) would give the following approximate thresholds of detection of spatial variations.

| | |
|---|---|
| Porosity | 1.6% |
| Shaliness | 6.5% |
| Sandstone/limestone fraction | 28% |
| Overpressure in shales | 11% |
| Gas/oil or oil/water saturation | 90% (i.e. borderline) |
| Overpressure in reservoirs | undetectable. |

Note that the pressure results are not directly comparable with the others, being in a different dimension.

Whichever formation variable (excepting fluid saturation) changes, the velocity and density respond in roughly similar proportions. Since these are the main acoustic parameters determined from seismic results, the need is indicated for auxiliary relationships, such as porosity/shaliness correlations, in the derivation of reservoir parameters from seismic data. A fluid saturation change, however, may be distinguished by exhibiting a disproportionately large density change relative to the velocity change. In practice, of course, a fluid interface reflection's unconformability with structural reflections is usually a stronger diagnostic.

### Resolution

Figure 5 is a N-S seismic reflectivity section with an effective bandwidth of approximately 400 Hz. It clearly shows a separate reflection from each of the nine interfaces within each of the three hypothetical reservoirs of the model, except of course where there was no acoustic impedance contrast across the interface (e.g. the top zones of the lowest reservoir). This indicates that if 400 Hz surface seismic data could be recovered from 3000 m

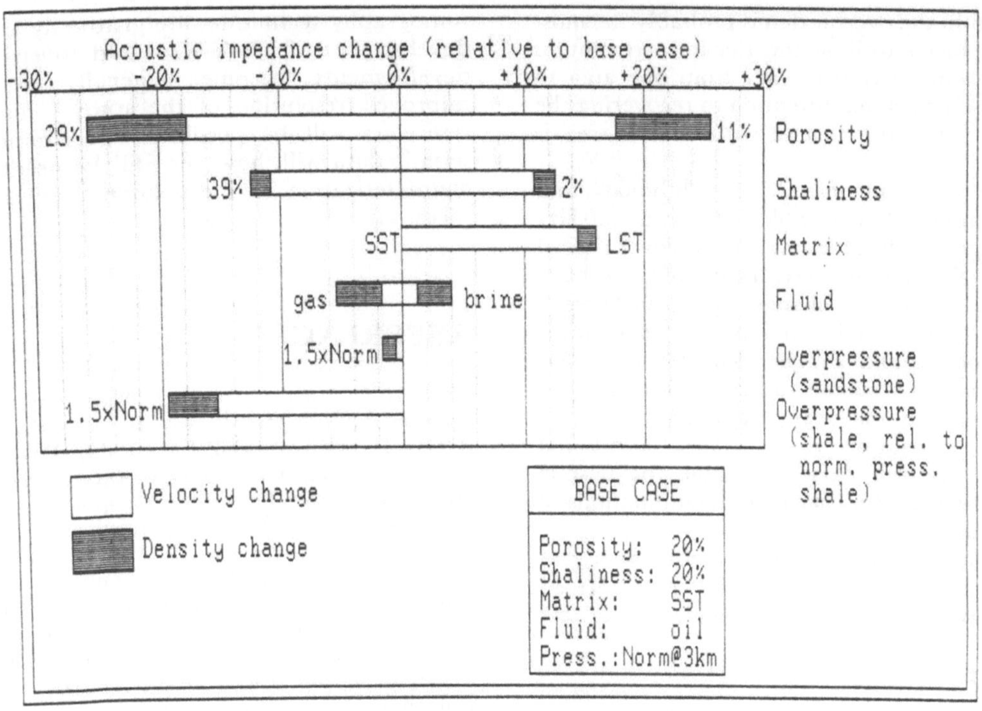


**Fig. 9.** Changes of acoustic impedance (partitioned into velocity and density contributions) that would be expected to result from changes of individual formation parameters to extreme values, relative to a base case.

(MSL), there would be no difficulty in estimating average porosity in 10 m thick zones.

On Fig. 6, which shows 30–200 Hz filtered data, the 10 m resolution would still be possible, although the seismic inversion stage would need to cope with the multilobed character of the wavelet. Information theory states that 200 Hz waves may contain independent data at a density of no more than 400 values per second on average, which at the seismic velocity of some 4000 m/sec, would give a value every 10 m of two-way travel, i.e. 5 m depth. These 200 Hz data could therefore yield two values per 10 m zone, enough to identify the slab-like nature of the reservoir as well as provide an estimate of reservoir quality.

The 15–100 Hz band of Fig. 7 clearly shows a peak-trough or trough–peak pair for each 20 m reservoir unit, and in some cases finer subdivisions of the waveform are visible, corresponding to the boundaries of the 10 m zones. Thus one could say that this bandwidth is marginal for detecting a succession of 10 m zones. This is in agreement with expectations from information theory.

The 7.5–50 Hz band shown in Fig. 8, although not very promising at first sight, does exhibit four or five extrema over each 80 m thick reservoir. These waveforms are the seismic responses averaged over approximately 20 m successive thicknesses of formation (with some overlap because of the wavelet shape). However, with this model, it is not possible to identify a simple interpretation because different parameters vary in an uncorrelated manner in successive 10 m zones. An algorithm based on interpolation between and extrapolation from control points (borehole information) would be needed.

### Interaction between resolution and sensitivity

In order to interpret a detailed distribution of porosity (or other) variations in 10 m thick zones from 50 Hz seismic data, it appears that two conditions are required. Firstly, the interpreter must invoke external information regarding the proportion of the seismic response that may be attributed to porosity rather than to other aspects of lithology. Secondly, the vertical distribution of porosity between each 10 m target zone and its immediate neighbours requires additional information (such as the observation of a consistent, coarsening-upwards trend within a formation).

It is with regard to this second point that the interaction between resolution and sensitivity occurs, in a type of "uncertainty principle". To a first approximation, the seismic response at a given level is the aggregated void volume over a thickness of rock equivalent to a quarter-wavelength (one-way travel) of the highest seismic frequency recovered. That is, for the example North Sea condition of 50 Hz recovered from 4000 m/sec rocks under 3000 m of $Q=100$ overburden, the thickness is 20 m.

## CONCLUSIONS

This study has been restricted to a limited range of geology and depths. Also the synthetic seismic reflectivity logs were based on one-dimensional modelling, and no account has been taken of complexities, such as the imaging and reflection amplitude recovery difficulties experienced in seismic processing over structurally complex areas with strong multiple reflections.

Within those limits, it appears that reservoir characterization using surface seismic data could be of value to reservoir engineers, particularly in its ability to predict variations in "average" porosity and shaliness at a thickness scale throughout a reservoir that is appropriate for reservoir simulation modelling.

In other circumstances, additional or different parameters may be interpreted. For example, in a shallow, poorly compacted reservoir, greatly increased sensitivity to pressure may well exist, and considerable sensitivity to saturating fluid type might be anticipated.

Absorption within the overburden is probably the most important factor in controlling the thickness resolution achievable. This will, of course, vary from one area to another, and it is obvious that attention to recovering the widest frequency band practicable within each area is necessary.

From the expected seismic sensitivities to lithology, and the current limitations of achievable bandwidth (50 Hz) and dynamic range (25 dB) of seismic sections over typical North Sea reservoirs that we have estimated, we have shown that lateral or vertical changes of 2% in average porosity or 7% in average shale fraction between adjacent 20 m thick grid blocks should often be resolvable. However, since seismic reflectivity responds to both of these and (to lesser degrees) to other lithological parameters, there is an ambiguity in interpretation that requires additional information, such as well data, to enable the seismic variations to be attributed to particular combinations of lithological variations.

Resolution of such variations in 10 m thick or finer grid blocks can be achieved by including borehole-derived systematic variations within 20 m intervals.

## Future developments

Considerable progress is currently being made (Cornish and King, 1988) in the area of spatial extension of borehole information, and integration with seismic data. These efforts are possibly the key to obtaining maximum information extraction and usage, and in overcoming the ambiguity and limited resolution of seismic responses.

Improvements in seismic sensitivity to lithology are mainly directed towards moving from the acoustic model of the Earth, the concern of this paper, to elastic or viscoelastic models, where shear-wave propagation and perhaps absorption are investigated. These should not only improve the sensitivity to certain parameters such as fluid saturation, and perhaps even permeability, but also increase the independence of the determined parameters. The most promising approaches involve "elastic inversion" replacing the seismic acoustic inversion. In this technique, prestack P-wave sourced seismic data can yield information on shear waves contained in the variations of received amplitudes versus source-receiver offset. In land surveys, S-wave sourced surveys may be used as well. Converted waves (P-S-P) are also being considered for use offshore, but there are problems—for example, extremely long receiver streamers seem to be required.

The attainment of a markedly greater bandwidth appears to be impeded mainly by the absorptivity of the overburden, so the main hopes for improvement in this area involve placing the sources, receivers, or both, closer to the target formations. The vertical seismic profile techniques (VSP) using sources at the surface and receivers in the borehole are of course widely used today, and can provide relatively detailed information near the boreholes.

One of the possibilities for a significant increase in bandwidth to the kilohertz range, and permitting trans-borehole ranges of up to around a kilometre, is associated with research taking place into the development of safe and effective downhole sources and multicomponent, multi-element downhole receivers. At the same time, cross-hole tomography techniques are gradually giving way to the development of whole-wavefield inversion. When these developments become generally available, greatly improved resolution of the spatial characterization of reservoirs will be possible. Three-dimensional surface seismic data will, however, still be needed to extend the characterization to the volumes between the interwell slices.

## REFERENCES


Angeleri, G.P. and Carpi, R. (1982). Porosity prediction from seismic data. *Geophys. Prospect.* **30**, 580–607.

Bilgeri, D. and Ademeno, E.B. (1982). Predicting abnormally pressured sedimentary rocks. *Geophys. Prospect.* **30**, 608–621.

Birtles, R. (1986). The seismic flatspot and the discovery and delineation of the Troll Field. In: Spencer, A.M. *et al.* (eds.), *Habitat of Hydrocarbons on the Norwegian Continental Shelf*, pp. 207–215. Graham & Trotman, London.

Cornish, B.E.. and King, G.A. (1988). Combined interactive analysis and stochastic inversion for high-resolution reservoir modelling. Presented at the *European Association of Exploration Geophysicists' 50th Annual Meeting, The Hague.*

Domenico, S.N. (1984). Rock lithology and porosity determination from shear and compressional wave velocity. *Geophysics* **49**, 1188–1195.

Farr, J.B. (1978). High resolution seismic as a reservoir analysis technique. Presented at *53rd SPE Annual Meeting, Houston*, paper SPE 7440.

Fillipone, W.R. (1982). Estimation of formation parameters and the prediction of overpressures from seismic data. *Research Symposium on Geopressure Studies, Society of Exploration Geophysicists' 52nd Annual Meeting, Dallas.*

Gardner, G.H.F., Gardner, L.W. and Gregory, A.R. (1974). Formation velocity and density—The diagnostic basics for stratigraphic traps. *Geophysics* **39**, 770–780.

Hamilton, E.L. (1972). Compressional wave attenuation in marine sediments. *Geophysics* **37**, 620–646.

Han, D., Nur, A. and Morgan, D. (1986). Effects of porosity and clay content on wave velocities in sandstones. *Geophysics* **51**, 2093–2107.

Jones, T.D. (1986). Pore fluids and frequency-dependent wave propagation in rocks. *Geophysics* **51**, 1939–1953.

King, M.S. (1966). Wave velocities in rocks as a function of changes in overburden pressure and pore fluid saturants. *Geophysics* **31**, 50–73.

Martinez, R.D. (1985). Deterministic estimation of porosity and formation pressure from seismic data. *Proceedings of the Society of Exploration Geophysicists' 55th Annual Meeting, Washington, D.C.*

Schlumberger (1986). *Log Interpretation Charts.* Schlumberger Well Services, U.S.A.

Shannon, C.E. (1949). Communication in the presence of noise. *Proc. Inst. Radio Eng.* **37**, 11.

Stainsby, S.D. and Worthington, M.H. (1985). Q estimation from vertical seismic profile data and anomalous variations in the central North Sea. *Geophysics* **50**, 615–626.

Toksoz, M.N., Cheng, C.H. and Timur, A. (1976). Velocities of seismic waves in porous rocks. *Geophysics* **41**, 621–645.

Wyllie, M.R.J., Gregory, A.R. and Gardner, G.H.F. (1958). An experimental investigation of factors affecting elastic wave velocities in porous media. *Geophysics* **23**, 459–493.

Wyllie, M.R.J., Gregory, A.R. and Gardner, L.W. (1956). Elastic wave velocities in heterogeneous and porous media. *Geophysics* **21**, 41–70.

# 4

# Stratigraphic Processing of 3D Seismic by Concurrent Analysis of Surface and Borehole Data

*P.S. Schultz*[a], *R. Gir*[b], *D. Pajot*[c] *and S.S.C. Wu*[d]

[a]GECO Geophysical Company Ltd, Orpington, UK
[b]Schlumberger Overseas S.A., Kuala Lumpur, Malaysia
[c]Schlumberger Austin Systems Center, Texas, USA
[d]Chinese Petroleum Corporation

When seismic data are available in production areas, it is often desirable to use them for stratigraphic interpretation to estimate the extent of existing reservoirs. Three-dimensional (3D) seismic data that contain the producing wells within their volume are particularly suitable for such an exercise. We present an interpretative processing procedure, supported by a case study, where surface and borehole data are used in cooperation to achieve maximum reliability of the final surface seismic inversion.

The study addresses a series of producing sand layers from a single well to be analysed for reservoir extent. The processing procedure begins with suitably processed 3D seismic and borehole data, from which are extracted: (1) a *deviated trace*, interpolated along the well trajectory; (2) *general time surfaces*, a set of mathematical surfaces approximating horizons, and (3) a *corrugated section*, a generalized 2D section coincident with the deviated well. It continues with corrections and calibrations to the residual wavelet, computed using the borehole data and applied to the 3D surface seismic data along structure. The procedure concludes with a bandlimited inversion of the seismic data, calibrated by the borehole data, applied to a generalized 2D section, and extrapolated along structure away from the borehole. With the aid of the final calibrated inversion, a refined estimate of the current reservoir volume was possible.

## INTRODUCTION

This study focused on a single structure within an active and producing offshore gas field. A large and impermeable listric fault controls closure for several producing sands on the upthrown side (underside) of the fault. Several production wells have been drilled and logged in the area. We studied a production zone of one of those wells, comprising a series of sand sheets between 2.0 and 3.0 seconds on the section. We concentrated on a formation consisting of a series of producing layers in a 250 ms interval. The two-way travel times in this study are given only approximately to protect the confidentiality of the data.

Figure 1 is a simplified structure map of the producing formation, showing the controlling listric fault and the location of three of the many wells drilled into the formation. Two producing areas can be seen on the map. A subset of the full 3D data was chosen for this case study, shown as the small (1 km$^2$) square enclosing the eastern producing area. The well used in this study is deviated. It intersects the fault plane about 500 ms above the formation under study. Thereafter, it remains almost parallel to it within the upthrown block, so as to intersect all possible gas accumulations.

The purpose of the study was to provide, by careful and concurrent analysis of surface and borehole data, independent verification of the existing estimates of reservoir volume, which had been obtained from a well-to-well correlation analysis of several wells penetrating the formation (not shown in the figure).

We shall describe a stratigraphic processing sequence that was used in this study to produce a calibrated inversion to acoustic impedance, from which an estimate of the reservoir extent was made. The processing procedure is applied post-stack, and is composed of three major steps: data preparation, residual wavelet processing, and calibrated inversion.

We shall describe all the main components of the stratigraphic processing procedure in the context of the present case study. Inasmuch as 3D seismic data were used for the study, all descriptions of the processing procedure will refer to 3D data, although the approach can be adapted to 2D data as well.

*North Sea Oil and Gas Reservoirs—II*

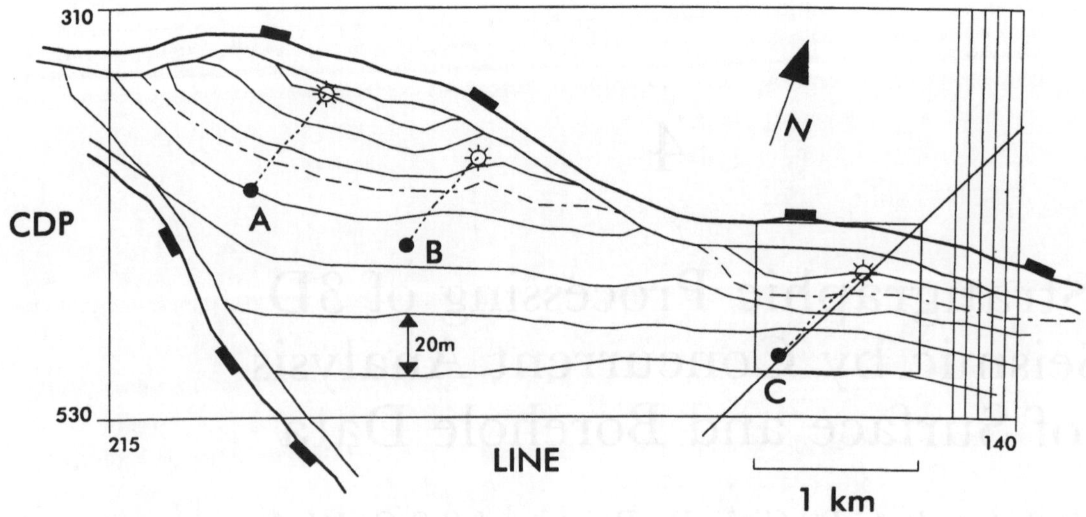


**Fig. 1.** A simplified structure map of the gas-producing formation. A large controlling listric fault trends E – W. Two reservoirs are shown, with the gas – water contact indicated by a dashed line. Many wells have been drilled into these reservoirs, but only three of them are shown. The large rectangular area shows the extent of the full 3D survey, while the small square (1 km²) enclosing deviated well "C" is the data subset used in this study. The straight line drawn through the platform location of "C" indicates the azimuth of the *corrugated section* extracted for the inversion.

## DATA PREPARATION

A vertical raypath VSP, a large suite of logs, and 3D seismic data were available. Each of these data types was processed in the normal fashion. The 3D seismic data were processed to a migrated volume. The logs were edited and processed, and the sonic (calibrated using the VSP) and density logs were combined to produce an acoustic impedance log. The VSP was processed to produce a corridor stack.

All concurrent surface and borehole analyses were performed in the time domain, and the logs were converted to the time domain using the time–depth curve from the VSP. The time domain was chosen because the logs could be accurately converted to time using the VSP, whereas the seismic data volume could not be converted to depth with equivalent accuracy.

Although the record of processing history was scrutinized and found to be acceptable, assumptions had to be made regarding the suitability of prestack processing applied to the data. In general, suitable processing of the seismic data for these purposes would ideally include the following non-exhaustive list of procedures to be followed carefully at all stages of processing: proper handling of relative amplitudes, wavelet compression, attenuation of multiples, suppression of other coherent noise in the data, and 3D migration with accurate velocities. Although not available for this dataset, a zero-offset stack from an amplitude-versus-offset analysis is appropriate when signal-to-noise ratio warrants it, since quantitative comparisons will be made with vertical raypath VSPs.

We have used three uncommon data and interpretation constructs that proved convenient during the course of this study. These constructs are (1) *the deviated trace*, which is a trace interpolated from the 3D data along the trajectory of the deviated well; (2) *general time surfaces*, which are a set of mathematical surfaces approximating geological horizons; and (3) *the corrugated section*, which is a generalized 2D section of arbitrary azimuth extracted from the 3D volume in such a way that the deviated well is coincident at all times. We have taken the liberty of giving these constructs names because they did not appear commonly in the geophysical vocabulary.

### The deviated trace

When analysing borehole and seismic data concurrently, it is convenient to be able to study a single seismic trace representing data at the well. Extracting such a trace from the 3D volume when the well is deviated represents a non-standard data representation. We used such a data representation in this study, and have named it the *deviated trace*. Its extraction is described below.

The initial step is to convert the well deviation data into the time domain (two-way travel time). This was done using the time–depth curve from the vertical raypath VSP. After the seismic data are 3D migrated, the 3D data volume is interpolated along the deviation of the well (as shown in Fig. 2) to extract the deviated trace, $T_D(t)$, which is defined as follows: If $P(x,y,t)$ is the 3D seismic data volume, and $(X_W(t), Y_W(t))$ is the deviated well trajectory expressed as a function of time, then

$$T_D(t) \equiv P(X_W(t), Y_W(t), t) \tag{1}$$

The sampling of this trace requires spatial interpolation only, as can be seen from the equation. The time sampling is unchanged, thus retaining the integrity of wavelet shapes. This trace will be used to calibrate the surface seismic data at the well.

The spatial interpolation was done in a straightforward manner, using $(x,y)$ spatial interpolation filters. No accommodation was made for the treatment of aliased steeply dipping events during the interpolation step. For these data, there was no need, since dips were small. For other data, a more elaborate interpolation scheme might be necessary to ensure the preservation of the integrity of amplitudes of events of any dip.

### General time surfaces

Approximate structural time surfaces, or *general time surfaces*, were then extracted from the data. These surfaces, which resemble horizons, are mathematical constructs designed to follow structure approximately. They have the properties that they:

- *represent local dip*—local tangents to the surfaces are parallel to the local dip in the data;

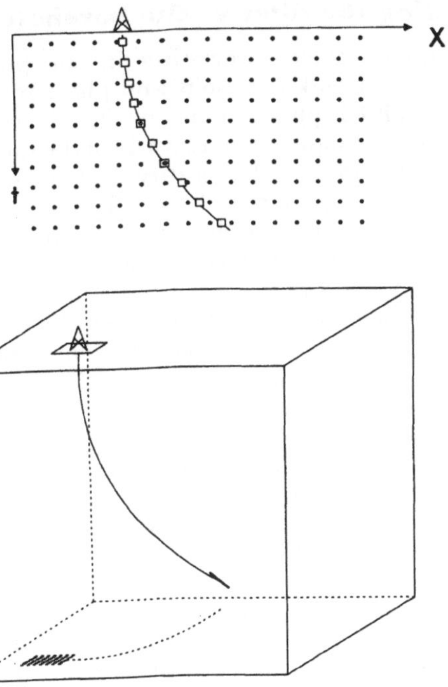


**Fig. 2.** The *deviated trace* extraction. The 3D seismic data volume is interpolated along the well trajectory using the time-based deviation information. Interpolation is performed spatially only; the uniform time sampling is maintained, thus preserving wavelet shapes.

- *are single-valued and continuous*—they are not double-valued in the presence of reverse faults, nor subject to gaps in the presence of normal faults;
- *span the entire area*—they do not terminate as do horizons in the presence of pinchouts or erosional unconformities.
- *are conformable*—they do not intersect or truncate into one another;
- *can be interpolated*—because they are conformable and continuous, they can be interpolated such that a time surface can be constructed to pass through any desired point.

These surfaces will be used to compute, automatically, spatially varying time gates that follow structure for wavelet correction filtering. They will also be used to extrapolate the low-frequency constraints in the inversion step.

Figure 3 shows a vertical section from the 3D data volume through the reservoir containing the controlling listric fault. Intersections with the set of general time surfaces are superimposed. It can be seen that they follow the structural trends in the data without actually being a proper geological horizon. Geological horizons can be used globally for these purposes only if they meet the five criteria given above, which in this case, and in most cases, they cannot.

## Corrugated section

The inclusion of constraints from borehole data in the inversion to acoustic impedance required the extraction of a generalized 2D section, or *corrugated section*, from the 3D volume. It is a section of arbitrary azimuth, on which the deviated well is coincident at all times.

The corrugated section is defined by a horizontal generating line of fixed azimuth, which cuts the 3D volume while intersecting the well at all times. We define a

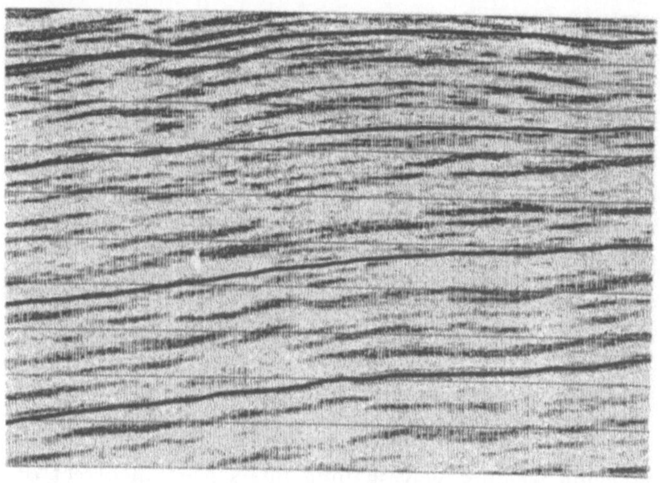

**Fig. 3.** *General time surfaces* are surfaces approximating horizons, and are computed automatically. Four surfaces and the large listric fault can be seen in this figure. Because of their special mathematical features (see text), they can be used for extrapolating time gates and low-frequency components away from the borehole.

corrugated section, $P_C(x, t)$, to be

$$P_C(x, t) \equiv P[x, y = Y_W(t) + (x - X_W(t) \tan\alpha, t] \qquad (2)$$

for $\alpha \neq \pi/2$, or,

$$P_C(y, t) = P[x = X_W(t) + (y - Y_W(t)) \cot\alpha, y, t] \qquad (3)$$

for $\alpha \neq 0$, where $\alpha$ is the angle of the generating line with the $x$-axis. Figure 4 shows a schematic diagram of corrugated sections for azimuths of 0° and 90°. Figure 1 shows the azimuth of the corrugated section extracted for this study.

## Residual wavelet processing

Because most standard wavelet-processing techniques in surface seismic processing are statistical in nature, there

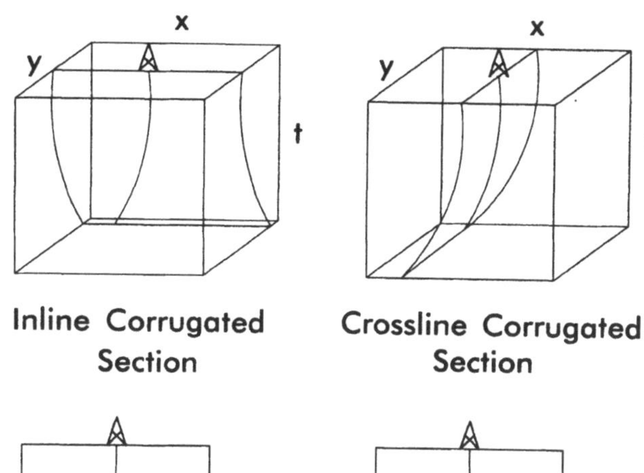


**Fig. 4.** The *corrugated section* is a generalized 2D section constructed with a fixed azimuth, and with the property that its surface is coincident with the well. This section is extracted from the 3D data volume using the azimuth shown in Fig. 1. All subsequent seismic section figures show the corrugated section.

will commonly be a residual distortion of the final wavelet from the desired zero-phase ideal. Commonly used statistical deconvolution techniques (e.g., whitening deconvolution, predictive deconvolution, MED and various parsimonious deconvolution methods) are applied in the hope of correcting wavelet-distorting effects both in the earth and in processing.

In the physical world, sources of wavelet distortion include short-period near-surface reverberations, short-period interbed multiples, angle-dependent reflection coefficients, attenuation/dispersion, and source and receiver array responses. In the processing world, final wavelets are affected by the choice of processing parameters in these and other common processing steps: mute pattern selection (especially when the wavelet changes with offset), prestack and post-stack deconvolution, stacking velocity selection, stacking weights, waveshaping filters from incorrect or incomplete knowledge of the source wavelet, and multichannel filtering operations.

*Residual wavelet processing* attempts to address this distortion after processing through the use of borehole seismic data. In the residual wavelet processing procedure, borehole seismic data are used to compute and to apply correction filters to the processed surface seismic data, shaping the residual seismic wavelet at the borehole. This is done by determining and applying a filter that converts the surface seismic deviated trace into the borehole corridor stack trace (or synthetic seismogram) within selected time gates. The filters are then extrapolated away from the borehole within spatially varing time gates that follow structure into the 3D data volume.

An alternative approach is one we call the *earth filter method*, in which borehole seismic data are used to calculate the earth transmission response for subsequent use in the surface seismic processing. This method was not used for several reasons, but chief among them is that it is not a residual processing method, and requires a full reprocessing of the surface seismic, beginning with prestack data. Appendix A describes these two methods mathematically, and compares them.

The underlying assumptions in estimating and applying a filter poststack from comparisons with borehole data is that the convolutional model is valid for the processed 3D data, and that the reflectivity is common between the surface seismic (deviated trace) and borehole seismic (corridor stack or synthetic seismogram) data. Furthermore, we assume that the borehole seismic trace is a zero-phase bandlimited reflectivity series.

We also assume that the surface seismic data have been sufficiently well processed so that: (1) The data are properly imaged in time (migrated). Although depth migration in principle can be used when overburden velocities vary laterally, the residual wavelet processing analysis should be done in the time domain because large velocity gradients can cause distortions of long wavelets during depth conversion. (2) Long period multiples and other coherent noise have been attenuated. (3) Any remaining deviations from the surface seismic being a zero-phase bandlimited reflectivity series can be described mathematically by a time-varying wavelet distortion.

The mechanism by which these filters are estimated can vary. For this study, the analysis was performed in an interactive environment. The borehole seismic data were a corridor stack from a vertical raypath VSP (i.e. where the surface source was moved to be always directly over the receiver in a deviated well); a synthetic seismogram was available and could have been used in the absence of a VSP. The surface seismic trace used in this analysis was the deviated trace, as described above.

## Estimating the filter at the borehole

The residual wavelet processing in the present study involved only a bulk time shift and phase rotation filters (i.e. time-shifted phase rotation filters), since it was decided that the signal-to-noise ratio of the surface seismic data was not high enough to justify applying more gener: shaping filters. The bulk shift was first estimated over large time gate. Then, within smaller time gates, ph: rotation filters were estimated to match the deviated tra to the corridor stack.

Figure 5 shows an image taken from the screen duri an interactive session to estimate the correction filter a shallow time gate, and shows the effectiveness of t time-shifted phase rotation filter in enhancing the mu interpretability of the surface and borehole data. For estimation of a bulk time shift and phase rotation filter envelope of the cross-correlation is a useful tool. The time shift is chosen from the location of the peak of the cross-correlation envelope, while the phase rotation filter angle is chosen to make the cross-correlation symmetrical within the envelope. When a phase rotation filter is applied to one of the traces in the cross-correlation, the envelope will remain invariant, although the cross-correlation itself undergoes changes (see Appendix B). This effect can be seen clearly in Figs. 5 and 6, which show the cross-correlation and its envelope before and after the shaping filter is applied.

In this study, the residual processing analysis estimated a bulk time shift of −26.60 ms, which was then applied to the surface seismic data. Although this time shift is larger than one might hope for, we have observed such shifts commonly in practice. The possible presence of large bulk time shifts underscores the need to include an analysis such as this one whenever a quantitative comparison between surface and borehole seismic is planned. Analysing a bulk time shift along a sizable length of the trace (Fig. 5 shows the window we used for the time shift calculation) when the well is deviated can be awkward and error-prone without the use of a construct like the deviated trace.

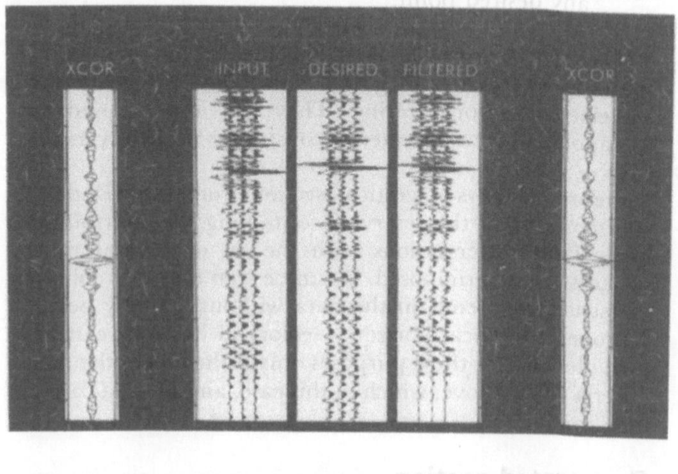


**Fig. 5.** The residual processing step involved computation and application of a time-shifted phase rotation filter. This figure shows the deviated trace (labelled, "Input"), the VSP corridor stack ("Desired"), and the filtered deviated trace ("Filtered"). The traces are repeated three times. The cross-correlation and its envelope are also shown. The left-hand cross-correlation trace is between the Input and Desired traces, and the right-hand one is between the Desired and Filtered. The corrective effect of the −26.6 ms time shift can be seen by comparing the cross-correlations.

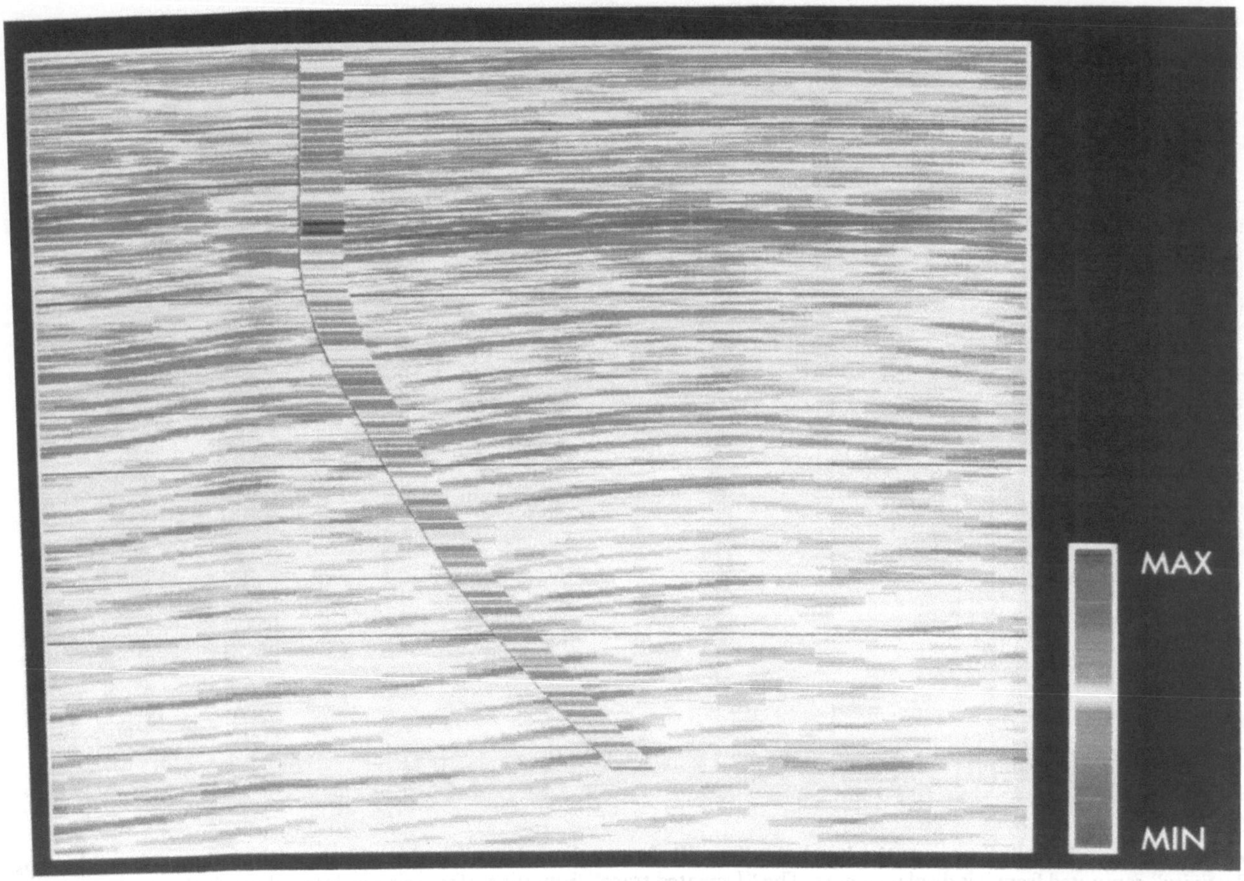


**Plate 17.** The *corrugated section* before the residual processing filter has been applied. The well is fully coincident with this section, and the trajectory is shown with the VSP corridor stack plotted alongside it to the right. The colour scale is common, but the amplitude normalization is different between the two data types, accounting for the colour disparities deeper in the section.

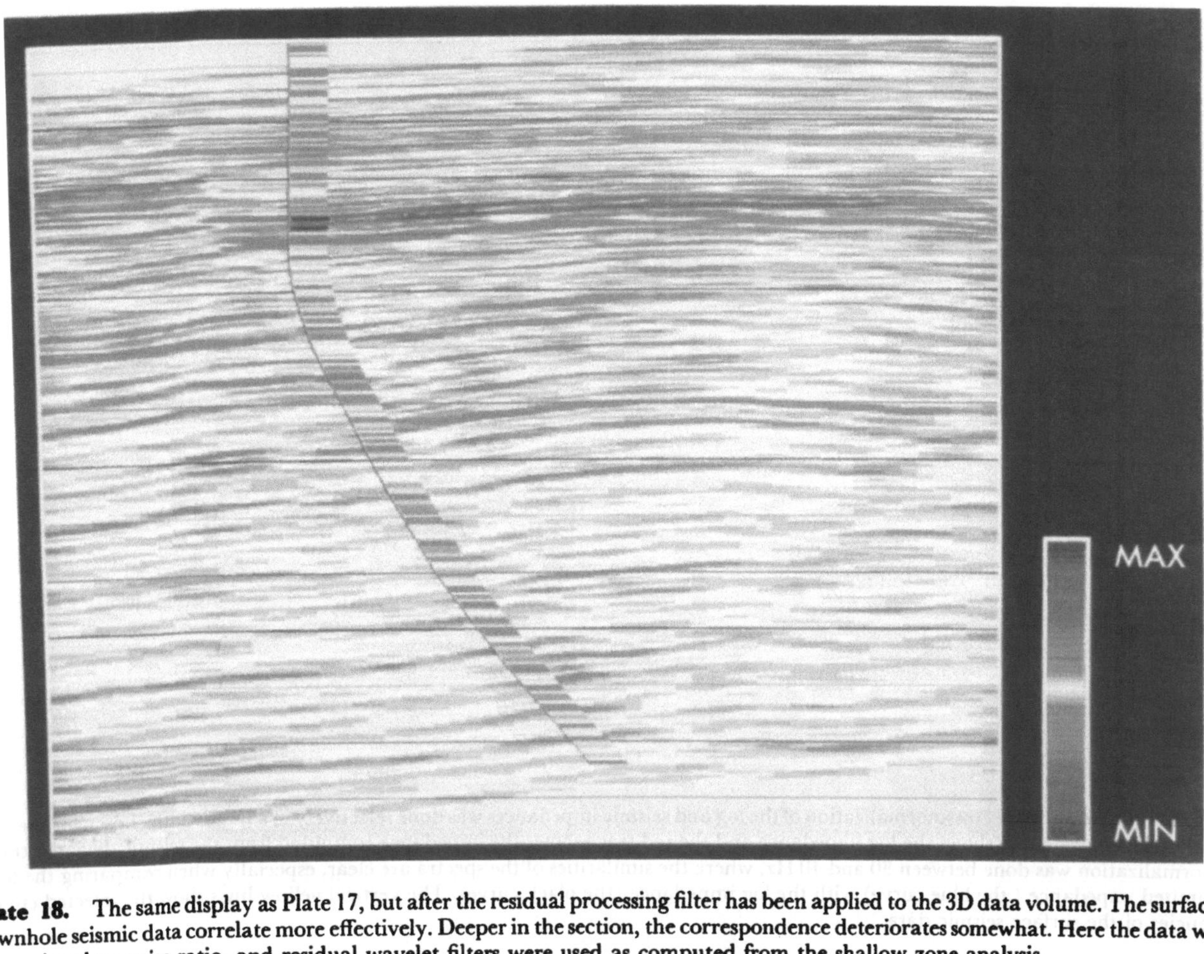


**Plate 18.** The same display as Plate 17, but after the residual processing filter has been applied to the 3D data volume. The surface and downhole seismic data correlate more effectively. Deeper in the section, the correspondence deteriorates somewhat. Here the data were of lower signal-to-noise ratio, and residual wavelet filters were used as computed from the shallow zone analysis.

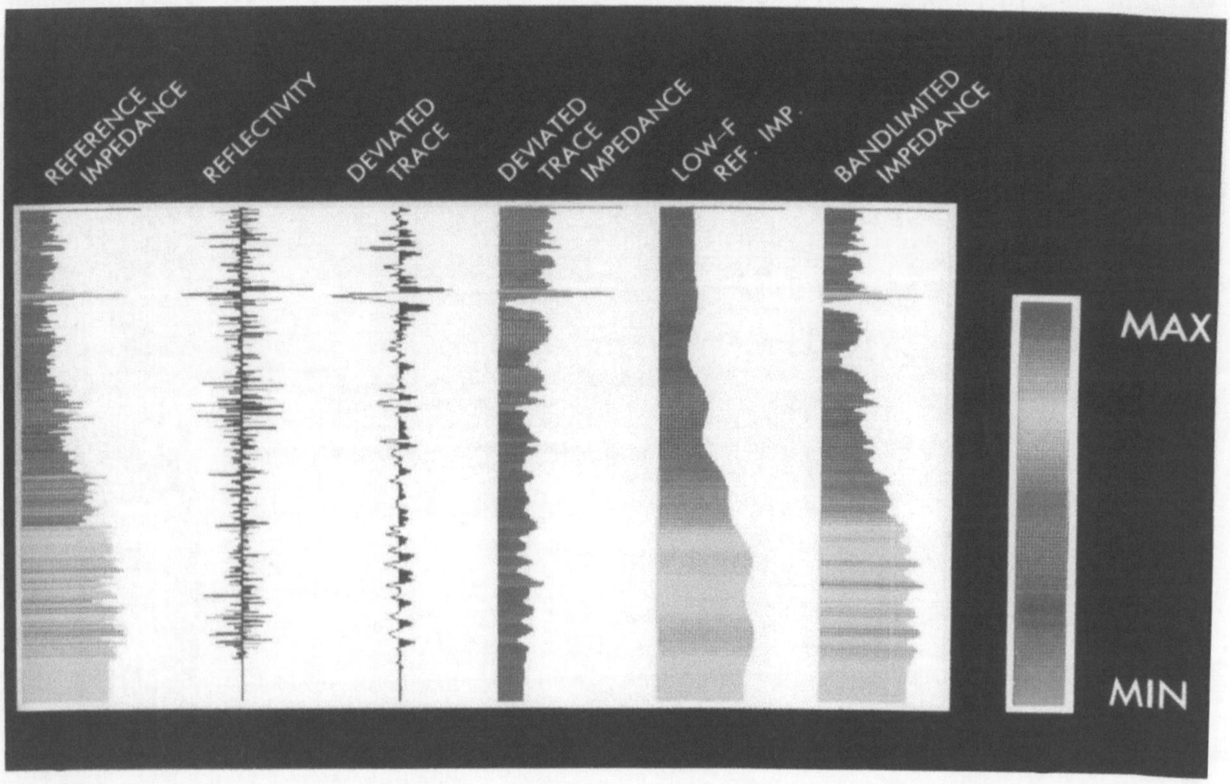


**Plate 19.** The key inversion steps at the borehole. The "Reference impedance" is the wide-band impedance from the logs; the "Reflectivity" computed from it is displayed next. The "Deviated trace" from the surface seismic data volume is the third trace, with the bandlimited impedance shown next, computed from the reflectivity equation. The Reference impedance is filtered to pass low frequencies only (those below the seismic band), and is shown as the fifth trace. The fourth and fifth traces are then amplitude-normalized (see Plate 20) and summed to form the final trace, labelled "Bandlimited impedance."

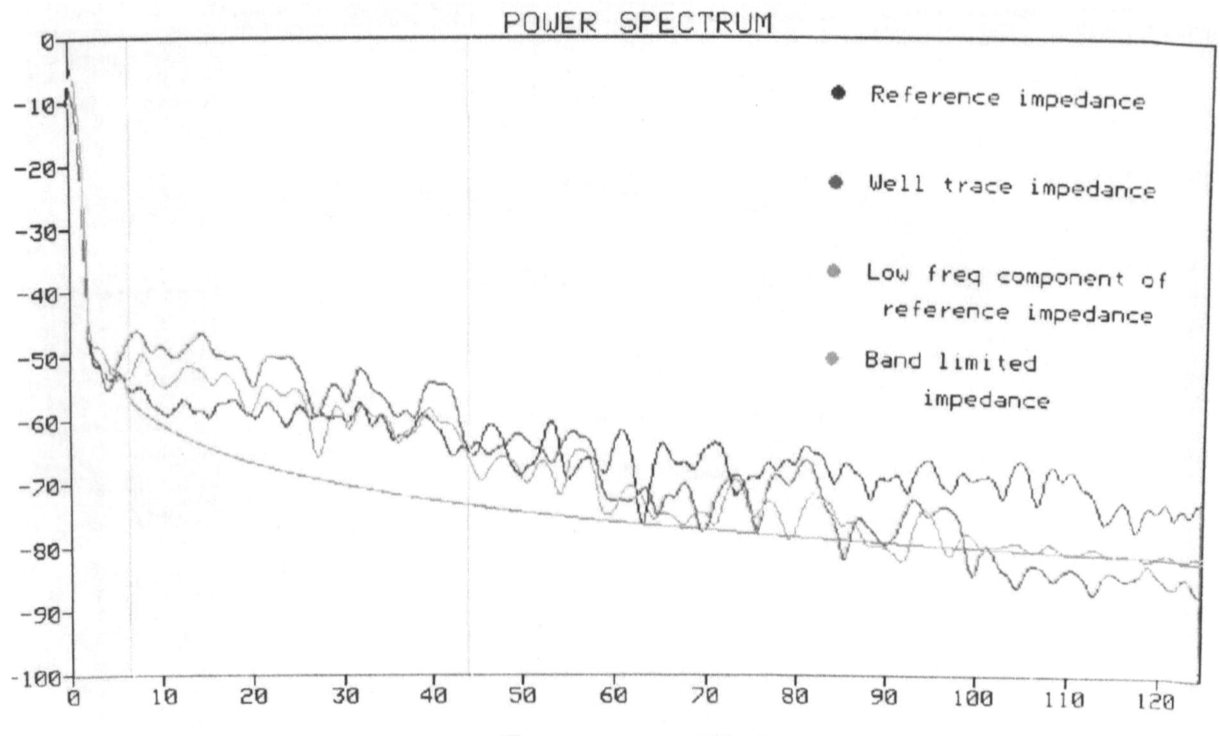


**Plate 20.** The amplitude cross-normalization of the log and seismic impedances was done selectively in a bandwidth chosen for spectral similarity. The black curve shows the log impedance and the red trace shows the impedance computed from the seismic deviated trace. The normalization was done between 30 and 40 Hz, where the similarities of the spectra are clear, especially when comparing the final bandlimited impedance ( the blue curve) with the log impedance (the black curve). The vertical yellow lines show the selected cut off frequencies of the surface seismic data.

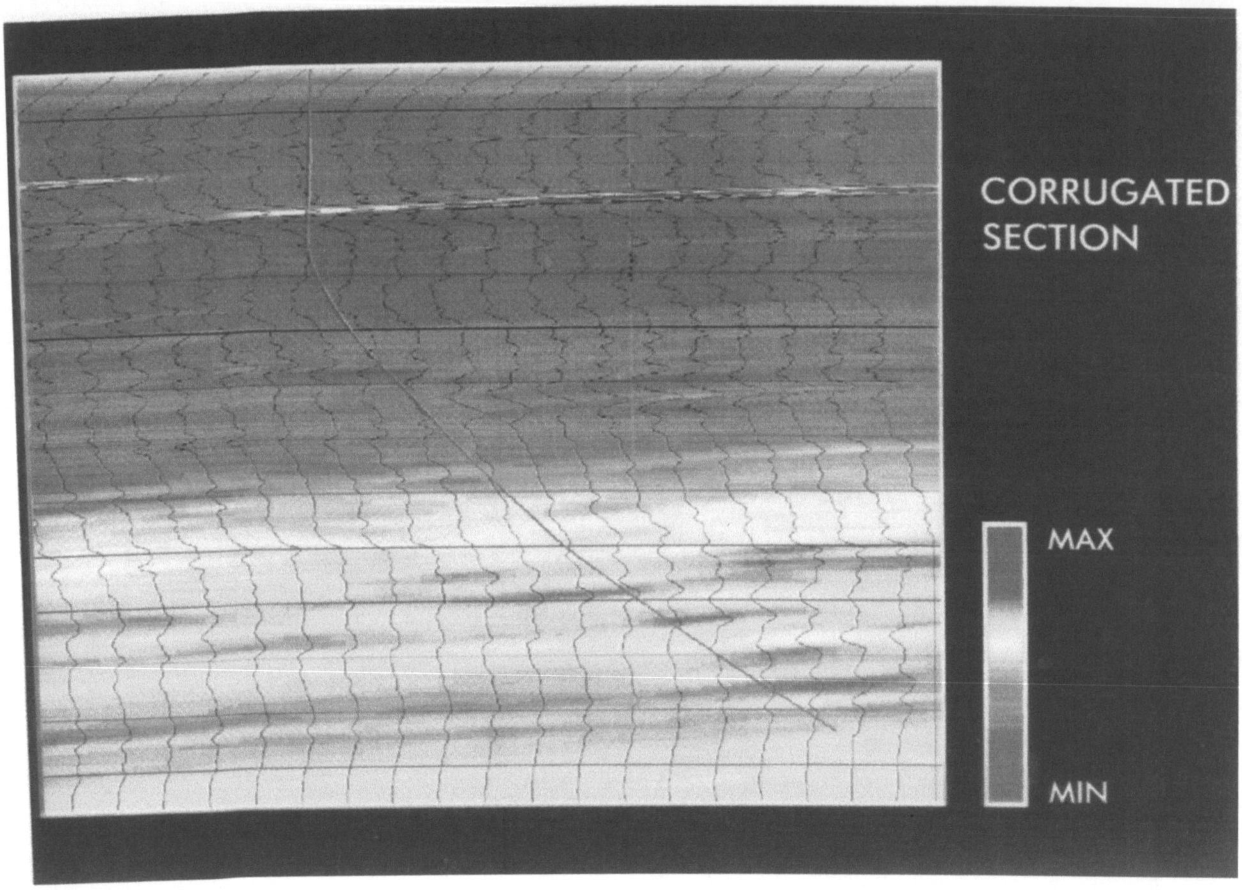


**Plate 21.** The *corrugated section* showing the final calibrated inversion. Two reservoirs can be seen as yellow zones near the bottom of the well. The colour scale was chosen to show a yellow/green contrast at the reservoirs. The lateral extent of these reservoir formations was estimated from this display and that of Plate 22.

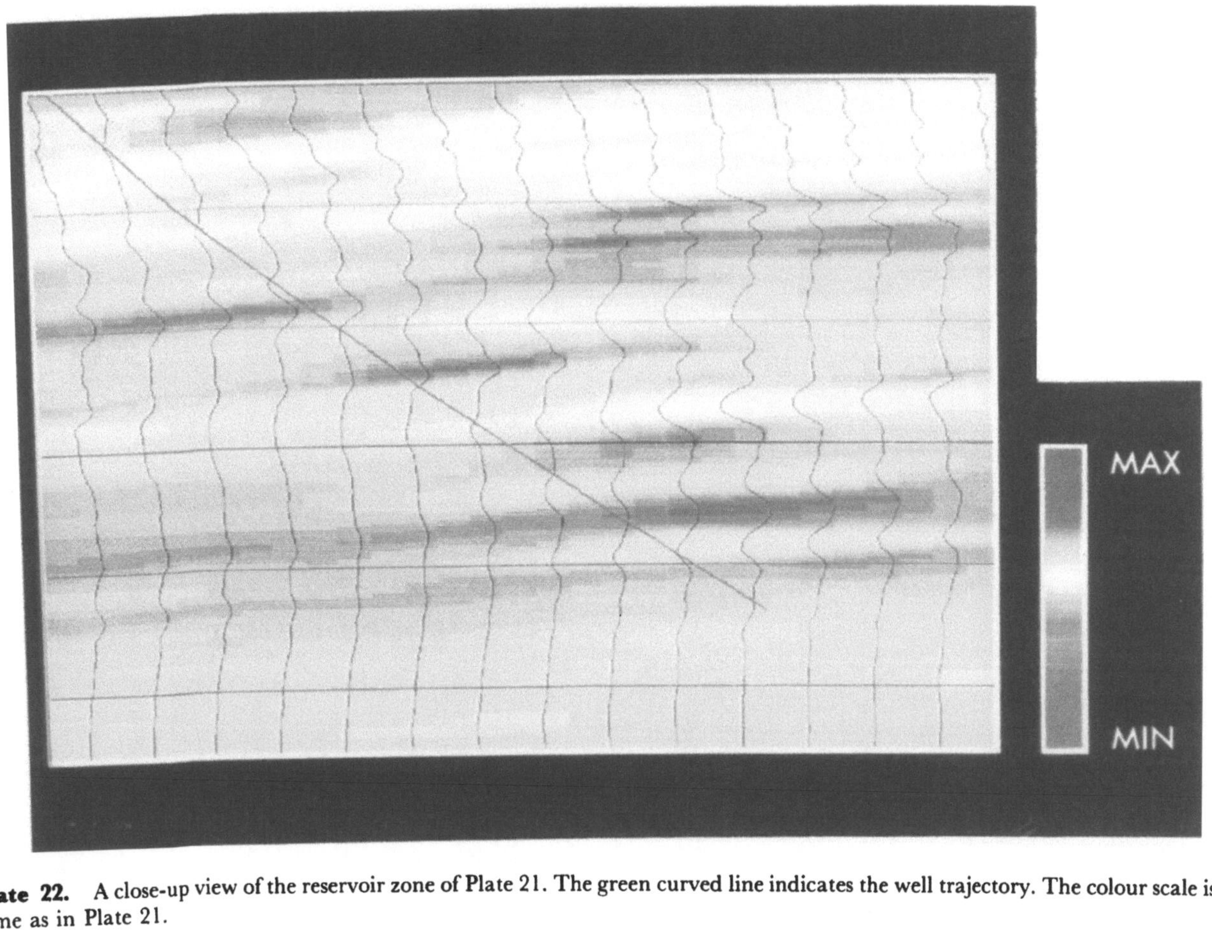


**Plate 22.** A close-up view of the reservoir zone of Plate 21. The green curved line indicates the well trajectory. The colour scale is the same as in Plate 21.

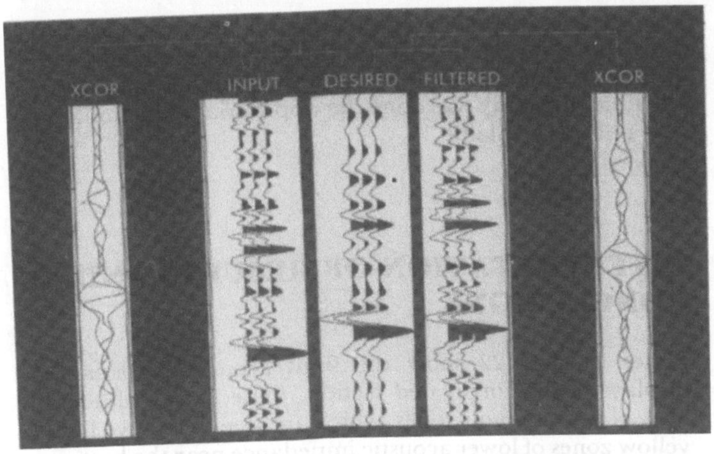


**Fig. 6.** Same as Fig. 5, but showing a close-up view of a strong event in the computation window. It can be seen here that the phase rotation filter has made the cross-correlation symmetrical within its envelope. The time-shifted phase rotation filter is a "soft" matching filter that does not change the fundamental information content of the surface seismic trace, yet brings it into closer correspondence with the VSP corridor stack.

### Extrapolating the filter away from the borehole

The phase rotation filters were computed in time gates at the borehole, and all involved approximately 50° of rotation. Those time gates were then extrapolated away from the borehole along structure in the following manner. The points on the deviated trace marking each time gate boundary were used as seed points for the computation of interpolated time surfaces, using the set of general time surfaces created earlier. This set of interpolated time surfaces form the boundaries of the spatially varying time gates.

The general time surfaces are interpolated as follows. If $(X_1, Y_1, t_1)$ is the seed point, which must lie on the surface to be interpolated, and it lies between two general time surfaces, $T_i(X, Y)$ and $T_{i+1}(X, Y)$, then the interpolated time surface, $I_1(X, Y)$, is given by,

$$I_1(X, Y) = aT_i(X, Y) + (1-a)T_{i+1}(X, Y) \tag{4}$$

where

$$a = \frac{T_{i+1}(X_1, Y_1) - t_1}{T_{i+1}(X_1, Y_1) - T_i(X_1, Y_1)} \tag{5}$$

Plates 17 and 18 show the corrugated section before and after filtering, with the corridor stack plotted along the well trajectory. This display shows the improved correlation between the surface and borehole data resulting from the residual wavelet processing step, especially in the shallow section.

Deeper in the section, the correlation between the two datasets is not perfect. Filters were not independently computed for time gates deeper in the section because there the data were not of sufficiently high signal-to-noise ratio in either the surface or borehole seismic. For these time gates, the phase rotation filters computed in shallow time gates were used.

Although the data could have been made to match more closely by designing "hard match" filters (see Appendix A), we chose the cautious route of terminating the estimate of new phase rotation filters when the signal-to-noise ratio began to degrade, and applying the filter from the time gate above. It is clear that we cannot claim to have completely compensated for residual wavelet effects deeper in the section, but we assume that the filter that was applied amounted to a partial correction to zero phase.

## CALIBRATED INVERSION

The third stage of the processing scenario is bandlimited inversion of the seismic data to acoustic impedance, where a calibrated inversion at the borehole is used to guide the inversion of traces away from the borehole. We inverted traces away from the borehole contributing to a single corrugated section of selected azimuth (see Fig. 1) in order to estimate the extent of the reservoir from the listric faults. For our purposes, it was not necessary to generate a complete acoustic impedance volume.

The final acoustic impedance traces will contain frequencies from approximately zero to the seismic maximum. The frequencies below the seismic band will come from the log impedance, the seismic band frequencies will come from the seismic data, and the frequencies above the seismic band will not be retained. Extrapolating these ultrahigh frequencies away from the borehole involves interpretative algorithms, such as sparse spike deconvolution or event detection, and the output tends to be highly dependent on the choice of algorithm, since these frequencies are not contained in the surface seismic data (see, for example, Oldenburg *et al.*, 1983).

### At the well

The inversion was performed first at the well location using the deviated trace and the acoustic impedance from borehole measurements. The acoustic impedance log (computed from the processed sonic and density logs) was used to establish the low-frequency component at the well location below the seismic band. Seismic band frequencies were computed from only the deviated trace by applying recursively the reflectivity equation for each sample. We assumed a zero-phase wavelet since we have performed the earlier residual wavelet processing step to ensure the compatibility of the surface and borehole data. The acoustic impedance computed from the seismic trace was then amplitude-normalized against the acoustic impedance log using a shared frequency band. The low-frequency component of the acoustic impedance from the log was then added to the impedance from the seismic trace to produce the final bandlimited impedance at the well.

Plate 19 shows key steps in the inversion at the well. The first trace, labelled "reference impedance", is the acoustic impedance from logs. The second trace is the reflectivity series computed from it, which was used only for visual quality control, and the third trace is the deviated trace. The fourth trace is the bandlimited acoustic impedance computed from the deviated trace only by recursively applying the reflectivity equation (Lindseth, 1979),

$$Z_i = Z_{i+1} \frac{1 - R_{i,i+1}}{1 + R_{i,i+1}} \tag{6}$$

where $Z_i$ is the impedance in layer $i$, and $R_{i,i+1}$ is the reflection coefficient at the interface between layers $i$ and $i+1$. This trace is shown biased from zero to enable a colour display.

Since the seismic and borehole data are separate measurements, they must be cross-calibrated before they can be combined. The seismic impedance trace is amplitude-normalized against the borehole impedance trace (the reference impedance) in the common frequency band, where a single scale factor is computed to apply to the surface seismic data. Plate 20 shows the spectra for several of the traces from Plate 19. The scale factor was computed from that part of the seismic band where the spectra of the reference impedance and the seismic impedance showed a similarity of character. Ideally, one hopes to see the seismic and log impedance match over the entire seismic band. However, if the match were exact, the deviated trace would be identical to the synthetic seismogram, a case not often found. In practice, one finds that the two impedance traces will match over some subset of the seismic band. In this study, the calibration was done over the range of approximately 30 – 40 Hz (compare the blue curve to the black curve in Plate 20).

When the scale factor is determined, it is applied to the seismic impedance trace. The seismic and borehole impedance are now ready to be merged. The fifth trace in Plate 19 is the low-frequency component of the reference impedance, where the frequency cutoff was selected to match the onset of the seismic energy. The sixth and final trace, labelled bandlimited impedance, is the sum of the low-frequency reference impedance with the normalized seismic impedance from the deviated trace.

This bandlimited impedance trace shows comforting similarities to the reference impedance. The yellow and orange layering toward the bottom of the trace corresponds to the location of the producing sand/shale layers.

### Extrapolation away from the well

For any trace away from the well, the acoustic impedance trace is the sum of the low-frequency component and the seismic band component. The seismic band component is computed for each subsequent trace in the same manner as for the original deviated trace inversion at the well, where the same normalization scale factor was used. The low-frequency component away from the well was computed by mapping the low-frequency component, as seen in the fifth trace of Plate 19, into the 3D volume using the general time surfaces so that the mapping follows structure.

The mapping procedure for the low-frequency component can be achieved in a variety of ways. The method used for this study is as follows. First, all minima and maxima were located on the low-frequency impedance trace. Then, interpolated time surfaces were computed from those points. Next, the extrapolation of each of those points was mapped to each output trace location. The low-frequency trace was then reproduced by a stretching transformation applied to the original low-frequency trace, using the mapped maxima and minima at each output trace location.

A more exact method for extrapolation of the low-frequency component would involve extrapolation of the full-band log impedance trace to each output trace location, and then applying the same low-frequency cut off filter. However, because the reservoir under study was small, over its limited lateral extent from the well the structural surfaces remained very close to parallel. Thus, there was virtually no stretching and compressing of the low-frequency component over the extent of the reservoir. The method we used was therefore equivalent, but more economical.

The effectiveness of a method such as the one we have used in this study can be further enhanced by the use of multiple wells to obtain low-frequency constraints that vary spatially over the geological surfaces. Instead of extending the low-frequency log impedance from a single well as a constant value over geological surfaces, the low-frequency component of log impedances from several wells within the producing reservoir are mapped over a set of closely spaced geological surfaces, and used as spatially varying low-frequency constraints on the extrapolation. This approach was not used in the present study, although there were multiple wells through the reservoir, because at the time this study was done the appropriate functionality was not yet available.

## INTERPRETATION OF RESULTS AND CONCLUSIONS

Plates 21 and 22 show the final acoustic impedance displays of the corrugated section. Plate 22 is a close-up view of two reservoir zones, which are seen as the two light yellow zones of lower acoustic impedance near the bottom of the well. These reservoirs are complex sand/shale sequences, as can be seen from the reference impedance in Plate 19, where producing sands are much thinner than the low-impedance zones of the bandlimited impedance, which gives a composite response. This composite response was interpreted for lateral extent, which was found to be limited to about 350 m from the well in the major sands in the direction parallel to the corrugated section (see Fig. 1).

This method may be extended further by using the calibrated impedance to infer and to map quantities such as porosity away from the borehole (des Ligneris, 1988, Doyen, 1988). In our case, reservoir extent was effectively estimated from the impedance itself.

Our estimate of reservoir extent is in essential agreement with the interpreted extent of the gas–water contact from the analysis of a multiple well correlation analysis. It verifies that the careful stratigraphic processing analysis yielded, in this case, a high-quality assessment of reservoir extent independently from a multiple well statistical correlation analysis. We conclude that this type of analysis can substitute extensive multiple well correlation analysis when a large number of wells are not available in the producing zone.

## ACKNOWLEDGEMENTS

We would like to express our appreciation to the Chinese Petroleum Corporation for their permission to show the data. We also are grateful to the following contributors who assisted in the development of the software used in this study: T. Endo, M. Hattori, H. Imai, M. Ishida, J. Kasahara, H. Mikada, H. Nishida, T. Ogawa, O. Osawa, N. Saito, and H. Yamamoto.

## REFERENCES

Claerbout, J.F., (1976). *Fundamentals of Geophysical Data Processing*. McGraw-Hill, New York.

Des Ligneris, S. (1988). Application of log and seismic data integration techniques to delineation of porosity in a carbonate reservoir. *58th Annual International SEG Meeting, Expanded Abstracts*, pp. 745–748.

Doyen, P.M. (1988). Porosity from seismic data: A geostatistical approach. *Geophysics* **53**, 1263–1275.

Lindseth, R.O. (1979). Synthetic sonic logs – A process for stratigraphic interpretation. *Geophysics* **44**, pp. 3–26.

Oldenburg, D.W., Scheurer, T. and Levy, S. (1983). Recovery of the acoustic impedance from reflection seismograms. *Geophysics*. **48**, 1318–1337.

# APPENDIX A

## A DESCRIPTION AND COMPARISON OF THE RESIDUAL PROCESSING AND THE EARTH FILTER METHODS FOR BOREHOLE CALIBRATION OF SURFACE SEISMIC WAVELETS

In the text, we mentioned two competing methods for the calibration of surface seismic wavelets using borehole data: the residual processing method an the earth filter method. These approaches are representative of two competing ways of performing wavelet calibration from borehole seismic (i.e. post-stack and prestack), and a quick look at them will be informative. However, they are not the only alternatives.

We present first a mathematical description of each, and then offer a comparison of their relative advantages and disadvantages.

### The residual processing method

We begin with the convolutional model of the seismogram and write equations for the surface seismic deviated trace and the borehole corridor stack trace. In $\mathcal{Z}$-transform notation,

$$S_s(\mathcal{Z}) = W_s(\mathcal{Z})R(\mathcal{Z}) + \mathcal{N}_s(\mathcal{Z}) \tag{A1}$$

and

$$S_v(\mathcal{Z}) = W_v(\mathcal{Z})R(\mathcal{Z}) + \mathcal{N}_v(\mathcal{Z}) \tag{A2}$$

where $S$ is the seismic trace, $W$ is the wavelet, $R$ is the common reflectivity, and $\mathcal{N}$ is the additive noise. Subscripts $S$ and $V$ refer to deviated trace and corridor stack data respectively. We assume that $W_v(\mathcal{Z})$ is the desired bandlimited zero-phase wavelet. So, we wish to find a filter, $M(\mathcal{Z})$, such that

$$W_v(\mathcal{Z}) = M(\mathcal{Z})W_s(\mathcal{Z}) \tag{A3}$$

or

$$W_v(\mathcal{Z})R(\mathcal{Z}) = M(\mathcal{Z})W_s(\mathcal{Z})R(\mathcal{Z})$$

and, from (A1) and (A2),

$$S_v(\mathcal{Z}) - \mathcal{N}_v(\mathcal{Z}) = M(\mathcal{Z})(S_s(\mathcal{Z}) - \mathcal{N}_s(\mathcal{Z}))$$

yielding,

$$S_v(\mathcal{Z}) = M(\mathcal{Z})S_s(\mathcal{Z}) \times [\mathcal{N}_v(\mathcal{Z}) - M(\mathcal{Z})\mathcal{N}_s(\mathcal{Z})] \tag{A4}$$

where (A4) shows the approximation involved in estimating $M$ from only the seismic traces themselves, $S_s$ and $S_v$.

The simplified procedure, then, for the estimation of $M$ is to find the approximate filter, $\hat{M}$, which satisfies the equation,

$$S_v(\mathcal{Z}) = \hat{M}(\mathcal{Z})S_s(\mathcal{Z}) \tag{A5}$$

The bracketed expression in (A4) indicates the error term due to additive noise. As this expression is the sum of two noise terms, we would expect the additive noise component to be amplified by $\sqrt{2}$ after application of the filter, $\hat{M}$, derived from (A5).

The procedure we have followed in the text is to find $\hat{M}$ such that (A5) is only approximately satisfied, by constraining $\hat{M}$ to be a time-shifted phase rotation filter. Such a filter would be an example of a "soft match". In practice, finding $\hat{M}$ such that (A5) is exactly satisfied (i.e. a "hard match") results in a filter that will progressively degrade the data as one gets increasingly distant from the borehole.

### The earth filter method

We begin with an expression for the downgoing wavefield at some depth $z_i$ from a surface source.

$$D_i(\mathcal{Z}) = W(\mathcal{Z})E_i(\mathcal{Z}) \tag{A6}$$

where $D$ is the downgoing wavefield, $W$ is the source wavelet, and $E$ is the earth filter, all at depth, $z_i$.

Now, if $E_i(\mathcal{Z})$ is the earth filter for the downgoing wavefield from the surface, $z_0$, to depth, $z_i$, then the earth filter response for a wave reflecting at depth, $z_i$, and returning to the surface is $E_i(\mathcal{Z})E_i(\mathcal{Z})$, the autoconvolution. In other words, the two-way travel path sees the earth filter twice.

In a VSP geometry, we often have a near-surface monitoring geophone. If we are able to record the full signature at that phone, then we have an estimate of $W(\mathcal{Z})$, leaving $E_i(\mathcal{Z})$ as the only unknown in (A6). Once $E_i(\mathcal{Z})$ is known at any depth, then a filter, $M(\mathcal{Z})$, to compensate for the earth transmission will satisfy the following relation:

$$M_i(\mathcal{Z})E_i(\mathcal{Z})E_i(\mathcal{Z}) = 1 \tag{A7}$$

In practice, $E$ would be estimated at several depths from the VSP downgoing wavefield, and $M$ would be estimated from (A7) for each of several time gates.

### A comparison of the two methods

The advantages of the residual processing method are, first, that it can be applied in a post-stack residual processing step, and second, that potentially it can correct for most prestack wavelet distorting effects mentioned in the text, even those not associated with earth transmission, such as angle-dependent reflection coefficients and mute patterns. A disadvantage of this particular method is that it requires both surface and borehole seismic data to be of relatively good quality.

The advantages of the earth filter method are that it gives an accurate estimate of the bandlimited transmission correction filter, and that it requires only relatively good quality VSP. The disadvantages are that it incorrectly assumes that the surface seismic raypaths are the same as those for the VSP, and also that the method must be used in a full reprocessing sequence, which is often inconvenient for 3D data.

# APPENDIX B

## THE CROSS-CORRELATION ENVELOPE AND PHASE ROTATION FILTERS

We show that the envelope of the cross-correlation function does not change when a phase rotation filter is applied to the surface seismic trace. Using $\mathcal{Z}$-transform notation, we first review the relationship between a time series and its envelope. Let us begin with the cross-correlation, $C(\mathcal{Z})$, between the surface seismic trace, $S(\mathcal{Z})$, and the VSP corridor trace, $V(\mathcal{Z})$, defined to be

$$C(\mathcal{Z}) = V\left(\frac{1}{\mathcal{Z}}\right)S(\mathcal{Z}) \tag{B1}$$

since the coefficients of $V$, $S$, and therefore $C$, are real. To compute the envelope of $C(\mathcal{Z})$, we first compute the analytic signal of $C(\mathcal{Z})$, call it $A(\mathcal{Z})$, which is defined as follows:

$$A(\mathcal{Z}) = C(\mathcal{Z}) + iQ(\mathcal{Z})C(\mathcal{Z}) \tag{B2}$$

where, expressed in the frequency domain,

$$Q(\omega) = \frac{-i\omega}{|\omega|} = -i\,\mathrm{sgn}(\omega), \qquad \text{(B3)}$$

and is the quadrature filter. $Q(Z)C(Z)$ is the Hilbert transform of $C(Z)$ (Claerbout, 1976, p. 20). The envelope of $C(Z)$ is then the modulus of the complex analytic trace, $A(Z)$.

Now because $A(Z)$ is a complex trace, we can perform a phase rotation by the complex multiply operation

$$A'(Z) = A(Z)e^{-i\phi} \qquad \text{(B4)}$$

where $A'(Z)$ is the phase-rotated version of $A(Z)$. Clearly, the modulus of $A'$ is the same as the modulus of $A$. If we can now show that convolving the seismic trace by a real phase rotation filter, $E_\phi(Z)$, of a form yet to be determined, results in the above change to $A(Z)$, then we will have demonstrated the invariance of the envelope of the cross-correlation to a phase rotation filter applied to the seismic trace.

It is easy to see that a phase rotation filter applied to one of the traces in the cross-correlation is equivalent to applying the same phase rotation filter to the cross-correlation itself, since

$$S'(Z) \equiv S(Z)R_\phi(Z) \qquad \text{(B5)}$$

and

$$C'(Z) \equiv C(Z)R_\phi(Z) = V\left(\frac{1}{Z}\right)S'(Z) \qquad \text{(B6)}$$

We now combine equations (B2) and (B4) and see what filter must be applied to the real seismic trace so that the complex analytic trace of the cross-correlation undergoes the phase rotation given in (B4).

$$\begin{aligned} A'(Z) &= A(Z)(\cos\phi - i\sin\phi) \qquad \text{(B7)}\\ &= (C(Z) + iQ(Z)C(Z))(\cos\phi - i\sin\phi)\\ &= C(Z)(\cos\phi + Q(Z)\sin\phi)\\ &\quad + iQ(Z)C(Z)(\cos\phi + Q(Z)\sin\phi)\\ &= C(Z)R_\phi(Z) + iQ(Z)C(Z)R_\phi(Z)\\ &= C'(Z) + iQ(Z)C'(Z) \end{aligned}$$

where we have used the identity

$$Q^{-1}(Z) = -Q(Z) \qquad \text{(B8)}$$

which is easily derived from equation (B3). Additionally, we have identified the general form of the real phase rotation operator,

$$R_\phi(Z) = \cos\phi + Q(Z)\sin\phi \qquad \text{(B9)}$$

So, the equation (B7) shows that there is no change to the envelope of the cross-correlation function when we apply the real phase rotation filter, $R_\phi(Z)$, in a form given by equation (B9), to the seismic trace, $S(Z)$.

# 5

# Frigg Field Reservoir Management

*R. Maritvold*

Elf Aquitaine Norge A.S., Stavanger, Norway

**A re-appraisal programme consisting of new appraisal wells, extensive coring and new 3D seismic has been necessary in order to evaluate the dynamic behaviour of the Frigg field. Having proved a strong water drive and a heterogeneous water entry into the gas zone, studies were focused on parameters important to sweep efficiency and final recovery: initial gas-in-place, residual gas saturation, heterogeneities and permeability barriers. The delineation of the permeability barriers and attempts to monitor the water rise outside the well control area required close cooperation between geophysicists, geologists and reservoir engineers. Some of the techniques applied came close to their limit of feasibility.**

**The use of seismic for monitoring the rise of the waterfront and control of sweep efficiency has been successful from a qualitative point of view. If these methods are to be used quantitatively, they require high-resolution seismic, optimized for this purpose in terms of aquisition and processing, and good control on stratigraphical variations. Modelling of the effects of geometry (dip, thickness) and geology are of course necessary for determining the reliability of the results.**

**Overall, the re-appraisal of the field has led to an improved assessment of the volumetric sweep efficiency and has allowed a better forecast of field potential in the decline phase. The remaining unswept zones have been located and their potential for further development evaluated. This could lead to the drilling of a highly deviated well(s) from one of the existing platforms in 1989 in order to improve the final recovery.**

## MAIN OBJECTIVES AND ORGANIZATION

The contractual constraints associated with the exploitation of a gas field require an accurate reservoir model capable of predicting the production potential evolution as a field enters the decline phase, as the water drive implies that the final reserves, being sensitive to the abandonment pressure, are affected by the chosen production profile. In addition, heterogeneous water entry could lead to the formation of pockets of unswept gas, possibly with the potential for further field development.

In order to achieve the required precision and to optimize recovery from the field, some key elements in the reservoir model obviously needed updating.

(1) A more complete geological model was required, taking into account:
- additional core and log data from the improved well coverage;
- palynological information;
- stratigraphic interpretation of the seismic facilitated by the new 3D lines and well seismic;
- information on the dynamic behaviour of the field;
- the possibility of mapping potential barriers using a seismic approach.

(2) Updating of structural maps that, together with new well information, would lead to an updating of the OGIP.
(3) Improved estimations of residual gas saturations using more representative core material and in-situ measurements.
(4) The use of direct measurements and seismic interpretation for fluid contact monitoring.
(5) Information on the behaviour of the satellite fields.
(6) Regional mapping and characterization of the Palaeocene aquifers.

The large amount of data interpreted within a relatively short time and often in parallel required a tight coordination in order to maintain consistency of the results, to assure interaction between the different studies and, last but not least, to respect deadlines. Owing to the specialization of some of the techniques and the geographical spread of different experts within an international oil company, the interpretation team was organized in two entities, applying exploration and reservoir engineering techniques respectively. In some cases, e.g. the monitoring of fluid movement by means of seismic, the techniques applied came close to being research. The coordination of the work was assumed by a reservoir engineer together with a production geologist. This provided the necessary link between the exploration and reservoir engineering entities.

Simulation by mathematical models was performed continuously, not only to give updated production forecasts but also to synthesize the interpretations as new well data

*North Sea Oil and Gas Reservoirs—II*

became available. This proved useful for focusing on areas needing closer attention, and in particular for demonstrating the need for further control wells to distinguish between possible hypotheses and for deciding on the need for re-completion of wells after water breakthrough.

## INTRODUCTION

Discovered in 1971, the Frigg field straddles the Norwegian – UK border and is situated in block 25/1 on the Norwegian side and block 10/1 on the British side (Fig. 1). It is now estimated that the original gas-in-place was in the range of $235 \times 10^9$ Sm³, of which $158 \times 10^9$ Sm³ have been produced from the start-up in 1977 until October 1988.

The reservoir consists of high porosity/permeability sands of lower Eocene and Palaeocene age, probably deposited as a channellized submarine fan. The maximum gas column was about 160 m, underlain by a 9 m thick oil leg covering an area of about 115 km². Four satellite fields, with a total gas-in-place in the order of $75 \times 10^9$ Sm³, are connected to the same limited aquifer and influence the regional pressure evolution. The Frigg Formation is in turn connected to large Palaeocene aquifers (Sele and Lista Formations) through the heterogeneous Balder Formation, which comprises interbedded shales, sand and tuff.

Based on information from a seismic survey acquired in 1973 (on a 1 × 1 km grid), ten exploration/appraisal wells together with five wells drilled on the satellite fields, the field was developed with four clusters of twelve wells each. Being centrally located on top of the structure and only penetrating about 35% of the total gas height (as a precaution in case of an active aquifer), the production wells yielded little new geological information.

Initially, the main uncertainties were linked to aquifer performance and, in particular, to the transmissibility of the tuff member of the Balder Formation which was believed to be a barrier to water movements. For this reason, one of the cluster wells, well 25/1-A22, was designed as an observation well deviated north-eastwards from the cluster and perforated in the Palaeocene aquifer (Fig. 2). The pressure evolution of the aquifer was monitored and the water encroachment into the gas pool was followed by cased hole logging. The shales within the Frigg sands were not assumed to have sufficient extents for creating barriers and a heterogeneous sweep, and were modelled by the production geologists using a statistical approach. The difficulties encountered in shale correlation are explained by a well coverage of 1 well per 12 km², and by the fact that the 1973 seismic data did not possess optimal resolution for detailed lithological interpretations. It should also be mentioned that core recovery with conventional core barrels during the exploration phase was poor.

By 1980, it was clear that the aquifer was active and that the water drive was very strong, but that the water entry was blocked locally by a shale layer in the observation well. Simulation of several scenarios of aquifer communication through the Balder Formation (Fig. 3) led to the deepening of another production well, 10/1-A25. The well results did not confirm the assumed geological model, as the main barriers were related to the shales rather than to the tuff.

A re-appraisal programme was undertaken and consisted of the deepening of two other production wells, 25/1-A14 and 10/1-A26, the drilling of three control wells, 10/1-5, 25/1-7 and 25/1-8, together with the acquisition of 4600 km of 3D high resolution seismic lines and an extensive study on about 1000 m of cores. Later, yet another production well (25/1-A3) was deepened into the Palaeocene aquifer for monitoring purposes, and the initial observation well, 25/1-A22, was re-converted to a producer thereby enabling pressure measurements across the shale barrier during the operation (Fig. 4).

This was the situation when a team of reservoir engineers, production geologists and geophysicists initiated the interpretation of the acquired data in order to produce a more

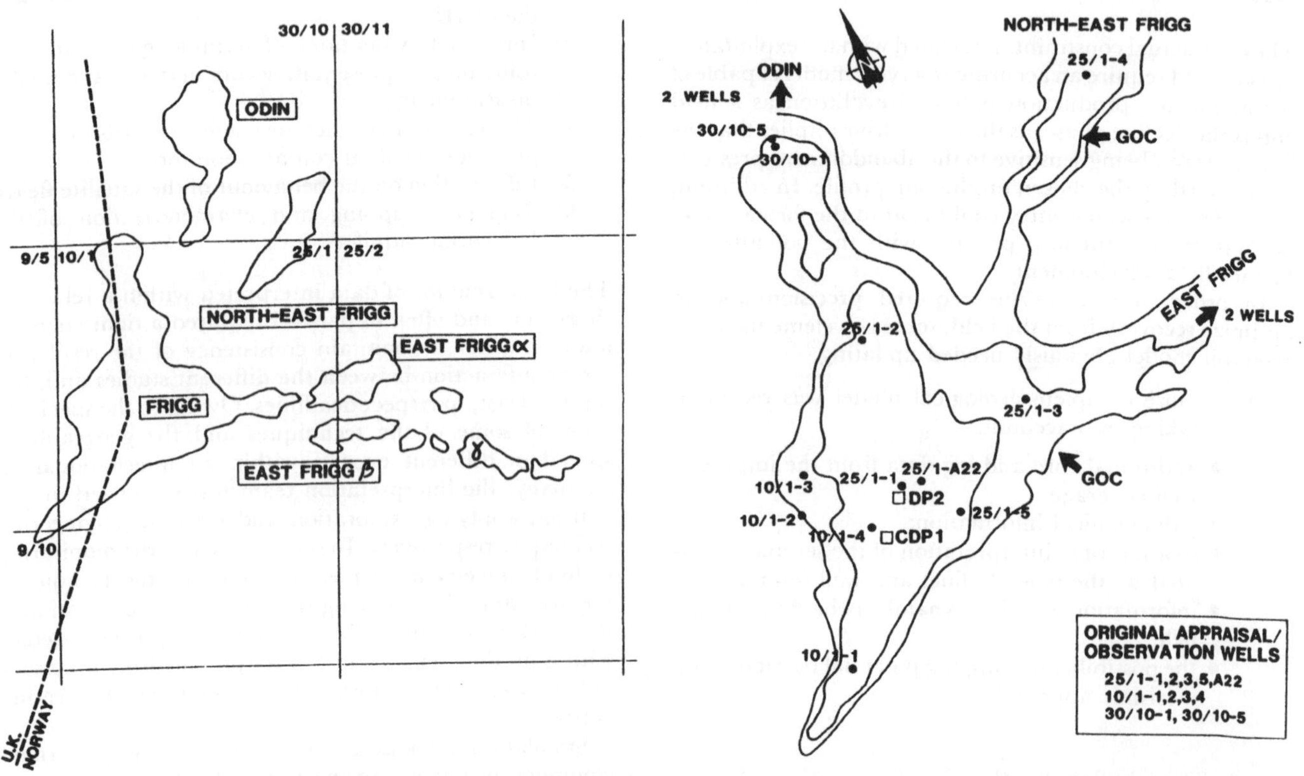


**Fig. 1.** Location of the Frigg field and its satellites (left) and the initial well coverage (right).

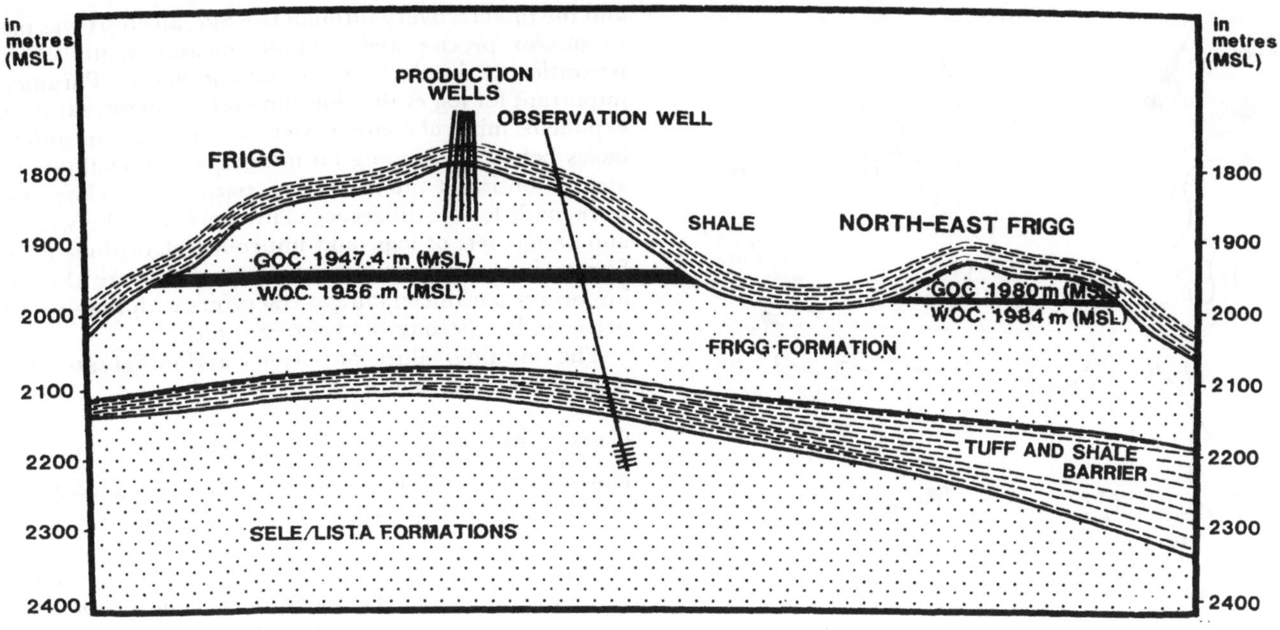


**Fig. 2.** The first geological model (simplified).

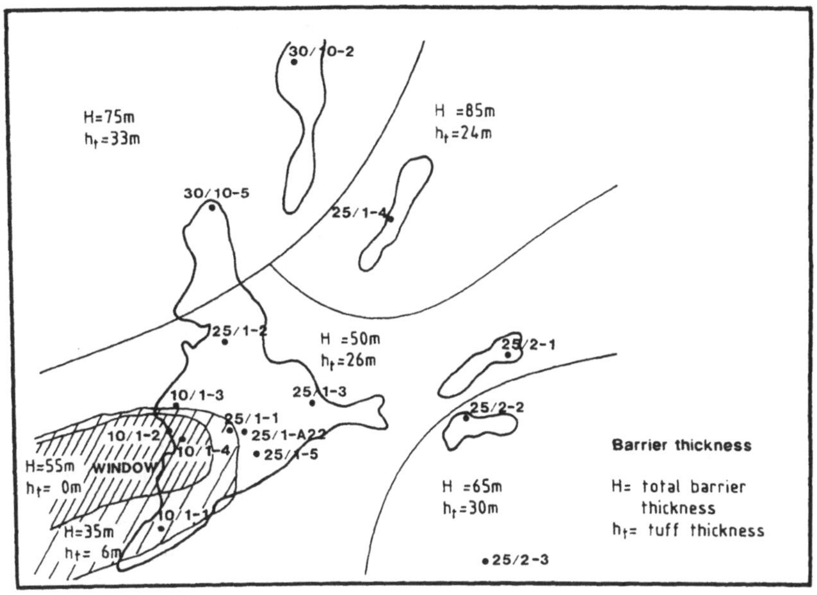


**Fig. 3.** Geological model prior to well 10/1-A25. Barrier between Frigg and the main aquifers.

precise forecast of pressure evolution and volumetric sweep efficiency.

Further details of the reservoir's geological setting and development up to 1985 are given by Brewster and Jeangeot (1987) and De Leebeck (1987).

# RESULTS

## Geological model

With access to a large amount of core data and improved well coverage, the complexity of the geological model increased considerably. Starting with a relatively homogeneous sand package with randomly distributed shale layers, a heterogeneous system composed of several succeeding channels (eroding and overlapping each other), shales and zones of debris flow (breccia) evolved.

The core data permitted the identification of 12 lithofacies with several subcategories. Having determined a sedimentological scheme and possible depositional sequences, a stratigraphic zonation of the wells was performed using techniques such as petrography, X-ray diffraction and isotope analysis, supplemented by palynological data (Fig. 5). Extrapolation of this zonation to most of the uncored wells was achieved using dipmeter microresistivity responses. The geometry of the units was modelled using seismic stratigraphy.

Pressure gradients measured in the appraisal wells proved the existence of several barriers to water encroachment. As the extension and location of these barriers are decisive for the final sweep efficiency, an effort to map potential barriers using stratigraphic modelling and seismic facies analysis was undertaken:

- identification of the main barriers and calibration of

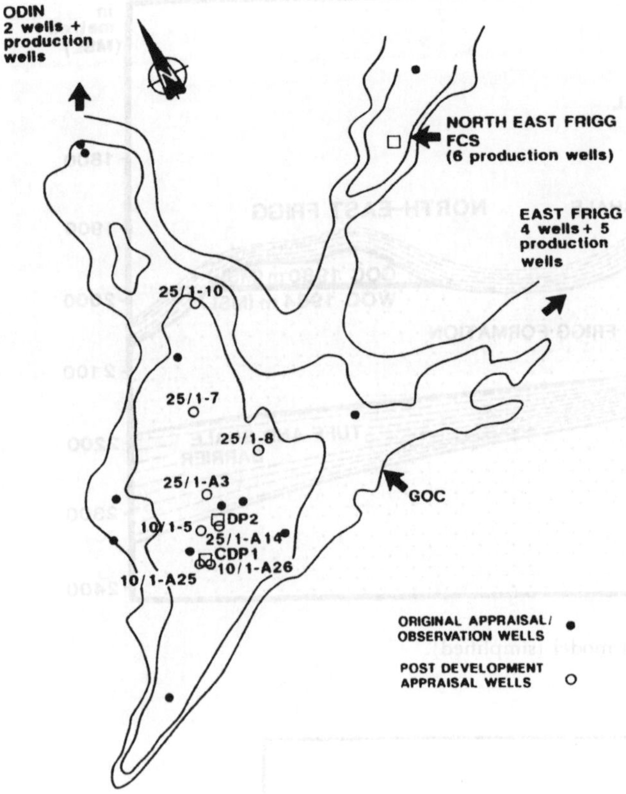


**Fig. 4.** Current well coverage.

well data on a seismic line using synthetic seismograms;

- modelling of the variation in seismic response with lithological changes in the barrier;
- comparison with seismic lines and mapping of the first interpretation;
- modelling of different possible scenarios of barrier configuration (continuous, overlapping, discontinuous) and re-comparison with the seismic data.

Finally, the study concluded with depth and extension maps of facies with barrier potential. The most important of these facies are:

- continuous tuff overlain by shales;
- heterolithic "breccia" facies;
- shale and interbedded sand/shale laminations within the upper part of the Frigg Formation (ntIIb).

The introduction of these barriers into the mathematical model had an important impact on the geometry of the swept zone. The main zone of unswept gas at platform shut-in is created by the dynamic conditions and the stratigraphical representation of the area, and its volume and extent are determined by the barrier continuity.

## Petrophysical measurements/log interpretation

Given the importance of the residual gas saturation (Sgr) on final recovery, a major part of the petrophysical measurements addressed this parameter by studying the effect of several variables: porosity; initial water saturation; depletion; shaliness; facies (laminated/homogeneous sands); diffusion effects; and measurement methodology. A relationship between residual gas saturation, porosity and initial water saturation was obtained and confirmed by open-hole log measurements in swept zones.

Since the porosity affects both the gas-in-place estimate and the final recovery through the Sgr, an effort was made to obtain precise and reliable measurements of core porosities at simulated reservoir conditions. Parameters important for log evaluation (formation factor, saturation exponent, mineral content) were also studied for different facies before performing an integrated log evaluation to assure consistent interpretation parameters. Corrections were made for the difference of interpretation between old and recent sets of logs, and improved algorithms for the filtrate invasion were used because these are the most important parameters in gas reservoirs with high vertical permeability (strong fluid segregation).

The resulting values for porosity and initial saturations were included in the updating of gas-in-place.

## Gas-in-place

Given the updated petrophysical data, a new geological model and the acquired 3D seismic, an updating of the initial gas-in-place was undertaken using a geostatistical approach. In short, the method consists of mapping all variables by means of statistical analysis and an assessment of their regional trend. In the case of Frigg, this is particularly important when determining the rock-volume because the seismic marker is situated at a variable distance from the first gas-bearing sand. The estimate of original gas-in-place has been reduced from the official evaluation in 1976 of $265 \times 10^9\,Sm^3$ to a 1988 figure of $235 \pm 14 \times 10^9\,Sm^3$.

## Geophysical monitoring

The central location of the producers led to poor control of the sweep over most parts of the field. Even if the control wells gave some specific information, the spatial coverage was poor. In particular, the sweep of the northern arm was an open question and the mathematical model, given the uncertainties of gas-in-place, residual gas saturation and gas entry from the satellite fields, did not have sufficient precision to delineate the size of the possible unswept zone.

The first attempts at seismic monitoring were performed by studying the possible migration of gas from the satellite fields towards Frigg. Seismic lines where shot in 1979, 1981 and 1983 and an expansion of the gas from North-East Frigg was detected and later confirmed by pressure measurements. These results and the visualization of the new gas–liquid contact by VSPs in appraisal wells showed that the sweep efficiency could perhaps be approached by studying changes of seismic response with production. Different methods were suggested.

(a) *Variation of flat spot amplitude vs. rise of gas–liquid contact* (*GLC*)

Several wells were modelled in order to deduce the seismic impedance from logs. The changes of acoustic impedance related to the rise of the GLC were computed using the GASSMANN model (Gassmann, 1951):

$$\text{Velocity} = f(\phi, \rho_{ma}, \rho_f, C_{ma}, C_f, \mu, \sigma, Z)$$

where $\phi$ = porosity, $\rho$ = density of matrix and fluid respectively, $C$ = compressibilities of the same; $\mu$ = Poisson's ratio; $\sigma$ = spherical component of the stress tensor; $z$ = depth. The sensitivity of the acoustic impedance to gas saturation is illustrated in Fig. 6.

The conclusion reached was that the acoustic response appears to be homogeneous and that the rising liquid level induces an attenuation of the amplitude of the flat spot (from 12 to 40%). The relative attenuation goes through a maximum value, decreases and then stabilizes. This behaviour is related to the interference between the old

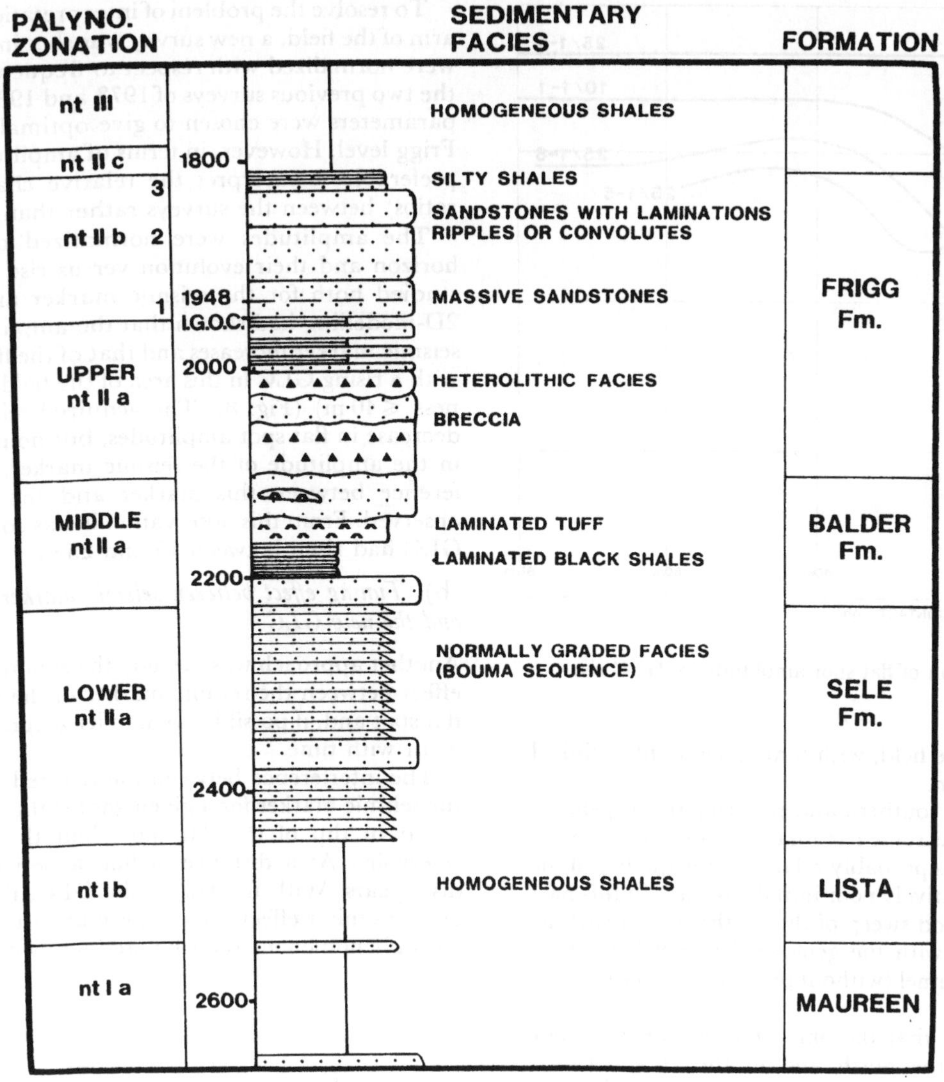


**Fig. 5.** Stratigraphical zonation.

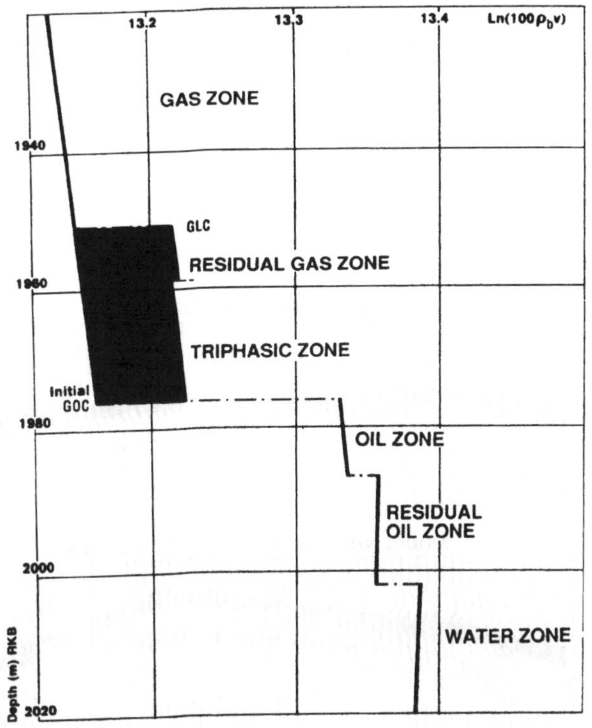


**Fig. 6.** Acoustic impedance vs. hydrocarbon saturation.

and the new gas–liquid contact. The well responses in Fig. 7 show a consistent trend, with some effects of lithology interference. The atypical response of well 25/1-3 is caused by a limited reservoir thickness (< 20 m).

Important changes in the seismic expression of the reservoir between the seismic marker and the flat spot were observed in the models and on the seismic lines. However, it was not feasible in practice to correlate these events directly with the rise of the liquid level owing to the complexity of the system (interference between old/new GLC, lithological changes and the Frigg seismic marker) and the frequency content of the data.

Before interpreting the seismic surveys, they were all normalized on a 1.5 × 1.5 km grid (370 km of lines). This was done to reduce as far as possible the differences caused by acquisition parameters and initial processing, and to optimize the processing with respect to resolution at the Frigg sand level. The quality of this normalization was verified by controlling the stability of the amplitude of the Frigg seismic marker, particularily outside the field. It was concluded that 60–70% of the normalized data were of good quality.

The interpretation is synthesized in Plate 23 which shows the attenuation of the flat spot amplitude over the field. This, together with maps of the stability of Frigg seismic marker, reinforcement of intermediate events, and changes at the seismic marker/flat spot intersection qualitatively suggests that the gas–liquid contact has

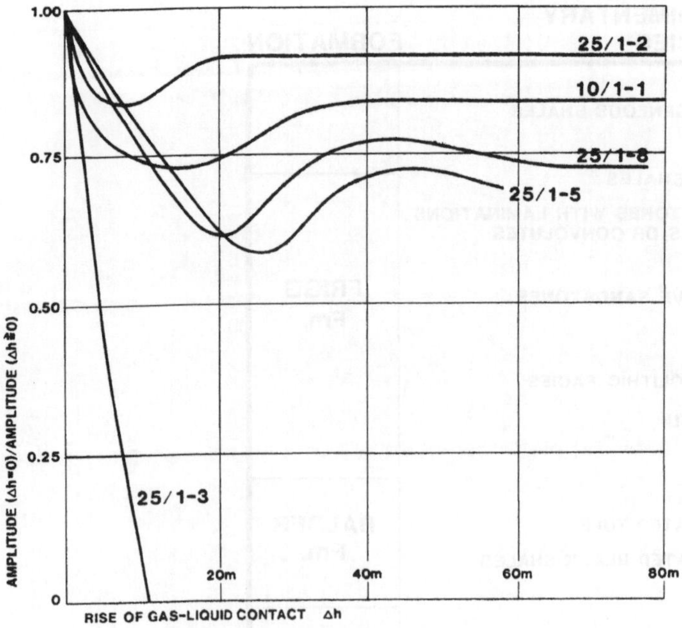


**Fig. 7.** Variation of flat spot amplitude vs. GLC rise.

risen over the whole field, with some uncertainties linked to the northern arm.

The sweep of the southern and eastern parts appears to be uniform, whereas some irregularities could exist in the western part. This is probably related to the quality of the data and to the relatively poor initial coverage in this part of the field. The good sweep of the south/west–north/east axis is compatible with the geological model having an important sand channel (without permeability barriers) in the east.

The study shows that the only area of potential poor sweep is towards the north and in the area between 10/1-3, 25/1-2 and 25/1-8, which appears less regular and more complicated with respect to the intermediate seismic events.

An attempt to quantify the rise of liquid level from the flat-spot attenuation (Fig. 7) was made in four wells:

| *Well* | *Initial gas height (m)* | *Estimated rise of gas–liquid contact (1985) (m)* |
|---|---|---|
| 25/1-2 | 47 | 10–30 |
| 25/1-3 | 20 | >10 |
| 25/1-5 | 64 | 25-45 |
| 10/1-1 | 95 | 35–60 |

It should be noted that a higher attenuation of the flat spot was generally observed on the seismic data (up to 80%) than indicated by the model, and that quantitative answers outside the well regions could not be considered reliable.

In order to investigate this difference, the validity of the acoustic behaviour model was checked by replacing the parameters used for velocity calculations with more precise values (improved corrections for pressure and the effect of filtrate on density and sonic logs). This exercise did not improve the acoustic sensitivity of the models, and the difference in attenuation of the flat spot data should probably be sought in the processing (e.g., amplitude preservation or possible variations of the amplitude vs. offset) or in the simplification of assuming a 1-dimensional model and normal reflection (energy dispersion due to the dipping gas–liquid contact).

To resolve the problem of interpretation in the northern arm of the field, a new survey was shot in 1987. These data were normalized with respect to frequency spectrum with the two previous surveys of 1973 and 1985, and processing parameters were chosen to give optimal resolution at the Frigg level. However, in terms of amplitude it still seemed preferable to interpret the relative changes (amplitude ratios) between the surveys rather than absolute values.

The amplitudes were normalized on a Cretaceous horizon and their evolution versus rise of the GLC was studied both for the seismic marker and the flat spot. 2D-modelling had shown that the amplitude of the Frigg seismic marker increases and that of the flat spot decreases with a rising GLC in this area of the field (reservoir thickness $\leqslant 40$ m) (Fig. 8). The acquired seismic data had a decrease in flat spot amplitudes, but no notable variation in the amplitude of the seismic marker. Thus, no interference between this marker and the new GLC was observed. From this observation it was concluded that the GLC had risen between 15 and 25 m.

(b) *Tuning effect between seismic marker, initial flat spot and the new GLC*

Another approach was to study the evolution of the tuning effect between the seismic marker of the top pay and the flat spot and, if possible, to map its geographical displacement with time.

The interference between the two reflectors attenuates the seismic marker for a given gas height and produces an apparent dip of the flat spot (half the dip of the top reservoir). At a distance of half a period, the flat spot disappears. With the rise of the GLC, the new flat spot gives a similar effect even if it is more attenuated (Fig. 9). This phenomenon could be used for two purposes.

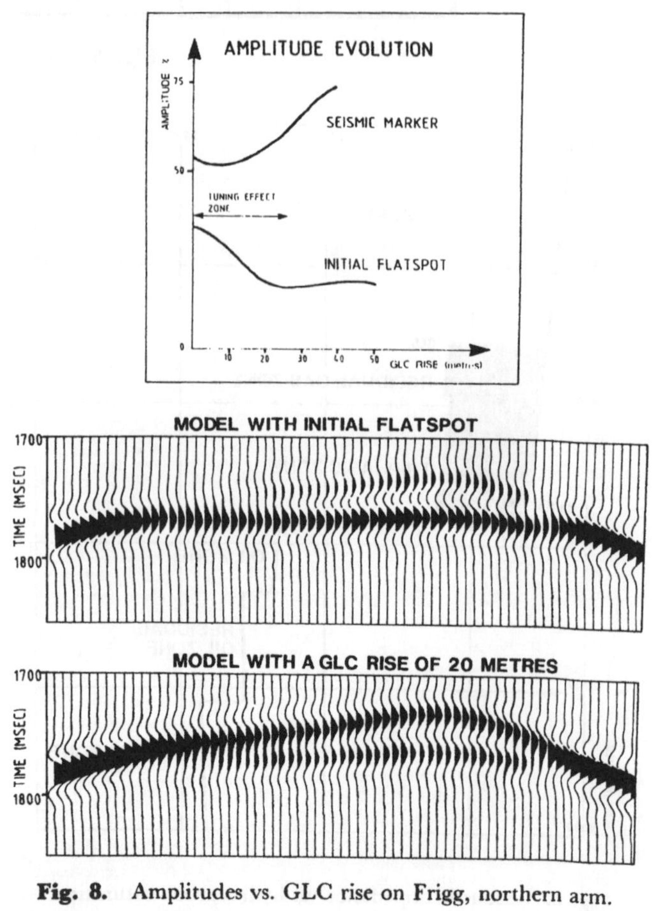


**Fig. 8.** Amplitudes vs. GLC rise on Frigg, northern arm.

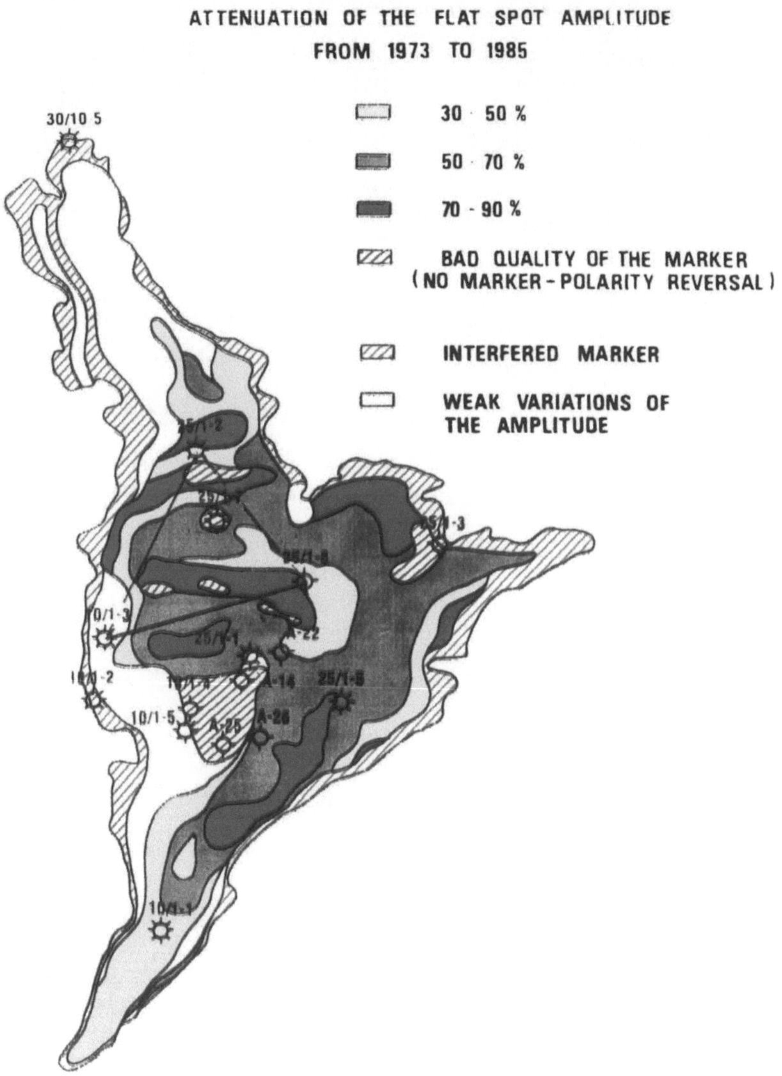


**Plate 23.** Map of flat spot amplitude attenuation.

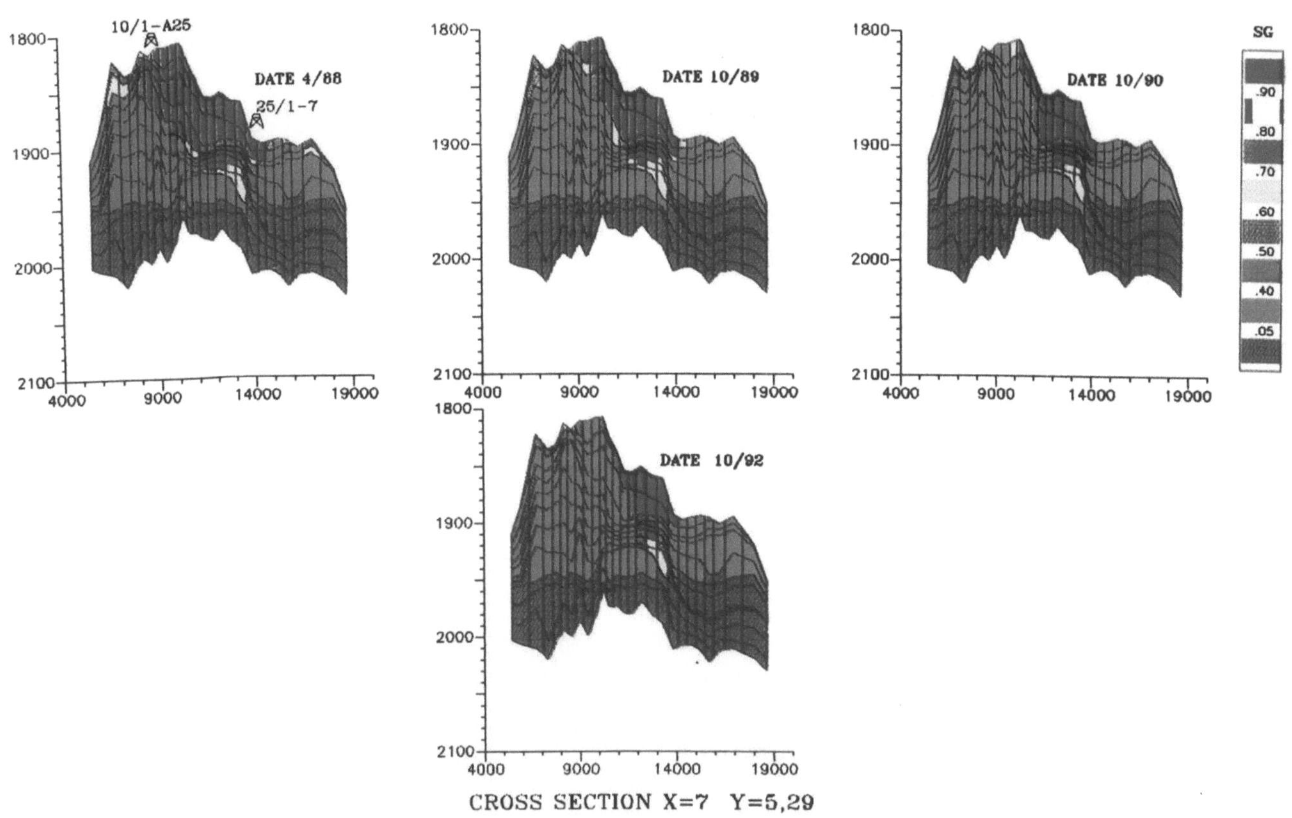


**Plate 24.** Unswept zones from simulations.

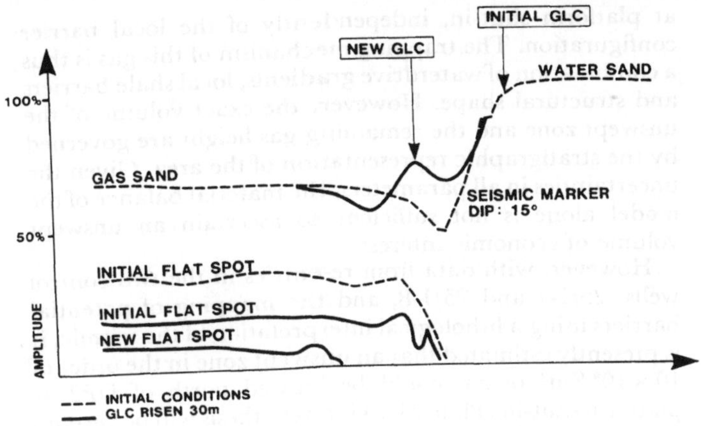


**Fig. 9.** Modelling the tuning effect between the Frigg seismic marker and the flat spot.

- A map of the maximum amplitude could be used to determine the exact edge of the field. The precision when mapping the initial GLC using the flat spot is a function of the dip, and a maximum amplitude map is particularly useful at low dips.
- The tuning effect of the new GLC can possibly be mapped, thus giving the location of the waterfront as a function of time. When it interferes with the initial GLC, this tuning effect is very weak and the lateral resolution is strongly dependent on the dip of top reservoir (200 m for 15° and 500 m for 5° dip). The vertical resolution is about 25 m at Frigg conditions.

The feasibility of the method is thus dependent on the configuration of the GLC rise and structural dip. For the time being (September 1988), the only attempt to apply the method has been on the northern arm of the field (dip around 3°), where a weak tuning effect occurs at a depth of -1925 m MSL (initial contact = -1948 m MSL). This depth also corresponds to the structural saddle between this part of the field and the platform area, and sweeping up to this depth seems logical. However, it is clear that the method is sensitive to the quality of the seismic data as the amplitude of the phenomenon is of the same order of magnitude as the seismic resolution.

(c) *Amplitude versus offset*

According to theory, a maximum amplitude could occur at an angle of incidence of 25° for a gas sand, and one idea was to see whether this maximum value could be followed on the different surveys. However, the testing of the method on a line failed and no conclusive results could be obtained. This could be due to a variation of amplitude laterally and/or to too low a signal/noise ratio.

*Summary* Qualitatively, seismic monitoring has been useful on Frigg. The methods are very sensitive to variations in acquisition parameters and processing (e.g. positioning of the lines, frequencies, bandwidth, etc.) and high-resolution seismic data are always required.

Schematically, the techniques applicable to semi-quantitative approach in the case of Frigg are:

| *Location* | *Central Field* | *Edge* |
|---|---|---|
| Rise of GLC < 25 m | Decrease of initial flat spot amplitude | Seismic monitoring difficult |
| Rise of GLC > 25 m | Seismic monitoring difficult; flat spot amplitude stable | Depending on dip and frequency spectrum, the tuning effect between the seismic marker and the new GLC could be used |

As always, the feasibility of seismic monitoring depends on the local geometry and geology.

## Other forms of monitoring

*Pressure measurements*

The field pressure is regularly monitored in the production wells, thereby giving an important input to the mathematical model. Particularly useful was the use of pressure gradient measurements in appraisal wells, and during recompletion allowing identification of permeability barriers.

*Liquid level*

The liquid level is continuously controlled by cased hole measurements in observation wells and direct sampling in some production wells (Fig. 10). As of today, water is found in most wells at the southern platform (CDP1) and coming close to the bottom of the northern wells (DP2).

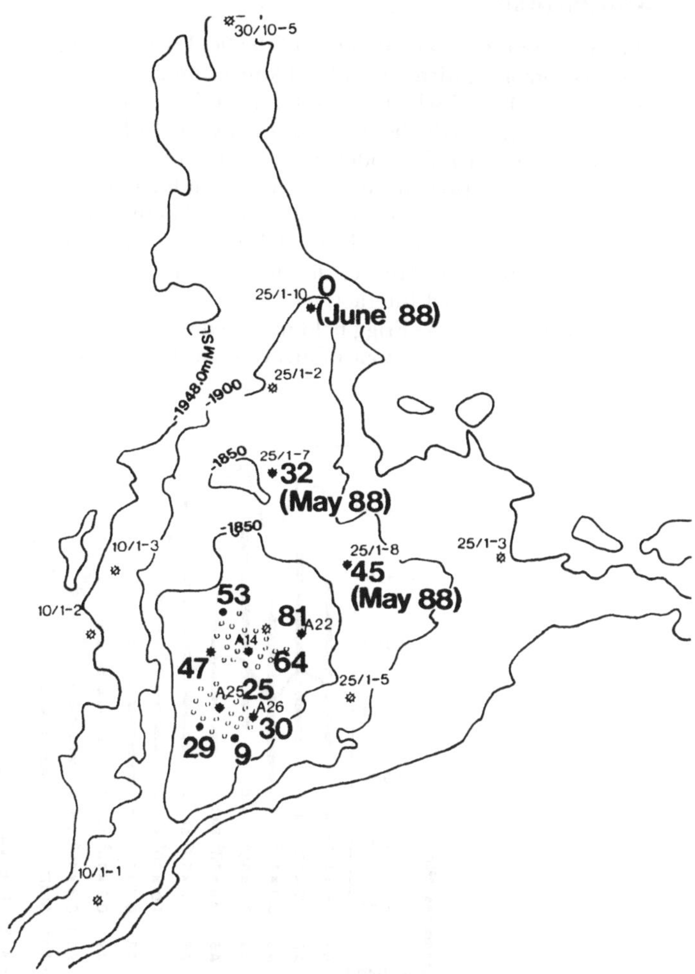


**Fig. 10.** Remaining gas height from well monitoring (October, 1988).

The sweep north of DP2, where seismic monitoring indicated an irregular water encroachment, is controlled by the two appraisal wells, 25/1–7 and 25/1–8. These wells were temporarily abandoned in 1985 and have since been re-entered twice for cased hole logging. The direct measurements obtained in the two wells have been very important matching points in the reservoir model and have confirmed barriers of a certain areal extent. These barriers have been mapped in detail by the latest lithological seismic interpretation. The result of a third re-entry foreseen during the spring of 1989 will be decisive in deciding whether or not to drill wells at high deviation towards this potentially unswept gas.

*Hydrated production*

Simulation and testing of the observation/production well 25/1-A14 (designed for the purpose of testing gas production with water in the well) and, later, wells on the CDP1 platform, have shown that the wells can be produced at reasonable flow rates even after water breakthrough. The usefulness of recompletion has been dismissed, as the low drawdown of the Frigg sands reduces coning within the wellbore. The communication through the sand screen is good enough to allow water to seep into the formation and thus the wells can be produced at very low flow-rates without liquid loading of condensed water. Figure 11 gives the position of the water within the production wells and the corresponding water free gas rate at end of October 1988. The possibility of producing in this manner gives a very good sweep efficiency in the platform area.

## Simulations

The present mathematical model of the field is quite large: it covers a geographical area of 961 km² using 29 × 38 × 16 cells (> 17 600, of which 10 500 are active). All three phases are simulated; the initial gas zone by 13 layers.

In addition, specific models of the satellite fields have been built using pressure data from a simplified greater Frigg area model for edge conditions. These specific models are more precise for the modelling of the expansion of gas towards Frigg. Hence, the different models are to a certain degree used interactively.

All the models of the Frigg field matched with respect to water rise have shown a zone of unswept gas north of DP2 at platform shut-in, independently of the local barrier configuration. The trapping mechanism of this gas is thus a combination of waterdrive gradients, local shale barriers and structural shape. However, the exact volume of the unswept zone and the remaining gas height are governed by the stratigraphic representation of the area. Given the uncertainties in all parameters, the material balance of the model alone is not sufficient to ascertain an unswept volume of economic interest.

However, with data from re-entries in the two control wells, 25/1-7 and 25/1-8, and the mapping of potential barriers using a lithological interpretation of the seismic, it is presently estimated that an unswept zone in the order of $10 \times 10^9$ Sm³ of gas could be located north of DP2 at platform shut-in (Plate 24). This hypothesis will be verified by new re-entries in the remote control wells early in 1989, which will possibly lead to the drilling of highly deviated wells (60–70°) from the DP2 platform to drain this zone. The other zones of unswept gas are related to structural tops/saddles, but are smaller in volume (totally $6–10 \times 10^9$ Sm³) and in remaining gas height. They are therefore not recoverable from a practical point of view.

At present, the best estimate of recovery is $176 \times 10^9$ Sm³ without additional wells, corresponding to a recovery factor of 72–74% (depending on the estimated leakage from the satellites and amount of liberation of gas from the oil zone). The volumetric sweep efficiency is good; around 89% with the present wells and perhaps above 90% with additional deviated wells.

## CONCLUSION

The strong waterdrive in the Frigg field necessitated a re-appraisal programme quite similar to the initial appraisal. It consisted of new wells, extensive coring and new seismic acquisition, but was focused on heterogeneities and barriers to flow that are important to the sweep efficiency. In the case of Frigg, with a large areal extent and barriers of unknown dimensions, this information could not be obtained under static conditions, and a certain production history was required. The stratigraphic interpretation of the seismic when combined with the information from new appraisal wells proved very useful when mapping the potential barriers.

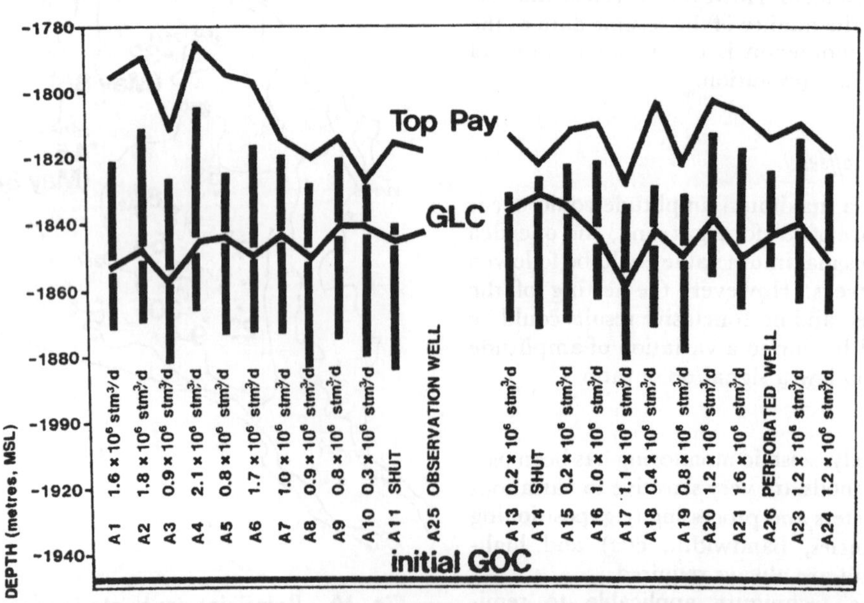


**Fig. 11.** Water level and production rates at the CDP1 platform (October 1988).

The direct information on water-entry and the seismic response under static and dynamic conditions allowed matching and confirmation of the mathematical models. This was possible even if the initial seismic was not very well adapted for quantitative evaluations. Except for a potentially unswept zone north of the platform area that could be drained using highly deviated wells, this interpretation has ruled out the existence of large unswept zones. The feasibility of seismic monitoring depends on the effects of geometry and geology on the seismic response. The studies performed on the Frigg seismic have also demonstrated the need for normalization in terms of the frequencies of the surveys to be compared, and that very good positioning of the lines is important.

## ACKNOWLEDGEMENTS

The author wishes to thank the management of Elf Aquitaine Norge A/S and the Frigg Unit Partners (Norsk Hydro A/S, Statoil, Total Marine Norsk A/S, Total Oil Marine and Elf UK,) for permission to publish this paper.

The text on the geophysical monitoring is based on reports for the Frigg Unit by J. Jestin, J.P. Genest and L. Latour of SNEA(P).

## REFERENCES

Brewster, J. and Jeangeot, G. (1987). The production geology of the Frigg Field. In: Kleppe *et al.* (eds.), *North Sea Oil and Gas Reservoirs*, pp. 75–89. Graham & Trotman, London.

De Leebeck, A. (1987). Frigg Field reservoir: characteristics and performance. In: Kleppe *et al.* (eds.), *North Sea Oil and Gas Reservoirs*, pp. 89–100. Graham & Trotman, London.

Gassmann, F. (1951). Elastic waves through a packing of spheres. *Geophysics* **15**, 673–685.

Thousand, A., Brewster, J. and Mure, E. (1986). Frigg Field. Mid life appraisal programme. *Proceedings of the Offshore Northern Seas Conference, Stavanger, August 1986.*

# 6

# Analysis of Formation Water Sampled from Core

*M.L. Coleman, M.R.O. Jones and M.A. Cox*

BP Research Centre, Sunbury-on-Thames, Middlesex, UK

Formation water analysis is valuable for prediction of chemical scale production (incompatibility with injected water) and for calibration of resistivity logs, and its variation within a field indicates extent of reservoir or zone connectivity in aquifer beneath a field. Water is not always produced, or is not produced long enough to ensure absence of contamination with drilling mud filtrate. Use of tritium tracer in mud allows correction for contamination. This work describes application of that approach to obtain composition of waters from core, where contamination is severe, but would be equally applicable to RFT (repeat formation tester) samples. Alternative analytical procedures were compared. Successive techniques could be applied to the same sample allowing samples as small as 0.2 ml to be analysed for chemical species Na, K, Mg, Ca, Sr, Ba, Cl and S and stable isotopes $\delta^2$D and $\delta^{18}$O. Dean and Stark distillation, as well as producing an accurate measure of water content, also gave a sample that could be analysed: $^3$T values of the distillate were accurate, but stable isotope compositions, $\delta^2$D and $\delta^{18}$O, are too negative by up to 60‰ and 10‰, respectively, about 50 times greater than analytical error. Leaching of the residual core gave chemical composition that could not be recognized as similar to formation water values. Samples ultracentrifuged from transition zone core gave consistent analyses comparable to produced water samples, after correction for contaminant mud-filtrate that formed up to 50% of the analysed sample. Cl compositions could be calculated to $\pm 1\%$, relative. K, Ba and S gave least-reproducible results, while Na, Ca, Sr, $\delta^2$D and $\delta^{18}$O gave consistent answers; better than 10% relative for the chemical compositions and 0.9‰ and 0.2‰ for $\delta^2$D and $\delta^{18}$O, respectively. Attempts were made to distinguish variations in mud infiltration between inner and outer portions of core but none were detected. Routine applications of these techniques would reduce water sampling cost whenever core is cut and could easily be developed for use on RFT samples.

A glossary of technical abbreviations is included in the text.

## INTRODUCTION

### Need for formation water analyses

Formation water composition may vary dramatically between different formations or even between different wells intersecting the same formation. A large amount of work has been carried out on problems associated with the salinity of formation waters (e.g. Morrow, 1971). It is important to be able to measure the chemical composition for the reasons listed below.

*Precipitation of chemical scale*

When a reservoir is produced by water injection, the injected water may be incompatible with the formation water and result in precipitation of chemical scale. Usually, damaging effects can be avoided by addition of scale inhibitor to the injected water. However, the inhibitor must be tailored to suit the particular combination of chemical compositions.

*Water saturation calibration*

Resistivity logs are used to determine initial water saturation in reservoirs ($S_w$). Resistivity is controlled by the ionic concentration of salts in solution and can be calculated precisely from chemical composition if a sample is not actually available for conductivity measurement.

*Reservoir description*

Variation of formation water composition can indicate isolation of separate reservoir units, with obvious implications for production planning, aquifer support and injection schemes. In addition, reservoir quality related to diagenesis depends on the mineral precipitation reactions occurring in formation water in pore space.

### Sample availability, validity and economics

Although accurate formation water compositions are necessary, samples of satisfactory quality are not always taken, despite considerable efforts to do so. Often because of cost or lack of awareness of their value, no sampling at all is performed.

*North Sea Oil and Gas Reservoirs—II*

*Sample types and their problems*

The main cause of poor quality samples is contamination by aqueous phase of drilling mud (mud filtrate contamination—MFC). The best-quality samples are those that are taken during flow testing after sufficient volume has been produced to flush out MFC. Ideally, samples should be taken at bottomhole pressure to preserve gases in solution, which are otherwise lost. Carbon dioxide and hydrogen sulphide, in particular, have an important effect on pH and, in some cases, on those features of the total chemical composition that are relevant to the topics mentioned above. However, the expense of taking downhole pressurized samples precludes their routine use.

Much valuable information can be gained from drill stem test (DST) waters that have been sampled at the surface. Validation of sample quality is often performed by making analyses of chloride (Cl) content on a series of samples until successive values are constant. This approach is satisfactory in most cases but fails if the MFC is similar to formation water for Cl but differs in other significant, minor components, for example barium (Ba), or sulphate ($SO_4$).

Other potential sources of formation water samples that are usually rejected because of high levels of MFC, are RFT samples and samples of core. RFT chambers can provide pressurized samples but MFC may contribute more than 50% of the composition. Core samples will be invaded by MFC to an extent that depends mainly on rate of penetration, formation permeability, mud solids contents and mud weight. Accurate estimation of MFC in core is a requirement when $S_w$ is measured on selected pieces of reservoir core, to calibrate resistivity logs. Standard preserved core samples are taken and total pore water is estimated by distillation (Dean and Stark method). Exaggerated estimates of $S_w$ result if correction is not made for petroleum displaced by MFC.

In summary, water sampling follows the same economic rules that apply almost universally. The best-quality samples cost a lot to take. Samples that offer various extents of reduced quality can be taken at lower cost and thus are more readily available. This chapter describes an approach that allows more and better-quality data to be obtained from poorer-quality samples by correcting for MFC contribution.

*Use of mud tracer*

Any sample can yield estimates of uncontaminated formation water composition even if there is a significant, but not overwhelming, content of MFC. In order to extract the true composition from the analysis of contaminated water, it is necessary to know only the composition and proportion of MFC. The former is achieved easily by sampling mud at regular intervals during drilling: mud composition changes by admixture with formation waters and from engineered changes to the mud. A tracer, a component not present in the formation water, can be used to monitor the amount of MFC content: tracer dilution in the mixture is a direct measure of content.

The work described here used tritium ($^3T$), the radioactive isotope of hydrogen, as a tracer. It has a half-life of just over 10 years, does not occur naturally in formation waters, is added in the form of tritiated water which is totally miscible with any aqueous fluid, and can be detected at low concentrations. These factors make it an economical and effective tracer.

## Experimental test conditions

This work, to estimate formation water composition, was combined with a routine use of tritiated aqueous mud to estimate $S_w$ from core. The initial test here was limited in scope and endeavoured to establish the feasibility of the methods used. Samples were taken from a reservoir zone with $S_w$ lower than ideal for testing water extraction from core, the transition zone between aquifer and oil zone in one of the Forties Field development wells. Aquifer core would have given more water. In some other respects, this choice was not ideal since good recent samples of formation water were not available. Forties has been produced by water injection for a long time and there must be doubt about the waters sampled from producing wells. Nevertheless, careful experimental design allowed controls to be included and direct comparison could be made of different methods of obtaining formation water analyses from core samples. The subsequent sections give details of the experiment and show that one technique, ultracentrifuge extraction, was extremely successful for determining chemical composition. A similar technique could be applied even more easily to obtaining formation water compositions from RFT samples.

# METHODS

## Analytical requirements

In many parts of the procedure, alternative methods were available to achieve what should be the same result. Direct comparisons could be made. In addition, a comparison of inner and outer portions of the same core was made. This gave paired samples from the same core, from which independent analyses could be made in which formation water compositions should be identical but MFC content might vary. Invasion of mud filtrate should be greater in outer portions provided that subsequent storage time of the whole core is not so great that diffusion causes homogenization of compositions.

*Extraction of water*

For $S_w$ purposes, total extraction of water from core is needed and the Dean and Stark distillation procedure (Rall and Taliaferro, 1946; Ward and Barnwell, 1980) is one of the few methods that accomplishes this quantitatively. This technique has the disadvantage that dissolved salts are left behind in the core. However, they can be recovered, with varying degrees of success, by leaching the dry core with distilled water. An alternative approach is ultracentrifuge extraction of water, which produces a sample representative of the chemical composition but only extracts part of the pore water. This makes it unsuitable for $S_w$ determination.

Water can be obtained from drilling mud by Dean and Stark distillation with the disadvantages for chemical analysis noted above. However, samples for chemical composition determination can be obtained by pressure filtering.

In both these cases, if water composition were the sole requirement then quantitative determination of water by Dean and Stark distillation could have been avoided. However, in this experiment $S_w$ was to be determined and this is likely to be the case in any practical application where maximum information should be obtained from expensively recovered samples.

*Chemical and isotopic analyses*

In addition to analyses for $^3T$ and chemical composition, analysis for oxygen and hydrogen isotopic composition of water was made. Natural variation in abundance of trace, stable (non-radioactive) isotopes deuterium ($^2D$) and $^{18}O$ can be used to characterize formation waters in addition to, but independent of, chemical compositions. The stable isotope ratios are expressed as parts per thousand or per mil (‰) variation of $^2D/^1H$ and $^{18}O/^{16}O$ ratio, relative to an internationally accepted Standard Mean Ocean Water (SMOW). These variations are expressed as $\delta^2D$ or $\delta^{18}O$ values whose sign (−or+) shows depletion or enrichment, respectively, of the trace isotope relative to abundance in the standard. Stable isotope compositions thus provide a further check on sample quality.

Because it was expected that only small volumes of water could be extracted by ultracentrifuge, procedures were devised to make successive analyses on the *same* small sample. Test samples were analysed and it could be shown that $^3T$, $\delta^{18}O$, $\delta^2D$ and chemical analyses could be made on the same sample, rather than separate aliquots. The introduction of errors by *successive* analyses, rather than separate portions, is discussed below, together with a comparison of apparently equivalent methods. However, a brief description of each analytical procedure is given first.

## Brief descriptions of analytical methods

*Core samples* — inner and outer

Inner and outer samples of preserved core were taken using a hammer and chisel. To ensure enough material for analysis and a representative sample, each core sample was quartered: split horizontally and vertically. Diagonally opposite (e.g. top left and bottom right) fractions of both inner and outer portions were combined to give two inner and two outer samples per core sample.

*Water extraction*

Dean and Stark distillation techniques were used to quantify mud and core water contents.

For ultracentrifugation, core samples were roughly pulverized with a hammer and 25 g samples were ultracentrifuged at $27 \times 10^3$ G for 24 hours. Mud samples were pressure filtered using nitrogen at 100 psi to produce clear filtrate.

*Isotopic analysis*

Samples for $\delta^2D$ and $d^{18}O$ analysis need to be purified first by distillation. Complete recovery is needed to avoid isotopic fractionation and therefore samples are distilled in vacuum using liquid nitrogen (−180°C) to condense the vapour and a final heating temperature of 140°C to ensure thorough dehydration of residual salts.

A 4 μl aliquot of water was used for $\delta^2D$ analysis. Hydrogen gas for mass spectrometric analysis was prepared by reducing water with Zn metal (Coleman *et al.*, 1982). $\delta^{18}O$ analysis is performed by mass spectrometric analysis of $CO_2$ that has equilibrated isotopically with a 1 ml sample of water (Epstein and Mayeda, 1953), a non-destructive process.

$^3T$ activity is measured by mixing a known quantity of water with a scintillant cocktail (Horrocks, 1974). A Quantulus scintillation counter records activity as disintegrations per minute, which is converted into $^3T$ concentration units/unit volume.

*Chemical analysis*

The major cations expected in formation waters were measured by inductive coupled plasma atomic emission spectrometry (ICPAES, or just ICP). The sample is aspirated into an argon plasma and the emitted visible light radiated is spectrally analysed to give simultaneous measurements for all relevant cations. Chloride was determined by ion chromatography.

*Sample dilution*

Since many of the methods could tolerate a degraded analytical precision but still yield significant results, gravimetric sample dilution was used to allow samples as small as 0.2 ml total to be analysed, if necessary. Precision is degraded by introducing the errors associated with weighing samples for dilution, a relatively minor effect.

## Experimental design

The description of methods shows that some techniques are non-destructive, or at least do not interfere with a selected subsequent measurement. Table I shows which successive applications of techiques were attempted, in some cases only in a preliminary test. A qualitative indication of success is also given in Table I, with more details shown in the next section. It can be seen that $^3T$ can be counted successfully after $\delta^{18}O$ measurement. Chemical analysis after distillation could be performed by redissolving the residue left after distilling water for isotopic analysis. Although major-species analyses gave satisfactory results, analyses of some minor or trace components behaved very poorly. Thus, the results could be used to make a fair estimate of total salinity but might be misleading with respect to trace components that are significant for chemical scale precipitation. The final experimental design is indicated in Fig. 1. All possible combinations of analysis are shown, but those parts of the programme which Table I indicates should be excluded are shown in broken-line boxes.

# RESULTS AND CALCULATED COMPOSITIONS

## Analytical errors

Firstly, the methods which, in combination are identified in Table I as being less than excellent are discussed below.

*Dean and Stark distillation and stable isotope analyses*

Analysis of water recovered after distillation showed that it was isotopically different from produced water. It was more negative in $\delta^2D$ by 10‰ to 60‰ and in $\delta^{18}O$ by 1‰ to 10‰; usual precision is 1‰ and 0.2‰, respectively. Depletion in $^2D$ correlated with depletion in $^{18}O$ and suggests less than 100% recovery of the water. To obtain accurate stable-isotope analyses, recoveries of better than 99% must be achieved. The under-recovery does not affect $^3T$ to a measurable extent.

*Chemical compositions of cores leached with water after Dean and Stark distillation*

A known volume of water was used for leaching and concentrations were calculated using measured volumes of water extracted. Initial results were disappointing and the full set of analyses was not completed. Sodium and chloride gave values approximately half the correct ones and calcium gave variable results, up to twice the true value.

*Chemical compositions of residues after distillation of water for isotopic analysis*

Measured formation water compositions are compared with the redissolved residues after distillation in Table II and shown graphically in Fig. 2. It can be seen that Ba analysis is not satisfactory and K only barely so. Normal precision for all cations is about 1%.

**Table I.** Success in applying successive methods to samples of core

| *Prior procedure (and use)* | *Subsequent procedure* | *Success* |
|---|---|---|
| Dean and Stark distillation (quantitative extraction of water) | $^3$T analysis | Excellent |
| | $\delta^2$D and $\delta^{18}$O analysis | Poor |
| | Chemistry from leached core | Poor |
| Ultracentrifuging (extraction of water) | Water content | Unusable |
| | $^3$T analysis | Excellent |
| | $\delta^2$D and $\delta^{18}$O analysis | Excellent |
| | Chemistry | Excellent |
| Distillation (sample purification for $\delta^2$D and $\delta^{18}$O analysis) | $^3$T analysis | Excellent |
| | $\delta$D and $\delta^{18}$O analysis | Excellent |
| | Chemistry on residue | Good |

## Analytical results

Using the preferred analytical procedures given in Fig. 1, analyses of inner and outer core samples for four depths are given in Table III, together with an analysis of produced water taken from aquifer a number of years earlier. Mud samples were taken immediately before and after the cored interval and their analyses are also shown in Table III.

Sulphate, as such, was not measured in the extracted samples. The field is free of hydrogen sulphide and the assumption is made that total sulphur is present as sulphate. This allowed S to be measured by ICP together with the cations, rather than the much more time-consuming (and expensive) ion chromatography.

There are significant values of tritium content in the

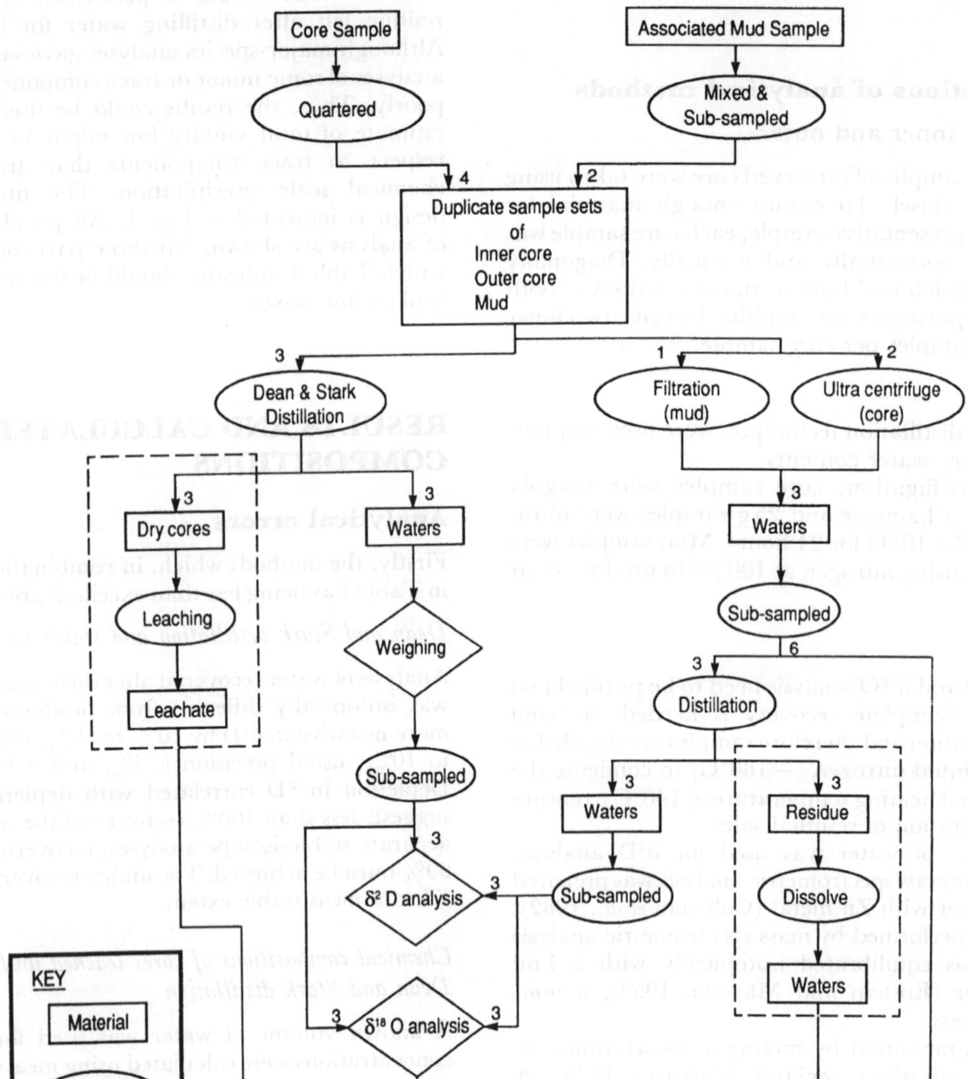


**Fig. 1.** Experimental design.

**Table II.** Measured formation-water composition[b] compared with redissolved residues after distillation

| *Sample*[a] | | *Na*[b] | *K* | *Mg* | *Ca* | *Sr* | *Ba* |
|---|---|---|---|---|---|---|---|
| 49 I | Meas. | 22140 | 7180 | 1250 | 2770 | 92 | 4.0 |
| | Resid. | 19290 | 6860 | 1075 | 2295 | 78 | 0.0 |
| 49 O | Meas. | 20450 | 6670 | 1130 | 2640 | 86 | 3.0 |
| | Resid. | 18680 | 8190 | 1165 | 2735 | 88 | 0.0 |
| 50 I | Meas. | 22460 | 5570 | 1220 | 2690 | 98 | 3.0 |
| | Resid. | 23420 | 6950 | 1295 | 3015 | 104 | 0.0 |
| 50 O | Meas. | 24080 | 4790 | 1290 | 3110 | 104 | 4.0 |
| | Resid. | 25070 | 6520 | 1265 | 2955 | 104 | 10.0 |
| 52 I | Meas. | 23000 | 3270 | 605 | 2650 | 301 | 5.0 |
| | Resid. | 27770 | 5470 | 675 | 2885 | 332 | 10.0 |
| 52 O | Meas. | 23940 | 3170 | 625 | 2710 | 308 | 5.0 |
| | Resid. | 26590 | 5250 | 670 | 2835 | 331 | 10.0 |
| 93 I | Meas. | 19620 | 6600 | 469 | 2190 | 213 | 4.0 |
| | Resid. | 21190 | 7470 | 450 | 2015 | 202 | 10.0 |
| 93 O | Meas. | 21230 | 7360 | 510 | 2380 | 230 | 4.0 |
| | Resid. | 21540 | 8460 | 495 | 2185 | 212 | 0.0 |
| Average % difference | | 3.5 | 30.0 | 0.6 | −1.1 | −0.2 | 12.5 |
| Standard Deviation | | 10.0 | 23.6 | 7.5 | 9.3 | 8.1 | 113.9 |

[a] I or O after the sample number indicates inner or outer subsample, respectively.
[b] All concentrations in ppm.

core waters, up to about half the mud value. This implies a major MFC component, also confirmed by the diference of the results form those of the produced water. The tritium measurements can be used to correct for MFC as shown below.

## Corrected results for water from core

The analysed samples are clearly mixtures of MFC and formation water. The tritium content of the mixture gives a measure of the MFC content. The calculation depends on knowing the $^3$T content of the relevant drilling mud. Drilling mud varies in composition during drilling as the mud is diluted with formation water. In order to compensate for this dilution, the mud composition is adjusted by addition of salts and also $^3$T tracer. Table III shows that $^3$T analyses of drilling muds, which bracket the relevant core depths, show some variation. For correction of MFC, a linear variation of $^3$T has been assumed. The basis of correction for each analysed component requires the concentration of that component in the mud and the fraction of MFC in the analysed mixture, $f_{mud}$. The analysed concentration, $a$, for each component, $i$, yields the concentration in the formation water, $w$, from the mud analysis, $m$, through the expression

$$a_i = m_i \times f_{mud} + w_i \times (1 - f_{mud})$$

## Discussion of results

### *Comparison of inner and outer portions*

The results for $^3$T concentration in Table III have been converted to $f_{mud}$ in Table IV. The correction process is shown for sodium and chloride in Fig. 3. It is slightly disappointing to see no real variation in mud infiltration between inners and outers. The extent of flushing was so great that invasion was to the same extent throughout the core. Although $f_{mud}$ is constant for any one depth, it varies between depths. This is not surprising as it would depend on formation permeability, mud weight and rate of penetration, among other things over the relevant interval. Despite these variations, corrected analyses give extremely consistent results.

### *Major components*

Examination of Table IV shows very encouraging results. The mean corrected values for Na and Cl are, within error, the same as produced water values. However, comparisons of inners and outers shown even greater consistency. It must be assumed that formation water is constant in composition over the distance of a single core sample. Similar homogeneity need not exist over the vertical distance encompassed by the core samples. The mean variations between pairs of inners and outers of the same depth, (7.9%, for Na and 1.0% for Cl) is similar to average

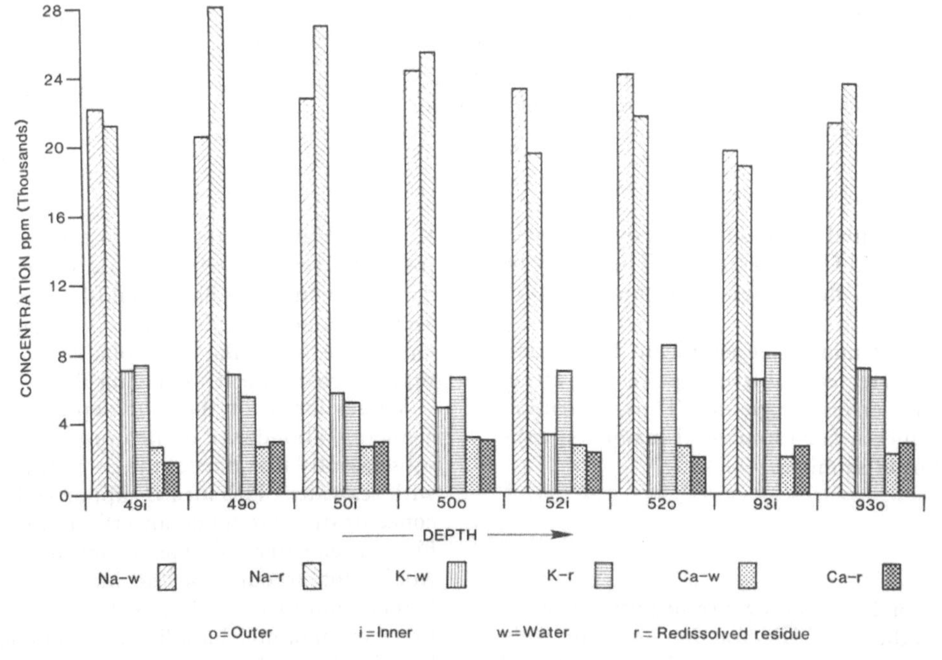


**Fig. 2.** Comparison of formation water composition measured directly and after dissolutions of distillation residue.

**Table III.** Chemical[b] and isotopic analyses of waters extracted from core, drilling mud and formation water.

| Sample[a] | Depth m | Na[b] | K | Mg | Ca | Sr | Ba | S | Cl | Tritium nCi/cm³ | $\delta^2 D$‰ | $\delta^{18} O$‰ |
|---|---|---|---|---|---|---|---|---|---|---|---|---|
| 49 I | 2448.95 | 22140 | 7180 | 1250 | 2770 | 92 | 4.0 | 3420 | 45600 | 0.82 | −16.4 | 0.4 |
| 49 O | 2448.95 | 20450 | 6670 | 1130 | 2640 | 86 | 3.0 | 3140 | 44900 | 0.84 | −16.2 | 0.2 |
| 50 I | 2449.20 | 22460 | 5570 | 1220 | 2690 | 98 | 3.0 | 3050 | 48300 | 0.69 | −16.7 | 0.5 |
| 50 O | 2449.20 | 24080 | 4790 | 1290 | 3110 | 104 | 4.0 | 3050 | 48700 | 0.69 | −17.6 | 0.7 |
| 52 I | 2453.05 | 23000 | 3270 | 605 | 2650 | 301 | 5.0 | 359 | 48600 | 0.57 | −19.6 | 0.7 |
| 52 O | 2453.05 | 23940 | 3170 | 625 | 2710 | 308 | 5.0 | 442 | 47700 | 0.59 | −18.6 | 1.1 |
| 93 I | 2483.80 | 19620 | 6600 | 469 | 2190 | 213 | 4.0 | 447 | 43500 | 0.92 | −18.1 | 0.3 |
| 93 O | 2483.80 | 21230 | 7360 | 510 | 2380 | 230 | 4.0 | 485 | 43700 | 0.91 | −16.5 | 0.6 |
| Drillmud 1 | 2442.00 | 9400 | 12300 | 35 | 1420 | 6 | 12.0 | 1120 | 31000 | 1.73 | −16.4 | −1.8 |
| Drillmud 2 | 2490.00 | 7900 | 12200 | 13 | 1470 | 6 | 18.0 | 995 | 27700 | 1.83 | −15.2 | −2.7 |
| Produced water | | 30990 | 472 | 522 | 3520 | 660 | 275 | 27 | 56700 | 0 | N/A[c] | N/A |

[a] I or O after the sample number indicates inner or outer subsample, respectively.
[b] All concentrations in ppm for non-isotopic species.
[c] N/A, not analysed.

**Table IV.** Analytical results[b] corrected for MFC

| Sample[a] | Depth (m) | Na[b] | K | Mg | Ca | Sr | Ba | S | Cl | Fraction MFC | $\delta^2 D$‰ | $\delta^{18} O$‰ |
|---|---|---|---|---|---|---|---|---|---|---|---|---|
| 49 I | 2448.95 | 22140 | 7180 | 1250 | 2770 | 92 | 4.0 | 3420 | 45600 | 0.47 | −16.4 | 0.4 |
| mud | | 9183 | 12286 | 32 | 1427 | 6 | 12.9 | 1102 | 30522 | | −16.2 | −1.9 |
| corr. val. | | 33633 | 2651 | 2331 | 3961 | 168 | −3.9 | 5476 | 58974 | | −16.6 | 2.4 |
| 49 O | 2448.95 | 20450 | 6670 | 1130 | 2640 | 86 | 3.0 | 3140 | 44900 | 0.48 | −16.2 | 0.2 |
| mud | | 9183 | 12286 | 32 | 1427 | 6 | 12.9 | 1102 | 30522 | | −16.2 | −1.9 |
| corr. val. | | 30914 | 1455 | 2150 | 3766 | 160 | −6.2 | 5033 | 58253 | | −16.2 | 2.2 |
| 50 I | 2449.20 | 22460 | 5570 | 1220 | 2690 | 98 | 3.0 | 3050 | 48300 | 0.40 | −16.7 | 0.5 |
| mud | | 9175 | 12285 | 32 | 1427 | 6 | 12.9 | 1101 | 30505 | | −16.2 | −1.9 |
| corr. val. | | 31149 | 1178 | 1997 | 3516 | 158 | −3.5 | 4325 | 59938 | | −17.0 | 2.1 |
| 50 O | 2449.20 | 24080 | 4790 | 1290 | 3110 | 104 | 4.0 | 3050 | 48700 | 0.40 | −17.6 | 0.7 |
| mud | | 9175 | 12285 | 32 | 1427 | 6 | 12.9 | 1101 | 30505 | | −16.2 | −1.9 |
| corr. val. | | 33828 | −112 | 2113 | 4210 | 168 | −1.8 | 4325 | 60600 | | −18.5 | 2.4 |
| 52 I | 2453.05 | 23000 | 3270 | 605 | 2650 | 301 | 5.0 | 359 | 48600 | 0.33 | −19.6 | 0.7 |
| mud | | 9055 | 12277 | 30 | 1432 | 6 | 13.4 | 1091 | 30240 | | −16.1 | −2.0 |
| corr. val. | | 29719 | −1070 | 882 | 3237 | 443 | 1.0 | 6 | 57446 | | −21.3 | 2.0 |
| 52 O | 2453.05 | 23940 | 3170 | 625 | 2710 | 308 | 5.0 | 442 | 47700 | 0.34 | −18.6 | 1.1 |
| mud | | 9055 | 12277 | 30 | 1432 | 6 | 13.4 | 1091 | 30240 | | −16.1 | −2.0 |
| corr. val. | | 31491 | −1450 | 927 | 3359 | 461 | 0.7 | 113 | 56557 | | −19.9 | 2.7 |
| 93 I | 2483.80 | 19620 | 6600 | 469 | 2190 | 213 | 4.0 | 447 | 43500 | 0.51 | −18.1 | 0.3 |
| mud | | 8094 | 12213 | 16 | 1464 | 6 | 17.2 | 1011 | 28126 | | −15.4 | −2.5 |
| corr. val. | | 31441 | 844 | 934 | 2935 | 425 | −9.6 | −132 | 59266 | | −20.9 | 3.2 |
| 93 O | 2483.80 | 21230 | 7360 | 510 | 2380 | 230 | 4.0 | 485 | 43700 | 0.50 | −16.5 | 0.6 |
| mud | | 8094 | 12213 | 16 | 1464 | 6 | 17.2 | 1011 | 28126 | | −15.4 | −2.5 |
| corr. val. | | 34408 | 2491 | 1006 | 3299 | 455 | −9.3 | −43 | 59324 | | −17.6 | 3.8 |

[a] I or O after the sample number indicates inner or outer subsample, respectively.
[b] All concentrations in ppm for non-isotopic species.

variation of each individual analysis from the overall mean (2.3% and 1.7% Na and Cl, respectively). There may be as small but significant fluctuation of formation water salinity with depth.

A similar pattern of minor but significant fluctuation can be seen in the corrected stable-isotope results. Here, however, the relative errors are greater. Thus, correction for MFC can produce a fair estimate of major ion concentration and thus of salinity.

*Minor ions*

Corrected values for K, Ba and S are negative in some cases. This implies that the MFC fraction, or more likely the analysis of the mud, was erroneous. Possible causes of error are cation exchange on clay minerals, giving a higher K concentration in the mud, or the reverse process lowering K by reaction with naturally occurring clays in the core. The latter problem could be overcome by ensuring analysis is performed as rapidly as possible after core is cut. In the case of Ba, it is probable that sulphate from the mud reacted with it, removing Ba as insoluble $BaSO_4$.

Values for S for the four depths show much greater elemental abundance in the upper two. It was discovered subsequently that part of the core may have been washed with seawater! This idea is supported by the fact that Sr concentration are much lower in those samples that have high S; seawater only has 8 ppm Sr. However, even this would not account for the high S concentations seen; higher than in normal seawater. It is important to note that care in sample handling is of paramount importance, but this lesson had not been learned at the time these samples were taken.

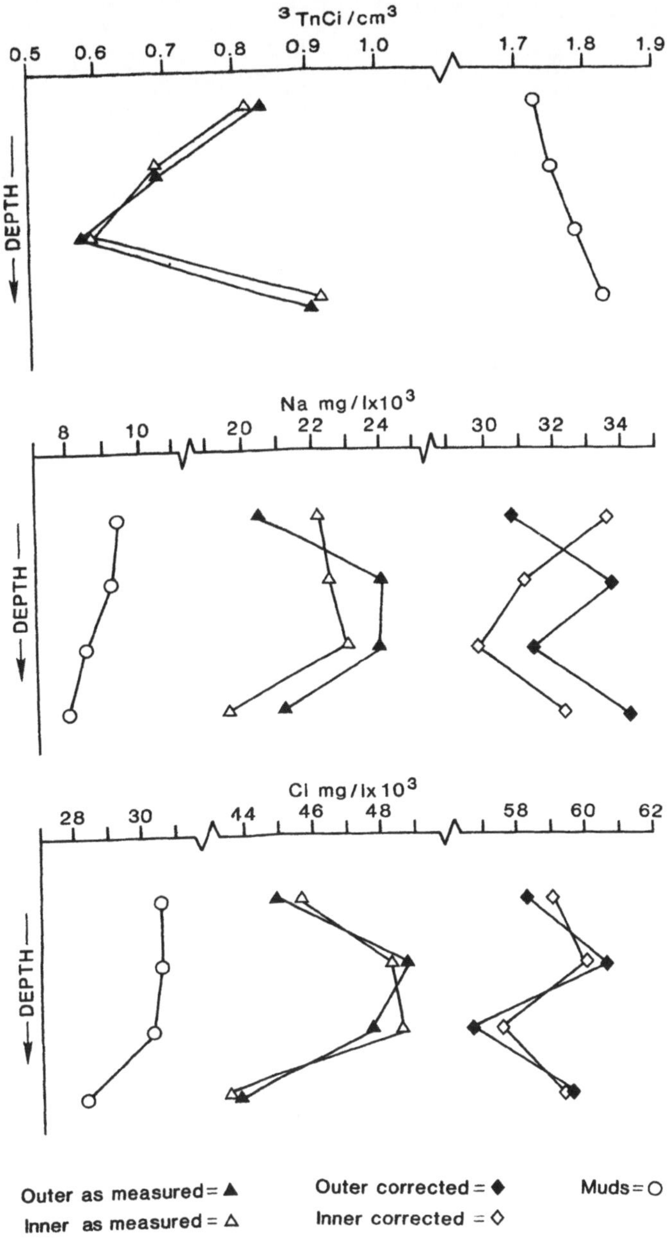


**Fig. 3.** Corrections for MFC for Na and Cl analyses.

## CONCLUSIONS

(1) Ultracentrifugation of core drilled with tritiated mud can provide a corrected chemical major-ion analysis of formation water of sufficient quality to calculate $R_w$.
(2) Dean and Stark distillation is essential for complete recovery of water from core for determination of true $S_w$.
(3) Leaching core after Dean and Stark extraction did not produce satisfactory chemical analyses of formation water.
(4) Minor ions that may be significant for prediction of chemical scale may not produce valid corrected analyses because of reaction with incompatible components in the mud.
(5) Use of this method would allow formation water salinity to be determined routinely whenever core is cut, without the expense of open-hole testing.
(6) Further development of the method should allow its application to deriving formation water compositions from recovered RFT samples.


## ACKNOWLEDGEMENTS

We are grateful to many colleagues in BP Research Centre for their help in running this project and getting it written. However, we would like to thank Jan Harte, Ted Baah and John Petler in particular. Colleagues in BP Exploration assisted greatly in discussion of the results but we would like to thank Dick Woodhouse especially for his helpful and instructive comments.


## GLOSSARY

| | |
|---|---|
| $\delta$ | Expression for $\left[\frac{R_{\text{sample}}}{R_{\text{standard}}} - 1\right] \times 1000$, where $r$ = ratio of trace isotope with respect to major isotope e.g. D/H |
| ‰ | Parts per thousand |
| $^2$D | Deuterium; stable isotope of hydrogen |
| DST | Drill stem test |
| $f_{\text{mud}}$ | Fraction of mud filtrate contamination in the analysed mixture of mud and formation water |
| ICPAES (ICP) | Inductively coupled plasma atomic emission spectrometry |
| MFC | Mud filtrate contamination |
| [nCi/cm$^3$] | Nanocuries per cubic centimetre 1 curie = $3.7 \times 10^{10}$ Becquerel |
| $^{18}$O | Heavy isotope of oxygen |
| RFT | Repeat formation test |
| $R_w$ | Resistivity of formation water |
| SMOW | Standard Mean Ocean Water; International Stable Isotope Standard |
| $S_w$ | Initial water saturation |
| $^3$T | Tritium; radioactive isotope of hydrogen |

## REFERENCES


Coleman, M.L., Shepherd, T.J., Durham, J.J., Rouse, J.E. and Moore, G.R. (1982). Reduction of water with zinc for hydrogen isotope analysis. *Analyt. Chem.* **54**, 993–995.

Epstein, S. and Mayeda, T.K. (1953). Variations of the O$^{18}$/O$^{16}$ ratio in natural waters. *Geochim. Cosmochim. Acta* **4**, 213.

Horrocks, D.L. (1974). *Applications of Liquid Scintillation Counting*. Academic Press, New York.

Morrow, N.R. (1971). The retention of connate water in hydrocarbon reservoirs. Pt. 2. Environment and properties of connate water. *J. Can. Petroleum Technol.* **10** (January–March), 47–55.

Rall, C.G. and Taliaferro, D.B. (1946). A method for determining simultaneously the oil and water saturations of oil sands. *U.S. Bureau of Mines Report of Investigations 4004*, December.

Ward, C.E. and Barnwell, J.L. (1980). Industrial survey of core handling, storage and analysis procedures for RHC measurements. Final report. *U.S. Department of Energy Report DOE/BC/10022*, October.

# 7

# Establishment of a Geological Fracture Model for Dual Porosity Simulations on the Ekofisk Field

*A. Fritsen*[a] *and T. Corrigan*[b]

[a]Norsk Hydro A.S., Stabekk, Norway
[b]Corrigan Associates, Ditchling, Sussex, UK

In order to construct a geologic fracture model for use in dual porosity simulation of the Ekofisk Field, all cored wells on the field were examined, and both lithological and fracture data were assembled. Fracture density distribution in the reservoir was mapped on the basis of estimated tectonic matrix block dimensions in the cored walls. Observed relationships between fracture density, lithology and structural positions formed the basis for statistical evaluation of logs in non-cored wells to assess the fracture density in areas outside the control of cored wells. Different logfacies types, defined by cluster analysis on log data, were quantified with respect to fracture density, and the tectonic fracture densities were predicted and mapped per reservoir layer. Equations for calculating fracture porosity and permeability were subsequently established, and the results contoured.

The geological fracture model was thereafter adjusted to well-test data, and the porosity and permeability for the stylolite fracture network was calculated separately and added to the tectonic fracture maps.

## INTRODUCTION

Over the past 10 years there has been a recognition that fractured reservoirs cannot adequately be simulated by single porosity simulators, and complex dual porosity/dual permeability reservoir performance simulators have been developed. This, however, raised the need for input parameters to describe the fracture systems. The complexity of fracture origins, and the resulting variability in morphologies and distributions of natural fracture systems require unique quantitative geological data for any fractured reservoir modelling.

The Ekofisk Field is a dome shaped, fractured chalk reservoir located at 3200 m depth in the Central Graben of the North Sea. The field consists of two main reservoirs, the Tor Formation of Maastrichtian age and the Ekofisk Formation of Danian age. The two reservoirs are separated by a tight argillaceous layer.

As part of Norsk Hydro's reservoir modelling project of the Ekofisk Field, a fracture model based on geological data was developed. This chapter follows the development of this model and discusses the procedures used and results obtained, as well as precautions and assumptions involved in the work.

*North Sea Oil and Gas Reservoirs—II*


It is recognized that the numerical fracture model that was eventually adapted to the simulator will never reflect completely the considerable geological variability found in the Ekofisk fracture network. However, the study has shown that, using core and log data, a model can be developed that successfully matches the fracture permeability independently assessed from well test data. Critically, for reservoir simulation, the model provides fracture data in all layers throughout the reservoir.

## CORE OBSERVATIONS AND CONCLUSIONS

Considerable effort was spent on core studies to provide a sound basis for the model. Cores from 13 wells with a total of 990 m of core were described in detail (Fig. 1), and both fracture systems and sedimentological facies were described.

### Stratigraphic and structural distribution of fractures

In all wells two main fracture types with different stratigraphic and structural distributions occur, as illustrated in Fig. 2.

The *tectonic fractures* can be classified as subvertical, small-scale normal faults arranged in parallel and conjugate sets cutting through the stylolites. These fractures are seen

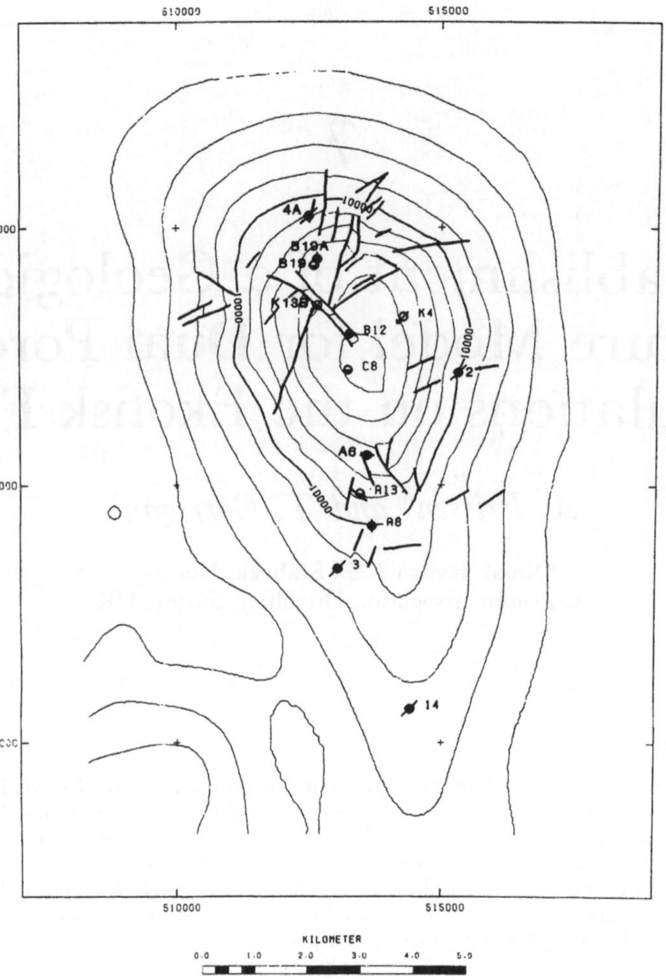


**Fig. 1.** Top Ekofisk Formation depth map showing location of cored wells.

to be interconnected, forming an effective network of flow channels, and are responsible for most of the fluid flow within the reservoir.

*Stylolite fractures* are closely associated with stylolites, forming vertical fracture planes usually normal to the stylolite surface and terminating against the tips of the stylolite teeth (Watts, 1983). They are rarely observed to cross or offset any stylolite surfaces. Their average vertical length is 10 cm and the spacing between the fractures is often less than 5 cm. They are generally not mineralized and they may represent highly permeable zones parallel to the stylolites, thus linking together the longer-spaced tectonic fractures.

The stylolite fractures are typically found in the Tor Formation, but in some downflank wells they also occur in the lower part of the Ekofisk Formation. In contrast, tectonic fractures are typically found in the Ekofisk Formation, but are also common in the upper Tor Formation, mainly in crestal wells or where the matrix porosity is high. Figure 3 shows schematically the distribution of the fracture types.

Whereas stylolite fractures in the Tor Formation seem to occur in all wells, independently of structural position, the presence of tectonic fractures appears to depend strongly on a structural control (the structural factor). In all layers, the tectonic fractures are most closely spaced in the crestal and fault-dominated areas, with spacing increasing markedly downflank.

## Matrix block dimensions

A matrix block is defined in engineering terms as a volume of chalk matrix completely surrounded by open tectonic fractures that disrupt capillary continuity. In this study, fracture density was therefore expressed as the average size of matrix blocks, defined by height, width and length (Fig. 4).

The approach used was to measure two of these dimensions (block height and width), directely on the core, producing results as shown in Figs. 6 and 7. However, estimation of block sizes from cores was complicated by two factors: one is that approximately 60% of all tectonic fractures seem to terminate at fracture intersections, thus forming polygons of variable shapes and sizes rather than a geometrically well-defined pattern. The second is the fact that since the core width is a maximum of 10 cm, only a fraction of these matrix block polygons can actually be seen. Consequently, judgement had to be applied in averaging interval block sizes. The upper limits of block sizes feasible to define by core examination was determined to be approximately 50 cm. In general, the block sizes so defined represent the minimum block sizes within the given interval. However, when dynamic conditions are considered, the block sizes represent a minimum height of capillary continuity only in sequences of massive chalk with no stylolites, clay seams or other bed-parallel disruptions. Elsewhere, the stylolite spacings or the spacing

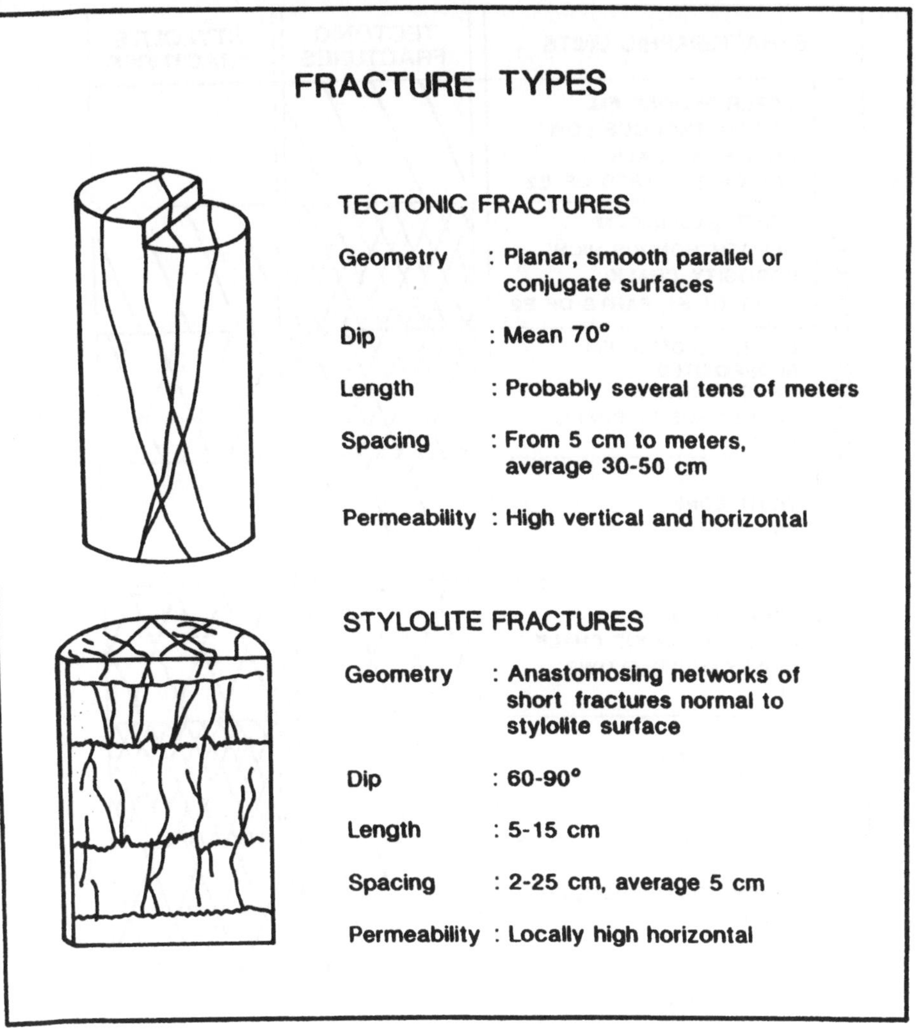


**Fig. 2.** Characteristics of the fracture types in the Ekofisk Field.

between tight laminae will define the "true" dynamic block heights, as illustrated in Fig. 5.

### Distribution of fracture density

From the core observations, two clear relationships of fracture density distribution were seen. First, the fracture density in a well is controlled by the position of that well on the structure. Wells on the flanks of the structure generally show lower fracture densities and therefore have larger matrix blocks than the crestal wells.

Second, the matrix block sizes are controlled by the lithological variations within the reservoir. When the matrix block sizes were plotted, together with well logs and core plug data, a stratigraphic variation in fracturing clearly occurred (Figs. 6 and 7). This variation is controlled by two major litological factors: clay size/insoluble residue content and matrix porosity.

In all wells, increasing shaliness in the chalk matrix reduces the number of open fractures. This is most clearly seen in the argillaceous Ekofisk Tight Zone, which contains no open fractures, except in the most-crestal wells. Similarly, the upper part of the Ekofisk Formation, where increasing shaliness marks the transition to the overlying Maureen Formation, is far less fractured than the average formation. The clay content appears to give the rock a greater plasticity, reducing the number of fractures.

In addition to this argillaceous control on fracturing, matrix porosity also seems to cause variation in fracture density. In A-8 and C-8 it is seen that the most heavily fractured intervals consist of allochthonous (redeposited) sediments exhibiting extreme values of core plug porosity (above 40%) (Figs. 6 and 7).

### Fracture pattern

No conclusive models for the possible origin and timing of the fracturing in the Ekofisk Field have to this date been published. The main reasons for this are the complexity of the fracturing and the problem of explaining the discrepancy in the fracture orientations in the Tor Formation compared to the Ekofisk Formation, and the fact that orientation data is only available from eight wells.

However, in its simplest form, the orientations suggest a radial fracture pattern that is modified by a regional NE–SW trend and by local faulting effects. This model, illustrated in Fig. 8, was used as the basic tectonic fracture network anisotropy.

## FRACTURE DENSITY IN NON-CORED WELLS

Mapping of fracture density/matrix block dimensions over the structure, solely on basis of the cored wells, was

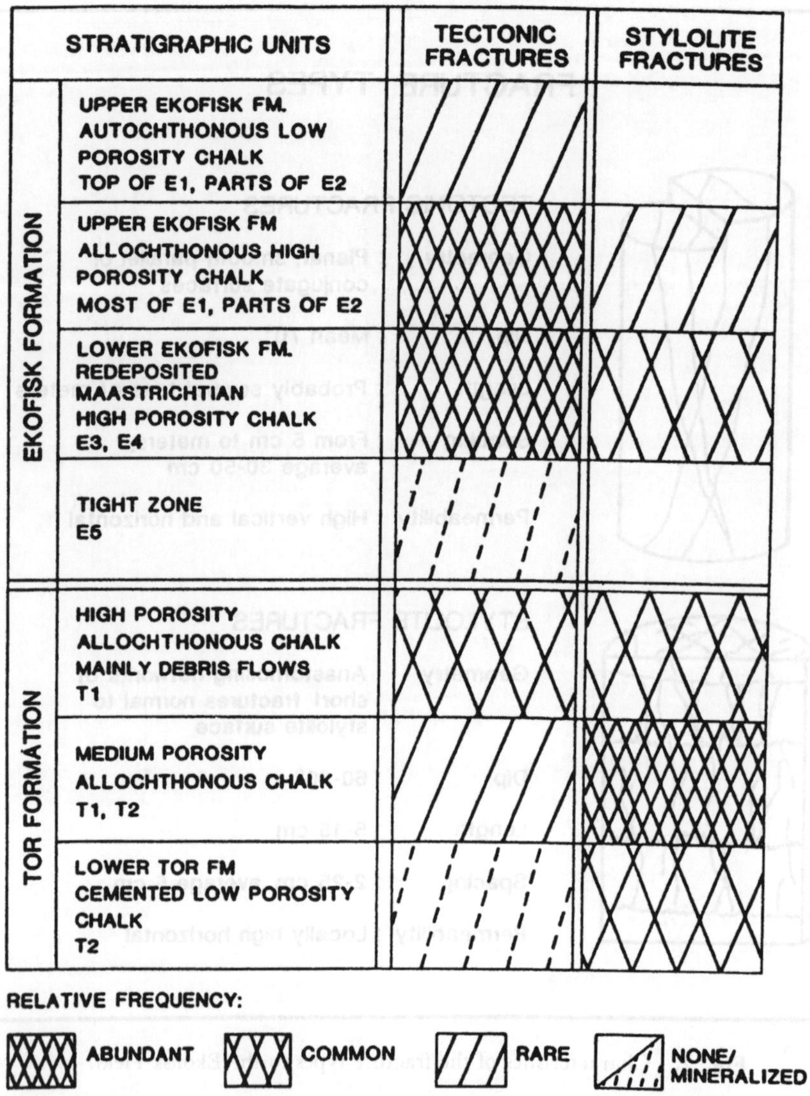


**Fig. 3.** Simplified stratigraphic distribution of tectonic and stylolite fractures.

considered unsatisfactory, both because of their limited number compared to total wells drilled and because of the location of the cored wells mainly along the N-S axis of the structure. It was therefore necessary to develop a method through which non-cored wells and non-cored intervals in cored wells could be evaluated for fracture density.

The objective was to establish, through statistical methods, a set of fracture facies types, each with a defined fracture density, that could be automatically recognized from conventional logs in wells in any structural position on the field.

The computer software used for the task was the in-house developed CREVAS programs (Compiled Rock Evaluation Software package) built on the SAS statistical systems package.

Initially, all log, core and fracture data from the oil zones in wells C-8 and A-8 were entered into the data base. These two wells were chosen as reference wells because they both have detailed core descriptions and good core condition, and they represent two different structural positions, one crestal and one downflank.

## Logfacies classification of reference wells

Classification of the chalk into different logfacies types was made through cluster analysis and discriminant analysis on the gamma-ray, FDC, CNL and sonic logs on A-8 and C-8 in combination. A total of eight logfacies types were defined and the classification rule was saved to be used on other wells. Simple statistics were then run on the logfacies groups to describe their characteristics, as summarized in Table I.

*Logfacies A* This is a small group characterized by very high gamma-ray and high formation density (Plate 25), low porosity and permeability, and low to medium sonic transit time. This group has the highest stability and the longest distance to the other clusters and was always the first cluster that separated out in the data set during the statistical runs. It consists solely of autochthonous pelagic chalk and turbidites. When its distribution in each well (Plate 26) is examined, it is found to occur in the uppermost part of the Ekofisk Formation, marking the argillaceous transition to the Maureen Formation, and in the Ekofisk Tight Zone. Mineralogical analyses by X-ray diffractometry show that this facies contains between 10% and 50% clay and mica and between 10% and 30% quartz. Very small pore throat sizes (0.1–0.25 μm) are also recorded, and in thin section this facies has a clay-rich matrix and is completely cemented by calcite.

*Logfacies B* This group has very low gamma-ray and consists of calcite-cemented thin-bedded debris flows.

**Table I.** Summary of statistics for each logfacies, wells A-8 and C-8

| | | *Logfacies* | | | | | | | | |
|---|---|---|---|---|---|---|---|---|---|---|
| | | *A* | *B* | *C* | *D* | *E* | *F* | *G* | *H* | *For all facies* |
| GR | Mean | 18.069 | 5.902 | 8.009 | 9.829 | 7.898 | 7.522 | 8.993 | 7.672 | 8.749 |
| | SD | 2.604 | 0.506 | 1.281 | 1.442 | 0.843 | 0.724 | 0.919 | 0.865 | 2.41 |
| | *n*[a] | 60 | 51 | 388 | 325 | 107 | 289 | 259 | 88 | 1567 |
| FDC | Mean | 2.291 | 2.224 | 2.302 | 2.130 | 2.174 | 2.061 | 2.039 | 1.954 | 2.147 |
| | SD | 0.064 | 0.071 | 0.058 | 0.038 | 0.043 | 0.034 | 0.037 | 0.036 | 0.12 |
| | *n* | 60 | 51 | 388 | 325 | 107 | 289 | 259 | 88 | 1567 |
| CNL | Mean | 26.665 | 25.875 | 23.332 | 33.552 | 28.558 | 33.485 | 38.228 | 37.301 | 31.138 |
| | SD | 3.517 | 3.133 | 2.769 | 2.165 | 2.017 | 1.911 | 2.443 | 1.691 | 5.97 |
| | *n* | 60 | 51 | 388 | 325 | 107 | 289 | 259 | 88 | 1567 |
| Sonic | Mean | 97.872 | 86.177 | 81.264 | 101.34 | 96.358 | 112.275 | 115.37 | 131.799 | 101.440 |
| | SD | 5.337 | 8.225 | 5.418 | 5.332 | 4.441 | 4.943 | 5.015 | 4.803 | 15.61 |
| | *n* | 60 | 51 | 388 | 325 | 107 | 289 | 259 | 88 | 1567 |
| Core porosity (%) | Mean | 25.121 | 31.253 | 24.994 | 34.911 | 32.214 | 39.349 | 40.525 | 42.127 | 33.637 |
| | SD | 4.167 | 5.244 | 5.112 | 4.496 | 3.145 | 3.249 | 3.263 | 3.773 | 7.46 |
| | *n* | 42 | 19 | 178 | 209 | 37 | 86 | 166 | 11 | 748 |
| Core $K_M$ (md) | Mean | 0.382 | 2.252 | 1.616 | 3.009 | 3.815 | 5.272 | 5.552 | 7.255 | 3.438 |
| | SD | 0.549 | 1.183 | 5.500 | 2.658 | 4.827 | 6.904 | 7.183 | 9.278 | 5.59 |
| | *n* | 42 | 19 | 178 | 209 | 37 | 86 | 166 | 11 | 784 |
| Block width (cm) | Mean | 33.579 | 12.175 | 14.610 | 16.422 | 12.129 | 6.933 | 10.727 | 4.149 | 13.824 |
| | SD | 16.821 | 4.920 | 9.776 | 12.572 | 10.348 | 3.041 | 6.933 | 1.580 | 11.40 |
| | *n* | 49 | 20 | 218 | 247 | 50 | 116 | 207 | 23 | 930 |
| Block height (cm) | Mean | 43.335 | 33.575 | 37.051 | 32.869 | 30.088 | 20.486 | 23.182 | 12.138 | 30.053 |
| | SD | 12.666 | 10.754 | 12.419 | 15.098 | 15.81 | 11.591 | 14.598 | 8.034 | 15.45 |
| | *n* | 49 | 20 | 218 | 247 | 50 | 116 | 207 | 23 | 930 |
| Quartz (%) | Mean | 12 | — | 17 | 17 | 3 | 5 | 9 | — | 8 |
| | *n* | 15 | 0 | 4 | 9 | 1 | 6 | 6 | 0 | 41 |
| Clay (%) | Mean | 27 | — | 3 | 7 | 2 | 0.8 | 3 | — | 4 |
| | *n* | 15 | 0 | 4 | 9 | 1 | 6 | 6 | 0 | 41 |
| Pore throat (μm) | Mean | 0.19 | — | 0.38 | 0.54 | 0.51 | 0.66 | 0.72 | 0.43 | 0.54 |
| | *n* | 4 | 0 | 20 | 14 | 1 | 9 | 12 | 1 | 61 |

[a] *n* = number of observations.

*Logfacies C* This group is characterized by very low porosity and permeability, low gamma-ray, very low sonic and very high formation density (Plate 25). The group consists mainly of slumps and debris flows and it occurs associated with thin chert-cemented layers in the Ekofisk Formation and within calcite-cemented parts of the Tor Formation. The group has high statistical stability and is well-separated from the other groups. Mineralogical analyses show only minor amounts of clay, but an average quartz content of 17%. In thin section this facies shows a high degree of calcite cementation of both matrix and macropores.

*Logfacies D* This group is one of the most frequent and has medium gamma-ray, porosity and permeability. It is separated from the other porous groups only by its slightly higher GR readings, indicating a less-clean chalk.

*Logfacies E, F and G* All these groups consist of high-porosity allochthonous chalk, with low gamma-ray, high sonic time, and low formation density. "E" represents the transition between the cemented "C" and the porous "F" and "D" types.

*Logfacies H* Although a very small group, its characteristics are remarkable. It has very low gamma-ray, very high porosity and permeability, extremely high sonic transit time and extremely low formation density (Plate 25). It consists of all allochthonous sedimentary facies, and is found only in C-8, mainly in the lower Ekofisk Formation. The group is easily separated and has a high stability.

It is felt that the validity of the statistical analysis performed is supported by the ease with which the logfacies could be related to distinct geological parameters.

## Relationship between logfacies and fracture density

To study the fracture density of these different logfacies, the arithmetic-average matrix block height and width, defined in the core study, was plotted against core porosity for each facies (Fig. 9). A clear relationship appeared in the plot.

- Logfacies A shows the largest average matrix block sizes, with a mean block height of 43 cm. Considering the upper "cutoff" of 50 cm on block heights, it is obvious that this lithofacies rarely contains open fractures. This agrees with the core observations that the shaly pelagic facies on top of the chalk and in the Tight Zone has very few open fractures.
- Logfacies C, although having an average porosity similar to that of "A", still has a higher average level of fracturing, with matrix block sizes between 15 and 37 cm. Apparently, the cementation of this logfacies may have resulted in relatively brittle behaviour,

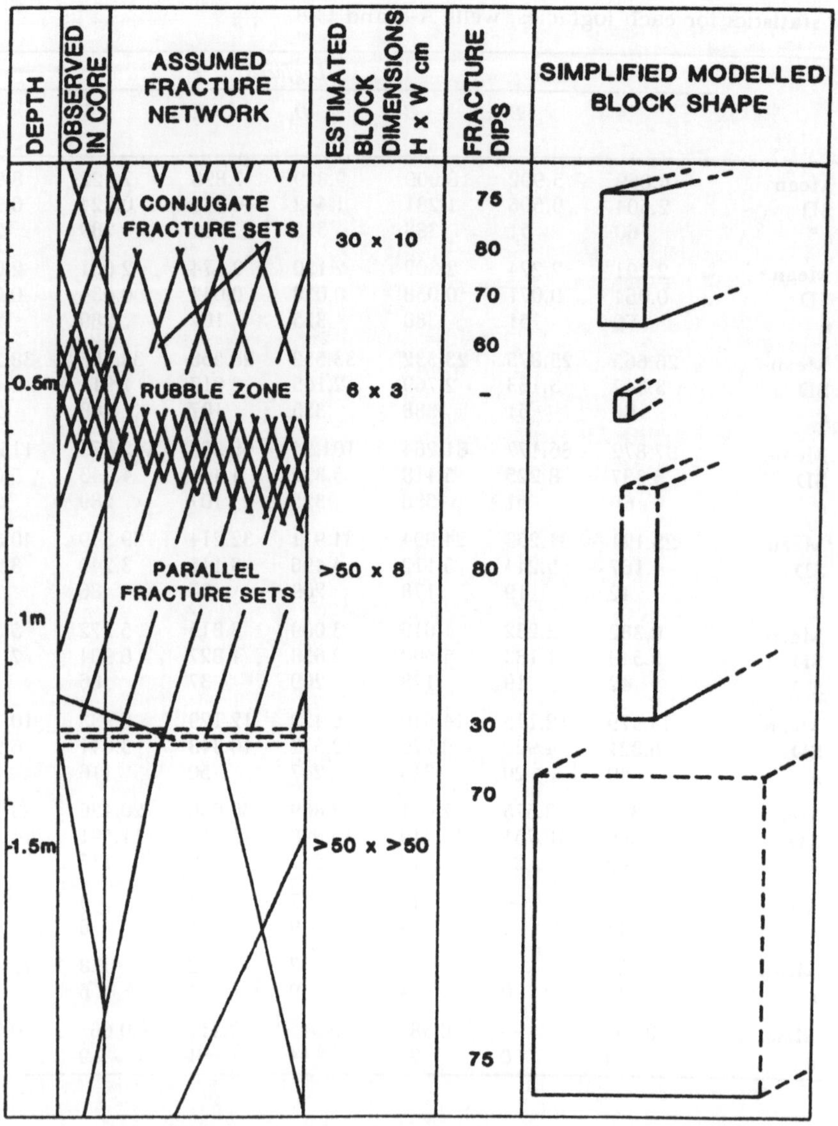


**Fig. 4.** Definition of matrix block dimensions by core description.

which shows that using matrix porosity only for fracture indication would be erroneous.

- Logfacies H is the most highly fractured, with an average matrix block height of 12 cm, indicating that most depths classified as "H" appear in the cores as rubble zones.

## Quantification of fracture density per logfacies

Following this initial step of logfacies classification using wells A-8 and C-8, which produced satisfactory results, the other cored wells were prepared for classification using the saved rules and the same procedures. The average values of matrix block height and width for each logfacies in these wells could, however, not be compared since the wells represent different locations on the structure, and therefore have different degrees of fracturing. In order to quantify this structural factor, the assumption was made that the difference in fracturing for one logfacies type between two wells is caused by the different structural positions of those two wells. The ratios of matrix block dimensions in each logfacies between C-8, which was chosen as a reference well, and the other cored wells were consequently plotted on a base map and contoured. The contouring was based on the structural model set up using core data and fracture orientations, as discussed earlier. Also, dipmeter data and well-performance data were used to assist the contouring process.

Using these maps, the matrix block dimensions for each logfacies in all cored wells were transformed and averaged to a basic set of "minimum crestal values" of matrix block height and width in the Ekofisk and Tor Formations separately.

## Analyses of non-cored wells

Having defined a set of basic values for matrix block dimensions for each logfacies, the procedure followed for the uncored wells was first to run a discriminant analysis on the log data based on the saved classification rules. Then the basic values of fracture density for each logfacies were entered and average block sizes per reservoir zone were calculated. Finally, this was multiplied by the structural factor for each well and the resulting values were mapped.

A total of 39 wells were classified and matrix block dimensions were calculated. These wells were chosen deliberately to give an optimized distribution of data points in each reservoir layer.

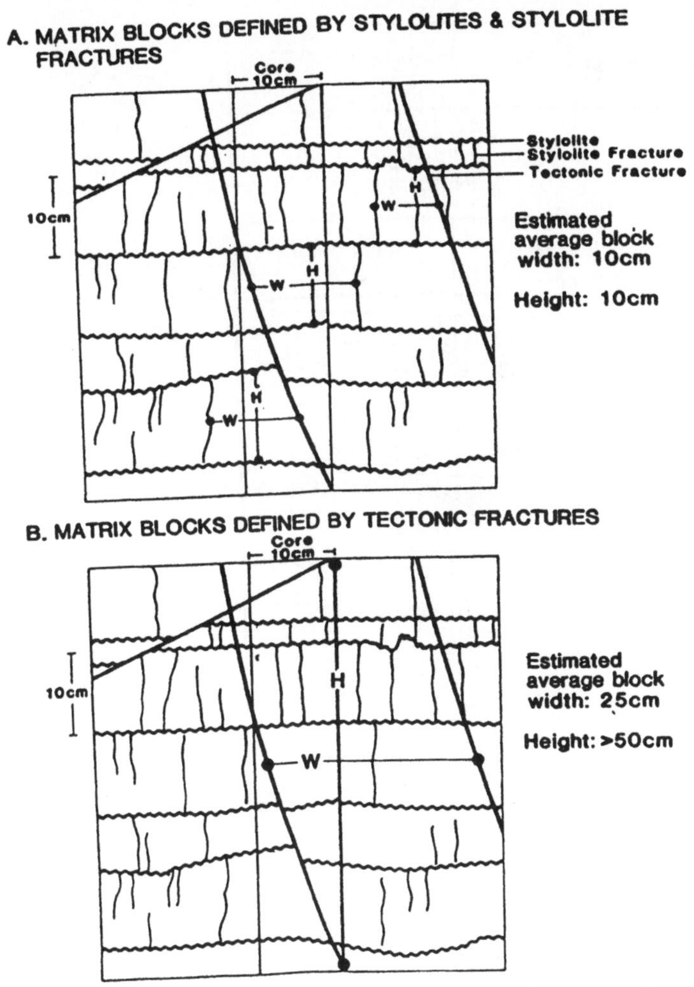


**Fig. 5.** Definition of matrix block dimensions within stylolite zones.

### Matrix block lengths

It is only possible to estimate and measure two dimensions of the matrix blocks from core. The longer of the two horizontal dimensions, the block length, was therefore defined indirectely, controlled by fracture orientation data.

Rose diagrams for each reservoir layer from the wells with oriented cores show that the range of fracture strike differs between the wells. Some wells exhibit a very small variance in orientation of the fractures, indicating that most fractures trend parallel and very few cut in different directions. This will result in very long and narrow matrix blocks. In contrast to this, other wells have more than one main fracture direction, and represent an orthogonal fracture network that will give rise to short matrix blocks. In this way, the rose diagrams indicate the ratios between matrix block widths and lengths. The frequency of fracture orientations in each layer was projected to a horizontal plane, and the width/length ratios could be defined. These ratios were plotted on maps and contoured, and maps of block length were then constructed on the basis of the previously prepared block width maps.

## FRACTURE POROSITY AND PERMEABILITY

A number of techniques for estimation of these parameters exist in the literature. The approach used in this study was to apply the results of core measurements, as far as possible, and supplement these with published data.

### Width of fracture opening

This important parameter is difficult to assess in core because fractures are not under reservoir stress and are usually broken open. However, studies of unbroken-open, and of cemented fractures provided a useful guide. This was supplemented with published information (Weber and Bakker, 1981; Nelson, 1985; Watts, 1983; Van Golf-Racht, 1982; Corbett, *et al.*, 1987). On the basis of these data, fracture widths were estimated as being 0.01 mm down-flank increasing to 0.1 mm in the crestal area, and maps were drawn on this basis.

### Fracture porosity

Since the matrix block dimensions had been calculated, and a distribution of fracture width had been defined, the fracture porosity could be calculated by the computer by introducing equation (1) in Table II to the map grids.

**Table II.** Equations for fracture porosity and permeability calculations

| | |
|---|---|
| **Fracture porosity** | |
| (1) $\phi_F = b\left(\frac{W\cdot H + W\cdot L + H\cdot L}{W\cdot H\cdot L}\right)\times 100$ | $b$ = fracture width<br>$W$ = matrix block width<br>$H$ = matrix block height<br>$L$ = matrix block length |
| **Conventional fracture permeability** | |
| (2) $K_F \frac{b^3}{12H}$ | $H$ = distance between infinite, parallel regularly spaced fractures |
| **Fracture permeability for inclined fractures** | |
| (3) $K_F = \frac{b^3}{12H}\cos^3\alpha$ | $\alpha$ = the angle between the axis of the pressure gradient and the fracture planes |
| **Fracture permeability for multiple fracture sets** | |
| (4) $K_{fy} = \frac{b^3}{12L}\sin^2\beta_w + \left(\frac{b^3}{12W} + \frac{b^3}{12H}\right)\sin^2\beta_L$ | $\beta_W$ = average angle of matrix block width direction<br>$\beta_L$ = average angle of matrix block length direction |

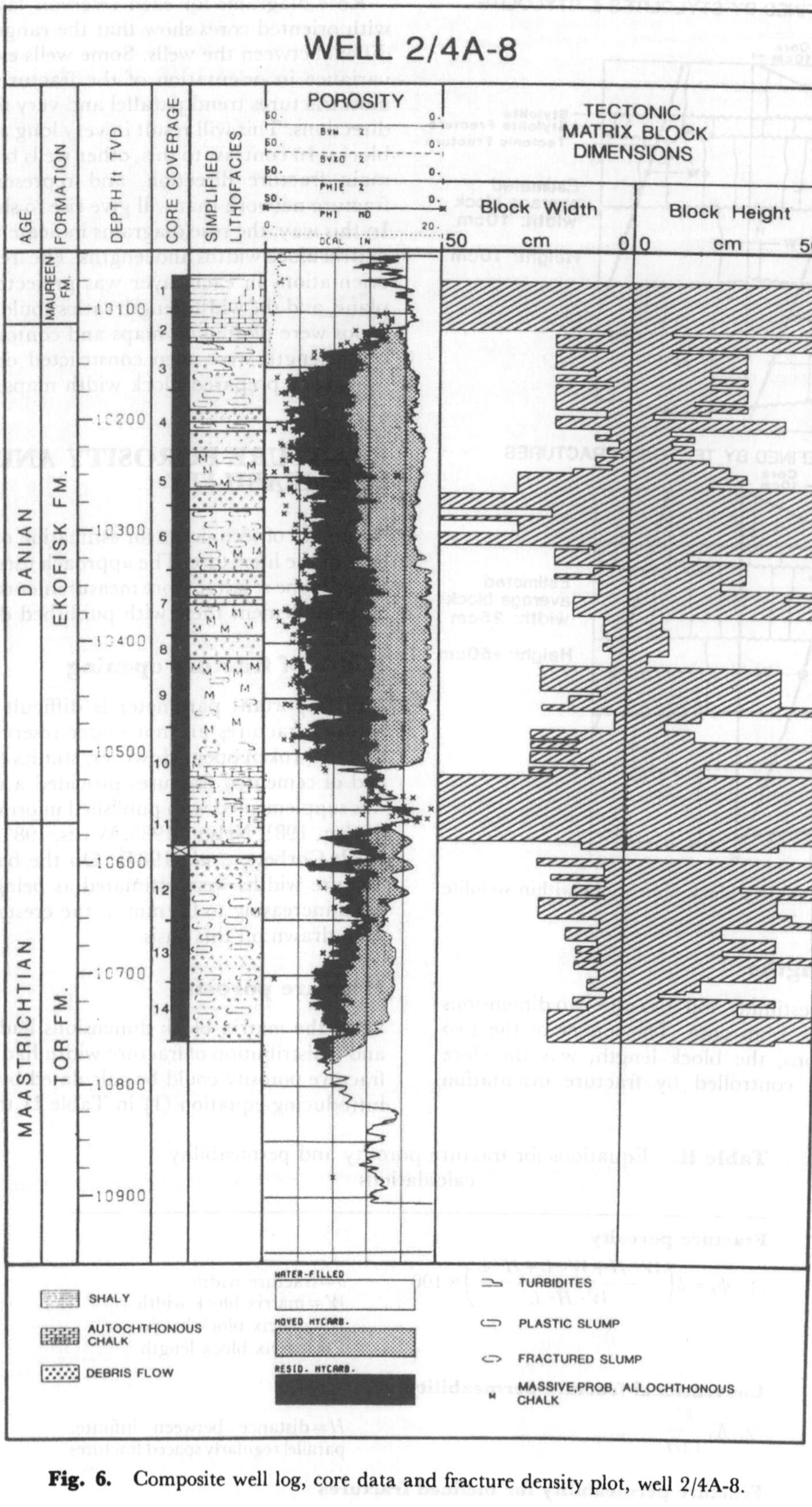


**Fig. 6.** Composite well log, core data and fracture density plot, well 2/4A-8.

The resulting values vary from less than 0.02% downflank to 0.2% on the crest in the Ekofisk Formation (except in the Tight Zone), and between 0.01% and 0.1% in the Tor Formation. These values show good agreement with published data (Weber and Bakker, 1981). Note that this is the porosity of the tectonic fracture network only. The additional porosity provided by the stylolite fractures is discussed later.

## Fracture permeability

Having quantified the matrix block sizes and fracture widths, the fracture permeability in such a fracture network could then be calculated.

Discussion of methods for calculating fracture permeability can be found in Van Golf-Racht (1982) and Nelson (1985). The conventional equation for fracture perme-

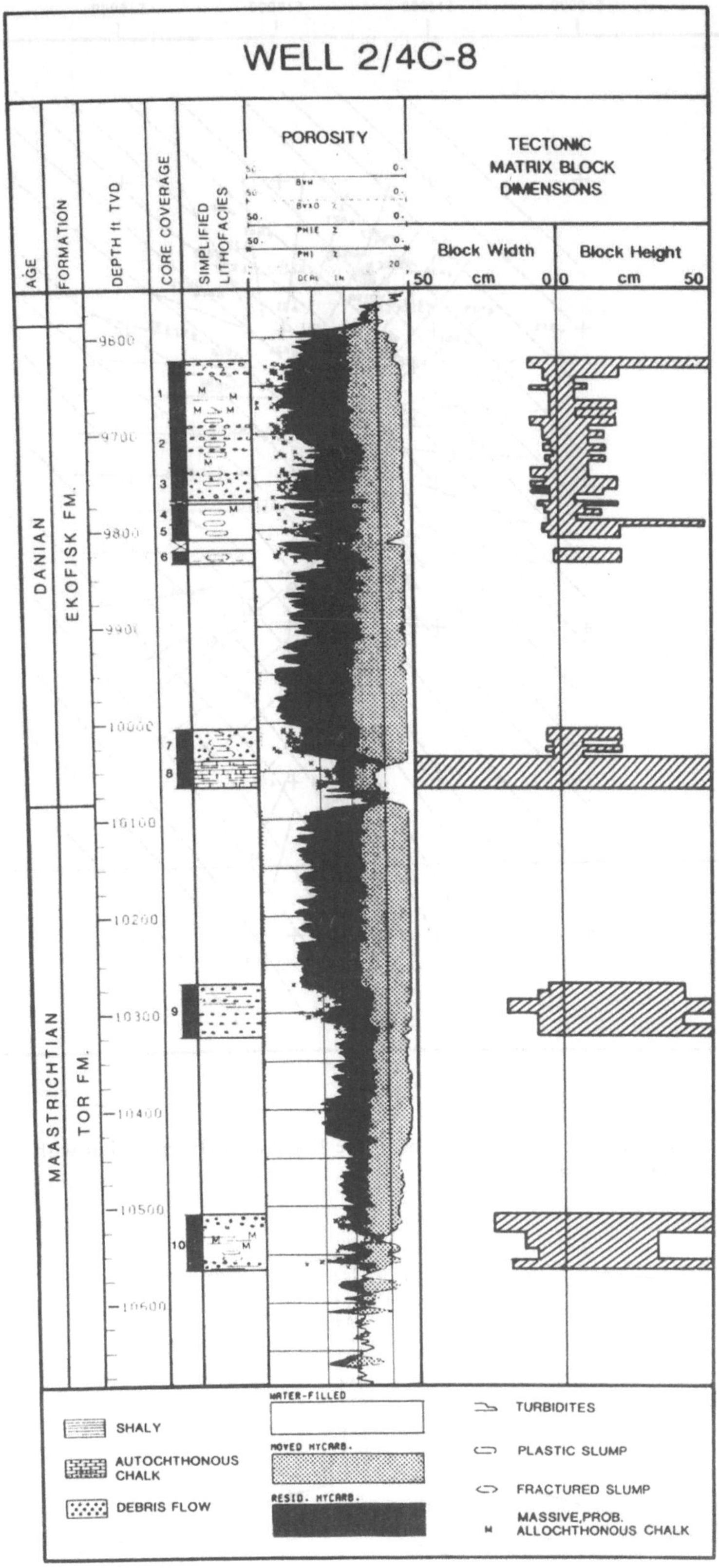


**Fig. 7.** Composite well log, core data and fracture density plot, well 2/4 C-8.

ability, (equation (2) in Table II) was not sufficient to represent the observed complexity of fracture distribution. This equation can be modified to allow for fracture dip and orientation, as shown in equation (3) in Table II. However, where several fracture sets with different angles and spacings occur, the fracture permeability is expressed by the total contribution of each set (equation (4) in Table II). This equation was used, and the actual and modelled flow paths are illustrated in Fig. 10. Both vertical and E–W and N–S horizontal fracture permeabilities were calculated with respect to the simulation grid. The angles between the fracture planes and the N–S and E–W simulation grid orientation were defined from rose diagrams and mapped for each reservoir layer. The average fracture dip found in core studies and used in the equation was 70°. Thus, having quantified and mapped the factors that influence

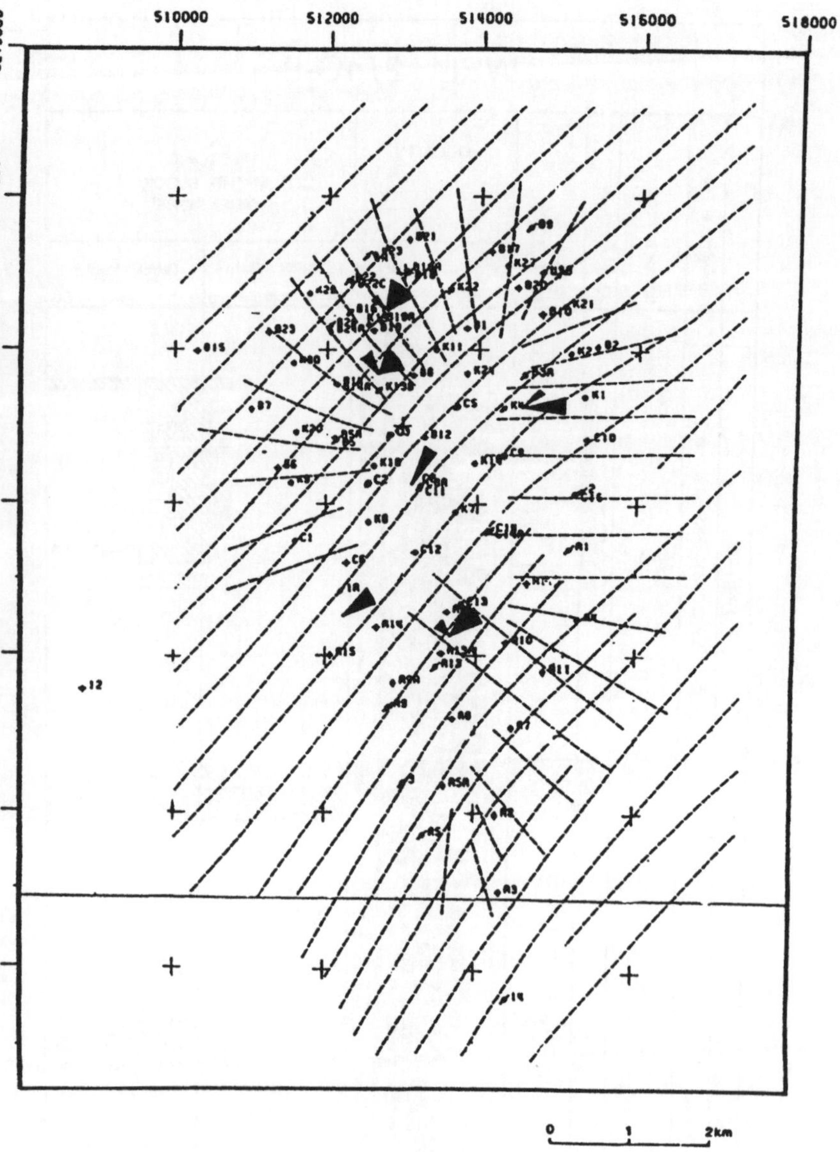

**Fig. 8.** Generalized outline for tectonic fracture orientations in the Ekofisk Field; semiradial with regional NNE–SSW overprint. Measured orientations in six wells are marked.

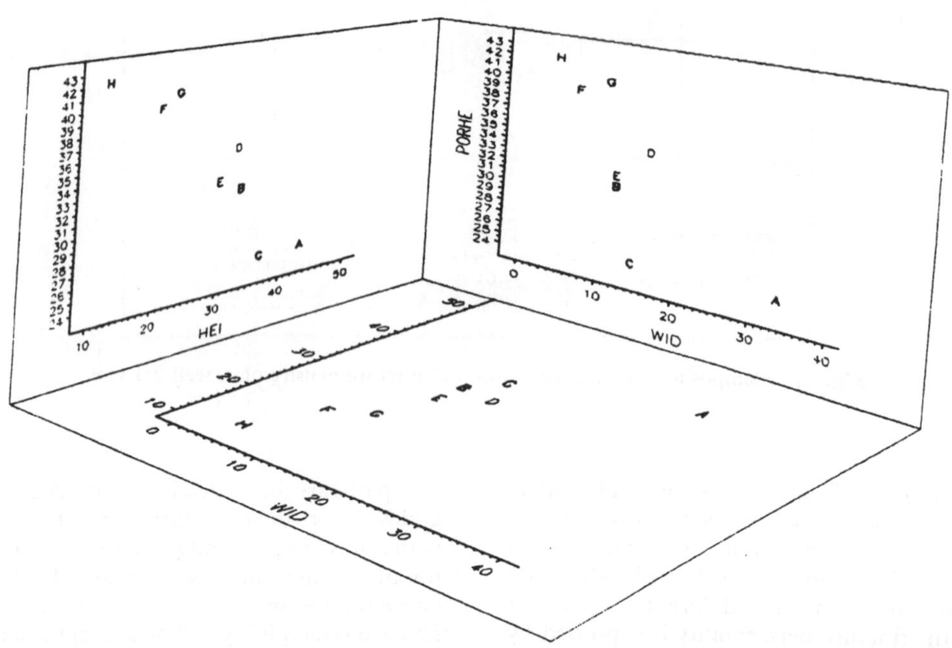


**Fig. 9.** Average (mean) core porosity, matrix block width and matrix block height plotted for each log facies, wells 2/4 A-8 and 2/4 C-8.

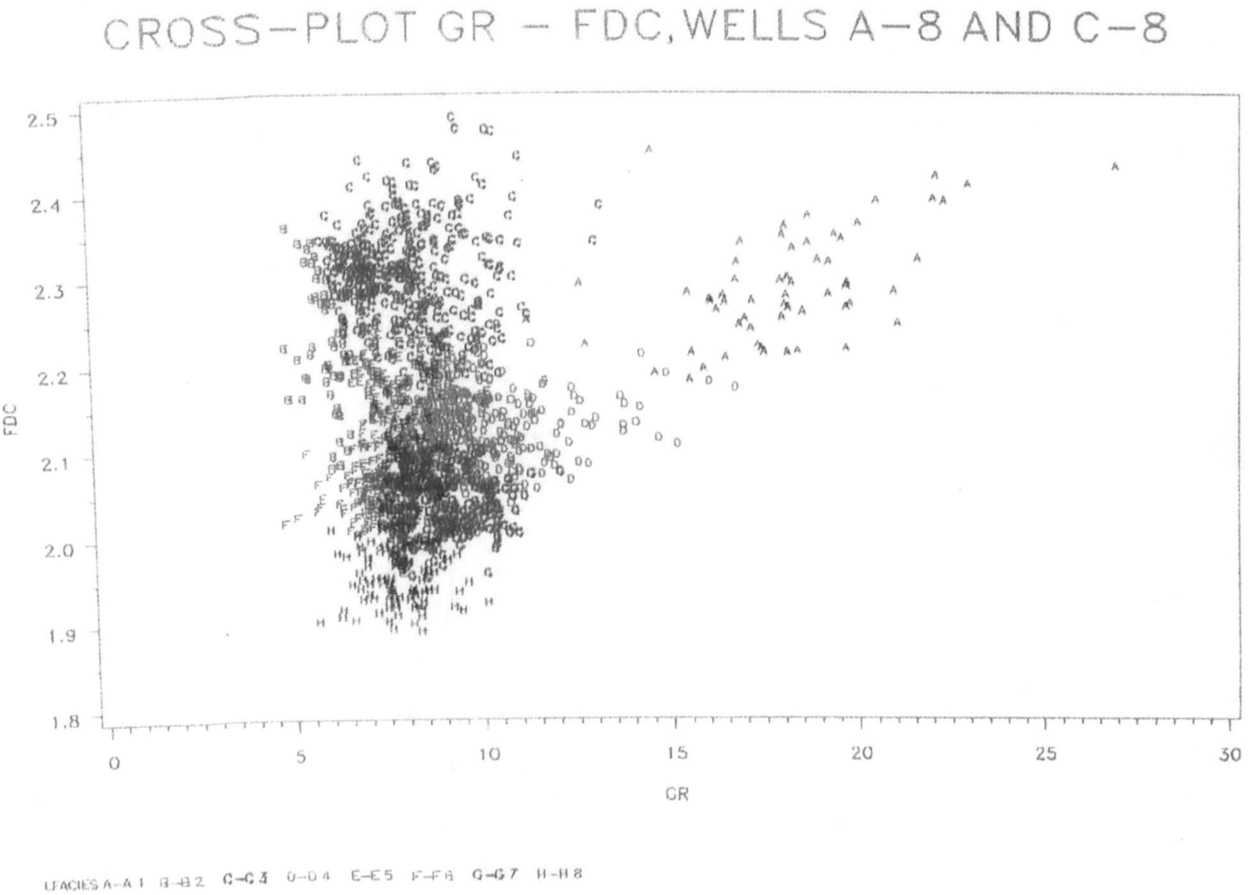


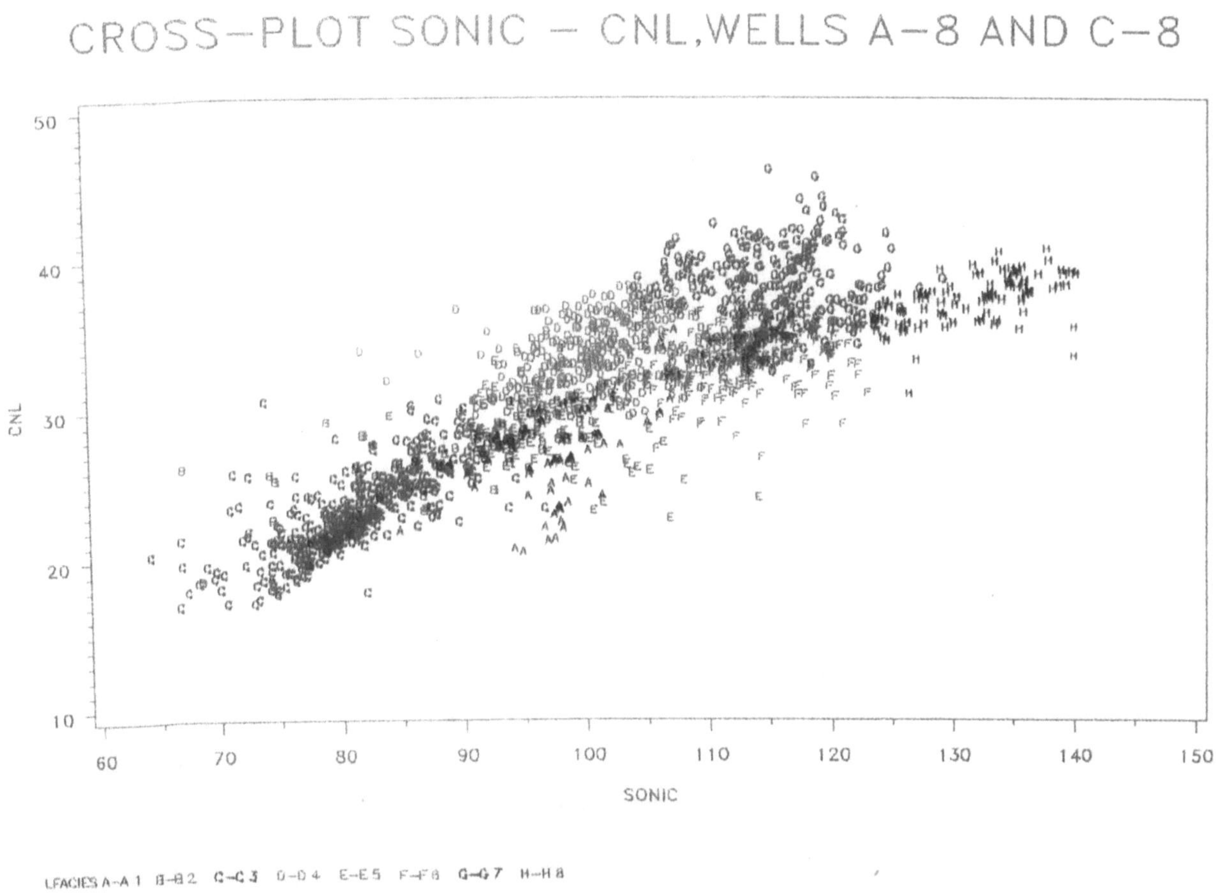


**Plate 25.** Cross-plots of well log readings showing the clustering of data from wells 2/4 A-8 and 2/4 C-8 into different logfacies.

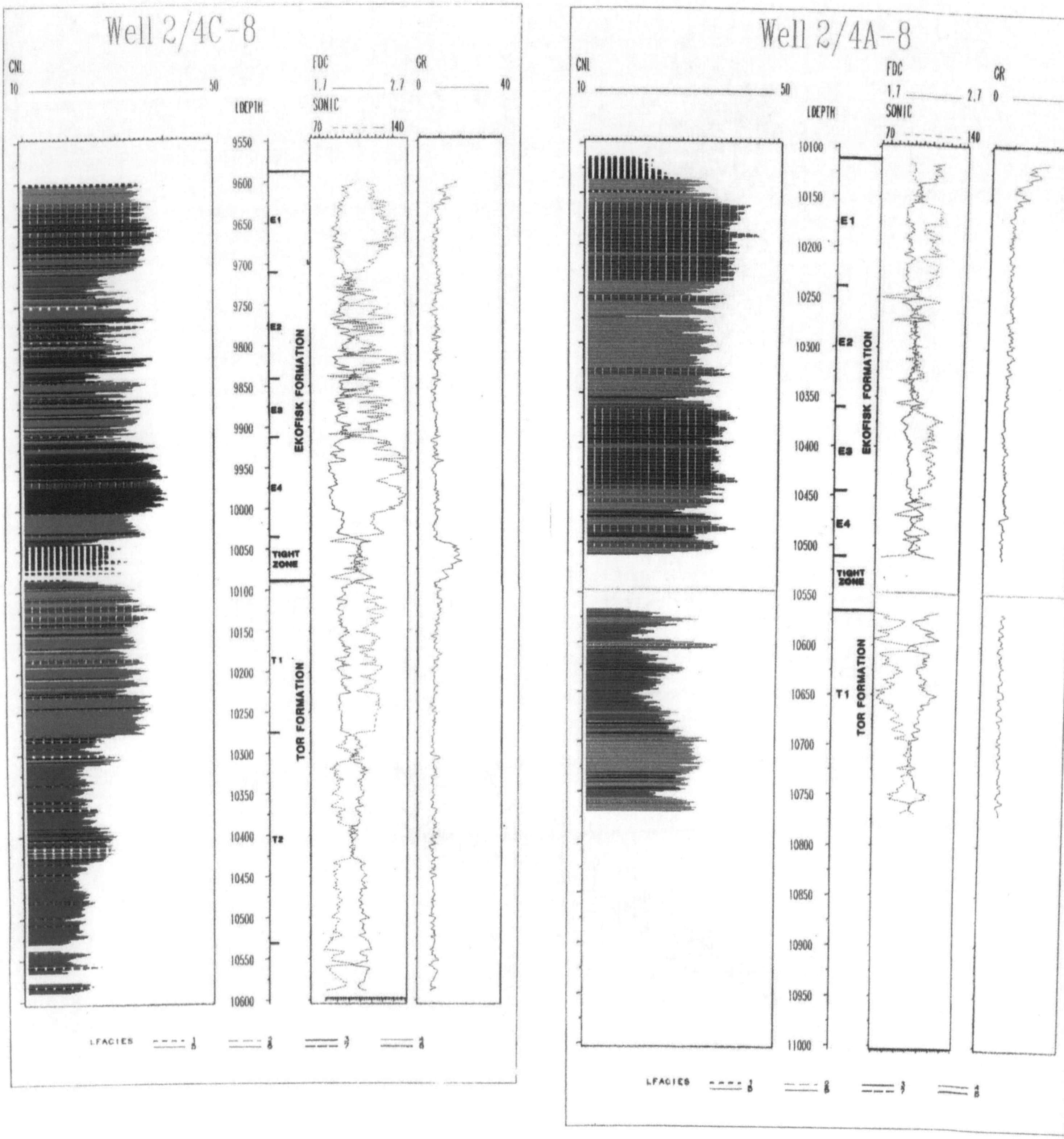


**Plate 26.** Logfacies classification, wells 2/4 A-8 and 2/4 C-8.

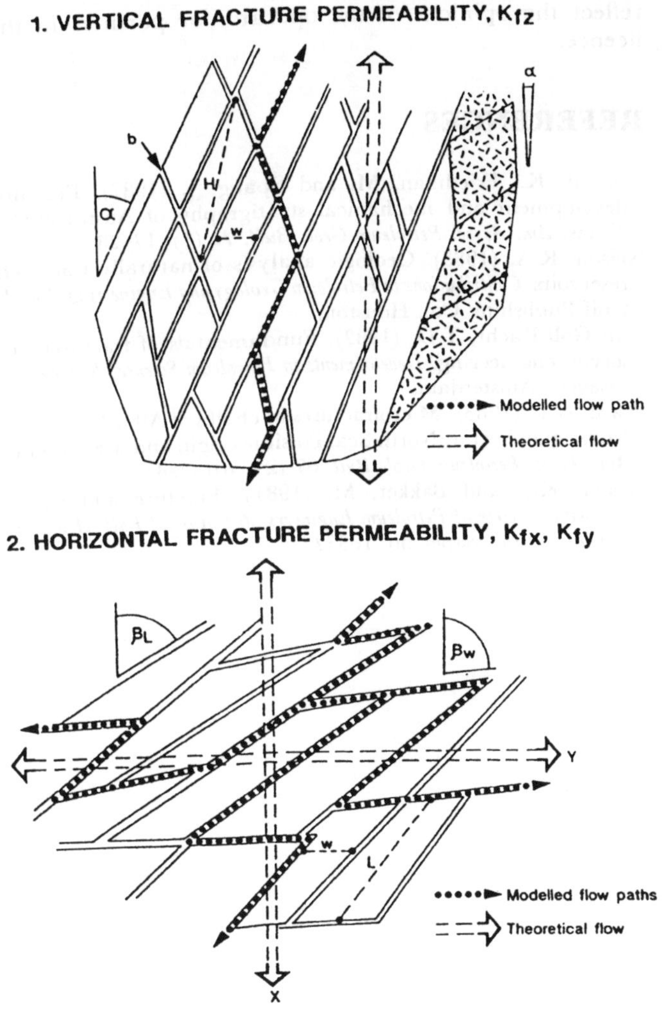


**Fig. 10.** Modelled and theoretical fluid flow paths through a fractured reservoir.

fracture permeability, the calculations were run in the computer and fracture permeability maps were constructed.

## DISCUSSION AND IMPLEMENTATION OF RESULTS

The process of establishing a fracture model, up to this point, embraced simplifications and assumptions of the natural fracture network of significant importance to the reservoir behaviour. This was, however, necessary if any quantification of the fractures was to be developed for numerical simulations. The quantified model describes grid blocks containing an orthogonal fracture network with equally spaced matrix blocks, and flow through planar, infinite, vertical and horizontal tectonic fractures. Undoubtedly, this produces far too optimistic values of fluid flow through the fractures.

In core, fractures terminate against other fractures and bed surfaces, especially those marking lithological boundaries. In order to quantify this effect on fracture permeability, a study was made of the well-test performance data from the field, because it is clear, by comparison with matrix permeabilities, that the major part of the flow rates measured in tests is the response of the fracture system. These permeabilities were then compared with the vectorized modelled fracture permeability maps. The ratio of predicted permeabilities to well-test permeabilities was found to be approximately 1.5. Based on this, it may be assumed that finite fractures make up approximately 1/3 of the fracture network, and that one out of three fractures is either a dead end or closed by mineralization. All fracture permeability maps were consequently multiplied by a value of 0.666. The resulting permeabilities vary between less than 10 mD downflank to several hundred millidarcies in crestal areas.

Up to this point, all maps constructed represented the tectonic fractures only. However, integration of core and well-performance studies indicate that the stylotite fractures play an important role in the productivity of the reservoir. Since they increase the drainage of the matrix and carry fluids horizontally to the larger-scaled tectonic fracture network, the stylolite fractures will represent a significant part of the horizontal fracture permeability, especially in the Tor Formation. Stylolite fractures also contribute to the total fracture porosity in the reservoir. It was therefore decided to incorporate stylolite fractures in the total fracture porosity and permeability maps.

On the basis of core study, the model that was adopted describes vertical stylolite fractures as 100 mm high, 500 mm long and 0.04 mm thick. From core examination it was seen that the stylolite fracture density is generally higher downflank than in crestal areas. The horizontal spacings between the vertical stylolite fracture plates were therefore varied over the field between 200 and 600 mm. The stylolite fracture porosity and permeability in N–S and E–W directions of each layer were calculated using modifications of equations (1) and (4) in Table II. The calculated values were then contoured and added to the previously constructed tectonic fracture porosity and horizontal fracture permeability maps.

After running a cross-sectional simulation study on this fracture model, the vertical fracture permeabilities were adjusted in order to match the field reservoir pressure and GOR history. This was achieved by introducing heterogeneously distributed horizontal unfractured layers, together with a generally increased portion of finite fractures, compared to infinite ones, in the vertical dimension.

## CONCLUSIONS

The fracture model described here is based on two major assumptions, both of which have been independently verified: the relationship between fracturing and lithological facies by extensive core study, and the relationship between fracturing and field structure by core study and well-test analysis. Furthermore, the model produced fracture permeabilities, by layer, that compared well with those computed from test data. As the model is logical and internally consistent, it can be upgraded whenever improved data become available.

However, the model also contains a number of assumptions that are not independently verifiable. One of the principal problems is the fact that true reservoir block heights, as defined by capillary continuity, are impossible to measure directly in core. A related problem is the definition of block width, since, except in zones of very dense fracturing, the bed-parallel distance between fractures is greater than the core width.

The presence of different scales of fracture networks, like stylolite fractures, complicates the fracture modelling. As it is not known to what degree these multiple sets of fracture networks are interconnected, prediction of the permeability contribution of each set of fractures is very difficult.

However, the model successfully demonstrates that, using the approach described, realistic fracture properties can be assigned throughout a highly complex reservoir where both core and well-test data are somewhat limited. In particular, the ability of the model to provide fracture properties for every reservoir layer is critical in fulfilling the requirements for simulation.

## ACKNOWLEDGEMENTS

The authors wish to thank Phillips, Petrofina, Agip, Total, Elf Aquitaine and Statoil for their permission to publish this paper. In particular, H. Farrell of Phillips and colleagues in Norsk Hydro, Petroleum Technology, are thanked for contributions to the work. The opinions expressed are those of the authors and do not necessarily reflect the opinions of the operator or partners in the licence.

## REFERENCES

Corbett, K., Friedman, M. and Spang, J. (1987). Fracture development and mechanical stratigraphy of Austin chalk, Texas. *Am. Assoc. Petroleum Geol. Bull.* **71** (1), 17–28.

Nelson, R.A. (1985). Geologic analysis of naturally fractured reservoirs. *Contributions in Petroleum Geology and Engineering, Vol. 1.* Gulf Publishing Co., Houston.

Van Golf-Racht, T.D. (1982). Fundamentals of fractured reservoir engineering. *Developments in Petroleum Science, Number 12.* Elsevier, Amsterdam.

Watts, N.L. (1983). Microfractures in chalks of Albuskjell Field, Norwegian Sector, North Sea: Possible origin and distribution. *Am. Assoc. Petroleum Geol. Bull.* **67** (2), 201–234.

Weber, K.J. and Bakker, M. (1981). Fracture and vuggy porosity. *Society of Petroleum Engineers 56th Annual Fall Mtg., San Antonio, Texas*, paper no. 10332.

# 8

# A New Approach to Production Log Analysis

*G. Nerby*

READ Petroleum Engineering Services A.S., Sandvika, Norway

This contribution presents a new approach to production log analysis. It employs a line search optimizing procedure to minimize the variance between log survey measurements and the equivalent predictions from theoretical physical models. The main features are:

- *Transfer functions* coupling the actual specific properties of the fluid elements to the sensor responses taking into consideration the constitutive physical relations and the physical rheology of the current flow pattern and fluid system in exposure to the sensor surroundings.
- *Dynamic flow models* are coupled in a continuity principle for mono- and multi-phase flow. For a given thermodynamic state and assumed flow flux, the phase velocities, flow regime, pressure gradient, velocity and distribution slip together with longitudinal and lateral heat transfer in the flow environments for the log survey are computed as a complete set of result parameters. The physical models are based on general continuity, constitutive relations and documentated correlations.
- Correlations for the *physical fluid properties*, concerning density, viscosity and surface tension based on one or more references at definite thermodynamical states.
- *Statistical modules* for filtering, variance estimates, curve fits and regression analysis between general properties and parameters in the multiphase flow hiarchy.
- A *line search optimizing module* based on golden section unidimensional search. The search directions are based on a conjugated direction propagation principle. This line optimization splices together the operational log measurements and the theoretical predictions. The search is made with respect to the multiphase flow fluxes as free variables.

Applicability of the system goes beyond conventional PLT analysis in that the uncertainty of each sensor is brought into consideration in the analysis. The determination of a given flow parameter can hence be performed in a flexible manner with a quality-weighted utilization of the downhole readings. This approach is general in nature and can easily accommodate new logging and interpretation techniques as well as improvement in flow analysis.

## INTRODUCTION

Production logging (PL) is one of the few tools available for direct monitoring of the dynamic reservoir and well inflow performance. Especially in the mature stages of field production, it has proved vital to the understanding of reservoir depletion, cross-flow, breakthrough and coning problems or recompletion requirements. Traditionally, however, PL-data has been used mainly for qualitative analysis, the reason most probably being the unreliability of existing models in describing the complex multiphase well inflow behaviour. Consequently a vast amount of information collected in expensive PL operations is rarely actively used for quantitative reservoir-management purposes.

*North Sea Oil and Gas Reservoirs-II*


This has been the motivation for the development of a new approach to production log analysis. In the following, the background and the principles of the system are described together with two examples that illustrate the performance of this concept.

## REVIEW OF PRODUCTION LOG ANALYSIS (PLA)

Some of the most advanced current PLA models utilize flow regime maps that are based on observations at ideal surface conditions for small pipe diameters. The sensor responses require explicit correlation charts based on a variety of assumed parameters. An empirical slip and pressure-gradient function is then found from these maps and the explicit corrected measurements. The flow rates are derived from this. These principles have many drawbacks. A flow map referred to laboratory conditions with definite fluid properties cannot be extrapolated directly to downhole conditions. Any flow regime prediction needs to be closely related to downhole conditions. More elaborate principles of statistical tool behaviour and flow physics needs to be given higher status than in present PLA models. Explicit chart corrections for the PL sensors should be replaced by implicit computations based on the in-situ physical and dynamical fluid behaviour. This has been the motivation for this approach to analysis of PL measurements. The genesis of this principle of analysis was described by Nuland and Nerby (1986), where the OLGA (oil and gas pipeline simulator) system was connected to optimization logic to perform production log analysis.

## MODEL DESCRIPTION

The system input comprises sensor readings, physical fluid properties, well deviation, internal diameter and assumed roughness of the well/toolstring-body (The latter can also be computed internally.) The required relative accuracies for the different sensors downhole are needed as input for the numerical flow analysis. Optional inputs are the surface oil, gas and water production rates, and the relative certainties for these different measurements.

The main results from the system are the phase velocities, the cross-sectional phase distribution, the areal-averaged phase fractions and the different contributors to the total pressure gradient such as fluid/pipewall friction, fluid/toolbody friction, gravitational and inertial acceleration. The following flow regimes are presently handled by the system: oil/gas and water/gas—mist, bubble/slug flow; water/oil—dispersed, emulsion and churn flow. Monophase flows are treated as limits of the flow patterns mentioned. In three-phase flow of gas, oil and water, the oil and water are treated as completely mixed and interacting with free gas as the second homogeneous phase. This is a simplification of a complicated three-phase flow regime, but is an acceptable approach when one liquid phase volumetrically dominates the other liquid.

The system enables analysis of PL measurements at shut-in or flow conditions, from a deep water/oil zone in monophase conditions, through dispersion of oil in water, and up to liquid/liquid churn or emulsified flows and eventually into free gas/liquid mixtures in slug, bubble or mist distributions. The whole analysis is performed in one continuous chain of computations.

The system contains four main modules: PL measurement precomputations (Mod1), computation of physical fluid properties (Mod2), numerical multi- or monophase flow analysis (Mod3) and data management, display and plotting (Mod4).

## MEASUREMENT PRE-COMPUTATIONS (Mod1)

The objective of the different precomputations is to correct implicitly for local fluid-mechanical effects on the sensor responses prior to utilization in the later flow analysis. The measurements are based on inline or fullbore spinner, densitometer, strain or quartz gauge readings, temperature, and optionally liquid hold-up capacitance sensors. Even if density data are not available, a complete analysis can still be performed if the pressure data are stable and of good quality. A typical tool string is illustrated in Fig. 1.

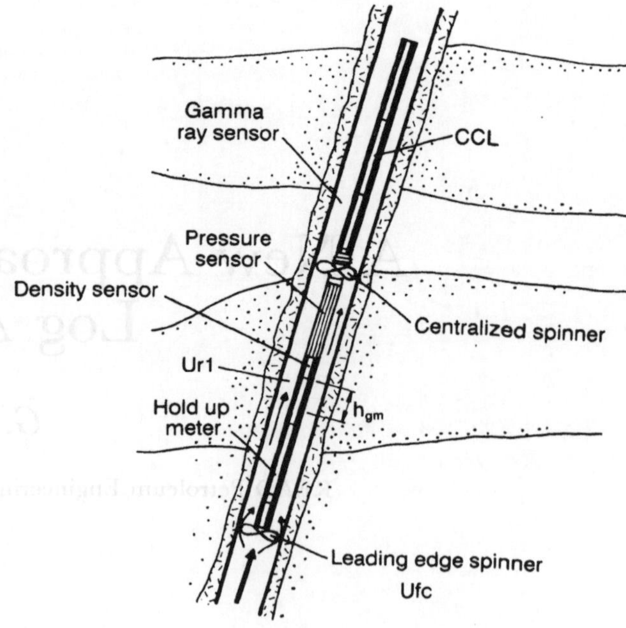


**Fig. 1.** Typical production log tool string

### Spinner sensor

The aim of this subprocess is to convert spinner rotations into an equivalent mass-averaged fluid velocity at or close to the centre of the pipe. This velocity is a complex parameter including results of slip and phase-distribution effects. The following well-known relation is utilized independently of spinner type:

$$\omega = S_1 \cdot U_{fmc} + O_f \tag{1}$$

However, the total fluid viscosity and density of the fluid gives a non-linear correction on the parameters $S_1$ and $O_f$ in Eq. (1). In addition, three-dimensional eddies in the sensor environments can locally influence the $S_1$ factor especially around the inflow zones of a well. A dimensional analysis based on density, viscosity, angle of attach and a spinner with defined pitch, diameter, mass and bearing system gives:

$$\omega = \text{const} \cdot \left(\frac{\rho_f}{\mu_f}\right)^{0.5} \cdot \tan\alpha \tag{2}$$

This equation shows how three important variables influence the spinner response: the angle of attack between the spinner axis and the local resultant fluid vector, and the total density and viscosity of the well fluid. The computed slope ($S_1$) at each depth increment is corrected for the estimated fluid viscosity, denisty and fluid angle of attack based on a nominal reference slope from a static zone of the well. In this procedure, a function is defined:

$$F_{s1} = (S_{1c} - S_{1\text{ref}})^2 \tag{3}$$

Assume that $N$ passes up and down are available. An elimination of spinner passes in the slope computation of Eq. (1) is done until the best fit is obtained between the current slope ($S_{1c}$) and the reference slope ($S_{1\text{ref}}$) corrected by Eq. (2). This procedure ensures a correction for local eddies and fluid rheology effects on the spinner in the estimation of $U_{fmc}$. The algorithm is outlined in Fig. 2.

If the polarity of the spinner is unknown, the computation of the averaged fluid velocity is performed in two stages. First, a fluid velocity, VPOL is estimated at each depth level defined by the fluid velocity, implying a change of

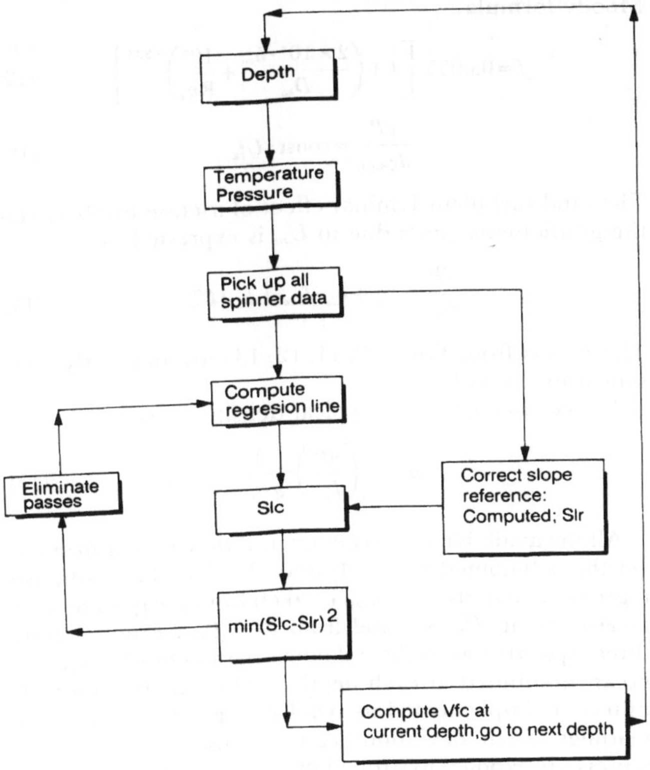


**Fig. 2.** Algorithm for estimating mass averaged fluid velocity

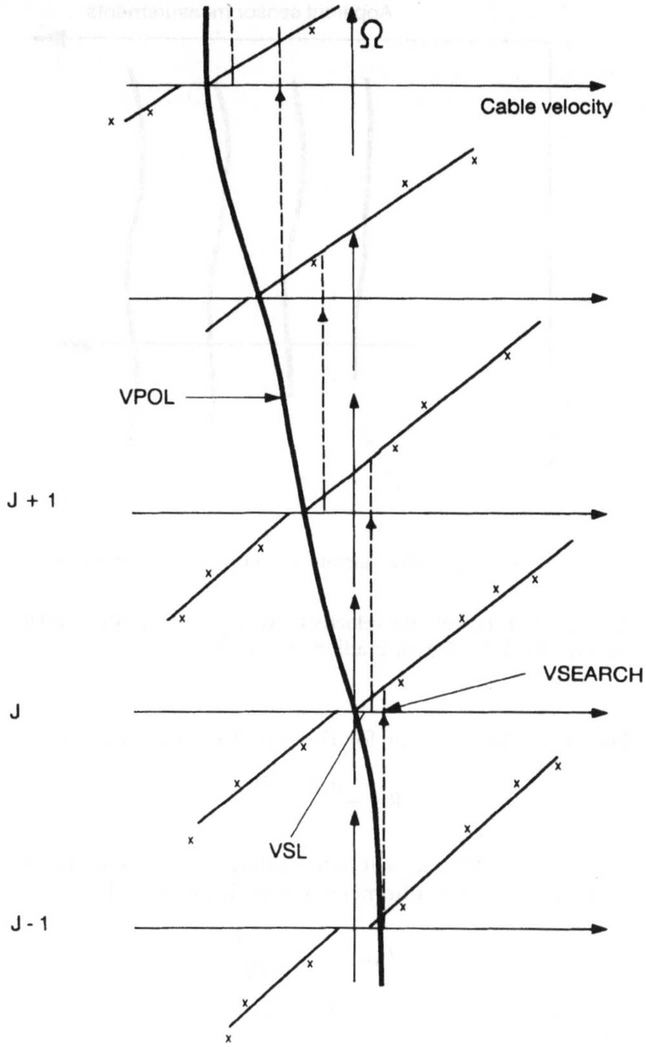


**Fig. 3.** Principle for detection of spinner polarity

rotation direction of a stationary propeller. At any depth, a cable velocity above VPOL is defined as applying to passes related to the positive polarity of rotation, and vice versa. To allow for local backflow, a search velocity for the next downstream depth level is defined by Eq. (4) utilizing a slide velocity (VSL), and VPOL at the upstream depth gives VSEARCH. All the cable passes higher than VSEARCH are used to compute the VPOL for the next depth level. Equation (4) gives the link between neighbour depths $j$ and $j+1$ in the computation of VPOL.

$$\mathrm{VSEARCH}_{j+1} = \mathrm{VPOL}_j - \mathrm{VSL}_j \tag{4}$$

Assuming VPOL close to zero initially at bottom depth, the system computes independent VSEARCH and VPOL in the complete depth interval. The principle of polarity detection is displayed in Fig. (3). After prediction of VPOL, $U_{fmr}$ can be computed just as if the senor readings detected the direction of the rotations.

## Pressure measurements

The system uses the absolute pressure of the strain and quartz gauges as input to the computations of fluid properties, together with pressure gradient versus depth. A local linear pressure function is assumed valid locally in a small depth interval (window):

$$P(z) = k_{1p} + k_{2p} \cdot z \tag{5}$$

The constants in the expression $(k_{1p}, k_{2p})$ are found by a least-squares method. The length of the sliding window can be set depending on the log quality and flow situation. A filter limit is applied in the regression leading to Eq. (5), implying that pressure data are avoided in the function when the local quadratic distance is greater than 90% of the total variance of the local pressure computed from Eq. (5). The pressure gradient versus depth is found directly by differentiation of Eq. (5):

$$\frac{dP(z)}{dz} = k_{2p} \tag{6}$$

Occasionally, the pressure gradient can indicate presence of liquid slugs and slug lengths in gas/liquid flow. Combined with the equivalent effects from the spinner sensor and total gas rates, data that previously has been interpreted as "noise" can now be analysed as intermittent slug flow.

## Density measurements

Three types of density sensor principles can be handled: gradiometer with and without internal buffer fluid, buoyancy cells, and gamma densitometers. Gradiometers are based on a differential-pressure principle. Turbulent and viscous shear influences the gradiometer readings at flow conditions in two ways: energy dissipation related to relative movement between well fluid and the tool body, and dissipation as a consequence of the fluid flow itself during the log survey. The first effect is displayed in Fig. 4. *These two effects have to be considered in any interpretation of these sensor readings.* The system treats the two dissipation effects in separate stages. The first-mentioned effect is eliminated after $U_{fmr}$ in Eq. (1) has been computed. The hydraulic diameter close to the gradiometer sensor-ports is

$$D_{ha} = \frac{D_p^2 - D_b^2}{D_p + D_b} \tag{7}$$

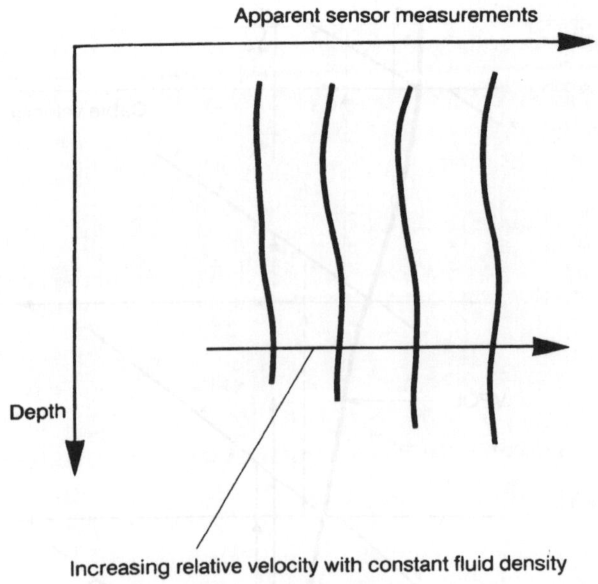


**Fig. 4.** Effect of relative velocity between toolstring and fluid for constant fluid density on gradiometer sensor

The Reynolds number for the pipe/toolbody annulus

$$\mathrm{Re}_a = \frac{\rho_f \cdot U_{ra} \cdot D_{ha}}{\mu_f} \quad (8)$$

where the local mass averaged relative fluid velocity between fluid and gradiometer sensor is estimated

$$U_{ra} = (U_{fmc} - U_{cab}) \cdot \frac{D_p^2}{(D_p^2 - D_b^2)} \quad (9a)$$

$U_{ra}$ will be different for each log-pass owing to changing cable velocity ($U_{cab}$).

The pressure and gradiometer sensors are exposed to the fluid in the annulus between the pipe and toolstring. The dissipation and other friction effects estimated from the sensor readings are induced from these annulus velocities, which are dependent on the cable speed in the different log-passes, as given by Eq. (9a). An average of the annulus phase velocity with respect to the log passes up and down is consequently used in the expressions for viscous and turbulent shear in the numerical models, apart from the expressions for fluid interphase friction. When the phase velocities in undisturbed pipe ($U_k$) are known or assumed, the annulus velocity for phase $k$ is expressed by

$$U_{ka} = \frac{1}{N_{pass}} \cdot \sum_{i=1}^{N_{pass}} \left\{ \frac{D_p^2}{(D_p^2 - D_b^2)} \cdot U_k + U_{cabi} \cdot \left[ 1 - \frac{D_p^2}{(D_p^2 - D_b^2)} \right] \right\} \quad (9b)$$

The index $a$ implies velocity referred to the pipe/toolbody annulus; $k = g, o, w$ depending on the fluid system. This equation is used later in the analysis when dissipation effects are estimated.

First, the relative friction effect induced by $U_{ra}$ is computed. For laminar conditions, the friction factor is generally given by

$$f = \frac{64}{\mathrm{Re}_a} \quad (10)$$

The viscous pressure loss between the gradiometer ports due to relative motion is given by $U_{ra}$:

$$\frac{dP}{dz_{rlam}} = \mathrm{const} \cdot U_{ra} \quad (11)$$

Turbulent flow gives equivalent effects according to the Moody formula:

$$f = 0.0055 \cdot \left[ 1 + \left( \frac{2 \times 10^4 \cdot h_{ro}}{D_{ha}} + \frac{10^6}{\mathrm{Re}_a} \right)^{0.333} \right] \quad (12)$$

$$\frac{dP}{dz_{rturb}} = \mathrm{const} \cdot U_{ra}^2 \quad (13)$$

The total turbulent/laminar effect of friction loss between the gradiometer ports due to $U_{ra}$ is expressed as

$$\frac{dP}{dz_{rfr}} = k_{1d} + k_{2d} \cdot U_{ra} + k_{3d} \cdot U_{ra}^2 \quad (14)$$

The effects from Eqs. (10, 11, 12, 13) are implicitly contained in Eq. (14).

Converting $dP/dz_{rfr}$ to typical density units:

$$\rho_{raw} = \left( \frac{dP}{dz_{rfr}} \right) \cdot \frac{1}{g \cdot h_{gm}} \quad (15)$$

All the available log-passes up and down are utilized to get the polynomial Eqs. (14) and (15) based on ordinary regression analysis. The $\rho_{raw}$ is taken from the gradiometer measurement. $U_{ra}$ is found from Eq. (9a). From at least three separate passes the unknown parameters $k_{1d}$, $k_{2d}$ and $k_{3d}$ are computed at each depth level of the log interval. Filters are implemented to avoid strongly deviated sensor readings. Such deviations can be observed where the sensors are exposed to strong inflow zones. The $k_{1d}$ factor in Eq. (14) is stored as the resultant tool reading for further flow analysis as $\rho_{sen}$. This value is interpreted as a sensor response equivalent to a differential pressure detector that ideally follows the averaged fluid flow. Often, at low flow rates, this dissipation effect is not significant. The total roughness factor for the pipe and tool constellation is estimated from the turbulent contributor $k_{3d}$ in Eq. (14). Combining Eqs. (12), (13) and (14):

$$h_{ro} = \left[ \left( \frac{k_{3d} \cdot D_{ha}}{\rho_f \cdot 0.0027} - 1 \right)^3 - \frac{10^6}{\mathrm{Re}_a} \right] \cdot \frac{D_{ha}}{2 \times 10^4} \quad (16)$$

This roughness estimate is used as input or a verification of any given or assumed pipe-roughness parameter in further steps of the analysis. To take care of dissipation concerning the fluid flow itself, all the contributors to the response differences between the two gradiometer sensor ports have to be considered. In density units, the readings of a gradiometer without internal buffer fluid, when the above-mentioned relative friction effects are eliminated, includes

$$\rho_{sen} = \rho_f \cdot \cos\varphi + \frac{dP}{dz_{fr}} \cdot \frac{1}{g \cdot h_{gm}} + \frac{dP}{dz_{acc}} \cdot \frac{1}{g \cdot h_{gm}} \quad (17a)$$

Note that Eq. (17a) contains the total pressure drop, while Eqs. (10) – (16) consider a relative pressure drop due to relative annulus velocity at the local depth levels. In the same units, the gradiometer with an internal buffer fluid of density $\rho_{sb}$ at well conditions gives the equivalent equation:

$$\rho_{sen} = (\rho_f - \rho_{sb}) \cdot \cos\varphi + \rho_{sb} + \frac{dP}{dz_{fr}} \cdot \frac{1}{g \cdot h_{gm}} + \frac{dP}{dz_{acc}} \cdot \frac{1}{g \cdot h_{gm}} \quad (17b)$$

The frictional pressure drop and convective pressure loss in Eqs. (17a,b) are unknown quantities at this stage. The numerical flow equations has to be solved for both multi- and monophase flow systems to eliminate these contributors to Eqs. (17a,b). This means that the absolute friction effect on the sensor has to be solved implicitly in the numerical flow equations. The readings from the other two densitometers are defined as a one-sensor principle and are therefore not corrected for any dynamic dissipation or acceleration pressure drop. The measurements are only volumetrically phase-dependent, and are used directly in the following analysis:

$$\rho_{sen} = \rho_f \tag{17c}$$

A correction for pressure and temperature is sometimes needed for some sensor principles in Eq. (17c).

### Liquid hold-up measurements

When these sensor readings are available for the analysis, a linear transfer are assumed between frequency of a capacitance sensor probe and the fluid fraction—Water fraction in oil, water fraction in gas, or oil fraction in gas:

$$y_{lm} = \frac{j_m - j_k}{j_l - j_k} \quad \begin{cases} l = \text{oil or water} \\ k = \text{gas in gas/water}, k = \text{oil in oil/water.} \end{cases} \tag{18}$$

In churn flow, the core of oil is assumed to be exposed to the tool sensor. In intermittent bubble/slug gas/liquid flow, the liquid hold-up measurements are assumed to be a resultant of exposure only to liquid in the slugs and the gas pockets but not to the liquid beneath the gas pocket. The linear relationship between phase fraction and frequency in Eq. (18) is not well verified. No correction for salinity versus reference frequencies is performed.

### The complete set of measurements

In summary, the precalculations performed on the measured spinner density and pressure data result in the following set of variables: *Spinnerdata*: a mass-averaged fluid velocity, corrected for swirl, changing density and viscosity at the spinner surface; $U_{fmc}$ (m/s). *Density measurements*: for gradiometer principles, a resultant sensor reading that virtually follows the averaged fluid flow is computed as a gradient versus depth in density equivalent units; $\rho_{sen}$ (g/cm³). *Pressure measurements*: $P_m$ (bar). Gradients from quartz and strain sensors are computed:

$$\frac{dP}{dz_{m,\,\text{quartz}}}, \quad \frac{dP}{dz_{m,\,\text{strain}}} \quad \text{(Pa/m)}$$

The *liquid hold-up measurements* from a capacitance sensor is converted to an area-based void fraction of oil or water: $y_{lm}$. The temperature data are used directly in the further analysis: $T_m$(°C).

## PHYSICAL FLUID PROPERTIES (Mod2)

Fluid properties are either taken directly from laboratory measurements or generated internally by means of the correlations described in the following: The Glasø (1980) correlations is utilized for saturated parameters and densities of oil and water. In this case, the surface properties of the fluid together with the measured pressure and temperature are used to predict the equivalent downhole properties. The gas density is computed from a known or estimated gas $Z$ factor following Standing (1947). The estimates of the pure phase viscosities are based on the following: oil; Beggs and Robinson (1975); gas; Lee *et al.* (1966); water; Haywood (1966). Linear volumetric averaging of pure oil and water viscosity is used in the estimation of effective viscosity of oil/water emulsions. The surface tensions for oil/gas, water/gas and oil/water are computed following Firoozabadi and Ramey (1988) and Firoozabadi *et al.* (1985). Both surface and bubble-point conditions can be used as reference for the state equations.

## NUMERICAL FLUID FLOW MODELS (Mod3)

Numerical fluid flow models serve as a key element in obtaining the phase velocities and distributions together with phase fractions and densities at each depth increment of the logged well. Numerical models are grouped into different submodels, depending on the actual fluid phases and phase distribution. The different submodels are based on general conservation laws and constructive relations that handle the interphase coupling and shear relations. The main difference between the submodels is the geometric fluid distribution across the pipe area and the momentum coupling between two interfaces. The computation of the wetted-phase perimeters and hydraulic diameters in the different constitutive relations is the main difficulty. The flow models are classified into liquid/liquid and gas/liquid systems. Further subdivisions are made according to the different flow patterns. For all the models, the input parameters include the heavy and light phase fluxes, the fluid properties and pipe geometry. The different flow regimes are illustrated in Fig. 5.

### Liquid/liquid systems

The expected flow patterns for water/oil systems are; *bubble flow*, where oil droplets are dispersed in a continuous water phase; *churn flow*, where oil moves centred in the pipe and the water appear as an annulus, therefore wetting the pipewall; *emulsion flow*, where water is continuously distributed in the oil (high oil fraction). The bubble and emulsified systems have a homogeneous appearance in the pipe cross-section and the local relative velocities are assumed to be small and constant in the radial direction. Under such homogeneous conditions a radial velocity profile factor is estimated depending on the turbulent intensity and the fluid properties. The principle is closely related to the Prandtl 1/7 law described by Wallis (1969). The slip velocity is connected to the droplet spectrum and shape. The small slip for this regime is in agreement with Tulsa measurements described by Mukherjee *et al.* (1981). Churn flow implies a coupling of oil and water at a definite interphase. Presently it is assumed that no water is dispersed in the churn of the oil.

The interphase momentum interaction is assumed to be proportional to the relative velocity between the oil and water, the ripples or waves at the interphase and the physical fluid properties. The slip velocity for churn flow will vary much more than for the dispersed flow regime. No radial velocity profile is computed for churn flow, because of possible discontinuity in the phase velocities across the pipe area.

Momentum balances for both fluid phases are set up independently of the flow regime. Each fluid phase is exposed to the other phase, to the toolstring and to the pipe wall. The following balance is valid for all the fluid systems.

MOMENTUM BALANCE OF THE LIGHT PHASE (OIL)

$$\frac{d}{dz}(y_o \cdot \rho_o \cdot U_{oa}^2) = -y_o \cdot \frac{dP}{dz_c} + f_o \cdot \frac{1}{2} \cdot \rho_o \cdot U_{oa}^2 \cdot \frac{S_o}{4A_o}$$

$$+ f_i \cdot \frac{1}{2} \cdot \rho_o \cdot (U_r)^2 \cdot \frac{S_{io}}{4 \cdot A_o} + y_o \cdot g \cdot \rho_o \cdot \cos\varphi \tag{19a}$$

$$U_o = S_d(U_{sc} + U_r) \tag{19b}$$

For dispersed or emulsified flow:

$$S_d = \frac{1 - y_o}{K_a - y_o} \tag{19c}$$

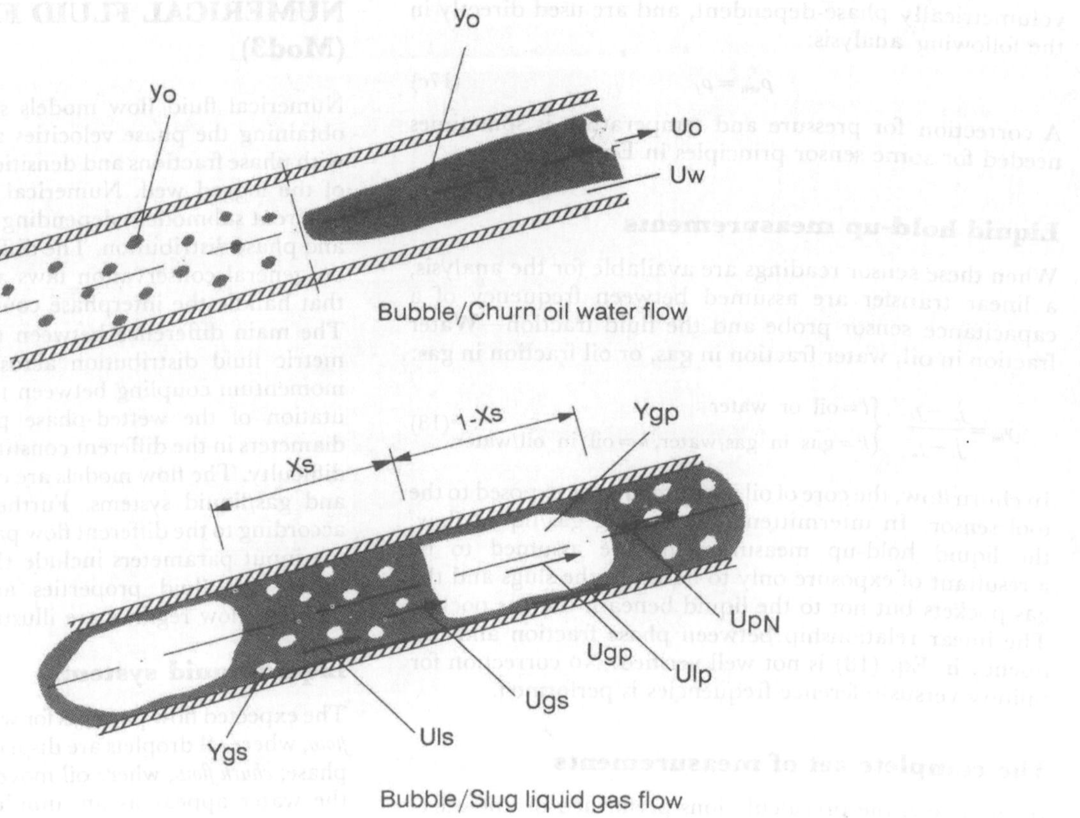


**Fig. 5.** Bubble/Churn oil water flow — Bubble/Slug liquid gas flow

$K_a$ = a function of velocity profile and fluid system
For churn flow, $S_d = 1$

MOMENTUM EQUATION FOR THE HEAVY PHASE (WATER)

$$\frac{d}{dz}(y_w\cdot \rho_w\cdot U_{wa}^2) = -y_w\cdot \frac{dP}{dz_c} + f_w\cdot \frac{1}{2}\cdot \rho_w\cdot \frac{U_{wa}^2\cdot S_w}{4\cdot A_w} - f_i\cdot \frac{1}{2}\cdot \rho_o\cdot (U_r)^2\cdot \frac{S_{io}}{4\cdot A_o} + y_w\cdot g\cdot \rho_w\cdot \cos\varphi \quad (20)$$

The annulus phase velocities are related to the undisturbed velocities in Eq. (9). A drag coefficient between the two phases gives the dynamic link between these:

$$f_i = f_i'\left(1 + \frac{k_1}{[(\sigma_{ow}/g\cdot(\rho_w - \rho_o))]^{0.5}}\cdot\frac{h_{wp}}{D_p}\right) \quad (21a)$$

$h_{wp}$ is close to zero for bubble/emulsion flow. $k_1$ is the local constant in Eq. (21a).

$$h_{wp} = \text{const}\cdot\frac{U_r^2}{g\cdot D_p\cdot (y_0)^{0.5}} \quad (21b)$$

$h_{wp}$ indicates the fluid interphase wave height from the Froudé number. For disperse or emulsified flow, Eqs. (19a,b,c) give the oil and gas phase velocities.

For churn flow, Eqs. (21a,b) give the slip distribution in such cases instead of the distribution slip factor, Eqs. (19b,c).

The shear force due to viscous dissipation and pipe/tool roughness is expressed by the Moody equivalent formula for each of the phases:

$$f_k = 0.0055\left[1 + \left(\frac{2\times 10^4\cdot h_{ro}}{D_{hk}} + \frac{10^6}{\mathrm{Re}_k}\right)^{0.333}\right],\ k = \text{oil, water} \quad (22)$$

The hydraulic diameter and Reynolds number are defined by Eq. (25). It is emphasized, that Eqs. (19) – (22) are valid for all the actual flow regimes of oil and water. However, the hydraulic phase diameters and wetted perimeter together with the formulation for the phase interaction imply a distinction between different flow patterns. The phase interaction for dispersed or emulsified flow imply water or oil as continuous phase and the other phase as the discontinuous phase, respectively. For small dispergents, less than 0.05 mm, the drag factor is given by Stokes' hypothesis described by Wallis (1969):

$$f_o = \frac{9}{\mathrm{Re}'},\qquad \mathrm{Re} = \left(\frac{U_r\cdot R_d\cdot \rho_c}{\mu_c}\right)$$

$c$ = continuous phase $\quad d$ = dispersed phase (23a)

For deformed droplets larger than 0.05 mm, the drag factor is given by Peebels and Garber as described by Malnes (1979):

$$f_o = \frac{3}{8}\cdot\left(\frac{g\cdot R_d\cdot U_r\cdot \rho_c^3}{\sigma_{uo}^3}\right) \quad (23b)$$

This expression is utilized directly in Eq. (21).

The Nukiama Tanasawa correlations in Wallis (1969) are utilized for the size of droplets to get an estimated averaged dispergent diameter. The phase that is assumed as continuous is dependent on the total oil fraction. For small oil fractions a bubble flow will be estimated. At higher fractions, either churn or emulsion flow can be assumed. At this stage of the discussion, the oil can be assumed to be a stable continuous phase in the bubble/emulsion equations. The velocity envelope defined by centre/average velocity of the cross-section at each depth interval is estimated according to Bendiksen (1984):

$$\mathrm{Re}=\frac{U_{fmc}\cdot\rho_f\cdot D_p}{\mu_c} \tag{24a}$$

$$R_r=\frac{U_{cc}}{U_c}=1-0.5\cdot\frac{D_c}{2}[3-\log(\mathrm{Re})]\exp\left(\frac{1}{x_c}\right) \tag{24b}$$

$$x_o=\frac{4\cdot\sigma_{ou}}{g\cdot\rho_f\cdot D_p^2}\quad\text{(Eötwod number)} \tag{24c}$$

For dispersed/emulsion flow the hydraulic diameter and wetted perimeter is directly related to the pipe dimension for the present continuous phase. The wetted perimeter and hydraulic diameter for the discontinuous phase is set to zero. For churn flow, two independent continuous phases are defined in the cross-section of the pipe. The interphase momentum coupling is given by Eq. (21), as for bubble flow: The following wetted perimeters, phase areas, and phase hydraulic diameter are valid for churn flow:

$$S_o=\Pi\cdot(y_o)^{0.5}\cdot D_p,\qquad S_i=S_o \tag{25a}$$

$$S_u=\Pi\cdot D_p[1+(y_o)^{0.5}] \tag{25b}$$

$$A_u=\frac{\Pi}{4}\cdot(D_p-D_o)^2 \tag{25c}$$

$$A_o=\frac{\Pi}{4}\cdot y_o\cdot D_p^2 \tag{25d}$$

$$D_{ho}=\frac{4\cdot A_o}{S_o} \tag{25e}$$

$$D_{hu}=\frac{4\cdot A_u}{S_u} \tag{25f}$$

$$\mathrm{Re}_k=\frac{\rho_k\cdot U_k\cdot D_{hk}}{\mu_k} \tag{25g}$$

Concerning the interphase friction Eq. (21), small fractions of oil will give an expression of the interphase friction similar to Eq. (23) for bubble flow, since the relations in Eqs. (21a,b) are applied as a general interphase correlation. This provides a continuous transition from bubble to churn flow in the numerical model, since the hydraulic diameters and wetted perimeters for slug and churn flow will be similar for small oil fractions.

The total set of equations is solved by first assuming a set of superficial velocities of oil and water. The pressure and temperature are known from the well measurements. The physical fluid properties are estimated and utilized directly in the flow equations. Since the churn flow model is equivalent to bubble flow for small oil fractions, the churn model is initially assumed. The Eqs. (9b), (19) and (25) are combined and solved with respect to the oil fraction. When the equations have converged to a solution, the oil fraction, phase velocities and averaged densities are found for the assumed flow rate. In order to compare the computed parameters directly with the measured equivalents, the average velocity and density referred to the average of the cross-section to the centre of area is applied. For churnflow, and the oil core is larger than the spinner diameter:

CENTRE MASS AVERAGED VELOCITY

$$U_{cc}=U_o \tag{26}$$

WATER HOLD-UP AT CENTRE

$$y_w=0 \tag{27}$$

For dispersion/emulsion flow, the computed centre velocity close to the spinner detection zone is

$$U_{cc}=R_r\cdot\left(\frac{U_{so}\cdot\rho_o+U_{su}\cdot\rho_u}{\rho_o\cdot y_o+\rho_u\cdot y_u}\right) \tag{28}$$

The $R_r$ factor transforms the area-averaged velocity to the area centre equivalent velocity:

$$y_u=1-y_o \tag{29}$$

The cross-sectional averaged computed density is, independent of flow regime,

$$\rho_c=\rho_o\cdot y_o+\rho_u\cdot y_u \tag{30}$$

The computed pressure gradient is assumed homogeneous in the pipe cross-section, and the centre and average related pressure drop is

$$\frac{dP}{dz_c}=(\rho_o\cdot y_o+\rho_u\cdot y_u)\cdot g\cdot\cos\varphi\quad\text{(gravitational contributor)}$$

$$+\frac{1}{2}\cdot\rho_o\cdot f_o\cdot\frac{U_{oa}^2}{D_{ho}}+\frac{1}{2}\cdot\rho_u\cdot f_u\cdot\frac{U_{ua}^2}{D_{hu}}\quad\text{(friction of phase } k\text{)}$$

$$\frac{d}{dz}(\rho_o\cdot y_o\cdot U_o^2+\rho_u\cdot y_u\cdot U_u^2)\quad\text{(convective acceleration)} \tag{31}$$

The convective contributor to Eq. (31) is found from the present phase velocities and the phase velocities at the depth level below the current one. When increasing the oil fraction from stable bubble flow, transition into either churn or emulsion flow is physically possible. Both the fluid rheology and the local flow dynamics are key parameters in which of the two flow patterns to be predicted. In general, the condition that implies the smallest free energy, considering the chemical, surface and mass contributors to the total energy of the fluid system, will determine the flow regime for the given fluid fluxes. However, this would be a complicated balance, owing to surface interactions, chemical emulsifiers, energy dissipation, etc. The selection between churn and emulsion flow is based on comparison of the computed parameters, Eqs. (26)–(29), with the equivalents measured at several depth levels. The condition implying best fit between the measurements and computations at each depth level predicts the local dynamic fluid behaviour.

## Gas/liquid systems

Bubble/slug and mist flow models are implemented in the system. The slug model concept is based on principles from Bendiksen (1984). Transition from slug to mist flow is numerically considered as a limit of the bubble/slug model. Equations (19), (20), (21), (22) and (23) are directly applied for gas/liquid concerning the momentum interactions between the pipe wall and the present fluid phases, independent of flow regime. However, the indexes for oil/water have to be changed to gas/liquid in the abovementioned equations. "Liquid" symbolizes water, oil or a complete mixture of these liquids. The gas/liquid model treats the bubble/slug transition as a continuous process where the local gas flow is a key parameter. At low gas fluxes, bubble flow will be predicted. At increasing gas rates, the small gas bubbles will agglomerate or coalesce into large-scale pockets moving relative to the liquid in the pipe. The liquid will appear as slugs intermittently bridging the liquid annulus in the gas pockets. In addition, some gas will be dispersed in the liquid slugs. A unit volume of the total fluid is divided into gas pocket (P) and liquid slug (S) sections defined for each depth level in the log interval. Each of these sections contains gas of different quantities and distribution geometries. The gas and liquid velocities are also different in the two sections of the unit volume. Therefore, six unknowns are to be solved: gas and liquid velocity and gas fraction in the slug and pocket, respectively. To find these, other parameters have to be defined, implying further constitutive equations and unknowns, which all have to be solved implicitly with the six main unknowns. The gas fraction in the slug is assumed proportional to the liquid velocity difference in the pocket and the slug section, pipe deviation and fluid properties. The gas fraction in slugs is estimated from

$$y_{gs}=\frac{k_g\cdot(U_{ls}-U_{lp})}{U_{ls}}\cdot\frac{U_{sg}}{U_s} \tag{32}$$

$k_g$ = a local constant for Eq. (32). The index $l$ is used for phase "liquid" in liquid/gas flow.

Drag effects on the droplets, including slip effects based on distribution and local velocity differences can be expressed as given by Malnes (1979) and Wallis (1969) for the slug section:

$$U_{gs}=Sd(U_{ls}+U_{0s}) \tag{33a}$$

$$Sd=\frac{1-y_{gs}}{K_a-y_{gs}} \tag{33b}$$

$$U_{0s}=1.4\cdot\left(\frac{g\cdot\sigma_{gl}\cdot(\rho_l-\rho_g)}{\rho_w^2}\right)^{0.25}\cdot(1-y_{gs})^2 \tag{34}$$

This expression takes into consideration the radial phase distribution in the liquid slug. The pocket's nose velocity is (Wallis, 1969)

$$U_{pn}=Ca\cdot U_{ls}+U_{0p} \tag{35}$$

According to measurements and analysis of Bendiksen (1984), the well deviation and Froudé number are connected to this expression:

$$Ca=\frac{\overline{(y_g\cdot U_s)}}{\overline{y_g}\cdot\overline{U_s}},\qquad \mathrm{Fr}=\frac{U_{ls}}{(gD)^{0.5}}\quad\text{(Froudé number)} \tag{36a}$$

$$Ca=1.05+0.15\cos^2\varphi,\qquad \mathrm{Fr}<3.5 \tag{36b}$$

$$=1.2,\qquad \mathrm{Fr}>3.5$$

$$U_{0p}=-0.54\sin\varphi+0.35\cos\varphi\cdot\left[\left(1-\frac{\rho_g}{\rho_l}\right)\cdot g\cdot D_p\right]^{0.5},\qquad \mathrm{Fr}<3.5 \tag{36c}$$

$$U_{0p}=0.35\cos\varphi\cdot\left[\left(1-\frac{\rho_g}{\rho_l}\right)\cdot g\cdot D_p\right]^{0.5},\qquad \mathrm{Fr}<3.5 \tag{36d}$$

The mass continuity referred to the pocket front velocity gives the averaged velocity of the gas in the pocket:

$$U_{gp}=U_{pn}-(U_{pn}-U_{gs})\cdot\frac{y_{gs}}{y_{gp}} \tag{37}$$

In the pocket, there is a connection between the liquid and the gas due to the interphase shear:

$$f_i=0.02\cdot\left(1+\frac{h_{lp}}{D_p}\right) \tag{38a}$$

$$h_{lp}\sim\frac{(U_{gp}-U_{lp})^2}{g\cdot D_p\cdot(y_{gp})^{0.5}} \tag{38b}$$

The liquid layer beneath the gas pocket is related to the local gas velocity:

$$U_{lp}=U_{ls}\cdot\frac{1-y_{gs}}{1-y_{gp}}-U_{pn}\cdot\frac{y_{gp}-y_{gs}}{1-y_{gp}} \tag{39}$$

The force along the pipewall due to viscous shear and pipe roughness is given by Eq. (22) with correct hydraulic diameters for liquid and gas. The wetted perimeters and hydraulic diameters are defined equivalently to Eq. (25).

An internal momentum equation for the pocket section links $U_{lp}$ with $U_{gp}$:

$$f_i\cdot\frac{1}{2}\rho_g\cdot(U_{gp}-U_{lp})^2\cdot\frac{S_i}{4A_g}+(1-y_{gp})\cdot\frac{1}{2}\cdot f_g\cdot\rho_g\cdot U_{gp}^2\cdot\frac{S_{gp}}{4A_{gp}} \tag{40}$$

$$-y_{gp}\cdot\frac{1}{2}\cdot f_l\cdot\rho_l\cdot U_{lp}^2\cdot\frac{S_{lp}}{4A_{lp}}-y_{gp}\cdot(1-y_{gp})\cdot g\cdot(\rho_l-\rho_g)\cdot\cos\varphi=0$$

A fraction of liquid slug in a pipe section unit is defined by

$$X_s=\frac{\text{length of slug}}{\text{slug}+\text{pocket length}}=\frac{(U_{gp}\cdot y_{gp}-U_{sg})}{U_{pn}\cdot(y_{gp}-y_{gs})} \tag{41}$$

$X_s$ can be defined as a liquid slug frequency by a stationary observer.

The expression (41) is based on the interaction between gas velocities in the slug and pocket including mass balances. $X_s$ is 1 for clean slug flow, which is equivalent to pure bubble flow. $X_s$ will be 0 for clean pocket flow, which is equivalent to annular or mist flow.

The mist flow appears when the flooding velocity of gas has been reached and all the liquid has been entrained into the gas phase. A limited gas fraction at which slug flow can appear is estimated comparing the slip relation in the pocket section. If there is an increase of gas above this limit, mist flow is definitely assumed. This limit is found from the combination of Eqs. (33), (35) and (36) giving the transition from slug to mist. As the limit velocity approaches infinity, a complete decoupling indicating an annular or mist flow occurs:

$$\mathrm{Slip}=\frac{U_{pn}\cdot(1-y_{gp})}{U_{sl}}=\frac{1-y_{gp}}{U_s-y_{gp}\cdot U_{gp}},\qquad U_s=U_{sl}+U_{sg} \tag{42}$$

which can be expressed as Eq. (43), resulting in Eq. (44):

$$\mathrm{Slip}=\frac{1-y_{gp}}{\dfrac{1}{Ca+U_{0p}/U_s}-y_{gp}} \tag{43}$$

$$\mathrm{Slip}=\frac{1-y_{pn}}{\dfrac{1}{Ca}-y_{pn}} \tag{44}$$

Slip unlimited, hence $y_{gp\,\mathrm{lim}}=0.83$.

Practical experiments of Wallis (1969) have indicated the fractional gas limit for the slug flow regime to be between 0.8 and 0.86, thus confirming the limit expressed in Eq. (44). The void limit in this expression implies a change from slug to mist flow, since all liquid is assumed to be entrained in the gas. Independent of this criterion, the difference of the measured and the equivalent computed pressure gradient indicates the accuracy of the assumed flow regime. For a given set of phase fluxes, different pressure gradients will be computed for slug and mist phase distributions. In the case of slug flow, the gravity of the liquid film at the pipe wall will nearly balance the wall and interfacial shear force in the pocket section, indicating a lower pressure drop. In mist flow, there will be a definite shear force at the wall because the gas with dispersed liquid is exposed to the wall and no water film is assumed to compensate for the wall friction. All the water will be dispersed in the gas, implying a higher gravity and total pressure gradient. The above equations for slug/bubble flow will then be simplified for mist flow since no liquid film is assur:ed on the pipewall and no intermittent slugs or gas pockets are assumed. The following relations are valid for the mist flow:

$$X_s = 0, \qquad D_{hg} = d_p, \qquad D_{hl} = 0 \tag{45}$$

$$U_g = \frac{1 - y_g}{K_a - y_g} \cdot (U_l + U_{os}) \tag{46}$$

The profile factor ($R_v$), Eq. (24), is computed equivalently to the bubble liquid/liquid flow, since a homogeneous distribution of the phases is assumed. The mass-averaged computed velocity at the centre of the cross-section of the pipe is:

MIST FLOW

$$U_{cc} = R_v \cdot \left( \frac{U_{sg} \cdot \rho_g + U_{sl} \cdot \rho_l}{\rho_g \cdot y_g + \rho_l \cdot y_l} \right) \tag{47}$$

SLUG/BUBBLE FLOW

$$\begin{aligned} U_{cc} = \{ & U_{gp} \cdot (1 - X_{sl}) \cdot \rho_g \cdot y_{gp} + U_{gs} \cdot X_s \cdot \rho_g \cdot y_{gs} \\ & + U_{ls} \cdot X_s \cdot \rho_l \cdot y_{ls} \} / \{ R_v \cdot [(1 - X_{sl} \cdot \rho_g \cdot y_{gp}) \\ & + X_s \cdot (\rho_g \cdot y_{gs} + \rho_l \cdot y_{ls})] \} \end{aligned} \tag{48}$$

The averaged density

MIST FLOW

$$\rho_c = \rho_g \cdot y_g + \rho_l \cdot y_l \tag{49}$$

$$y_{lc} = 1 - y_g \tag{50}$$

SLUG/BUBBLE FLOW

$$\rho_c = (\rho_g \cdot y_{gs} + \rho_l \cdot y_{ls}) \cdot X_s + (\rho_g \cdot y_{gb} + \rho_l \cdot y_{lb}) \cdot (1 - X_s) \tag{51}$$

$$y_{lc} = 1 - (y_{gp} \cdot (1 - X_s) + y_{gs} \cdot X_s) \tag{52}$$

THE AVERAGED PRESSURE GRADIENT FOR ALL GAS/LIQUID FLOWREGIMES

$$\begin{aligned} \frac{dP}{dz_c} = & \left[ (\rho_g \cdot y_{gs} + \rho_l \cdot y_{ls}) \cdot g \cdot \cos\varphi + f_1 \cdot 0.5 \cdot \rho_l \cdot |U_{lsa}| \cdot U_{lsa} \cdot \frac{1}{D_p} \right] \\ & + (1 - X_s) \cdot \left[ (\rho_g \cdot y_{gp} + \rho_l \cdot y_{lp}) \cdot g \cdot \cos\varphi \right. \\ & \left. + f_1 \cdot 0.5 \cdot \rho_l \cdot |U_{lpa}| \cdot U_{lpa} \cdot \frac{1}{D_{hlp}} \right] \\ & + (1 - X_s) \cdot \rho_l \cdot (U_{ls} - U_{lp})^2 \end{aligned} \tag{53}$$

where $D_{hlp}$ is hydraulic diameter of liquid beneath the gas pocket.

## Connection between measurements and computation

To obtain the correct set of phase fluxes at the different depth levels for all the fluid systems, an error function with the superficial velocities as free variables is defined. This function links the measurements and the physical fluid model. For liquid/liquid and gas/liquid systems:

$$\begin{aligned} \text{Fert} = & W_{sp} \cdot (U_{fmc} - U_{cc})^2 && \text{(Spinner link)} \\ & + W_{gm} \cdot (\rho_{sen} - \rho_{ceq})^2 && \text{(Density link)} \\ & + W_{pr} \cdot \left( \frac{dP}{dz_m} - \frac{dP}{dz_c} \right)^2 && \text{(Pressure gradient link)} \\ & + W_{\text{cap}} \cdot (y_{lm} - y_{lc})^2 && \text{(Liquid hold-up link)} \end{aligned} \tag{54a}$$

If gradiometer sensors without buffer fluid are applied:

$$\rho_{\text{ceq}} = \rho_c \cdot \cos\varphi + \left( \frac{dP}{dz_{\text{fr}}} + \frac{dP}{dz_{\text{acc}}} \right) \cdot \frac{1}{g \cdot h_{gm}} \tag{54b}$$

If gradiometer sensor with buffer fluid is applied:

$$\rho_{\text{ceq}} = (\rho_c - \rho_{sb}) \cdot \cos\varphi + \rho_{sb} + \left( \frac{dP}{dz_{\text{fr}}} + \frac{dP}{dz_{\text{acc}}} \right) \cdot \frac{1}{g \cdot h_{gm}} \tag{54c}$$

Or if buoyancy or gamma sensor is applied for density detection:

$$\rho_{\text{ceq}} = \rho_c \tag{54d}$$

An iteration with respect to the flow fluxes is performed until "Fert", Eq. (54), has reached a minimum value. Each depth level has one unique solution concerning phase velocities, pressure drop and phase fractions defined by minimum of Eq. (54). The multiphase flow analysis begins at the bottom of the log interval, where the precomputed data are picked up for use in the error function. From the precomputed pressure gradient, a liquid/liquid or a gas/liquid submodel is initially selected. When convergence is achieved, it is checked whether eventually another flow pattern may occur with the current set of measurements. In such cases, the flow regime implying the smallest error function is selected. The result for the current depth level is saved and used as a first guess for the phase fluxes at the next depth. The principle for the procedure is illustrated in Fig. 6. The interactions are performed by a numerical line search method. The search vectors are defined as the superficial velocities. The vector search is based on "conjugated directional search", according to Powell (1964). The search accuracy is based on user-defined limits. The iteration at each depth stops if the search distances for any of the two vectors are below a given minimum, or if the change in error function is close to zero close to the converged vector combination. This basic principle of line

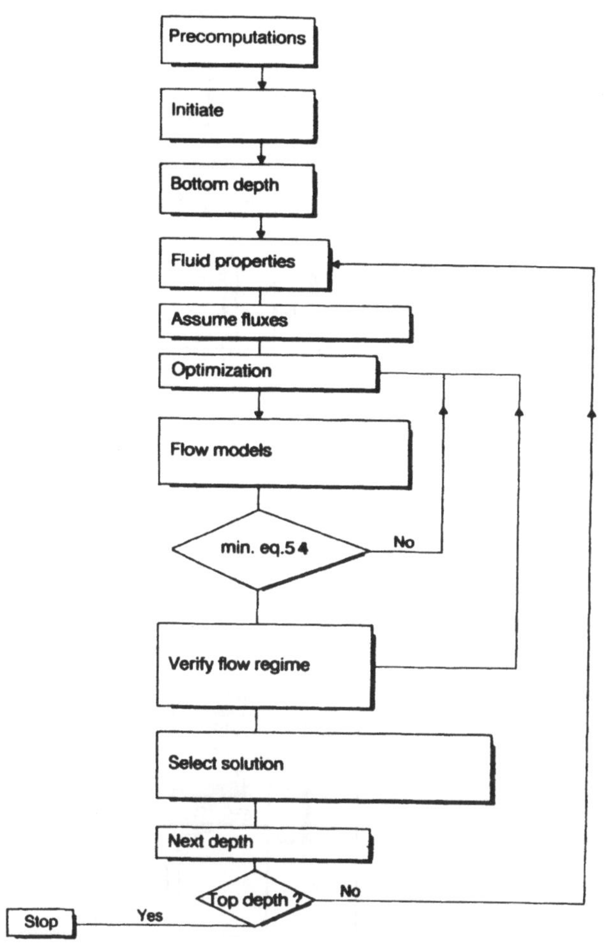


**Fig. 6.** Algorithm for the numerical flow analysis

search was utilized in flow analysis in the project system described by Nerby (1986) and Nuland and Nerby (1986).

# RESULTS FROM ANALYSIS OF WELL MEASUREMENTS

To illustrate how the outlined theory works, measurements from one gas/liquid and one liquid/liquid well from the North Sea are treated. The system has been implemented into a computer system, called "DYNALOG", which is here utilized for the analysis of these two wells. The gradiometer log data are displayed in Figs. 7 and 9. The results from both precomputations and the complete two-phase analysis are displayed in Figs. 8 and 10.

## The gas/liquid well

The aim of this analysis was to obtain a measure of the performance of the gas/oil model data from the North Sea well and computing the phase fluxes and pressure drops. The oil and gas rates predicted above the upper perforation were found to be about 10% higher than the measurements performed on the test separator. Since the PL measurements were of very poor quality, the result of the analysis is considered highly satisfactory and shows that the system can handle PL measurements that are on the lower limit of acceptance.

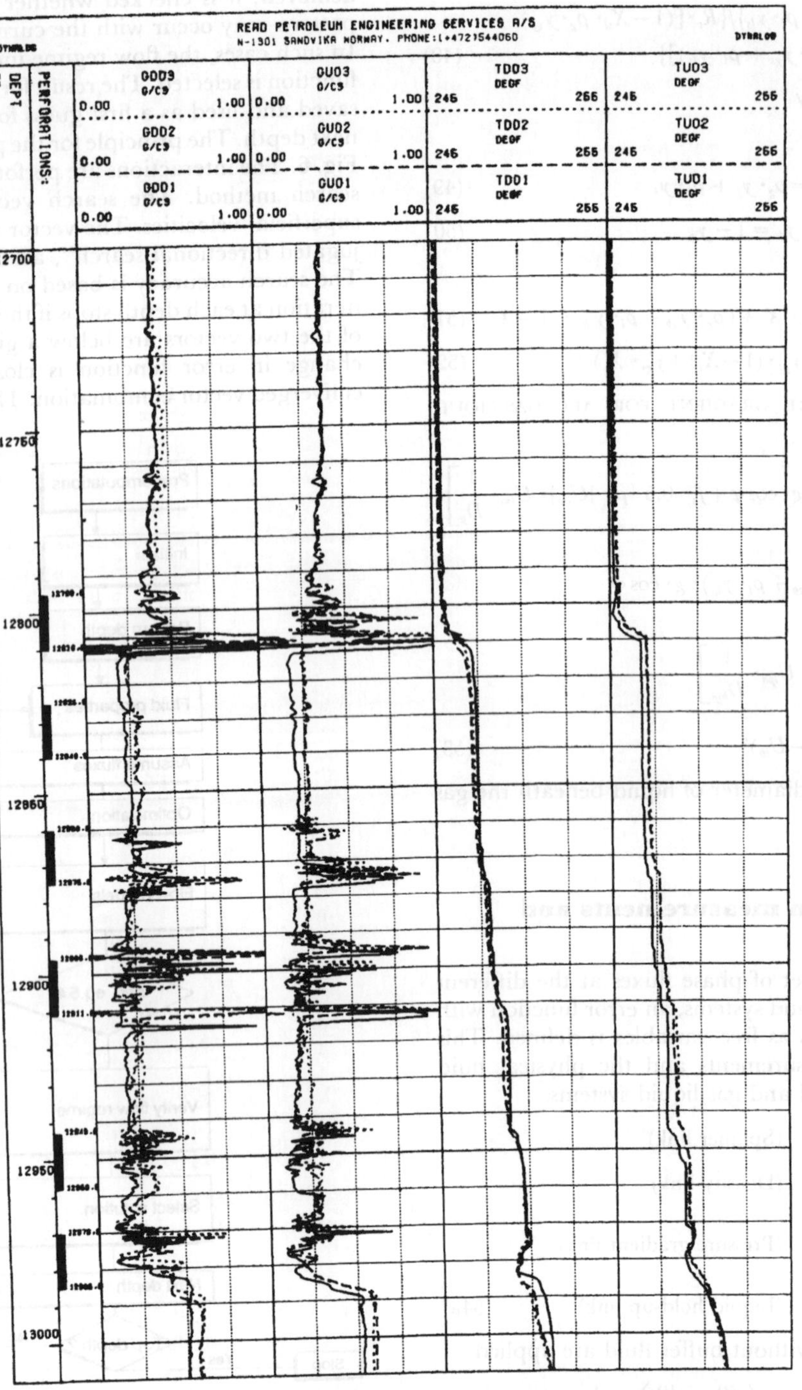


**Fig. 7.** Gradiometer measurements from flow passes

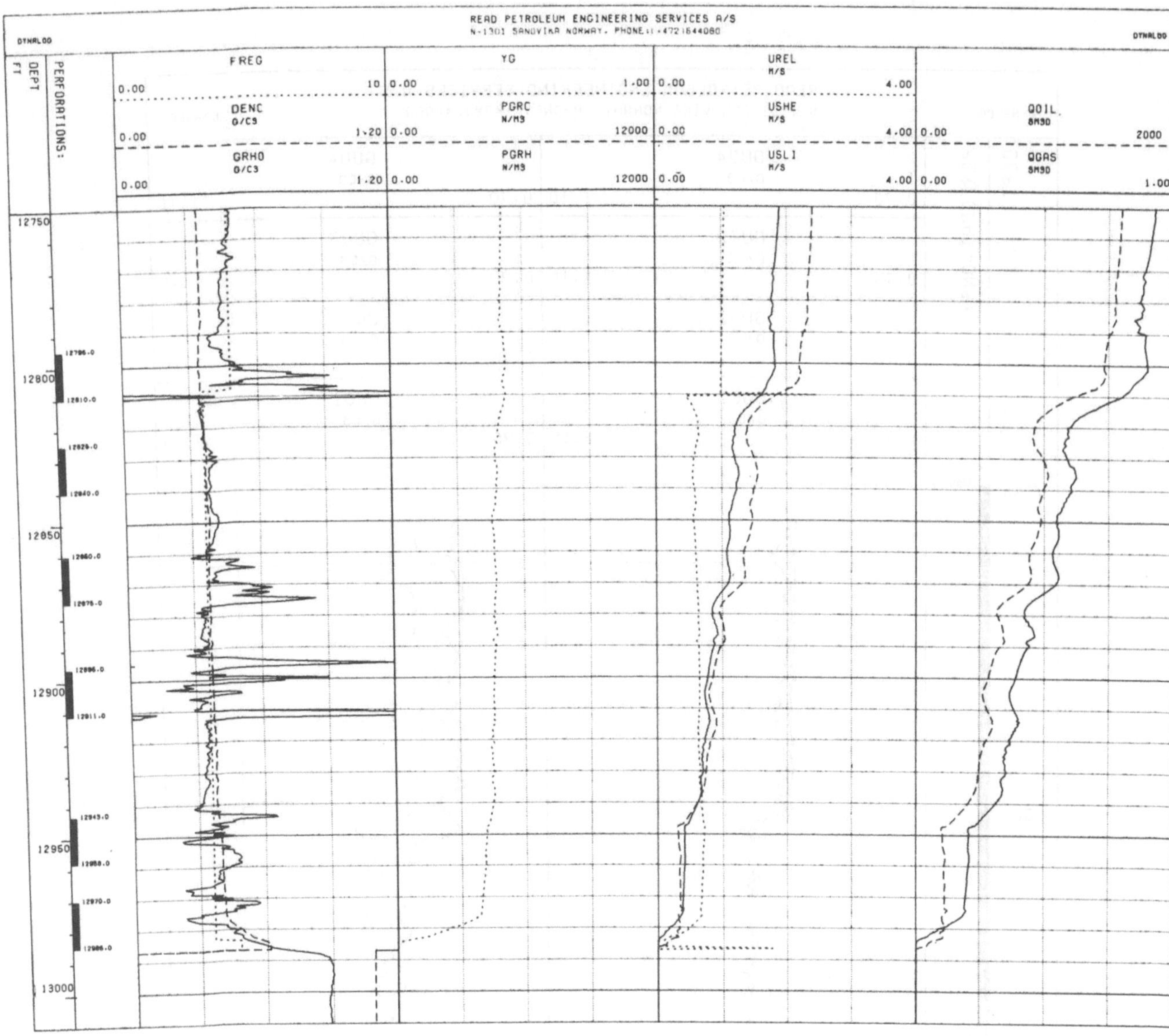


**Fig. 8.** PLA results from flowing well

## The liquid/liquid well

The goal for this analysis was to quantify the production of water and oil in the perforation zones. In addition, the level for the water production was to be accurately detected for further well management to depress the water production. Precomputations and two-phase flow analysis was performed utilizing the different oil/water submodels. The roughness factor of the inner pipe was estimated at about 90 $\mu$m, in the typical range of corroded steel pipes. A transitional bubble/churn flow regime was predicted in the entire interval. The consequence of this was a high slip velocity compared to the phase velocity difference in homogeneous bubble flow. The churn flow regime is a reasonable prediction considering the actual 45° well deviation. In this well, emulsion of water/oil could occur because of the high oil fraction. However, comparing the predicted fluxes of oil and gas using the emulsion flow model did not fit with the rates at the test separator. The deviation from the separator measurements for bubble/churn flow was close to 7% for oil, and 5% for the water. From this analysis, the production of both water and oil could be predicted quite accurately.

## CONCLUSION

The outlined system has successfully merged state-of-the-art numerical multiphase flow modelling and sensor measurement techniques into one implicit analysis. The system treats the different observed flow patterns selectively to get the most appropriate transfer functions between the sensor readings and the actual properties describing the fluid system. The system avoids ambiguous explicit corrections on the sensor readings by implicitly applying in-situ measurements. The system is modular, for easily incorporation of new sensor techniques and numerical models including physicochemical fluid properties. Examples from the analysis of two different wells provide results with acceptable accuracy when compared with the separator rates at the surface.

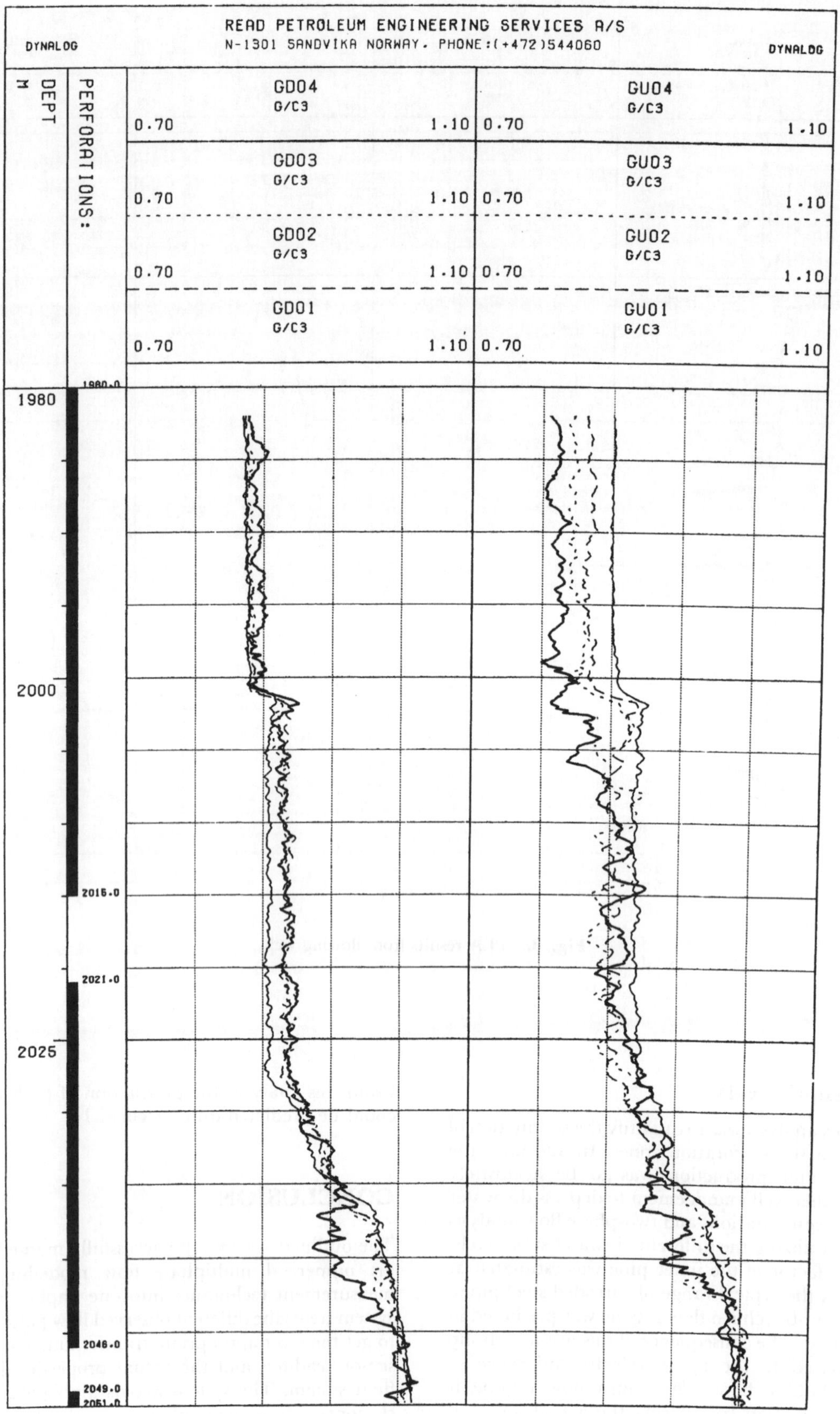


**Fig. 9.** Gradiometer measurements from flowing period

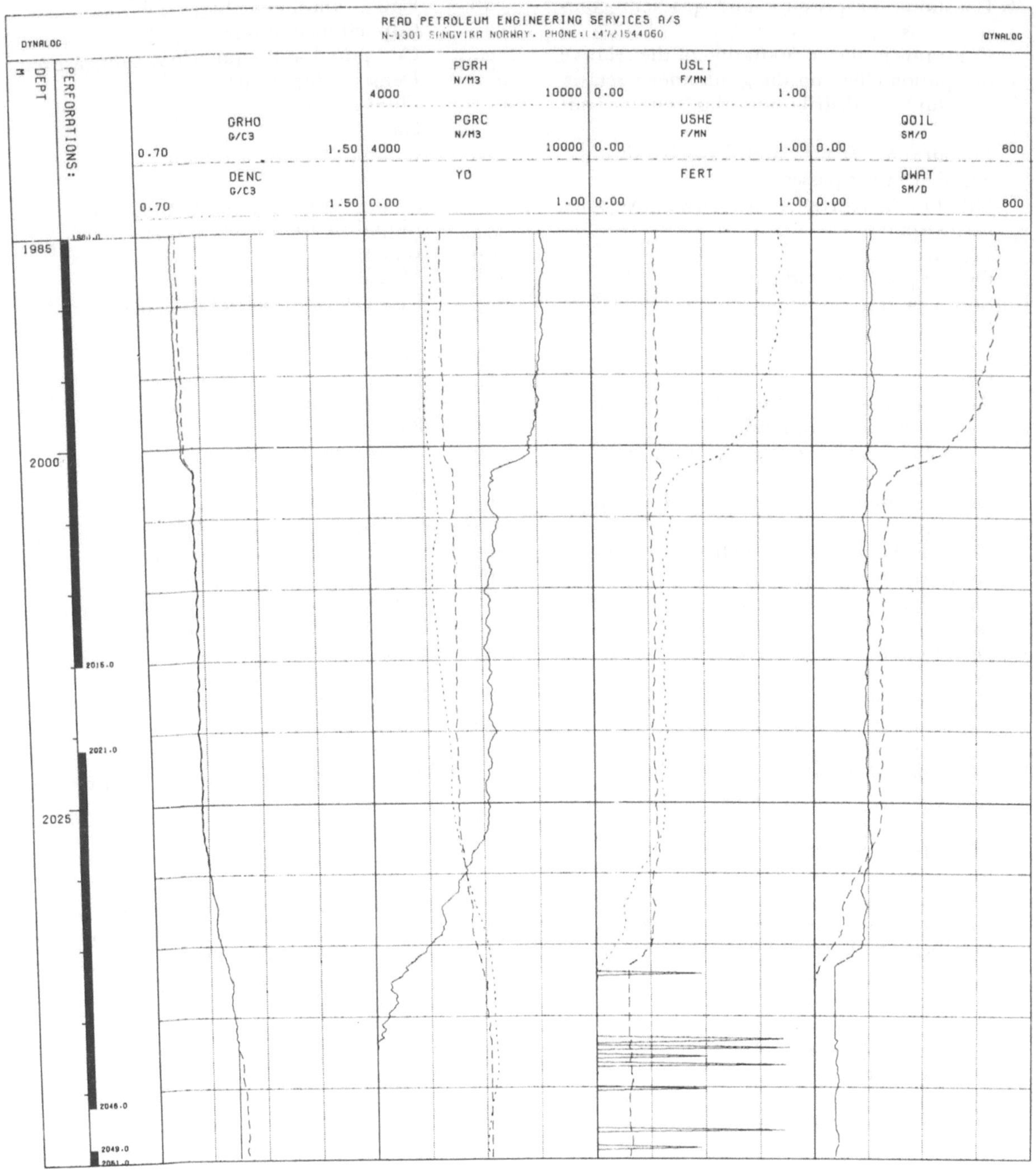


**Fig. 10.** PLA results from the liquid liquid well

## NOMENCLATURE

Symbols used in this paper are listed in alphabetical order.

| | |
|---|---|
| $A_k$ | Area of phase $k$ |
| $C_a$ | Constant in distribution slip relation gas/liquid |
| $\frac{dp}{dz_m}$ | Gradient from the strain or quartz sensors |
| $\frac{dP}{dz_c}$ | Computed pressure gradient |
| $\frac{dP}{dz_{fr}}$ | Total friction pressure drop |
| $\frac{dP}{dz_{acc}}$ | Total acceleration pressure drop |
| DENS | Computed fluid density |
| $D_p$, $D_b$ | Pipe internal diameter, toolbody outer diameter, respectively. |
| $D_{hk}$ | Hydraulic diameter of phase $k$. |
| $D_{ha}$ | Hydraulic diameter related to annulus between toolbody and pipe wall |
| Fert | Total error function |
| $f$ | Friction/dissipation factor, in general |
| $f_k$ | Friction/dissipation at wall of phase $k$ |
| $f_i$ | Interphase friction |
| $f_o$ | Drag factor between dispersent and bulk phase |
| $Fr$ | Froudé number |
| $g$ | Acceleration due to gravity |
| GRHO | Gradiometer measurement |
| $h_{wp}$ | Ripples on smooth/wavy interphase between two fluid phases |
| $h_{ro}$ | Pipe roughness |
| $h_{gm}$ | Distance between gradiometer sensor ports |
| $j_k$ | Hold-up frequency of capacitance sensor, fluid phase $k$ |

$k_{1p}, k_{2p}, k_{1d}, k_{2d}$ Constants for pressure and gradient computations.
$k_{3d}$ Constants in the computation of the relative dissipation effect on the gradiometer sensor
$K_a$ Constant in distribution slip relation oil/water
MPK Pressure, $K-H$(HP) and $K=S$ (strain)
$N_{pass}$ Number of log passes
$O_f$ Offset of the regression line between velocity versus spinner
$P$ Pressure
PGRH Quartz pressure gradient
PGRS Strain pressure gradient
$Q_{wat}$ Water rate
$Q_{oil}$ Oil rate
$Q_{gas}$ Gas rate
$R_v$ Relation between centre and mass-averaged velocity in dispersed flow
Re Reynolds number related to pipe without tool sensor
$Re_a$ Reynolds number related to toolbody/pipe annulus
$S_l$ Slope in the regression line between velocity versus rotations of spinner
Slip Slip relation defined by gas/liquid velocity ratio
$Sd$ Distribution – slip relation
$S_k$ Wetted perimeter, phase $k$
$T$ Temperature
$U_s$ Total superficial velocity
$U_{sk}$ Superficial velocity, phase $k$
$U_{0p}$ Rise velocity, Nicklin pockets
$U_{0s}$ Rise velocity, dispersed fluid
$U_{pn}$ Pocket nose velocity
$U_{fmc}$ Measured, average fluid velocity, estimated from spinner sensor
$U_r$ Relative velocity between the fluid phases
$U_{ra}$ Relative velocity between cable velocity and fluid velocity detected by the spinner sensor in toll body/pipe annulus
$U_k$ Velocity, phase $k$
$U_{ks}$ Velocity, phase $k$ in slug
$U_{kp}$ Velocity, phase $k$ in pocket
$U_{cab}$ Cable velocity
$U_{ka}$ Velocity of phase $k$ referred to tool/pipe annulus averaged from all log-passes up and down.
$U_c$ Computed mass averaged velocity referred to total cross-sectional area.
$U_{cc}$ Centered computed mass averaged velocity
$U_{0s}, U_{0p}$ Rise velocity of dispersed fluid, and Nicklin pocket, respectively.
VTAV Spinner related fluid velocity
$W_{sp}, W_{gm}, W_{pr}, W_{cap}$ Weight factors on spinner, density, pressure and hold-up sensors respectively
$X_s$ Slug fraction of total pipe length
$y_{kp}$ Fraction of phase $k$ in pocket
$y_{ks}$ Fraction of phase $k$ in slug
$y_{kc}, y_k$ Fraction of phase $k$, computed
$y_{lm}$ Fraction of liquid phase $l$, measured
$z$ Coordinate along pipe axis as depth reference

SUBSCRIPTS

$a$ Annulus tool/pipe
acc Acceleration
$b$ Body of tool sensor
$c$ Computed
$c$ Centre of pipe
cab Cable (wireline-cable)
cap Capacitance sensor
ceq Computed and equivalent to measured value
$d$ Density; dispersent
$f$ Fluid mixture
$g$ Gas
$gm$ Gradiomanometer sensor
$h$ Hydraulic
$k$ Fluid phase $k$; $k$=oil, water, gas
$l$ Liquid phase, $l$=oil or water
lam Laminar
$m$ Measured
0 Rise velocity due to buoyancy
$o$ Oil
$p$ Pressure, pipe
$pn$ Pocket nose
$pr$ Pressure sensor
$p$ Pocket
raw Uncorrected measurement
$r$ Relative
$rfr$ Relative friction
$r$lam, $r$turb, $r$tot Laminar, turbulent and total friction effect due to relative motion between fluid and tool body
ro Roughness of pipe/tool constellation.
$s$ Liquid slug, superficial velocity
sen Sensor
$sb$ Silicone buffer fluid
$sp$ Spinner sensor
turb Turbulent
$w$ Water
$wp$ Waves, ripples between two fluid interfaces

GREEK

$\alpha$ Angle between spinner blade and fluid velocity vector
$\rho_{raw}$ Uncorrected gradiometer sensor readings
$\rho_{ceq}$ Computed density equivalent to the measured value. Internal buffer fluid, friction and deviation are taken into consideration
$\rho_{sen}$ Measured densitometer value corrected for relative friction effects
$\rho_f$ Total fluid density
$\rho_k$ Density of phase $k$
$\rho_{sb}$ Buffer fluid density
$\rho_c$ Computed fluid density
$\varphi$ Pipe deviation measured from vertical axis.
$\sigma_{kp}$ Surface tension between phase $k$ and $p$
$\mu_f$ Total fluid viscosity
$\mu_k$ Viscosity of phase $k$
$\omega$ Rotations of spinner

## REFERENCES

Beggs, H.D. and Robinson, H.R. (1975). Estimating the Viscosity of crude oil systems. *J. Petroleum Technol.* 1140–1141.

Bendiksen, K.H. (1984). An experimental investigation of the Motion of the long bubbles in inclined tubes. *Int. J. Multiphase Flow* **10** (4), 467.

Firoozabadi, A. and Ramey , H.J. (1988). Surface tension of water hydrocarbon systems at reservoir conditions. *J. Can. Petroleum Technol.* 27–41.

Firoozabadi, A., Katz, D.L., Soroosh, H. and Sajjadian, V.A. (1985). Surface tension of reservoir crude oil-gas systems Recognizing the asphalt in the heavy crudes, SPE 13826.

Glasø, Ø. (1980). Generalized PVT correlations. *J. Petroleum Technol.* 785–790.

Haywood, R.W. (1966). Sixth International Conference on the

Properties of Steam. Supplement on Transport Properties. (Journal of Engineering for Power—Trans. ASME) pp. 63–66.

Lee, A.L., Gonzalez, M.H., Eakin, B.E. (1966). The viscosity of natural gases. *J. Petroleum Technol.* 997–1000.

Malnes, D. (1979). Slip relations and momentum equations in two phase flow. *Institute for Energiteknikk,* (IFE) *report EST-1.*

Mukherjee, H., Brill, J.P. and Begg, H.D. (1981). Experimental study of oil–water flow in inclined pipes. *J. Energy Resources Techn.*, 56–66.

Nerby, G. (1986). *PLING Program Code Description and User Manual.* IFE/KR/F-86/170.

Nuland, S. and Nerby, G. (1986). *Production Logging Interpretation Guide – PLING-principles and Equations.* IFE/KR/F-86/051.

Powell, M.J.D. (1964). An efficient method for finding the minimum of a function of several variables without calculating derivatives. *Computer J.* **7**, 39.

Standing, M.B. (1947). A pressure–volume–temperature correlation for California oil and gases. *Am. Petroleum Institute Drill. Prod. Pract.*, 275.

Wallis, G.B. (1969). *One-dimensional Two-phase Flow.* McGraw Hill, New York.

# 9

# The Geotechnical Characteristics of Weak North Sea Reservoir Rocks

*M.E. Jones[a], M.J. Leddra[a], A. Goldsmith[a], O.P. Berget[b] and I. Tappel[b]*

[a]Department of Geological Sciences, University College London, UK
[b]Oljedirektoratet, Stavanger, Norway

Laboratory investigations of the deformation characterstics of chalks and weakly cemented sandstones from a number of North Sea reservoirs, and from outcrops, demonstrate that these materials are elastic under initial loading, but exhibit strongly non-linear deformation behaviour at higher stresses. This behaviour can be described by a characteristic failure surface in stress/pore volume coordinates. This paper presents a description of this behaviour. When reservoir rocks of this type are loaded, maintaining a uniaxial strain condition, the material, undergoes ductile yield at relatively low stress followed by significant porosity decrease at almost constant stress (pore collapse). The onset of pore collapse depends on the rock type, its pre-deformational porosity and the extent and nature of the cementing material. Eventually a hardening occurs that is a response to the increasing compaction and porosity decline in the sample. Reservoir rocks will mobilize maximum compaction drive if they yield in this manner, but may also suffer a permeability loss due to the decrease in porosity. Other detrimental aspects of reservoir compaction (well casing collapse and surface subsidence) may also occur. When subject to large shear stresses in the presence of high pore pressures, the materials investigated are observed to deform readily in shear. They may approach a liquefact stat. This aspect of the mechanical behaviour is interpreted as a possible mechanism for the rapid transfer of substantial volumes of reservoir solids in a well. This phenomenon has been reported to occur in a number of fields. generally, following rapid drawdown of previously "shut-in" well.

## INTRODUCTION

A number of geotechnical problems have developed in connection with production of hydrocarbons from reservoirs in the Norwegian sector of the North Sea. The most publicized example has been the compaction of the Ekofisk reservoir and associated seafloor subsidence (Aam, 1988; Potts *et al.* 1988), and other cases of reservoir compaction have been reported (Ruddy *et al.*, 1988). Other problems occur in the near-wellbore region and include damage to well casings, the production of significant quantities of the reservoir rock, problems encountered during drilling and poor core recovery (Ruddy *et al*, 1988; Pattillo and Smith, 1982; Risnes, *et al.*, Bratli and Risnes, 1981). Reservoir geotechnical problems in the Norwegian sector of the North Sea have proved to be very expensive, have required costly and imaginative solutions, and where severe, have strongly influenced the production strategy for the field.

The occurrence of these problems of reservoir rock deformation have prompted a number of detailed geotechnical investigations to be undertaken. These investigations have concentrated on both measuring the mechanical behaviour of the reservoir rocks (Addis, 1987; Johnson and Rhett, 1987; Ruddy *et al.*, 1988) and modelling the various deformation processes (Potts *et al.*, 1988; Barton *et al*, 1988; Ruddy *et al.*, 1988). It has been a common observation that major deformations of the reservoir only occur when the reservoir rock exhibits non-linear deformation characteristics (Jones *et al.*, 1987). North Sea reservoirs located in rocks that are elastic (which in practical terms means rocks that are of low compressibility) throughout the entire stress regime experienced during production are not reported to have exhibited any serious geotechnical problems.

Deformations of the reservoir rock occur in response to changes in the effective stresses acting upon the reservoir during production (Jones *et al.*, 1987; Terzarghi, 1936, 1943). Reservoir-scale deformations such as compaction are a consequence of reservoir-scale effective stress change due to pressure depletion (Potts *et al.*, 1988). Deformations restricted to the near-wellbore region are normally a consequence of local pressure reduction, or drilling activity, although they can also be a secondary consequence of compaction (Jones and Leddra, 1987; Leddra and Jones, 1989).

Reservoir geotechnical problems generally have an adverse effect on the economics of the field. In most cases

*North Sea Oil and Gas Reservoirs—II*

this is because the problmes were not predicted and only recognized after the field had been put into production. Assessment of geotechnical risk would, in some cases, have avoided expensive remedial action. For example, had the compaction and seafloor subsidence at Ekofisk been predicted at the start of production, elevation of the production complex (Aam, 1988) could have been avoided.

Once a geotechnical problem has been recognized, its investigation has often been hampered by poor description of the mechanical behaviour of the reservoir rock. This is because the majority of the world's hydrocarbons occur in reservoirs that behave in an elastic manner throughout the effective stress change imposed during production. The laboratory measurement of the mechanical behaviour of reservoir rocks is often restricted to determining a compressibility under isotropic stress conditions. From this, linear elastic theory is used to define a uniaxial compressibility (Teeuw, 1971). Generally, a detailed experimental investigation (which might reveal non-linear and inelastic behaviour) is only adopted after a geotechnical problem has been identified and, although theories for the behaviour of bonded materials exist (see for example Vaughan, 1985), they have not been widely used.

Studies undertaken by the authors have shown that the non-linear behaviour exhibited by weak reservoir rocks follows a characteristic form irrespective of the rock type. This behaviour is also exhibited by other cemented (bonded) porous materials (Jones *et al.*, 1987; Addis, 1987; Vaughan, 1985; Uriel and Serrano, 1973) and has two important properties. Firstly, compactional strains can develop at nearly constant stress for an interval of strain immediately following the onset of yield. Secondly, under large shear stresses, the materials can deform at constant pore volume and constant stress (or even decreasing stress) to large strains. The authors propose that the former of these properties will be important in determining the extent of reservoir compaction whilst the latter will be important in promoting large-scale solids production.

## EXPERIMENTAL RESULTS

An extensive programme of rock deformation experiments has been undertaken to determine the nature of the deformation behaviour of porous sedimentary rocks. These experiments have been conducted on chalks and sandstones (see Yassir, 1989, for a treatment of mudrock behaviour). The experimental study has been conducted using core samples recovered from a number of North Sea reservoirs, and analogue materials from outcrops in Southern England and Denmark.

In the experiments, axial stress, confining stress, axial displacement, radial strain, top and bottom sample pore fluid pressure and, in drained experiments, volume of expelled pore fluid, were monitored. The experiments were conducted using high-pressure triaxial apparatus of the designs described by Jones (1988) and Jones *et al.* (1987).

### The experimental investigation

To investigate the compaction and shear characteristics of the sedimentary rocks under study here, two different types of experiment have been conducted using the triaxial cell. These are uniaxial strain ($K_0$) and triaxial tests. Triaxial tests have been conducted under fully drained and undrained pore pressure conditions (Bishop and Henkel, 1962). Native North Sea hydrocarbons, brines and pure water have been used as pore fluids. A limited number of isotropic stress (hydrostatic) compression experiments have also been undertaken.

Representative stress paths for the uniaxial strain experiments conducted on chalks and sandstones are illustrated in Fig. 1 in the axes of mean effective stress ($P'$) and deviatoric stress ($q$). These show a systematic variation in both cases, with the material exhibiting a high rigidity at low stress, a substantially reduced rigidity (increased restraining stress) at intermediate stress levels and an increase in rigidity as the applied stresses are further increased. This can be most clearly illustrated by considering the ratio between the rate of change of the least and greatest principal effective stresses (incremental $K_0$) defined by each section of these curves. These values are marked on Fig. 1.

The stress-strain curves that correspond to the stress paths illustrated in Fig. 1 are shown in Fig. 2. As with the stress paths, these also exhibit a characteristic form. At low stresses the stress increases rapidly with strain, and the strain is fully recoverable on unloading (Leddra *et al.*, 1989). Above a particular stress level the curves become flatter as strain increases substantially with minimal increase in stress, and at larger strains the curves exhibit work hardening. All strains developed after the initial steep section of the curves are permanent.

If the expelled volume of pore fluid is measured, as the sample compacts, in these drained experiments, and providing sample pore pressure is monitored to ensure that no excess pressures are developing, it is possible to determine the change in pore volume and hence change in porosity or voids ratio (Lambe and Whitman, 1979) directly. The relationship between porosity and the mean effective stress for the materials whose behaviour is illustrated in Figs. 1 and 2 is shown in Fig. 3. As with the stress paths and stress-strain curves, these data also exhibit three distinct regimes of behaviour. At low stress, the pore volume is relatively insensitive to increasing stress. Then, above a critical stress, a substantial reduction in pore volume occurs over a narrow stress interval. At still larger effective stresses the rate of pore volume change is seen to decrease with further stress increase.

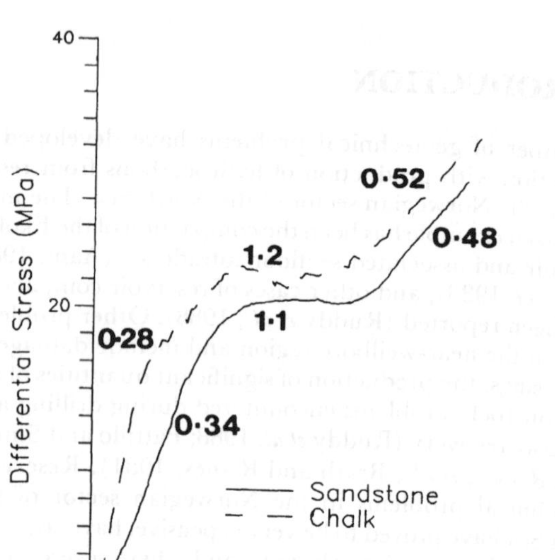


**Fig. 1.** Stress path diagram in the coordinates of differential stress-mean effective stress, showing the typical behaviour of a chalk and a weak sandstone sample deforming under uniaxial strain conditions. The curves are labelled with the stress ratio ($K_0$) values.

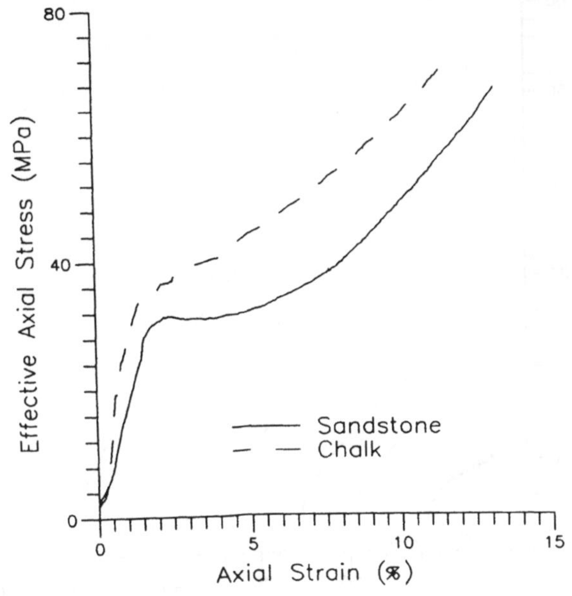


**Fig. 2.** Axial stress–axial strain curves for the two experiments described by the stress paths in Fig. 1.

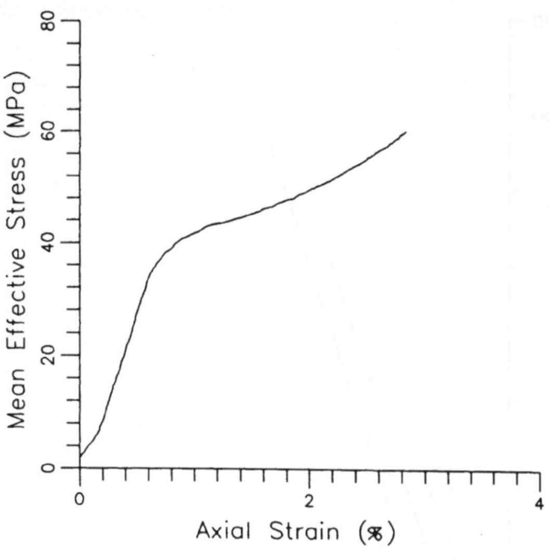


**Fig. 4.** Mean effective stress (hydrostatic stress)–axial strain recorded during an isotropic deformation experiment on a North Sea Chalk.

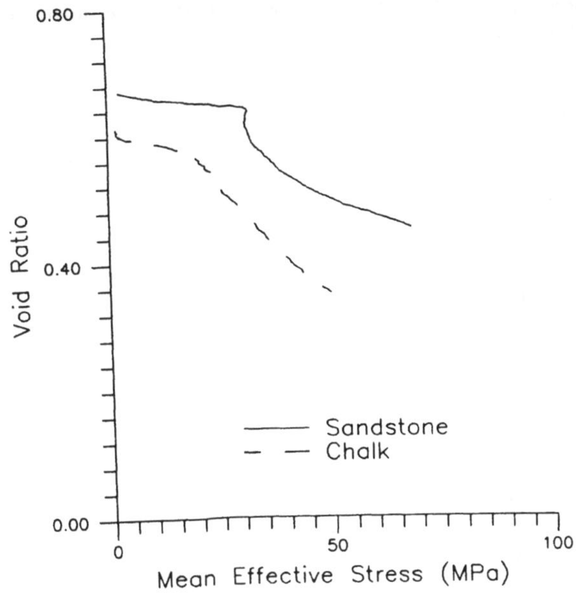


**Fig. 3.** The change in voids ratio (voids ratio = porosity/(100 − porosity)) with mean effective stress recorded during the two experiments described in Figs 1 and 2.

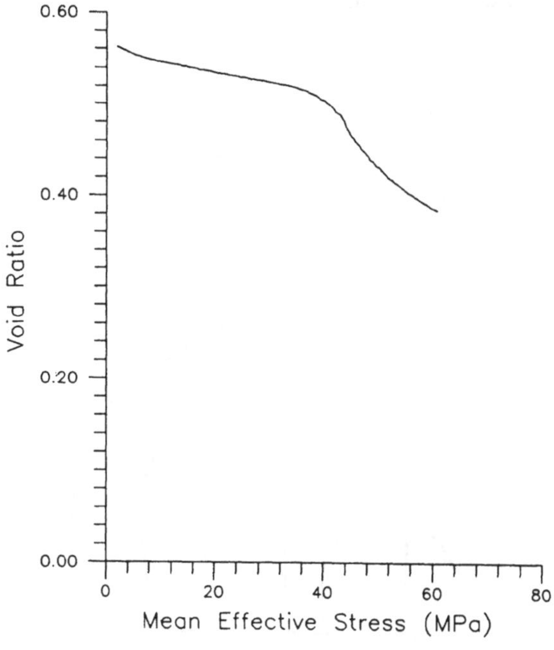


**Fig. 5.** Change in voids ratio during isotropic loading of the chalk reported in Fig. 4.

Figures 4 and 5 show the results of an isotropic stress compression experiment conducted on a chalk sample. The effective stress–axial strain plot in Fig. 4 has a similar form to the stress–strain curves in Fig. 2, and the porosity–effective stress plot (Fig. 5) is similar in form to those shown in Fig. 3.

Triaxial experiments were conducted following the method of Bishop and Henkel (1962) in which the sample was allowed to consolidate to an isotropic or $K_0$ stress state, and was then loaded axially with constant confining stress. Two different types of triaxial test have been undertaken during this study, drained and undrained. Drained experiments have been conducted on weak sandstones and undrained experiments on chalks. The stress paths for each data set are illustrated in Figs. 6 and 7 respectively. It is important to note the difference between the drained and undrained triaxial test data. In drained triaxial tests, the sample undergoes a reduction in pore volume until a critical shear condition is attained at which the pore volume remains constant. In the undrained experiments, the pore volume remains unchanged during the loading. The pore pressure increases substantially in the undrained experiments. This can be seen if sample pore pressure is plotted against the mean effective stress for the undrained chalk, triaxial experiments (Fig. 8).

Differential stress–strain curves for the chalk and sandstone triaxial experiments are shown in Figs. 9 and 10. The strain data in these curves have been normalized to zero strain at the onset of triaxial loading (all consolidation strains have been subtracted). These curves show a yield point after which deformation is nominally steady-state.

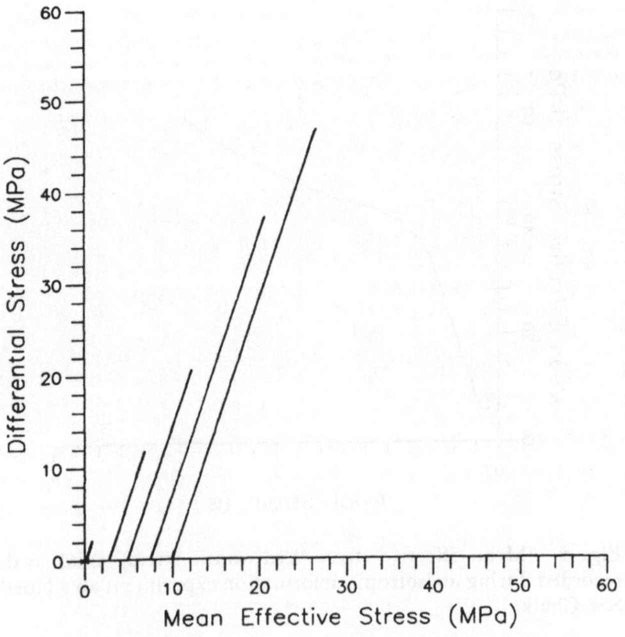


**Fig. 6.** Stress paths followed during drained triaxial deformation of a weak sandstone.

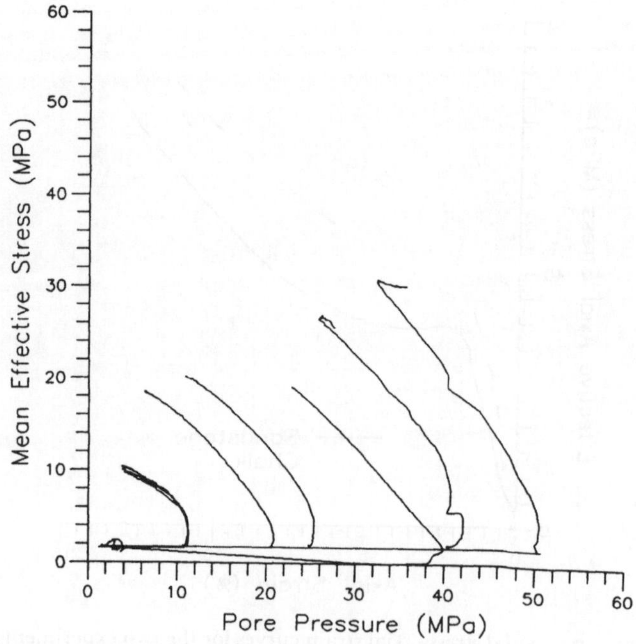


**Fig. 8.** The pore pressure response of the 37% porosity chalk during undrained triaxial loading (Fig. 7). The pore pressure in the sandstone samples (Fig. 6) remained constant throughout the experiment at 1.0 MPa, which was equal to the downstream backpressure.

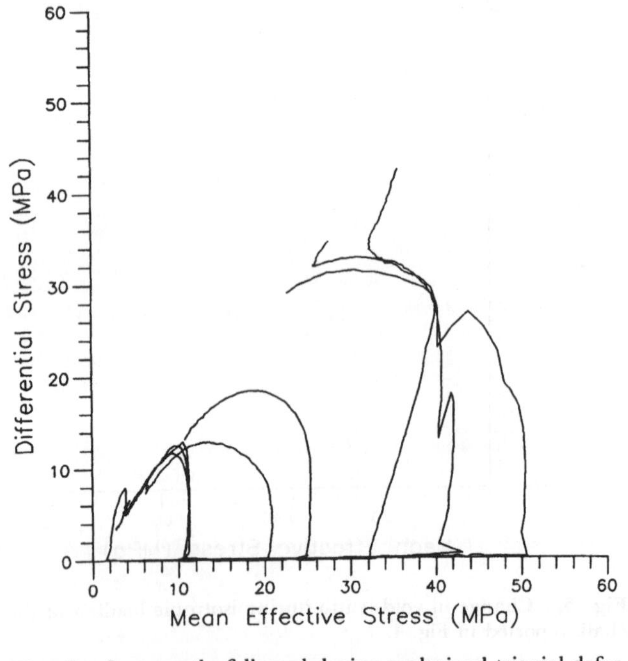


**Fig. 7.** Stress paths followed during undrained triaxial deformation of a 37% porosity chalk.

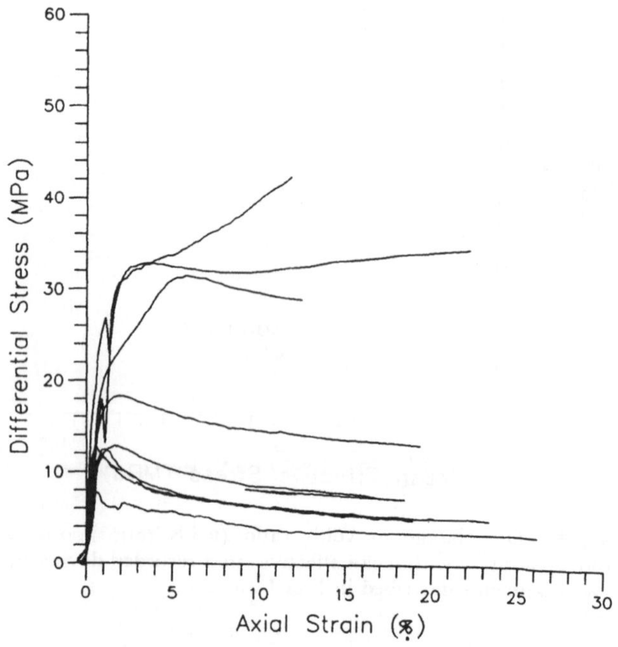


**Fig. 9.** Differential stress–axial strain curves for the 37% porosity chalk deformed under undrained triaxial conditions (Fig. 7).

Large strains are achieved, but at nominally constant pore volume—Figs. 11 and 12.

Comparison of the data from the uniaxial strain, isotropic and triaxial experiments indicates that although similarities exist between the isotropic and uniaxial strain experiments, the triaxial experiments differ in a fundamental manner. This is discussed in the following section.

## Interpretation of the experimental results

Deformations occur in response to the action of both shear and normal stresses. If a porous sedimentary rock is subjected only to the action of normal stress (as in a hydrostatic rock deformation test), the resulting deformation will be volumetric, with either the generation of an excess pore pressure, or, if a flow path is available, the flow of pore fluid from the rock. If the same sedimentary rock is subjected only to shear stress, then it will respond by changing its shape, but generally this shape change will not be accompanied by changes in pore fluid pressure or pore volume.

The two stress–deformation conditions outlined above are end members of a spectrum of stress states that a porous sedimentary rock might experience. It therefore follows that by far the majority of deformations will contain

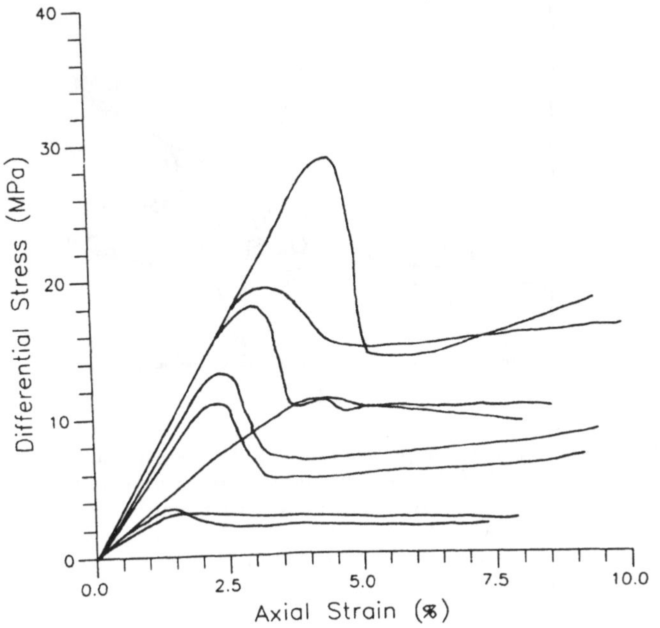


**Fig. 10.** Differential stress–strain curves for the triaxial experiments on sandstones (Fig. 6)

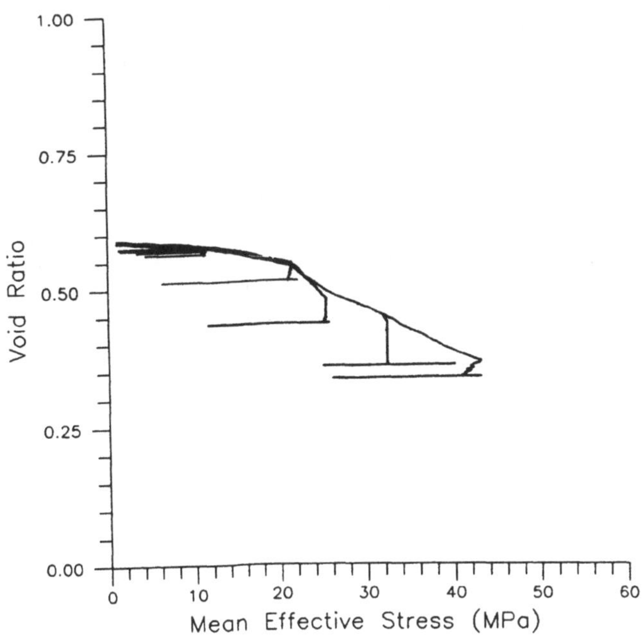


**Fig. 11.** Change in voids ratio with mean effective stress in the undrained triaxial experiments on 37% porosity chalk.

elements of volume change and distortion, because the applied stress system will contain both normal and shear stress components. The standard Mohr circle diagram (Farmer, 1983) provides a possible method of representing these different stress paths, but a more powerful alternative is to plot the mean effective stress acting upon the material against the difference between the greatest and least principal effective stresses (the differential or deviatoric stress). The paths followed during the uniaxial strain ($K_0$) and triaxial deformation experiments are shown in Figs. 1, 6 and 7. As there is no shear stress developed in hydrostatic experiments, this stress path cannot usefully be represented in either Mohr or stress-path coordinates.

The hydrostatic stress path corresponds to the condition in which the sediment experiences a equal all-round stress

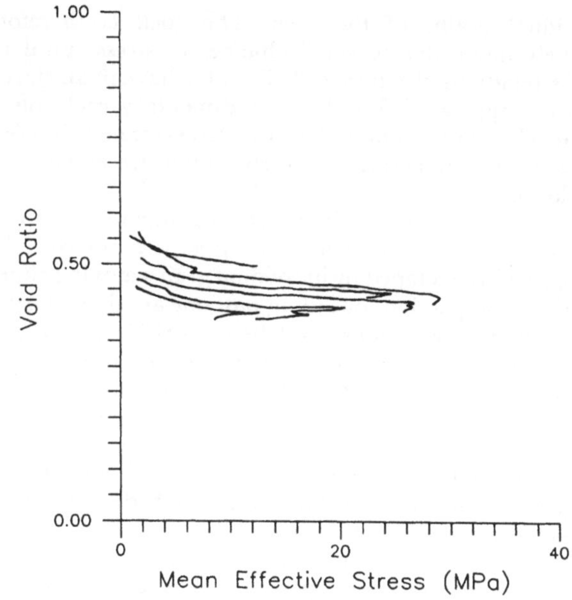


**Fig. 12.** Change in void ratio with mean effective stess for the drained triaxial experiments on sandstones.

increase. No shear stresses are developed. Under uniaxial strain ($K_0$) conditions, the greatest and least principal effective stresses are increased to permit shortening of the sample while maintaining a condition of no lateral strain. This type of deformation is always accompanied by an increase in shear stress with increasing principal stress. The stress ratio ($K_0$) in a uniaxial strain deformation is normally a constant in particulate materials, but varies in a predictable manner (Figs. 1 and 2) for bonded materials (Vaughan, 1985; Addis, 1987). The triaxial stress path (Jones *et al.*, 1987; Atkinson and Bransby, 1978) leads to failure of the test material under the maximum attainable shear stress conditions.

In Figs. 6 and 7, two different paths are shown for triaxial deformation experiments. This is because these diagrams are constructed in terms of effective stress that depends on the magnitude of the pore pressure in the sample. If the pore pressure increases, the mean effective stress decreases. The shear stress is dependent upon the difference between the greatest and least principal effective stress; its magnitude is therefore independent of pore pressure. During an undrained triaxial test, values of the mean effective stress will therefore decrease as the shear stress increases (Fig. 7), whereas in a drained triaxial test values of mean effective stress will increase as the shear stress increases (Fig. 6).

The triaxial stress paths illustrated in Figs. 6 and 7 are seen to end about a common line. This is a failure envelope, termed the critical state line in soil mechanics, where it is regarded as a state boundary between possible and impossible stress states of the material. If a stress path intersects the critical state line, it is obliged either to remain on that line or to return into allowable stress space. Any attempt to force the sample to attain stress states outside of the critical state are always resisted through deformation of the sample, which relieves the stress. Deformations of particulate materials at the critical state condition tend to be serious, and normally occur at constant stress with little or no change in pore pressure or pore volume.

The data presented in the preceding section illustrate that the reservoir rocks studied are bonded materials. This means that natural geological processes have caused mineral material to form cement bonds between the individual grains of the rock. The rock is therefore relatvely insensitive to small changes in stress (until its yield strength has been exceeded) and behave in an elastic manner (Figs. 2, 4, 9 and 10). Sedimentary rocks often remain elastic throughout the entire stress range of interest to reservoir engineering, if cement bonds are extensively developed.

While the rock is elastic, all strains imposed by stress increases are recovered when the stress is removed. The strain itself is developed by individual grains moving closer together (compaction) or sliding past one another (shear), so the elastic distortions (and hence strain energy) are concentrated in the cement bonds between grains (Fig. 13). In laboratory experiments, it is generally seen that a sample will deform elastically by no more than 2% of its original length. During compaction, this elastic shortening occurs through pore-volume reduction, either due to compression of the pore fluid or through dissipation of excess plore pressures (Figs. 2 and 3).

The distortional strain energy associated with elastic deformation of a cemented porous rock is taken up in the cement bonds betwen grains (Fig. 13). With increasing effective load, these bonds will fail. As the stress is increased, progressive failure of grain-to-grain bonds occurs, leading to a deviation from the elastic stress–strain relationship. Failure of the cemented structure occurs on all stress paths associated with increasing effective stress (including hydrostatic loading). This progressive collapse of the cemented mineral skeleton is often termed "yield" and by determining the stress conditions of the onset of yield on different stress paths, an elastic yield envelope is defined that has a characteristic form (Fig. 14).

After yield, the cemented structure of the rock is substantially destroyed. The post-yield behaviour should then be similar to the behaviour of a particulate material, i.e. deformation will be represented by a combination of pore pressure generation, pore-volume reduction and grain-boundary sliding depending on the applied stress condition (Atkinson and Bransby, 1978; Lambe and Whitman, 1979; Janbu, 1985).

The destruction of the elastic mineral skeleton in response to an increase in normal stress (even in the presence of a shear stress) occurs over a finite stress–strain interval (Figs. 1 and 2). This reflects a special form of yield

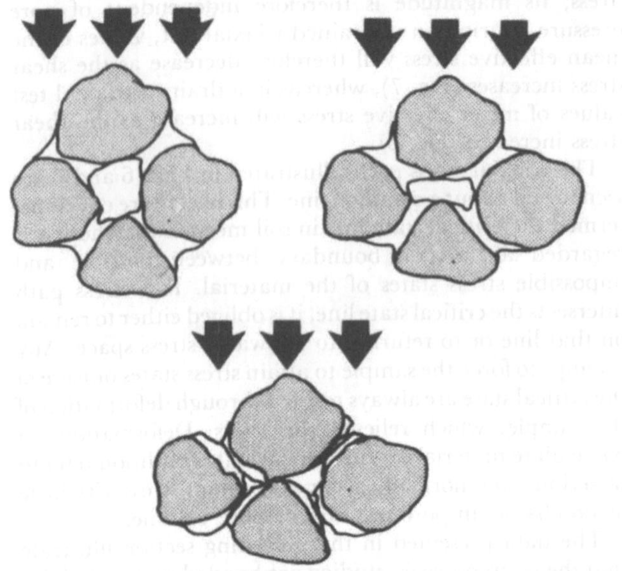

**Fig. 13.** Schematic representation of the elastic shortening and subsequent failure of cement bonds due to increasing load.

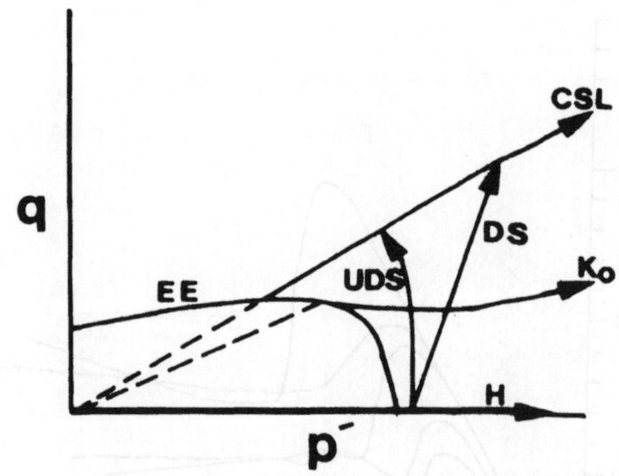


**Fig. 14.** Schematic representation of the yield surface for a bonded material. This is a projection on the differential stress ($q$)–mean effective stress ($p'$) plane. EE = elastic envelope; CSL = critical – state line; $K_0$ = uniaxial strain path; H = isotropic (hydrostatic) loading; UDS = undrained shear (triaxial); DS = drained shear (triaxial).

behaviour during which failure spreads through the material by the expansion of failed regions at the expense of the elastic rock. If the stress–strain curve and stress path are compared for a uniaxial strain experiment (Figs. 1 and 2) it is seen that the incremental $K_0$ ratio increases before the stress–strain curve flattens, so the inflexion points in these two curves do not correspond. The inflexion of the stress path reflects the onset of cement bond failure, whereas the inflexion in the stress-strain curve corresponds to the condition in which a majority of the cement bonds in the material have been severed. This latter condition is accompanied by the rapid decrease in pore volume (Figs. 2 and 3) often termed pore collapse. Yield therefore occurs over a finite stress–strain interval, which corresponds to the horizontal section of the $K_0$ stress path. The extent of this interval depends on the extent of the cementation (Fig. 15) such that samples with the lowest predeformation porosity have the highest yield strength and the greatest yield stress interval. Further compaction under higher stresses results in a decrease in the rate of pore volume change.

Only in highly permeable materials will effective stress change, and the corresponding strain, be achieved rapidly. In experiments conducted on low-permeability sediments, a finite time is necessary for excess pore pressures, created by the deformation, to dissipate. This process is termed primary consolidation (Atkinson and Bransby, 1978; Lambe and Whitman, 1979) and is well known for particulate materials. It should be noted that complete primary consolidation need not lead to an equilibrium strain condition because the majority of natural materials undergo additional creep deformations (termed secondary consolidation) (Janbu, 1985). The recognition of porosity/effective stress relationships for bonded materials (Vaughan, 1985; Addis, 1987) has led to the concept of a porosity trend line (Smits *et al.*, 1986) in which otherwise identical samples of different predeformational porosity compact to become asymptotic to a common line (Fig. 16). This approach provides a convenient way of treating non-linear deformation in reservoir compaction studies (Johnson *et al.*, 1988) and is similar to the void ratio–log mean effective stress plots of soil mechanics (Lambe and Whitman, 1979). As the quality of experimental data has improved, this concept is seen to be oversimplistic (Ruddy *et al.*, 1988). It is probable that for materials that exhibit any degree of sample heterogeneity, the porosity trend line is an approx-

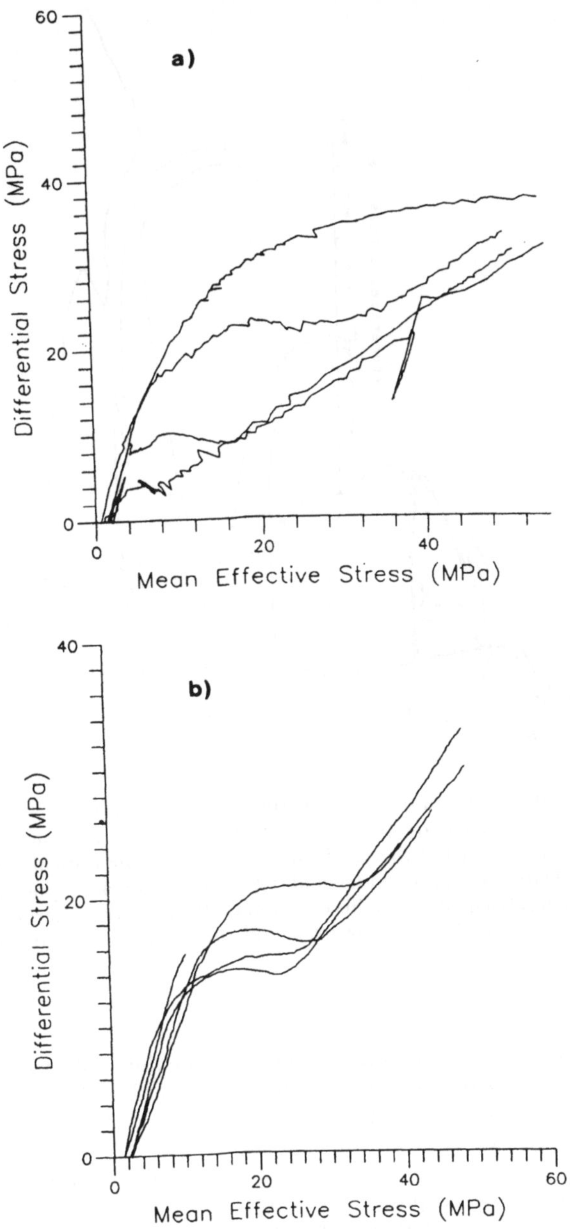


**Fig. 15.**(a) A compilation of $K_0$ stress paths for North Sea chalk, showing the limiting elastic and consolidation paths for $K_0$ deformation, a consequence of variation in predeformational porosity. (b). A similar compilation of $K_0$ stress paths for a weak North Sea sandstone.

imation. This approach to using stress-strain data is therefore only valid if the samples are strictly identical in all ways except for their predeformational porosity. Such a condition is unlikely to be widely realized by any natural bonded material.

## The yield surface

Terzarghi (1936, 1943) identified that changes in effective stress in soil were always accompanied by changes in pore volume. This concept has been refined to provide theories of yield and failure for soil materials such as the critical-state theory developed by the Cambridge Soil Mechanics Group (Atkinson and Bransby, 1978) in which failure is defined not only by a state of shear and normal stress but also by a specific pore volume.

As pore volume is dependent upon effective stress, the stress path diagram (Fig. 14) can be further refined. The

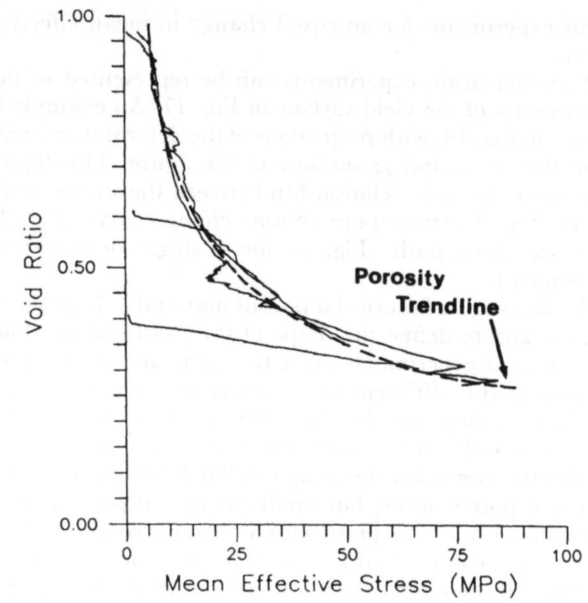


**Fig. 16.** Porosity trendline defined by four deformation experiments conducted on samples of North sea chalk that exhibit different predeformational porosities.

diagram as presented is a projection of a more complex figure with the axes of mean effective stress/shear (deviatoric) stress/pore volume onto a mean effective stress/shear stress plane. This relationship is illustrated in Fig. 17, which defines a yield surface of complex geometry separating possible from impossible stress states.

The shape of the yield surface illustrated in Fig. 17 indicates that for normal compactions (which occur on the end cap), the magnitude of the pore volume change will also depend on the shear stress. Large shear stresses overcome frictional resistance to grain-boundary sliding promoting larger pore volume changes for a given increase in mean effective stress. Loss of pore volume is therefore greater in uniaxial strain experiments than in isotropic

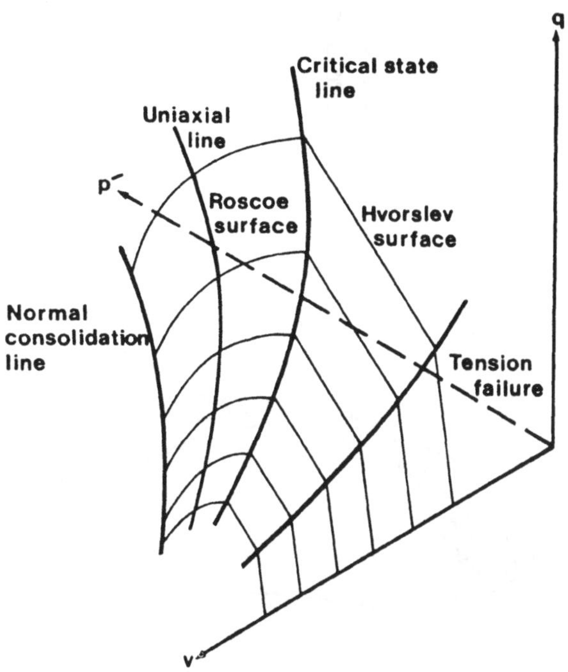


**Fig. 17.** Schematic representation of yield surface diagram (after Atkinson and Bransby, 1978). The axes are differential stress ($q$), mean effective stress ($p'$) and pore volume ($v$).

stress experiments for an equal change in mean effective stress.

Uniaxial strain experiments can be represented in the coordinates of the yield surface in Fig. 17. An example is shown in Fig. 18, with projections of the deformaton curve onto the *xy*, *xz* and *yz* surfaces of the figure. This figure illustrates the basic relationship between the stress–strain curve (Fig. 2), stress–pore volume change curve (Fig. 3) and the stress path (Fig. 1) for a single deformation experiment.

To investigate failure of a porous material at high shear stresses and to define the shape of the yield surface (Fig. 17), triaxial experiments must be conducted on samples consolidated to different effective stresses prior to triaxial loading (Bishop and Henkel, 1962). These experiments terminate at the stress conditions defining the critical state. Under this condition the sample will deform at nominally constant pore volume, but small changes in pore volume due to pore fluid compressibility or sample dilatancy will cause the shear stress at failure to increase or decrease in a consistent manner (strain-softening or strain-hardening responses) (Fig. 7). In either case, deformation at the critical state will lead to large strains at constant, or nearly constant, stress with little, or no, change in porosity. The deformation mechanism active at the critical state is frictional sliding of grains, either pervasively or within a discrete failure plane in the sample.

The experiments illustrated by the stress paths shown in Fig. 7 can also be represented in the coordinates of the yield surface (Fig. 19). This diagram shows the geometry of the failure surface, and also, if compared with Fig. 18, the major differences in the deformation paths followed during uniaxial strain and undrained triaxial experiments. The most significant difference pertains to the fact that the deformation plot for a uniaxial strain experiment contains the stress–strain curve, because the strain is accomplished by decrease in porosity whereas the deformation plot for the undrained triaxial experiments gives no indication of the strain magnitude. (The deformation occurring at nominally constant pore volume.) Despite these differences, it is also clear that the data in Figs. 18 and 19 belong to a common dataset and are simply following different paths on or below the same yield surface.

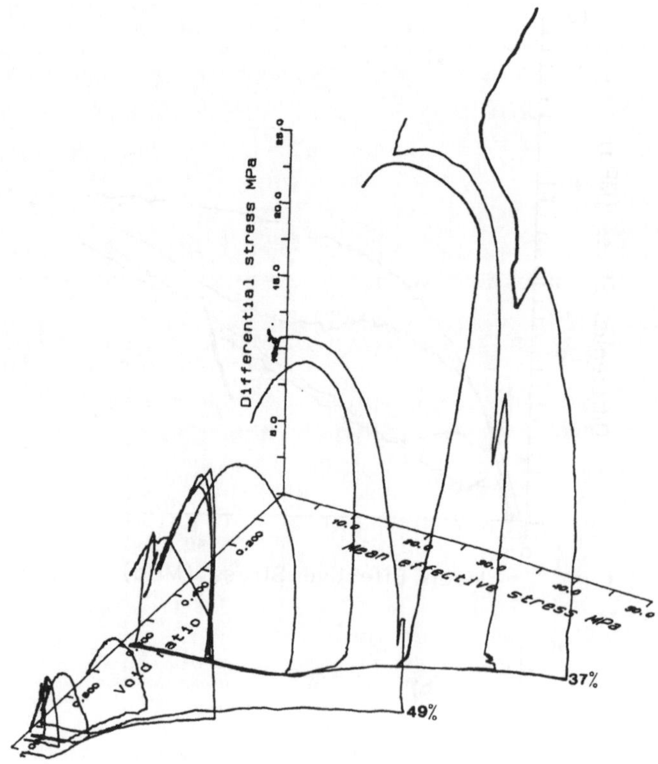


**Fig. 19.** Deformation plot of the undrained triaxial experiments conducted on the 37% porosity chalk. This shows the geometry of part of the yield surface, includng the critical-state line. Data for undrained triaxial deformation of a 49% porosity chalk has been added to this diagram to illustrate the divergence of the yield surfaces at low stress and the common critical-state line.

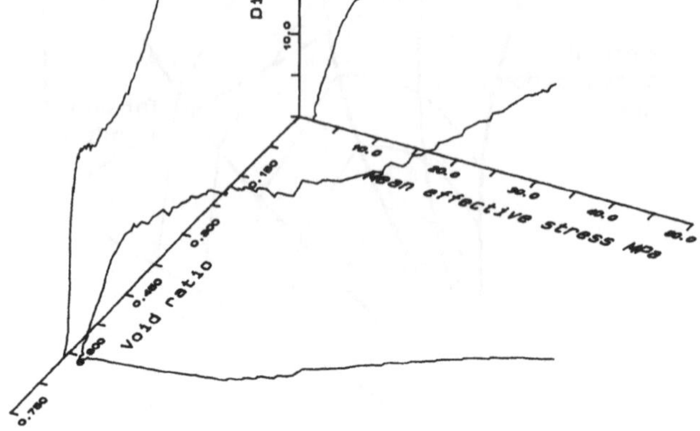


**Fig. 18.** Deformation plot of a uniaxial strain experiment conducted on a uniaxial strain experiment conducted on a North Sea chalk. The deformation curve is projected onto the three axial surfaces of this diagram.

The yield surface (Figs. 17, 18 and 19) is a state boundary surface between possible and impossible stress states. All deformations are therefore constrained to occur on this surface or within the space it encloses (Fig. 17). All deformations that are accompanied by changes in either pore pressure or pore volume, and all post-yield compactions, occur within this figure, or, more normally, on, or near, the Roscoe Surface (end cap). (Deformations of extremely overconsolidated materials may occur on the Hvorslev Surface, but a treatment of this is beyond the scope of this present contribution.) Major shear deformations occur only if the material is loaded in such a manner that it attains the critical state. Here materials show a great sensitivity to stress change and may deform under conditions of steady-state stress to large strains. Such a situation can lead to run away catastrophic failure of the type associated with liquefaction.

The yield surface diagram (Fig. 17) can be defined by taking a number of identical samples and deforming them on different stress paths. It therefore reflects the possible stress states that a material with a single predeformational porosity can attain. It is also the state boundary for that material in its post-yield particulate state. Two important aspects of reservoir rocks have therefore been ignored in the preceding discussion. Firstly the fact that they are elastic whilst their cement bonds are intact, and secondly that they exhibit a range of predeformational porosities so that a yield surface, defined for a material with a single predeformational porosity is of limited practical use. This is discussed in the following section.

## MATERIAL BEHAVIOUR

Materials such as the sandstones and chalks investigated during this study show variation in material properties both in response to differences in the extent of cementation (predeformational porosity (Figs. 14 and 16), and owing to changes in porosity caused by the applied stress system (Figs. 18 and 19). The behaviour is also influenced by the presence of the cement bonds and the change in state of the material that accompanies yield and the destruction of cementation. Any unified model of material behaviour for reservoir rocks must take into account these different aspects of material behaviour.

In Fig. 19, the results of undrained shear experiments conducted on two different chalks (predeformation porosities of 37% and 49%) are illustrated. These show that the yield surfaces for the two materials are separate, but that they share the same critical state line. The explanation for this is simply that at low stress the materials are elastic and exhibit quite different strength characteristics. After yield they are still different insofar as they support different pore volumes, but they are now essentially the same particulate material. As shear stress increases, they must therefore both fail on the same critical state line. Similarly, their consolidation paths under $K_0$ conditions should converge (Figs. 15 and 16).

The influence of the low-stress elasticity can be seen in the data in Fig. 20. In this diagram, the undrained triaxial test stress paths from Fig. 7 have been replotted with the stress paths from two $K_0$ experiments that were conducted on other samples of the same chalk. The positions of the elastic limit and the critical state line are also shown. In this diagram it is seen that in, both the lightly consolidated undrained triaxial tests, and the $K_0$ experiments, the material is taken to stress states outside of the critical state.

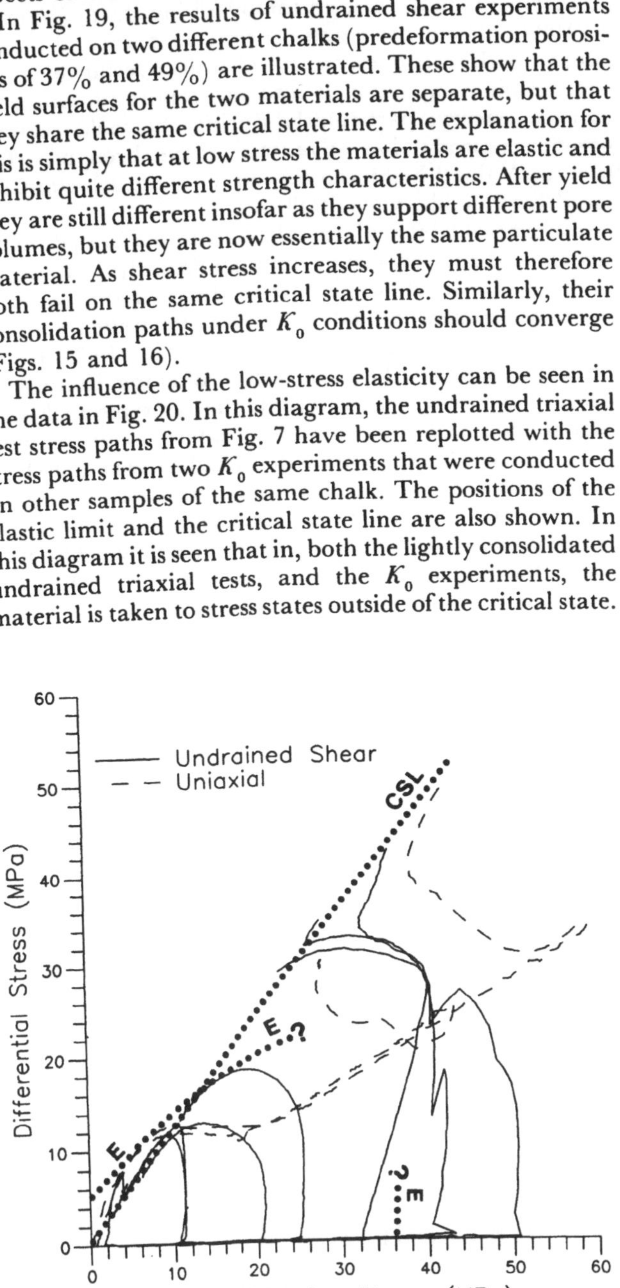


**Fig. 20.** Relationship between the $K_0$ stress paths and the undrained triaxial stress paths for the 37% porosity chalk. Both $K_0$ experiments were terminated by undrained triaxial loading. A section of the elastic envelope and the critical-state line have been marked on this drawing.

In the case of the $K_0$ experiments, the path returns into permissible stress space for the post-failure material, crossing the critical-state line at a point close to yield, and then continues in the linear manner typical of a particulate material. In the case of the triaxially loaded material, after attaining the elastic limit, the stresses drop back to the critical state, and all subsequent deformation is then constrained by stress conditions corresponding to the critical state line.

The data in Fig. 20 indicate that where a reservoir rock yields elastically under high shear stresses and low normal effective stresses, outside of the critical state, it will deform in a dramatic manner to relieve stress and attain the critical state condition. At the moment of yield, it will become extremely weak and a "runaway" failure of a liquefact type will occur. The same material under uniaxial strain conditions will respond to a change in effective stress (such as a decrease in pore fluid pressure) by compacting under stress/pore volume conditions that cause it to pass extremely close to the critical state condition during yield. This is also serious, because a small stress excursion such as may occur near an injector well, or a well that is being drawndown after a shut-in period, or local variations in pore volume, could promote a critical-state strain-softening failure in the material. It has been our experience that during the incremental $K_0 = 1.0$ regime of compaction behaviour it is very easy for a failure to develop in an experimental sample; this we believe is a consequence of the proximity of $K_0$ and critical-state conditions during yield. Compacting reservoir rocks may then experience serious mechanical instabilities when fluid pressure drawdown takes them through yield in the $K_0$ deformation regime. Promotion of elastic failure of such rocks, at high shear stresses, in the presence of high pore fluid pressures (low normal effective stresses), would promote an even more serious failure.

The samples in the two $K_0$ deformation experiments reported in Fig. 20 were both subject to undrained shear following $K_0$ compaction. These two stress paths both end on the critical-state line, indicating that the critical state is a material property that is independent of the route used to attain it. Once the critical state was attained, both of the samples strain hardened. This is typical of all of the chalks that were consolidated to stresses beyong their elastic limit prior to loading under undrained triaxial conditions.

## DISCUSSION AND CONCLUSIONS

### Material behaviour

The material deformation model described above identifies three regimes of deformation in response to increase in effective stress, elastic shortening, yield (pore collapse) and compaction. The model also identifies that a limit exists to allowable effective stress states. This limit is defined by the onset of deformation that is largely through shear. The phenomenological expression of critical state shear depends on the stiffness of the sample. If the sample is weak (low mean effective stresses and large pore volume), a high density of ductile shear zones will be established and the sample will behave in a liquefact manner (Leddra and Jones, 1989). Alternatively, if it is strong (large mean effective stress and small pore volume), the critical state condition may be realized by the development of a plastic behaviour. This behaviour is most pronounced if the sample has reached the elastic limit during the initial hydrostatic or $K_0$ loading. The form of the shear failure will therefore be dependent upon either the magnitude of the predeformational porosity or the magnitude of the compactional strain prior to shearing. In low-porosity materials under very low mean effective stresses, and low total stresses, the critical state may be represented by a single shear fracture.

The major geomechanical conclusion of this study can be summarized as follows. Reservoir chalks and porous sandstones behave in a similar manner, and both deform in a manner that can be taken as typical of bonded materials. The major similarities in behaviour can be summarized as follows

(1) The materials are bonded and behave in an elastic manner at low stresses.
(2) The material will yield at a stress that is dependent upon porosity (assuming that data for samples of the same composition deformed on the same stress path are compared).
(3) The data also show that the compressibility of the deforming samples increase substantially during yield.
(4) For samples deforming under hydrostatic or uniaxial strain conditions, the increased compressibility is associated with a breakdown of the cemented structure of the sediment.
(5) The data show that in hydrostatic and uniaxial strain experiments, three distinct regimes of stress–strain behaviour are evident. The three regimes of behaviour can be described as elastic, yield (pore collapse) and compaction.
(6) The elastic and compaction regimes of behaviour are fundamental properties of the material in its pre-yield and post-yield states. The post-elasticity strain equilibration regime is the transfer of the strain state between these two different material conditions.
(7) The materials exhibit an elastic yield envelope that closes to the zero shear stress condition with increasing mean effective stress. At low stresses, this yield envelop may be considered as constituting part of the boundary between permissible and impossible stress states. At higher stresses, this state boundary is replaced by the critical-state line.
(8) Post-yield behaviour is constrained to occur on, or inside, a figure in shear stress–mean effective stress–pore volume space. This figure need have no relationship with the elastic yield surface although they have a similar geometry in the same stress–pore volume co-ordinates.

### Applications to reservoir engineering

If production of hydrocarbons from a reservoir leads to a reduction in the fluid pressure, then the reservoir will deform. This deformation will generally be laterally constrained so that as a first approximation reservoir compaction can be regarded as occurring on the $K_0$ stress path. If the reservoir rock is of sufficient strength that it remains elastic throughout this deformation, the strains will tend to be small, and serious problems should not develop. However,if the reservoir rock enters into the yield regime, strains will become much larger and more serious problems of reservoir compaction may occur. The deformation mechanism responsible for this compaction is primarily pore volume reduction, so that compaction may be accompanied by permeability reduction. (Some of the weaker Brent Sandstones show pronounced permeability loss as they pass through yield (Goldsmith, 1989), and a similar but less pronounced effect has been reported for the chalk (Leddra *et al.*, 1989). Compaction in reservoirs where the rock has been loaded beyond its elastic limit will therefore lead to possible problems of permeability loss, and, if the Ekofisk experience is repeated, sea floor subsidence.

As the reservoir rock yields, in areas of stress variation (such as might occur in the vicinity of wells undergoing drawdown, dilatant fractures and injection wells) there is a real possibility that the reservoir rock may attain the critical state. The effect of this will be that wells and fractures may be invaded by the sediment flowing into available spaces, and, where such spaces are not available, the rock may remain in a weak condition until penetrated by another well. Here the weak state of the rock will lead to poor core recovery and instability of the bore holewalls.

Even if the reservoir rock does not attain the critical state, vertical compactional strain developing in a reservoir will cause a complicated distribution of strains and displacements to develop in the overlying strata (Jones *et al.*, 1989). These deformations may lead to well-casing failures, both in the reservoir and in its overburden.

Reservoir compaction is also an important source of reservoir energy, particularly during the yield regime. Significant production advantages have been reported from the Valhall Field as a consequence of such compaction drive (Ruddy *et al.*, 1988).

In near-wellbore regions, adjacent to perforated intervals, the stress system can deviate significantly from the $K_0$ condition. Under these conditions, the reservoir rock will experience stresses equivalent to either the critical state of the corresponding elastic limit (Risnes *et al.*, 1982a,b). If the pore volume is large, a strain-softening (liquefact) failure may occur. In this event, the reservoir rock will become mobilized and flow to an environment of lower stress (the wellbore). This type of failure will therefore manifest itself by large-scale solids production. Liquefact failure of the reservoir rock is most likely to occur when a previously shut-in well is rapidly drawn down. This will change the tangential and radial total stresses acting upon the adjacent reservoir, increase the shear stress acting in the rock adjacent to a perforation, and may lead to a pressure different between the wellbore and the reservoir pore fluid. This is an important mechanism for substantial solids production in both the Chalk and Brent Sandstone reservoirs (Leddra and Jones, 1989). It should be noted that the data in Figs. 18, 19 and 20 indicate that a reservoir rock prone to solids production is also a string candidate for compaction. Permeability loss, well-casing failures and even surface subsidence could follow if a horizon prone to sand production is also subject to widespread pore-pressure reduction.

The examples given in the preceding paragraphs indicate that detailed analysis of the geomechanics of the reservoir rock as part of the production planning for a reservoir has major economic advantages. The data find important application in petrophysical interpretation, reservoir simulation, the design of offshore structures to mitigate subsidence effects and in the design of well completions in weak reservoir rocks.

## ACKNOWLEDGEMENTS

The authors wish to thank NPD for sponsorship of a number of projects that have led us to develop the material model presented here and for their sponsorship of the geotechnical database that has been central to this work. Thanks are also due to Imperial College Soil Mechanics for providing some of the experimental facilities. Valuable discussions with Dave Edwards, Najwa Yassir,

Doug Rhett, Christian Schroder and Knut Pederstad are also acknowledged. Finally we thank Tony Addis whose work and constructive discussions so influenced attitudes and understanding in our research group. Two experimental results form his unpublished thesis have been used in Fig. 20.

## REFERENCES

Addis, M.A. (1987). Mechanisms of sediment compaction responsible for oil field subsidence. Unpublished Ph.D Thesis, University of London.

Aam, K. (1988). Ekofisk subsidence: The problem, the solution, and the future. In: *Proceedings of the International Conference on Behaviour of Offshore Structures, Trondheim, Norway, June* 1988, p. 19–40.

Atkinson, J.H. and Bransby, P.L. (1978). *The Mechanics of Soils: An Introduction to Critical State Soil Mechanics*, McGraw-Hill, London.

Barton, N., Makurat, A., Harvik, L., Vik, G., Bandis, S., Christianson, M. and Addis A. (1988). The discontinuum approach to compaction and subsidence modelling as applied to Ekofisk. In: *Proceeding of the International Conference on Behaviour of Offshore Structures, Trondheim, Norway, June 1988*, pp. 129–142.

Bishop, A.W. and Henkel, D.J. (1962). *The Measurement of Soil Properties in the Triaxial Test*. Edward Arnold, London.

Bratli. R.K. and Risnes, R. (1981) Stability and failure of sand arches. *Soc. Petroleum Eng.* (April), 236–248.

Farmer, I (1983) *Engineering Behaviour of Rocks*, 2nd ed. Chapman and Hall, London.

Goldsmith, A.S. (1989). Permeability decline and compressibility in sandstone reservoir rocks. In: V. Maury and D. Fourmaintraux (eds.), *Rock at Great Depth*, (Vol. 2), Balkema, Rotterdam, pp. 923–928.

Janbu, N. (1985). Soil models in offshore engineering. 1985 Rankine Lecture. *Geotechnique*, **35** (3), 241–281.

Johnson, J.P. and Rhett, D.W. (1987). Compaction behaviour of Ekofisk Chalk as a function of stress. SPE paper 15872, *SPE European Petroleum Conference, London, 20–22 October, 1986*.

Johnson, J.P., Rhett, D.W. and Seimers, W.T. (1988) Rock mechanics of the Ekofisk Reservoir in the evaluation of subsidence. *Offshore Technology Conference, Houston, Texas, OTC5621*, pp. 39–50.

Jones, M.E. (1988) Determination of the Mechanical Properties of Reservoir Rocks using the Triaxial Test: Experimental Guidelines. *Norwegian Petroleum Directorate, YA-524, Stavanger*.

Jones, M.E. and Leddra, M.J. (1987). Ground motions and seismicity due to fluid production from subsurface reservoirs. *Mem. Geol. Soc. China, N.9.* (December), 465–494.

Jones, M.E., Leddra, M.J. and Addis, M.A. (1987). *Reservoir Compaction and Sea floor Subsidence Due to Hydrocarbon Extraction.* Offshore Technology Report, OTH 87 276. HMSO, London.

Jones, M.E. Leddra, M.J. and Potts, D. (1989). Ground motions due to hydrocarbon production from the chalk. *International Chalk Symposium*, September. Preprint No. 59, Thomas Telford, London, pp. 341–347.

Lambe, T.W. and Whitman, R.V. (1979). *Soil Mechanics, S.I. Version*, 2nd edn. Wiley, New York.

Leddra, M.J. and Jones, M.E. (1989). Steady-state flow during undrained loading of the Chalk. *International Chalk Symposium*. Preprint No. 18, Thomas Telford, London, pp. 117–124.

Leddra, M.J., Pederstad, K., Lønøy, A. and Jones, M.E. (1989). The influence of increasing effective stress on the permeability of chalks under hydrocarbon reservoir conditions. *International Chalk Symposium*, September. Preprint No. 19, Thomas Telford, London, pp. 125–131.

Patillo, P.D. and Smith, M.B. (1982). The effect of formation flow on the integrity of perforated casing. *SPE 11123, presented at the SPE Fall Meeting, New Orleans, September*, 1982.

Potts, D.M., Jones, M.E. and Berget, O.P. (1988) Subsidence above the Ekofisk Oil Reservoirs. *Proceedings of the International Conference on Behaviour of Offshore Structures, Trondheim, Norway, June 1988*, p. 113–128.

Risnes, R., Bratli, R.K. and Horsrud, P. (1982a). Sand stresses around a wellbore. *Soc. Petroleum Eng. J.* (December), 883–898.

Risnes, R., Bratli, R.K. and Horsrud, P., (1982b). Sand arching – a case study. *European Petroleum Conference, London October, 1982.* Paper EUR 310.

Ruddy, I., Anderson, M.A., Pattillo, P.O., Bishlawi, M. and Foged, N. (1988) Rock compressibility, compaction, and subsidence in a high-porosity chalk reservoir: A case study of Valhall Field. *SPE paper 18278, SPE 63rd Annual Technical Conference, Houston, October*, 1988, pp. 179–186.

Smits, R.M.M., De Waal, A., and Van Kooten, J.F.C. (1986). Prediction of abrupt reservoir compaction and surface subsidence due to pore collapse in carbonates. *SPE paper 15642, SPE 61st Annual Technical Conference, New Orleans, October, 1986*, pp. 1–11.

Teeuw, D., (1971). Prediction of formation compaction from laboratory compressibility data. *Soc. Petroleum Eng. J.* (Sept.), 263–271.

Terzarghi, K. (1936). The shearing resistance of saturated soils and the angle between the planes of shear. *International Conference on Soil Mechanics and Foundation Engineering 1, Cambridge, Massachusetts, Proceedings*, Vol. 1, pp. 54–56.

Terzarghi, K., (1943). *Theoretical Soil Mechanics*. Wiley, New York.

Uriel, S. and Serrano, A.A. (1973) Geotechnical properties of two collapsible volcanic soils of low bulk density at the site of two dams on the Canary Islands *Proceedings of the 8th International Conference on Soil Mechanics and Foundation Engineering, Moscow, 1973*, pp. 251–264.

Vaughan, P.R. (1985). Mechanical and hydraulic properties of in situ residual soils. *1st International Conference in Geomechanics of Tropical and Saprolitic Soils, Brazilia, February*, 1985.

Yassir, N. (1989). Mud volcanoes and the behaviour of overpressured clays and silts. Unpublished PhD Thesis, Universty of London.

# 10

# The Effect of Low-Permeability Layers on Oil Production from Vertical and Horizontal Wells in the Troll Field

*L. Sundal*[a], *C.A. Kossack*[b] *and J. Kleppe*[b]

[a]Esso Norge a.s., Forus, Norway
[b]Norwegian Institute of Technology, Trondheim, Norway

The production of oil from the Troll Field in the North Sea has been studied numerically by Kossack, Kleppe and Aasen (1987). The reservoir description used in that study contained low-permeability layers of cemented sandstone at the gas–oil contact and below the water–oil contact. The lateral extent and continuity of these cemented layers and the exact value of their vertical permeability is unknown. As a result, the comparisons between the vertical and the horizontal well performance in the Troll Field as presented by Kossack *et al.* (1987) is only valid for the particular description used.

This contribution investigates the effect these calcite-cemented sandstone layers near horizontal and vertical wells have on the coning of gas and water, during oil production from thin oil zones. The vertical permeability of the calcite-cemented layers was varied from 0 mD (totally sealing) to 1800 mD (not present). The coning situation was studied at various oil production rates and with four different well configurations. Under certain conditions a secondary gas cap forms under the low-permeability layer at the gas–oil contact. At low and intermediate production rates the effect of this secondary gas cap on the coning result is surprising. A complicated interaction between gas coning from this secondary gas cap and the flow of primary gas cap gas through the layer was discovered as the production rate and vertical permeability of the layer were varied. An extensive analysis of these results explains the physical processes involved.

This phenomenon is of general interest in coning problems where low-permeability layers are present. The results can be applied to fields other than the Troll Field and to production situations involving both vertical and horizontal wells.

## INTRODUCTION

The thin oil rim found in the Troll Field contains considerable amounts of oil, 5700 Mstb* ($\approx$900 $MSm^3$) owing to its large areal extent. The oil overlays an active aquifer, and is in contact with a large gas cap. Owing to a thickness of less than 92 ft (28 m), the critical rate to coning is low considering the offshore environment. Producing the field with conventional wells is most likely not economical because of the gas coning. Experiences with horizontal wells have shown good results in thin oil zones, particularly when coning is a problem (Chaperon, 1987; Darley *et al.*, 1983; Ekrann, 1987; Johansen, 1986; Karcher *et al.*, 1986; Kossack *et al.*, 1987; Reiss, 1985). The relatively lower pressure drawdown required to produce a certain rate from a horizontal well compared to a vertical will reduce the tendency to coning.

In the paper of Kossack *et al.* (1987), some reservoir-management aspects of oil production from the Troll Field are numerically investigated. Production from different well configurations of horizontal and vertical wells are compared.

Their reservoir description contained two thin, continuous low-permeability layers of cemented sandstone. Laboratory studies of these layers estimate the vertical permeability to be of the order of 0.1 mD, but their exact permeability and their lateral extent are unknown. This geological feature is a dominant factor in the well's performance, thus, the results calculated by Kossack *et al.* (1987) are valid only for the assumed description. Their results show that a horizontal well will perform better than 2 or 3 vertical wells when low-permeability layers ($k_z$ = 0.1 mD) exist at the gas–oil contact (GOC) and below the water–oil contact (WOC). This result is also dependent on the production rate and the length of the horizontal well.

This paper presents the results of evaluation of the effect

* Prefix M implies millions.

*North Sea Oil and Gas Reservoirs—II*

of these impermeable or low-permeability cemented layers on the near-well coning of gas and water into vertical and horizontal wells in Troll Field. Since the existence, lateral extent, and vertical permeability of these layers is uncertain, in the analysis the vertical permeability of the layers was given the values of 0 mD (totally sealing), 0.1 mD, 1.0 mD, and 1800 mD (not present).

## REVIEW OF PREVIOUS STUDIES OF HORIZONTAL WELLS IN THE TROLL FIELD

An extensive study has been published recently by Kossack *et al.*, (1987) that compares oil production from vertical and horizontal wells in the Troll Field. The reservoir model used in this study contained two thin (1.6 ft), low-permeability layers (LPL) ($k_H = 10$ mD, $k_z = 0.1$ mD). These cemented sandstone layers were located at the gas–oil contact and 6.6 ft (2 m) below the water-oil contact. The results of their study, summarized in this section, were highly dependent on the low vertical permeability of these two layers. This dependence is extensively investigated in the remainder of this paper. The basic rock properties, PVT data and reservoir description are given in their paper.

The basic objective of Kossack *et al.*'s (1987) study was to compare the oil production performance of 1, 2 and 3 vertical wells and a large-diameter horizontal well in a sector of a circular drainage area (full pattern). This configuration is most easily explained using Fig. 1, where six 60° sectors are shown in the full pattern. In their numerical study a 3-dimensional, finite-difference grid was constructed around 1, 2 or 3 vertical wells or a horizontal well in a symmetry element of the sector; see Fig. 2. In most of the simulations, the oil production rate for a sector was held constant with time. They investigated sector production rates from 4000 to 24 000 stb of oil per day. Note that the sector production rate is the sum of oil produced by all wells in a sector. The oil recovery results at shut-in gas–oil ratio (GOR) limit of 5000 scf/stb are summarized in Fig. 3. At low sector production rates, the

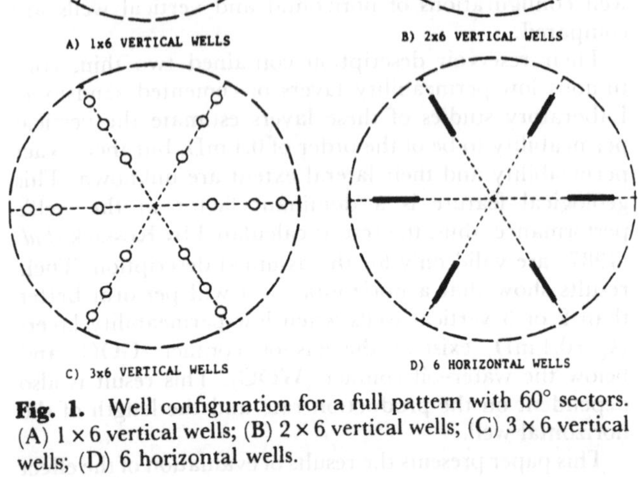


**Fig. 1.** Well configuration for a full pattern with 60° sectors. (A) 1 × 6 vertical wells; (B) 2 × 6 vertical wells; (C) 3 × 6 vertical wells; (D) 6 horizontal wells.

1500 ft (460 m) and 2000 ft (610 m) long horizontal wells produced much more oil than the 1, 2, or 3 vertical wells. As the sector production rate increased to 8000 stb/d, the recovery from the shorter horizontal well became less than from 3 vertical wells. Eventually, as the rate increased further, the performance of the shorter horizontal well equalled that of 2 vertical wells. The 2000 ft horizontal well performed much better than all the vertical well configurations for all production rates studied.

During these simulations the average water cut for all well configurations was between 0.36 and 0.40. The authors evaluated the sensitivity to the sector angle. The results are given in their paper. Economics, i.e. horizontal well drilling costs, operating costs, and oil price, would specify which sector angle has the highest present value.

In their final simulations Kossack *et al.* (1987) produced the horizontal wells (60° sector, 1500 ft length) in a more standard way, i.e. by reducing the required oil production rate by 50% every time the GOR rose above the solution gas value. These results are shown in Fig. 4, where schedule A is a vertical well producing 1000 stb/d of oil, schedule B is a horizontal well at 1000 stb/d, schedule C is a horizontal well producing in the sequence 8000, 4000, 2000, 1000 stb/d, and schedule D is a horizontal well with the sequence 4000, 2000, 1000 stb/d. Schedules B, C and D recover approximately 30% of the oil in the sector.

The conclusions of Kossack *et al.* (1987) were that a horizontal well in the Troll Field would replace 2 or 3 vertical wells, but the results are dependent on production rate and the horizontal well's length. Also, by a gradual reduction of production rate, a horizontal well can recover 30% of the initial oil in place.

## EVALUATION OF EFFECT OF LOW-PERMEABILITY LAYERS (LPL) ON OIL PRODUCTION

The effect of the near-wellbore reservoir description on oil production can be evaluated by comparing the results of simulations from various scenarios. These scenarios should involve alterations in the reservoir description for the various well configurations. From these results, the importance of the LPL's permeability and extent will be shown in the differences in the fluid flow and the production GORs. To follow the lead given by Kossack *et al.* (1987), the numerical simulations were run with 1, 2 or 3 vertical wells or a large-diameter 1500 ft horizontal well.

The following two studies were performed and comparisons were made between these four well configurations.

(1) At a constant production rate and a constant vertical permeability of the calcite-cemented layers, the LPL at both the GOC and below the WOC were removed sequentially from the description. This yields four cases, which will be discribed in a later section.
(2) A sensitivity study of the vertical permeability of LPL at the GOC in the absence of the layer below the WOC was made. An additional parameter, the sector oil production rate, was varied from 4000 to 24 000 stb/day for all descriptions.

## THE NUMERICAL MODELS

### Reservoir description

In order to simulate both gas and water coning, the 3-dimensional, 3-phase black oil simulation model

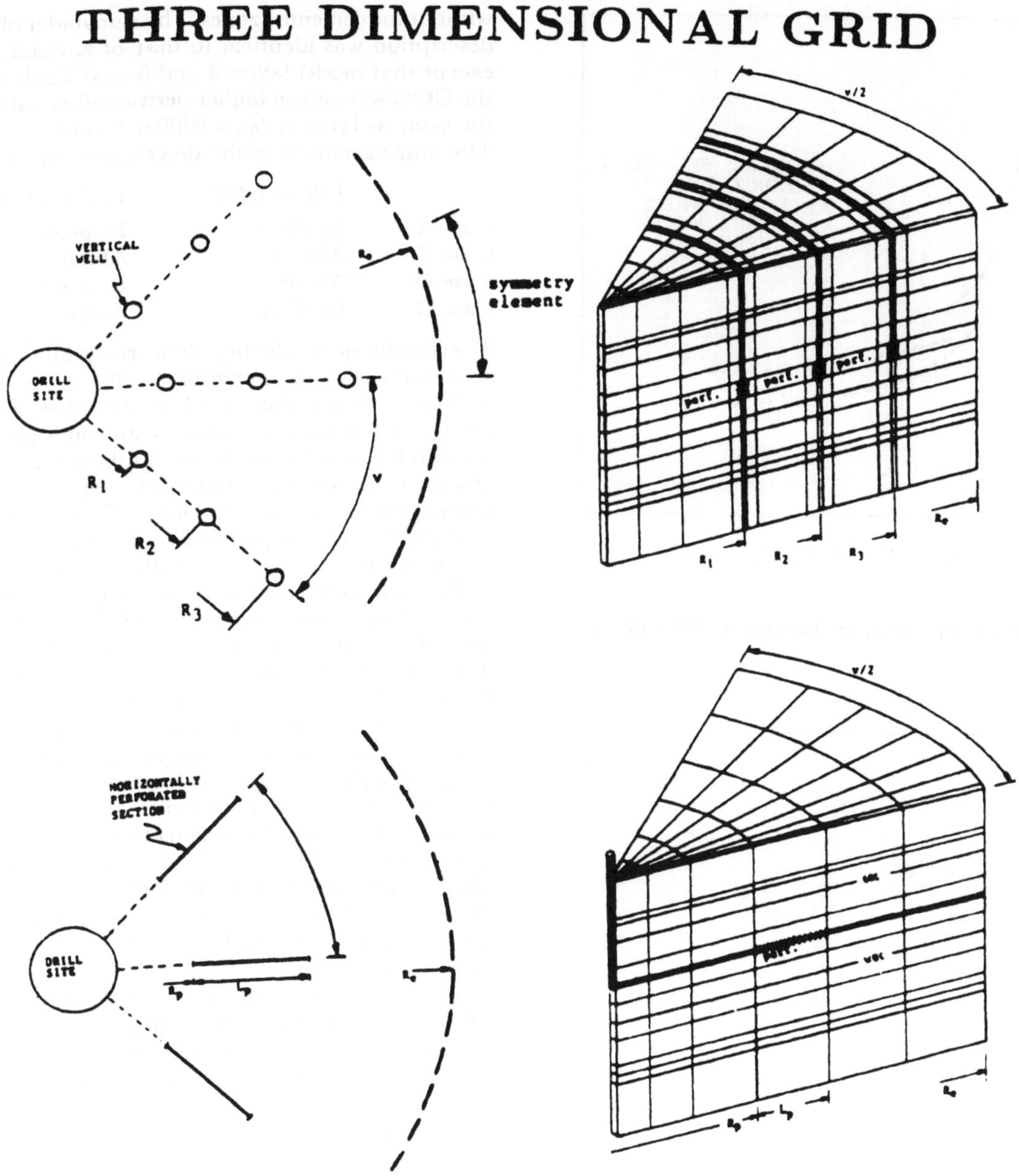


**Fig. 2.** Typical 3-dimensional grid for numerical simulations: (a) Vertical wells; (b) horizontal wells.

**Table I.** Description of numerical layering (from Kossack *et al.*, 1987) with modifications for this study given in parentheses.

| *Model layer* | *Thickness (ft)* | *Porosity (%)* | *Permeability (mD)* | | *Initial saturation (%)* | |
|---|---|---|---|---|---|---|
| | | | *Hor.* | *Vert.* | $S_w$ | $S_g$ |
| 1 | 144 | 29 | 1500 | 300 | 15 | 85 |
| 2 | 29.5 | 30 | 4550 | 920 | 8 | 92 |
| 3 | 1.6 | 25 | 10 | 0.1 | 20 | 0 |
| 4 | 4.9 | 28 | 150(9000) | 10(1800) | 20(8) | 0 |
| 5 | 6.6 | 28 | 30(9000) | 1(1800) | 23(8) | 0 |
| 6 | 9.0 | 31 | 9000 | 1800 | 8 | 0 |
| 7 | 9.0 | 31 | 10 000 | 2000 | 8 | 0 |
| 8 | 9.0 | 31 | 10 000 | 2000 | 8 | 0 |
| 9 | 2.6 | 31 | 9000 | 1800 | 18 | 0 |
| 10 | 6.6 | 31 | 9000 | 1800 | 85 | 0 |
| 11 | 1.6 | 25 | 10 | 0.1 | 85 | 0 |
| 12 | 18.0 | 27 | 2200 | 440 | 85 | 0 |
| 13 | 1181.1 | 30 | 100 | 20 | 100 | 0 |

ECLIPSE* was used. A short summary of the layers in the numerical model used by Kossack *et al.* (1987) and, with a slight modification, also used in this study, is given in Table I. The remainder of the reservoir, rock, fluid, and grid data are given in detail by Kossack *et al.* (1987). The gas, oil, and aquifer are 170 ft (52 m), 42 ft (13 m), and 1100 ft (335 m) thick, respectively. Permeabilities in the oil zone are high; approximately 9D in the horizontal plane. The drainage radius is 5000 feet (1525 m). In the first study, two calcite-cemented layers were placed in the same locations as discussed by Kossack *et al.* (1987), that is, at the GOC and 6.6 feet (2 m) below the WOC. In the second study, only the low-permeability layer at the GOC was considered.

## Sector description and well representation

The sector size used in these simulations was the same as Kossack *et al.s* (1987) base case, that is, 60°: see Fig. 1 for

*Trademark of Exploration Consultants Ltd, England.

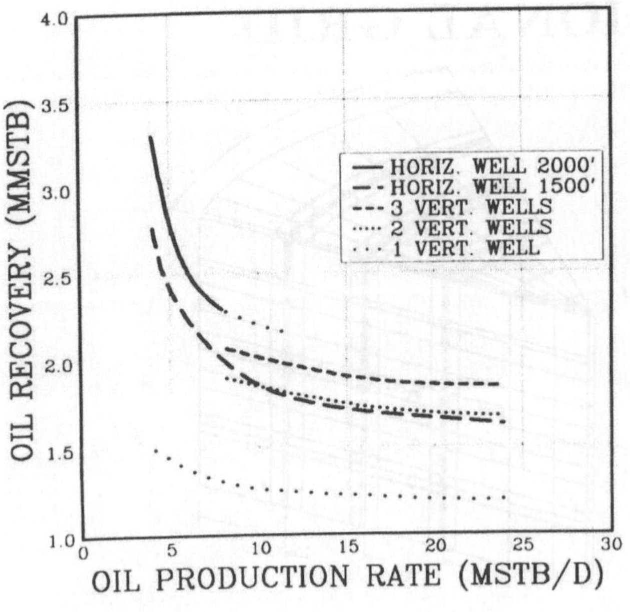


**Fig. 3.** Oil recovery vs. sector production rate. (From Kossack *et al.*, 1987.

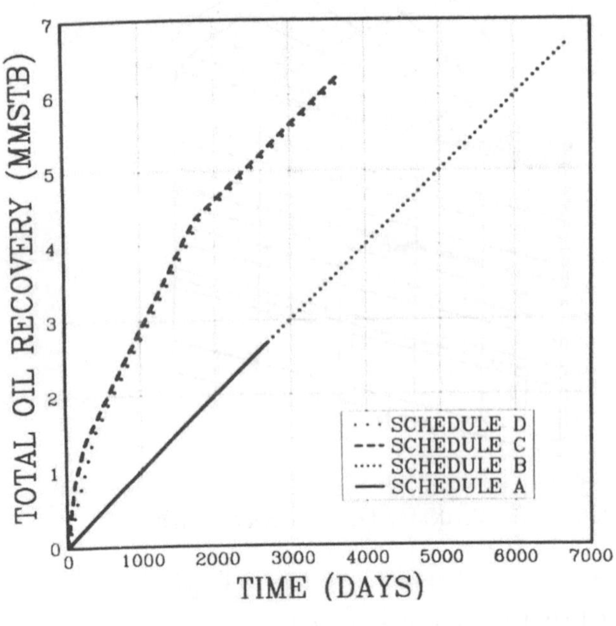


**Fig. 4.** Optimized oil recovery vs. time. (From Kossack *et al.*, 1987).

a sketch of the situation. The well was represented as a very high-permeability, high-porosity row of grid blocks as previously described by Kossack *et al.* (1987). The pressure drop along this large-diameter, horizontal well was of the order of a few psi. Additional sensitivity studies have been carried out on this well representation (Sundal, 1987), but will not be discussed here.

## EFFECT OF LOW-PERMEABILITY LAYERS AT 8000 stb/d OIL PRODUCTION

To evaluate the effect that the two low-permeability layers (LPL) had on the published results of Kossack *et al.* (1987), the case with $Q_o = 8000$ stb/day was simulated with and without the cemented zones. The remainder of the reservoir description was identical to that of Kossack *et al.* (1987) except that model layers 4 and 5 (see Table I) just below the GOC were given higher permeability values that were the same as layer 6, ($k_H = 9000$ mD and $k_z = 1800$ mDD). The four variations of the description are as follows:

| | *LPL at GOC* | *LPL below WOC* |
|---|---|---|
| Case A | In place | In place |
| Case B | Absent | Absent |
| Case C | Absent | In place |
| Case D | In place | Absent |

The results of producing these reservoirs with 1, 2, or 3 vertical wells and a horizontal well, are compared in Fig. 5. The wells are shut-in when the GOR reaches 5000 scf/stb. In this figure the time to shut-in is plotted against the number of wells where the numbers 1, 2 and 3 (on the abscissa) are for the number of vertical wells and the 4 stands for the horizontal well case. The results of Kossack *et al.* (1987) are also plotted in the figure for comparison. The points for the five cases (A, B, C, D, and Kossack *et al.* (1987)) are connected with line segments even though intermediate values have no meaning. These "curves" do show the trends as the number of vertical wells increases. Actually, over the entire range of abscissa values (1 to 4) the vertical pressure gradient near the well(s) is decreasing (as the abscissa value increases). This pressure gradient controls the rate of gas coning. As the gradient decreases, the time to shut-in generally increases. Also, the effect of the reservoir description for a given well configuration can be seen along vertical line segments of the grid. For one and two vertical wells the lack of a LPL at the GOC is disastrous. On the other hand, for three vertical wells and the horizontal well, the descriptions without the LPL actually yields better results than the cases with LPL at the GOC. The explanation of this unexpected behaviour will be given in a later section. The description with both LPLs present always yields a higher recovery than the simulations using the description with the LPL at the GOC only. The Kossack *et al.* (1987) results are better than all of the vertical well cases, since the moderate permeability layers in numerical layers 4 and 5 provide additional protection from gas coning. For the horizontal well, the Kossack *et al.* case shows a decline in well life compared to three vertical wells, since horizontal wells generally show a relative decrease in performance as the heterogeneity of

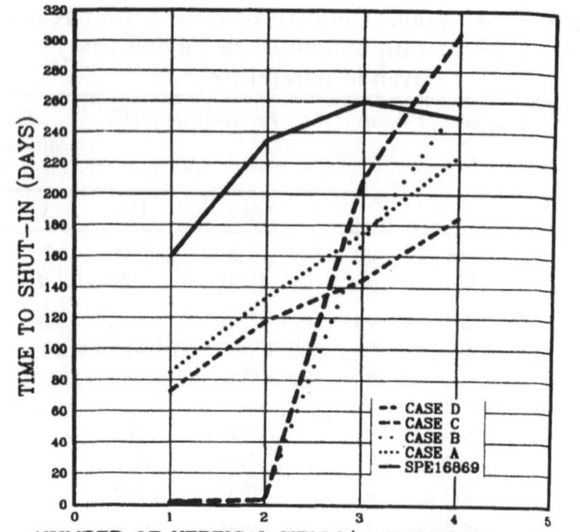


**Fig. 5.** Effect of removing low-permeability layers from 1, 2, and 3 vertical wells and from a horizontal well.

the description increases. The watercut results, not shown in detail, vary from a low of 0.38 to a high of 0.50 when the LPL at the WOC is absent.

## EFFECT OF A LOW-PERMEABILITY LAYER AT THE GOC AT VARIOUS PRODUCTION RATES

The previous section showed that the presence of a LPL with a vertical permeability of 0.1 mD at the GOC had a substantial and, in some cases, an unexpected effect on the gas coning results. An extensive study is described in this section in which the LPL at the GOC is assigned a range of values of permeability. The LPL below the OWC that predominately affects the water production is not present in the reservoir description used for these simulations. In a 60° sector, the oil production rates for the 1, 2 or 3 vertical wells or one 1500 ft horizontal well are varied from 4000 to 24 000 stb/d.

The vertical permeability of the LPL used in the previous section (0.1 mD) was just an estimated value, since the actual permeability is unknown and will most likely also vary laterally. To investigate a full range of possible conditions of the LPL, permeabilities were assigned to the LPL as follows:

| | |
|---|---|
| 0.0 mD | Totally sealing barrier |
| 0.1 mD | Estimated value from geologists (used by Kossack *et al.* (1987) |
| 1.0 mD | Moderate permeability layer |
| 1800 mD | No layer present (same permeability as neighboring layers below) |

A complete factorial study was performed on the three variables as follows:

| | |
|---|---|
| Well configuration $k_z$ of LPL | 1, 2, 3 vertical, 1 horizontal 0, 0.1, 1.0, 1800 mD |
| Sector oil production rate $Q_0$ | 4000, 8000, 16 000, 24 000 stb/d |

The results from the simulations consist of water cut (WC), gas–oil ratio (GOR) values (until the GOR reaches a value of 5.0 Mscf/stb and the well is shut-in), and time to shut-in. The water cut results vary from a low of 0.47 to a high of 0.70. A summary of the water cut results shows the following:

- At a given rate the horizontal well has a slightly higher WC than the vertical well cases.
- WC increases as $k_z$ of the LPL decreases. When $k_z = 0$, WC is in the range of 0.6 to 0.7.
- Simulations that produce for a longer time, say more than 250 days, have slightly higher water cuts.

Several mechanisms are involved in controlling the gas flow into the vertical and horizontal wells. A partial understanding of the mechanisms can be gained by analysing the GOR-vs.-time results. Figures 6 and 7 provide some insight into the processes. At sector oil production rate of 4000 stb/d, (Fig. 6) two descriptions are shown, e.g. $k_z = 0$, and $k_z = 1800$ mD. In both parts of Fig. 6, two types of GOR responses are observed; (1) curves that have a finite slope as free gas breakthrough occurs (1, 2, 3 vertical wells at $k_z = 0$ and 2 and 3 vertical wells at $k_z = 1800$) and, (2) curves that have a nearly infinite slope (the horizontal well at $k_z = 0$, and 1 vertical well and the horizontal well at $k_z = 1800$). The same situations exist at $Q_0 = 16\,000$ stb/d, (Fig. 7). At $k_z = 0$ and at 1800 mD,

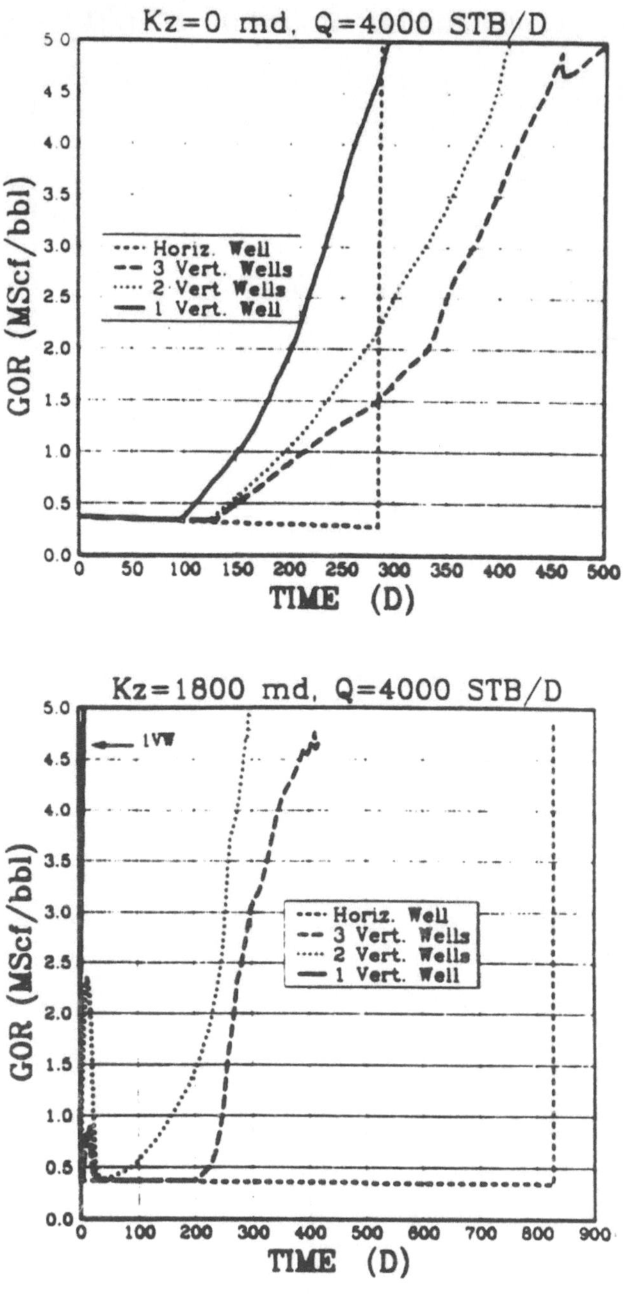


**Fig. 6.** GOR profiles vs. time at 4000 stb/d of oil production for (upper) $k_z = 0$ mD and (lower) $k_z = 1800$ mD.

the 1, 2, 3 vertical well cases have finite slope, while the horizontal well cases both have infinite slopes, initially. The difference in these two situations comes from the condition of the gas being produced. When a large bubble of free gas suddenly cones into a well, the GOR jumps from the solution gas value to a larger value with nearly an infinite slope. If this gas bubble is of limited size (see upper part of Fig. 7, $k_z = 0$, horizontal well) the jump has a limited height. On the other hand, when the well produces solution gas that has become free, before this gas can rise to the top of the oil column, the GOR curve rises above the solution gas value with a finite slope.

To display the vast quantity of results from the factorial sensitivity study, the shut-in time was plotted against sector oil production rate with the permeability of the LPL as a parameter. Four plots, one for each well configuration, show all the results (Figs. 8–11).

The relationship between shut-in time and production rate for one vertical well (Fig. 8) is as one would expect.

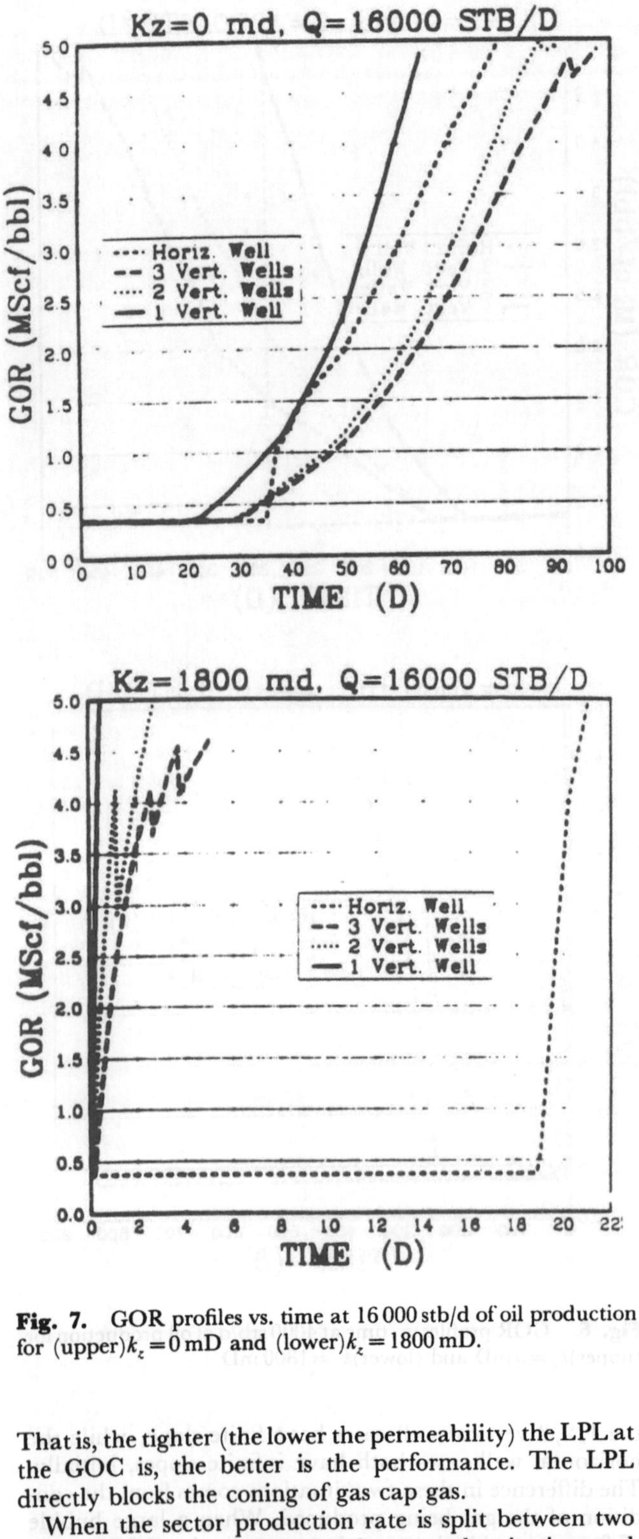


**Fig. 7.** GOR profiles vs. time at 16 000 stb/d of oil production for (upper) $k_z = 0$ mD and (lower) $k_z = 1800$ mD.

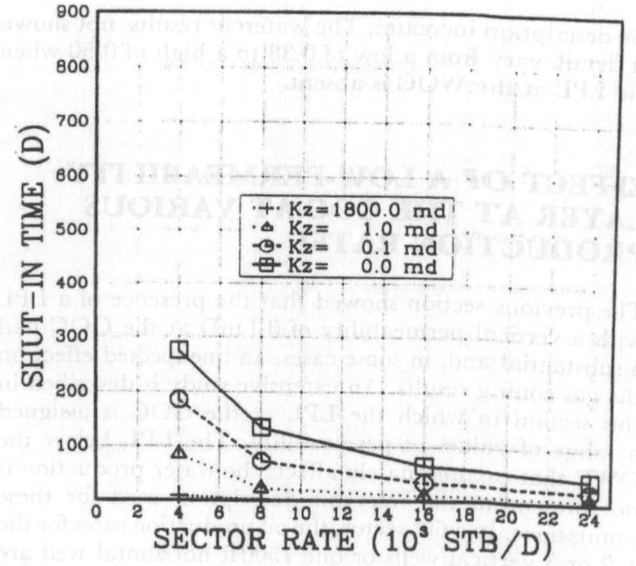


**Fig. 8.** Shut-in time vs. sector production rate for one vertical well.

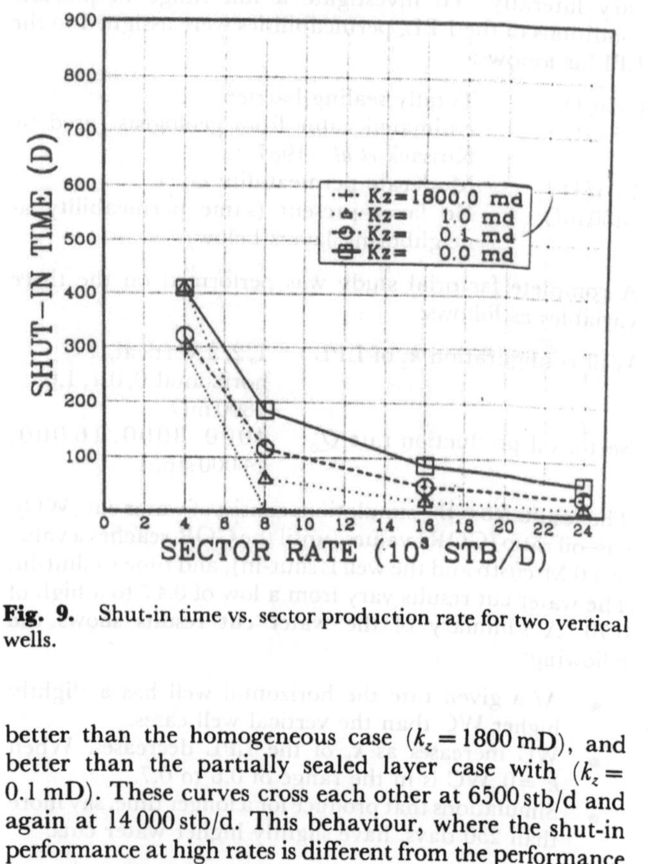


**Fig. 9.** Shut-in time vs. sector production rate for two vertical wells.

That is, the tighter (the lower the permeability) the LPL at the GOC is, the better is the performance. The LPL directly blocks the coning of gas cap gas.

When the sector production rate is split between two wells (Fig. 9), the relationship between shut-in time and production rate is as expected at rates above 8000 stb/d. At 4000 stb/d an *a priori* unexpected result can be seen when $k_z$ varies between 0.0 and 1.0 mD. For $k_z$ values of 0.0 and 1.0 mD the wells shut in at 410 days, while the simulation with the intermediate permeability value, ($k_z = 0.1$ mD), shuts in in less than 320 days. This might seem contradictory or perhaps indicate an error, but, as will be explained later, two coning mechanisms are competing in these simulations and as $k_z$ changes from 0.0 to 1800 mD the dominating mechanism shifts.

Spreading the production further with three vertical wells (Fig. 10) has interesting results at sector production rates less than 6000 stb/d. The LPL with $k_z = 1.0$ mD performs better than the fully sealed case ($k_z = 0.0$ mD), better than the homogeneous case ($k_z = 1800$ mD), and better than the partially sealed layer case with ($k_z = 0.1$ mD). These curves cross each other at 6500 stb/d and again at 14 000 stb/d. This behaviour, where the shut-in performance at high rates is different from the performance at low rates as the vertical permeability rises (i.e. the curves cross), is called "crossover" in the following. At high rates the sealed case gives the best results. When production is obtained from a 1500 ft horizontal well this "crossover" behaviour is even more evident (Fig. 11). At high production rates the presence of a tight zone is favourable, but at rates below 10–12 000 stb/d the picture is the opposite. In fact, when the layer is sealing ($k_z = 0$) the horizontal well shuts in at the same time as one vertical well at all production rates. At low rates (below 10 000 stb/d) it is surprising to see the best horizontal well performance in the homogeneous reservoir. Any permeability reduction at the GOC causes the horizontal well operating at these rates to shut in earlier. In the homogeneous reservoir the

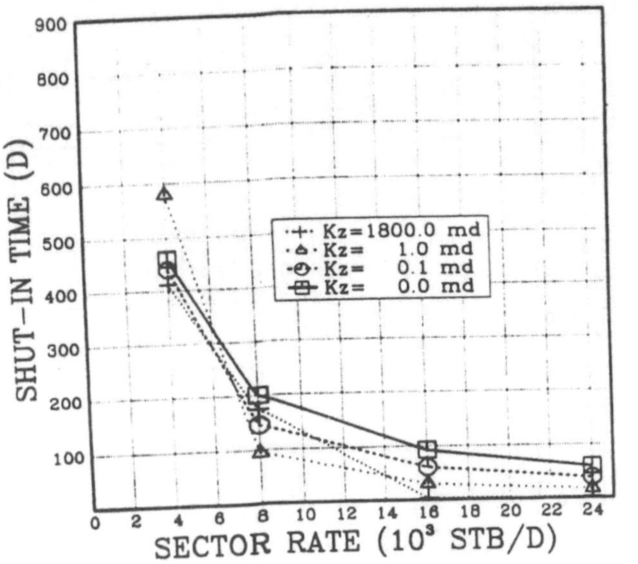


**Fig. 10.** Shut-in time vs. sector production rate for three vertical wells.

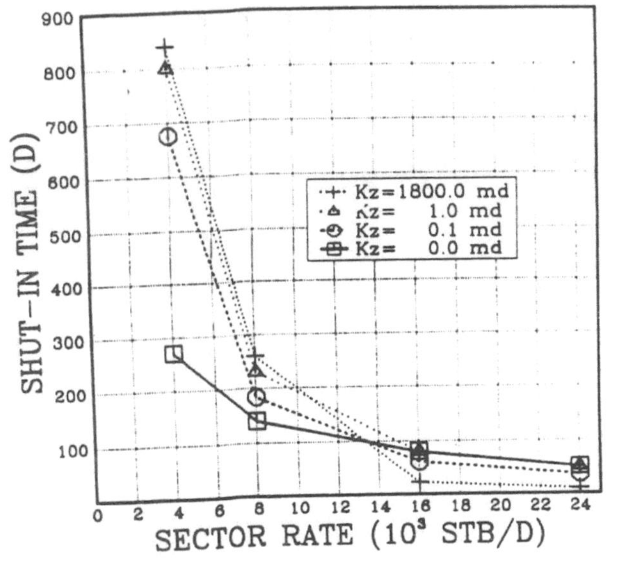


**Fig. 11.** Shut-in time vs. sector production rate for a 1500-ft horizontal well.

horizontal well performs very well. In fact, at 4000 stb/d the horizontal well has improvement factors (that is, time to shut-in for a horizontal well divided by the time to shut-in for a vertical well configuration) in shut-in time of 105, 2.8 and 2.0 for 1, 2 and 3 vertical wells, respectively. At 8000 stb/d the improvement factors are 144, 74 and 1.5, respectively.

The cause of this "crossover" behaviour, i.e. where the sealing barrier gives the best results at high rates but the worst results at low rates, is a complicated interaction between sector production rate; well configuration and interaction between wells; primary and secondary gas cap coning; pressure support from the gas cap; oil zone permeability. Two different situations occur below and above critical gas coning rate. If the well production rate is below the critical gas coning rate, a simple depletion process takes place for a period of time. Depending on the amount of pressure support from the gas cap, a certain quantity of liberated solution gas will evolve near the well. This gas will either rise towards the top by buoyancy or be pulled into the well. If the gas rises, it may join the primary gas cap or form a secondary gas cap, depending on how tight the LPL is.

At 4000 stb/d of oil production with a sealing barrier, no gas cap pressure support is possible. With 1, 2 and 3 vertical wells the evolved solution gas is *not* allowed to rise and form a secondary gas cap. As time increases the well(s) pull in more and more free gas with the oil. This is evident from the slopes of the GOR vs. time curves for 1, 2, and 3 vertical wells (Fig. 6, upper). As described earlier, a finite positive slope shows that the well is producing free gas evolved near the well. With the horizontal well, the pressure drawdown near the perforations is low, and the evolved gas rises to form a secondary gas cap. At 270 days this relatively large secondary gas cap cones into the well and shuts it in. Note the infinite slope of this curve as a high-saturation gas bubble reaches the well (Fig. 6). The process is shown in the three sequential sketches of Fig. 12.

On the other hand, in the homogeneous reservoir (no LPL), $Q_0 = 4000$ stb/d is above the critical coning rate for one vertical well and it shuts in after 8 days. The other well configurations with higher critical coning rates receive pressure support from the primary gas cap and do not immediately cone gas, except for a transient pulse of free gas for the 2 and 3 vertical well cases that lasts 10–20 days (Fig. 6, lower). These two configurations (2 and 3 vertical wells) produce mainly free solution gas, as described earlier when the GOR vs. time curve has a finite slope. The horizontal well with a lower drawndown allows the evolved solution gas to migrate up to the primary gas cap, and then at 840 days the horizontal well suddenly cones the gas cap gas (infinite slope in the GOR plot). This is

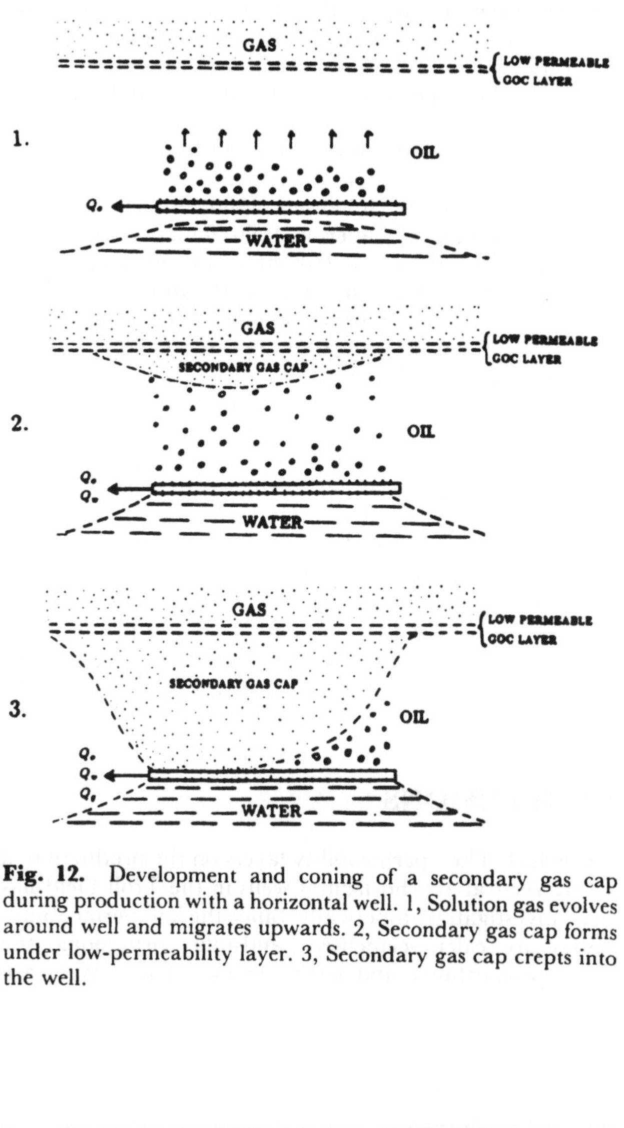


**Fig. 12.** Development and coning of a secondary gas cap during production with a horizontal well. 1, Solution gas evolves around well and migrates upwards. 2, Secondary gas cap forms under low-permeability layer. 3, Secondary gas cap crepts into the well.

because the downward movement of the gas cap (for pressure support as the oil is depleted) has reduced the distance between the gas cap and the well. In summary, when the production is subcritical and the LPL at the GOC has a vertical permeability less than 1800 mD, the pressure support from the gas cap is reduced, but then also the ability of evolved solution gas to escape to the primary gas cap is reduced. The results at $k_z = 0.1$ and 1.0 mD are results of the combination of these two mechanisms.

Above the critical coning rate, the situations have similar aspects. At $Q_0 = 16\,000$ stb/d and $k_z = 0$ mD (Fig. 7, upper), without pressure support from the gas cap, all the wells produce the evolved solution gas at increasingly higher GORs. The horizontal well allows a small amount of evolved solution gas to form a secondary gas cap for a few days. After 35 days this small secondary gas cap is coned into the well. Note the very steep slope in this curve at that point of time. When $k_z = 1800$ mD and $Q_0 = 16\,000$ stb/d (Fig. 7, lower), the vertical wells all cone gas within a few days. Note the discontinuous behaviour in these curves when an individual well in the sector reaches 5000 scf/stb and is shut-in. The horizontal well is producing close to the critical coning rate and operates for only 21 days.

In summary, this crossover behaviour, seen most clearly in the horizontal well's shut-in times (Fig. 11), occurs when a volatile oil is produced at subcritical rates in the presence of a sealing or partially sealing, layer at the GOC. The pressure support from the primary gas cap and the migration of evolved solution gas (1) to the well, (2) to a secondary gas cap or (3) to the primary gas cap are controlled by the vertical permeability of this LPL. The physics that reduces the performance of the horizontal well in the presence of a LPL is also what makes the well perform exceptionally well in a homogeneous reservoir; that is, the low vertical drawdown (pressure gradient) at any position along the completion interval. This physical situation allows the gas to migrate up and join the primary gas.

The critical coning rates estimated from this study for both vertical and horizontal wells agree with the approximate value given by Johansen (1986) for a vertical well and the equation of Chaperon (1986) for a horizontal well. The calculated values of 2300 stb/d for one vertical well and 7500 stb/d for a horizontal well support the observations seen in the homogeneous case. These approximate formulae can not be applied in the case when heterogeneities (such as the LPL) exist.

Two important results observed in these simulations are (1) that the horizontal well behaves differently from 1, 2 and 3 vertical wells in most of these cases, and (2) the horizontal well performs better than vertical wells, at low intermediate sector production rates in a *homogeneous* reservoir. In other words, the horizontal well's relative performance decreases substantially in the presence of a low permeability layer. This result is also illustrated in Fig. 13–16, where the cumulative oil production from a sector at various rates is plotted for the 4-well configuration at the values of $k_z = 0.0$, 0.1, 1.0 and 1800 mD, respectively.

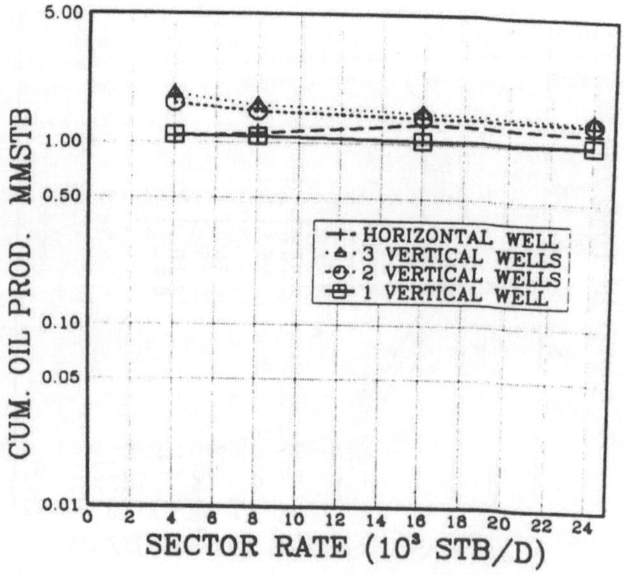


**Fig. 13.** Cumulative oil production vs. sector production rate when $k_z = 0$ mD.

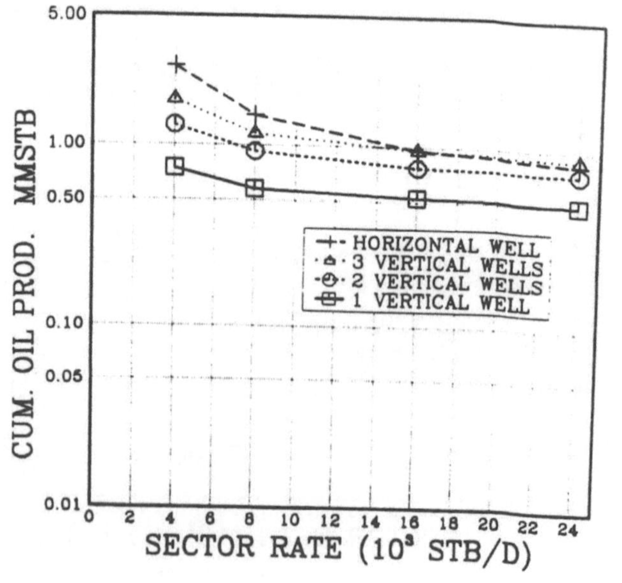


**Fig. 14.** Cumulative oil production vs. sector production rate when $k_z = 0.1$ mD.

## CONCLUSIONS

The effect of low-permeability layers on the production of oil by vertical and horizontal wells in the Troll Field has been investigated numerically. Since the low-permeability layers are calcite-cemented sandstone formations, the exact permeability and lateral extent of the layers are unknown. Thus, to evaluate how these layers, present or not, would affect oil production, their permeability was varied as a parameter in the study.

The model assumes a total reservoir drainage area with a radius of 5000 ft, divided in sectors by no-flow boundaries. Each section is drained by 1, 2, or 3 vertical or a 1500 ft horizontal well.

The following conclusions can be drawn:

(1) The presence of the low-permeability cemented sandstone layers found in the Troll Field and presented in Kossack *et al.*'s (1987) reservoir description has a substantial effect on the performance of vertical and horizontal wells. Without a partially sealing low-permeability layer at the GOC ($k_z = 0.1$ mD), the 1 and 2 vertical well cases cone gas in only a few days at moderate and high sector production rates. The effect of spreading out the sector production with three vertical walls or a hor-

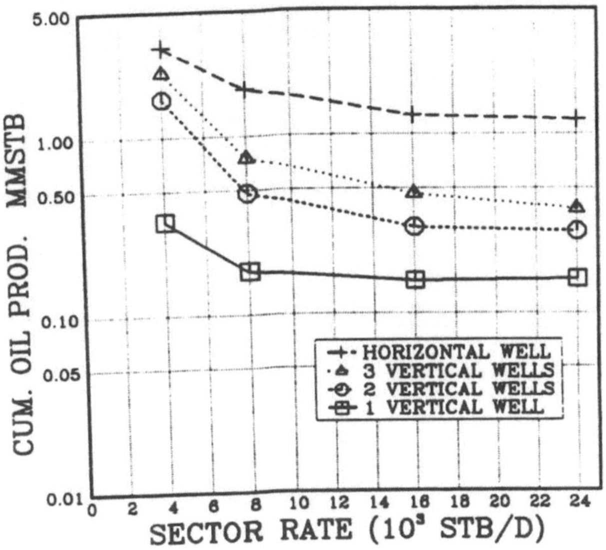


**Fig. 15.** Cumulative oil production vs. sector production rate when $k_z = 1.0$ mD.

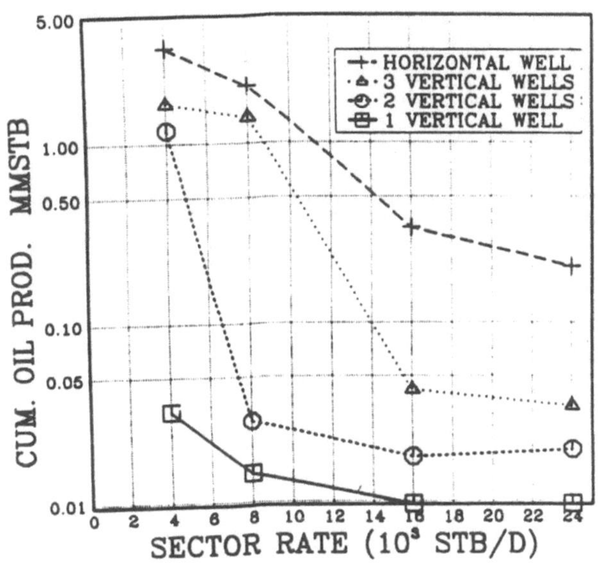


**Fig. 16.** Cumulative oil production vs. sector production rate when $k_z = 1800$ mD.

izontal well makes these configurations less dependent on partially sealing low-permeable layers at the GOC at moderate production rates, and their performance is good.

(2) The horizontal well exhibits very unusual behaviour in reservoirs with low-permeability layers at the GOC. At high production rates the sealing barrier ($k_z = 0$ mD) results in the longest production, and the homogeneous layer ($k_z = 1800$ mD) produces the shortest period of time. At a sector production rate between 10 000 and 12 000 stb/d of oil, the curves cross over. Below these rates the horizontal well produces the most oil in the homogeneous reservoir and the least in the sealing barrier case. This crossover phenomena is a result of the interaction of the horizontal well production, the low-permeability layer, the primary gas cap movement, and the development of a secondary gas cap.

(3) The placement of horizontal wells in a gas coning situation should take into account the presence of low-permeability layers that would tend to reduce the communication between the primary gas cap and the oil zone. If a layer near the GOC has a very low vertical permeability, conventional wells may perform as good as or better than a horizontal well. Otherwise a horizontal well still performs better than vertical wells.

(4) In the Troll Field, the use of horizontal wells as calculated by Kossack *et al.* (1987) still looks very attractive. In fact, if the low-permeability layer at the GOC is not continuous over a large area, horizontal wells are the only proven technology that may economically recover Troll oil.

## NOMENCLATURE

| | |
|---|---|
| GOC | Gas–oil contact |
| GOR | Gas–oil ratio |
| $k_H$ | Horizontal permeability |
| $k_z$ | Vertical permeability |
| LPL | Low-permeability layer(s) |
| M | Million ($10^6$) |
| $Q_o$ | Oil production rate (stb/d) |
| WC | Water cut |
| WOC | Water–oil contact |

## REFERENCES


Chaperon, I. (1986). Theoretical study of coning toward horizontal and vertical wells in anisotropic formations: subcritical and critical rates. Paper SPE 15377.

Darley, J.C., Røssland, L., van Golf-Racht, T. and Krakstad, O. (1983). Reservoir simulation of the Troll Field. *Offshore North Seas Advanced Projects Conference, Stavanger, Norway.*

Ekrann, S. (1987). Production from thin oil zones. *SPOR report no. SPT T-6/87.* Rogalandsforskning, Stavanger, Norway.

Johansen, G.K. (1986). Coning in a high permeability environment (the Troll Field). Thesis, Norwegian Institute of Technology, Trontheim, Norway.

Kossack, C.A., Kleppe, J. and Aasen, T. (1987). Oil production from the Troll Field: A comparison of horizontal and vertical wells. Papers SPE 16869.

Karcher, B.J., Giger, F.M. and Combe, J. (1986). Some practical formulas to predict horizontal well behavior. Paper SPE 15430.

Reiss, L.H. (1985). Horizontal wells production after five years. Paper SPE 14338.

Sundal, L. (1987). Diploma Thesis. The Norwegian Institute of Technology, Division of Petroleum Engineering and Applied Geophysics, Trondheim, Norway.

# 11

# Statistical Approach to the Computer Modelling of Hydraulic Fracturing in Hydrocarbon Reservoirs

*R. Romagnoli and R. Varvelli*

Dipartimento Georisorse e Territorio, Politecnico di Torino, Torino, Italy

This paper presents the first results of research in progress at Turin Politechnico, concerning the optimization of the techniques for improvement of the formation permeability around oil and gas wells in hydrocarbon reservoirs.

The paper examines the potential of implementing existing simulators by introducing sections that are able to treat the variables statistically, and parallel sections where the influence of the single variables is evaluated by the response surface method.

An integrated simulation system has been determined in which different approaches can be applied in order to perform an integrated study of the hydraulic fracturing phenomena. The integrated use of these methodologies and techniques allows wide ranges of requirements to be faced and provides a reliable tool both for reservoir-engineering and production personnel.

## INTRODUCTION

During the last few years both oil companies and research institutions have started numerous scientific research programmes to improve understanding of the mechanics of the hydraulic fracturing processes. So far, a number of hydraulic fracturing methods have already been studied extensively and applied to the secondary recovery of the in-place hydrocarbons. A group of quite interesting connections have resulted, such as those that relate to the fracturing-pressure response and the pressure transient analysis currently adopted in reservoir engineering. Subsequent studies have been carried out in order to extend the application of hydraulic fracturing methods to other fields, (e.g. geothermal energy extraction, radioactive waste disposal).

The design of fracturing stimulation processes usually includes three stages: (1) fracture geometry prediction, (2) fracture clean-up prediction, (3) fractured well performance prediction. Many simulators already exist that are able to predict all three stages of the hydraulic fracturing process. Hagoort, *et al.* (1980) and Settari (1980) made use of sequential methods of solution in order to solve the reservoir flow equations, the fracture flow equation and the fracture geometry equation; Nghiem, *et al.* (1984) described a fully implicit hydraulic fracture model able to carry out the said three stages by solving all the equations simultaneously, discussing only the vertical fracture pattern described by Geertsma and de Klerk's modelling.

The first models still only serve a purpose when fracture treatments are small and the fracture length is short; so a good deal of new and original pseudo-three-dimensional modelling techniques have been optimized in order to describe the evolution of the geometry of three-dimensional (3D) hydraulic fracture systems caused by fluid injection (Settari and Cleary, 1986). This work has also set up suitable computation sections for incorporation purposes inside 3D fracturing simulators. In other words, for massive hydraulic fracturing operations today, necessary to rely upon models that are able to compute all the fracture dimensions, rather than guessing some of them and using mass-conservation equations to estimate the others.

Nowadays, one of the most advanced approaches to the complete characterization of the problem is fracturing-pressure analysis and interpretation. This has been carefully studied and evaluated in a number of significant technical papers (Nolte, 1988, above all). It is worth remarking that hydraulic fracturing is often complicated, because of many different variable factors or conditions that heavily influence and modify the results and the costs. Efforts have to be made to evaluate and demonstrate the influence of the most significant parameters that govern the complex subject of "hydraulic fracturing operations", and to establish the real value of the new complex models vs. the simpler (and more subjective) ones.

*North Sea Oil and Gas Reservoirs—II*

## INTRODUCTION TO THE PROBLEM AND TO THE RELATED MODELLING

Hydraulic fracturing is the process of creating fracture systems in porous media by injecting fluids under pressure through a wellbore in order to overcome the existing stresses and to cause material failures and cracks. The main theories of the mechanics of failure of rock subjected to internal fluid pressure are still the basis of study. Most work is focused on the results of hydraulic fracturing operations.

Although there is disagreement about the orientation of fractures, most authors assume that when pressure increases in the formation breakage occurs in the plane that is perpendicular to the direction of the least compressive stress. The determination of fracture geometry requires consideration of three events simultaneously: (1) the fracture opening, (2) the fluid flow inside the fracture, (3) the leaking of the fracturing fluids into the formation strata.

One of the most typical phenomena that is the immediate result of well stimulation by hydraulic fracturing is the curvilinear growth of fracture systems, with subsequent interaction.

Fracture paths in reservoir structures are known to be associated with changes in stress loading in both time and space. Curved paths are often caused by the conformation of stress fields, while multiple fractures in the same wellbore and linked fracture networks provide a maximization of the communication of the considered wells with the formation.

This chapter points out the importance of the implementation of existing simulators by the introduction of sections that are able to treat the data statistically (by probability distributions of the input variables instead of single-valued functions). Second, it indicates the effectiveness of the evaluation of the influence of single parameters by the response surface method in an advanced, reliable operational manner.

Existing simulators have been implemented according to these aims. One particular simulator has been taken as a starting point. This has been assembled on the basis of that described and formulated by Settari and Cleary (1984). The simulator is able to couple 3D 2-phase flow in the reservoir with 3D fracture modelling in vertical planes, proppant transport and heat propagation. Furthermore, it includes in the overall formulation the 3D 2-phase flow inside the reservoir; 2-phase non-Newtonian flow in the fracture; heat transfer in the fracture system and in the reservoir; propagation of the crack in the space; proppant transport; wellbore hydraulics, and the PVT behaviour of the reservoir fluids and the fracturing fluids. The turbulence effects, which are not negligible for gas flow at high velocity, are also taken in account by the generally non-linear Forchheimer's equation (from Settari and Cleary, 1984). The identification of fracture geometry and the subsequent pressure distribution computations can be performed directly from the force balance and flow equations. In the implementation stage the continuous crack-growth process has always been approximated by stepwise sequences in the finite-element method (FEM) approach. For any growth step, the fluid volume in the total crack cavity is assumed to remain constant. So far, the modelling work has not included the possible additional effect(s) of proppants on the fracture geometry.

The implementation stage of the simulator has led to the introduction of sets of probability distributions. For any of the considered variables of the problem inside chosen ranges it is possible to define sets of values generated according to any statistical law (e.g. fixed standard deviation, defined stochastic distributions, uncontrolled variations inside fixed ranges). This has been done in order to simulate the uncertainty that is so often derived from the attempt to evaluate the considered variables and unknowns. The need to quantify the related influence of any parameter on the results of the modelling and forecasting stages has led to the development of a parallel approach to system identification. This has made it possible to study problems where a response variable changes in a pattern according to the different levels of solicitation of more stimulus variables. This choice has not been made because of the lack of consolidated theory, rather the need for an operative tool for use in connection with the already available deterministic and statistical ones.

The first attempt by the surface response method has been made by introducing inverse polynomial response surface in $k$ variables of input (with levels: $X_1, X_2, \ldots, X_k$), defined as

$$P(x) = \frac{\prod_{i=1,k} (X_i)}{y} \quad (1)$$

where $y$ is the corresponding mean response, and $P(x)$ is a polynomial in the variables: $X_1, X_2, \ldots, X_k$. Here each two parameter form in a single stimulus variable gives an asymptotic curve, while the three parameter form will have simple non-symmetrical optima. Since the use of the ordinary polynomial response surfaces does not allow the existence of non-symmetrical or asymptotic responses the use of the inverse polynomial response surfaces has been preferred. There is nevertheless a need for careful extrapolation to the edges of the regions under examination. The choice of inverse polynomial response surface gives the following advantages:

(1) It is possible to obtain asymptotic forms in two variables.
(2) It is possible to achieve a marked lack of symmetry—about the optimum—in the placing of the contours.
(3) The contours themselves can show quite appreciable differences from the elliptical contours of the general bivariate quadratic.

Developing the $y$ contained in Eq. (1) in the form:

$$y_j = \sum_p \left(\frac{\beta_p}{U_{pj}}\right) \quad (2)$$

where the $U_{pj}$ are the cross-products/ratios of the $X$ levels (e.g. $X_1 \times X_z$, $X_2/X_3 \times X_5$), and the $\beta_p$ are the numerical coefficients of the $X$ levels, when log–normal distributions are assumed for the $y$ (AAVV, 1984), then weighted least squares can be successfully used as a starting point for iterative maximum likelihood fitting processes. If $E(y_j) = y_j$ and $\mathrm{Var}\,(y_j) = \sigma^2 y_j^2$, then an initial vector $b$ of parameter estimates can be obtained by solving the system: $Ab = z$ where

$$A_{pq} = \sum_j (U_{pj} U_{qj} y_j^2) \quad \text{and} \quad z_p = \sum_j (U_{pj} y_j)$$

Further corrections $\delta\beta$ can be obtained by solving the system: $C\delta\beta = w$, where

$$C_{pq} = \sum_j (U_{pj} U_{qj} \hat{y}_j^2) \quad \text{and} \quad w_p = \sum_j (U_{pj} \hat{y}_j \ln [\hat{y}_j/y_j])$$

where the asymptotic variance–covariance matrix of the parameter estimates $\beta$ is given by $\sigma^2/C$.

## SOLUTION TECHNIQUES AND ANALYSIS OF THE RESULTS

Today's date processing systems allow computer programs to be run that are able to solve highly complex structures. FEM modelling gives the opportunity of assigning the constitutive laws of any material, choosing the most adequate element type from a wide range, discretizing complex domains with the desired detail, and solving linear equation systems through adequate storage memories by iterative, direct front, or band-solving methods (Dhatt and Touzot, 1984; Bathe and Wilson, 1976; Cheung and Zienkiewicz, 1967).

The finite-element method has been adopted in this context mainly in order to perform detailed analyses of spatial correlations, interactions and changes. FEM analyses go through the formulation of the matrices; numerical integrations for the matrix computation term-to-term; the treatment of single element matrices up to the assembly of global ones representing the discretized domains; and the solution of the systems of equations that describe the physical problem.

Variations in time have been studied by conventional finite-difference techniques. As far as flow problems are concerned, the required integral formulation (for FEM application) has been obtained from the differential formulation by weighted residual methods. The method has led to the use of Galerkin's scheme by selecting the weight functions on the basis of the opportune functions that are able to approximate the variable field behaviour.

As expected, 2D and 3D FEM analyses have greatly emphasized the fact that the input data are much smaller causes of uncertainty and numerical instability for the fixed-height geometry models.

As far as the 3D FEM modelling is concerned, trouble may also come from the need to assign the variables describing the physical properties of the confining layers and the stress fields acting inside the entire masses. The 3D computation of crack extension also requires detailed knowledge of the values of the bottomhole pressure during hydraulic fracturing jobs. Since the behaviour of pressure while fracturing is in progress is directly influenced by the vertical growth of the fractures, the post-fracturing evaluation study also requires the performance of back-analysis operations. The pressure declines after the treatments can supply reliable quantitative information both for interpretation and design purposes if the height of the fracture has been kept constant or has shown stable growth during the treatments and if the extension has not been restricted.

As pointed out by Nolte (1988), in cases of non-stable height growth and/or restricted extensions it is possible to have reliable qualitative information. Thus, fracturing-pressure analysis, with pressure transient analysis and curve-matching solution techniques, seems to be one of the most powerful tools of scientific investigation. The propped fracture conductivity is one of the parameters that directly determines the production rate values after the closure; it depends on the history of the stimulation job and on the position of fractures vs. pay, since only the part of the channel that is positioned directly against the pay really contributes to the productivity gain.

Figure 1 shows that an approximation of the value of the propped fracture conductivity in the range 200 md $\pm$ 100 md can lead to predicted gains in the well production rate that may differ more than 10–12 times. A moving-boundary modelling analysis stage has been performed by checking the fracture tip position at the end of every time step, generating new fracture nodes and repeating the computation step whenever the tip has come to the interest of new grid-blocks. Good stability in the procedure has been obtained by zeroing the fluid loss term from the fracture section to the (other) new blocks when the tip retreats to the block boundary.

A grid sensitivity analysis stage has been performed in order to check the convergence of the computed solutions to the theoretical ones while refining the grid mesh in different directions. Over 250 two-phase cases, the global injected rate varied in the range 300–6000 $m^3$/day at an initial pressure of 3300 psi. It has been observed that mesh

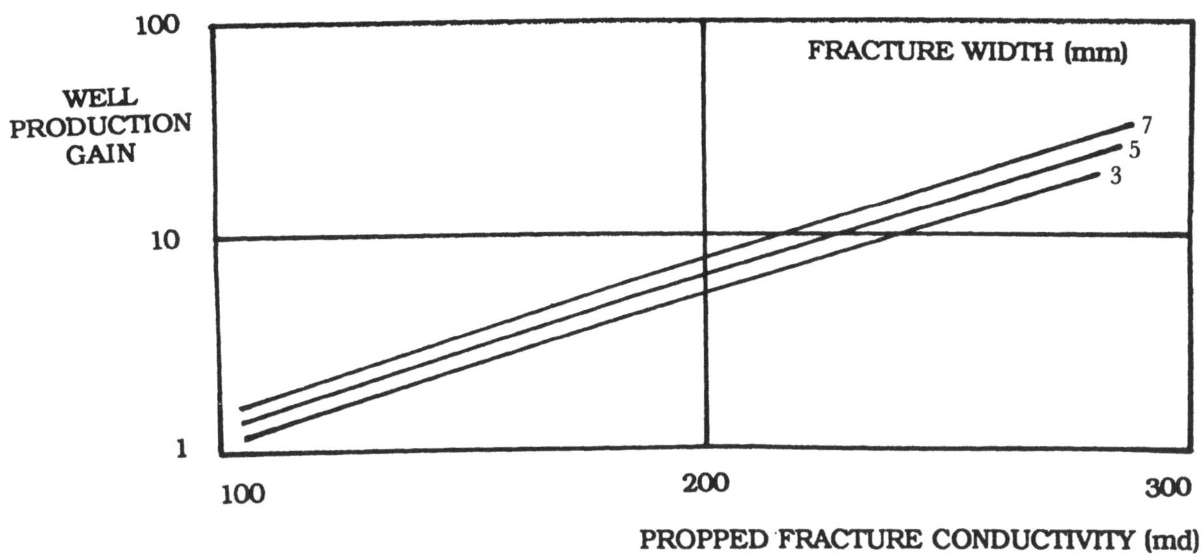


**Fig. 1.** Analysis of the influence of the propped fracture conductivity on the well production gain.

| | |
|---|---|
| Formation permeability | 1 md |
| Oil viscosity | 1 cp |
| Fracturing fluid viscosity | 3 cp |
| Propped fracture width | 4 mm $\pm$ 1 mm |
| Propped fracture permeability | 500 md $\pm$ 100 md |
| Bottom hole pressure range | 2800–3800 psia |
| Intermediate strength proppant | 12/20 ISP type |

refinement perpendicular to the fracture propagation direction leads to much better results than in any other spatial direction. The average ratio of precision in the results obtained for the perpendicular case vs. the parallel direction mesh refinement, determined on the basis of the relative errors, is 5.5 to 1. Obviously such an analysis stage requires a great deal of computation time.

Further developments enable some useful information to be reached to allow back-analysis study. The time step discretization has involved negligible errors whenever the time steps are designed according to geometrical series with ratios lower than 1.120 and initial values lower than 0.002 days.

The development of sections in the simulator, where it is possible to introduce the probability distribution of a variable instead of single values, has immediately shown the possibility of getting good quality characterizations of the geomechanical properties of the confining rock layers. This problem is usually bypassed by 2D FEM modelling, which only considers the value of the fracture height. As far as 3D FEM modelling is concerned, the available data are inadequate for deterministic predictive applications, because of the frequency of insufficient information about the properties of the layers that surround the reservoir. Probabilistic studies of Young's modulus of the upper and lower boundary, with the use of spatial correlation spherical-type semivariograms and of sets of frequency hystograms referred to 5, 4, 3, and 2 m correlation ranges (in order to approximate Monte Carlo simulations, where the range tends to 0), have revealed the importance of such an approach.

Fig. 2 shows that asymptotic trends are reached in the majority of the cases over a wide range of different kinds of formations (the confining layers Young's modulus ranging up to $1–20 \times 10^6$ psia). Analogous results have also been obtained for other geomechanical parameters (cohesion, and Poisson's ratio in particular); this confirms that the statistical approach provides reliable results even when the analytic one is extremely impractical. Element dimensions that are much smaller than the correlation ranges have been chosen in order to avoid any interference with the impact of spatial correlations (this is a strong potential cause of disturbance).

The employment of the response surface method (RSM) has established which of the input variables really affects the response variable, thus integrating the simulator by introducing sections where simple smoothing functions are studied to get accurate representations of the true underlying surfaces.

The choice of the inverse polynomial response surfaces for the evaluation of the influence of the single parameters has sufficiently represented both marked non-symmetrical and asymptotic responses, also in pairs of variables. The construction of the inverse-type polynomial models (AAVV., 1984) has followed the criterion of searching axial plots using the centre point and the axial points for the considered variables, in order to obtain as much information as possible (all the other variables remaining at their known nominal values). After a rough introductory process of fitting an ordinary quadratic surface, in order to decide which powers should be used for each variable and approximating the possible interactions between the different variables, the inverse polynomial fitting itself includes the identification of (1) the term that is analogous to the general mean; (2) the terms that are analogous to the main effects; (3) the terms that are analogous to significant factor pair(s) interactions (from the established ordinary polynomials). Origins (for variables that need them) are chosen by the residual sum-of-squares method, non-significant terms are eliminated by backward methods; then forward selection allows focusing and includes the terms corresponding to two factor interactions between important variables.

An application example has been developed, considering the response variable $y$ = production gain to be influenced by the four stimulus variables of propped fracture permeability, propped fracture width, bottomhole pressure, and injected rate. Such an analysis originates a 25-point central composite design including a full $2 \times 4$ factorial pattern. Stepwise fits of ordinary quadratic response surfaces allow a first rough evaluation of the most adequate powers for the variables involved.

The axial plots analysis allows a quite careful identification of the investigated degrees of influence. Figure 3 depicts the trends of the four stimulus variables considered, with respect to their ranges of variation. In single detail the partial results have been:

(1) $X = (1000\ 1750\ 2500)$ $y = (25\ 33\ 01)$ probable powers (2 3)
(2) $X = (450\ 500\ 550)$ $y = (40\ 55\ 65)$ probable powers (1 2 3)
(3) $X = (3.5\ 4.0\ 4.5)$ $y = (15\ 25\ 42)$ probable powers (2 3)
(4) $X = (3550\ 3300\ 3050)$ $y = (49\ 21\ 16)$ probable powers (1 2)

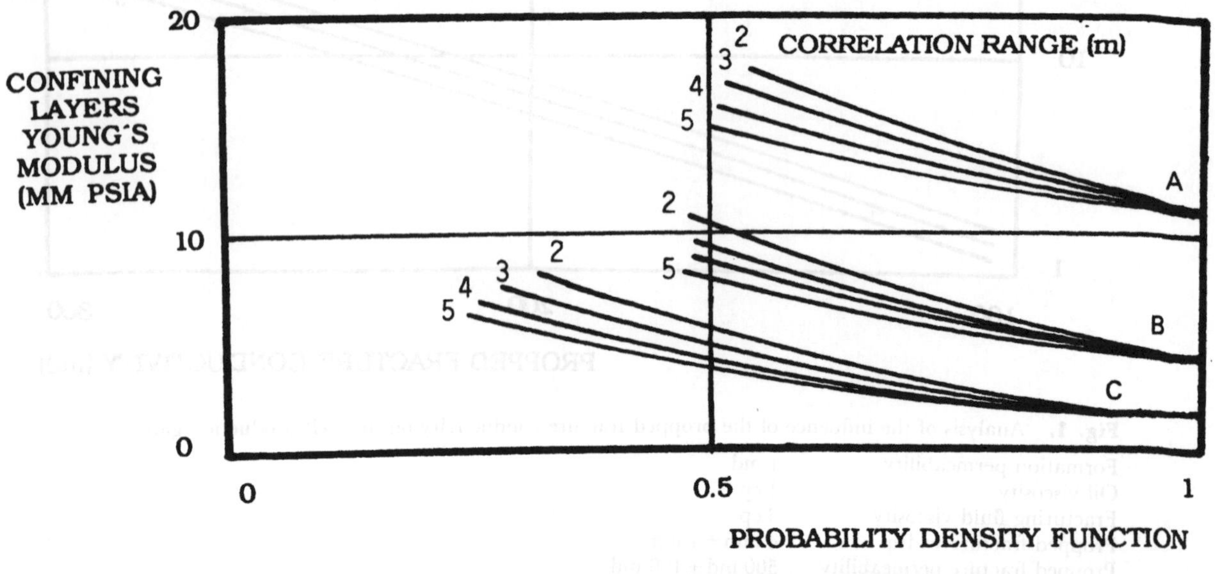


**Fig. 2.** Statistical simulation of Young's modulus of the confining layers (same data as Fig. 1), for three different formation cases (A, diorites; B, dolomite limestones; C, compact sandstones).

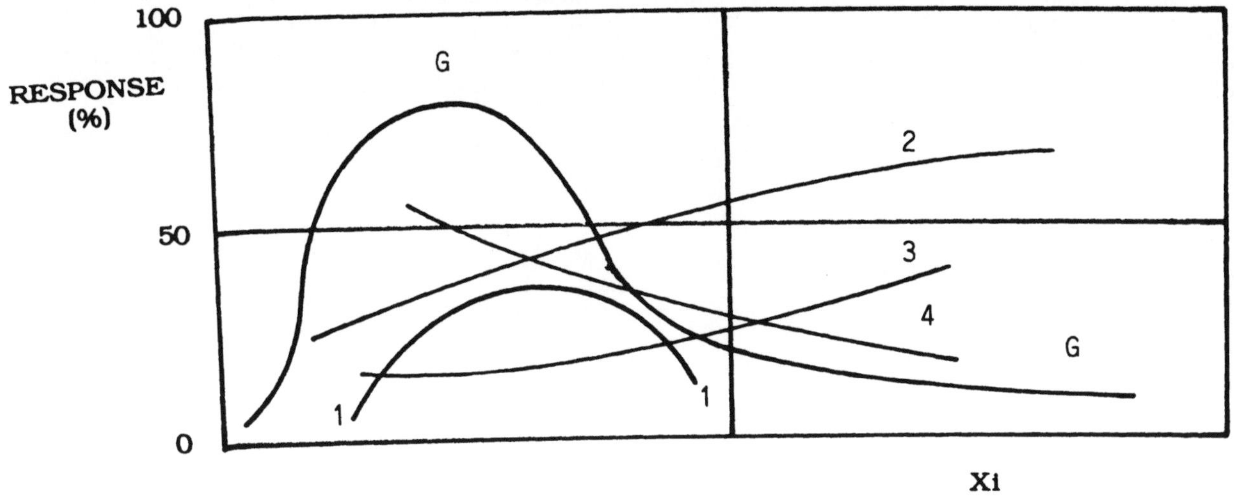


**Fig. 3.** Comparison of the effects produced by four different stimulus variables, using the response surface method.
1, injected rate (range = 500–5000 m³/day)
2, propped fracture permeability (range = 400–600 mD)
3, propped fracture width (range = 3–5 mm)
4, bottom hole pressure (range = 3800–2800 psia)
G, inverse polynomial resulting curve

It can be shown that, for the probable sets of powers given above (referring to the pivot points obtained), the initial set of terms — according to the possible choices — has the following meanings:

| $X_1$ | $X_2$ | $X_3$ | $X_4$ | |
|---|---|---|---|---|
| 1 | 1 | 1 | 1 | GENERAL UNDEFINED MEANING |
| 2 | 1 | 1 | 1 | $X_1$ HAS THE MAIN EFFECT |
| 1 | 3 | 1 | 1 | $X_2$ HAS THE MAIN EFFECT |
| 1 | 1 | 2 | 1 | $X_3$ HAS THE MAIN EFFECT |
| 1 | 1 | 1 | 2 | $X_4$ HAS THE MAIN EFFECT |
| 1 | 2 | 2 | 1 | INTERACTION EXISTS $X_2 \times X_3$ |
| 2 | 1 | 2 | 1 | INTERACTION EXISTS $X_1 \times X_3$ |
| 1 | 2 | 1 | 2 | INTERACTION EXISTS $X_2 \times X_4$ |
| 2 | 2 | 1 | 1 | INTERACTION EXISTS $X_1 \times X_2$ |

The performance of the search of the origin corresponding to $X_1$ allows the direct elimination of the corresponding non-significant terms. The fitting process for the inverse polynomial response surfaces, applied as described in the previous section, has involved the assumption of log–normal distributions for $y$; non-zero origins have been fitted by carrying out grid searches to find which origin values give the minimum value of the residual sum of squares. The data coding process has also been adopted in order to check whether the effect of rounding errors will produce any kind of disturbance. A simple assignment like $X_i' = (X_i/\bar{X}_j)y' = y/\bar{y}$ (before the fitting) has meant that all the noise coming from the existing difference of scales, which is not negligible, can be avoided.

As a final remark, as shown in Fig. 3, one of the most influential variables is the injected rate: both because of the similarity of the $X_i$–$G$ curves, and the obtained mathematical fits, results confirm that it is possible to optimize fracturing jobs by carefully studying man-made operations. As Settari and Cleary have stated in many publications, the success of fracturing jobs is largely due to accurate preliminary design.

## CONCLUSION

This paper has examined the potential of implementing existing simulators by introducing sections, that are able to treat the variables statistically, and parallel sections where the influence of the single variables is evaluated by the response surface method. An integrated simulation system has been determined in which different approaches can be applied in order to perform an integrated study of the evolution of the hydraulic fracturing phenomena that are the result of well stimulation operations and the energization in the hydrocarbon reservoirs.

FEM modelling has been adopted in order to study fracturing phenomena in space, coupled with finite-difference discretization schemes in the time domain.

The integrated use of these methodologies and techniques allows wide ranges of requirements to be faced and provides a reliable tool both for reservoir-engineering and production-engineering personnel.

## REFERENCES

AAVV (1984). *Response Surface Methodology Handbook For Nuclear Reactor Safety.* Directorate General Science, Research and Development Joint Research Centre, Commission of the European Communities, ECSE – EEC – EAEC, Brussels-Luxembourg.

Bathe, K.J. and Wilson, E.L. (1976). *Numerical Methods in Finite Element Analysis.* Prentice-Hall, Englewood Cliffs, N.J.

Cheung, Y.K. and Zienkiewicz, O.C. (1967). *The Finite Element Method in Structural and Continuum Mechanics.* McGraw-Hill, London.

Dhatt, G. and Touzot, G. (1984). *The Finite Element Method Displayed.* New York.

Hagoort, J., Weatherill, B.D. and Settari, A. (1980). Modeling the propagation of waterflood-induced hydraulic fractures. *Soc. Petroleum Eng. J.* **20**, (4), 293–303.

Nghiem, L.X., Forsyth, P.A. Jr. and Behie, A. (1984). A fully implicit hydraulic fracture model. *J. Petroleum Technol.* **36** (8), 1191–1198.

Nolte, K.G. (1988). Principles for fracture design based on pressure analysis. *Soc. Petroleum Eng. Prod. Eng. J.* **3** (1), 22–30.

Settari, A. (1980). Simulation of hydraulic fracturing processes. *Soc. Petroleum Eng. J.* **20** (6), 487–500.

Settari, A. and Cleary, M.P. (1984). Three-dimensional simulation of hydraulic fracturing. *J. Petroleum Technol.* **36** (8), 1177–1190.

Settari, A. and Cleary, M.P. (1986). Development and testing of a pseudo-three-dimensional model of hydraulic fracture geometry. *Soc. Petroleum Eng. Prod. Eng. J.* **1** (6), 449–466.

Keynote Address 2

# Quantitative Determination of Pore Structure and Application to Fluid Displacement in Reservoir Rocks

*N.C. Wardlaw*

Department of Geology and Geophysics, University of Calgary, Calgary, Canada

The flow of one or more fluids in a porous rock and the displacement of one fluid by another is affected by pore structure. In order to explain observed differences in fluid flow it is necessary to have quantitative models of pore structure. Five methods of obtaining information about pore structure are considered; (1) mercury capillary pressure curves; (2) petrographic image analysis using binary images of thin sections from impregnated rocks; (3) dispersion phenomena during displacement of one miscible fluid by another; (4) inertial coefficient evaluated by pressure transient flow test; and (5) serial sectioning of rock samples and computer reconstruction of pore system in three dimensions. A combination of porosimetry to obtain essential throat characteristics and two-dimensional image analysis to determine pore characteristics provides an effective means of obtaining quantitative information about pore structures that is better than either technique alone. Potential uses of quantitative pore-structure data are in predicting, under specified conditions, the effects of pore properties on permeability, relative permeability, dispersion, capillary pressure–saturation relations, the critical end points of residual wetting and non-wetting phase saturation, the efficiency with which one fluid displaces another, and electrical conductivity.

## INTRODUCTION

The flow of one or more fluids in a porous rock and the displacement of one fluid by another is affected by pore structure. In order to explain observed differences in fluid flow it is necessary to have quantitative models of pore structure. The major objective of this presentation is to outline and discuss several methods by which this can be achieved. Potential uses of quantitative pore-structure data are in predicting, under specified conditions, such properties as permeability, relative permeability, dispersion, capillary pressure–saturation relations, the critical end points of residual wetting and non-wetting phase saturation and the efficiency with which one fluid displaces another.

These properties are not controlled by pore structure alone but also by fluid properties (viscosity, density, interfacial tension, wetting and saturation history) and by the effects of several forces (capillary, viscous, gravity and inertial). These relationships can be depicted as

Pore structure + fluid properties + forces = fluid flow and displacement

*North Sea Oil and Gas Reservoirs–II*


Oil and gas reservoir systems include a wide range of pore structures, fluid properties and relative magnitudes of forces. It is impractical to undertake experimental tests to evaluate every set of conditions or production options and there is an incentive to understand the displacement mechanism and to make predictions about fluid flow and displacement in reservoir rocks. While these predictions should be possible from adequate knowledge of pore structure, fluid properties and forces, it is also true that pore structure may be evaluated by analysis of selected fluid displacement tests in combination with known fluid properties and forces. It would be advantageous if simpler types of fluid displacement tests could be used, in combination with other information about a given system, to predict flow in some more complex circumstances where experimentation is more difficult. Successes in this area have been limited.

Five methods of obtaining information about pore structure are considered here:

(1) determination from mercury capillary pressure curves (van Brakel, 1981; Li *et al.*, 1986; Wardlaw and Taylor, 1976);
(2) petrographic image analysis using binary images of thin sections from impregnated rocks (Ehrlich *et al.*, 1984; McCreesh *et al.*, in press; Etris *et al.*, 1988);
(3) dispersion phenomena during displacement of one

miscible fluid by another (Bretz *et al.*, 1988; Perkins and Johnston, 1963; Batycky *et al.*, 1982);
(4) inertial coefficient evaluated by pressure-transient flow tests (Jones 1987);
(5) serial sectioning of rock samples and computer reconstruction of pore system in three dimensions (Lin and Cohen, 1982); Macdonald *et al.*, 1986a,b).

A combination of mercury porosimetry and image analysis appears to provide the most useful method of obtaining quantitative information about pore structure for purposes of predicting fluid flow and displacement of one fluid by another under specified conditions.

## DETERMINATION OF PORE STRUCTURE FROM CAPILLARY PRESSURE CURVES

### Outline of method

Drainage and imbibition capillary pressure curves are highly reproducible for a given sample or lithology but differ widely in form among samples with different pore structures. This indicates that curve form is sensitive to pore structure (Wardlaw and Taylor, 1976). Although capillary pressure diagrams have long been used to interpret properties of porous media (van Brakel, 1981), a full evaluation of the information that could be derived from such diagrams has been lacking. Li *et al.* (1986) laid the ground work for such an evaluation. The objective in this section is to show how pore structure can be evaluated from a combination of porosimetry and image analysis.

A large number of different pore–throat structures could all generate similar drainage curves (non-wetting phase displacing wetting phase) but the number of structures that would also generate a specific imbibition curve (wetting phase displacing non-wetting phase) is lower. If additional secondary drainage curves and scanning loops also are considered, the number of structures that satisfy the entire suite of curves is even further reduced. In fact, the pore structure may become uniquely defined in terms of pore and throat connectivity, size and arrangement.

Quantitative information about pore structure can be obtained from capillary pressure curves by (1) defining the physical principles ("rules") that govern interface motions; (2) incorporating these "rules" into computer simulations of drainage and imbibition; (3) generating capillary pressure curves for three-dimensional networks with known pore–throat properties; (4) assessing the sensitivity of capillary pressure curve shape to known pore–throat variables (Li *et al.*, 1986); (5) using (1)–(4) for interpretation of pore–throat variables for samples of unknown pore structure.

The method has been tested by comparing interpretations of selected pore variables made from capillary pressure curves with independent measurements of the same pore variables obtained by direct observation (Wardlaw *et al.*, 1987).

### Pore variables

A porous medium can be considered to consist of larger spaces, referred to as pores, that are connected by smaller spaces or constrictions, referred to as throats. Pores are by definition larger than directly connected throats.

Throat size can be defined in terms of a plane of minimum cross-section between two adjacent pores. Pore size can be defined as a plane of maximum cross-section within a pore and also as the volume between two or more throats. Thus, throats are defined in terms of diameter and pores in terms of diameter and volume.

Properties of interest include the size frequency distribution, shape and connectivity of pores and throats, the size correlation of directly connected throats and pores and the spatial arrangement of pores and throats (Fig. 1). The composition and roughness of surfaces also are significant to the distribution, movement and trapping of fluids.

The arrangement of pores and throats of various sizes relative to each other is described as pore–throat correlation. Where there is no relationship between the sizes of throats and the pores connected by them, the structure is described as uncorrelated (Fig. 1(C)). In correlated structures, larger pores are connected by larger throats and smaller pores by smaller throats (Fig. 1(D)). In correlated pore–throat structures, the largest throat is not necessarily smaller than the smallest pore.

The connectivity of a pore-throat system is defined in terms of numbers of throats that connect with pores (Fig. 1(B)). A system may be described in terms of a range, as well as an average, connectivity. Connectivity may be related to pore size.

The arrangement of pores of various sizes relative to each other is described as the spatial order of pores. If there is no relationship between the size of a pore and the size of neighboring pores the system is disordered (Fig. 1(E)). If large pores tend to occur in proximity to large pores or smaller pores in proximity to other small pores the system has a degree of spatial order (Fig. 1(F)). Since a throat, by definition, is smaller than the pores that it connects, spatial order of pores may also imply a degree of pore–throat size correlation.

Throats and/or pores may have preferred orientations that confer preferred directions of flow within a pore system. A further pore property is that of irregularities or roughness of the walls of pores, which then have increased surface areas. The quantification of preferred orientation and surface roughness is not considered here.

### Relationship of pore and throat size to capillary pressure

The relationship between diameter ($d$) and equilibrium capillary pressure ($P_c$) is given by the Young–Laplace equation of capillarity, which, for cylindrical tubes is

$$P_c = P_{Hg} - P_v = \frac{4\gamma \cos\theta}{d} \qquad (1)$$

**Fig. 1.** Properties of pore systems; (A) pore shape; (B) connectivity or number of throats that connect with each pore (3 and 6 illustrated); (C) uncorrelated pore–throat structure; (D) correlated pore–throat structure; (E) spatially disordered pores (not clustered); (F) Spatially ordered pores (clustered). (From Li and Wardlaw, in press.)

where $P_{Hg}$ and $P_v$ are pressures measured in mercury and partial vacuum respectively. $P_v$ is a very small pressure and can be neglected, so that $P_c$ is equal to $P_{Hg}$; $\gamma$ is air–mercury surface tension; and $\theta$ is the contact angle measured in mercury.

Pore systems in rocks are characterized by spaces of complex geometry that are not cylindrical tubes, and fluid interfaces that, although minimal surfaces, may not be spherical. Lenormand *et al.* (1983) show that an effective diameter ($d_e$) for a meniscus can be obtained

$$d_e = \frac{2}{(1/x) + (1/y)} \qquad (2)$$

where $x$ and $y$ are the two principal diameters of the meniscus and provide a measure of the effective diameter of the pore or throat occupied by that meniscus.

The contact angle for mercury advancing ($\theta_A$) is not likely to be the same as for mercury receding ($\theta_R$), and $\theta_A$ and $\theta_R$ are required parameters in calculating the effective diameters of throats and pores respectively (Wardlaw *et al.*, 1988).

If the applied pressure ($P_a$) is larger than the critical capillary pressure for a throat ($P_{ct}$), penetration of a throat of effective diameter ($d_{et}$) by mercury occurs and the pore behind the throat will be filled with mercury. That is, for intrusion;

$$P_a > P_{ct} = \frac{4\gamma \cos\theta_A}{d_{et}} \qquad (3)$$

An interface will vacate a pore (effective diameter $d_{ep}$) if the applied pressure is less than the critical capillary pressure of the interface in the pore ($P_{cp}$). That is, for extrusion:

$$P_a < P_{cp} = \frac{4\gamma \cos\theta_R}{d_{ep}} \qquad (4)$$

Thus, whereas throats are the deciding barriers during mercury intrusion, pores are the deciding barriers for mercury extrusion or withdrawal. As defined previously, all of the volume is associated with the pores, and throats are planes of minimum effective diameter separating adjacent pores.

## Effect of pore and throat variables on shapes of capillary pressure curves

Computer simulation of capillary pressure curves were performed to determine the effects of pore–throat correlation and connectivity, spatial order or clustering of pores and wetting phase compressibility on: (a) non-wetting phase (nwp) breakthrough in drainage; (b) residual wetting phase (wp) saturation ($S_{wi}$); (c) residual nwp saturation ($S_{nwr}$); and (d) displacement efficiency ($D_E$). The relative effects of several variables are summarized in Tables I and II.

Pore–throat size correlation has larger effects on displacement than do variations in either pore–throat connectivity, in the range 4 to 8, or whether the wetting phase is compressible or incompressible.

The degree and type of clustering of pores affects both the displacement of wp or nwp in the clusters and in the "matrix" between the clusters. Computer simulations of the effects of clustering are reported elsewhere (Li *et al.*, 1986; Li and Wardlaw, in press). Discontinuous clusters of small pores can contribute to large $S_{wi}$ and discontinuous clusters of large pores or vugs can contribute to large $S_{nwr}$ (Morrow, 1971; Chatzis *et al.*, 1983; Wardlaw and Cassan, 1978; Wardlaw, 1980). Clearly, the spatial order of pores has large effects on fluid displacement, as do the size frequency and shapes of pores.

The effects of several pore–throat variables on curve form are given in Li *et al.* (1986) and Li and Wardlaw (in press) and are summarized here in Fig. 2. The curves presented (with the exception of 2(D), (G) and (H), which are diagrammatic) are based on computer simulations of 3D network structures the details of which are given elsewhere (Li *et al.*, 1986). All systems have uniform size distributions of pores and throats for uniformity and ease of comparison but can easily be transformed to any specific size frequency distribution. Nwp saturation is based on number occupancy of pores and the capillary pressure scale is based on the number fraction of throats (in drainage) smaller than a certain size and pores (in imbibition) larger than a certain size that are occupied by nwp (or wp). Since the purpose is to indicate relative effects, the absolute scales are not given but can be found in Li *et al.* (1986).

All simulations are for systems with six throats connecting with each pore ($Z=6$) except for Figs. 2(E) and (F), where systems with $Z=4$ and $Z=8$ are shown.

*Correlated and uncorrelated P–T structures*

Computer simulations reveal several differences in the form of capillary pressure curves for uncorrelated and correlated *P–T* structures.

(1) The relative positions of the end points ($S_{nwr}$) of two or more imbibition capillary pressure curves initiated at different nwp saturations (Figs. 2(A) and (B), $I_1$ and $I_2$).
(2) The relative positions of the breakthrough pressures for primary and secondary drainage curves (Fig. 2(A) and (B), $B_1$ and $B_2$).
(3) The shapes of scanning curves differ for the two models (Figs. 2(A) and (B)).
(4) Nwp breakthrough saturation (pore occupancy) in primary drainage is larger for uncorrelated than correlated structures (Fig. 2(A) and (B), bt).
(5) $D_E$ from $S_{wi}$ is much larger for correlated than for uncorrelated structures (Figs. 2(A) and (B)) (Table II).
(6) For the correlated model, the difference in saturation at the end points of two imbibitions is equal to the difference in saturation between the starting point of the first imbibition curve and the point on the second secondary drainage curve at the same pressure. This is not the case for the uncorrelated model (Figs. 2(A) and (B)).

The relationships (1), (2) and (3) should be apparent regardless of pore and throat size frequency distributions.

Relationship (1) can be explained in the following way. For a model with perfect pore–throat correlation, the domain of pores and throats that is occupied by nwp at breakthrough in drainage persists intact during imbibition until the moment when nwp continuity is finally broken. This means that imbibition curves initiated at different saturations ($d_1$ and $d_2$) will end at the same pressure as is seen for points $I_1$ ad $I_2$ on Fig. 2(B).

The essence of the correlated pore–throat structure is that the filling of the larger throats with nwp in drainage is accompanied by the filling of the larger pores and, conversely, the emptying of the larger pores is accompanied by nwp withdrawal from the larger throats.

This is not the case in the uncorrelated pore-throat model, in which there is no relationship between the sizes of throats and the sizes of the pores which they connect. Thus, during all stages of drainage, pores of all sizes are

**Table I.** Comparison of effects of variables on critical points. Numbers are differences in the percentage frequency of pores occupied by nwp or wp as indicated. Breakthrough in drainage (BT), discontinuous wp saturation in drainage ($S_{wi}$), discontinuous nwp saturation in imbibition ($S_{nwr}$) and nwp displacement efficiency initiated at $S_{wi}$ ($D_E$). Unless otherwise indicated, $Z=6$ and wp is compressible. Modified after Li and Wardlaw (in press)

| | | 1 | 2 | 3 | 4 | 5 |
|---|---|---|---|---|---|---|
| | | *Correlated P–T minus* | *$Z=8$ minus $Z=4$* | | *Compressible minus incompressible wp* | |
| | | *Uncorrelated P–T* | *Uncorrelated P–T* | *Correlated P–T* | *Uncorrelated P–T* | *Correlated P–T* |
| BT | nwp saturation | −17 | −21 | −15 | 0 | 0 |
| | pressure | −13 | −21 | −13 | 0 | 0 |
| $S_{wi}$ | wp saturation | −7 | 7 | −10 | 0 | −5 |
| | pressure | +11 | −19 | +6 | +3 | −2 |
| $S_{nwr}$ | nwp saturation | −36 | 0 | −12 | −3 | −1 |
| | pressure | −51 | −14 | −13 | −7 | 0 |
| $D_E$ | $\frac{100-S_{wi}-S_{nwr}}{100-S_{wi}}$ | +42 | +8 | +24 | +8 | +5 |

**Table II.** Displacement efficiency for displacements initiated at 90% nwp saturation and 40% nwp saturation are $D_{E(90)}$ and $D_{E(40)}$ respectively. $Z=6$ and wp is compressible

| *Uncorrelated P–T* | | | *Correlated P–T* | | |
|---|---|---|---|---|---|
| $D_{E(90)}$ | $D_{E(40)}$ | $D_{E(90)}-D_{E(40)}$ | $D_{E(90)}$ | $D_{E(40)}$ | $D_{E(90)}-D_{E(40)}$ |
| 18 | 6 | 12 | 57 | 27 | 30 |

filled in a random manner. Although the total number of pores occupied by nwp is greater at $d_2$ than at $d_1$ (Fig. 2(A)), at both $d_1$ and $d_2$ the relative frequency of pores occupied by nwp in all size categories is equal. However, nwp withdraws in an order of increasing pore size and, for imbibition commencing at a higher saturation ($d_2$), a larger fraction of pores occupied by nwp can be emptied before all nwp is finally disconnected. It follows that the pore size emptying at the imbibition end point ($I_2$) will be larger than at the imbibition end point $I_1$ (Fig. 2(A)). For this reason, imbibition initiated at higher nwp ($d_2$), in uncorrelated pore–throat models, causes an imbibition end point ($I_2$) that is at a lower imbibition pressure.

Relationship (2) can be explained as follows. In the correlated pore–throat model, breakthrough pressures are similar for the primary and for the secondary drainage curves (Fig. 2(B)). Breakthrough pressures for secondary drainage curves are not affected by residual nwp saturation because the domain occupied by nwp that exists just prior to the imbibition end point resembles the domain occupied by nwp at breakthrough in drainage.

This is not the case for the uncorrelated pore–throat model, in which the residual nwp at the end of imbibition contributes to the reconnection of nwp in secondary drainage. In this case, nwp continuity is established at a lower pressure in secondary drainage than in primary drainage (Fig. 2(A)) (Wardlaw *et al.*, 1987).

Relationships 1 and 2 have been observed for capillary pressure curves (Figs. 3 and 4) from rock samples in which it was possible to verify independently by direct visual measurements the sizes of pores and associated throats (Wardlaw *et al.*, 1987).

Direct measurements of throat and pore sizes invaded by Wood's metal show that Indiana Limestone has a strong component of pore–throat size correlation, with correlation coefficients ($R^2$) of 0.87, 0.94, and 0.93 at three successively higher capillary pressures. Similar measurements for Berea Sandstone show no obvious correlation. Correlation coefficients ($R^2$) of 0.14, 0.14 and 0.18 at three different capillary pressures indicate the prevailing disordered arrangement of throats with respect to pores. These results are consistent with the observation from computer simulations that pore–throat size correlation is a property that can be evaluated from the relative pressures of the end points of imbibition curves initiated at different wp saturations.

### *Pore shape*

Scanning curves lie between imbibition curves and secondary drainage curves. Scanning curve shape, as well as being affected by pore–throat correlation, is sensitive to pore shape.

An assumption used in the computer simulations is that when a pore empties at a pressure related to its diameter it does so completely. In fact, pores may be of highly irregular shape and accommodate interfaces with a range of curvatures such that nwp withdrawal occurs over a range of capillary pressures. Large pores in which nwp is trapped at the termination of imbibition may have partially emptied before nwp continuity was lost. Such partial withdrawal reduces $S_{nwr}$ and affects the shape of imbibition curves. Similarly, the shapes of drainage curves may be affected by partial filling events.

Some information about partial withdrawal and pore shape can be obtained from the ascending portions of scaning curves. Figures 2(C) and (D) illustrate an ascending scanning curve initiated from a point $S_1$ on an imbibition curve (I) before the $S_{nwr}$ end point was reached. If pores are of spherical shape and empty completely at a specific capillary pressure, the scanning curve should ascend vertically to a capillary pressure related to the size of the largest throat accessing a pore occupied by wp (Fig. 2(C), curve $S_1$–$S_2$). Further increase of capillary pressure would result in nwp invasion and departure from parallelism with the pressure axis. However, if there are pores of irregular shapes that empty over a range of capillary pressures (Fig. 2(D) inset), nwp can enter partially emptied pores that maintain nwp continuity to the nwp sink. There is no threshold capillary pressure associated with a throat to be overcome and nwp invasion proceeds immediately with increase in pressure (Fig. 2(D), curve $S_1$–$S_2$). The larger the angle $S_1$–$S_2$ makes with the vertical, the greater the

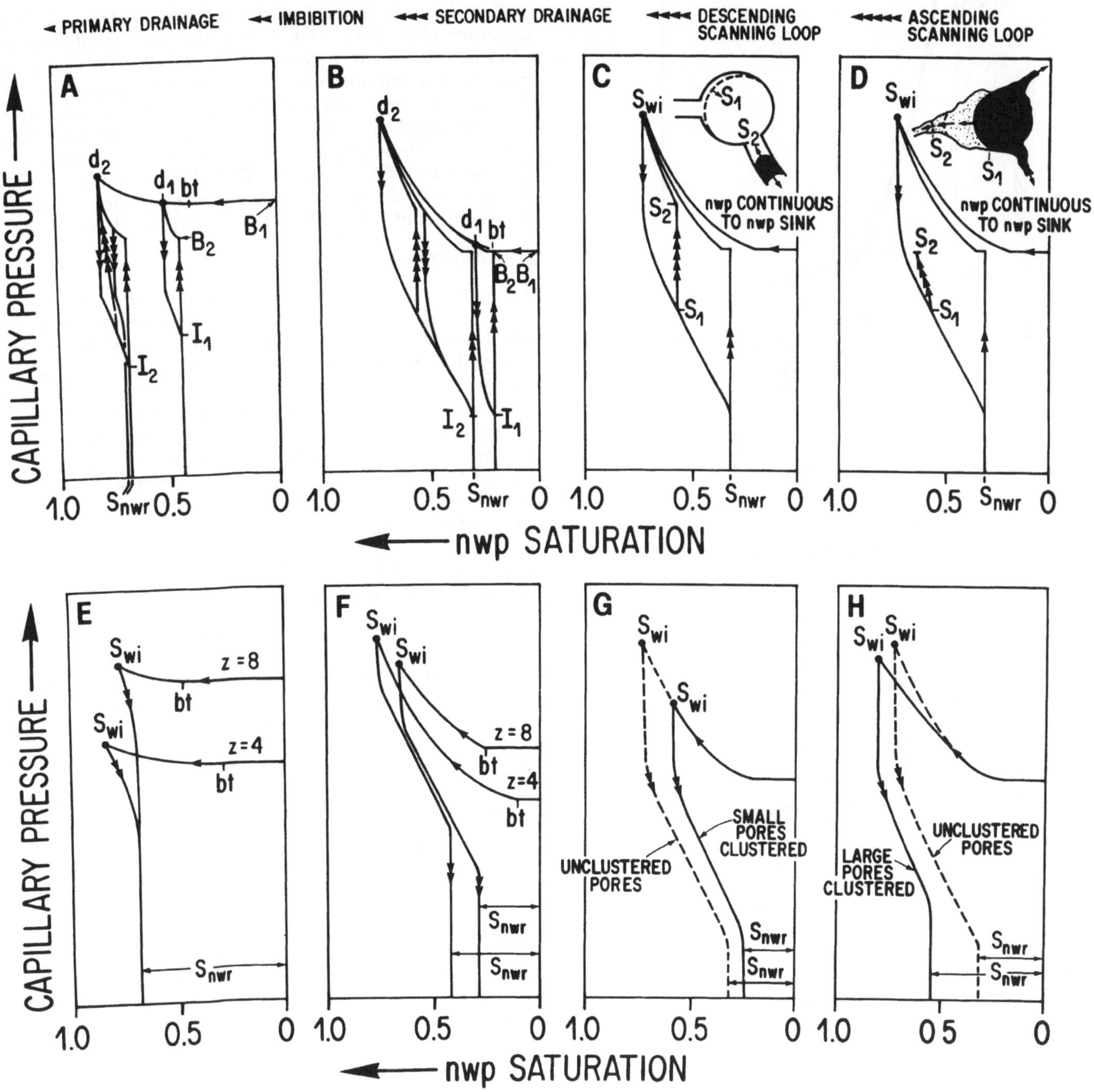


**Fig. 2.** Capillary pressure diagrams to illustrate the effects of several pore–throat variables on curve form. (A and B) uncorrelated and correlated pore-throat structures; (C and D) spherical and non-spherical pore shapes; (E and F) pore–throat connectivities of 4 and 8 for uncorrelated and correlated pore–throat structures; (G and H) clustering of small pores and clustering of large pores.

(A) Capillary pressure diagram for an uncorrelated pore–throat model illustrating the breakthrough pressure ($B_1$) and saturation (bt) in drainage, imbibition curves initiated at two different saturations $d_1$ and $d_2$ ($d_2$ equivalent to $S_{wi}$), and the end point pressures and saturations on the imbibition curves $I_1$ and $I_2$. The breakthrough pressure on the secondary drainage curve is $B_2$.

(B) As for (A) but for a correlated pore–throat model.

(C) Arrows as for (A) and (B); correlated pore–throat system. Inset shows interface in spherical pore (broken line) just prior to complete emptying at pressure $S_1$. Pore is re-invaded when threshold pressure for throat is exceeded at capillary presure $S_2$.

(D) Inset shows nwp partially withdrawn from non-spherical pore at capillary pressure $S_1$ and subsequently advancing as capillary pressure is increased to $S_2$. Unlike the case in (*C*) re-invasion of pore does not require that threshold pressure of a connecting throat be exceeded.

(E) Primary drainage and imbibition curves for uncorrelated pore–throat systems with pore–throat connectivity $\mathcal{Z}=4$ and $\mathcal{Z}=8$.

(F) Primary drainage and imbibition curves for correlated pore–throat systems with pore–throat connectivity $\mathcal{Z}=4$ and $\mathcal{Z}=8$.

(G) Drainage and imbibition curves for correlated pore–throat model with no clustering of pores (broken lines) and with some clustering of small pores that has increased $S_{wi}$ by an arbitrary amount (solid lines).

(H) Drainage and imbibition curves for correlated pore–throat model with no clustering of pores (broken lines) and with some clustering of large pores that has increased $S_{nwr}$ by an arbitrary amount (solid lines).

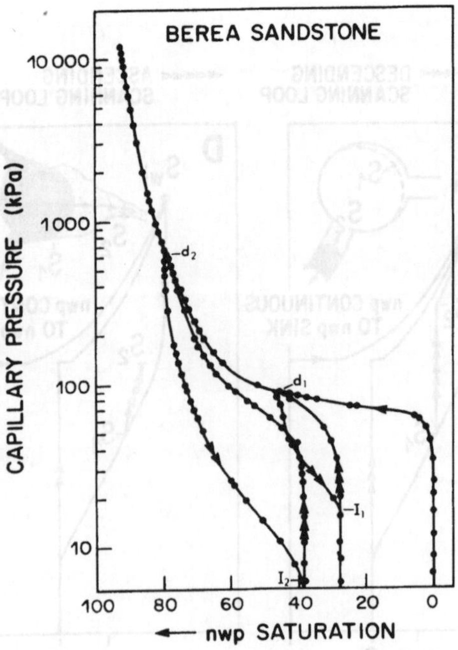


**Fig. 3.** Experimental mercury capillary pressure curves for Berea Sandstone. Arrows as in caption for Fig. 2. (From Wardlaw *et al.*, 1987.)

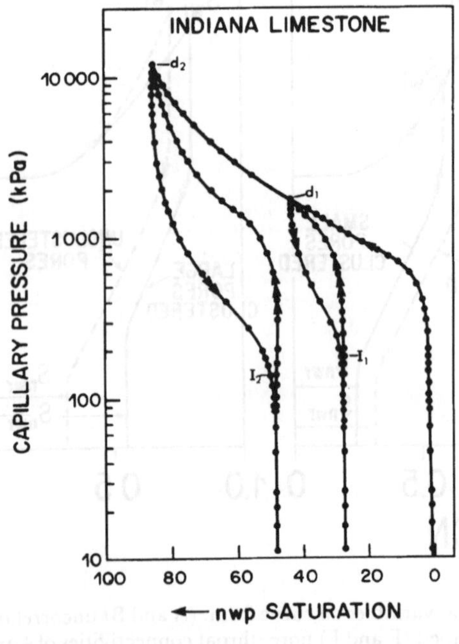


**Fig. 4.** Experimental mercury capillary pressure curves for Indiana Limestone. Arrows as in caption for Fig. 2. (From Wardlaw *et al.*, 1987.)

departure of pore shape from spherical form. The imbibition pressure $S_1$ is related to pore size and by initiating ascending scanning curves at different pressures it should be possible to obtain information about the relative shapes of pores of different sizes.

*Pore–throat connectivity*

For an uncorrelated P–T model, the fraction of pores occupied by nwp at $S_{wi}$ for $Z=4$ (4 throats connect with each pore) is 77% compared with 84% for $Z=8$. The fraction of pores occupied by nwp at $S_{nwp}$ is 70% for both $Z=4$ and $Z=8$. Thus, changes from $Z=4$ to $Z=8$ have only small effects on nwp displacement efficiency (Fig. 2(E) (Li *et al.*, 1986).

The situation is much different for a correlated P–T structure (Fig. 2(F)), for which the nwp displacement efficiency is much higher where P–T connectivity is greater. The fraction of pores occupied by nwp at $S_{wi}$ for $Z=4$ is 66% compared with 76% for $Z=8$. The fraction of pores occupied by nwp at $S_{nwp}$ is 42% compared with 30% for $Z=8$. Thus, displacement efficiency $D_E=36\%$ for $Z=4$ and $D_E=60\%$ for $Z=8$. The above examples are for the compressible wp case and for nwp withdrawal from $S_{wi}$ (see Li *et al.*, 1986, Table 4, for further details).

*Pore clustering*

Throats are by definition smaller than the pores that they connect and clustering of small pores implies clustering of small throats and, therfore, a degree of pore–throat size correlation. Clustering of isolated domains of fine pores may increase $S_{wi}$ and of large pores may increase $S_{nwr}$ (Figs. 2(G) and (H)) and the shapes of drainage and imbibition curves are affected. Pore clustering affects displacement efficiency ($D_E$) in the intervening matrix, as is discussed in Li and Wardlaw (in press), but these effects on displacement efficiency are probably small compared with those of the clusters themselves.

## Pore structure determination by porosimetry and image analysis

The computer simulations indicate that all of the variables (Fig. 1) affect all of the capillary pressure curves (primary drainage, imbibition, secondary drainage and scanning curves) to some degree. This causes difficulties in approaching a unique solution for a pore structure based on capillary pressure curves alone.

The interpretation can be simplified if independent information on some aspects of the structure is available. Information about pore size frequency and pore clustering can be obtained by image analysis of pores intersected on random planes (Thompson *et al.*, 1987a), but throat size information, pore–throat correlation and pore–throat connectivity—all properties strongly affecting permeability and displacement efficiency—cannot be obtained by direct observation from 2D sections. These properties, however, can be evaluated from capillary pressure curves if pore size distribution and clustering are known.

The complexity of the task is a function of the nature of the pore structure and is simplest for systems with uncorrelated pore–throat structures and pores that are spatially disordered (not clustered). Initially, the following assumptions are made; $Z$ is constant for all pores, pore shape is independent of size, volume of throats is negligible, pores are not clustered, snap-off does not occur in throats during imbibition, and advancing and receding contact angles and interfacial tension are known.

The analysis proceeds in the following way.

(1) Pore size distribution and degree and type of pore clustering are obtained by image analysis of 2D sections.
(2) Degree of pore–throat size correlation is estimated from:
  (a) relative positions of end points ($S_{nwr}$) of two or more imbibition curves (Figs. 2(A) and (B), $I_1$ and $I_2$).
  (b) relative positions of breakthrough pressures for primary and secondary drainage curves (Figs. 2(A) and (B), $B_1$ and $B_2$).
(3) Connectivity ($Z$) is obtained, knowing 1 and 2, from the breakthrough pressure on a drainage curve and

from the end point of an imbibition curve ($S_{nwr}$) initiated at $S_{wi}$.

(4) Properties 1, 2 and 3 above are necessary in order to obtain throat size frequency distribution from a drainage curve.

(5) Pore shape information may be obtained from the slope of the ascending scanning curve.

(6) If pores are clustered, fine pore clusters may increase $S_{wi}$ and large pore clusters increase $S_{nwr}$, but precise analysis is complex because the presence of clusters also affects displacement in the "matrix" between clusters.

In conclusion, the shapes of capillary pressure curves are sensitive to pore structure and provide the possibility of obtaining quantitative information about pore structure that is difficult to obtain by other means. For example, information about the size, frequency distribution of throats and the degree of pore–throat correlation, which have large effects on permeability, relative permeability and displacement efficiency, can be obtained. However, a combination of capillary pressure data and image analysis provides a better means of determining quantitative information about pore structure than either method alone. Porosimetry provides the best method of obtaining information about throats and image analysis the best method of getting pore properties.

## PORE STRUCTURE AND PETROPHYSICAL PROPERTIES OBTAINED FROM 2D IMAGES

A pore system is a 3D network of interconnected spaces that in a 2D slice or thin section consists of discrete areas that have been referred to as porosity elements or porels by Ehrlich in Etris *et al.* (1988). The term has no implication that an area is either pore or throat. The size, shape and arrangement of porels is related to the form of the 3D network and, provided the 3D structure is composed of randomly oriented elements, the area of porels in a random 2D slice will be in proportion to the volume of the 3D pore system. Thus, porosity may be obtained from a random slice.

Petrophysical properties such as permeability, relative permeability, electrical conductivity (formation factor) and drainage capillary pressure are strongly influenced by the sizes of throats, which are constrictions of minimum cross-sectional area between pores. However, the probability that a random plane of section will intercept a throat is almost nil. Throats are virtually absent from the plane of section and, if present, are impossible to recognize. Surprisingly, it has been shown that information from 2D slices through pore systems is useful in predicting permeability, electrical conductivity and drainage capillary pressure (Brumfield, 1988; McCreesh *et al.*, in press), because predictable relationships exist between pores and throats.

### Porel measurement

A pore system is impregnated with coloured epoxy, which enables it to be distinguished from surrounding rock, and a thin section is prepared. From a microscopic field of view, a digital image is obtained of all porels. The porels are then subjected to iterative erosion–dilation cycles that probe their sizes and geometric complexities (Ehrlich *et al.*, 1984). The first erosion–dilation cycle strips the outermost pixels from a porel (one pixel erosion) and then subsequently adds one pixel to the diminished porel (one pixel dilation). A final operation subtracts the number of porel pixels after the erosion–dilation cycle from those present in the original porel. Dilation following erosion usually will not restore the shape of the original porel because surface roughness, one or two pixels wide, vanishes without leaving traces that could allow restoration during the subsequent dilation. A second erosion–dilation cycle erodes and dilates a two-pixel increment during which larger-scale roughness features are lost. Subsequent erosion cycles progressively "eat away" surface roughness until a "smooth component" of the porel is obtained which, upon dilation, maintains a smooth shape. Ultimately there is a final erosion cycle during which the porel disappears completely leaving no nucleus to be expanded by subsequent dilation. This allows definition of the size of a porel. Also, the pixels lost during successive cycles of erosion–dilation differencing represent the frequency distribution of roughness elements of various sizes. The size of the smooth component and of the roughness elements can be partitioned into two separate spectra and their absolute sizes calibrated from the known size of a pixel (Fig. 5).

The frequency distributions from all porels were summed for each field of view and those for several fields of view can be summed for a thin section that represents a rock sample.

Perimeters and areas of each porel can be made during each cycle. Porels with constrictions fission during erosion and the size of the constrictions can be evaluated from the number of pixels it is necessary to erode in order to cause fissioning (Fig. 6). This provides a measure of "2D throat size" that is different from, but perhaps related to, throats as previously defined, that is, as minimum cross-sectional areas between pores. The number of daughter elements into which a porel fissions also provides a measure of the interconnectedness of the pore system.

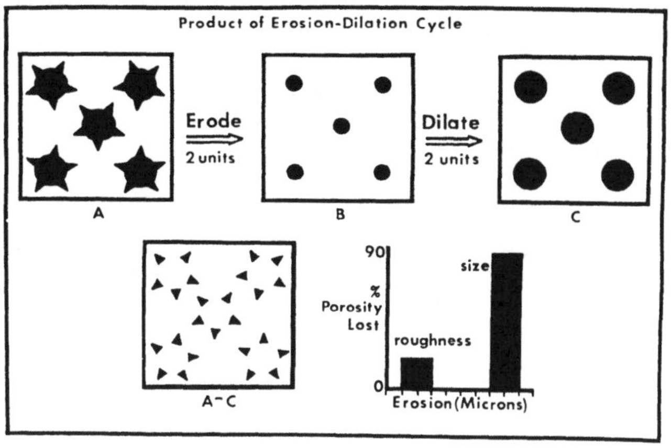


**Fig. 5.** Erosion progressively removes roughness elements producing a smooth pore that, upon dilation, retains a smooth shape. (After Ehrlich *et al.*, 1984.)

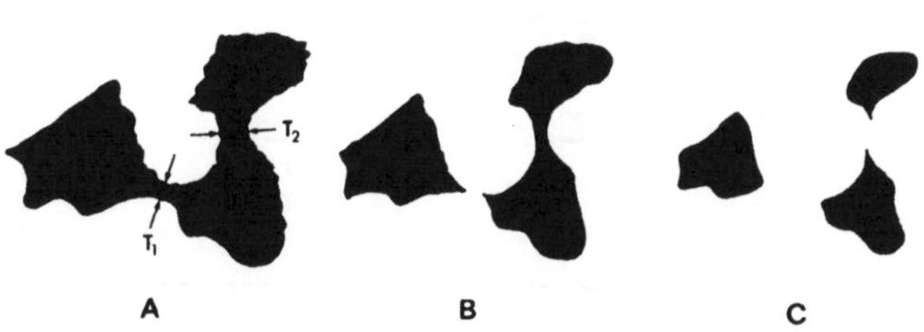


**Fig. 6.** Porel fissions into three daughter porels during successive erosion cycles, which provides a measure of 2D throat size ($T_1$ and $T_2$).

## Pore classification

Frequency distributions for the rough and smooth components obtained by erosion–dilation differencing can be used to deduce the number, size and volumetric proportions of pore types present. This is done using the pattern-recognition/pore classification algorithm SAWVEC (Ehrlich and Full, 1988) derived from the programs EXTENDED CABFAC (Klovan and Miesch, 1976) and the QMODEL algorithms (Klovan and Miesch, 1976; Full *et al.*, 1981, 1982). It is assumed that the relative proportions of pore types vary between samples and that the number of samples is larger than the number of pore types (Ehrlich and Full, 1988).

Although classification into pore types has been accomplished mainly using components of the rough and smooth spectra, other spectra related to 2D throat size and to the fissioning of porels during successive erosion cycles might also be used.

Pore types that emerge from the classification typically have different sizes along with differences in other characteristics (Fig. 7).

The variance in size of a pore type can be large or small. In the pore-typing procedure it is recognized that a 3D pore of irregular shape will appear as a variety of shapes and sizes of 2D porels in section. However, the variance of porel sizes in 2D will be related to the variance of pore sizes in 3D.

## Correlation of pore types with drainage capillary pressure curves

A mercury injection capillary pressure curve provides information about the volume of pores that are invadable via throats within specified size ranges but does not allow deductions concerning the sizes, shapes or numbers of pores.

McCreesh *et al.* (in press) obtained impregnated thin sections and mercury injection capillary pressure curves for suites of rock samples from various formations. They classified the pore types present in a suite of samples using the methods referenced in the previous section and implemented a regression-based procedure to determine the proportion of each pore type that filled during invasion of successive segments of the drainage curve, i.e. at successive pressure and saturation increments.

The regression showed that, for a particular suite of samples, the proportion of the several pore types present could vary substantially from sample to sample but that the proportion of a given pore type that filled during successive pressure increments did not vary much. Figure 8 provides an example of relationships obtained, by the regression procedure, between pore-type proportion and throat size accessibility.

In the sandstone illustrated in Fig. 7, five distinct porosity types were classified. Six equations were generated to estimate mercury saturation in six pressure intervals based on the percentage of porosity contributed by each pore type in individual samples. The results are displayed in Fig. 8. Coefficients of correlation ($R^2$) range between 0.80 and 0.87 indicating that the equations account for between 80 and 87% of the variability in data. A predicted mercury injection curve can be generated for every sample from its pore type compositions using these equations. The proportion of porosity associated with each pore type is close to the ideal 100% but may fall short because some pores are unfilled at the highest injection pressure or some pores may be too small to be resolved during imaging.

McCreesh et al. defined an "entropy" parameter that provides a measure of the strength of pore–throat association. Entropy measures the degree of spread of throat sizes associated with each pore type. If large pores are accessible via large throats and small pores via small throats, there is

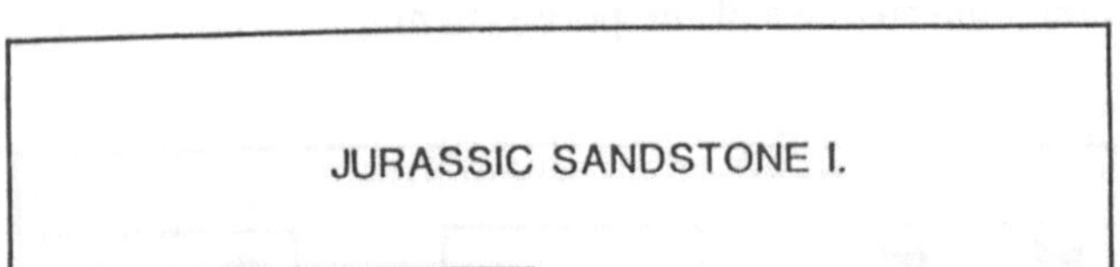


**Fig. 7.** Binary plots of porosity representing each of the five pore types comprising a Jurassic Sandstone sample suite. Each sample can be quantified in terms of proportions of each type (From McCreesh *et al.*, in press.)

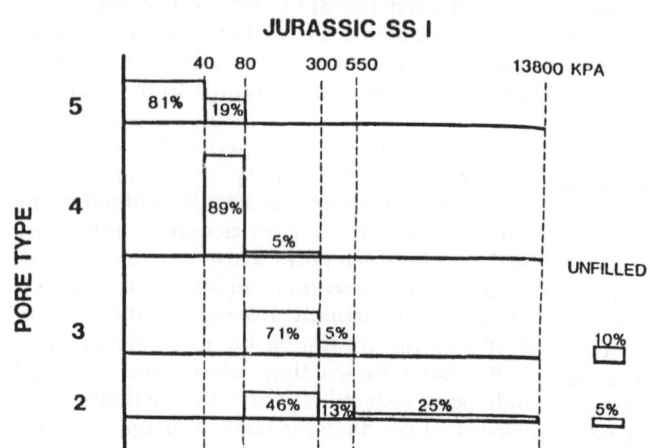


**Fig. 8.** The contribution of each pore type to saturation in each pressure interval in the Jurassic Sandstone sample suite. The pore types are numbered in order of size. There is a strong association between pore size and throat size. Each pore type fills in a restricted pressure interval and smaller pores fill at higher pressures. (From McCreesh *et al.*, in press.)

a high degree of pore–throat association and the entropy of pore types will be low. This is the case illustrated in Fig. 8. If large pores are accessible via throats of all sizes, then pore–throat correlation is a minimum and entropy a maximum for the pore type. Surprisingly, the ability successfully to predict the form of drainage curves from thin-section images does not appear to be related to the degree of pore–throat correlation or entropy.

In summary, for a suite of samples from a sedimentary formation, pore throat sizes as probed by drainage capillary pressure curves show consistent relationships with the several pore types that may be present. Since the pore types are determined from images of impregnated thin sections, information from these images provides a means of predicting drainage capillary pressure curves. However, the technique is not likely to be successful when applied to samples that are not genetically related.

## Correlation of pore types with permeability and formation factor

Etris *et al.* (1988) report very high coefficients of correlation ($R^2 = 99\%$) for the prediction of permeability by petrographic image analysis (PIA). A suite of 10 thin sections of micritized packstones and grainstones from the Middle East were analysed with accompanying mercury injection capillary pressure curves and measurements of core porosity and permeability.

Three pore types were defined in the thin sections based on the pattern-recognition/pore classification algorithms and the relative proportions of the pore types in each sample. The procedures developed by McCreesh *et al.* were applied and regression equations obtained to predict the shapes of mercury injection capillary pressure curves. Thus, relationships were obtained that indicate the size ranges of throats that access the different pore types as illustrated on Fig. 8.

Using the sizes of the pore types, the amount of porosity associated with them and their corresponding throat sizes, a Hagen–Poiseuille version of Darcy's Law was used to provide a simple capillary tube model of flow through a porous medium (Scheidegger, 1974; Brumfield, 1988). In this model, a flow path is treated as a capillary tube with a radius equal to that of the throat of a given pore type, and the pore system is treated as a bundle of capillary tubes of various radii.

The number of pores of a given type in a sample can be estimated from the size of a given pore type, its relative proportion and the known porosity of the sample. It is assumed that the number of throats aligned in the direction of flow is proportional to the number of pores of a given type and that the number of throats is obtained by multiplying the number of pores by a constant. It is assumed that this constant applies for all pore types within a given rock. The constant can be obtained by comparing the calculated values of permeability (assuming the constant = 1) with the actual values.

Brumfield (1988) reported that for three different suites of sandstone samples the number of "effective" throats per pore, calculated using the above methods, ranged from 2.3 to 3.3. In an ideal hexagonal structure there would be 6 throats per pore and two of these could be considered as oriented in each axial direction. With less ideal packing of grains the number of throats would be somewhat larger.

A fractional permeability is calculated for each pore type and the sum of these provides an estimate of the total permeability of the sample. For a genetically related suite of samples, the success of the method in predicting permeability from the frequency of various pore types determined from 2D images appears to depend on two other relationships; firstly, a given pore type is accessible via throats within predictable ranges of sizes and this relationship is not affected significantly by the abundance of a given pore type in a sample; secondly, the number of throats accessing pores is constant and independent of the sizes of the pores.

The mercury injection capillary pressure curve is responsive to the accessibility of pores via throats of certain sizes and the relationship between pore type and accessing throat size is made as if all pore types and accessing throat sizes were in parallel. The interpretation also uses a capillary tube model in inferring throat size. Accessibility via throats in parallel and a capillary tube model are both used in the permeability calculation and the precision of the prediction of permeability then depends mainly on the assumptions given in the previous paragraph. The success of the prediction does not depend on any particular correlation between porosity and permeability within a sample suite.

Brumfield (1988) and Etris *et al.* (1988) showed that a simple capillary tube model is sufficient to estimate Archie formation factors ($m$) for the suites of sandstone and carbonate samples studied. The higher the $m$ value the more conductive the porosity, and a relationship might be anticipated between a pore type that enhances permeability and one that enhances electrical conductivity. The methods used for the prediction of formation factor are similar to those used for the prediction of permeability. The average coefficients of proportionality between measured formation factor values and predicted formation factor values were relatively constant within a given data set for a genetically related suite of samples (i.e. samples from the same formation). However, the physical significance of the coefficient of proportionality is not understood.

The conclusion is that a thin section in combination with a mercury injection capillary pressure curve may provide sufficient information for the prediction of formation factor and that the Archie exponent ($m$) is likely to vary as pore–throat relationships vary.

## Pore systems in sedimentary rocks

The findings of Brumfield (1988) and Etris *et al.* (1988) that permeability and Archie's formation factor can be predicted using a combination of pore images from impregnated thin sections in combination with mercury injection capillary pressure curves provides support for the findings of McCreesh *et al.* (in press). That is, within a suite of genetically related rock samples, the geometric–topological aspects of the pore systems are not infinitely varied. Although the proportion of various pore types may vary from sample to sample within a suite of samples from a particular formation, the accessibility of pores of given types by throats of particular sizes does not appear to vary much. It is these relationships that allow pore characteristics to be used to predict permeability, electrical conductivity and drainage capillary pressure although these properties depend on throats more than pores.

It might be argued that permeability and electrical conductivity or formation factor are more easily obtained by direct measurement than by the methods outlined above, but where only small sample cuttings are available direct measurement are not possible by conventional methods. Perhaps a greater value of these results is in a better understanding of pore systems and their variability and the hope that the models devised may be useful in predicting other aspects of flow, such as relative permeability, which are much more difficult to obtain experimentally.

## PORE STRUCTURE AND DISPERSION

Dispersion in a porous rock is the phenomenon of mixing of two miscible fluids, one displacing the other. Adjacent flow paths of different sizes and different resistances to flow separate and rejoin, causing variations in both local pressure gradients and flow rates. A displacing phase entering a system as a slug or pulse becomes hydrodynamically mixed, and dispersion provides a measure of the extent of the zone of mixing.

A longitudinal dispersion coefficient ($K_l$) has been defined (Perkins and Johnson, 1963):

$$K_l = UL\left(\frac{\lambda_{90}-\lambda_{10}}{C}\right)^2 \qquad (5)$$

where

$\lambda_{90} = \dfrac{(V/V_p)-1}{\sqrt{V/V_p}}$ when effluent contains 90% displacing fluid

$\lambda_{10}$ = as above when effluent contains 10% displacing fluid

$C$ = constant

$L$ = length of sample (cm)

$U$ = average interstitial velocity

$V$ = volume of displacing phase injected

$V_p$ = total pore volume

and longitudinal dispersivity $\alpha_d$ is defined as

$$\alpha_d = \frac{K_l}{U} \qquad (6)$$

Thus, the extent of the spread of a concentration front by hydrodynamic mixing depends upon the dispersivity of the porous medium. However, as flow velocity decreases, the effects of molecular diffusion become increasingly important as a mixing mechanism.

In the following section, the effects of pore structure on dispersion at the scale of core samples are discussed.

### Pore structure and effluent profiles

Since the main interest here is in the effects of pore structure on dispersion, other factors that may affect dispersion are considered as being minimized. These include differences in viscosity and density of displacing and displaced phases; molecular diffusion; adsorption–desorption; connate water saturation; displacement rate; and sample size and shape.

Coats and Smith (1964) proposed a dispersion–capacitance model with a single average axial velocity based on the displaceable pore volume. Two distinct types of pore spaces are assumed. One type, the flowing fraction ($f$), conducts fluid at the average axial velocity, and the remaining portion ($1-f$) is assumed to be stagnant. Mass transfer is assumed between the stagnant and flowing fluids. The dispersion coefficient ($K_l$) contains the cumulative effects of various mechanisms causing axial dispersion in the flowing fraction. The rate constant for mass transfer ($K_m$) may modify the effects of dispersion by transferring components between flowing and stagnant portions. Mass transfer may occur by local eddy mixing as well as by molecular diffusion, so that both $f$ and $K_m$ can be rate dependent.

Some aspects of pore structure can be interpreted from responses of effluent concentration ($C$) at the core exit to step changes in ejected concentration at the core inlet from zero to $C_0$. Figure 9, modified after Batycky *et al.* (1982), illustrates profiles obtained by finite-difference solution of the dispersion–capacitance equations of Coats and Smith (1964).

If there were no dispersion ($K_l = 0$), a step change would be observed at the outlet identical to the one imposed at the inlet (Fig. 9(A)). As the dispersion coefficient ($K_l$) increases, the length of the zone of mixing increases and injected phase breaks through at smaller pore volumes of input. However, where there is no stagnant fraction ($f=1$), the effluent concentration ($C/C_0$) is always 0.5 of the injected concentration at one pore volume injected and the concentration gradient is mirror-symmetric about one pore volume of injection (Fig. 9(A)). The effluent concentration is $C$ at the core exit and $C_0$ at the core inlet.

For a system with random arrangement of pores and throats and $f=1$, an increase in the variability of pore and throat sizes caused increased dispersion (Fig. 9(A)). Bretz *et al.* (1988) have argued that earlier breakthrough (Fig. 9(B)) has the added requirement of preferential flow paths that imply some spatial ordering of throats. Spatial ordering of throats may be, but is not necessarily, related to spatial ordering of pores.

As the flowing fraction is decreased ($f<1$), the effluent concentration curve is displaced to the left and breakthrough occurs earlier because a portion of the pore system is not conductive (Fig. 9(C)). The magnitude of the flowing fraction can be estimated from the fraction of the pore volume corresponding to an effluent concentration ($C/C_0$) of 0.5. The symmetry of the profile is maintained provided there is no mass transfer between flowing and stagnant fractions ($K_m=0$).

The effect of mass transfer ($K_m>0$) is to skew the effluent profile, with increasing values of mass transfer increasing skewness (Fig. 9(D)). This asymmetric tailing increases the value of injected pore volume where $C/C_0=0.5$.

$K_m$ is dependent on viscosities, diffusion coefficient and the occurrence of eddies (Batycky *et al.*, 1982) and, as it increases further, the behaviour begins to approach that corresponding to larger values of $f$ and smaller values of $K_m$ (Fig. 9(D), $K_{m3}$). Thus, solution of the Coats and Smith (1964) differential equation may not be unique and interpretation using "best fit" parameters may be subjective.

Baker (1977) reports that the dispersion coefficient ($K_l$) and transfer coefficient ($K_m$) are velocity dependent, but that the flowing fractions ($f$) are relatively independent of velocity.

Bretz *et al.* (1988) correlate thin-section observations of pore size distributions and spatial clustering of pores with the form of effluent profiles. Berea standstone and two other sandstone samples were modelled well by flowing fractions close to 1 and had symmetric profiles with 0.5 normalized concentration arriving at about one pore volume injection. The pore size distributions were found to be relatively narrow and unimodal. There was no evidence of spatial ordering of pores for two of the samples and only limited ordering for the third. Thus, the pore structure data from thin sections are qualitatively consistent with the form of the effluent profiles.

Effluent profiles for carbonate samples tested showed early breakthrough, much lower flowing fractions and correspondingly wider pore size distributions and much more pore size correlation or clustering. Again, the observations from thin sections appeared to be consistent with the form of the effluent profiles.

Measurements by others of dispersion coefficients for bead packs indicate that the dispersion coefficient increases

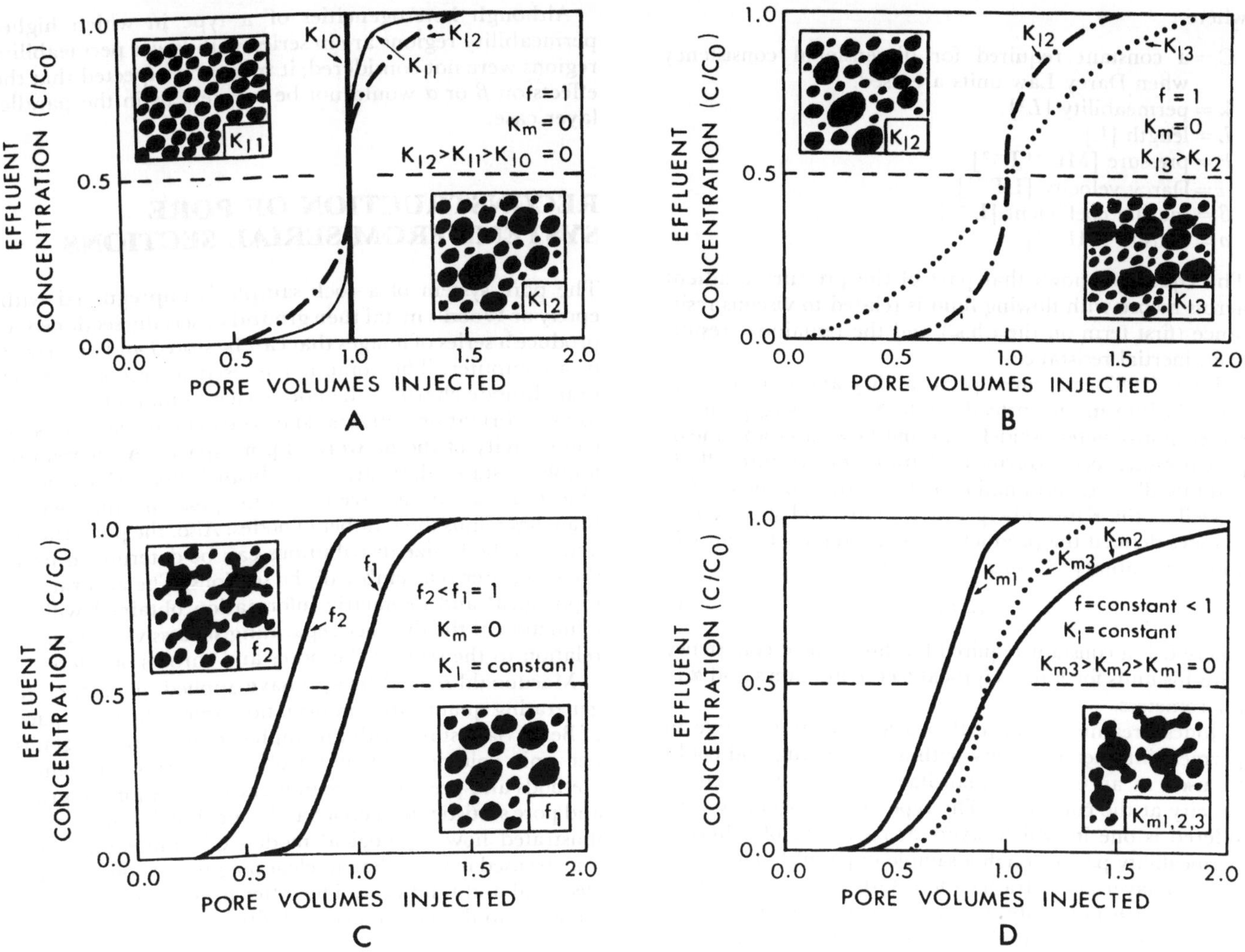


**Fig. 9.** Effluent concentration profiles from the Coats–Smith capacitance–dispersion model: (A) pores and throats are random but $K_{11}$ has less size variance that $K_{12}$; (B) $K_{12}$ has pores and throats randomly arranged whereas $K_{13}$ has "channels" of higher conductance traversing sample; (C) $f_1$ has no stagnant pore volume whereas $f_2$ has; (D) illustrates the effects of variable mass transfer between flowing and stagnant fractions. (Modified from Batycky *et al.*, 1982, and Bretz *et al.*, 1988.)

as the width of the particle size distribution increases (Perkins and Johnston, 1963). Spence and Watkins (1980) found that cores with wide pore size distributions had higher residual oil saturations following $CO_2$ displacements.

Ericson (1988) described a gaseous tracer technique for characterizing pore structures in macroscopically uniform granular materials. The histories of tracer concentrations at inlet and outlet are compared and both non-adsorbing and adsorbing tracers are used in a carrier gas. The properties of the porous structure that are evaluated are the fractions of flowing and stagnant porosity; the tortuosity of the flow porosity; the specific surface area of the pore space; and the fraction of the specific surface area that is associated with the flow porosity. Agreement between known and experimentally determined parameters was good in the granular materials tested.

### Relationships amongst dispersion, permeability and capillary pressure

Menzie (1987) investigated relationships between dispersivity and other core characteristics. The most conspicuous relationship existed between the permeability and the dispersion coefficient. For Berea sandstone cores the dispersion coefficient $K_l$ and the dispersivity $\alpha_d$ increased with the permeability. Similar relationships have been reported by Dutta (1984), Shamir and Harleman (1967) and Patel and Greaves (1987).

Menzie (1987) also showed that as the dispersion coefficient for Berea sandstone decreased, the plateau of the drainage capillary pressure curve increased, indicating smaller average throat sizes, which also corresponded with samples having lower core permeabilities.

Menzie (1987) also reported that Koval's (1963) heterogeneity factor ($H$) was higher for rocks with higher dispersivity. Since these rocks also had higher permeabilities, there were positive correlations of heterogeneity with permeability and dispersivity.

## INERTIAL COEFFICIENT AS A MEASURE OF HETEROGENEITY

Jones (1987) has presented evidence that the inertial coefficient ($\beta$) is strongly influenced by the degree of permeability heterogeneity in reservoir rocks. The inertial coefficient ($\beta$) is defined in the Forchheimer equation, which for one-dimensional flow is

$$-\frac{dP}{dL}=\frac{\mu v}{k}+C\beta\rho v^2 \qquad (7)$$

where

C = a constant required for dimensional consistency when Darcy Law units are used.
k = permeability $[L^2]$
$L$ = length [L]
$P$ = pressure $[ML^{-1}T^{-2}]$
$v$ = Darcy velocity $[LT^{-1}]$
$\beta$ = inertial coefficient $[L^{-1}]$
$\rho$ = density $[ML^{-3}]$

The equation shows that part of the pressure gradient across a core with flowing fluid is related to viscous resistance (first term on the r.h.s.) and the remainder results from inertial resistance.

Jones (1987) measured permeability and pore volume using helium in an unsteady-state Klinkenberg permeameter–porosimeter. Widely varying flow rates and mean pore pressures occur during each transient-pressure fall-off and this allows determination of the inertial coefficient ($\beta$) as well as the Klinkenberg permeability and slip factor.

Jones defined the product of the inertial coefficient ($\beta$) and permeability ($k$) as $\alpha$;

$$\alpha = C\beta k \qquad (8)$$

where $C$ is a constant required for dimensional consistency with his units for $\beta$ in reciprocal feet, $k$ in darcys and $\alpha$ in micrometres.

Jones presents evidence that $\alpha$ and $\beta$ are sensitive to permeability heterogeneity within a core plug and could be used as an effective quantitative measure of heterogeneity at the core scale. The type of heterogeneity considered is one in which layers or pathways of differing permeability pass through a sample in parallel. This is the case for layer-cake stratification. Other types of heterogeneity with permeability blocks in series or randomly arranged were not considered. The non-linear dependency of the inertial resistance on the square of velocity results in an unexpectedly high coefficient when it is computed from the average flow rate through such a non-homogeneous core plug.

When the pressure gradient and velocity of flow are low, the inertial resistance in the most permeable layer is fairly low and the flow rate in each layer can be approximated from the permeability. Jones calculated the effective inertial coefficients for composite systems with layers of differing permeability and showed that these coefficients were much higher than the coefficients of the layers making up the system. For a system with two layers, it was shown that the maximum effective coefficient occurs where about half of the total flow is in each of the layers.

Jones argues that at high pressure gradients, inertial resistance is high in the most permeable layers where velocity is high but is low in tight layers where velocity is lower. This increased resistance in high permeability layers reduces the flow in these layers. The same flow reduction could be accounted for with a permeability reduction of these layers. This would reduce the apparent permeability contrast and reduce the effective value of $\alpha_{eff}$. At lower pressure gradients, the apparent permeability reduction due to inertial resistance in the high-permeability layers is less. Thus, the apparent permeability contrast is greater and the resultant $\alpha_{eff}$ is larger than it was at the higher pressure gradients. The maximum $\alpha_{eff}$ for a given system occurs when the inertial resistance in the most permeable layer is just barely observable.

Jones concludes that determination of permeability heterogeneity from the measurement of $\alpha$ or $\beta$ is considerably easier and more sensitive than any other method available.

Although heterogeneities of a type in which higher permeability regions are in series with lower permeability regions were not considered, it might be expected that the effects on $\beta$ or $\alpha$ would not be as great as in the parallel layer case.

## RECONSTRUCTION OF PORE SYSTEMS FROM SERIAL SECTIONS

The pore system of a rock sample is impregnated with epoxy or Wood's metal then ground or serially sectioned to produce a series of images that can be stored and processed in a computer. The computer is used to reconstruct 3D branch-node charts of the pore space, commonly referred to as deformation retracts, and to calculate the genus or connectivity of the network of pore spaces. A theorem of topology states that, given the branch-node chart of an object, its genus is given by one plus the number of branches minus the number of nodes. Also, the pore–throat and pore-body size distributions can be determined from the computer reassembly of the sections. The accuracy of topological and geometric information obtained will be a function of the distance separating successive sections in relation to the sizes of the pores and throats of interest.

Macdonald *et al.* (1986a,b) have applied the method to generating quantitative information above the pore space of Berea Sandstone with an emphasis on obtaining genus per unit volume. Lin and Cohen (1982) used similar methods and presented preliminary observations of grain and pore statistics for Berea Sandstone. Furthermore, they illustrated how topological models of sedimentary rock may be used in modelling nuclear magnetic resonance and electrical and fluid flow. The latter provides the basis for going on to derive the permeability of the rock.

## SUMMARY AND DISCUSSION

Table III presents a summary of the methods considered and their use in determining the properties of pore systems. Also summarized are the fluid flow and displacement properties that can be assessed directly by these methods.

### Porosimetry

As mercury (nwp) penetrates in drainage, the fractions of the pore volume that are accessible via throats of various sizes can be measured (Wardlaw and Taylor, 1976). However, the size frequency distribution of throats is much more difficult to obtain because this requires additional information about the average numbers of throats connecting with pores, pore–throat size correlation and pore size distribution.

Pore size distribution can be derived from mercury withdrawal (imbibition) curves but only for those pores from which mercury withdraws. Mercury in the larger pores becomes disconnected and trapped before their withdrawal pressures are reached and their sizes are, therefore, not probed. Only a partial pore size distribution is obtained. Also, the physical processes that occur during imbibition are more varied and complex than during drainage (Li and Wardlaw, 1986; Chen and Koplik, 1985), making interpretation of imbibition curves more difficult.

As an alternative to the pressure-controlled porosimetry discussed in the preceding, both throat and pore sizes can be resolved by a procedure in which mercury is injected at very slow, constant rates such that small pressure drops can

**Table III.** Methods considered and their use in determining the properties of pore systems. Also summarized are the fluid flow and displacement properties that can be assessed directly by these methods. X indicates a strong relationship and x a weaker relationship

| | *Properties of pore system* | | | | | | | | | | | *Flow and displacement* | | | |
|---|---|---|---|---|---|---|---|---|---|---|---|---|---|---|---|
| *Method* | *Throat Size* | *Pore Size* | *Pore-size Clustering* | *Pore-throat Correlation* | *Pore-throat Connectivity* | *Surface Roughness* | *Dead-end and flowing Pore Fractions* | *Capillary Pressure-Saturation* | *Intrinsic Permeability* | *Relative Permeability* | *Electrical Conductance* | $S_{wi}$ | $S_{or}$ | $D_E$ | *Dispersivity* |
| Porosimetry | X | x | x | X | x | X | x | X | x | | x | X | X | X | |
| 2D image analysis | | X | X | | x | X | | | | | | | | | |
| Dispersion experiment | x | | x | | | | X | | | | | | | | X |
| Inertial coefficient experiment | | | X | | | | | | | | | | | | |
| 3D serial sections | x | X | X | x | X | x | x | | | | | | | | |

be associated with the filling of pores (Yuan and Swanson, 1986). However, the sizes of very small pores cannot be resolved by this method.

Interpretation of pore structure is simplified if the size, shape and arrangement of pores are obtained independently by image analysis and then combined with throat size frequency, pore–throat connectivity and degree of pore–throat correlation obtained from capillary pressure curves. Pore–throat correlation can be obtained from imbibition and secondary drainage curves and is a property that is difficult to obtain by other means but which affects relative permeability and displacement efficiency.

The critical end point $S_{or}$ and displacement efficiency can be obtained directly from capillary pressure curves, and several methods of determining permeability have been reported (Purcell, 1949; Swanson, 1978; Swanson, 1981; Thompson *et al.*, 1987a). Using these methods, permeability could be estimated from drill cuttings not suitable for conventional measurements.

## Image analysis

Information about the size, shape and arrangement of pores can be acquired from 2D slices of impregnated rock by image analysis. However, the sizes of throats, defined as minimum effective diameters between pores, cannot be obtained from 2D images.

However, geometric information obtained by image analysis has been correlated with data from drainage capillary pressure curves and it has been shown that predictable relationships commonly exist between the characteristics of pores and throats for suites of genetically related rock samples. With these relationships, it has been possible to use 2D pore characteristics to predict properties, such as drainage capillary pressure curves, permeability and electrical conductivity, that are related to throat sizes rather than pore sizes (Etris *et al.*, 1988; McCreesh *et al.*, in press).

The reasons for these relationships are not yet understood but a conclusion is that pore structures in rocks are not infinitely variable and display some common characteristics where they have been subjected to similar depositional and diagenetic histories. The method requires that the petrographic image analysis system be "trained" on a geologically related group of samples, for which independent measurements of the property of interest have been measured. Then, estimates of the property can be made for "unknown" but related samples. What constitutes a "related group of samples" has not yet been defined.

Advantages of the method are that it is not necessary to understand the complex physical events that occur during flow and displacement in order to make predictions of these properties. Since it is possible to identify the types of pores within a pore system that contribute the most to permeability, there is the potential for a better understanding of flow.

## Dispersion

Dispersion of miscible fluids flowing in core is influenced by the variability of pore and throat sizes, the presence of preferred flow paths and heterogeneity, all of which may be assessed to some degree by porosimetry and image analysis. However, dispersion also is strongly affected by the fraction of "dead end" pore volume in relation to the "flowing" pore volume and this property is best assessed from the structure of the effluent profile in a miscible displacement experiment. "Flowing fraction" has also been measured in an adaptation of mercury porosimetry in which, following breakthrough of mercury, mercury with a radioactive tracer is injected. From the volume injected at tracer breakthrough, in relation to the total volume of mercury present, the "flowing fraction" can be estimated (Chatzis and Dullien, 1985).

## Inertial coefficient

The effects of small-scale stratification (layers in parallel) on flow have been assessed through their effects on the inertial coefficient during unsteady-state flow tests. The extent to which quantitative information about other types of small-scale heterogeneity might be obtained from similar tests has yet to be explored.

## Serial sectioning and 3D computer reconstruction

This is the only method by which all aspects of pore structure can be resolved by direct observation. The disadvantages are the time and expense of the method and the relatively small sample volume that can be processed. If the determination of pore structure is not to be an end in itself, it is still necessary to interpret the effects of structure on flow and displacement.

# CONCLUSIONS

Quantitative models of pore structure are required to explain observed differences in fluid flow.

Potential uses of quantitative pore structure data are in predicting, under specified conditions, such properties as permeability, relative permeability, dispersion, capillary pressure–saturation relations, the critical end points of residual wetting and non-wetting phase saturation, the efficiency with which one fluid displaces another, and electrical conductivity.

Most of the properties of pore systems can be derived from a combination of mercury porosimetry, used in a constant injection rate mode as well as the more usual pressure-controlled mode, and with the addition of radioactive mercury miscibly displacing normal mercury. However, a more efficient method would be to use a combination of pressure-controlled porosimetry and image analysis in which essential throat characteristics derived by porosimetry are correlated with pore characteristics obtained by image analysis of 2D sections.

## NOMENCLATURE

| | |
|---|---|
| $C$ | Concentration $[ML^{-3}]$ |
| $d$ | Diameter $[L]$ |
| $d_e$ | Effective diameter $[L]$ |
| $D_e$ | Displacement efficiency |
| $f$ | Flowing fraction |
| $k$ | Permeability $[L^2]$ |
| $K_l$ | Longitudinal dispersion coefficient $[L^2T^{-1}]$ |
| $K_m$ | Rate constant for mass transfer $[T^{-1}]$ |
| $L$ | Length of sample $[L]$ |
| nwp | Non-wetting phase |
| $P$ | Pressure $[MT^{-2}L^{-1}]$ |
| $P_c$ | Capillary pressure $[MT^{-2}L^{-1}]$ |
| $S$ | Saturation |
| $U$ | Average interstitial velocity $[LT^{-1}]$ |
| $v$ | Darcy velocity $[LT^{-1}]$ |
| V | Volume $[L^3]$ |
| $V_p$ | Pore volume $[L^3]$ |
| wp | Wetting phase |
| $x$ | Principal diameter of a meniscus $[L]$ |
| $y$ | Principal diameter of a meniscus $[L]$ |
| $Z$ | Number of throats connecting with each pore |

SUBSCRIPTS

| | |
|---|---|
| $a$ | Applied |
| $A$ | Advancing |
| $C$ | Capillary |
| $e$ | Effective |
| Hg | Mercury |
| $i$ | Irreducible |
| $l$ | Longitudinal |
| $nw$ | Non-wetting |
| $p$ | Pore |
| $r$ | Residual |
| $R$ | Receding |
| $t$ | Throat |
| $v$ | Vacuum |
| $w$ | Water |

GREEK

| | |
|---|---|
| $\alpha =$ | $C\beta k$, where $\beta$ is in reciprocal feet, $k$ is darcys and $\alpha$ in micrometres. $C$ is a constant required for dimensional consistency. |
| $\alpha_d$ | Longitudinal dispersivity $[L]$ |
| $\beta$ | Inertial coefficient $[L^{-1}]$ |
| $\gamma$ | Surface tension $[MT^{-2}]$ |
| $\theta$ | Contact angle (degrees) |
| $\lambda_{90}$ | $(V/V_p) - 1/\sqrt{V/V_p}$ when effluent contains 90% displacing fluid |

## REFERENCES


Baker, L.E. (1977). Effects of dispersion and dead-end pore volume in miscible flooding. *Soc. Petroleum Eng. J.* **263** 219–227.

Batycky, J.P., Maini, B.B. and Fisher, D.B. (1982). Simulation of miscible displacement on full diameter carbonate cores. *Soc. Petroleum Eng. J.* **22**, 647–657.

Bretz, R.E., Specter, R.M. and Orr, F.M. (1988). Effect of pore structure on miscible displacement in laboratory cores. *Soc. Petroleum Eng. Reservoir Eng.* **3**, 857–866.

Brumfield, D.S. (1988). Derivation of an empirical model consistent with a variety of physical properties. Unpublished Master's thesis, The University of South Carolina.

Chatzis, I. and Dullien, F.A.L. (1985). The modeling of mercury porosimetry and the relative permeability of mercury in sandstones using percolation theory. *Int. Chem. Eng.* **25**, 47–66.

Chatzis, I., Morrow, N.R. and Lim, H.T. (1983). Magnitude and detailed structure of residual oil saturation. *Soc. Petroleum Eng. J.* **23**, 311–326.

Chen, J.D. and Koplik, J. (1985). Immiscible fluid displacement in small networks. *J. Colloid Interface Sci.* **108**, 304–330.

Coats, K.H. and Smith, B.D. (1964). Dead end pore volume and dispersion in porous media. *Soc. Petroleum Eng. J.* **231**, 73–84.

Dawe, R.A., Wheat, M.R. and Wright, R.J. (1985). Effects of heterogeneities on miscible flow processes in EOR. *3rd European Meeting on Improved Oil Recovery, Rome, April.*

Dutta, S. (1984). Dispersivity as an oil reservoir rock property. M.S. Thesis, Universty of Oklahoma, Norman.

Dykstra, H. and Parsons, R.L. (1950). The prediction of oil recovery by waterflood. In *Secondary Recovery of Oil in the United States*, 2nd edn., pp. 160–174. American Petroleum Institute.

Ehrlich, R. and Full, W.E. (1988). Sorting out geology—unmixing mixtures. *In*: Mann, J. McCammon, R., and Jones, T. (eds.). *Use and Abuse of Statistical Methods of the Earth Sciences.* Oxford Universty Press. (In press.)

Ehrlich, R., Crabtree, S.J., Kennedy, S.K. and Cannon, R.L. (1984). Petrographic image analysis I-analysis of reservoir pore complexes: *J. Sediment. Petrol.* **54** (4), 1365–1378.

Ehrlich, R., Horkowitz, K.O., Horkowitz, J.P. and Crabtree, S.J., Jr. (1988). Functional and objective classification of reservoir porosity: relating petrographic image analysis to petrophysics II: *Am. Assoc. Petroleum Geol. Bull.* (In press)

Ericson, K.L. (1988). Gaseous tracer techniques for pore structure characterization of granular materials. United States Department of Energy Contract DE-AC04-76DP 00789, Sandia National Laboratories, sand 86-2585C.

Etris, E.L., Brumfield, D.S., Ehrlich, R. and Crabtree, S.J. (1988). Relations between pores, throats and permeability: a petrographic/physical analysis of some carbonate grainstones and packstones. *Carbonates and Evaporites* **3**, 17–32.

Full, W.E., Ehrlich, R. and Klovan, J.E. (1981). Extended Qmodel—objective definition of external end members in the analysis of mixtures. *Math. Geol.* **13** (4), 331–344.

Full, W.E., Ehrlich, R. and Bezdek, J.C. (1982). Fuzzy Qmodel—a new approach for linear unmixing. *Math. Geol.* **14**, 259–270.

Jones, S.C. (1987). Using the inertial coefficient, $\beta$, to characterize heterogeneity in reservoir rock. *Society of Petroleum Engineers 62nd Annual Technical Conference, Dallas, Texas, September*, pp. 165–174, SPE 16949.

Klovan, J.E. and Miesch, A.T. (1976). Extended CABFAC and Qmodel computer programs for Q-mode factor analysis of compositional data. *Computers and Geosciences* **1**, 161–178.

Koval, E.J. (1963). A method for predicting the performance of unstable miscible displacement in heterogeneous media. *Soc. Petroleum Eng. J.* **228**, 145–154.

Lenormand, R., Zarcone, C. and Sarr, A. (1983). Mechanisms of the displacement of one fluid by another in a network of capillary ducts. *J. Fluid Mech.* **135**, 337–353.

Li, Y. and Wardlaw, N.C. (1986). Mechanisms of non-wetting phase trapping during imbibition at slow rates. *J. Colloid Interface Sci.* **109**, 473–486.

Li, Y. and Wardlaw, N.C. (in press). Quantitative determination of pore structure from capillary pressure curves. In: N. Morrow (ed.), *Interfacial Phenomena in Petroleum Recovery*. Marcel Dekker, New York.

Li, Y., Laidlaw, W.G., Wardlaw, N.C. (1986). Sensitivity of drainage and imbibition to pore structures as revealed by computer simulation of displacement process. *Adv. Colloid Interface Sci.* **26**, 1–68.

Lin, C. and Cohen, M.H. (1982). Quantitative methods of microgeometric modelling. *J. Appl. Phys.* **53**, 4152–4165.

Macdonald, I.F., Kaufmann, P. and Dullien, F.A.L. (1986a). Quantitative image analysis of finite porous media 1. Development of genus and pore map software. *J. Microsc.* **144**, 277–296.

Macdonald, I.F., Kaufmann, P. and Dullien, F.A.L. (1986b). Quantitative image analysis of finite porous media 2. Specific genus of cubic lattice models and Berea sandstone. *J. Microsc.* **144**, 297–316.

McCreesh, C.A., Ehrlich, R. and Crabtree, S.J., Jr. (in press). Estimation of mercury-injection curves from thin-section porosity measurements: Relating petrographic image analysis to petrophysics I. *Am. Assoc. Petroleum Geol. Bull.*

Melrose, J.C. (1988a). Valid capillary pressure data at low wetting phase saturations. *Society of Petroleum Engineers 63rd Annual Conference, Houston, Texas*, pp. 713–718, SPE 18331.

Melrose, J.C. (1988b). Use of water-vapor desorption data in the determination of capillary pressures at low water saturations. *Soc. Petroleum Eng. Reservoir Eng.* **3**, 913–918.

Menzie, D.E. (1987). Dispersivity as an oil reservoir rock characteristic. *U.S. Department of Energy Progress Review* (for quarter ending Sept. 30) pp. 111–113.

Morrow, N.R. (1971). Small scale packing heterogeneities in porous sedimentary rocks. *Am. Assoc. Petroleum Geologists Bull.* **55**, 514–522.

Patel, K.M. and Greaves, M. (1987). Surfactant dispersion in porous media. *Chem. Eng. Res. Design* **65**, 12–22.

Perkins, T.K. and Johnston, O.C. (1963). A review of diffusion and dispersion in porous media. *Soc. Petroleum Eng. J.* **227**, 70–84.

Purcell, W.R. (1949). Capillary pressures—their measurement using mercury and the calculation of permeability therefrom. *Trans. Am. Inst. Mining Eng.* **186**, 39–48.

Scheidegger, A.E. (1974). *The Physics of Flow Through Porous Media*, 3rd edn. University of Toronto Press.

Shamir, U.Y. and Harleman, D.R.F. (1967). Dispersion in layered porous media. *J. Hydraulics Division, Proc. Am. Soc. Civil Eng* **93** (No. Hy5), 237–260.

Spence, A.P. and Watkins, R.W. (1980). The effect of microscopic core heterogeneity on miscible flood residual oil saturation. *Society of Petroleum Engineers of AIME, 55th Annual Technical Conference, Dallas, Texas*, pp. 1–17, SPE 9229.

Swanson, B.F. (1978). A simple correlation between air permeabilities and stressed brine permeabilities with mercury capillary pressures. *Society of Petroleum Engineers 53rd Technical Conference, Houston, Texas*, pp. 1–8, SPE 8234.

Swanson, B.F. (1981). A simple correlation between permeabilities and capillary pressures. *J. Petroleum Technol.* **33**, 2498–2504.

Thompson, A.H., Katz, A.J. and Krohn, C.E. (1987a). The microgeometry and transport properties of sedimentary rock. *Adv. Phys.* **36**, 625–694.

Thompson, A.H., Katz, A.J. and Raschke, R.A. (1987b). Estimation of absolute permeability from capillary pressure measurements. *Society of Petroleum Engineers 62nd Annual Technical Conference, Dallas, Texas*, pp. 475–481, SPE 16794.

van Brakel, J. (1981). A special issue devoted to mercury porosimetry. *Powder Technol.* **29**, 1–208.

Wardlaw, N.C. (1980). The effects of pore structure on displacement efficiency in reservoir rocks and in glass micromodels. *Society of Petroleum Engineers First Joint SPE/DOE Symposium on Enhanced Oil Recovery, Tulsa, Oklahoma*, pp. 345–352, SPE 8843.

Wardlaw, N.C. and Cassan, J.P. (1978). Estimation of recovery efficiency by visual observation of pore systems in reservoir rocks. *Bull. Can. Petroleum Geol.* **26**, 572–585.

Wardlaw, N.C. and Taylor, R.P. (1976). Mercury capillary pressure curves and the interpretation of pore structure and capillary behaviour in reservoir rocks. *Bull. Can. Petroleum Geol.* **24**, 225–262.

Wardlaw, N.C., Li, Y. and Forbes, D. (1987). Pore–throat size correlation of capillary pressure curves. *Transport in Porous Media* **2**, 597–614.

Wardlaw, N.C., McKellar, M. and Li, Y. (1988). Pore and throat size distributions determined by mercury porosimetry and by direct observation. *Carbonates and Evaporites* **3**, 1–15.

Yuan, H.H. and Swanson, B.F. (1986). Resolving pore space characteristics by rate controlled porosimetry. *Society of Petroleum Engineers 5th Symposium on Enhanced Oil Recovery, Tulsa, Oklahoma*, pp. 167–180, SPE 14892.

# 12

# Comparison between Laboratory Experiments and Detailed Simulations of Unstable Miscible Displacement Influenced by Gravity

*M.A. Christie, A.D.W. Jones and A.H. Muggeridge*

BP Research, Sunbury-on-Thames, Middlesex, UK

This paper presents a study of oil displacement by a less-dense, less-viscous solvent. A comparison is made between laboratory observations and high-resolution simulations in order to test the predictive capability of the numerical model. Observations are made of linear displacement in a Ballotini pack with the principal flow at right angles to the direction of gravity. The pack is sufficiently thin that flow is approximately two-dimensional. Flow is visualized using a dye tracer and recorded on video; subsequent image processing of the video film enables the concentration distribution to be determined for any chosen time. The effluent is analysed by colorimetry. The numerical model used to simulate the experiment has a simplified physical description (two-dimensional displacement of oil and solvent). High-order finite-difference techniques in conjunction with extremely fine grids provide detailed resolution of the flow.

We obtain good agreement between simulation and experimentally observed outlet concentration and production curves. The agreement was obtained using only measured laboratory data together with a random mechanism for initiating fingers in the simulation. The visual appearance of the simulation is close to the observations, exhibiting fingering and gravity override and comparisons of concentration profiles show good agreement between simulation and experiment. The simulations correctly predict the transition between gravity override and fingering as a function of rate.

## INTRODUCTION

In any field-scale miscible gas project, viscous fingering and gravity override of the lighter more mobile solvent are likely to occur. It is important to have accurate predictions of breakthrough time and sweep efficiency to establish the design and to assess the economic risks in implementing such a project.

Computer simulation of the effects of viscous fingering can in principle predict breakthrough times and sweep efficiencies (assuming that physical and reservoir properties are known) provided that (i) the physics of the flow is adequately described, (ii) grid resolution is fine enough, and (iii) the mechanism for triggering the fingers models reality. To assess whether these conditions are achievable, we need to compare numerical simulation with a well-characterized experiment.

Blackwell's experiments (Blackwell *et al.*, 1959) provide good, well-characterized data on unstable miscible flow in a thin, horizontal bead-peak. Blackwell investigated viscous fingering in a linear displacement; other experiments reported results on fingering in a quarter five-spot (Habermann, 1960); Lacey *et al.*, 1961), or fingering influenced by gravity (Dumoré, 1964; van der Poel, 1962; Crane *et al.*, 1963). Peaceman and Rachford (1962) reported early studies of numerical simulation of fingering on a 40 × 20 grid. Three-dimensional (3D) miscible fingering experiments were recently reported by Withjack and Akervoll (1988).

Christie and Bond (1987) obtained good agreement between high-resolution simulations and Blackwell's experiments. They used the measured experimental data as input to a computer code that featured accurate numerical methods on a fine grid. They used a random initial concentration fluctuation to simulate experiments at mobility ratios of 5 and 86, and obtained excellent predictive agreement with the recovery curves reported by Blackwell. Using the same high-resolution code, Christie and Jones (1987) found good agreement between simulated and observed production rates and average concentrations for a horizontal bead-pack.

The effect of gravity on unstable miscible flow has not generally been considered in published simulations. Fayers and Newley (1988) reported Christie's comparison of high-resolution simulations with van der Poel's (1962) experiments, which were on a bead-pack of dimensions

*North Sea Oil and Gas Reservoirs–II*

80 cm × 4.8 cm × 1.8 cm. The rate at which the behaviour switched from gravity override to fingering was underpredicted by a significant factor. However, good agreement was obtained with the experimental results for the low- and high-rate limits—see Fig. 1. A possible reason for this lack of agreement between the measured and predicted behaviour over the transition rates is the use of 2D code to model a 3D experiment. Fayers and Muggeridge (1989) carried out simulations of experiments reported by Crane *et al.* (1963) that support this idea, but a detailed comparison with a well-characterized experiment is required to confirm this assumption.

The purpose of this paper is to report a detailed comparison between simulation and experiment for unstable miscible flow in a thin bead-pack with flow in a vertical plane and with gravity acting at right angles to the principal direction of flow. The aim is to compare simulated and experimental results over a range of rates and density differences, so that the transition from gravity-dominated to viscous-dominated flow can be characterized.

## EXPERIMENT

### Design and techniques

A pack of 355 ± 65 μm diameter Ballotini beads was chosen as the porous medium bcause it enabled a relatively homogeneous sample to be constructed and simple flow-visualization techniques to be used. The pack, which was made from Perspex, was 60 cm long, 9 cm wide and 0.65 cm thick with a rectangular cross-section. The packing technique was based on a method described by Wygal (1963). With the enclosure held vertically, the beads were poured in through a stack of five eight-mesh wire screens spaced 2.5 cm apart, with the lowest screen 38 cm above the end of the pack. The model was then vibrated for 30–60 s, in order to pack down the beads; more beads were added and the process was repeated until the model was full.

The pack thickness was determined by the requirement that the flow be essentially two-dimensional so that direct comparison with 2D numerical simulations could be made. The wavelength (given by $2\pi/k$, where $k$ is the wavenumber) of the smallest unstable mode was calculated from (see eq. (4) of Christie and Bond, 1987)

$$k^2_{\max} = \frac{v}{\lambda D_T} \frac{d\lambda}{dc} \frac{\partial c}{\partial x} \qquad (1)$$

to be of order 0.2 cm. In Eq. (1), $v$ is the velocity, $\lambda$ is the mobility (permeability/viscosity), $c$ is the solvent concentration and $D_T$ is the transverse dispersion coefficient. The concentration gradient $\partial c/\partial x$ was estimated by assuming that the concentration went from 1 to 0 over a length scale of approximately 8 beads or 0.3 cm. The maximum growth rate of the allowable unstable modes in the third dimension can be calculated as $\omega = 8.4 \times 10^{-4}\,s^{-1}$ (or, in units of pore volumes injected, $\omega \simeq 3pv^{-1}$). This growth rate is sufficiently small that fingering in the third dimension will not be important, and no 3D effects were seen in the experiment.

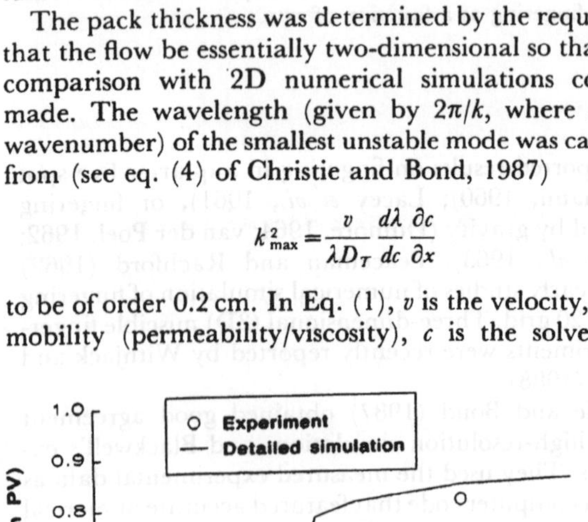


**Fig. 1.** Dependence of recovery on flow-rate for the van der Poel experiments ($M = 7.9$).

To establish uniform flow at the inlet, fluid was injected into a 1 cm wide reservoir that was separated from the pack by a wire gauze. The reservoir served to smooth out pressure variations. At the outlet there was a 0.2 cm wide reservoir that was divided into three sections with an outlet port in the centre of each section.

In the displacement experiments, the model rested on one edge with the gravity vector perpendicular to the principal flow direction and fluid was injected at constant rate and produced at constant pressure (Fig. 2). Before conducting a displacement, the model was first turned so that the inlet pointed upwards. One fluid was then injected through the inlet and the other through the outlet and both were produced through bleed holes 5 cm downstream of the inlet. The model was then turned back onto its edge and the displacement was started. In this way a sharp, vertical interface was obtained giving well-defined initial conditions.

Dye was used as a tracer in the displaced phase, which enabled the effluent to be analysed by colorimetry. From this data, outlet concentration and production were determined with an accuracy of ±2%. The displacement was recorded on video and individual frames were subsequently digitized using an image processor to give quantitative luminance maps. Calibration curves relating luminance to concentration and position (to account for the possibility of non-uniform illumination of the pack) were obtained by analysing images of the pack filled with known fluid concentrations. Using these calibration curves, a concentration map could be found for any chosen time during displacement. Concentration profiles across the pack and mean concentration as a function of distance from the inlet were found for selected times. The point concentration measurements were accurate to ±10% and the mean profiles were accurate to ±5%.

### Pack properties

The Ballotini pack had a 39% porosity and a permeability of 115 D. These figures are typical of values reported for homogeneous packs (Wheat, 1984). The dispersion characteristics were measured at flow rates between 0.15 cm s$^{-1}$ and 0.060 cm s$^{-1}$, which was the range used in the displacement experiments. The measurements of the longitudinal dispersion coefficient were made at a favourable mobility ratio of 0.26 and a least-squares fit to the data gave

$$D_L = 0.045(v/\phi) \quad \text{cm}^2\,\text{s}^{-1} \qquad (2)$$

The transverse dispersion coefficient was determined from production data from two parallel matched viscosity streams (Blackwell, 1962). A least-squares fit to the data gave

$$D_T = 0.0015(v/\phi) \quad \text{cm}^2\,\text{s}^{-1} \qquad (3)$$

These dispersion coefficients are at the lower end of the ranges reported in the literature for homogeneous packs (Blackwell, 1962; Brigham *et al.*, 1961) and indicate that the model packing was uniform. This was confirmed by the

**Observed flow**

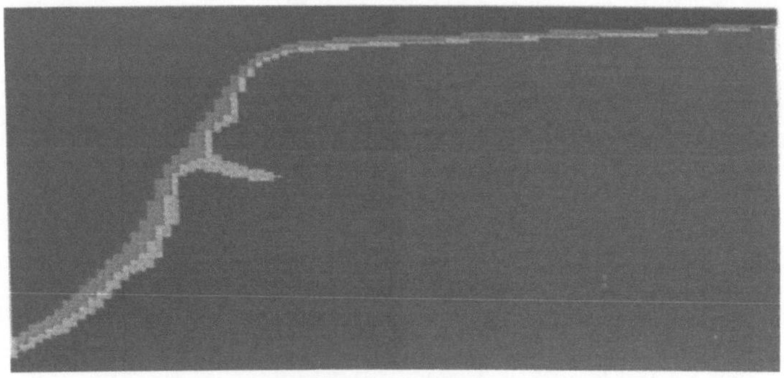

**Simulated flow**

**Plate 27.** Detail of flow near inlet at 0.1 pvi for experiment 1.

**Observed flow**

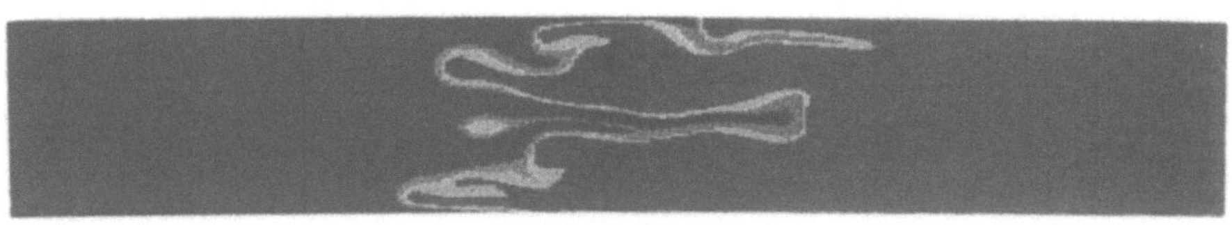

**Simulated flow**

**Plate 28.** Observed and predicted behaviour at 0.5 pvi for experiment 3.

**Observed flow**

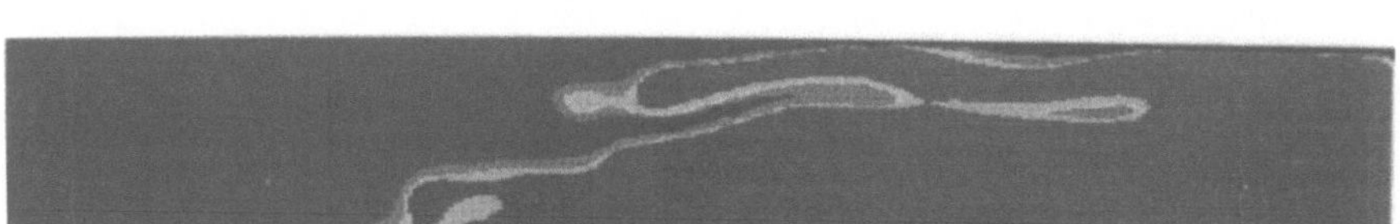

**Simulated flow**

**Plate 29.** Observed and predicted behaviour at 0.5 pvi for experiment 6.

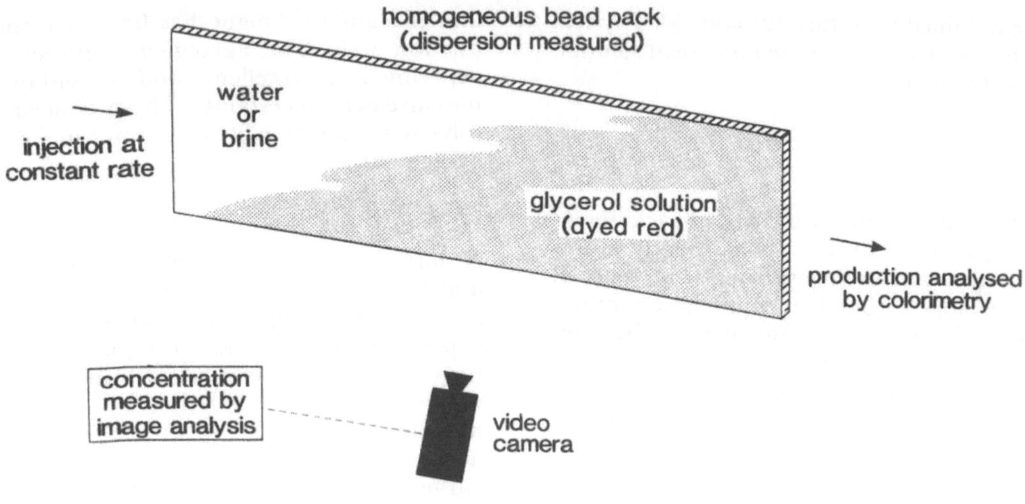


**Fig. 2.** Schematic diagram of the experiment.

uniformity of the front in a unit mobility ratio displacement.

## Experimental conditions

Six different experiments were carried out using two fluid pairs (water/glycerol and brine/glycerol) and three different rates. The variation of viscosity with concentration was measured for each fluid pair and it was found that the data were accurately represented by the quarter-power mixing rule. This curve lay within 3% of the water/glycerol data, and within 1.5% of the brine/glycerol data. The relevant experimental data are summarized in Table I.

The behaviour of the displacement can be characterized by the ratio of the time taken to inject one pore volume of fluid ($L/v$) to the time taken for solvent to rise to the top of the pack under gravity ($\Delta\rho\, gk/\mu_s H$). A suitable dimensionless number to characterize the displacement is given by

$$N_g = \frac{v\mu_s}{\Delta\rho\, gk}\frac{H}{L}(M-1) \tag{4}$$

where v is the Darcy velocity, $H$ is the pack height, $L$ is the pack length, $M$ is the mobility ratio, $\Delta\rho$ is the density difference, $\mu_s$ is the solvent viscosity, $g$ is the acceleration due to gravity and $k$ is the permeability. The $M-1$ factor ensures that $N_g$ has the correct behaviour for mobility ratios close to 1 (see Fayers and Muggeridge, 1989). The experiments lay in the range $N_g=0.35$ to $N_g=7.78$ and covered the transition from gravity override to viscous fingering.

**Table I.** Summary of measured experimental data

| *Expt.* | $\mu_s(cP)$ | $\Delta\rho(g\,cm^{-3})$ | $v/\phi(cm\,s^{-1})$ | $M$ | $N_g$ |
|---|---|---|---|---|---|
| 1 | 0.951 | 0.136±0.002 | 0.015 | 6.8±0.3 | 0.35 |
| 2 | 0.951 | 0.136±0.002 | 0.030 | 6.8±0.3 | 0.70 |
| 3 | 0.951 | 0.136±0.002 | 0.059 | 6.8±0.3 | 1.37 |
| 4 | 1.280 | 0.023±0.002 | 0.015 | 5.0±0.2 | 1.91 |
| 5 | 1.280 | 0.023±0.002 | 0.030 | 5.0±0.2 | 3.83 |
| 6 | 1.280 | 0.023±0.002 | 0.061 | 5.0±0.2 | 7.78 |

# NUMERICAL SIMULATION

The design aims for the computer code to model fingering were to obtain both accuracy and speed so that the full range of unstable wavelengths could be modelled. The accuracy requirements were met by using an explicit finite-difference code based on flux corrected transport (FCT) on a fine grid, and the speed was obtained by vectorization of the code for use on a CRAY computer. A description of the model and its validation against Blackwell's experiments is given by Christie and Bond (1987). For completeness, a brief review of the assumptions and equations is given below. In order to obtain the highest resolution of fingering behaviour we used a formulation containing only the essential physics to describe unstable flow, and made the following assumptions:

(i) 2-phase Darcy flow (aqueous and oleic phases);
(ii) 3 components—water, oil, solvent;
(iii) all components incompressible;
(iv) oil and solvent first contact miscible;
(v) ideal mixing for oil and solvent densities;
(vi) specified mixing rule for viscosities.

The conservation equations describing the flow were written using the total velocity formulation described by Peaceman (1977), and are described in detail by Christie (1989). Only the equations governing solvent–oil flow are reproduced below:

PRESSURE EQUATION

$$\nabla\cdot\lambda\nabla p + Q = \nabla\cdot\gamma(c)\mathbf{g} \tag{5}$$

where

$p$ is the pressure
$c$ is the concentration
$Q$ is the source/sink term
$\mathbf{g}$ $=(g_x, g_y)$ is the (normalized) gravity term
$\lambda(c)$ $=k(x,y)/\mu_o(c)$ is the oil phase mobility
$\gamma(c)$ $=c\rho_s+(1-c)\rho_1$ is the specific gravity

TOTAL VELOCITY

$$\mathbf{v}_t = -\lambda\nabla p + \gamma(c)\mathbf{g} \tag{6}$$

CONSERVATION OF SOLVENT

$$\frac{\partial c}{\partial t} + \nabla\cdot\mathbf{v}_t c = \nabla\cdot D\nabla c \tag{7}$$

where $D$ is the dispersion tensor. For all the cases studied we used the simplified form:

$$D=\begin{pmatrix} D_L & 0 \\ 0 & D_T \end{pmatrix} \tag{8}$$

$D_L$ and $D_T$ were obtained from Eqs. (2) and (3). Physical dispersion effects were included in the numerical solution (see Christie and Bond, 1987).

## RESULTS

Experimental observation of fingering in bead packs shows that recovery as a function of time is consistently reproducible. In simulating viscous fingering in the presence of gravity, we would hope to be able to predict quantities such as average concentration, average finger width, and concentration in the fingers. We would not expect to be able to predict exact locations of individual fingers, since we do not know the details of how those fingers are triggered. However, by comparing these measured average quantities with the equivalent values predicted by simulation, we can assess how well the mathematical model predicts the overall flow behaviour.

### Input data and initial conditions

All the input data used in the simulation was taken from experimental measurements, and is shown in Table I. The simulation started with a vertical solvent–oil interface at the position of the bleed valves. All the simulations reported were carried out on a 300 × 120 grid. At this level of resolution each grid block has dimensions 0.18 cm × 0.075 cm, which is approximately 5 bead diameters by 2 bead diameters. These grid block sizes are still too large to resolve fully the measured levels of physical dispersion. To some extent this limits the resolution of the fingers by excluding small-wavelength modes with positive growth rates. Refining the grid further was not considered worthwhile because the assumptions used to derive Darcy's law break down on these length scales, and details such as mixing due to the tortuous flow paths as well as fingering triggered by correlations in the positions of individual beads become increasingly important.

### Comparison of simulation and experiment

*Low rate*

At very low rates, buoyancy terms dominate viscous forces and a gravity tongue is formed. Any fingers that are triggered rise to the top of the pack. Plate 27 shows the development of the gravity tongue after 0.1 pv for experiment 1. Both the simulation and experiment show significant gravity override, and a small finger can be seen just over half-way up the pack. Both the simulation and experiment show that the foot of the initial interface moves backward at early times. The location of the leading edge of the gravity tongue is also accurately predicted. Figure 3 shows a plot of the 50% concentration contour from the simulation after 0.5 pv injected, with the experimentally determined height of the 50% concentration level plotted at 20 points. The finger visible in Plate 27 has disappeared, and the gravity tongue has broken through to the production well. The agreement between simulation and experiment is excellent, and is within the bounds of measurement uncertainty. The produced concentration of solvent as a function of time, shown in Fig. 4, is also in good agreement.

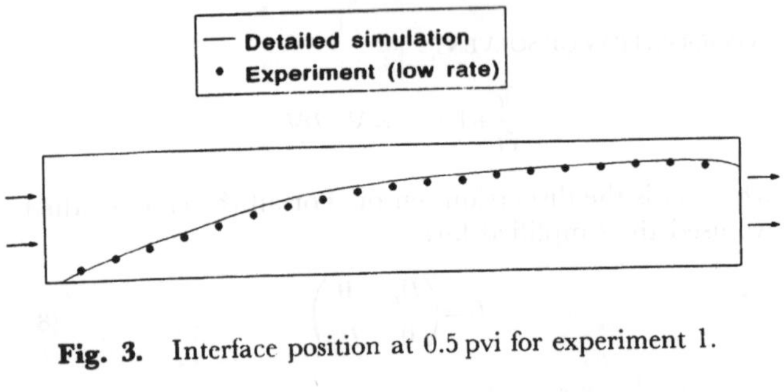


**Fig. 3.** Interface position at 0.5 pvi for experiment 1.

*High rate*

At high rates, viscous forces dominate over gravity forces and the displacement is dominated by fingering. Figure 5 shows the average concentration $\hat{c} = \int_0^1 c(x,y)\,dy$ for both experiment 6 and the corresponding simulation. The fingering gives rise to a broad transition zone that is well represented by the simulation. The simulated curve shows fluctuations arising from averaging over individual fingers. These fluctuations are not apparent in the experimental curve under these conditions. Horizontal experiments at higher mobility ratios carried out by Christie and Jones (1987) showed fluctuations in the equivalent experimental curve. However, both horizontal and vertical experiments show consistently lower levels of fluctuations than the equivalent simulations, and it is not clear why the discrepancy occurs. The solvent production curves for both experiment and simulation shown in Fig. 6 are in good agreement.

To compare the range of finger wavelengths seen in the simulations and in the experiment, we computed discrete Fourier transforms of the concentration fluctuations at a particular distance down the core. We computed

$$z_k(x) = \sum_{j=1}^{NY} c(x, y_j) \exp\left(-i\frac{2\pi(j-1)k}{NY}\right) \quad (9)$$

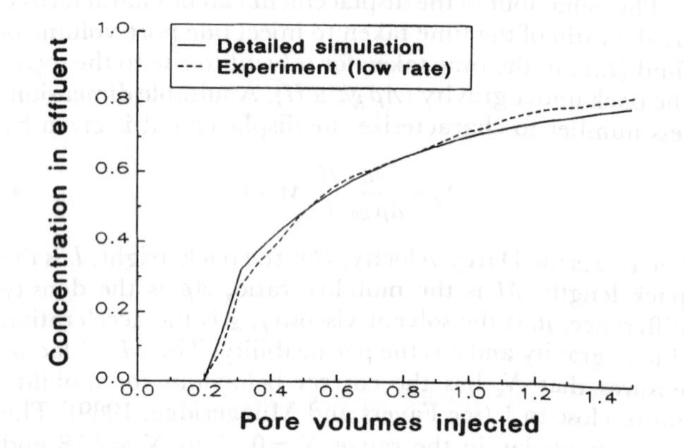


**Fig. 4.** Comparison of solvent production curves for experiment 1.

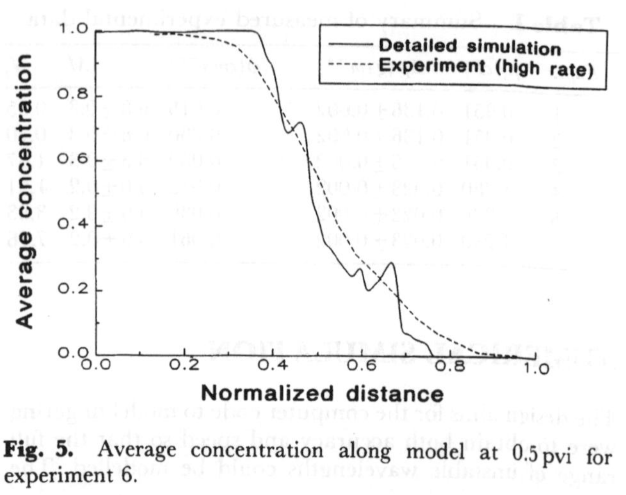


**Fig. 5.** Average concentration along model at 0.5 pvi for experiment 6.

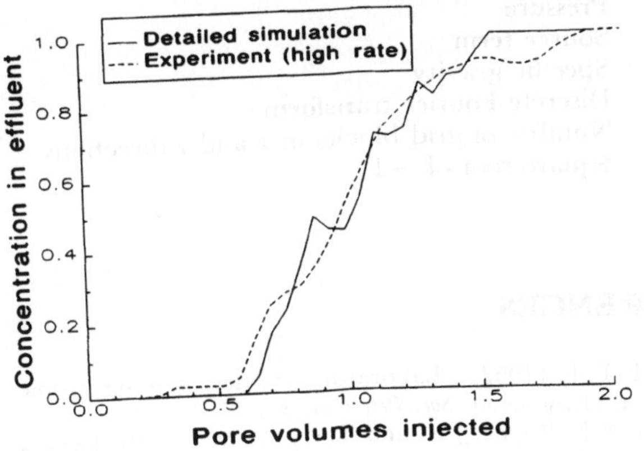


**Fig. 6.** Comparison of solvent production curves for experiment 6.

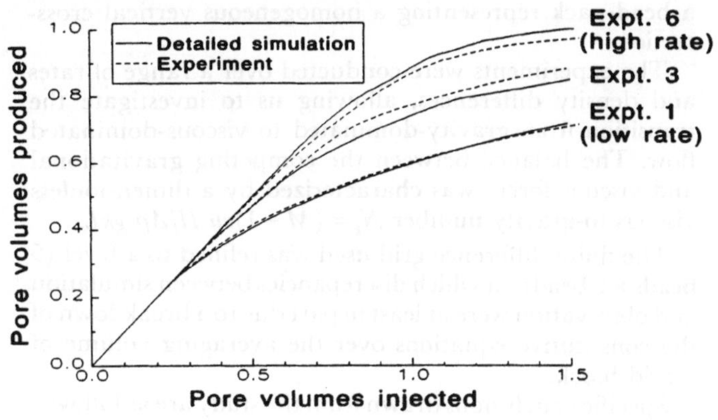


**Fig. 8.** Comparison of oil recovery curves for experiments 1, 3, and 6.

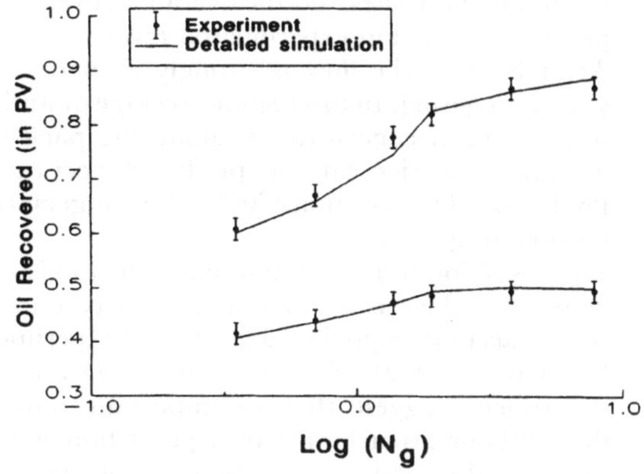


**Fig. 9.** Dependence of recovery on flow rate.

Figure 7 shows a plot of the magnitude of $z_k(0.5)$ against $k$ for both simulation and experiment for experiment 6. We can see the same behaviour in that the amplitude increases at small wavenumber (long wavelength) for both simulation and experiment. The experiment shows more evidence of short-wavelength fluctuations than the simulations. There are also some differences at long wavelengths, but since both plots represent a single slice along $x=0.5$ at $t=0.5$ from one realization, the agreement can be considered reasonable. Plate 29 shows a comparison of the experimental and simulated fingering at 0.5 pv injected. There is broad agreement on the large-scale features, but the experiment shows more fine-scale detail than the simulation; this is consistent with the Fourier transform results.

*Transition region*

The variation of pore volumes of oil recovered against time are plotted in Fig. 8 for experiments 1, 3, and 6. For low rates (experiment 1), a single finger forms and recovery is low. At high rates, viscous fingering dominates and a greater number of fingers gives rise to higher recoveries. Experiment 3 lies between the two extremes of behaviour and gives intermediate recoveries. Plate 28 shows a comparison of simulated and observed concentrations after 0.5 pv injected. The general level of agreement is good, although the fine details are different.

Figure 9 shows recoveries at 0.5 and 1 pv plotted against $\mathcal{N}_g$ for all the experiments simulated. The experimentally measured values fall on the predicted line within the limits of experimental uncertainty. The recovery at 1 pv injected for experiment 3 lies slightly above the predicted value even allowing for the $\pm 2\%$ measurement accuracy.

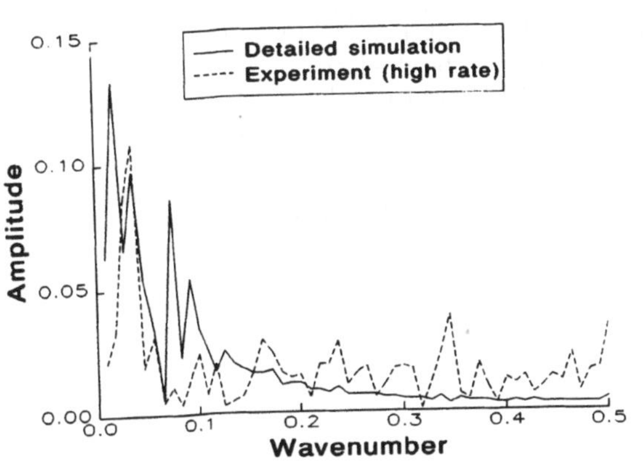


**Fig. 7.** Amplitude spectrum for experiment 6 at 0.5 pvi.

Studies reported by Fayers and Muggeridge (1989) show that recoveries in the transition region are affected by the initial locations of the fingers; we have not carried out further simulations to assess the sensitivity of this case to random number seed.

The results presented above allow us to draw some conclusions about the importance of 3D effects. The agreement obtained between simulation and the 2D experiment for recovery at 1 pore volume against $\mathcal{N}_g$ was very good. The earlier simulations of the 3D van der Poel experiments reported by Fayers and Newley (1988) showed a systematic error in prediction of the critical rate at which recovery changes from gravity-dominated to viscous-dominated. The source of this discrepancy is likely to be circulation terms that become important in 3D flow but which are excluded in a 2D simulation. If we write the velocity as $\mathbf{v} = -\lambda \nabla p$, the vorticity can be written as

$$\nabla \times \mathbf{v} = -\nabla \lambda \times \nabla p \qquad (10)$$

Thus, a pressure gradient in the $z$ direction crossed with a mobility gradient in the $y$ direction will give rise to a rotation in the $y$–$z$ plane. This will tend to promote gravity override and delay the onset of viscous fingering, leading to exactly the kind of discrepancy observed.

## CONCLUSIONS

We have compared the results of high-resolution numerical simulation of unstable flow with detailed observations of a well-characterized laboratory experiment. The particular process studied was unstable miscible displacement in a bead-pack representing a homogeneous vertical cross-section.

The experiments were conducted over a range of rates and density differences, allowing us to investigate the transition from gravity-dominated to viscous-dominated flow. The balance between the competing gravitational and viscous forces was characterized by a dimensionless viscous-to-gravity number $N_g = (M-1)v\mu_s H/\Delta\rho\, gkL$.

The finite-difference grid used was refined to a level (5 beads × 2 beads) at which discrepancies between simulation and observation were at least in part due to a breakdown of the constitutive equations over the averaging volume of a grid block.

Specific conclusions drawn from the study are as follows.

(a) At low rates, the displacement is dominated by the formation of a gravity tongue and the flow behaviour can be predicted accurately. For the lowest-rate experiment simulated, the interface position, the production rates and the overall recovery were predicted to within experimental error.

(b) At high rates, the flow is strongly influenced by viscous fingering. In this regime, average quantities such as mean concentration along the pack and overall recoveries can be predicted accurately. Prediction of the location of individual fingers is not possible in general.

(c) The transition from gravity-dominated to viscous-dominated flow as a function of gravity number $N_g$ was accurately predicted for this 2D experiment. Comparison of 2D simulation with 2D and 3D experiments suggests that it is important to model the third dimension for accurate prediction of flows close to the critical transition region between gravity override and viscous fingering

## ACKNOWLEDGEMENT

Permission to publish this work has been given by the British Petroleum Company plc.

## NOMENCLATURE

| | |
|---|---|
| $k$ | Wavenumber |
| $v$ | Velocity |
| $\lambda$ | Mobility |
| $D_L, D_T$ | Longitudinal and transverse dispersion coefficients |
| $c$ | Concentration |
| $\omega$ | Linear growth rate |
| $\phi$ | Porosity |
| $\Delta\rho$ | Density difference |
| $\mu_s, \mu_0$ | Solvent, oil viscosities |
| $M$ | Mobility ratio |
| $N_g$ | Gravity number (see Eq. (4)) |
| $\hat{c}$ | Average concentration |
| $\mathbf{g}$ | Gravity vector |
| $H$ | Pack height |
| $L$ | Pack length |
| $p$ | Pressure |
| $Q$ | Source term |
| $\gamma(c)$ | Specific gravity |
| $z_k$ | Discrete Fourier transform |
| $NX, NY$ | Number of grid blocks in $x$ and $y$ directions |
| $i$ | Square root of $-1$ |

## REFERENCES


Blackwell, R.J. (1962). Laboratory studies of miscroscopic dispersion phenomena. *Soc. Petroleum Eng. J.* **2**, 1–8.

Blackwell, R.J., Rayne, J.R. and Terry, W.M. (1959). Factors influencing the efficiency of miscible displacement. *Trans. Am. Inst. Mech. Eng.* **216**, 1–8.

Brigham, W.E., Reed, P.W. and Dew, J.N. (1961). Experiments on mixing during miscible displacement in porous media. *Soc. Petroleum Eng. J.* **1**, 1–8.

Christie, M.A. (1989). High resolution simulation of unstable flows in porous media. *Soc. Petroleum Eng. Reservoir Eng.* **4** (3), 297–303.

Christie, M.A. and Bond, D.J. (1987). Detailed simulation of unstable processes in miscible flooding. *Soc. Petroleum Eng. Reservoir Eng.* **2**, 514–522.

Christie, M.A. and Jones, A.D.W. (1987). Comparison between laboratory experiments and detailed simulation of miscible viscous fingering. Presented at *4th European Symposium on Enhanced Oil Recovery, Hamburg*.

Crane, F.E., Kendall, H.A. and Gardner, G.H.F. (1963). Some experiments in the flow of miscible fluids of unequal density through porous media. *Soc. Petroleum Eng. J.* 277–280.

Dumoré, J.M. (1964). Stability considerations in downward miscible displacements. *Trans. Am. Inst. Mech. Eng.* **231**, 356–361.

Fayers, F.J. and Muggeridge, A.H. (1989). Extensions to Dietz theory and behaviour of gravity tongues in slightly tilted reservoirs. Presented at the *Society of Petroleum Engineer Symposium on Reservoir Simulation, Houston, Texas*, SPE 18438.

Fayers, F.J. and Newley, T.M.J. (1988). Detailed validation of an empirical model for viscous fingering with gravity effects. *Soc. Petroleum Eng. Reservoir Eng.* **3**, 542–550.

Habermann, B. (1960). The efficiency of miscible displacement as a function of mobility ratio. *Trans. Am. Inst. Mech. Eng.* **219**, 264–272.

Lacey, J.W., Faris, J.E. and Brinkman, F.H. (1961). Effect of bank size on oil recovery in the high pressure gas-driven LPG-bank process. *J. Petroleum Technol.* 806–810.

Peaceman, D.W. (1977). *Fundamentals of Numerical Reservoir Simulation*. Elsevier.

Peaceman, D.W. and Rachford, H.H. (1962). Numerical calculation of multidimensional miscible displacement. *Soc. Petroleum Eng. J.* **2**, 327–339.

van der Poel, C. (1962). Effect of lateral diffusivity on miscible displacement in horizontal reservoirs. *Trans. Am. Inst. Mech. Eng.* **225**, 317–326.

Wheat, M.R. (1984). The role of heterogeneities in oil recovery. PhD Thesis, Imperial College, London.

Withjack, E.M. and Akervoll, I. (1988). Computed tomography studies of 3-D miscible displacement behaviour in a laboratory five-spot model. Presented at the *63rd Annual Technical Conference and Exhibition of the Society of Petroleum Engineers, Houston, Texas*, SPE 18096.

Wygal, R.J. (1963). Construction of models that simulate oil reservoirs. *Soc. Petroleum Eng. J.* **3**, 281–286.

# 13

# Relationship between Azimuths of Flood Anisotropy and Local Earth Stresses in Oil Reservoirs

*K.J. Heffer*[a] *and A.B. Dowokpor*[b]

[a]BP Exploration Company Ltd, London, UK
[b]Imperial College of Science and Technology, London, UK

Theoretical models have recently opened up the possibility that extensive induced rock fracturing can accompany many normal oilfield waterfloods, particularly in some North Sea reservoirs. Additionally, multiple sets of natural fractures in flooded reservoirs are likely to have anisotropy in conductivity governed partly by local earth stresses. If fracturing were implicitly involved in the processes of fluid movement through a reservoir, then one could anticipate a relationship between the preferred direction of water movement in a flood and the azimuth of local maximum horizontal principal stress. Modern-day earth stress orientation data are sparse, but coherent trends are apparent from published measurements across North America and a growing number of measurements in North Sea wells.

Information on flood breakthrough patterns, as well as oil production responses and anisotropy indicated by pulse or interference testing, has been examined in 36 case histories, and a correlation between flood and horizontal stress axes has been found. Fracturing—natural (including faults, microcracks, or syn-sedimentary fault control on depositional trends), or induced—is inferred to be a prime factor in the directionality of floods, albeit the type and scale of fractures is as yet uncertain. The correlation suggests that particular configurations of injectors and producers with respect to the orientations of earth stresses can delay the occurrence of breakthrough and improve oil recoveries in a large variety of fields.

## INTRODUCTION

One of the more important reservoir parameters upon which reservoir engineering attention is often focused is anisotropy in horizontal permeability; that is, the condition under which lateral fluid flow across a region of the reservoir, for given pressure gradient, is not independent of azimuthal direction. Lateral anisotropy is of obvious importance to areal sweep efficiencies in secondary or tertiary floods of any type.

Causes of anisotropy have been discussed by Alexander (1986), Evans (1987), Lake (1988a,b), Rajan (1988), and Goggin *et al.* (1988). These discussions have been in terms of depositional and diagenetic influences; they highlight the importance of the scale of the region over which directional permeability is defined.

It has long been recognized that fractures and faults can also provide preferred directions of fluid movement during waterflooding operations: the Spraberry Trend, West Texas (e.g. Dyes and Johnston, 1953; Barfield *et al.*, 1959; Elkins and Skov, 1962; Elkins *et al.*, 1968) provides a classic example of the influence of a dominant natural fracture set. Carnes (1966) has reviewed the effects of natural fractures on directionality in several waterfloods in Oklahoma and Texas. The influences of faults and fractures upon injectivities and waterflood directionality in the Ventura field, California have been discussed by Schneider (1972). Craig, 1971 recognized that fractures and planes of weakness are not uncommon in reservoirs but dismissed the general importance of directional (matrix) permeability. Another, more recent, example from the North Sea is a pilot waterflood in the Ekofisk field, which has also demonstrated anisotropy attributed to natural fractures, as observed in interference testing and water breakthrough behaviour (Thomas *et al.*, 1987).

Natural fractures, particularly joints (fractures along which there has been little or no displacement) are widespread at all scales in all types of rocks. The styles, intensities and conductivities of fractures depend upon the lithology, the histories of stress and diagenesis (e.g. Jaeger and Cook, 1979, Chap. 1), and, perhaps most importantly for conductivities, the modern-day local stress state (e.g. Walsh, 1981). In general, however, even where conductive natural fractures are known to exist in a reservoir, their influence upon flood directionality is not always obvious, nor is it necessarily anticipated before the flood is initiated.

Fractures can also be induced during flooding operations, either deliberately (for stimulation) or inadvertently.

*North Sea Oil and Gas Reservoirs–II*

It is becoming apparent from careful monitoring of injection pressures that many injectors have fractures induced in the reservoir rock under normal waterflooding operations, particularly in the North Sea. An important influence upon fracturing conditions is the cooling of the formation by the injected water: within the cooled zone, which in general is expected to grow to significant lateral dimensions, thermal contraction of the rock can lead to large reductions in local earth stresses. Pore pressure increase also acts to reduce local effective stresses and bring rock closer to a failure condition (e.g. Dusseault, 1985). There have been some recent attempts to model induced fracture propagation incorporating these influences (Perkins and Gonzalez, 1984, 1985; Koning, 1985; Koning and Niko, 1985; Morales *et al.*, 1985; Dikken and Niko, 1987; Garon *et al.*, 1988). However, these simulations have treated the rock as a continuous medium with constant properties and assumed linear elasticity in fracture mechanics and rock deformation. There are several reasons (outlined later under Discussion of Results and Possible Explanations) to believe that rock deformation is more complex. There is also uncertainty in the ability of conventional methods of determining fracture length (from pressure fall-off tests and step-rate test matching) to provide reliable results in the presence of a large, and possibly changing, fracture storage term. Another complicating factor arises from the possible existence of an extra resistance to flow on the faces of fracture walls due to deposits from injected water of poor quality. The extent and configuration of induced fractures are, therefore, in general, uncertain. Several instances of induced fractures extending to interwell distances have been reported in the literature (e.g. the El Dorado field reported in Szpakiewicz *et al.*, 1987, and the East Velma West Block reported in Bolling, 1985).

The above considerations imply the possibility that fracturing plays a significant role in many oilfield floods, although the role cannot always be expected to be obvious.

## Objective

The objective of this study was to determine whether fractures, natural or induced, could be shown from field data on flood performance histories, to have a widespread, significant influence on the anisotropy of reservoir flow properties.

Since the study was intended to take a purely empirical approach, and since also fracture influence upon extensive subsurface fluid movements is difficult to infer directly except in the most obvious of cases, some other observable parameter, strongly associated with fracturing, had to be used to fulfill the objective indirectly: the parameter chosen was orientation of horizontal earth stresses.

## Method

The control on the general orientation of fractures during creation is the contemporaneous local earth stress tensor. Also, in a network of existing natural fracture sets, any anisotropy in the modern-day local earth stresses will be an influence on the distribution with orientation of the fracture widths, either *in situ* or during fluid injection; hence stresses can govern the anisotropy in fluid conductivity in a system of natural fractures. Perhaps the best direct demonstration of this latter principle has been obtained from observations of microseismicity during water injection into jointed granite geothermal reservoirs (Pine and Batchelor, 1984; Fehler *et al.*, 1987). In projection onto horizontal planes, the clouds of microseismicity, arising from small shearing events on the joints and marking the progress of the injected water banks, had long axes that were aligned close to the local modern-day maximum horizontal principal stress axes. This directional fluid progress was reproduced in the cited studies, and also in Harper and Last (1989), with distinct element modelling for a discontinuous medium.

Given that the stress state is a strong control on fracturing and fracture conductivity, the method chosen to achieve our study objective was to assess the relationship, over many different flooding projects, between the azimuths of the preferred directions of horizontal fluid flow and of the orientation of the local modern-day earth stress axes.

If fractures, natural or induced, are conductive over extensive areas of a project, over which the local modern-day principal stress axis is (sub)horizontal, then we might anticipate that the preferred flooding directions would tend to be at small angles to the orientation of the local modern-day major horizontal principal stress axis. This latter axis will be referred to as SH for brevity. The small angles would be symptomatic of the involvement of extension, shear, or hybrid shear fractures, or a mixture of these fracture types.

As an example already explicitly noted in the literature, the waterflood of the Pembina oilfield, western Alberta, shows a strong NE–SW directionality in its progress. Bell and Gough (1981) have noted the correspondence of this direction with the local azimuth of SH and have inferred the influence of vertical induced fractures aligned with SH on permeability anisotropy.

It was recognized in the general study that other sources of directionality could of course confuse any relationship with stress orientation. These were the variations, for any one injector, in individual distances to nearest-neighbour producers; and, probably more importantly, in the offtake rates at the producers. Also, any large-scale pressure gradient from production or injection external to the study areas, or general formation dip, could distort the apparent anisotropy. Without full history-matching of each field through detailed simulation—not a practical task within the scope of this study—such influences upon flood directionality could not be nullified. It was probable that these influences were in directions that were, over the total population of injection cells, reasonably uniformly distributed around the azimuths, and uncorrelated with SH directions; nevertheless, it was recognized that for some individual cases, particularly those with small populations, such an assumption might not have been valid. Conventional principles of reservoir engineering were thus, of necessity, put aside in order to test the hypothesis of fracturing influence: the validity of this abeyance could only be judged by the success or failure of the sought-for correlation. If success were found, the existence of "extraneous" and potentially conflicting influences on directionality could only strengthen the inference that stress orientation is a prime indicator for flooding directionality.

In contrast, there are more probable links between sedimentological or diagenetic influences on anisotropy and the orientation of the stress state, given that little stress rotation has occurred to the present-day. Some of these influences would tend to be in an opposite sense to that of fracturing. For example, pressure dissolution seams (stylolites), the most obvious potential link between diagenesis and stress orientation, form on surfaces that are orthogonal to the maximum compression: therefore, in compressional tectonic regimes where these surfaces can tend towards vertical and can influence horizontal permeabilities, the implied anisotropy is in a sense opposite to that of fracturing. Again, proximal paleoslopes, and therefore alluvial channel axes would, in a broad sense, trend normal to the major faults of extensional basins and therefore orthogonal to SH. However, axial channels in rift basins and syn-sedimentary fault control on depositional trends are examples where preferred reservoir continuity might be expected to be subparallel to the paleo-orientation of SH (eg. Budding *et al.*, 1988). Even in these latter examples, the question of whether maximum horizontal permeability is parallel or orthogonal to the channel axis might depend upon the well spacing relative to the channel width. It will require a separate study to distinguish the relative influences of depositional trends controlled by faulting and of post-sedimentary natural fractures or induced fracturing.

## DATA AVAILABILITY AND COLLATION

### Information on earth stresses

Data on modern-day stresses have been rapidly accumulating recently, particularly on orientations. Wellbore ellipticity analyses from multiarm caliper logs have provided major contributions to this information. It is becoming apparent that there are geographical regional trends in orientations of horizontal stress axes, associated strongly with the motions of plate tectonics, albeit the exact mechanism linking them is as yet uncertain. Maps have been published for stress orientations across Western Europe (Klein and Barr, 1986; Janot *et al.*, 1988), the Continental United States (Zoback and Zoback, 1987) and Western Canada (Bell and Babcock, 1986). Orientations of SH used in this study have generally been read from these maps, unless more detailed information was available for a particular case.

Across the southern and central North Sea, SH is generally aligned NW–SE or NNW–SSE, apparently following the broad alignment of the major basin-bounding faults. However, in the northern North Sea there is evidence that SH tends more to E–W (Janot *et al.*, 1988), in line with an E–W trend in SH onshore southern Scandinavia (Klein and Barr, 1986).

For much of the United States and Canada east of the Rocky Mountains, SH is oriented NE–SW (Zoback and Zoback, 1987; Bell and Babock, 1986).

The above trends are average only: there is considerable local variation of stress direction, even within particular structures. In general there is an uncertainty in azimuth of SH measurements in these maps of from 20° to 30°, except that the more concentrated data in particular areas of the United States has allowed this uncertainty to be reduced to about 15° (Zoback and Zoback, 1987). Interpolation between published data points, and variability caused by local structure, will further degrade the accuracy in estimated orientation of SH. For this reason, the attempted correlation of this study needed to be based on a large population of cases, such that a statistical relationship could be sought despite individual data errors.

### Information on flood anisotropies

Unfortunately, data available from fields in the North Sea were insufficient to produce a reliable correlation. Although the effects being studied were anticipated as being of large importance to the waterfloods in the North Sea, the developments were in general not mature enough to provide sufficient evidence of their influence. Only three cases were available for the correlation from the North Sea area and one from onshore United Kingdom.

Therefore, a greater quantity of data was culled from the literature. Even this required an intensive (although not exhaustive) search for unambiguous indicators of preferred flooding directions: breakthrough of injected fluid to producing wells cannot always be attributed uniquely to a "culprit" injector well, leaving the path through the reservoir indeterminate in many cases. The use of tracers in the injected fluid overcomes this problem; data from interference or pulse tests (after initiation of flooding) and oil production responses to start-up of injection at a particular well also allow unique indication of directionality in flooding. Careful reservoir monitoring of this type has been more common in enhanced oil-recovery projects.

### Database

A database of 36 case histories from the literature and from data owned by the BP Exploration Company was gathered. It should be emphasized that no selection of cases was made during this cull: the only criterion for including a case in the data base as it was encountered was whether directions for both flood and stress anisotropies could be assessed.

Owing to the desirability of including as much data as possible in the study, information of several different types and degrees of completeness was gathered on flood directionality. In order to provide reasonable consistency and objectivity in the study, the following data for each case was stored in the database.

(1) A series of values of azimuthal angles for the directions joining pairs of injector and producer wells between which fluid breakthrough had, or potentially could have, occurred. Only nearest-neighbour pairs were included for potential paths of breakthrough not yet actually taken, but all directions in which breakthrough had actually occurred were included, even where the producers concerned were beyond the rank nearest to the "culprit" injector. In a few cases, only a general direction of early breakthrough had been indicated and only a rough estimate of the number of wells was available. The term "breakthrough direction", actual or potential, is abbreviated to BTDIR in the following text.

(2) A weighting factor for each value of BTDIR. This in general took the value unity if BTDIR was associated, either with actual breakthrough of injection fluid or tracer, or with oil production response to injection start-up. The weight was given a value zero if no breakthrough or response had occurred along BTDIR. If quantitative data could be assessed from the case history that indicated the severity of breakthrough (from the relative time delays, from the percentages of injected tracer detected in each surrounding producer, or from the magnitude of oil production response), or if an interference test gave quantitative assessment of directional permeability across a project, then these data, normalized, were used for the weight. The normalization procedure caused the weights for any one case to sum to the number of actual BTDIRs for that case: this provided consistency between these cases and those where only unity weights were assigned. Reference will be made below to a direction in the database with a non-zero weight as a "taken BTDIR"; and to a general direction in the database, including any with zero weight, as an "available BTDIR".

(3) One value for the local orientation of SH.

(4) Other characteristics concerning the reservoir that might, where available, have bearing on the degree of influence of fracturing on flood anisotropies are location, age, depth, lithology, depositional environment, thickness, permeability, porosity, oil gravity, oil viscosity, gas–oil

**Table I.** Characteristics of cases studied

| CASE | FIELD NAME | LOCATION | LITH | NF | A | DEPTH FT | K | H | F | SH | REF. |
|---|---|---|---|---|---|---|---|---|---|---|---|
| 1 | NPR-3 | Wyoming | S | Y | T | 500 | 63 | 120 | P | –25 | SPE 14918 |
| 2 | Eliasville | N. Texas | L | Y | I | 3200 | 11 | 40 | P | 76 | SPE 12637 |
| 3 | 8 Fields | W. Texas | L/D | Y | B/T | 6000 | – | – | G/W | 90 | JPT JAN 73 |
| 4 | Velma | S. Oklahoma | S | Y | B/T | 6500 | 70 | 300 | W | 70 | JPT AUG 85 |
| 5 | Glennpool | Oklahoma | S | Y | T | 1500 | 150 | 88 | SU/P | 70 | SPE 15551 |
| 6 | Jay–Lec | NW Florida | L/D | N | T | 15400 | 35 | 95 | W | –60 | SPE 9476 |
| 7 | W.Burkburnett | N. Texas | S/SHL | N | R | 1700 | 75 | 16 | SU | 65 | SPE 10162 |
| 8 | Hitts Lake | E. Texas | S | N | B | 7250 | 300 | – | P | 70 | SPE 13123 |
| 9 | Salem Unit | Illinois | S | N | T | 1750 | 125 | 50 | SU/P | 65 | SPE 14442 |
| 10 | Kelsey | S. Texas | S/SHL | N | I | – | – | – | W | 9 | JPT AUG 79 |
| 11 | Pembina | W. Alberta | S | N? | B | 5000 | 40 | 28 | M/W | 42 | B+G 81 |
| 12 | W. Sumatra | Montana | S/SHL/L | N | T | 4500 | 900 | – | W | 45 | SPE 15161 |
| 13 | Spraberry | W. Texas | S | Y | B/I | 6800 | 16 | 30 | W | 86 | JPT SEP 68 |
| 14 | Langston–KL | N. Texas | S | N | B | 3550 | 47 | 10 | W | 65 | AIME,244,55 |
| 15 | Fairway | E. Texas | L | N | T | 9550 | 20 | 70 | WAG | 65 | JPT OCT 70 |
| 16 | Lick Creek | Arkansas | S | N | B | 2550 | 1200 | 8 | P/W | 70 | SPE 14958 |
| 17 | N.Burbank | Oklahoma | S | Y | I/T | – | 50 | 42 | SU/P | 65 | SPE 12695 |
| 18 | Midale | S.E. Saskat. | L/D | Y | B | – | 25 | 54 | W | 45 | SPE 15635 |
| 19 | Claymore | C.N. Sea | S | N | B | 8000 | 650 | 1000 | W | –15 | SPE 16556 |
| 20 | Maljamar | New Mexico | S | Y | I/T | 4000 | 18 | 27 | C | –65 | SPE 14940 |
| 21 | Levelland | W. Texas | – | N | T | – | – | – | M | –60 | JPT NOV 77 |
| 22 | L.Buffalo | Wyoming | – | Y | T | – | – | – | W | 0 | JPT NOV 77 |
| 23 | S.Swan Hills | W.C. Alberta | L | N | T | 8500 | 50 | 74 | WAG | 45 | SPE 15547 |
| 24 | BP | – | S | Y | T | 9000 | 100 | 500 | W | –48 | – |
| 25 | Horseshoe G. | NW New Mex. | S/SHL | Y | B/T | 1200 | 60 | 8 | W | –60 | JPT OCT 64 |
| 26 | Clear Ck. | SW Wyoming | S | Y | B | 11500 | 250 | 1000 | G | –25 | SPE 15528 |
| 27 | Robertson | Texas | D | Y | B/I | 6500 | 1 | 300 | W | –60 | SPE 15568 |
| 28 | Bradford | Pennsylvania | S | Y | B/T/I | 1300 | 200 | 20 | W/P | 72 | SPE 15550 |
| 29 | Ekofisk | C.N. Sea | L | Y | B/I | 10400 | 100 | 590 | W | –45 | JPT FEB 87 |
| 30 | Judy Creek | C. Alberta | L | N | B | – | 70 | – | W | 45 | JPT JUL 77 |
| 31 | Alvord S. | N. Texas | CON | – | B | 5700 | 60 | 25 | C | 60 | SPE 14309 |
| 32 | BP | – | S/SLT | Y | T | 6500 | 10 | 80 | W | –5 | – |
| 33 | BP | S.N. Sea | S | N | B | 6500 | 250 | 90 | W | –45 | – |
| 34 | BP | UK | S | Y | B/T I/R | 3000 | 10 | 50 | W/SU/C | –31 | – |
| 35 | Higgs Unit | W. Texas | S | N | T | 1870 | 500 | 14 | W/P/O | 68 | JPT JAN 73 |
| 36 | BP | – | S | Y | T | 9000 | 250 | 500 | WAG | –34 | – |

| KEY | |
|---|---|
| A | – Anisotropy Indicator |
| B | – Breakthrough of Fluid |
| B+G | – Bell + Gough (1981) |
| BP | – BP Proprietary Data |
| C | – $CO_2$ Flood |
| CON | – Conglomerate |
| F | – Fluid injected |
| G | – Gas injection |
| H | – Average Gross Thickness, feet |
| I | – Interference Test |
| K | – Average Permeability, MD |
| L | – Limestone |
| LITH | – Reservoir Lithology |
| M | – Miscible Flood |
| NF | – Naturally Fractured? |
| N | – Not Fractured |
| O | – Oil Injection |
| P | – Polymer Flood |
| R | – Rate Response |
| REF | – Latest Reference Used |
| S | – Sandstone |
| SH | – Max. Horiz. Stress Direction |
| SHL | – Shale |
| SLT | – Siltstone |
| SU | – Surfactant Flood |
| T | – Tracer |
| W | – Waterflood |
| WAG | – Water-Alternating-Gas |
| Y | – Yes, Naturally Factured |

ratio, reservoir temperature, well spacing, tectonic regime (extensional or compressional) and also whether fractures are reported as a prominent feature of the reservoir.

Table I lists the 36 cases studied together with their characteristics. The wide variations in permeabilities, depths and thicknesses should be noted. The large majority of cases involved flooding with water or water-based fluids, but a few concerned just gaseous injection. There was almost equal division of cases between sandstone and carbonate reservoirs, and between compressional and extensional tectonic regimes. Figure 1 shows the frequency distribution of azimuths of SH for the 36 cases. Although there is a strong frequency in the NE–SW direction, deriving from the prevalence of this orientation of major compression in North American cases, there is nevertheless a wide distribution of SH amongst other azimuths.

# TREATMENT OF DATA

Two further problems hindered the interpretation of preferred flood directions: firstly, production wells only sparsely sample the azimuths from any one injector well; and, secondly, breakthrough of injected fluid is not expected to be uniquely in the direction of major horizontal permeability except early in the life of a highly anisotropic flood. It was for these reasons that a straightforward regression of taken BTDIRs on SH was only partially informative: linear regression plots indicated significant positive correlation but not unit slopes. The plots could not take into account the lack of uniformity in the distribution of available BTDIRs over the range in SH.

Two ways of treating the data were used to counter these problems. The first was to aggregate all of the data from all of the cases in the database relative to the local orientation of SH. This aggregation provided a significant number of available BTDIRs in any 10° sector relative to SH, and so allowed equitable assessment of the quantity of taken BTDIRs in an individual sector, accounting for the number of available BTDIRs in that sector.

Since the results of this aggregation may have been unduly influenced by a minority of the cases in which fracturing is particularly prominent, a second method treated the data on the basis of individual cases. In order to do this despite the sparseness of the data, the concept of an azimuthally dependent function for the "probability of flood breakthrough" was introduced. This concept, although heuristic, was nevertheless felt to provide a consistent, objective means of assessing directionality in any one flood.

As a check, the compatibility between these two methods of treating the data was demonstrated. The two methods and their results will now be discussed in detail.

## Analysis of aggregated relative orientational data

The simple basis of this method was to combine, in an aggregated rose diagram, values of BTDIR relative to the local orientations of SH (that is, values of BTDIR – SH), from all cases of the database. In Fig. 2 is plotted the available distribution of relative orientations as the outer rose diagram, and the taken distribution as the inner rose diagram. The total population of available directions was 1168, of which 700 were taken; i.e. an overall breakthrough proportion of approximately 60%. The available frequencies show peaks at $-45°$, $0°$, $+45°$ and $+90°$: these peaks reflect prominences of available BTDIRs at multiples of roughly 45°, deriving from the common deployment of regular five-spot or nine-spot flooding patterns in North America, combined with the regional NE–SW trend in SH east of the Rocky Mountains. The taken frequencies incorporate the weights associated with the BTDIRs. The taken distribution shows considerable peaks and troughs, and, although there are strong frequencies in the range $-50°$ to $0°$ (relative to SH), the picture is not clear-cut.

In order to account for the availability of directions in assessing the taken distribution, the frequency of taken directions in each 10° sector was normalized by dividing by the frequency of available directions in that sector. The resulting normalized function, shown in Fig. 3, represents the proportion of available directions in any sector that have been taken. The following will be observed from Fig. 3.

(1) The function shows "noise" of much less amplitude than does Fig. 2, and reveals the existence of a consistent underlying trend.

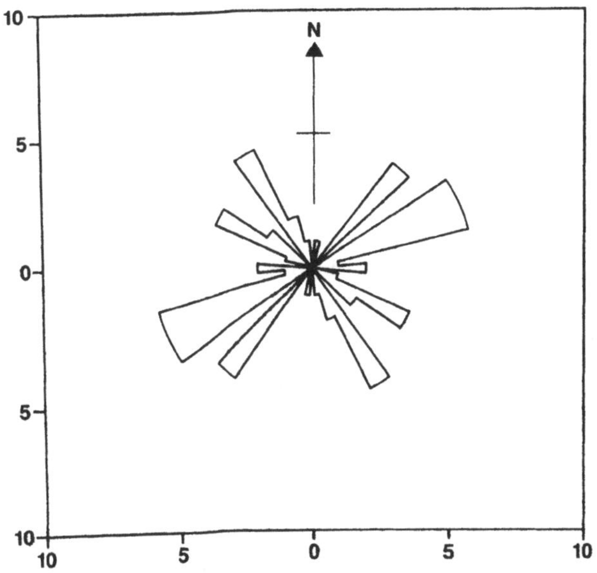


**Fig. 1.** Azimuthal frequency distribution of major horizontal principal stress axes from all cases, relative to geographic North.

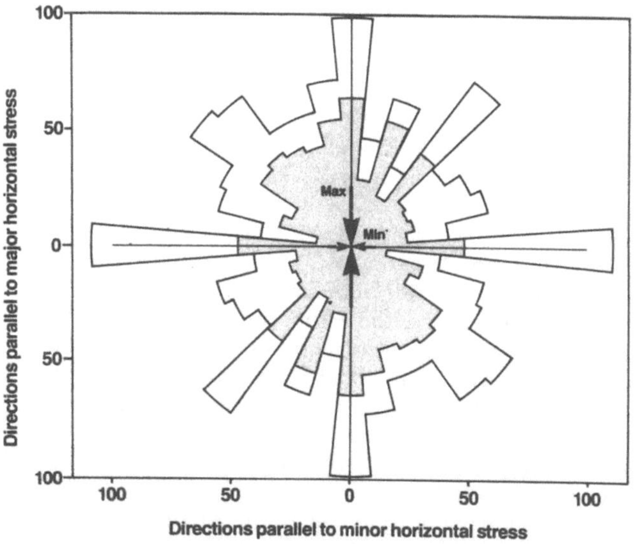


**Fig. 2.** Frequency distributions of available (outer rose diagram) and taken (inner rose diagram; shaded) breakthrough directions from all cases relative to the horizontal principal stress axes.

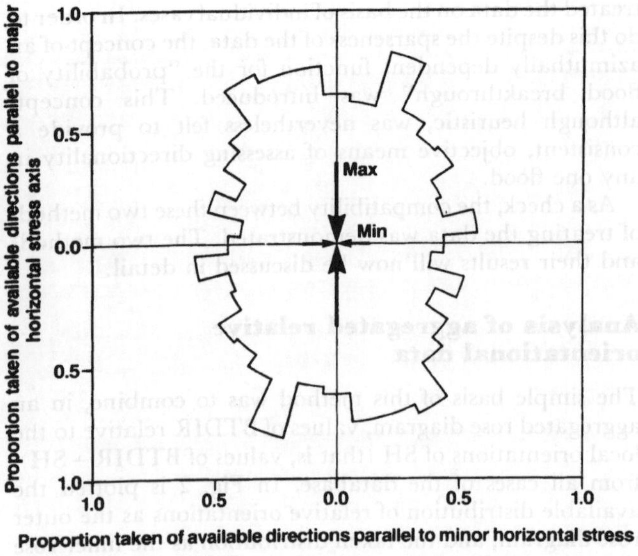


**Fig. 3.** Proportion of available breakthrough directions from all cases that are taken, as a function of orientation relative to horizontal principal stress axes.

(2) The area within the normalized function is approximately equal on either side of the 0–180° axis.

(3) The proportion of available directions parallel to SH, which have been taken, is larger than the proportion orthogonal to SH. The degree of anisotropy is about 3:2 in favour of SH. In other words, fluid breakthrough has occurred to producers that were positioned, relative to the injector, parallel to SH with 50% more likelihood than for producers in the orthogonal direction.

It should be noted that the normalized function is not a probability density function of angle, but rather it describes for each sector the mean outcome of a series of "experiments", each with only two possible results: available BTDIR is either taken or not. The mean outcome in each sector approaches, for large populations, the underlying probability of breakthrough for that particular orientation relative to SH. As an illustration of the significance of the variations, the likelihoods of the actual observations were calculated under the assumption that the underlying probability of breakthrough was not variable with orientation relative to SH, but, instead, had a constant value of 0.60 for all orientations. The deviation of the observed total number of taken BTDIRs in the range of angles 45° either side of SH allowed the hypothesis to be rejected at the 99.9% confidence level. A similar rejection resulted from the deviation of the observations in the range of angles 45° either side of the minor horizontal principal stress axis. Therefore, the actual observations of proportional breakthrough were extremely unlikely to have been random variations about an orientation-independent mean. It is difficult to deduce further statistics for the function because an exact model for the variations of breakthrough probability with angle relative to SH is not obvious.

At this point in the argument, it still remains a possibility that these results were influenced by a minority of cases; perhaps, for example, those where natural fractures are a prominent feature of the reservoir description. In order to examine this issue, analysis by individual case was necessary.

## Analysis by individual case

In some of the cases, the anisotropy in fluid flow was obvious: either from statements in the reports of the case or from strong consistency in directionality of fluid breakthrough. In other cases, the preferred directions were more ambiguous and possibly confused by the availability of breakthrough directions. In order to allow extraction of a preferred direction of flooding for all cases, a consistent objective procedure was required.

The heuristic method chosen was to define a "probability of breakthrough function" for each case. This concept was devised to represent the probability that the flood front of injected fluid at a particular orientation from any injector had reached an imaginary reference circle centred on that injector. The ellipse was chosen to describe the variation of the breakthrough probability with orientation because it is a simple geometry by which to represent axial asymmetry; and because, also, the aggregated normalized breakthrough distribution of Fig. 3 was roughly elliptical. For most injectors, the radius of the reference circle was implicitly associated with the average distance to nearest-neighbour producing wells surrounding that injector. The actual observations of (non-)breakthrough to nearest neighbours could then be used as data on which regression would yield the parameters of the probability ellipse. A probability ellipse was deemed to be common for all injectors in a particular case, implying that an average directionality over the pattern cells of that case was being sought. Note that it was only directionality that was of interest to this study, so that the separate magnitudes of the ellipses' axes were immaterial: only their orientation and length ratio were important. Therefore, the diversities between injectors of injection rates and times, average nearest-neighbour distances and floodfront dispersions were not of concern.

The fact that actual breakthroughs *beyond* nearest-neighbour producers were included in the database did not invalidate the procedure, since a "long-range" BTDIR of that type implied an imaginary BTDIR to a point at a similar azimuth projected onto the reference circle for the injector concerned.

In cases where non-unit weights were available, then the number of (pseudo-)producers for each BTDIR was put equal to its weight, despite it being non-integral: this procedure implied the concept of a group of essentially coincident pseudo-observations of injection fluid passing a particular point on the reference circle. This device was considered to give a reasonable reflection of the "weight" of one actual breakthrough, derived from various types of information, in the degree of anisotropy "observed". Weights derived from oil production response or interference tests, although not necessarily associated with actual breakthrough, were used in the same way, to imply a non-integral group of pseudo-observations of breakthrough: this was also felt to reasonably reproduce the degree and direction of observed anisotropy in transmissibility as it might affect flooding progress.

The parameters of the breakthrough probability ellipses were determined individually for each case by the method of maximum likelihood. The likelihood function for the set of (pseudo-)observations of (non-)breakthrough was taken to be

$$L(\Phi, a, b; B, N) = \prod_{i \in B} [P(\theta_i; \Phi, a, b)]^{W_i} \times \prod_{j \in N} [1 - P(\theta_j; \Phi, a, b)] \quad (1)$$

where

$\Phi, a, b$ are the parameters of the probability of breakthough ellipse as depicted in Fig. 4. ($\Phi$ is the orientation of its major axis relative to

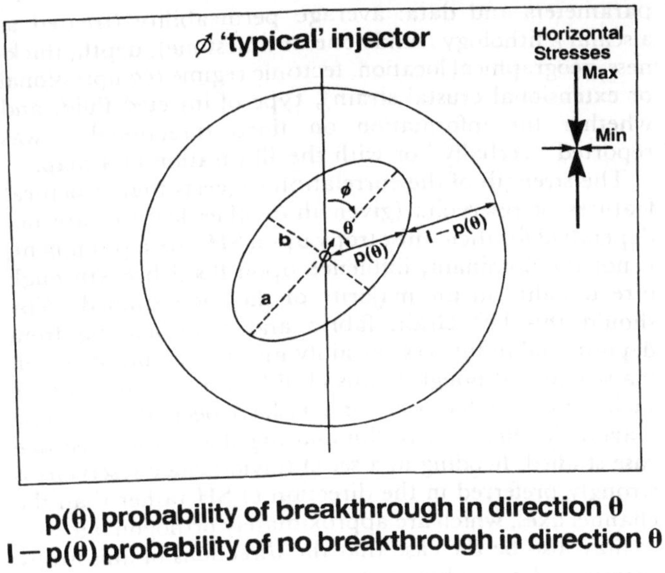


**Fig. 4.** Example of probability of breakthrough $p(\theta)$ as an elliptical function of orientation relative to horizontal principal stress axes, for use in individual case analysis.

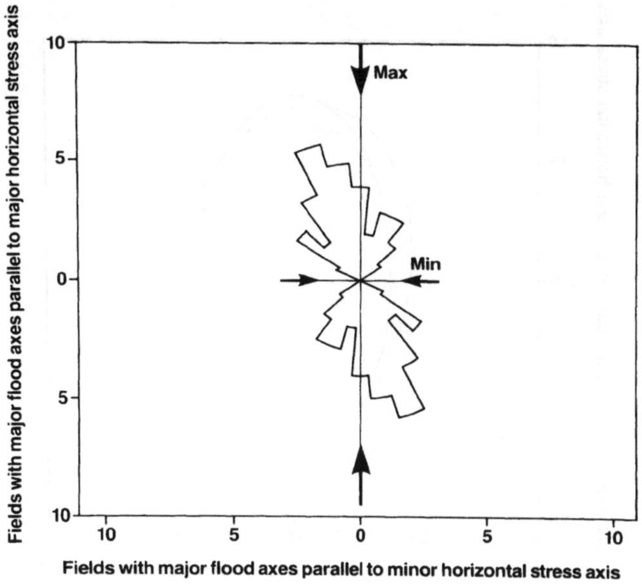


**Fig. 5.** Frequency distribution of orientations of major flood axes for individual cases, relative to horizontal principal stress axes.

local SH; $a$ and $b$ are the lengths of its major and minor semi-axes respectively)

$B$ is the set of (pseudo-)observations of breakthrough for the case

$\mathcal{N}$ is the set of observations of non-breakthrough for the case

$\theta_i$ is the orientation of observation $i$ relative to SH

$W_i$ is the weight of observation $i$

$P(\theta; \Phi, a, b)$ is the probability of observing breakthrough to the reference circle at relative orientation $\theta$, given the parameters of the ellipse, $\Phi, a, b$

$[1-P(\ )]$ is the probability of observing no breakthrough under the same circumstances

The model for the breakthrough probability variation with orientation was the ellipse:

$$P(\theta; \Phi, a, b) = \frac{1}{\sqrt{[\cos^2(\theta-\Phi)/a^2 + \sin^2(\theta-\Phi)/b^2]}} \tag{2}$$

The likelihood function (or rather its logarithm) was maximized for each case by variation of the parameters of the ellipse, $\Phi, a, b$. The maximization was performed by explicit calculation of $L$ over the complete ranges of parameter values (in steps discretized so as to give adequate accuracy: in particular, 10° steps in $\Phi$ were used). This procedure avoided the potential problem that iterative methods might have encountered, of convergence on a local, as opposed to global, maximum.

The complete set of probability ellipses for the 36 cases displays a tendency for the major axes to be close to SH. Some ellipses have low eccentricity (i.e. are almost circles): such instances may have derived either from a low degree of anisotropy in the flood, or, alternatively, from the report of the case being at a late stage in the flood history, without quantitative data on breakthroughs being available to allow non-unit weighting of the taken BTDIRs.

Figure 5, the frequency distribution of the angles, $\Phi$, between the ellipse major axes and SH, emphasizes the clustering of the preferred directions of flooding close to SH. In 31 cases (86%) $|\Phi|$ is less than or equal to 45°. In only one case is $|\Phi|$ greater than 55°. The mean value of $\Phi$ over the range −90° to +90° is −8°, with 90% confidence limits of −16° and 0°. The mode (peak) of the distribution is seen to be biased approximately 20° anticlockwise from SH. These results will be discussed further below.

## Consistency of the two methods

As a check on the results of the individual fits of probability ellipses, their consistency with the aggregated analysis was demonstrated. The individual probability ellipses were averaged by two different methods: the first was to average the probabilities over all cases in each sector of 1°; the second was to form arithmetic averages of the parameters, $\Phi, a, b$, of the individual ellipses, with which to form an overall average ellipse. In addition, an overall probability ellipse was fitted by the maximum-likelihood method to the aggregated database. The results of these three averaging processes are shown in Fig. 6. The ellipse fitted to the aggregated database has a major axis that is oriented 10° anticlockwise from the SH axis, and has an eccentricity of 0.77, corresponding to a ratio in lengths of its axes of 1.56.

The three averaged ellipses are reasonably similar to one another and also to the normalized function of Fig. 3, implying the required consistency.

## Separate analysis of "unfractured" reservoirs

As pointed out above, in about half of the cases, the reported reservoir description had specifically mentioned the presence of fractures, natural, induced deliberately or induced inadvertently: in the remainder of the cases, no special mention had been made. It might be surmised that the relationship between BTDIR and SH derived above could be attributed mainly to the former set of cases, whilst the latter set would show no, or only a weak, relationship. To test this idea, the same analysis steps were performed separately on "fractured" and "unfractured" cases.

In fact, the same general picture emerged for each set of cases. Both aggregated normalized functions (corresponding to Fig. 3) for the two sets showed more accentuated peaks and troughs, owing to the smaller populations; but both also possessed generally larger values close to SH. The degree of anisotropy, determined from fitting an aggregated

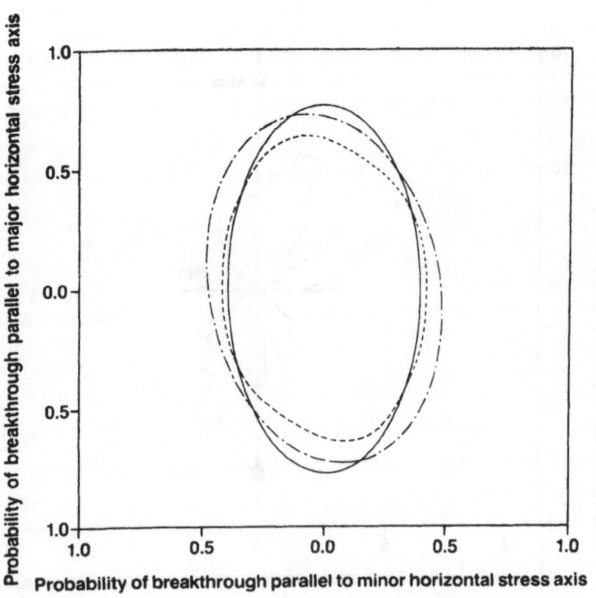


**Fig. 6.** Averaged ellipses describing the probability of flood breakthrough in individual cases as a function of orientation relative to horizontal principal stress axes. (1) Individual ellipses averaged in each sector (broken curve). (2) Ellipse using average of individual elliptical parameters (solid curve). (3) Ellipse determined from aggregated data from all cases (chain curve).

probability ellipse, was again approximately 3 : 2 in favour of SH in either case. The distributions of relative orientations of probability ellipse major axes (corresponding to Fig. 5) for each set again clustered close to zero: the distribution for the "unfractured" reservoirs showed, at a relative orientation of $-10°$, a more marked peak than for the "fractured" reservoirs.

It was concluded that there was no substantial difference between the strength of the correlations for reservoirs with obvious fracturing and those deemed to be "unfractured". This may be because all reservoirs contain a significant density of fractures at some scale, which might not be obvious in core or from well performance.

## DISCUSSION OF RESULTS AND POSSIBLE EXPLANATIONS

Considering the uncertainties associated with the SH values, with the heuristic method of eliciting preferred flooding directions, and the several other potential causes of anisotropic flooding noted under Method above, the degree of clustering of Fig. 5 is remarkable. Various changes to weighting factors and repeated passes through the data correcting minor errors have indicated that this picture is reasonably robust: a few cases switched preferred directions with these changes but not sufficiently to alter the clustering of $\Phi$ values close to zero. Part of the success of the correlation probably stems from the likelihood that an orientation of SH drawn from published maps approximates to an *average* of local variations across an individual field, and that this was compared in each case with the preferred flooding direction computed from fieldwide data, with the major flooding axis representing again an *average* of these data.

No significant categorization of cases could be deduced from examination of the values of $\Phi$ (i.e. degree of match between major flooding and stress axes) for individual fields with respect to the following associated reservoir parameters and data: average permeability (treated as a scalar), lithology (sandstone or limestone), depth, thickness, geographical location, tectonic regime (compressional or extensional crustal strain), type of injected fluid, and whether the information on flood directionality was reported "verbally" or with the illustration of a map.

The strength of the correlation suggests that structural features of reservoirs (given that other features are not dependent for their anisotropy upon SH) are a prominent, if not the dominant, influence upon flood breakthrough directionality in the majority of the cases studied. Why should this be? Grain fabric and cross-bedding from depositional influences certainly give rise to anisotropy in many cases, although Evans (1987) cites instances where such anisotropy has been, or may have been, destroyed by diagenetic effects. It is illuminating that, in at least one case studied, flooding in a set of fluvio-deltaic reservoirs is strongly preferred in the direction of SH rather than the channel axes, which are approximately orthogonal to SH. It may also be the case that the directions of anisotropy determined by sedimentological influences are not consistent on a sufficiently large scale to show up in the overall flooding trends for a field. On the other hand, stress fields can have the degree of lateral coherence in average orientation (albeit with local variability) necessary to cause the reservoirwide trends. This is not to say that the anisotropy will be strongly defined in most cases: that the data have required careful processing to pick out preferred directions in many instances is indication that the results are statistical and easily masked in small samples by local variations. Local variations may derive from lithological influences, local pressure gradients or dips, or from variations in natural fracture trends or stress orientations.

The possible structural features in a reservoir that may be responsible for the correlation are as follows:

(1) Natural fractures (joints or faults) that are naturally conductive *in situ*: the anisotropy may derive either from the dominant intensity or continuity of a particular set of fractures aligned close to SH; or from the influence of the stress state in causing fractures striking close to SH to be more conductive than for those fractures at large angles to SH.

(2) Natural fractures whose conductivity is initiated or significantly enhanced by the stress changes introduced into the reservoir rock by the fluid injection. The conductivity may be due to just unloading of the normal stress on fractures without significant extension, to extensional parting of fracture walls at locations of tensile effective stress, or to shear movement with dilation on fractures under appropriate conditions. The scale of fracture on which these changes might occur could range from those of microcrack to major fault. Mismatch of thermal contractions or poroelastic expansions could play a significant role at any scale, from that of individual grains, upwards through various sizes of heterogeneity, each of which might undergo differential cooling and/or have variable expansion coefficients. Although induced fractures might be oriented stochastically, their strikes are likely to be biased on average towards SH, since this configuration would be associated with lower strain energies in the rock.

(3) Natural fractures or faults that are sealing, or that have low conductivity relative to the matrix permeability, and that are aligned close to SH. The sense of anisotropy in this situation will be the same as for that caused by highly conductive fractures. This explanation again requires that there has been little rotation of SH from the time of fracture genesis. Furthermore, it is not uncommon for

faults to be non-sealing, and indeed some cases in this study have indicated a direct path of tracer breakthrough that crosses known faults: sealing fractures are therefore not thought to be a dominant cause behind the correlation. Strain-hardening faults, as are commonly found in porous sandstones (e.g. Aydin and Johnson, 1983), could also provide aligned surfaces of low permeability. It is difficult to predict whether these features would retain their low conductivity if effective stresses were substantially reduced in their locality.

(4) Fresh fractures created by the injection process itself. Whilst a hydraulic fracture induced close to the injection well could clearly cause fresh breaks in the rock, it is open to question whether such fractures could extend to large distances without being modified by natural fractures. Observations and analysis of such interactions have been reported by Warpinski and Teufel (1987). New shear fractures may also be possible but, again, it is arguable that pre-existing discontinuities would be followed as in (2) above. An exception might be where the natural fractures were stronger than the undeformed rock, as for the strain-hardening faults found in porous sandstones (e.g. Aydin and Johnson, 1983).

(5) Deliberately induced stimulation fractures from either injector or producer wells that were reported in a minority of cases. Those from injector wells may have propagated during injection.

(6) Extensive dilatancy anisotropy (e.g. Crampin and Atkinson, 1985). Widely occurring natural microcracks (or anisotropic pore openings) aligned with SH have been inferred from observations of shear-wave splitting in many seismic surveys. Natural realignment of these microcracks or pores is inferred to be by subcritical crack growth, involving (possibly very slow) corrosion of the rock at the stress concentrations, and redeposition of corrosion products at locations of lower stress intensity. If in sufficient density, which may be articially promoted by the stress changes due to injection, such features may also provide the basis for the observed anisotropy in reservoir permeability.

(7) Following from the earlier discussion under method, it cannot be ruled out that flow anisotropy is derived from depositional trends controlled by syn-sedimentary faults striking subparallel to the paleo-orientation of SH.

While no categorical selection of universal mechanism may be made from the evidence examined in this study, it is apparent that, in some cases at least, openings for enhanced flow occur along the discontinuities of natural fractures; it also seems likely at this stage that the stress changes that injection induces in the reservoir are an important influence in initiating or widening openings.

### Implications of the correlation

Whatever the true cause behind the correlation, it provides a strong indicator for the most efficient orientation of waterflood configurations. SH can be predetermined by a variety of methods, of which breakout analysis is the easiest (although sometimes, by itself, less than definitive). Given no other overriding reservoir engineering grounds for alternative configurations, a line drive with producer–injector axes aligned orthogonally to SH will, on the basis of these statistics alone, provide for the smallest probability of early breakthroughs. However, there may be significant variations in orientation of local SH, which might be predictable from structural considerations: alternatively, a dominant fracture set may be observed to strike at acute but significant angles to SH, and may provide a more likely preferred direction for flood progress.

We have studied here relative flood breakthrough occurrences that can be associated with areal sweep efficiencies to breakthrough; the link between these and ultimate sweep efficiencies is complicated. Nevertheless, there is a strong likelihood that effects on the former have similar, albeit generally less pronounced, implications for the latter (eg. Landrum and Crawford, 1960; Mortada and Nabor, 1961). The effect of the role of fracturing on vertical sweep efficiencies has not been considered here, but may be just as important.

Obviously there is a strong incentive to improve both the basis for the correlation, through more detailed information and a greater population of cases, and also our understanding of its causes.

## CONCLUSIONS

(1) The empirical probability of breakthrough of injected fluid having occurred to neighbouring producer wells at intermediate times during flooding oil reservoirs has been demonstrated to have a strong consistency in its azimuthal dependence, such that breakthrough from an injector in the direction parallel to the major horizontal principal axis of the local modern-day stress (SH) had more than 50% greater probability of occurring than did breakthrough parallel to the minor principal stress axis. The study population for this deduction came from 36 separate cases and comprised 700 instances of breakthrough from out of 1168 available injector–producer pairs.

(2) Further analysis of the data indicated that the strong relationship between preferred breakthrough directions and SH also held individually for the large majority of cases studied. In 86% of the cases the most probable breakthrough direction was determined to be within 45° of SH. The mean relative orientation of the most probable breakthrough directions from all the cases was 8° anticlockwise from SH.

(3) No characteristic of the flood or the reservoir was found to have significant bearing on this relationship: in particular, approximately the same relationship was found to hold for the subset of reservoirs in which natural fractures were not a prominent characteristic as for those reservoirs deemed to be "naturally fractured".

(4) It is inferred that fracturing—natural, induced, or more probably a combination of both—plays a fundamental role in determining the areal anisotropy of floods over a wide selection of reservoirs. The type, scale and mechanism of fracturing (under which term are potentially included faults, microcracks and syn-sedimentary faults controlling depositional trends) cannot be specified from this study, but understanding of these processes provides a goal of importance to improved planning of floods, and prediction of their performance.

## REFERENCES

Alexander, W.G. (1986). Geologic controls on producibility at Clear Creek field, Uinta County, Wyoming. Presented at the *61st Annual Meeting of the Society of Petroleum Engineers, New Orleans*, October 1986, SPE 15528.

Aydin, A. and Johnson, A.M. (1983). Analysis of faulting in porous sandstones. *Struct. Geol.* **5** (1), 19–31.

Barfield, E.C., Jordan, J.K. and Moore, W.D. (1959). An

analysis of large-scale flooding in the fractured Spraberry Trend area reservoir. *J. Petroleum Technol.* **11** (4) 15–19.

Bell, J.S. and Babcock, E.A. (1986). The stress regime of the Western Canadian basin and implications for hydrocarbon production. *Bull. Can. Petroleum Geol.* **34** (3), 364–378.

Bell, J.S. and Gough, D.I. (1981). Intraplate stress orientations from Alberta oil-wells. In: O'Connell, R.J. and Fyfe, W.S. (eds), *Evolution of the Earth.* American Geophysical Union and Geological Society of America, Geodynamics Series, Vol. 5, pp. 94–104.

Bolling, J.D. (1985). A full field model study of the East Velma West Block Sims Sand Unit reservoir. *J. Petroleum Technol.* **37** (9), pp. 1429–1440.

Budding, M.C. *et al.* (1988). Probabilistic modelling of discontinuous reservoirs. *Proceedings Indonesian Petroleum Association Seventeenth Annual Convention, October 1988*, paper IPA 88–21.04.

Carnes, P.S. (1966). Effects of natural fractures or directional permeability in water flooding. *Society of Petroleum Engineers of American Institute of Mining, Metallurgical, and Petroleum Engineers, North Texas Section seventh biennial secondary recovery symposium,* preprints pp. 17–24, paper SPE1423.

Craig, F.F., Jr. (1971). The reservoir engineering aspects of waterflooding. *Society of Petroleum Engineers American Institute of Mining, Metallurgical, and Petroleum Engineers, Monographs*, Vol. 3, *Henry L. Doherty series.*

Crampin, S. and Atkinson, B.K. (1985). Microcracks in the Earth's crust. *First Break* **3**, 16–20.

Dikken, B.J. and Niko, H. (1987). Waterflood-induced fractures: a simulation study of their propagation and effects on waterflood sweep efficiency. Presented at *Offshore Europe '87, Aberdeen, September*, 1987, SPE 16551.

Dusseault, M.B. (1985). Fluid injection design to preclude induced shear failure of formations. Presented at the *36th Annual Technical Meeting of the Petroleum Society of Canadian Institute of Mining, Edmonton*, June 1985, paper 85–36–6.

Dyes, A.B. and Johnston, O.C. (1953). Spraberry permeability from build-up curve analyses. *Petroleum Trans. Am. Inst. Mining, Metall. Petroleum Eng.* **198**, 135–138.

Elkins, L.F. and Skov, A.M. (1962). Large scale waterflood performance, Sprayberry Field, West Texas. Presented at the *37th Annual Meeting of the Society of Petroleum Engineers of the American Institute of Mining, Metallurgical and Petroleum Engineers in Los Angeles, October*, 1962, SPE 405.

Elkins, L.F., Skov, A.M. and Gould, R.C. (1968). Progress report on Spraberry waterflood-reservoir performance, well stimulation and water treating and handling. *J. Petroleum Technol.* **20** 1039–1049.

Evans, R.C. (1987). An investigation into the influence of common sedimentary structures and diagenesis on permeability, heterogeneity and anisotropy in selected sands and sandstones. Society of Petroleum Engineers unsolicited paper number SPE 17130.

Fehler, M., House, L. and Kaieda, H. (1987). Determining planes along which earthquakes occur: method and application to earthquakes accompanying hydraulic fracturing. *J. Geophys. Res.*, **92** (B9), 9407–9414.

Garon, A.M., Lin, C.Y. and Dunayevsky, V.A. (1988). Simulation of thermally-induced waterflood fracturing in Prudhoe Bay. Presented at the *Society of Petroleum Engineers California Regional Meeting, Long Beach, California, March 1988*, SPE 17417.

Goggin, D.J. *et al.* (1988). Patterns of permeability in eolian deposits: Page sandstone (Jurassic), Northeastern Arizona. *SPE Formation Evaluation* **3** (2), 297–306.

Harper, T.R. and Last, N.C. (1989). Interpretation by numerical modelling of changes of fracture system hydraulic conductivity induced by fluid injection. *Geotechnique* **39** (1), 1–11.

Jaeger, J.C. and Cook, N.G.W. (1979). *Fundamentals of Rock Mechanics* (3rd edn.). Chapman and Hall, London.

Janot, P., Gauer, P. and Gross, E. (1988). Orientation de la contrainte tectonique dans l'Europe de l'Ouest a partir des ovalisations de trous de forages. *Revue de l'Institute Francais du Petrole* **43** (4), 517–521.

Klein, R.J. and Barr, M.V. (1986). Regional state of stress in Western Europe. In: *Proceedings of the International Symposium on Rock Stress and Rock Stress Measurements, Stockholm, September, 1986*, pp. 33–44.

Koning, E.J.L. (1985). Fractured water wells—analytical modelling of fracture propagation. Paper SPE 14684.

Koning, E.J.L. and Niko, H. (1985). The effect of reservoir pressure and temperature variations on fracturing conditions during a waterflood operation. *Proceedings of the Third European Meeting on Improved Oil Recovery, Rome, April*, 1985.

Lake, L.W. (1988a). The origins of anisotropy. *J. Petroleum Technol.* **40** (4) 395–396.

Lake, L.W. (1988b). Author's reply to discussion of The origins of anisotropy. *J. Petroleum Technol.* **40** (7), 905.

Landrum, B.L. and Crawford, P.B. (1960). Effect of directional permeability on sweep efficiency and production capacity. *Trans. Am. Inst. Mining, Metall. Petroleum Eng.* **219**, 407–411.

Morales, R.H. *et al.* (1985). Detection of formation fracture in a waterflooding experiment. Paper SPE 13747 Presented at the *Society of Petroleum Engineers 1985 Middle East Oil Technical Conference and Exhibition, Bahrain, March*, 1985, SPE 13747.

Mortada, M. and Nabor, G.W. (1961). An approximate method for determining areal sweep efficiency and flow capacity in formations with anisotropic permeability. *Soc. Petroleum Eng. J.* **1**, 277–286.

Perkins, T.K. and Gonzalez, J.A. (1984). Changes in earth stresses around a wellbore caused by radially symmetrical pressure and temperature gradients. *Soc. Petroleum Eng. J.* **24** (2), 129–140.

Perkins, T.K. and Gonzalez, J.A. (1985). The effect of thermoelastic stresses on injection well fracturing. *Soc. Petroleum Eng. J.* **25** (1), 78–88.

Pine, R.J. and Batchelor, A.S. (1984). Downward migration of shearing in jointed rock during hydraulic injections. *Int. J. Rock Mech. Mining Sci. Geomech. Abstr.* **21** (5) 249–263.

Rajan, V.S.V. (1988). Discussion of the origins of anisotropy. *J. Petroleum Technol.* **40** (7), 905.

Schneider, J.J. (1972). Geological factors on the design and surveillance of waterfloods in the thick structurally complex reservoirs in the Ventura field, California. *Paper SPE 4049*, Presented at the *47th Annual Fall Meeting of the Society of Petroleum Engineers of the American Institute of Mining, Metallurgical and Petroleum Engineers in San Antonio, Texas, October 1972*, SPE 4049.

Szpakiewicz, M.J., McGee, K. and Sharma, B. (1987). Geologic problems related to characterization of clastic reservoirs for EOR. *SPE Formation Evaluation* **2** (4), 449–460.

Thomas, L.K. *et al.* (1987). Ekofisk waterflood pilot. *J. Petroleum Technol.* **39** (2), 221–232.

Walsh, J.B. (1981). Effect of pore pressure and confining pressure on fracture permeability. *Int. J. Rock Mech., Mining Sci. Geomech. Abstr.* **18**, 429–435.

Warpinski, N.R. and Teufel, L.W. (1987). Influence of geologic discontinuities on hydraulic fracture propagation. *J. Petroleum Technol.* **39** (2), 209–220.

Zoback, M.L. and Zoback, M.D. (1987). Tectonic stress field of the Continental United States. In: Pakiser, L. and Mooney, W. (eds), *Geophysical Framework of the Continental United States.* Geological Society of America Memoir.

# 14

# Tracer Tests in Oil Appraisal and Reservoir Evaluation: State of the Art

*T. Bjørnstad, K. Garder, I. Hundere, and O.B. Michelsen*

Institute for Energy Technology, Kjeller, Norway

This chapter describes the use of tracers in oil and gas reservoir exploration and evaluation. The tracers may be radioactive atoms or molecules that are detected by nuclear measuring techniques, naturally occurring stable isotopic ratios detected by electromagnetic isotope separation, or chemical tracers that may be measured by more or less conventional chemical techniques, e.g. gas chromatography. The topics selected for a closer description include:

(1) tracers during drilling and completion of a well (mud infiltration studies, study of the "near wellbore" chemical reactions, control of mud circulation, measurement of injectivity profiles, location of cementing level, detection of casing leaks, selective acidizing and treatment of fractures);
(2) tracers in reservoir evaluation (well-to-well flow studies for water and gas flooding);
(3) tracers for EOR-chemicals (the study of surfactant and polymer behaviour under relevant reservoir conditions);
(4) tracers for the determination of residual oil saturation (partitioning tracers in single well or in well-to-well experiments).

In addition, we provide critical comments on past and present practices in tracer application and give some viewpoints concerning future needs and the direction of development.

## INTRODUCTION

In general, a tracer is a substance (solid, liquid, gas) that is introduced into a process in order to describe the development of the process and to study its mechanism. It is a requirement that the injection and presence of the tracer do not perturb the process in any way, nor must the process itself perturb the tracer behaviour differently from the medium to be traced.

In some cases, however, one may want the tracer to behave "non-ideally" and to undergo some form of physicochemical change. In such cases, it is the degree of this physicochemical change that constitutes the measure of the process or parameter to be studied.

In order to maintain the proper conditions for a tracer experiment, the amount of tracer material applied must normally be kept very low. Furthermore, the detectability of the tracer has to be correspondingly good. These are the main reasons why radioactive atoms and molecules have been, and still are, the only realistic tracers in many cases, especially where high dilution factors are involved. However, chemical substances, like the perfluorocarbons, with nearly comparable detectability (by gas chromatography in combination with electron capture detection) are now receiving increasingly more attention (Kerrin, 1981; Lovelock and Ferber, 1982; Dietz and Senum, 1984; Janota-Bzowski, 1985).

Tracers constitute an irreplaceable experimental tool in most branches of natural science and technology today, including oil and gas reservoir exploration and evaluation, and the potential is greater than most reservoir engineers are aware. Below, we outline in some detail selected types of tracer experiments that are either performed routinely in the offshore oil industry or that may be performed on a regular basis after a modest amount of development. We also provide some critical comments on past and present experimental practice, express some views on improvements needed and point to some future directions of development.

## TRACERS IN EXPLORATION

There are a number of ways in which tracers may be applied during the process of well drilling, completion and

*North Sea Oil and Gas Reservoirs–II*

treatment (Clayton, 1967), and the first publications date back to the early forties (Buckley, 1940; Teichman, 1944). Some of these areas are briefly treated below.

## Application during well drilling

### *Measurement of mud water infiltration*

During drilling and completion of a well, the mud filtrate, the completion fluids and particles from the drilling process will normally infiltrate the near-wellbore zones. The amount of infiltration depends on several factors, two of the most important being the quality of the mud cake formed and the permeability of the formation material. Here, we will concentrate on the mud filtrate invasion, and especially treat water-based mud. This infiltration may result in chemical processes occurring in the liquid phase (precipitation, complexation, etc.) when the intruder liquid encounters the formation liquid of different chemical composition. Chemical reactions may also occur between the liquids and the formation minerals (ion exchange, dissolution, etc.). These interactions may result in changes of important physical parameters like permeability, solid composition in the invaded zone, electrical conductivity, etc. Such changes may again affect the response of electrical and nuclear logs (Phelps *et al.*, 1984) (also for oil-based mud), and the chemical composition of samples taken from the formation water, either by formation tester equipment or in the course of regular water production tests. Good knowledge of the true composition of the formation water is important for evaluation of the reservoir.

How is knowledge of the mud infiltration depth obtained? One way is to dope the mud with a dissolvable constituent that experience has demonstrated is not present in the formation water and that will not undergo any measurable reaction with the formation minerals or other constituents of the formation water or the adjacent oil. This constituent will then perform as a tracer for the mud water.

Chemical species like nitrate, bromide and iodide have been used, and are still, to some extent, in use today. However, the background concentration is not always negligible, nor can chemical or biological reactions in the invaded zone be excluded. Bromide seems to be the least affected of the three tracers mentioned (Davis *et al.*, 1980; Bowman, 1984). Iodine may be lost under aerobic field conditions while nitrate is subject to a number of chemical and biological transformations that may take place under both aerobic and anaerobic conditions. In addition, the analysis may be cumbersome considering the small amounts of additives that are applied.

In order to overcome these difficulties, the use of tritiated water, HTO, has been proposed (Millar and Buckles, 1975) as a tracer for the mud water. Tritium is the heaviest existing hydrogen atom. It is radioactive and emits very low-energy $\beta$-particles with a half-life of 12.32 years. The chemical and physical behaviour of HTO is not distinguishable from the reactions of the normal water molecule in this kind of application. Moreover, the analysis of HTO in very small quantities is readily performed by nuclear detection techniques. This method has been further developed (Dahl, 1987) and appears today as the most reliable method for assessing the amount of mud filtrate in water samples taken from the formation. In brief, it consists of the following steps.

- A few tens of metres before the bit enters the zone of interest (the oil-bearing formation), the drilling mud is doped with HTO to a homogeneous concentration (100–500 Bq/ml).
- The HTO content is kept constant ($\pm 2\%$) when drilling through the oil zone and into the underlying water zone by frequent analysis of the circulating mud (at least one analysis every hour). The concentration of HTO is adjusted when necessary by addition from an adjustment reservoir (see Fig. 1).
- FMT (RFT) water samples taken can readily be analysed with respect to the mud filtrate content. If this content is denoted by R, the relationship is simply

$$R=\frac{A_{SW}}{A_{MW}} \tag{1}$$

where $A_{SW}$ is the corrected counting rate in the sample taken, and $A_{MW}$ is the corrected counting rate in the pure mud filtrate. By using FMT samples with different content of mud filtrate as determined by the HTO content, one may by ordinary chemical analysis of the collected samples determine the concentration of ions like $Na^+$ and $Cl^-$ in the formation water, as illustrated in Fig. 2.
- By following the HTO concentration in samples taken from a water production test as a function of produced water volume, it is possible to estimate the approximate depth of infiltration. However, research remains to be carried out on the determination of invasion *profile*.
- By analysing the consecutive water samples with respect to the content of ions like $Ba^{2+}$, $Sr^{2+}$, $Ca^{2+}$, $SO_4^{2-}$, $CO_3^{2-}$, etc., and correlating the results with

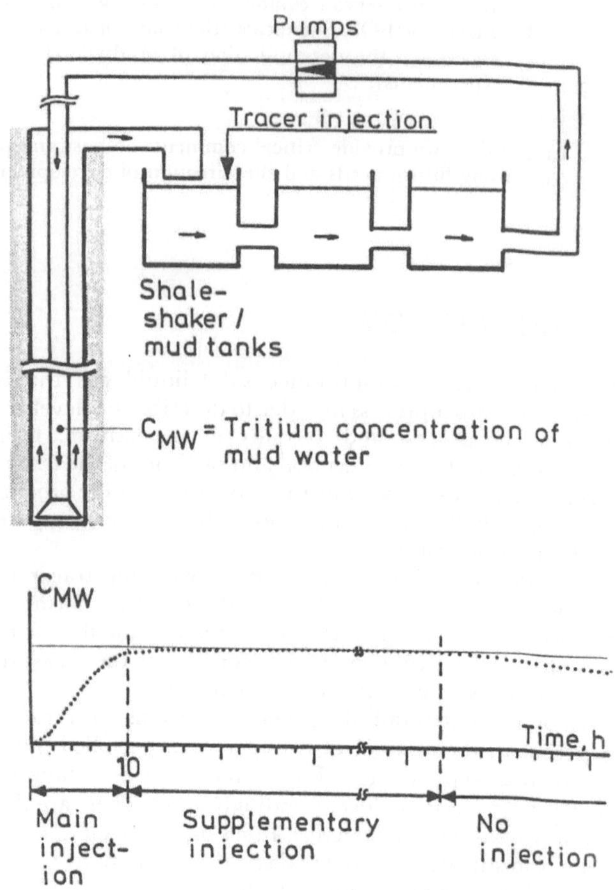


**Fig. 1.** Use of tritiated water, HTO, as a tracer for the mud filtrate during exploratory drilling through the geological structures of interest.

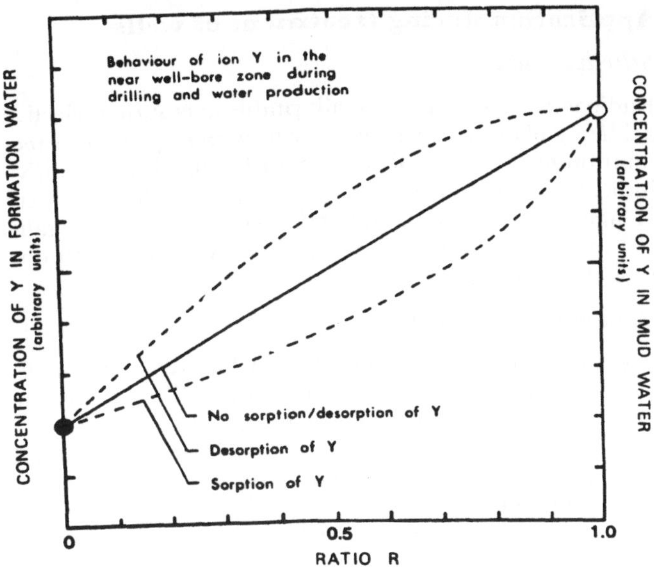


**Fig. 2.** Principal behaviour of an ion Y (sorbing, desorbing or non-interacting as derived from its measured concentration in produced water as a function of $R$ (Eq. 1):
(——) derived concentration of Y in formation water;
(– – –) measured concentration of Y in mud filtrate.

the values of $R$, one may estimate the degree to which chemical reactions have occurred in the near-well bore zone*. The method is at presently under development, and preliminary results have been obtained. However, further research, is needed in order to quantify the chemical reactions and estimate their damaging effects.

The radiotracer method for the measurement of mud infiltration for oil based mud has not yet been developed, but we hope to be able to start work on this shortly. Quantification of the mud oil or mud emulsion infiltration is strongly required.

*Control of drilling fluid circulation, location of porous zones and measurement of injectivity profiles*

In normal drilling operations the ratio of drilling fluid entering the borehole to the drilling fluid leaving the borehole should be constant. If the fluid output exceeds the input, a potentially hazardous condition develops. If the input exceeds the output, the difference is lost to the formation, either through high-permeability zones or into macroscopic fractures.

The simplest way of locating high-permeability zones is by incorporating a $\gamma$-emitting tracer into the drilling mud. After a period of circulation, the well is cleaned and a $\gamma$-logging tool is lowered to register zones of increased radioactivity, corresponding to the depths where the mud was lost into the formation (Flagg *et al.*, 1955). Among the suitable, relatively short-lived radiotracers are $^{198}Au$ ($t_{1/2}$ = 2.69 d), which in solution precipitates onto the mud particles, and $^{131}I^-$ ($t_{1/2}$ = 8.02 d) or $^{82}Br^-$ ($t_{1/2}$ = 35.34 h) which remain as dissolved species in the mud water. In cases where longer-lived tracers are desirable, $^{192}Ir$ ($t_{1/2}$ = 74.0 d) and $^{46}Sc$ ($t_{1/2}$ = 43.8 d) in various chemical forms may be used.

The so-called "flow rate method" is a more elaborate technique for the location of high-permeability zones, measurement of drilling fluid circulation and evaluation of injectivity profiles (Ford, 1962; Gore, 1963; Walker *et al.*, 1964; Kelldorf, 1970; Wiley and Cocanower, 1975; Hill and Solares, 1985). The principle is illustrated in Fig. 3. An instrument containing a bank of tracer solution and one or two $\gamma$-ray detectors is lowered into the wellbore. A tracer slug is injected, either into the flowing mud or into the injection water, and the time needed for transport to the detector, or between the two detectors, is measured. The tool is lowered to successively new positions and the operation is repeated. Thus, the liquid travel time in the various sections of the wellbore can be calculated and the differential fluid loss along the wellbore determined.

Only a small amount of radiotracer is required for each measurement, and commonly $^{131}I^-$ is used as a tracer. Instead of using a bank of radiotracer solution, one could apply radionuclide generator systems with long-lived mothers and shorter-lived daughters. Table I gives some examples of such generation systems. Here, the daughter nuclides are eluted off the generator when needed, and the generator grows into saturation when not in use. Although

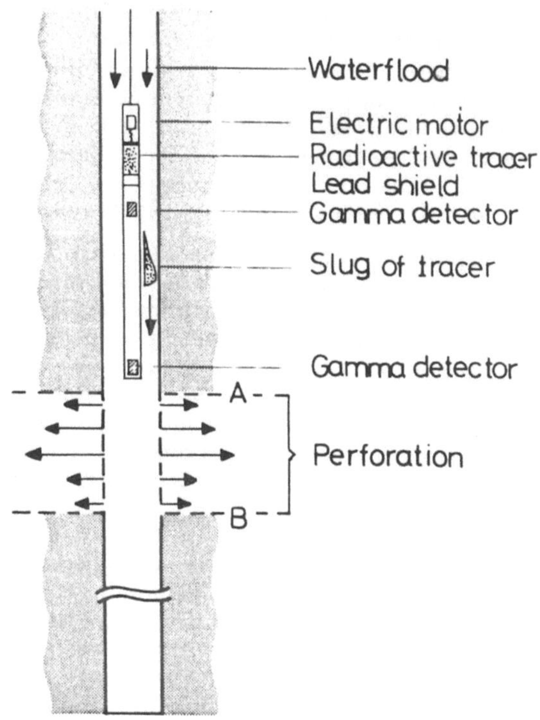


**Fig. 3.** Radioactive tracer log for measurement of mud circulation rate (or water injection/production rate).

**Table I.** Radionuclide generator systems applicable in well-logging tools

| *Radionuclide of interest* | *Generator system* | *Principal γ-rays* |
|---|---|---|
| $^{106}Rh$ (30 s) | $^{106}Ru$ (368 d) → $^{106}Rh$ | 512,622 keV |
| $^{137m}Ba$ (2.55 m) | $^{137}Cs$ (30.17 y) → $^{137m}Ba$ | 611 keV |
| $^{68}Ga$ (68.3 m) | $^{68}Ge$ (288 d) → $^{68}Ga$ | 511,1077 keV |
| $^{113m}In$ (99.5 m) | $^{113}Sn$ (115.1 d) → $^{113m}In$ | 392 keV |
| $^{99m}Tc$ (6.0 h) | $^{99}Mo$ (66 h) → $^{99m}Tc$ | 141 keV |

* Bjørnstad, T., Dahl, J.B., Hvidsten, J., Kleven, R., Qvenild, C. and Tollan, O. On the use of tracers in mud filtrate and completion fluids invasion studies. To be published as IFE/KR/F report in 1989.

there could be large advantages in using generator systems instead of tracer solutions, they have not, to our best knowledge, been used for this purpose previously. A research and development period would be necessary in order to make such a system operable.

## Application during completion of wells

Well completion practices differ greatly, and vary with such factors as the depth of the producing zone and the general stratigraphic conditions encountered in a well. However, some general methods for the use of radiotracers can be mentioned.

*Location of cementing level behind casing*

When cement containing an insoluble $\gamma$-emitting tracer like $^{46}Sc_2O_3$ is squeezed behind the casing, the "cement top" can be detected with a $\gamma$-ray log lowered into position (Flagg *et al.*, 1955; Evans *et al.*, 1980). In this way the location of the cement top can be followed during the whole procedure. One should ensure that the tracer used actually follows the cement. It should not therefore, be dissolvable in water. It is also possible to use various constituents of the cement itself as tracer which can be made radioactive, preferably by irradiation in a thermal nuclear reactor. One example of such a tracer, dispersed inside the cement grains after neutron activation, is $^{24}Na$ ($t_{1/2} = 15.0$ h). A modest amount of cement is radioactivated and mixed with the first part of the cement slug pumped into the wellbore. If a slightly longer-lived radiotracer is desired, $^{198}Au$ may be precipitated on the outer surfaces of cement grains in a small cement slug, which, in turn, is mixed with the first cement entering the wellbore.

*Location of casing leaks and channels behind casing*

A common problem encountered in production operations is that a potential producing zone does not respond as anticipated from core data, drill-stem tests and well logs. This malfunction may be due to a faulty cementing job, thus giving rise to the movement of water in channels within the cement annulus behind the casing. Such channels provide undesirable communication between various formation layers of different pressures. These kind of leaks can be detected by two kinds of nuclear techniques as follows.

(1) A tool containing a 14 MeV neutron generator in one end, and a $\gamma$-ray detector in the other end (Paap *et al.*, 1977), is lowered into the wellbore. The flowing water behind the casing is radioactivated by the reaction $^{16}O(n, p)^{16}N$ ($t_{1/2} = 7.13$ s) and the resulting high-energy $\gamma$-radiation of 6.1 and 7.1 MeV is detected with the $\gamma$-detector. These energies are considerably higher than the background radiation from the formation minerals and may therefore be registered without disturbance at relatively small intensities.

(2) A radioactive solution containing a non-sorbable $\gamma$-emitting tracer like $^{131}I^-$ or $^{82}Br^-$ is injected (and it is supposed that logs of the natural $\gamma$-ray activity from the formation exist) (Flagg *et al.*, 1955; Gore and Terry, 1956; Charyulu and Devrajan, 1977). The tracer solution is squeezed through the perforation, and the wellbore is cleaned with an overflush of inactive water. During this stage, a series of logs is taken and compared with the original logs. Leakages and fluid directions are readily detected, and remedial action can be taken.

## Application during treatment of wells

*Selective acidizing*

Radioactive tracers are also adaptable in ensuring selective acidizing of oil-bearing zones, both with respect to correct positioning of the acid and in order to obtain positive indications about which zones are benefiting from the acid (Flagg *et al.*, 1955; Williams and McCarthy, 1987). Again, $^{131}I$ in the form of HI can be applied by direct dissolution into the treating acid.

*Treatment of fractures*

Well fractures may be treated with high-viscosity fluids, sometimes loaded with an abrasive sand. To measure the effect of the treatment, radiotracers may be applied either dissolved in the liquid or as a constituent of the sand (Pearce, 1977; Lopus *et al.*, 1987). For successive treatment cycles, different radiotracers are preferred. Ordinary $\gamma$-ray logging tools are usually applied for detection of the radiation.

Often, the use of tracers has aided decisively in the performance of successful fracture simulation. Sometimes, however, the tracer tests reported in the literature have not given the desired result, largely because of the choice of the wrong tracer. In addition, it seems that the radiation detection and data evaluation techniques have been too primitive.

# TRACERS FOR RESERVOIR EVALUATION

## Well-to-well tracer test

Well-to-well tracer testing is an increasingly used tool for reservoir characterization. Its usefulness stems from the fact that tracer tests provide information concerning the flow of fluid on the *reservoir scale* (Trantham *et al.*, 1980; Beier and Sheely, 1988; McIntyre and Polkowski, 1985). Tracers are used in reservoir studies to trace the movement of the injected fluid through the reservoir and to monitor reservoir performance. Both radioactive and non-radioactive (chemical or isotopic ratio) tracers may be used. Generally the injection fluid is labelled with a chemical or radioactive tracer and the produced fluid from the well(s) of interest is sampled and analysed to determine the tracer response curve. The analysis of the tracer response curve (concentration of tracers vs. time) can provide important information about the character of the reservoir and makes it possible to improve production strategy.

Specifically, tracer results are used to obtain information on:

- flow patterns and rate of movement of injected fluid;
- accurate breakthrough times and source of breakthrough;
- heterogeneities, fractures, permeability variations;
- variation of sweep improvement treatments, etc.

Information of this kind can be used to improve sweep efficiency, and to provide a guide to remedial action when problems arise. Conventional and improved waterfloods are perhaps the most documented assisted recovery methods for tracer application. However, tracer tests are used for any type of improved oil-recovery process, including thermal and chemical processes and all types of gas injections.

Design and planning of the tracer programme is extremely important. It usually involves the choice of tracers, tracer injection well patterns, the amount of tracer injection strategy, the sampling programme and interpretation. In this type of tracer study, "ideal tracers" are used. The "ideality" refers mainly to the chemical and physical behaviour of the tracer, which should exhibit the following properties.

(1) It should move at a velocity very close to the velocity of the injected fluid, and it should not be sorbed onto the geological material or react with the reservoir fluids.
(2) The tracer must have good chemical and thermal stability, and should not be degradable by sulphate-reducing bacteria.

The following more general tracer requirements also apply.

(3) The detection limit should, in general, be as low as possible in order to minimize the amount of tracer needed.
(4) The tracer should be of low toxicity to allow safe usage.
(5) The availability and cost of the tracer and the cost of analysis should be reasonable.

## Tracers for water injection

Radioactive species are probably the most used tracers in well-to-well tests. The main advantage of radioactive tracers is their ease of detection in extremely small quantities. A large number of tests involving such tracers have been reported in the literature (Wagner *et al.*, 1974; D'Hooge *et al.*, 1981; Garder and Michelsen, 1988). However, there are few available data on systematic testing of the tracer material, and in many projects this has led to a very low recovery efficiency of the injected tracers.

At IFE we have carried out systematic investigations to test various water tracers for use in Norwegian reservoirs, both within the SPOR programme and for actual field applications. The laboratory testing of tracers has involved both static adsorption studies and dynamic studies under reservoir conditions (Garder and Hundere, 1988). As a result of these investigations, the following water tracers may now be regarded as suitable for application in well-to-well tests: tritiated water, $^{14}$C-labelled thiocyanate, $^{22}$Na and $^{36}$Cl. $^{35}$S-labelled thiocyanate and $^{125}$I are also suitable tracers, but their shorter half-lives (87 days and 60 days) may, in some instances, be a limiting factor.

Various chemical substances have also been reported as useful for tracing water movement in reservoirs. They include compounds such as thiocyanate, iodides, various alcohols, etc. These tracers might be applied in small reservoirs; they would not, however, be suitable in the large reservoirs of the North Sea. The main limiting factor in the use of chemical tracers in large reservoirs is the high least-detectable concentration (ppb to ppm range). In addition, they are often affected by chemical and biodegradation. Consequently, it may be necessary to compensate for this by injecting large amounts of material.

Certain perfluorinated chemicals offer the possibility of exceptionally low detection limits. These compounds are chemically very stable, non-toxic and should be suitable for large-scale investigations. At present, compounds such as perfluorodecalin and fluorinated organic acids etc. have been studied and used as tracers in hydrological and oceanographic investigations (Ledwell *et al.*, 1986; Watson and Ledwell, 1988). At IFE we are now testing perfluorinated compounds for use as tracers in petroleum reservoirs. This is part of a comprehensive tracer testing programme that involves both field studies and laboratory investigations (Garder and Michelsen, 1988; Garder and Hundere, 1988).

## Natural isotopic tracers

The natural isotopic composition of a water and its solutes provides an inherent "fingerprint" that can be used to identify the origin of the water (e.g. meteoric/marine, nature of water–rock interaction), and to quantify mixing processes between different volumes of water. Isotopic tracers that are potentially important as tracers for water injection are $\delta$D and $\delta^{18}$O values—which reflect the $^2$H/$^1$H and $^{18}$O/$^{16}$O isotope ratios of the water molecules—and $^{87}$Sr/$^{86}$Sr values, which reflect the isotopic composition of dissolved strontium. In terms of these isotope systems, modern seawater has a homogeneous composition ($\delta D = 0‰$, $\delta^{18}O = 0‰$, $^{87}Sr/^{86}Sr = 0.70920$). Formation waters, on the other hand, can have a wide range of compositions, which reflect: (a) their ultimate origin (meteoric/marine); (b) different seawater compositions in the past, particularly important for $^{87}$Sr/$^{86}$Sr; and (c) water–rock interaction since deposition. Because of these factors it would be an extreme coincidence if the formation water in a North Sea reservoir had the same isotopic composition as modern seawater. It is the difference between seawater and formation waters that forms the basis for using these tracers for seawater injection.

A prerequisite for such a tracer study is that the isotopic composition of the formation water has been well characterized. If production waters are subsequently monitored for their isotopic composition, breakthrough of injected seawater will naturally be recognized by a change in the isotopic composition of the production water towards that of modern seawater. Just how soon breakthrough may be detected is dependent upon the absolute difference between the seawater and formation water compositions, the reproducibility of the isotopic analysis and the homogeneity of the formation water.

## Tracers for gas injections

An ideal gas tracer should have the same thermodynamic properties as the injected gas. Hence, the most obvious way of selecting a tracer for a gas flood experiment is to label the actual injected gas with $\beta$-emitting atoms where physically possible. Thus a $CO_2$ flood could be labelled with $^{14}CO_2$ and a hydrocarbon gas flood could be traced with tritated or $^{14}$C-labelled hydrocarbons. The tracers most frequently used as gas tracers have been radioactive tracers, mainly tritiated or $^{14}$C-labelled hydrocarbons and $^{85}$Kr (Tinker, 1973; Omoregie *et al.*, 1987). More recently, the use of chemical tracers in the form of halogen compounds (freons) as tracers for gas injection has been reported (Omoregie *et al.*, 1987). At IFE we are investigating perfluorocarbons (PFC) as tracers for gas injections. PFC compounds are made by substitution of fluorine for hydrogen in hydrocarbons. They are known to be chemically very stable up to temperatures of 400–500°C. They are physically inert and have low toxicity. Owing to the high affinity of their reaction with electrons, the cyclic PFCs are among the most sensitive compounds in the electron capture detection (ECD) system. At present, we are testing perfluoromethylcyclopentane, perfluorodimethylcyclohexane and perfluoromethylcyclohexane. The, testing involves both laboratory investigations and field studies.

## The interpretation of a tracer test

The interpretation of the tracer reponse curve is an important part of the tracer test. The tracers' arrival times at the producer constitute a useful tool for evaluating a heterogeneous reservoir. Primarily, the results indicate the continuity or discontinuity of the reservoir. The tracer response curve gives information on the permeability distribution and fluid transmissibility within the reservoir. When non-communicating layers with different permeabilities exist, the tracer response curve is the summation of effects from each layer, as shown in Fig. 4. When different tracers are used in a multiwell water-injection programme, the tracer response curves for the differents tracers show sources of breakthrough and transmissibility (Fig. 5). Simple models are available for calculation of the volumetric sweep efficiency, reservoir permeability and permeability distribution. Usually, computer simulation is used to match the tracer response curves, and tracer modelling has lately been a field of development. This may be exemplified by:

- the development of a reservoir layering model (Abbaszaden-Dehghani and Brigham, 1984);
- the use of a stochastic approach to analyse tracer response from a heterogeneous reservoir (Smith and Brown, 1984);
- the development of the UTCHEM simulator, which includes tracer adsorption and patitioning (Pope and Sepehrnoori, 1988);

and the ongoing tracer modelling work at IFE*.

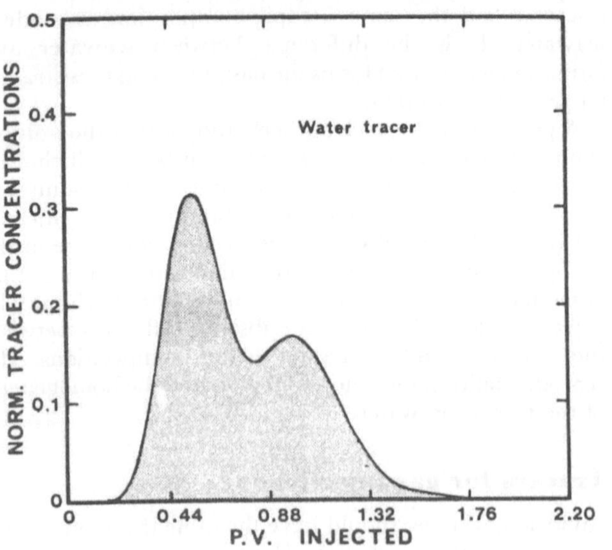


**Fig. 4.** Idealized tracer response curve for reservoirs with two permeability layers.

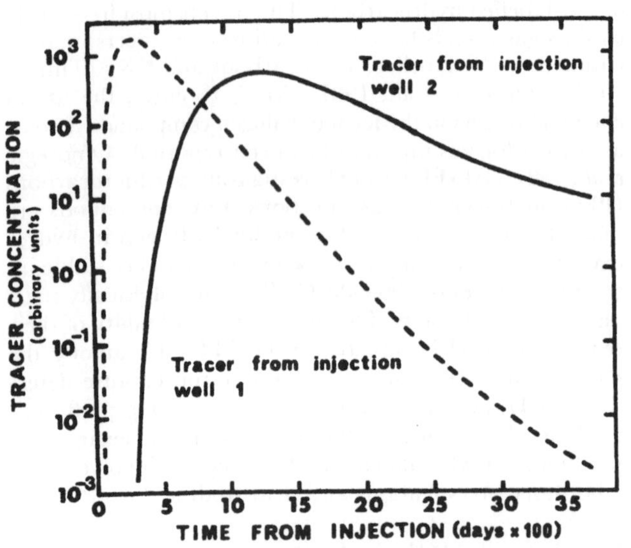


**Fig. 5.** Typical tracer response curves in a producing well with breakthrough of tracers from two injection wells.

## TRACERS FOR EOR CHEMICALS

Enhanced oil recovery (EOR) is being used more and more as the need for making the best use of existing resources is recognized; and, as field test are proving, the use of EOR methods is economically feasible.

A simplified definition of EOR is the recovery of oil by the injection of materials not normally present in the reservoir. The definition comprises all modes of oil recovery processes (drive, push-pull, and well treatments), and covers a large number of oil-recovery agents (Lake, 1988).

Two types of processes in particular are expected to be of great importance and require large quantities of chemicals: polymer flooding and surfactant (micellar-polymer) flooding. In polymer flooding, the chemical is dissolved in injected brine to increase its viscosity, leading to oil displacement of a more favourable mobility ratio. Surfactant flooding is a process that results in very low residual saturations in swept zones by reducing the oil–water interfacial tension by several orders of magnitude (Chemicals in the Oil Industry Symposium, 1985).

### Why use (radio)tracers?

Like a number of other materials used in the oil industry, EOR chemicals need to be tested under field conditions for stability, possible interaction with the environment, behaviour during migration through the reservoir formation, and so on. Conventional analytical methods are, in many cases, unsatisfactory for the observation of some of these properties. For example, laboratory testing often implies that core samples are enclosed in steel cylinders under high pressure and temperature while a liquid containing the compound under study is forced through the core. If this compound is labelled with a radionuclide capable of emitting $\gamma$-radiation of sufficient energy, the movement of the liquid, or, more precisely, of the compound, can be followed directly and continously without interrupting the experiment, which cannot be achieved by conventional methods.

Also, the migration of the compounds in the reservoir itself can be studied favourably, since the identification of radiolabelled compounds is simple and the sensitivity of detection is extremely high. In this case, both $\gamma$- and $\beta$-active radionuclides are of interest as labels for EOR chemical molecules.

### Polymer flooding

A number of polymers have been considered for polymer flooding, such as:

xanthan gum
scleroglucan
hydrolysed polyacrylamide (HPAM)
copolymers of acrylic acid and acrylamide

* Johansen, T., Antonsen, B. and Hidle, S. Tracer testing for improved reservoir description. Talk given at the SPOR seminar, Norwegian Petroleum Directorate, Stavanger 1988.

copolymers of acrylamide and 2-acrylamide-2-methyl-propane sulphate (AM/AMPS)
hydroxyethylcellulose (HEC)
carboxymethylhydroxyethylcellulose (CMHEC)
polyacrylamide (PAM)
polyacrylic acid
glucan
dextran polyethyleneoxide (PEO)
polyvinylalcohol

Only the first four have actually been field tested. However, the list demonstrates that there are many potentially suitable chemicals, some of which may prove to be more effective than those currently used. It is not possible here to go through the list in detail, and therefore we find it convenient to pick out xanthan as an example, firstly because it is one of the most promising and widely used compounds, and, secondly, because at IFE we have examined some of the possibilities of preparing xanthan spiked with a suitable radio-nuclide. For laboratory investigations, $^{131}I$ is such a nuclide. We have tried various ways to introduce iodine in the xanthan molecule, which chemically is a polysaccharide produced by bacterial strains of the genus xanthomonas. Direct methods have so far proved unsuccessful, but it seems very likely that the objective can be achieved by cultivating the bacteria on a substrate containing iodine-labelled glucose, which is one of the building stones of xanthan. $^{14}C$-labelled xanthan has already been prepared by this process and possibly also the tritium-labelled compound. Polyvalent ions such as $Cr^{3+}$ and $Al^{3+}$ are frequently employed in combination with xanthan and other polymers in order to produce in-situ gel formation in reservoir rocks through cross-linking (Chyi-Gang Huang *et al.*, 1984; Dovan and Hutchins, 1984; Prud'homme and Uhl, 1984; Sohail Aslam and Willhite, 1984; Abda *et al.*, 1984; Hubbard *et al.*, 1986). In the case of $Cr^{3+}$, certain aspects of the gelation reaction can be studied using radioanalytical methods, since this element has an isotope $^{51}Cr$ ($t_{1/2}$= 27.7 d) with favourable tracer properties (Hubbard *et al.*, 1986).

### Surfactant flooding

There are four types of surface-active materials (some say five, including a group of water-insoluble emulsifying agents). Table II lists the four types with examples showing the chemical form.

A recent literature search gave many examples of the application of tracers in connection with EOR (Michelsen *et al.*, 1987; Garder and Michelsen, 1988), but none that demonstrated the use of labelled surfactants to enable us to study these specifically. We know, however, that a demand for such tracers exists. Since most surfactants are made synthetically, incorporation of, for instance, radiotracers in such compounds may most conveniently be done in the process of their synthesis. This should not present any great difficulties in the case of $^{14}C$ and tritium, but when it comes to γ-emitting tracer nuclides, problems may be encountered because these, in most cases, represent elements not normally found in the molecule. The problem may be a synthetic one, or it may turn out that the tracer molecule behaves anomalously as compared to the "true" one. For example, since sulphur is a common constituent of many important surface-active agents, we have considered replacing some of the sulphur atoms by selenium, which can easily be radioactivated. However, it remains to be seen whether a possible selenium-containing molecule will exhibit the same properties as the sulphur one. Radioactive iodine will be much easier to introduce from a synthetic point of view, but it may turn out that the "new" molecules will differ more from their originals than in the case of selenium.

## TRACER TECHNIQUES FOR DETERMINATION OF OIL SATURATION

In the tracer studies reported in the preceding sections, the tracers have been assumed to show no chromatographic effects (i.e. no adsorption and partitioning).

### Partitioning tracers

The basis of the tracer technique for the measurement of oil saturation is the partition of a chemical, called a tracer, between the oil and water phases in the pore space of a reservoir rock. This partitioning is identical to that observed in laboratory chromatographic separations, and was first proposed by Cooke (1971) as a potential method for SOR measurement. The method is based on the injection of brine containing two tracers into one well and monitoring the arrival time of these tracers at a second well some distance away. Cooke showed that for two tracers having different partition coefficients ($K$-values) between oil and water, their separation could be related to the oil saturation. The fundamental equation for the tracer method is:

$$V_i = \frac{(1-S_o)V_w + K_i S_o V_o}{(1-S_o)+K_i S_o} \quad (2)$$

which quantitatively relates the velocities $V_w$ and $V_o$, for water and oil, respectively, the partition coefficient $K_i$ and the oil saturation $S_o$. So if $K_i$ is measured in the laboratory, and the velocities are measured in a field experiment, the saturation $S_o$ can be obtained.

In our work we have used the UTCHEM (Pope and Sepehrnoori, 1988) simulator to study the effect of partitioning of tracers between the oil and water phases on a reservoir scale. This was done in combination with laboratory experiments for the determination of partition coefficients for chemical tracers. With this approach we tried to determine the number, type and amount of tracers

**Table II.** Classification of surfactants and examples

| Anionics | Cationics | Amphoterics | Non-ionics |
|---|---|---|---|
| Sulphonates<br>Sulphates<br>Carboxylates<br>Phosphates<br>etc. | Quaternary ammonium<br>Pyridinum<br>Imidazolinium<br>Piperidinium<br>Sulphonium Compounds<br>Compounds<br>Others | Aminocarboxylic acids<br>Others | Alkyl-, Alkyl-, Aryl-, Acyl, Acylamido-, Acylaminepoly-glycol ethers<br>Polyol ethers<br>Alkanolamides<br>Others |

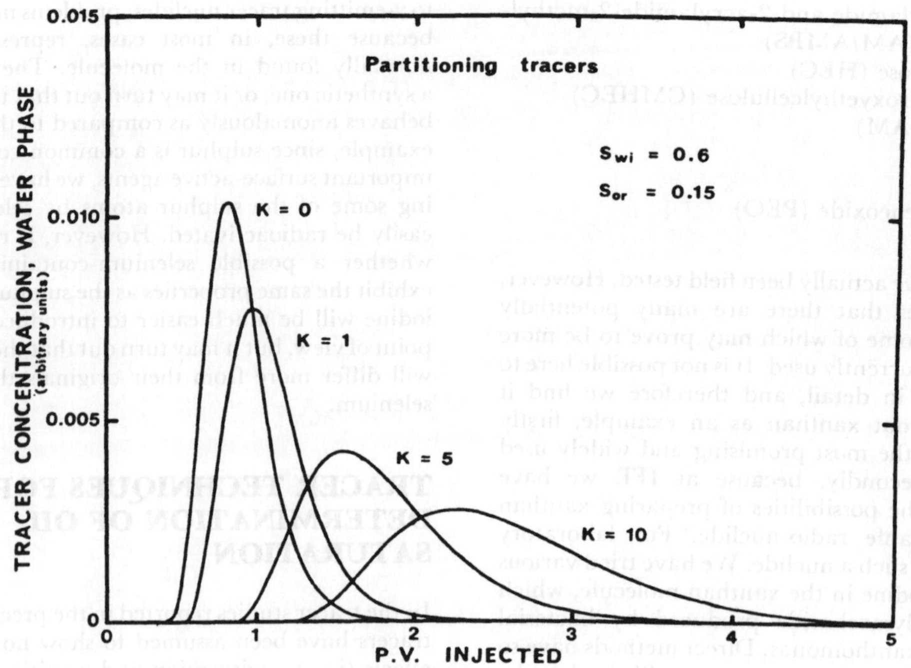


**Fig. 6.** Tracer response curves for tracers with varying degree of partitioning to the oil phase.

that should be injected in a field study in order to gain as much information as possible from the reservoir.

The purpose of the study with UTCHEM was to examine the behaviour of tracers in a reservoir by injecting different types of tracers, and selecting different reservoir oil saturations, in order to see the effect on tracer histories. The case selected for this study was a homogeneous reservoir, 165 × 15 × 15 feet. The porosity was 30%, the horizontal permeability was 1000 mD, and the vertical permeability 100 mD.

An example of a tracer response curve for four different tracers is shown in Fig. 6. In this case the residual oil saturation was 15%. As can be seen from the figure, the tracers show different behaviours as expected. The "ideal water tracer" ($K=0$) breaks through and reaches peak concentration at an earlier time, and has a higher normalised peak concentration than those that are partitioning.

These differences are caused by the fact that fractions of the partitioning tracers stay for part of the time in the oil that resides in the reservoir, which in turn results in a delay of the tracer movement. With a higher value of the partition coefficient ($K$), the tracer movement is more delayed.

In other experiments it has been shown how an increasing residual oil saturation leads to an increased delay of tracer movement. In this project we are now studying the partitioning of various perfluorinated chemicals in order to develop methods to obtain information on the internal structure and associated fluid saturation of the petroleum reservoirs.

## Single well tracer test

The most common application of tracer tests for the measurement of oil saturation has been the single well tracer test, carried out by injection and production of reservoir brine containing tracers in the same well.

In a single well tracer test the reversibility of the flow pattern destroys the chromatographic separation between two tracers, the separation that occurs during injection is essentially removed as the tracers return to the test well. This reversibility is avoided by using a chemically reactive tracer.

The tracer bank is pushed into the reservoir to the desired depth of investigation. Flow is then stopped long enough for a detectable amount of secondary tracers (alcohols) to be formed by hydrolysis. Then the flow is reversed and the produced fluids are sampled and analysed to determine the arrival times of original and secondary tracers. Residual oil saturation can than be related to the separation of the two tracers as shown in the equation above. Frequently, a water tracer, such as tritiated water, is used to establish a material balance.

The single well technique has found wide application in the oil industry. During the last 10 years the method has also been refined in several ways. Improvements have been made with regard to:

- development of a wider selection of tracers;
- better laboratory procedures for providing essential test data (Deans, 1969; Tomich *et al.*, 1973; Charlisle and Kapoor, 1982; Knaepen *et al.*, 1988);
- computer models for test interpretation (Deans *et al.*, 1980).

Most of the single well tracer tests have been carried out with esters (methyl, ethyl, propyl, acetate). Recently, however, a variety of tracers have been described in the patent literature. These include chemicals such as ketoesters, carbon dioxide, alcohols, etc. (Keller, 1974; Deans, 1976; Sheely and Baldwin, 1978; Sheely, 1979; Richardson, 1986, 1987). The chemicals may also be tagged with a radioactive isotope such as $^{14}C$. Single well tracer tests have also been used to evaluate EOR processes. The method utilized a multitracer version of the single well tracer test for determining residual oil saturation by measuring the oil displaced by an EOR slug process. Methyl acetate, ethyl acetate and isopropyl acetate were used as tracers.

* Johansen, T., Antonsen, B. and Hidle, S. Tracer testing for improved reservoir description. Talk given at the SPOR seminar, Norwegian Petroleum Directorate, Stavanger 1988.

## CONCLUSION

Tracer techniques have proved their usefulness in the petroleum industry for more than 40 years. In particular, the use of radiotracers has added a new dimension to the exploration and evaluation possibilities of large reservoirs owing to their very good detectability.

However, in spite of the inherent power of the technique, it has not yet gained the broad application that it deserves. There may be several reasons contributing to this: insufficient knowledge of its existence, the need for specially trained and skilled personnel who are often unavailable within the companies, and perhaps the fact that sometimes the tracer tests have not given the desired clarification. The last is an argument that frequently comes as a result of unprofessionally conducted experiments. Apparently, the tracers used have been chosen with the half-life and $\gamma$-ray energy as the only important criteria. Two other criteria of overriding importance have been left unconsidered: the chemical and physical properties of the tracers under the conditions in which they are meant to operate. And of course it is of vital importance that the radiotracers do trace the material under study.

A general problem in this kind of tracer application has been that laboratory research of the tracer's behaviour has been insufficient. We are happy to note, however, that there now seems to be a growing awareness of the need for such preparative investigations, including studies of a basic nature. The increasing interest in the application of tracers has also triggered investigation of chemical tracers of a new kind in connection with reservoir exploration, namely the perfluorocarbons.

We believe the development is now heading in the right direction, and are convinced that results will be produced that will not only benefit oil exploration and production, but also in the end the community as a whole.

## ACKNOWLEDGEMENTS

The reserach work at IFE referred to in this paper has partly been carried out within the SPOR programme (SPOR: Statlig FoU-Program innen økt Oljeutvinning og Reservoarteknikk) and partly in cooperation with the oil companies Statoil, Saga Petroleum, and Phillips Petroleum Company, Norway. The authors wish to express their gratitude to all of them for their invaluable support.

## REFERENCES

Abbaszaden-Dehghani, M. and Brigham, W.E. (1984). Analysis of well-to-well flow to determine reservoir layering. *J. Petroleum Technol.* (Oct.), 1752–1762.

Abda, M.K., Chung, H.S., Phelps, C.H. and Klaric, T.M. (1984). Field experience with floodwater diversion by complexed biopolymers. Paper no. SPE/DOE 12642.

Beier, R.A. and Sheely, C.Q. (1988). Tracer survey to identify channels for remedial work prior to $CO_2$ injection at MCA unit, New Mexico. Paper no. SPE/DOE 17371.

Bjørnstad, T. *et al.* (1989). On the use of tracers in mud filtrate and completion fluid invasion studies. To be published as IFE research report.

Bowman, R.S. (1984). Evaluation of some new tracers for soil water studies. *Soil Sci. Soc. Am. J.* **48**, 987.

Buckley, S.E. (1940). Method of locating detectable cement in a borehole US Patent Number 2.220.205.

Charlisle, T. and Kapoor, S. (1982). Development of a rapid and accurate method for determining partition coefficients of chemical tracers between oils and brines. Paper no. DOE/BC/10100–4.

Charyulu, K.S. and Devrajan, N.R. (1977). Role of radioisotopes in oil well investigations. *Proceedings National Symposium on Isotopes Applications in Industry. Trombay, Bombay.* 1977, p. 102.

*Chemicals in the Oil Industry.* Proceedings from a symposium organised by the Royal Society of Chemical Industry, Division North West Region, Owens Park, University of Manchester, March 1985.

Chyi-Gang Huang, Green, D.W. and Willhite, G.P. (1984). An experimental study of the in-situ gelation of chromium (+3)-polyamide polymer in porous media. Paper no. SPE/DOE 12638.

Clayton, C.G. (1967). A survey of the application of radiation techniques in oil and mineral boreholes. Paper no. AERE-R 5368.

Cooke, C.E., Jr. (1971). Method of determining residual oil saturation in reservoirs. US Patent Number 3.590.923.

Dahl, J.B. (1987). Tracer techniques in near bore-hole chemistry. In Hagebø, E. and Salbu, B. (eds.), "Aspects of Nuclear Science" (in Honour of Prof. A.C. Pappas), p. 142. Norwegian University Press.

Davis, S.N. *et al.* (1980). Ground water tracers—a short review. *Groundwater* **18**, 14.

Deans, H.A. (1969). Method of determining fluid saturation in reservoirs. US Patent Number 3.623.842.

Deans, H.A. (1976). Method of determining the relation between fractional flow and saturation of oil. US Patent Number 3.990.298.

Deans, H.A., Majores, S. and Heemstra, R.J. (1980). Single well chemical tracer test simulation handbook: "Improved accuracy and interpretation for residual oil saturation measurement. DOE/BC/20006–14. US Department of Energy.

Dietz, R.N. and Senum, G.I. Capabilities, needs and applications of gaseous tracers. *Presented at the Atmospheric Tracer Workshop, Santa Fe,* May 1984, pp. 123–174. Paper no. LA-10301-C.

D'Hooge, Sheely, C.G. and Williams, B.J. (1981). Interwell tracers—an effective reservoir evaluation tool. West Sumatra field results. *J. Petroleum Technol.* **33**, 779.

Dovan, H.T. and Hutchins, R.D. (1984) Development of a new aluminium-polymer gel system for permeability adjustment. Paper no. SPE/DOE 12641.

Evans, G.V., Parsons, T.V. and Wallace, M.R.G. (1980). Nuclear grout monitoring on offshore platforms. *The 12th Annual Offshore Technology Conference, Houston, Texas, May, 1980,* paper no. OTC-3791.

Flagg, A.H., Myers, J.P., Campell, J.L.P., Terry, J.M. and Mardock, E.S. (1955). Radioactive tracers in oil production problems. *J. Petroleum Technol.* **7**, 1.

Ford, W.O. Jr. (1962). How new injectivity profiling method works. *World Oil* 1 August 43.

Garder, K. and Hundere, I.A. (1988). SPOR-OPT tracer methods for reservoir studies. Institute for Energy Technology, paper no. IFE/KR/F-88/098.

Garder, K. and Michelsen, O.B. (1988). SPOR-OPT tracer studies for improved reservoir description. Institute for Energy Technology, paper no. IFE/KR/F-88/079.

Gore, G.L. (1963). Relative injectivity profiles of irregular size holes. *J. Petroleum Technol.* **15**, 922.

Gore, G.L. and Terry, L.L. (1956). Radioactive tracer techniques. *J.* Petroleum Technol. **8**, 12.

Hill, A.D. and Solares, J.R. (1985). Improved analysis methods for radioactive tracer injection logging. *J. Petroleum Technol.* **37**, 511.

Hubbard, S., Roberts, L.J. and Sorbie, K.S. (1986). Experimental and theoretical investigation of time-setting polymer gels in porous media. Paper no. SPE/DOE 14959.

Janota-Bzowski, J. (1985). Identification of air pollution sources by making use of tracers. *Environ. Prof. Eng.* **11**(2) 27.

Johansen, T., Antonsen, B. and Hidle, S. (1988). Tracer testing for improved reservoir description. *SPOR Seminar, Norwegian Petroleum Directorate, Stavanger.*

Johansen, T., Antonsen, B. and Hidle, S. (1989). A theory of multicomponent chromatography in two phase flow. *Fifth European Symposium on Improved Oil Recovery, Budapest, April 1989.*

Kelldorf, W.F.N. (1970). Radioactive tracer surveying—a comprehensive report. *J. Petroleum Technol.* **22**, 661.

Keller, E. (1974). Dual temperature tracer method for determining fluid saturation in petroleum reservoirs. US Patent Number 3.847.548.

Kerrin, S.L. (1981). Perfluorocarbons as hydrological tracers. Paper no. NSF/CPE-81015.

Knaepen, W.A.I., Tijssen, R. and van der Bergen, E.A. (1988). Experimental aspects of the partitioning tracer test for residual oil saturation determination using FIA-based laboratory equipment. *SPE European Petroleum Conference, London, October 1988*, paper no. SPE 18387.

Lake, L.W. (1988). *Enhanced Oil Recovery Manual*. Department of Petroleum Engineering, The University of Texas at Austin.

Ledwell, J.R., Watson, A.J. and Broccher, W.S. (1986). A deliberate tracer experiment in Santa Monica basin. *Nature* **323**, 322–326.

Lopus, T.A., Seifert, D.J. and Schein, G.W. (1987). Production improvement through identification of conductive natural fractures utilizing multiple radioactive isotope tochnology. Paper no. SPE-16192.

Lovelock, J.E. and Ferber, G.J. (1982). Exotic tracers for atmospheric studies. *Atmos. Environ.* **16**, 1467.

McIntyre, F.J. and Polkowski, G. (1985). Radioactive tracer application to monitoring solvent spreading in the Rainbow Keg River B pool vertical hydrocarbon mischible flood. Paper no. SPE 14440.

Michelsen, O.B., Garder, K. and Bjørnstad, T. (1987). Application of radioactive tracers for improved oil recovery. Radiotracer evaluation—literature survey. Paper no. IFE/KR/F-87/167.

Millar, R.H.G. and Buckles, R.S. (1975). Tritiated water as a drilling mud tracer in Beaufort exploration wells. *J. Can. Petroleum Technol.* **14**, 42.

Omoregie, Z.S. *et al.* (1987). Monitoring the Mitsue hydrocarbon mischible flood program design. Implementation and preliminary results. *The 38th Annual Technical Meeting of the Petroleum Society of CIM, Calgary 1987*, paper no. 87-38-06.

Paap, H.J., Arnold, D.M. and Peelman, H.E. (1977). Behind casing water flow detection using continuous oxygen activation. US Patent Number 4.032.780.

Pearce, R.M. (1977). Evaluation of fracture treatment using tracer and temperature survey. Paper no. SPE 7910.

Phelps, G.D., Stewart, G. and Peden, J.M. (1984). The analysis of the invaded zone characteristics and their influence on wireline log and well-test interpretation. Paper no. SPE 13287.

Pope, G.A. and Sepehrnoori, K. (1988). Field tracer studies using a compositional simulator. *Fifth Annual Report, Center for Enhanced Oil and Gas Recovery Research*, University of Texas.

Prud'homme, R.K. and Uhl, J.T. (1984). Kinetics of polymer/metalion gelation. Paper no. SPE/DOE 12640.

Richardson, E.A. (1986). Determining residual oil saturation by injecting $CO_2$ and base generating reactant. US Patent Number 4.617.994.

Richardson, E.A. (1987). Determining residual oil saturation by injecting salts of carbonic and halocarboxylic acids. US Patent Number 4.646.832.

Sheely, C.G. (1979). Single well surfactant test to evaluate surfactant floods using multi tracer method. US Patent Number 4.168. 746.

Sheely, C.Q., Jr. and Baldwin, D.E., Jr. (1978). Single well tracer method to evaluate enhanced recovery. US Patent Number 4.099.565.

Smith, P.J. and Brown, C.E. (1984). The interpretation of tracer test response from heterogeneous reservoirs. Paper no. SPE 13262.

Søhail Aslam and Willhite, G.P. (1984). Viscosimetric measurement of chromium (III)-polyacrylamide gels by Weissenberg rheogoniometer. Paper no. SPE/DOE 12639.

Teichman, C.F. (1944). Method of determining the porosity and location of permeable formations in oil wells. US Patent Number 2.358.945.

Tinker, G.E. (1973). Gas injection with radioactive tracer to determine reservoir continuity—East Coalings field, California. *J. Petroleum Technol.* (Nov.), 1251–1254.

Tømich *et al.* (1973). Single well tracer method to measure residual oil saturation. *J. Petroleum Technol.* (Feb.), 211–218.

Trantham, J.C., Threlkeld; C.B. and Patterson, H.L., Jr. (1980). Reservoir description for a surfactant/polymer pilot in a fractured, oil-wet reservoir. North Burbank Unit Tract 97. *J. Petroleum Technol.* (Sept.) 1647.

Wagner, O.R., Baker, L.E. and Scott, G.R. (1974). The design and implementation of multiple tracer program for multifluid, multiwell injection projects. Paper no. SPE 5125.

Walker, T., Sherwood, J., Sumner, C.R. and Marshall, R. (1964). The fluid travel log. *J. Petroleum Technol.* **16**, 851.

Watson, A.J. and Ledwell, J.R. (1988). Purposefully released tracers. *Tracers in the Ocean*, The Royal Society, London.

Wiley, R. and Cocanower, R.D. (1975). A quantitative technique for determining injectivity profiles using radioactive tracers. *Paper no. SPE 5513.*

Williams, R.L. and McCarthy, J.T. (1987). Using multiple radioactive tracers to optimize stimulation designs. *Paper no. SPE 16383.*

# 15

# Immiscible Displacement in a Porous Medium Simulated by a Statistical Model

*D. Zhou and E.H. Stenby*

Instituttet for Kemiteknik, Technical University of Denmark, Lyngby, Denmark

Invasion percolation theory is here applied to the simulation of immiscible displacement of two phases in well-defined porous media. A model has been developed that takes not only the influence of the capillary number into account but also the wettability of the system and the viscosity ratio of the immiscible phases. The porous medium in this model is described by a network of pores of equal length. Their size (radius) follows a selected distribution function. Four different distribution functions have been tested in this work: the even distribution function $f(r)=c$, the $\chi^2$ function (with $n$ equal to 4 and 10) and a function $f(r)=2r\cdot\exp(-r^2)$, which was also used by Heiba *et al.* (1986). The influence of the distribution functions on the simulation results was investigated. A method to generate the corresponding network model based on the properties of a given medium is also proposed. The simulation results show that the displacement results cannot be described by the capillary number alone, and the influence of the viscosity ratio between the two phases varies with the capillary number. The larger the capillary number, the larger the influence. The wettability of the system also changes significantly the dependence of the residual oil saturation (both for the oil-wet and the water-wet cases) on the capillary number. This is most dominant when the capillary number is relatively small. The critical capillary number is determined for porous media with different wettabilities. The simulated results are compared with experimental results published by other investigators and this laboratory.

## INTRODUCTION

To optimize the performance of water, polymer and surfactant flooding in oil-recovery processes, it is essential to understand the dynamics of multiphase flow through a porous medium and the connection between the macroscopic behaviour and the microscopic structure of the medium. The injection rate, the viscosities of the injected and the displaced phase (we limit this study to two-phase displacement), the interfacial tension between the two phases, the densities of the two phases, and the properties of the reservoir rock all contribute to the final result. Experimental and theoretical work has been carried out during the last hundred years. Most commonly, the above-mentioned factors have been divided into four groups: the capillary number, $N_c$, which has been defined in several ways (Chatzis and Morrow, 1984), the viscosity ratio between the two phases, the bond number, which expresses the competition of the capillary force with the buoyancy force caused by the density difference between the two phases, and finally, the properties of the reservoir. In this paper a micromodel has been developed. It is based on an attempt to portray the properties of a porous medium and predict the influence of the capillary number, the viscosity ratio, and the wettability of the medium. The effects caused by the density difference could have been included in the simulation without any major changes. The reason why no density influence is included in this work is the lack of experimental data available to compare it with. Several pore-level models have already been developed. Models such as the well-known capillary tube models (with either equal diameters or changing diameters) are very successful in describing single-phase flow through a porous medium.

However, they are unable to model multiphase flow through a porous medium, since they ignore the throats connecting the pores. The network models overcome this disability of the capillary tube models, although they are still much simpler than the true porous media. Network models have earlier been generated in the following ways.

(1) Pores of equal size and different pore throat diameters are regularly arranged with a given node number, which is the number of pores directly connected to the node.

(2) Pores of equal or unequal size, but with different pore throat diameters, are first regularly arranged with one node number, then a fraction of the pores

*North Sea Oil and Gas Reservoirs—II*

are randomly blocked to generate a lower average node number. This makes the model more realistic in simulating the randomness of the medium.

(3) The pore body diameters and the pore throat diameters follow certain distribution functions, and the pore throat volume is so small that it can be neglected when calculating the void volume of the medium. The pores are regularly arranged with a certain node number.

These procedures focus on the representation of the randomness and the interconnections of the pores. No relationship between the generation of the medium and the macroscopic properties of the medium has previously been proposed for a pore-level model. Still, some data were recommended in the model by Gun Oh and Slatter (1973), such that the ratio of a pore body diameter to the pore throat diameter should be 2.5 and the ratio of the pore length of a pore to the pore throat diameter should be 5.5 to 6, for simulation of the mobilization of the residual oil.

The introduction of percolation theory in the simulation of multiphase flow at the pore level opened up new possibilities. It led to the simulation of displacement processes at capillary numbers so low that the capillary forces dominate the process. It is based on the assumption that the wetting phase always enters the smallest accessible pores, and that the non-wetting phase enters the biggest pores (Chandler *et al.*, 1982). This is only a crucial point when we are considering EOR processes. For these processes we can change the viscous and capillary forces by choosing the injection conditions and the injected fluid. In all simulations it was assumed that the non-wetting phase could not be displaced once it was isolated by the wetting phase. The residual oil saturation is calculated as the isolated non-wetting phase. The entrapment of the non-wetting phase is a result of the bypass process, which is caused by the difference in the flow rate in different flow paths.

Mohanty and co-workers (1987) analysed the physics of the wetting phase displacing the non-wetting phase, generally referred to as the imbibition process, and developed a procedure for simulating the influence of the capillary number on the recovery efficiency at low to moderate capillary numbers. Darcy's law was applied and two of the three main mechanisms of entrapment of the non-wetting phase were taken into account, i.e. the by-pass and the snap-off. The criterion for the occurrence of snap-off in their work was whether or not the ratio of a pore body diameter to the pore throat diameter was larger than 3.0. If the ratio were larger than 3.0, snap-off would occur when the wetting phase flowed through it. The simulated value of the residual "oil" (non-wetting phase) saturation in a two-dimensional model was larger than the experimental results. This was explained by the different dimensions.

Koplik and Lasseter (1982) treated their network as an electrical analogue of constraints. Bypassing was assumed to be the only mechanism of entrapping the non-wetting phase. This approach seems more theoretically reasonable than most others, but the practical use of this model is limited because too many variables have to be evaluated in each step of the simulation. It can, therefore, only be used for a very limited number of pores.

Most previous simulation works have focused on the discussion of the capillary number, which, of course, is the most sensitive of the four important factors: capillary number, viscosity ratio, bond number and the properties of the reservoir rock.

The dependence of residual oil saturations on capillary numbers for both connected and disconnected phase was reviewed by Larson *et al.* (1981). As mentioned, the variation of the normalized residual oil saturation with the capillary number is coincident for the Berea sandstones studied. The absolute values are different for different cores, even for the same type of medium. Foster (1973) reported his correlation of the residual oil saturation and the capillary number as a range of expected values instead of a single curve.

Abrams (1975) conducted more than 100 displacement experiments on sandstone cores with different flow rates, viscosity ratios, and interfacial tensions. He correlated his results with a dimensionless number, $F$, defined as

$$F=\left(\frac{v\mu_w}{\sigma}\right)\left(\frac{\mu_o}{\mu_w}\right)^{0.4} \tag{1}$$

which is a combination of the capillary number and the viscosity ratio. The results indicated that the larger the viscosity ratio between the displacing phase and the displaced phase, the more efficient was the displacement. Mungan (1966) investigated the variation in the final recovery and the recovery at breakthrough with the viscosity ratio. The results showed that the recovery of the displaced phase was most sensitive to the capillary number when the viscosity ratio was close to unity. We find the same trend in our simulation results. The influence of wettability of a medium on residual oil saturation was recently reviewed by Anderson (1987). It was concluded that the recovery efficiency is higher for water-wet reservoirs than for oil-wet reservoirs at similar conditions (flow rates, viscosity ratio, etc.). The work by Gupta and Trushenski (1979) shows that the critical capillary number for getting significant oil recovery depends strongly on the wettability of the medium. The displacement of the wetting phase by the non-wetting phase, generally referred to as the drainage process, requires a capillary number approximately 10 times higher than that required for the imbibition process (the capillary number was calculated with the viscosity of water both for water displacing oil and oil displacing water). Apparently the displacement of oil by water was carried out with a disconnected oil phase. In that case, a higher capillary number is needed to obtain a significantly higher recovery than for a connected oil phase, as reviewed by Larson *et al.* (1981). In that study, some of the wettability effects on oil recovery were shown. We conducted a series of displacement experiments (Stenby and Zhou, in prep.) and found that there is a significant difference between the imbibition process and the drainage process.

The simulations presented in this paper show very good agreement with those results.

## DESCRIPTION OF A POROUS MEDIUM

It is impossible to give a realistic picture of a porous medium at the pore level, owing to the complexity of the pore structures. But even very simplified models are helpful for understanding porous media made by nature. The following rules were used for generating networks as models of porous media.

(1) The void space of the medium consists of a number of pores arranged in a regular lattice.
(2) The pores creating the void space of the medium have the same length.
(3) The shape of the individual pores must be defined in such a way that we can define the relationship between the dimensions.

(4) The pore body and throat diameters follow given distribution functions.
(5) All pores are well interconnected with a node number of 4 for the two-dimensional model.
(6) The nodes, where the pores meet, are assumed to have no volume, and no viscous resistance.

The ratio between the average pore body diameter, $D_{ab}$, and the average pore throat diameter, $D_{at}$, is assumed to be equal to a constant, $A$, for a given medium:

$$\frac{D_{ab}}{D_{at}}=A \tag{2}$$

The ratio between the pore length of a medium, $L$, and $D_{ab}$ is assumed to be a constant, equal to $B$:

$$\frac{L}{D_{ab}}=B \tag{3}$$

$A$ and $B$ are characteristic constants for a given medium.

The properties of the porous medium (porosity, $\phi$, and permeability, $K$) can be expressed as in Eqs (4) and (5), where

$$\phi=\frac{3\pi}{4L^3}D_{av}^2L \tag{4}$$

and

$$K=c\cdot\phi\cdot D_{af}^2 \tag{5}$$

Here $c$ is a constant for a given medium. The average flowing pore diameter, $D_{af}$, and the average volume pore diameter, $D_{av}$, can be calculated from the distribution functions and the pore shape functions. Equation (5) can be rewritten as

$$K=c\cdot\frac{3\cdot\pi}{4}\cdot\frac{D_{af}^2\cdot D_{av}^2}{L^2} \tag{6}$$

Having derived these expressions, we can define the average pore length if we know the pore size distribution function. A common procedure for determining the pore size distribution is mercury injection as a drainage process. Photomicrographic techniques have also been used to determine pore size distribution (Dullien *et al.*, 1972).

## AN ANALYIS OF TWO-PHASE DISPLACEMENT IN A POROUS MEDIUM

### Competition between the viscous and capillary forces

First we consider a capillary tube with a varying diameter along the axial direction, $d(x)$. The wetting phase displaces the non-wetting phase, and the interfacial tension is $\sigma$. Fig. 1 shows this, and the capillary pressure, $P_c$, is given as

$$P_c=\frac{4\cdot\sigma}{d(x)}\cdot\cos(\theta) \tag{7}$$

where $\theta$ is the contact angle of the displacing phase. If the fluids are Newtonian and incompressible, the pressure drop $P_a$, caused by the viscosities of these two fluids, is

$$P_a=\frac{32\cdot v\cdot L}{d(x)^2}[x\cdot\mu_w+(1-x)\cdot\mu_o] \tag{8}$$

and $x=X/L$. The pressure difference over this tube is

$$P=P_a-P_c \tag{9}$$

that is

$$p=-\frac{4\cdot\sigma}{d(x)}\cdot\cos(\theta)+\frac{32\cdot v\cdot L}{d^2(x)}[x\cdot\mu_w+(1-x)\cdot\mu_o] \tag{10}$$

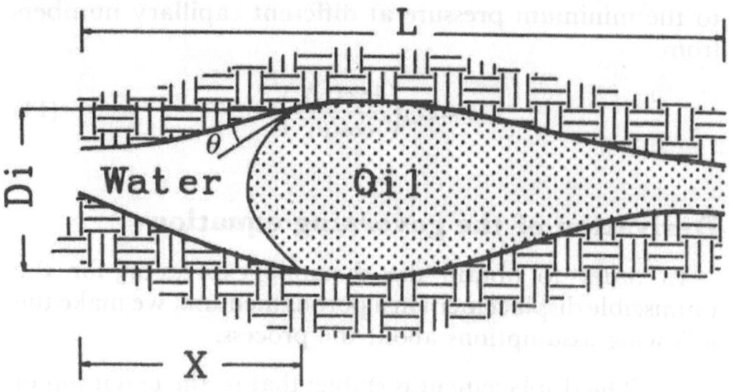


**Fig. 1.** Two-phase immiscible displacement in a capillary tube with varying diameter along the flow direction. $L$ is the length of the tube, $X$ is the distance from the inlet to the interface, $\theta$ is the contact angle and $D_i$ is the flow diameter.

Next we consider several tubes of the same length but with different diameters. The average diameter of the tubes is $D_a$, and the average flow rate in each tube is $v$. So the linear flow rate in tube $i$ with a diameter $d_i$ is:

$$v_i=\frac{D_a^2}{(d_i(x))^2}\cdot v \tag{11}$$

Combined with Eq. (11), Eq. (10) can be rewritten as

$$P=-\frac{4\cdot\sigma}{d_i}\cdot\cos(\theta)+\frac{32\cdot D_a^2\cdot v\cdot L}{d_i^4}[x\cdot\mu_w+(1-x)\cdot\mu_o] \tag{12}$$

When the flow rate is relatively high and the interfacial tension is low, it can be seen that for the imbibition process the minimum $P$ value does not occur for the smallest $d_i$ because of the second term on the right-hand side of Eq. (12).

If we use Eq. (12) for a system of pores instead of tubes, it can be rewritten as

$$\frac{P\cdot D_{ab}}{4\cdot\sigma}=-\frac{D_{ab}}{d_i}\cdot\cos(\theta)+\frac{D_{ab}^4}{d_i^4}N_c\cdot m \tag{13}$$

where $m$ is a constant depending on the dimensions of the pores, and $d_i$ is the diameter of pore $i$.

For a given medium, the average pore body diameter is a constant; if we assume that the interfacial tension is also a constant, we can find the pore diameters corresponding

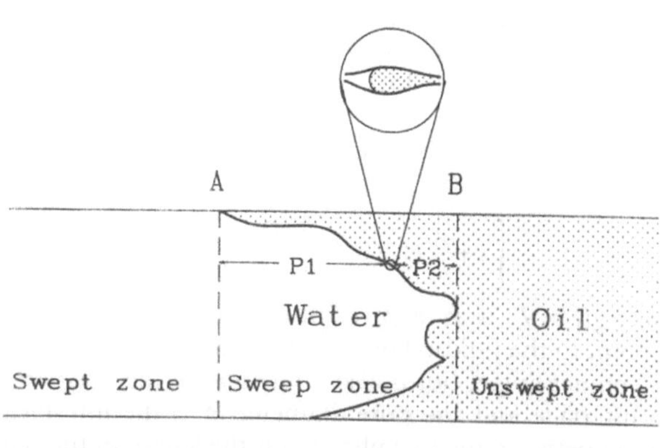


**Fig. 2.** Visualizing the displacement process. The three zones are, from the left: the *swept zone*, where the trapped phase is immobile in the medium; the *sweep zone*, where the stable interface passes and the displacement takes place; and the *unswept zone* with the initial saturations. The pressure drop across the selected pore is illustrated by the pressure $P1$ from the swept zone and $P2$ to the unswept zone, plus the capillary pressure $P_c$ in the pore.

to the minimum pressure at different capillary numbers from

$$d_i = D_{ab}\left(\frac{4 \cdot m \cdot \mathcal{N}_c}{\cos(\theta)}\right)^{1/3} \tag{14}$$

## Derivation of the governing equation

In order to obtain the dynamic expression for the immiscible displacement in a porous medium, we make the following assumptions about the process.

(1) The displacement is stable; that is, the extension of the sweep zone (see Fig. 2) stays constant during the displacement after it has been fully developed.
(2) Bypassing is the only procedure for entrapping the displaced phase.
(3) The clusters of the displaced phase, isolated by the displacing phase, cannot be moved and form the residual saturation. In earlier works, this assumption was made when the capillary number was reasonably small. It was assumed that the isolated displaced phase could be removed at high capillary number. But the higher the capillary number, the smaller are the isolated clusters of the displaced phase, and the more difficult it will be to move them. We use this assumption for all values of the capillary number.
(4) The flow in the pores is viscous Newtonian flow.
(5) The fluids are incompressible.
(6) Darcy's law applies.

Consider a displacement process as shown in Fig. 2. The pressure in the displacing phase should be the same at any point of section A; so should the pressure in the displaced phase in section B. The pressure difference between sections A and B through all paths should be the same and can be written as:

$$P = P_1 + P_2 - P_{i,\mathrm{int}} + P_{i,\mathrm{vis}} \tag{15}$$

where $P$ is the pressure difference between sections A and B, $P_1$ is the pressure drop caused by the flow of the displacing phase through $n_1$ pores, $P_2$ is the pressure drop caused by the flow of the displaced phase through $n_2$ pores, $P_{i,\mathrm{int}}$ is the capillary pressure due to the interface between the two phases in pore $i$, and $P_{i,\mathrm{vis}}$ is the pressure drop caused by the viscous flow through the pore containing the interface.

From our assumptions, $P_1$ and $P_2$ can be calculated by using Darcy's law, $P_{\mathrm{int}}$ by using the Laplace equation, and $P_{i,\mathrm{vis}}$ by using the Hage–Poiseuille equation. Therefore the pressure difference $P$ can then be expressed (as shown below) by using Eq. (15):

$$P = \frac{4 \cdot \sigma}{D_{ab}}\left\{-\frac{D_{ab}}{D_{ti}} \cdot g(i, x, d_i) + \mathcal{N}_c(8 \cdot B \cdot \left(\frac{D_{ab}^2 \cdot D_{af}^2}{D_{if}^4}\right)\right. \\ \left.\left[x + \frac{\mu_o}{\mu_w}(1-x) + \frac{B \cdot D_{ab}^2}{4 \cdot c \cdot D_{af}^2}\left(\frac{n_1}{K_{rw}} + \frac{n_2\mu_o}{K_{ro}\mu_w}\right)\right]\right\} \tag{16}$$

where $g(i, x, d_i)$ is an expression for the wettability of the medium to the fluids. This cannot be a simple function of the contact angle, but is also a function of the position of the interface and the pore diameter. $\sigma$ is the interfacial tension between the two phases; $x$ is the length of the pore occupied by the displaced phase divided by the length of the pore; $D_{ti}$ is the pore throat diameter of pore $i$; $D_{if}$ is the flowing pore diameter of pore $i$; $K_{rw}$ is the relative permeability of the displacing phase; $K_{ro}$ is the relative permeability of the displaced phase, $n_1$ is the number of pores behind the interface; $n_2$ is the number of pores in front of the interface; $\mathcal{N}_c$ is the capillary number (as defined in Eq. (17)) and $\phi$ is the porosity of the medium.

$$\mathcal{N}_c = \frac{v\mu_w}{\sigma\phi} \tag{17}$$

When one phase is injected in order to displace the second phase, the pressure difference we apply should overcome the biggest pressure drop created by the injected phase flowing through pore $i$. It is approximated in the following way:

$$P_{\max} \approx P_1 + P_2 + P_i(\mathrm{vis})_{\max} - P_i(\mathrm{int})_{\min} \tag{18}$$

$$P_{\max} \approx \frac{4 \cdot \sigma}{D_{ab}}\left\{-\frac{D_{ab}}{D_{ti}} \cdot g(i, x, d_i)_{\min} + B_n\left[32 \cdot c \cdot \frac{D_{af}^4}{D_{if}^4}\left(x + \frac{\mu_o}{\mu_w}(1-x)\right)_{\max} \right.\right. \\ \left.\left. + \left(\frac{n_1}{K_{rw}} + \frac{n_2\mu_o}{K_{ro}\mu_w}\right)\right]\right\} \tag{19}$$

and

$$B_n = \frac{D_{ab}^2}{D_{af}^2} \cdot \frac{B}{4 \cdot c} \cdot \mathcal{N}_c \tag{20}$$

where $P_{\max}$ is the maximum pressure needed to displace the second phase from a pore. The value is hypothetical, since $P_i(\mathrm{vis})_{\max}$ and $P_i(\mathrm{int})_{\max}$ cannot reach their maximum values for the same $x$ value. $B_n$ is a dimensionless number that combines the capillary number and the relative dimensions of the medium. The determination of $g(i, x, d_i)_{\min}$ was discussed by Lin and Slatter (1982); they concluded that when the wetting phase displaces the non-wetting phase

$$g_{\min} = \frac{\cos(\theta_1)}{A_i} \tag{21}$$

and when the non-wetting phase displaces the wetting phase

$$g_{\min} = \cos(\theta_2) \tag{22}$$

where $A_i$ is the ratio of the pore body diameter to the pore throat diameter of pore $i$, $\theta_1$ is the contact angle for the imbibition, and $\theta_2$ is the contact angle for the drainage process.

# SIMULATION

According to the approach proposed above, we can generate the medium and conduct our simulations provided that we have the pore body and the pore throat size distribution functions. But for most of the porous media from which the displacement data are obtained, these functions are unavailable. It is a rather complicated process to determine these functions. Three distribution functions are used in this work to show the influence of the distribution function on the final results. The medium is considered to be either completely water-wet or oil-wet in the simulations.

## Relative permeability

To find the relation between relative permeability and saturation, we employed the procedure developed by Lin and Slatter (1982):

$$K_{ri} = \frac{\Sigma D_{if}^4}{\Sigma D_{If}^4} \tag{23}$$

where the relative permeability, $K_{ri}$, for phase $i$ is calculated as the ratio of the summed flow diameter of the pores filled with phase $i$ to the fourth power, and the total sum of all diameters, also to the fourth power, indicated by $j$.

The flow diameter was defined as the average of the pore throat and the pore body diameters.

### Distribution functions

We have not been able to extract any specific distribution functions from the experimental data reported in the literature. The even distribution function was first chosen for its simplicity, and later we tested the function used by Heiba *et al.* (1986), from which they obtained good simulated results when predicting relative permeabilities. To show the relation between the results from the even distribution function and the function mentioned above, we used $\chi^2$ functions with two different values of $n$. The $\chi^2$ function is defined as:

$$\chi^2 = X_1^2 + X_2^2 + X_3^2 + \cdots + X_n^2 \tag{24}$$

where $X_i$ is a random number following the even distribution function (0, 1).

Figure 3 shows the distribution functions generated for our simulations.

## RESULTS AND DISCUSSION

The computer programs were written in Pascal and the simulations were carried out on an HP9000/350 microcomputer. Figure 4 shows examples of the simulations. It can be seen that at a very low capillary number the size of the trapped oil clusters is bigger than at a higher capillary number. The difference between Figs. 4(b) and 4(c) is in the viscosity ratio, which is 1.0 and 0.3 respectively. The population of oil clusters in Fig. 4(b) is larger than in 4(c), but in both cases the sizes of the clusters are small.

### Dependence on the distribution function

In Fig. 5, the simulated normalized saturation is shown as a function of the capillary number for the different distribution functions. It is interesting to see that these curves are rather similar to each other for both imbibition and drainage processes. This gives a possible explanation of why Larson *et al.* (1981) reached their surprising conclusions after analysing a large collection of experimental data. As mentioned earlier, they stated that the dependence of the normalized residual oil saturation on the capillary number could be coincident for data from Berea sandstones. To make a close comparison, we selected one curve from each distribution function with a viscosity ratio of 0.3 and plotted them in Fig. 6. The shape of the curves shows a qualitative likeness, but the quantitative difference is of the same order of magnitude as the scatter in experimental data reported in the literature. This indicates that information about the distribution function can reduce the uncertainty of simulations. Another important result is that the curves from different distribution functions start to decline dramatically from almost the same $B_n$ value. That may imply that the critical $B_n$ value is universal for all of the media. These values will be used in the determination of the critical capillary number for both types of displacement. The critical $B_n$ value is $3.0 \times 10^{-4}$ for the imbibition process, and $6.0 \times 10^{-3}$ for the drainage process. We may note that the relations simulated for the drainage processes scatter at low capillary number. This is an effect of working with a small model ($20 \times 100$) where the effect of a single pore can be seen for low recoveries.

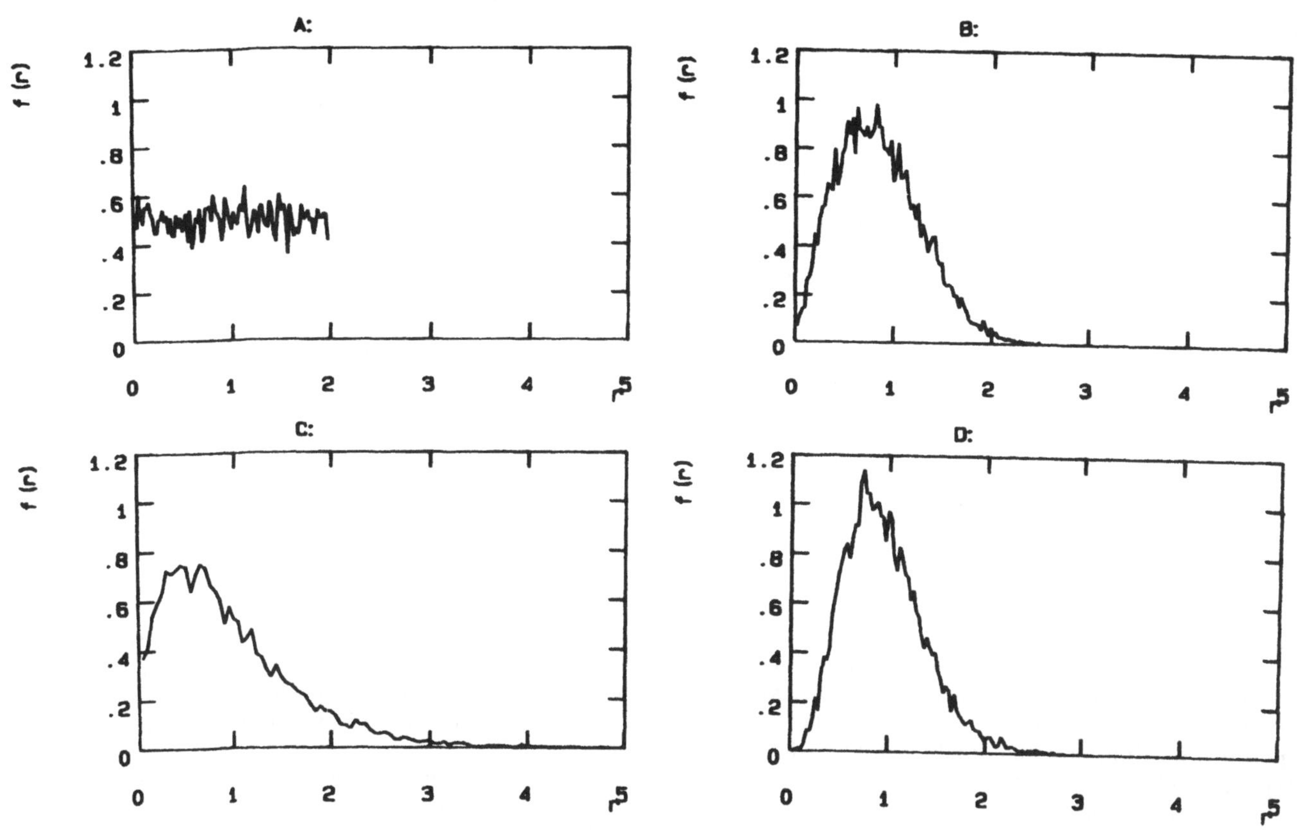


**Fig. 3.** The distribution functions used in this work. (A) $f_1$, the even distribution function; (B) $f_2 = 2r\exp(-r^2)$; (C) $f_3 = \chi^2$ with $n = 4$; (D) $f_4 = \chi^2$ with $n = 10$.

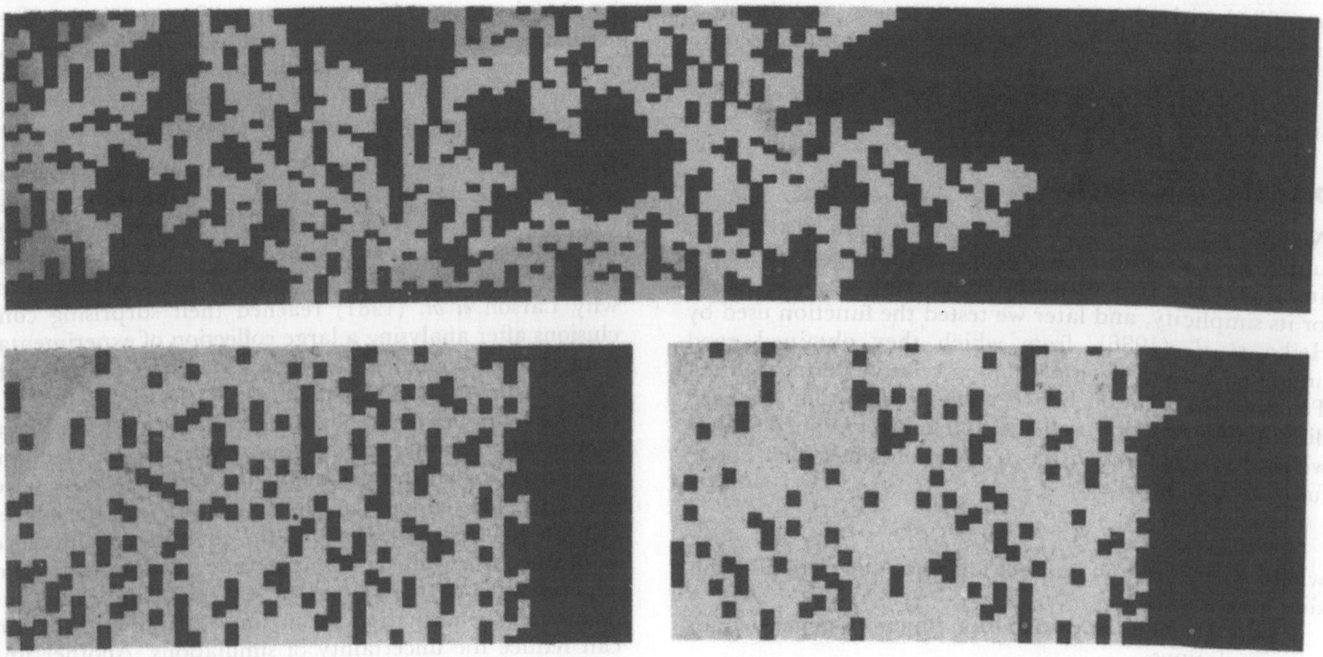

**Fig. 4.** Three examples of the simulated displacement processes after 0.35 pore volumes injected. The capillary number $N_c$, and the viscosity ratio, $U_R = \mu_o/\mu_w$, are the variable parameters. (A) $N_c = 10^{-6}$ and $U_r = 1.0$ or 0.0, the result is independent of the value of $U_r$ in this case; (B) $N_c = 10^{-4}$ and $U_r = 1.0$; (C) $N_c = 10^{-4}$ and $U_r = 0.0$.

## The effect of the Viscosity Ratio

Figure 7 shows a comparison between the experimental data and the simulated results using the distribution function from Heiba *et al.* (1986) for imbibition processes. The experimental data are from sample 799 of Abrams (1975); the viscosity of the displacing phase is 32.3 cP, and the viscosities of the displaced oil phases are 1.26 cP and 30.5 cP. The squares represent the data with "unity" viscosity ratios (30.5/32.3) and the crosses the data with "zero" viscosity ratios (1.26/32.3). The dashed curve corresponds to the first case, the solid curve to the second. Good agreement is obtained, especially in view of the simplicity of the model. From Figs. 5 and 6, it can be seen that the viscosity ratio influences the results only as the capillary number reaches a value larger than the critical capillary number. In the range with a significant viscosity effect, the recovery efficiency is most sensitive to the capillary number for viscosity ratios around unity.

## The effect of wettability

In Eq. (18), the value of $B_n$ depends only on the properties of the medium if the capillary number, $N_c$, is constant. The only difference between the imbibition process and the drainage process is the calculation of the $g$ value. The controlling diameter for the first case is the pore body diameter, and for the second case the pore throat diameter. This shifts the curve towards higher capillary numbers for the drainage process. Another contribution to the difference between imbibition and drainage comes from the calculation of the capillary pressure. Because the wetting phase enters the pore with the smallest pore diameter (which is much smaller than the average), the capillary pressure becomes much larger and then makes a more important contribution to the total pressure in the drainage process. The experimental results by Stenby and Zhou are shown in Fig. 8 together with the simulated results. The ratio of the average pore body diameter to the average pore throat diameter is $A = 3$; the ratio of the average pore length to the average pore body diameter, $B = 3$; and the constant defined by Eq. (5), $c = 0.02$. The experimental work was conducted on a small artificial glass core (length 5.0 cm and diameter 2.54 cm) at 25°C. The wetting phase was water (1.0 cP) or aqueous surfactant solutions, and the non-wetting phase was heptane (0.4 cP) or hexadecane (3.0 cP).

## Determination of the critical capillary number

The critical $B_n$ values, for different distribution functions are almost identical in our simulations. On this basis we can determine the critical capillary pressure for a given medium. From Fig. 6, $3.0 \times 10^{-4}$ is found as the critical $B_n$ value for the imbibition process, and $6.0 \times 10^{-3}$ for the drainage process. So

$$(N_c)_c \cdot \frac{1}{4c} \cdot \frac{D_{ab}^2}{D_{af}^2} = 3.0 \times 10^{-4} \tag{25}$$

where $(N_c)_c$ is the critical capillary number. In order to make a comparison with earlier works, Eq. (25) can be rearranged as

$$(N_c)_c \left(\frac{\phi}{K}\right) = C_a \left(\frac{K}{\phi}\right)^{-1} \tag{26}$$

where

$$C_a = \frac{4c}{1} \cdot \frac{D_{af}^2}{D_{ab}^2} \cdot 3.0 \times 10^{-4} \tag{27}$$

If the relative dimensions $A$ and $B$, and constat $c$ of the medium only depend on the material forming the medium, the critical capillary numbers will be unique for a given material, independently of the variation in permeability and porosity.

The same procedure could be used for the displacement of the wetting phase by the non-wetting phase.

Chatzis and Morrow (1984) investigated the relation between capillary number and the displacement of a disconnected oil phase for sandstones. They found that the

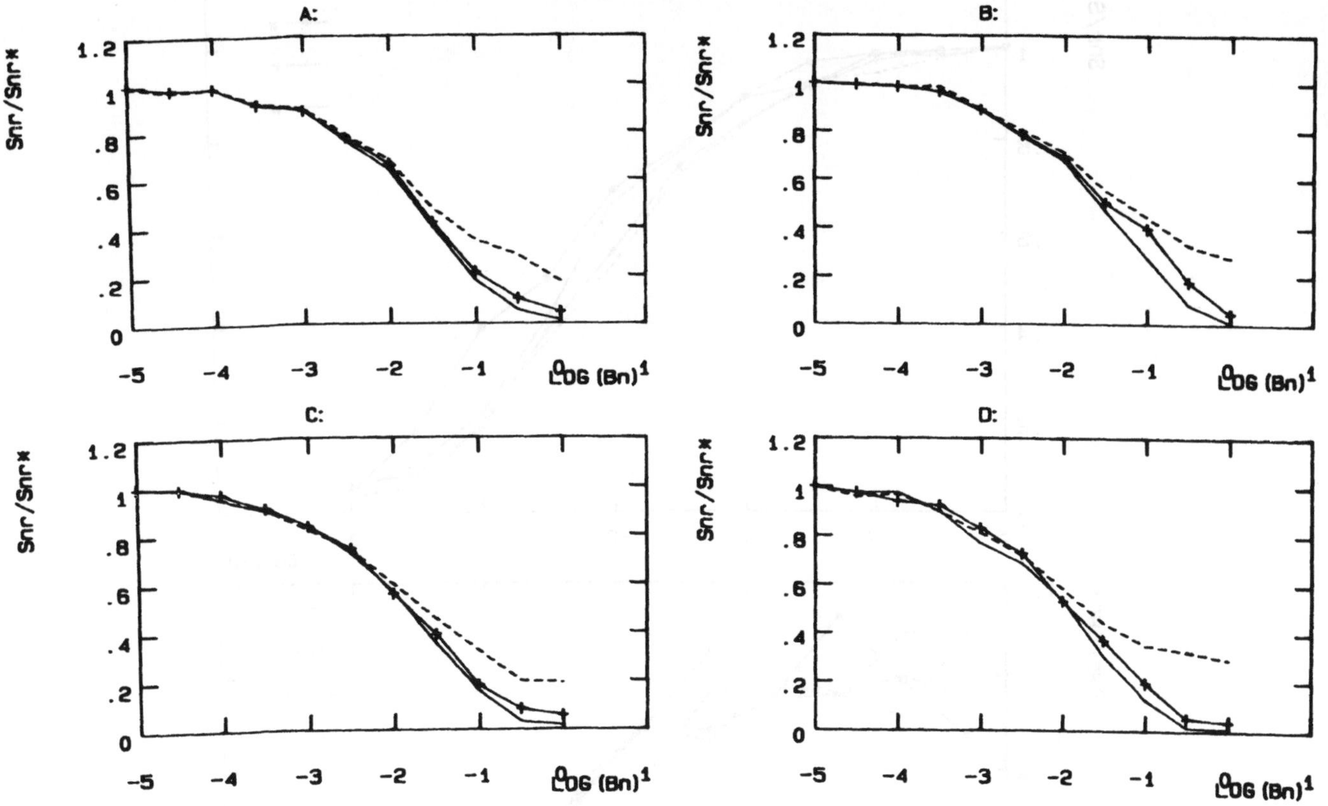


**Fig. 5a.** The simulated relationship between the normalized residual non-wetting phase saturation and the $B_n$ values (defined in Eq. (22)) using different distribution functions. (A), (B), (C), and (D) are from distribution functions $f_1, f_2, f_3$ and $f_4$ respectively. The dashed curves represent the cases with unity viscosity ratio, the curves with crosses a viscosity ratio of 0.3, and the continuous curves a viscosity ratio of 0.0.

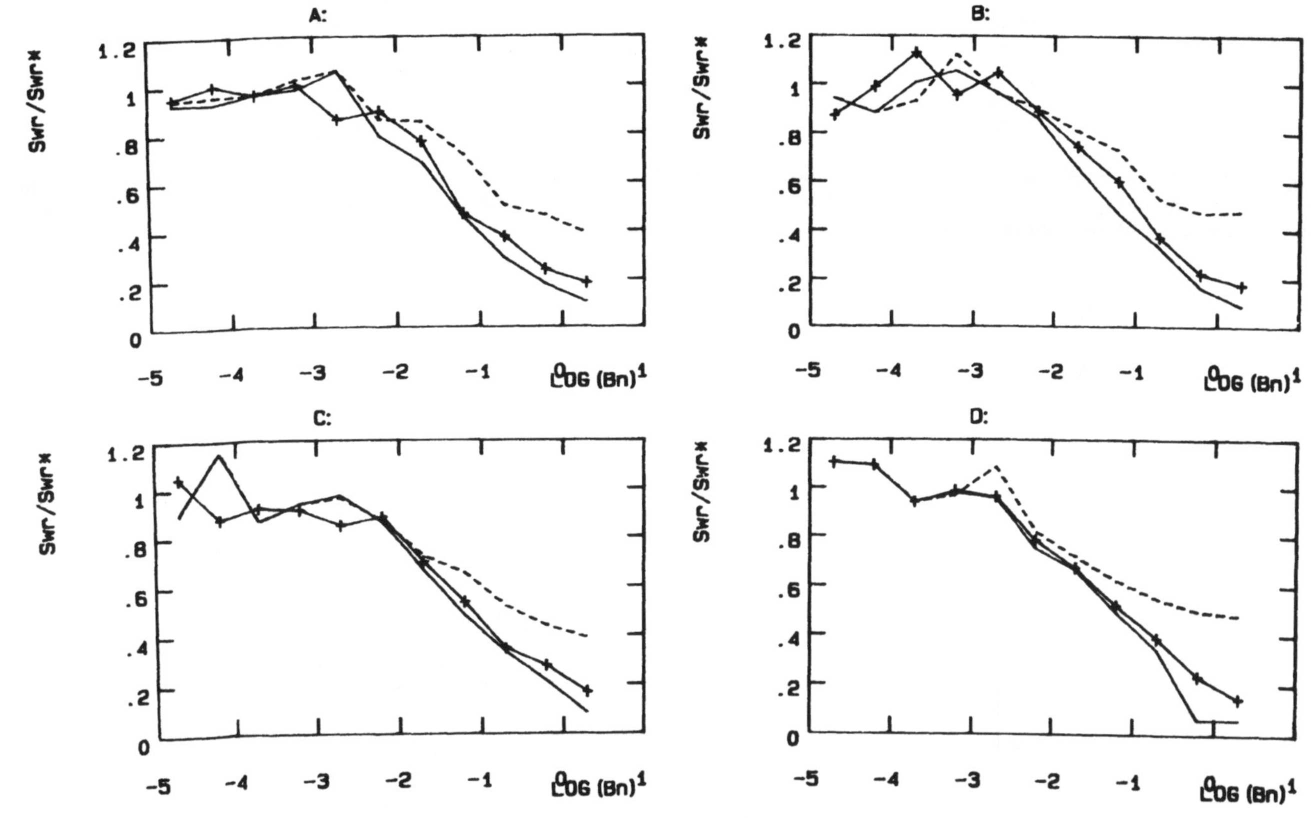


**Fig. 5b.** The simulated relationships between the normalized residual wetting phase saturations and the $B_n$ values using different distribution functions. (A), (B), (C) and (D) are from distribution functions $f_1, f_2, f_3$ and $f_4$ respectively. The dashed curves represent the cases with unity viscosity ratio, the curves with crosses a viscosity ratio of 0.3 and, the continuous curves a viscosity ratio of 0.0.

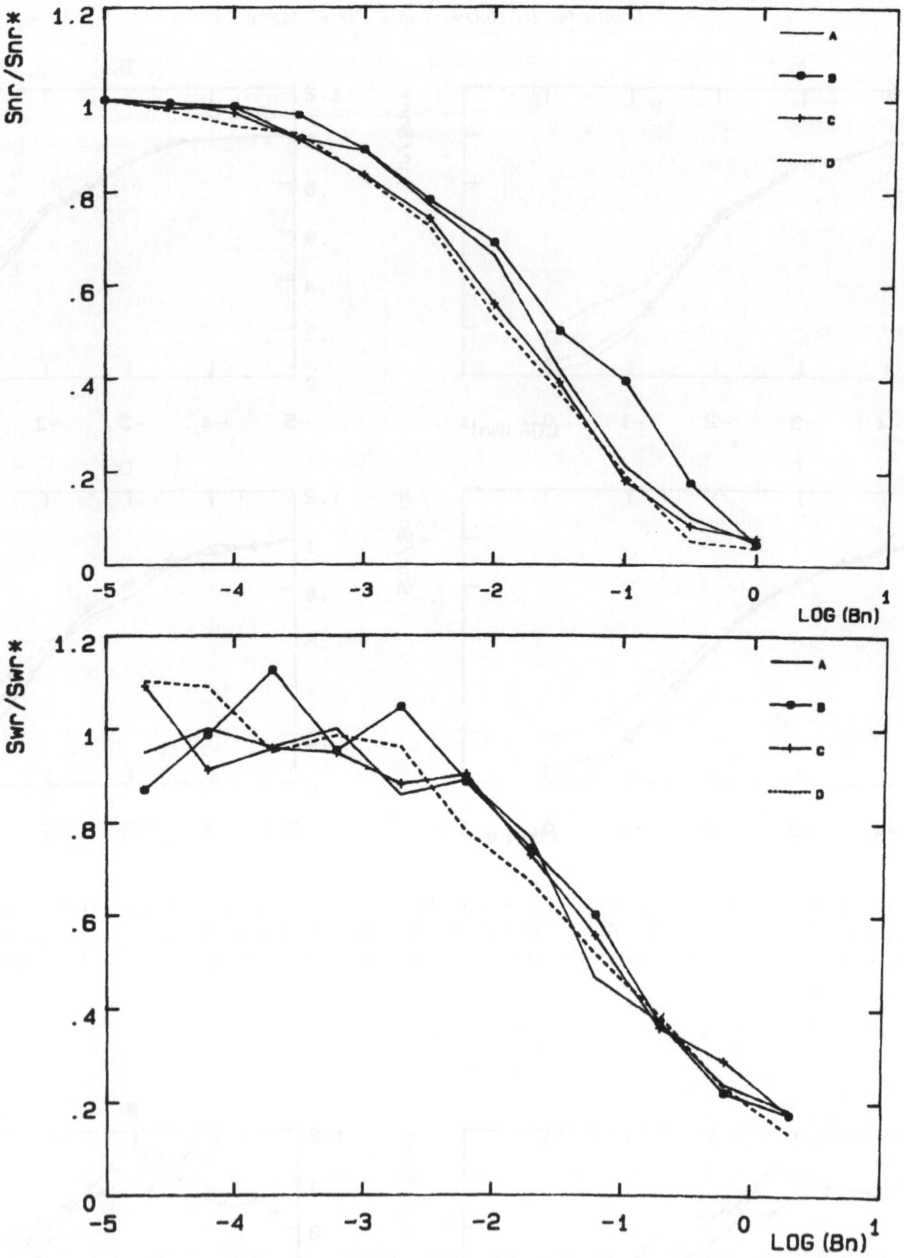


**Fig. 6.** Simulation results with different distribution functions describing the porous medium. Curves A, B, C and D are from distribution functons $f_1, f_2, f_3$ and $f_4$, respectively. In both (a) and (b) the viscosity ratio, $U_r$, is 0.3. The results in (a) are for imbibition processes, and in (b) for drainage processes.

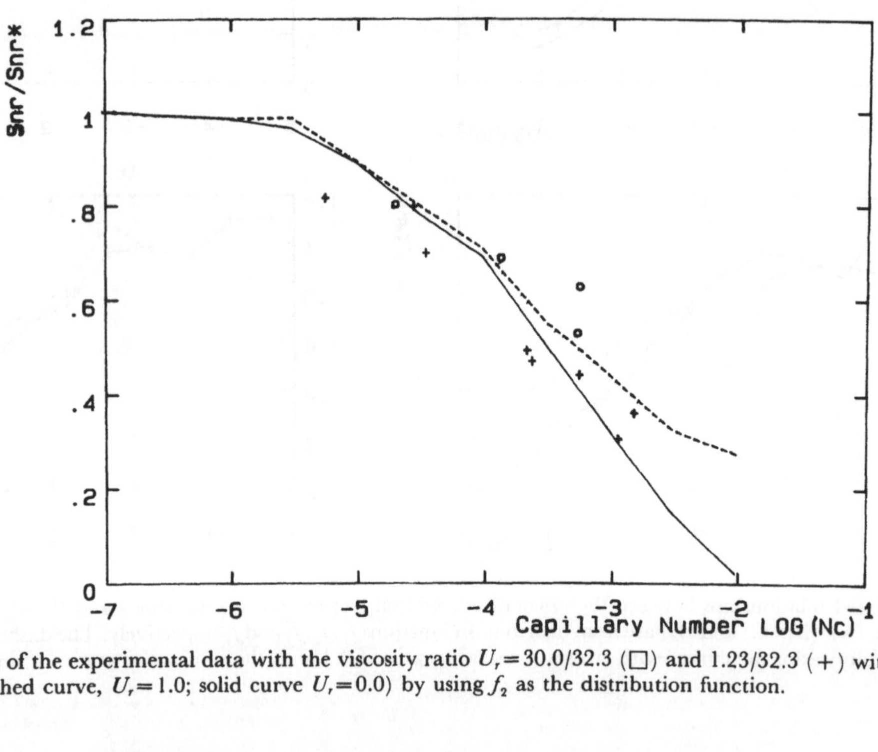


**Fig. 7.** Comparison of the experimental data with the viscosity ratio $U_r = 30.0/32.3$ (□) and 1.23/32.3 (+) with the corresponding simulated results (dashed curve, $U_r = 1.0$; solid curve $U_r = 0.0$) by using $f_2$ as the distribution function.

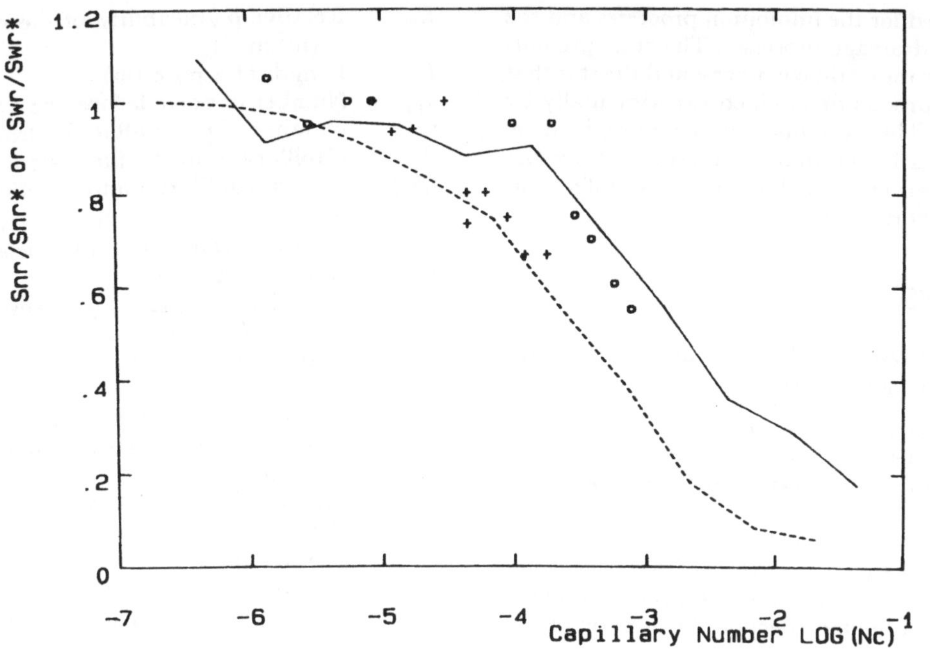


**Fig. 8.** Comparison of the experimental data for different wettabilities with the predicted results by using $f_3$ as the distribution function. The boxes (□) show the experimental data from drainage processes and the crosses (+) those from imbibition processes. The viscosity ratio is 0.3 in both cases. (The experimental data are from Stenby and Zhou, in prep.)

critical capillary number could be correlated to the medium properties by the following equation:

$$(N_c)_c \cdot \left(\frac{\phi}{K_a}\right) = C \cdot K_{rw} \left(\frac{K_a}{\phi}\right)^{-1.115} \tag{28}$$

where $C$ is a constant, and $K_a$ is the permeability of the medium to air. Since relative permeability is a function of saturation, the saturation value does not change very much around the critical capillary number.

The product of $C$ and $K_{rw}$ in Eq. (28) is a constant. Equations (26) and (28) are clearly similar. But the two equations are derived for the displacement of disconnected and connected phases respectively. The relation between these two cases is not clear.

Figure 9 shows the experimental results and the predicted critical capillary numbers obtained using Eq. (25). The parameters $A$, $B$, and $c$ have the same values Fig. 8. The boxes are experimental data for the imbibition processes carried out by Chatzis and Morrow (1984). The

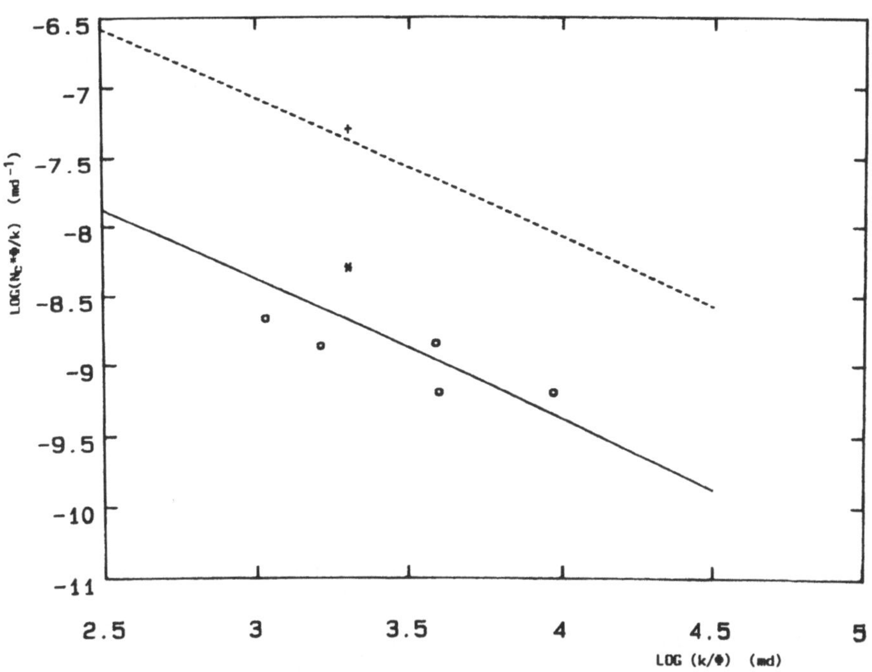


**Fig. 9.** Comparison of experimental and predicted values of the critical capillary number for significant recovery. The boxes (□) show the experimental data of Chatzis and Morrow (1984) for imbibition processes, and the solid line shows the predicted results. The dashed line is predicted for drainage processes, and the cross (+) and star (*) are the experimental data of Stenby and Zhou (in prep.) for drainage and imbibition processes respectively.

solid line is predicted for the imbibition processes and the dashed line for the drainage processes. The cross presents the critical value for the drainage process and the star that for the imbibition process determined experimentally by Stenby and Zhou. The experimental value obtained by Stenby and Zhou is larger than that predicted by the model. This can be explained by the end effects in experiments with short cores.

## CONCLUSIONS

From the simulated results and the discussion above, we can draw the following conclusions.

(1) This modified percolation invasion theory is successful in modelling both imbibition and drainage processes with viscosity ratios less than or equal to unity.
(2) The four different distribution functions tested show some influence on the shape of the curve of the normalized residual saturation versus the capillary number, and a limited effect on the position of the curve.
(3) The relative dimensions of the pores in a given medium specify the critical capillary numbers for both imbibition and drainage processes.
(4) The viscosity ratio can alter the dependence of the normalized residual saturation on the capillary number. The effect is largest for viscosity ratios close to unity.
(5) The wettability of a medium to the flowing fluids significantly affects the dependence of the normalized residual saturation on the capillary number. The drainage process needs a capillary number 20 times larger than for the imbibition process to obtain a significant improvement in recovery.

## ACKNOWLEDGEMENT

The Danish Technical Reseach Council (STVF) is acknowledged for its financial support.

## NOMENCLATURE

| | |
|---|---|
| $A$ | Ratio of the average pore body diameter to the average pore throat diameter ($m\,m^{-1}$) |
| $A_i$ | Ratio of the pore body diameter to the throat diameter of pore $i$ ($m\,m^{-1}$) |
| $B$ | Ratio of the average pore length to the average pore body diameter ($m\,m^{-1}$) |
| $B_n$ | A constant for a given medium (dimensionless) |
| $c$ | A constant for a given medium (dimensionless) |
| $C$ | A constant for a given medium (dimensionless) |
| $C_a$ | A constant for a given medium (dimensionless) |
| $D_{ab}$ | Average pore body diameter (m) |
| $D_{af}$ | Average flow diameter (m) |
| $D_{av}$ | Average volume diameter (m) |
| $D_{at}$ | Average pore throat diameter (m) |
| $D_{ib}$ | Body diameter of pore $i$ (m) |
| $D_{ti}$ | Throat diameter of pore $i$ (m) |
| $F$ | Dimensionless number |
| $g$ | A function of wettability of a medium (dimensionless) |
| $g_{max}$ | The maximum value of $g$ in a pore |
| $K$ | Permeability ($m^2$) |
| $K_{rw}$ | Relative permeability of the wetting phase ($m^2\,m^{-2}$) |
| $K_{ro}$ | Relative permeability of the non-wetting phase ($m^2\,m^{-2}$) |
| $L$ | Length of a pore (m) |
| $n_1$ | Number of pores before the front (dimensionless) |
| $n_2$ | Number of pores after the front (dimensionless) |
| $\mathcal{N}_c$ | Capillary number(dimensionless) |
| $(\mathcal{N}_c)_c$ | Critical capillary number (dimensionless) |
| $P$ | Pressure drop over a pore or several pores ($N\,m^{-2}$) |
| $P_a$ | Pressure drop caused by viscosities ($N\,m^{-2}$) |
| $r$ | Radius of a pore (m) |
| $U_r$ | Viscosity ratio, $\mu_o/\mu_w$ (dimensionless) |
| $v$ | Volume flow rate ($m\,s^{-1}$) |
| $v_i$ | Volume flow rate through pore $i$ ($m\,s^{-1}$) |
| $x$ | Ratio of the length of the pore occupied by the displacing phase to the length of the pore ($m\,m^{-1}$) |
| $X$ | Distance from the pore throat to the interface in the pore |
| $\sigma$ | Interfacial tension ($N\,m^{-1}$) |
| $\theta$ | Contact angle (degrees) |
| $\phi$ | Porosity ($m^3\,m^{-3}$) |
| $\mu_w$ | Viscosity of the displacing phase ($N\,m^{-1}\,s^{-1}$) |
| $\mu_0$ | Viscosity of the displaced phase ($N\,m^{-1}\,s^{-1}$) |

## REFERENCES


Abrams, A. (1975). The influence of fluid viscosity, interfacial tension, and flow velocity on residual oil saturation left by waterflood. *Soc. Petroleum Eng. J.*, (Oct.), 437–447.

Anderson, W.G. (1987). Wettability literature survey—Part 6: The effects of wettability on waterflooding. *SPE Reservoir Eng.* (Dec.), 1605–1622.

Chandler, R., Koplik, J., Lerman, K. and Willemsen, J. (1982). Capillary displacement and percolation in porous media. *J. Fluid Mechanism* **119**, 249–267.

Chatzis, I. and Morrow, N.R., (1984). Correlation of capillary number relationships for sandstone. *Soc. Petroleum Eng. J.* (Oct.), 555–562.

Dullien, F.A.L., Dkawan, G.K., Gurak, N. and Babjak, L. (1972). A relationship between pore structure and residual oil saturation in tertiary surfactant floods. *Soc. Petroleum Eng. J.* (Aug.), 25–53.

Foster, W.A., (1973). A low-tension waterflooding process. *J. Petroleum Technol.* No. 25, 205–210.

Gun Oh, S. and Slatter, J. (1973). Interfacial tension required for significant displacement of residual oil. *Soc. Petroleum Eng. J.* (April), 83–95.

Gupta, S.P. and Trushenski, S.P. (1979). Micellar flooding—Compositional effects on oil displacement. *Soc. Petroleum Eng. J.* (April).

Heiba, A.A., Jerauld, G.R., Davis, H.T., and Scriven, L.E. (1986). Mechanism-based simulation of oil recovery processes. Paper no. SPE 15593.

Koplik, J. and Lasseter, T.J. (1982). One- and two-phase flow in network models of porous media. *Chem. Eng. Commun.* **26**, 285–295.

Larson, R.G., Scriven, L.E. and Davies, H.T. (1981). Displacement of residual nonwetting fluid from porous media. *Chem. Eng. Sci.* **36**, 75–85.

Lin, C.H. and Slatter, J.C. (1982). Three-dimensional randomized network model for two phase flow through porous media. *Am. Ind. Chem. Eng. J.* **28** (2), 311–324.

Mohanty, K.K., Davies, H.T. and Scriven, L.E. (1987). Physics of oil entrapment in water-wet rock. *SPE Reservoir Eng.* (Feb.), 113–128.

Mungan, J. (1966). Interfacial effects in immiscible liquid–liquid displacement in porous media. *Soc. Petroleum Eng. J.* (Sept.), 247–253.

Stenby, E.H. and Zhou D.E. (in prep.). A Laboratory and theoretical study on immiscible displacement processes in artificial cores. Submitted to SPE/DOE Seventh Symposium on Enhanced Oil Recovery, Tulsa, Oklahoma, April, 1990.

# 16

# Ekofisk Field Enhanced Recovery

*R.M. Sulak, G.R. Nossa and D.A. Thompson*

Phillips Petroleum Company Norway, Stavanger, Norway

Ekofisk Field, discovered in late 1969, was the first commercial discovery in the Norwegian Sector of the North Sea. The field has produced some 856 million barrels of oil and 2.6 trillion standard cubic feet of gas through 1988. Almost all of this has been primary production with solution gas drive and compaction playing major roles.

In June of 1986, the 2/4 K Waterflood Platform was installed at Ekofisk to enhance the recovery from the Tor Formation. The platform, weighing 38 000 tons, has slots for 30 injection wells and is equipped to inject 375 000 barrels of water per day. The platform was installed on the basis of results of laboratory imbibition tests and the Tor Formation Waterflood Pilot conducted from April 1981 to July 1984. These results indicated spontaneous imbibition of 48% could be expected in the Tor Formation. Simply stated, this meant that if seawater were introduced to the Tor Formation fracture system, it would enter the matrix, without benefit of a pressure differential, and displace, on average, hydrocarbons from 48% of the pore volume. Water injection was initiated in November 1987.

Laboratory tests on the Lower Ekofisk chalk indicate that spontaneous imbibition values are in the 25% range, but considerably greater changes in water saturation are achieved by viscous displacement with relatively small pressure differentials. The Lower Ekofisk Waterflood Pilot was initiated in July 1986 to determine whether sufficient water saturation changes could be achieved in the field, through a combination of imbibition and viscous displacement, to make waterflooding the Lower Ekofisk Formation commercial. The results of the pilot are positive and the expansion of the waterflood to the Lower Ekofisk as well as the south is in progress. Innovative modifications of existing platforms are being made to minimize the cost of project implementation.

Nitrogen injection appears to be an attractive method for displacing hydrocarbons, primarily natural gas, from the Upper Ekofisk Formation. A thorough evaluation is almost complete and a decision concerning nitrogen injection is anticipated by November 1989.

## INTRODUCTION

The Ekofisk Field, located in the Central Graben in the Norwegian sector of the North Sea, originally contained more than 8 billion barrels of oil equivalents. Initially, only about 20% was thought to be recoverable through primary depletion. Reservoir compaction, which was discovered in late 1984, increased primary recovery estimates to 24%. The challenge that the Phillips Norway Group has faced and continues to face is how economically to enhance recovery from the fractured chalk reservoirs located in the middle of the hostile North Sea.

Early laboratory testing and a field waterflood pilot resulted in the decision to waterflood the northern two-thirds of the Tor Formation. A second waterflood pilot was established to determine the feasibility of waterflooding the Lower Ekofisk Formation. Positive results coupled with innovative concepts for platform modifications to accommodate additional producing wells led to expansion of the waterflood to the Lower Ekofisk Formation and to the southern portion of the field.

*North Sea Oil and Gas Reservoirs–II*


Nitrogen injection can improve recovery by molecular diffusion into the matrix, vaporization of solution gas from residual oil, and overall pressure maintenance. A decision on the project is expected in November 1989.

## RESERVOIR DESCRIPTION

The Ekofisk Field is an elongated anticline with the major axis running N–S (Fig. 1) and covers approximately 12 000 acres (4860/ha).

### Ekofisk Formation

The upper producing horizon of Ekofisk Field is the Ekofisk Formation of Danian age in the Paleocene Period. It is located at a depth of about 9600 ft and ranges in thickness from 330–500 ft. The natural fractures in this chalk reservoir enhance matrix permeability values, typically 1–5–mD, to an effective permeability approaching 100 mD. Porosity ranges from 30% to 48%.

### Tight Zone

The Tight Zone is the bottom portion of the Ekofisk Formation and forms an impermeable barrier between the

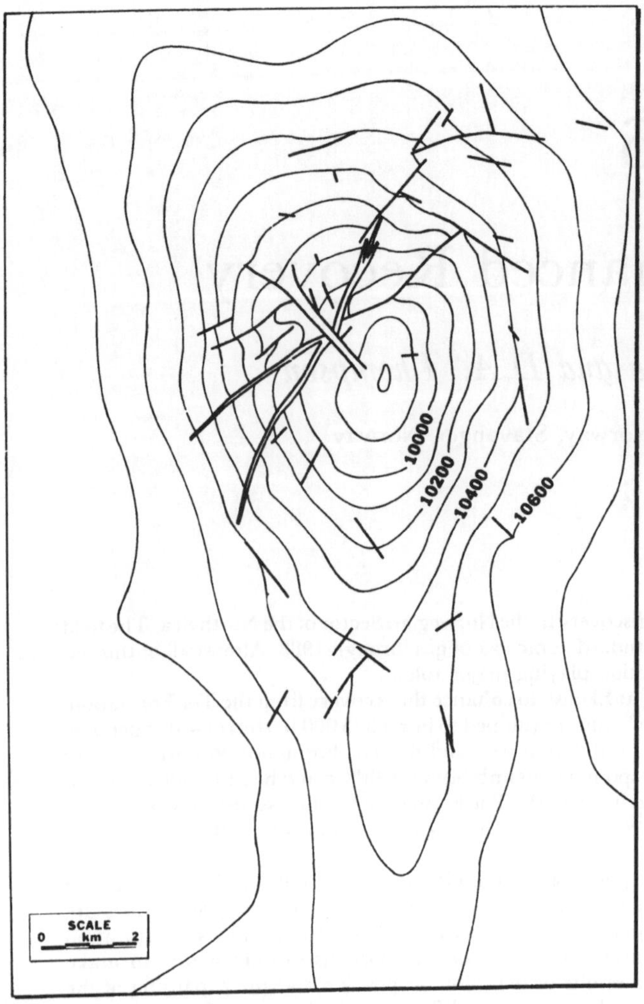


**Fig. 1.** Structure map of Ekofisk Field, Top Tor Formation.

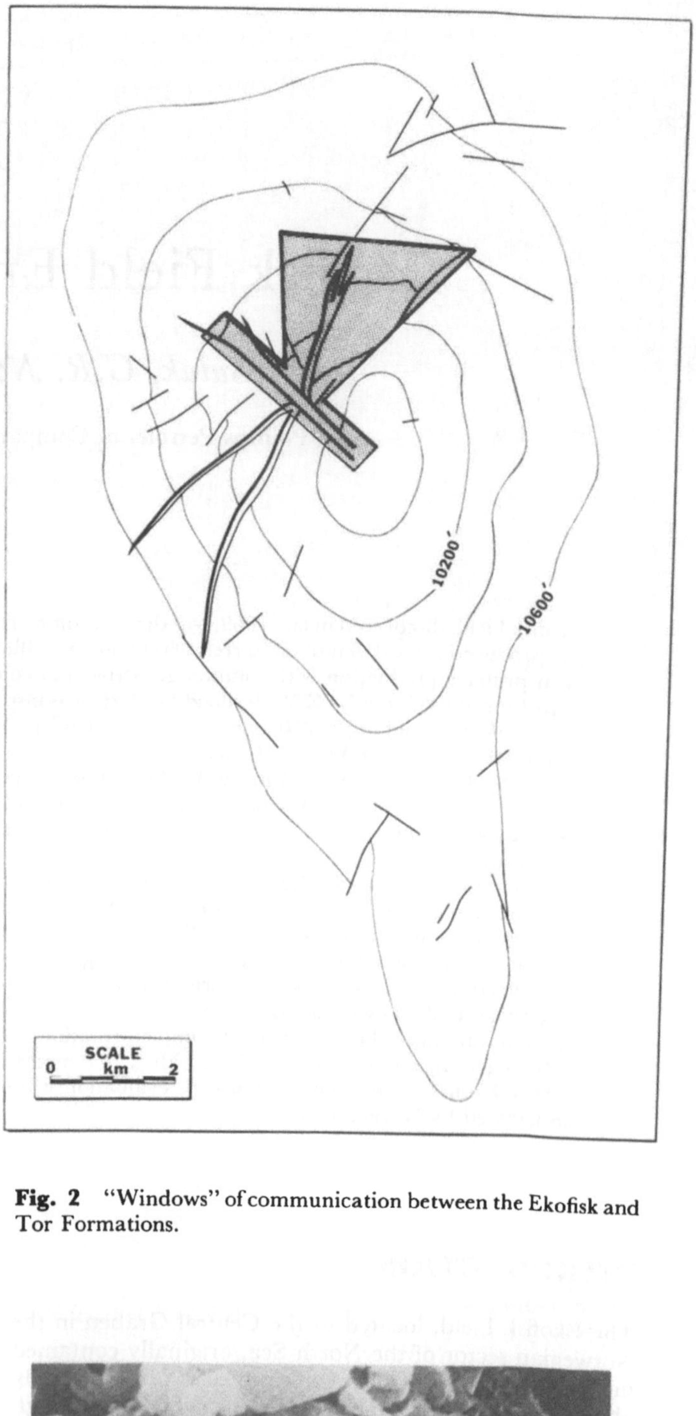


**Fig. 2** "Windows" of communication between the Ekofisk and Tor Formations.

Ekofisk and Tor Formations. It is approximately 50 ft thick and has a porosity in the 10–20% range. The Tight Zone has been faulted out in two areas of the field forming "windows" between the two reservoirs (Fig. 2).

## Tor Formation

The lower reservoir is the Tor Formation of Maastrichtian age in the Cretaceous period. It is also a fractured chalk reservoir ranging in thickness from 250 to 500 ft. Its porosity ranges from 30% to 40% + and effective permeability reaches 100 mD.

## Mineralogy

The mineralogy of both formations is similar, being a fine-grained limestone composed of skeletal debris of coccolithophorides, a pelagic unicellular alga. The calcareous exoskeletons of these algae are called coccospheres, comprised of wheel-shaped coccoliths. The great majority of the matrix is composed of the calcite crystal component of a coccolith called a coccolith platelet. Whole coccoliths are relatively common but preserved coccospheres are rare (Fig. 3).

Chalk in the Lower Ekofisk and Tor Formations is relatively clean with generally less than 5% non-carbonate impurities. The Upper Ekofisk, however, contains as much as 20% impurities, mainly silica.

**Fig. 3.** SEM micrograph of a preserved "coccosphere".

### Fluid properties

Ekofisk was initially an undersaturated volatile oil reservoir. Initial reservoir pressure was 7135 psia at datum (−10 400 ft) and bubble-point pressure was 5443 psia. The 36°API gravity crude had a solution gas–oil ratio (GOR) of 1547. Reservoir temperature is 131°C (268°F) at datum.

### Overburden

The overburden is approximately 9300 ft thick and consists primarily of clays and shales with interbedded shale streaks. Below approximately 4500 ft the overburden is over-pressured.

## FIELD DEVELOPMENT—PRIMARY RECOVERY

The Ekofisk Field was discovered in late 1969. It was the first major discovery in the North Sea.

### Development history

The magnitude of the reservoir was understood following the drilling of three appraisal wells, but the producibility of this fractured chalk reservoir was in question. Early on it was feared that depletion of reservoir pressure would lead to a closure of the natural fractures that were essential for commercial flow rates from the reservoir. It was reasoned that if compaction occurred, fracture closure and reduced permeability could be expected.

To answer the producibility question, the discovery well and three appraisal wells were completed subsea and tied into the converted jack-up platform *Gulftide* for an extended production test. From July 1971 to May 1974, 28 million barrels of oil were produced. Productivity did not decline and the decision to develop Ekofisk with permanent structures was made in 1972. (Boyce, 1972; van den Bark and Thomas, 1980).

Development required the installation of three drilling/production platforms, a field terminal platform and living quarters. Since sea conditions often prevent offshore loading in the North Sea, a one-million-barrel concrete storage tank was also installed. Production through the permanent facilities was initiated in May 1974. In October 1975 a 220-mile oil pipeline to Teesside, England, became operational. Following installation of compressors in February 1975 all produced gas was injected until a 274-mile gas pipeline to Emden, Germany became operational in September 1977. After that, "swing gas"—gas in excess of sales—was injected. The exception was the period August 1985 to October 1986 when an average of 324 mmscf/d* was injected to mitigate subsidence. Following the discovery of subsidence in November 1984, 9 months were required to divert gas from sales to injection. This lag time was rather minimal considering the long-term nature of the gas sales agreements and was possible only with a high degree of cooperation from the purchasers of Ekofisk gas. After the subsidence remedy (platform elevation) was determined in January 1986, another 9 months were required to divert gas from injection to sales.

### Drive mechanisms

Until the discovery of subsidence in November 1984, solution gas drive was considered to be the predominant drive mechanism. Reservoir compaction, which is responsible for subsidence, is now also considered as one of the main drive mechanisms (Johnson *et al.*, 1988; Johnson and Rhett, 1986; Sulak and Danielsen, 1988). Oil expansion played a role in the early days of Ekofisk. Water influx in the Tor Formation is also significant.

Consider a typical piece of chalk from the Lower Ekofisk or Tor Formation that had an original porosity of 38%. Reservoir pressure reduction from initial to current pressure has resulted in a reduction in porosity to 32.9% (Fig. 4), based on trendlines developed for Ekofisk chalk. Taking into account the 7.6% reduction in thickness that must accompany this porosity reduction (assuming incompressible rock matrix), the pore volume reduction in this example is 20%. Under constant pressure conditions, pore fluids that once occupied these now non-existent pore volumes were produced. Reducing this pore volume change to account for the proper volumetric distribution of oil and gas, formation volume factors of the oil and gas, solution gas in the oil and the liquids in the gas, it is apparent that compaction drive by itself is responsible for a recovery factor of about 10% in this example.

*$m=10^3$; $mm=10^6$

## FIELD DEVELOPMENT—ENHANCED RECOVERY

### Dry gas injection

Upper Ekofisk dry gas injection was introduced in February 1975 with five crestal injectors providing up to 500 mmscf/d injection. Initially, all produced gas was reinjected until the gas sales line to Emden became operational in September 1977. Then, gas in excess of sales was reinjected. As of 1 November 1988, over 900 billion scf dry gas has been reinjected, providing partial pressure maintenance and increased hydrocarbon liquids production. Excess gas will continue to be injected at Ekofisk with potential expansion to Eldfisk upon implementation of Ekofisk nitrogen injection.

### Tor waterflood

Waterflooding is the most promising and practical enhanced oil-recovery (EOR) process for hydrocarbon recovery in the Tor Formation. This conclusion was reached after extensive screening of all available EOR processes, i.e. gas flooding (miscible and immiscible), chemical flooding (micellar, polymer, caustic, etc.) and thermal methods. Selection criteria were dominated by the need to find a process that could be economically applied to a naturally fractured chalk reservoir offshore.

Fractured reservoirs have been studied in detail with respect to pressure transient analysis, but there is less understanding of displacement mechanisms and associated oil recovery. Two displacement mechanisms operate simultaneously and independently in waterflooding of a naturally fractured chalk reservoir. The dominant process, either spontaneous imbibition or forced imbibition, depends on fracture intensity.

When fracture intensity is low, a pressure gradient is created such that, when a threshold pressure of 1–2 psi is exceeded, water will migrate from the fractures into the matrix blocks and displace oil as in a conventional waterflood. When fracture intensity is high, as is the case for Ekofisk Field, water will preferentially travel through the fracture system, displacing oil from the fractures. In this situation, conventional waterflood theory does not apply and the controlling factor of the countercurrent process (water moving from fracture to matrix and oil

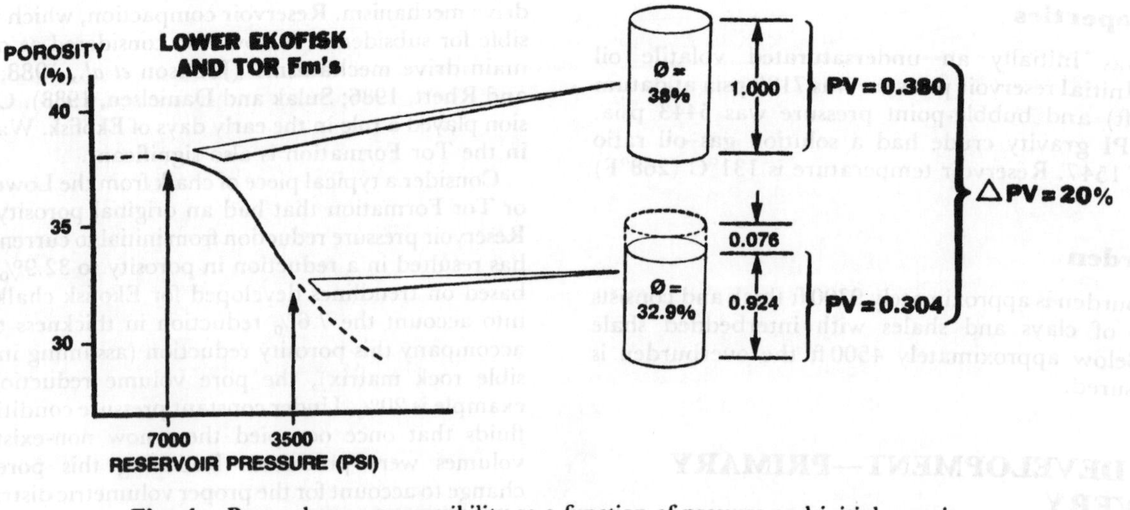


**Fig. 4.** Pore volume compressibility as a function of pressure and initial porosity.

moving from matrix to fracture) will be capillary imbibition.

*Laboratory imbibition tests*

Laboratory experiments have been run for both spontaneous and forced imbibition. Results of the forced imbibition tests will be discussed in more detail under Laboratory Tests—Imbibition for the Lower Ekofisk Formation.

Spontaneous imbibition tests have been run both at room and elevated temperatures and the results have been published in several papers. The imbibition behaviour of the Tor Formation samples indicates an excellent correlation of initial water saturation ($S_{wi}$), water imbibition ($\Delta S_w$) and final water saturation ($S_{wf}$) with porosity (Fig. 5) (Gunlikson and Wier, 1985).

The results indicate the Tor Formation is strongly water-wet and exhibits high water imbibition values (48% on average).

Imbibition experiments were systematically varied to identify potential sources of laboratory-induced bias of the results. Sensitivities evaluated Ekofisk crude and synthetic oil, room and elevated temperatures, and preserved and extracted cores. Core preparation and oil type did not significantly affect results. An imbibition increase of 10% was obtained at 132°C (270°F) compared to that at room temperature.

Laboratory tests were also performed to evaluate waterflooding performance in a depleted reservoir. Gas-saturated samples were subjected to oil imbibition followed by water imbibition. The experiments indicate that oil imbibition occurs until virtually all of the gas has been displaced. Water imbibition will then displace the oil and any remaining gas. The experiments did not indicate any gas being trapped after water imbibition.

*Tor Pilot Waterflood*

The encouraging laboratory imbibition results prompted a field test to evaluate imbibition characteristics. In addition to imbibition, it was necessary to establish whether water could be injected at significant rates, to test rock integrity and to establish reservoir anisotropy. A single injection well pilot project with three producing wells (Fig. 6), confined vertically but not areally (Fig. 7), was implemented.

Water injection commenced in April 1981 into the Tor Formation of Well B-16. The project demonstrated that water can be successfully injected into Tor Formation. Injection rates of up to 32 000 bw/d were obtained with average injection rates of over 25 000 bw/d during the $3\frac{1}{2}$ year project life.

Water injection was terminated on 16 July 1984 after injection of 22.8 million barrels of water. Water breakthrough occurred in April 1982 in Well B-22 (Fig. 8) and in July 1983 for Well B-24. Water breakthrough was not observed in Well B-19. Positive waterflood response was observed shortly after the start of the water injection by a sharp GOR decrease (Fig. 9–11), typical of the imbibition displacement mechanism.

Figure 12 shows the watercut behaviour of Well B-22. This behaviour, typical of fractured reservoirs, shows an immediate response to reduced water injection in the form of decreased water cut accompanied by increased oil production and GORs. During injection, water is transported through the fracture system and imbibed into the matrix. Once the injection ceases, water transport through the fractures ceases and only imbibition continues. This allows water in the fractures to be imbibed into the matrix, displacing oil into the fractures. This results in the sharp drop in water cuts and increase in oil production. GOR increases are the result of decreased reservoir pressure around the producers. This phenomenon provides a tool for effective waterflood management when watercuts become excessive.

Pilot performance was matched using a dual porosity 3-phase black-oil numerical model that was discussed by Thomas *et al.* in 1987. The model confirmed imbibition values developed in the laboratory.

From the Tor Pilot Waterflood, it was concluded that water can be injected at significant rates without significant effects on rock integrity. A definite N–S major permeability trend was defined by pulse testing. However, no rapid water breakthrough to the producers was observed. Most importantly, the good imbibition characteristics of the Tor formation were confirmed.

*2/4 K Waterflood Project*

With favourable imbibition results from both laboratory tests and the pilot waterflood, it was decided in 1983 to implement a waterflood project in the Tor Formation covering the northern two-thirds of the field. To accomplish this, a new water-injection platform (2/4K) was designed, built and, in 1986, installed alongside the existing 2/4 B platform.

The maximum water injection rate is 375 000 bw/d at 5000 psi. Figure 13 shows the water-injection well locations

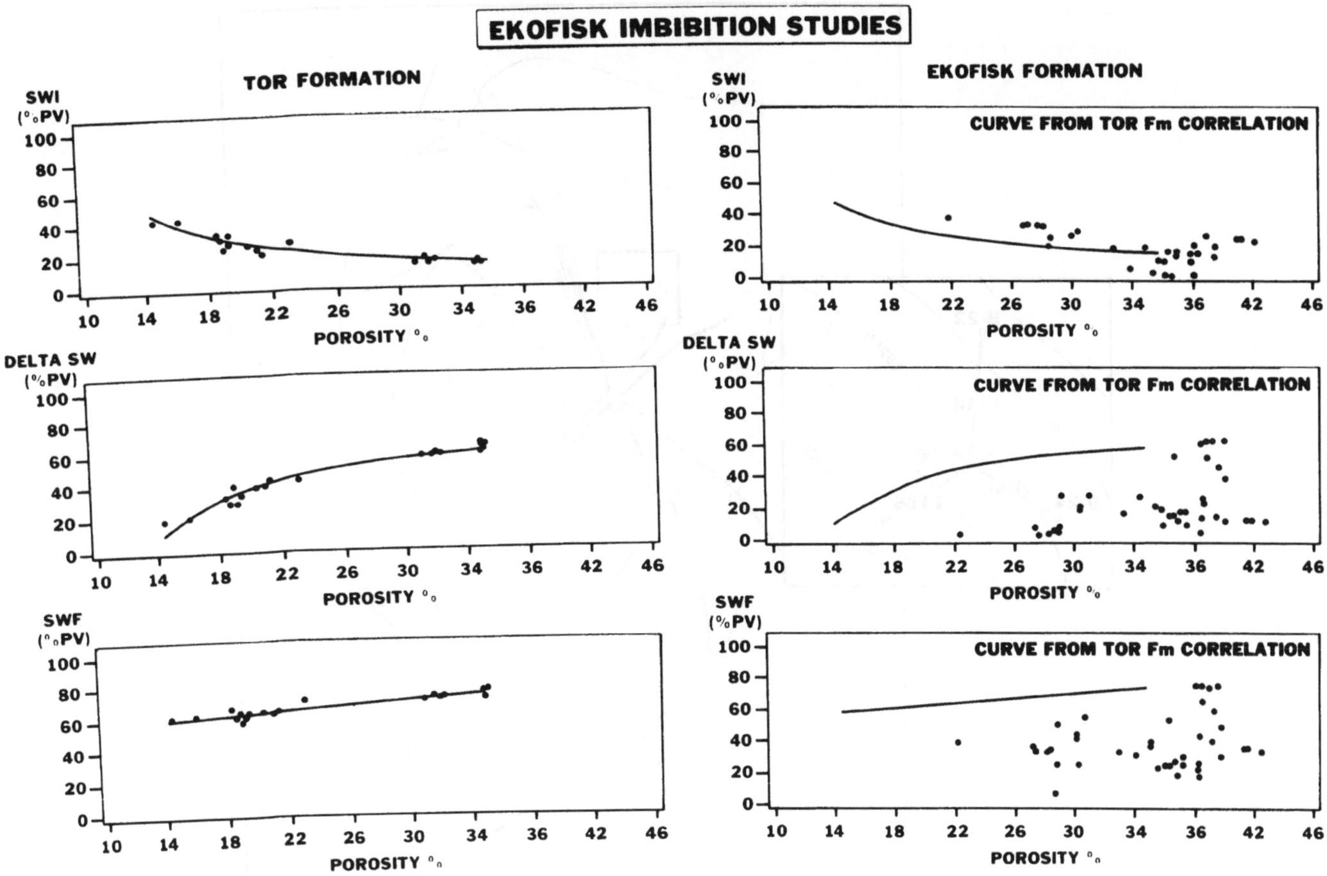


**Fig. 5.** Initial water saturations, changes due to imbibition, and final water saturations for the Tor and Ekofisk imbibition studies.

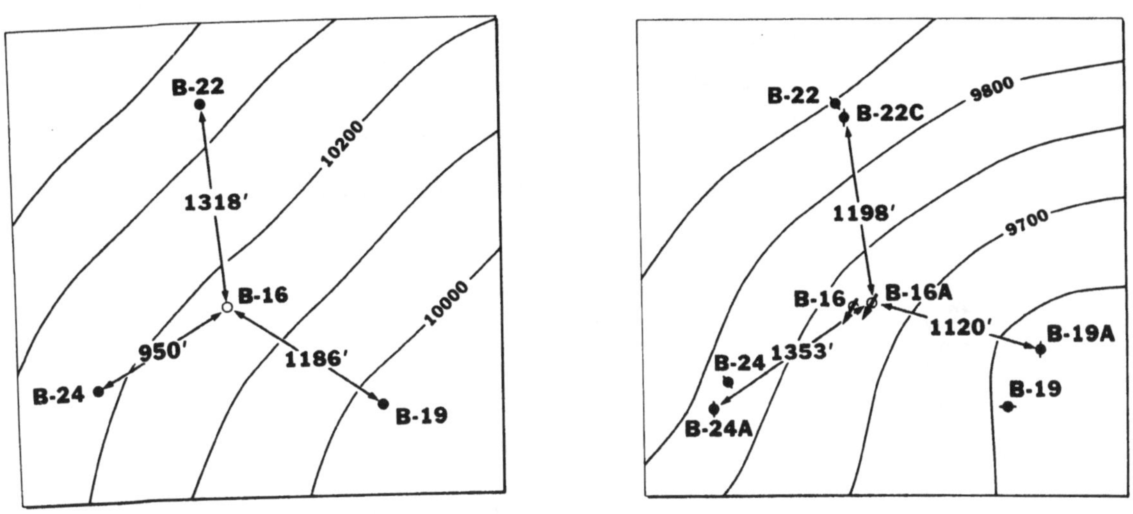


**Fig. 6.** Tor Pilot Waterflood area.

aligned and spaced N–S to form a line drive pattern taking advantage of anisotropy trends in the field.

Reservoir performance has been modeled using a three-dimensional (3D), 3-phase modified black-oil simulator. Historical GOR and pressure performance, from the start of production until the present, has been matched. Laboratory imbibition results and field pilot results have been incorporated in the full-field model by use of pseudorelative permeability concepts. This model has then been used to predict oil, gas and water production rates as well as recoveries expected under the different conditions examined.

To quantify 2/4 K waterflood incremental recoveries, simulation runs were made with and without the 2/4 K project. The results indicate an incremental recovery of 160 mmboe. Figure 14 shows predicted waterflood performance in terms of barrels of oil equivalent. This represents an incremental recovery of 1.9% of the original hydrocarbons in place.

Water injection from 2/4 K began in November 1987.

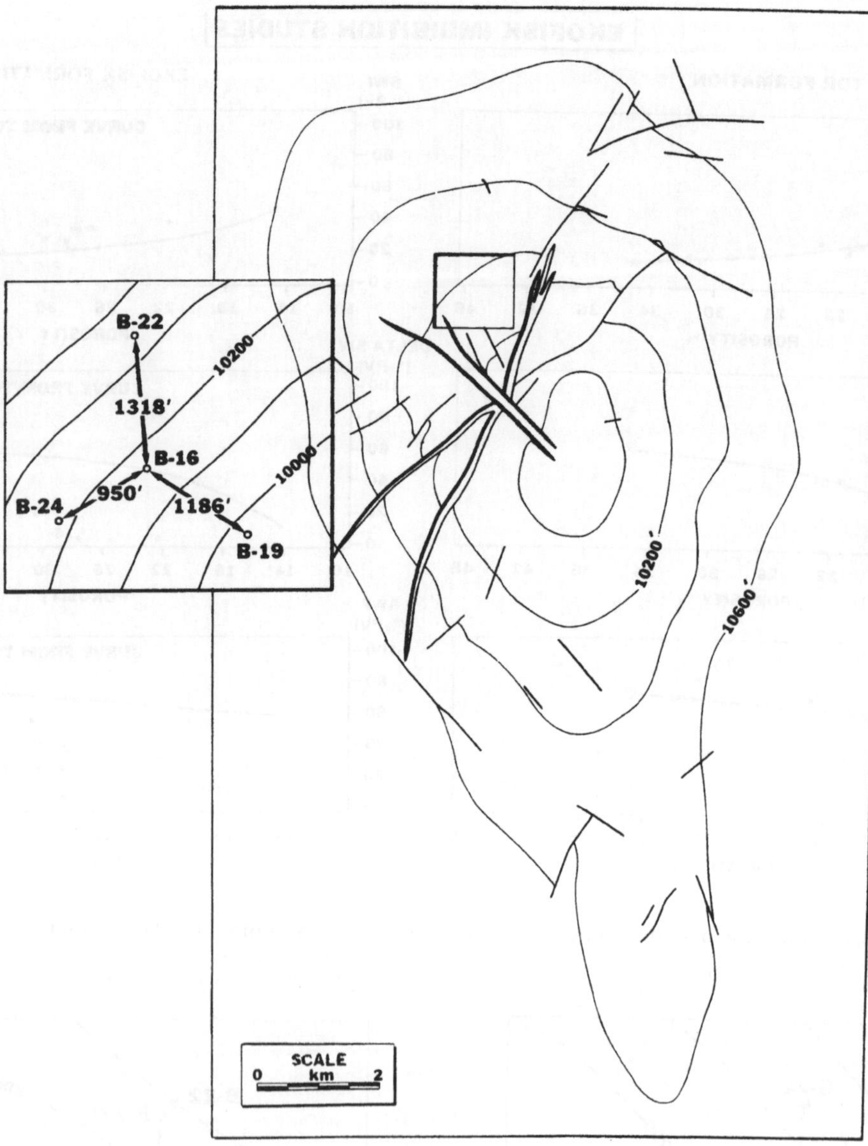


**Fig. 7.** Field location of the Tor Pilot Waterflood.

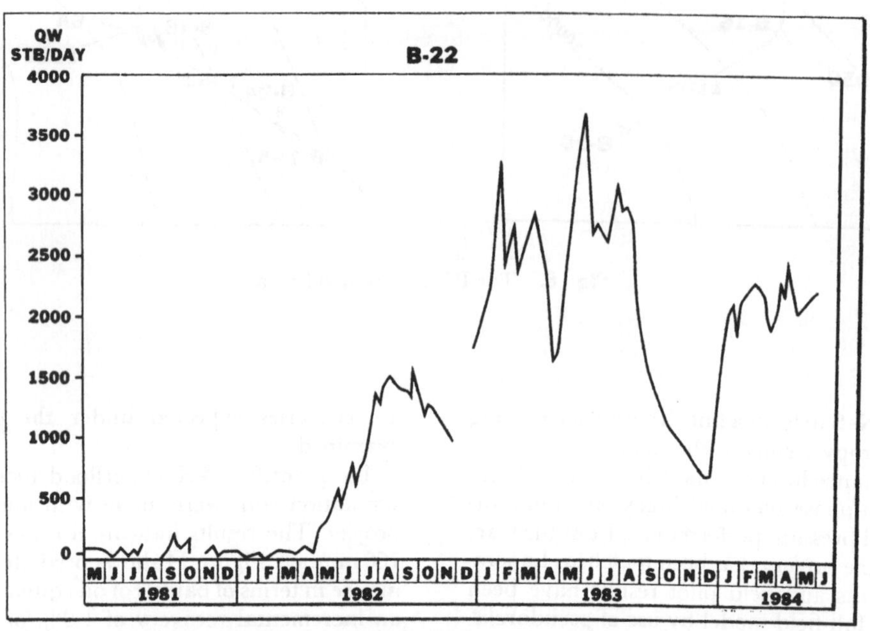


**Fig. 8.** Water breakthrough in Tor Pilot Well B-22.

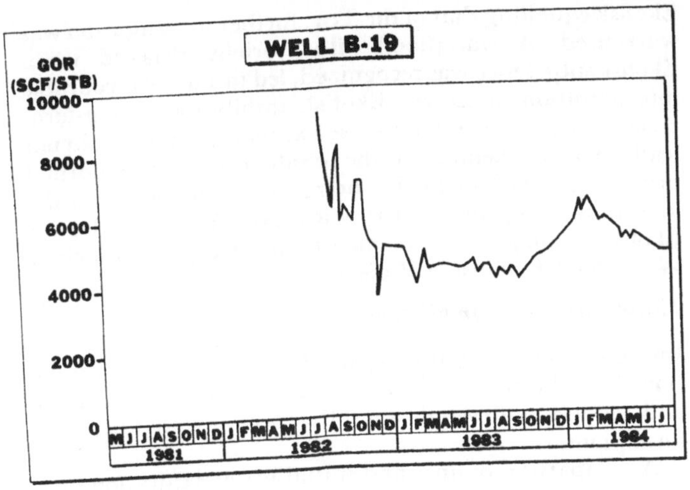


**Fig. 9.** Observed gas–oil ratios from Tor Pilot Well B-19.

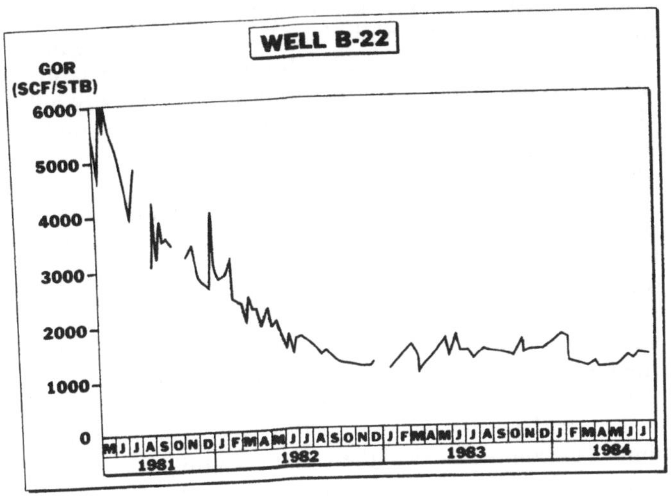


**Fig. 10.** Observed gas–oil ratios from Tor Pilot Well B-22.

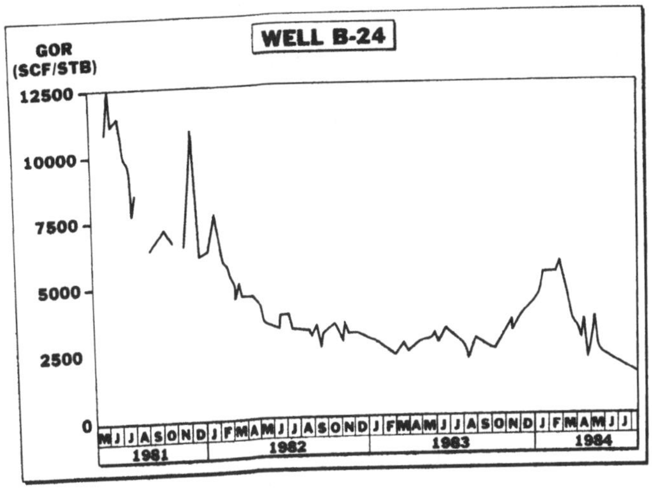


**Fig. 11.** Observed gas–oil ratios from Tor Pilot Well B-24.

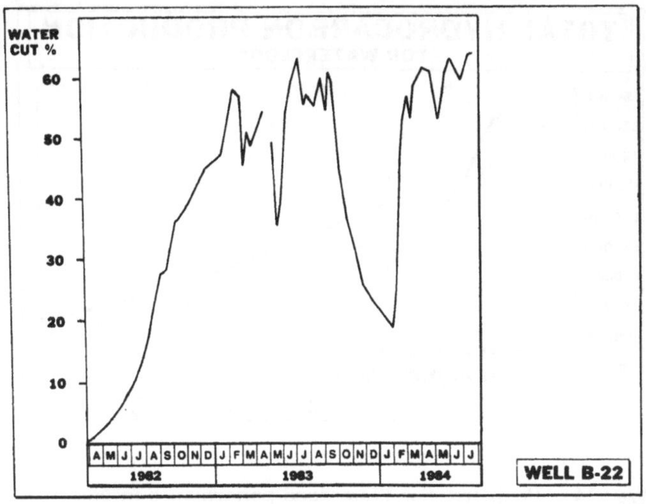


**Fig. 12.** Watercut from Tor Pilot Well B-22

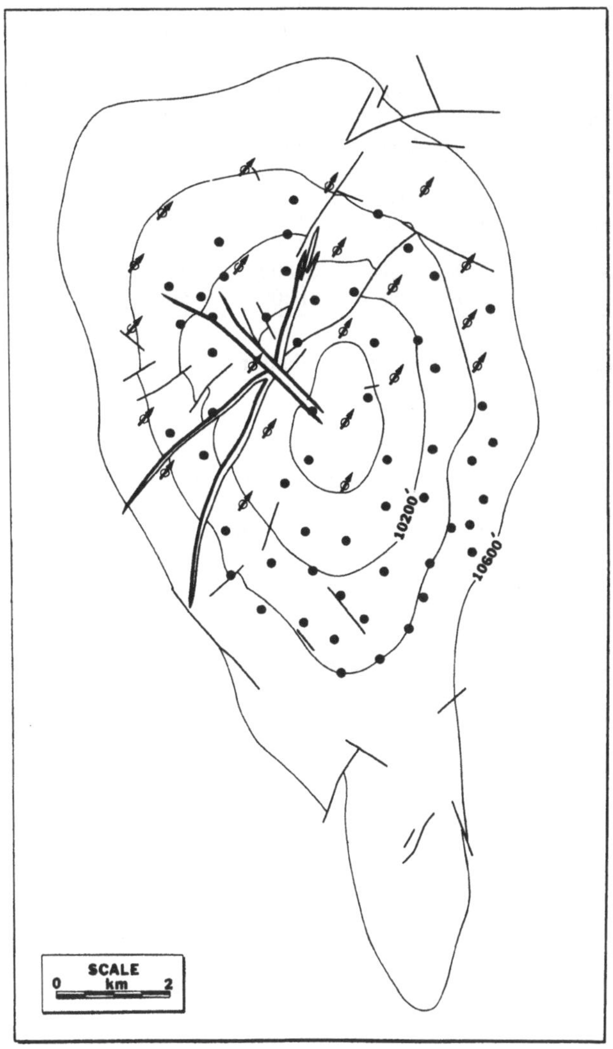


**Fig. 13.** Tor Waterflood pattern of injectors and producers.

The project includes injection into 21 wells, nine of which have been drilled and put on injection as of 1 November 1988. Individual injector performance is tabulated in Fig. 15.

Figure 16 shows performance curves for a typical injector. After one year of partial injection, waterflood response has been observed in the GOR behaviour of wells B-3, B-8, B-18, B-20; oil rate has increased in wells B-3 and B-8 and watercut has increased in wells B-3 and B-20.

Initial indications are as expected. There has been a decrease in GORs, no significant water production, directional movement of water in a N–S trend and toward

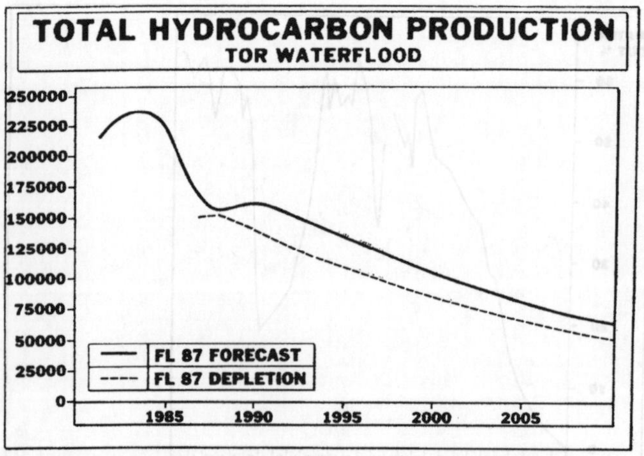


**Fig. 14.** Predicted Tor Waterflood performance in barrels of oil equivalent.

| *Well* | *Start date* | *Injection rate (bw/d)* | *Injection pressure (psig)* | *Cumulative injection (mbbls)* |
|---|---|---|---|---|
| K-30 | 28 Dec. 1987 | 8 500 | 3 000 | 2 781 |
| K-4 | 29 Dec. 1987 | 23 800 | 2 700 | 6 740 |
| K-22 | 21 Jan. 1988 | 23 100 | 2 600 | 6 513 |
| K-12 | 21 Feb. 1988 | 17 800 | 2 600 | 4 516 |
| K-13 | 11 Mar. 1988 | 20 500 | 3 000 | 4 327 |
| K-27 | 16 July 1988 | 15 000 | 3 000 | 1 572 |
| K-28 | 2 Aug. 1988 | 18 000 | 3 000 | 1 639 |
| K-9 | 7 Oct. 1988 | 24 800 | 2 700 | 6 448 |
| K-11 | 15 Oct. 1988 | 21 800 | 3 000 | 3 133 |

**Fig. 15.** Performance of individual injectors as of 1 November 1988.

high permeability areas. Additionally, waterflood response across faults has been observed.

Figure 17 shows response of well B-8 to water injection. Implementation of the full-field waterflood will continue with the goal of having the waterflood operating at full capacity by early 1990.

## Lower Ekofisk waterflood

Early laboratory tests yielded low imbibition values for the Lower Ekofisk but, with the in-place oil in the Lower Ekofisk equalling that of the Tor, further investigation was warranted. A waterflood pilot, briefly delayed when Ekofisk subsidence was recognized, led to a more favourable interpretation of Lower Ekofisk imbibition. Long-term creep tests demonstrated that seawater injection would not lead to a weakening of the chalk during the normal producing life of the field. The results of the pilot and the long-term creep tests led to waterflooding of the Lower Ekofisk, a major component of the recently authorized Waterflood Expansion Project.

### *Laboratory tests—imbibition*

Imbibition characteristics of the Ekofisk and Tor Formations have been extensively studied by Phillips Research and Services, with significant differences observed between formations.

A comparison of final and initial water saturation and the implied spontaneous imbibition as a function of core porosity are provided in Fig. 5 for the Tor and Ekofisk Formations (Gunlikson and Wier, 1985). A strong correlation between porosity and imbibition exists in the Tor, indicating an average imbibition of 48%. Overlaying these correlations on Ekofisk samples shows initial water saturations are similar, but Ekofisk imbibition is generally only 20–25% and does not correlate with porosity. Some higher imbibition values for the Ekofisk Formation have been observed, particularly in the south.

Petrographic analysis found no significant differences in pore geometry between the formations that would account for imbibition variances (Weseloh, 1985). Analysis of surface properties and core wettability found that solvent extraction improved poor imbibers while having no effect on strong imbibers, indicating that the core's surface chemistry dominated imbibition performance. Analysis of the extracted material identified an organic coating that had been deposited on the rock surfaces over geological time. The compositions of these materials differed between formations and effectively determined the reservoir's surface properties. Highly oxidized material exhibited water-wet characteristics, while less-oxidized organics tended to be more oil-wet. The organic surface, seen by scanning electron microscope (Fig. 18), shows the organic coating on a core surface before cleaning and a similar surface after solvent extraction. Differences in spontaneous imbibition between the Ekofisk and Tor Formations appear to be naturally occurring phenomena. Work continues to identify factors affecting the laboratory imbibition results and methods to correlate Ekofisk imbibition.

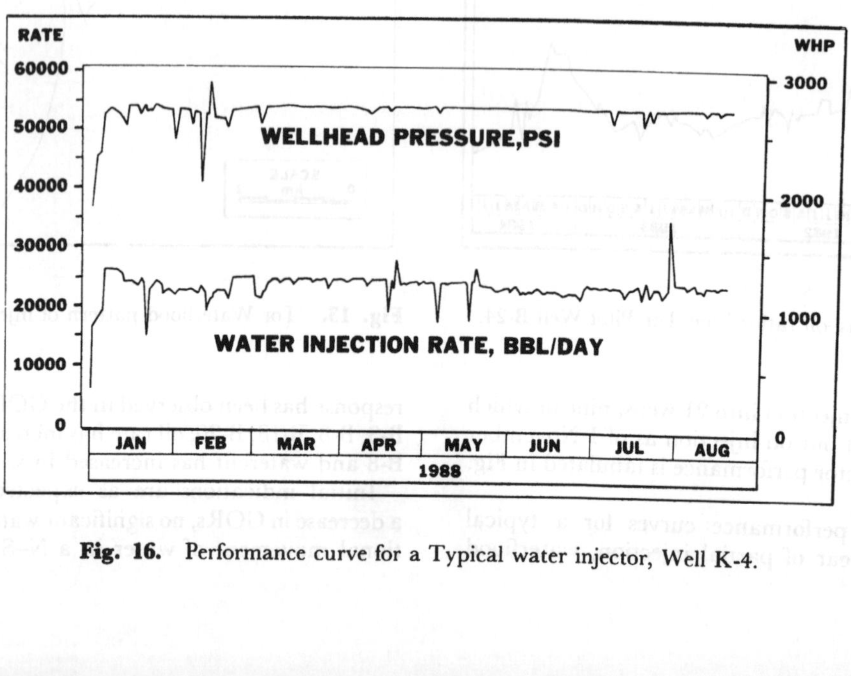


**Fig. 16.** Performance curve for a Typical water injector, Well K-4.

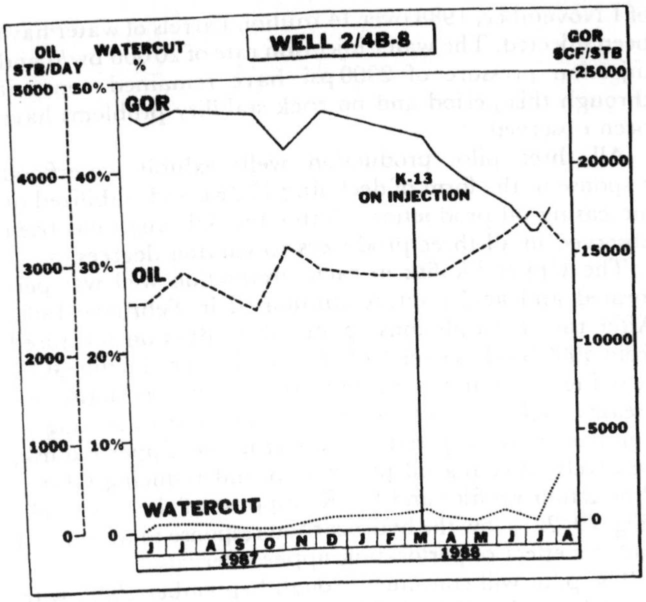


**Fig. 17.** Response of Well B-8 to full-field Tor water injection.

### *Laboratory tests—rock strength*

Before a commitment to waterflooding of the Lower Ekofisk Formation, the effect, if any, of seawater injection on the mechanical properties of the chalk had to be determined. Phillips and Shell independently studied this question in 1985. Both concluded from these early studies that seawater injected at room temperature would have no short-term effect on the strength of the chalk. (Smits *et al.*, 1986).

In 1986, Phillips performed hydrostatic compressibility tests at reservoir temperature (Johnson *et al.*, 1988). Adverse effects of seawater injection were not detected even at reservoir temperature, but the long-term effects were still unknown.

To investigate the long-term effects of seawater injection, horizontal companion core plugs from the Lower Ekofisk in Well 2/4 A-8 were tested in hydrostatic test cells starting in the Spring of 1987. Both plugs were brought up to 6000 psi confining pressure at reservoir temperature in steps of 1000 psi. At 4000 psi confining pressure, 10 pore volumes of seawater were injected through one of the plugs. Both plugs stopped compacting after about 2 weeks and neither has reinitiated compaction during the test period, which is up to 18 months at the time of writing.

The preservation of higher porosity in the hydrocarbon-filled pores of the reservoir suggests that fluid content has an effect on chalk strength that is detectable over geological time. The purpose of the long-term test was to identify any differences in long-term creep behaviour between the waterflooded and a control plug, that would suggest a weakening of the waterflooded chalk that could be expected during the life of the Ekofisk Field.

### *Lower Ekofisk Pilot Waterflood*

Despite poor laboratory imbibition results, it was recognized that additional mechanisms could crontribute to a successful Lower Ekofisk waterflood. This prompted installation of the Lower Ekofisk pilot waterflood to investigate recovery mechanisms, injectivity performance and rock stability in the presence of seawater (Sylte *et al.*, 1988).

The Lower Ekofisk Pilot consists of a single water injector, B-16, surrounded by three producers B-19, B-22 and B-24, as shown in Fig. 19. Injection facilities and wells from the Tor pilot were utilized. The wells were sidetracked to within 500 ft of the original bottomhole location, perforated, and acid-fractured in the Lower Ekofisk.

A water injectivity test was conducted on well B-16 before the sidetrack to determine Lower Ekofisk injectivity and rock strength. Water injection averaged 25 000 bw/d

**Fig. 18.** SEM micrographs of chalk surfaces shown prior to cleaning and after cleaning.

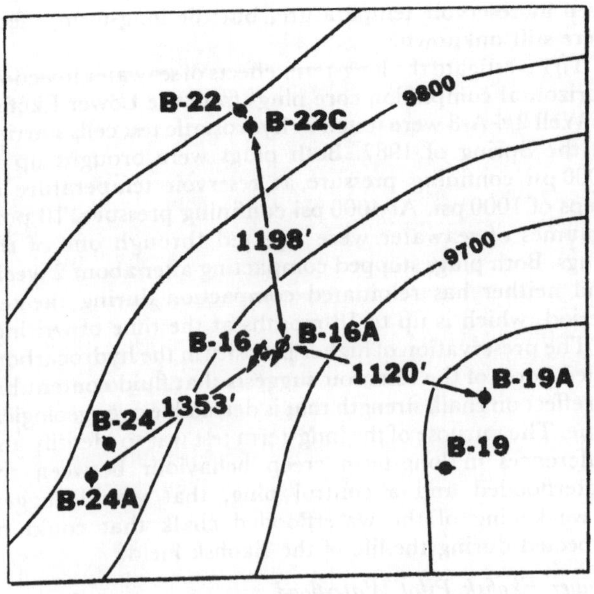


**Fig. 19.** Ekofisk Formation Pilot Waterflood area.

for a 12-day period with no observed decline in injectivity. The flowback was performed over the next 3 months without significant solids production. These results show sufficient rates can be injected into the Lower Ekofisk Formation and rock strength, at least on a short-term basis, is not adversely affected by seawater injection.

B-16 was then sidetracked 150 ft from the previous location and completed in the Lower Ekofisk. A comparison of CPI logs from the original B-16 and the sidetracked location showed a water-saturation increase of over 50%. The water-saturation increase in the sidetrack indicates that pilot area imbibition values were much higher than observed in the laboratory for the Lower Ekofisk. Continuous water injection into B-16 began in July 1986 and as of 1 November, 1988 over 14 million barrels of water have been injected. The water injection rate of 20 000 bw/d and injection pressure of 2900 psi have remained constant through this period and no rock stability problems have been observed.

All three pilot production wells exhibit waterflood response in the form of declining GORs and stabilized or increasing oil production. Water breakthrough has been observed in all three producers to varying degrees.

The Upper Ekofisk in each production well was perforated and acid fracture stimulated in February 1988. After the recompletions, pilot oil production increased from 5000 bo/d to over 10 400 bo/d. The producing GOR remained essentially constant and water production increased slightly. These results indicate that water migration has provided pressure support to the Upper Ekofisk, positively affecting oil production and reducing GORs. Production profiles and GOR response of the pilot wells (Fig. 20–22) show the hydrocarbon response to waterflood and the effect of perforating upper intervals.

The pilot will continue through September 1989, with an additional well drilled in the pilot area in early 1989. Openhole logs from this well will help quantify waterflood residual saturations in the Lower Ekofisk Formation.

The Lower Ekofisk pilot has been successful in increasing oil rates in pilot production wells while maintaining low water production. Additionally, water injectivity has been maintained with no adverse effects observed in rock strength. These favourable results provide a technical basis for expanding the existing Tor waterflood to the Lower Ekofisk Formation.

*Waterflood Expansion Project*

Tor waterflood facilities on platform 2/4 K have injection capacity of 375 000 bw/d and slots for 30 water-injection wells. Peak Tor injection is forecast at 300 000 bw/d but on average will utilize only 125 000 bw/d. Additionally, only 21 of the 30 slots are used. Because of excess capacity and Lower Ekofisk reserves potential, expansion of the Tor

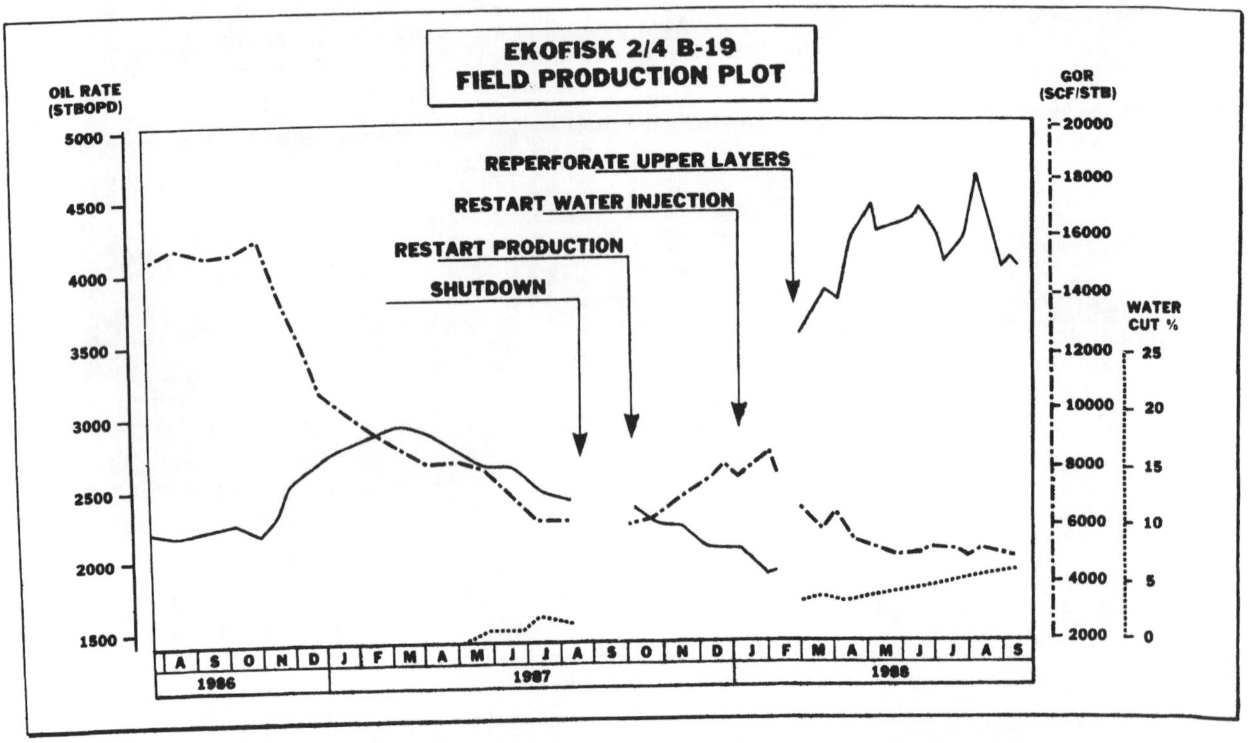


**Fig. 20.** Oil Production, GOR and watercut behaviour of Ekofisk Pilot Well B-19.

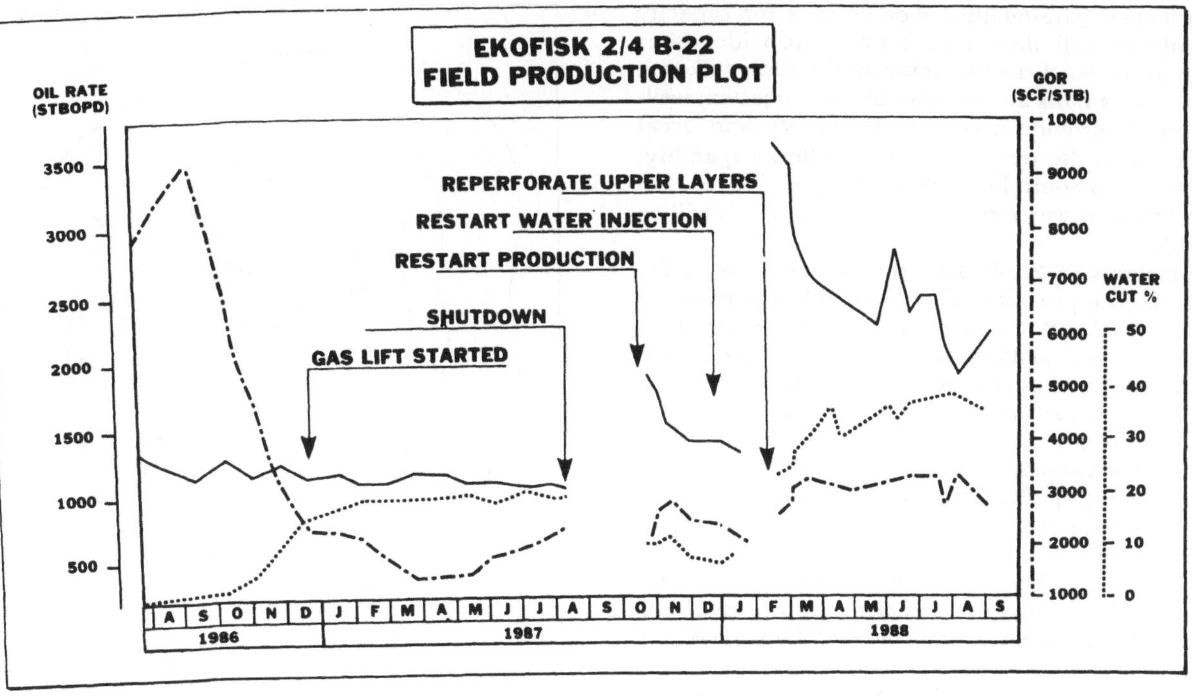


**Fig. 21.** Oil production, GOR and watercut behaviour of Ekofisk Pilot Well B-22.

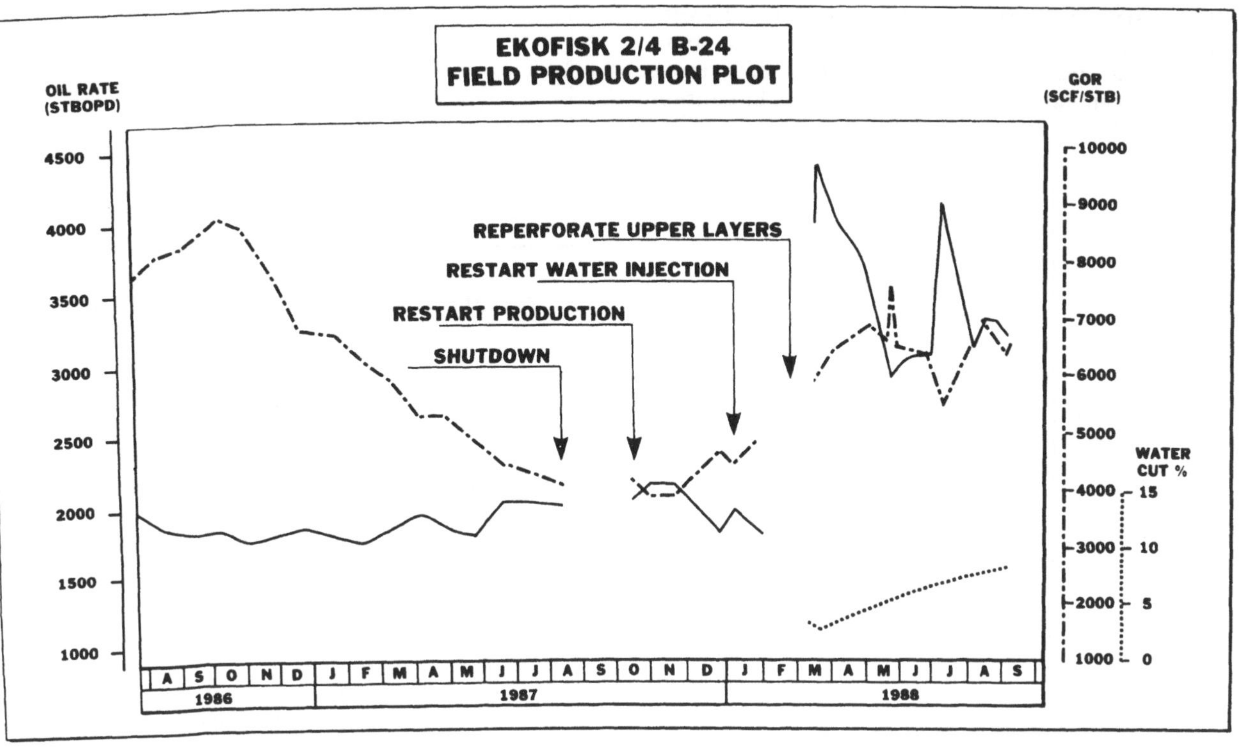


**Fig. 22.** Oil production, GOR and watercut behaviour of Ekofisk Pilot Well B-24.

waterflood to the southern Tor and Lower Ekofisk fieldwide is a high priority.

Waterflood expansion to the southern Tor was first investigated in Fall 1986. Project economics were hindered owing to the limited waterflood project area and the need for a new platform. The project scope was restricted to the southern Tor, with Lower Ekofisk expansion not possible because of apparently negative laboratory imbibition results.

Continued engineering efforts overcame problems encountered in the initial expansion study. Favourable Ekofisk pilot results indicated higher waterflood displacement than that suggested by laboratory tests, allowing expansion of the potential waterflood area to include the Lower Ekofisk. Innovative design modifications to platforms 2/4 A, 2/4 C and the South Flare Bridge Support II provide 24 additional slots for injectors and producers. The slot distribution allows project expansion to the central and southern areas of the field, eliminating the need for new platforms and additional drilling rigs.

Waterflood expansion utilized excess injection capacity and available well slots from 2/4 K to provide water injection to the northern two-thirds of the Lower Ekofisk Formation. Five flank and two crestal water injection wells are placed to provide maximum injectivity and areal coverage within the reaches of 2/4 K drilling capability. Modifications to South Flare Bridge Support II (Fig. 23) provide 8 slots for water injection wells, 7 of which will be commingled.

Modifications to 2/4 A platform (8 slots) and 2/4 C platform (8 slots) provide additional withdrawal points in the Tor and Ekofisk Formations. Bottomhole locations targeted from these platforms are selected in areas of high mobile oil saturation and high permeability with preference given to commingled wells. The resulting pattern of producers and injectors is a staggered line drive, oriented N–S to take advantage of favourable anisotropy in that direction. (Fig. 24). Platform modifications are possible because of existing wellhead spacing, which allows additional conductors to be located in the diagonals of existing wellheads. The wellhead configuration in Fig. 25 shows the original 15 slots and the platform modifications that result in 8 additional slots (numbered 16–23) on 2/4 A and C platforms.

Simulation results indicate Waterflood Expansion will increase recovery by 118 mmbo and 434 bscf gas for a total hydrocarbon recovery of 190 mmboe. Recovery comes primarily from the Lower Ekofisk (71%), with the remainder attributed to Tor expansion. The high liquids recovery is consistent with the goal of enhancing liquids production, leaving Ekofisk gas recovery as the target of a more efficient displacement process. The Waterflood Expansion Project will increase hydrocarbon recovery from Ekofisk by 2.3%.

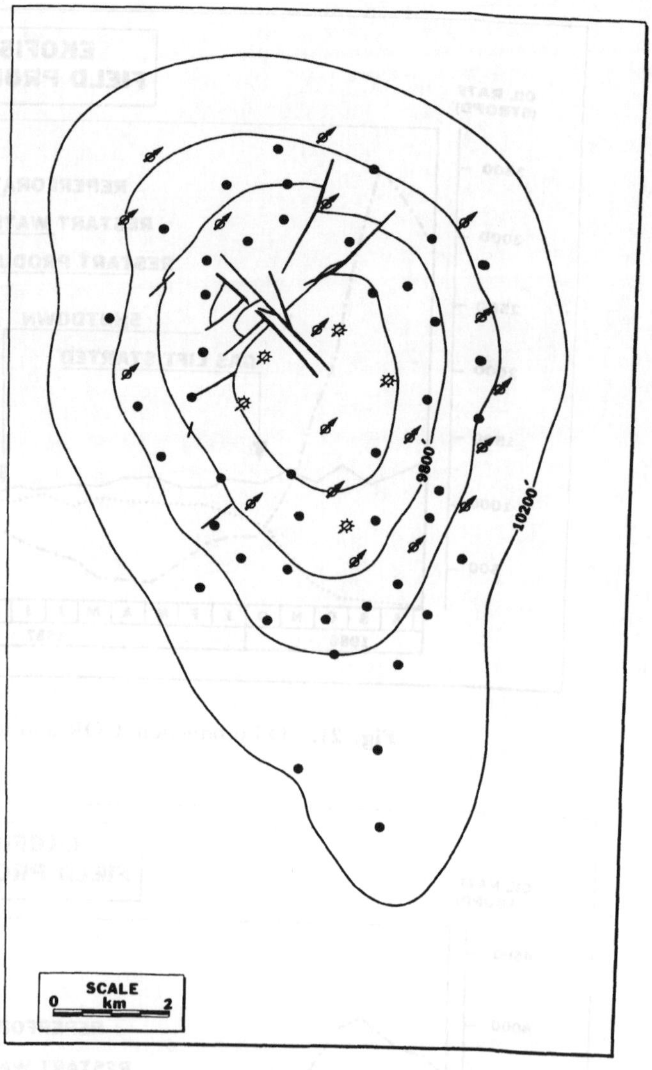


**Fig. 24.** Ekofisk Formation water injection pattern.

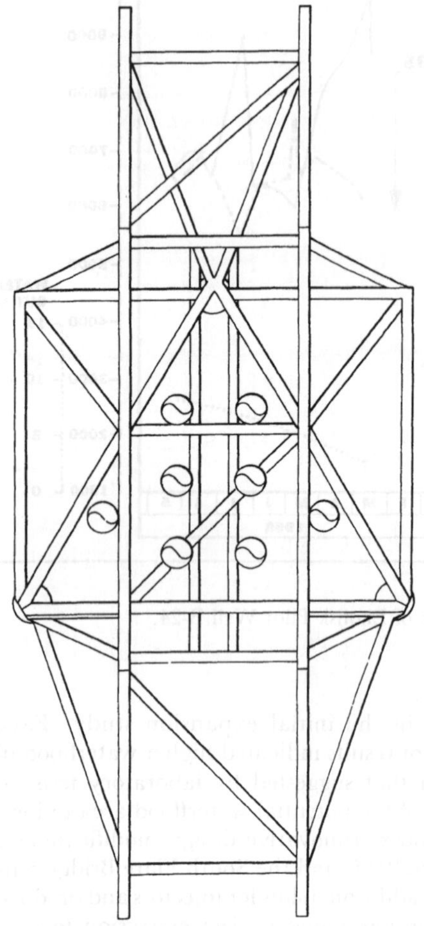

**Fig. 23.** Modification of bridge support no. 2, providing eight slots for water-injection wells.

## Upper Ekofisk Nitrogen Injection

To complete improved reco ery at Ekofisk, a Nitrogen Injection Project has been proposed for the Upper Ekofisk Formation. Initially, the Upper Ekofisk contained only undersaturated oil and connate water. Over 900 bscf dry gas has already been injected into the Upper Ekofisk. With continued injection of swing gas, almost one pore volume of dry gas will have been injected into the crest of the Upper Ekofisk before the start-up of the Nitrogen Injection Project, currently scheduled for October 1993. Coupled with the gas liberated since the bubble point was reached in 1978, the free gas saturation in the crest of the Upper Ekofisk is now aproximately 50%. Oil saturation is at its residual to gas, 40% in this case, and connate water saturation remains at 10%. The initial saturation conditions for the crestal Upper Ekofisk as well as those at project start are depicted in Fig. 26. Nitrogen will displace the free gas as well as vaporize gas from the residual oil, increasing gas sales while maintaining reservoir pressure.

*Recovery mechanisms*

Three recovery mechanisms play a role in the Upper Ekofisk Nitrogen Injection Project, namely molecular diffusion, vaporization and pressure maintenance. The

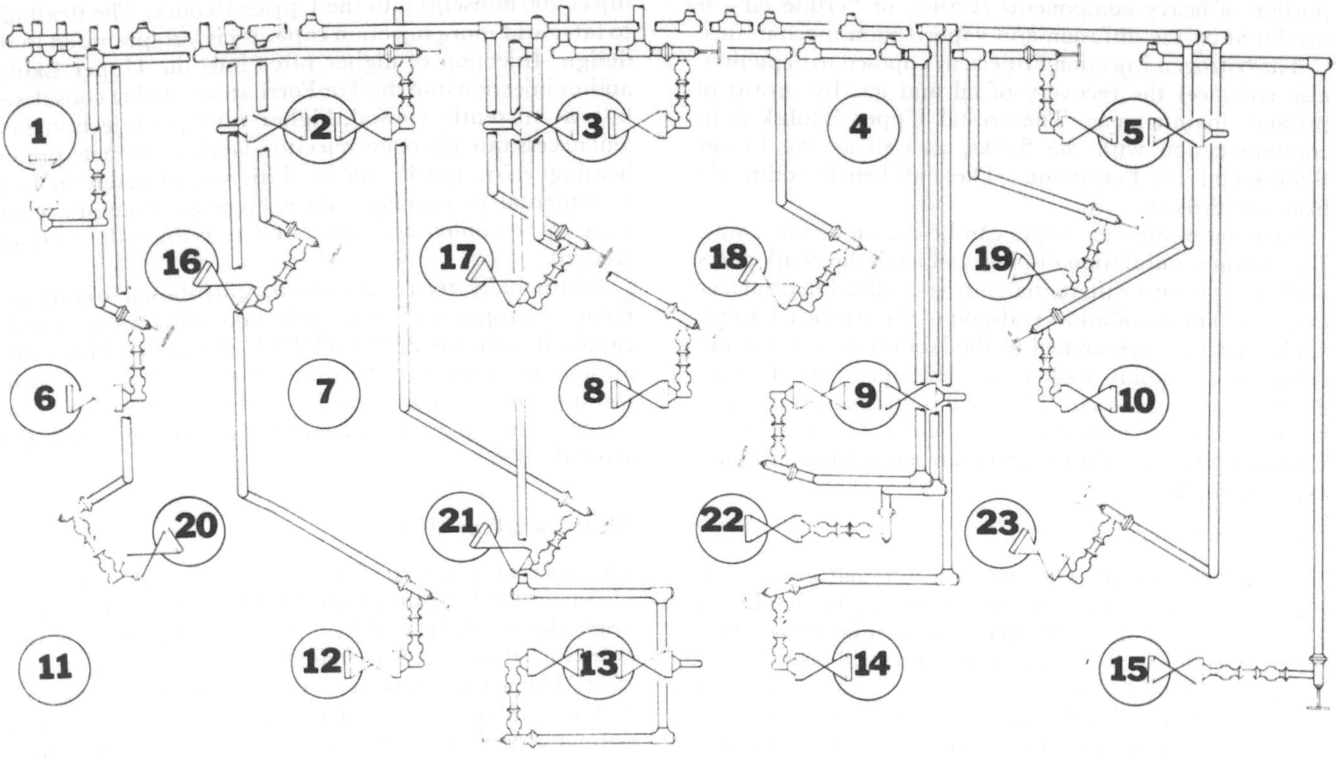


**Fig. 25.** Wellhead configuration showing original slots (1–15) and the added slots (16–23) on platforms Alpha and Charlie.

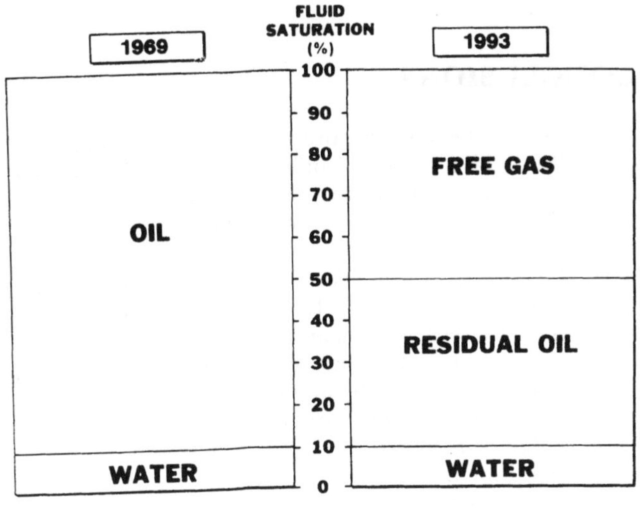


**Fig. 26.** Schematic of Upper Ekofisk Formation saturation conditions prior to nitrogen injection.

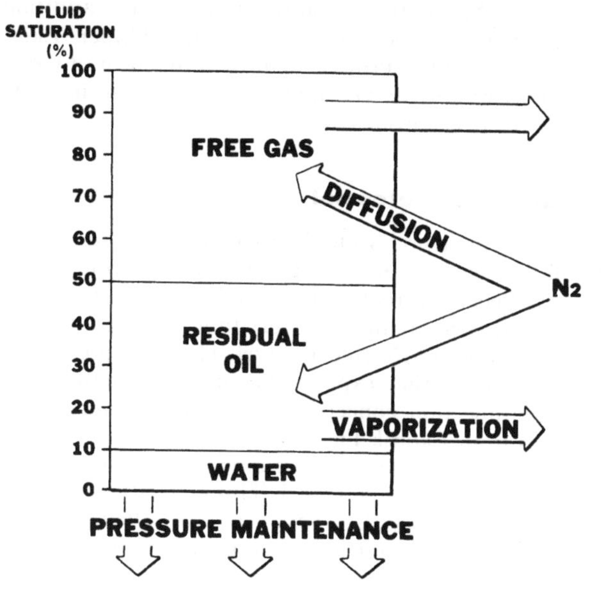


**Fig. 27.** Dominant hydrocarbon displacement mechanisms of the Upper Ekofisk Nitrogen Injection Project.

first two enhance gas recovery almost exclusively, while pressure maintenance enhances both oil and gas recoveries.

Figure 27 shows the diffusion mechanisms at work under the conditions at project start. Nitrogen flowing through the fractures diffuses into the matrix, displacing natural gas. This process continues until thermodynamic equilibrium is reached. Transport of nitrogen into the matrix and gas out is governed by the diffusion coefficient and component gradient between fracture and matrix. With pure nitrogen in the fracture system, the process will continue until all free gas is displaced. Laboratory experiments conducted by Phillips Research and Services show that 90% of the natural gas in the matrix will be displaced by nitrogen diffusion in a few days where fracture spacing is less than 1 metre. This fracture spacing is representative of the Upper Ekofisk Formation. Diffusion experiments carried out by Winfrith under the direction of Fina Exploration Norway, Inc., has yielded similar results. Approximately 70% of the crestal Upper Ekofisk gas that is recovered is the result of molecular diffusion.

Also shown in Fig. 27 is the vaporization mechanism at work. Nitrogen, which enters the matrix by molecular diffusion, contacts residual oil and vaporizes the lighter components. Almost all of the solution gas will be vaporized and displaced into the fracture system by additional diffusion, i.e. the other 30% gas recovery results from vaporization combined with diffusion. Only a small portion of heavy components (C5+), or "crude oil", is produced by the diffusion and vaporization mechanisms.

The Nitrogen Injection Project, as opposed to depletion, also enhances the recovery of oil and gas by means of pressure maintenance. The crestal Upper Ekofisk is in communication with the flanks, as well as the Lower Ekofisk and Tor Formations. These all benefit indirectly from the Project.

Some interesting analogies can be drawn at this point. Just as water imbibition displaces oil from the chalk pores without a pressure differential, nitrogen diffusion displaces free gas. The imbibition end-points for various Ekofisk chalks has been researched in the laboratories, while the diffusion end-point had been developed by thermodynamicists decades ago. The imbibition rate was inferred to be satisfactory from the results of the Tor and Lower Ekofisk pilots, while the diffusion rate has been determined experimentally.

*Project description*

The current concept is to install a nitrogen generation platform east of 2/4 FTP. Following start-up in October of 1993, 200 mmscf/d of nitrogen will be injected in two crestal injection wells. Swing gas injection will probably be diverted to Eldfisk. Reducing nitrogen breakthrough to under 100 mmscf/d will allow the entire produced gas stream to be sent through the pipeline to Emden, Germany, within contract specifications. Nitrogen breakthrough can be controlled by selective recompletion of high-breakthrough wells or by limiting production of these wells and/or diverting the gas stream of these wells to Eldfisk injection. Simulation studies indicate an additional 165 mmboe will be recovered by the Upper Ekofisk Nitrogen Injection Project, mainly in the form of natural gas. This represents yet another 2% increase in hydrocarbon recovery from Ekofisk.

## Future EOR potential

After the fieldwide waterfloods of the Tor and Lower Ekofisk reservoirs and nitrogen injection into the Upper Ekofisk become a reality, there is little more that can be done to further enhance recovery from Ekofisk if the price of oil remains at $15–18 per barrel. There are possibilities, however, to further expand the current waterflood and nitrogen-injection concepts to recover yet more hydrocarbons from Ekofisk Field.

The 2/4 K platform, as mentioned earlier, has an injection capacity of 375 000 bw/d. This capacity results from the three turbines that have been installed. A fourth turbine could be installed, increasing injection capacity to 500 000 bw/d, if warranted. Expansion to the higher injection rate is being evaluated today and hinges on several factors. First of all, fieldwide well injectivity must be verified. Secondly, fieldwide imbibition in the Tor and Lower Ekofisk Formations must be verified by flooding. Also, satisfactory imbibition rates in these formations must be verified. Finally, higher Upper Ekofisk imbibition values than currently assumed will probably be required. Although water is injected into the Lower Ekofisk, water will migrate to the Upper Ekofisk. Additional recoveries from the Lower Ekofisk will most likely be required to support justification of the expansion to a higher water-injection rate. When water-injection well K-23 is drilled in the waterflood pilot area in Spring 1989, openhole logs will help in further evaluating the in-situ imbibition values for the Upper Ekofisk as well as for the Lower Ekofisk and Tor Formations.

The Nitrogen Injection Project is being designed to inject 200 mmscf/d into the Upper Ekofisk. The flexibility to later add more injection capacity is also premised in the design. Injection of higher rates into the Upper Ekofisk and/or injection into the Tor Formation is being considered, but, as currently viewed, higher nitrogen-injection rates will necessitate nitrogen-rejection facilities to meet present heating value specifications. This would result in large incremental investments, since huge gas volumes would have to be chilled and processed to reduce the nitrogen content.

Additionally, recoveries currently attributed to enhanced recovery projects in Ekofisk only take into account gas-lift capabilities on the 2/4 B and 2/4 C platforms. These were authorized as part of the original 2/4 K Project. Gas-lift facilities on 2/4 A will further enhance recoveries from Ekofisk when they are authorized for timely installation around 1995.

## Ekofisk after 2011

The current production licence for Ekofisk expires on 31 August 2011. By then some 30–32% of the hydrocarbons originally in place will have been produced. It is now estimated that an additional 2% will be produced after 2011. The actual amount will depend on operating costs and the value of oil and gas at that time. Ekofisk is estimated to produce to 2030 or beyond. The platforms will then be 50 years old. Economic lifetime will come when the cost of repairing the platforms and/or processing equipment will exceed the net present value of additional hydrocarbons that could otherwise be recovered.

# CONCLUSIONS

(1) When the current production license expires in 2011, 24% of the 8.4 billion boe originally in place in Ekofisk will have been produced by primary drive mechanisms—mainly solution gas drive and reservoir compaction.

(2) The 2/4 K Project, Tor Formation Waterflood, will enhance recovery by 1.9% or 160 mmboe.

(3) The Waterflood Expansion Project, expansion to the southern Tor and to the Lower Ekofisk, will enhance recovery by another 2.3% or 190 mmboe.

(4) The Upper Ekofisk Nitrogen Injection should further enhance recovery by 2.0% or 165 mmboe.

(5) Additional enhanced recovery potential by means of increased water and/or nitrogen injection capacity is currently being evaluated. This could result in additional recoveries up to approximately 200 mmboe.

(6) Further enhancement would require significant increase in the price of oil.

# ACKNOWLEDGEMENTS

The authors thank Phillips Petroleum Company and Ekofisk partners, including Fina Exploration Norway, Inc., Norsk Agip A/S, Elf Aquitaine Norge A/S, Norsk Hydro a.s., Total Marine Norsk A/S and Den Norske Stats Oljeselskap a.s. for permission to publish this paper. Special thanks to J.E. Sylte for valuable technical contribution to this paper and to Anne Mette Rimestad for typing this and several other papers published by the Reservoir Engineering Department of Phillips Petroleum Company Norway over the past few years.

## REFERENCES

Boyce, B.M. (1972). Ekofisk Phase 1 production looks good. *Oil Gas J.* (June 12), 87–88.

Gunlikson, R.D. and Wier, D.R. (1985). Laboratory studies on water imbibition in Ekofisk. *North Sea Chalk Symposium—Chalk Research Monograms*, Book III, paper no. 5.

Johnson, J.P. and Rhett, D.W. (1986). Compaction behaviour of Ekofisk chalk as a function of applied stress. *European Petroleum Conference Proceedings*, 1986, pp. 221–231.

Johnson, J.P., Rhett, D.W. and Siemers, W.T. (1988). Rock mechanics of the Ekofisk reservoir in the evaluation of subsidence. *Proceedings of the Offshore Technology Conference, Houston, 1988*, pp. 39–50.

Smits, R.M.M., de Waal, J.A., and van Kooten, J.F.C. (1986). Prediction of abrupt reservoir compaction and surface subsidence due to pore collapse in carbonates. *Proceedings, SPE Annual Meeting*, 1986.

Sulak, R.M. and Danielsen, J. (1988). Reservoir aspects of Ekofisk subsidence. *Proceedings of the Offshore Technology Conference, Houston, 1988*, pp. 9–21.

Sylte, J.E., Hallenbeck, L.D., and Thomas, L.K. (1988). Ekofisk formation pilot waterflood. *Proceedings, SPE Annual Meeting Reservoir Engineering, 1988*, pp. 153–168.

Thomas, L.K., Dixon, T.N., Evans, C.E. and Vienot, M.E. (1987). Ekofisk waterflood pilot. *J. Petroleum Technol.* (February), 221–232.

Van den Bark, E. and Thomas, O.D. (1980). Ekofisk. First of the giant oil fields in Western Europe. *Am. Assoc. Petroleum Geol. Bull.* **65**, 2341–2362.

Weseloh, C.J. (1985). Improved accuracy in capillary pressure measurements. *North Sea Chalk Symposium—Chalk Research Monograms*, Book III, paper no. 6.

## Keynote Address 3

# Pseudofunctions – The Key to Practical Use of Reservoir Description

*L.W. Lake, E. Kasap and M. Shook*

Department of Petroleum Engineering, University of Texas at Austin, USA

Any transport property that does not have a molecular analogue must be averaged (*pseudoized*) or scale-adjusted so that it can be used in making reservoir performance predictions. In modelling flow through permeable media, typical properties so treated are absolute permeability, relative permeability, capillary pressure, dispersion coefficients, fingering factors, and equilibrium constants. These properties are rendered scale-dependent by the average. We review and justify the philosophy behind pseudofunction generation based on the notion of *vertical equilibrium*, a necessary condition for time-independent pseudofunctions. After discussing the traditional approaches to scale adjustment, we give some interesting special cases and illustrate areas of current research. The central theme, however, is that pseudofunctions represent the only practical method of combining fluid and transport properties with spatially varying reservoir properties.

## INTRODUCTION

Pseudofunctions are transport relations or functions that describe flow through permeable media on a scale that is larger than that which can be measured directly. Their primary purpose is to relate fluid flow rate (flux) to an imposed potential difference on a practical scale. Virtually all permeable media properties are subject to being "pseudoized": absolute permeability, capillary pressure, relative permeability, dispersion coefficients, fingering factors and equilibrium constants. Permeability, capillary pressure, relative permeability and dispersivity are *explicit* pseudos because they appear in the original local flow equations; fingering factors and equilibrium constants are *implicit* because they do not.

Pseudo properties do exist outside the field of flow through permeable media, mainly in turbulent mass and energy transfer. Scale dependency exists here also, but it is less pronounced than in reservoir engineering because of the tremendous difference between local and large (field) scales (factors of $10^8$ in some cases). We are all aware of the difficulties that the scale dependence brings: difficulty in deriving pseudofunctions analytically, obscuring and overlapping of mechanisms, and an inability to validate directly the properties.

However, pseudofunctions can actually contain quite significant advantages because it is through them that we can insert properly weighted and scale-adjusted transport properties into reservoir calculations. They make it possible to lump a great deal of information (local absolute and relative permeabilities, capillary pressures, viscosities, densities, dip and heterogeneity) in a form that can be used easily. In any event, *all* permeable media properties are in some sense pseudos because all measurements deal with a finite volume of rock. In this sense, then, the issue of pseudoizing is the same as the age-old issue of adjusting measurements on one scale to accommodate an application on another.

Basically, there are two ways to generate pseudofunctions: through averaging a refined simulation and through analytic means. The first procedure is extremely common in practice. One simply makes several detailed simulation runs, usually of a cross-section or even of a pattern, and then captures the flux–potential gradient relationship from these runs in permeability, relative permeability functions and/or dispersion functions for use on a larger scale. Examples of this procedure are in the works of Kortekaas (1983), Lasseter *et al.* (1986) and Arya *et al.* (1988). Analytic pseudofunctions can be derived by a variety of methods, many of which we discuss here.

Whatever the procedure, all pseudofunctions must be independent of time for practical use, although they can depend on position and must depend on the averaging scale.

## VERTICAL EQUILIBRIUM

A very useful procedure for generating pseudofunctions is to invoke the assumption of vertical equilibrium (VE)

*North Sea Oil and Gas Reservoirs – II*

across the cross-section of the reservoir in which the displacement is taking place. When VE applies, it is possible to combine vertical and displacement sweep efficiencies into a pseudodisplacement sweep, which may then be estimated by a lower dimensional procedure. This gain is not without its price, however, which we attempt to spell out in this paper.

Formally, vertical equilibrium is a condition in which the sum of all the fluid flow driving forces in the direction perpendicular to the direction of bulk fluid flow is zero. For dipping reservoirs, the vertical and transverse directions do not coincide; therefore, the condition discussed here should really be "transverse" equilibrium. The terminology "vertical" is well imbedded in the petroleum literature, however, and we retain it for consistency. A consequence of the VE assumption is that it represents a state of maximum transverse fluid movement or *cross-flow*. Calculations based on VE, therefore are of use in estimating the tendency of cross-flow to affect displacements when compared to the no-cross-flow calculations. More importantly, however, VE must hold in order for the pseudofunctions to be independent of time and position (Lake, 1989).

Maximum vertical fluid movement seems highly restrictive because we are continually upsetting the transverse driving forces by injecting and producing through wells. So we must go to some length to see that it is actually fairly common.

Figure 1 shows cross-sections of reservoirs into which a single-phase fluid is being injected only through the interval shown on the left. The contours are non-dimensionalized flow potentials at steady state calculated from an analytical solution (Lake and Zapata, 1987). The boundary on the right is an axis of symmetry where the dimensionless flow potential is held constant at zero. Figure 1(a) clearly shows the effects of the limited injection, as the contours are highly distorted around the inlet. The contours in Fig. 1(b), however, act as though the partial injection is not present, because the potential contours are essentially vertical lines throughout the cross-section. The second panel is clearly in VE because the vertical potential gradient, the only driving force in the problem, is zero nearly everywhere.

The only difference between the pressures in the two panels in Fig. 1 is in the value of the effective length–thickness ratio $R_L$ defined below:

$$R_L=\left(\frac{L}{H}\right)\left(\frac{k_v}{k}\right)^{1/2} \qquad (1a)$$

$L$ can be the injector–producer distance, the length of an average streamline or any other characteristic length that expresses the well-to-well residence time of the fluid, and $k$ and $k_v$ are the horizontal and vertical permeabilities. $H$ is either the net or gross pay depending on how $k_v$ is calculated.

$R_L$ may be regarded as a ratio of a characteristic time for fluid to cross the reservoir in the longitudinal or $x$ direction to that in the transverse or $z$ direction. If $R_L$ is large, saturation, concentration, or pressure fluctuations in the

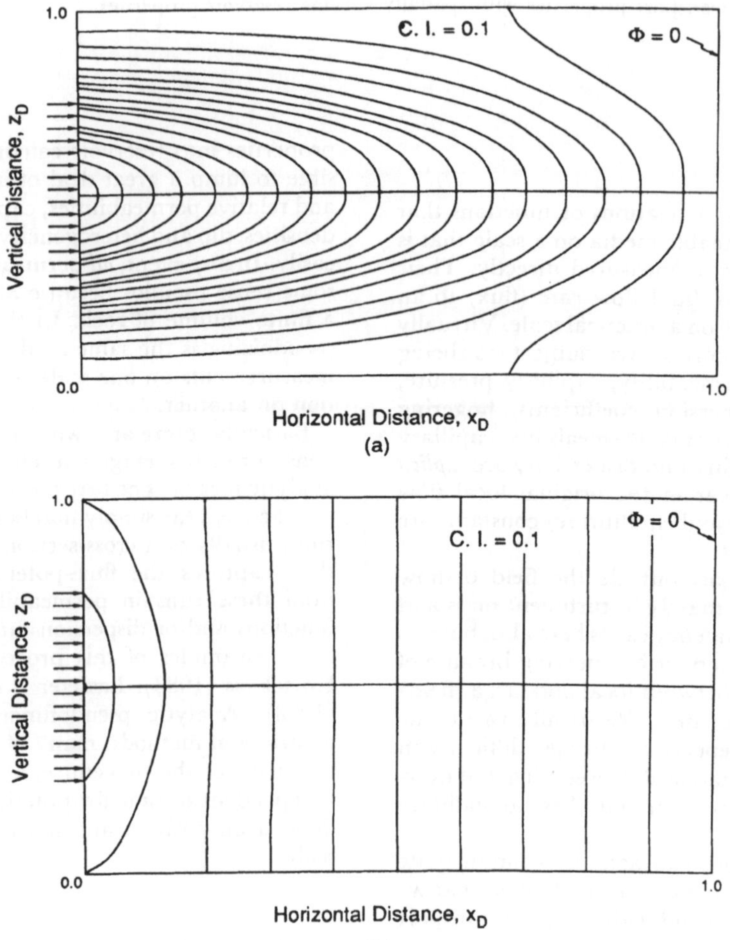


**Fig. 1.** Isopotential contours for two cases of steady-state flow involving partial injection into the right side of the field: (a) $R_L=0.2$; (b) $R_L=2.0$.

$z$-direction decay much faster than those in the $x$ direction. We, therefore, neglect the $z$ direction perturbations as a consequence. Thus, when we say that the VE assumption "applies", or that the subject reservoir is "in vertical equilibrium", we are saying that for the bulk of the reservoir $z$ direction fluctuations are negligible and the $z$ direction gradients are small.

The permeabilities in Eq. (1a) are themselves averages. In the simple case shown in Fig. 1, there is no ambiguity in what they mean. In layered reservoirs, the following formulae do a reasonable, though approximate job:

$$k=\frac{\Sigma k_i h_i}{H}, \qquad \frac{H}{k_v}=\Sigma h_i k_{vi} \tag{1b}$$

where $H=\Sigma h_i$ and all sums run over all transverse layers. By taking the $k_v$ to be a harmonic average over the reservoir interval, it is clear that $k_v=R_L=0$ if there are one or more impermeable barriers (e.g., continuous shale layers) within the interval $H$. Clearly, the VE assumption will not apply in this case. However, the pseudodisplacement sweep efficiency of the intervals between the barriers may be estimated on the basis of VE. The combined response of all such intervals follows from methods for non-communicating layers.

It is not hard to see that $R_L$ is typically large, as shown by the following example. If we take the well spacing to be $L=1000$ m, reservoir thickness to be $H=60$ m, and $k_v=0.1k$, values not unreasonable in many North Sea reservoirs, then we have $R_L=5.3$. This value of $R_L$ is obviously large enough for VE to apply in the second example in Fig. 1 (lower panel). The situation is essentially the same for numerical simulation wherein $L$ and $H$ are grid-block dimensions. Table I gives typical grid-block sizes. Of course we rarely know the permeability ratio, $k_v/k$. But in this case the idea of $R_L$ can be used to find the ratio required for VE to apply.

Vertical equilibrium also applies to flow in anisotropic media, but now it depends on boundary conditions and the nature of the local permeability tensor. Figure 2 shows steady-state potential contours in an anisotropic medium, which has its major and minor axes oriented as shown, and into which fluid is being injected over the entire vertical interval. The orientation of the permeability tensor is to simulate cross-bedding in geological strata (Kasap and Lake, 1989). Figure 2(a) has impermeable transverse boundaries and Fig. 2(b) has flow potential changing linearly along the transverse boundaries.

The presence of both the injection and production wells is clearly present in Fig. 2(a), but the effect persists only short distances into the medium. For the bulk of the flow, the isopotential lines are straight and oriented at a constant angle determined by the directions of the major and minor permeability axes. Thus, VE in anisotropic media with no-flow transverse boundary conditions means that the potential gradients are perpendicular to the bedding planes. This conclusion can be radically altered if the transverse boundary condition changes to constant pressure gradient along the boundary as in Fig. 2(b). Now the isopotential lines are straight and vertical, without a hint of the cross-bedding, but the velocity vectors are oriented along the bedding plane. The vertical isopotential lines (a type of enforced VE) are of great assistance in calculating pseudopermeabilities.

The observation about $R_L$ determining when VE applies evidently holds for flow fields in layered media (Zapata and Lake, 1981) and those having randomly heterogeneous media (Arya *et al.*, 1988). The specific value of $R_L$ required for VE to apply is different for these cases, as would be true for processes other than waterflooding. Other fluid driving processes, such as dispersion and capillary pressure, might require different dimensionless groups, as the following example shows.

Figure 3 shows cross-sections of reservoirs into which a fluid is being injected from the left. This fluid is single-phase, and has constant viscosity and density. Although injection occurs along the entire cross-section, a tracer of unit concentration is also being injected in a subinterval as shown on the left. The contours drawn are steady-state isoconcentration lines that were obtained from an analytical solution of the two-dimensional convection-diffusion equation (Shook, 1990).

The driving "force" here is the transverse concentration gradient caused by the non-uniform injection condition. The only difference between Figs. 3(a) and 3(b) is the magnitude of the transverse dispersion number $\mathcal{N}_{TD}$

$$\mathcal{N}_{TD}=\frac{L}{H^2}\alpha_t \tag{2}$$

where $\alpha_t$ is the transverse dispersivity.

In Fig. 3(a) the concentration contours exhibit substantial effects of the partial injection. In Fig. 3(b), on the other hand, the transverse concentration gradient is quickly damped, and the effect of the partial injection dies out a short distance into the reservoir. This is another example of VE, since the transverse driving force is zero nearly everywhere.

It is tempting to imagine that there is a connection between the VE indicators $R_L$ and $\mathcal{N}_{TD}$. $R_L$ indicates VE with respect to viscous forces and $\mathcal{N}_{TD}$ with respect to dispersion. That they are different arises, in part, because of the differences in the physical processes they represent; however, it is unlikely that a reservoir with a zero vertical permeability would have a non-zero transverse dispersivity. Locally, permeability is related to the mean of the grain size distribution and dispersivity to the variance. Because grain size is always non-zero it follows that the variance must approach zero, as does the mean, so that $R_L$ should be zero when $\mathcal{N}_{TD}$ is also. However, there is no definitive relationship between vertical permeability and transverse dispersivity.

When VE applies, the pseudorepresentation is time independent. The advantages brought by this feature accounts for our dwelling on it here. However, when VE does not apply, the material properties may change so slowly with time that a pseudorepresentation is still possible, at least approximately. A good example of this is in immiscible displacements in non-communicating, stratified media (Waggoner, 1985). Here, averaged fractional flow curves changed so slowly that a single representation was possible. Other material properties, such as dispersivities, evidently change much more dramatically (Arya *et al.*, 1988).

**Table I.** Typical block sizes in simulation[a]

| *Process* | *L/H* | |
|---|---|---|
| | *Minimum* | *Maximum* |
| Undersaturated oil reservoirs | 53 | 71 |
| Gas reservoir | 6.7 | 21 |
| Well test | 4 | 110 |
| Retrograde gas | 3.9 | 14 |
| Solution gas drive | 33 | 200 |

[a] From Haldorsen (1983).

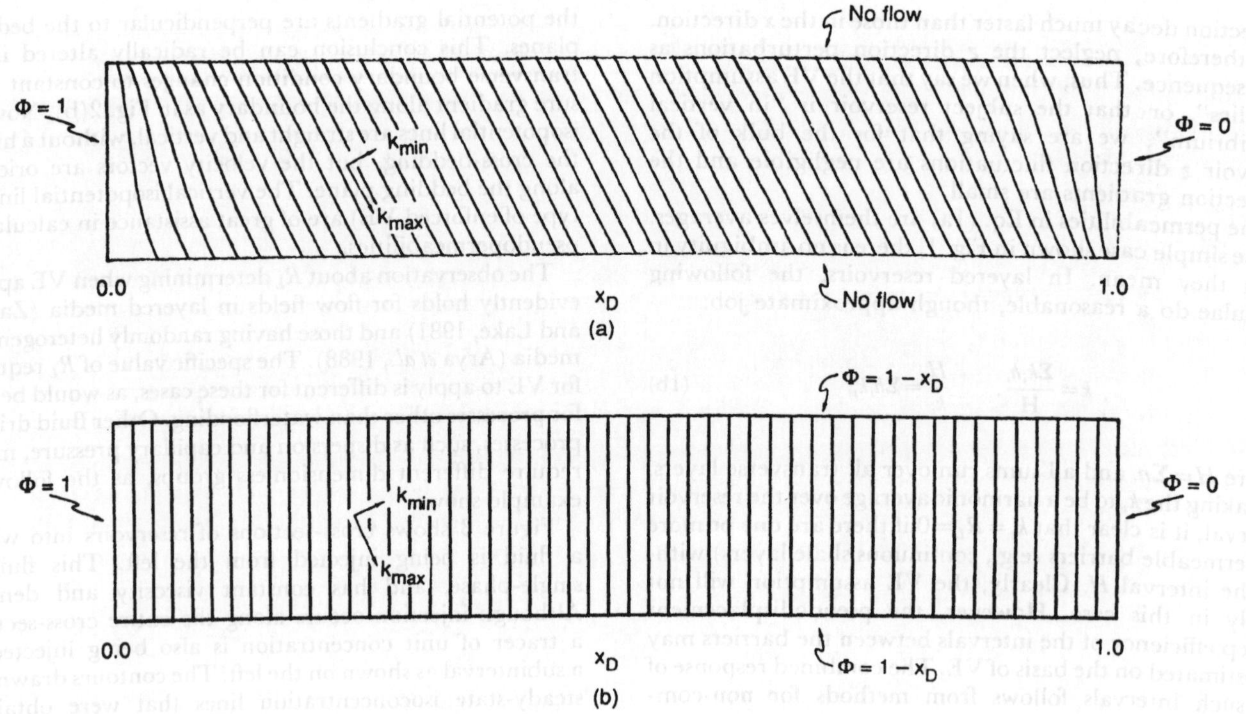


**Fig. 2.** **Isopotential contours for two cases of steady-state flow in an anisotropic permeable media: (a) no flow transverse boundaries; (b) linear potential change on boundary. (From Kasap, 1990.)**

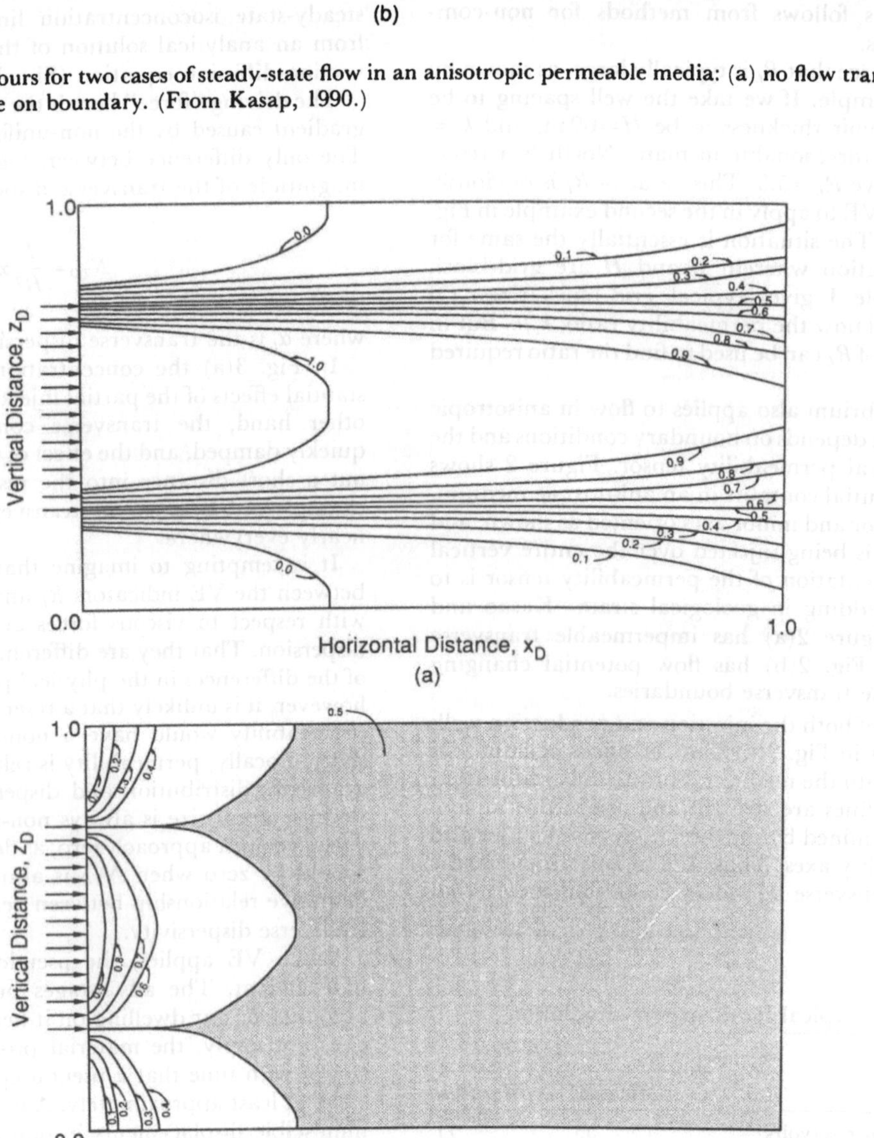


**Fig. 3.** **Isoconcentration contours for two cases of a constant viscosity and density miscible steady-state cross-section: (a) $N_{TD} = 0.01$; (b) $N_{TD} = 0.5$. Shaded area has constant concentration. (From Shook, 1990.)**

## PSEUDOPERMEABILITY

Permeability has been one of the two most studied pseudoproperties, probably because it can be analysed in single-phase, steady-state flow. Vertical equilibrium for this case implies that the flow potential is constant in the transverse direction. For incompressible flow in a reservoir of constant dip angle $\alpha$, this is

$$\Phi = P + \rho g \cos(\alpha) = f(x) \text{ only} \tag{3}$$

In this equation, $P$ and $\rho$ are the fluid pressure and density, respectively, and $g$ is the gravitation acceleration constant.

There are two limiting cases of pseudoized permeability applicable to layered media. If the layers are parallel to the flow, the thickness-weighted arithmetic average applies:

$$\bar{k} = \frac{\Sigma h_i k_i}{H} \tag{4a}$$

If the layers are perpendicular to the flow, the thickness-weighted harmonic average applies:

$$\bar{k} = \frac{H}{\Sigma h_i / k_i} \tag{4b}$$

These two well-known formulae are rarely thought of as pseudorepresentations, but such is obviously true, because their derivation requires the transverse potential gradients to be zero in both cases, as required by Eq. (3).

Unfortunately, the layers in real reservoirs, if they exist, are not homogeneous and isotropic as required by Eqs. (4). For anisotropic layers, the pseudopermeability will be a tensor whose components depend on the local anisotropy and on the heterogeneity of the layers. Figure 4 shows that VE may be imposed (through the boundary condition in Fig. 2(b)) on parallel and serial isotropic layer configurations so that the components of the pseudopermeability tensor may be calculated (Kasap and Lake, 1989). These are

$$\tilde{k}_{xy} = \frac{h k_{xy2} k_{xx1} + k_{xy1} k_{xx2} (L-h)}{h k_{xx1} + (L-h) k_{xx2}}$$

$$\tilde{k}_{yy} = \frac{1}{L}\left((L-h) k_{yy1} + h k_{yy2} + \frac{L-h}{k_{xx1}}[(\tilde{k}_{xy} - k_{xy1})(k_{xy1} - k_{xy2})]\right) \tag{5}$$

$$\tilde{k}_{xx} = \frac{L k_{xx1} k_{xx2}}{(L-h) k_{xx2} + h k_{xx1}}$$

where the subscripts 1 and 2 refer to the local permeability tensor components and the lengths, $L$ and $h$, are defined in Fig. 4. These formulae collapse to the appropriate limits of Eq. (4) upon the imposition of local permeability isotropy ($k_{xy1} = k_{xy2} = 0$; $k_{x1} = k_{y1} = k_1$; $k_{x2} = k_{y2} = k_2$). Kasap and Lake (1989) extended the results in Eq. (5) to account for partial layering and acute-angle orientations by approximations. More general treatments of local anisotropy and layering follow from the volume-averaging approach of Quintard and Whitaker (1988), but the VE assumption is still required. Effective medium theory (King, 1987) allows for rather complex arrangements of permeability, but the permeability must be locally isotropic.

Both numerical and analytical methods will apply to stochastically generated permeability fields. The most common heterogeneity investigated in this manner has been that of impermeable shales imbedded in a homogeneous flow field. The effective or averaged permeability of individual realizations can now be determined by numerical methods (Haldorsen, 1983), analytical methods (Begg and King, 1985) and subsequently applied to transient flow behaviour (Haldorsen *et al.*, 1987). The shales themselves may be randomly situated or correlated

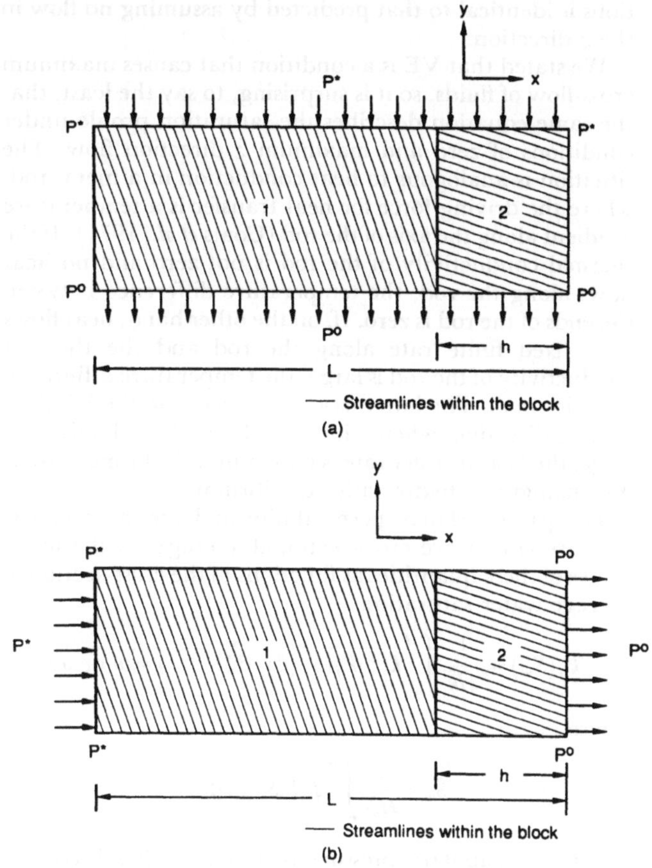


**Fig. 4.** Schematics of parallel and serial flow in two-layered media consisting of anisotropic layers: (a) parallel flow; (b) serial flow. (From Kasap and Lake (1989). Reproduced by permission of SPE-AIME; from SPE paper no. 18434: *An Analytical Method to Calculate the Effective Permeability Tensor of a Grid Block and its Application in an Outcrop Study.*)

(Desbarats, 1987) and the pseudopermeability can have a tensorial character (White and Horne, 1987). An analytical method that is well suited to stochastic fields is the renormalization group method (King, 1988). In this method, averaging takes place over successively larger fields until the variance of the derived permeability approaches zero, at which point the mean value of the distribution becomes the pseudopermeability.

## PSEUDORELATIVE PERMEABILITY AND CAPILLARY PRESSURE

Relative permeability is no more difficult to average than absolute permeability. However, pseudorelative permeability depends on an average phase saturation, which cannot be calculated analytically without assuming VE.

For a two-phase waterflood with constant phase compositions, the incompressible VE condition reduces to

$$P_c - P_c|_{z=0} = \Delta\rho g z \cos(\alpha) \tag{6}$$

where $\Delta\rho = \rho_w - \rho_o$ is the density difference between the immiscible phases water ($w$) and oil ($o$), and we have used the definition for oil–water capillary pressure $P_c = P_o - P_w$ in this equation. Equation (6) describes implicitly the water saturation profile in the $z$ direction since $P_c$ is a known function of water saturation. But this saturation distribution is just what would be observed in the transition zone between oil and water under static conditions. Hence, the $z$ direction saturation profile under VE conditions is identical to that predicted by assuming no flow in the $z$ direction.

We stated that VE is a condition that causes maximum cross-flow of fluids, so it is surprising, to say the least, that the same equation describes the saturation profile under conditions of zero and maximum $z$ direction flow. The situation is analogous to heat conduction in a metal rod, where the driving force for heat transfer is a temperature gradient along the axis of the rod (Coats *et al.*, 1971). If the thermal conductivity of the rod is not zero and no heat flows along the rod, the temperature difference between the ends of the rod is zero. If, on the other hand, heat flows at a fixed finite rate along the rod and the thermal conductivity of the rod is large, the temperature difference is again small. The latter case is analogous to the VE flow in the $z$ direction, where, since the thermal conductivity is large, the heat transfer rate is maximum; the former case is the analogue of hydrostatic equilibrium.

The pseudorelative permeability and pseudocapillary pressure curves are cross-sectional averages of the local values of these quantities as functions of the cross-sectional average water saturation.

$$\tilde{k}_{rj}(\bar{S}_w) = \frac{1}{H\bar{k}} \int_0^H k(\xi) k_{rj}(S_w(\xi))\,d\xi, \qquad j = w \text{ or } o \quad (7a)$$

where

$$\bar{S}_w = \frac{1}{H\bar{\phi}} \int_0^H \phi(\xi) S_w(\xi)\,d\xi \quad (7b)$$

The pseudocapillary pressure is somewhat less involved because of the VE assumption.

$$\tilde{P}_c = P_c(\bar{S}_w) \quad (7c)$$

The $z$ direction saturation profile gives us all the information we need to determine these quantities according to Eqs. (7) (see Lake (1989) for a derivation of these equations).

In a sense, Eqs. (6) and (7) are the basic tools of reservoir characterization. The pseudo curves, $\tilde{k}_{rw}$, $\tilde{k}_{ro}$ and $\tilde{P}_c$ versus $\bar{S}_w$, may be used directly in numerical simulation. They embody nearly all of what we believe to be important about flow through permeable media. The pore size distribution, wettability and fluid properties are present through the local properties $k_{rj}$, $P_c$ and the fluid density. The dimensions of the medium are present through $H$ and dip angle $\alpha$, and the geological heterogeneity is present through the variation of permeability and porosity with depth. Most importantly, these quantities are entered into the pseudo properties with the appropriate weight for the flow under consideration.

Figure 5(a) gives some calculated pseudorelative permeability curves from Dake (1978), who has an excellent discussion on the topic for immiscible two-phase displacements. This figure shows the extremes of diffuse flow, an essentially constant saturation in the vertical direction, and of segregated flow. Real cases usually fall somewhere in between these limits.

Figure 5(b), also from Dake (1978), shows the effect of layer ordering. These are two curves from a four-layer reservoir with the highest permeability at either the top or the base. Because the gravity effect through Eq. (6) is always downwards, there is a pronounced effect of the ordering of the layers. The displacement is much less efficient when the high permeability layer is on the bottom because gravity accentuates the water channelling in this case. Note also that pseudorelative permeability curves can take on the ordinarily non-physical shape of being concave downwards.

## SPECIAL CASES OF VERTICAL EQUILIBRIUM IN IMMISCIBLE DISPLACEMENTS

Though the VE procedure is quite general, being restricted only to reservoirs having constant properties in the $x$ direction and having a large $R_L$ (or its equivalent), there are several VE flows that are special cases of Eqs. (6) and (7). We review these special cases in this section and show how they follow from the general theory.

### Gravity tonguing

Easily the most celebrated of the VE theories is the theory of gravity tonguing or under-running originally proposed by Dietz (1954). This theory was first proposed as an alternative to the Buckley–Leverett theory for waterflooding, but it is actually a special case of the VE theory because a finite time is required for the conditions underlying the theory to apply. Since the publication of the original Dietz paper, variations of the theory have been applied to miscible gas process, steam flooding and *in-situ* combustion. We restrict ourselves to the case of water displacing oil, though the over-running case can be developed in quite a similar manner.

The key assumption in the Dietz theory is the absence of a transition zone, or *segregated* flow. This condition can only be true for conditions where the capillary pressure is small (well-sorted and/or high permeability media) or the reservoir thickness is very large compared to the capillary transition zone. The sharp transition zone or macroscopic interface resulting from this condition suggest that the theory is applicable to any displacement, miscible or 'immiscible, where simultaneous flow of more than one component or phase is absent at any point in the reservoir. Equation (6) cannot be satisfied at any point in the reservoir, since the oil and water densities are not, in general, equal. The resolution of this contradiction is to allow the water hydrostatic equation to apply in regions flowing only water, and the analogous equation for oil to apply to zones flowing only oil. The pseudorelative permeabilities for this case are linear functions of the average water saturation.

### Viscous cross-flow

Next we consider a uniformly layered medium with no gravity or capillary pressure. For this case the permeable medium consists of $N_L$ parallel, isotropic layers each of contrasting thickness, permeability, and porosity. Despite the simplicity of these restrictions, this case leads to some remarkable insights about flow and mixing in viscously dominated flows.

Since neither gravity nor capillary pressure is present, Eq. (6) is trivially satisfied, and there is no constraint on the saturations in the $z$ direction. To resolve this, Hearn (1971) assumed segregated flow within a layer. The relative permeabilities in the averaging formulae become merely the end-point values

$$\begin{aligned} \tilde{k}_{rw} &= \frac{1}{H\bar{k}} \sum_{i=n+1}^{N_L} (khk^0_{rw})_i \\ \tilde{k}_{ro} &= \frac{1}{H\bar{k}} \sum_{i=1}^{n} (khk^0_{ro})_i \\ \bar{S}_w &= \frac{1}{\bar{\phi}H} \left\{ \sum_{i=1}^{n} [\phi h(1-S_{or})]_i + \sum_{i=n+1}^{N_L} (\phi h S_{wr})_i \right\} \end{aligned} \quad (8)$$

where $n$ is the number of the slowest layer (smallest $k/\phi$) flowing water at a given cross-section. The average water

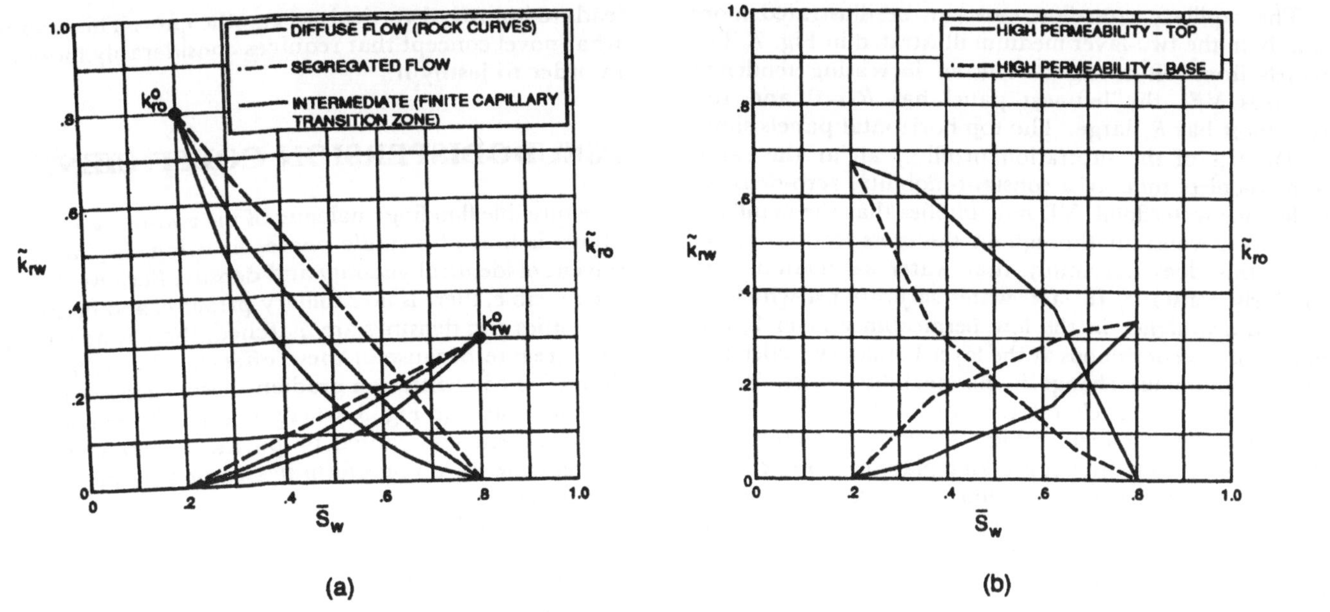


**Fig. 5.** Pseudorelative permeability curves which show (a) the effect of diffuse versus segregated flow, and (b) layer order. (Reproduced by permission of Elsevier Science Publishers; from Dake, L.P. (1978), *Fundamentals of Reservoir Engineering*, Elsevier.)

saturation and pseudorelative permeabilities are, therefore, functions of $n$ parametrically, and can be regarded as functions of each other in this way.

Assuming that segregated flow is correct if the end-point mobility ratio ($M^o = \lambda^o_{rw}/\lambda^o_{ro}$) of the displacement is favourable ($M^o < 1$), the directions of this *viscous* cross-flow are such that only a pure fluid (that is, either oil or water) flows from their respective layers. This flow causes the front positions to come closer together with a net increase in vertical sweep efficiency (Fig. 6(a)). Thus, we see that lowering the mobility ratio increases oil recovery by improving fluid distribution at the injection face (as would occur in the no-cross-flow case), and through viscous cross-flow. In fact, if VE holds it is possible for $M^o$ to be so low that the effect of the heterogeneities is entirely suppressed (Zapata and Lake, 1981).

When the displacement is unfavourable ($M^o > 1$), however, the viscous forces cause a mixing zone to develop between the front in the fastest layer and that in the slowest layer. This mixing zone causes the vertical sweep efficiency actually to be greater than the corresponding no-cross-flow case, since the mixing zone attenuates the effect of the unfavourable mobility ratio (Fig. 6(b)). That longitudinally non-segregated flow can occur in VE displacements in the absence of capillary pressure is a major revelation in the understanding of such processes. Such flow has implications in the study of viscous fingering and the assignment of pseudofingering factors.

## Capillary equilibrium

When capillary forces are restored to the above problem and transverse viscous forces are removed by setting the mobility ratio to 1, the resulting cross-flow is now capillary cross-flow. There are no gravity forces to counteract the $z$ direction imbibition, and the $z$ direction water saturation profile is uniform within each layer. Because of the variable properties in the $z$ direction, however, the $P_c - \bar{S}_w$ function changes. The capillary pressure (not the capillary pressure function, which scales with the square root of permeability) is a constant through any cross-section.

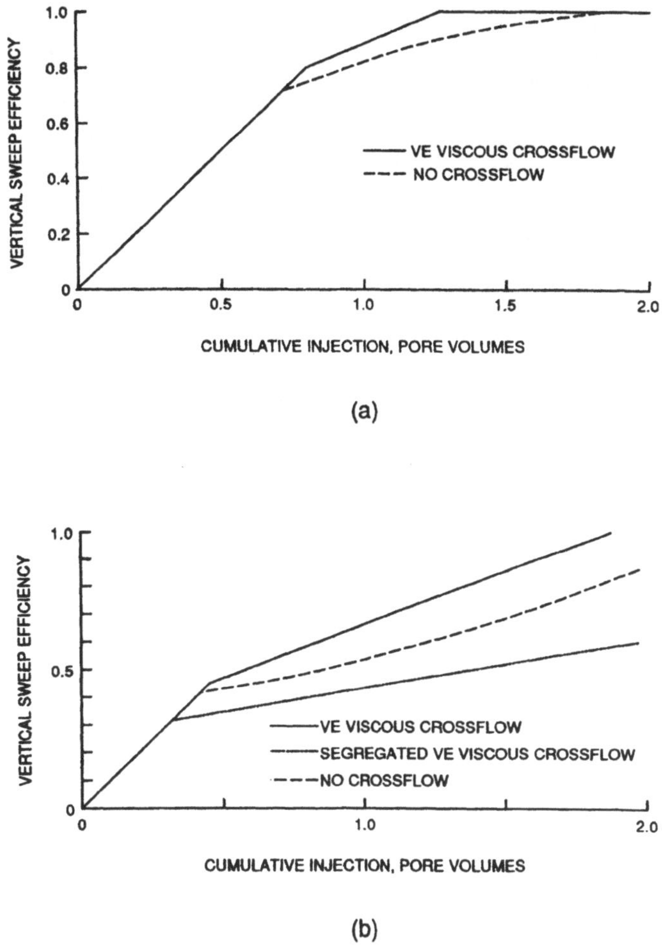


**Fig. 6.** Vertical sweep efficiencies for viscously-dominated, piston-like displacements in two-layered permeable media: (a) Favourable mobility ratio ($M^o < 1$); (b) unfavourable mobility ratio ($M^o > 1$). (From Zapata and Lake, 1981. Reproduced by permission of SPE-AIME; from paper number 10111: *A Theoretical Analysis of Viscous Crossflow*.)

The capillary cross-flow case can be illustrated more simply in the two-layer medium illustrated in Fig. 7. The panels from left to right show an increasing tendency towards VE; the leftmost panel has $R_L=0$ and the rightmost has $R_L$ large. The top horizontal panels show schematics of the saturation profiles, all at the same dimensionless time, of a constant-mobility, zero-density-difference waterflood. VE now implies that the capillary pressure is constant throughout the cross-section. However, this does not imply that water saturations are similarly constant. In fact, water saturations, which in general are higher in the low permeability layer 2, are inevitably discontinuous at the layer boundary, reflecting the discontinuous change in permeability there. When $R_L=0$, the two-layer character of the flood is evident in the water cut (lower panels of Fig. 7); when $R_L$ is large, the flood behaves as though it is taking place in a single-layer reservoir that has a larger capillary pressure range (larger pore size distribution). The capillary pressure function needed to model the behaviour in the large $R_L$ case is the pseudocapillary pressure.

Additional complexity is added when capillary pressure is added to a displacement whose mobility ratio is not unity. In this situation, both viscous cross-flow and capillary pressure-induced cross-flow occur. Preliminary results indicate that, for favourable mobility ratios, a small mixing zone develops because of pressure differences in the oil bank in the "slow" layer and the brine in the "fast" layer. Since the viscous cross-flow is such that the displacement fronts move closer together, this mixing zone remains small and does not appear greatly to influence the displacement.

For an unfavourable mobility ratio, the cross-flow because of capillary pressure $P_c$ influences the displacement much more. In this case, viscous cross-flow causes a mixing zone to develop between the displacement front in the fastest and slowest layers. When capillary pressure is introduced, however, the subsequent cross-flow is such that the mixing zone diminishes substantially. This suggests that, for the cases studied thus far, the capillary pressure cross-flow dominates the viscous cross-flow, and leads to improved vertical sweep efficiency. This is a somewhat novel concept that requires considerably more work in order to justify it.

## PSEUDODISPERSION COEFFICIENT

The miscible flooding analogue of the two-phase case given above is that of a single-phase oil being displaced by a solvent of identical viscosity and density. Because the fluids are miscible, there is no capillary pressure, and because the viscosities and densities are matched, there is no tendency to segregate because of these effects. The only driving "force" is concentration gradients. We encountered such a displacement in Fig. 3, where a non-uniform concentration gradient was caused by partial injection. Here we concentrate on non-uniformities caused by permeability variations.

The general theory for a miscible displacement in a strictly layered medium has been worked out by Van den Broeck and Mazo (1983) and Gelhar *et al.*, (1979). We will reproduce a qualitative description restricted to media with two isotropic layers (Lake and Hirasaki, 1981).

Let a two-layer non-communicating ($N_{TD}=0$) reservoir be as shown in Fig. 8. The left panel shows the displacement (and its mixing) well confined to the individual layers. The effluent concentration shows definite evidence of the two layers with intralayer mixing superimposed on the much larger two-layer response. If $N_{TD}$ is an intermediate value, the isoconcentration lines extend entirely across the reservoir (left panel of Fig. 8) and the breakthrough curve (right panel) shows only slight evidence of the layering. If $N_{TD}$ is large enough for the displacement to be in VE, we have the case shown in the bottom panel. Now the isoconcentration lines are perfectly straight in the $z$ direction as required by VE (see Fig. 3). The effluent is exactly the same as that of a single-layer system except that the mixing is substantially larger. We could easily represent the long-term response of this system with a pseudodispersion coefficient. Lake and Hirasaki (1981) show how this idea can be extended approximately to multiple layers

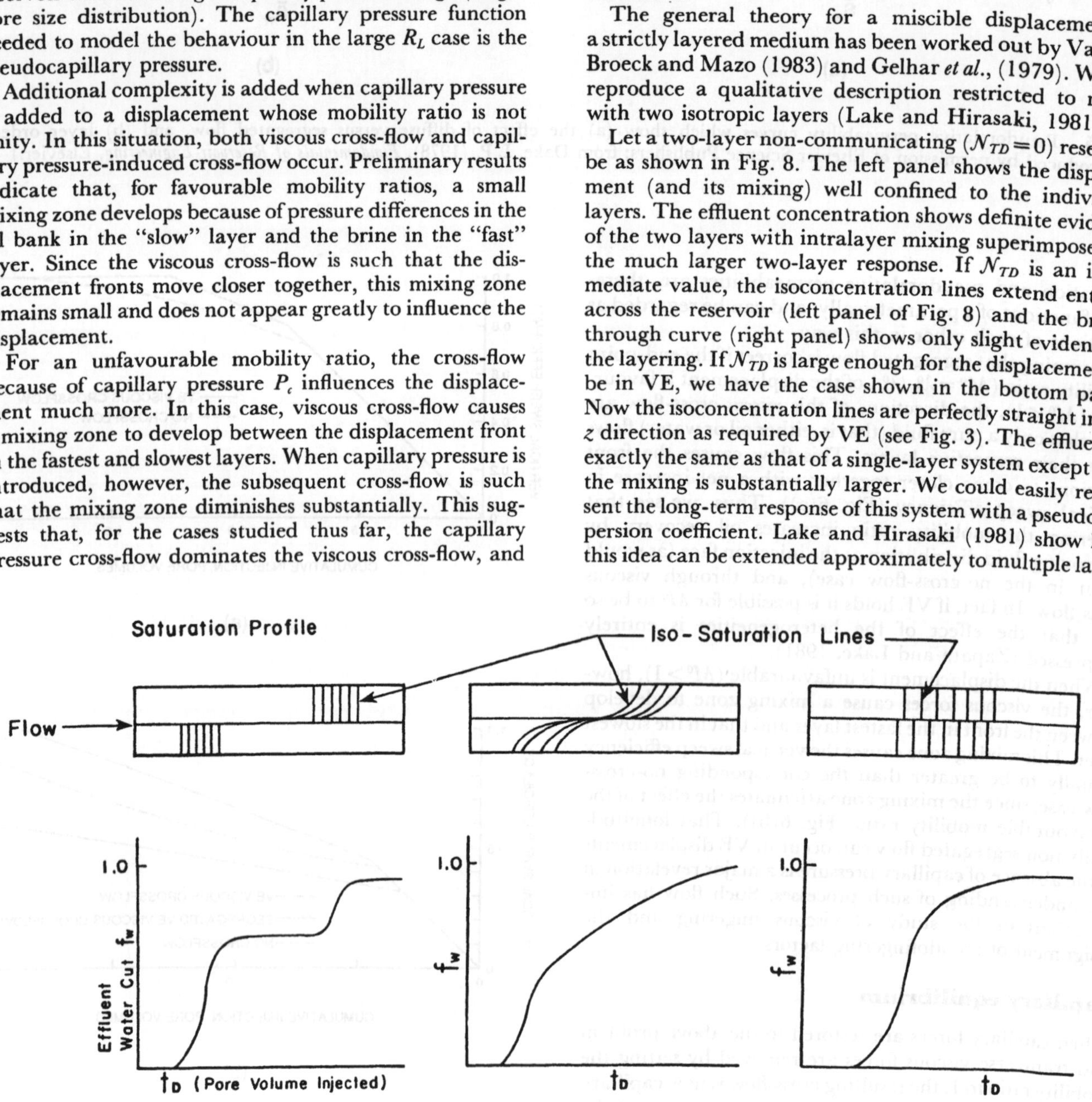


**Fig. 7.** Schematic isosaturation profiles and effluent water fractional flow for a waterflood in various stages of vertical equilibrium. (From Yokoyama, 1981.)

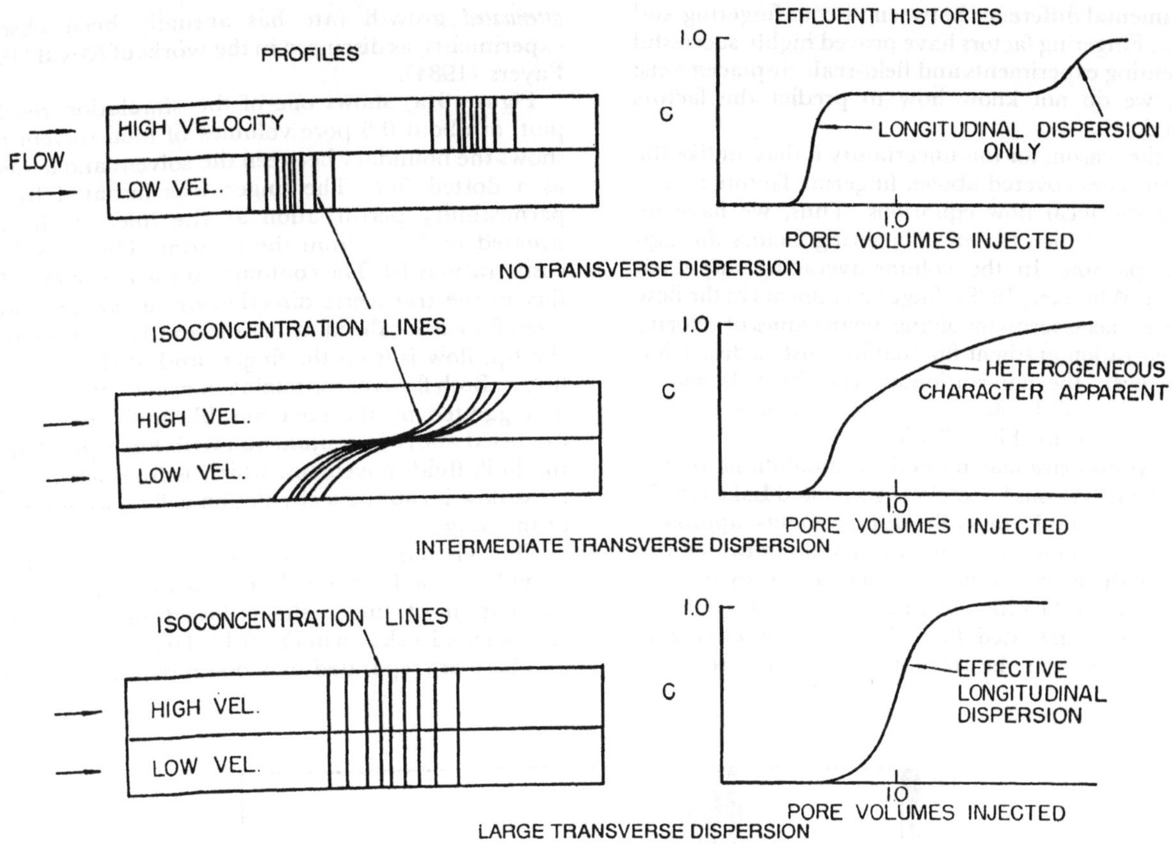


**Fig. 8.** Schematic isoconcentration profiles and effluent concentration for an equal density and viscosity miscible displacement in various stages of vertical equilibrium. (From Lake, L.W. and Hirasaki (1981). Reproduced by permission of SPE-AIME; Taylor's Dispersion in Stratified Porous Media, *Soc. Petroleum Eng. J.*, **21**, 459–468.)

in which extension partially explains the increase in apparent dispersion with scale observed in field displacements.

By far the most intensive effort on pseudodispersion coefficients, however, has been directed to flows through stochastically-generated fields. Some typical references in this regard are Gelhar and Axness (1983), Dagan (1988), and Arya *et al.* (1988). Part of the emphasis on stochastic dispersion coefficients is the tacit acknowledgement that it is impossible to describe local velocity fields deterministically because the local permeability field is unknown. Therefore, we find dispersion coefficients developed for a wide variety of stochastic permeability fields ranging from totally uncorrelated (Arya *et al.*, 1988) to correlated on all scales (Wheatcraft and Tyler, 1988). A second reason for the emphasis on stochastic investigations is that the molecular analogue of dispersion – diffusion – is the consequence of statistically uncorrelated fluctuations about a mean position. But the fluctuations in local velocity during flow through permeable media are not in general uncorrelated; hence, the analogue is probably closer to mixing in turbulent flow rather than to diffusion (Hewett, 1986). Indeed, the presence of correlation generally renders the pseudodispersion coefficient time dependent, a feature which limits the quantity's utility and indicates that the physical premise—dispersive flux being proportional to concentration gradient—is questionable.

Probably the most physically evident exposition of the consequences of stochastically derived dispersion coefficients lies again in the volume-averaging approach (Plumb and Whitaker, 1988). This approach says that the dispersion coefficient of a mean concentration in a stochastic field is actually the correlation coefficient between fluctuations in local velocity and concentration gradient. Given the heterogeneous nature of permeable media, the first is relatively easy to visualize, since it can be related to the variations in local permeability—at least in single-phase, constant-mobility flow (Yang, 1990). Fluctuations in local concentration gradient are much harder to deal with because the notion of a true mean concentration is unclear in an inherently unsteady displacement. The best approach may be to take the mean concentration to be given by the solution to the averaged transport equation, but it is not clear that this approach will give the fluctuations properties that will make them statistically benevolent. Plumb and Whitaker (1988) partially solve this problem by invoking VE as in Fig. 8.

## FINGERING FACTORS

The most important barrier to complete oil recovery in miscible or miscible-tending displacements is viscous fingering. This phenomenon occurs when there is a tendency for the displacing agent to move through the medium faster than the fluid being displaced. A good deal is known about when and how viscous fingers form, but relatively little is known about how to represent them by pseudofunctions, particularly in their interrelationship between heterogeneity and dispersion. For a good review of this subject, see Fayers (1987).

One way to represent viscous fingers is to introduce so-called fingering factors into a fractional flow representation. Such representations generally lead to mixing zones that grow in proportion to time, rather than with the square root of time as does dispersion. This distinction is the fundamental difference between viscous fingering and dispersion. Fingering factors have proved highly successful in representing experiments and field-scale displacements; however, we do not know how to predict the factors beforehand.

One of the reasons for this uncertainty is that, unlike the pseudoproperties covered above, fingering factors do not appear in the local flow equations. Thus, we have no analogues with which to define fingering factors through direct comparison. In the volume-averaging approach (Plumb and Whitaker, 1988), fingering appears in the flow equations as excess terms involving mean values of velocity and concentration gradient fluctuations just as dispersion does. However, the non-stationary velocity field makes these terms very difficult to evaluate. VE may provide a way to circumvent this difficulty.

Using two-dimensional numerical simulations in the presence of only viscous forces, Waggoner and Lake (1987) noticed that unstable miscible displacements approach VE. This was evidenced by the linear growth rate of the cross-sectionally averaged mixing zone. The growth rate, however, was significantly less than that which would be expected from segregated flow; hence, local mixing is taking place between the solvent and the oil. Such an *attenuated* growth rate has actually been observed in experiments, as discussed in the works of Koval (1963) and Fayers (1984).

Figure 9(a) shows one of the simulation results. This plot, at about 0.5 pore volumes of total solvent injected, shows the boundary between the solvent and a miscible oil as a dotted line. The finger was initiated by a small permeability perturbation at the inlet of the medium situated half-way from the bottom. The oil–solvent viscosity ratio is 10. The contours on the figure are of solvent flux in the transverse direction or *viscous* cross-flow. The cross-flow is localized at the tip and base of the finger; at the tip, flow is from the finger, and at the base, into the finger. Both flows cause mixing in exactly the same manner as suggested by the unfavourable mobility ratio mixing discussed above in conjunction with Fig. 6(b). Taken with the bulk fluid movement, which is from left to right, the cross-flow is causing a circulation cell to occur on the sides of the finger.

This viscous mixing is the cause of the attenuated growth rate of the finger. To demonstrate this, consider the calculation of finger shape based on the method of characteristics shown in Fig. 9(b). Here the blunt finger tip and base are separated by a no-cross-flow curved bound-

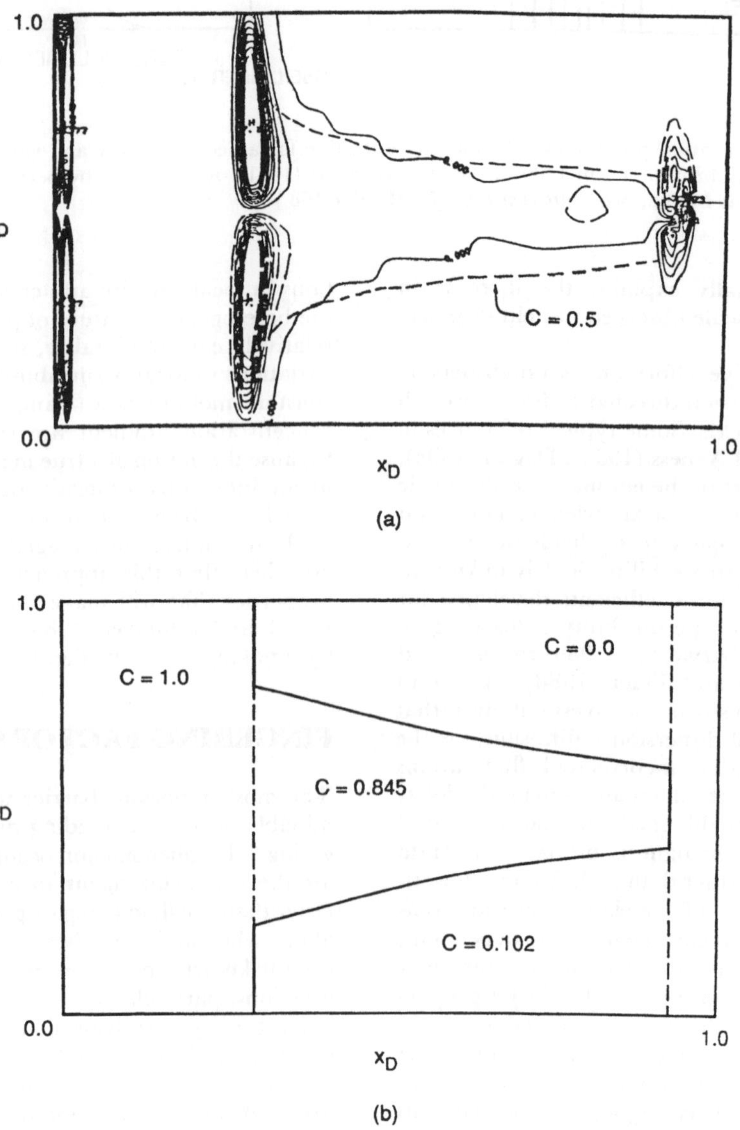


**Fig. 9.** Isocontours of transverse flux around the tip and base of a single viscous finger at $t_D = 0.5$ based on numerical simulation: (a) Under VE, the cross-flow suggests an analytic model that gives the finger shape shown in (b). (From Waggoner *et al.*, in preparation.)

ary (also suggested by simulation) between the solvent and oil. The advantage of solution by the method of characteristics is that it is entirely dispersion-free, which means that mixing so induced *must* come from viscous forces. The concentrations within and outside the finger are shown on the figure; these agree fairly well with what is required to produce the desired growth rate. What we conclude from this calculation is that viscous mixing plays a large role in unstable displacements.

Real displacements, of course, do have local mixing caused by dispersion and diffusion. Transverse dispersion, which itself may play a part in the attenuated finger growth described above, actually suppresses fingering by causing the fingers to damp out before they have grown too much (Heller, 1963). This damping is much like that depicted in Fig. 8, except that the origin of the velocity fluctuations, and the difference in front positions are now mobility inequalities rather than heterogeneities. After the damping has occurred, the process proceeds as a stable miscible displacement, but with a large dispersion coefficient. This effect is undoubtedly predominant in small-diameter experiments as would be present in slim tube experiments. It probably accounts for the apparent stabilization of unstable displacements reported by Young (1986).

## CONCLUDING REMARKS

Much of this paper has involved a discussion of vertical equilibrium (VE) as it applies to developing pseudofunctions for integrating reservoir description and fluid mechanics into reservoir predictions. VE is a tool for developing analytical pseudofunctions, even in rather complicated flows, and it thereby eliminates much of the expense and time of detailed simulations, the alternative to assuming VE.

VE is also a necessary condition for the pseudoproperties to be independent of time. Pseudofunctions must be independent of time in order for them to have practical use. To see this, imagine a reservoir simulation involving injection and production into and from several wells, to the blocks of which we assign time-dependent pseudorepresentations. If $t=t_1$ corresponds to the initiation of injection into the first well, we can easily alter the pseudofunctions of all the reservoir blocks as functions of time $(t-t_1)$. But there are actually several time scales, because other wells can be initiated at $t=t_2$, $t=t_3$, etc., and any block in the reservoir influenced by more than one well has several different time scales $(t=t_1, t=t_2, t=t_3$, etc.) associated with it. Thus, even if the time-dependency of the pseudofunctions were known perfectly, it would still be unclear as to which time to use in calculating the pseudofunctions of most of the blocks.

Using time-invariant pseudofunctions, on the other hand, is not extremely restrictive, particularly in reservoirs with large well spacings. The functions are the focal point of reservoir description because they represent the merging of all the diverse disciplines associated with this highly interesting and difficult endeavour.

## NOMENCLATURE

| | |
|---|---|
| [=] | Means has units of length ($L$), time ($t$), force ($F$) or mass ($m$) |
| $C$ | Concentration |
| $g$ | Gravitational acceleration constant [=] $L/t^2$ |
| $h$ | Layer thickness [=] $L$ |
| $H$ | Characteristic thickness [=] $L$ |
| $k$ | Permeability [=] $L^2$ |
| $k_v$ | Vertical permeability [=] $L^2$ |
| $k_r$ | Relative permeability |
| $L$ | Characteristic length [=] $L$ |
| $M$ | Mobility ratio |
| $n$ | Layer index |
| $N_L$ | Total number of layers ($n=1, \ldots, N_L$) |
| $N_{TD}$ | Transverse dispersion number |
| $P$ | Pressure [=] $F/L^2$ |
| $P_c$ | Capillary pressure [=] $F/L^2$ |
| $R_L$ | Effective length–thickness ratio |
| $S$ | Saturation |
| $t$ | Time [=] $t$ |
| VE | Vertical equilibrium |
| $x$ | Coordinate direction parallel to bulk flow [=] $L$ |
| $z$ | Transverse ("vertical") coordinate [=] $L$ |

**GREEK**

| | |
|---|---|
| $\alpha$ | Dip angle |
| $\alpha_t$ | Transverse dispersivity [=] $L$ |
| $\Delta$ | Change |
| $\phi$ | Porosity |
| $\Phi$ | Fluid potential [=] $F/L^2$ |
| $\lambda_r$ | Relative mobility [=] $F \cdot t/L^2$ |
| $\rho$ | Fluid density [=] $m/L^3$ |

**SUPERSCRIPT**

| | |
|---|---|
| ~ | Pseudoproperty |
| – | Average |
| 0 | end-point |

**SUBSCRIPT**

| | |
|---|---|
| 1, 2 | Refers to regions in Fig. 4 |
| $i$ | Index on layers |
| $j$ | Phase index |
| $w, o$ | Water and oil |
| $xx, yy$ | Diagonal terms in pseudopermeability tensor |
| $xy, yx$ | Off-diagonal terms in pseudopermeability tensor |

## ACKNOWLEDGEMENTS

Support for this work has come from a variety of sources: US Department of Energy, Cray Research, the State of Texas Advanced Technology Program, and the Center for Enhanced Oil and Gas Recovery Research at The University of Texas at Austin. Larry W. Lake holds the Shell Distinguished Chair in Petroleum Engineering.

## REFERENCES

Arya, A., Hewett, T.A., Larson, R.G. and Lake, L.W. (1988). Dispersion and reservoir heterogeneity. *Soc. Petroleum Eng. Reservoir Eng.*, 139–148.

Begg, S.H. and King, P.R. (1985). Modelling the effects of shales on reservoir performance: Calculation of effective vertical permeability. Presented at the *Society of Petroleum Engineers Reservoir Simulation Symposium, Dallas*, paper no. SPE 13529.

Coats, K.R., Dempsey, J.R. and Henderson, J.H. (1971). The use of vertical equilibrium in two-dimensional simulation of three-dimensional reservoir performance. *Soc. Petroleum Eng. J.* **11**, 63–71.

Dagan, Gedeon (1988). Time-dependent macrodispersion for solute transport in anisotropic heterogeneous aquifers. *Water Resources Res.* **24**, 1491–1500.

Dake, L.P. (1978). *Fundamentals of Reservoir Engineering*, pp. 372–412. Elsevier, New York.

Desbarats, A.J. (1987). Numerical estimation of effective permeability in sand-shale formations. *Water Resources Res.* **23**, 272–286.

Dietz, D.N. (1954). A theoretical approach to the problem of encroaching and by-passing edge water. *Akad. van Wetenschappen, Amsterdam, Proc.* **56-B**, 83.

Fayers, F.J. (1984). An approximate model with physically interpretable parameters for representing viscous fingering. Presented at the *59th Annual Society of Petroleum Engineers Fall Technical Conference and Exhibition, Houston*, paper no. SPE13166.

Fayers, F.J. (1987). A review of the status of computation for evaluating miscible flood performance. In: Kleppe, J. *et al.* (eds.), *North Sea Oil and Gas Reservoirs*, pp. 231–244. The Norwegian Institute of Technology, Graham & Trotman, London.

Gelhar, L.W. and Axness, C.L. (1983). Three-dimensional stochastic analysis of macrodispersion in aquifers. *Water Resources Res.* **19**, 161–180.

Gelhar, L.W., Gutjahr, A.L. and Naff, R.L. (1979). Stochastic analysis of macrodispersion in a stratified aquifer. *Water Resources Res.* **15**, 1387–1397.

Haldorsen, H.H. (1983). Reservoir characterization procedures for numerical simulation. PhD Dissertation, The University of Texas at Austin.

Haldorsen, H.H., Chang, D.M. and Begg, S.H. (1987). Discontinuous vertical permeability barriers: A challenge to engineers and geologists. In: Kleppe, J. *et al.* (eds.), *North Sea Oil and Gas Reservoirs*, pp. 127–151. The Norwegian Institute of Technology, Graham & Trotman, Ltd, London.

Hearn, C.L. (1971). Simulation of stratified waterflooding by pseudo-relative permeability curves. *J. Petroleum Technol.* 805–813.

Heller, J.P. (1963). The interpretation of model experiments for the displacement of fluids through porous media. *Am. Inst. Chem. Eng. J.* **9**(4), 452–459.

Hewett, T.A. (1986). Fractal distributions of reservoir heterogeneity and their influence on fluid transport. Presented at the *61st Annual Technical Conference and Exhibition of the Society of Petroleum Engineers, New Orleans*, paper no. SPE 15386.

Kasap, E. (1990). PhD Dissertation, The University of Texas at Austin.

Kasap, E. and Lake, L.W. (1989). An analytical method to calculate the effective permeability tensor of a grid block and its application in an outcrop study. Presented at the *Numerical Simulation Symposium of the Society of Petroleum Engineers, Houston, Texas*, paper no. SPE 18434.

King, P.R. (1987). The use of field theoretic methods for the study of flow in a heterogeneous porous medium. *J. Phys. A: Math. Gen.* **20**, 3935–3947.

King, P.R. (1988). The use of renormalization for calculating effective permeability. *Transport in Porous Media*, 643–644.

Kortekaas, T.F.M. (1983). Water/oil displacement characteristics in cross-bedded reservoir zones. Presented at the *58th Annual Technical Conference and Exhibition of the Society of Petroleum Engineers, San Francisco*, paper no. SPE 12112.

Koval, E.J. (1963). A method for predicting the performance of unstable miscible displacement in heterogeneous media. *Soc. Petroleum Eng. J.* **3**, 145–155.

Lake, L.W. (1989). *Fundamentals of Enhanced Oil Recovery.* Prentice-Hall, Englewood Cliffs, N.J.

Lake, L.W. and Hirasaki, G.J. (1981). Taylor's dispersion in stratified porous media. *Soc. Petroleum Eng. J.* **21**, 459–468

Lake, L.W. and Zapata, V.J. (1987). Estimating the vertical flaring in in situ leaching. *In Situ* **11**(1), 39–62.

Lasseter, T.J., Waggoner, J.R. and Lake, L.W. (1986). Reservoir heterogeneities and their influence on ultimate recovery. In: Lake L.W. and Carroll, H.B. (eds.), *Reservoir Characterization*, pp. 545–560. Academic Press, New York.

Plumb, O.A. and Whitaker, S. (1988). Dispersion in heterogeneous porous media I. Local volume averaging and large-scale averaging. *Water Resources Res.* **24**, 913–926.

Quintard, M. and Whitaker, S. (1988). Two-phase flow in heterogeneous porous media: The method of large-scale averaging. *Transport in Porous Media* **3**, 357–413.

Shook, M. (1990). PhD Dissertation. The University of Texas at Austin.

Van den Broeck, C. and Mazo, R.M. (1983). Exact results for the asymptotic dispersion of particles in n-layer systems. *Phys. Rev. Lett.* **51**(15), 1309–1312.

Waggoner, J.R. (1985). Waterflood behaviour in heterogeneous sands. MS Thesis, The University of Texas of Austin.

Waggoner, J.R. and Lake, L.W. (1987). A detailed look at simple viscous fingering. Presented at the *II Simposio Internacional sobre Recuperacion Mejorada de Crudo Trabajos Tecnicos, Maracaibo, Venezuela*, pp. 395–428.

Waggoner, J.R., Zapata, V.J. and Lake, L.W. (in prep). Viscous mixing in flows through permeable media.

Wheatcraft, S.W. and Tyler, S.W. (1988). An explanation of scale-dependent dispersivity in heterogeneous aquifers using concepts of fractal geometry. *Water Resources Res.* **24**, 566–578.

White, C.D. and Horne, R.N. (1987). Computing absolute transmissibility in the presence of fine-scale heterogeneity. Presented at the *Society of Petroleum Engineers 9th Symposium on Reservoir Simulation, San Antonio*, pp. 209–220, paper no. SPE 16011.

Yang, An-Ping (1990). PhD Dissertation, The University of Texas at Austin.

Yokoyama, Y. (1981). The effects of capillary pressure on displacements in stratified porous media. MS Thesis, The University of Texas at Austin.

Young, L.C. (1986). The use of dispersion relationships to model adverse mobility ratio miscible displacements. Presented at the *Society of Petroleum Engineers/United States Department of Energy Fifth Symposium on Enhanced Oil Recovery, Tulsa, Oklahoma*, paper no. SPE/DOE 14899.

Zapata, V.J. and Lake, L.W. (1981). A theoretical analysis of viscous crossflow. Presented at the *Society of Petroleum Engineers Fall Technical Conference and Exhibition, San Antonio, Texas*, paper no. SPE 10111.

# 17

# A Method for Estimation of the Density of Fault Displacements below the Limits of Seismic Resolution in Reservoir Formations

*C. Childs, J.J. Walsh and J. Watterson*

Department of Earth Sciences, University of Liverpool, Liverpool, UK

The fracture density per unit volume in a reservoir formation can be measured directly from seismic data for fractures with displacements greater than the limit of seismic resolution. Plots of displacement size vs. cumulative number are produced from seismic interpretations digitized and processed using software developed for the analysis of fault geometries and displacements.

Data from several North Sea oilfields have been analysed and show that cumulative fault displacement distributions are non-linear but systematic and are a function of (i) the variation in displacement on individual faults, which varies with fault size, and (ii) the size population of faults within the measured volume. On logarithmic plots of displacement size vs. cumulative number, the oilfield data distribution curves have straight central segments with slopes between −0.5 and −1.0. The differences in slope probably represent real differences in the type of fault population. Straight segments of the population curves can be extrapolated from the lower limit of seismic resolution, about 10 ms or about 15 m, to lower displacement values for estimating the densities of fractures with displacements too small to be resolved seismically. Extrapolation to displacements as small as 1 cm is justified by numerically modelled population curves and by the empirical demonstration that fault data, from coal mines and other sources, show systematic distributions for displacements from 50 m down to 1 cm. Fault displacement distributions appear to be similar in extensional and compressional regimes. Fault displacement distribution plots can be used to determine the limits of seismic resolution and could form the basis of a method for testing the quality of seismic interpretations.

The theoretical model developed for fracture density distribution is based on theoretical displacement distributions on single faults, combined with a "dead" fault population model that is based on earthquake populations and a fault growth model.

## INTRODUCTION

One of the problems of characterizing reservoir formations in terms of their hydraulic properties is the determination of the density of faults that are on a scale below the limit of seismic resolution, but on a scale above that adequately represented in well or core data. A method has been developed for measuring the density of fault displacements above the limit of seismic resolution that allows extrapolation of the measured population curve into the field representing smaller fault displacements. The method deals only with fractures on which there is shear displacement. It is anticipated that further experience of the method, with a wider variety of data than has been examined so far, will allow development of a procedure that can be routinely applied to seismic interpretations. Fault displacement population curves derived from what are considered to be good data sets have characteristic shapes similar to one another and to those of theoretical curves. The theoretical basis for the systematics of fault and fault displacement populations, and a description of the model used for plotting theoretical displacement populations are given in Walsh and Watterson (in preparation). Population curves derived from what are considered to be poor data sets are significantly different from the theoretical curves and it may prove possible to test the quality of an interpretation by examination of the fault displacement population curves. An unexpected benefit of population curves is that they give a clear indication of the effective

*North Sea Oil and Gas Reservoirs—II*

limit of seismic resolution of fault displacement for a given data set.

Displacements can be measured and plotted manually for small data sets, but with the large data sets commonly available for offshore developments a computer system is needed. Those displacement plots illustrated that are derived from oilfield data have been plotted directly from seismic data sets digitized and processed using Fault Analysis Projection System (FAPS) software developed by the Fault Analysis Group at Liverpool.

## CHARACTERIZATION OF FAULT DENSITY

The minimum requirements of a method of characterizing fault density, or a fault population, are that (i) objective measurements should be obtainable from standard data sets, and (ii) the characterization should allow quantitative comparisons of fault populations in different areas or rock volumes.

Any method of characterization of fault populations must take account of the observation that displacement varies over each fault surface: in ideal simple faults, the displacement varies systematically but non-linearly from a maximum at the centre to zero at the tip-line (Barnett *et al.*, 1987; Walsh and Watterson, 1987). Characterization of the fault population by reference to numbers of faults of specified maximum displacement, either intersecting a 2-dimensional surface or within a rock volume, is impractical because points of maximum displacement of faults will not all lie within either a single surface or a specified volume and because fault branching may not permit objective identification of individual faults.

For practical purposes the numbers of faults of different size, where size is defined as the maximum displacement of a fault, is of less interest than either the numbers of displacements of particular sizes encountered along a linear pathway, of specified orientation, or the area of fault surface per unit volume. In the latter case the total area of fault surface can be subdivided according to displacement value and this is the basis of the method used here to characterize fault displacement populations. For some purposes, e.g. determining the relative numbers of displacements of different sizes, the shapes of the displacement distribution curves are sufficient and absolute values of fault surface area per unit volume are not needed.

## REPRESENTATION

Populations of fault displacements, either on a surface or within a volume, are expressed as plots of $\log d$ vs. $\log N$, where $d$ = measured displacement and $N$ = cumulative number of displacements greater than $d$ (Fig. 1). The plot represents a *fault displacement population* rather than a *fault population*. If required, $N$ can be converted to length/unit area (2D) or area/unit volume according to the spacing and length of sampling lines. *Fault displacement population* plots allow simple visual comparison between displacement populations and are similar to the conventional method of representing earthquake populations ($m$ vs $\log N$, where $m$ = earthquake magnitude). Displacements measured and plotted can be either the total dip–slip displacement or either the vertical component or horizontal component of dip–slip displacement. All plots illustrated are of the vertical component of displacement, i.e. fault throw, and displacement values are expressed either in metres or in two-way travel time.

## MEASUREMENT

A displacement population is measured by sampling the displacements at all points of intersection between sampling lines and fault surfaces. A single fault surface is therefore sampled as many times as it is intersected by sampling lines. When seismic sections are approximately normal to the strike of all faults in a field, the sampling lines are the picked reflectors on all interpreted sections. A statistically ideal sample would require equally spaced sampling lines, and all plots presented are of this type, but this requirement is not crucial. When differently oriented fault systems occur, each system is measured separately; FAPS has a provision for setting up synthetic shotlines in the appropriate orientation for each fault direction. Displacement population plots are derived from interpreted data digitized by FAPS, which requires each individual fault and fault array to be labelled in a hierarchical manner, allowing data to be selected by any combination of shotlines, reflectors or fault labels.

The displacement population of a reflector can also be represented in terms of fault trace length (for a given range of displacements) per unit area of the reflector (Fig. 1), by taking account of the spacing between seismic lines and the relative orientations of seismic lines and fault strike. When seismic lines are approximately normal to fault strike, as is the case for the oilfield data presented, each displacement is taken to represent a fault trace length equal to the spacing between the sampled seismic lines. The area of the reflector is simply determined from the lengths and spacings of the sampled lines. The potentially most useful plot is that representing fault surface area per unit volume of rock (see Fig. 7). For the volume between two reflectors, each displacement reading is taken to represent a fault surface area calculated from the line spacing, the stratigraphic interval between the two reflectors and the fault dips between the two reflectors. For volumes containing more than two reflectors, the fault surface area per unit volume is obtained by weighting the values for each reflector pair according to the stratigraphic interval between them.

## THEORETICAL DISPLACEMENT POPULATIONS

Theoretical population curves can be calculated by the method given by Walsh and Watterson (in preparation). The method is based on (i) the well established, but empirical, magnitude–frequency relationship for populations of tectonic earthquakes; (ii) a derived relationship between an active earthquake population and the total fault population existing at the end of a tectonic episode; and (iii) the displacement distribution on individual fault surfaces and its variation with fault size.

The magnitude–frequency behaviour of earthquakes satisfies the empirical relation

$$\log N = a - bm \qquad (1)$$

where $N$ is the total number of shocks of magnitude $m$ or greater (Ishimoto and Iida, 1939; Gutenberg and Richter, 1949) and holds not only for natural earthquakes but also for mining-induced seismicity (Kusznir *et al.*, 1984) and microfractures in laboratory compression of rock specimens (Mogi, 1962; Scholz, 1968). The value of $a$ is determined only by the number of events, but the value of $b$, the negative of the slope of the straight line distribution, is close to 1 and is rarely less than 0.7 or greater than 1.3 (King, 1983). The concept of the "characteristic earthquake"

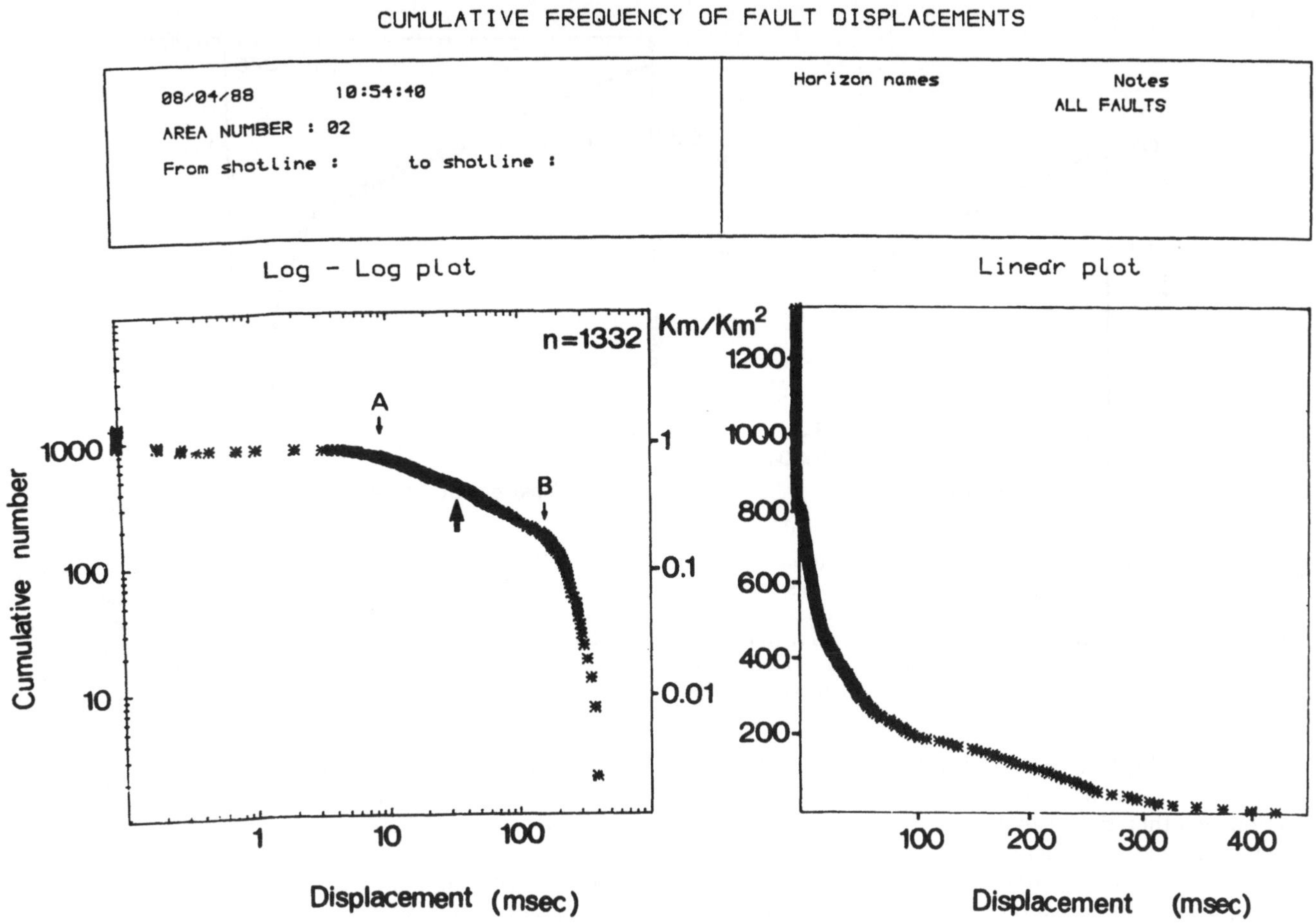


**Fig. 1.** Standard output for a fault displacement population, with a logarithmic plot (*left*) and arithmetic plot (*right*) of the same data. Data ($n = 1332$) are for all faults intersecting a single reflector in one offshore field. The limits of the approximately straight segment are arrowed (*A* and *B*) and the slope is $-0.54$, spanning 1.2 orders of magnitude. A small kink in the straight segment is also shown.

(Schwartz and Coppersmith, 1984; Wesnousky *et al.*, 1983) stipulates that successive earthquakes on the same fault are of the same magnitude and that the magnitude is related to the dimension of the fault surface ($M_s \propto R^3$, where $M_s$ = surface wave magnitude and $R$ = radius of a circular fault). As a fault grows with time, the characteristic earthquake of the fault will increase in magnitude, but for a small number of successive events the change in earthquake magnitude will be imperceptible. An earthquake population is therefore related to the active fault population in the period during which the earthquake population is measured. The physical basis for the magnitude–frequency relation and the reason for the variation in *b* value are not understood but the relation is an indication of the fractal nature of faults (King, 1983).

The theoretical model for fault displacement populations is based on earthquake population data because these represent the most objective and consistent data on fault size distribution, but applications are not restricted to seismic faults. The total fault population existing at the end of a tectonic episode, the "dead" fault population, is derived from an active earthquake fault population using a fault growth model described by Watterson (1986). The numbers and sizes of faults in the total population are combined with the displacement distributions on faults of each size (Walsh and Watterson, 1987, 1988) to produce the numbers of displacements of each size in the total population.

A theoretical displacement population (Fig. 2, left) shows the population curve to consist of two distinct parts. The right-hand segment of the curve intersects the abscissa at the largest measured displacement in the population and curves steeply upwards to join a straight segment. The extent and shape of the right-hand segment of the curve vary with the number of readings in the sample, and with the difference between the maximum displacement on the largest fault in the population (only a part of which may be in the sampled volume) and the largest displacement in the sample. The significant part of the curve is the straight segment, the slope of which varies according to the input parameters but is not affected by the number of displacement readings. Of the input parameters, the most significant is the slope of the straight line describing the "dead" fault population on a plot of $\log D$ vs. $\log \mathcal{N}$, where $D$ = maximum displacement on a fault and $\mathcal{N}$ is the cumulative number of faults of displacement greater than $D$. Numerically derived "dead" fault populations are described by the following expression (Walsh and Watterson, in preparation):

$$\log \mathcal{N} = a - B \log D \qquad (2)$$

where $a$ varies with the number of faults in the population and $-B$ is the slope of the straight line distribution. The value of $B$ varies from 1.3 to 2 for realistic values of input parameters, the most significant of which is $b$ (see Eq. 1).

The other significant parameters for a model fault displacement population are the size of the largest fault in the population, expressed as the maximum displacement on the fault, and the size of the largest displacement in the measured population, which will usually be less than

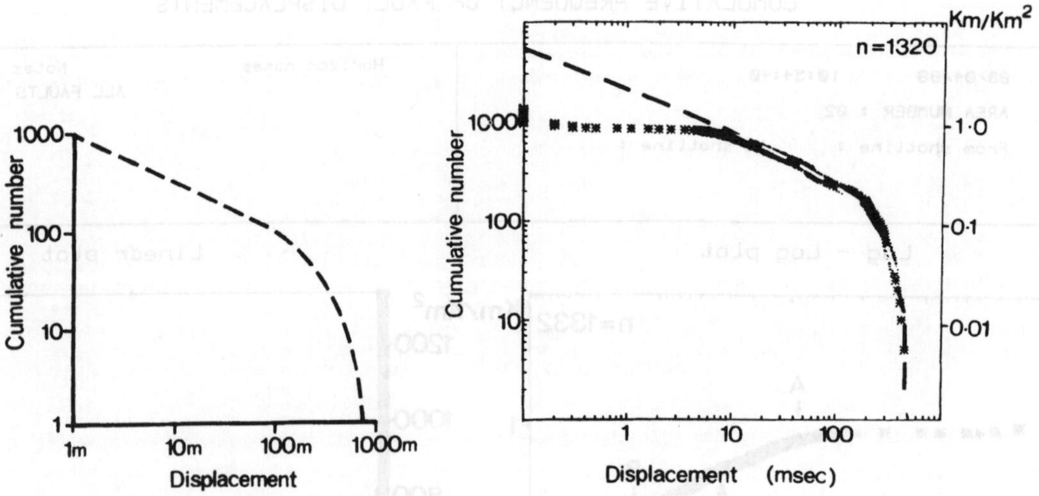


**Fig. 2.** *Left*: Theoretical fault displacement population for a fault population ($B=1.5$) with a maximum displacement of 1000 m on the largest fault and 1 m on the smallest faults. Maximum displacement in the sample = 725 m; minimum displacement = 1 m; total number of displacements = 1000. *Right*: Population plot for all data on a single reflector in an offshore field: slope of straight segment of curve is −0.45. Theoretical fault displacement population curve (dashed line) for a fault population ($B=1.41$) with a maximum displacement of 1500 ms on the largest fault and 0.1 ms on the smallest faults. Maximum displacement in the sample = 450 ms; minimum displacement = 0.1 ms; total number of displacement values = 5300.

the maximum displacement on the largest fault. Many measured displacement populations derived from seismic reflection or other data can be precisely matched with theoretical curves by judicious adjustment of the input parameters (Fig. 2); it is well to remember that precise matching of data by a theoretical curve may be possible even when the processes incorporated in the theoretical model differ from the actual processes. Validation of the theoretical population characteristics can come only from data sets for different size ranges of faults and from a variety of tectonic environments.

The crucial conclusion to be drawn from the theoretical curves is that the straight-line segment persists to displacement values smaller than those actually measured in any real data set.

# DISPLACEMENT POPULATION DATA

## Population data

Population data have been collected from British coalfield mine plans and from seismic reflection data from individual oilfields. Additional sources provide population data from a variety of tectonic environments. A feature common to all data sets is the deterioration of data quality at low displacement values.

## Oilfield data

The displacement population curve shown in Fig. 1 for a single reflector in an offshore field is derived from a total of ca. 900 km of seismic lines and is typical of populations derived from what are judged to be reasonably good data. The curve is comparable with the theoretical curve (Fig. 2) except that a third element intervenes between the straight segment (arrowed A and B, Fig. 1) and the ordinate. The third segment joins the straight segment at the limit of seismic resolution, about 10 ms, and represents an invalid portion of the curve where the data do not adequately represent the population. The invalid segment of the curve will not be plotted if all displacements less than the supposed limit of resolution are excluded but, in practice, it is useful to plot this segment and so identify the point at which the straight line segment degenerates. The theoretical curves indicate that the true displacement population can be defined by extrapolating the straight segment of the curve, thus estimating the population of displacements that are less than the limit of resolution (Fig. 2). A cumulative arithmetic curve (Fig. 1) is plotted routinely to test for random data that would produce a straight line on this plot, the slope varying with the total number of displacement measurements.

The robustness of the data has been tested by plotting populations for decreasing proportions of the data set, i.e. by including only every 2nd, 3rd, 4th, 5th or 10th seismic line (Fig. 3). The straight-line segment persists, with no change in slope, but has begun to degenerate where every 10th line has been sampled and only 137 measurements are included.

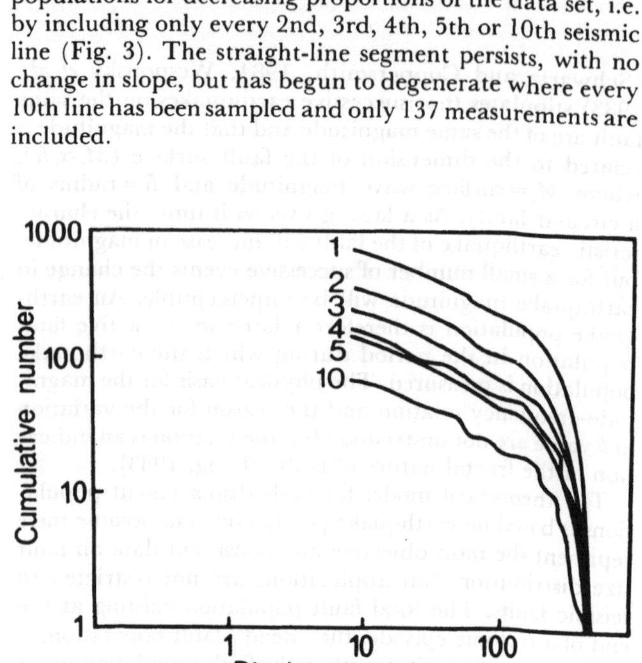


**Fig. 3.** Population plots for all data (1) on a single reflector in a field (same data as Fig. 1), and for data from every 2nd, 3rd, 4th, 5th and 10th seismic line. Number of displacement readings are 1320 (1), 661 (2), 431 (3), 327 (4), 266 (5) and 137 (10). Displacements of less than 10 ms are omitted from each data curve. The slope of the straight segment of all of the curves is ca. −0.49.

Differences between the displacement populations of individual reflectors within a single field are shown in Fig. 4. Reflectors A, B and C pre-date the faulting and have almost identical populations. Reflectors D, E and F represent progressively younger syn-faulting horizons, each with a different displacement population in which the straight segment of the curve progressively disappears and the displacement population approaches (Fig. 4, line F) that expected for a single fault (Fig. 5). The seismic sections, from which Fig. 4 is derived, show that only the master fault of the fault system propagated through the syn-faulting part of the sequence, possibly as a consequence of compaction, and that displacements of reflector F are mostly on this one fault.

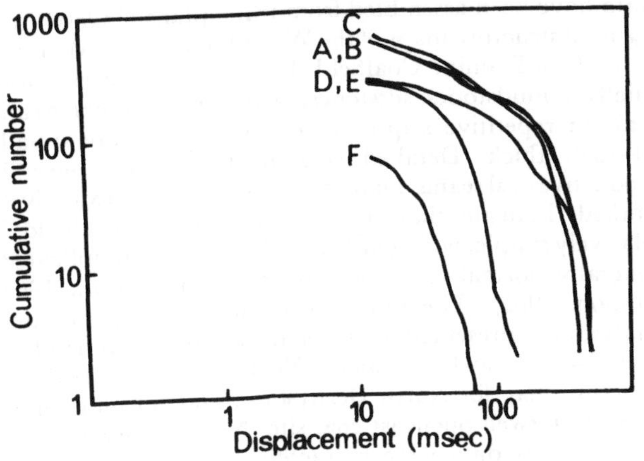


**Fig. 4.** Displacement populations of each of six reflectors A–F from an offshore field. A–C pre-date the faulting and D–F are progressively younger and within the syn-faulting part of the sequence. Slopes of the straight segment of population curves of pre-faulting reflectors range from −0.49 to −0.54.

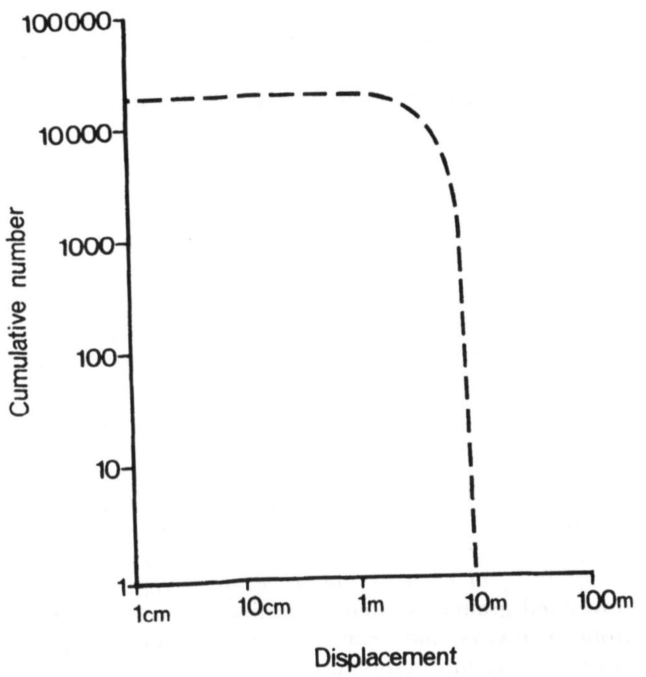


**Fig. 5.** Theoretical displacement population plot for a single fault with maximum displacement 10 m; maximum displacement reading 9.5 m; minimum displacement reading 0.01 m; and 19000 readings.

A plot of displacement data for a single fault on three reflectors (Fig. 6) has no central straight segment and is similar to the theoretical curve for a single fault (Fig. 5). Where a field is bounded by a major fault that is included in the data set, the areal extent of the data set may not be sufficient to encompass a fault population representative of the faulted area (the data set may be dominated by the major fault); ideally, the area should extend across strike half-way to the next major fault. In such cases the population curve is a composite product of a curve representing a single large fault and another representing a standard population in which the maximum fault size is less than that of the major fault. The composite curve will show a kink in the otherwise straight segment where the numbers of large displacement readings do not match the numbers of smaller readings; a kink of this type is evident in the population shown in Fig. 1. It is anticipated that for a given maximum size of fault in a population there will be a corresponding minimum area that must be sampled to obtain a standard displacement population. The relationship between maximum fault size and minimum sampling area, which will vary with the fault density, has not yet been established.

Although population plots from what are judged to be good data are similar to one another, those from what are judged to be poor data are more varied. These judgements of data are made on the basis of the quality of the seismic data and of the interpretation and are made before the fault displacement populations are examined. A typical population curve derived from good data is shown in Fig. 7 but has, however, two small kinks (arrowed) in what is otherwise a straight segment. The straight segment on this curve has a higher slope (−0.97) than that shown in Fig. 1 and this difference is discussed in a later section. Examples of plots from poor data are shown in Figs 8 and 9. In Fig. 8 there is no straight segment, or possibly a short and poorly defined one, nor a significant kink. The curve shown in Fig. 9 has three straight segments separated by two sharp kinks, but the central segment spans only half an order of magnitude of displacement, comprises several minor steps and has a very high slope (−1.66). Populations derived from good data characteristically have straight

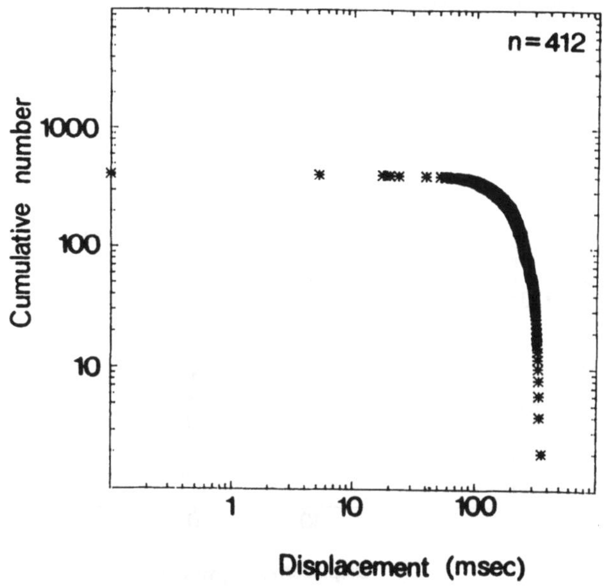


**Fig. 6.** Displacement population plot of a single large fault on three reflectors from an offshore field.

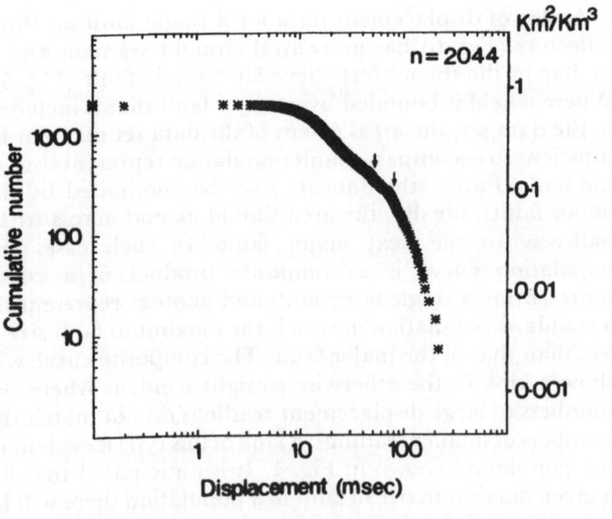


**Fig. 7.** Displacement population plot for all data on six pre-faulting reflectors from an offshore field. Two small kinks in the straight segment of the population curve are arrowed. The slope of the straight segment of the population curve is −0.97.

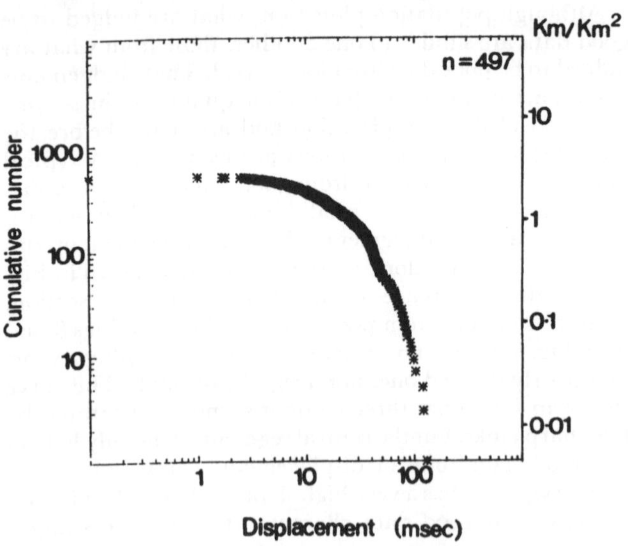


**Fig. 8.** Displacement population plot from what are judged to be poor data from an offshore field. Note the absence of an obvious straight central segment and the sharp kink.

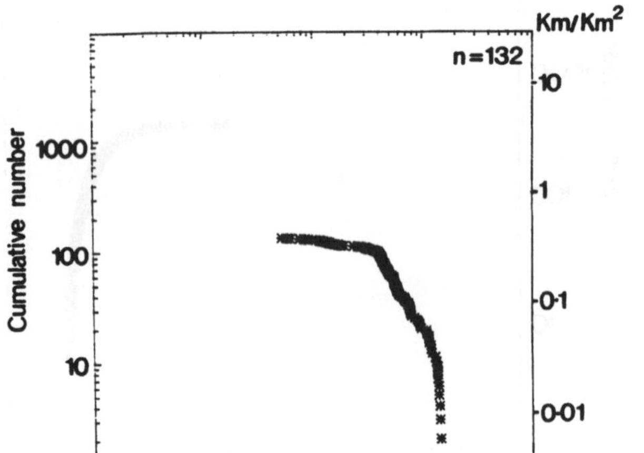


**Fig. 9.** Displacement population plot from what are judged to be poor data from an offshore field, showing straight segments separated by two kinks.

segments spanning at least one order of magnitude of displacement.

The displacement populations illustrated, and others that have been constructed from oilfield data, are similar to one another and to the theoretical populations. Before the straight segments of the data curves can confidently be extrapolated to displacement values less than those measured, as the theoretical work would suggest, confirmation is needed that populations of fault displacements less than ca. 30 m behave in a similar way to the larger ones. Populations of displacements between 30 m and 10 cm, derived from coalfield data, have been investigated to provide the required information about this size range.

## Coalfield data

Data have been assembled from mine plans (1:2500) and regional structure maps of the Westphalian Coal Measures of the East Pennine Coalfield, UK. The measures consist chiefly of mudstones, siltstones, sandstones, seatearths and coals in repetitive sequences or cycles that are usually 10–15 m thick. Detailed structural data for over 25 productive coal seams, many of which have been extensively worked, indicate that the structure of the coalfield is relatively simple, with gentle bed dips and uncomplicated, generally normal, tectonic faulting (Smith *et al.*, 1967; Rippon, 1985). The results of displacement population analyses are presented for three data sets: (a) mine plans from North Gawber colliery, Yorkshire; (b) a regional structure map for North Derbyshire; and (c) mine plans from Butterwell opencast coal site, Northumberland. All displacement data were extracted manually from these data sets.

### *North Gawber colliery, Yorkshire*

The most extensively worked seam is the Lidgett for which mine plans (1:2500) recording structure contours, fault traces and measured fault throws, are available in an intensively worked area of about 9 km². Data are also available for overlying and underlying seams, the Low Haigh Moor (10 m above) and Thorncliffe (40 m below), in intensively worked areas of 7 km² and 9.4 km² respectively. Two principal fault sets are distinguished on the basis of fault strike: (a) NE–SW strike, NW and SE dipping, and (b) NW–SE strike, NE and SW dipping. Almost 1500 throw (vertical displacement) readings, mostly ranging between 10 cm and 50 m, are available. NE–SW striking faults contribute ca. 60% of the fault displacement data and 100% of the displacements greater than 10 m.

Line sampling requires standardization of procedures to ensure objectivity and reproducibility. Displacement at a fault trace/sampling line intersection is found by linear interpolation between two displacement readings on the fault trace. Displacement at the point of divergence, or branch-point, of two splays from a single fault is determined by linear interpolation between a displacement on the single fault and the sum of the displacements ($A$ and $B$) on the two splays; at intervening points, displacement on each splay is determined by the ratio $A$:$B$. Workings cover about 90% of each of the three areas and, although unworked ground is often determined by faulting, data from roadways and seam contours usually permit a reasonably accurate estimate of displacement in unworked ground. Other procedures used where areas of unworked ground occur include: (a) tip-points of faults that terminate within unworked ground are determined by linear extrapolation of throw readings, and (b) branch-points of splays are located by extrapolation of a splay to a main fault.

Line sampling of the Lidgett Seam was carried out at 250 m, 125 m and 25 m line spacings. The curves for the data from 125 m and 25 m samplings are of similar shape (Fig. 10) and, as the 125 m data adequately describe the displacement population, this spacing was used for the Thorncliffe and Low Haigh Moor seams. Population curves are similar for the three seams studied (Figs. 11 and 12) and show the same pattern as the populations derived from seismic data. The range of displacements of the straight-line segments of the population curves exceeds 1.5 orders of magnitude (Figs. 11 and 12). The lower limit of resolution, as defined by the junction between the straight-line and left-hand segment of the curves, is at 30 cm displacement (Fig. 11). Fault displacements on mine plans are sometimes recorded down to 10 cm (formerly 6 inches) but this is not usually the case and, because fault tips (zero displacement) often cannot be positioned precisely, displacement values of less than ca. 30 cm are expected to be under-represented. Displacement population curves have been plotted for each fault set (NW and NE strikes) and for all faults on a single seam or combination of seams (Figs. 10–13). Although all curves show similar patterns and demonstrate that displacements down to 30 cm conform with the theoretical curves, the slopes of the straight segments of the population curves for NW and NE striking data sets are different (Fig. 13). This difference is discussed in a later section.

*Examples of poor data*

The displacement population curve for a 40 km² area containing Bolsover Colliery, North Derbyshire and derived from a regional structure map (scale 1:95000, British Geological Survey Sheet 112 Chesterfield; Smith *et al.*, 1967) is shown in Fig. 14. The lower limit of acceptable displacement data is not known, but the shape of the population curve suggests that displacements of less than about 10 m are under-represented. If so, a large proportion (ca. 50%) of the data must be disregarded and the range of displacements of the straight-line segment of the population curve is less that one order of magnitude. Furthermore, the slope of the straight segment of the population curve is significantly steeper than for straight segments on population curves produced from oilfield and mine plan data. The accuracy of the regional structure map was checked by reference to mine plans (1:2500) and the quality of recorded data, particularly below 10 m, is poor.

The displacement population curve derived from a seam plan (1 : 1250), covering an area of 0.26 km², from Butterwell opencast coal site, Northumberland, is also shown in Fig. 14. Structural analysis of underlying and overlying seams indicates that the quality of displacement data is poor, with fault heaves greatly exaggerated. The irregular form of the displacement population curve reflects these inaccuracies (Fig. 14).

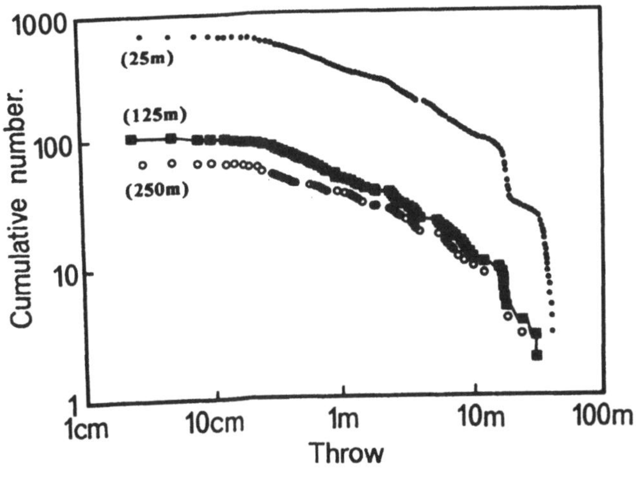


**Fig. 10.** Displacement population plot for a NE striking fault set, from mine plan data of Lidgett seam, North Gawber colliery, Yorkshire. Data curves are for line spacings of 250 m (open circles), 125 m (solid squares) and 25 m (solid circles). Data points for the 125 m line spacing curve are joined by a solid line. Slopes of straight segments of curves are ca. −0.50.

## Data from other sources

Data from published sources have been assembled and plotted on logarithmic plots of displacement size vs. cumulative number. For some data sets the sampling methods are uncertain. Displacement values that appear to be below the limit of easily resolved displacements have been excluded in each case. The principal data sets are described below and are for normal faults unless stated otherwise.

(a) Throw measurements ($n=39$), from 64 m to 0.25 m, in a 5 km development roadway in Cotgrave Colliery, Nottinghamshire, UK (Rippon and Raine, 1984). The sampling method used is similar to a line sampling technique. The population curve is shown in Fig. 15.

(b) Two data sets ($n=380$ and 76) with throws ranging from 0.6 m to 3 m from two areas (4 km² and 0.6 km²) in Granville Colliery, Shropshire, UK (Buck, 1977). The sample appears to represent the population of maximum fault throws (Fig. 15).

(c) Three data sets with throws ranging from 1 m to 25 m from collieries in the Kent Coalfield, UK (Warren, 1977). Sampled areas are 6 km², 6 km² and 8 km² with 82, 116 and 365 readings respectively. The sample appears to represent the population of maximum fault throws (Fig. 15).

(d) Displacements, referred to as offsets, which are assumed to be throws, ranging from 10 cm to 20 m from La Houve Colliery, Lorraine Coalfield, France (Villemin *et al.*, 1987). The area sampled is 5 km², within which 1600 readings were measured. The sampling technique is not described (Fig. 16).

(e) Two data sets with displacements ranging from 1.5 cm to 48 m from two thrust sheets ($n=135$ and 279) in the Appalachian Coalfield (Wojtal, 1986). Data curves are derived from profiles at given structural levels above major thrusts and are probably comparable with data curves derived by line sampling (Fig. 16).

(f) Eight data sets of throw populations from faults within Tertiary/Quaternary sediments in the S. Kanto District, Japan (Kakimi, 1980). Throws range from 1 cm to 4 m and the numbers of throw readings range from 80 to 500. One data set is for reverse faults. One data set, which includes throws down to 1 cm, was collected using a line sampling method, but the sampling technique for other data sets is not known. It is demonstrated that displacement population curves for different areas have similar forms and that straight-line segments of data curves have a limited range of slope. By detailed analysis of one area, Kakimi (1980) also demonstrated that displacement populations are systematic down to throws of 1 cm and suggested that the straight segments of other displacement curves could be extrapolated to 1 cm (Fig. 17).

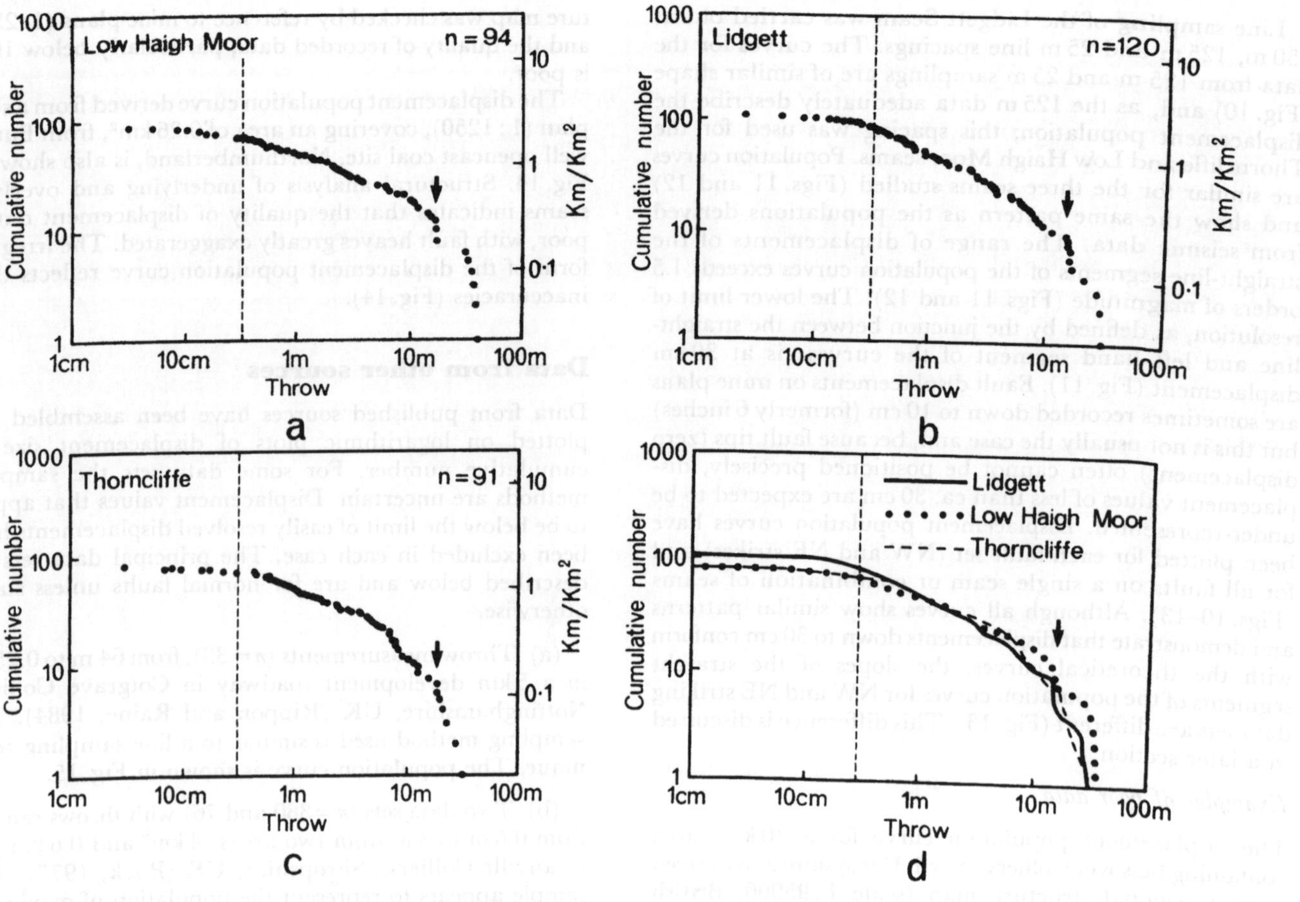


**Fig. 11.** Displacement population plot for NE striking fault set, from mine plan data of three seams, North Gawber colliery, Yorkshire. The quality of data is poor for throws less than 30 cm (broken line). Arrows show the positions at which North Gawber displacement population curves steepen owing to sampling effects (ca. 15 m). Slopes of straight segments of population curves are −0.49 (Lidgett seam), −0.37 (Low Haigh Moor seam) and ca. −0.404 (Thorncliffe seam).

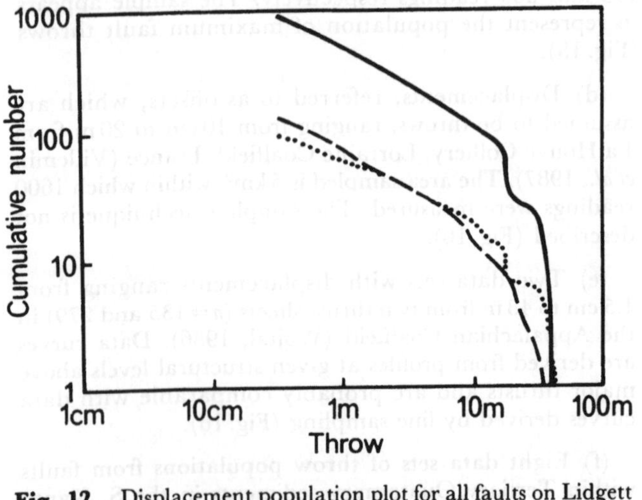


**Fig. 12.** Displacement population plot for all faults on Lidgett (solid line), Thorncliffe (broken line) and Low Haigh Moor (dotted line) seams, North Gawber colliery, Yorkshire. Line spacing is 25 m for Lidgett and 125 m for Thorncliffe and Low Haigh Moor. Slopes of straight segments of population curves are −0.60 (Lidgett seam), −0.58 (Thorncliffe seam) and −0.47 (Low Haigh Moor seam).

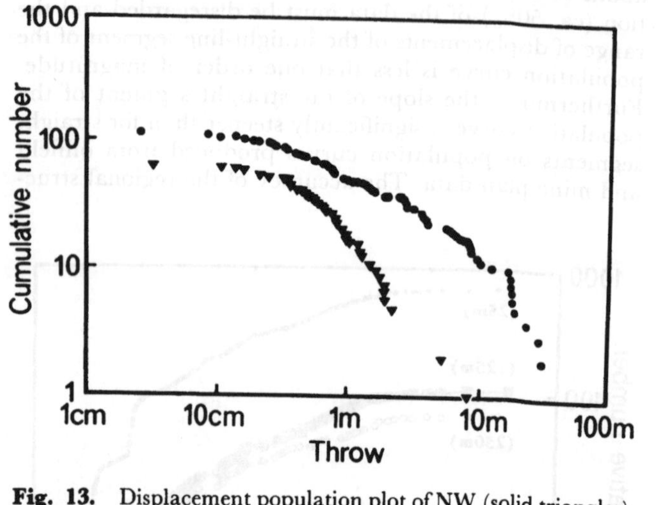


**Fig. 13.** Displacement population plot of NW (solid triangles) and NE (solid circles) striking faults intersecting the Lidgett seam in North Gawber colliery, Yorkshire. Line sampling at 125 m spacing curves are shown. The quality of throw data is poor below 30 cm. Slopes of straight segments of curves are −1.04 (NW) and −0.49 (NE).

## Conclusions from coalfield and additional data

Population curves derived from these data are comparable with those derived from seismic data but cover a different range of displacements. A straight segment is characteristic of curves derived from good data and in many cases displacements on the straight segments span ca. 1.5 orders of magnitude. The combined range of displacement for all straight segments on coalfield and additional data curves is 3.5 orders of magnitude (1 cm to ca. 50 m) demonstrating that displacements well below the limit of seismic resolution

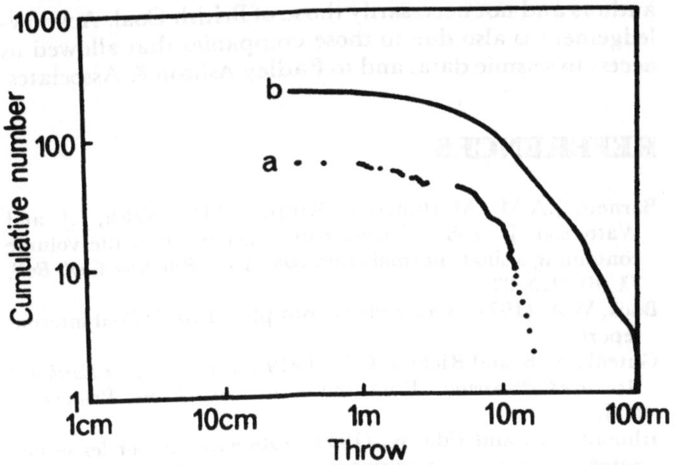


**Fig. 14.** Displacement population curves for data from (a) mine plans of Butterwell opencast coal site, Northumberland and (b) North Derbyshire regional structure map (Smith *et al.*, 1967). Slope of North Derbyshire population curve is −1.48. See text for details.

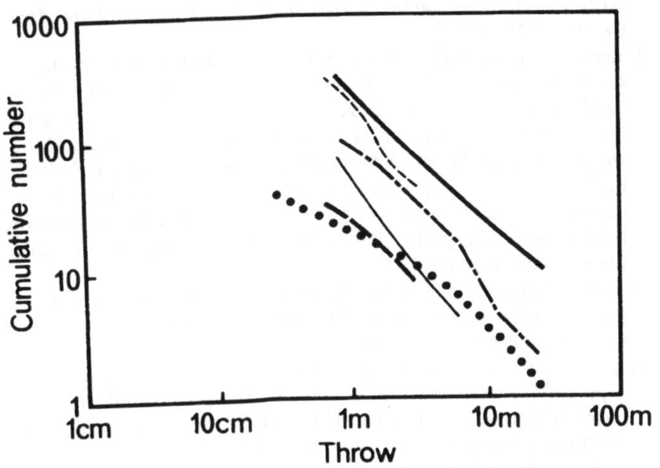


**Fig. 15.** Population curves for data from Cotgrave colliery, Nottinghamshire (dotted line; Rippon and Raine, 1984); Kent coalfield (solid lines; Warren, 1977); and Granville colliery, Shropshire (broken lines; Buck, 1977). Mean slopes of population curves are: −0.53 (Nottinghamshire), −1.43 to −1.00 (Kent) and −1.00 to ca. −1.40 (Shropshire).

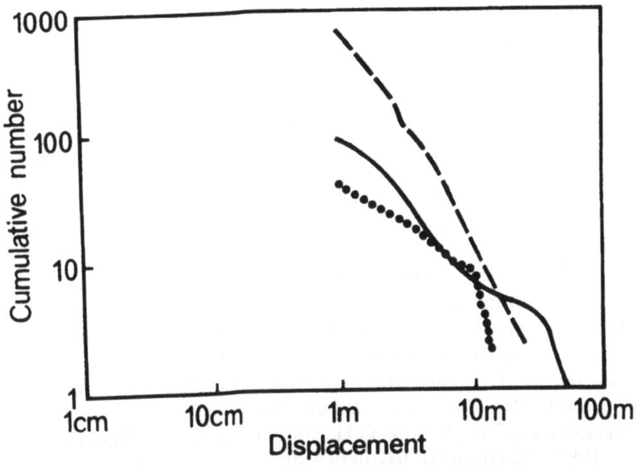


**Fig. 16.** Displacement population curves for data from reverse faults (solid and dotted lines) in Appalachian coalfields (Wojtal, 1986) and normal faults (broken line) from Lorraine coafield (Villemin *et al.*, 1987). Mean slopes of population curves are: −1.73 (Lorraine Coalfield) and −1.00 to −0.63 (Appalachian Coalfield).

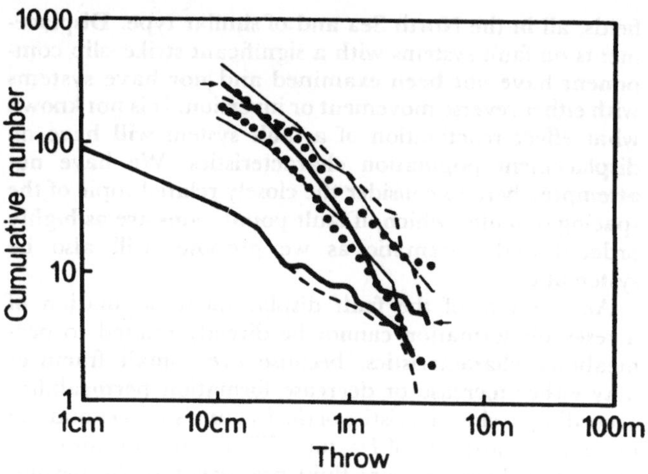


**Fig. 17.** Displacement population curves for eight data sets from the South Kanto district, Japan (Kakimi, 1980). All data sets are for normal faults, except one, which is for reversed faults (arrowed). Slopes of population curves range from −0.5 to −1.3.

conform with the theoretical model, and that extrapolation of oilfield data is justified down to displacements of as little as 1 cm.

## DISCUSSION

A relationship between earthquake populations and those of "dead" fault populations has previously been proposed by Kakimi (1980) and by Villemin *et al.*, (1987) on the basis of the straight segments of fault displacement population curves. Kakimi (1980) and Wojtal (1986) independently recognized that the form of displacement population curves is a product of incremental growth of faults by which, with time, an increasing proportion of displacement is accommodated on larger faults. Wojtal (1986) also interpreted population data as indicating a progressive increase in displacement rate on individual faults. The principal purpose of this chapter is to demonstrate empirical relationships, and the theoretical aspects are fully discussed elsewhere (Walsh and Watterson, in preparation).

It is evident that the straight segments of displacement population curves derived from good data differ significantly in slope, but the reason for this is uncertain. An obvious explanation, and one consistent with theoretical modelling, is that the different slopes represent genuine differences in the type of fault population. These differences would correspond to the genuine, but unexplained, differences between earthquake populations indicated by the range of *b* values (Eq. (1)). However, insufficient data have been examined to demonstrate conclusively that the different slopes are not artefacts. We have not yet been able to investigate populations derived from different interpretations of the same seismic data; different styles of interpretation undoubtedly exist, but we see no positive indication that two interpretations of the same seismic data could give rise to different slopes. The most convincing evidence that different slopes reflect genuine differences between fault systems are the data from the North Gawber colliery, which show that fault sets with different strike have straight segments of significantly different slope. In this case at least, slope difference is neither an artefact of the data collection nor due to lithological control of the fault populations.

A shortcoming of the investigation so far is the limited range of seismic data examined, which are from only five fields, all in the North Sea and of similar type. Displacements on fault systems with a significant strike–slip component have not been examined and nor have systems with either reverse movement or inversion. It is not known what effect reactivation of a fault system will have on displacement population characteristics. We have not attempted here to consider the closely related topic of the spacing of faults, which, if fault populations are as highly ordered and systematic as we propose, will also be systematic.

An estimate of the fault displacement population of a reservoir formation cannot be directly related to permeability characteristics, because even small fractures may either increase or decrease formation permeability according to characteristics of the fracture surfaces and the presence or absence of fracture fillings, and because the interaction between pore fluid pressure and the permeability of fractures of different orientations must be taken into account. The immediate aims of this work are to correlate data on the different scales represented by seismic and core data, and to establish empirical relationships between fault displacement populations derived from seismic data and formation properties determined by other means.

## CONCLUSIONS

(i) Fault displacement populations represented by plots of $\log N$ vs $\log d$ can be objectively derived from seismic interpretations.

(ii) Populations derived from good data have a characteristic form and have a straight line segment spanning at least one order of magnitude of displacement.

(iii) Theoretical modelling suggests that the straight segments can be extrapolated to estimate populations of displacements down to as little as 1 cm.

(iv) Fault displacement population plots derived from coalfield and other data show straight segments down to displacements as little as 1 cm, and provide empirical justification for the extrapolation of population curves derived from seismic data.

(v) Fault displacement plots can be used to determine the limit of seismic resolution of a data set and, with further experience, may afford a simple means of testing the quality of seismic interpretations.

(vi) The difference in slope between straight segments of fault displacement curves probably reflects a real difference between the fault populations.

(vii) Empirical relationships should be sought between fault displacement populations and other reservoir parameters.

## ACKNOWLEDGEMENTS

We thank colleagues in the Fault Analysis Group, John Gibson, Chris Lavers and Nick Lindsay, for their help carrying out this work. British Coal Chief Geologist Mike Allen is thanked for his encouragement of the project, which is partly funded by British Coal contract YCE.30/20214. The views expressed are those of the authors and not necessarily those of British Coal. Acknowledgement is also due to those companies that allowed us access to seismic data, and to Badley Ashton & Associates.

## REFERENCES

Barnett, J.A.M., Mortimer, J., Rippon, J.H., Walsh, J.J. and Watterson, J. (1987). Displacement geometry in the volume containing a single normal fault. *Am. Assoc. Petroleum Geol. Bull.* **71** (8), 925–937.

Buck, W.A. (1977). Granville Geosimplan. British Coal internal report.

Gutenberg, B. and Richter, C.F. (1949). *Seismicity of the Earth and Associated Phenomena.* Princeton University Press, Princeton, N.J.

Ishimoto, M. and Iida, K. (1939). Observations sur les seisms enregistres par le microseismograph construit dernierement (1). *Bull. Earthquake Res. Inst. Univ. Tokyo* **17**, 443–478.

Kakimi, T. (1980). Magnitude–frequency relation for displacement of minor faults and its significance in crustal deformation. *Bull. Geol. Surv. Japan*, **31**, 467–487.

King, G. (1983). The accommodation of large strains in the upper lithosphere of the earth and other solids by self-similar fault systems: the geometrical origin of *b*-value. *Pure Appl. Geophys.* **121**, 761–815.

Kusznir, N.J., Al-Saigh, N.H. and Ashwin, D.P. (1984). Induced seismicity generated by longwall coal mining in the North Staffordshire Coal-field, U.K. In: Gay, N.C. and Wainwright, E.H. (eds.), *Proceedings of the 1st International Congress on Rockbursts and Seismicity in Mines, Johannesburg, South Africa*, pp. 153–160.

Mogi, K. (1962). Magnitude frequency relation for elastic shocks caused by the fracture of heterogeneous materials and its relation to earthquake phenomena. *Bull. Earthquake Res. Inst. Univ. Tokyo* **40**, 125–173.

Rippon, J.H. (1985). Contoured patterns of the throw and hade of normal faults in the Coal Measures (Westphalian) of north-east Derbyshire. *Proc. Yorkshire Geol. Soc.* **45**, 147–161.

Rippon, J.H. and Raine, J.D. (1984). Fault forecasting and exploration techniques in the East Midlands Coalfield. Unpublished presentation to the Midlands Branch of the Institution of Mining Engineers, 14 November 1984.

Scholz, C.H. (1968). The frequency magnitude relation of microfracturing in rock and its relation to earthquakes. *Bull. Seismol. Soc. Am.* **58**, 399–415.

Schwartz, D.P. and Coppersmith, K.J. (1984). Fault behaviour and characteristic earthquakes: Examples from the Wasatch and San Andreas fault zones. *J. Geophys. Res.* **89**, 5681–5698.

Smith, E.G., Rhys, G.H. and Eden, R.A. (1967). Geology of the country around Chesterfield, Matlock and Mansfield. *Mem. Geolog. Surv. Great Britain.*

Villemin, T., Angelier, J. and Sunwoo, C. (1987). Distribution of fault-fracture patterns: orientation, density, offset, length and spacing. *Tectonic Studies Group Meeting, Palaeostresses and Fault Systems, Swansea*, (abstract).

Walsh, J.J. and Watterson, J. (1987). Distributions of cumulative displacement and seismic slip on a single normal fault surface. *J. Struct. Geol.* **9**, 1039–1046.

Walsh, J.J. and Watterson, J. (1988). Analysis of the relationship between displacements and dimensions of faults. *J. Struct. Geol.* **10**, 239–247.

Walsh, J.J. and Watterson, J. (in preparation) Populations of active and inactive faults.

Warren, S.T. (1977). A consideration of fault associations in the Kent Coalfield. British Coal internal report.

Watterson, J. (1986). Fault dimensions, displacements and growth. *Pure Appl. Geophys.* **124**, 365–373.

Wesnousky, S.G., Scholz, C.H., Shimazaki, K. and Matsuda, T. (1983). Earthquake frequency distribution and mechanics of faulting. *J. Geophys. Res.* **88**, 9331–9340.

Wojtal, S. (1986). Deformation within foreland thrust sheets by populations of minor faults. *J. Struct. Geol.* **8**, 341–360.

# 18

# Determination of Petroleum Accumulation Histories: Examples from the Ula Field, Central Graben, Norwegian North Sea

*S.R. Larter[a], K.O. Bjørlykke[a], D.A. Karlsen[a], T. Nedkvitne[a], T. Eglinton[a], P.E. Johansen[a], D. Leythaeuser[a], P.C. Mason[b], A.W. Mitchell[b] and G.A. Newcombe[b]*

[a]Department of Geology, University of Oslo, Oslo, Norway
[b]BP Petroleum Development Norway A/S, Forus, Norway

The application of coupled geological and geochemical studies of oilfields can provide information on the maturity of the source rocks that fed the field, the directions of filling and estimates on the timing of accumulation development. In this ongoing study, detailed three-dimensional compositional mapping of the Ula Field permitted the selection of samples for more detailed molecular analysis, providing data on both source facies and maturity. It was observed from these data that, in addition to the expected lateral variations in chemical composition, vertical variations are also evident. Furthermore, local polar compound concentrations (small tar mats that are chemically relatable to the accumulation) are definable on a stratigraphic basis. Despite these chemical compositional variations and maturity differences, the petroleum column in the Ula Field appears mechanically stable. Indeed, the observed chemical compositional variations and the mechanical properties of the field are consistent with Late Tertiary filling as indicated by maturation modelling of maturity parameters.

## INTRODUCTION

Although large-scale compositional heterogeneity in petroleum accumulations has been known for some time (Sage and Lacey, 1939; Espach and Fry, 1951; McCord, 1953; Schulte, 1980), it is only recently that these have been interpreted using petroleum geochemical techniques. In addition to recognizing biodegradation and water-washing effects, compositional variations in petroleum accumulations may also be interpreted as arising from source facies and maturity variations in the petroleum charge feeding the accumulation. Through time, the variable composition of the charge to the field is integrated and preserved as compositional variations in the petroleum column due to incomplete mixing.

England and Mackenzie (1989) have suggested that, while diffusive or density-driven mixing can eliminate inherited gross compositional variations in a petroleum column (ca. 100 m) within the time scale of reservoir filling (ca. 10 Ma), lateral compositional gradients across a field are not equilibrated (cf. also Hirschberg, 1984; England *et al.*, 1987). These compositional variations may be interpreted to indicate field-filling directions and place limits on the probable locations and types of source rocks filling a field (England and Mackenzie, 1989; Leythaeuser and Rückheim, in press; Karlsen and Larter, 1989).

This area of geochemical study may prove to be one of the most powerful applications of petroleum geochemistry to basin evaluation problems. Even in extensively drilled provinces such as the North Sea, the vast majority of well locations are targeted at structural highs such that the mature petroleum source rocks relevant to known petroleum accumulations are only rarely sampled. Thus, a systematic study of petroleum column heterogeneity is not of only academic interest, but is also directly applicable to determining the location, type and maturity of the source rocks responsible for an accumulation. Furthermore, by directly determining the location of field fill points, it may prove possible to quantify more accurately the petroleum carrier system and identify further exploration potential. Additionally, an understanding of the mixing processes that occur may also allow the distributions of petroleum variations within fields to be used to tune production programmes.

*North Sea Oil and Gas Reservoirs—II*

In this contribution we explore some of the methodologies and applications of reservoir petroleum geochemistry and provide some guidelines for such studies using general examples from an integrated geochemical/geological study of the Ula Field in the Norwegian Central Graben.

## THE ULA FIELD

The Ula Field lies some 280 km SW of Stavanger and about 65 km NNW of the Ekofisk chalk fields (Fig. 1). The 7/12-2 Ula discovery well was drilled by BP in 1976, encountering significant oil in marine sandstones of Late Jurassic age. The field is an anticlinal structure located at the eastern margin of the Central Graben. The structure developed as a result of regional mid-Campanian and Tertiary inversion episodes superimposed on a Permo-Triassic salt feature. Crestally, the field is dissected by a N–W striking fault system and the field is closed to the north against the Hidra Fault System.

The Late Jurassic reservoir sandstone, the Ula Formation, was deposited on a storm-dominated shallow marine shelf and unconformably overlies eroded Middle Jurassic and Triassic fluvial sediments (Home, 1987). It is conformably overlain by Late Jurassic mudstones including those of the Mandal Formation that are the source of the petroleums. The Ula reservoir sandstone can be broadly subdivided into an upper and lower member, with the upper unit containing the best-quality reservoir (Fig. 1). Early appraisal drilling additionally confirmed a deeper oilpool in the subcropping fluvial sandstones of the Triassic Skagerrak Formation. However, the productivity of these sands is low owing to poor reservoir quality, and this oil is not included in the Ula Field recoverable reserves estimate of 330 million barrels.

## GEOCHEMICAL APPROACH

A filling model for any petroleum accumulation should comprise the following information:

- Definition of the location, organic facies, maturity level, expulsion efficiency and spatial extent of the source rocks contributing petroleums to the field.
- Definition of the location, orientation, connectivity and residual petroleum saturation of the regional carrier system transporting petroleum from source basin to trap.
- Determination of the times at which various portions of the field reached significant petroleum saturation levels and how this distribution of petroleum is related to the sedimentology and diagenesis of the reservoir.
- Determination of the extent of post-filling mixing of petroleum within the field by diffusive and advective processes such as mechanical stirring.

An initial step in the determination of the filling history is the three-dimensional compositional mapping of the petroleum accumulation (Karlsen and Larter, 1989) and the definition of the maturity structure (the spatial relationships between petroleums of differing source maturities). The latter may be determined using one of the multitude of biomarker-based (Mackenzie, 1984; Cornford *et al.* 1986) or alkylaromatic hydrocarbon-based maturity parameters (Radke, 1988), or in terms of a gross maturity-related parameter such as solution gas–oil ratio (GOR) or saturation pressure/bubble point (England and Mackenzie, 1989; Karlsen and Larter, 1989).

A filling direction for a single-reservoir field can then usually be ascertained in terms of a gradient of maturity within the field; the most mature petroleums being nearest the filling points of the field (cf. Karlsen and Larter, 1989; Leythaeuser and Rückheim, in press; England and Mackenzie, 1989). However, we do not discuss filling directions in this preliminary report but concentrate instead on illustrating the general approach to reservoir geochemistry by discussing the principles involved in mapping of the petroleum accumulation.

## PETROLEUM COMPOSITIONAL HETEROGENEITY

An understanding of the mechanisms of field filling must be based on a high-resolution petroleum population distribution map of the reservoir (where the term petroleum population refers to a petroleum zone of definable bulk composition with similar maturation and/or source characteristics). A cost-effective approach to population mapping of petroleum reservoirs on a metre scale using solvent extraction techniques has previously been described by Karlsen and Larter (1989). The resulting gross compositional data (i.e. concentrations of saturated hydrocarbons, aromatic hydrocarbon ratios, etc.) are discussed in turn below with examples from the Ula Field.

### Saturated hydrocarbons

The concentration versus depth of saturated hydrocarbons in petroleums from extracted core material for six wells from the Ula Field are illustrated in Fig. 2. In most wells, these are relatively uniform with depth and values are typically in the range of 62–74% in the upper part of the reservoir, decreasing systematically with depth in wells 3A and 4. Well A01 stands out from this set, with saturated percentages ranging from 80–90%. However, representative gas chromatograms (Fig. 2) suggest that these high saturated hydrocarbon concentrations may in part be due to the presence of drilling-mud contamination. The gas chromatograms for the other wells show comparable n-alkane/isoprenoid alkane distributions, implying a common source with generally similar maturity characteristics for the petroleums.

### Aromatic hydrocarbons

In Fig. 3, the aromatic hydrocarbon compositions versus depth plots show that wells 6, A18 and 4, which are located along the NW–SE crestal axis of the field, typically have mono- and di-aromatic hydrocarbon contents in the range of 70–80% of the total aromatic hydrocarbon concentration of the petroleum. The low concentrations measured in the lower part of well 4 are attributable to interbedded coals in the Middle Jurassic section of this well. In contrast, the flank wells A01, A03 and 3A show significant differences in the concentration of these aromatic hydrocarbons. Well A03 has uniformly the lowest concentrations with values of ca. 60%, whereas wells A01 and 3A display large variations with several zones containing only traces of polyaromatic hydrocarbons relative to the mono- and di-aromatic hydrocarbons. Although in the case of A01 these low concentrations may be a function of contamination with oil-based mud, well 3A was cored with a water-based mud system.

These data, together with biomarker analyses and solution gas data (see below), suggest that while the petroleums in the crestal wells are closely related in terms of maturity and source, wells on the flanks of the structure

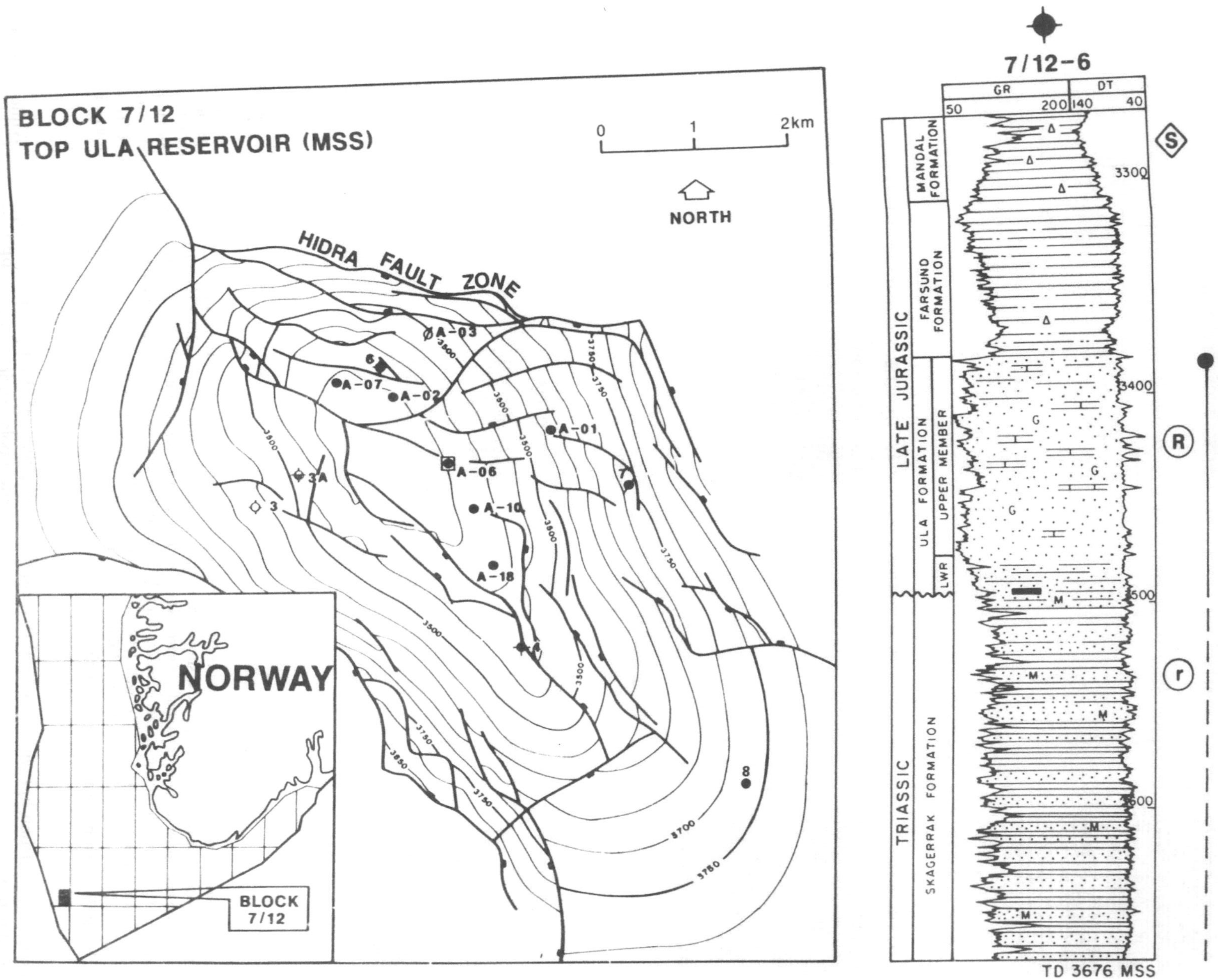


**Fig. 1.** Locations of wells and major elements of the Ula Field.

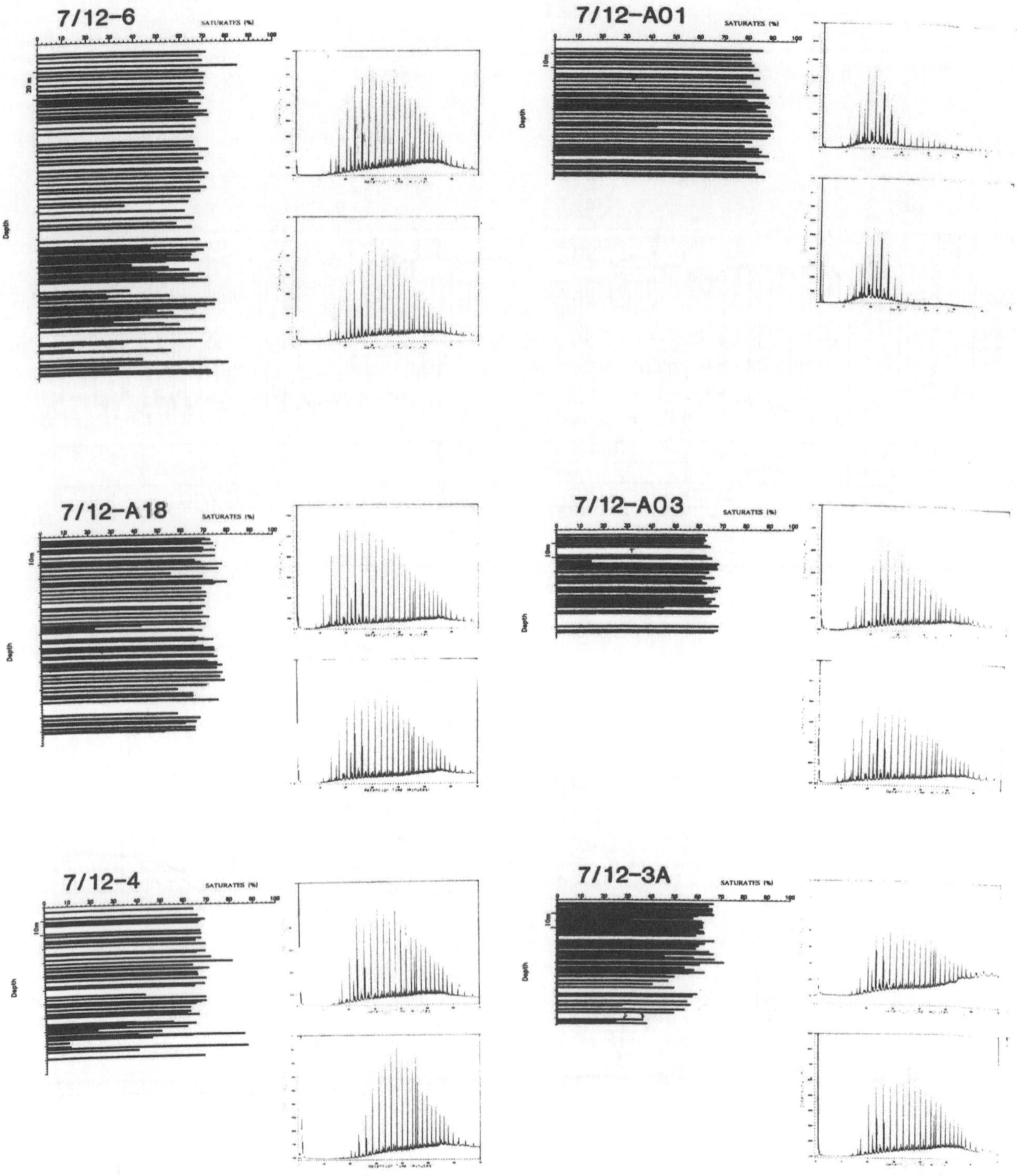


**Fig. 2.** Depth variations in percentage saturated hydrocarbons (determined by the Iatroscan TLC/FID technique) for six cores through the reservoir zone of the Ula Field. Also shown for each core are two representative GC–FID traces of the saturated fractions of core extracts from the upper and lower part of the core. Note the shift from mostly around 66–72% saturated hydrocarbons in the upper part of core 7/12-6, to a much higher spread in values (mostly around 30–72%) in the lower part of this core. While most cores display percentage saturates values around 66–72%, well 7/12-A01 has significantly more saturates (mostly 80–88%). The GC–FID traces are, except for well 7/12-A01, very similar for all wells (differences in low-boiling range are due to evaporative loss during coring and storing). Well A01 has suffered oil-based mud contamination.

contain related petroleums but with some local facies and maturity influences.

## Polar compounds

Figure 4 shows a vertical profile through part of the field in which large variations in the gross composition of the petroleum are evident, with the lower portion of the accumulation in particular containing petroleums enriched in polar (NSO + asphaltene) components. While some of these polar-enriched zones (PEZ) are associated with low absolute concentrations of petroleum in carbonate-cemented intervals, several PEZ appear associated with viable, or formerly viable, reservoir intervals and are systematically associated with the bases of coarsening-upwards sequences.

Furthermore, the composition of the asphaltenes in these discrete zones is very similar to the composition of the

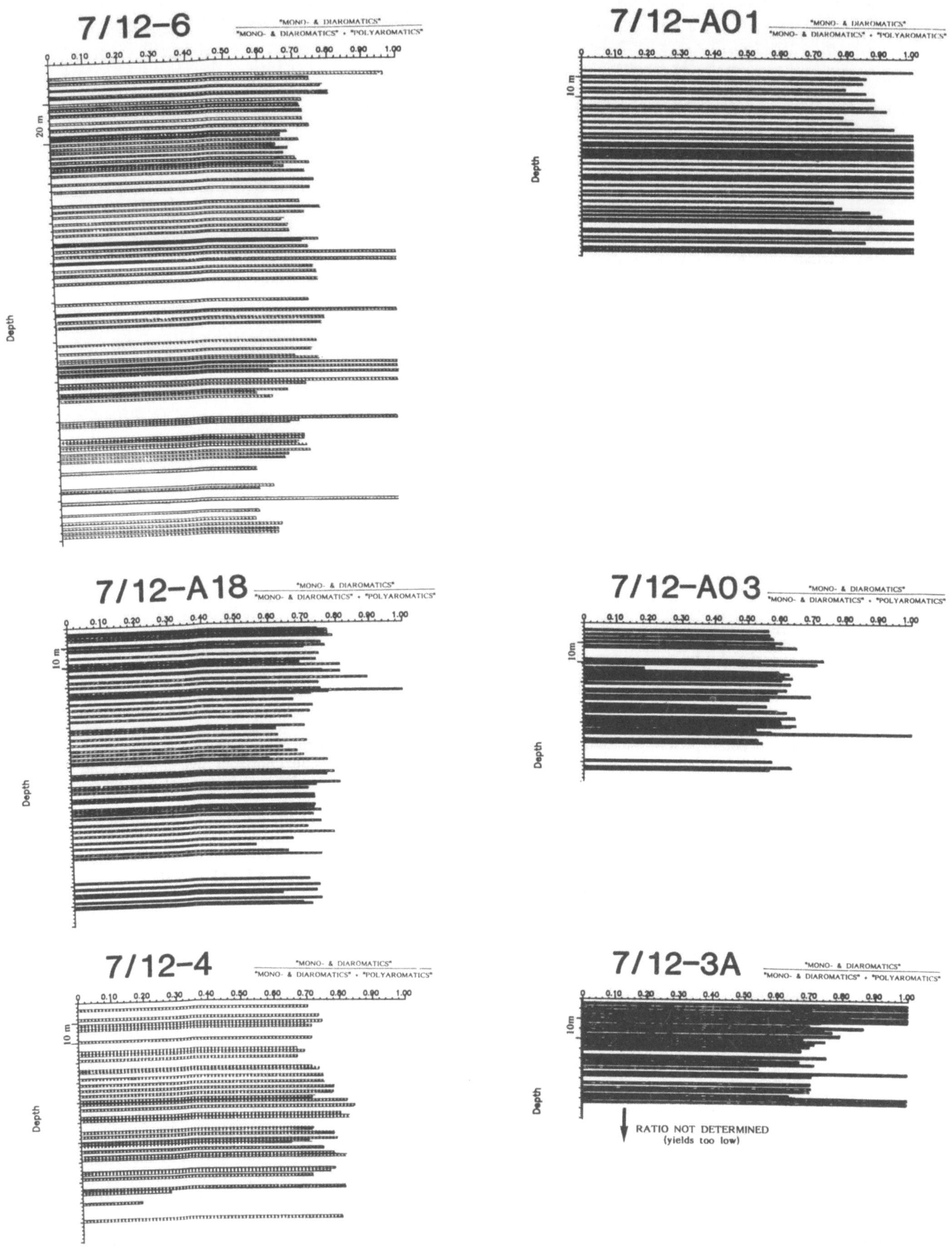


**Fig. 3.** Depth variations in the amount of mono- and diaromatic hydrocarbons relative to polyaromatic hydrocarbons as determined by the Iatroscan technique. Variation in this ratio can be used to trace differences in petroleum composition that are related to variations in source facies and maturity. The low concentrations of polyaromatic hydrocarbons in parts of A01 are characteristic of oil-based drilling muds.

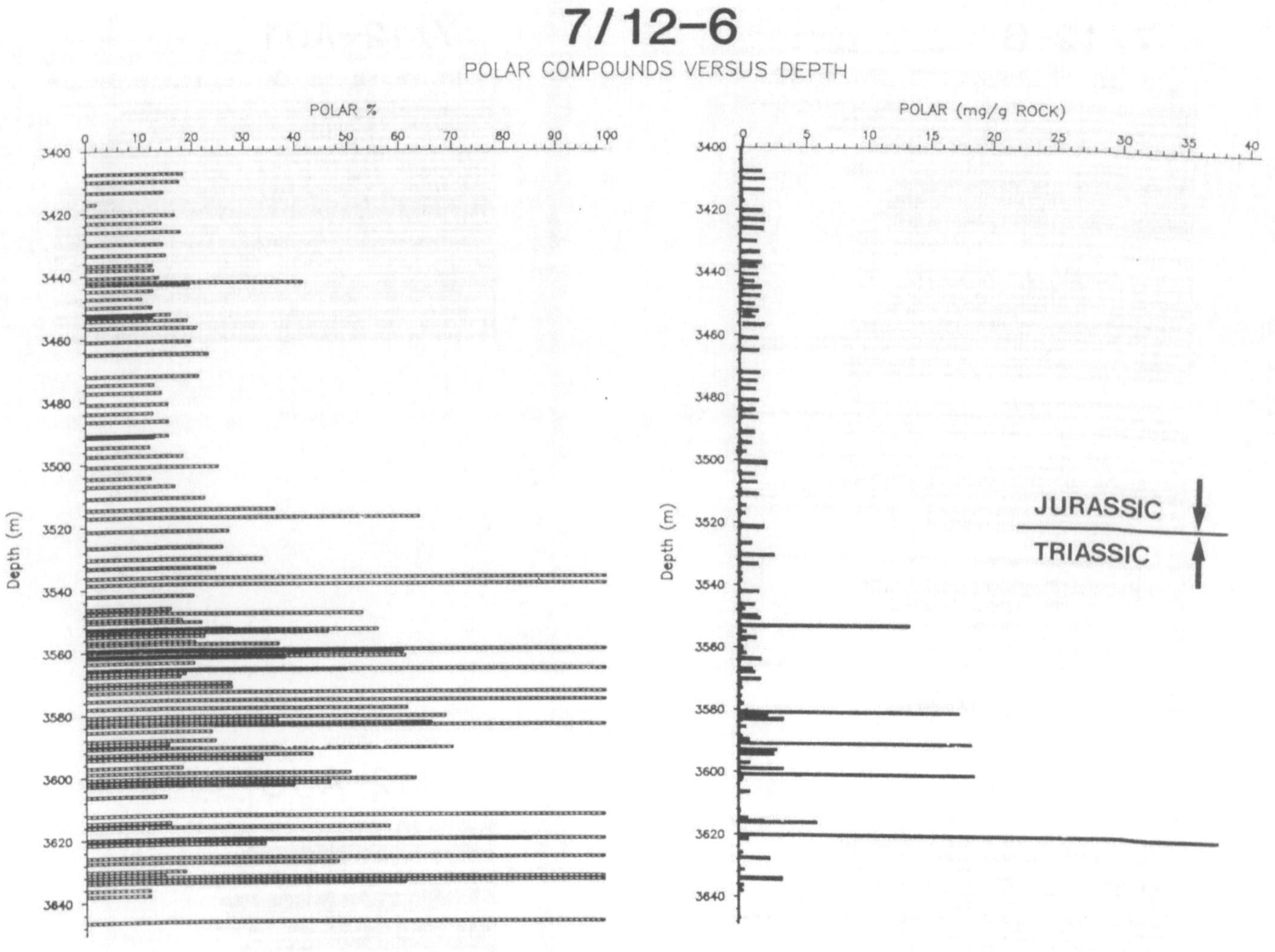


**Fig. 4.** Depth variations of polar species (resins plus asphaltenes) in well 7/12-6. Note the increase in polars in the Triassic reservoir. While most of the samples high in relative amounts of polar compounds are carbonate-cemented horizons, five samples stand out with high absolute amounts of polar components. Microscopy of these samples shows that these polar-enriched petroleums fill the porosity in these intervals.

asphaltenes in drill stem tests from related larger test intervals in the well. Thus, a comparison using quantitative pyrolysis–mass spectrometry (Eglinton *et al.*, in prep.) shows that the asphaltenes from the Triassic reservoir, for example, are preferentially enriched in aromatic pyrolysates (Fig. 5).

These observations suggest that these horizons, up to 1 metre in thickness, may represent zones of excess asphaltene precipitation associated with the main oil column and are not local in-situ phenomena. The origin of these horizons is unclear at this time, but it may be speculated that they may be associated with local conduits or mixing boundaries within the reservoir.

Figure 6 shows pyrolysis–mass spectra of asphaltenes and kerogens from a sample of the Mandal Formation source rock at a depth of about 4 km and at a maturity equivalent to ca. 0.8% $R_0$. The pyrograms from kerogens and asphaltenes confirm that asphaltenes are valid oil-soluble kerogen analogues, carrying both source and maturity information (Eglinton *et al.*, in prep.). The general fingerprint and quantitative similarities between the Mandal Formation asphaltenes and the Ula oil asphaltenes (not shown) implies that the former is the dominant source rock to the Ula system and thus confirms previous oil–source rock correlation studies (Spencer *et al.*, 1986).

## PETROLEUM MATURITIES

The observed lateral and vertical gradients in gross petroleum composition can also be followed in terms of maturity variations in the field. These variations reveal much about the internal architecture of the accumulation and indicate the level of maturity in the source kitchen.

### Solution gases

Compositional and isotopic data from separated petroleum solution gases from several production wells (Ula Formation) are shown in Table I and the isotopic data are plotted in Fig. 7 on some of the commonly used diagrams for gas genetic and maturity interpretations. These plots show that the gases may be classified as being thermogenic, oil-associated in origin.

Using the isotopic equilibrium approach of Sundberg and Bennet (1981), the data imply generation temperatures ranging from 116 to 155°C with the lowest temperatures indicated for the methane–ethane pairs. Yet applying the maturity model of Clayton (in press) suggests that the solution gases lie at the higher end of this temperature range and are associated with source rocks near the end of phase of labile kerogen conversion to petroleum. This

**Table I.** Gas composition and stable isotope data

| *WELL* | $C_1$% | $C_2$% | $C_3$% | i-$C_4$% | n-$C_4$% | i-$C_5$% | n-$C_5$% | $CO_2$% | $C_1$(%)HC | $C_{2+}$(%)HC | i-$C_4$/n-$C_4$ | $C_1/(C_2+C_3)$ |
|---|---|---|---|---|---|---|---|---|---|---|---|---|
| 7/12-A01 | 74.35 | 14.69 | 6.30 | 0.66 | 1.14 | 0.08 | 0.08 | 2.7 | 76.4 | 23.6 | 0.58 | 3.5 |
| 7/12-A02 | 70.70 | 11.90 | 8.50 | 1.27 | 2.90 | 0.46 | 0.45 | 3.9 | 73.6 | 25.4 | 0.44 | 3.5 |
| 7/12-A06 | 73.50 | 12.10 | 7.50 | 0.96 | 2.17 | 0.30 | 0.29 | 3.2 | 75.9 | 24.1 | 0.44 | 3.8 |
| 7/12-A07 | 68.50 | 12.30 | 9.20 | 1.48 | 3.36 | 0.47 | 0.42 | 4.3 | 71.6 | 28.4 | 0.44 | 3.2 |
| 7/12-A10 | 71.20 | 12.50 | 7.30 | 1.07 | 2.41 | 0.37 | 0.31 | 4.8 | 74.8 | 25.2 | 0.44 | 3.6 |
| 7/12-A18 | 69.30 | 12.30 | 8.50 | 1.35 | 3.03 | 0.44 | 0.36 | 4.7 | 72.7 | 27.3 | 0.45 | 3.3 |
| 7/12-7 | 70.10 | 15.80 | 9.00 | 0.91 | 1.77 | 0.10 | 0.12 | 2.2 | 71.7 | 28.3 | 0.51 | 2.8 |
| 7/12-8 DST 1 | 65.68 | 16.90 | 10.10 | 1.41 | 3.19 | 0.25 | 0.27 | 2.2 | 67.2 | 32.8 | 0.44 | 2.4 |

| *WELL* | $\delta^{13}C_1$ | $\delta^2H_{C_1}$ | $\delta^{13}C_2$ | $\delta^{13}C_3$ | $\delta^{13}C_{i\text{-}C_4}$ | $\delta^{13}C_{n\text{-}C_4}$ | $\delta^{13}C_{CO_2}$ | $\delta^{18}O_{CO_2}$ | $\Delta\delta^{13}C_2$-$C_1$ | $\Delta\delta^{13}C_3$-$C_1$ | $\Delta\delta^{13}C_3$-$C_2$ |
|---|---|---|---|---|---|---|---|---|---|---|---|
| 7/12-A01 | −47.4 | −245 | −33.9 | −30.4 | −30.1 | −29.6 | −13.5 | −6.9 | 13.5 | 17.0 | 3.5 |
| 7/12-A02 | −47.3 | −238 | −33.8 | −30.4 | −29.8 | −29.8 | −8.2 | −9.3 | 13.5 | 16.9 | 3.4 |
| 7/12-A06 | −47.3 | −230 | −33.7 | −30.0 | −29.8 | −29.6 | −8.2 | −7.6 | 13.6 | 17.3 | 3.7 |
| 7/12-A07 | −47.5 | −253 | −33.8 | −30.6 | −30.2 | −29.8 | −8.2 | −12.0 | 13.7 | 16.9 | 3.2 |
| 7/12-A10 | −47.5 | −233 | −33.7 | −30.5 | −29.9 | −29.7 | −7.8 | −6.3 | 13.8 | 17.0 | 3.2 |
| 7/12-A18 | −47.3 | −238 | −33.6 | −30.4 | −29.2 | −29.6 | −7.6 | −7.0 | 13.7 | 16.9 | 3.2 |
| 7/12-7 | −48.3 | −262 | −34.5 | −30.3 | −30.6 | −30.4 | −7.8 | −2.2 | 13.8 | 18.0 | 4.2 |
| 7/12-8 DST 1 | −47.7 | −272 | −34.0 | −29.7 | −29.9 | −29.9 | −10.4 | −14.4 | 13.7 | 18.0 | 4.3 |

would imply that generation occurred at temperatures of around 140–150°C, equivalent to vitrinite reflectances of ca. 0.8–0.9%$R_0$. The spread of temperatures from the isotopic equilibration approach is believed to be related to non-homogeneous isotope compositions in the precursor source kerogen (Karlsen *et al.*, in prep.) and to mixing problems associated with additional influx of methane.

The latter is indicated by the difference in the isotopic ratios of the methane and ethane components. For all the samples, this is slightly larger than that usually observed for cogenetic gases and thus implies that the gases are to some extent mixtures. This is supported by statistical analysis of the compositional data, which indicate a trend (at the 5–10% significance level) of increasingly light carbon isotopic values for ethane, the butanes and carbon dioxide along the crest of the field in a SE to NW direction. These isotopically light gases in the north and east of the field are therefore thought to be most probably associated with local facies variations in the source section charging these areas.

## Oils

Table II shows summary petroleum column maturity data for a vertical profile through the field. These parameters are derived either from discrete molecular species present in low concentrations in the petroleum (e.g. $C_{27}$ hopane isomer ratios ($T_s/T_m$) and sterane isomerization at $C_{20}$; Mackenzie, 1984), or from the distribution of abundant components of the oils in the gasoline range that can be related empirically to source rock temperatures (Thompson, 1979; Schaefer *et al.*, 1984). These data indicate that the

**Table II.** Maturity and bulk property summary-vertical profile, Ula Field[a]

| | 1 | 2 | 3 | 4 | 5 | 6 | 7 | 8 |
|---|---|---|---|---|---|---|---|---|
| Ula Formation | 1.5 | 4.0 | 1.4 | 41 | 672 | 29 | 2330 | 693 |
| Skagerrak Formation | 2.4 | 3.0 | 2.0 | 50 | 487 | 21 | 1600 | 699 |

[a] (1) 1st Parrafin Index modified after Thompson, 1979).
(2) *Trans/cis* dimethylcyclopentane (Schaeffer *et al.*, 1984).
(3) $T_s/T_m$.
(4) %20S C29 steranes (cf. Mackenzie, 1984).
(5) Gas–oil ratio (scf/bbl).
(6) Mol% methane in petroleum.
(7) Saturation pressure (psi).
(8) Subsurface bulk density (kg/m³).

oils were expelled from the source rocks at temperatures of 140–150°C, equivalent to vitrinite reflectances of 0.8–0.95%$R_0$. Notably, the Triassic petroleum is interpreted to be more mature than that in the Jurassic reservoir.

The degree of $C_{29}$ sterane isomerization has been kinetically modelled (cf. Mackenzie and McKenzie, 1983) assuming an average geothermal gradient of 35°C/km for the source basin and a conservatively average heating rate of 4°C/Ma for the rapid Late Tertiary burial (since 40 Ma) phase. The results indicate that the current levels of isomerization were only achieved in the Late Neogene, suggesting a recent petroleum filling event for the field. Indeed, the temperatures derived from the data in Table II approximately equate to the present-day burial depths of the source kitchen.

The higher maturity of the Triassic petroleum might suggest that extensive, high petroleum saturations were achieved in the Jurassic reservoir before the Triassic reservoir was filled with the current petroleum charge. However, the Triassic petroleum has a lower saturation pressure and GOR than the Jurassic petroleum, and yet the subsurface densities of the two petroleums are nearly identical. Indeed, it is probable, from the available data, that the subsurface densities in the field are everywhere sub-equal, confirming the proposition of England and Mackenzie (1989) that most oilfields are rapidly density-stabilized by mechanical processes.

Although the GOR of a petroleum is controlled by source facies and temperature, by source heating rate and significantly by source rock expulsion efficiency (Mason and Larter—unpublished data), the GOR of petroleum expelled from a given source rock will typically increase with maturity. This is supported by laboratory pyrolysis results (Lewan *et al.*, 1979; England and Mackenzie, 1989), field data (Mason, unpublished) and computer models of the petroleum-generation process (Mackenzie and Quigley, 1988; Larter, in press). It is therefore reasonable to speculate that the Triassic reservoir oil charge was initially of higher GOR than that of the initial petroleum charge in the Jurassic reservoir.

However, the influx of additional methane (see parameter 5, Table II) to the Jurassic reservoir has resulted in equalization of subsurface densities in the petroleum columns and, apparently, attainment of the current situation with regard to saturation pressure and GOR. Indeed, the vertical mole per cent methane variations in the field are larger than those expected purely by gravita-

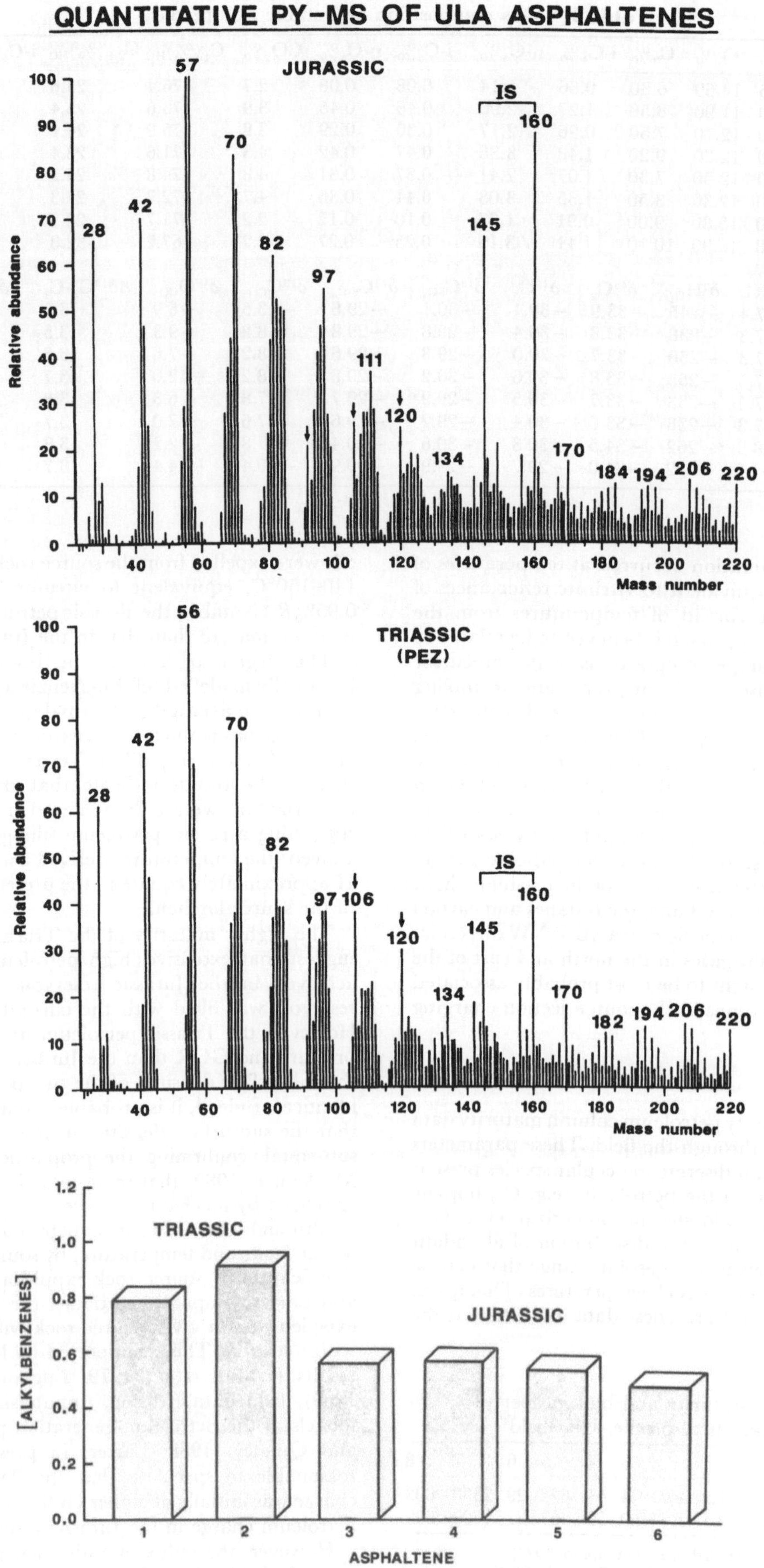


**Fig. 5.** Pyrolysis–mass spectra of asphaltenes from petroleums in the Jurassic and Triassic reservoirs of well 7/12-6. The marked peaks are fragments due to alkylbenzenes, the internal standard peaks are marked IS. The data were obtained on separated asphaltenes admixed with a poly-(butylstyrene) internal standard and pyrolysed at 770°C for 1 second on a FOM AUTO PYMS, the ion source operating at 16 eV. This analysis allows a clear discrimination, the Triassic asphaltenes pyrolysates being more aromatic. The histogram in the lower part of the diagram shows absolute concentrations of alkylbenzenes in asphaltene pyrolysates from the Ula and Skagerrak Formations core extracts and DST oils.

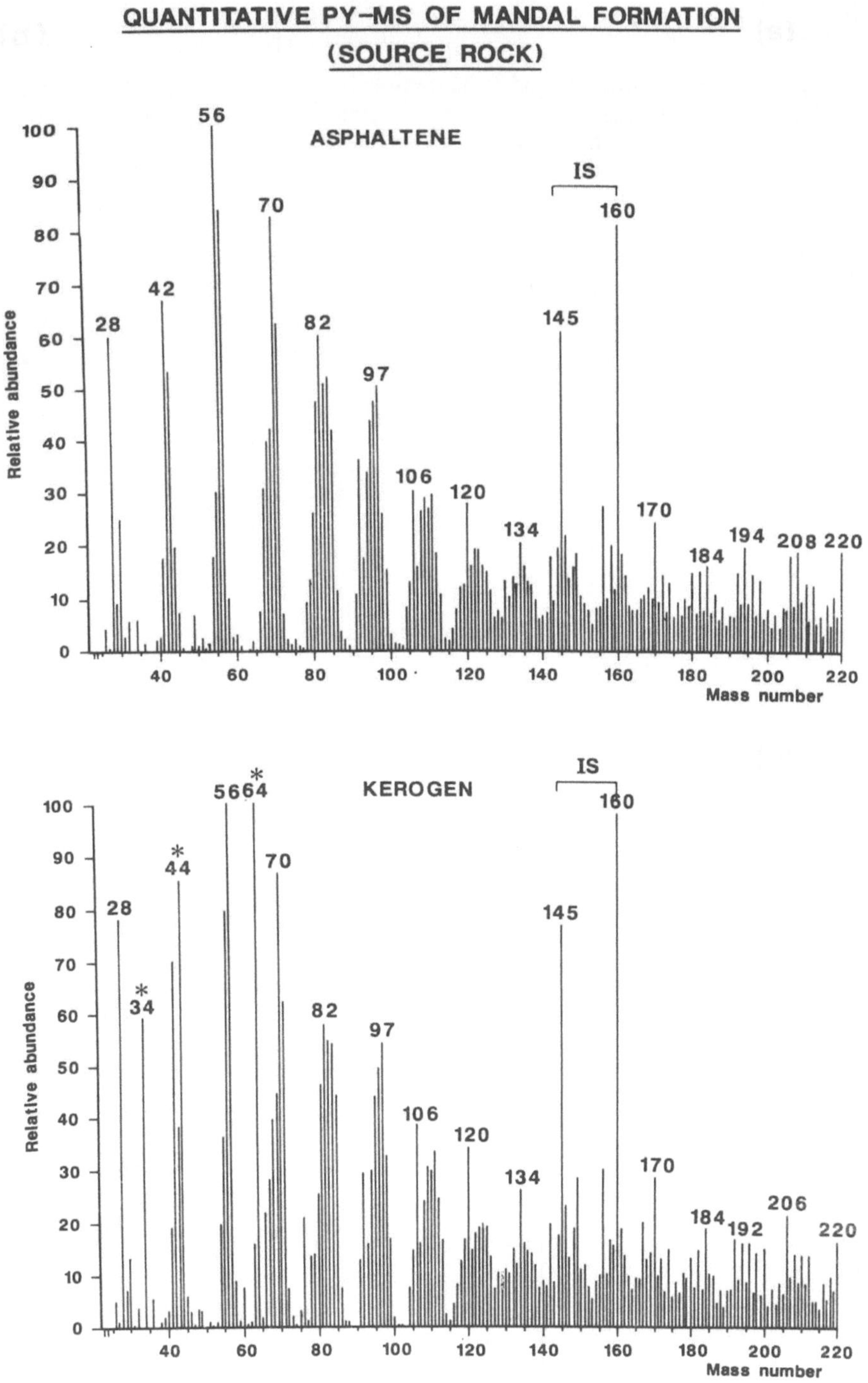


**Fig. 6.** Pyrolysis–mass spectra of asphaltenes and kerogens from a mature Mandal Formation sequence. PY–MS conditions as for Fig. 5. Note the similarity between source rock asphaltene and kerogen pyrograms and between the Mandal Formation and Ula Field petroleum asphaltenes (Fig. 5).

tional or thermal segregation in ideal fluids (Hirschberg, 1984; Schulte, 1980), and similar gradients are not observed in the other gas or heavier petroleum components.

## DISCUSSION

England and Mackenzie (1989) have suggested that diffusion and density-driven mixing processes are the main processes of achieving mechanical and chemical equilibrium in petroleum reservoirs. They claim that most natural reservoirs are in mechanical equilibrium (i.e. petroleum densities are equal at a given depth and remain constant or increase with depth). They also suggest that, in typical reservoirs, chemical equilibrium is attained rapidly (in ca. 1 Ma) in a vertical sense by diffusion on a 100 m scale, but that lateral compositional gradients persist on a field length scale (of kilometres) owing to the inability of lateral diffusion, with a (length)$^2$ time dependency, to achieve chemical equilibrium in a typical field fill-time.

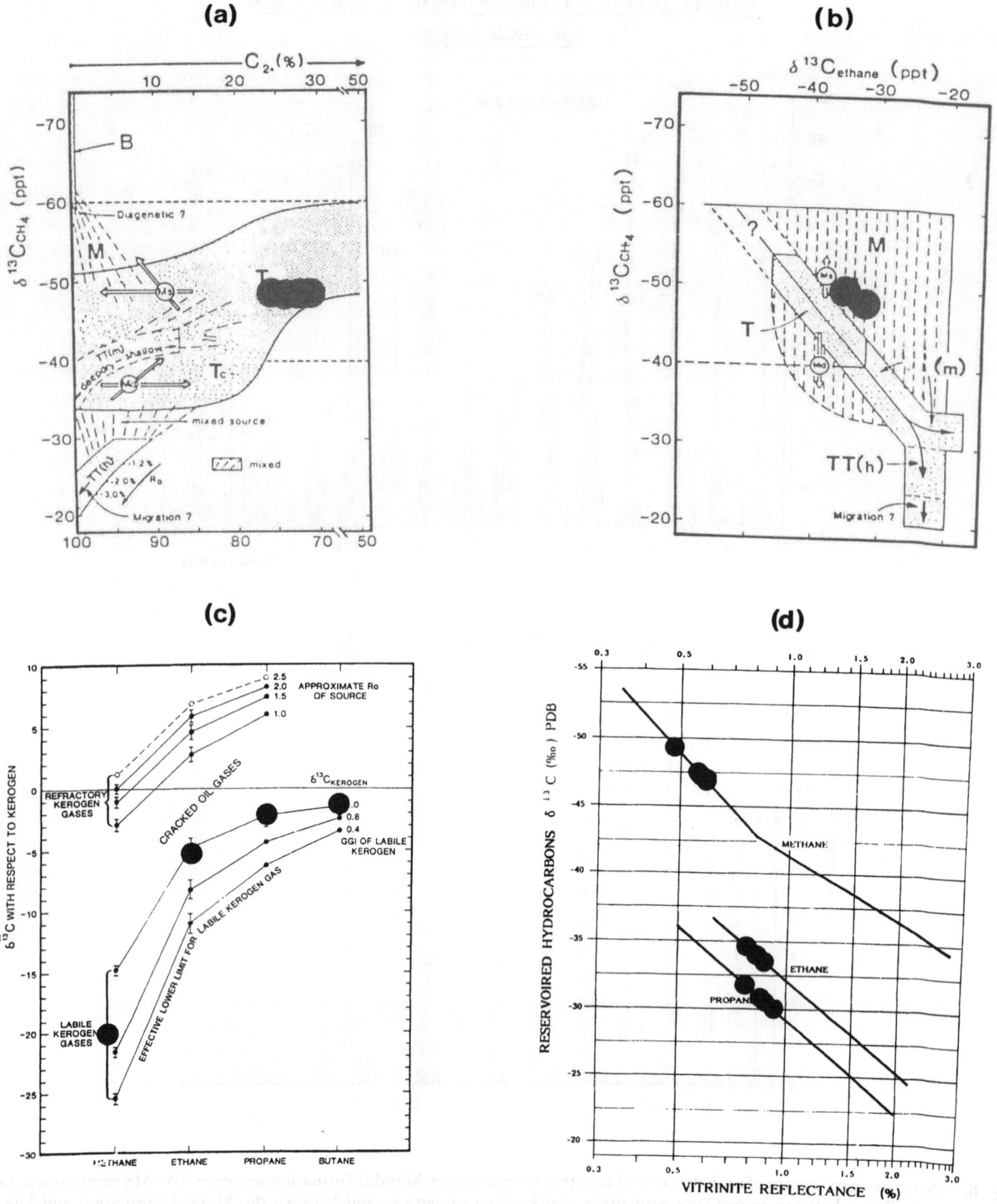


**Fig. 7.** Stable carbon isotope and gas composition data of production solution gases from several wells in the Ula Field plotted in commonly used interpretative diagrams. The two upper diagrams (a, b) are from Schoell (1983); of the lower diagrams, (c) is from Bundesanstalt fur Geowissenschaften und Rohstoffe (BGR, West Germany) and (d) from Clayton (in press). See text for discussion.

Although in the high-quality upper Ula reservoir zone petroleum compositions do appear quite homogeneous (Figs 2 and 3), it is clear from the compositional data (Figs 2–4) that full vertical chemical equilibrium in the field has not been established, except perhaps for mobile components such as methane (modified by gravitational effects). Within local portions of the reservoir, petroleum compositions are quite variable on a metre scale, particularly for polar species (Fig. 4). This indicates that simple diffusion models cannot completely account for the observed variation in petroleum column compositions.

It may be speculated that surface adsorbtion effects on minerals (cf. Cuiec, 1984) may control local petroleum compositions in polar components such that diffusive processes may be unable to homogenize gross petroleum compositions on even a submetre scale in clay-rich reservoirs (cf. the lower reservoir zone). In this respect, the diagenesis of the reservoirs has been a critical element in controlling petroleum distribution. For example, bitumen-stained stylolites are commonly observed in the Ula Formation (Fig. 8). These occur mainly in the argillaceous sandstones and commonly along burrow-walls in exten-

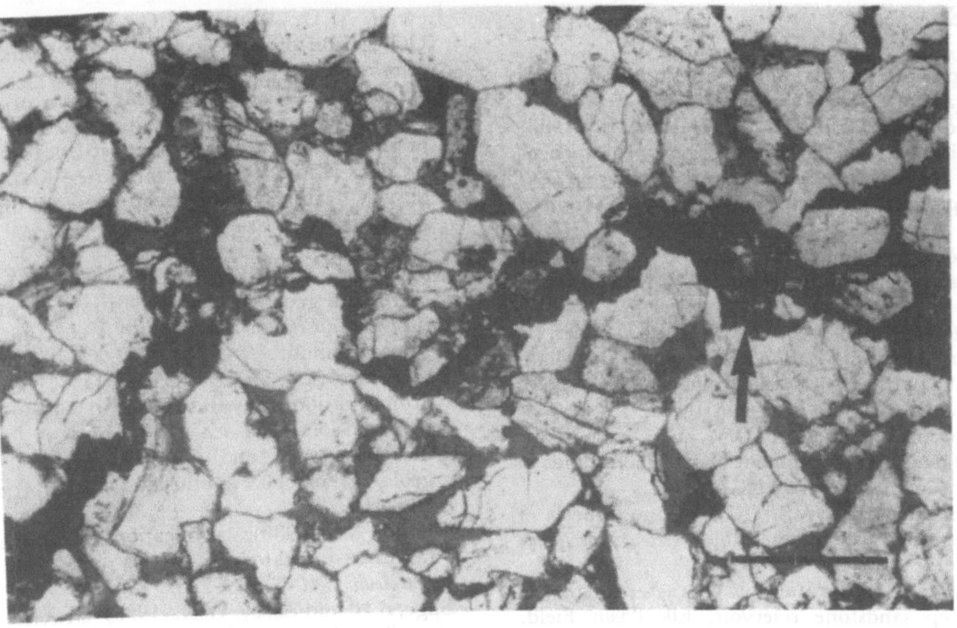

**Fig. 8.** A stylolite in the Ula Formation reservoir. The scale bar represents 500 μm.

sively bioturbated portions of the formation. Clay and mica seem to have a promotional effect on pressure solution such that the stylolites are probably first developed in these argillaceous sandstones. In turn, the silica released from pressure solution probably represents a significant source of authigenic quartz in the reservoir, which has a major effect on the porosity and permeability of the sandstone.

While the Ula Field petroleum column has not been chemically homogenized, except in the upper part, the density profile suggests that mechanical equilibrium has probably been attained in the field such that densities in the upper and lower reservoir zones are subequal. This is possible because petroleum is a multicomponent system and a given subsurface density can be achieved by many individual compositions. Indeed, the order-of-magnitude estimates of the rates of vertical diffusive mixing and mechanical mixing in reservoirs proposed by England and Mackenzie (1989), show that for reservoirs with high lateral permeabilities and significant vertical tortuosity (filled with thick columns of low-viscosity oils), mechanical mixing to attain mechanical equilibrium will occur more rapidly than attainment of vertical chemical equilibrium by diffusion. Thus, in the example of the Ula Field, where the filling event is relatively late, it might be expected that density-driven overturning of the field may dominate over diffusive equilibration processes in the petroleum column.

## SUMMARY

In this study a detailed three-dimensional compositional mapping of the accumulation permitted the selection of samples for more detailed molecular analysis, providing data on source facies and maturity. It was observed that:

- Within any portion of the field, the petroleum column of the upper Ula Formation is chemically fairly homogeneous vertically, but lateral variations across the field are evident in the solution gas and molecular data.
- The petroleum columns of the lower Ula Formation and Triassic reservoirs are chemically variable on both large and small scales, and local polar compound concentrations (small tar mats), chemically relatable to the accumulation, are definable on a stratigraphic basis in the lower portion of the reservoir. These may represent polar compound "lags" in potential carrier sands.
- It is noted that, in studies of field filling processes, traditional visual data analysis methods may need to be supported by numerical and statistical analyses of data sets.
- Despite chemical compositional variations and maturity differences in the field, the petroleum column in the Ula Field appears mechanically stable. This is attributed to methane transport within the field.
- The observed chemical compositional variations and the mechanical properties of the field are consistent with a Neogene filling as indicated by maturation modelling.

## ACKNOWLEDGEMENTS

This project is part of an ongoing joint study between the Department of Geology at the University of Oslo and BP Petroleum Development Norway A/S. The authors are grateful to BP for extensive technical and financial support. William England (BP Sunbury) and David Nash (BPPDN) provided many useful suggestions and comments. Bjørg Andresen (IFE) and Rainer Schaefer (KFA) performed the isotopic and light-hydrocarbon analyses respectively. We are also grateful to the ULA licence partners: Statoil, Norske Conoco A/S, K/S A/S Pelican & Co., Svenska Petroleum A/S, for access to sample material and permission to publish.

Much of the data reported here represents doctoral studies by Tor Nedkvitne and Dag Karlsen, who are supported by the Nordic Ministeråd and NAVF respectively. Steve Larter is supported by the Nordic Ministeråd Energy Research Program.

## REFERENCES


Clayton, C.J. (in press). Carbon isotope fractionation during natural gas generation. *Marine Petroleum Geol.*

Cornford, C., Needham, C.E.J. and De Walque, L. (1986). Geochemical habitat of North Sea oils and gases. In: Spencer, A.M. *et al.* (eds.), *Habitat of Hydrocarbons on the Norwegian Shelf*, pp. 39–54, Norwegian Petroleum Society, Graham & Trotman, London.

Cuiec, L. (1984). Rock/crude-oil interactions and wettability: an attempt to understand their interrelation. *Soc. Petroleum Eng. AIME*, 13211.

Eglinton, T.I., Larter, S.R., Karlsen, D.A., Brandt-De Boer, B., Boon, J.J. (in prep.). Characterisation of kerogens and asphaltenes by quantitative pyrolysis-mass spectrometry (Q-PY-MS): Applications to source rock and reservoir studies.

England, W.A., Mackenzie, A.S., Mann D.M. and Quigley T.M. (1987). The movement and entrapment of petroleum fluids in the subsurface. *J. Geol. Soc. London* **144**, 327–347.

England, W.A. and Mackenzie, A.S. (1989). Some aspects of the organic geochemistry of petroleum fluids. *Geologische Rundschau* **78** (1).

Espach, R.H. and Fry, J. (1951). Variable characteristics of the oil in the Tensleep sandstone reservoir, Elk Basin Field, Wyoming and Montana. *Petroleum Trans., AIME* **192**, 75–83.

Hirschberg, A. (1984). The role of asphaltenes in compositional grading of a reservoir's fluid column. *Soc. Petroleum Eng. AIME*, 13171.

Home, P.C. (1987). The Ula Field, block 7/12, Norway. In: Spencer A.M. *et al.* (eds), *Geology of the Norwegian Oil and Gas Fields*, pp. 143–151. Norwegian Petroleum Society, Graham & Trotman, London.

Karlsen, D. and Larter, S.R. (1989). A rapid correlation method for petroleum population mapping within individual petroleum reservoirs – application to petroleum reservoir description. In: Collinson, J. (ed.), *Correlation in Hydrocarbon Exploration* Norwegian Petroleum Society, Graham & Trotman, London, pp. 77–85.

Karlsen, D., Leythaeuser, D. and Larter, S.R. (in prep.). Gas isotope and light hydrocarbon maturity/source parameter, a question of sample and method quality. Submitted to *Organic Geochem.*

Larter, S.R. (in press). The molecular characterization of kerogen—applications to primary and secondary migration studies and to maturation modelling. *Proceedings of the 7th World Palynological Congress, Symposium on Kerogen Analysis in Hydrocarbon Exploration. Geologische Rundschau* **78** (1).

Lewan, M.D., Winters, J.C., and McDonald, J.H. (1979). Generation of oil-like pyrolysates from organic rich shales. *Science* **203**, 897–899.

Leythaeuser, D. and Rückheim, J. (in press). Heterogeneity of petroleum composition within a reservoir as a reflection of the migration and accumulation history. *Geochim. Cosmochim. Acta.*

McCord, D.R. (1953). Performance predictions incorporating gravity drainage and gas cap pressure maintenance—LL-370 Area, Bolivar Coastal Field. *Petroleum Trans, AIME* **198**, 231–248.

Mackenzie, A.S. (1984). Applications of biological markers in petroleum geochemistry. In: Brooks, J. and Welte, D. (eds.), *Advances in Petroleum Geochemistry* Vol. 1, pp. 115–214. Academic Press, London.

Mackenzie, A.S. and McKenzie, D. (1983). Isomerisation and aromatisation of hydrocarbons in sedimentary basins formed by extension. *Geol. Mag.* **120**, 417–528.

Mackenzie, A.S. and Quigley, T.M. (1988). Principles of geological prospect appraisal. *Bull. Am. Assoc. Petroleum Geol.* **72**, 399–415.

Radke, M. (1988). Application of aromatic compounds as maturity indicators in source rocks and crude oils. *Marine Petroleum Geol.* **5**, 224–226.

Sage, B.H. and Lacey, W.N. (1939). Gravitational concentration gradients in static columns of hydrocarbon fluids. *Trans. AIME* **132**, 120–131.

Schoell, M. (1983). Genetic characterization of natural gases. *Bull. Am. Assoc. Petroleum Geol.* **67**, 2225–2238.

Schulte, A.M. (1980). Compositional variations within a hydrocarbon column due to gravity. *Soc. Petroleum Eng., AIME*, paper no. 9235.

Schaefer, R.G., Welte, D.H. and Pooch, H. (1984). Geochemistry of low molecular weight hydrocarbons in two exploration wells of the Elmworth gas field (Western Canada Basin). *Organic Geochem.* **6**, 695–701.

Spencer, A.M., Home, P.C. and Wiik, V. (1986). Habitat of hydrocarbons in the Jurassic Ula Trend, Central Graben, Norway. In: Spencer, A.M. *et al.* (eds), *Habitat of Hydrocarbons on the Norwegian Continental Shelf* pp. 111–127. Norwegian Petroleum Society, Graham & Trotman, London.

Sundberg, K.R. and Bennet, C.R. 1981. Carbon isotope paleothermometry of natural gas. In: Bjorøy, M. *et al.* (ed.), *Advances in Organic Geochemistry 1981*, pp. 769–774. Wiley, New York.

Thompson, K.F.M. (1979). Light hydrocarbons in subsurface sediments. *Geochim. Cosmochim. Acta* **43**, 657–672.

# 19

# Lateral Extent of Calcite-cemented Zones in Shallow Marine Sandstones

*P.A. Bjørkum and O. Walderhaug*

Rogaland Research Institute, Stavanger, Norway

Calcite cementation in shallow marine sandstones occurs as continuously cemented layers, as stratabound concretions and as scattered concretions. These geometrically different types of calcite cementation can all be explained by diffusion-controlled growth of calcite cement at the expense of biogenic carbonate contained within the sandstones. When biogenic carbonate is concentrated in layers, stratabound concretions or continuously cemented layers form because calcite-cement nuclei may then be concentrated within the biogenic carbonate-rich layers. Continuously cemented layers are thought to form by the merging of stratabound concretions; stratabound concretions form when the biogenic carbonate within a layer is exhausted prior to merging of the concretions. Scattered concretions probably form when biogenic carbonate is homogeneously distributed within a sandstone as preferred levels of nucleation may be absent. Calcite-cemented layers formed from biogenic carbonate inherit the geometry of the biogenic carbonate-rich beds. Constraints may therefore be placed on the probable lateral extents of cored calcite-cemented layers by identifying the type of biogenic carbonate-rich beds within which the calcite cemented layers were formed. Cored calcite-cemented zones belonging to scattered concretions can in some cases be identified by the physical sedimentary structures preserved within them, and by not being associated with surfaces of erosion or non-deposition.

## INTRODUCTION

Shallow marine sandstones on the Norwegian shelf commonly contain calcite-cemented layers with thicknesses of up to a few metres (Bailey *et al.*, 1981; Olaussen *et al.*, 1984; Kantorowicz *et al.*, 1987; Saigal and Bjørlykke, 1987; Walderhaug *et al.*, 1989). Within such layers practically all porosity is filled by calcite cement, and permeability is insignificant. Similar calcite-cemented layers in shallow marine sandstones exposed onshore (Fig. 1(a)) may have lateral extends of several kilometres (Davies, 1967; Brenner and Davies, 1973; Bryant *et al.*, 1988), and correlation studies suggest that calcite-cemented layers in shallow marine sandstones on the Norwegian shelf may have lateral extents of at least 6 km (Walderhaug *et al.*, 1989).

In this chapter we present a model for the formation of calcite-cemented layers and discuss the prediction of their lateral extent when encountered in wells. The model is based on a study of calcite-cemented layers in Jurassic sandstones from the North Sea, the Haltenbanken area of the Norwegian Sea and the Barents Sea, combined with outcrop studies of the Jurassic Bridport Sands and Bencliff Grit in Dorset, southern England.

*North Sea Oil and Gas Reservoirs–II*


## SOURCES OF CALCITE CEMENT

Carbonate fossils are common in many recent shallow marine sands (e.g. Kumar and Sanders, 1976; Aigner and Reineck, 1982), and calcite-cemented zones in ancient shallow marine sandstones are often associated with concentrations of carbonate fossils (e.g. Brenner and Davies, 1973; Fursich, 1982; Bryant *et al.*, 1988). Diagenetic dissolution of carbonate fossils within shallow marine sandstones has also been observed (Knox *et al.*, 1982; Hudson and Andrews, 1987; Walderhaug *et al.*, 1989), and in some cases the isotopic composition of calcite cement clearly indicates that biogenic carbonate has been its dominant source (Fig. 2). Commonly, however, isotopic analysis shows that calcite cement contains carbon derived from the decomposition of organic matter (Irwin *et al.*, 1977; Curtis, 1978), but an influx of $CO_2$ cannot lead to the precipitation of more calcite than the amount that could form from the biogenic carbonate already within the sandstone, unless some other source of $Ca^{2+}$ is present (Saigal and Bjørlykke, 1987). $Ca^{2+}$ might be derived from plagioclase, but several sandstones on the Norwegian shelf contain calcite-cemented intervals and little evidence of plagioclase dissolution (Saigal and Bjørlykke, 1987; Walderhaug *et al.*, 1989). It is therefore thought that plagioclase is probably not an important source of $Ca^{2+}$ for calcite cement in sandstones from the Norwegian shelf.

The transport of dissolved calcite from underlying sediments is probably not significant either, as there is

**(a)**

**(b)**

normally little upward pore water flow through sandstones on the Norwegian shelf (Haszeldine *et al.*, 1984). In addition, the solubility of calcite does not necessarily increase with depth (Bjørlykke *et al.*, 1988). Fresh water flushing may increase the cumulative water flow through a sandstone, but in many cases isotopic analysis indicates that calcite cement has precipitated at advanced burial depths beyond the influence of fresh water flushing (Bjørlykke *et al.*, 1986; Lønøy, *et al.*, 1986; Saigal and Bjørlykke, 1987; Walderhaug *et al.*, 1989), or close to the seafloor prior to any fresh water flushing (Kantorowicz *et al.*, 1987). Furthermore, fresh water entering a sandstone may well be undersaturated with regard to $CaCO_3$.

When carbonate cementation takes place at the seafloor

(c)

**Fig. 1.** (a) Laterally extensive calcite-cemented layers in the coastal exposures of the Jurassic Bridport Sands south of Burton Bradstock in Dorset (England). The calcite-cemented layers stand out in relief owning to greater resistance to weathering, and can be traced laterally for 3 km. (b) Close-up view of calcite cementation in the Bridport Sands, Burton Cliff, Dorset. The photograph illustrates that calcite cementation occurs both as continuously cemented layers (top and bottom layers) and as layers of stratabound concretions (two middle layers). The upper continuously cemented layer consists of a first generation of calcite cement forming evenly spaced concretions (light colour) and later precipitated cement of a darker colour. Scale is 2 m long. (c) Large calcite-cemented concretion from the Jurassic Bencliff Grit, Osmington Mills, Dorset. The Bencliff Grit contains calcite-cemented concretions scattered throughout the sandstone. Many of the concretions at this locality, including the one shown in the photograph, contain perfectly preserved hummocky lamination and are not associated with surfaces of erosion or non-deposition. Cored calcite-cemented intervals in the hummocky-laminated shoreface sndstones of the Jurassic Rannoch Formation (North Sea) may be of the same type. Scale = 1.1 m.

(hardgrounds), high-Mg calcite or aragonite may possibly be precipitated from seawater (Bathurst, 1975; Kennedy and Garrison, 1975).

## FORMATION OF CALCITE CEMENTED LAYERS—A GROWTH MODEL

Homogeneous nucleation of calcite cement is dependent on the spatial variation in the degree of supersaturation within the pore water, whereas heterogeneous nucleation is governed by the distribution of favourable nucleation substrates. Biogenic carbonate can be a favourable substrate for calcite nucleation (Bathurst, 1975), but as many carbonate fossils consist of aragonite or high-Mg calcite they may not be favourable nucleation substrates for calcite. Poisoning of carbonate fossil surfaces by various inhibitors may also prevent nucleation (Berner, 1980, p. 189). Sites of calcite-cement nucleation are therefore probably controlled by the spatial variation in concentration of dissolved $CaCO_3$. If it is accepted that biogenic carbonate is the dominant source of dissolved $CaCO_3$ within shallow marine sandstones, the critical supersaturations necessary to form calcite cement nuclei will first be achieved within layers rich in biogenic carbonate.

The formation of calcite-cement nuclei lowers the concentration of dissolved calcite in the surrounding pore water, and dissolved calcite will diffuse inwards and precipitate at the surfaces of the nuclei. This will in turn cause further dissolution of clastic carbonate within the range of influence of each nucleus (i.e. the radius of the volume where dissolved $CaCO_3$ concentrations are reduced

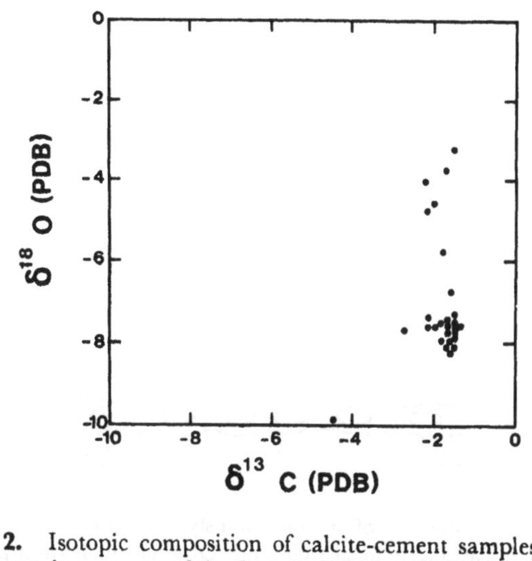


**Fig. 2.** Isotopic composition of calcite-cement samples from the coastal exposures of the Jurassic Bridport Sands (England). $\delta^{13}C$-values are close to zero, which strongly indicates that biogenic carbonate was the dominant source of calcite cement.

owing to the growth of a nucleus). Through this feedback mechanism a nucelus will grow into a calcite-cemented concretion at the expense of the clastic carbonate in the surrounding sediment. For pore water flow rates typical of compacting sedimentary basins, i.e. 0.01–0.001 cm/year (Bjørlykke *et al.*, 1988), the growth process is unaffected by fluid flow, and transport of dissolved calcite takes place by diffusion (Berner, 1980, p. 116). Since a nucleus lowers the concentration of dissolved calcite in its vicinity, nucleation of new nuclei will be unlikely within the range of influence of earlier-formed nuclei. This growth model can explain both the semiregular spacing observed between stratabound concretions and the fact that calcite-cemented layers may be continuously cemented or consist of stratabound concretions (Fig. 1(b)). Continuously cemented layers are regarded as a more advanced growth stage of stratabound concretions (Fig. 3). Whether continuously cemented layers or stratabound concretions form is a consequence of the amount of biogenic carbonate originally present within the carbonate fossil-rich layer. If the biogenic carbonate is exhausted before concretions merge into a continuous layer, stratabound concretions form. (For a more detailed discussion of the growth model, see Bjørkum and Walderhaug (in press)). The model may not apply to calcite-cemented layers formed at the seafloor or within thin, vertically confined zones characterized by specific microbiological reactions (Raiswell, 1987). The location of calcite-cement nuclei may then be controlled by factors other than the spatial variation in the content of biogenic carbonate.

## LATERAL EXTENT OF CALCITE CEMENTED LAYERS

According to the growth model presented here, a calcite-cemented layer formed from biogenic carbonate will inherit the geometry of the host biogenic carbonate layer. Predicting the lateral extent of calcite-cemented layers is thus largely a sedimentological problem, and data regarding their lateral extent can be obtained by studying carbonate fossil-rich/calcite-cemented layers in modern shallow marine sediments and in ancient analogues.

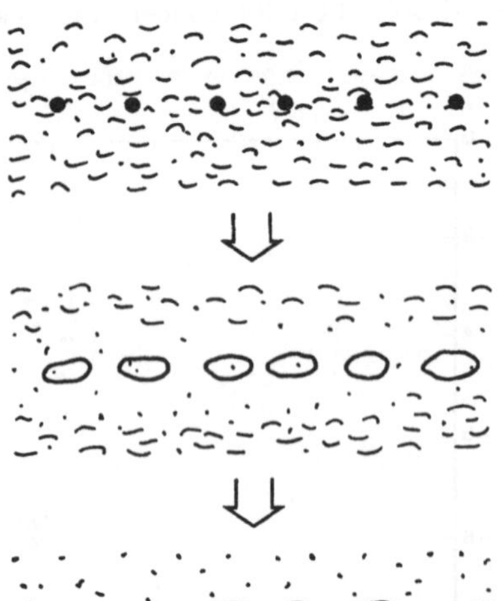

**Fig. 3.** Formation of a continuously cemented layer from a layer of carbonate fossil-rich sand. Original nucleation points grow into concretions as the clastic carbonate in the layer dissolves, and eventually the concretions merge and form a continuously cemented layer.

Several types of carbonate fossil-rich layers are present in modern and ancient shallow marine sediments. These include the following.

Type 1—layers rich in biogenic carbonate formed by the accumulation *in situ* of carbonate fossils during periods of low clastic input. The layers may experience short-term reworking *in-situ* and winnowing (Fursich, 1982; Davies, 1967)

Type 2—carbonate fossil lags at the base of storm-deposited layers (e.g. Kumar and Sanders, 1976; Aigner and Reineck, 1982; DeCelles, 1987)

Type 3—carbonate fossil lags located at the base of submarine channels (Brenner and Davies, 1973)

Type 4—carbonate fossil lags located above submarine erosion surfaces (Nummedal and Swift, 1987)

The Fensfjord Formation of the Brage Field contains calcite-cemented layers thought to have formed from type 1 accumulations. Correlation between wells suggests that the most extensive layers have lateral extents greater than 6 km (Walderhaug *et al.*, 1989). Davies (1967) considered that the laterally extensive (>3 km) calcite-cemented layers in the coastal exposures of the Bridport Sands (Fig. 1(a)) were formed from type 1 accumulations, although Bryant *et al.*, (1988) suggested that the same calcite-cemented layers formed from allochthonous biogenic carbonate transported during storms. We consider the interpretation presented by Davies (1967) to be the most likely. A study of recent shallow marine sands seawards from Sapelo Island has shown that biogenic carbonate-rich layers formed by storm reworking and winnowing can have lateral extents of 15 km (Howard and Reineck, 1972). Little quantitative data regarding the lateral extent of type 2 layers are available, but they have been reported to have lateral extents ranging from 1 metre to a few hundred metres (Brenner and Davies, 1973; DeCelles, 1987). Laterally amalgamated lags of type 3 have been reported to have lateral extents of more than 1 km (Brenner and Davies, 1973), whereas lateral extents of transgressive lags of type 4 are uncertain.

Calcite-cemented layers formed from type 1 accumulations can often be recognized in cores by extensive bioturbation, by absence of a coarse siliciclastic lag, by absence of fossils transported from other environments, and by not being located above an erosion surface. Type 2 layers may be recognized by their location at the base of a storm-deposited layer (as described by Kumar and Sanders, 1976), by their containing a coarse siliciclastic lag, and sometimes by their containing fossils transported from other environments (Aigner and Reineck, 1982). Layers of types 3 and 4 are located above erosion surfaces and are coarser grained than the underlying sediments.

Biogenic carbonate may also be homogeneously distributed within a sandstone instead of being concentrated in certain layers. In such cases the growth model predicts that no preferred levels of nucleation exist. Nucleation points will consequently be uniformly distributed in three dimensions, and calcite cementation will occur as scattered concretions (Bjørkum and Walderhaug, in press). This type of calcite cementation is well-developed in the Bencliff Grit (Fig. 1(c)). In a non-bioturbated sequence, such calcite-cemented concretions can probably be recognized in cores by not being associated with surfaces of erosion or non-deposition (Fig. 1(c)).

Carbonate-cemented layers formed at the seafloor (hardgrounds) may have considerable lateral extents, as they form in response to periods of non-deposition (Kennedy and Garrison, 1975) which can affect very large areas. Hardgrounds may often be recognized by fibrous cement textures, borings, encrusting fauna, cemented intraformational clasts (Bathurst, 1975), abundant glauconite or phosphate (Kennedy and Garrison, 1975) and by $\delta^{18}O$-values close to zero (Nelson and Lawrence, 1984). Calcite-cemented layers formed in the sulphate-reducing zone, largely from organically derived $CO_2$, may also have considerable lateral extents, since they also form in response to periods of non-deposition (Raiswell, 1987; Kantorowicz *et al.*, 1987). This type of calcite cementation may be characterized by very negative $\delta^{13}C$ values ($-20$ to $-50$), $\delta^{18}O$-values close to zero, and low iron content in the calcite cement (Kantorowicz *et al.*, 1987).

If several wells penetrate the same sequence, correlation allows constraints to be placed on the lateral extents of calcite-cemented layers. If calcite-cemented intervals encountered in different wells belong to the same layer, the cored zones should belong to the same genetic type of calcite-cemented layer. Other points in favour of correlating the cored calcite-cemented intervals include similar thickness, the same grain size of the siliciclastic material, the same type of fossils (and possibly trace fossils), the same degree of fossil fragmentation and similar types and amounts of early diagenetic minerals such as glauconite, phosphate or pyrite (Walderhaug *et al.*, 1989).

Additional information on the lateral extent of calcite-cemented layers could probably be obtained in some cases from pressure tests and by the use of tracers. Calcite-cemented layers are too thin to be recognizable on standard seismic, but further advances in geophysical methods might possibly allow the lateral extent of the thickest calcite-cemented layers to be determined (Berg *et al.*, this volume).

If continuous layers do in fact form by the merging of concretions, any evidence within a cored calcite-cemented interval that indicates merging of neighbouring concretions probably also indicates the presence of a continuous layer. One sign of a continuous layer is concave upper and lower boundaries as shown in Fig. 4. If closely spaced samples along a horizontal transect of the cored calcite-cemented zone reveal a symmetrical variation in geochemical parameters (e.g. $\delta^{13}C$, $\delta^{18}O$, $^{87}Sr/^{86}Sr$ or trace element content) (Fig. 4), this would also indicate the presence of two merged concretions, since radial geochemical variation in concretions commonly occurs (e.g. Coleman and Raiswell, 1981; Curtis and Coleman, 1986). The thickness of a calcite-cemented interval in a core is probably not a reliable indicator of continuity, since concretions in the Bencliff Grit may be 1 metre thick (Fig. 1(c)), whereas continuous (3 km) layers in the Bridport Sands may be only a few decimetres thick (Fig. 1(b)).

Even if it cannot be determined whether a given calcite-cemented layer is continuous or not, the growth model presented here allows predictions to be made about the possible distribution of uncemented "holes" within a layer. Since original nucleation points are thought to be located with a semiregular spacing, the distribution of uncemented holes will also be semiregular. The variation in the diameters of any holes present is expected to be small owing to the uniformity of the growth process. If cementation has gone so far that the neighbouring concretions are in contact with each other, any remaining uncemented holes will be smaller than the concretions, i.e. a few decimetres or less in diameter.

## CONCLUSIONS

Calcite cement in shallow marine sandstones is largely derived from biogenic carbonate contained within the sandstones, but $CO_2$ produced by microbiological decomposition of organic matter is also an important source of $CO_3^{2-}$.

Calcite cement occurs as continuously cemented layers, as stratabound concretions, and as scattered concretions. This can be explained by assuming diffusion-controlled growth of calcite cement at the expense of biogenic carbonate contained within the sandstones. Cemented layers and stratabound concretions are thought to form when clastic carbonate is concentrated in layers. Concentrations of dissolved $CaCO_3$ are then highest within these layers, and calcite-cement nuclei are consequently concentrated in their interiors. Continuously cemented layers are regarded as a more advanced growth stage of stratabound concretions, and form when the amount of biogenic carbonate within the bed is high enough to allow concretions to merge. Scattered concretions form when biogenic carbonate is homogeneously distributed within a sandstone because preferred levels of nucleation may be absent in such a system.

Detailed core description, petrographic study and geochemical analysis can enable cored layers to be assigned to different genetic classes. Since lateral extent seems to vary from class to class, some constraints can be placed on the probable extent of cored calcite-cemented layers. Lateral extents of calcite-cemented layers can be further constrained by correlation between wells, and scattered concretions may in some cases be identified by their content of preserved sedimentary structures and their lack of association with surfaces of erosion or non-deposition.

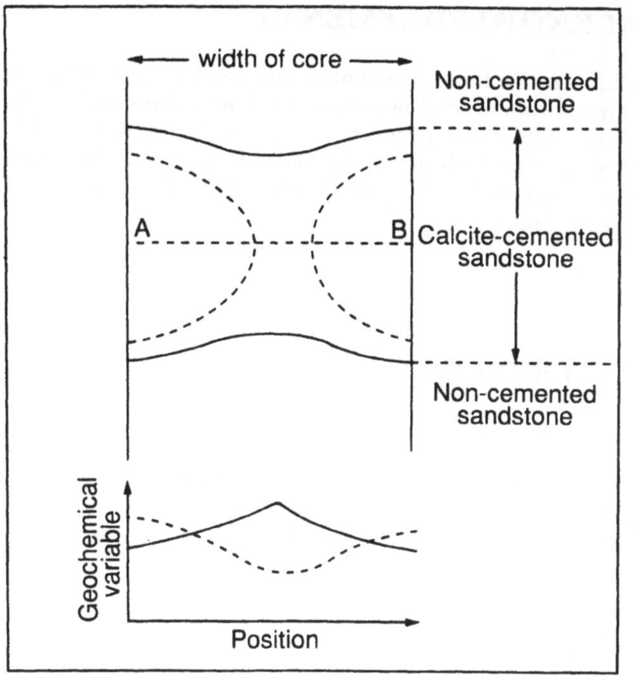


**Fig. 4.** Recognizing a continuously cemented layer in a core. If the upper and lower boundaries of the layer are concave, this may indicate that two concretions have merged and that the cored interval is part of a continuously cemented layer. If geochemical analyses are performed along a horizontal transect of the layer and a symmetrical variation in geochemical parameters is found, this will also indicate that the cored interval contains two merged concretions, as concretions commonly show a radial variation in geochemical composition.

## ACKNOWLEDGEMENTS

This contribution is based on results from a study of calcite cementation in shallow marine sandstones funded by Esso Norge A/S. Their permission to publish results from this project is gratefully acknowledged. The paper has benefited from reviews by Knut Bjørlykke, Steve Ehrenberg and Girish Saigal.


## REFERENCES


Aigner, T. and Reineck, H.-E. (1982). Proximality trends in modern storm sands from the Helgoland Bight (North Sea) and their implications for basin analysis. *Senkenbergiana Maritima*, **14**, 183–215.

Bailey, C.C., Price, I. and Spencer, A.M. (1981). The Ula oil field, block 7/12, Norway. In: *Norwegian Symposium on Exploration*, article 18, pp. 1–26.

Bathurst, R.G.C. 1975. *Carbonate Sediments and their Diagenesis*, 2nd Edn. Elsevier, Amsterdam.

Berg, E., Brevik, I. and Buller, A.T. Experiences gained using a seismic inversion method for detailed reservoir studies. (*This volume*)

Berner, R.A. 1980. Early Diagenesis. Princeton University Press, Princeton, 241 p.

Bjørkum, P.A. and Walderhaug, O. (in press). Geometrical arrangement of calcite cementation in shallow marine sandstones. In: *Proceedings from a Conference on Geochemical Self-Organization, Santa Barbara, California, June 1988.* Earth Science Reviews.

Bjørlykke, K., Aagaard, P., Dypvik, H., Hastings, D.S. and Harper, A.S. (1986). Diagenesis and reservoir properties of Jurassic sandstones from the Haltenbanken area offshore mid Norway. In: Spencer, A.M. *et al.*, (eds.), *Habitat of Hydrocarbons on the Norwegian Continental Shelf* pp. 275–286. Norwegian Petroleum Society, Graham & Trotman, London.

Bjørlykke, K., Mo, A. and Palm, E. (1988). Modelling of thermal convection in sedimentary basins and its relevance to diagenetic reactions. *Marine Petroleum Geol.* **5**, 338–351.

Brenner, R.L. and Davies, D.K. (1973). Storm generated coquinoid sandstone: genesis of high-energy marine sediments from the Upper Jurassic of Wyoming and Montana. *Geol. Soc. Am. Bull.*, **84**, 1685–1698.

Bryant, I.D., Kantorowicz, J.D. and Love, C.F. (1988). The origin and recognition of laterally continuous carbonate-cemented horizons in the Upper Lias sands of southern England. *Marine Petroleum Geol.* **5**, 108–133.

Coleman, M.L. and Raiswell, R. (1981). Carbon, oxygen and sulphur isotope variations in concretions from the Upper Lias of N.E. England. *Geochim. Cosmochim. Acta* **45**, 329–340.

Curtis, C.D. and Coleman, M.L. (1986). Controls on the precipitation of early diagenetic calcite, dolomite and siderite concretions in complex depositional sequences. In: Gautier, D.L. (ed.), *Roles of Organic Matter in Sedimentary Diagenesis*, pp. 23–35. Special Publication No. 38. Society of Economic Paleontologists and Mineralogists, Tulsa.
D.L. (ed.), *Roles of Organic Matter in Sedimentary Diagenesis*, pp. 23–25. Special Publication No. 38. Society of Economic Paleontologists and Mineralogists, Tulsa.

Davies, D.K. (1967). Origin of friable sandstone—calcareous sandstone rhythms in the Upper Lias of England. *J. Sediment. Petrol.* **37**, 1179–1188.

DeCelles, P.G. (1987). Variable preservation of Middle Tertiary coarse-grained, nearshore to outer shelf storm deposits in southern California. *J. Sediment. Petrol.* **57**, 250–264.

Fursich, F.T. (1982). Rhythmic bedding and shell bed formation in the Upper Jurassic of east Greenland. In: Einsele, Grand Seilacher, A. (eds.), *Cyclic and Event Stratification*, pp. 208–222. Springer, Berlin.

Haszeldine, R.S., Samson, I.M. and Cornford, C. (1984). Quartz diagenesis and convective fluid movement: Beatrice oil field, U.K., North Sea. *Clay Minerals* **19**, 391–402.

Howard, J.D. and Reineck, H.-E. (1972). Georgia coastal region, Sapelo Island, U.S.A.: Sedimentology and biology. IV. Physical and biogenic sedimentary structures of the nearshore shelf. *Senkenbergiana Maritima*, **4**, 47–79.

Hudson, J.D. and Andrews, W.J. (1987). The diagenesis of the Great Estuarine Group, Middle Jurassic, Inner Hebrides, Scotland. In: Marshall, J.D. (ed.), *Diagenesis of Sedimentary Sequences*, pp. 259–276. Geological Society Special Publication N. 36. Blackwell, Oxford.

Irwin, H., Curtis, C. and Coleman, M. (1977). Isotopic evidence for source of diagenetic carbonates formed during burial of organic-rich sediments. *Nature* **269**, 209–213.

Kantorowicz, J.D., Bryant, I.D. and Dawans, J.M. (1987). Controls on geometry and distribution of carbonate cements in Jurassic sandstones: Bridport Sands, southern England and the Viking Group, Troll Field, Norway. In: Marshall, J.D. (ed.), *Diagenesis of Sedimentary Sequences*, pp. 103–118. Geological Society Special Publication No. 36. Blackwell, Oxford.

Kennedy, W.J. and Garrison, R.E., (1975). Morphology and genesis of nodular chalks and hardgrounds in the Upper Cretaceous of southern England. *Sedimentology* **22**, 311–386.

Knox, R.W.O'B., Morton, A.C. and Lott, G.K. (1982). Petrology of the Bridport Sands in the Winterborne Kingston borehole, Dorset. In: Rhys, G.H., Lott, G.K. and Calver, M.A. (eds.), *The Winterborne Kingston Borehole, Dorset, England*, pp. 107–121. Reports of the Institute of Geological Sciences, Leeds, N. 81/3.

Kumar, N. and Sanders, J.E. (1976). Characteristics of shoreface storm deposits: modern and ancient examples. *J. Sediment. Petrol.* **46**, 145–162.

Lønøy, A., Akselsen, J. and Rønning, K. (1986). Diagenesis of a deeply buried sandstone reservoir: Hild Field, northern North Sea. *Clay Minerals* **21**, 497–511.

Nelson, C.S. and Lawrence, M.F. (1984). Methane-derived high-Mg-calcite submarine cement in Holocene nodules from the Fraser delta, British Columbia, Canada, *Sedimentology* **31**, 645–654.

Nummedal, D. and Swift, D.J.P. (1987). Transgressive stratigraphy at sequence-bounding unconformities: some principles derived from Holocene and Cretaceous examples. In: Nummedal, D., Pilkey, O.H. and Howard, J.D. (eds.), *Sea-level Fluctuation and Coastal Evolution* pp. 241–260. Special Publication No. 41. Society of Economic Paleontologists and Mineralogists, Tulsa.

Olaussen, S., Dalland, A., Gloppen, T.G. and Johannessen, E. (1984). Depositional environment and diagenesis of Jurassic reservoir sandstones in the eastern part of the Troms I area. In: Spencer, A.M. *et al.* (eds.), *Petroleum Geology of the North European Margin*, pp. 61–79. Norwegian Petroleum Society, Graham & Trotman, London.

Raiswell, R. (1987). Non-steady state microbiological diagenesis and the origin of concretions and nodular limestones. In: Marshall, J.D. (ed.), *Diagenesis of Sedimentary Sequences*, pp. 41–54. Geological Society Special Publication No. 36. Blackwell, Oxford.

Saigal, G. and Bjørlykke, K. (1987). Carbonate cements in clastic reservoir rocks from offshore mid Norway – relationship between isotopic composition, textural development and burial depth. In: Marshall, J.D. (ed.), *Diagenesis of Sedimentary Sequences*, pp. 313–324. Geological Society Special Publication No. 36. Blackwell, Oxford.

Walderhaug, O., Bjørkum, P.A. and Bolås, H.M.N. (1989). Correlation of calcite-cemented layers in shallow-marine sandstones of the Fensfjord Formation in the Brage Field. In: Collinson, J.D. (ed.), *Correlation in Hydrocarbon Exploration* pp. 367–375. Norwegian Petroleum Society, Graham & Trotman, London.

# 20

# Sand Body Dimensions and Infill Sequences of Stable, Humid-Climate Delta Plain Channels

*T. Dreyer*

Norsk Hydro Research Centre, Bergen, Norway

Sediments deposited in delta plain settings are common in the Jurassic sequences of the Norwegian continental shelf (e.g. the Ness Formation). The most prominent reservoir units in this type of deposit are distributary channel sand bodies. These mainly appear to be laterally discontinuous in a direction normal to the palaeoslope, and probably reflect cut-and-fill processes in stable distributaries. Based on well-data alone, it is difficult to determine vital reservoir parameters such as the geometry and interconnectedness of these channel sand bodies. Studies of well-exposed possible analogues like the Saltwick Formation of Yorkshire (England) and the Aspelintoppen Formation of Spitsbergen were therefore undertaken. This study indicates that single-storey sand bodies deposited in stable cut-and-fill channels have ribbon-like geometries, whereas multistorey channel bodies of the same type tend to display more sheet-like dimensions. A curve relating the width and thickness of these sand bodies has been constructed. Humid climate cut-and-fill sand bodies deposited in stabilized distributaries can be recognized by an infill sequence that, when complete, reflects three stages of infilling (active, partly active and passive phase fills). Active phase infills frequently have the best reservoir quality at the top, owing to high-discharge events prior to (partial) abandonment. In addition to palaeo-river morphology, sandbody dimensions and the degree of interconnectedness are governed by the pattern of differential subsidence along depositional strike, and by the ratio between sediment supply and accommodation.

## INTRODUCTION

In recent years, much attention has been focused on studies concerned with increasing recovery from proven hydrocarbon accumulations. Within the field of geology, this interest in reservoir optimization has led to a growing need for quantified geological models where the architecture of the reservoir can be specified according to geostatistical methods (e.g. Lake and Carroll, 1986; Gundesø and Egeland, this volume). One of the key elements in geological quantification is the concept of reservoir "building blocks" (Fig. 1). Briefly, a reservoir building block is a body of rock distinct from other bodies of rock on the basis of specific petrophysical and geometrical properties. The first step towards defining building blocks is the breakdown of the reservoir into large-scale permeability classes (Dreyer *et al.*, in press). The next steps involve dimensional analyses of the various building blocks (fieldwork studies of analogues), coupled with studies of their interrelationships.

The quantification of geological parameters is especially important in reservoirs with a high degree of heterogeneity. The Jurassic Ness Formation, which consists of delta plain deposits (e.g. Ryseth, in press), is one of the most heterogeneous reservoir types on the Norwegian shelf. The present study is concerned with the main reservoir building block of such delta plain sequences; the distributary channel sand bodies. These usually represent high-permeability "pipelines" running through sequences that tend to be dominated by fine-grained material. The main aim of this study is to present data concerning the geometry and infill sequences of these delta plain channel sand bodies. Detailed comparative studies between exposed and subsurface delta plain deposits are outside the scope of this contribution. The dimensional data given here can be used in geological simulations of delta plain reservoirs (Gundesø and Egeland, this volume), or as a basis for detailed correlations in delta plain sequences (e.g. Ryseth, in press).

*North Sea Oil and Gas Reservoirs–II*

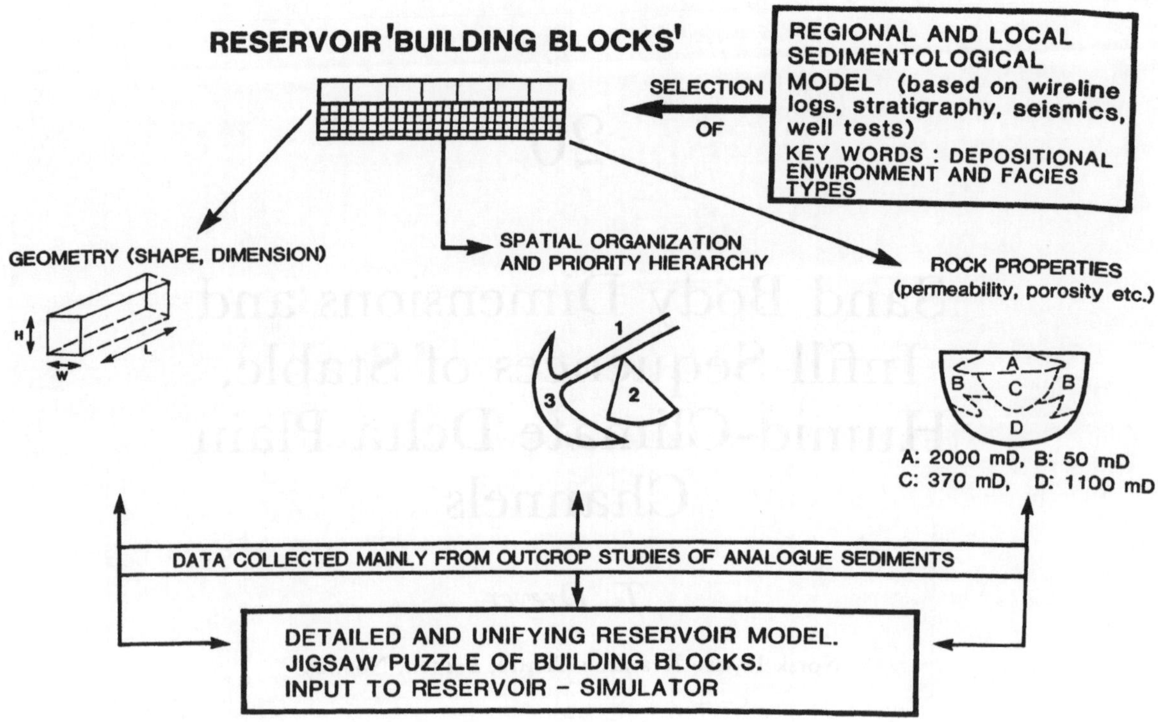


**Fig. 1.** Organization of studies concerned with quantification of geological reservoir parameters.

## COMMENTS ON DELTA PLAIN CHANNEL DEPOSITS IN THE NESS FORMATION, NORTHERN NORTH SEA

The Middle Jurassic Brent Group has long been the primary reservoir target in North Sea hydrocarbon exploration. Until recently, it has been customary to produce only from the sandy delta front/shallow marine intervals (Rannoch, Etive, and Tarbert Formations) of this reservoir sequence. The reservoir potential of the heterogeneous delta plain sediments of the Ness Formation has largely been neglected. However, the growing need for reservoir optimization has lately caused a re-evaluation of the production strategy for the Brent Group. In several North Sea fields, a significant amount of the reserves in the Ness Formation is presently changing status from "in place" to "recoverable".

Quantification of sand body geometries in the Ness Formation requires an understanding of the fundamental processes which controlled deposition in the Viking Graben during the Middle Jurassic. It is firmly established that the Ness Formation represents the delta/coastal plain part of the Brent deltaic system (e.g. Graue *et al.*, 1987). At maximum progradation (Fig. 2), the fluvial/wave-dominated delta covered an extensive area, and upper delta plain conditions were present up to 61° 30′N. Sedimentation was influenced by synsedimentary tectonics, especially in the form of increased rates of subsidence on the downthrown side of fault segments, and a relatively humid climate prevailed (Ryseth, in press).

Detailed investigations of core material from the Ness Formation channel sandstones have been carried out by Dreyer (unpublished Norsk Hydro data) and Ryseth (in press). Interpreted vertical core sequences from the Ness Formation are shown in Fig. 3. Although no definite conclusions regarding channel type and sand body extents can be drawn from studies of borehole data (e.g. Bridge, 1985), there are some indications that Ness channel bodies for the most part are laterally restricted sand units that formed in stabilized (non-migrating) rivers. These indications can be summarized as follows.

(1) A compilation of Ness Formation channel sand vs. floodplain fines ratios from northern North Sea wells gives an average ratio of 0.34. This suggests that the delta plain was dominated by interchannel sedimentation. It is thus unlikely that sheet-like sand bodies produced by highly mobile channels or channel-belts (e.g. braided streams) are common in the Ness Formation (Allen, 1978; Bridge and Leeder, 1979).

(2) Lateral accretion is the primary sandbody-formation mechanism in high-sinuosity meandering streams. In cores, this process can be recognized by a more or less well-developed fining-upwards sequence, accompanied by a structural assemblage corresponding to point-bar migration (e.g. Allen, 1970; Jackson, 1978). The infill sequences in Ness channels rarely conform to these meandering stream characteristics, but tend to reflect vertical accretion and a stepwise decrease in flow-energy. This will be discussed in detail in a separate section.

(3) With the closer well spacing obtained in the production phase of North Sea fields, correlations normal to the main channel flow direction may give considerable insight into the geometry of the channel sand bodies. In the Oseberg field, channel bodies can usually not be correlated even between well-pairs a few hundred metres apart in sections normal to palaeocurrent (e.g. Fig. 4). This suggests a dominance of laterally restricted sand bodies (usually less than a few hundred metres wide) in the Ness Formation. However, as discussed below, vertical stacking together with splitting and joining of channel bodies in the horizontal plane may prevent the laterally restricted channel bodies from becoming isolated.

(4) Ryseth (in press) shows that Ness channels seem to be characterized by strong incision into the surrounding

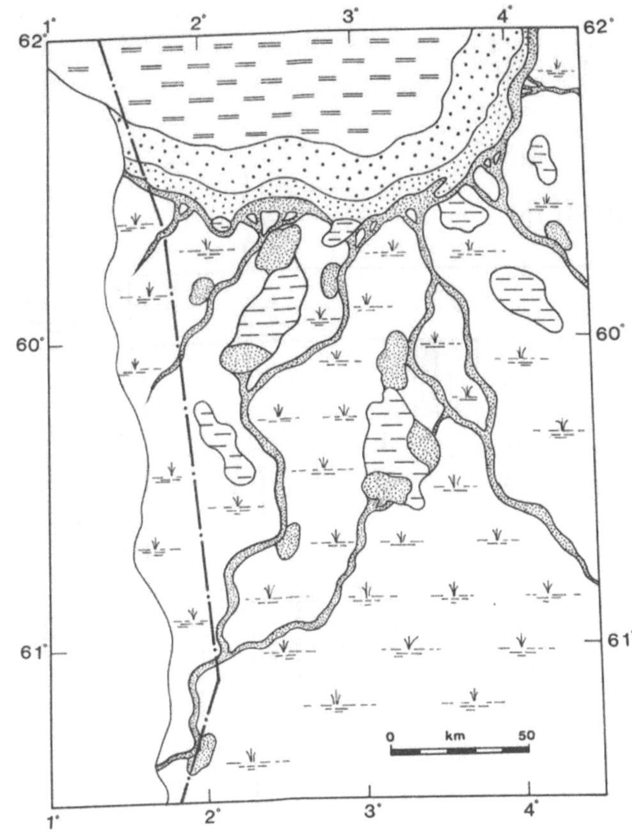


**Fig. 2.** Northern North Sea palaeogeography during maximum extension of the Brent delta (Bajocian). (Modified from Graue *et al.*, 1987.)

fines. This has also been documented for distributary channels in a similar delta plain setting by Hopkins (1985). Deep incision and stable banks will fix the positions of channels, thus preventing substantial lateral migration (e.g. Coleman and Prior, 1982). As a consequence, ribbon-shaped sand bodies will tend to form.

(5) Single-storey channels in the Ness Formation often exhibit thicknesses exceeding 5 m, whereas multistorey sand bodies (2 or 3 storeys) may be more than 20 m thick (Fig. 3). This implies the presence of relatively deep channels. The absence of sequences relatable to point-bar deposition (above) suggests that these channels were filled mainly by vertical accretion in a non-migrating river.

The palaeochannel-pattern of the non-migrating (stable) rivers that produced the laterally restricted sand bodies in the Ness Formation might have varied considerably. According to Schumm (1977) and Rust (1978), stable river courses may be either of the low-sinuousity, single-channel type (*straight*) or of the multiple-channel, high- to variable-sinuosity type (*anastomosed*). Both types are common in settings with very low slopes, such as an extensive delta plain (cf. the Ness Formation, Graue *et al.*, 1987). Channels of the former type are commonly dominated by rapid vertical infilling following a phase of incision into surrounding fine-grained sediments (Kelling and George, 1971; Meckel, 1972; Land and Weimer, 1978; Coleman and Prior, 1982; Haszeldine, 1983; Okolo, 1983; Hopkins, 1985). In some cases, lateral accretion associated with fluvial bars may also contribute to the infilling process (e.g. Haszeldine, 1983). The straight channels tend to shift position frequently in response to changes in gradients on the delta plain, but remain stable in their positions between shifts (fixed channels, Friend, 1983). Anastomosed rivers are also dominated by rapid vertical infilling, but commonly display a branching and joining pattern of relatively deep and stable channels (e.g. Smith and Smith, 1980; Smith, 1986). Channel bodies formed in this river type tend to be quite thick and often vertically stacked. This is probably related to the long residence times of individual channels or generations of channels in fixed positions on the floodplain (Smith, 1986). Settings where a moderate to high subsidence rate is balanced by a moderate to high rate of sediment supply promote this kind of vertical aggradation (Smith, 1986, Fig. 14).

As stated by Bridge (1985), it is doubtful whether an accurate reconstruction of paleochannel patterns can be made from subsurface data. Hence, it is very difficult to determine whether the Ness channel bodies were formed in straight or anastomosed river reaches. Perhaps the only way to make this distinction is to construct palaeogeographic maps along time-lines within the reservoir. These maps show the number and locations of co-existing channel bodies, thus facilitating the partial reconstruction of Ness palaeochannel patterns. This method is currently being tested on the Oseberg field (Dreyer, unpublished data), and it is strongly dependent on both well-spacing and the ability to define time-lines. Preliminary results suggest that vertical changes from straight to anastomosed palaeochannel patterns take place within the Ness Formation.

In conclusion, it is believed that the channels in the Ness Formation formed on an extensive delta plain wetland area characterized by low slopes, and probably reflect deposition in both straight and anastomosing river reaches. On the basis of subsurface data alone, it is very difficult to distinguish between these river types because the individual channels in both systems can be characterized as "stable (non-migrating), vertically aggraded cut-and-fill channels". As a consequence, all discussions concerning channel-bodies in the remainder of this chapter refer to channels conforming to this descriptive term.

## GEOLOGICAL SETTING OF SALTWICK AND ASPELINTOPPEN FORMATIONS

Quantitative information regarding humid-climate delta plain channel bodies is scarce in the literature (Fielding and Crane, 1987). To obtain the data necessary for constructing reservoir geological models for delta plain reservoirs (e.g. the Ness Formation), field studies in deposits of this kind were undertaken in the Eocene of Spitsbergen (the Aspelintoppen Formation, Fig. 5) and the Middle Jurassic of England (the Saltwick Formation, Figs. 6 and 8). These formations were both deposited on a humid climate delta plain characterized by discrete channel bodies set in a matrix of coals, impermeable floodplain mudstones, and muddy sandstones/sandy mudstones of bay-fill or crevasse-splay origin (Figs. 5 and 6). As for the Ness Formation, the sand/shale ratio in these sequences varies locally, but mostly stays within the 0.25–0.45 range.

The Saltwick Formation is 30–55 m thick, and represents the first episode of non-marine sedimentation in the Middle Jurassic sequence in Yorkshire (Livera and Leeder, 1981; Hancock and Fisher, 1981). It is progradational in its lower part, and shows signs of delta lobe abandonment towards the top (Livera and Leeder, 1981). Sand bodies deposited by both low-and high-sinuosity channels occur in the Saltwick Formation (Hemingway, 1974; Hancock

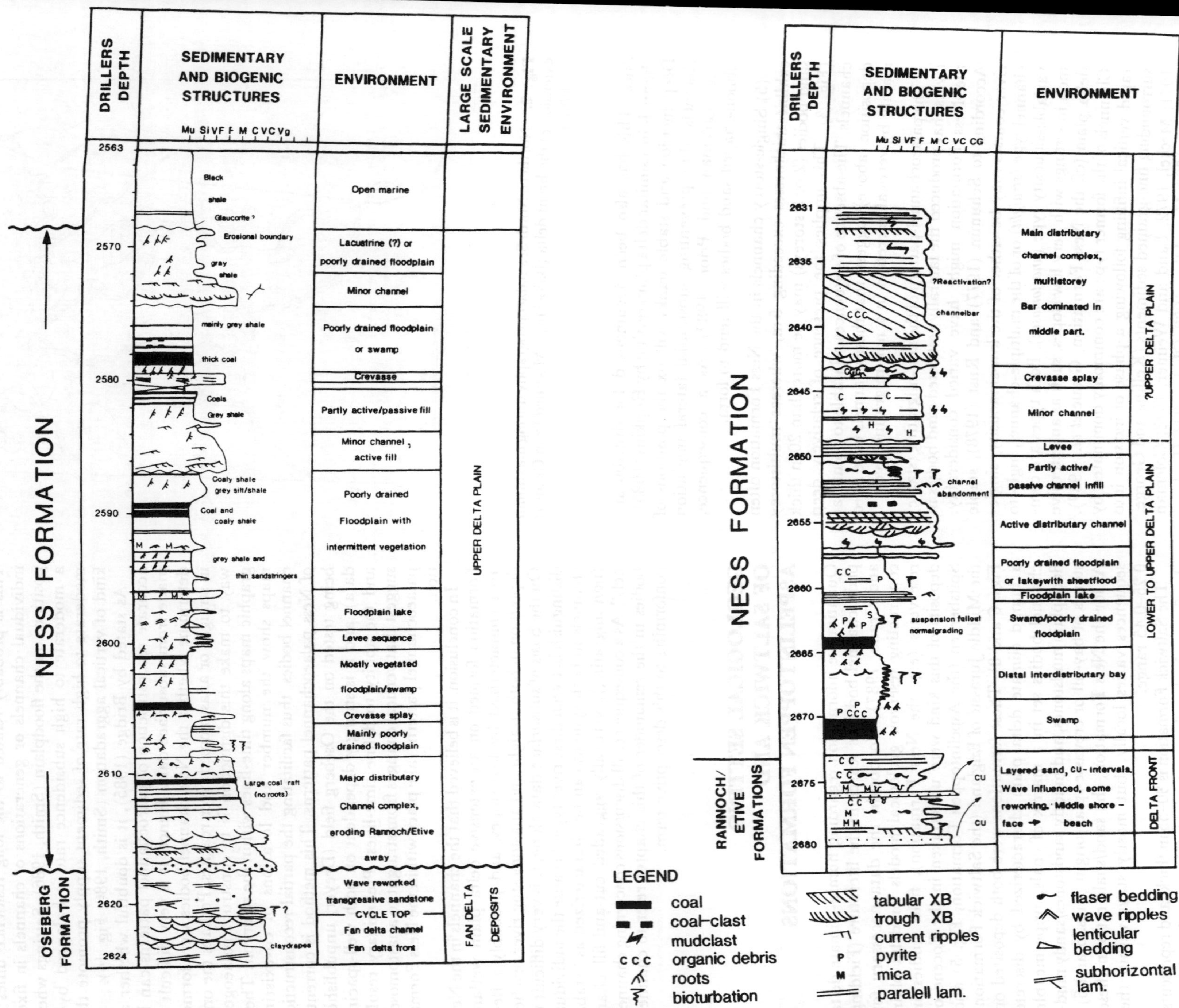


**Fig. 3.** Interpreted vertical profiles through delta plain sequences of the Ness Formation in two wells fromthe Oseberg field. Note the variation in amounts and vertical positions of channel sand bodies.

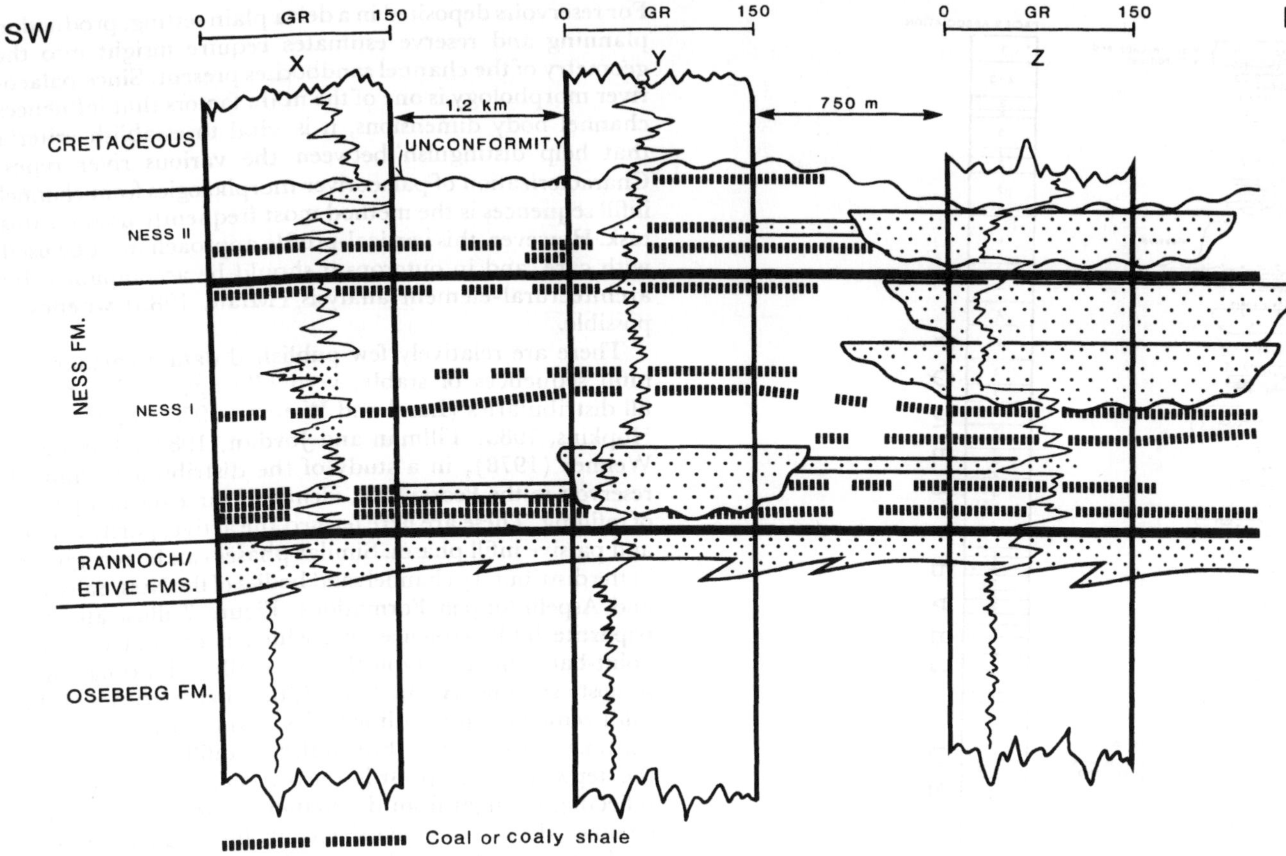


**Fig. 4.** Reservoir-scale correlation of the Ness Formation in three wells from the Oseberg field. Note the discontinuous channel sand bodies. The correlation is approximately normal to the main palaeocurrent direction.

and Fisher, 1981). In this study, we investigated only those channel bodies in which signs of lateral migration were not evident. They were distinguished by their concave bases, their lack of well-developed clay plugs and lateral accretion bedding, their tripartite infill sequences (see below), and their higher content of coarse-grained material. Moreover, in the cases where channel bodies could be traced for a considerable distance parallel with the palaeoslope, the low spread in paleocurrents (Fig. 11) suggests deposition in rather straight channels (e.g. Okolo, 1983; Haszeldine, 1983).

The distribution of channel sandstones in the Saltwick Formation seems to be partly dependent on the pattern of differential subsidence that existed during deposition (Alexander, 1986). It is likely that similar syn-sedimentary effects influenced Ness Formation sedimentation, as witnessed by the considerable thickening of this formation on the downthrown side of faults (e.g. Graue *et al.*, 1987). No such effects have been documented for the more than 900 m thick Aspelintoppen Formation (Steel *et al.*, 1981). This extremely thick delta plain sequence is characterized by vertical aggradation, related to a near balance between sediment supply and the space available for sediment infilling (accommodation; Posamentier and Vail, 1988). In the Ness Formation, a similar near-balance situation existed in the period between maximum progradation (Fig. 2) and the first major transgressive pulse (Helland-Hansen *et al.*, in press). The Aspelintoppen Formation was deposited proximally in a foreland basin, and overlies the cyclic delta front sediments of the Battfjellet Formation (Helland-Hansen, 1985). The cyclicity in these delta-front deposits is related by Helland-Hansen (1985) to frequent autocyclic avulsions of the Aspelintoppen distributary channels. Immature sandstone textures, a dominance of massive beds and an abundance of soft-sediment deformation suggest rapid deposition in the distributaries. Palaeocurrent data from the Aspelintoppen Formation (Fig. 11) show that the within-channel spread in transport directions is low, whereas the between-channel spread is significant. The implication of this latter feature is that the mean flow direction varied through time, which might indicate that the stable delta plain channels of the Aspelintoppen Formation were of the anastomosed type. This tentative suggestion is supported by the high amount of bank-stabilizing material in the formation (Fig. 5) and the inferred high rates of subsidence and sedimentation (Smith, 1986). However, features like the limited channel thicknesses (usually between 2 and 5 m, very rarely exceeding 10 m, Fig. 10), the unstable depositional surface (as witnessed by the abundant soft-sediment deformation structures), and the scarcity of laterally adjacent (perhaps branching) channel bodies at any level within the formation, argue against an anastomosing channel pattern. Until more work has been done on the Aspelintoppen Formation, the channel bodies in this sequence should thus be referred to by the general descriptive term for stable delta plain distributaries introduced in the previous section.

## CHARACTERISTIC INFILL SEQUENCES OF STABLE, VERTICALLY AGGRADED CUT-AND-FILL DISTRIBUTARIES

In reservoir studies, it is of crucial importance to determine the dimensions of the sedimentary building blocks (Fig. 1).

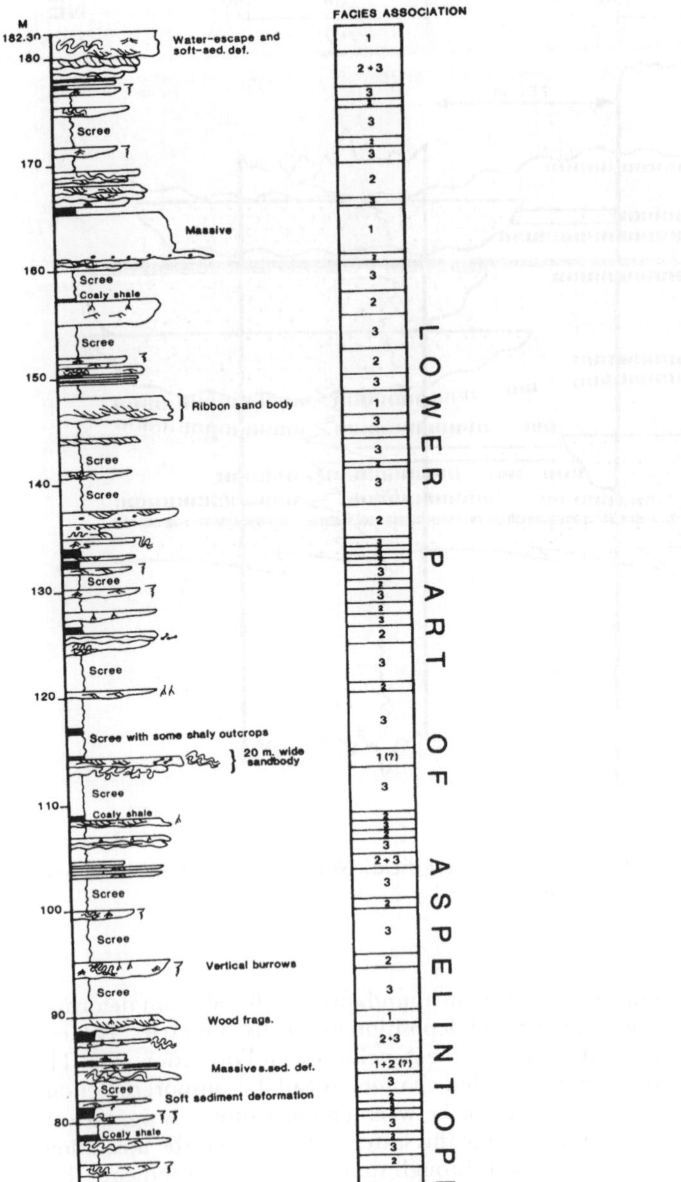


**Fig. 5.** Sedimentological log through the lower part of the Aspelintoppen Formation at Ishøgda, Nordenskioldland, Spitzbergen. The numbered facies associations are: 1, stable, vertically aggraded cut-and-fill channel deposits; 2, sandy interchannel sediments; 3, fine-grained interchannel sediments.

For reservoirs deposited in a delta plain setting, production planning and reserve estimates require insight into the geometry of the channel sandbodies present. Since palaeoriver morphology is one of the main factors that influences channel body dimensions, it is vital to establish criteria that help distinguish between the various river types. Characterization of palaeoriver morphologies from channel infill sequences is the method most frequently used for this task. However, this vertical-profile approach must be used with care, and in outcrops it should be accompanied by architectural-element analysis (Miall, 1985) whenever possible.

There are relatively few published data regarding the infill sequences of stable, vertically aggraded cut-and-fill distributaries (Land and Weimer, 1978; Okolo, 1983; Hopkins, 1985; Tillman and Jordan, 1987). Land and Weimer (1978), in a study of the distributary channel reservoir in the Peoria field, identified three distinct phases of infilling. These are here termed the active, partly active, and passive infill phases. Similar phases can be recognized in the distributary channel sand bodies of the Ness, Saltwick, and Aspelintoppen Formations. Figure 7 illustrates this tripartite infill sequence (B), which is compared to the point-bar sequence typical of laterally migrating high-sinuosty streams (A). In Table I, the characteristics of this infill sequence are outlined. Its main features are the contrasts between the three different infill types (Fig. 7), the stepwise fining upwards, the presence of internal scours reflecting changes from deposition to erosion within the channel, the common massive/convolute bedded intervals in the active fill, and the strongly undirectional within-channel palaeocurrent indicators. These features will be discussed below. It should be emphasized that in theory this tripartite infill sequence could form in any channel undergoing abandonment. However, in river systems where mobile channel belts are developed (braided and meandering streams, Friend, 1983), the effects of (point) bar migration and frequent local shifts in channel courses would tend to dominate vertical sequences. An example of the stepwise deterioration of reservoir quality associated with this vertical infilling is shown in Fig. 7(C).

The stepwise fining up (Figs. 7 and 8) associated with stable cut-and-fill channels can be explained by sequential avulsion (stage 3 in Fig. 9). After incision, the cut-and-fill distributary enters its active fill phase (stage 2 on Fig. 9). In humid climate settings, discharge is often consistently high throughout this phase (Coleman and Prior, 1982; Okolo, 1983), and sandy bedload material is the dominant infill material (Tillman and Jordan, 1987). If the discharge is fluctuating, finer-grained low-stage deposits, similar in appearance to the partly active fill (Fig. 7), may be interbedded with the sandy active phase fill. This kind of interbedding is occasionally observed both in the Saltwick and Ness Formation channels.

Internal scour surfaces (Figs. 7 and 8) become progressively more common towards the top of the active phase fill (stage 3A in Fig. 9). These scour surfaces are planar features which in the Saltwick Formation can be followed for 33–375 m along depositional dip. In a direction normal to the palaeocurrent, they tend to be present only in the central portion of the channel bodies. The erosional relief associated with these surfaces never exceeds 10 cm. Sediments overlying the scours seem to have been deposited by

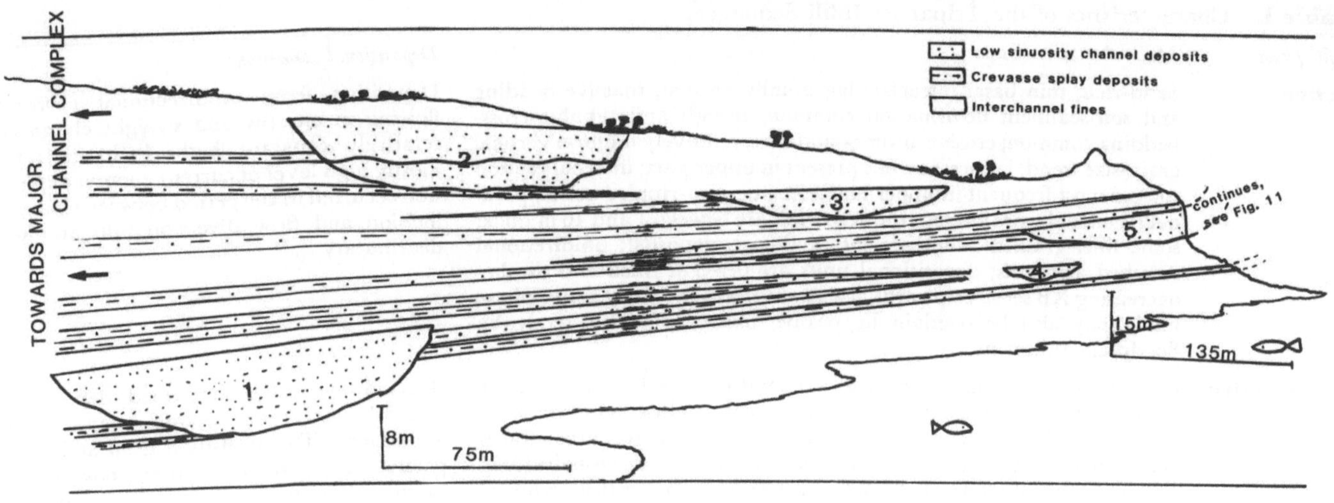


**Fig. 6.** Outcrop sketch of the Saltwick Formation, north of High Hawsker, Yorkshire, England. The section is approximately normal to the main palaeocurrent direction. Note the low $W/T$–ratios of the channel sand bodies.

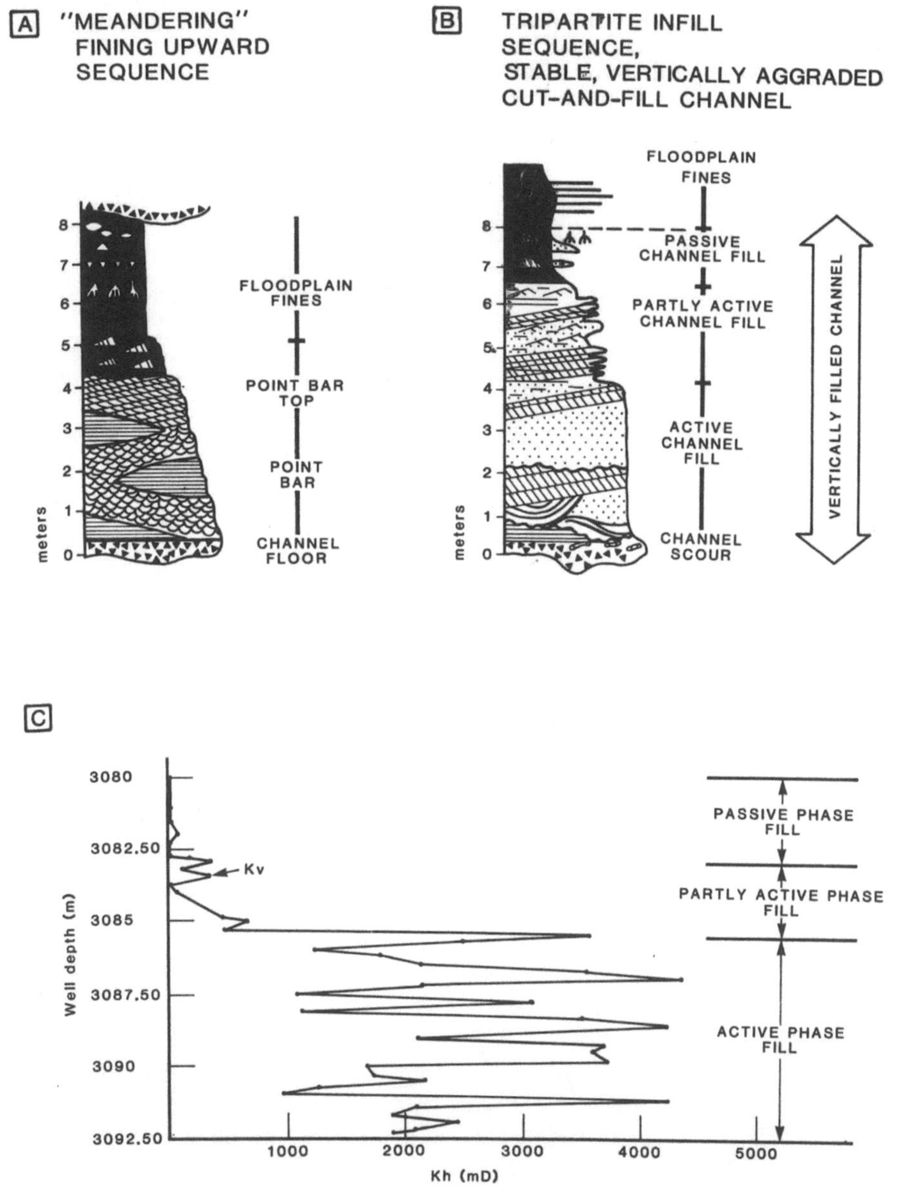


**Fig. 7.** Comparative infill sequences of "ideal" meandering streams (A; Allen, 1970) and stable, vertically aggraded cut-and-fill channels (B, see Table I for characteristics). In (C), the stepwise vertical deterioration in reservoir quality in a Ness channel body from the Oseberg field is related to a tripartite cut-and-fill channel infill sequence.

**Table I.** Characteristics of the Tripartite Infill Sequence.

| *Infill phase* | *Main characteristics* | *Depositional conditions* |
|---|---|---|
| **Active** | Sand-rich, thin basal intraclast lag usually present; massive bedding and soft-sediment deformation common; trough and tabular cross-bedding common; erosive lower boundaries; relatively uniform vertical grain-size trend; large scour-fills present in upper part; internal erosion sufaces most frequent in upper half; some current-rippled zones, plant debris often drapes foresets; the active fills are beween 1 and 16 m thick; some finer-grained zones present (low-stage deposits); unidiretional crossbed azimuths; depositional units are 0.1–3 m thick and display decreasing XB set size upwards; usually overlain by partly active phase infill; may also be overlain by passive phase infill or directly by floodplain sediments | Deposition from unidirectional currents flowing in narrow and straight channels; relatively constant high discharge rate; mostly high level of current energy; deposition occurred in the period between channel incision and flow diversion into another distributary |
| **Partly active** | Consists of alternating sandstone and mudstone beds, very variable grain size; beds usually thin (1–30 cm); plant debris very common; current ripples is the most common sedimentary structure; root-traces penetrate many beds; parallel lamination and small-scale cross-bedding also occur; grain-size and bed thickness either show no vertical trends or fining/thinning upwards; these units are 0.5–7 m thick, always underlain by active fill, overlain by passive fill or floodplain sediments; flaser-bedding common; sand-beds are usually normal-graded and have erosive lower boundaries | Bedload and suspended load deposition in a channel that became progressively abandoned. The transition from active to partly active sedimentation took place rapidly. Sandbeds formed in the partly abandoned channel mainly during floods in the river system (pulse-like sedimentation) |
| **Passive** | Dark to light grey mudstone; organic-rich intervals; thin lenses or beds of fine- to very fine-grained sandstone, commonly capped by a coalbed or coaly shale; root-traces occur; beds often display normal grading from very fine-grained sand to clay; some burrows; units from 0.1 to 3 m thick overlie active or partly active phase fills with an unscoured boundary; the mudstone may be massive or parallel to subparallel laminated | Suspension fallout in completely abandoned channel. Stagnant water. Fluctuating water depth (flood-controlled?) |

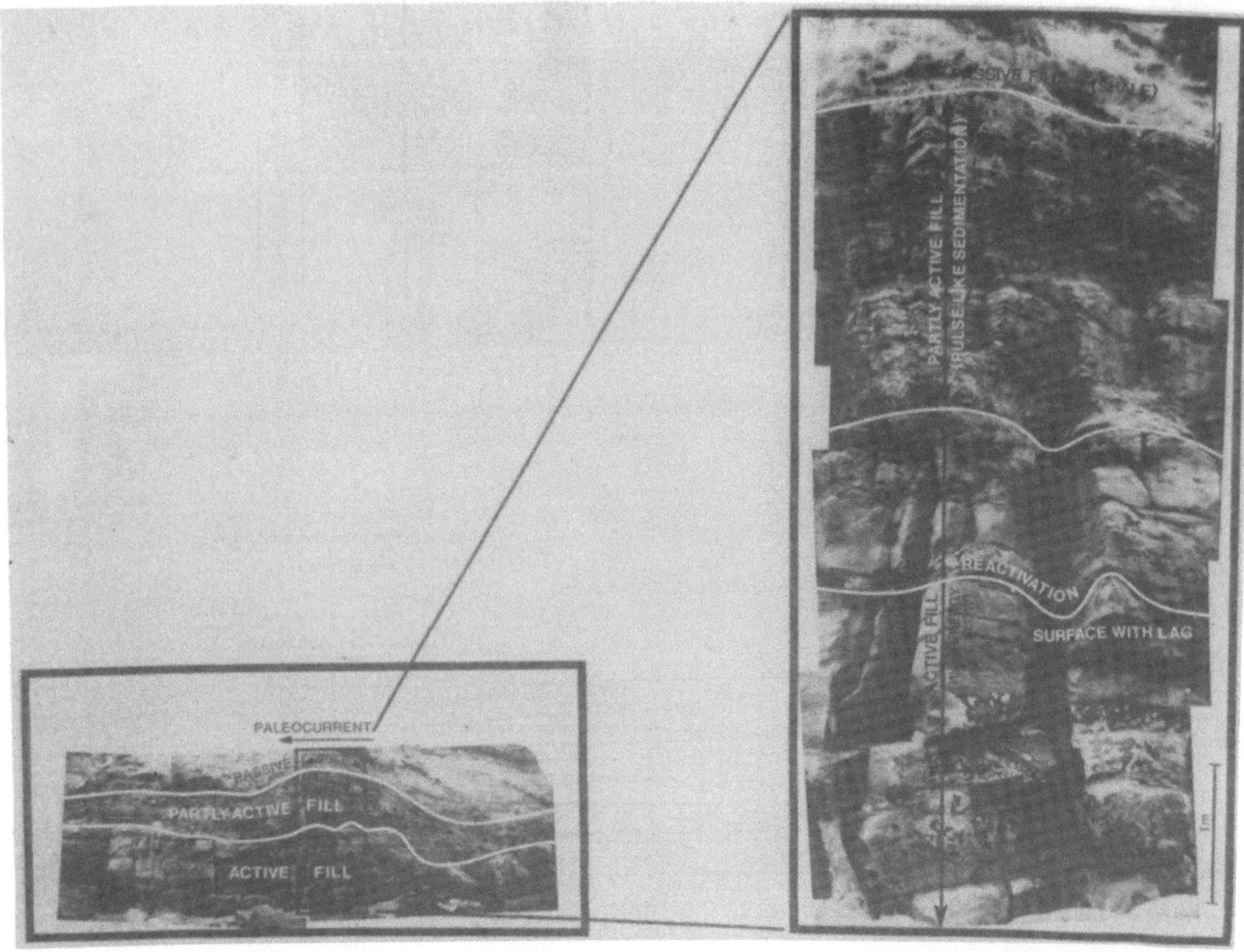


**Fig. 8.** Close-up photos of a cut-and-fill channel deposit at Hayburn Wyke, Yorkshire, England. Note the characteristic tripartite infill sequence (Table I).

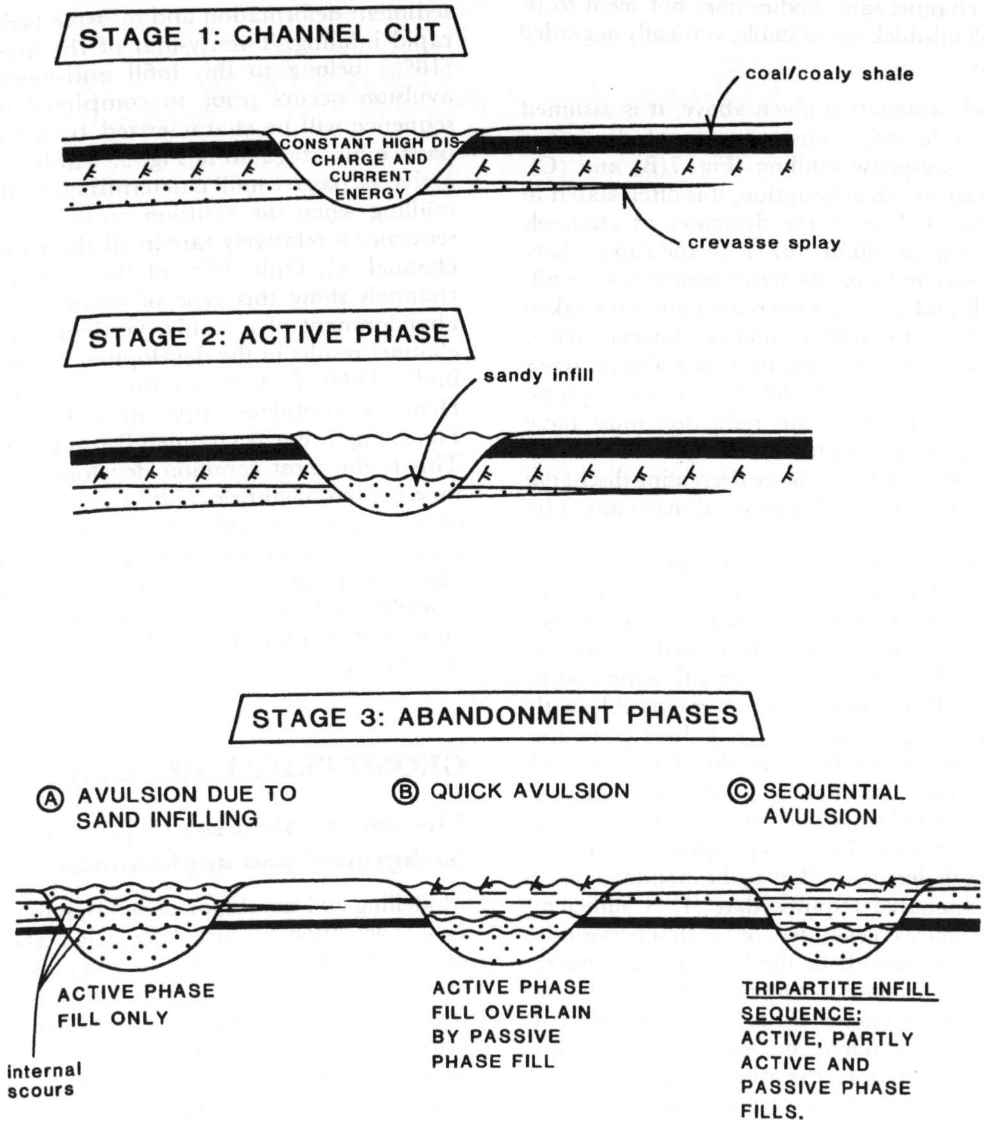


**Fig. 9.** Cut-and-fill processes and products in a stable cut-and-fill distributary channel. The alternative developments in the abandonment phase largely determine the nature of the channel infill. The sequential avulsion process (stage 3C) is thought to be responsible for the generation of the complete tripartite infill sequence.

more vigorous currents than those below (coarser grain size, less clay in the grain framework, massive bedding or large-scale cross-bedding replacing current ripples or small-scale cross-bedding). Clay or siderite clasts are usually strewn on the scour surface. At the terminations of these surfaces, gentle upward curvature was observed. On the basis of the above, it seems unlikely that these scour surfaces represent boundaries between independent channel deposits in a multistorey channel complex. Instead, these numerous, discontinuous and strictly intra sand body scour surfaces probably formed within single-channel tracts as the threshold of deposition was exceeded during active phase vertical aggradation (stage 3A in Fig. 9). This is supported by their dominance near the top of the active infill. There are at least two ways of explaining the generation of these scour surfaces:

(1). The threshold of deposition may have been exceeded as a result of sudden increases in discharge, i.e. flood events. However, this explanation seems to be at odds with the consistent increase in scour surfaces upwards in the active fills (Fig. 9, stage 3).

(2). If water is flowing at a relatively constant rate through a channel that is shallowing (and narrowing) due to infilling, it can be derived from the Shields' criterion (Blatt *et al.*, 1980) that the shear stress on the bottom increases as the flow depth decreases. Eventually, a point will be reached where the shear stress promotes erosion instead of settling of grains. Erosion/non-deposition continues until the shear stress along the bottom again falls below the threshold of deposition (loss of flow energy or increased flow depth). Several cycles of deposition and erosion may take place during vertical channel aggradation, and the likelihood of generating scour surfaces increases as the channel depth decreases. An important implication of this development is that the *upper part* of the active fill sand body becomes dominated by high-permeability sand (Dreyer *et al.*, in press). In reservoir modelling of distributary channel sandstones, it is common to regard only the sandy active phase fill as the "channel sediments", whereas the partly active and passive infill is considered non-reservoir (Fig. 7(C)). Since the active phase fill in this case has the most permeable zone at the top, the upward decrease in reservoir quality commonly associated with channel sand bodies does not seem to be valid over the full fill-thickness of stable, vertically aggraded channel deposits.

For the second explanation given above, it is assumed that there is a relatively constant rate of discharge throughout the *active*-phase infilling (Fig. 7(B) and (C), Fig. 9). In contrast to this assumption, it is often stated in the literature that the discharge decreases as channels become infilled (e.g. Schumm, 1977). In the cases where the active fill is overlain by partly active and/or passive fills (Fig. 9, stages 3B and 3C), this decrease may have taken place during the latter two infilling phases. However, if one postulates a reduction in discharge over time also for those channel bodies that are entirely filled with active-phase sediments (Fig. 9, stage 3A), this reduction must have taken place during the active phase. Hence, if it is true that channels of this type always experience decreasing discharge during infilling, the second explanation (above) cannot be valid.

There are, however, some observations pertinent to the stable, cut-and-fill channels that imply that a decrease in discharge during infilling does not always have to take place. For instance, the weakly developed to absent fining-upwards tendency in these "active fill" sand bodies (Fig. 10, channels B, C, and D) is not compatible with a decrease in discharge. Moreover, it follows from the assumption of semiconstant discharge that the volume of water passing through the channel must at all times be roughly the same. Similarly, the space available for this "constant" volume of water becomes progressively smaller as the channel depth decreases. Hence, the maintenance of a semiconstant discharge requires that (1) a sufficient amount of "new" space is generated, or (2) that previously unoccupied space is utilized. In the first case, the water could increase the in-channel accommodation by eroding previously deposited channel sediments. The internal scours described above might be the product of this process. In the second case, the river could widen its channel, or allow more of its discharge to enter the floodplain as overbank floods.

The effects of such events can be observed in the Saltwick and Aspelintoppen Formations. In the Saltwick Formation, Mjøs and Walderhaug (1989) observed that upwards along the margins of channel bodies the number and thickness of laterally adjacent crevasse-splay deposits usually increase. This suggests that overbank flooding took place more frequently towards the end of channel infilling periods. In both the Aspelintoppen and Saltwick Formations, most of the laterally restricted channel bodies can be seen to have maximum widths along a horizontal plane that is parallel with the top of the active fill. This might imply that the width of the channel was increased in the last part of the active infilling phase. Some active-fill channels have well-developed, erosively based wings (e.g. Friend *et al.*, 1979) at their tops, a feature that may indicate extensive crevasse-splay development or an extreme widening of the channel immediately prior to abandonment.

Changes in floodplain gradients, and/or loss of discharge due to flow diversion through a favoured distributary (Hopkins, 1985), eventually lead to the abandonment of channels. This abandonment stage can be viewed in terms of three end-members. Avulsion may occur after complete infilling while the channel remains in its active phase (stage 3A in Fig. 9). In this case, a channel body containing only sandstone-dominated active phase infill (Table I) will form. In the Aspelintoppen Formation, this development is commonly seen (Fig. 10, channels B, C and D). The resulting sand bodies contain abundant soft-sediment deformation and massive bedding, indicative of rapid infilling. Twenty-four of the investigated channels (18%) belong to this infill end-member. If a sudden avulsion occurs prior to completed infilling, the infill sequence will be characterized by active fill overlain by passive fill (stage 3B in Fig. 9, Table I). The thickness of active vs. passive infill is determined by the completeness of infilling when the avulsion occurred. This type of infill sequence is relatively rare in all three formations (Fig. 10, channel A). Only 10% of the investigated cut-and-fill channels show this type of vertical profile. Progressive abandonment of a stable, vertically infilled cut-and-fill channel results in the development of partly active phase infill (Table I) between the active and passive parts. Hence, a "complete" tripartite infill sequence (stage 3C in Fig. 9, Fig. 7, Fig. 8, channels E and F in Fig. 10) is formed. This is the most common development, with 95 of the investigated channels (72%) in the three formations displaying this profile. The gradual loss of discharge associated with this infill sequence may be due to progressive blocking of the channel entrance, diversion of flow into a newly cut distributary, changes in floodplain gradients, an overall decrease in fluvial discharge, or a combination of these factors.

## GEOMETRICAL RELATIONSHIPS

### The width/thickness (*W*/*T*)-plot: background and applications

Modelling and simulation of delta-plain reservoirs like the Ness Formation require knowledge about the channel sand body dimensions. Widths and maximum thicknesses of stable, humid-climate cut-and-fill sand bodies were registered in the Aspelintoppen and Saltwick Formations. Care was taken to ensure that the widths were measured normal to the channel axis. The lengths of these channel bodies were not measured, since this dimension may be considered infinitely large in most reservoir simulations. For sand bodies consisting of more than one palaeochannel deposit (Fig. 12), the dimensions of individual channel bodies were noted in addition to the width and maximum thickness of the composite body. Note that within composite sand bodies, multistorey channel erosion often reduces the original dimensions of individual sand bodies (Fig. 12). Hence, the widths and thicknesses recorded in these cases are minimum values. In this study, component channel bodies that appeared to be significantly affected by erosion were not measured.

The author has also collected width and thickness data for cut-and-fill sand bodies in the Scalby and Cloughton Formations (Middle Jurassic, Yorkshire—see Nami and Leeder, 1978; Hancock and Fisher, 1981) and the Dakota Group (Upper Cretaceous, Denver area, USA—see Weimer and Land, 1972). Literature data in Okolo (1983) were also relevant. The results are presented as *W*/*T*-plots in Fig. 13.

It is important to realize that the data in Fig. 13 refer to sand body dimensions, not paleochannel dimensions (Fielding and Crane, 1987). In reservoir modelling, we are primarily interested in the dimensions of units with distinctly different reservoir qualities (Fig. 1). Hence, the sandstone-dominated active and partly active phase infills could not be grouped together with the shaly passive phase infill (Fig. 7). Furthermore, it is easier to recognize channel sand bodies than entire channel fills from subsurface data because of the problems associated with separating a passive phase fill from non-channel fines. In the field, we therefore

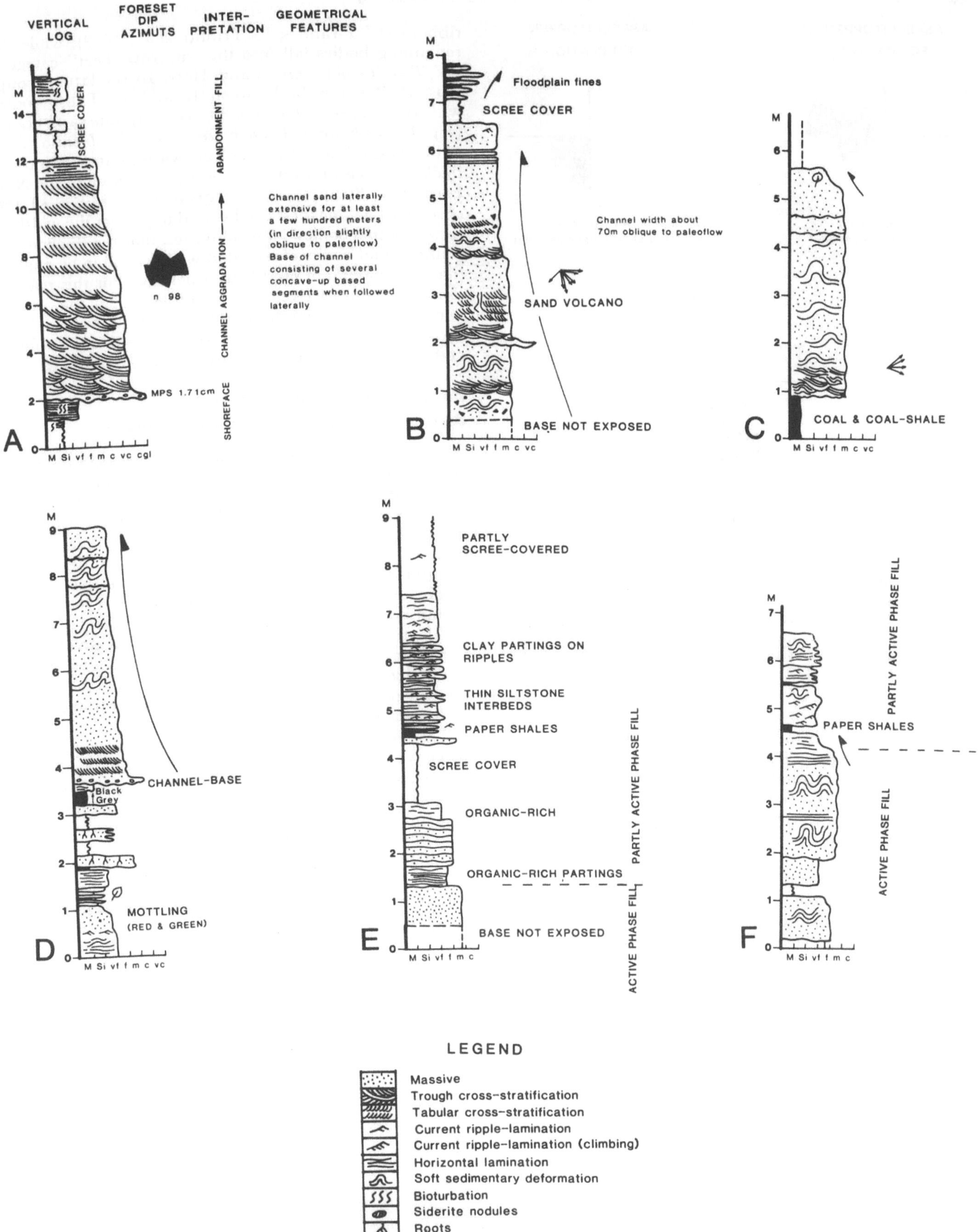


**Fig. 10.** Examples of different types of infill sequences from stable, vertically aggraded channels in the Aspelintoppen Formation, Spitzbergen. (A) channel-sandstone incised into shoreface sands; (B) channel-fill sequence about 50 m above base Aspelintoppen Formation; (C) example of extensively soft sedimentary deformed channel-fill; (D) channel sequence from uppermost part of Aspelintoppen formation (about 800 m above base); (E–F) channel-fill sequences displaying frequent ripple-lamination in upper parts (A-C: Liljevalchfjellet; D-F: Wahlenbergfjellet).

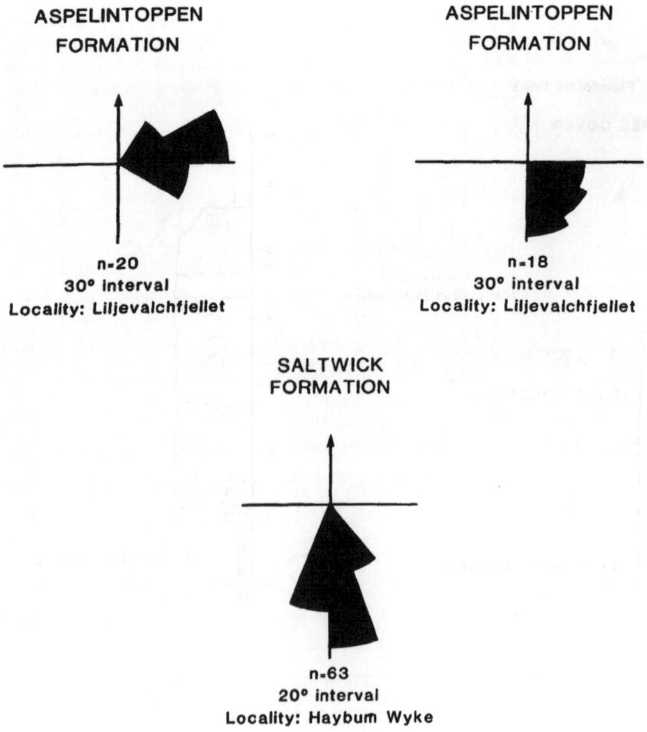


**Fig. 11.** Paleocurrent rose diagrams for three stable, humid climate cut-and-fill channels. Note the low within-channel spread in palaeocurrent directions, and the significant between-channel spread in the Aspelintoppen Formation. The palaeocurrent data from these formation were derived from flute casts and cross-beds.

measured the width and thickness of the active and partly active phase fills only (Fig. 8). In the lateral dimension, the difference between sand body and palaeochannel width is probably small for this channel type. For the thickness measurements, however, the difference may be considerable, as indicated by Fielding and Crane (1987).

The data in Fig. 13 suggest that there is a crude relationship between thickness and width in stable, vertically accreted cut-and-fill channel sand bodies. In general, it seems that the width of this sand body type increases with increasing thickness. The regression line on Fig. 13 is a linear function with a correlation coefficient ($r$) of 0.71. More than 50% of the sand bodies can be classified as ribbons ($W/T$-ratio $<15$; Friend, *et al.*, 1979), while the remaining bodies fall into the "narrow sheet" category ($W/T$-ratios between 15 and 100). In the latter group, most $W/T$-values lie between 15 and 35. Those bodies with $W/T$ ratios in excess of 35 are composite sand bodies (Fig. 13, see below). These have a mean $W/T$-ratio of 54.

In wells, it is a fairly straightforward matter to find the thickness of cut-and-fill sand-bodies like those in the Ness Formation. By applying the regression line in Fig. 13, the observed well thickness can be used to predict the width of these channel bodies. Note, however, that the thickness of channel sand bodies observed in wells will often be smaller than the maximum sand body thickness. For this reason, the width values derived from an application of Fig. 13 will in many cases be minimum values. Acquisition of this kind of data is fundamental for reservoir-communication studies and estimates of recoverable hydrocarbon volumes.

## Factors governing channel body dimensions and interconnectedness

The ribbon-like geometry of these channel sand bodies is partly due to the morphology and infilling mechanism of the stable cut-and-fill distributaries. This type of river system is usually characterized by deeply incised, rather narrow channels of varying sinuosity (e.g. Meckel, 1972; Smith, 1986; Coleman and Prior, 1982; Shuster and Steidtmann, 1987). As discussed above, vertical accretion dominates thereby preventing the channel from generating a sand body significantly wider than the initial channel width. In systems like this, the surface area covered by distributaries may be small compared to the total extent of the floodplain. Hence, this type of fluvial system has the potential of generating numerous ribbon-like sand bodies set in a matrix of floodplain fines.

The density and interconnectedness of fluvial channels is strongly influenced by external factors acting on a basin-wide scale. Allen (1978) and Bridge and Leeder (1979) list avulsion frequency, floodplain width, climate/discharge regime, subsidence rate and sedimentation rate among the most important. Assuming a system of ribbon-like cut-and-fill channels, a humid climate, a constant and uniform avulsion frequency, and a large floodplain width, the alluvial architecture of the Ness, Saltwick and Aspelintoppen Formations can be studied as a function of variations in the ratio between rate of sediment supply and rate of accommodation (Fig. 14). This model also assumes a uniform and constant rate of subsidence normal to the

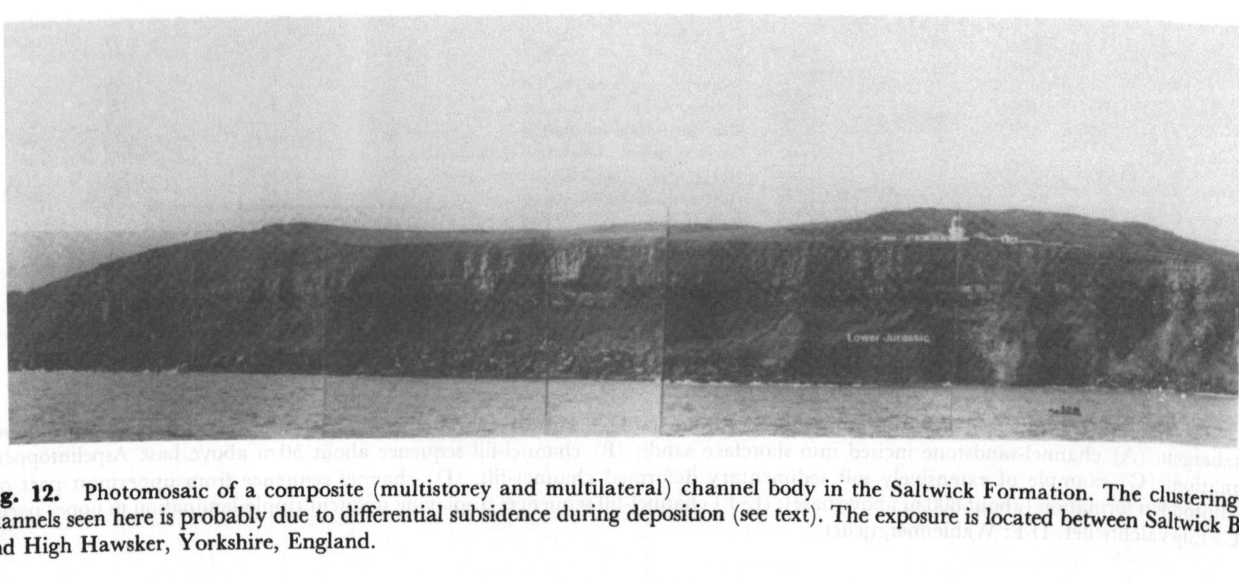


**Fig. 12.** Photomosaic of a composite (multistorey and multilateral) channel body in the Saltwick Formation. The clustering of channels seen here is probably due to differential subsidence during deposition (see text). The exposure is located between Saltwick Bay and High Hawsker, Yorkshire, England.

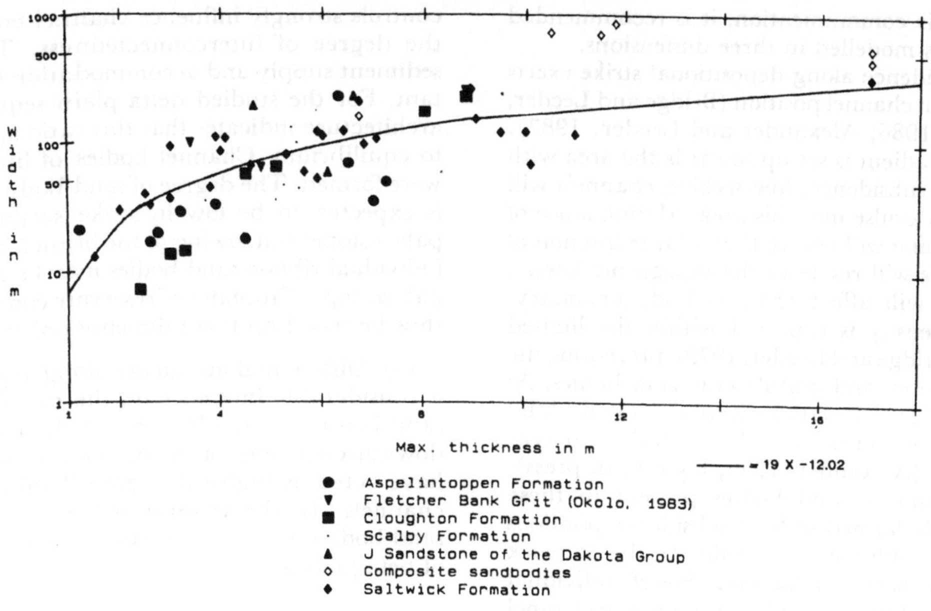


**Fig. 13.** *W/T*-relationships for humid climate cut-and-fill channel sandbodies. The regression line has been constructed for data from single-storey channel deposits and individual channels in multistorey deposits (all data points except the open diamonds). Correlation coefficient ($r$) = 0.71. Note the high percentage of ribbon-like sandbodies (*W/T* ratio of 15 or less).

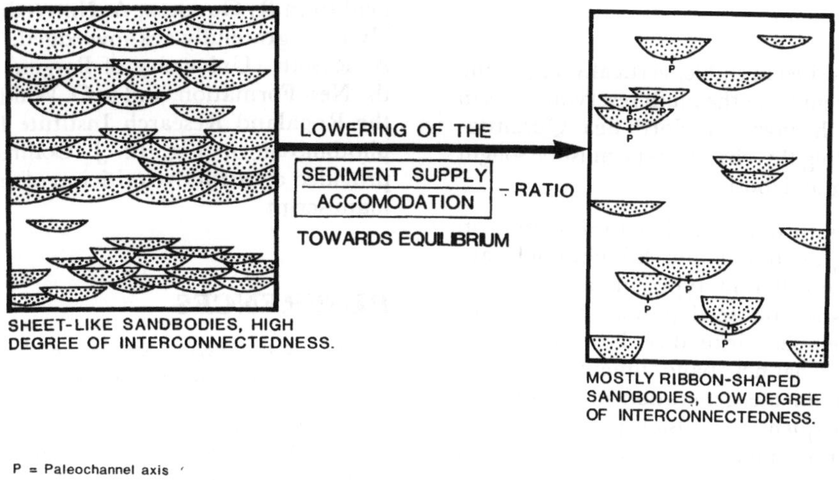


**Fig. 14.** The effects of variations in the sediment supply vs. accommodation ratio on sand body interconnectedness. (P = palaeochannel axis.)

main palaeocurrent direction (see below). If the ratio between sediment supply and accommodation is high (Fig. 14, left), each increment of space generated in the basin will contain a volumetrically large percentage of channel deposits. Younger channels will tend to rework older channel deposits and interchannel fines as the system progrades. This obviously leads to a high degree of interconnectedness with numerous sheet-like sand bodies. The most sand-rich portions of the Ness Formation on the Oseberg field may have accumulated under these circumstances.

A lowering of the ratio between sediment supply and accommodation towards 1 (equilibrium) will alter the sand body density (Fig. 14, right). (Note that if the ratio falls below 1, a transgression is initiated in the most distal part of the system). In this setting, a considerable amount of accommodation is generated between each avulsive jump. Owing to the restricted lateral extent of the vertically aggrading cut-and-fill channels, the majority of this newly created space will be occupied by vertically accreting interchannel sediments. Each increment of space generated in the basin will thus have a moderate to low channel density (Shuster and Steidtmann, 1987). The alluvial architectures of the Aspelintoppen Formation, parts of the Saltwick Formation and probably most of the Ness Formation suggest deposition under these circumstances. The degree of sand body interconnectedness will be low in sections normal to the palaeocurrent (Fig. 14). However, stable cut-and-fill channel systems commonly branch or bifurcate rather extensively (Land and Weimer, 1978; Smith and Smith, 1980; Coleman and Prior, 1982; Fielding, 1986). Thus, individual channels may join up in the dip direction (along the palaeoslope). In the dip-section, communication between reservoir sand bodies may also be established where a younger channel erosively crosses the path of an older one. To obtain a realistic estimate of reservoir communication, it is recommended that the reservoir is modelled in three dimensions.

Differential subsidence along depositional strike exerts a strong influence on channel position (Bridge and Leeder, 1979; Alexander, 1986; Alexander and Leeder, 1987). Assuming that a gradient is set up towards the area with the highest rate of subsidence, low-seeking channels will have a tendency to avulse into this area. Maintenance of the gradient over time will ensure that a large fraction of the fluvial channels will reside in the topograpic low(s). This confinement will affect channel body geometry. A high channel density is expected within the limited "lowlying" area (Bridge and Leeder, 1979), promoting the creation of multistorey and multilateral sand bodies. As the Saltwick and Ness Formations were deposited, differential subsidence patterns existed in relation to syn-sedimentary faults (Alexander, 1986; Ryseth, in press). Many of the composite sand bodies present in these formations probably formed in high-subsidence positions on the downthrown sides of these faults. In the Saltwick Formation, it can be observed that even though individual channel bodies are ribbon-shaped, the composite channel sand bodies may attain sheet-like geometries (Figs. 12 and 13). In reservoir studies, the effects of differential subsidence need to be taken into account when constructing the geological model.

## CONCLUSIONS

(1) Sand bodies deposited in stable, vertically aggrading cut-and-fill channels seem to be the main reservoir units in the Ness Formation of the northern North Sea. Quantitative information regarding this channel type must be sought in well-exposed analogue deposits.

(2) Vertical profiles through sand bodies formed in stable, vertically aggraded cut-and-fill channels are characterized by three different types of infill (active, partly active and passive phase deposits). Within this tripartite sequence, the reservoir quality shows a stepwise upwards deterioration. Channel abandonment by sequential avulsion may explain the generation of a tripartite infill sequence. If the sequential avulsion process is interrupted before completion, more simple vertical profiles form.

(3) Internal scour surfaces in the upper part of the active phase infills were probably formed as the threshold of deposition was exceeded during vertical aggradation. These surfaces should not be confused with the bounding surfaces created by erosion of one channel into another. High-permeability sediments are deposited between the scour surfaces. Hence, the active phase fills in this type of sand body do not commonly exhibit the upwards decrease in permeability normally associated with fluvial channel deposits.

(4) In the subsurface, knowledge about the lateral extent of sand bodies deposited in stable cut-and-fill channels is vital for production planning. The *W*/*T*-plot for stable, humid-climate cut-and-fill channel sand bodies (Fig. 13) allows the prediction of sand body dimensions from estimated sand body thicknesses. Outcrop measurements indicate that single-storey cut-and-fill channels mostly have ribbon-like geometries (*W*/*T*-ratios usually between 8 and 30). Multistorey (composite) channel bodies of this type are more sheet-like, with an average *W*/*T*-ratio of 54.

(5) In addition to palaeoriver morphology, external controls strongly influence channel body dimensions and the degree of interconnectedness. The ratio between sediment supply and accommodation is expecially important. For the studied delta plain sequences, the alluvial architecture indicates that this ratio apparently was close to equilibrium. Channel bodies of limited lateral extent were formed. The degree of sand body interconnectedness is expected to be low in strike sections (normal to the palaeoslope) but, owing to the branching channel pattern, individual ribbon sand bodies may be connected along the palaeoslope. Estimates of reservoir communication should thus be based on three dimensional modelling.

(6) Differential subsidence along depositional strike has a considerable influence on the distribution of channel sand bodies. In the Ness and Saltwick Formations, the downthrown sides of major syn-sedimentary faults may have acted as high-subsidence "traps" for distributary channels. In the vicinity of these "traps", composite sand bodies with a sheet-like geometry had the potential of being formed.

## ACKNOWLEDGEMENTS

The author would like to thank W. Helland-Hansen (Norsk Hydro) for participation in the fieldwork on the Aspelintoppen Formation. In Yorkshire, assistance in the field from R. Gundesø, A. Nøttvedt and R. Steel of Norsk Hydro is gratefully acknowledged. Thanks are also due to A. Ryseth (University of Bergen) for discussions about the Ness Formation, and to O. Walderhaug and R. Mjøs of the Rogaland Research Institute for compilation of the photomosaic in Fig. 12. J. Collinson and J.P. Nystuen provided constructive criticism of an earlier version of this manuscript.

## REFERENCES

Allen, J.R.L. (1970). Studies in fluviatile sedimentation: A comparison of fining-upwards cyclothems with special reference to coarse-member composition and interpretation. *J. Sediment. Petrol.* **40**, 298–323.

Allen, J.R.L. (1978). Studies in fluviatile sedimentation: An exploratory quantitative model for the architecture of avulsion-controlled alluvial suites. *Sediment. Geol.* **21**, 129–147.

Alexander, J. (1986). Idealized flow models to predict alluvial sandstone body distribution in the Middle Jurassic Yorkshire Basin. *Marine Petroleum Geol.* **3**, 298–305.

Alexander, J. and Leeder, M.R. (1987). Active tectonic control on alluvial architecture. In: Ethridge, F.G., Flores, R.M. and Harvey, M.D. (eds.). *Recent Developments in Fluvial Sedimentology* pp. 243–252. Society of Economic Paleontologists and Mineralogists Special Publication No. 39.

Blatt, H., Middleton, G. and Murray, R. (1980). *Origin of Sedimentary Rocks.* Prentice-Hall, Englewood Cliffs, N.J.

Bridge, J.S. (1985). Paleochannel patterns inferred from alluvial deposits: a critical evaluation. *J. Sediment. Petrol.* **55**, 579–389.

Bridge, J.S. and Leeder, M.R. (1979). A simulation model of alluvial stratigraphy. *Sedimentology* **26**, 617–644.

Coleman, J.M. and Prior, D.B. (1982). Deltaic environments of deposition. In: Scholle, P.A. and Spearing, D. (eds.) *Sandstone Depositional Environments*, pp. 139–179. American Association of Petroleum Geologists Memoir 31.

Dreyer, T., Scheie, Å. and Walderhaug, O. (in press). A mini-permeameter-based study of permeabilty trends in channel sandbodies. *Am. Assoc. Petroleum Geol. Bull.*

Fielding, C.F. (1986). Fluvial channel and overbank deposits from the Westphalian of the Durham Coalfield, NE England. *Sedimentology* **33**, 119–140.

Fielding, C.F. and Crane, R.C. (1987). An application of

statistical modelling to the prediction of hydrocarbon recovery factors in fluvial reservoir sequences. In: Etheridge, F.G., Flores, R.M. and Harvey, M.D. (eds.), *Recent Developments in Fluvial Sedimentology*, 321–328. Society of Economic Paleontologists and Mineralogists Special Publication No. 39.

Friend, P.F. (1983). Towards the field classification of alluvial architecture or sequence. In: Collinson, J.D. and Lewin, J. (eds.), *Modern and Ancient Fluvial Systems* pp. 345–354. Special Publication No. 6 of the International Association of Sedimentologists.

Friend, P.F., Slater, M.J. and Williams, R.C. (1979). Vertical and lateral building of river sandstone bodies, Ebro Basin, Spain. *J. Geol. Soc. London* **136**, 39–46.

Graue, E., Helland-Hansen, W., Johnsen, J., Lømo, L., Nøttvedt, A., Rønning, K., Ryseth, A. and Steel, R. (1987). Advance and retreat of Brent Delta system, Norwegian North Sea. In: Brooks, J. and Glennie, K.W. (eds.), *Petroleum Geology of North West Europe*, Graham and Trotman, London, pp. 915–938.

Gundesø, R. and Egeland, O. SESIMIRA—a new tool for 3D modelling of heterogeneous reservoirs. *This volume.*

Hancock, N.J. and Fisher, M.J. (1981). Middle Jurassic North Sea deltas, with particular reference to Yorkshire. In: Illing, L.V. and Hobson, G.D. (eds.), *Petroleum Geology of the Continental Shelf of NW Europe*, pp. 186–195. Heyden, London.

Haszeldine, R.S. (1983). Fluvial bars reconstructed from a deep, straight channel, Upper Carboniferous Coalfield of North-east England. *J. Sediment. Petrol.* **53**, 1233–1247.

Helland-Hansen, W. (1985). Sedimentology of the Battfjellet Formation (Paleogene) in Nordenskiold Land, Spitsbergen. Unpublished Cand.Scient. thesis, University of Bergen.

Helland-Hansen, W., Steel, R., Nakayama, K. and Kendall, C.G. St. (in press). Review and computer modelling of the Brent Group stratigraphy. In: Whateley, K.G. and Pickering, K.T. (eds.), *Deltas: Sites and Traps for Fossil Fuels.* Geological Society Special Publication.

Hemingway, J.E. (1974). Jurassic. In: Rayner D.H. and Hemmingway, J.E. (eds.), *The Geology and Mineral Resources of Yorkshire*, eds. pp. 161–224. Yorkshire Geological Society.

Hopkins, J.C. (1985). Channel-fill deposits formed by aggradation in deeply scoured, superimposed distributaties of the Lower Kootenai Formation (Cretaceous). *J. Sediment. Petrol.* **55**, 42–52.

Jackson, R.G. (1978). Preliminary evaluation of lithofacies model for meandering alluvial streams. In: Miall, A.D. (ed.), *Fluvial Sedimentology*, pp. 543–576. Canadian Society of Petroleum Geologists Memoir 5.

Kelling, G. and George, G.T. (1971). Upper Carboniferous sedimentation in the Pembrokeshire Coalfield. In: Bassett, D.A. and Bassett, M.G. (eds.), Geological Excursions in South Wales and Forest of Dean, pp. 240–259.

Lake, L.W. and Carroll, H.B. Jr. (eds.) (1986). *Reservoir Characterization.* Academic Press, London.

Land, C.B. and Weimer, R.J. (1978). Peoria field, Denver Basin, Colorado—sandstone distributary channel reservoir. In: Pruit, J.D. and Cottin, P.E. (Eds.), *Energy Resources of the Denver Basin*, pp. 81–104.

Livera, S.E. and Leeder, M.R. (1981). The Middle Jurassic Ravenscar Group ("Deltaic Series") of Yorkshire: recent sedimentological studies as demonstrated during a field meeting, 2–3 May 1980. *Proc. Geol. Assoc.* **92**, 241–250.

Meckel, L.D. (1972). Anatomy of distributary channel-fill deposits in recent mud-deltas (abstract). *Am. Assoc. Petroleum Geol. Bull.* **56**, 639.

Miall, A.D. (1985). Architectural-element analysis: a new method of facies analysis applied to fluvial deposits. *Earth Sci. Rev.* **22**, 261–308.

Mjøs, R. and Walderhaug, O. (1989). Sandstone geometry in fluviodeltaic sediments of the Ravenscar Group, Yorkshire. SPOR Water Report, Rogaland Research Institute, Stavanger.

Nami, M. and Leeder, M.R. (1978). Changing channel morphology and magnitude in the Scalby Formation (M. Jurassic) of Yorkshire, England. In: Miall, A.D. (ed.), *Fluvial Sedimentology* pp. 431–440. Canadian Society of Petroleum Geologists Memoir 5.

Okolo, S.A. (1983). Fluvial distributary channels in the Fletcher Bank Grit (Namurian R2b), at Ramsbottom, Lancashire, England. In: Collinson, J.D. and Lewin, J. (eds.), *Modern and Ancient Fluvial Systems* pp. 421–433. Special Publication No. 6 of the International Association of Sedimentologists.

Posamentier, H. and Vail, P. (1988). Eustatic controls on clastic deposition. In: Wilgus, C. (ed.), *Sea Level Changes—An Integrated Approach* pp. 125–154. Society of Economic Paleontologists and Mineralogists Special Publication No 42.

Rust, B.R. (1978). A classification of alluvial channel systems. Miall, A.D. (ed.), *Fluvial Sedimentology*, pp. 187–198. Canadian Society of Petroleum Geologists Memoir 5.

Ryseth, A. (in press). Correlation of depositional patterns in the Ness Formation, Oseberg area. In: Collinson, J.D. (ed.), *Correlation in Hydrocarbon Exploration.* Norwegian Petroleum Society, Graham and Trotman, London.

Schumm, S.A. (1977). *The Fluvial System.* Wiley, New York.

Shuster, M.W. and Steidtmann, J.R. (1987). Fluvial-sandstone architecture and thrust-induced subsidence, Northern Green River Basin, Wyoming. In: Ethridge, F.G., Flores, R.M. and Harvey, M.D. (eds.), *Recent Developments in Fluvial Sedimentology*, pp. 279–286. Society of Economic Paleontologists and Mineralogists Special Publication No. 39.

Smith, D.G. (1986). Anastomosing river deposits, sedimentation rates and basin subsidence, Magdalena River, northwestern Colombia, South America. *Sediment. Geol.* **46**, 177–196.

Smith, D.G. and Smith, N.D. (1980). Sedimentation in anastomosed river systems: examples from alluvial valleys near Banff, Alberta. *J. Sediment. Petrol.* **50**, 157–164.

Steel, R.J., Dalland, A., Kalgraff, K. and Larsen, V. (1981). The cental Tertiary basin of Spitsbergen—sedimentary development of a sheared-margin basin. In: Kerr, J.W. and Ferguson, A.J. (eds.), *Geology of the North Atlantic Borderlands*, 647–664. Canadian Society of Petroleum Geologists Memoir 7.

Tillman, R.W. and Jordan, D.W. (1987). Sedimentology and subsurface geology of deltaic facies: Admire 650' Sandstone, El Dorado field, Kansas. In: Tillman, R.W. and Weber, K.J. (eds.), *Reservoir Sedimentology*, pp. 221–292. Society of Economic Paleontologists and Mineralogists Special Publication No. 40.

Weimer, R.J. and Land, C.B. (1972). Field guide to the Dakota Group (Cretaceous) stratigraphy, Golden-Morrison area. *Mountain Geologist* **9**, 240–267.

# 21

# The Connectivity and Conductivity of Overlapping Sand Bodies

*P.R. King*

BP Research Centre, Sunbury-on-Thames, Middlesex, UK

Many oil reservoirs have a highly complicated geometry owing to the complex environment in which the sands were deposited. There is usually a mixture of good (that is, high-permeability) sandstones and poorer (low-permeability) siltstones, mudstones and shales. The main flow of oil during recovery is through the sandstones, the flow rates through the poorer rocks being too slow to be of economic consequence. Clearly, the connectivity of the sandstones is of prime importance in determining how much oil is potentially recoverable with a particular well configuration The overall shape of the connected sandstone is also important in determining the flow rates in the reservoir, through tortuosity effects, for example.

The aim of this contribution is to show that it is possible to explain the generic behaviour of connectivity and conductivity of uncorrelated, overlapping sand bodies in terms of some simple theory. With this general understanding we then hope to be able to predict the behaviour of a specific system that is of use in an engineering application. Thus, for the purpose of this paper we will consider a simple model of the sands. We model them as randomly placed geometrical shapes embedded in an impermeable background rock. Although this is not geologically very realistic, it does give an insight into the behaviour of such systems. This same theory will then be used to study real systems. We use the methods of percolation theory to analyse the results. This shows that there is a distinct threshold in net-to-gross below which there is no overall connectivity. Above this threshold the connectivity increases rapidly, whereas the effective permeability only increases slowly. Finite size scaling within percolation theory is used to show how the connectivity and conductivity decrease as a function of well spacing.

## INTRODUCTION

Many hydrocarbon reservoirs are the outcome of complex depositional processes that have led to a complicated distribution of sands. Broadly speaking, these sands may (usually) be divided into good (high-permeability) and bad. It is common now to speak of "flow units" or "hydraulic units" within reservoirs (Haldorsen *et al.*, 1988). These flow units may be connected in a complicated manner determined by the sedimentary processes (Bridge and Leeder, 1979). Throughout this paper we will refer to sand bodies, but we really mean flow units. Consider the specific example of sand bodies deposited by a meandering river system (Fig. 1). A meandering river deposits sand to form a sand body covering the meander belt. At some time an event occurs upstream which causes the river to change its path. This causes a new sand body to be laid down that may intersect a previous body. This process is continued until a system of sand bodies is formed embedded in an impermeable background. Of course, in the real world the situation is far more complicated. There may be crevasse splays linking together channel sands via low-permeability paths. The sand bodies may be in physical contact but not in hydraulic contact because of mud drapes. Post-depositional events, such as faulting, will also alter this simple model. However, we use this as a basic model of overlapping sand bodies that may be elaborated at a later stage when more reliable data are available.

We shall assume that the bodies may be represented as rectangles in two dimensions or cuboids in three. Further, we shall assume that the bodies are distributed independently in space. Finally, we assume perfect hydraulic contact between the bodies. These assumptions may not be very realistic but they will allow us to develop an understanding of the way that overlapping bodies connect and form flow pathways. These details do not materially affect the style of analysis presented here. Also some of the assumptions will not strongly affect the quantitative results. For example, we will provide evidence that it is not important whether the sands are rectangular but only that they are convex. Also, it is unlikely to be important that there is a stacking sequence of the sands (Collinson, 1978; Bridge and Leeder, 1979) provided the system is larger than the stacking sizes. The stacking sequence is a correlation between the centres of the sand bodies. The existence of a correlation (which would make the sands appear to be "attractive" or "repulsive") will not affect the nature of the calculations reported here, although it may affect some of the numerical values. Hence, in this paper we shall only discuss uncorrelated sand bodies. This is to be taken as read when mention is made of overlapp-

*North Sea Oil and Gas Reservoirs--II*

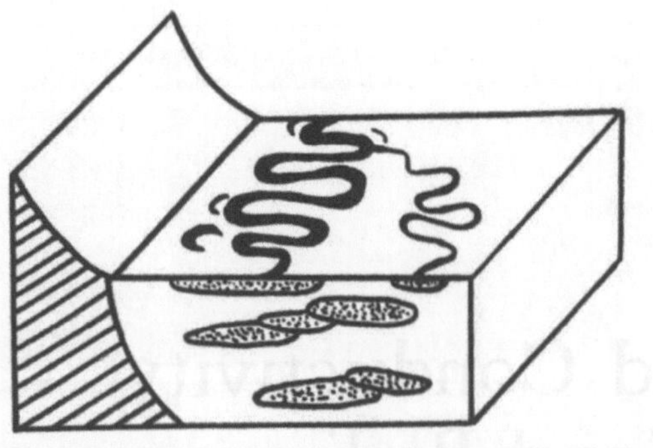

**Fig. 1.** Overlapping sand bodies deposited by a meandering river system.

ing sands. All of these assumptions may be investigated and the model may be made more realistic in a number of ways. However, the purpose of this paper is to understand the general behaviour of overlapping bodies in terms of a simple model. The exact details of the model will be described in the next section.

We wish to determine the probability that wells are in connection via the good sand. Also, the fraction of the total sand (and hence oil or gas) that is in contact with these wells, at what rate the oil will flow and what fraction is recoverable are important parameters to study. The behaviour of this model is simple to visualize. At a small number density, the sand bodies will be isolated. As more bodies are added they will overlap to form clusters until eventually they form a large cluster that spans the region. This type of behaviour is well described by percolation theory, which we use to describe the connectivity and conductivity of the overlapping sand bodies.

## THE MODEL

The main method that has been used previously to model overlapping sand bodies has been a ballistic or Boolean algorithm (Haldorsen *et al.*, 1988); i.e. the Cartesian coordinates of the centre of each body are chosen at random and then the dimensions of the body are chosen. Overlap with other bodies is then tested for. This is not a very efficient algorithm for large numbers of bodies. To generate the realizations we use a method previously used in the study of continuum percolation of discs (Gawlinski and Stanley, 1981). We consider the sand bodies to be rectangles of size $a \times b$ in a square region (arbitrarily set to $100 \times 100$). This region is divided into a coarse grid, each grid-block having the same size as the sand bodies. Any sand bodies whose centres lie within the same grid-block must then overlap. Also sands cannot overlap with those whose centres do not lie within the nine blocks surrounding the current one.

We now use the fact that the rectangles have a Poisson probability distribution to occupy one of these grid-blocks. A grid-block is occupied if the centre of the sand lies within it. The probability $P_k$ that a grid-block has $k$ sand bodies inside it is

$$P_k = \frac{x^k e^{-x}}{k!} \qquad (1)$$

where $x = Nab/A$. Here $N$ is the total number of sand bodies and $A$ is the area of the region ($= 10^4$ here). We generate the bodies by sweeping sequentially through the grid. In a given grid block we generate a number, $y$, uniformly from the interval [0,1]. We then generate the cumulative of the Poisson distribution Eq. (1), $C_k = \Sigma_{i=1}^{k} P_i$. The number of sands to be placed within this grid-block is then $k$ such that $C_k < y < C_{k+1}$. These $k$ bodies are then distributed independently and uniformly within the grid block. As we put each body down, we give it a label, $i$ for the $i$th body and so on. If a body overlaps with a previous one in an adjacent grid-block, we change the label of the current body to that of the one with which it overlaps. In this way we build up clusters of overlapping sand bodies. Each cluster is identified by a unique label. For the two-dimensional simulations we can check which clusters intersect the left- and right-hand ends of the region and whether there are any clusters of overlapping sand bodies. Each cluster is ident- i.e. they form a cluster that links two wells located at the boundaries of the region.

To calculate the fraction of sand (or net-to-gross $p$) we put a fine grid over the system and simply count the fraction of points that fall within sand bodies. Likewise, we can count the fraction of these occupied points that lie within the spanning cluster. We call this the connected sand fraction ($P$). We find that the connection between the connected sand fraction and the net-to-gross ratio is well explained by percolation theory. So before discussing the results of these simulations in detail we give a brief review of percolation theory.

## PERCOLATION THEORY

There are a number of good reviews of percolation theory (Essam, 1980; Stauffer, 1979, 1985), so here we only review that material which is relevant to the problem of overlapping sand bodies. The subject was first developed by Broadbent and Hammersley (1957). The simplest version may be stated as follows. Consider the sites of a network (the squares in Fig. 2). These are occupied with probability $p$ (the black squares in Fig. 2). Neighbouring sites are in the same cluster if they are both occupied. Percolation theory is essentially interested in the size and shape of these clusters as a function of $p$.

At low $p$ (Fig. 2) the clusters are small (generally they are single sites) and well spaced. As $p$ increases the clusters begin to join up to form larger clusters. The main mechanism is by bridging between clusters. The clusters become dominated by a single large cluster (shown in black in Fig. 3 whilst the others are grey). At one value of $p$ (about 0.59725 in this case) this largest cluster spans the whole region. This is the percolation threshold (denoted $p_c$). Above this value more and more of the clusters become absorbed by the percolating cluster leaving only a few, isolated, occupied sites (Fig. 4).

This transition from disconnected clusters to a single spanning cluster is the main feature of percolation. It is only a sharp transition for infinitely large systems. We will discuss the effects of finite boundaries later. We have only described lattice-based percolation. We could do away with the lattice altogether. This system of continuum percolation (relevant to overlapping sand bodies) shares the same qualitative behaviour and indeed many of the quantitative features are the same. The one value that is closely associated with the system under consideration is the percolation threshold.

We now define a few terms relevant to our discussion of percolation theory. We have already defined the probability $p$ that a site is occupied. We define the percolation probability ($P_\infty(p)$) to be the fraction of occupied sites belonging to the infinite percolating network. Alternatively, this is the probability that an occupied site belongs to the infinite cluster. $P_\infty$ is zero below $p_c$ and it rises rapidly for

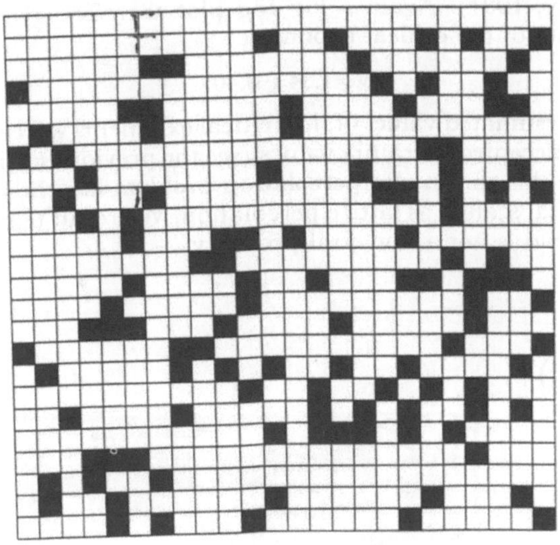

**Fig. 2.** Lattice percolation: p=0.2.

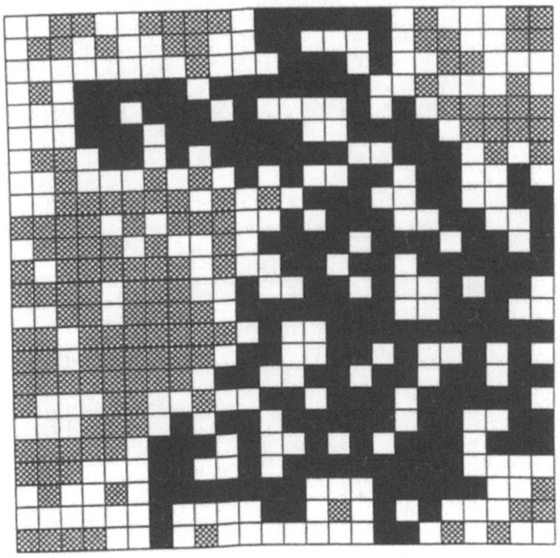

**Fig. 3.** Lattice percolation: p=0.6.

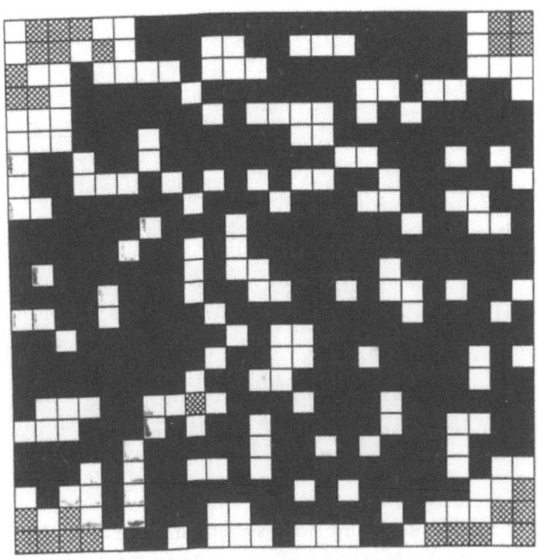

**Fig. 4.** Lattice percolation: p=0.7.

$p$ above $p_c$. Considerable numerical studies indicate that this rise may be described by a power law:

$$P_\infty \propto (p-p_c)^\beta \tag{2}$$

This defines the exponent $\beta$, which is known as a critical exponent, the connectivity exponent. This is the first of many critical exponents used in percolation theory. We shall only need a small number of them.

The next property we need to define is the correlation length or connectedness length. This is related to the size of the largest cluster. The correlation function $g(r)$ is the probability that a site at a distance $r$ from an occupied site in a finite cluster is also occupied and belongs to the same cluster. The connectivity length $\xi$ is defined as

$$\xi^2 = \sum_r r^2 g(r) \Big/ \sum_r g(r) \tag{3}$$

This takes the scaling form

$$\xi \propto |p-p_c|^{-\nu} \tag{4}$$

where $\nu$ is a new critical exponent, the correlation length exponent. The correlation length decreases above the threshold because the infinite cluster is excluded from the sum shown in Eq. (3).

Finally, we are interested in the conductivity ($\Sigma$) of the system. Imagine that the occupied sites are conducting and the unoccupied sites are insulating. Then the conductivity of the lattice is measured by placing a potential difference across the network and measuring the current flow. Clearly, below the threshold there is no overall connectivity so the conductance is zero. As $p$ increases, the conductance increases with a power law.

$$\Sigma \propto (p-p_c)^\mu \tag{5}$$

These three power laws define three critical exponents; the connectivity exponent $\beta$, the correlation length exponent $\nu$ and the conductivity exponent $\mu$. These exponents appear to be independent of the underlying lattice, unlike the percolation threshold. This property is known as universality. The underlying principle behind universality is that the large-scale properties of the percolating system should be insensitive to the small-scale structure of the problem. Numerical experiments indicate that this is true. They do, however, depend on the number of dimensions of the simulation. In this paper we will be mostly interested in two dimensions, but we give some results for three dimensions at the end. The generally accepted values from the literature (Stauffer, 1985) are (in two dimensions)

$$\beta = 5/36 \approx 0.1389; \qquad \nu = 4/3 \approx 1.3333; \qquad \mu \approx 1.3$$

So far we have only discussed infinite systems; in reality all systems are finite. The effect of the finite boundaries is to smear out the percolation transition; i.e. the average connectivity may be finite below the threshold. The average here is an average over many realizations. There will be fluctuations about this mean behaviour that decrease as the system size increases. Within percolation theory, finite size scaling theory gives a way of relating the mean behaviour at different sizes. Finite size scaling theory is based on the principle that everything depends only on the ratio of the two length scales involved. Those are the system size $L$ and the size of the dominant cluster $\xi$; $L/\xi$ is the effective, dimensionless length scale of the system. Strictly, both of these lengths should be much larger than the sand size; however, we find empirically that they only need to be greater than around 5 times the sand length. If $L$ is very much larger than $\xi$ we would expect the system to behave as though it were infinite. For $L/\xi < 1$ we would expect the system to notice the boundaries. Hence, if we look at the probability $P(p, L)$ that an occupied site belongs to the largest cluster at a filling fraction $p$, ($P(p, L)$ is the finite size analogue of $P_\infty$). We expect this to be a function of $L/\xi$ only. $\xi$ scales with $p-p_c$ as in Eq. (4) and for convenience we use $(L/\xi)^{1/\nu}$. So we write

$$P(p, L)=f[(p-p_c)L^{1/\nu}] \quad (6)$$

where $f$ is some function. We now consider how the behaviour of the system changes under an arbitrary change in length. Let $L\rightarrow\kappa L$. Under this change of scale we expect the essential percolation behaviour to be unaltered. That is, $P(p, \kappa L)\rightarrow c(\kappa)P(p, L)$. The only function of $L$ that has this property is the power. Hence we write

$$P(p, L)=L^A F[(p-p_c)L^{1/\nu}] \quad (7)$$

where $F$ is a universal function. As $L\rightarrow\infty$ we must recover the power law Eq. (2). Hence, $F[z]\rightarrow z^\beta$ for large $L$. To cancel the $L$ dependence in Eq. (7) we must have $A=-\beta/\nu$. Hence, the finite size scaling law for the connectivity can be written

$$P(p, L)=L^{-\beta/\nu}F[(p-p_c)L^{1/\nu}] \quad (8)$$

So if we plot $L^{\beta/\nu}P$ against $(p-p_c)L^{1/\nu}$ we would expect to see a universal function independent of the system size. We shall see this when we examine the results for the overlapping sand bodies.

## CONNECTIVITY

We use the algorithm described earlier to generate realizations of overlapping sand bodies. Three typical examples are shown in Fig. 5 for three different net-to-gross ratios. It is clear that the same kind of percolative behaviour is present as was described in the previous section; i.e. at low net-to-gross the sand body clusters are small and isolated. They grow until one cluster spans the region. This cluster absorbs mass until only a small number of bodies are isolated. First, we will consider square sand bodies embedded in a square region, because rectangular bodies introduce a further complication, which we will deal with later. If we plot the connected sand fraction against the net-to-gross we get the typical percolation behaviour of rapid increase in connectivity above threshold (Fig. 6). There is a considerable spread in results arising from finite size effects. However, if we average the data (averaged at fixed net-to-gross) we obtain the curves shown in Fig. 7. The dependence on system size is clear (the system size $L$ is defined as the ratio of the system length to the sand size; $100/a$, for sands of size $a\times a$ in a region $100\times 100$). The averages were made for around 15 000 realizations at each size ($L$).

First we must determine the percolation threshold. This is done by estimating the threshold at each size ($L$) and then extrapolating to the infinite system. We used several different estimates of the threshold. We found

$$p_c=0.668\pm 0.003 \quad (9)$$

This is in good agreement with previous work (Pike and Seager, 1974; Balberg *et al.*, 1984). The errors refer to the variance found from using different estimates of $p_c$. We then checked the universality of the exponents. To estimate $\beta$ we took a log–log plot of $P$ against $(p-p_c)$ to give

$$\beta=0.14\pm 0.02 \quad (10)$$

To estimate $\nu$ we first looked at the probability distribution that the system will percolate at net to gross $p$. The variance in this distribution scales like $L^{1/\nu}$ (Gawlinski and Stanley, 1981). Again a log–log plot therefore gives an estimate of the critical exponent

$$\nu=1.34\pm 0.40 \quad (11)$$

These estimated values of the critical exponents were close to the accepted literature values (see the previous section). Hence this continuum percolation problem has the same universal scaling as lattice percolation. We recall that the threshold is not a universal property.

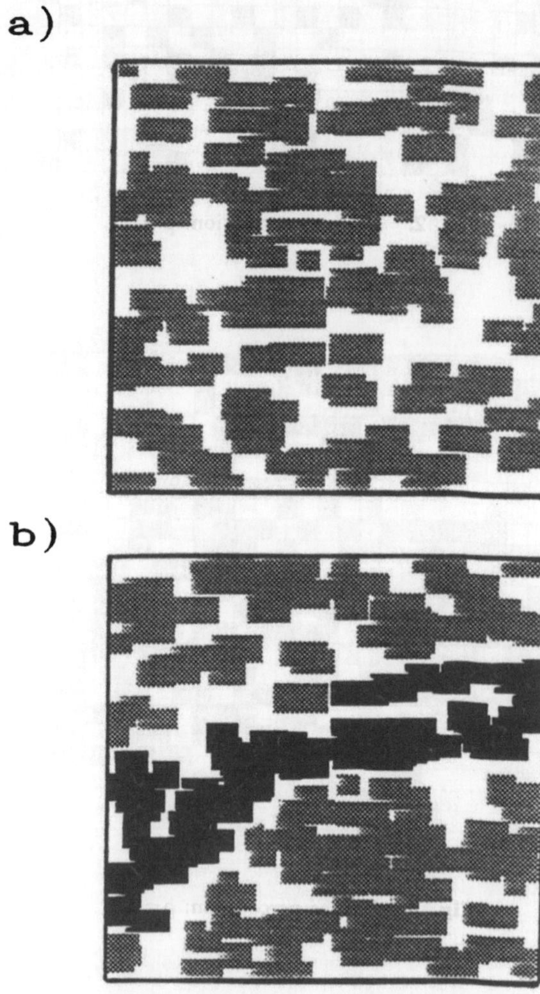


**Fig. 5.** Overlapping sand body system at three different net to gross ratios. (a) *Net to gross 61.53%, connected sand fraction 0%*. (b) *Net to gross 67.19%, connected sand fraction 30%*. (c) *Net to gross 78.10%, connected sand fraction 93%*.

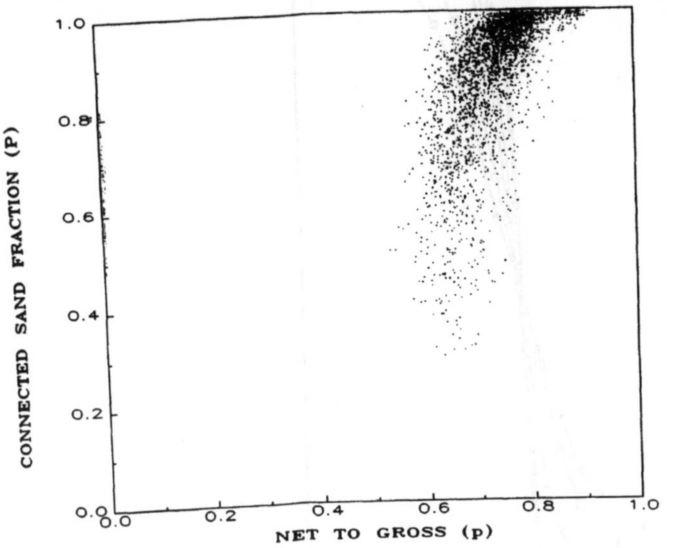


**Fig. 6.** Connectivity of square sand bodies.

We now apply finite size scaling. We plot $L^{\beta/\nu}P$ against $(p-p_c)L^{1/\nu}$ Fig. 8). The empirically determined values of the critical exponents were used here. It can be seen that all of the curves lie nearly on top of each other. There is a certain amount of noise and the behaviour is not universal as $p$ approaches unity. The power $z^{\beta}$ is shown as the dotted line to show the asymptotic behaviour. Hence, from the universal curve we could reconstruct the curves of Fig. 7 without having to perform the many realizations that were originally required.

We must now consider the case of rectangular sand bodies. First it is necessary to recall that it is the ratio of lengths that is important in determining the aspect ratio of the system. Consider a real case of a region 1000 m × 25 m in which there are sand bodies 100 m × 5 m. The aspect ratio of the sands is 20 : 1. However, the aspect ratio of the region size to the sand size in each direction is only 2 : 1. It is this ratio that is important for the percolation behaviour. For this case of rectangular bodies there is obviously a difference between the horizontal and vertical connectivity (for these purposes we define the horizontal direction to be parallel to the long axis of the bodies). Finite size scaling theory says nothing about the shape of the bodies. So we will proceed to replot the connectivities using the rescaled axes as before (Fig. 9) ($\omega$ is the aspect ratio $a/b$ of the rectangles). It can be seen that it is easier to connect in the horizontal direction than in the vertical direction. Also, above an aspect ratio of about 4 : 1 the horizontal connectivity appears to be independent of the aspect ratio. The vertical connectivity, however, is still sensitive about this value. It is not yet understood how to rescale the curves for the rectangular bodies onto the isotropic case, so we use the different curves of Fig. 9.

With these rules we can calculate the connectivity as a function of well spacing. This is effectively achieved by simply inverting the finite size scaling laws that have just been presented. As a concrete example, we consider a layer 10 m thick with sand bodies 50 m long by 1 m thick. Then consider a 500 m well spacing. In terms of sand body units the vertical thickness is 10 and the horizontal distance is also 10; this corresponds to the isotropic case (squares). Hence, to calculate the connectivity at a particular net-to-gross ($p$), we simply invert Eq. 8 using $L=10$ and the numerical universal function $F$ from Fig. 8. At a well spacing of 1000 m we now have a system which (in sand body units) is 20 × 10. This is equivalent to vertical flow in a system of aspect ratio 2 : 1 with $L=10$. So we use the appropriate curve from Fig. 9 and the scaling equation, Eq. (8). We continue this process to determine the connectivity as a function of well spacing for various net-to-gross ratios. This is given in Fig. 10. This is an example of the general behaviour of the overlapping sand bodies that we are seeking. From a few simple arguments and understanding of the universal behaviour of percolating systems, we have shown that it is possible to deduce much about the behaviour of a real system. The result given here is for the fraction of sand connected to two wells and hence the fraction of original oil in place that is in contact with the wells. However, this is not necessarily the same as the recoverable oil from these wells. In the next section we will describe the effective permeability of the percolating system and the implications for the recoverable fraction of oil.

## EFFECTIVE PERMEABILITY

We shall now consider a simple model for the effective permeability of the sand body system. We shall give all the sands the same permeability (which without loss of generality we may set equal to unity). The background material we shall set to zero permeability. We shall assume that there is perfect connectivity between sands (no impermeable mud drapes and so on). We can then define the effective single-phase permeability between two wells as that permeability which gives the same overall flow rate for a fixed pressure drop as in the heterogeneous system. We could calculate this permeability by solving the single-phase pressure equation ($\nabla \cdot K\nabla P=0$) within the overlapping sands, calculating the total flow rate and equating the effective permeability to the ratio of this total flow to the overall pressure gradient (Begg and King, 1985). This, however, requires a lot of computer time and storage and becomes prohibitively expensive when many realizations are required to determine the mean behaviour. Instead we use a computationally efficient approximation known as real space renormalization (King, 1988). The renormalization technique works on the principle of calculating the effective permeability of a small region of space to define a coarser grid. We then repeat this procedure on the coarse grid and so on until a fixed value is found. To do this for the continuum overlapping sand body problem we first place a fine grid of points over the model and assign grid-point permeabilities of 0 or 1 depending on whether the point lies within shale or sand respectively. As with the connectivity there is a scatter in the results from finite size effects. Here, we shall only discuss the mean behaviour.

It might be expected that the sand conductivity ($\Sigma$) is simply proportional to the amount of sand in the connecting cluster ($\Sigma(p) \propto P(p)$). However, simple experiments (Last and Thouless, 1971) and the results of these simulations showed that this is not the case. Figure 11 shows the results of a large size system. It can clearly be seen that the effective permeability (solid line) rises much more slowly than the connectivity (dotted line). The reason for this difference is that the percolating cluster has a tortuous conducting path with many dead ends. In Plate 30 we show the result of introducing tracer particles into the left-hand side of a simulation of the flow on a sand body system. The different colours represent different times. It can be seen that there is almost no flow in the dead ends. These dead ends contribute to the fraction of sand that is connected, but not to the conductivity. Likewise these regions will not be swept in secondary displacement

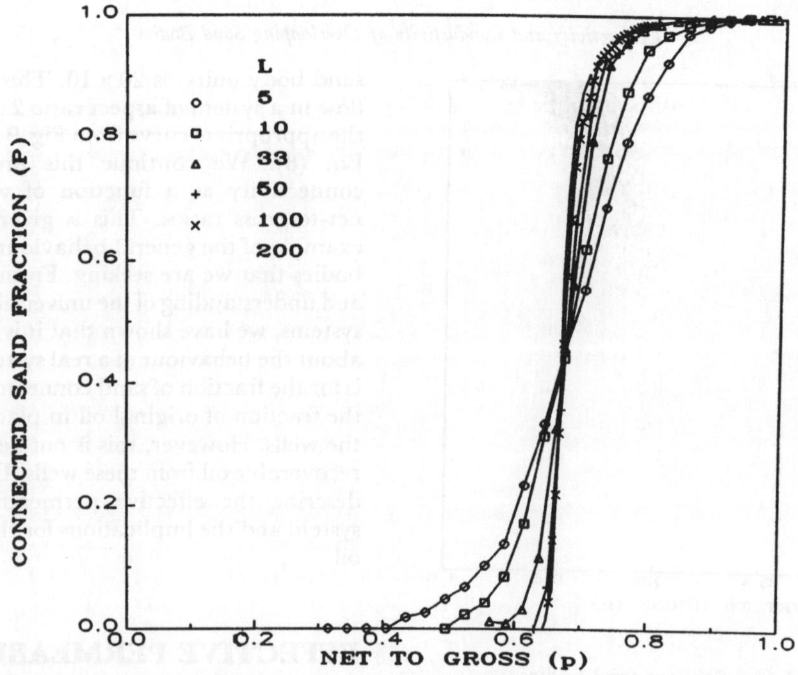


**Fig. 7.** Average connectivity of square sand bodies as a function of system size.

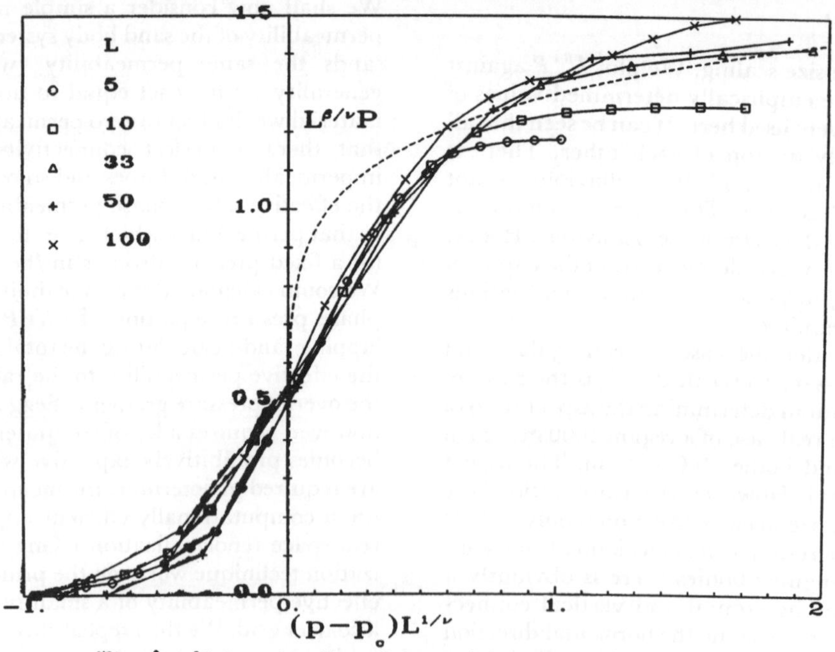


**Fig. 8.** Universal scaling of connectivity to account for finite size.

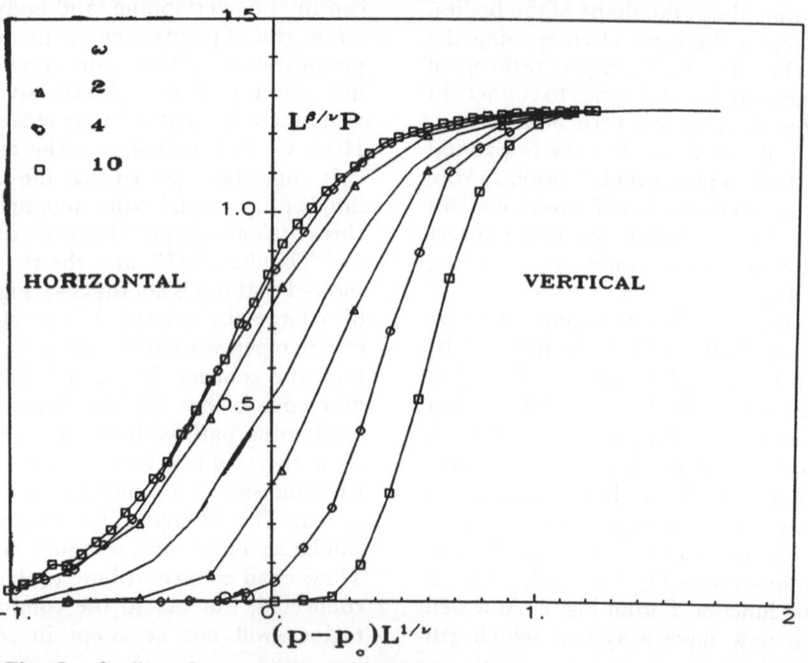


**Fig. 9.** Scaling of connectivity of rectangular sand bodies, effect of aspect ratio.

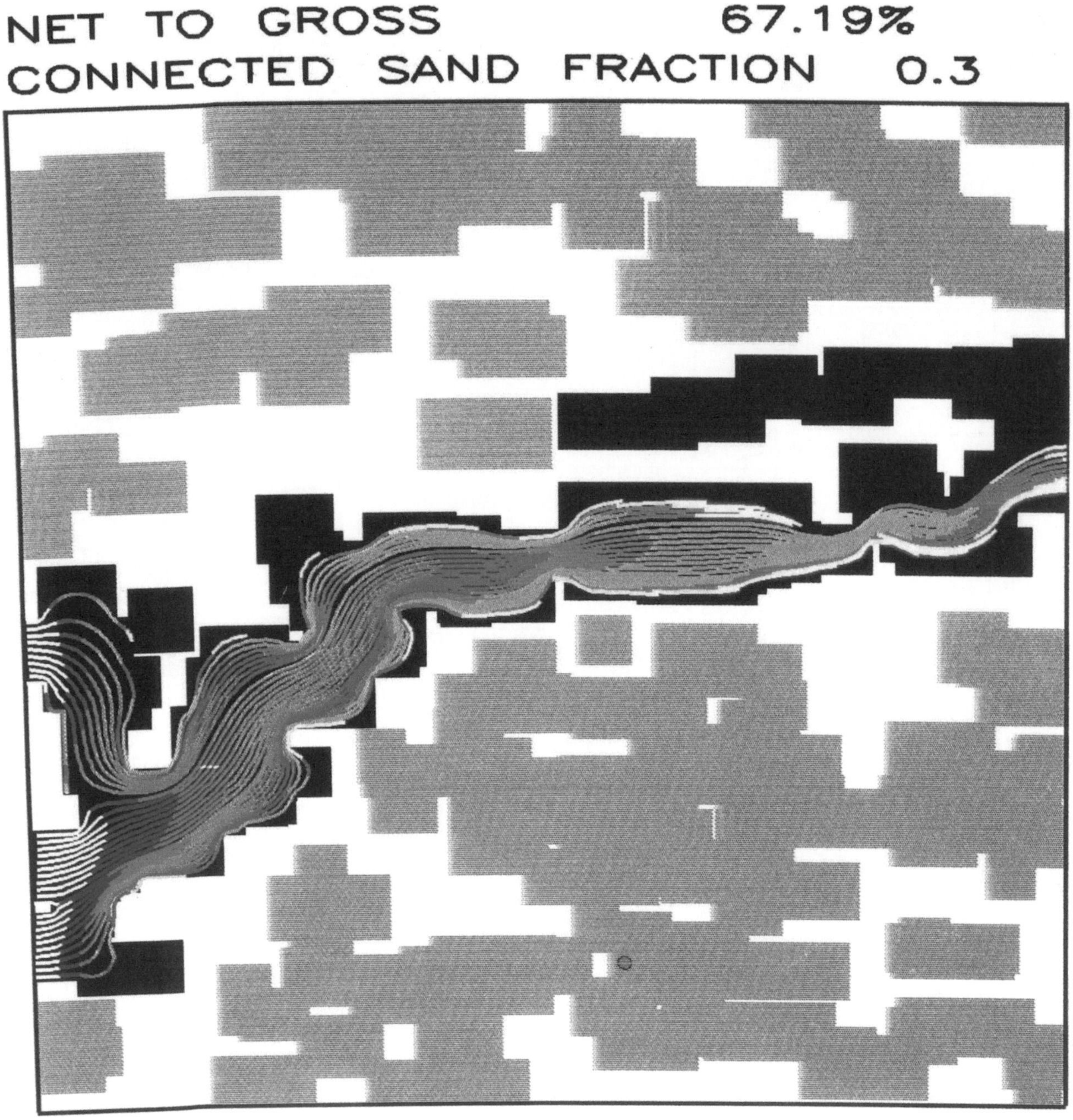


**Plate 30.** Streamlines in a sand body cluster showing non-swept dead ends.

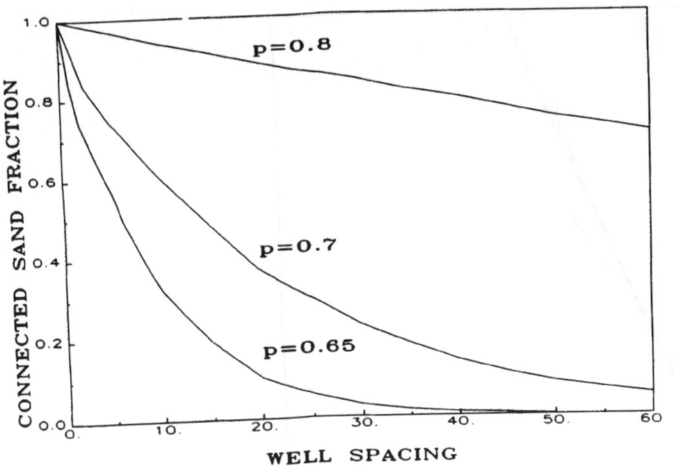


**Fig. 10.** Connectivity as a function of well spacing at different net to gross ratios.

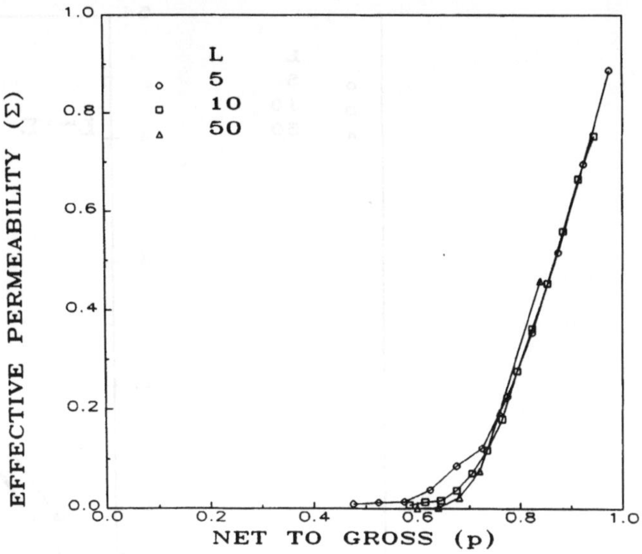


**Fig. 12.** Average effective permeability as a function of system size.

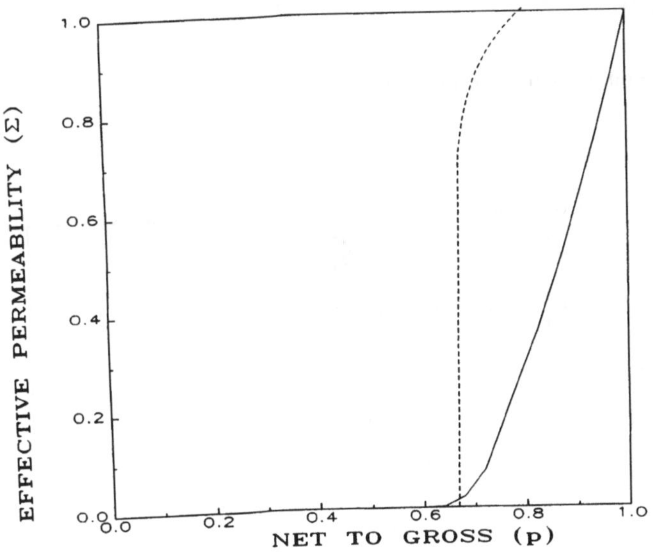


**Fig. 11.** Contrast between connectivity and effective permeability.

processes. Hence, the recoverable oil fraction will increase slowly like the effective permeability.

We have previously shown that the percolation theory of connectivity gives rise to a power law analogous to that for the conductivity. The best numerical experiments on lattices give rise to the exponent $\mu \approx 1.3$. We were unable to perform a sufficient number of realizations to obtain a good numerical estimate in the case of the sand body problem and so we used the literature value (Stauffer, 1985). In Figure 12 we show the effective permeability as a function of net to gross for a variety of system sizes in the case of square bodies in a square region. Finite size arguments analogous to those given previously give

$$\Sigma(p, L) = L^{-\mu/\nu} G[(p-p_c)L^{1/\nu}] \tag{12}$$

where $G$ is a universal function with the asymptotic behaviour $G[z] \rightarrow z^{\mu}$ as $z \rightarrow \infty$. Rescaling the results of Fig. 12 gives this universal behaviour (Fig. 13), the asymptotic behaviour is shown by the dashed line.

For rectangular bodies we apply the finite size theory Eq. (12). As with the connectivity, the rescaled curves depend on the aspect ratio. Using this it is possible to construct curves of effective permeability against well spacing for a variety of net-to-gross ratios in the same way as before (Fig. 14). It can be seen that the effective permeability drops off much more rapidly than the connectivity. We might expect that the recoverable oil fraction would drop off in a similar fashion. This is the subject of future work.

These results for the connectivity and conductivity are important in that they give the value of infill drilling in terms of the increase in recoverable oil and the flow rates as a function of the well spacing. However, some of these results are sensitive to details of the model. The main problem with the model as presented here is that it is two-dimensional. Real reservoirs are, of course, three-dimensional. In the next section we shall investigate the effects of the third dimension.

## THREE DIMENSIONS

The essential details of the three-dimensional model are the same as for the two-dimensional one. The same generation technique as the one previously described is used, except that a cuboidal coarse grid is used in which the sand bodies are generated. The high cost of the numerical simulation has so far prohibited the same level of study that has been achieved in the two-dimensional case. For example, in three dimensions the wells should be represented as a pair of lines. There is then a difficult question as to the effects of alignment of the location of these lines with the principal axes of the sands. This is certainly of importance in the finite size behaviour, especially with non-isotropic sands. All we have studied here is the basic percolation behaviour. To do this, the exact boundaries for connectivity are not important, although they will affect the actual connectivity in an application. Hence, we have considered the connectivity between two faces of a cube. From this we found the percolation threshold to be

$$p_c = 0.276 \pm 0.004 \tag{13}$$

We have not done a sufficient number of simulations to determine the exponents with any accuracy. However,

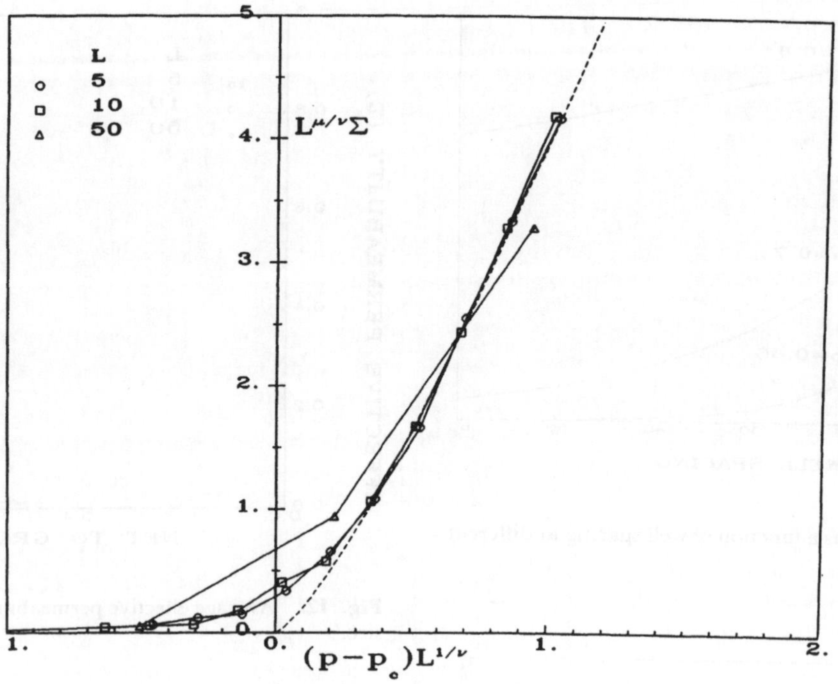


**Fig. 13.** Scaling of effective permeability to account for finite size.

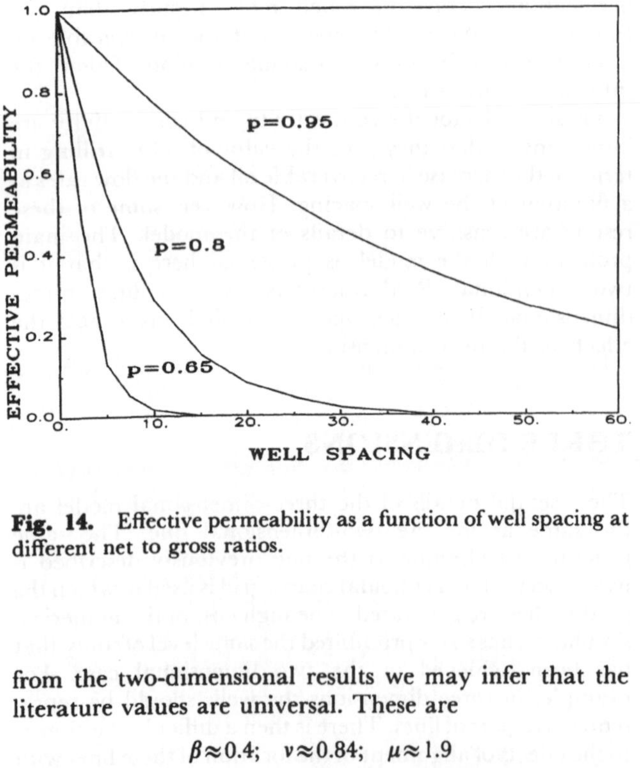


**Fig. 14.** Effective permeability as a function of well spacing at different net to gross ratios.

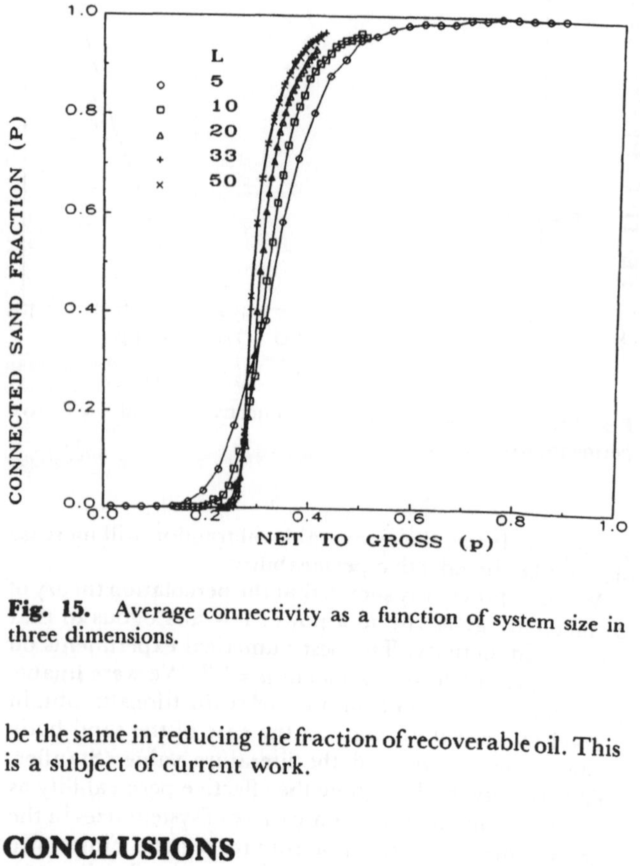


**Fig. 15.** Average connectivity as a function of system size in three dimensions.

from the two-dimensional results we may infer that the literature values are universal. These are

$$\beta \approx 0.4; \quad \nu \approx 0.84; \quad \mu \approx 1.9$$

The finite size behaviour is given in Fig. 15. These results are for cubic sands in a cubic region. The finite size scaling given by Eq. 8 is identical in three dimensions (with the appropriate exponents, of course). This is exhibited in Fig. 16. Further studies are required to infer the behaviour of rectangular bodies.

The main consequence of this small amount of three-dimensional work is that the threshold is very much reduced. Hence, at a given net-to-gross the connectivity is much higher for the three-dimensional system than for two dimensions. We would expect this to be true for the conductivity as well. However, the effects of dead ends will be the same in reducing the fraction of recoverable oil. This is a subject of current work.

## CONCLUSIONS

We set out to explain the connectivity and conductivity behaviour of overlapping sand bodies in terms of some simple theoretical model. In particular we wanted to see whether there was any kind of universal behaviour that could be used to help produce results of interest to the engineer without having to perform a large number of numerical realizations for each new reservoir. We have demonstrated that for a simple model of sand bodies it is

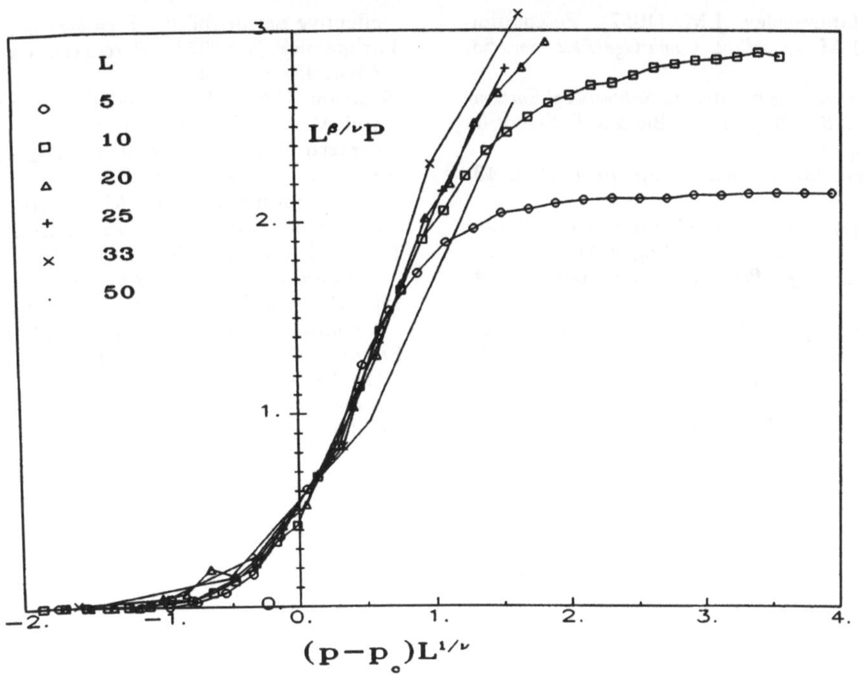


**Fig. 16.** Finite size scaling in three dimensions.

possible to recover the behaviour at any scale by using percolation theory. This theory is easy to apply in that it involves nothing more complicated than powers. We have shown that the application of this theory, along with a few numerical results, has given curves for connected sand fraction and effective permeability (which will be related to recoverable oil fraction) as a function of well spacing and net to gross.

Whilst these results are very encouraging, it should be pointed out that there are a few limitations to the theory as well as to the model. From the theoretical point of view, percolation theory is only strictly accurate near the threshold. However, from the results presented here, it would appear that the region of applicability covers most of the region of interest. Also in the region away from the threshold (high net-to-gross of order unity) it is easy to do exact calculations using mean-field theory (Kirkpatrick, 1973). This region is not very broad and is not of much interest in field applications. Also, percolation theory, even with the finite size corrections, applies to large systems. Again from the results presented here, "large" seems to mean greater than about five sand body sizes. However, all statistical approaches have difficulties with small numbers. Also, if there are only a very small number of bodies it is possible to do a large number of realizations directly and it should not be necessary to resort to theoretical models. There are also unresolved difficulties associated with the finite size effects, in particular in relation to rectangular bodies. These are not fully understood and are the subject of current research.

It is recognized that the sand body model presented here is very simple. Real sands are not rectangles imbedded independently in an impermeable background with perfect connection between the bodies. However, the purpose of this model was to demonstrate the theoretical methods that can be used to analyse the connectivity and conductivity. It is possible to play games with the basic model to make it more realistic. However, this will require more data about the disposition and dimensions of sand bodies. It would also be possible to use a model based on sedimentary processes (Allen, 1978) or on field observations (Knutson *et al.*, 1971; Knutson, 1976).

## NOMENCLATURE

| | |
|---|---|
| $a$, $b$ | Linear dimensions of the sand bodies |
| $F$ | Universal scaling function for connectivity |
| $G$ | Universal scaling function for conductivity |
| $g(r)$ | Correlation function |
| $L$ | Linear system size |
| $P$ | Connected sand fraction |
| $p$ | Net to gross ratio |
| $p_c$ | Percolation threshold |
| $z$ | Reduced net to gross $=(p-p_c)L^{1/\nu}$ |
| $\beta$ | Connectivity exponent |
| $\mu$ | Conductivity (permeability) exponent |
| $\nu$ | Correlation length exponent |
| $\xi$ | Correlation length |
| $\Sigma$ | Effective permeability (conductivity) |
| $\omega$ | Aspect ratio of the sands $=a/b$ |

## ACKNOWLEDGEMENT

The author would like to thank the British Petroleum Company plc for permission to publish this paper.

## REFERENCES

Allen, J.R.L. (1978). Studies in fluviatile sedimentation. *Sediment. Geol.* **21**, 129–147.

Balberg, I., Anderson, C.H., Alexander, S. and Wagner, N. (1984). Excluded volume and its relation to the onset of percolation. *Phys. Rev.* **B**, **30**, 3933–3943.

Begg, S.H. and King, P.R. (1985). Modelling the effects of shales on reservoir performance: Calculation of effective vertical permeability. Paper no. *SPE* 13529.

Bridge, J.S. and Leeder, M.R. (1979). A simulation model of alluvial stratigraphy. *Sedimentology* **26**, 617–644.

Broadbent, S.R. and Hammersley, J.M. (1957). Percolation processes: Crystals and Masers. *Proc. Cambridge Phil. Soc.* **53**, 629–641.

Collinson, J.D. 1978. Alluvial sediments. In: *Sedimentary Environments and Facies*, (H.G. Reading, Ed.), Blackwell Scientific Publication, Oxford, pp. 15–60.

Essam, J.W. (1980). Percolation theory. *Rep. Prog. Phys.* **43**, 833–912.

Gawlinski, E.T. and Stanley, H.E. (1981). Continuum percolation in two dimensions: Monte Carlo tests of scaling and universality for non-interacting discs. *J. Phys. A: Math. General*, **14**, L291–L299.

Haldorsen, H.H., Brand, P.J. and Macdonald, C.J. (1988). Review of the stochastic nature of reservoirs. In: Edwards, S.F. and King, P.R. (eds.), *Mathematics in Oil Production*, IMA Conference Series no. 18. Oxford University Press.

King, P.R. (1988). The use of renormalization for calculating effective permeability. *Transport in Porous Media*, **4**, 37–58.

Kirkpatrick, S. (1973). Percolation and conduction. *Rev. Mod. Phys.*, **45**, 574–588.

Knutson, C.F. (1976). Modeling of noncontinuous Fort Union and Mesaverde sandstone reservoirs, Piceance Basin, Northwestern Colorado. *SPE J.* **16**, 175–188.

Knutson, C.F., Maxwell, E.L. and Millheim, K. (1971). Sandstone continuity in the Mesaverde formation, Rulison field area, Colorado. *J. Petroleum Technol.* (Aug.) 911–919.

Last, B.J. and Thouless, D.J. (1971). Percolation theory and electrical conductivity. *Phys. Rev. Lett.* **27**, 1719–1721.

Pike, G.E. and Seager, C.H. (1974). Percolation and conductivity: A computer study. *Phys. Rev. B* **10**, 1421–1434.

Stauffer, D. (1979). Scaling theory of percolation clusters. *Phys. Rep.* **54**, 1–74.

Stauffer, D. (1985). *Introduction to Percolation Theory*. Taylor and Francis, London.

# 22

# SESIMIRA—A New Geological Tool for 3D Modelling of Heterogeneous Reservoirs

*R. Gundesø*[a] *and O. Egeland*[b]

[a]Norsk Hydro AS, Stabekk, Norway
[b]Geomatic AS, Gjettum, Norway

Set against a background of large uncertainties connected with reservoir description and subsequent recovery evaluation of offshore hydrocarbon fields, a requirement for improved reservoir modelling techniques has become evident. This paper presents a computer program, SESIMIRA, that uses bit-matrices to create high-resolution three-dimensional grids for the three-dimensional modelling of heterogeneous reservoirs. The main objective of the program is to generate reservoir models that clearly preserve a realistic heterogeneity and adequately represent the distribution of important reservoir parameters. The reservoir modelling procedure is applicable to any type of geological setting. The concept is based on the integration of sedimentological, structural and petrophysical data that can be treated both deterministically and stochastically. The user has the option of including any number of facies types, and can require that different conditions (rules) are fulfilled during the modelling process. The current version of SESIMIRA utilizes a Boolean approach to create facies bodies in the reservoir model.

The program was tested on data from the Middle Jurassic Ness Formation in a North Sea field. Data from 15 wells were included, and a sedimentological model containing channel and crevasse-splay sandstones encased in non-permeable overbank mudstones was reproduced. A number of realizations were made, and reservoir volumes connected to specific production wells were calculated directly from the SESIMIRA-generated models.

Two different methods of transferring the detailed geological output from SESIMIRA to reservoir simulators have been developed for the purpose of predicting hydrocarbon recovery. One is based on a hexagonal grid system using non-neighbour connections to describe additional fluid flow directions in a standard Cartesian simulator; the other method involves finite-element techniques for displacement-front tracking simulation. Both methods have been successfully used for fluid displacement simulation of two-dimensional areal models.

## INTRODUCTION

Although reservoir simulation has become a standard tool for evaluating field development options for North Sea oilfields (e.g. Chauvin *et al.*, 1979; Nadir, 1980; Stiles and Bobek, 1982), the production history of many fields has demonstrated that simulation predictions are often unreliable and have been significantly in error (e.g. Bayat and Tehrani, 1985; Van Oert, 1988; Corrigan, 1988).

As a result of high costs in offshore areas like the North Sea, only a minimum of exploration and appraisal wells can be justified before important field development decisions are made. The use of oversimplified geological models based on data from a limited number of widely spaced wells is one of the most important reasons for the failure of simulators to predict field performance. Oversimplification and the use of unrealistic geological models are partly due to the paucity of well data, but also to the inappropriate use of the available data. Experience from densely drilled fields and outcrop studies shows, for example, that the normal approach of linear interpolation of reservoir quality between wells some kilometres apart may often produce unrealistic models of petrophysical variation. These interpolation techniques may give a reasonable average value for volume-in-place calculations, but they do not provide a realistic representation of heterogeneity required for the prediction of fluid displacement.

The use of stochastic techniques as a way of improving geological reservoir modelling has grown rapidly during the last few years (e.g. Haldorsen and Lake, 1984; Skaug and Gundesø, 1986; Haldorsen and Macdonald 1987; Augedal *et al.*, 1986). The objective of these techniques is to generate probabilistic representations of reservoir heterogeneity to compensate for a lack of directly measured field data. These models should provide an improved basis for recovery predictions. Additionally, the uncertainty and

*North Sea Oil and Gas Reservoirs—II*

risk associated with different development options can be better quantified using such methods.

A new concept for computer reservoir modelling involving some use of stochastic techniques will be presented here. A program, SESIMIRA (*se*dimentological *sim*ulation in *irap*), has been developed. This is based on a general mapping system called IRAP (Egeland, 1987), but has been extended to handle three-dimensional (3D) grids. The tool allows sedimentological, structural and petrophysical information to be integrated into consistent reservoir models. Deterministic data are honoured and the required number of sedimentological facies can be included. It allows the user to define rules (conditions) to make the program perform in a geologically sound manner. A case study of the delta-plain sediments of the Middle Jurassic Ness Formation in the North Sea is presented to illustrate the modelling procedure.

A major challenge has been the transferrence of complex geological models into practical reservoir simulation models without losing important details. Two innovative methods have been developed and utilized to simulate flow through complex 2D areal models.

## GEOLOGICAL INPUT DATA

The extent of a reservoir database will vary from fields in the appraisal phase with a few wells to producing fields with perhaps more than 100 wells. In the North Sea, well spacing of between 2 and 10 km is typical at the end of the appraisal phase; it then decreases to between a few hundred metres and 2 km during the development phase. In each case, maximum utilization of the available data should occur in order to produce the best possible reservoir model. The most important data input requirements are summarized below.

### Qualitative data

Sedimentological core description provides a classification of a sequence into sedimentary facies and thus an interpretation of depositional environments. A correct interpretation of channel type, for example, has implications with respect to reservoir dimensions and connectivity (Fig. 1, upper). As more wells become available, well correlations give indications of sand continuity, variation in formation thicknesses and the lateral distribution of sedimentary facies. Use of data from outcrops analogous to the reservoir can often improve this understanding.

The result of a conventional facies analysis of a subsurface reservoir is usually a conceptual model of the general geological setting during deposition; this may be illustrated by schematic block diagrams, palaeogeographic maps and cross-sections. The aim of SESIMIRA is to reproduce similar models, but in a systematic, efficient and quantitative manner. The resultant models should be faithful to the available well data and in a form suitable for use in a reservoir simulator.

Both deterministic and stochastic data (Haldorsen and Lake, 1984) are used to produce the models. Isochore maps, large-scale faults, well data and beds correlatable between several wells are examples of data that can be used deterministically, while sand body geometry, orientation and location in the interwell volume are variables that often are associated with significant uncertainty and require a stochastic treatment.

A convenient set of facies types or facies associations are selected as building blocks. Since the main objective of reservoir modelling is to quantify fluid displacement, the subdivision of the reservoir into building blocks should primarily reflect differences in permeability and other reservoir properties. A study of the typical spatial relationships between the building blocks (hereafter called facies) is also required. An example of reservoir architecture in a deltaic setting is given in Fig. 1 (lower).

### Quantitative input

For each facies type used in the model, SESIMIRA requires the following set of parameters to be specified:

- thickness
- width
- length
- basic shape
- volumetric proportion of the total reservoir volume
- orientation (azimuth)
- rotation (dip)
- lateral trends
- vertical trends.

Many of these parameters are normally difficult to assess, and as such are quantified in terms of probabilistic distributions (except volumetric proportions). In SESIMIRA, any type of statistical function can be handled, but in most cases fairly simple distributions are used because of the often limited number of observations on which they are based.

Thicknesses can be derived quite easily from well data, whereas lateral dimensions are more difficult to determine. Results from well correlations, outcrop studies and Drill Stem Test (DST) analysis are most frequently used. These interpretations and observations are generally not very accurate but they can give a reasonable range of values (Fielding and Crane, 1987; Miall, 1988). A basic form for each facies type must be selected. Standard geometrical forms (e.g. disks, ellipsoids, rectangular boxes), combinations of these or digitized forms can be chosen. The basic form is scaled to its stochastic geometry during the modelling process. The volumetric content that each facies occupies relative to the total rock volume is calculated from well data. This parameter is used as a "stop" criterion in the modelling.

In fluvial settings, palaeocurrent information and sandbody orientation play an important role in determining the reservoir architecture and reservoir communication. The main direction of palaeocurrent might be uncertain, and many different channel orientations might occur within one fluvial system. Therefore, a wide range of possibilities must often be used for this parameter. Also, certain lateral and vertical variations in the relative abundance of each facies can often be observed. In SESIMIRA these are formulated in probability curves. Lateral trends are based on maps contoured between wells, while sedimentological well profiles give input to the vertical distribution curves.

## BASIC CONCEPTS OF THE PROGRAM

### Algorithm

SESIMIRA is a general tool applicable to all types of sedimentary settings. The user can apply various options in the program to build an algorithm that effectively reproduces any reservoir architecture. The modelling is organized in several steps. First comes the construction of the deterministic matrices for parameters such as reservoir geometry, facies observed in the wells, faults and any fixed geological information that will be preserved during the

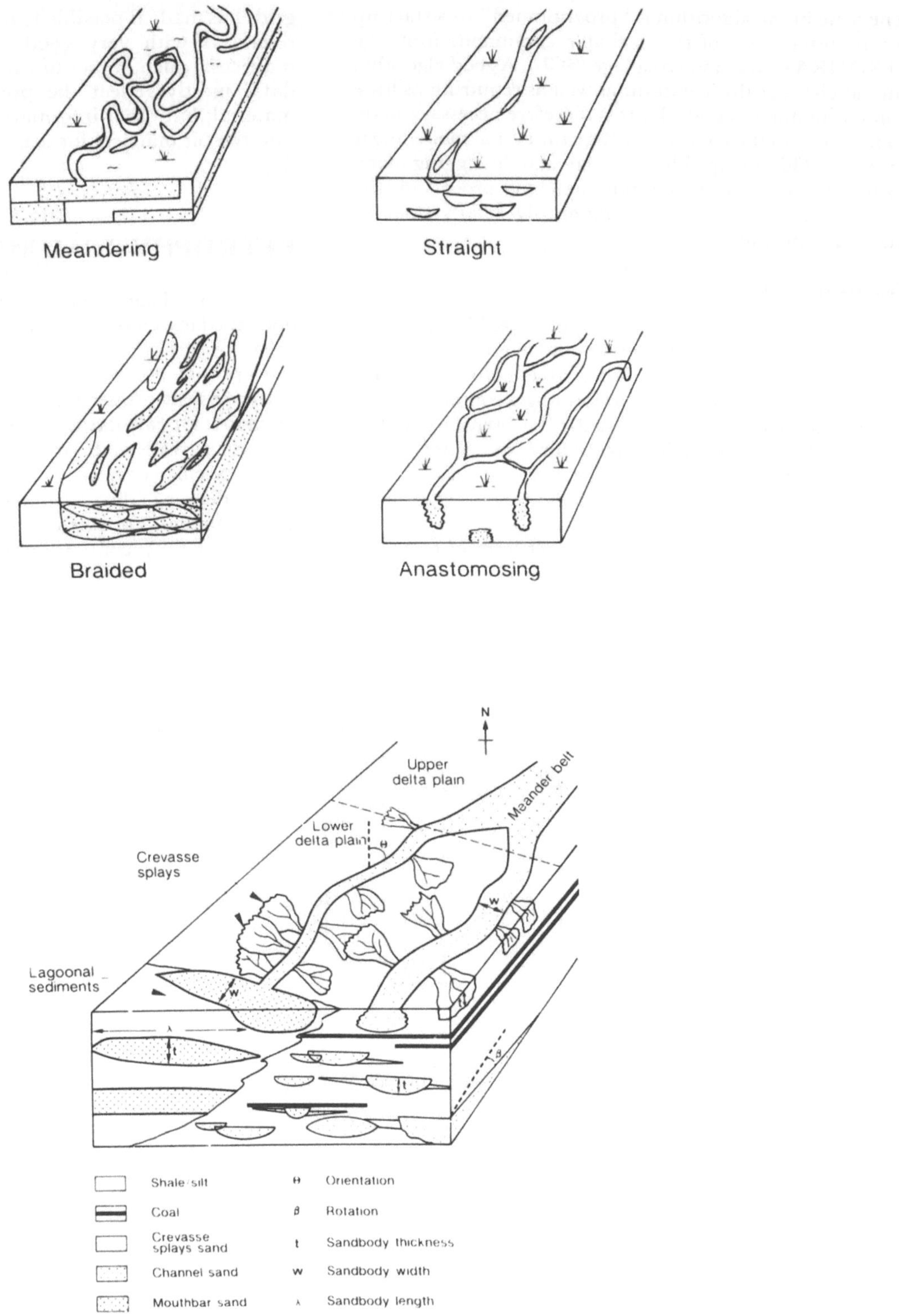


**Fig. 1.** Sedimentological illustrations: (*upper*) channel types; (*lower*) schematic model of a deltaic setting.

stochastic process. The remaining reservoir volume is then filled by stochastically assigned data. The generation and distribution of facies bodies are carried out in a traditional Boolean way (Delhomme and Giannesini, 1979). In other words, each body is given dimensions and an orientation from the probability distributions, and the centre point of the body is placed in the reservoir model according to probability functions for horizontal and vertical trends. The location of each body can be rejected or slightly modified when it is in conflict with deterministic data or other conditions. This is ensured by additional sets of bit matrices and control arrays. More and more bodies are defined in the model until the stop criterion (the correct volumetric proportion of facies) is reached.

The following factors must be specified in the algorithm:

- deterministic data
- spatial combination of facies (e.g. crevasse splays are always connected to channels)
- the priority when overlap occurs
- a sequence for creating the facies
- stop criteria

The logic in the algorithm is "programmed" in a start-up file by using a set of the available commands from the SESIMIRA Command Language (SCL). A good algorithm should give a realistic output, as well as requiring as little computing time as possible. It is therefore necessary to do some testing of the algorithm before the actual modelling is initiated. This is helped by a display and debugging mode so that the user can follow the modelling process on the screen and halt the process temporarily at any stage to make modifications.

### Bit matrices

The 3D grids are constructed as a set of IRAP grids (also called 2.5D grids) with regular and coinciding $x$ and $y$ positions of the grid nodes. The $z$ values in these grids are treated as columns of the 32 bits forming a computer word to form a single 3D grid (Fig. 2). To obtain a higher number of grid levels in the vertical direction, a new single 3D grid is stacked on top of the first one in order to represent an additional 32 levels. In this way the grid nodes of the 3D matrix are constructed in the $z$ direction as a multiple of the 32 bits. The compact form of these 3D grids has made it possible to model the facies generation of reservoirs with very good resolution. The bits in the matrices, which are set to 1 or 0 to represent some kind of data, mostly reflect the presence or absence of facies material; but other information can also be incorporated into the bit matrices for controlling the computer modelling.

## PETROPHYSICAL MODELLING

Petrophysical parameters are assigned to the reservoir rock once the facies modelling is complete. No fixed procedure exists for this in SESIMIRA. It is up to the user to choose the optimal procedure in each case. The modelling of petrophysical properties is carried out for two principal purposes: for calculation of hydrocarbons in place, and to provide a realistic basis for the reservoir simulation of recovery efficiency.

The degree of detail required for these two objectives is quite different. For calculations of hydrocarbons in place, conventional 2D contour maps between wells constructed

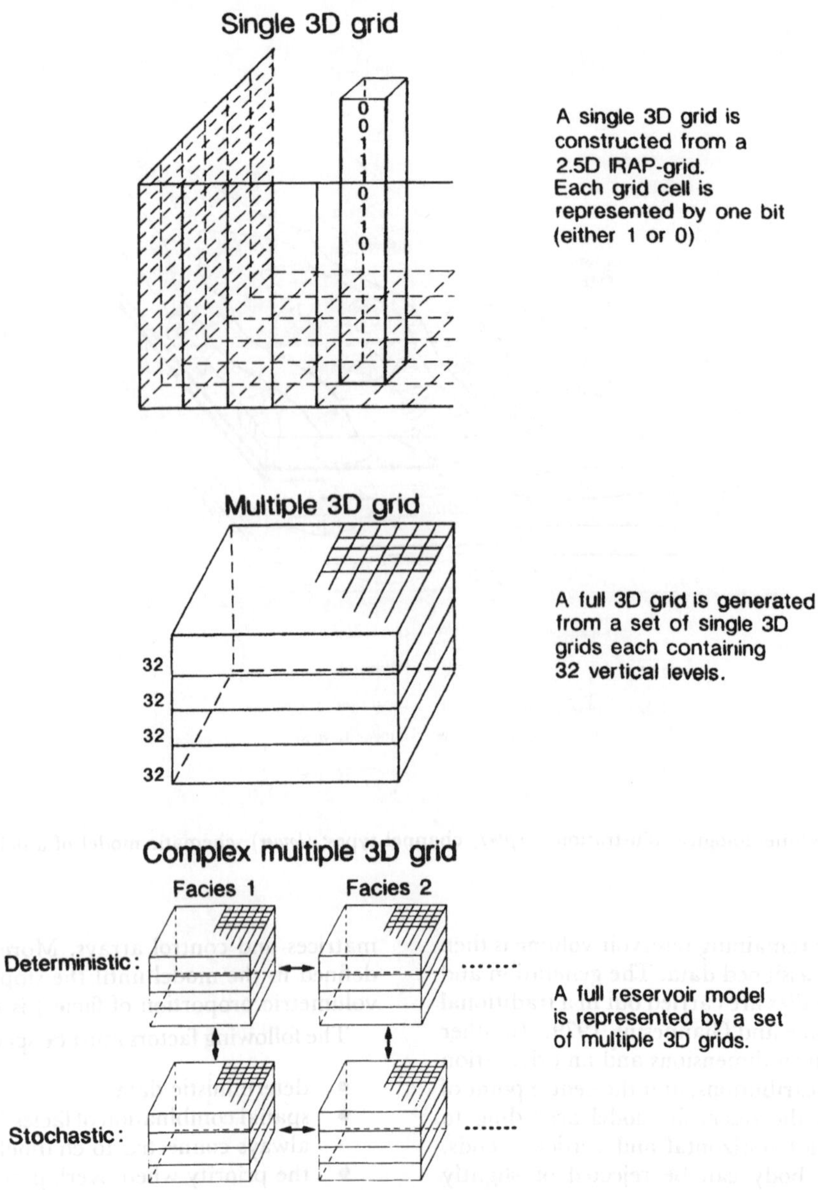


**Fig. 2.** Bit matrices as used in the SESIMIRA program. (For further explanation, see text.)

by interpolation techniques may be sufficient. Eventually a set of maps for each facies type can be generated. A Monte Carlo approach for volume determination is another available option.

On the other hand, an investigation of recovery efficiency requires detailed modelling of petrophysical heterogeneity and has an even greater requirement for statistical methods than facies modelling. Heterogeneities at all scales may have a significant effect on sweep efficiency and/or displacement efficiency (e.g. Hewett and Behrens, 1988; Lasseter *et al.*, 1986). Horizontal correlation lengths for permeability are normally only a few metres, and less than a metre vertically (Dreyer and Scheie, in press; Tomutsa *et al.*, 1986). However, generating field models with this degree of resolution is not a practical proposition because of the enormous computing resources required.

The modelling of petrophysical point values that takes into account small-scale heterogeneities ("numerical rock") can therefore only be carried out for smaller subvolumes within the reservoir (for instance, a simulation cell). Methods of scaling up the effect of these heterogeneities to field-scale models are described in Begg and Carter (1987) and Lasseter *et al.*, (1986).

In most actual cases the small-scale heterogeneities are neglected, with values for porosity and permeability being assigned directly to reservoir simulation cells. Within the SESIMIRA concept such parameters can be related to facies, location and sandbody geometry.

## CASE STUDY

A case study of the Middle Jurassic Ness Formation in a North Sea field was performed in order to test the program. The sequence modelled was originally 60 m thick, but has been reduced over large areas by post-depositional erosion. The sediments were deposited in a delta-plain environment consisting of channel sands, crevasse splays and fine-grained material desposited in interchannel areas during overbank flooding (Fig. 1, lower). The channels are interpreted as simple, straight, ribbon distributaries and anastomosing channels (Fig. 1, upper). Some of the channels have amalgamated laterally to produce sand bodies classified as ribbon-belts with lateral extensions larger than that of a single distributary. Numerous minor channels split and rejoin, producing a complex network of channels. The geometrical nature of the sedimentary facies and variable net/gross ratios (0.1–0.7) within this sequence indicate that a number of discontinuous sand bodies are present. This is confirmed by DST data and well correlations (Fig. 3).

Deterministic well data from 15 wells were included in the model. Based on an isochore map, a regular SESIMIRA 3D grid with 12.8 million (500 × 200 × 128) cells was generated giving a resolution of 50 m × 50 m × 0.5 m. The general model was composed of porous and permeable channel and crevasse reservoirs embedded in fine-grained non-reservoir overbank facies. The modelling algorithm used the following steps:

- initialization of the 3D grid
- assigning deterministic data to well locations
- extrapolation of channels and crevasses in wells
- assigning channels to deterministic crevasses
- generating stochastic channels
- assigning stochastic crevasses to channels
- modelling of petrophysical data
- calculation of volumes, interconnectedness, etc.
- transferring data to the reservoir simulator

The results of the facies modelling can be inspected in vertical and areal sections (Plates 31 and 32). This is an important option in order to check and eventually modify the model. In a case study, some 10–20 repetitions should normally be carried out in order to achieve a statistically representative spread in the realizations.

Fixed locations for 10 production wells were selected (along the fluid contact) in order to investigate the consequences of the modelling. Owing to variable channel orientation and the presence of crevasse splays, the majority of the sand bodies seem to be interconnected. However, any flow through the reservoir, for instance between an injector and a producer, will often have to follow fairly complicated and tortuous flow paths. It is considered unlikely that these large sand volumes are in full communication during production. Possible diagenetic effects on erosional surfaces, levées along the channels, uncertain communication through thin beds, etc., might form transmissibility reducers or barriers. To enable an investigation of these effects, two boundary conditions were introduced in order to calculate more realistic pore volumes assumed to be in effective communication with the 10 production wells (Table I):

- No vertical communication is assumed if the sand connectedness between two adjacent layers is less than (in this case) 15% of the total area, unless the area of connectedness is penetrated by a production well (Fig. 4).
- No horizontal communication is assumed if the sand body thickness is less than 0.5 m (1 grid cell).

These practical conditions are just examples of several options; different rules should normally be tested and sensitivity studies should be run. An application of these calculations is to plot expected drainage volume as a function of the number of production wells (Fig. 5). This can be used to investigate further optimal well spacing, assess the need for infill drilling and optimization of injection and production well patterns (Haldorsen and Macdonald, 1987). In the Ness Formation case study, most of the hydrocarbon pore volume seems to be connected to production wells after the drilling of 10 wells.

Figure 6 shows areal sections from the same level for the eight different runs. The differences represent the randomness in the stochastic process and serve as illustrations of the modelling uncertainty. Table I shows the effect this might have on the average net-to-gross ratio in the wells (i.e. the productivity per well) and the estimation of hydrocarbon pore volumes (HCPV).

**Table I.** Results from case study

| *Run no.* | *Sand content ch. sand* | *pr. well[a] cr. sand* | *Relative HCPV[b]* | *HCPV connected to wells* (%) |
|---|---|---|---|---|
| 1 | 0.26 | 0.13 | 100.0 | 85.9 |
| 2 | 0.31 | 0.16 | 105.6 | 92.4 |
| 3 | 0.19 | 0.11 | 94.2 | 82.3 |
| 4 | 0.28 | 0.14 | 97.7 | 84.5 |
| 5 | 0.34 | 0.18 | 107.1 | 95.8 |
| 6 | 0.32 | 0.16 | 103.5 | 90.1 |
| 7 | 0.31 | 0.14 | 102.4 | 87.1 |
| 8 | 0.24 | 0.16 | 95.9 | 83.4 |

[a] $\sum_{i=1}^{n} \sum_{j=1}^{m} h_j / \sum_{i=1}^{n} H_i$ when $H_i$ = total isopach in well $i$, $h_j$ = sand-body thickness, $n$ = number of wells (= 10), $m$ = number of sand-bodies in well $i$.

[b] Run 1: index 100.

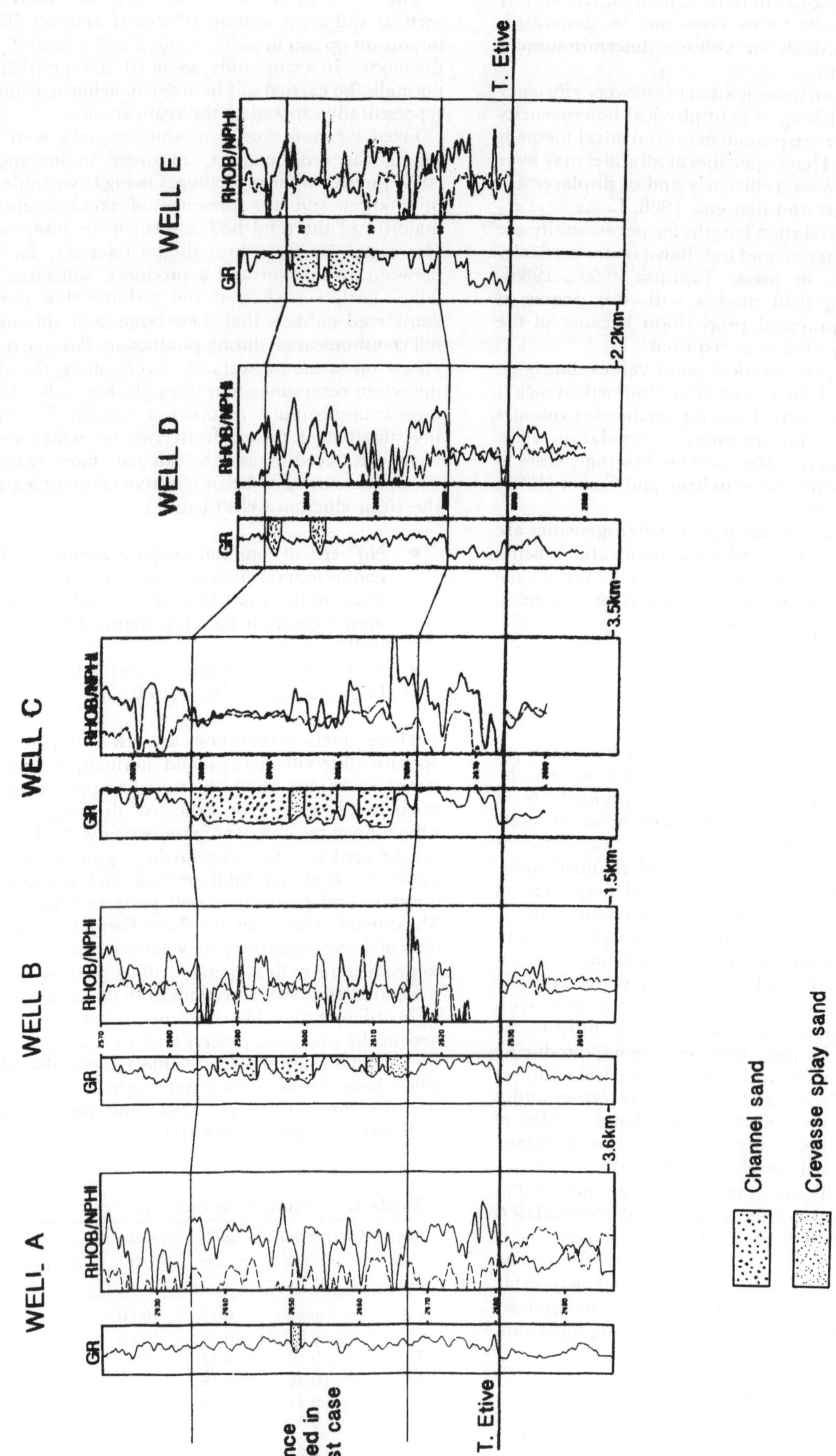


**Fig. 3.** Well correlation in Middle Jurassic Ness Formation. Gr, gamma log; RHOB, density log; NPHI, neutron log.

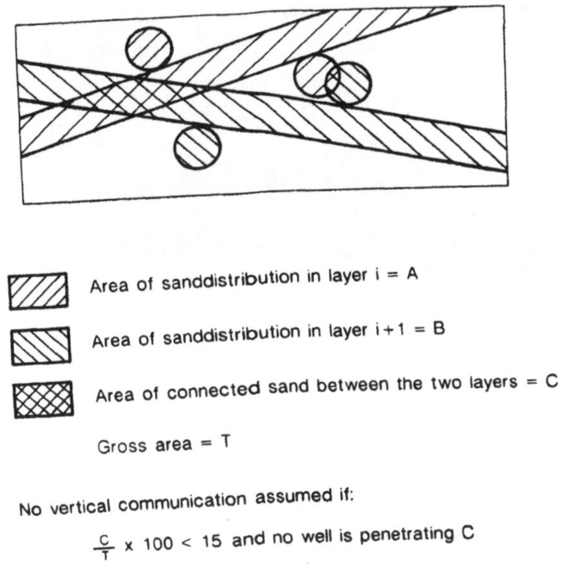


**Fig. 4.** Illustration explaining the cut-off criterion used for assuming vertical communication applied to the calculation of "effective communication".

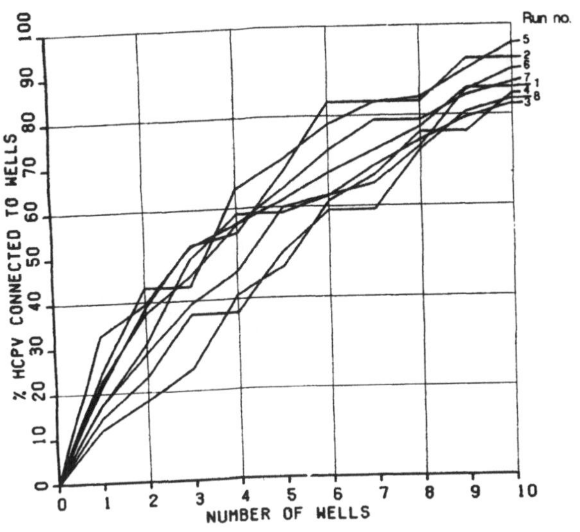


**Fig. 5.** Plot of the percentage of hydrocarbon porevolume (HCPV) in effective communication with production wells as a function of the number of wells (runs nos. 1–8).

## RESERVOIR SIMULATION

The ultimate objective of most reservoir studies is to derive a production profile for the prospect. These predictions are normally based on reservoir simulations. During reservoir simulation studies, the sensitivity of production performance to a variety of possible development options is investigated. To ensure that reservoir simulation studies are based on realistic geological input, a procedure to construct reservoir simulation models that accounts for complex geological geometries is required. This requirement meets two practical problems:

- Only a small number of model realizations can be simulated within practical cost constraints.
- The large number of grid cells applied in SESIMIRA must be significantly reduced owing to constraints in the computing capacity of the reservoir simulation.

The first problem can be reduced by ranking the geological model realizations on the basis of reservoir interconnectedness (qualitatively or quantitatively). Simulation results of a "good" case and of a "bad" case will give a fair range of the uncertainty. The problem of reducing the number of grid cells without losing vital geological complexity is more challenging. This, however, has been achieved successfully with the application of new reservoir-simulation technology. Two methods were tested, both of which gave encouraging results.

### Simulation using a hexagonal grid

This method utilizes parts of the HEX program described by Wadsley *et al.*, (this volume). HEX subdivides the plane into a hexagonal grid and uses non-neighbour connections to provide the extra intercell transmissibilities required as a consequence of an areal coordination number of 6, rather than 4 as in a rectangular grid. This enables the application of a conventional Cartesian simulator, and permits a more correct representation of the flow direction.

A representative 2D areal section from a SESIMIRA-generated model was selected. This model was automatically re-gridded into a hexagonal grid of 200 active cells, which sufficiently maintained the geological complexity (Plate 33, upper). The grid was compressed by a factor of 0.7 in order to enhance the resolution across the width of the channels. Different petrophysical properties were used for the two sandy facies, and the overbank deposits were assumed to be non-permeable. The model was then used for reservoir simulation with three oil-production and three water-injection wells for a period of 10 years. The results are shown in Plate 33 (lower). The system has lately been extended to 3D simulations by calculating effective permeabilities over the simulation cells based on the dense SESIMIRA mesh.

### Simulation by front tracking

Front tracking reservoir simulators are based on a different technology from that of traditional simulation tools. The solution for the differential equations describing pressure behaviour is obtained by means of a finite-element method, which gives high accuracy and flexibility with respect to grid geometry (Plate 34, upper). The saturation equation is a hyperbolic conservation law describing the velocity of saturation shock fronts. Besides utilizing a more flexible grid, the advantage with this technique is that it requires less computing time and excludes numerical dispersion.

A prototype of a front tracking program, FRONT (Bratvedt and Rian, 1988), was tested on the same input data as that used for the hexagonal grid. A grid generator developed as an interactive preprocessor was used to reduce the number of cells in the SESIMIRA output to a practical grid detailed enough to describe fluid flow properly. In general, both the geological details and the dynamic flow behaviour could be satisfactorily described by a 1000-cell system. The output data from this simulation are shown in Plate 34 (lower).

## DISCUSSION

The objective of using SESIMIRA is to generate realistic models of heterolithic and discontinuous reservoirs that should support more accurate predictions of hydrocarbon recovery and more reliable estimates of the uncertainty involved. The ability of the geologist and engineer to identify the depositional system and establish the input parameters is the most important constraint on how

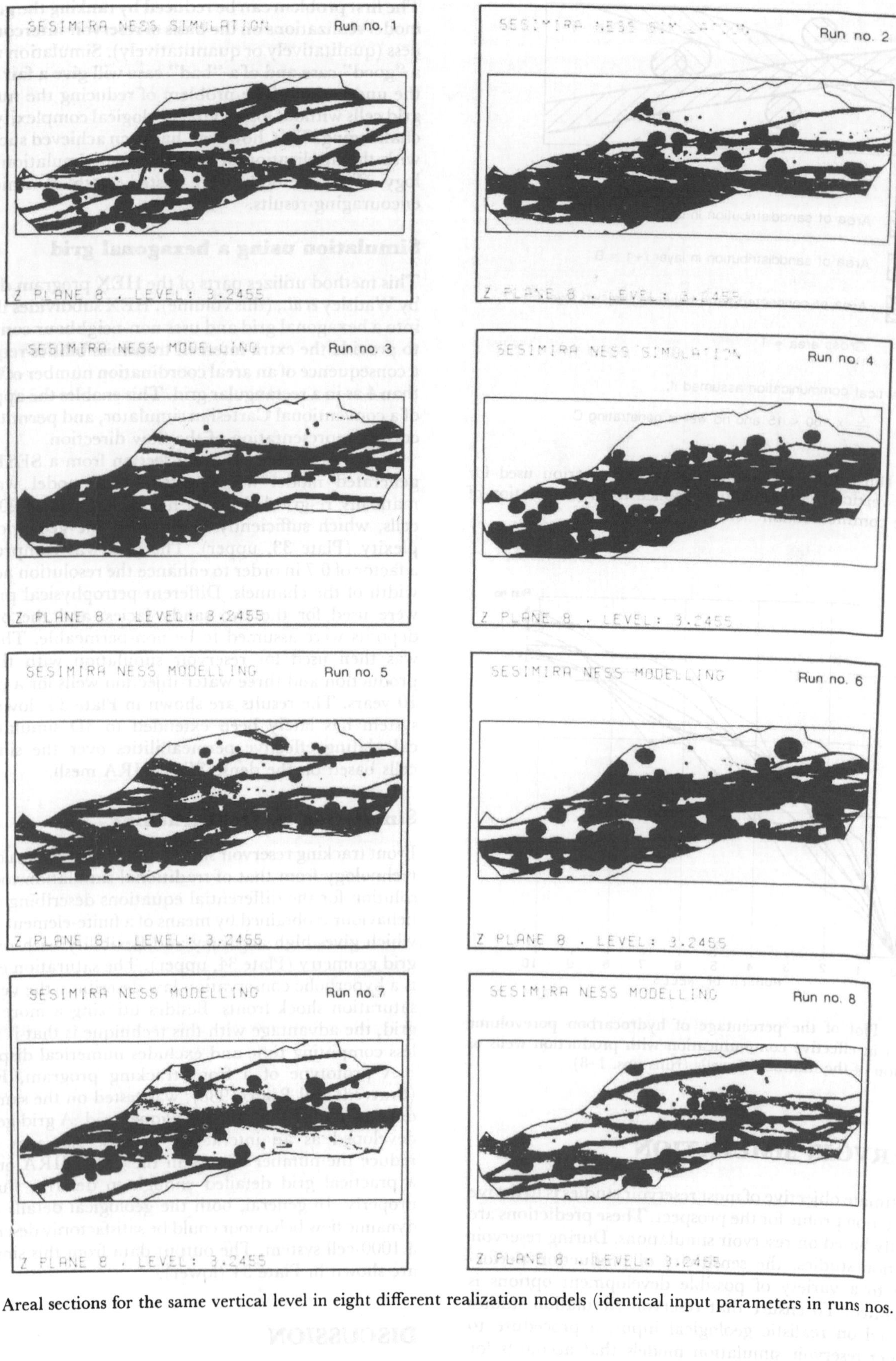


**Fig. 6.** Areal sections for the same vertical level in eight different realization models (identical input parameters in runs nos. 1–8).

accurately the reservoir can be described. Since the description may easily be erroneous, careful use of the results is essential. For example, stochastic modelling should not be relied upon to predict accurately the precise position of a specific discontinuous sand body. However, as the volume of available geological and engineering data increases and the data are integrated, the modelling uncertainty decreases. To meet the challenge of improving the estimation of reservoir parameters, numerous comparative sedimentological and minipermeameter studies have been carried out in parallel with this project (Dreyer, this volume; Dreyer and Scheie, in press).

A brief review of stochastic techniques for producing geological models of petroleum reservoirs is given by

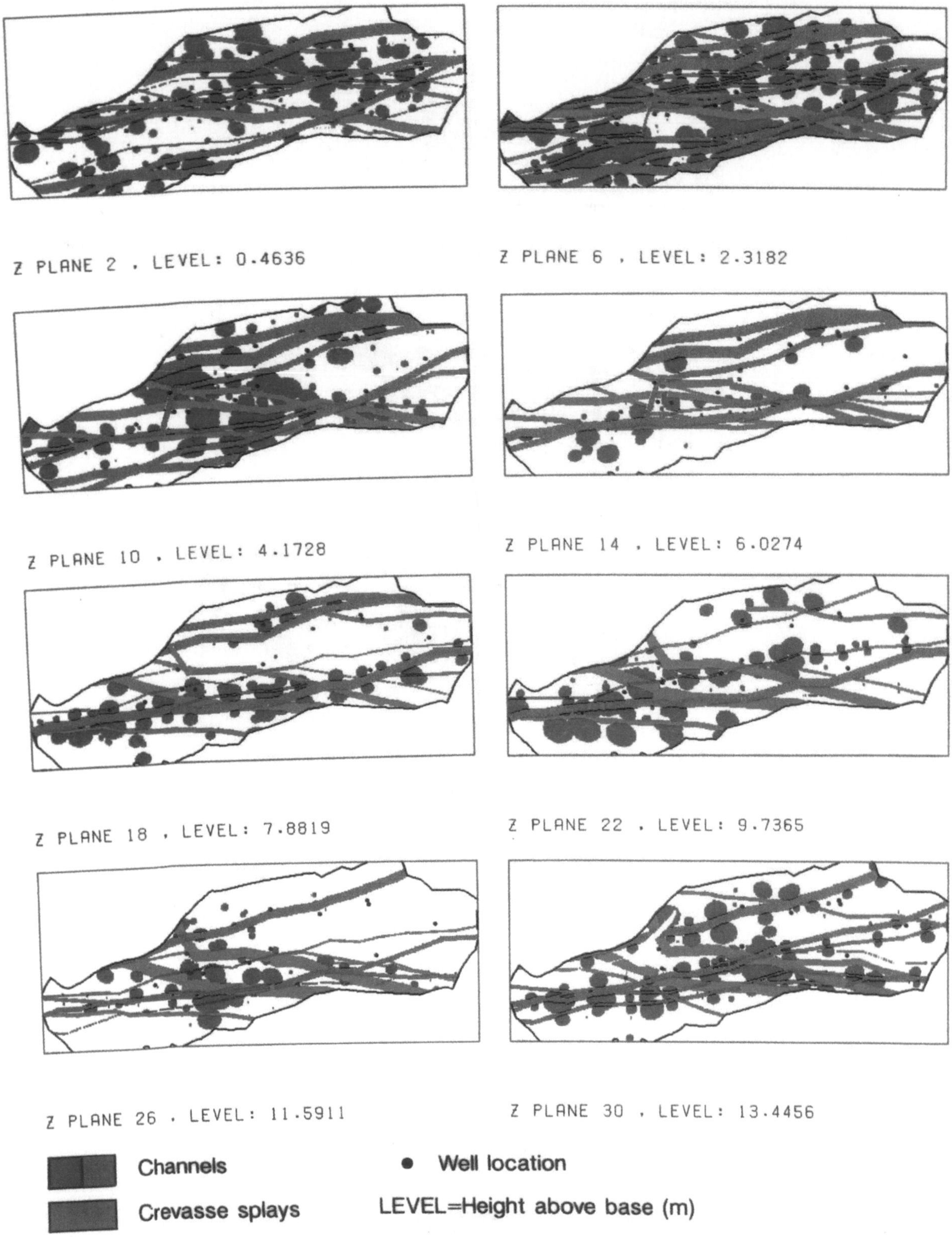


**Plate 31.** Areal sections of a test case model (run no. 1).

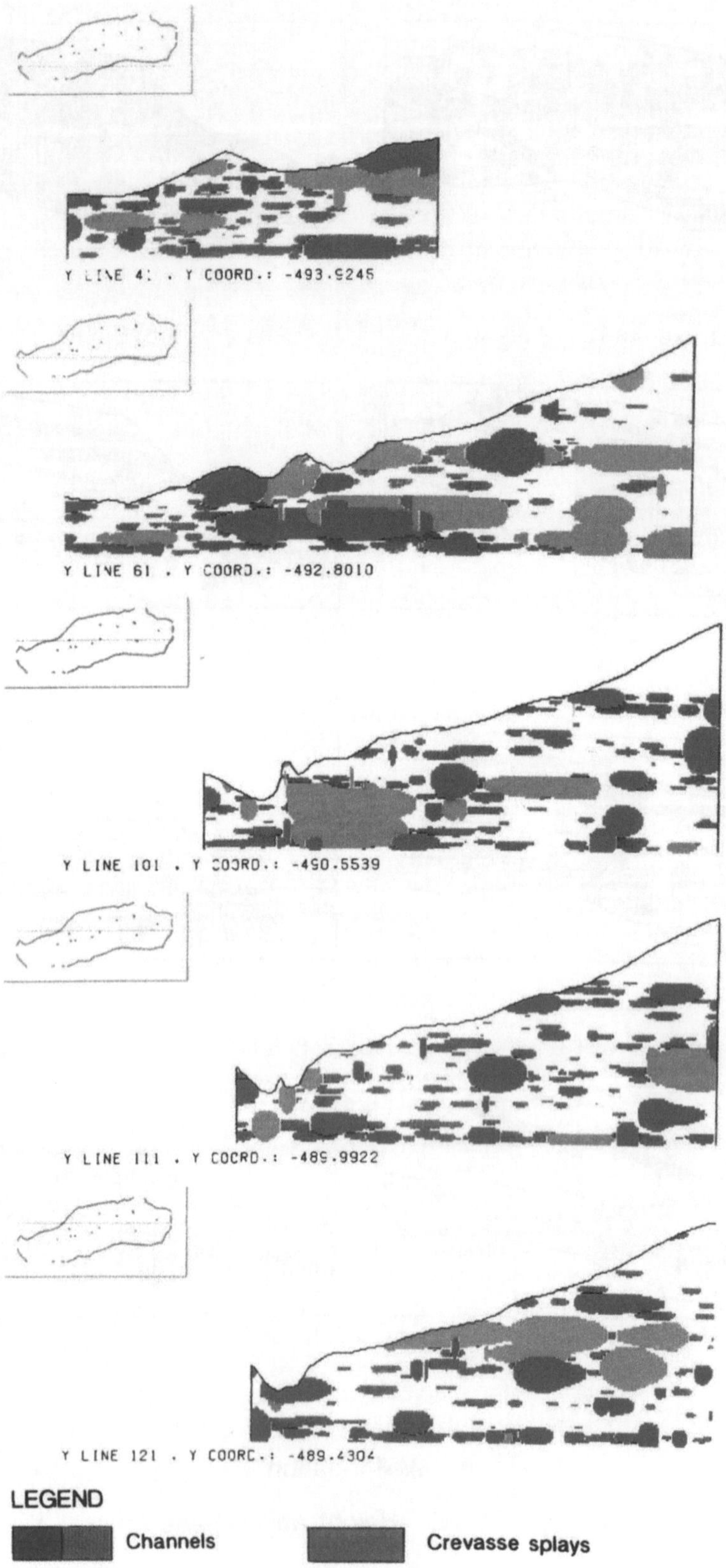


**Plate 32.** Cross-sections of a test case model (run no. 1).

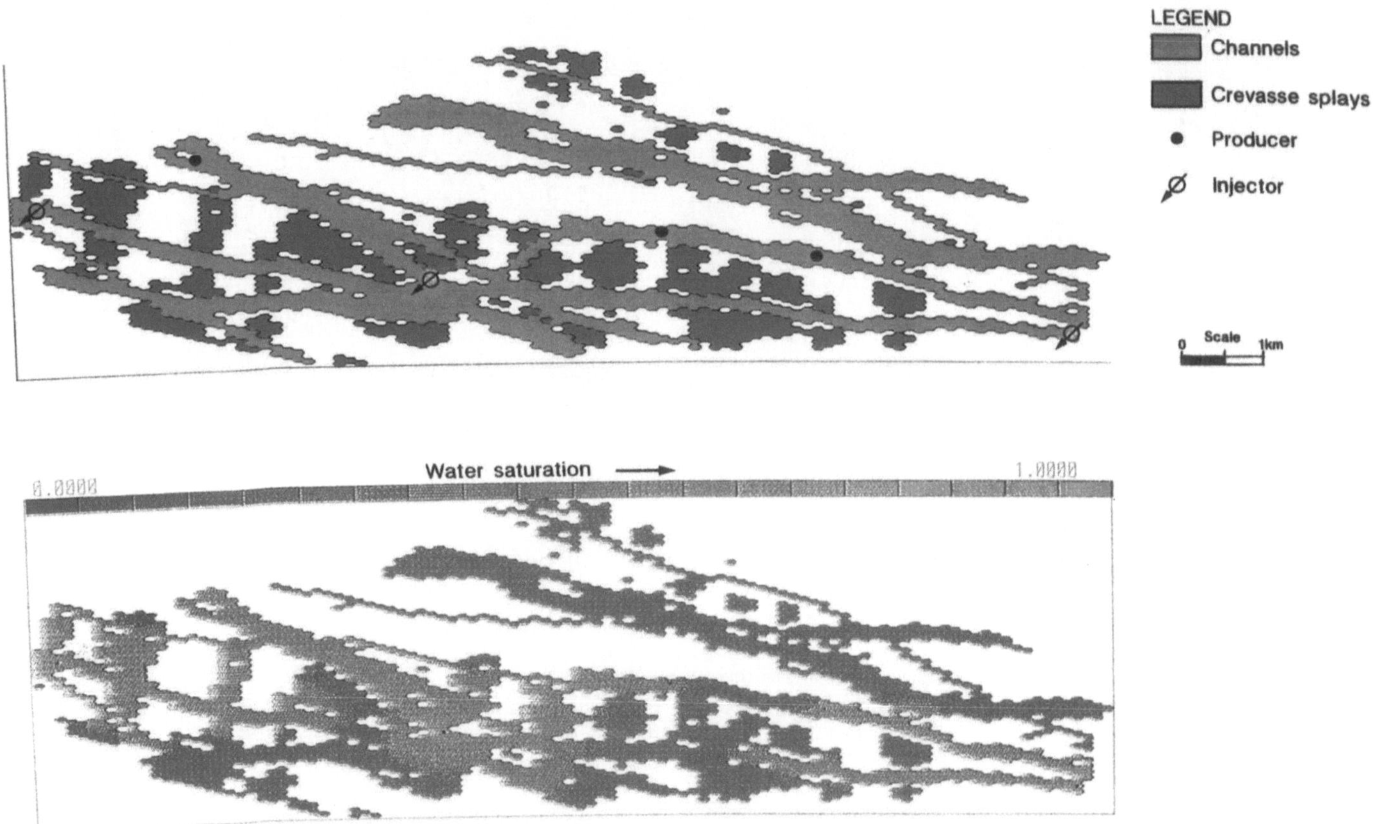


**Plate 33.** Reservoir simulation on a hexagonal grid: *Upper*: input model; *Lower*: simulation result after 10 years.

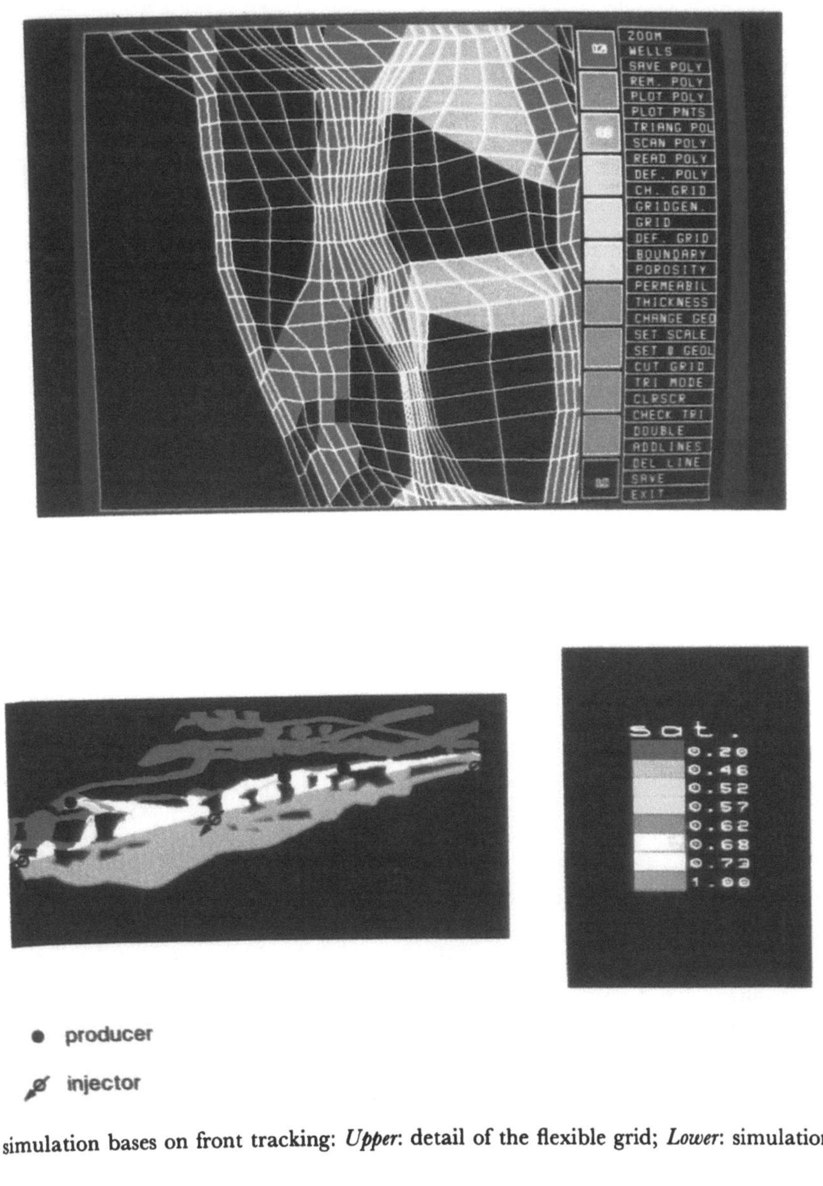


**Plate 34.** Reservoir simulation bases on front tracking: *Upper*: detail of the flexible grid; *Lower*: simulation result after 10 years.

Dubrule (1988). Dubrule states that the Boolean approach to stochastic modelling is most applicable in the early stage of a field's life when sparse well data are available. Our experience shows that the Boolean method may also be used successfully when a large number of constraints is required to reproduce a geological setting dominated by discontinuous facies bodies. However, in other reservoir types with more continuous sand bodies, the Boolean method has proved to be less effective.

Consequently in future program development, options for other methods such as variogram-based simulations (Journel, 1974; Matheron *et al.*, 1987), fractals (Hewett and Behrens, 1988) and Markov methods (Farmer, 1987) could be included. Since SESIMIRA already contains the basis functions for grid handling, fault modelling, statistical options, volumetrical calculations etc., a change of simulation technique will require only minor modifications.

## CONCLUSIONS

A large part of future North Sea field developments might be economically fairly marginal owing to large investments and the current low oil prices. This demands more accurate production forecasts than have been seen in the past in order to avoid financial disasters. To achieve this, an important requirement is to base recovery calculations on realistic geolological models. Such models should contain the required degree of heterogeneity important for the flow description, and should make use of all of the available data.

A new system has been described here which conditions deterministic and stochastic data in a flexible way thereby enabling the modelling of various types of reservoir architechture. An example has been shown in which a conceptual model of a delta-plain setting has been reproduced by the SESIMIRA program. Furthermore, uncertainties as to hydrocarbons in place and reservoir communication have been investigated, and calculations of recovery efficiency based on new techniques have also been carried out.

Reservoir computer modelling using some sort of stochastic technique is a practical way of generating quantitative 3D models of complex reservoir geometries. It also seems to have a considerable potential for improving reservoir calculations.

## ACKNOWLEDGEMENTS

We thank the management of Norsk Hydro for granting permission to publish this paper. Thanks are also given to Technical Software Consultants and Ian Ferguson for providing the results of the simulations. Finally, T. Corrigan, J.E. Evensen and H. Goemans are acknowledged for their review of earlier versions of the manuscript.

## REFERENCES

Augedal, H.O., Omre, H. and Stanley, K.O. (1986). SISABOSA, a program for stochastic modelling and evaluation of reservoir geology. Paper presented at the *IFE/SSI Seminar Reservoir Description and Simulation with Emphasis on EOR, Oslo, 1986.*

Bayat, M.G. and Tehrani, D.H. (1985). The Thistle field—analysis of its past performance and optimisation of its future development. Presented at *Ofshore Europe'85 Conference, Aberdeen, 1985*, SPE paper no. 13989.

Begg, S.H. and Carter, R.R. (1987). Assigning effective values to simulator grid-block parameters in heterogeneous reservoirs. *62nd Annual Technical Conference, Dallas, 1987*, SPE paper No. 16574.

Bratvedt, K. and Rian, D.T. (1988). Modelling and simulation of a fluvial geological system by the application of front tracking methods. *The 1st International Forum on Reservoir Simulation, Alpbach, 1988.*

Chauvin, A.L., Jackson, R.F., McGee, P.R. and McMichael, C.L. (1979). Development of the Statfjord Field using three-dimensional and areal reservoir simulation. Presented at *54th Annual Fall Technical Conference and Exhibition, Las Vegas, 1979*, SPE paper no. 8334.

Corrigan, T. (1988). Factors controlling successful reserve prediction; a cautionary tale from the U.K. North Sea. Presented at the *2nd. Conference on Reservoir Management in Field Development and Production, Stavanger, 1988.*

Delhomme, A.E.K. and Giannesini, J.F. (1979). New reservoir description techniques improve simulation results in Hassi-Messaoud field, Algeria. SPE paper no. 8435.

Dreyer, T. Sand body dimensions and infill sequences of stable, humid climate delta plain channels. *This volume.*

Dreyer, T. and Scheie, Å. (in press) A minipermeameter-based study of permeability trends in channel sandbodies. *AAPG.*

Dubrule, O. (1988). A review of stochastic models for petroleum reservoirs. Paper presented at the *BSRG Seminar Quantification of Sediment Body Geometries and their Internal Heterogeneities, London, 1988.*

Egeland, O. (1987). *IRAP-Interactive Reservoir Analysis Package.* User manual.

Farmer, C.L. (1987). The generation of stochastic fields of reservoir parameters with specified geostatistical distribution. Presented at the *IMA Conference, Cambridge, 1987.*

Fielding, C.R. and Crane, R.C. (1987). An application of statistical modelling to the prediction of the hydrocarbon recovery factors in fluvial reservoir sequences. In: Ethridge F.G., Flores R.M.and Harvey M.D. (eds), *Recent Developments in Fluvial Sedimentology*, pp. 321–327. SEPM Special Publication No. 39.

Haldorsen, H.H. and Lake, L.W. (1984). A new approach to shale management in field-scale models, *Soc. Petroleum Eng. J.* (August) 447–457.

Haldorsen, H.H. and Macdonald, C.J. (1987). Stochastic modelling of underground reservoir facies. *62nd Annual Technical Conference of SPE, Dallas, 1987*, paper no. SPE 16751.

Hewett, T.A. and Behrens R.A. (1988). Conditional simulation of reservoir heterogeneity with fractals. *63rd Annual Technical Conference Houston, 1988*, paper no. SPE 18326.

Journel, A.G. (1974). Simulation conditionelle, theorie et practique. These de docteur-ingenieur, Universite de Nancy I.

Lasseter, T.J. Waggoner J.R. and Lake L.W. (1986). Reservoir heterogeneities and their influence on ultimate recovery. In: Lake, L.W. and Corroll, H.B., Jr. (eds.), *Reservoir Characterization* pp. 545–559. Academic Press, Orlando.

Matheron, G., Beucher, H., de Fouget C., Galli, A., Guerillot, D. and Ravenne, C. (1987). Conditional simulation of the geometry of fluvio-deltaic reservoirs. *62nd Annual Technical Conference, Dallas, 1987*, SPE paper no. 16573.

Miall, A.D. (1988). Reservoir heterogeneties in fluvial sandstones: lessons from outcrop studies. *AAPG Bulletin*, **72**(6), 682–697.

Nadir, F.T. (1980). Thistle field development. Presented at *European Offshore Petroleum Conference and Exhibition, London, 1980*, EUR paper no. 165.

Skaug, M. and Gundesø, R. (1986). Geological modelling of the Frigg Field with special emphasis on shale mapping. *European Petroleum Conference of SPE, London, 1986*, SPE paper no. 15859.

Stiles, J.H. and Bobek, J.E. (1982). Fulmar pre-development simulation study. Presented at *European Petroleum Conference, London, 1982*, EUR paper no. 282.

Tomutsa, L., Jackson, S.R. and Szpakiewicz 1986. Geostatistical characterization and comparison of outcrop and subsurface facies: Shannon Shelf Sand Ridges. *California Regional Meeting of SPE, Oakland, 1986*, SPE paper no. 15127.

Van Oert, B. (1988). Lessons learned in North Sea oil field developments. *J. Can. Petroleum Technol.* **27**, 123–132.

Wadsley, A.W., Erlandsen S. and Goemans, H.W. Hex, a tool for integrated fluvial architechture modelling and numerical simulation of recovery processes. *This volume.*

# 23

# A Computer Program for Evaluation of Fluvial Reservoirs

*R. Clemetsen[a], A.R. Hurst[b], R. Knarud[c] and H. Omre[a]*

[a]Norwegian Computing Center, Oslo, Norway
[b]Statoil a.s., Stavanger, Norway
[c]Saga Petroleum a.s., Høvik, Norway

Heterogeneities create large problems for the production of fluvial petroleum reservoirs. The procedures for evaluating the potential of such reservoirs at an early stage of investigation must rely heavily on geological interpretation. FLUREMO is a program system that can support geologists and reservoir engineers in this evaluation. The system simulates possible reservoir descriptions in three dimensions. Avulsion and clustering effects in the river channels, sinuosity as well as varying channel sizes, can be modelled. Existing well observations can be reproduced. Various contact measures for the reservoir can be computed and input to reservoir production simulators can be prepared. The program system is presented and guidelines for its use are provided. An example of the inference of model parameters from realizations from Bridge and Leeder simulations (1979) is presented. Finally, a discussion of various measures of contact is given.

## INTRODUCTION

Heterogeneity is recognized as a major complication in the evaluation of petroleum reservoirs. In reservoirs of fluvial origin, heterogeneity constitutes the main characteristic; models in three dimensions are therefore required. The FLUREMO program system provides an opportunity for the three-dimensional modelling and evaluation of such reservoirs.

Some quantitative aspects of fluvial sedimentology, primarily sand body connectedness, have been presented by Allen (1978) and Bridge and Leeder (1979). The characteristics of petroleum reservoirs of fluvial origin are treated in Augedal *et al.* (1986), Haldorsen and MacDonald (1987) and Henriquez *et al.* (1988) among others. FLUREMO is a development of the SISABOSA system described in Augedal *et al.* (1986).

The FLUREMO program system is characterized as being descriptive. This entails that the characteristics of the reservoir as it presently appears are represented by statistical measures such as net-to-gross ratio, river channel sizes and interaction parameters. No time-dependent model that can represent the physical processes involved in the sedimentation is included. Examples of process-related models can be found in Bridge and Leeder (1979) and Scott (1986).

There are three main reasons for choosing a descriptive model as opposed to a process-related model. Firstly, conditioning on observations made in wells can be performed more simply. Secondly, process-related models tend to be strongly dependent on the specification of geological events, the magnitudes of which are poorly known. Thirdly, process-related models in three dimensions require complicated parameter input in order to produce realistic results. For descriptive models this is much simpler. Process-related models are definitely of interest, but it is expected that considerable research will be necessary before reliable models and simulation procedures can be obtained.

A brief introduction to the geological characteristics of fluvial reservoirs is presented in the next section. The model assumptions underlying the program and the simulation procedure are then specified, followed by an example of an application. The methodology is further specified in the appendices.

## GEOLOGICAL SETTING

Large reserves of hydrocarbons are found in resevoirs with fluvial affinities in the North Sea area. Examples are the Jurassic Ness and Statfjord Formations, in addition to the Triassic upper member of the Lunde formation in the Snorre Field. Evaluation of reservoir volumetrics and reservoir performance are traditionally difficult in fluvial reservoirs. This stems from the uncertainty involved in trying to correlate between sandstones located at similar levels in different wells. The complexity of this correlation is discussed by Nystuen *et al.* (1989). Rules of thumb

*North Sea Oil and Gas Reservoirs–II*

suggest that when the net-to-gross ratio is below 0.5 caution should be applied when correlating between wells (Yinan Qiu *et al.*, 1987). A better understanding of the interaction between reservoir characteristics and reservoir production performance is of great importance.

Sedimentation in fluvial systems is influenced by numerous physical processes that control the distribution of sediments both laterally and vertically within the basin. In a well the result of these processes is usually recognized by cyclic sequences of sediments that occur at various scales (Fig. 1). In general, large cycles reflect large-scale allocyclic processes, i.e. changes in tectonic style, sea level fluctuations, shifts of erosional base level or climatic changes. Shorter cycles reflect autocyclic processes, such as local isostatic compaction, rate of sedimentation, type of sediment supply, diurnal climatic changes or avulsion processes. Structural correlation of long cycles is in general simple to make within a given exploration area. Structural correlation within or between individual short cycles is usually very complicated owing to the internal complexity referred to above, and to well spacing.

Here a stochastic approach for the evaluation of short-cycle correlation is presented. Typical deposition of fluvial sand bodies consists of a series of deposits within a channel belt. The belts have an elongated geometry and are filled predominantly by sand. Beyond the channel belts fine-grained deposits gather subaerially or in lakes on the fluvial plain. Channel belts have finite lifetimes, known as periods of avulsion, after which the creation of a new channel belt occurs in a new location. New channel belts may erode into previous channel belts, so producing a complex sand body geometry (Fig. 1).

## MODEL ASSUMPTIONS AND PROGRAM SYSTEM

The program system is based on a set of model assumptions. The assumptions constitute a trade-off between realism in representing a reservoir of fluvial origin and feasibility from the point of view of processing. In order to structure the presentation, it is convenient to give an overview of the FLUREMO system first. The flow chart for the FLUREMO system is presented in Fig. 2. The challenge is to transform information collected during early stages of investigation into information relevant for decision making concerning the future strategy for the reservoir. The following modules are defined.

- *Parameter Estimation.* This module is not formalized as software and has to be performed manually. It should be based on available knowledge of fluvial environments, the actual reservoir and knowledge of the program system.
- *Reservoir Description Simulation.* This module constitutes the main part of the program system, and generates simulations of reservoirs of fluvial origin. The simulation is based on model assumptions, parameter values and conditioning data. Several equi-probable reservoir representations are generated and stored in a parametric form.
- *Reservoir Display.* This module provides a graphical display of the reservoir representations.
- *Reservoir Contact.* This module computes various measures of inter-connectedness and continuity in the reservoir representations. These measures are believed to be related to the production potential of the representations.
- *Reservoir Gridding.* This module converts the reservoir representations into a grid form suitable for input to reservoir production simulators.

The characteristics of, and the assumptions underlying the various modules are presented below.

### Reservoir Description Simulation

The module for Reservoir Description Simulation is based on the following model assumptions:

- *Dimensions.* The model is three-dimensional and a parametric storage of the reservoir representations is used.
- *Facies.* The reservoir comprises two facies:
  (1) the reservoir facies: a sand facies of fluvial origin with homogeneous porosity, water saturation and permeability characteristics;
  (2) the residual facies; a non-permeable fluvial plain facies that constitutes the background sedimentation.
- *Net-to-gross ratio.* The net-to-gross of the reservoir facies, which is equivalent to the sand-to-gross, is user-specified and may vary both horizontally and vertically (Fig. 3). A tolerance for the average net-to-gross must be specified.
- *Reservoir facies.* The sand bodies that represent sand deposited within single river channels are organized in a hierarchical structure. A family of sand bodies, or a channel belt, comprises a set of stacked sand bodies that form genetically related units deposited during separate periods between avulsions. Each period of accumulation is assumed to constitute reasonable stable sediment supply and climatic conditions. This creates a unique channel belt system on the fluvial plain. The various families, or channel belts, interact in a repulsive manner. A new family may occur as a result of autocyclic processes controlled by isostatic compaction, change in sedimentation rate or climatic extremes. A significant shift in channel position may occur as a response to avulsion.
  (i) River channel scale: Each river channel is represented by a horizontally located body with a rectangular cross-section. It is sinuous in the direction of the channel (Fig. 4). The dimensions of the rectangular cross-section and the sinuousity parameters are user-specified as probability distributions.
  (ii) Channel belt scale: The number of sand bodies in a family is user-specified from a probability distribution. Each family has a direction and a sinuousity attached to it. The probability distributions for these parameters are user-specified. Sand bodies in a specific family are attracted to each other. Implicit within this is a relatively great probability that bodies in the same family will overlap. This attraction is specified by a maximum for distances between body centres and an attraction function. This function makes it possible to encourage sand bodies to stack according to certain patterns (Fig. 5). The stacking pattern will influence the probability of contact between individual river channels and hence the production potential. The methodology used is presented in Appendix I.
  (iii) Reservoir scale: The various families interact in a repulsive manner through their family location

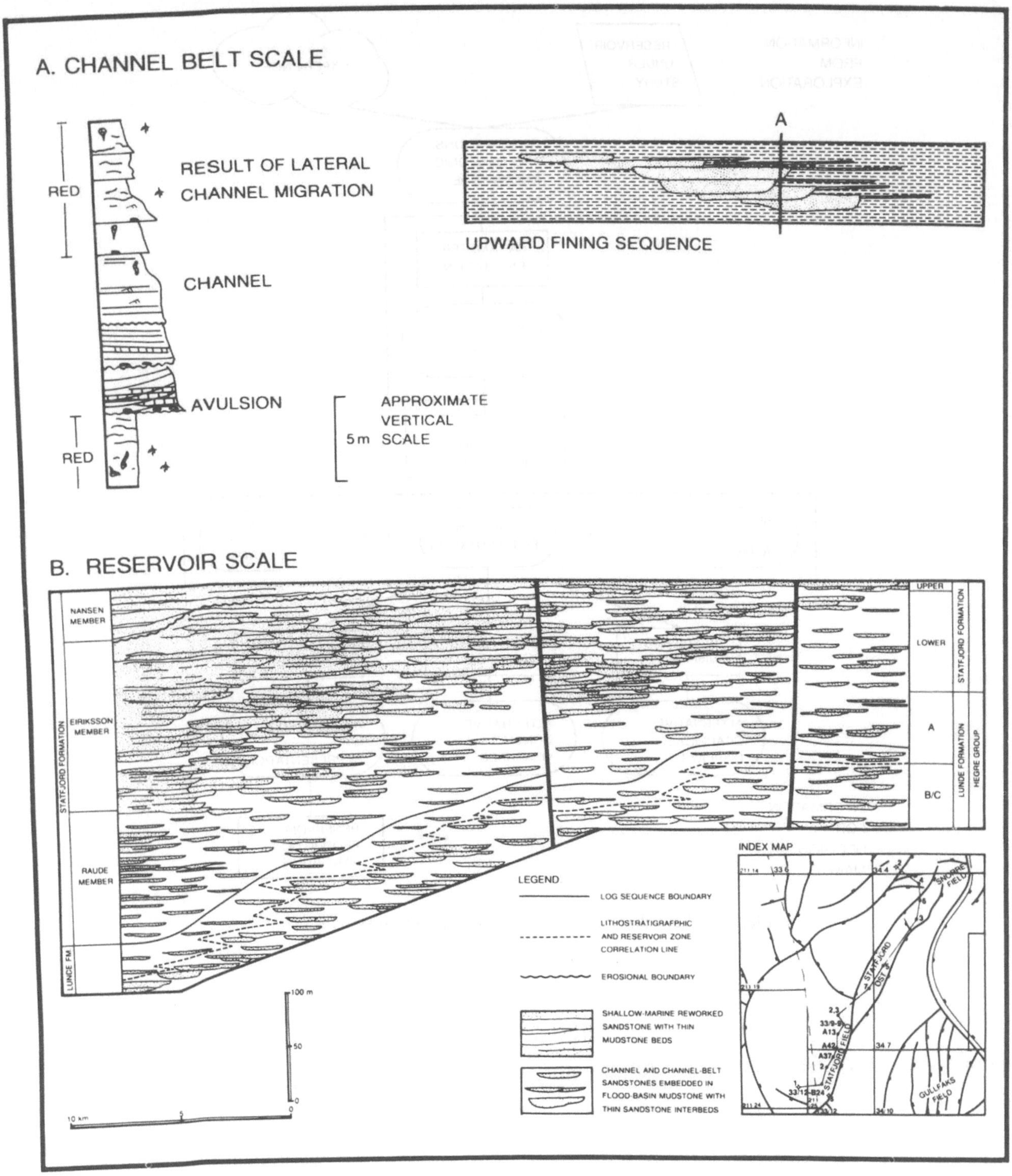


**Fig. 1.** Geological setting, fluvial environment. (Reproduced by permission of the Norwegian Petroleum Society, NPF; from Nystuen *et al.*, 1989.)

and direction. There is thus a tendency for sand bodies in different families not to overlap, (Fig. 6). Repulsion tends to disperse the occurrence of shale and hence influence the production potential. The repulsion function has to be specified by the user, and the methodology is based on the theory of fibre processes (see Appendix II). Attraction within families and repulsion between families are treated independently.

- *Data conditioning*. Consistency with sand bodies observed in the wells is ensured in the simulation procedure. The sand bodies observed and their family connections have to be specified by the user.

The simulation procedure in the Reservoir Description Simulation module contains the following steps:

(A) Estimation of the expected number of families to be present, given the actual parameter values.

(B) Generation of the locations and directions of families observed in wells. The locations and directions of the remaining families are generated without taking repulsion into consideration.

(C) Fulfillment of the repulsion requirements between families by an iterative procedure that follows the

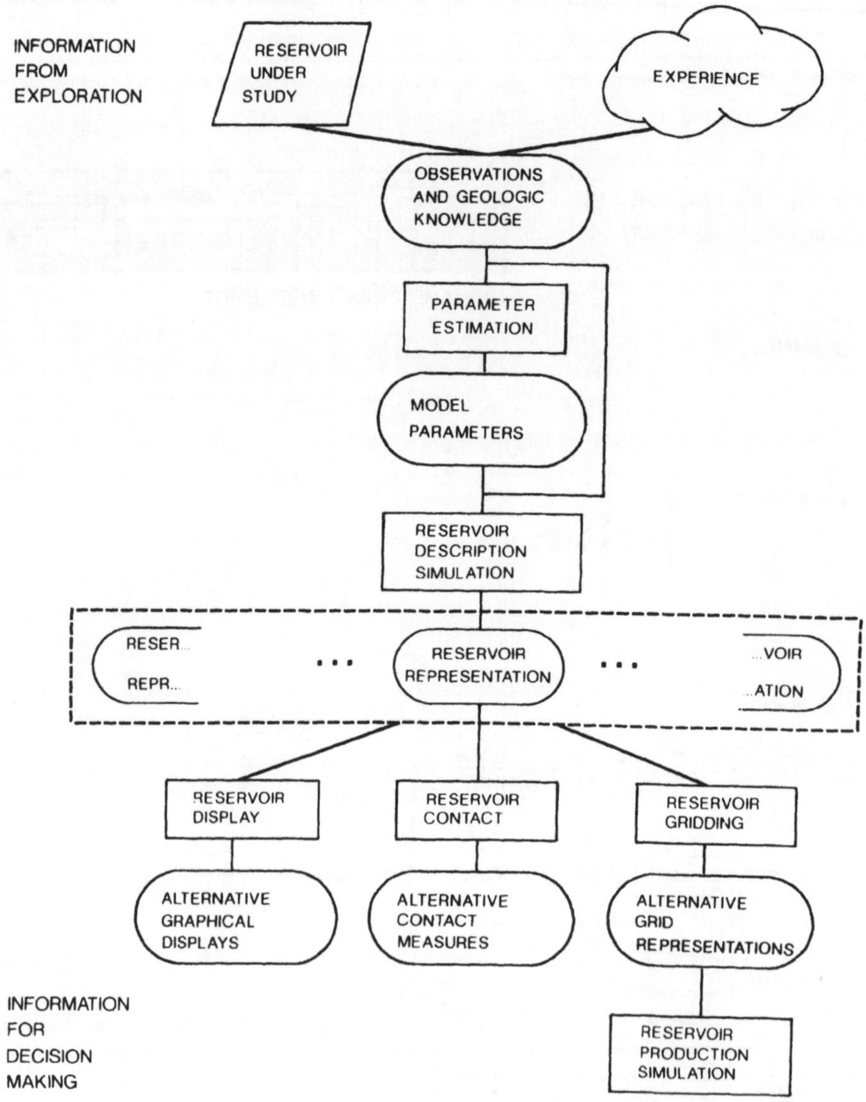


**Fig. 2.** Flow chart of the FLUREMO program system.

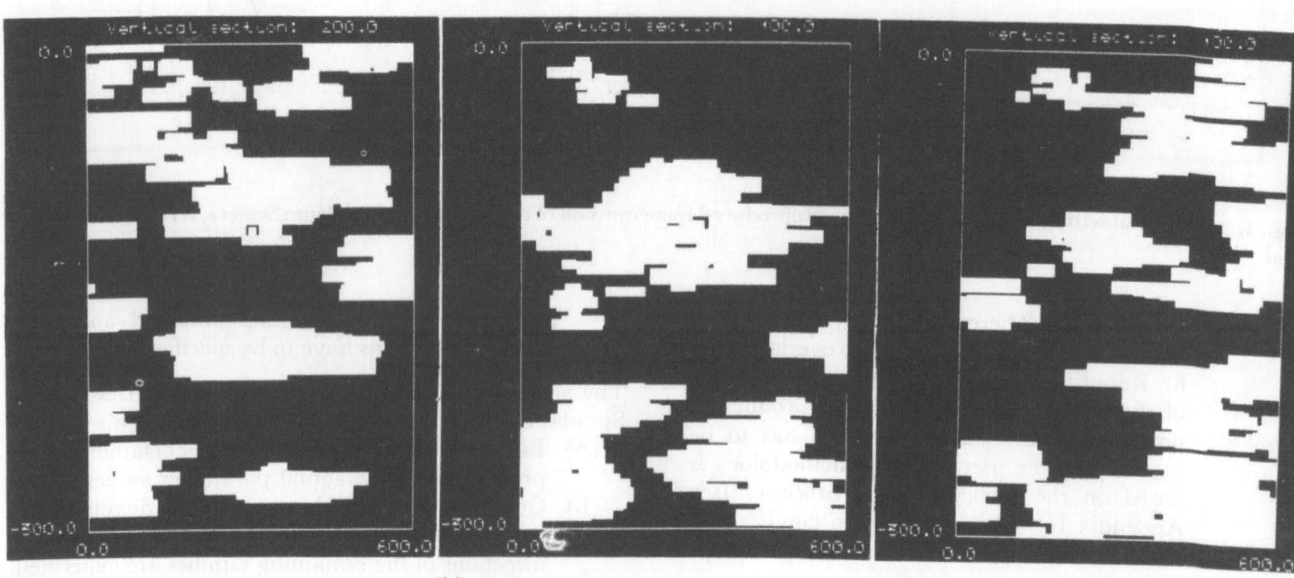


**Fig. 3.** Examples of cross sections with varying net-to-gross ratio: (a) uniform; (b) vertical variation; (c) vertical and lateral variation.

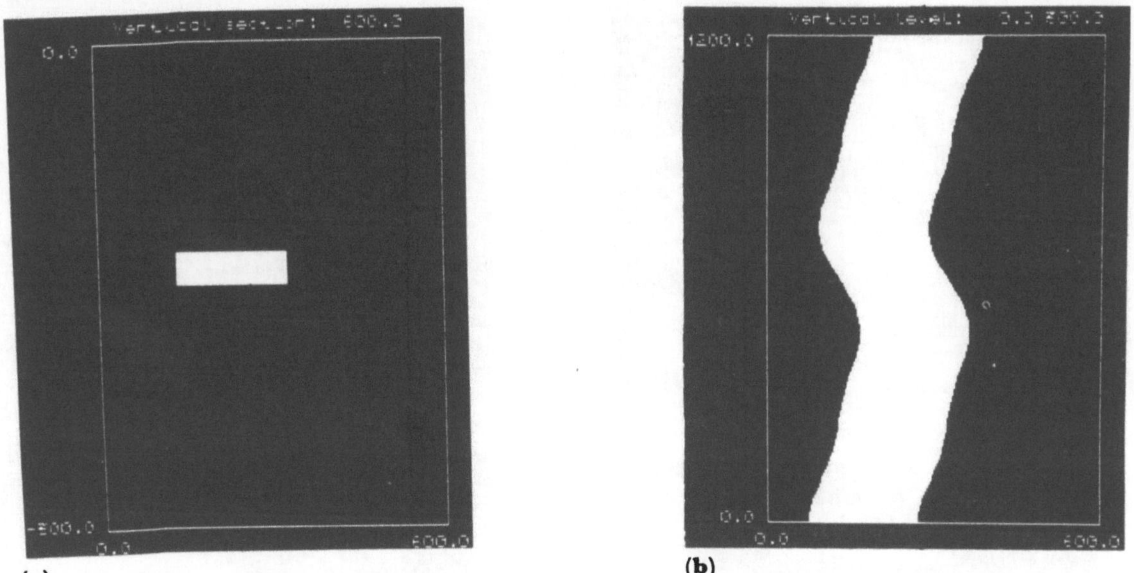


(a) (b)

**Fig. 4.** Example of one river channel: (a) cross section; (b) horizontal plane through channel.

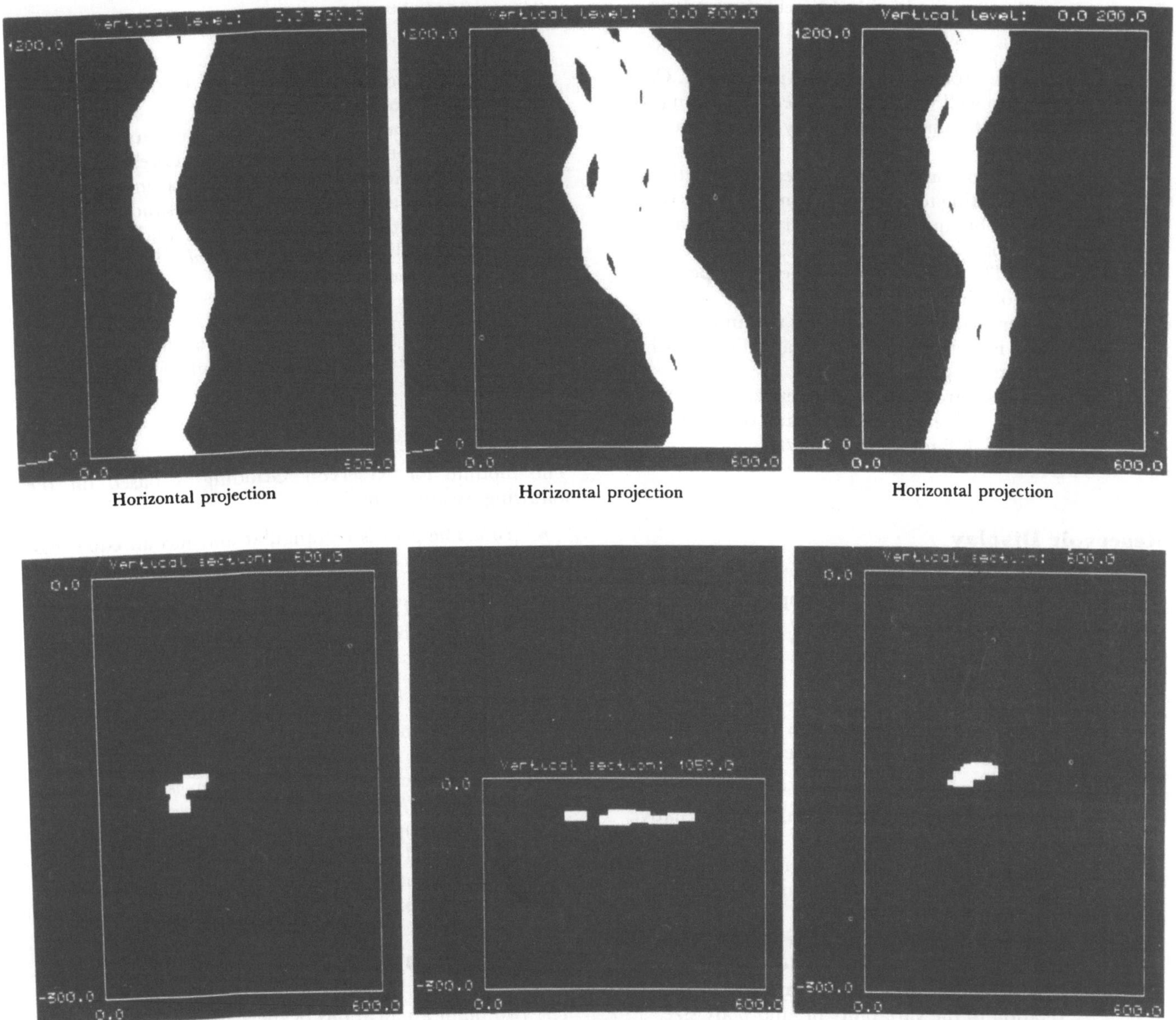


Horizontal projection

Horizontal projection

Horizontal projection

Cross-section *Vertical stacking*

Cross-section *Horizontal stacking*

Cross-section *Diagonal stack*

**Fig. 5.** Examples of stacking in channel belts.

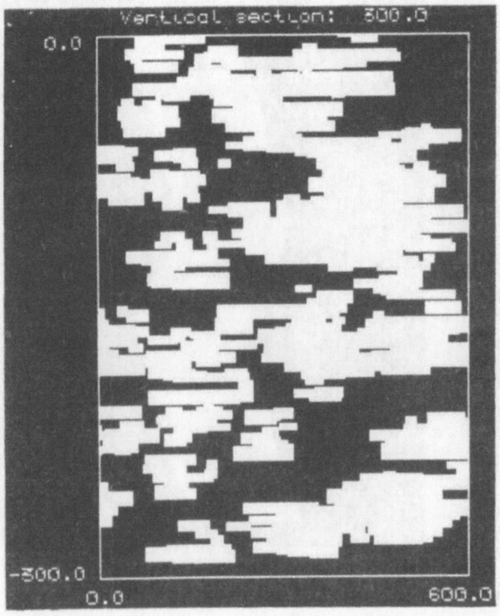


**(a)**

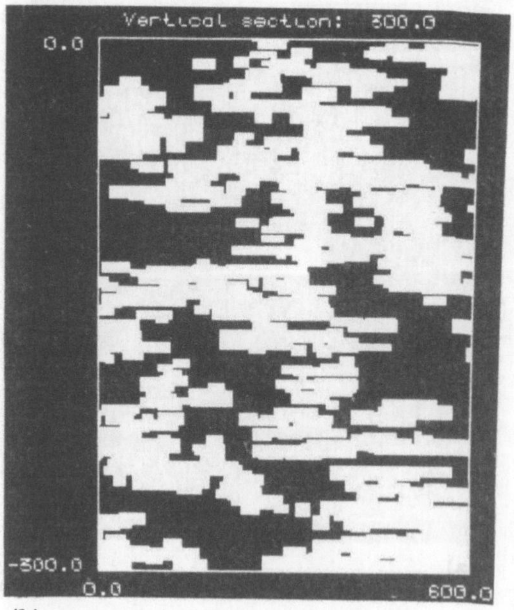


**(b)**

**Fig. 6.** Examples of reservoir with channel belts: (a) Uniform random; (b) repulsive random.

death/birth process outlined in Appendix II. Conditioning to well values is obtained by keeping the conditioned families alive. Finally, family sinuosity is generated.

(D) Generation of sand bodies in each conditioned family according to observations made in the wells. Generation of sand bodies in each of the remaining families, ensuring that no intersection occurs with the wells. The procedure is outlined in Appendix I.

(E) Testing of the net/gross ratio of the reservoir realizations to check whether it is within tolerance.
- If the net/gross ratio of the realization is within tolerance—go to F.
- If the net/gross ratio of the realization is NOT within tolerance—re-estimate the expected number of families and return to B.

(F) STOP.

## Reservoir Display

The module for Reservoir Display is based on the UNIRAS system and includes several graphical options. Examples are given in the next section.

## Reservoir Contact

The module for Reservoir Contact has two options for describing the reservoir potential of a representation.

(1) *Interconnectedness*: a measure of overlap of the sand bodies in a reservoir representation. Two alternative measures are implemented:
- Expected overlap ratio, defined as the expected fraction of the surface of a sand body that falls within other sand bodies. It is a generalization of the measure used by Allen (1978) and Bridge and Leeder (1979), see Fig. 7. This measure has a complicated relation to the variability of sand body size.
- Effective sand size, a function of direction and distance. It is defined as the probability of having sand connected further than a given distance in a given direction. This measure must be evaluated relative to the specified sand body dimensions.

2 *Continuity ratio*: a measure of the maximum production potential in a reservoir representation for given locations of injection and production wells. It is defined as the fraction of the sand volume in contact with both the injection and production wells (Fig. 8) (also see Augedal *et al.*, 1986; Haldorsen and McDonald, 1987).

Note that the interconnectedness is only reservoir-dependent, while the continuity ratio as defined is conditioned to the well locations. Care should be taken not to mix these terms.

## Reservoir Gridding

The module for Reservoir Gridding is based on the following assumptions:

- *Grid.* The grid is rectangular and has an equal size over a given volume. Both the grid size and volume are user-specified.
- *Sand ratio.* The fraction of sand is computed in each grid cell. The data when multiplied by porosity and water saturation provide input suitable for a reservoir production simulator.
- *Flow ability.* The flow characteristics are estimated in the three main directions between planes perpendicular to the flow direction at the centres of neighbouring cells. The algorithm is based on the arithmetic/harmonic average approach with a correction term for the possibility of barriers (see Appendix III). With the addition of a re-scaling based on sand permeability, these also provide input to a reservoir production simulator.

# EXAMPLE OF USE

The two main applications of the system are as follows.

(1) Evaluation of the accuracy and precision in the predicted production potential of a reservoir, where the uncertainty is due to dispersion and uncertainty present in the input parameters for the reservoir

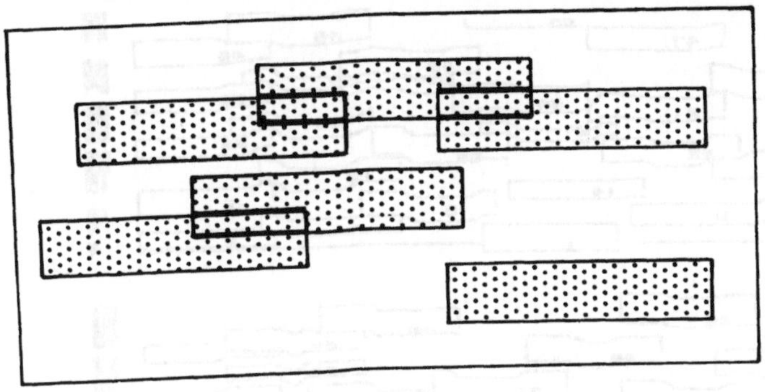

**Fig. 7.** Definition of interconnectedness. Length of the thicker lines over the total horizontal boundary.

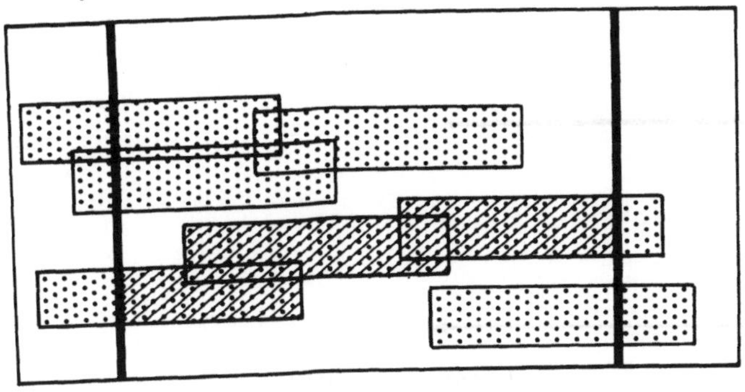


**Fig. 8.** Definition of continuity ratio. Hatched area over the total sand area between the wells.

production simulation. In particular, its sensitivity to individual parameters can be evaluated. The use of stochastic models in order to reproduce the inherent variability in the reservoir description is required if the uncertainty in the production potential is to be assessed. The conditioning of representations to observations in wells is not necessary unless an inference about a given reservoir is required. In the system, parameters like river channel dimensions, sinuosity, directional deviations of channel belts, channel belt stacking and avulsion characteristics can be evaluated. The result makes a quantitative evaluation of further development strategy possible.

(2) Quantification of the uncertainty of the production potential for a given reservoir. To obtain this, the system parameters must be specified such that both the inherent dispersion and the user's lack of knowledge about them for the specific reservoir are represented. Conditioning of the representations to observations in wells is required in order to ensure the generation of consistent representations for the reservoir, and to provide a reduction in uncertainty with increasing amounts of data.

It is worth noting that the former application of the system requires fewer assumptions than the latter. The results of the former are assumed to be valid for a class of reservoirs for which the model assumptions are considered valid. The results of the latter apply directly to a given reservoir, and hence are more complicated to verify.

Typical guidelines for the application of the system are as follows:

(A) Gather all of the relevant information about the reservoir to be studied and the class of reservoirs to which it belongs. Justify the use of the FLUREMO system from the main characteristics of the reservoir. A minimum requirement is that the reservoir is of a fluvial origin.

If the user feels confident about the sedimentological processes and variables, a process-related stochastic model may be used. From the reservoir representations of this type of model, system parameters can be estimated.

In an example, the repulsion function between channel belts is estimated from two-dimensional reservoir representations generated from a modified version of the program given by Bridge and Leeder (1979) and Bridge (1979). Some representations are presented in Fig. 9(a), and the corresponding estimate for the repulsion function is given in Fig. 9(b). The function describes the relative probability of two channel belt centres being a given distance apart (Appendix II).

Other system parameters are estimated from other available information. Such information may typically come from extensively drilled reservoirs with similar characteristics, comparable outcrops and studies of modern fluvial systems.

(B) Activate the Reservoir Description Simulation module and critically inspect the reservoir representations. Several of the system parameters are difficult to infer from the limited amount of data available. Thus, the final justification of the model should be made by letting experienced geologists verify cross-sections produced from the reservoir representations.

When a set of parameters valid for the reservoir under study is obtained, several reservoir representations are generated. The number needed is dependent upon the variability in the representations and the averaging effect of the computation in which they are later involved.

(C) The graphical options of the Reservoir Display module are used for presenting the reservoir representations (Fig. 10).

(D) The Reservoir Contact module computes various contact measures which are assumed to be related to the production potential of the representations. Even full-scale flow simulations performed with traditional production simulators appear to be unreliable for heterogeneous reservoirs. Thus, it is not obvious which prediction of production potential should be trusted. This problem requires further research.

Returning to the contact measures, a simple schematic example will be used to demonstrate that their characteristics are very different. Consider two cross-sections with an identical net-to-gross ratio (Fig. 11(a) (b)). In Fig. 11(a), the sand bodies have a uniform random location, i.e. no repulsion. In Fig. 11(b) they are located with maximum repulsion, which implies a regular location. The sand bodies are all of equal sizes. Now increase the net-to-gross ratio by increasing the size of the sand bodies. Note that in the cross-section without repulsion the total number of sand bodies must also be increased in order to match the net-to-gross ratio of the other.

In Fig. 12(a) and (b), the values of the contact measures, interconnectedness (expected overlap ratio) and continuity ratio are presented as a function of the net-to-gross ratio for each of the cross-sections. These measures were defined in the previous section. The graphs are not exact computations, but they do show the main characteristics for the two measures:

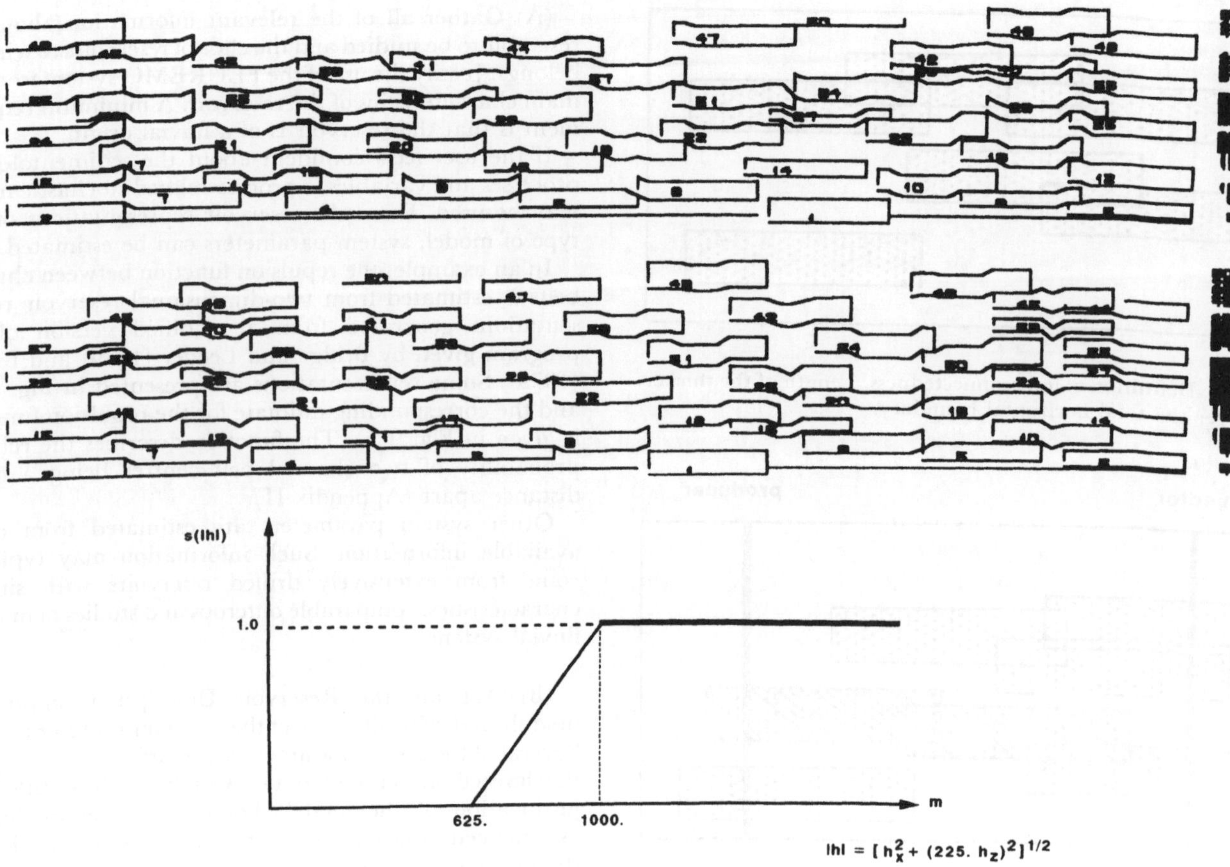


**Fig. 9.** Estimation of repulsion function from realizations produced using a modified version of Bridge's (1979) program. (a) Two realizations. Channel height ≈ 5 m; width ≈ 1000 m. (b) Estimated repulsion function.

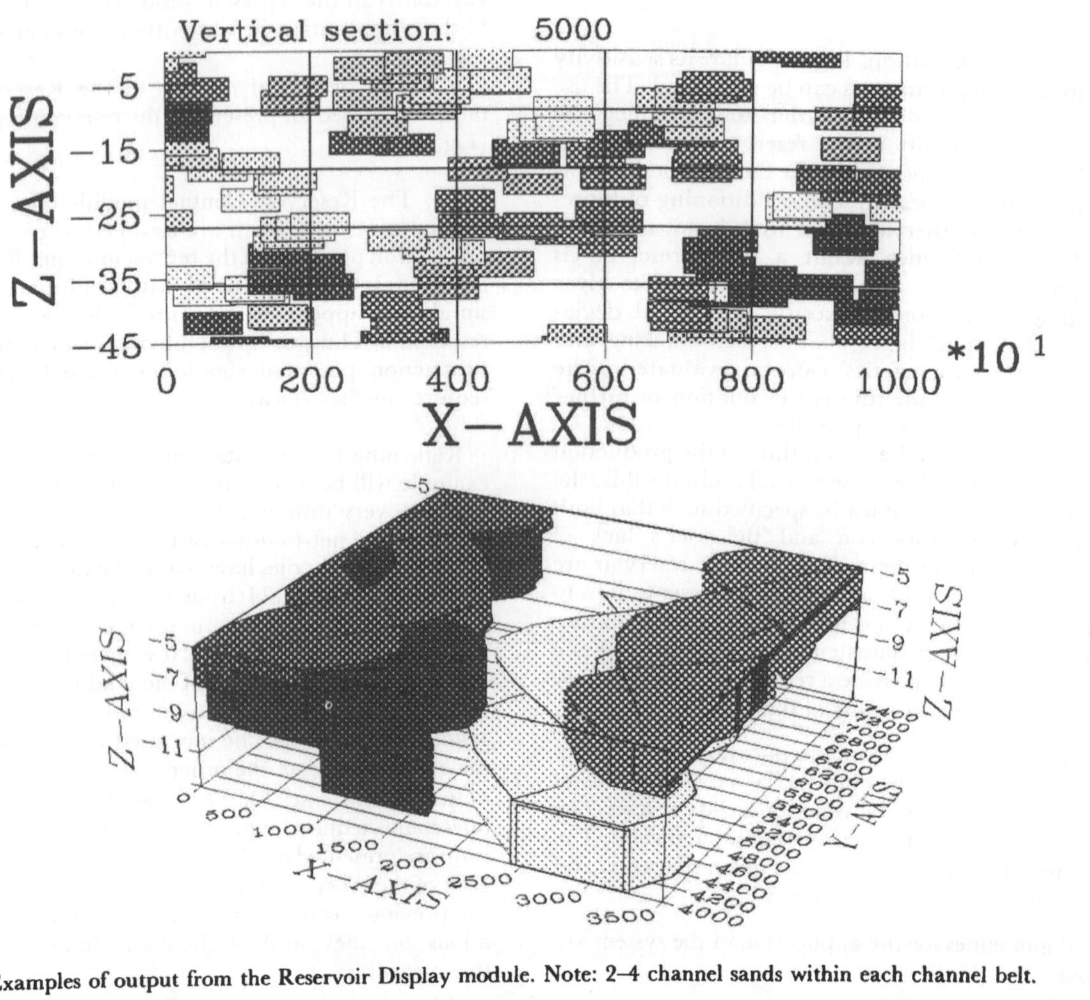


**Fig. 10.** Examples of output from the Reservoir Display module. Note: 2–4 channel sands within each channel belt.

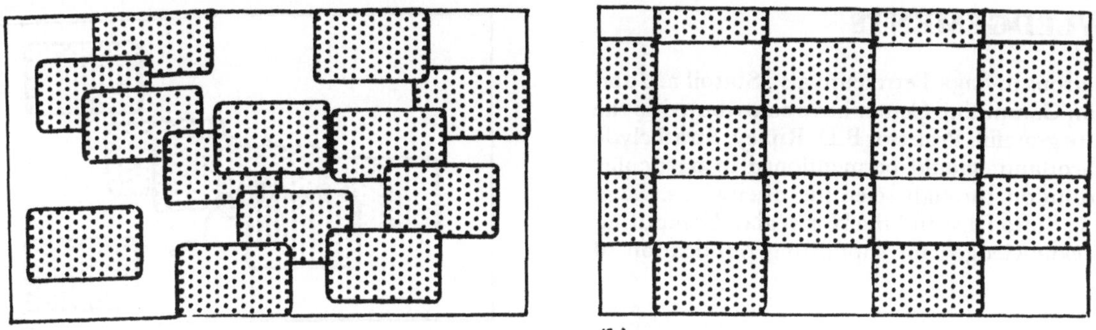


**Fig. 11.** Schematic cross sections: (a) without repulsion; (b) with repulsion.

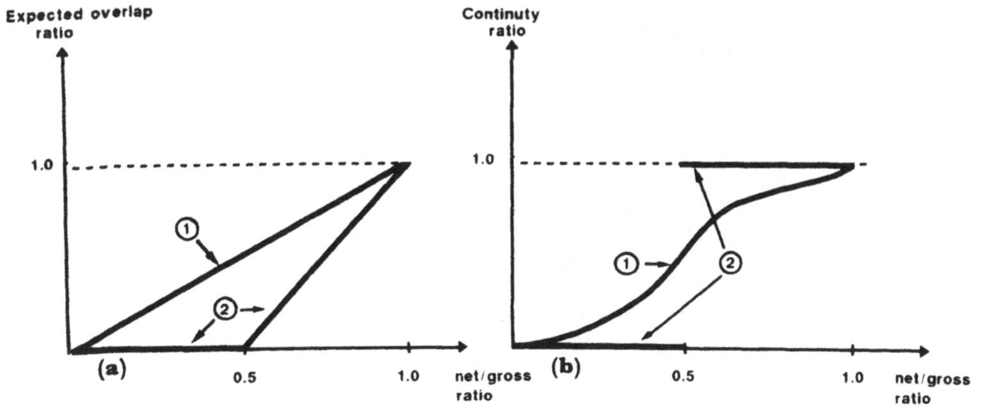


**Fig. 12.** Contact measures: (a) interconnectedness; (b) continuity ratio, 1, Cross-section without repulsion (see Fig. 11(a) M. 42.2, Cross-section with repulsion (see Fig. 11(b)).

- The measure of interconnectedness for the cross-section without repulsion is larger than that of the cross-section with large repulsion for all net-to-gross ratios. Hence, increasing repulsion between sand bodies tends to decrease the interconnectedness measure. Intuitively, this makes sense as repulsion is designed to avoid overlap. Note also that interconnectedness is independent of the dimension of the model.
- The continuity ratio has a more complex behaviour. For no repulsion, the continuity ratio grows steadily with increasing net-to-gross ratio. With repulsion, however, there is a discontinuous step at a net/gross ratio of 0.5. For a net-to-gross ratio less than 0.5, there is no continuity. For a net-to-gross ratio larger than 0.5, the continuity ratio is 1.0. This is intuitively reasonable because once two sand bodies touch, they all touch, owing to the regularity. Consequently, for the continuity ratio no simple ordering exists with respect to the repulsion of sand bodies. Furthermore, the continuity ratio tends to increase with increasing dimensions of the model.

Both the interconnectedness and the continuity ratio influence the true production potential. However, the manner in which they are related remains to be fully understood.

(E) The reservoir representations are transferred to a grid form suitable for input to reservoir production simulators. The production simulators provide production rates over time, which is an important variable for the evaluation of a development strategy. Unfortunately, traditional reservoir production simulators are poorly suited for the evaluation of heterogeneous reservoirs. The extent to which the results can be trusted is an open question. What is clear, however, is that production simulations require considerable computer resources. To avoid unnecessary computation, one may use one of the contact measures as a stratification variable. This entails using the contact measure to select the reservoir representation to be used in a later evaluation, such that it reflects the variability of the complete set of representations.

## CLOSING REMARKS

The program system FLUREMO is based on a descriptive, stochastic model of a reservoir. Reservoir characteristics are therefore represented by probability distributions and correlation measures. No time dependence is included in the model. The FLUREMO systems provides greater flexibility in describing a reservoir than other available program systems. The interactions between river channels defined by channel belts with pairwise repulsion is one example. This characteristic is expected to have a considerable impact on production. The use of process-related simulations (like the Bridge and Leeder procedure) in order to estimate system parameters for interaction is expected to have great potential. This, however, remains to be confirmed on real case studies.

In the near future the system will be used for evaluating the sensitivity of production to the various parameters. Case studies on real reservoir data will also be performed.

Future extensions of the program system may be in the directions of conditioning on a large number of wells with specified intercorrelations, defining more facies, or simulation of microvariability in the sand facies.

## ACKNOWLEDGEMENTS

Numerous persons in Saga Petroleum a.s., Statoil a.s. and the Norwegian Computing Center have contributed to the project. We are grateful. Professor B.D. Ripley, Strathclyde University, Scotland, should be mentioned in particular for his contribution through scientific discussions. Saga Petroleum a.s., Statoil a.s. and the Norwegian Computing Center are acknowledged for supporting publication of this paper.


## References


Allen, J.R.L. (1978). Studies in fluviatile sedimentation: an exploratory quantitative model for the architecture of avulsion-controlled alluvial suites. *Sediment. Geol.* **21**, 129–147.

Augedal, H.O., Stanley, K.O. and Omre, H. (1986). SISABOSA, a program for stochastic modelling and evaluation of reservoir geology. In Nitteberg, J. (ed.), *Proceedings from Conference on Reservoir Description and Simulation with Emphasis on EOR, Oslo, September 1986*. Institute for Energy Technology.

Bridge, J.S. 1979. A FORTRAN IV program to simulate alluvial stratigraphy. *Computers & Geosciences*, 5, pp. 335–348.

Bridge, J.S. and Leeder, M.R. (1979). A simulation model of alluvial stratigraphy. *Sedimentology* **26**, 617–644.

Haldorsen, H.H. and MacDonald, G.J. (1987). Stochastic modeling of underground reservoir facies (SMURF). *Annual Technical Conference and Exhibit of SPE, Dallas, Texas, September*, 1987, paper no. SPE 16751.

Henriquez, A., Tyler, K.J. and Hurst, A. (1988). Characterization of fluvial sedimentology for reservoir simulation modeling. *Annual Technical Conference and Exhibit of SPE, Houston, Texas, October* 1988, paper no. SPE 18323.

Nystuen, J.P., Knarud, R. and Jorde, K. (1989). Correlation of Triassic to Lower Jurassic sequences, Snorre Field and adjacent areas, northern North Sea. In: Collinson, J.D. (ed.) *Correlation in Hydrocarbon Exploration* pp. 273–289. Norwegian Petroleum Society, Graham and Trotman, London.

Ripley, B.D. (1987). *Stochastic Simulation.* Wiley, New York.

Scott, N. (1986). Modern vs. ancient braided-stream deposits: a comparison between simulated sedimentary deposits and the Ivishak Formation of the Prudhoe Bay field, Alaska. MSc. Thesis, Department of Applied Earth Sciences, Standford University, California.

Stoyan, D., Kendall, W.S. and Mecke, J. (1987). *Stochastic Geometry and its Applications.* Wiley, New York.

Yinan Qiu Q., Peihua Xue and Jingsiu Xia. (1987). Fluvial sandstone bodies as hydrocarbon reservoirs in lake basins. In: Ethridge, F.G., Flores, R.M. and Harvey, M.D. (eds.), *Recent Development in Fluvial Sedimentology.* Special Publication No. 39. Society of Economic Palaeontologists and Mineralogists, Tulsa.


## APPENDIX I. MODEL ON CHANNEL BELT SCALE

Each channel belt comprises a random number of river channels (sand bodies) that tend to attract each other.

Consider a cross-section perpendicular to the main flow direction of the fluvial system (Fig. A1). Such processes can be represented by marked point processes in a statistical setting (Stoyan *et al.* 1987; Ripley, 1987). Introduce the following notation:

$x_0$ The centre of the channel belt.

$D_{x_0}$ A domain centred at $x_0$ defining the maximum area within which the centres of the sand bodies of this channel belt can possibly be located. The domain is defined by the user. $|D|$ is the area of the domain.

$\mathcal{N}$ The number of sand bodies in the channel belt. A random number that is user-specified from the probability distribution $p_{\mathcal{N}}^i = \text{Prob}\{\mathcal{N}=i\}$.

$x$ Centre of an arbitrary sand body.

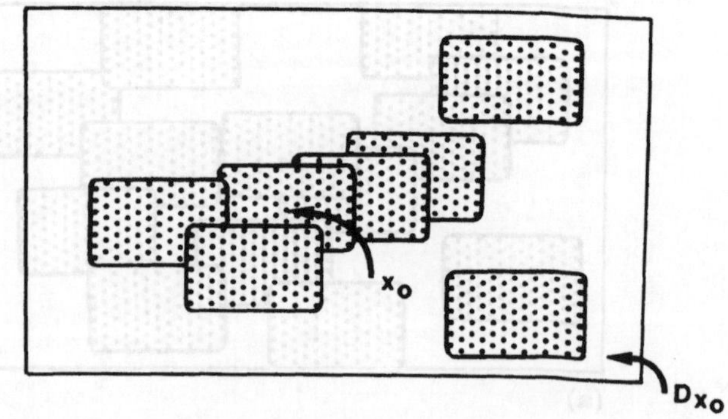


**Fig. A1.** Example of a cross-section containing one channel belt system.

The interaction model is defined as being conditioned on the number of sand bodies, $n$, and has the following form:

$$p_n(x_1, \ldots, x_n) = C_O \exp\left\{-\left[\sum_i^n b(x_i) + \sum_i^n \sum_j^n c(x_i - x_j)\right]\right\}$$

with

$p_n(x_1, \ldots, x_n)$ The joint probability density function for sand bodies to be located at exactly $(x_1, \ldots, x_n) \in D_{x_0}$.

$C_O$ A normalizing constant.

$b(\cdot), c(\cdot)$ Functions specifying the relative frequency and pairwise interaction. These functions are assumed to be independent of $n$.

This model is constrained to account for pairwise interaction only. It was first applied to physical processes with $c(\cdot)$ being related to the energy potential of particles. Familiar probabilistic relations provide the conditional distribution of $x$ given the location of $(x_1, \ldots, x_{n-1})$:

$$p_{n|n-1}(x|x_1, \ldots, x_{n-1}) = \exp\left\{-\left[b(x) + \sum_i^{n-1} c(x - x_i)\right]\right\}$$
$$= r(x) \prod_i^{n-1} s(x - x_i)$$

with

$$r(x) = \exp\{-b(x)\}$$
$$s(x' - x'') = \exp\{-c(x' - x'')\}$$

$r(x);\ x \in D_{x_0}$ A function over $D_{x_0}$ specifying the relative frequency of sand bodies. The function is specified by the user.

$s(x' - x'');\ x', x'' \in D_{x_0}$ A function specifying the pairwise interaction between two sand bodies located at $x'$ and $x''$ respectively. The function is specified by the user and $x' - x''$ will usually be vectorial.

Consider the following examples supported by Fig. A2 (A, B and C).

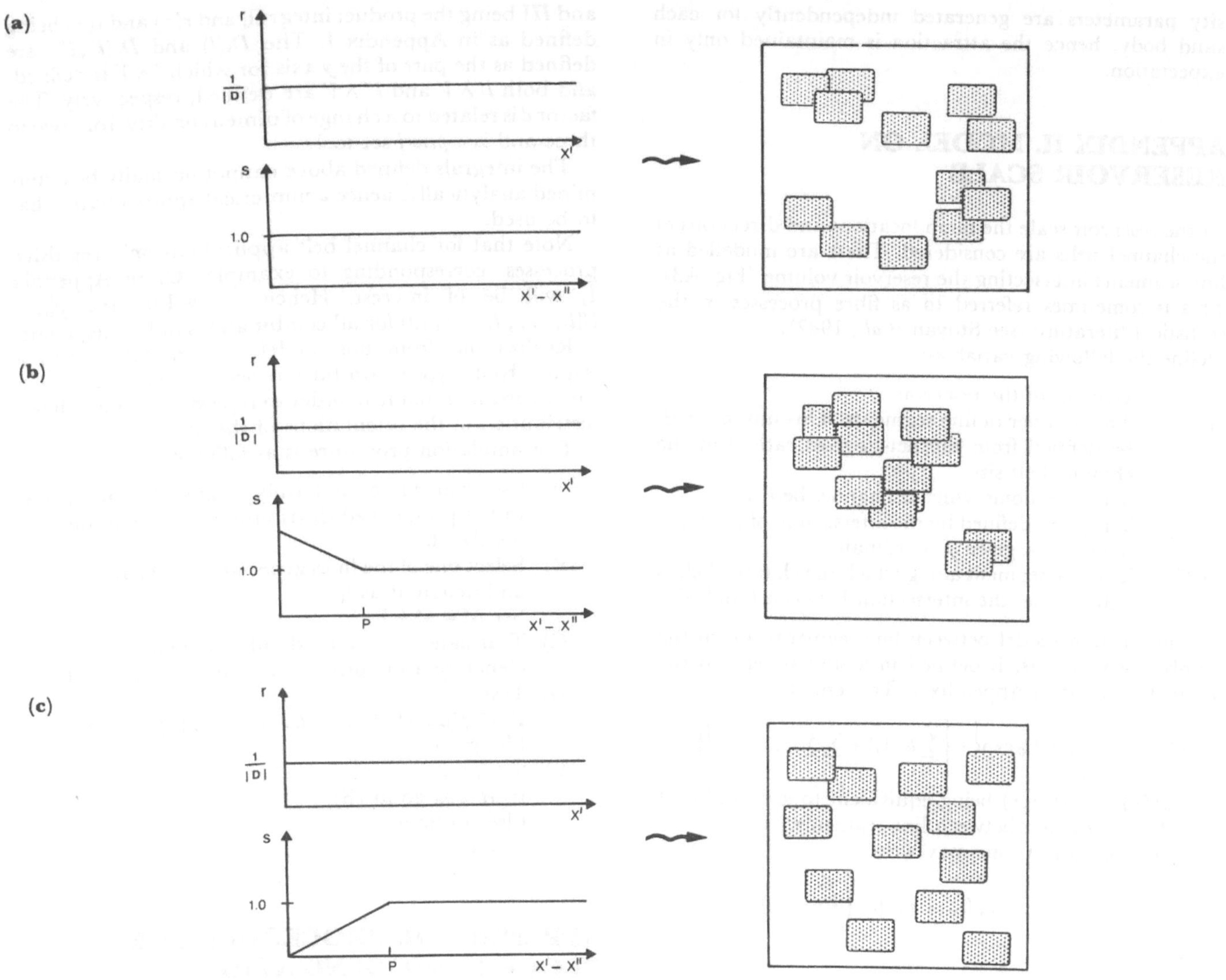


**Fig. A2.** Examples of various interaction processes. (a) Poisson process on *D*. (b) Attraction process on *D*. (c) Repulsive process on *D*.

A. ASSUME THE FOLLOWING PARAMETERS:

$$r(x) = \frac{1}{|D|}$$
$$s(x' - x'') = 1.0$$

Then the conditional probability is:

$$p_{n|n-1}(x|x_1, \ldots, x_{n-1}) = C'_0 \frac{1}{|D|}$$

which entails a Poisson process with uniform intensity over *D*.

B. ASSUME THE FOLLOWING PARAMETERS:

$$r(x) = \frac{1}{|D|}$$

$$s(x' - x'') = \begin{cases} q - \frac{1}{u}|x' - x''|; & |x' - x''| \leqslant u(q - 1.0); \quad q < 1 \\ 1.0 & \text{else} \end{cases}$$

Then sand bodies tend to attract each other, but a probability for a uniform random location with D still remains.

C. ASSUME THE FOLLOWING PARAMETERS:

$$r(x) = \frac{1}{|D|}$$

$$s(x' - x'') = \begin{cases} \frac{1}{u}|x' - x''|; & |x' - x''| \leqslant u \\ 1.0 & \text{else} \end{cases}$$

Then sand bodies exhibit a repulsive behaviour.

At a channel belt scale, processes showing attraction are of interest. Example B is therefore of greatest interest.

Realizations from such models are usually generated from Ripley–Kelly death/birth processes. It is recognized, however, that attraction processes may cause problems owing to larger computer time. In the present implementation, a much simpler simulation procedure is applied. Its properties are not studied thoroughly, but it does reproduce the main features of the model. The procedure is:

(A) Generate the number of sand bodies, $n$, from $p_i^N$.
(B) Do for $i = 1, \ldots, n$
Generate $x_i$ according to $p_{i|i-1}(x|x_1, \ldots, x_{i-1})$.

It is worth noting that sinuousity is later assigned to the sand bodies in the channel belt direction. The sinuousity parameters are generated independently for each sand body; hence the attraction is maintained only in expectation.

## APPENDIX II. MODEL ON RESERVOIR SCALE

At the reservoir scale the main locations and directions of the channel belts are considered. These are modelled as line segments intersecting the reservoir volume (Fig. A3). This is sometimes referred to as fibre processes in the statistical literature (see Stoyan *et al.*, 1987).

Define the following variables:

| | |
|---|---|
| $V$ | Domain of the reservoir. |
| $N$ | The number of line segments. This number will be defined from the net-to-gross ratio and the channel belt size. |
| $\mathrm{l}$ | The line along which a channel belt lies. |
| $l(x, z\|y)$ | The point defined by the intersection of l and the plane for a given $y$ coordinate. |
| $l \wedge V$ | The line segment along which the channel belt is defined, i.e. the intersection between $l$ and $V$. |

The interaction model between line segments, given the number of segments, is defined in a similar way to the model for points in Appendix I. This entails:

$$p_n^{\mathrm{l}}(\mathrm{l}_1, \ldots, \mathrm{l}_n) = C_O \exp\left\{-\left[\sum_i^n b^{\mathrm{l}}(\mathrm{l}_i) + \sum_i^n \sum_j^n c^{\mathrm{l}}(\mathrm{l}_i, \mathrm{l}_j)\right]\right\}$$

with $b^{\mathrm{l}}(\cdot)$ and $c^{\mathrm{l}}(\cdot,\cdot)$ being equivalent to $b(\cdot)$ and $c(\cdot)$ valid for interaction between line segments.

As in Appendix I, one may write:

$$p_{n|n-1}^{l}(l|l_1, \ldots, l_{n-1}) = r^l(l) \prod_i^{n-1} s^l(l, l_i)$$

with

$$r^l(l) = \prod_{D_y(\mathrm{l})} I\,[r(l(x, z|y))]^d\,dy$$

$$= \exp\left\{d \cdot \int_{D_y(l)} \ln\{r(l(x, z|y))\}\,dy\right\}$$

$$s^l(l', l'') = \prod I\ [s(l'(x, z|y) - l''(x, z|y))]^d\,dy$$

$$= \exp\left\{d \cdot \int_{D_y(l', l'')} \ln\{s(l'(x, z|y) - l''(x, z|y))\}\,dy\right\}$$

and $\Pi I$ being the product integral, and $r(\cdot)$ and $s(\cdot)$ being defined as in Appendix I. The $D_y(l)$ and $D_y(l', l'')$ are defined as the part of the $y$ axis for which $l \wedge V$ is defined, and both $l' \wedge V$ and $l'' \wedge V$ are defined, respectively. The factor $d$ is related to a change of dimensionality from two to three and is *a priori* set to 1.

The integrals defined above cannot normally be determined analytically: hence a numerical approximation has to be used.

Note that for channel belt applications only repulsive processes, corresponding to example (C) in Appendix I, will be of interest. Hence, $s(x) \leqslant 1.0$, and $p_{n|n-1}^l(l|l_1, \ldots, l_{n-1}) \leqslant 1.0$ for all combinations of line segments.

Realizations from this model are generated from a Ripley–Kelly type death/birth process. A slight change in the procedure is made in order to reproduce a pre-defined distribution of the orientations of the lines.

The simulation procedure is as follows:

(A) Generate $n$ lines uniformly located, but according to the pre-defined distribution of orientation.
Set $M = 0$.

(B) Select one of the lines generated in (A) at random, and denote it as $l_k$.
Set $M = M + 1$.

(C) Translate the line randomly to obtain $l_k^l$.
Generate a uniform random number $u \in [0, 1]$.

(D) Test
If $u \geqslant p_{n|n-1}^l(l_k^l|l_1, \ldots, l_{k-1}, l_{k+1}, l_n)$, go to (E).
Else go to (C).

(E) Set $l_k = l_k^l$.
If $M < 3n$ go to (B).
Else continue.

(F) STOP.

## APPENDIX III. PROCEDURE FOR BLOCK HOMOGENIZATION

The reservoir representations are stored in a parametric form. Reservoir production simulators require input on a grid block format. Each block has to have three values for the assigned effective permeability, one for each main direction. The procedure for obtaining effective permeability over a given grid block is as follows.

Define:

$$D_{xyz} = \text{domain of grid block}$$

$$i(x, y, z) = \begin{cases} 1 & \text{if sand in } (x, y, z) \\ 0 & \text{if shale,} \end{cases} \quad \text{indicator of sand}$$

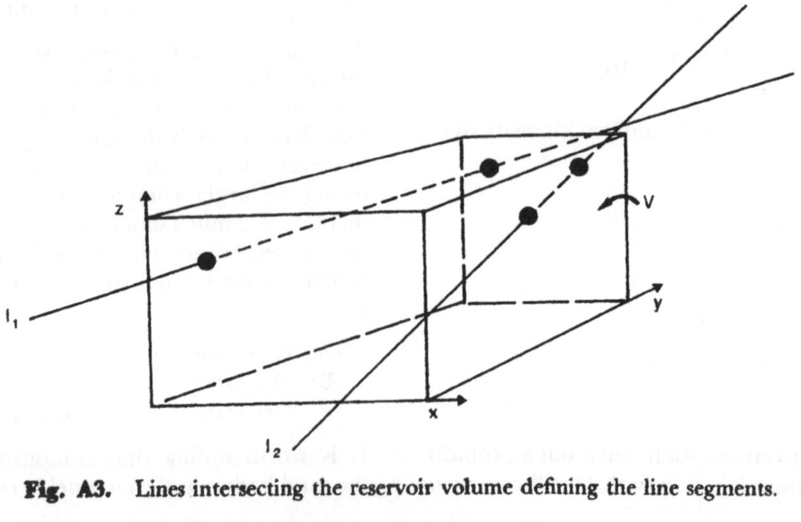


**Fig. A3.** Lines intersecting the reservoir volume defining the line segments.

The effective permeability in the $x$ direction is defined by

$$k_s \times \left[ \frac{1}{|D_x|} \int_{D_x} \left[ \frac{1}{|D_{yz}|} \int_{D_{yz}} i(y, z|x)\, dy\, dz \right]^{-1} dx \right]^{-1}$$

$$\times \left[ 1 - \prod_{D_{yz}} I \left(1 - \prod_{D_x} I\, i(x|y, z)\, dx\right) dy\, dz \right]$$

with $k_s$ being the permeability of the sand facies, and $\Pi I$ the product integral.

The rather complicated expression for effective permeability contains three terms separated by multiplication signs. The first term characterizes the quality of the sand. The second term is identical to the traditional arithmetic/harmonic average expression. The third term is a correction term implying that if all possible straight lines parallel to the $x$-axis through the block intersect a shale, then the block is impermeable. The latter correction term is needed in the situation sketched in Fig. A4.

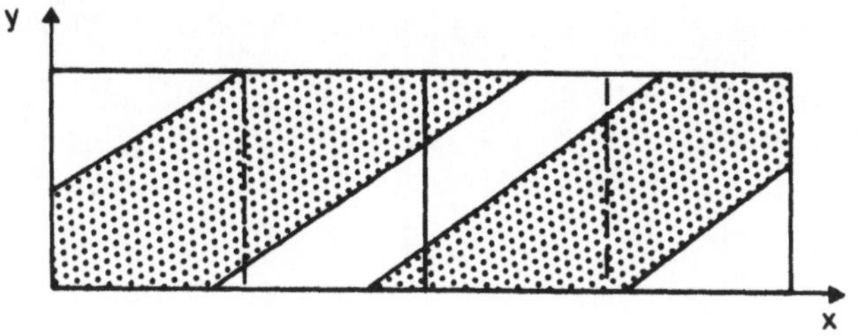


**Fig. A4.** Pair of blocks with no flow in the $x$ direction between centres.

Identical expressions are used for the $y$ and $z$ directions. In practice the expression has to be approximated numerically by point sampling in the block.

# 24

# HEX – A Tool for Integrated Fluvial Architecture Modelling and Numerical Simulation of Recovery Processes

*A.W. Wadsley, S. Erlandsen and H.W. Goemans*

Norsk Hydro AS, Oslo, Norway

This contribution describes HEX, a computer tool for integrated fluvial architecture modelling and simulation of recovery processes, using hexagonal grid systems. Stochastic modelling of a system of intersecting channel sands deposited in a braided stream environment is described. An hexagonal sub-division of the model plane is used to discretize fluvial sand bodies. The same hexagonal grid is used for subsequent reservoir simulation, and thus the simulation model produced is a clear representation of the geological model of the reservoir. The derivation of intercell transmissibilities is described together with the use of non-neighbour connections between grid cells. This conventiently enables the model to be run using a conventional (Cartesian) simulator. Examples are presented of simulation studies of water–oil displacement in a reservoir in which the effect of permeability heterogeneity in the reservoir is investigated, together with the effect of lateral barriers to flow, along the channel banks. An example of matching pressure behaviour of a long-term production and build-up test is given, using a stochastic reservoir model generated by HEX, in which the deterministic parameters of three wells were honoured. The use of a hexagonal grid overcomes the problems of grid-orientation effects and allows accurate modelling of fluid flow both diagonal and parallel to the principal grid directions. The technique is straightforward to use and enhances the capacity to plan the development of complex fluvial reservoirs.

## INTRODUCTION

Geological and engineering modelling of fluvial reservoir units is becoming increasingly important in North Sea Jurassic Fields (e.g. the Ness, Statfjord and Lunde formations). The complex architecture of interconnected sand bodies, which may range from narrow channels to areally extensive sheets presents a simulation modelling challenge owning to:

(1) the frequent lack of correlation between sands observed in adjacent wells, suggesting their discontinuous nature;
(2) the apparent randomness (within certain geological frameworks) of the distribution and sizes of channel sand bodies;
(3) uncertainty with regard to flow continuity between sand bodies, which is influenced, among other factors, by:
   - the degree of erosional intersection of channels;
   - presence of (eroded) channel bounds;
   - permeability variation between channels;
   - permeability variation within channels;

(4) the requirement to honour well and well test data that often describe various (radial and linear) flow regimes.

If a deterministic representation of the location, orientation and physical dimensions of the sand body architecture is not available, a stochastically derived model of the reservoir (based on sound geological principles) may be used to investigate its flow behaviour (Haldorsen and Macdonald, 1987).

We here describe the integrated geological modelling and reservoir simulation of a fluvial Jurassic sand under appraisal. Three units are investigated: the Upper, Middle and Lower reservoir units of the Statfjord Formation in a North Sea field sector. The objective of the study was to calculate the recovery efficiency of these reservoirs under water flood, and to investigate the sensitivity of recovery to variations in the assumed reservoir permeability, sand distribution and planned well pattern.

Results of representative simulation runs are presented. A principal concern is the accurate prediction of water-breakthrough at the producers, and the modelling of both

*North Sea Oil Gas Reservoirs—II*

areal and vertical sweep efficiencies. Because the precise sand distribution in the reservoir can not be defined uniquely by conventional means, stochastic methods were used to interpolate between wells. The simulation results in turn are to be probabilistically interpreted, and a range of possible recoveries from the reservoir is to be defined. However, it is beyond the scope of the work presented here to show results for a whole population of reservoir realizations.

There is little practical experience in the development of such reservoirs in the North Sea. Thus, the expected recovery can best be determined through simulation and cannot be independently verified. Several modelling approaches have been published earlier (Augedal *et al.*, 1986; Martin *et al.*, 1988). The HEX approach is an attempt to minimize any bias in the calculation of recovery and sweep that can be attributed to the modelling process itself, rather than to real physical reservoir effects. Therefore, a clear geometrical relation between the stochastically generated reservoir model and the simulation model should be maintained. To this end, a novel approach was adopted in which a hexagonal grid system was used to develop realizations of sand body distribution with the same grid being used as the basis for a conventional finite-difference simulation of oil and water displacement.

## GEOLOGICAL RESERVOIR MODELS

In the example discussed, it is assumed that only one permeable reservoir facies is present; that is, sand bodies resulting from the infill of river and stream channels in a moderate- to high-energy fluvial depositional environment. Each channel is thought to be of low sinuosity and is schematically represented as being straight, fully straddling the reservoir dimensions and with a rectangular cross-section. The channels are assumed to run parallel to the inferred palaeoslope of the reservoir. Channel width, thickness and orientation (and petrophysical properties) are stochastically generated. The erosion of channels into each other is modelled by the successive intersection of channels as they are generated in the model.

Schematic areal views of such channel systems are shown in Plate 35. This plate shows the change in sand body connectivity as the total sand content in the reservoir increases. The sand content is defined as the ratio of (porous) sand volume to gross rock volume. This definition distinguishes between reservoir rock (sand) and non-reservoir rock (non-pay or shale). At low sand content there is almost no connectivity transverse to the principal palaeoflow. As the sand content increases, transverse connectivity increases until at high sand contents the distribution of sand in the reservoir approaches that of a homogeneous sheet sand. The sand content of the reservoir units in the example study ranges from 32% to 80%.

The primary concern when simulating this type of reservoir is to preserve the major geological controls that affect fluid displacement. Apart from being affected by permeability distribution, the gross areal movement of oil and water will clearly be dominated by the orientation of the channels and the channel width unless there is sufficient variability in their orientations and sufficient channel intersection to provide effective lateral communication.

As an application example here, a reservoir section of the Jurassic Upper, Middle and Lower Statfjord units is used whereby the Upper and Middle units are separated by a fieldwide shale and the sand content in the Lower unit is reduced (See Fig. 6.)

## SIMULATION GRIDS

Correct numerical simulation of geological reservoir models is in practice constrained by the use of a standard (5-point) Cartesian grid system (Heinemann *et al.*, 1989). In such a system, flow into or out of a grid block can only occur along the directions parallel to the grid axes. This restriction can lead to a retardation of the flood front in directions diagonal to the grid axes. This phenomenon, known as the grid-orientation effect, is well documented (Todd *et al.*, 1972; Aziz and Settari, 1979). Pruess and Bodvarsson, 1983 proposed a uniform hexagonal grid pattern to improve grid-orientation effects.

In a Cartesian grid problems arise also if two approximately parallel sand bodies intersect the same grid block. An example of this is shown in Fig. 1. In this case, the porosity of the grid cell is derived from that of both sand bodies. This results in an "areal smearing" of the sand distribution. Moreover, a non-zero transmissibility may be calculated between the channels in the grid system, whereas physically they may not intersect. Sufficiently small grid blocks may be used to overcome this drawback, but in practice this often leads to unacceptable high usage of CPU time. These modelling problems arise if the simulation grid does not possess a clear geometrical relation to the sand body geometry.

In order to overcome the geometrical limitations of conventional quadrangular gridding, our approach was to use a hexagonal subdivision of the plane in order to create an areal discretization of each sand body. In this system, each sand body is composed of a connected set of areal hexagons that are smaller than the channel width to adequately reflect the areal geometry of the created sand body. The objectives of this approach are:

(1) to provide a numerical simulation model that is a clear representation of the geological model,

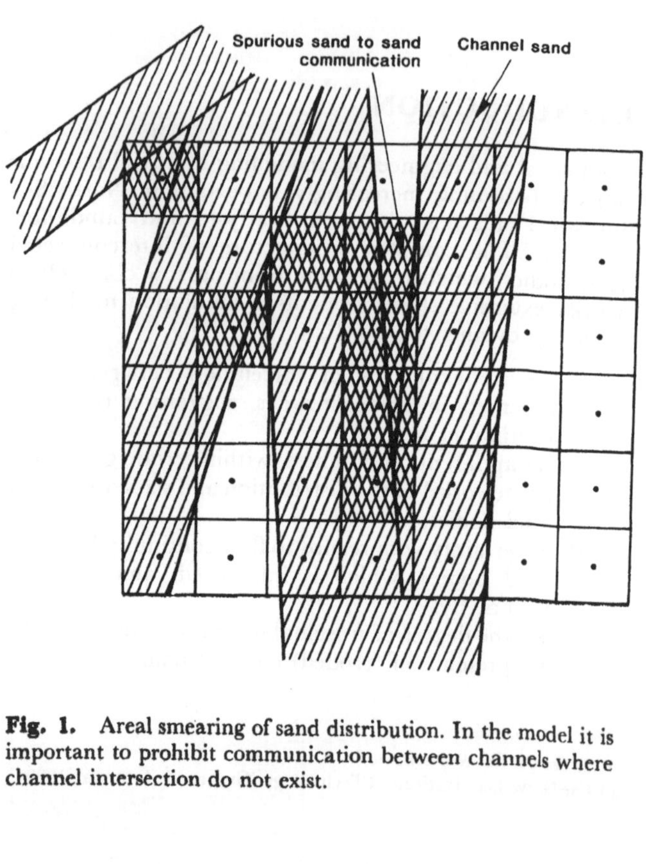


**Fig. 1.** Areal smearing of sand distribution. In the model it is important to prohibit communication between channels where channel intersection do not exist.

maintaining the sand–shale sequence observed in wells;

(2) to provide good channel continuity in all directions, both parallel and diagonal to the simulation grid orientation;

(3) to define lateral and vertical reservoir continuity through active (shale) barrier modelling in addition to the sand body generation.

## SAND CREATION

Sand bodies are created in sequence, with the cell properties of the grid system updated as each sand body is created. The sand creation sequence terminates whenever the sand content target for a reservoir region is met. The sand content is an input value that can be specified for each model layer. Shales are created in two ways:

(a) absence of sand;

(b) introduction of transmissibility barriers between cells.

Case (a) defines a "passive" (implicit) system of shale generation. Case (b) defines an "active" system of (shale) barrier creation, which can be used to associate lateral permeability barriers with the channel sands.

A sand-creation sequence begins with the stochastic generation of the top layer of the sand body. The vertical position of the top layer is selected at random from all available predefined layer position using a uniform frequency distribution. This is the layer below which the sand body is hung, as shown schematically in Fig. 2. The centre point of the channel is picked from distributions defining the central $x$ and $y$ coordinates of each channel, these distributions may be layer-dependent. The cumulative sand content of the region containing the start location, and the sand content of the reservoir layer corresponding to the top layer of the sand body, are checked to see whether they have reached their target or maximum sand content. If this is exceeded, another layer and location for the start of sandbody creation is tried.

The remaining sand body characteristics (width, thickness, direction, porosity, permeability, net-to-gross ratio and $K_v/K_h$ ratio) are selected by stochastic techniques (Augedal *et al.*, 1986) from a cumulative frequency distribution (which can be specified according to layer and reservoir region).

The central "thread" of the sand body is then determined. This "thread" is a connected series of grid cells that starts from the centre point of the sand body and extends in the (stochastically chosen) channel direction until it reaches the reservoir boundary in both orientations. An example of a channel "thread" is shown in Fig. 3. The construction of this "thread" ensures that the channel will consist of a connected sequence of cells, independent of the

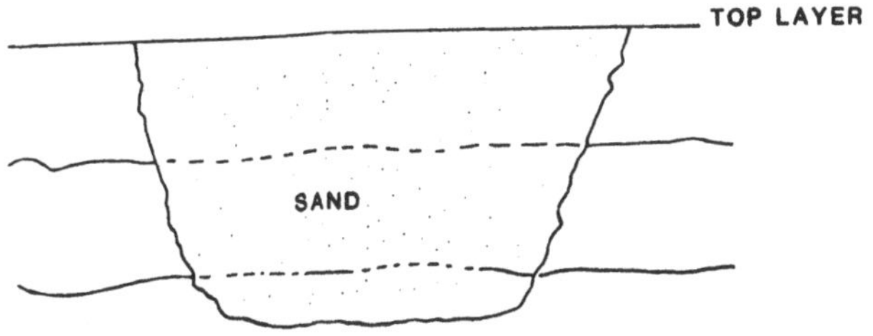


**Fig. 2.** A channel is hung off the top of a layer. The channel may fully penetrate several layers if they are thin compared to the sand-body thickness.

specified width of the channel. The practical effect is to force the smallest effective width for any channel to be 1.5 times the outside radius of each hexagonal cell.

Cells between the centre line of the sand body, as determined by its direction and its central location, and the parallel lines at distances equal to half the sand body width are then infilled. This method effectively constructs a rectangular channel. (Fig. 4).

Once the areal location of a sand body has been determined, its vertical extent is then calculated together with the corresponding grid-cell properties. In each vertical (cross-sectional) column where sand is present the fractional penetration of the reservoir grid layers is determined by the ratio of sand-body thickness to cumulative thickness of the reservoir layers, starting from the top layer of the sand body. In order to obtain a realistic distribution of shales beneath each channel, the layer thickness of each grid cell may be randomly generated from an input cumulative frequency distribution. This method avoids a rigid geometrical approach to reservoir layering by allowing for random variation of reservoir layer thicknesses. The relationship between the variation of reservoir thickness and sand-body thickness distributions determines the qualitative characteristics of the vertical sand–shale sequences in the reservoir.

Channel penetration into a lower reservoir layer is shown in Fig. 5. It can be seen that only parts of the channel base are in communication with this reservoir grid layer. A typical reservoir section is presented in Fig. 6.

If the top layer of a new sand channel is already occupied by a high previously defined sand body, then penetration of this column will only commence below the bottom of this layer. In turn, the newly created sand body will cut any sand bodies below it of older origin (Fig. 7). For partially penetrated layers, the updated block properties are calculated proportionately, based on the new penetration and the previous sand body penetration of the layer. Fully penetrated layers are given properties equal to the current channel properties. Thus sand-body creation adheres strictly to geological principles: channels created in high layers cut channels whose tops lie in layers lower down in the reservoir sequence.

### Deterministic wells, well conditioning

The reservoir modelling allows well "conditioning", where in a specific position deterministic grid-block parameters can be specified to maintain observed reservoir sequences in wells. Grid blocks defined as sand in these positions form the starting point for new channel "threads". When several deterministic wells are to be conditioned, channel sand bodies in any reservoir layer are created alternating from well to well until the specified sand content is met. If a stochastically generated thread meets another deterministic "conditioning" point, it will annul the "thread" and take an alternative direction.

### Lateral transmissibility

All grid cells in the same reservoir layer and belonging to the same sand body are assumed to be in communication. A transmissibility flag is set so as to prohibit flow between adjacent sand bodies. Thus, lateral communication between two adjacent cells in the grid system can only be effected if these cells have "belonged" to the same channel at some time during the reservoir history. This method ensures that communication between sand bodies that are adjacent but do not cut into each other ("smearing") does not occur. An additional option allows for the introduction of lateral transmissibility barriers along the boundaries of

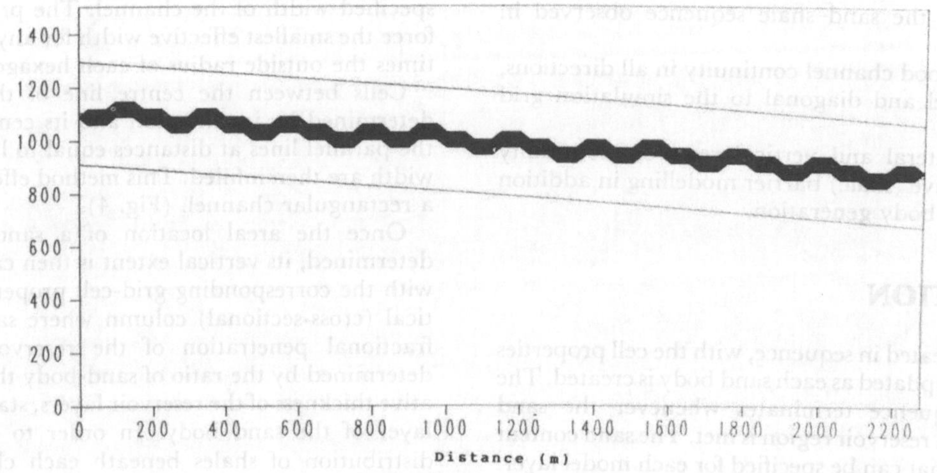


**Fig. 3.** Grid cells along the central line of the channel are filled with sand first ("channel thread").

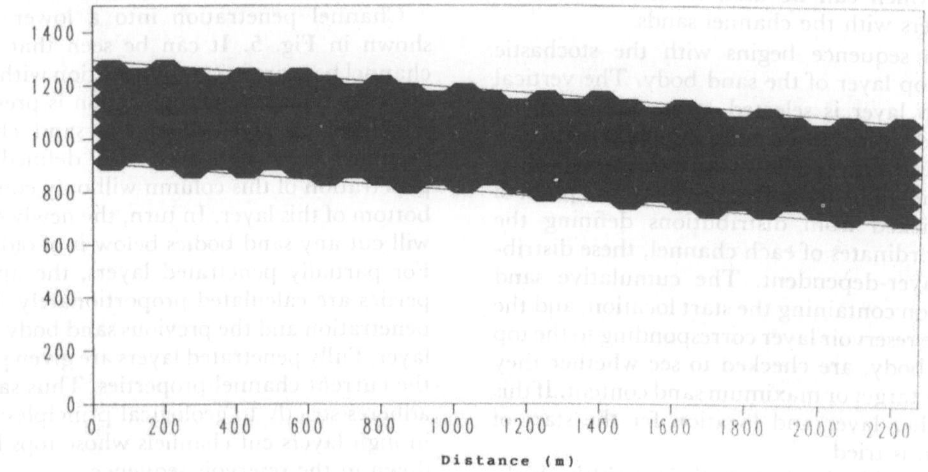


**Fig. 4.** Cells are filled with sand to complete the channel width.

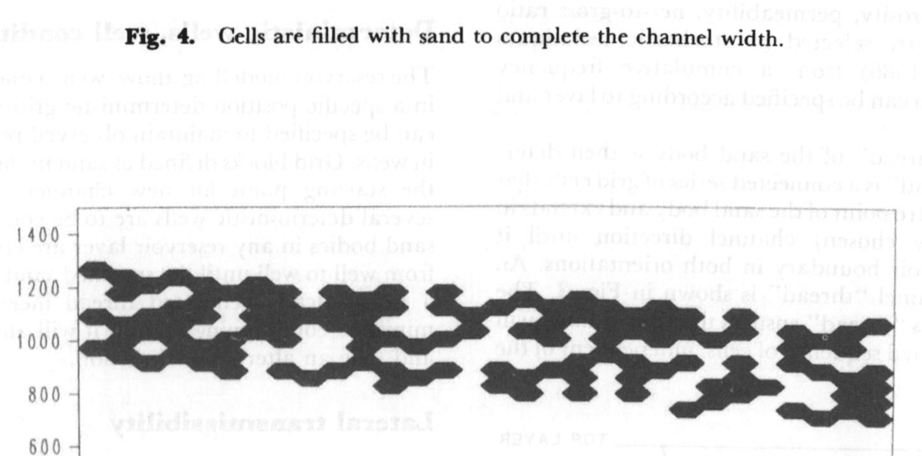


**Fig. 5.** Channel penetration into a lower layer. The occasional appearance of sand (black) is caused by superimposing a constant channel thickness on a varying reservoir layer thickness.

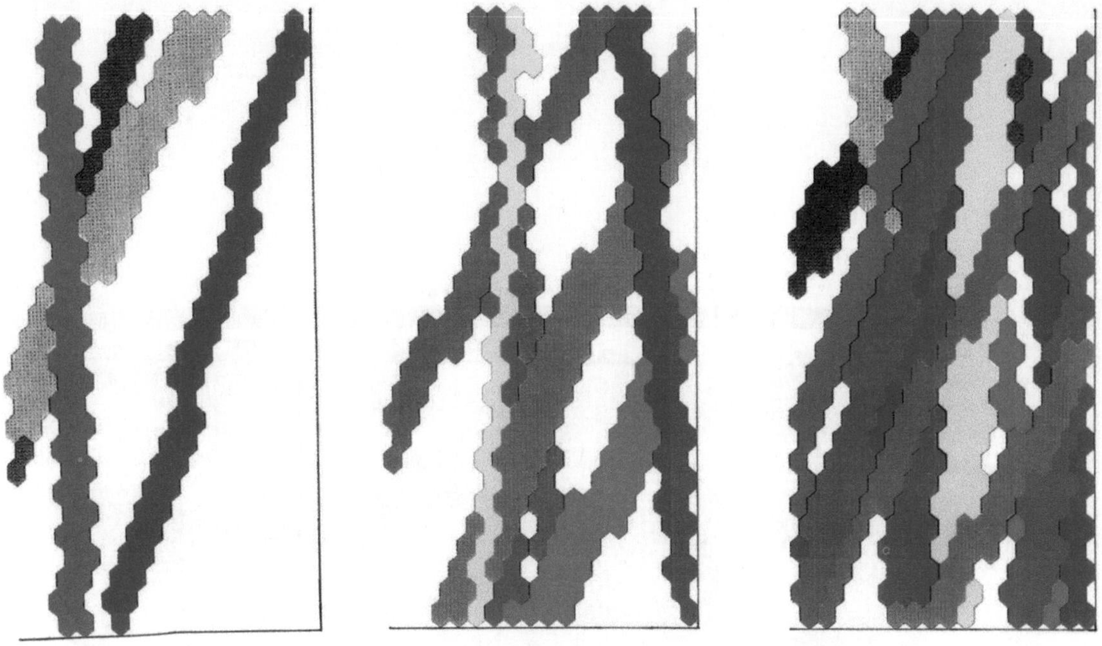

**Plate 35.** Schematic areal views of fluvial channel systems. Note that lateral continuity increases as the sand content increases.

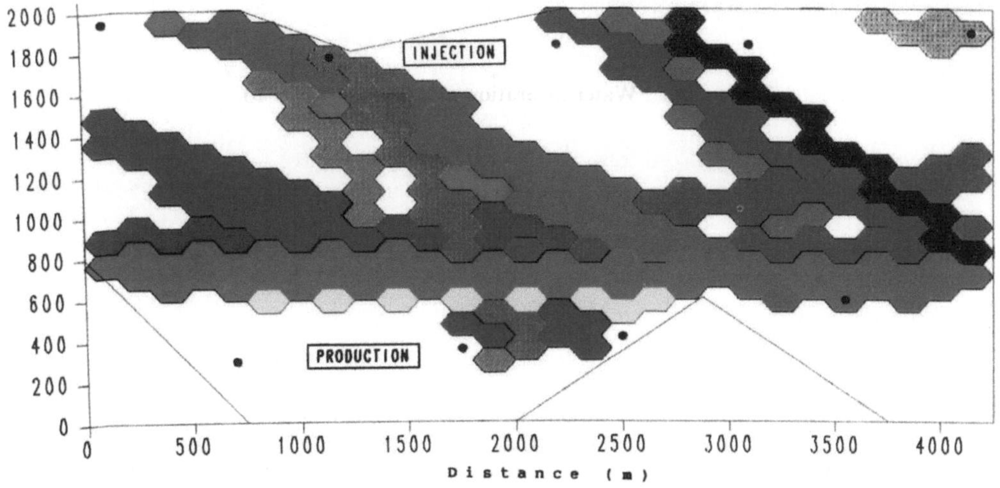


**Plate 36.** Plan view of the grid showing channels and well locations.

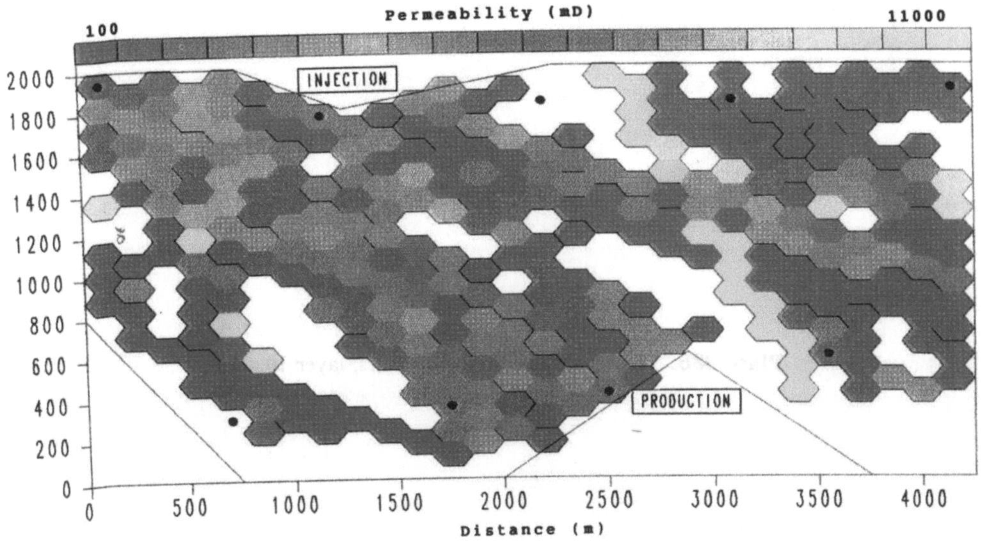


**Plate 37.** Permeability distribution in layer 10.

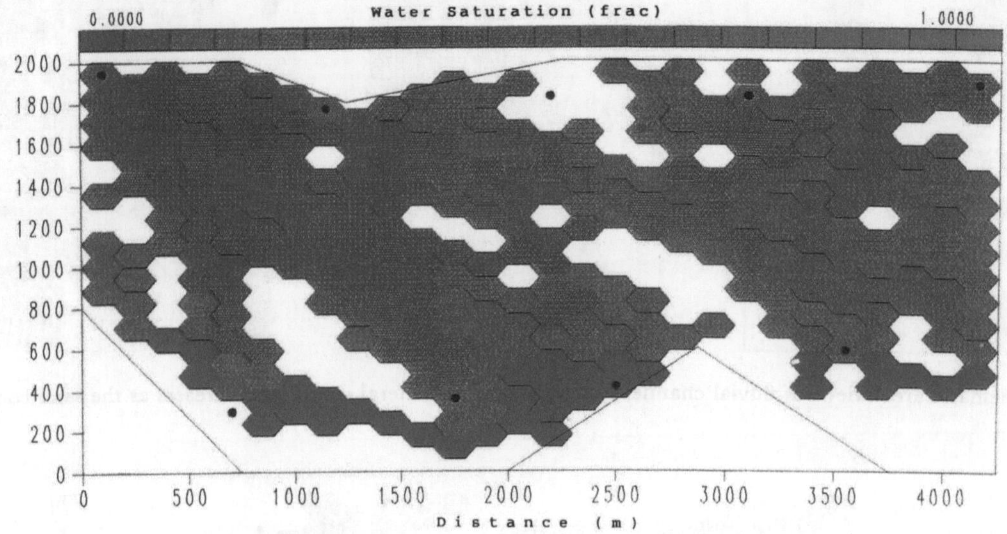


**Plate 38a.** Water saturation at 1.0 years, layer 10.

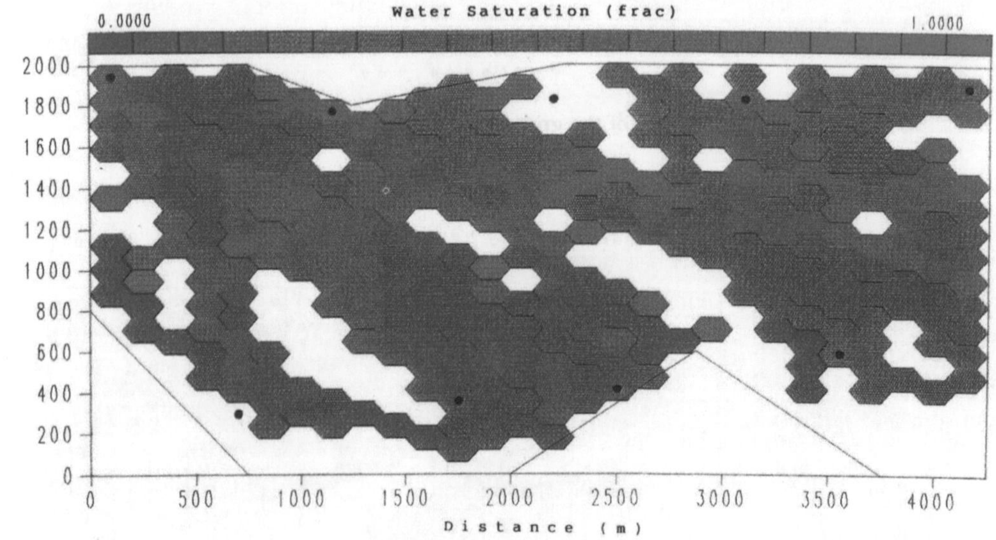


**Plate 38b.** Water saturation at 8.0 years, layer 10.

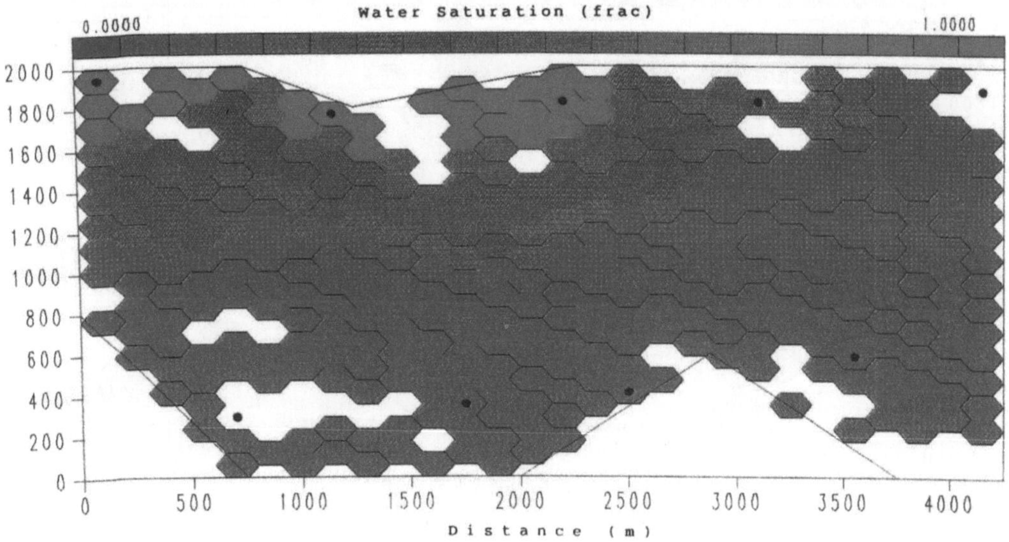


**Plate 39a.** Water saturation at 1.0 years, layer 18.

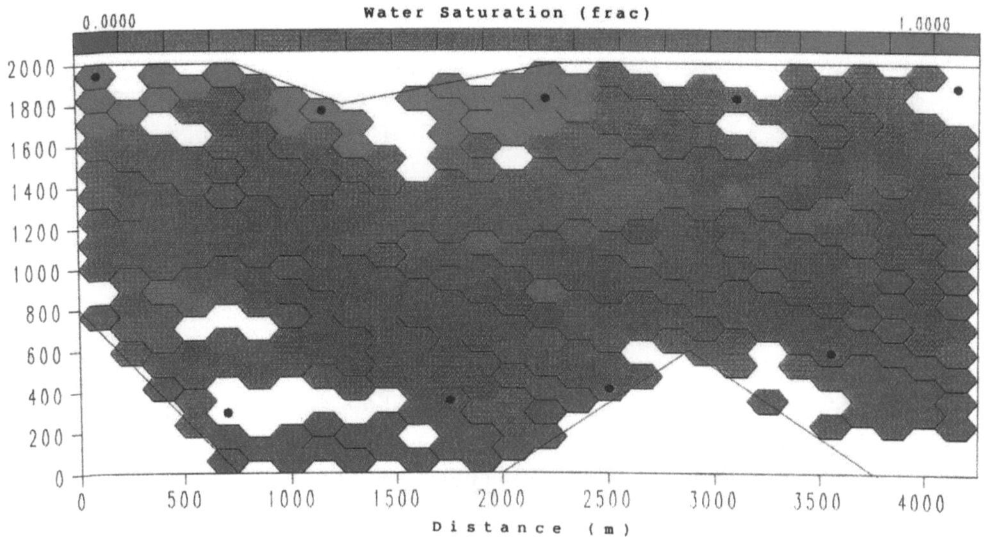


**Plate 39b.** Water saturation at 8.0 years, layer 18.

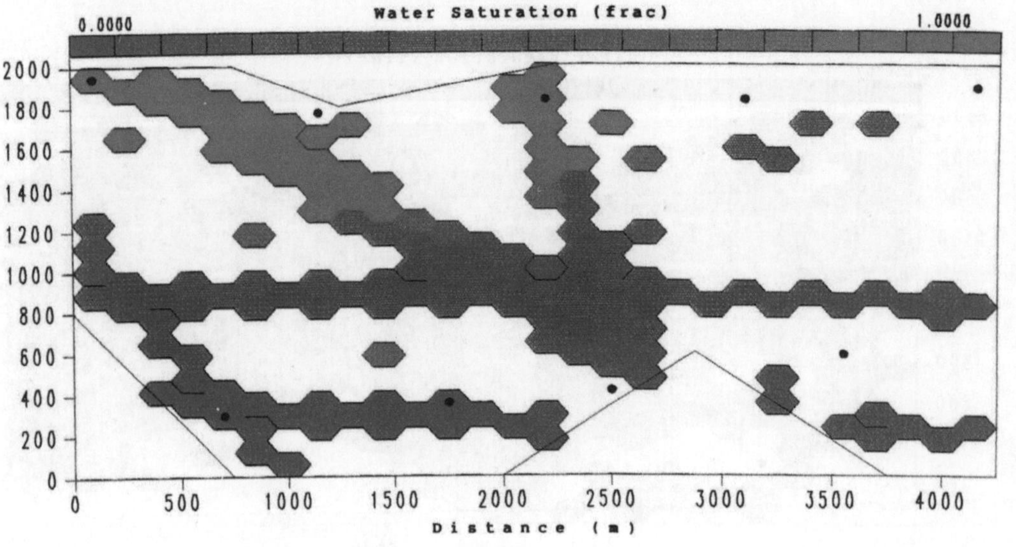


**Plate 40a.** Water saturation at 1.0 years, layer 25.

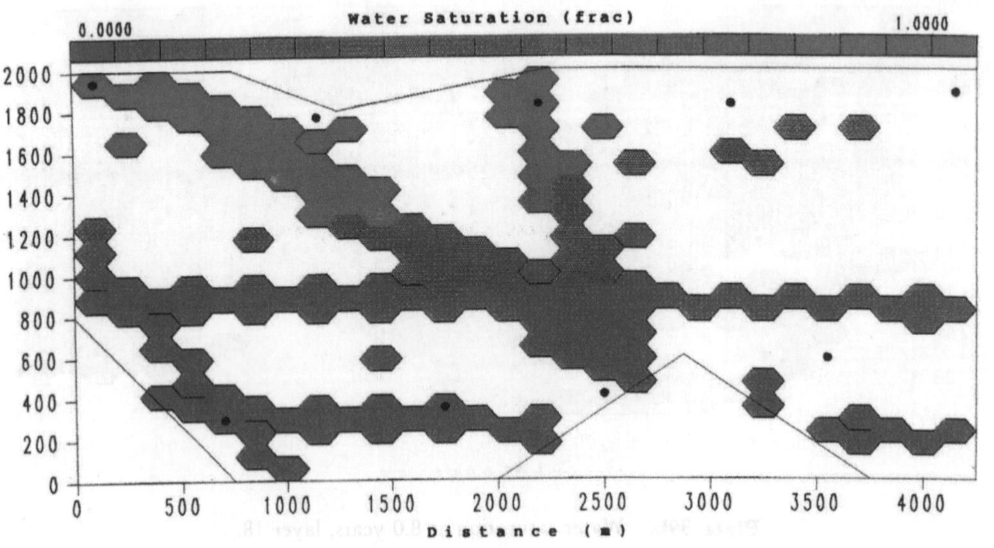


**Plate 40b.** Water saturation at 8.0 years, layer 25.

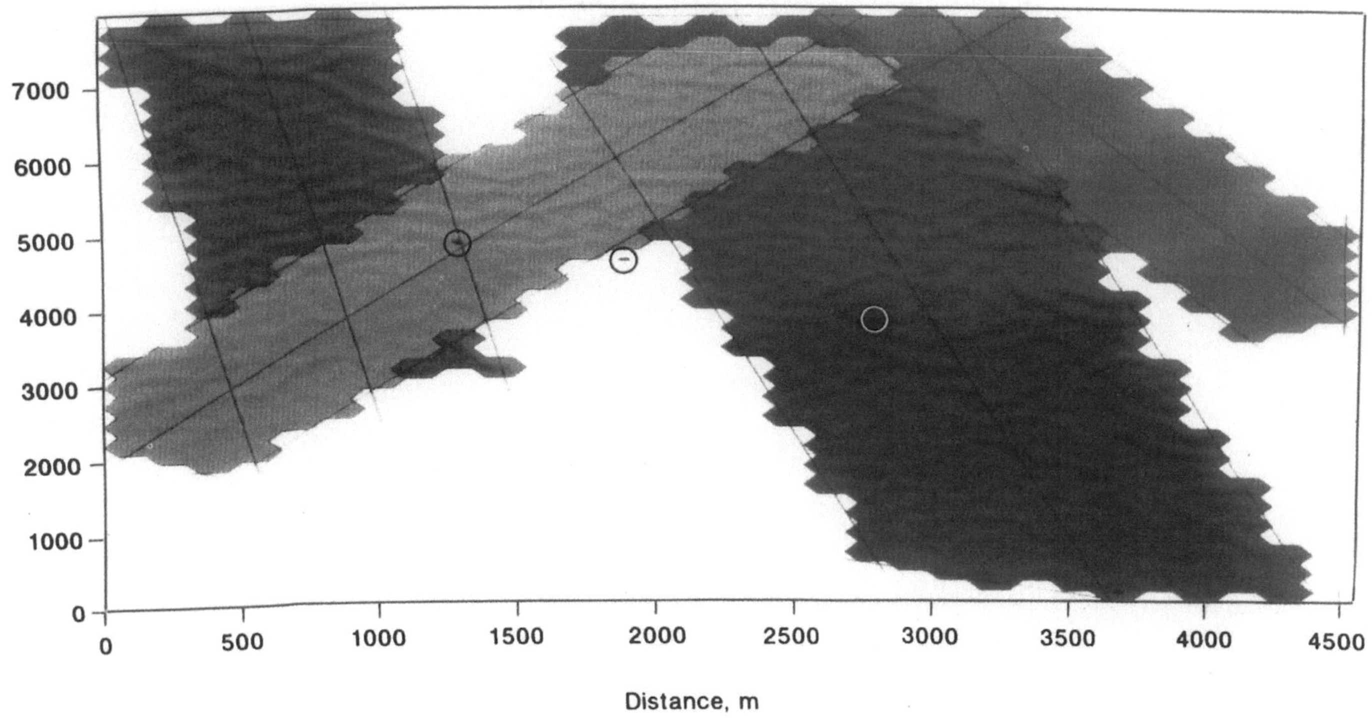


**Plate 41.** Channel model used in production test interpretation. Tested well indicated by yellow circle.

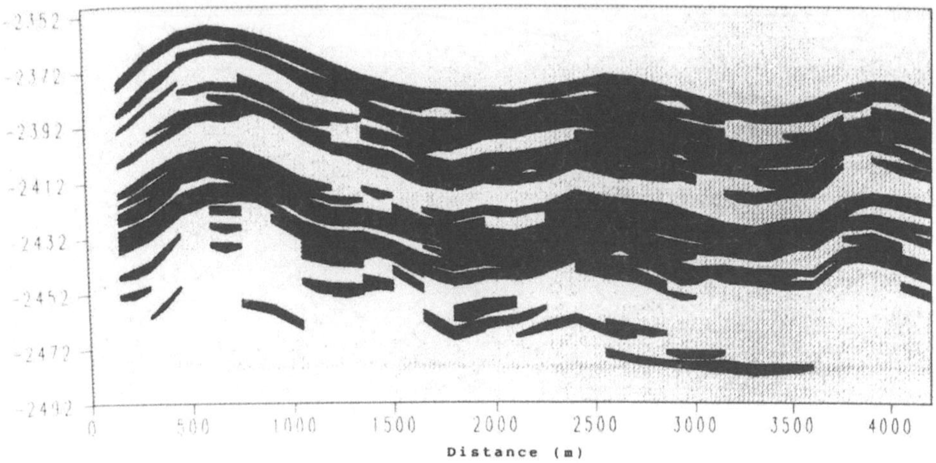


**Fig. 6.** Typical reservoir section, the sand is shown in black and the shale in shaded grey.

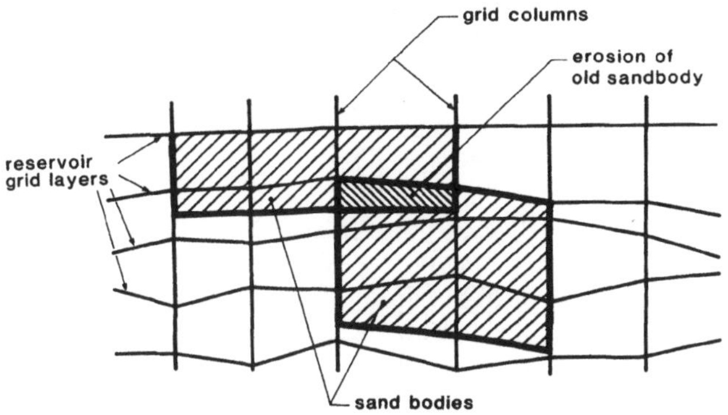


**Fig. 7.** Erosion of an older sand-body by a newly created one.

a channel without affecting vertical transmissibility. This option reduces lateral transmissibility at the sides of a channel. This reduction is in proportion to the ratio of new sand body penetration in the cell over the thickness of old sand in the cell. This reduces or destroys transmissibility created by earlier sand penetration. Figure 8a shows lateral barriers that are introduced when using this option. The corresponding layer without the introduction of lateral barriers is shown in Fig. 8(b).

If the fractional sand thickness of a grid cell is unity, then it is assumed to be in communication with the cell below (unless either cell has zero vertical permeability porosity).

# MATHEMATICAL SIMULATION MODEL

The hexagonal grid used for geological modelling of sand bodies is also used for numerical simulation of the water–oil displacement. The extra degree of direct communication available with the use of hexagonal systems significantly reduces errors in simulating the transport of fluids diagonally across the simulation grid. This was a primary consideration for choosing a hexagonal system for the modelling of fluvial channel sands. Narrow channels of width similar to the dimensions of the cells of the simulation grid are difficult to model accurately with the standard Cartesian grid system. A hexagonal grid can represent connected sand bodies in all directions of the plane.

A standard commercial black-oil reservoir simulation program was used. The program uses a standard Cartesian ordering of the grid cells with an option allowing for non-neighbour connections between cells. The mapping of the full hexagonal grid system onto the *I–J* ordering required for simulation input is shown in Fig. 9. In this figure it can be seen that each column of hexagons is parallel to the $y$-axis with constant $x$ coordinate. However, the rows of hexagons do not have constant $y$ coordinates; these vary depending whether the corresponding column is odd or even. The coordinates of the centres with respect to the coordinate system shown in Fig. 9 are:

$$x(i,j) = R[1 + 1.5(i-1)] \tag{1}$$

$$y(i,j) = R[\sqrt{3}j - \tfrac{1}{2}\sqrt{3} \bmod (i,2)] \tag{2}$$

where,

| | |
|---|---|
| $i,j$ | are grid cell locations |
| $x,y$ | are cell Cartesian coordinates |
| $R$ | is the hexagon outside radius |

$\bmod(i,2) = 0$, if $i$ is even
$= 1$, if $i$ is odd

A constant scaling factor, $F$, can also be applied in the $y$-direction to scale the $x$–$y$ dimensions of the simulation grid:

$$y(i,j) = F\{R[\sqrt{3}j - \tfrac{1}{2}\sqrt{3} \bmod (i,2)]\} \tag{3}$$

## Hexagonal cell connections

Figure 10 shows the two possible connections between adjacent hexagonal cells for odd and even columns respectively. Standard connections are equivalent to those in a Cartesian ordering of the grid cells. The "non-neighbour" connections signify connections that are not recognized by the standard Cartesian ordering scheme. The terminology is that of standard reservoir simulators. The connections are shown schematically below:

COLUMN WITH ODD $i$.

```
⟨standard connection⟩
                    i,j+1
  i−l,j               |              i+1,j   ⟨standard connection⟩
           \                  /
                     i,j
              /*            \*
i−1,j−1               |           i+1,j−1  *⟨non-neighbour
                    i,j−1                     connection⟩
```

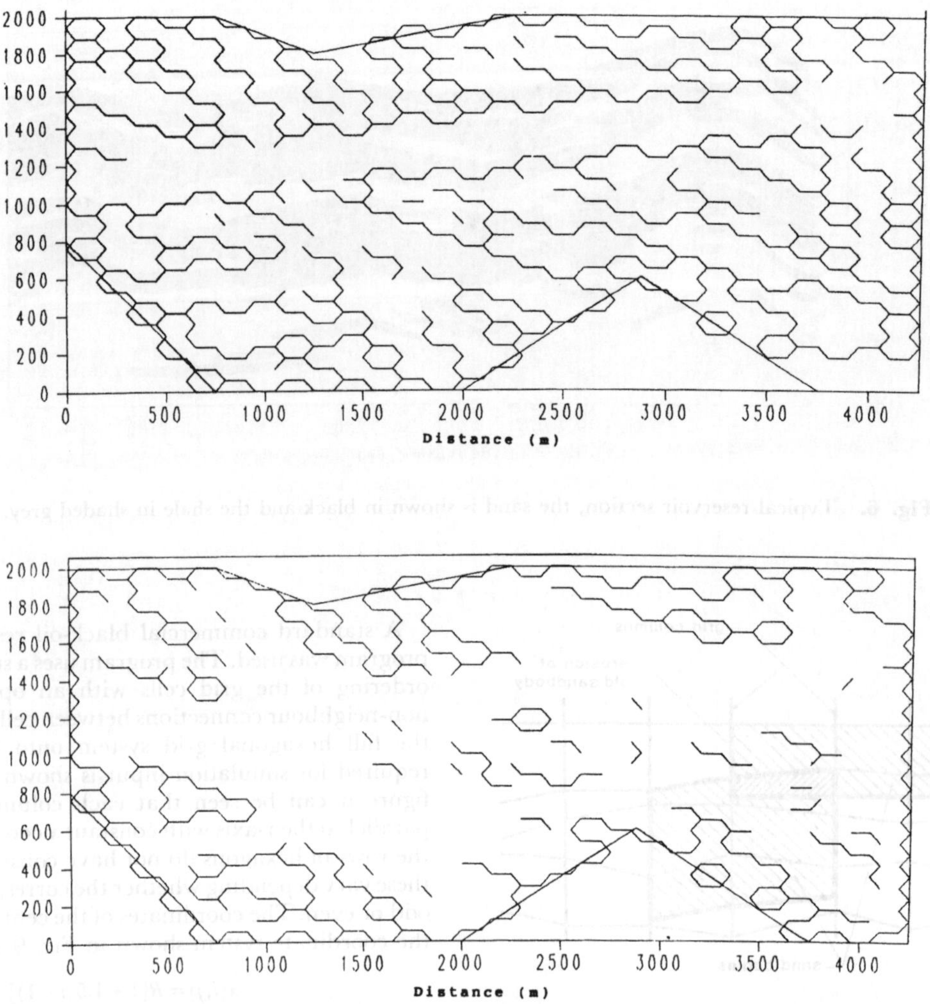


**Fig. 8**(a) Reservoir layer with lateral barriers. (b) Reservoir layer without lateral barriers.

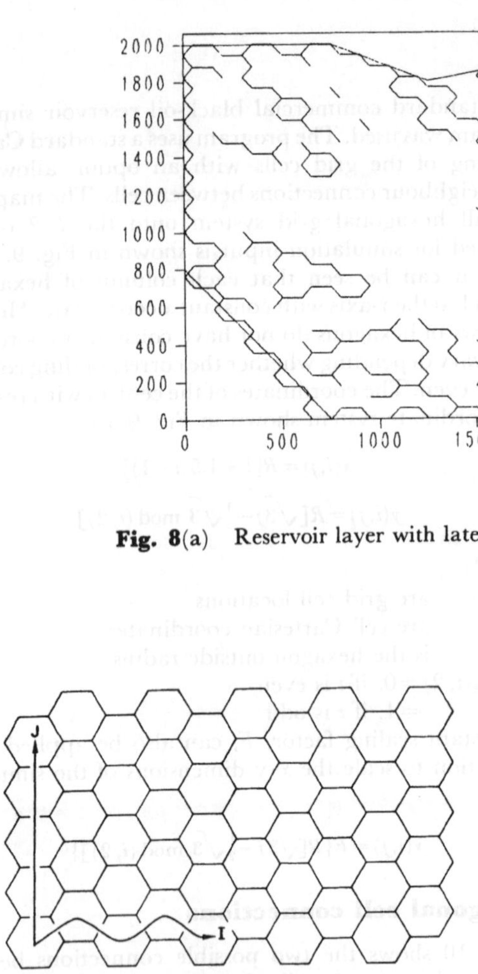


**Fig. 9.** Orientation of hexagonal grid.

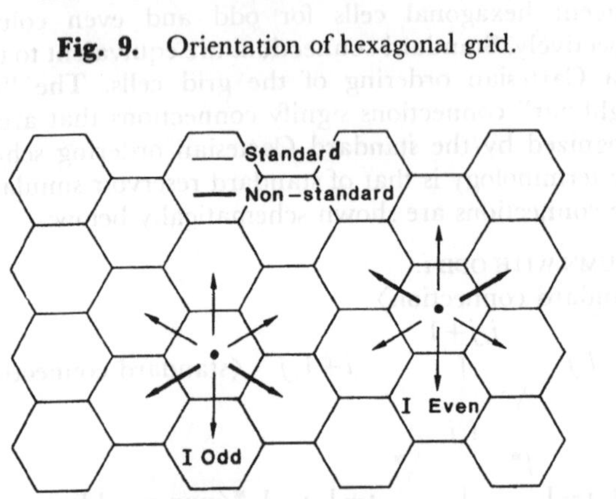


**Fig. 10.** Connections between adjacent hexagonal cells.

COLUMN WITH EVEN $i$.
⟨standard connection⟩

$i,j+1$

$i-1,j+1$ * | /* $i+1,j+1$ *⟨non-neighbour connection⟩

$i,j$

/ | \\

$i-1,j$ $i+1,j$ ⟨standard connection⟩

$i,j-1$

The connections in the $j$ direction are always standard. In the schematic connections as shown above, the connection at 60°, 120° and 240° to the vertical is either a standard or a non-neighbour connection depending upon whether $i$ is odd or even.

## Geometric hexagonal transmissibility

The calculation of the fluid flow transmissibility between adjacent hexagonal cells is based upon the finite-element method (Zienkiewicz, 1977; Lemonnier, 1979). The calculation given here is based upon a derivation of transmissibility formulae (Kinzelbach, 1986).

Figure 11(A) shows three mutually adjacent hexagonal cells whose centres define the corner points of a triangle. The transmissibility and area terms are independent of coordinate translations. The coordinate differences, $b1$, $b2$

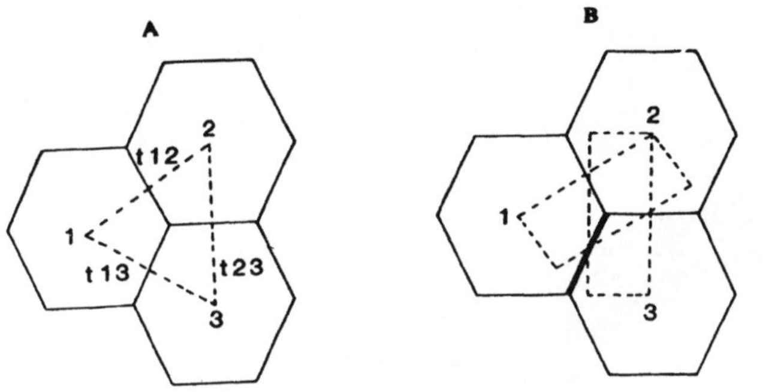


**Fig. 11.** Connections for transmissibility calculation.

$b3$, $c1$, $c2$, $c3$ and area factor $d$ are defined by:

$$b1 = y2 - y3 = \sqrt{3}FR; \qquad c1 = x3 - x2 = 0 \tag{4}$$

$$b2 = y3 - y1 = -\tfrac{1}{2}\sqrt{3}FR; \qquad c2 = x1 - x3 = -\tfrac{3}{2}R \tag{5}$$

$$b3 = y1 - y2 = \tfrac{1}{2}\sqrt{3}FR; \qquad c3 = x2 - x1 = \tfrac{3}{2}R \tag{6}$$

$$d = \text{area of hexagon} = \frac{3\sqrt{3}}{2}FR^2 \tag{7}$$

On the assumption of linear flow between hexagonal nodes, the transmissibility factors for flow between the hexagonal nodes become:

$$t12 = -\left(\frac{b1 \cdot b2 + c1 \cdot c2}{2d}\right) = \frac{F}{(2\sqrt{3})} \tag{8}$$

$$t13 = -\left(\frac{b1 \cdot b3 + c1 \cdot c3}{2d}\right) = \frac{F}{(2\sqrt{3})} \tag{9}$$

$$t23 = -\left(\frac{b2 \cdot b3 + c2 \cdot c3}{2d}\right) = \frac{1}{4\sqrt{3}}\left(\frac{3}{F} - F\right) \tag{10}$$

These factors exclude a unit-dependent constant and the *kh* factor.

The transmissibilities are independent of the outside hexagon radius $R$. In the case that the $y$ scale factor $F = 1$, they reduce to the common value, $\tfrac{1}{2}/\sqrt{3}$. This implies that the increasing distance between the centres of each hexagon, which could lead to a reduction in trasmissibility, is exactly offset by the increase in cross-sectional area to flow as the radius increases.

If the factor $F$ increases to values greater than $\sqrt{3}$, then the transmissibility factor t23 goes to zero and becomes negative. This factor corresponds to a $j$ direction transmissibility. This behaviour is mathematically correct; at the value $F = \sqrt{3}$, the angle between the triangular sides 12 and 13 becomes 90°. In this case, the mesh is equivalent to a rectangular system lying at 45° to the coordinate axes. The connectivity of this system is defined by the linear flow assumption to the triangle, which reduces the diagonal flow term to zero. Such a grid system is identical to a standard Cartesian (5-point) system. For values $F > \sqrt{3}$, the negative transmissibility compensates for the too-large transmissibilities being calculated for the factors $t12$ and $t13$. Use of such a distorted grid for multiphase fluid flow calculations may, however, lead to large numerical errors and is not recommended.

The above calculation assumes that all three cells are in communication. Figure 11(B) shows the similar system with a transmissibility barrier between two cells; in this case the transmissibility factor $t13$ is zero.

If the factors t12 and $t23$ calculated above were used for the remaining transmissibilities, low values would result as compared to the equivalent linear system. In this case, the transmissibility corresponding to the half-cell adjacent to the barrier can be calculated assuming flow through the equivalent rectangle as shown in Fig. 11(B). Further, the quarter-cell transmissibilities can be calculated on the basis of the presence or absence of a barrier. A harmonic summation is used to obtain transmissibility between to (half-) sides of adjacent cells, and arithmetic summation is used to obtain full-face transmissibility.

## MODEL APPLICATION

A simulation study of the Upper, Middle and Lower Statfjord reservoir units in the field sector under waterflood will now be discussed. The simulation model used a 28 × 17 areal grid. The base hexagon radius was 100 m, with a scale factor of 0.688 applied in the $y$ direction giving maximum grid extensions of 4250 m in the $x$ direction and 2061 m in the $y$ direction. The reservoir dips at approximately 8° at the top of the structure. Plate 36 shows a plan view of the grid giving injection and production well locations. The mean channel direction as modelled is some −30° to the $x$ axis direction, as shown in the same figure where the channels penetrating layer 1 of the Upper unit are shown. Figure 12 shows a cross-section for column 6($i$=6). Injection wells are located downdip of the reservoir.

### Reservoir unit description

A 27-layer model was built incroporating the Upper, Middle and Lower reservoir units:

| | |
|---|---|
| layers 1–10 | Upper unit |
| layer 11 | Reservoirwide shale |
| layers 12–18 | Middle unit |
| layers 19–27 | Lower unit |

Sand-content targets for the three reservoir units were 70%, 80% and 32% respectively. The mean layer thickness was 3.5 m. The actual grid block thicknesses were picked at random from a uniform distribution of thickness from 2 m to 5 m, similar to the sand-body distribution. This was chosen to give a reasonable shale distribution between sand bodies.

Channels were presumed to have their central starting position, or mid-points, located arbitrarily within the reservoir unit with uniform probability. Each channel was sufficiently long to completely span the areal dimensions of the simulation grid. Channel orientation was defined as +10° to −70° with respect to the $x$ axis. An independent distribution of chanel widths was assumed from 120 m to 300 m.

For sand bodies in the Upper unit, a constant porosity of 27% with constant effective permeability of 2000 mD was adopted. The vertical $K_v/K_h$ ratio in use was 0.7. In general, the Middle unit has poorer reservoir quality than the Upper unit (notwithstanding its higher sand content); the average permeability is 250 mD and the porosity is 25%. The permeability of the lower unit is 250 mD (the same as the Middle unit), with a porosity of 22%.

### Lateral transmissibility barriers

The Middle unit differs from the Upper unit also in a production test response in this unit. The Middle unit exhibits strong linear flow behaviour as compared with radial flow in the upper reservoir unit. This difference is difficult to reconcile, given the higher sand content

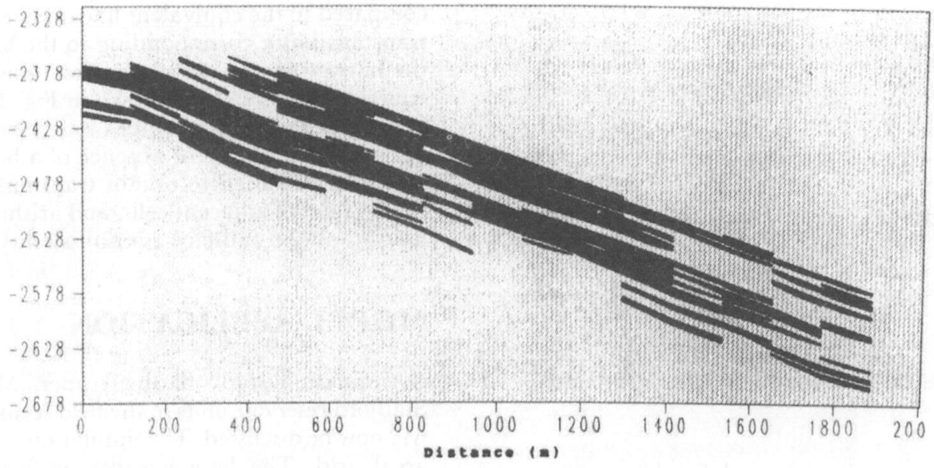


**Fig. 12.** Cross-section for **YZ** plane 6 ($i=6$).

assumed for the Middle unit, unless the channel development of this unit is different from that of the Upper unit. To account for linear pressure build-up sequences that were interpreted from production tests in the Middle unit, lateral transmissibility barriers were actively created along the boundaries of each sand body during their creation. The active creation of barriers accounts for high sand content coupled with linear inflow observed in well tests. Vertical communication between sand bodies resulting from erosion is unaffected.

## Wells

The well locations are shown in Plate 36. The maximum injection pressure was set to 428 bar; and the minimum bottomhole flowing pressure of the producers was fixed at 250 bar. This bottomhole pressure is equivalent to a flowing tubing-head pressure of 60 bar. A maximum reservoir voidage constraint of 8300 $m^3$/d (res. cond.) was imposed on all wells.

## Results

Three representative simulation runs are given as examples, the first (case 1) with constant permeability in each of the reservoir units, the second (case 2) with individual channel permeability stochastically generated from the distribution of permeabilities as seen in the wells. The distribution of permeability used ranged from 100 mD to 11 000 mD with an average (arithmetic) of 2000 mD in the Upper unit. The permeability distribution used for the Middle and Lower units was uniform between 100 and 400 mD. The third sensitivity (case 3) is similar to case 2 but without lateral barrriers along the channel boundaries.

Figure 13 presents a plot of oil rate and water-cut development against recovery for cases 1 and 2. It can be seen that the initial oil production plateau rate for each case is 22 000 $Sm^3$/d and 17 000 $Sm^3$/d respectively.

Case 2, reflecting increased permeability heterogeneity, shows a marked decrease in deliverability. This choking effect is due to lower effective permeability in the reservoir. In this case the effective permeability of the reservoir is given by the harmonic average of the permeability distribution, and not by the arithmetic average as assumed in case 1. Early water-breakthrough occurs because of the presence of a few channels of high permeability that act as water "pipelines" within the reservoir.

Case 3 shows the effect of lateral permeability barriers (Fig. 14). When lateral barriers are absent in the model, the water-cut development for the Middle unit is delayed.

## Qualitative areal sweep

A map of the permeability distribution for layer 10 (in the Upper unit) is shown in Plate 37. The permeability distribution is displayed on a linear scale from 100 mD to 11 000 mD. Each change of tone corresponds to a permeability increment of 500 mD. Of interest is the high proportion of blocks with permeabilities below 600 mD (the lowest permeability class displayed) and the very high permeability channel crossing the reservoir at an approximate $x$ coordinate of 3000 m. Also, an area of moderately high permeability (~5000 mD) can be seen following the main channel direction between 500 m and 1500 m along the $x$ axis.

A water-saturation map for the same layer after 1 year of water injection/production is shown in Plate 38a. This is the bottom layer of the Upper unit, so some water slumping can be expected. However, early water-breakthrough is seen to be associated with the two high-permeability zones described above, with rapid displacement of water in these areas of the reservoir. Additionally, water fow follows the preferred channel direction, with low volumes of water sweeping the north-eastern side of the reservoir ($x$ coordinate 4000 m).

As seen in Plate 38b, almost total flooding of this layer has taken place after 8 years. Even at this time, however, significant regions of the reservoir are poorly swept. Thus, the influence of both permeability heterogeneity and preferred channel directions leads to low areal sweep efficiency.

The effect of lateral barriers is clearly seen in Plates 39a and 39b, where water saturation in the bottom layer of the Middle unit is shown at 1 year and 8 years respectively. At 1 year, high water saturations are evident near injectors 1, 2 and 3 but the water front is constrained by the lateral barriers along the channels. In this case, most lateral movement of oil and water must also be accompanied by vertical flow through cannel erosional surfaces. As seen in Plate 39b, at 8 years, significant areas of this lowermost layer are unswept.

An example of extreme sweep variability is shown in Plates 40a and 40b for layer 25 situated in the Lower unit, which has a low sand content of only 32%. Here, both

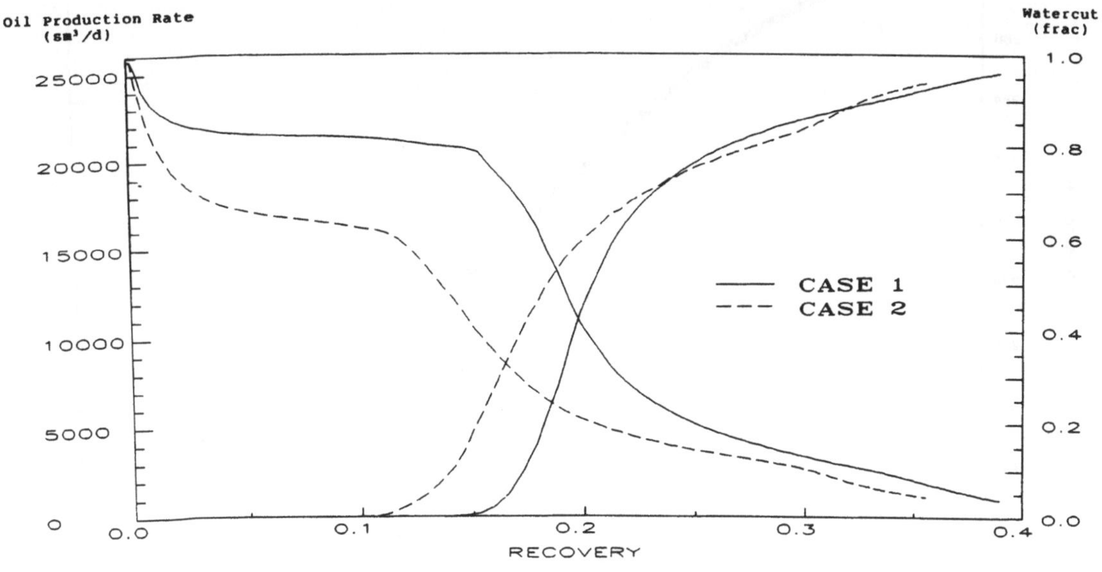


**Fig. 13.** Production performance of cases 1 and 2.

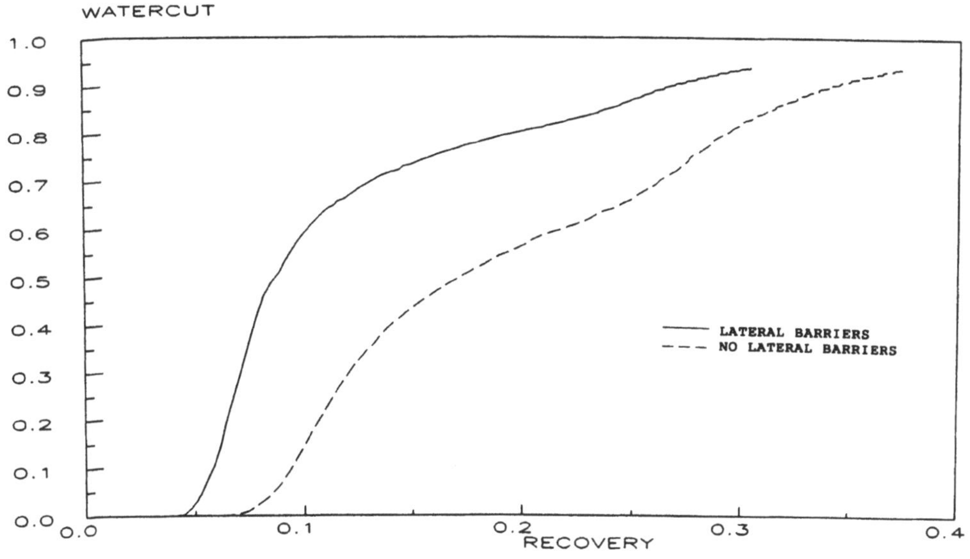


**Fig. 14.** Watercut development vs. recovery for the Middle unit for Case 2 (lateral barriers) and Case 3 (no lateral barriers).

lateral barriers and channel geometry tend to reduce areal sweep efficiency markedly.

## Well-test pressure response matching

A long-term production test of half a year's duration, followed by a main pressure build-up period of 3 months has been carried out in the Ness Formation, a member of the middle Jurassic Brent group, which has been interpreted as a fluvial channel deposit. The oil-bearing interval in the tested well is in communication with an aquifer and a gas-cap.

A superposition plot (generalized Horner plot) and a dimensionless pressure derivative plot of the pressure build up are shown in Figs 15 and 16. After a short period of radial flow, the pressure interpretation yields a linear pressure response that indicates a channel width of over 1000 m, followed by pressure support from intersecting channels and the gas cap.

A stochastic realization of the channel sands was modelled using the HEX tool. Deterministic parameters in the test well and two other wells penetrating the formation were honoured. One layer of the channel model is shown in Plate 41. The production test was then simulated. A comparison of measured and simulated pressure response during the main pressure build-up period is shown in Fig. 17.

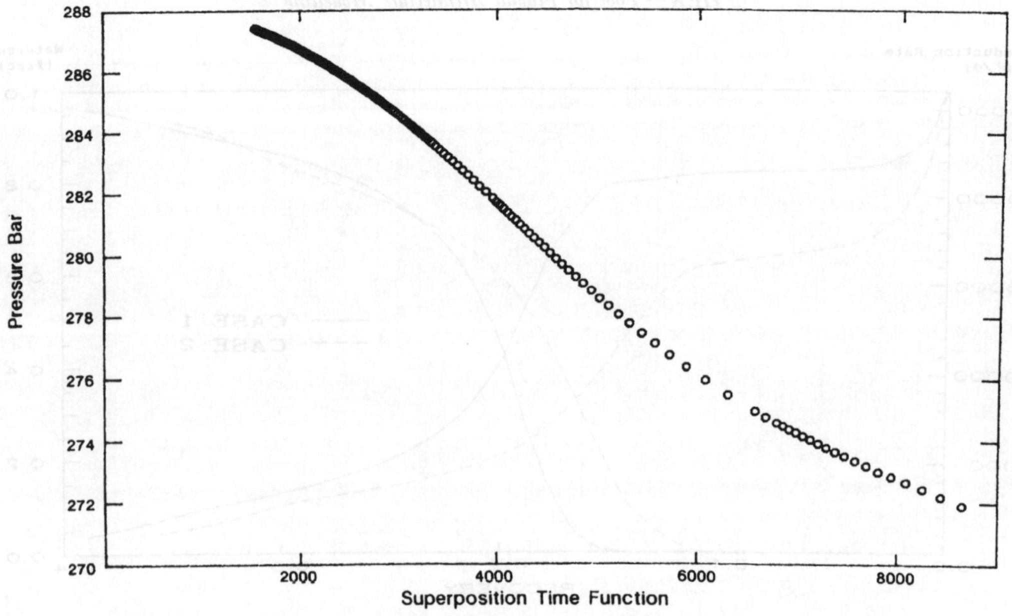


**Fig. 15.** Pressure build-up superposition plot (generalized Horner plot).

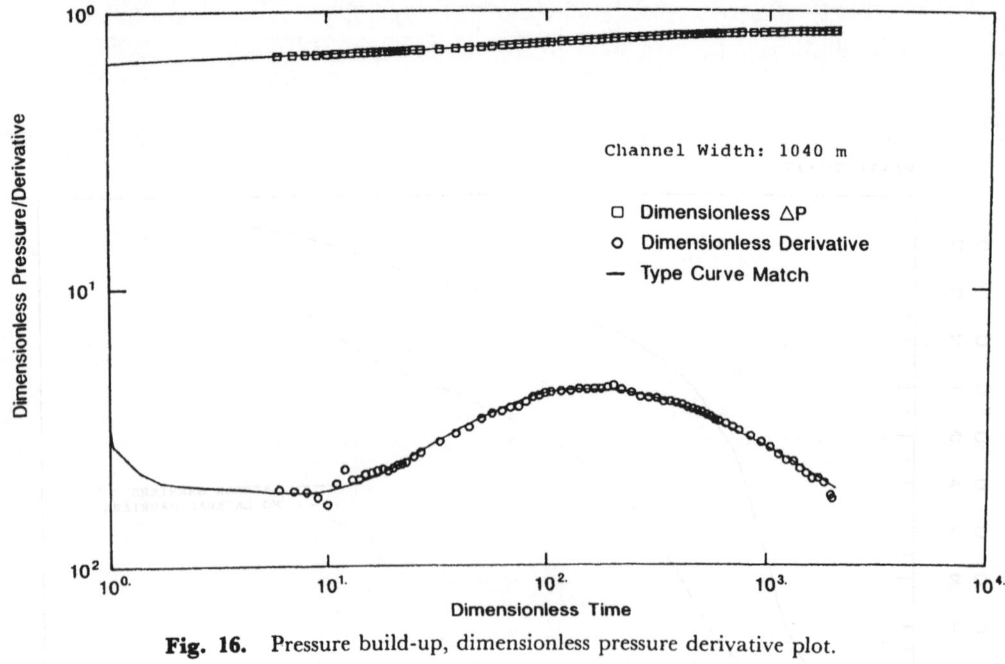


**Fig. 16.** Pressure build-up, dimensionless pressure derivative plot.

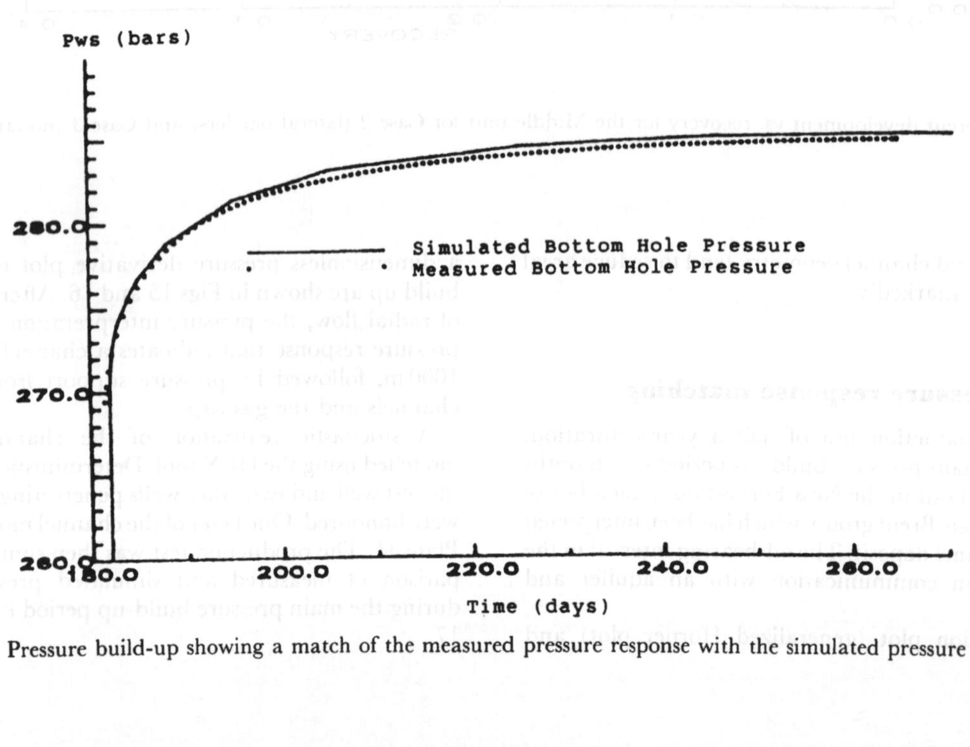


**Fig. 17.** Pressure build-up showing a match of the measured pressure response with the simulated pressure response.

## CONCLUSIONS

We have shown how a hexagonal grid system can be used both for the geometrical representation of sands in a fluvial reservoir, and for the simulation of water–oil displacement in such reservoirs.

The extra degree of communication inherent in the use of hexagonal grid systems reduces errors in simulating the transport of fluids diagonally across the simulation grid.

The hexagonal grid approach has been used for the appraisal of a fluvial reservoir in the North Sea. The system was easy to use and was found to use run CPU time comparable to conventional simulation, only increasing through the relatively high number of non-neighbour connections in the simulation grid. Generally some 15–20% higher CPU time was recorded in comparison with conventional models using a similar number of grid blocks.

Owing to the lack of artificial grid constraints on flow in the model, an accurate picture of probable areal sweep in the highly heterogeneous reservoir sands could be developed. The significance to sweep efficiency of lateral transmissibility barriers, required to honour linear flow conditions as interpreted from well-tests, is illustrated.

Pressure response during a long-term production and build-up test was matched, using a stochastically generated reservoir model in which the deterministic parameters of three wells were honoured.

The model significantly improved the possibility of quantifying principal flow mechanisms in the reservoir under scrutiny.

## NOMENCLATURE

| | |
|---|---|
| $i, j$ | Grid cell location |
| $x, y$ | Cell Cartesian coordinates |
| $R$ | Hexagon, outside radius |
| mod $(i, 2)$ | $=0$, if $i$ is even |
| | $=1$, if $i$ is odd |
| $F$ | Grid scaling factor in $y$ direction |

## ACKNOWLEDGEMENTS

We thank the management of Norsk Hydro A/S for support in publishing this paper. Thanks are also due to Helge Haldorsen for his timely comments and Paul Fjerstad for contributing well-testing work.

## REFERENCES

Augedal, H.O., Omre, H. and Stanley, K.O. (1986). SISABOSA, a program for stochastic modelling and evaluation for reservoir geology. *Proceedings. IFE/SSI Seminar on Reservoir Description and Simulation with Emphasis on EOR, Oslo, 1986.*

Aziz, K. and Settari, A. (1979). *Petroleum Reservoir Simulation*, p. 332. Applied Science, London.

Haldorsen, H.H. and Macdonald, C.J. (1987). Stochastic modelling of underground reservoir facies (SMURF). Paper no. SPE 16751.

Heinemann, Z.E. Brand C., Munka M and Chen, Y.M. (1989). Modelling reservoir geometry with irregular grids. Paper no. SPE 18412.

Kinzelbach, W. (1986). *Ground Water Modelling*, Elsevier, Amsterdam.

Lemonnier, P.A. (1979). Improvement in reservoir simulation by a triangular discontinuous finite element method. Paper no. SPE 8249.

Martin, J.H, Evans, A.J. and Raper, J.K. (1988). Reservoir Modelling of low sinuosity channel sads: A network approach. Paper no. SPE 18364.

Pruess, K. and Bodvarsson, G.S. (1983). A seven point finite difference method for improved grid orientation performance in pattern steamfloods. Paper no. SPE 12252.

Todd, M.R., O'Dell, P.R. and Hirasaki, G.J. (1972). Methods for increased accuracy in numerical reservoir simulators. *Trans. Soc. Petroleum Eng. of AIME*, **253**, 515–530.

Zienkiewicz, O.C. (1977). *The Finite Element Method*. McGraw-Hill, London.

# 25

# An Integrated Software for Stochastic Modelling of Reservoir Lithology and Property with an Example from the Yorkshire Middle Jurassic

*J.L. Rudkiewicz*[a], *D. Guérillot*[a], *A. Galli*[b] *and Group Heresi**

[a]Institut Français du Pétrole, Rueil-Malmaison Cédex, France
[b]Centre de Géostatistique, Fontainebleau, France

Direct observations on a reservoir based on core and log analysis provide high-resolution data on a centimetric or decimetric scale in a series of essentially one-dimensional (1D) slices. Reservoir fluid flow prediction however requires a 2D or 3D description of petrophysical variations using grids with decametric to hectometric blocks. We have developed an interactive and integrated software that simulates lithology and reservoir parameter variations in 2D at high resolution and scales them up to a scale suitable for conventional reservoir simulators. The modelling is carried out in three successive steps.

(1) High resolution (HR) lithological description of the reservoir with a geostatistical variogram-based method. Input data are the succession of lithofacies observed at the wells, the global proportions of each lithofacies and their variograms in both horizontal and vertical directions. Output data are a 2D lithologic grid.
(2) HR porosity and permeability description. Field observations provide empirical relations between each lithofacies and its porosity and permeability. On the basis of these relationships, numerical values of porosity and permeability are assigned to the HR grid.
(3) Scaling up HR porosity and permeability in order to produce a grid suitable for fluid flow prediction with conventional reservoir models. The scale averaging problem is solved in a CPU time-effective way using both algebraic fomula and fluid flow simulation. Output of this stage is a 2D low-resolution (LR) porosity and permeability description, suitable as input to a full-field reservoir simulator.

This software has been used to model lithofacies ad porosity/permeability variations of the fluvio-deltaic Yorkshire Middle Jurassic, an equivalent to the Brent Group in the North Sea.

## INTRODUCTION

### Building a reservoir model

Geological information on a field is usually available through well and seismic data. Well data are often logs, sometimes cores, providing information on a centimetric to decimetric scale, hereafter called high-resolution (HR). Reservoir geologists face the problem of extrapolating this essentially one-dimensional (1D) information onto a two- or three-dimensional object, in order to build a reservoir model, suitable for predictions of hydrocarbon fluid flow. However, the grid blocks used in fluid flow simulators range from a decametric to a hectometric size, called low-resolution (LR). According to the classification of Haldorsen (1986), the well data belong to the macroscale, whereas the grid blocks belong to the megascale.

The conventional way of building megascale reservoir models was to scale up information on each 1D well, then to extrapolate the averaged data on the complete grid. Heterogeneity, in this case, was only megascale heterogeneity. But, HR heterogeneity also influences hydrocarbon flow and should be taken into account when building the reservoir model. Modelling HR heterogeneity with deter-

*Group Heresi works on *he*terogeneous *re*servoir *si*mulation and is: J.D. Cartigny, B. Doligez, R. Eschard, D. Guérillot, P. Houel, Y. Mathieu, C. Ravenne, G. Renard, J.L. Rudkiewicz (Institute Français du Pétrole) and H. Beucher, C. de Fouquet and A. Galli (Ecole des Mines de Paris/Centre de Géostatistique).

*North Sea Oil Gas Reservoirs—II*

ministic physical laws seemed impossible, as too many unknown parameter would be involved. A stochastic approach seemed more fruitful. Interactive software named HERESIM (an acronym for heterogeneous reservoir simulation), which aims at reproducing the macroscale heterogeneity with stochastic methods, has therefore been developed.

## The HERESIM software

The main philosophy behind this software is that a correct reservoir characterization can only be achieved through a realistic geological description. In most siliciclastic reservoirs, permeability and porosity heterogeneity is mainly related to lithofacies heterogeneity. Furthermore, there is an important advantage in simulating first the lithofacies, then porosity and permeability, as stochastic lithofacies descriptions can be compared with conceptual sedimentological models. Their consistency with geological knowledge can be tested. Conceptual models of sedimentary environments exist and are well established (see Scholle and Spearing, 1982, for sandstone deposits), whereas only few conceptual models of porosity and permeability distributions exist. Thus, a drawback of direct porosity and/or permeability simulation is that such control of consistency is at least less easy, perhaps impossible.

Other workers used this kind of mixed approach, combining discrete lithofacies simulation with continuous petrophysical properties, especially to model North Sea reservoirs (Omre *et al.*, 1988, and others in this volume). They usually generate sand bodies, whose dimensions are taken from given distribution function, and whose spatial positions are independent, randomly placed into a shale background. This geostatistical method is known as *Boolean.* The geostatistical method presented here is *variogram-based,* which means that the lithofacies are *correlated*: their spatial positions are not independant, as in a Boolean model.

The flow chart of HERESIM (Fig. 1) involves five tasks:

(1) data input, choice of simulation grid and of the correlation horizons;
(2) high-resolution (HR) lithological description, at the decimetric scale and with a control on the sedimentary organization;

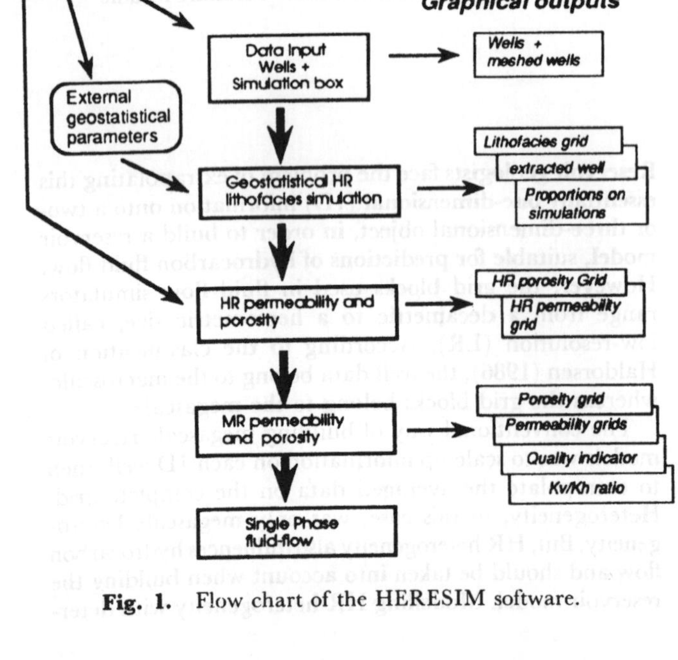


**Fig. 1.** Flow chart of the HERESIM software.

(3) HR porosity and permeability description, controlled mainly by the HR lithology;
(4) composition of HR porosity and permeability, in order to estimate parameters on an medium-resolution (MR) grid;
(5) single-phase fluid fow simulation through this MR grid to obtain effective permeability on the ultimate LR grid.

The user, either reservoir engineer or geologist, can easily step through each task, each time varying the parameters according to different assumptions. We scribe the software and illustrate an application on a fluviodeltaic reservoir. The first sets of parameters are based on data gathered in Yorkshire. There, the Middle Jurassic Ravenscar Group offers broad outcrops with geological characteristics similar to the Brent Group (Hancock and Fisher, 1981). The methodology used by our group to study these outcrops, focusing on the Scalby Formation, has already been presented (see Ravenne *et al.*, 1987, or Eschard *et al.*, 1987). It has resulted in a detailed sedimentary reconstruction.

# HIGH-RESOLUTION LITHO-FACIES DESCRIPTION OF A RESERVOIR

The reservoir is described by a regular grid composed of rectangular cells. In order to characterize the reservoir fully, we have to assign a single lithofacies to each cell. The reservoir is divided into one or several homogeneous units, deposited within the same environment. Inside each unit, lithofacies are identified by a geological study of the available well data.

The lithofacies are *conditionally simulated* between the wells, which means that:

(1) The reservoir description is not only an interpolation of lithofacies between wells, but the lithofacies are allowed to vary within a fixed range, given that they respect some parameters: in other words, they are simulated.
(2) The simulated lithofacies at the wells exactly fit the actual observed lithofacies: in other words, the simulation is conditioned by the wells.

## Geostatistical method used for conditional simulation

The reservoir is described using geostatistics, a mathematical method which assumes that geological phenomena can be described by random functions. These functions must respect a given correlation function (Matheron, 1971).

Lithofacies are coded as integers, thus, the reservoir is modelled as a random function that takes only integer values. We recap briefly the simulation method, which uses truncated Gaussian functions (see also Matheron *et al.*, 1987; De Fouquet *et al.*, 1988). All equations are written in a 2D horizontal–vertical section $X, Z$.

The main assumptions are that a sedimentary unit can be described by a random function *Lit* which has the following properties.

- It has a correlation structure following perpendicular and parallel lines to sedimentary horizons, i.e. whose covariance function is cov $(s, t) = f_1(s) f_2(t)$, where $s$ and $t$ are local coordinates and $f_1$ and $f_2$ are *ad hoc* functions.
- It is the transformation of a Gaussian function $G(0, 1)$ (i.e. of mean 0 and standard deviation 1),

according to Eq. (1). $G$ is called the *underlying Gaussian function.*

$$\mathrm{Lit}_{ij}=\sum_{k=1}^{n}\mathrm{Cod}\,(k)\,\mathrm{Ind}_{t_{k-1}<G_{ij}\leqslant t_k} \qquad (1)$$

where

$\mathrm{Lit}_{ij}$ is the lithofacies at cell $ij$
$\mathrm{Cod}(k)$ is the discrete code of lithofacies number $k$
$n$ is the total number of lithofacies
$\mathrm{Ind}_{a<G\leqslant b}$ is an *indicator* function with the values
1 if $a<G\leqslant b$
0 if $G\leqslant a$ or $G>b$
$t_k$ are the Gaussian thresholds (see Eq. (2))

The Gaussian function $G$ ranges from minus $\infty$ to $+\infty$. It has to be truncated into $n$ parts at threshold points (Fig. 2). The thresholds $t_k$ are computed with the proportions of each lithofacies along the correlation lines, solving:

$$\int_{-\infty}^{t_k}\frac{1}{\sqrt{2\pi}}e^{-x^2/2}\,dx=\sum_{l=1}^{k}\mathrm{Prop}(l) \qquad (2)$$

where Prop ($l$) is the ratio of lithofacies number $l$ along a correlation line. Thresholds computed with Eq. (2) ensure that the average proportion of lithology number $l$ in the simulated grid (i.e. $E(\mathrm{Ind}_{\mathrm{Lit}_{ij}=\mathrm{Cod}(k)})$ will be Prop ($l$).

The conditional simulation of the lithofacies is then based on the conditional simulation of the Gaussian function $G$. The spatial correlation of $G$ is deduced from the spatial correlation of function Lit. The simulation involves three steps (Fig. 3):

(1) transformation of the known lithofacies, coded as integers, into Gaussian values;
(2) simulation of Gaussian values between wells using Eq. (3);

$$G_{ij}=G^{*}_{\text{known points}}+\sigma U \qquad (3)$$

where

$G^*$ is the linear approximation of $G$ with all the points previously known, that minimizes $var(G^*-G)$ (called kriged value)

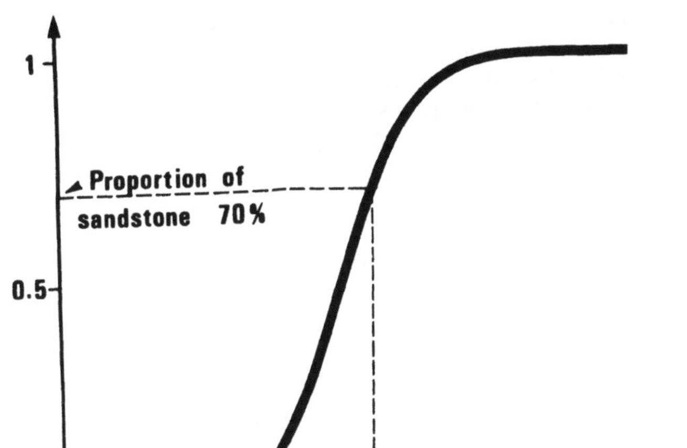


**Fig. 2.** Computation of Gaussian thresholds. The curve is the CDF function $F(x)=\int_{-\infty}^{x}1/\sqrt{2\pi}e^{-t^2/2}dt$. Supposing one has two lithofacies, sandstone and shale, with respective ratios of 70% and 30% ($Y$ axis), then the threshold $t_1$ is $F^{-1}$ (0.7).

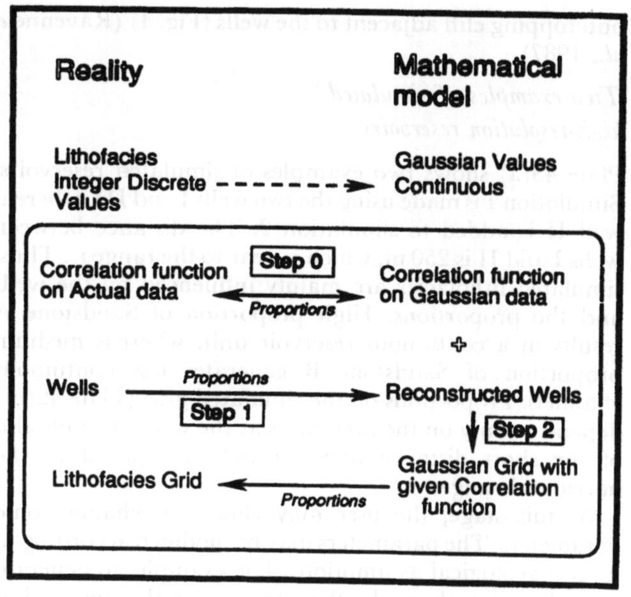


**Fig. 3.** Overview of the simulation method using Gaussian CDF and thresholds. The input parameters are horizontal proportions of lithofacies and correlation functions of the underlying Gaussian function. Both interact to give the correlation function of the lithofacies function.

$\sigma$ is the uncertainty on the approximation $G^*$
$U$ is a independent uniform Gaussian value

(3) transformation of generated Gaussian grid back into a lithofacies grid using Eq. (1).

The right-hand part of Eq. (3) is the sum of an interpolation term $G^*$ and of a random term $\sigma U$. As the interpolation $G^*$ is exact at the wells, the uncertainty $\sigma$ is zero. Thus, the computed value $G$ transformed with Eq. (1) gives the exact lithofacies and the simulation of *Lit* is conditional. The random term allows generation of various realizations of the same function $G$.

## A simulation based on actual data

### *The input data*

The lithofacies simulation is based on three wells I, II and II (Plate 42(a)) showing the Scalby Formation of the Ravenscar Group. The Scalby Formation has been divided into two sedimentary units (A and B), which are two fluviodeltaic sequences grading from sandstone to shale.

As the sediments form tabular units, the correlation lines have been chosen to be horizontal and vertical. In order to respect the vertical sequential evolution, facies proportions are allowed to vary vertically. They represent a global parameter that has to be computed from well data, or estimated from geological experience. The average horizontal proportions are computed from 8 out of the 36 available wells that have been drilled on this site (Plate 42(b)).

The analysis of correlation between lithofacies has shown that the covariance of the underlying function $G$ can be factorized into:

$$\mathrm{cov}(x,z)=\exp\left(-\frac{|x|}{r_x}\right)\exp\left(-\frac{|z|}{r_z}\right) \qquad (4)$$

Thus the ranges $r_x$ and $r_z$ are the only parameters used for the correlation function. These values represent the distance above which there is no correlation between lithofacies. They have been computed using available data on the outcropping cliff adjacent to the wells (Fig. 4) (Ravenne *et al.*, 1987).

*Two examples of simulated high-resolution reservoirs*

Plate 43(a) shows two examples of simulated reservoirs. Simulation 1 is made using the two wells I and III, whereas well II is added to simulation 2. The distance between wells I and II is 250 m, which is near to the range $r_x$. Thus, simulated lithofacies are mainly influenced by the wells and the proportions. High proportion of Sandstone A results in a continuous reservoir unit, whereas medium proportion of Sandstone B generates less continuous channels. Proportions on the simulated grid (Plate 42(c)) depend mainly on the lithofacies at the wells, a result also of the short distance between wells compared to the horizontal range.

At this stage, the user may choose to change some parameters. The parameters may be modified according to several geological assumptions. For example, to generate large but flat channels, the range $r_x$ of the correlation function may be increased, whereas $r_z$ is decreased. When the user is satisfied with his lithofacies description, he may go to the next step.

## HR POROSITY AND PERMEABILITY DESCRIPTION

### Porosity and permeability relationships as a function of lithology

Petrophysical properties are then generated on the previously defined lithofacies HR grid. Porosity is a scalar, so a single numerical value between 0 and 100% is generated for each cell. Permeability is a tensor, so in 2D, three numerical values representing *kx*, *kz* and *kxz* should be generated. Fortunately, field observations in Yorkshire have shown that the different types of sandy lithologies are homogeneous at the centimetric scale. At HR scale, the permeability tensor is therefore assumed to be diagonal and is modelled through a single value *k* that represents both *kx* and *kz*.

High-resolution porosity and permeability grids are built on the framework of the previously simulated lithofacies. The main assumption is that porosity and permeability depend strongle on the lithofacies of each cell. Porosity and permeability of each different lithofacies may also be modelled as stochastic functions, with a given correlation structure.

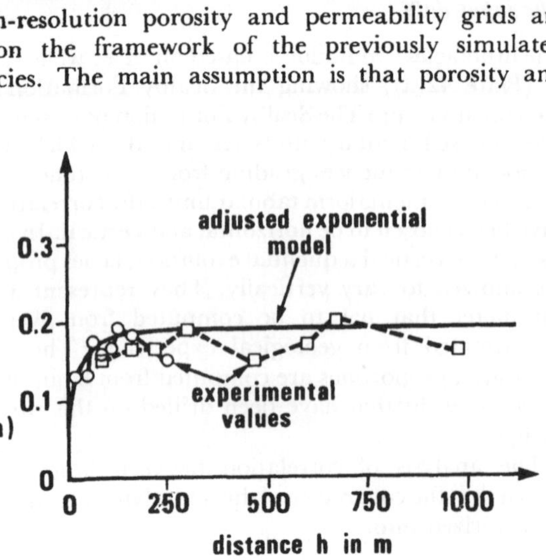


**Fig. 4.** Horizontal correlation of the integer function *Lit*. The full line represents the experimental correlation values, computed on two sets of values of Lit (squares: wide sampling, circles: close sampling); the dashed line represents the matched theoretical model. The range of the underlying covariance function is $r_x = 175$ m.

If the sedimentological description of the wells is accurate and discriminates sufficiently between several classes of lithologies, each lithofacies is homogeneous and has a unique porosity and permeability value. In this case, in a porosity–permeability plot, the data are clustered around separated permeability and porosity values, according to their lithofacies.

If there is significant variation of porosity and permeability within each lithofacies, spatially correlated porosity and permeability values are generated. If enough data are available, the software may compute the lithofacies specific porosity and permeability distribution functions and their covariance structure. A set of values based on these parameters can then be generated. This procedure is quite CPU time expensive and not really interactive. A more efficient method requires the further assumption that porosity distribution is normal and the permeability is log–normal. This assumption is based on field observation in Yorkshire.

### Simulation of porosity and permeability values

Porosity and permeability are efficiently simulated using the following method. A correlated Gaussian distribution $G_1$ (0, 1) is generated on the whole grid. Its covariance structure is also given by Eq. (4), but the ranges $r_x$ and $r_z$ may be different from the ranges taken for the lithofacies simulation. $G_1$ is then transformed using Eq. (5), in order to built the HR porosity grid.

$$\phi_{ij} = m(Lit_{ij}) + \phi(Lit_{ij})G_1 \qquad (5)$$

where

| | |
|---|---|
| $\phi_{ij}$ | is the porosity at cell $i\,j$ |
| $m(Lit_{ij})$ | is the mean porosity of the lithology of cell $i\,j$ |
| $\sigma(Lit_{ij})$ | is the standard deviation of the porosity of the lithology of cell $i\,j$ |

A second Gaussian distribution $G_2(0, 1)$ is then generated, with a correlation structure given by Eq. (4), though possibly with different ranges. In order to respect a given correlation between porosity and permeability, this distribution also satisfies Eq. (6):

$$\mathrm{cov}(G_1, G_2) = \frac{\mathrm{cov}(\phi_{ij}, \log k_{ij})}{\mathrm{var}_\phi \, \mathrm{var}_{\log k}} \qquad (6)$$

The right-hand side is computed experimentally from the field data, as are the means and standard deviations in Eq. (5). The value $G_2$ is then transformed, with an equation similar to 5, to get a permeability value.

### Two examples of simulations

The cores taken from the Scalby Member have been plugged every 0.25 m whenever possible. The air-permeability was then measured. In 7 out of 14 lithofacies, there were enough data to compute the standard statistical parameters (Fig. 5). The other facies were shaly and could not have been plugged easily. None-the-less, as their permeability is below 0.01 mD, unique values of 0.01 mD and 0% were given to each cell of these lithofacies.

Based on the lithofacies grids simulated above and on the means and standard deviations as shown in Fig. 5, HR porosity and permeability grids have been generated (Plate 43(b)). The covariance function used here is the

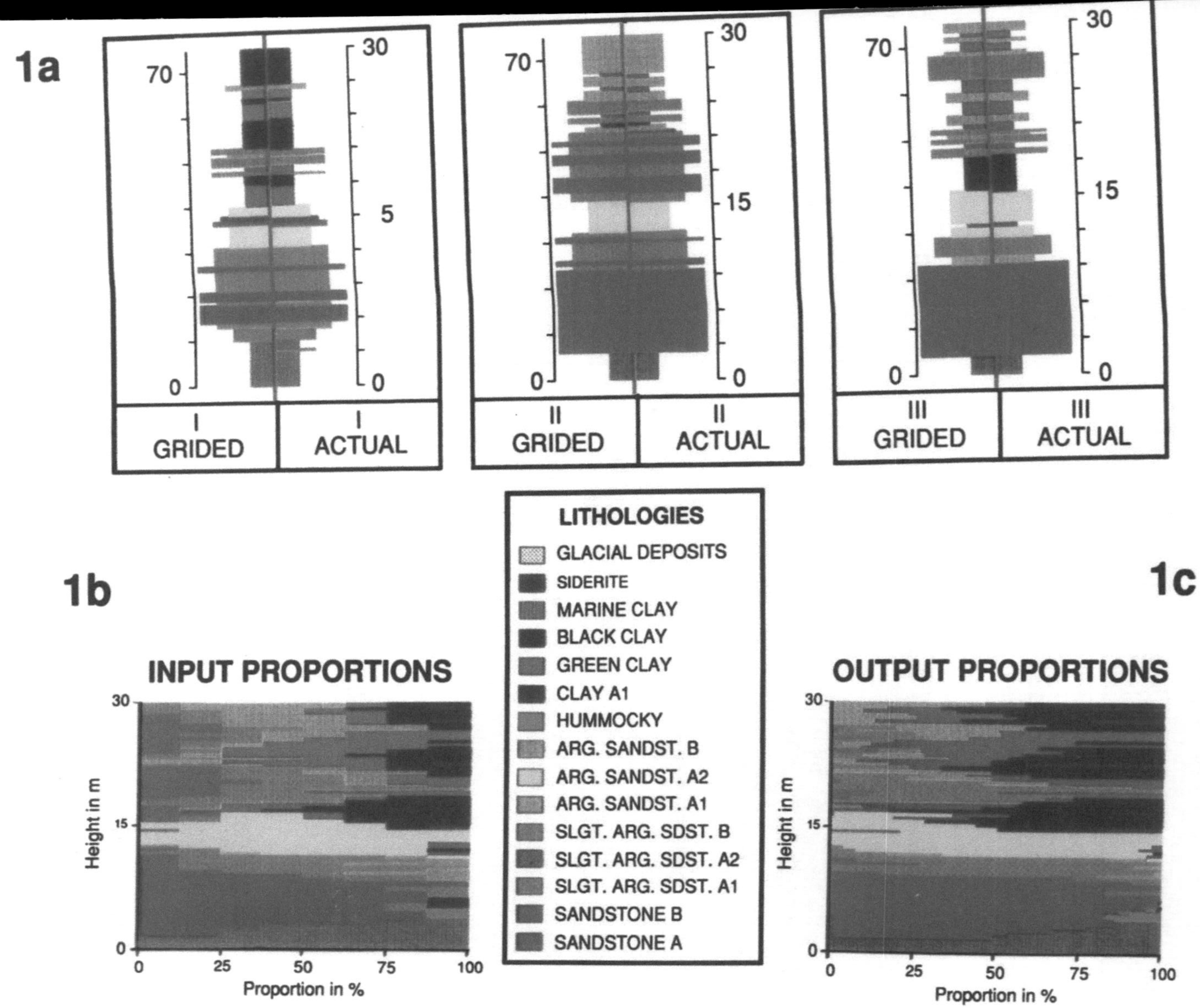


**Plate 42.** (a) **Input wells I, II and III:** On a total thickness of 30 m, the observations indicate two successive sedimentary sequences A and B. Both sequences grade from sandstone to shale; A is upper-estuarine and erodes the underlying marine Scarborough Formation, whereas B is lower-deltaic, with meandering channels. Depths are local to each well. (b) **Horizontal proportions of lithofacies along the correlation lines:** Given the lithofacies in each cell at the wells, their proportions are computed along the correlation lines. Here, these lines are horizontal, so along line number $p$, the ratio of lithofacies $l$ is:

$$\mathrm{Prop}(l) = \frac{1}{\mathcal{N}} \sum_{k=1}^{\mathcal{N}} Ind_{\mathrm{Lit}_{k,p}=l}$$

$\mathcal{N}$ being the number of cells in the $X$ direction and *Ind* the indicator function (see text). (c) **Lithofacies proportions calculated on a simulated grid**: These ratios are computed along the 250 horizontal cells of simulation 2 (Plate 43a). Compared to the input proportions, the sandy lithofacies ratios of unit B are reduced, owing to the absence of these facies in the wells.

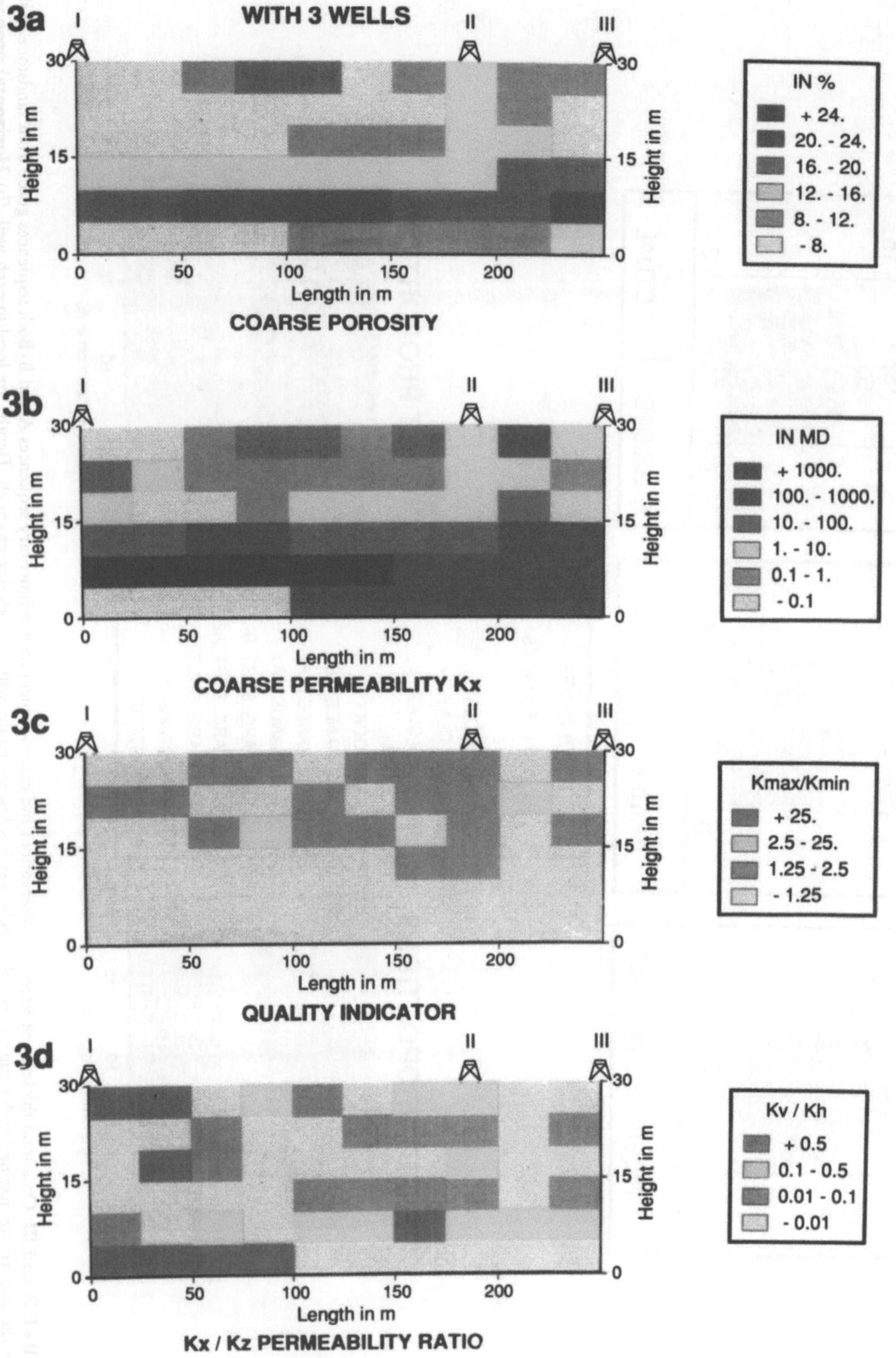


**Plate 44.** Results on the MR grid. In all pictures, the grid size is 10 cells in $X$ and 6 in $Z$, resulting in a cell size of 25 m × 5 m. Here, $SAF_1$ is greater than 300. (a) **Porosity:** Averaged porosity is easily calculated on the MR grid, as porosity is an additive scalar. (b) **Averaged permeability in *X* direction:** Algebraic formulae give an estimation of averaged permeability. Analogous results could be obtained for the vertical permeability. (c) **Quality indicator:** The quality indicator helps the HERESIM user to select the suitable MR grid. (d) **Anisotropy factor:** This factor is the ratio of the vertical averaged permeability versus the horizontal one.

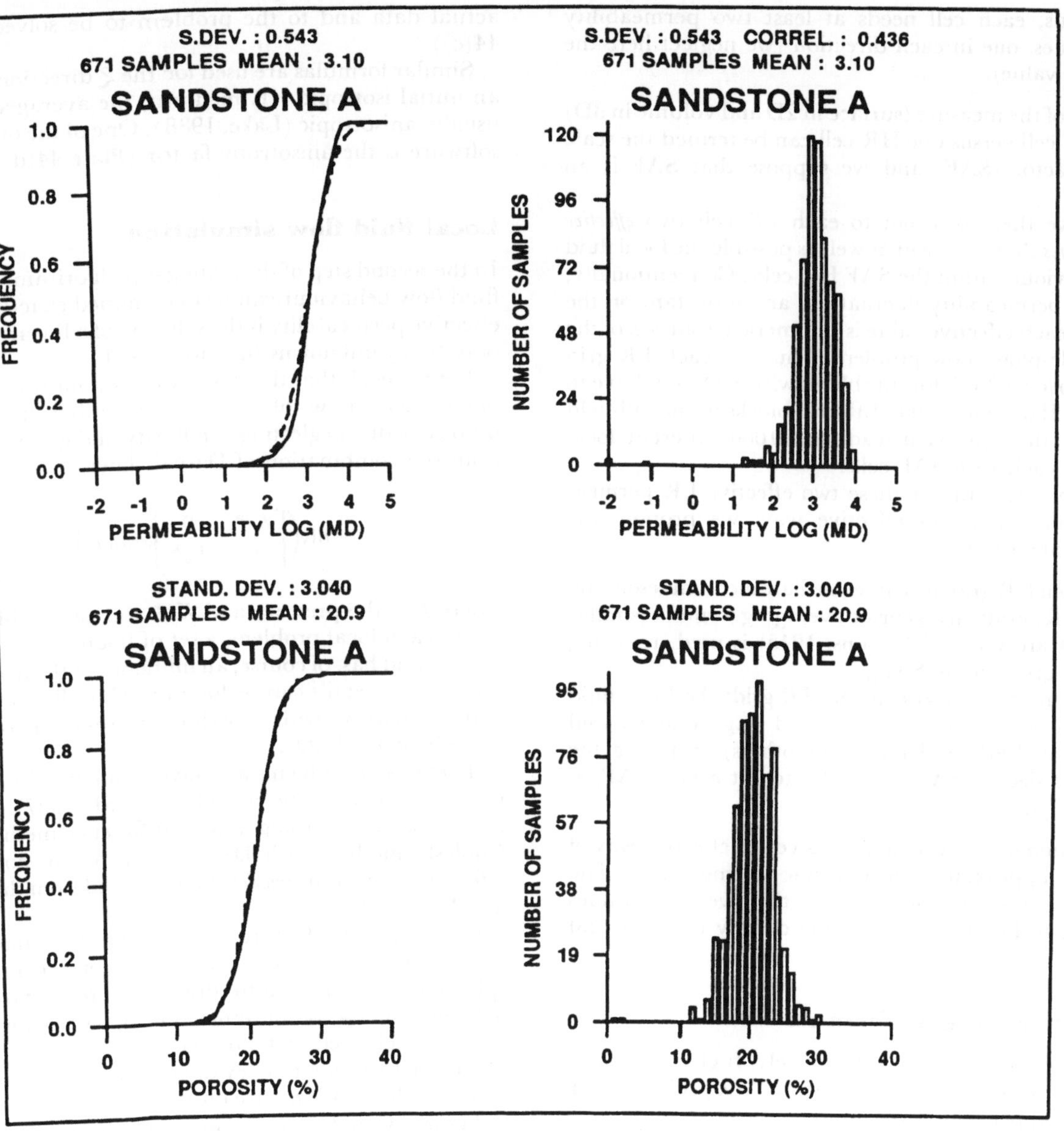


**Fig. 5.** CDF and histograms of porosities and permeabilities. Full lines are the cumulated distribution functions (CDF) for permeability and porosity in lithofacies Sandstone A. Dashed lines are the theoretical Gaussian CDF with means and standard deviations indicated above the charts. Porosities are normal and permeabilities log–normal.

same as the covariance function used for the lithofacies, because petrophysical properties are controlled by the depositional environment. Petrographic analysis shows that porosity and permeability depend mainly of the amount of syngenetic clay minerals trapped in the sandstones. The ranges would be different if diagenesis had a stronger and more variable influence on petrophysical properties. At this stage the total porous volume of each lithofacies is easy to compute, giving an estimate of the average reserves.

## SCALING UP HR PORHCITY AND PERMEABILITY

### Scale averaging problem

The previous step leads to a HR characterization of the reservoir. Unfortunately, existing computers cannot simulate fluid flow directly on HR grids, because they are not able to handle such a quantity of data. Even in 2D, with a cross-section of 1000 m length by 100 m height, and a HR cell of $1 \times 0.1$ m, the number of HR grid blocks reaches 1 million! So, for reservoir engineering purposes, it is fundamental to be able to reduce the HR infomation to obtain effective parameters useful for fluid flow reservoir modelling. Remember that the usual size of LR grid blocks in a full-field simulator is about $100 \times 100 \times 10$ m.

Reduction of grid size for additive values such as fluid volumes is straightforward (Plate 44(a)), but it is more complex for non-additive parameters such as absolute permeability, relative permeabilities and capillary pressure curves.

Here, only the absolute permeability averaging procedure will be considered, assuming that the fluid flow can be reduced to a 2D problem.

### Averaging the HR absolute permeability

Let us consider two grids at two different scales:

(1) the HR grid, where each cell has an isotropic permeability value;
(2) the reservoir simulator (LR) grid. At this scale, permeability cannot be assumed to be isotropic.

Thus, each cell needs at least two permeability values, one in each direction (we neglect here the $K_{xz}$ value).

The ratio of the measure (surface in 2D and volume in 3D) of one LR cell versus one HR cell can be termed the scale average factor (SAF) and we suppose that SAF is an integer.

We have then to assign to each LR cell two *effective permeabilities* that represent as well as possible the local fluid flow behaviour within the SAF HR cells. Conventionally, when the permeability fluctuations are important on the HR grid, each effective value is a numerical solution of the local single-phase flow problem. Thus, for each LR grid block, two local fluid flow problems with SAF cells have to be solved. If one has 10 000 LR grid blocks in the full-field simulator, this approach leads to 20 000 different local problems, each with SAF cells!

Therefore, to estimate these two effective LR permeability values in a cost effective way, we propose the following algorithm.

(1) From HR grid to an intermediate medium resolution (MR) grid: an algebraic averaging formula (similar to Cardwell and Parsons, 1945) is used, reducing the grid size by $SAF_1$;
(2) From this *MR* grid to the *LR* grid: the local fluid flow is numerically simulated (e.g. Warren and Price, 1961, and numerous others), reducing the grid size by $SAF_2$ (in order to get $SAF = SAF_1 \times SAF_2$).

Combination of the two methods gives an effective way of obtaining approximate megascopic permeability. The user can adjust the precision of the averaged values according to his reservoir and the quality of the actual data.

## Algebraic averaging formula

The first step is carried out interactively. It gives estimates of the effective $Kx$ and $Kz$ values on the MR grid, together with their upper and lower limits.

Let us consider a regular $m \times n(=SAF_1)$ square grid in $X$, $Z$ and $k_{ij} (i=1, m \text{ and } j=1, n)$ the HR permeability. Given some assumptions, the average permeability $Kx$ in this grid is between the two limits $Kx_{min}$ and $Kx_{max}$:

$$Kx_{min} = \frac{m}{n} \sum_{j=1}^{n} \left( \sum_{i=1}^{m} k_{ij}^{-1} \right)^{-1} \quad (7)$$

$$Kx_{max} = \frac{m}{n} \left[ \sum_{i=1}^{m} \left( \sum_{j=1}^{n} k_{ij} \right)^{-1} \right]^{-1} \quad (8)$$

The coarser the MR grid, the wider the range between these limits. The estimated average permeability $Kx^*$ in this grid is then between the two limits and taken as follows (plate 44(b)):

$$Kx^* = \sqrt{Kx_{min} Kx_{max}} \quad (9)$$

The estimate is correct for specific geometries of the LR cells (in series or in parallels) and can be extended in 3D with irregular cell sizes. For other geometries, its accuracy can be tested with the quality indicator.

The quality indicator $QUx(m, n)$ is defined as:

$$QUx(m,n) = \frac{Kx^*}{Kx_{min}} = \frac{Kx_{max}}{Kx^*} = \sqrt{\frac{Kx_{max}}{Kx_{min}}} \quad (10)$$

To find a suitable MR grid, the user can interactively try several reduction factors, until $QUx(m, n)$ and $QUz(m, n)$ are considered small enough according to the quality of the actual data and to the problem to be solved (see Plate 44(c)).

Similar formulas are used for the $Z$ direction. Even with an initial isotropic HR medium, the averaged LR one is usually anisotropic (Lake, 1988). One of the outputs of the software is the anisotropy factor (Plate 44(d)).

## Local fluid flow simulation

In the second step of the scaling-up algorithm, the average fluid flow behaviour can be determined exactly. Now, the effective permeability is the solution of a linear system with only $SAF_2$ unknowns instead of SAF.

Let us recall that the basic equation for the steady-state monophasic flow of an incompressible fluid through porous media, neglecting capillarity and gravity, is derived from the combination of Darcy's law:

$$-\text{div}\left( \begin{bmatrix} Kx^* & 0 \\ 0 & Kz^* \end{bmatrix} \text{grad } P \right) = 0 \quad (11)$$

where $P$ is the pressure and $Kx^*$ the permeability.

For each local problem, a set of boundary conditions is needed and has, of course, an influence of the average fluid flow. More details can be found in Guerillot *et al.* (1989), and comparison with experimental results is reported in Jacquin *et al.* (1989).

These local problems are solved numerically with a 2D reservoir stream tube model (Renard, 1987). Equation (11) is discretized using a finite-difference method (direct Choleski method with D4 ordering of the nodes). The effective permeabilities are then deduced from the pressure at all grid centres.

One important feature of the stream tube model is the automatic and precise determination of streamlines that split the flow area into drainage areas. Streamlines, defined as the loci of constant values of the stream function, are paths followed by fluid "particles" as they move from one place to another. They divide the total flow area into "stream channels" or stream tubes.

Figures 6 and 7 show two examples of stream tubes respectively obtained on the HR and MR grids. In the HERESIM software, fluid flow is usually calculated only on the MR grid (Fig. 7). The first fluid flow simulation has been done here only to compare it with the second one. Fluid goes mainly through the second tube (dotted part). As this tube looks similar on both figures, one does not lose much information using 60 cells instead of 18 750!

The LR grid can easily be used to simulate fluid flow of a full-field reservoir.

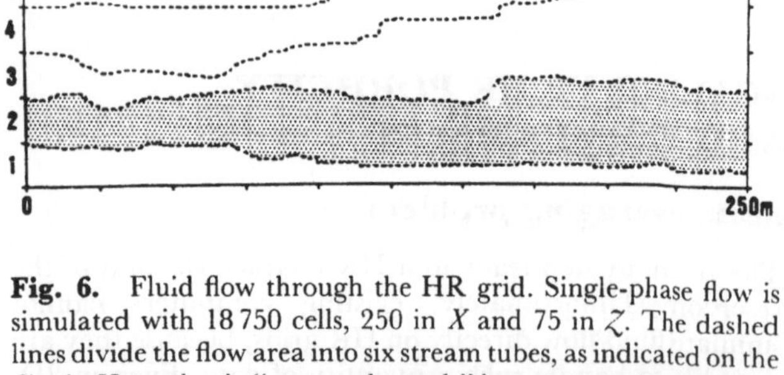


**Fig. 6.** Fluid flow through the HR grid. Single-phase flow is simulated with 18 750 cells, 250 in $X$ and 75 in $Z$. The dashed lines divide the flow area into six stream tubes, as indicated on the $Z$ axis. Here, the six "stream channels" have very contrasted flow rates. Fluid goes mainly through tube 2 (dotted area).

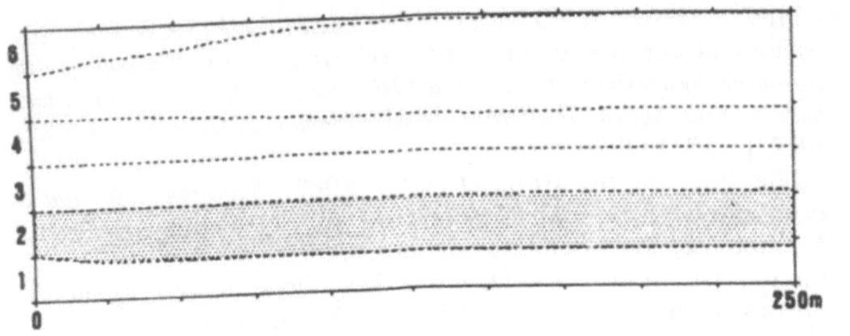


**Fig. 7.** Fluid flow through the MR grid. Single-phase flow is simulated here with only 60 cells, 10 in $X$ and 6 in $Z$. As in Fig. 6, fluid goes mainly through tube 2. The streamlines are less curved because we have neglected cross-terms $Kxz$ as usual in reservoir simulation.

## CONCLUSIONS

A precise characterization of reservoir heterogeneity is needed to perform correct fluid flow simulations. As a deterministic description of a reservoir seems impossible, owing to the high number of possible parameters involved, a stochastic description of the reservoir is a better solution. Based on this concept, an integrated software designed to characterize a reservoir has been developed and tested on experimental data from the Middle Jurassic Ravenscar Group in Yorkshire.

Here, the reservoir is described as one or more geologically homogeneous units (i.e. deposited in the same environment). Its stochastic parameters (correlation function and global lithofacies ratios) can be determined either from geological assumptions or from field data. The reservoir lithofacies is computed on a high-resolution (i.e. decimetric) grid. According to the previous parameters, a description is made with a geostatistical method, using truncated Gaussian functions. It is consistent with geological knowledge, depending on the depositional environment of the reservoir. High-resolution porosities and permeabilities related to the lithofacies are then produced. Averaged values of permeability, suitable as input data in reservoir fluid flow simulators, are computed in two steps. First, a medium resolution grid is computed, using algebraic approximations, in a CPU time-effective way. Second, computation of the monophasic fluid flow on this medium-resolution grid gives averaged permeabilities over larger grids.

Integrated software provides the advantage of being flexible and interdisciplinary. It is flexible, because each parameter can be easily modified, so that different assumptions, relying on different values of parameters, may be rapidly tested*. It is interdisciplinary, because the joint experience of geologists and reservoir engineers is needed in order to use it efficiently.

The stochastic techniques presented in this paper are based upon joint use of correlation functions and proportions. They allow rapid generation of different scales of heterogeneity, linked to the depositional environment of each reservoir unit. It a marine environment, the horizontal ranges, for example, will differ greatly from the ranges in a meandering fluviatile system. The required input parameters should be based on detailed sedimentological studies. Our stochastic methods provide new quantitative tools to the reservoir geologist: correlation functions and lithofacies proportions. They also help to put some quantitative of thinking into geological studies.

As with all stochastic simulation methods, HERESIM is only able to represent what reality could be, not what it is. In order to appreciate stochastic variations, several simulations should be run and analysed. In conventional statistical studies, the statistical parameters should be infered from large data sets. In reservoir studies, large data sets are usually available only when enough wells have been drilled. None-the-less, HERESIM can be applied in the apraisal stage of a field, in order to test different geological assumptions. For example, the hydrocarbon reserves may be computed as a function of the size of sandstone bodies, modelled with various given proportions and correlation length. But HERESIM can also be applied to a producing field, where the input parameters can be directly computed from the field data. It rapidly generates an updated full-size reservoir grid each time a new well is drilled.

HERESIM is at present an integrated 2D software. Fluid-fow behaviour is rarely a simple 2D problem, and 3D data should be processed. The stochastic methods are not restricted to 2D and have already been applied to simulate 3D lithology (De Fouquet *et al.*, 1988). However, storage requirements on computers limit the number of grid cells that can be simulated. Very heterogeneous reservoirs should be modelled through various 2D cross-sections. Porosity and permeability should then be scaled up and interpolated in the conventional way, until a full 3D software becomes available.

*On a 8650-VAX computer, building a 1000 × 200 HR grid takes less than 15 s of CPU- time.

## REFERENCES

Cardwell, W.T. and Parsons, R.L. (1945). Average permeabilities of heterogeneous oil sands, *Trans. Am. Inst. Min. Metall. Petroleum. Eng.* **160**, 34–42.

De Fouquet, C., Beucher, H., Galli, A. and Ravenne, C. (1988). Conditional simulation of radom sets. Application to an argilaceous sandstone reservoir, *Proceedings of the Third International Geostatistics Congress, Avignon.*

Eschard, R., Houel, P., Mathieu, Y., Ravenne, C. and Rudkiewicz, J.L. (1987). Géométrie de réservoirs fluvio-deltaiques dans les séries du Jurassique moyen du Yorkshire. *1er. Congrès Français de Sédimentologie, Paris.*

Guérillot, D., Rudkiewicz, J.L., Renard, G. and Galli, A. (1989). An integrated model for computer aided reservoir description: From outcrop study to fluid flow simulations. *Fifth European Symposium On Improved Oil Recovery, Budapest, Hungary, April, 1989.*

Hancock, N.J. and Fisher, M.J. (1981). Middle Jurassic North Sea Deltas with particular reference to Yorkshire. In: Illings, L.V. and Hobson, G.D. (ed.), *Petroleum Geology of the Continental Shelf of North West Europe*, pp. 186–195. Heyden, London.

Haldorsen, H.H. (1986) Simulator parameter assignment and the problem of scale in reservoir engineering. In: Lake L.W. and Carrol. M.B., Jr. (eds.), *Proceedings of the Reservoir Characterization Technical Conference, Dallas*, pp. 293–340. Academic Press, New York.

Haldorsen, H.H. and Macdonald, C.J. (1986). Stochastic modelling of Underground reservoir facies (SMURF). Paper no. SPE 16751.

Jacquin, Y., Henriette, A., Guérillot, D., Adler, P.M. (1989). Heterogeneity and effective permeability of porous rocks—experimental and numerical investigation. *Second International Reservoir Characterization Technical Conference, June, 1989, Dallas.*

Lake, L.W. (1988). The origins of anisotropy. *J. Petroleum Technol.* (April), 395–396.

Matheron, G. (1971. The theory of regionalized variables and its applications. *Les Cahiers du Centre de Morphologie Mathématique de Fontainebleau*, Ecole Nationale Supérieure des Mines de Paris, France.

Matheron, G., Beucher, H., de Fouquet, C., Galli, A., Guerillot,

D., Ravenne, C. (1987). Conditional simulation of the geometry of fluvio-deltaic reservoirs. Paper no. SPE 16753.

Omre, H., Halvorsen, K.B., Holden, L. and Høiberg S. (1988). Reservoir heterogeneity, geological description and effects on fluid flows. *Models for Heterogeneity; Proceedings from SPOR-Seminar.* Norwegian Petroleum Directorate, Stavanger, Norway.

Ravenne, C., Eschard, R., Galli, A., Mathieu, Y., Montadert, L., Rudkiewicz, J.L. (1987). Heterogeneities and geometry of sedimentary bodies in a fluvio- deltaic reservoir. Paper no. SPE 16752.

Renard G. (1987). A 2D Reservoir stream tube EOR model with periodical automatic generation of streamlines. *Proceedings 4th European Symposium on Enhanced Oil Recovery, Badische Anilin und Soda Fabrik Aktien Gesellschaft et al., Hamburg, Germany, October, 1987*, pp. 711–722.

Scholle, P.A. and Spearing, D. (ed.) (1982). *Sandstone depositional environments.* American Association of Petroleum Geologists Memoir no. 31.

Warren, J.E. and Price H.S. (1961). Flow in heterogeneous porous media, *Soc. Petroleum Eng. J.* **1**, 153–159.

# 26

# How Geostatistics Helps Decisions on Complementary Development

*O. Ricard*

ELF Aquitaine, Paris, La Défense, France

**The economic advantages of additional wells are strongly related to the remaining reserves and consequently to a precise knowledge of the range of OGIP. Direct measures of the influence of structural mapping and heterogeneity on the OGIP can be made using statistical or geostatistical concepts in a probabilistic evaluation. This knowledge helps in deciding development and study policy. There are two conventional approaches to the probabilistic evaluation of OGIP—Monte Carlo simulations on average values of parameters, or geostatistical conditional simulations. The first method gives pessimistic results when it is used on the subjective expectation curves of the main parameters. Conditional simulation needs powerful computer resources and gives results whose physical meaning is not easy to control. Both methods are global because they give only a global value of uncertainty without giving any information on the origin of this uncertainty. In particular, they do not compute directly the influence of heterogeneity on the OGIP uncertainty. Consequently, these methods are not well adapted for deciding what parameters must be specifically studied to reduce uncertainty. For these reasons, a third method has been developed by ELF for the Frigg field evaluation which is particularly adapted to the computation of uncertainty on developed fields. This method, which uses conventional notions of geostatistics, computes only the expected value and variance of Frigg OGIP, and the distribution of the various contributions to this variance. It has been shown that the global uncertainty on Frigg OGIP is small and that the influence of heterogeneities on the OGIP uncertainty is much smaller than the influence of structural maps and log interpretations.**

## INTRODUCTION

Two years ago, the "most probable" value of OGIP of Frigg was reduced owing to new seismic data, new interpretation of logs and new maps of petrophysical data. Before this period, the activity of a strong aquifer was determined by measuring the water rising in some wells and the corresponding pressure decline. In this type of problem, the remaining undeveloped reserves of gas are reserves that are going to be bypassed and trapped behind the water front. For a given position of the dynamic gas–oil contact (GOC) and a given value of residual saturations, the bigger the OGIP, the more important is the potential for trapped and undeveloped unswept reserves. Thus, the need for complementary development of the field is strongly dependent on the value of the OGIP. The validity of the decision for this development is closely linked to *the quality of the evaluation of the uncertainty of the most recent OGIP computation.*

*North Sea Oil and Gas Reservoirs-II*


## OBJECTIVE OR SUBJECTIVE EVALUATION OF UNCERTAINTY

The OGIP uncertainty evaluation is generally made before the main development of a field. As data are generally few and poor, this evaluation belongs to the subjective type, i.e. using either a *Monte Carlo* or a *probabilistic* aggregation of *subjective expectation curves* on each parameter of the OGIP computation. Elf's experience is that computed absolute range of the final OGIP is generally wide because the subjective estimates of expectation curves are always pessimistic. However, it can lead to a good decision because the *relative range of remaining reserves* $R=$ (possible–proven)/(probable) is relatively small.

As the gas is produced, the probable remaining reserves will decrease but the range of remaining reserves (possible–probable) will remain fairly constant as long as nothing is done to change the uncertainty evaluation method. Therefore, towards the end of the production, the ratio $R$ will increase and a high $R$ means that a decision for complementary development is hazardous.

The range of remaining reserves (possible–probable) cannot be sufficiently reduced by dynamic considerations: as usual, dynamic studies give very poor estimates of OGIP when an active—but unknown—aquifer exists.

Fortunately, after 10 years of production, new studies on complementary appreciation wells and seismic surveys give a lot of new data that allow an *objective* computation of uncertainty on OGIP, i.e. an evaluation based on *statistical* and *geostatistical* analyses of data.

## PRINCIPLES OF GEOSTATISTICS

The geostatistical approach uses two main steps: *structural analysis* and *kriging*.

- During the *structural analysis*, the structure of regional spot values of a parameter is studied with several statistical analysis tools, mainly the *experimental variogram* and the *automatic recognition of drift* (G. Matheron, 1973; D. Renard 1985)
- *Kriging* of spot data uses the *structure of data* given by the structural analysis to compute several "direct products" and "indirect products". Among direct products of kriging are:
  (1) the kriged map of the parameter;
  (2) the map of spot kriging variance (i.e. the map that gives, at any point, the variance of the difference between the kriged computed value and the unknown reality);
  (3) the kriged value of the average value of the parameter on a domain (for example, the gas-bearing zone of the field);
  (4) the kriging variance value of the average value of the parameter on a domain (i.e. the value of the variance of the difference between the kriged average value of the parameter on a domain and the unknown true average value on the same domain)

*Conditional simulation techniques* (G. Matheron 1973, P. Delfiner and J.P. Chiles 1977), are indirect products of kriging that use the kriging weights to generate realizations of the random function of each regional parameter.

## GEOSTATISTICS AND UNCERTAINTY COMPUTATION

The uncertainty due to mapping can be estimated either by the use of *kriging variance* or by the use of *conditional simulations*. In OGIP evaluations, the advantage of the conditional simulation approach was conventionally to take into account the ability of the hydrocarbon bearing surface to vary. Use of conditional simulation conventionally gives a mean value and a variance that take into account this ability. This technique has been used for more than 12 years by Elf, which developed the SIMPACK program with CGMM (Centre de Geomorphologie Mathematique de Fontainebleau).

However, according to Elf's experience, the conditional simulation technique has several practical drawbacks compared to the kriging variance technique as follows.

(1) Sometimes, the differnce between the median values given by the deterministic and the conditional simulation approaches cannot be explained only by the ability of the hydrocarbon-bearing zone to vary or by the correlations between parameters. This difference can be very large (on the Neocomian level of Elf Aquitaine's fields in the Paris basin, the deterministic value of OGIP was out of the range given by the conditional simulation method, in spite of the large number of wells drilled (166)). It can be due either to an inadequate use of geological information in the geostatistical approach or to methodological biases (like the use of hydrocarbon thicknesses—even after a Gaussian transformation—as the parameter to be simulated). When these differences exist, and as the deterministic value is generally chosen as the "right value", managers do not know how to use the expectation curve and the range given by the conditional simulation.

(2) If conditional simulation does not give the "physical meaning" of the mean value, then neither does it give the "physical meaning" of the computed uncertainty. It computes only a global value of uncertainty without providing any information on the origin of this uncertainty.

(These drawbacks exist not only for conditional simulation but also for all the methods based on Monte Carlo aggregation.)

(3) Conditional simulation on a reservoir corresponds to an extensive study generally performed after the deterministic computation and by people other than engineers involved in this deterministic computation. Owing to the generalization of the uncertainty computation, this computation must correspond to a small study, which has to be made at the same time as the deterministic estimation and by the same people.

Consequently, ELF Reservoir Department has tried to create a new method (PACEVAH: Procedure Analytique de Calcul de l'Esperance et de la Variance d'une Accumulation d'Hydrocarbure) which builds statistical results (like mean value or standard error) *around a* median value of OGIP that is easy to compare with the deterministic value. It attempts to improve the use of the direct products of kriging to estimate directly (1) the correction term to be added to the median value that will give the mean value, and (2) the variance on OGIP, without using conditional simulations or Monte Carlo aggregation. The principles of this method are given below, with the specific data available for the Frigg calculation.

Note that the PACEVAH method is not a general mathematical method: it uses certain approximations that are well adapted only for evaluating correctly appraised reservoirs.

## PRINCIPLES OF COMPUTATION OF MEAN VALUE AND VARIANCE OF OGIP IN THE PACEVAH METHOD

The program BLUEPACK® (D. RENARD, 1985) was used for all the kriging computation. The mean value of the OGIP and the standard deviation of this mean value was computed in two different steps.

### Computation of the mean value of OGIP

The mean value of OGIP is equal to the sum of a principal term OGIP1 and a corrective term $\Delta 1$:

$$E(\mathrm{OGIP}) = \mathrm{OGIP1} + \Delta 1 \qquad (1)$$

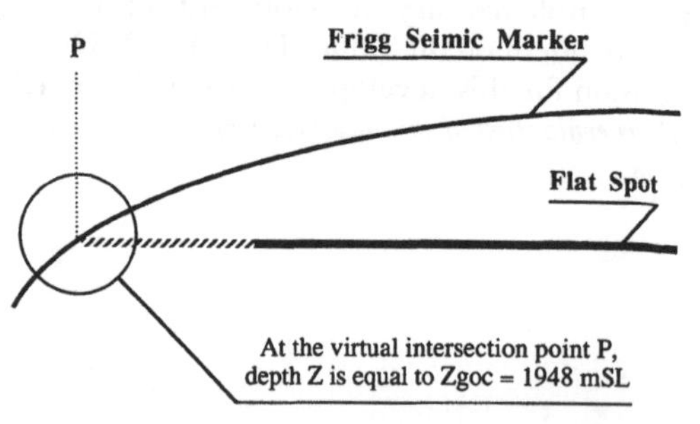


**Fig. 1.** Use of flat spot limit to match the FSM depth.

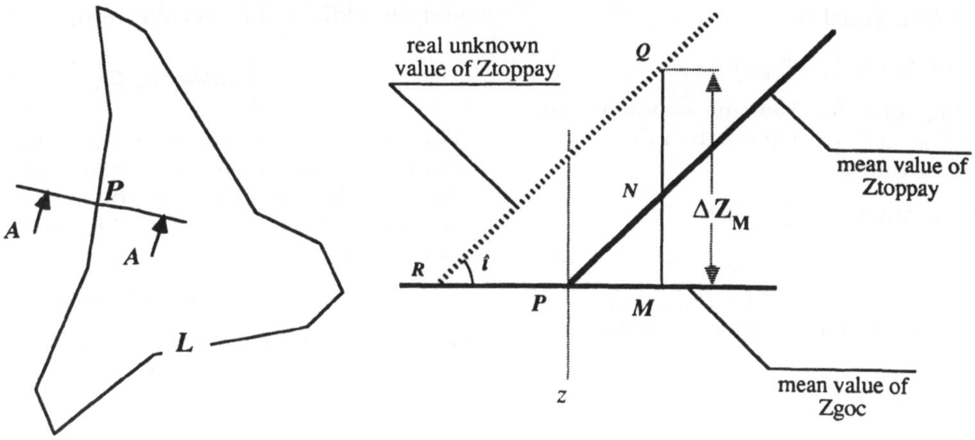


**Fig. 2.** Boundary of gas-bearing zone and overstepping correction.

*Main term computation*

The principal term OGIP1 is equal to the integral on the most probable gas bearing surface $\hat{S}$ of the product of the "mean maps" of *Ht*, *Alpha*, *Sg* and $\phi$:

$$\text{OGIP1} = \int_{\hat{S}} Ht_k \cdot Alpha_k \cdot \phi_k \cdot Sg_k \, ds \qquad (2)$$

$\hat{S}$ — is the most probable gas bearing surface inside the most probable boundary $\hat{L}$ defined as the intersection of the most probable depth of the top pay zone $Z_{tp}$ and the average depth of the gas–oil contact $\overline{Z_{GOC}}$.

$Alpha_k$ — is the map of the ratio:

$$\frac{\text{(Net thickness)}}{\text{(Gross thickness)}} = \frac{Hu}{Ht}$$

This map is computed by kriging the *Alpha* average values in the gas bearing zone at the 15 wells that have crossed this gas-bearing zone $(\{Alpha_i\}_{i=1,15})$, which can be written

$$Alpha_k = K(\{Alpha_i\}_{i=1,15}) \qquad (3)$$

$\phi_k$ — is the map of porosity of net thickness, which is computed by kriging the average values of $\phi$ in the gas-bearing zone at the 15 wells that have crossed this gas-bearing zone $(\{\phi_i\}_{i=1,15})$,

$$\phi_k = K(\{\phi_i\}_{i=1,15}) \qquad (4)$$

$Sg_k$ — is the map of gas saturation on net porous thickness, which is computed by kriging the average value of $S_g$ in the gas-bearing zone at the 15 wells that have crossed this gas-bearing zone $(\{Sg_i\}_{i=1,15})$,

$$Sg_k = K(\{Sg_i\}_{i=1,15}) \qquad (5)$$

$Ht_k$ — is equal to:

$$Ht_k = Z_{tpk} - Z_{GOC} \qquad (6)$$

$Z_{GOC}$ — is the *GOC* average depth, equal to 1948 m *SL*

$Z_{tpk}$ — is the depth map of the top pay zone, equal to

$$Z_{tpk} = T_{fsm} \cdot V_{fsm} + \Delta Z_{fsmk} + U0_k \qquad (7)$$

$T_{fsm}$ — isochron seismic map of the FSM (Frigg Seismic Marker), directly obtained from seismic interpretation (i.e. without correction at wells)

$V_{fsm}$ — isovelocity migration map of the FSM is a purely seismic datum deduced from the stacked velocities. The product $T_{fsm} \cdot V_{fsm}$ is a purely seismic depth because no well information at wells is used to compute it.

$U0_k$ — is the map of the "unit zero" level. This unit zero level separates the seismic marker from the top pay zone. This map is computed by kriging the *U*0 thicknesses at wells $\{U0_i\}_{i=1,34}$. The number of wells considered, which is equal to 34, is greater than the number of wells at Frigg because as *U*0 is a regional phenomenon, which is independent of the presence of gas, all the wells in the neighbourhood of Frigg have to be used.

$$U0_k = K(\{U0_i\}_{i=1,34}) \qquad (8)$$

$\Delta Z_{fsmk}$ — is the map of the corrections of the FSM seismic depth, which is computed by kriging the difference between the seismic depth and the true depth:

- at 18 real wells where complete sonic logs have been performed $\{DZ_{fsmt,i}\}_{i=1,18}$
- at 85 virtual depth control points at the boundary of the flat spot: $\{\Delta Z_{fsmv,i}\}_{i=1,85}$. As the original gas–oil contact is the same all over the field, the virtual intersection *p* of the Frigg Seismic Marker and the flat spot observed on a migrated seismic section corresponds to the depth $Z_{GOC} = 1948$ m *SL* (Fig. 1).

$$\Delta Z_{fsmk} = K(\{\Delta Z_{fsmt,i}\}_{i=1,18}, \{\Delta Z_{fsmv,i}\}_{i=1,85}) \qquad (9a)$$

The computed median value OGIP1 is about 2.3% greater than the value $\text{OGIP}_D$, computed by a totally deterministic approach using a complete geological description of the reservoir, based on a stratigraphical seismic interpretation. (This small difference is due to the fact that stratigraphical seismic interpretation shows some levels of laminated shales in zones without any wells).

*Corrective "overstepping term" computation*

It is possible to demonstrate that the value OGIP1, which is generally considered to be the most probable value of OGIP, is lower than the average value $\overline{\text{OGIP}}$ because of the boundary and overstepping effect. In the computation of OGIP, the value of the gross thickness *Ht* is

$$Ht = \text{MAX}(Z_{tp} - Z_{GOC}, 0) \qquad (9b)$$

The mean value of $Ht$ is equal to

$$\overline{Ht}=E(\text{MAX}(\mathcal{Z}_{tp}-\mathcal{Z}_{GOC},0)) \quad (9c)$$

In the gas-bearing zone, far from the boundary, the probability $P(\mathcal{Z}_{tp}<\mathcal{Z}_{GOC})$ of $\mathcal{Z}_{tp}$ being less than $\mathcal{Z}_{GOC}$ is very small, thus Eq. (9c) becomes

$$\overline{Ht}=E(\text{MAX}(\mathcal{Z}_{tp}-\mathcal{Z}_{GOC},0)) \approx E(\mathcal{Z}_{tp})-E(\mathcal{Z}_{GOC})=\overline{\mathcal{Z}_{tp}}-\overline{\mathcal{Z}_{GOC}} \quad (9d)$$

Near the boundary, the probability $P(\mathcal{Z}_{tp}<\mathcal{Z}_{GOC})$ of $\mathcal{Z}_{tp}$ being less than $\mathcal{Z}_{GOC}$ is not so small, thus Eq. (9c) becomes

$$\overline{Ht}=E(\text{MAX}(\mathcal{Z}_{tp}-\mathcal{Z}_{GOC},0)) > E(\mathcal{Z}_{tp})-E(\mathcal{Z}_{GOC})=\overline{\mathcal{Z}_{tp}}-\overline{\mathcal{Z}_{GOC}} \quad (9e)$$

As the computation of OGIP1 uses Eq. (6), which is similar to Eq. (9d), we are sure that, because of the boundary effects given by Eq. (9e), the true mean value $\overline{\text{OGIP}}$ will be greater than OGIP1.

The corrective term $\Delta 1$ can be estimated using the cross-section AA at any point P of the boundary of the gas-bearing zone. In Fig. 2, in this vertical plane AA, the Expected Value $E(SMQR)$ of the surface SMQR of the triangle $MQR$ (which corresponds to the computation of the expected value $\overline{\text{OGIP}}$) is equal to

$$E(\text{SMQR})=E\left(\frac{MO\cdot MR}{2}\right)=E\left(\frac{\Delta\mathcal{Z}_M\cdot\Delta\mathcal{Z}_M}{2\cdot tq(\hat{\imath})}\right) \quad (10a)$$

The dip $tg(\hat{\imath})$ is assumed to be constant on $MR$, and equal to the dip $tg(\hat{\imath}_p)$ at $P$. (This assumption is justified by the fact that the variogram of corrections to seismic $\mathcal{Z}_{fsm}$ shows a range larger than $\Delta\mathcal{Z}_M/tg(\hat{\imath})$. That means that the $\Delta\mathcal{Z}$ corrections are highly correlated between $M$ and $R$ and that $tg(\hat{\imath})\approx tg(\hat{\imath}_p)$).

According to the relation $E(X^2)=\sigma^2(X)+(E(X))^2$, Eq. (10a) can be written

$$E(SMQR)=\frac{E(\Delta\mathcal{Z}_M^2)}{2\cdot tg(\hat{\imath}_p)}=\frac{(E(\Delta\mathcal{Z}_M))^2}{2\cdot tg(\hat{\imath}_p)}+\frac{\sigma^2(\Delta\mathcal{Z}_M)}{2\cdot tg(\hat{\imath}_p)} \quad (10b)$$

But

$$\frac{(E(\Delta\mathcal{Z}_M))^2}{2\cdot tg(\hat{\imath}_p)}=SMNP \quad (10c)$$

So, Eq. (10b) becomes

$$E(SMQR)=SMNP+\frac{\sigma^2(\Delta\mathcal{Z}_M)}{2\cdot tg(\hat{\imath}_p)} \quad (10d)$$

$\sigma^2(\Delta\mathcal{Z}_M)$ is the variance of $\Delta\mathcal{Z}_M$ conditioned by all the data points. This variance can be written

$$\sigma^2(\Delta\mathcal{Z}_M)=\sigma_k^2(\Delta\mathcal{Z}_{fsm})+\sigma_k^2(U0)+\sigma^2(\mathcal{Z}_{GOC}) \quad (11)$$

where

- $\sigma_k^2(\Delta\mathcal{Z}_{fsm})$ is the spot kriging variance of $\Delta\mathcal{Z}_{fsm}$ at the point $M$
- $\sigma_k^2(U0)$ is the spot kriging variance of $U0$ at the point $M$
- $\sigma^2(\mathcal{Z}_{GOC})$ is the variance of initial gas–oil contact
- $SMNP$ is the surface that corresponds to the most probable value of OGIP, i.e. to the OGIP1 computation.

This gives an additional rock volume around the most probable boundary $\hat{L}$:

$$\Delta Rock\ V=\int_L \frac{\sigma^2(\Delta\mathcal{Z}_M)}{2\cdot tg(\hat{\imath}_p)}\cdot dl \quad (12)$$

and an additional corrective term to OGIP1

$$\Delta 1=\int_L Alpha_k\cdot\phi_k\cdot Sg_k\cdot\frac{\sigma^2(\Delta\mathcal{Z}_M)}{2\cdot tg(\hat{\imath}_p)}\cdot dl \quad (13)$$

This correction of the average value does not affect all the boundaries in the same way because it depends strongly on the dip $\hat{\imath}_p$: the smaller the dip, the bigger the boundary and the overstepping correction! On Frigg, this correction corresponds to:

$$\Delta 1=0.0114\cdot\text{OGIP1} \quad (14)$$

*Computation of the total mean value of OGIP*

The total mean value of OGIP is given by equation (1) which gives

$$\overline{\text{OGIP}}=\text{OGIP1}+0.0114\cdot\text{OGIP1}=1.0114\cdot\text{OGIP1}. \quad (15)$$

This means that the uncertainty on the boundary of the field introduces a distortion between the most probable value and the mean value: the mean value is 1.14% greater than the most probable value. To find this result, it is usually necessary to compute the expectation curve of OGIP by Monte Carlo simulations. The PACEVAH method introduces an improvement because the Monte Carlo simulation would only give a mean value, 2.3% + 1.14% = 3.44%, greater than the deterministic value $\text{OGIP}_D$, and the user would not have known whether this 3.44% was due to the boundary and overstepping correction or to some additional geological information in the deterministic approach.

## Computation of the standard deviation of OGIP

The variance of OGIP is equal to the sum of several groups of terms: the *main variance term group*, the *bias variance term group* and the *overstepping variance term*.

*Main variance term computation*

The main group of terms gives for all the regional parameters used in the OGIP computation ($\Delta\mathcal{Z}_{fsm}$, $U0$, $Alpha$, $\phi$ and $Sg$) the magnitude of the uncertainty due to the extension of the values at wells to an average value for the whole field. It measures the effects of the uncertainty of mapping on the uncertainty of this average value. The geostatistic concept which gives the uncertainty of the average value of a parameter $X$ on a domain is the estimation (or extension) variance of the average value on a domain. If the mapping is done using kriging methods, this uncertainty will be measured by the kriging variance $\sigma^2_{ks}$ of the average value on the domain $\hat{S}$.

$$\sigma^2_{ks}(X)=\sigma^2\left(\frac{1}{\hat{S}}\cdot\int_S X\cdot ds\right) \quad (16)$$

*Bias variance term computation*

The uncertainty of the extension is only part of the total uncertainty. In particular, if all the spot data at wells are affected by the same bias, the uncertainty due to this bias will not be computed by the extension variance because it cannot be shown in the structural analysis. (The variogram of a biased set of data is exactly the same as the variogram of the unbiased set of the same data!) This occurs especially for the petrophysical parameters resulting from a log interpretation: the results of all the wells are affected in a similar way by the choice of a common saturation formula, a mineralogical model or a gas-effect correction. It is therefore necessary to add a bias variance term $\sigma^2_{bi}$ to the extension variance term to compute the true uncertainty. Generally, these bias variances cannot be completely computed from a statistical analysis of data and they are partially subjectively estimated. However, classical statistical tools can be used to render these bias variance estimations more objective, as shown below.

In the Frigg study, bias variance terms have been used for the porosity $\phi$, the gas saturation *Sg* and for the gas–oil contact depth: $Z_{GOC}$.

- For porosity, the bias variance term has been estimated at $\sigma^2_{bi}(\phi) = (0.25\,\text{p.u.})^2$, assuming that the systematic bias on the porosity log interpretation corrected by cores corresponds to a range of $\pm 0.5$ p.u. This range has been estimated by comparing several log interpretations with different mineral models on the GLOBAL® SCHLUMBERGER interpretation program. As the interpretation with an incomplete mineral model differs from a more complete one by 1 p.u. in the gas zone, it has been assumed that the more complete one is good at $\pm 0.5$ p.u.

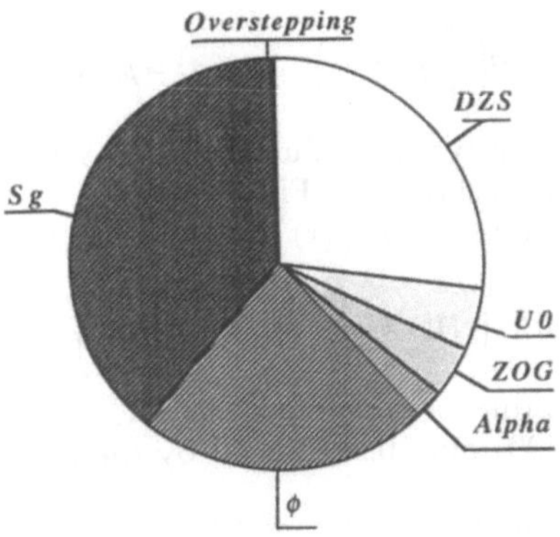


**Fig. 3.** Breakdown of variance of the OGIP of Frigg (parameters).

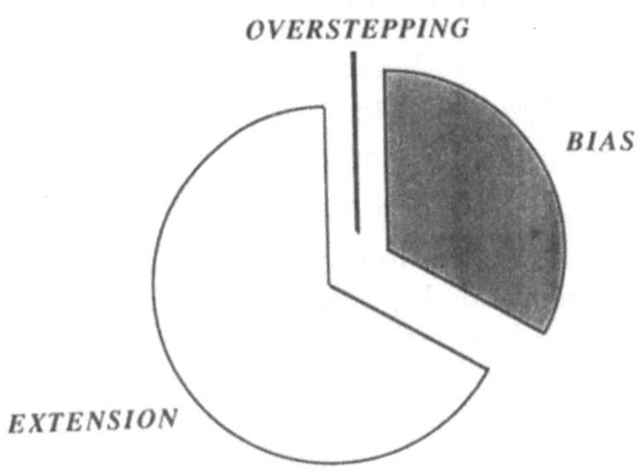


**Fig. 4.** Breakdown of variance of the OGIP of Frigg (extension/overstepping/bias).

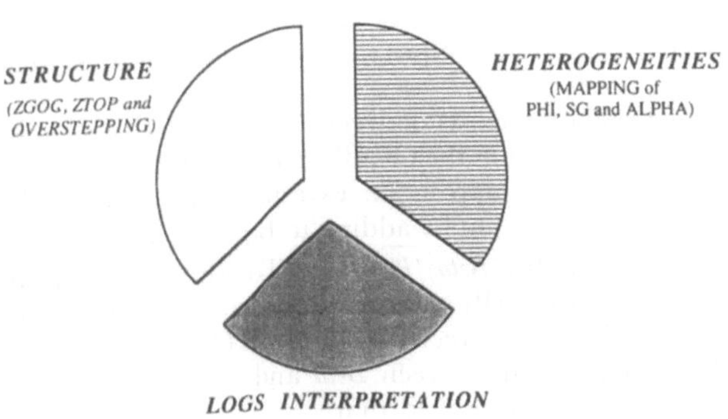


**Fig. 5.** Breakdown of variance of the OGIP of Frigg (heterogeneities/structure/logs interpretation).

- The uncertainty on $Z_{GOC}$ is analogous to a bias variance because it cannot be deduced from the structural analysis of data. The bias variance has been computed using the following method. A range of GOC depths was estimated *a priori* for each well crossing the GOC, according to the uncertainties on deviation, log depths, and presence of clay layers around the presumed GOC. A variance was computed for each range. An estimator $Z^*_{GOC}$ was defined as a linear combination of the $Z_{GOCi}$ of each well *i*:

$$Z^*_{GOC} = \sum_{i=1}^{N_{wells}} \beta_i \cdot Z_{GOCi} \tag{17}$$

with $N_{wells} = 15$.
The best estimator corresponds to values of $\beta_i$ that minimize the estimation variance of $(Z^*_{GOC} - Z\dagger_{GOC})$ where $Z\dagger_{GOC}$ is the true, unknown value of GOC depth. This best estimate of GOC was kept for all the OGIP computations and its variance was computed:

$$\sigma^2_{bi}(Z_{GOC}) = (0.34\,\text{m})^2 \tag{18}$$

- As usual, the estimation of the uncertainty of the gas saturation was the most subjective because it depends strongly on the water salinity in the gas zone, which it is impossible to know. After an estimation of a reasonable range for this salinity and of the corrections for laminated shales, this variance was estimated as

$$\sigma^2_{bi}(Sg) = (1.5\,\text{s.u.})^2 \tag{19}$$

No possible biases have been considered on the *U*0 thickness estimation, on the *Alpha* or on the $\Delta Z_{fsm}$ evaluation.

For *Alpha*, the choice of cut-off (cut-off on very high value of clay content) prohibits any bias due to log interpretation.

For *U*0, interpretation of sonic logs is assumed to be unbiased.

For $\Delta Z_{fsm}$ evaluation, virtual points are used to take into account the information on OGIP given by the flat spot extension. In this method (Fig. 1) there is an uncertainty $\pm \Delta Z(M)$ on the true depth of the point *P*:

(1) because the intersection between the flat spot and the FSM is only virtual, there is an uncertainty on the $(x, y)$ position of *P*;
(2) because the flat spot and the FSM do not have negligible thicknesses.

However, geophysicists consider that there is no systematic error on the *P* depth estimation: the mean value of $\Delta Z(P)$ is equal to zero around the most probable boundary $\hat{L}$. Therefore, uncertainty on $\Delta Z_{fsm}$ is entirely taken into account by the experimental variogram, and there is no need for an additional bias variance.

### *Overstepping variance term computation*

The last correction term corresponds to the uncertainty of the surface *S* of the gas-bearing zone. This uncertainty is given by its variance of overstepping, $\sigma^2_{overs}$. The overstepping variance $\sigma^2_{overs}$ can be computed if we consider that the overstepping rock volume term is equal to $\Delta R$ with

$$\Delta R = \int_L \frac{\Delta Z_M^2}{2 \cdot tg(\hat{\imath})} \cdot dl \tag{20}$$

It has already been seen that the mean value of $\Delta R$ is equal to

$$E(\Delta R) = \Delta Rock\ V = E\left(\int_L \frac{\Delta Z_M^2}{2 \cdot tg(\hat{\imath})} \cdot dl\right) = \int_L \frac{\sigma^2(Z_M)}{2 \cdot tg(\hat{\imath})} \cdot dl \tag{21}$$

The variance term is equal to

$$\sigma^2(\Delta R) = \sigma^2\left(\int_L \frac{\Delta Z_M^2}{2 \cdot tg(\hat{t})} \cdot dl\right) \tag{22a}$$

with

$$\Delta Z_M^2 = \Delta Z_{fsm}^2 + U0^2 + \Delta Z_{GOC^2} \tag{23}$$

Equation (22a) is estimated by

$$\sigma^2(\Delta R) \approx \frac{1}{4 \cdot tg^2(\hat{t}_1)} \cdot \sigma^2\left(\int_L \Delta Z_M^2 \cdot dl\right) \tag{22b}$$

where

$$\frac{1}{2 \cdot tg(\hat{t}_1)} = \int_L \frac{1}{2 \cdot tg(\hat{t})} \cdot dl$$

Thus

$$\sigma^2(\Delta R) \approx \frac{\hat{L}^2}{4 \cdot tg^2(\hat{t}_1)} (\sigma_{KL}^2(\Delta Z_{FSM^2}) + \sigma_{KL}^2(\Delta Z_{U0^2}) + \sigma^2(\Delta Z_{GOC^2})) \tag{24}$$

$\sigma_{KL}^2(X)$ means the kriging variance of the average value of $X$ around $\hat{L}$:

$$\sigma_{KL}^2(X) = \sigma^2\left(\frac{1}{\hat{L}} \cdot \int_L X \cdot dl\right) \tag{25}$$

To compute this kriging variance, it will be assumed, that if $s^2$ is the sill or the maximal experimental value of the variogram $\gamma_h(X)$ of the regional parameter $X$,

$$\gamma_h(X^2) \leqslant 4s^2 \cdot \gamma_h(X) \tag{26}$$

Furthermore, assuming that the kriging system $\{a_i\}$ of $X^2$ is roughly equivalent to the kriging system $\{b_i\}$ of $X$, then the kriging variance of $X^2$ i.e. $\sigma_{KL}^2(X^2)$ can be estimated by

$$\sigma_{KL}^2(X^2) \leqslant 4s^2 \cdot \sigma_{KL}^2(X) \tag{27}$$

Each term of the type $\sigma_{KL}^2(X)$ is computed by BLUEPACK in the neighbourhood of the average gas boundary zone $\hat{L}$ of Frigg.

Note that this variance is only estimated roughly because overstepping is a correction term, so the accuracy of overstepping variance does not need to be as great as the accuracy of the variance of the main term.

*Total variance computation*

The total variance of OGIP can be written as follows, using Eq. (1) and neglecting the covariance term:

$$\sigma^2(\text{OGIP}) = \sigma^2(\text{OGIP1} + \Delta 1) \approx \sigma^2(\text{OGIP1}) + \sigma^2(\Delta 1) \tag{28}$$

In view of the fact that $\sigma^2(\Delta 1)$ is the overstepping variance $\sigma_{\text{overs}}^2(\text{OGIP})$, the relative variance $\sigma^2(\text{OGIP})/\overline{\text{OGIP}^2}$ will be equal to

$$\frac{\sigma^2(\text{OGIP})}{\overline{\text{OGIP}^2}} = \frac{\sigma^2(\text{OGIP1})}{\overline{\text{OGIP}^2}} + \frac{\sigma_{\text{overs}}^2(\text{OGIP})}{\overline{\text{OGIP}^2}} \tag{29}$$

The term $\sigma^2(\text{OGIP1})/\overline{\text{OGIP}^2}$ can be deduced from Eq. (2):

$$\frac{\sigma^2(\text{OGIP1})}{\overline{\text{OGIP}^2}} = \frac{\sigma^2\left(\int_{\hat{S}} Ht \cdot Alpha \cdot \phi \cdot Sg\, ds\right)}{\overline{\text{OGIP}^2}} \tag{30}$$

This variance can be estimated using kriging variances of the stochastic integral $\sigma_{ks}^2(\int_S p \cdot Y\, ds)$ where $p$ is a function of $(x, y)$ and $Y$ a random function of $(x, y)$. $\sigma^2(\text{OGIP1})$ can be written (assuming no correlation between *Beta* and *Ht* and neglecting the second-order term):

$$\sigma^2(\text{OGIP1}) \approx \sigma_{ks}^2\left(\int_{\hat{S}} \text{Ht}_k \cdot Beta\, ds\right) + \sigma_{ks}^2\left(\int_{\hat{S}} Ht \cdot Beta_k\, ds\right) \tag{31}$$

*Beta*$(x)$ is a random function equal to

$$Beta = Alpha \cdot Sg \cdot \phi \tag{32}$$

*Beta*$_k(x)$ is a "mean map" equal to

$$Beta_k = (Alpha \cdot Sg \cdot \phi)_k \tag{33}$$

These variances of stochastic integrals cannot be computed directly by a normal kriging program such as BLUEPACK. However, the variances can be maximized. In particular, for a properly appraised field (i.e. a field in which most of the wells are drilled in the thickest part) the partial kriging variance is

$$\sigma^2\left(\int_{\hat{S}} Ht_k \cdot Beta\, ds\right) \leqslant \overline{Ht^2} \cdot \sigma^2\left(\int_{\hat{S}} Beta\, ds\right) \tag{34}$$

This relation has not been demonstrated but has been verified in some theoretical cases and is completely in accordance with common sense. (In Fig. 6, the $p(x)$ function is a cone and the variogram of the spherical type. In estimating

$$\sigma_t^2 = \sigma^2\left(\int_{\hat{S}} p \cdot Y\, ds\right) \quad \text{by} \quad \sigma_a^2 = p_m^2 \cdot \sigma^2\left(\int_{\hat{S}} Y\, ds\right)$$

where $p_m$ is the average value of $p(x)$ on $S$, the relative error $(\sigma_t^2 - \sigma_a^2)/\sigma_a^2$ is shown on Fig. 7.)

As variations of *Beta*$_k(x)$ are small on $\hat{S}$,

$$\sigma^2\left(\int_{\hat{S}} \text{Ht} \cdot Beta_k\, ds\right) \approx \overline{Beta^2} \cdot \sigma^2\left(\int_{\hat{S}} Ht\, ds\right) \tag{35}$$

Using the values obtained in Eqs. (30), (31), (33), (34) and (35), Eq. (29) may be replaced by the inequality

$$\frac{\sigma^2(\text{OGIP})}{\overline{\text{OGIP}^2}} \leqslant \frac{\sigma^2(Ht)}{\overline{Ht^2}} + \frac{\sigma^2(\phi)}{\overline{\phi^2}} + \frac{\sigma^2(Sg)}{\overline{\text{Sg}^2}} + \frac{\sigma^2(Alpha)}{\overline{Alpha^2}} + \frac{\sigma_{\text{overs}}^2(\text{OGIP})}{\overline{\text{OGIP}^2}} \tag{36}$$

$\overline{Ht}$, $\overline{\phi}$, ... are mean values over the field of $Ht$, $\phi$. ... The variances on the right-hand side are given by

$$\sigma^2(Ht) = \sigma_{ks}^2(\Delta Z_{fsm}) + \sigma_{ks}^2(U0) + \sigma_{bi}^2(Z_{GOC}) \tag{37}$$

$$\sigma^2(Alpha) = \sigma_{ks}^2(Alpha) \tag{38}$$

$$\sigma^2(\phi) = \sigma_{ks}^2(\phi) + \sigma_{bi}^2(\phi) \tag{39}$$

$$\sigma^2(Sg) = \sigma_{ks}^2(Sg) + \sigma_{bi}^2(Sg) \tag{40}$$

The variance computation approximation corresponding to Eq. (31) and finally Eq. (36) is sufficiently accurate if the relative standard deviation of each term is less than 10% (Fig. 8). This corresponds to the current values of relative standard deviations encountered in OGIP uncertainty computations at the end of a development.

In all these computations, the parameters are considered to be uncorrelated. If a correlation exists between petrophysical parameters, it is necessary to use the parameter *Beta* defined in Eq. (32) instead of *Alpha*, $\phi$ and *Sg* for the computation of OGIP1 (in kriging directly the *Beta* values at wells) and for the computation of the extension variance term.

If a correlation exists between *Beta* and *Ht*, it has to be taken into account in the OGIP1 computation (by kriging *Beta* with an *Ht* map as an external drift) and in the variance computation by adding to Eq. (36) a covariance term $2 \cdot \text{COV}(Ht, Beta)/\overline{Ht} \cdot \overline{Beta}$. To obtain the mean value from the OGIP1 value, it is theoretically necessary to add an additional corrective term $\Delta 2$ to OGIP1. If there is a linear regression between *Beta* and *Ht* such as $Beta = a + b \cdot Ht$, it can be proved that

$$\Delta 2 \approx b \cdot \int_S \sigma^2(Ht)\, ds \tag{41}$$

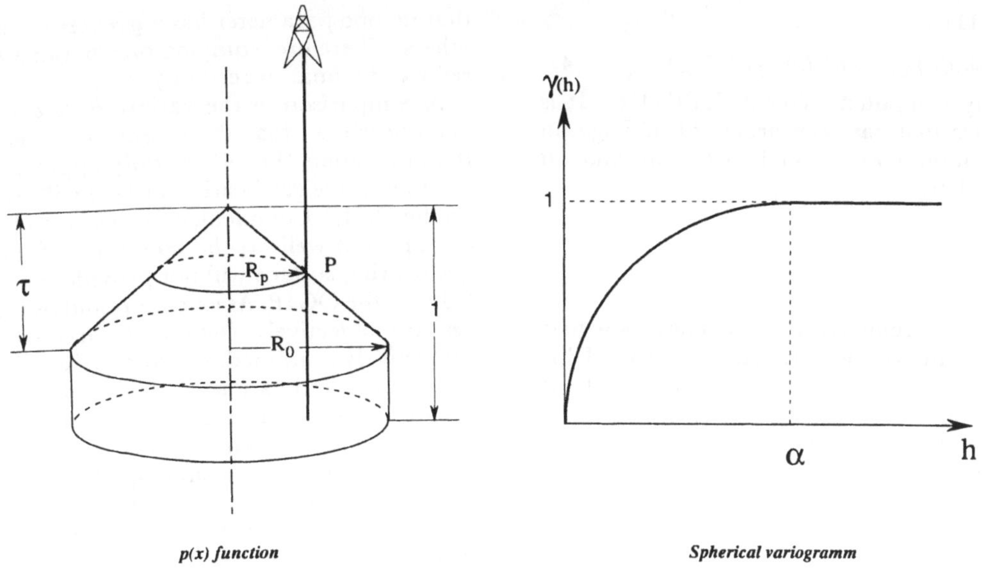


**Fig. 6.** Functions used in verification of inequality (34).

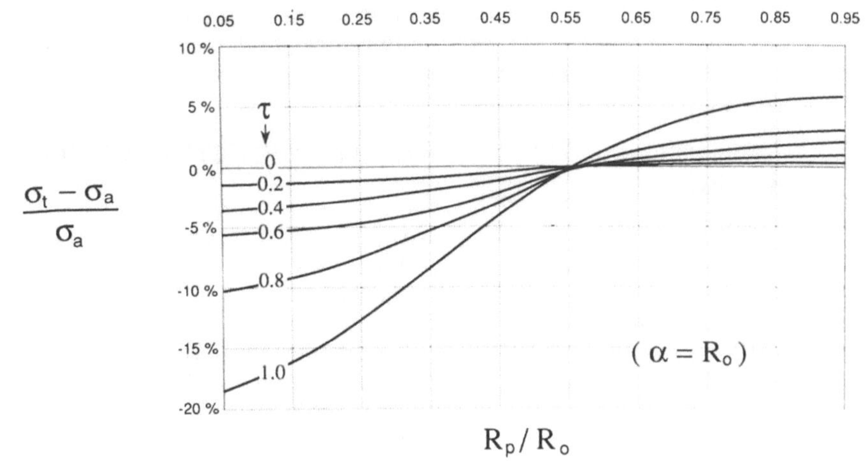


**Fig. 7.** Relative error in $\sigma_a = p_m \sigma(\int_S \Upsilon)$ as an approximation of $\sigma_t = \sigma(\int_S p \Upsilon)$.

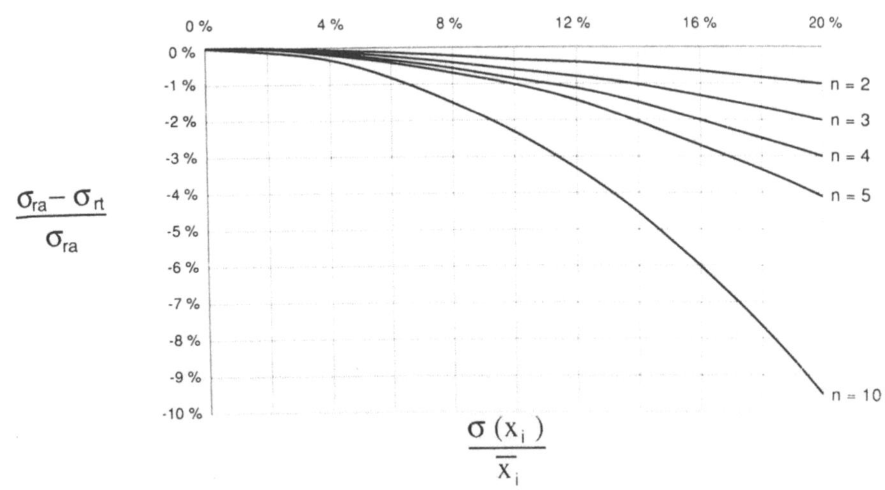


**Fig. 8.** Relative error in $\sigma_{ra} = (\Sigma_{i=1}^{n} \sigma^2(X_i)/\bar{X}_i^2)^{1/2}$ as an estimate of $\sigma_{rt} = \sigma(X_1, \ldots, X_n)\sqrt{(X_1, \ldots, X_n)}$.

where as in Eq. (11):

$$\sigma^2(Ht) = \sigma_k^2(\Delta Z_{fsm}) + \sigma_k^2(U0) + \sigma^2(Z_{GOC}) \qquad (42)$$

This term is easily computed using BLUEPACK. It is a second-order term that has been neglected in Frigg, in spite of a rather large correlation between *Sg* and *Ht* ($\Delta 2 = 0.1\%$ of OGIP1!)

## RESULTS

The main results are presented in Table I. They show that the relative uncertainty of the new estimate of OGIP is fairly small (with a relative standard deviation of about 3.5%). The volumes of unswept gas at shut-in of existing wells have been delineated, directing the complementary development policy.

**Table I.** Computation of standard deviation on OGIP

| *Variable* $X$ | *Kriging variance* $\sigma_{ks}^2(X)$ | *Bias variance* $\sigma_{bi}^2(X)$ | *Relative variance* $(\sigma^2(X)/X^2$ |
|---|---|---|---|
| $DZ_{fsm}$ | 0.7950(m²) | 0 | — |
| $U0$ | 0.1418(m²) | 0 | — |
| $Z_{GOC}$ | 0 | 0.1157(m²) | — |
| $Ht$ | | | $4.53 \times 10^{-4}$ |
| *Alpha* | 0.2872(%²) | 0 | $3.02 \times 10^{-5}$ |
| $\Phi$ | 0.1632(p.u.²) | 0.0625(p.u.²) | $2.87 \times 10^{-4}$ |
| $Sg$ | 1.6102(s.u.²) | 2.25 (s.u.²) | $4.86 \times 10^{-4}$ |
| Overstepping relative variance $\frac{\sigma^2_{\text{overs}}(\text{OGIP})}{\text{OGIP}^2}$ | | | $1.38 \times 10^{-6}$ |
| $\frac{\sigma^2(\text{OGIP})}{\text{OGIP}^2}$ | | | $1.258 \times 10^{-3}$ |
| Relative standard deviation on OGIP $\frac{\sigma(\text{OGIP})}{\text{OGIP}}$ | | | 3.54% |

## COMPLEMENTARY RESULTS GIVEN BY THE METHOD: THE ORIGIN OF UNCERTAINTY

The PACEVAH method gives results in addition to the mean value and global uncertainty of OGIP. It enables computation of the origins of the uncertainty, information that allows the technical manager to concentrate his funds on studying those parameters that are most important.

A breakdown of the variance is given in Fig. 3. A comparison of the variances of the parameters shows that no one parameter has a greater uncertainty than the others. Therefore, studying one parameter alone will not reduce the final uncertainty.

A comparison of the variances of *overstepping*, *extension* and *bias* shows that the extension uncertainty is greater than the others (Fig. 4). So only appraisal wells inside the most probable gas-bearing surface will reduce uncertainty on the OGIP. Complementary work on log interpretation or appraisal wells at the boundary of the most probable gas-bearing surface will not provide much greater knowledge of the OGIP. As extension and overstepping variances are objectively computed, and bias variances are subjectively computed, this comparison also shows that the uncertainty estimation of the OGIP is fairly objective because bias variance is only a quarter of the global.

A comparison of *heterogeneities*, *structure* and *logs* has been computed (Fig. 5) assuming that the summation of extension variances of petrophysical parameters is a direct measurement of the uncertainty due to heterogeneities. The summation of bias variances on these petrophysical parameters gives uncertainty on log interpretation. This comparison shows that the uncertainty of OGIP due to heterogeneities is lower than the uncertainty of structure and log interpretation, but similar to the uncertainty of OGIP due to the structural definition alone. It proves that for future OGIP estimations on Frigg, complementary information from stratigraphic seismic surveys have the same effect on the reduction of uncertainty as information from structural seismic.

## ACKNOWLEDGEMENTS

The author wishes to thank the management of Elf Aquitaine Norge A/S and the Frigg Unit partners: Norsk Hydro A/S, Statoil, Total Marine Norsk A/S, Total Oil Marine plc and Elf UK plc for permission to present this paper.

M. Alonso, P. Bitot, M. Guy, M. Mollier, B. Moulenes and L.H. Reiss of Elf Aquitaine, and R. Maritvold, F. Ravel and A. Thouand of Elf Aquitaine Norge A/S are also acknowledged.

## REFERENCES

Delfiner, P. and Chiles, J.P. (1977). *Conditional simulations: a new Monte Carlo approach to probabilistic evaluation of hydrocarbons in place*. Ecole des Mines de Paris, CGMM, Fontainebleau.

Matheron, G. (1972) *Théorie des Variables Régionalisées*. Informatique Géologique, Masson, Paris.

Matheron, G. (1973). The intrinsic random functions and their applications. *Advanced Applied Probabilities*, Vol. 5.

Renard, D. (1985). *Manuel d'utilisation de BLUEPACK®*. Ecole des Mines de Paris, CGMM, Fontainebleau.

# 27

# Hybrid Modelling of Heterogeneity to Improve 3D Simulation Results: Case Studies 1 and 2 on a North Sea Gas Field

*A.M. Haak and E.F.M. Elewaut*

Institute of Applied Geoscience, Delft, The Netherlands

A three-dimensional computer simulation model of a Rotliegend gas field, constructed to describe the fluid and reservoir behaviour under strong water drive, predicted certain problems in the gas field. A detailed characterization of the reservoir properties that influence the fluid movement was deemed necessary. Accordingly, a hybrid modelling technique was applied. The deterministic part of this method produced detailed maps of reservoir porosity and horizontal permeability based on a sedimentological model and a distinct facies correlation. The stochastic part of this method was developed to calculate how the presence of shale breaks affects fluid behaviour in terms of flow-path tortuosity and vertical effective permeability. Two different models based on different structure maps of the same gas reservoir were constructed. The geological models that resulted from applying the hybrid modelling technique provided a base for an adequate and simple matching procedure. The models were matched satisfactorily with the average reservoir pressure, water breakthrough in well 4 and the rise of the gas—water contact (GWC) in the field. Both models can predict the overall fluid movement within acceptable confidence levels.

## INTRODUCTION

Reservoir heterogeneity influences reservoir performance in terms of local and overall fluid movement. From preliminary simulation models of a Rotliegend gas field it was concluded that:

- Applying the rock data to the model on the assumption that the strata were homogeneous failed to model the fluid displacement processes accurately.
- Displacement of the gas by the underlying aquifer is stable; coning can only occur when the GWC is near the well perforations and occurs only locally.
- Displacement of the gas is mainly a result of vertical fluid movement.

It was therefore decided to construct a detailed production geology model.

A generally applicable hybrid modelling technique was developed that combined deterministic well data with a stochastic technique for calculating vertical permeability in the interwell area. The objective was to produce detailed descriptions of porosity, horizontal permeability and vertical permeability. The technique was applied in two different three-dimensional (3D) simulation models of the North Sea gas field. The results obtained from the first model have been presented earlier (Haak and Elewaut, 1988). Since then the method has been improved to be used in a second case study. The results of both case studies are compared in this paper.

## GEOLOGICAL FRAMEWORK

The Rotliegend in the North Sea is made up of four distinct facies associations that have been interpreted as the products of deposition in fluviatile (wadi), aeolian, sabkha and lacustrine environments. The Rotliegend is subdivided into three units:

(1) Lower Unit, 3–80 m, fluvial and minor aeolian sandstones;
(2) Middle Unit, 120–200 m, aeolian sandstone; the main gas reservoir;
(3) Upper Unit, 6–50 m, homogenized dunes and waterlaid sandstones.

The three units are separated by shaly zones. A sedimentological paradigm was prepared, so that any correlation between the different facies in the reservoir could be

*North Sea Oil Gas Reservoirs—II*

ascertained. This enabled the layers to be subdivided consistently on the basis of sedimentological knowledge and on the vertical facies detected in each of the wells.

## FIELD HISTORY

The gas field modelled in this study is situated in the Dutch North Sea sector. The Rotliegend sandstone reservoir consists mainly of aeolian and fluviatile deposits, with an occasional streak of lacustrine deposits. The field can be classified as a structural trap, bounded by faults on the south and west, crossed by N–S oriented faults and underlain by an aquifer. Table I summarizes the field data used for reservoir simulation. Initial pressure is 30 500 kPa, the GWC is located at 2804 m below sea level, some 130 m below the top of structure. The GIIP of the field was estimated to be $11–15 \times 10^9$ Sm$^3$ of gas. The field came on stream in January 1983. At present the field has six producing wells and a single platform with an average production of $3.5 \times 10^6$ m$^3$/day. The wells have been perforated both in the Upper Unit and in the Middle Unit.

Until February 1986 no water movement was seen on the pulsed neutron logs (PNL) of the wells penetrating the GWC. Since that date, the GWC seems to have risen by between 13 and 27 m. In March 1987 perforations in one well nearest to the GWC started to produce formation water. Another well was drilled updip, to take over the production of this partially plugged-off well.

**Table I.** Summarized field data

| *Parameter* | *Range* | *Units* |
|---|---|---|
| GIIP | $10–15 \times 10^9$ | Sm$^3$ |
| GWC | 2804 | m |
| Initial pressure | $305 \times 10^7$ | Pa |
| Temperature | 105 | °C |
| Initial saturation | | |
| gas | 0.65–0.30 | — |
| water | 0.35–0.70 | — |
| Viscosity | | |
| gas | 0.0000235 | Pa s |
| water | 0.000426 | Pa s |
| Depth | 2675–2804 | m ss. |
| Thickness | 0–129 | m |
| Permeability | | |
| aeolian | 0.05–1.000 | μm$^2$ |
| fluviatile | 0.001–0.005 | μm$^2$ |
| Porosity | | |
| aeolian | 14–25 | % |
| fluviatile | 7–9 | % |

## HYBRID MODELLING TECHNIQUE

Various authors (e.g. Begg *et al.*, 1985; Bifani *et al.*,1987; Kool, 1988) refer to block models that can be used to estimate the effective vertical permeability of a reservoir block volume that contains discontinuous permeability barriers. We modified the Haldorsen and Lake (1984) approach to serve as a basic procedure for relating vertical permeability to a known distribution of lateral permeability.

The outline of the method is illustrated in Fig. 1. The method involves using deterministic well data to define the lateral and vertical distribution of the porosity and the horizontal permeability, in accordance with the hypothesized geology of the Rotliegend reservoir. The effective vertical permeability was modelled using a stochastic technique to calculate the average tortuosity of the flow paths for a fluid particle in a discrete volume of the reservoir.

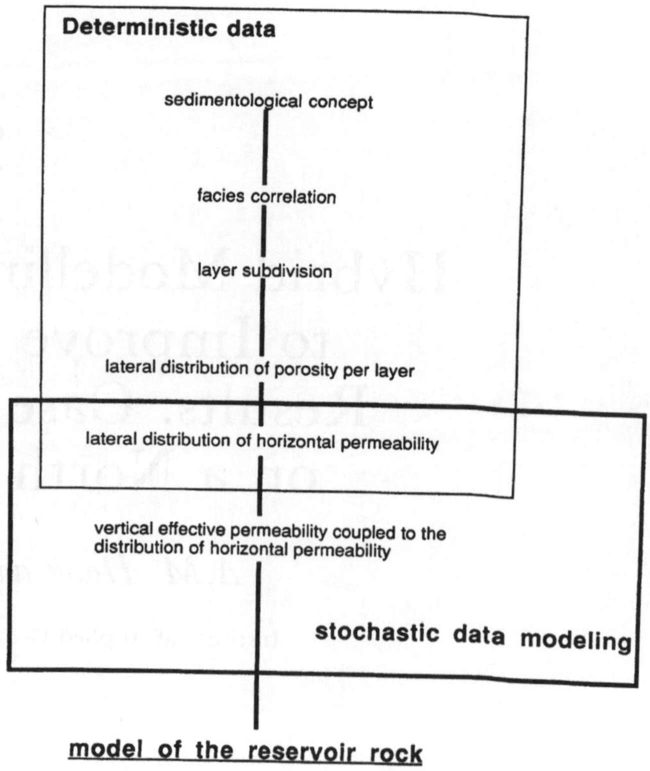


**Fig. 1.** Diagram of the hybrid modelling technique.

### Deterministic data (porosity, horizontal permeability)

*Sedimentology*

To model effectively the flow paths within the reservoir, one needs to know the lateral extent of the different media available for flow and of those preventing vertical flow. Therefore, a paradigm of the sedimentology of the Rotliegend reservoir was developed.

Sedimentological studies of the depositional environments of Rotliegend fields indicate that fluviatile sediments present within the dune units are probably the result of single flash floods in wadis, resulting from local showers over the dune fields. Because of the high infiltration and evaporation rate, these floods rapidly wane and the water stagnates, resulting in the deposition of clay drapes on top of the wadi sands (Rust, 1981). Studies in comparable sedimentary environments (Tunbridge, 1984), revealed that channels are mostly less than 60 m wide.

*Facies correlation*

In Fig. 2 the inferred palaeogeography of the Middle Unit demonstrates the predicted lateral facies distributions. The sedimentological paradigm was used to correlate the various facies of the reservoir in the interwell area. The aim was to construct a consistent subdivision of layers based on sedimentological knowledge and on the vertical facies detected in each of the wells. The flow chart of Fig. 3 illustrates how this was achieved.

First, a correlation between the core descriptions and the geophysical well logs was defined. For the Rotliegend deposits a good fit was obtained having a cut-off value of 90

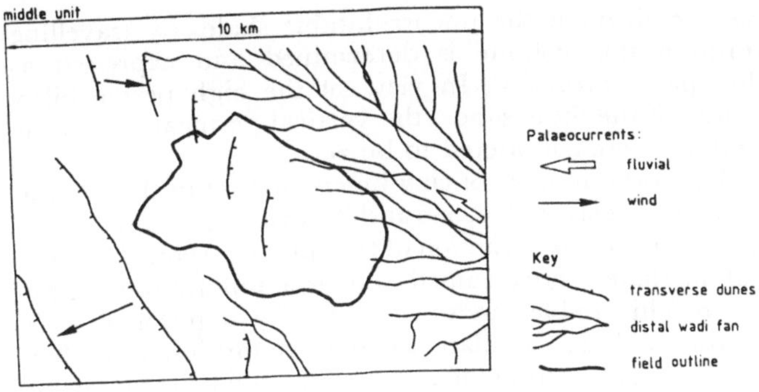


**Fig. 2.** Palaeogeography of the Middle Unit.

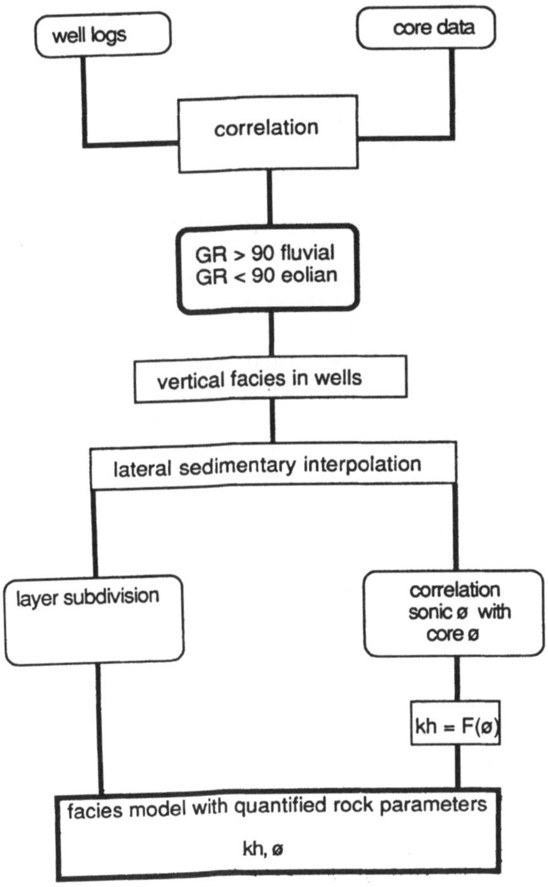


**Fig. 3.** Procedure for obtaining a facies correlation.

API on the gamma-ray and assigning the lower value to the aeolian facies. A facies correlation throughout the reservoir resulted in the gas-filled Rotliegend being subdivided into nine layers (Fig. 4).

*Rock properties*

Figure 4 illustrates the good correlation between the core and the acoustic (sonic) porosities for well 4. Based on the facies correlation in each of the other wells, the porosities can be calculated from the sonic logs according to

$$\phi_{ac} = \frac{dt_{log} - dt_{ma}}{dt_{fl} - dt_{ma}} \qquad (1)$$

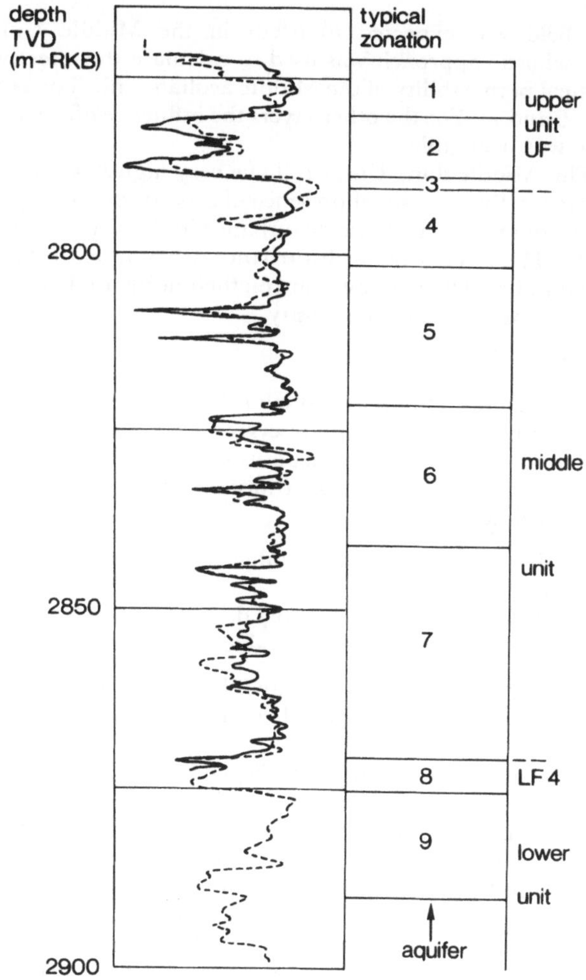


**Fig. 4.** Typical layer zonation. Well 4: ——, core porosity; – – –, sonic porosity.

Appropriate values for $dt_{ma}$, $dt_{fl}$ and $dt_{log}$ have been subtracted from the correlation in well 4 (Fig. 4).

The calculated porosities were averaged per layer. In conjunction with the sedimentological paradigm these values were used to construct porosity distribution maps for each layer. At this stage the porosity values of all nine maps were sampled according to the grid-block configuration of the simulation model.

In the literature several methods are used to calculate the permeability as a function of the porosity. Assuming that the data distribution is representative, one can find the relationship between permeability and porosity by cross-plotting core permeabilities versus core porosities. The permeability of each of the distinct geological zones (layers) was calculated from cross-plots, yielding average values for the horizontal permeabilities for all wells. Between the wells the values of horizontal permeability were interpolated in accordance with the sedimentological model.

Thus, on the basis of a sedimentological paradigm the distribution of rock properties that define simulation performance was represented in maps of porosity, permeability and layer thickness.

## Stochastic data (effective vertical permeability)

*Hypothesis*

In earlier studies it was found that the main vertical movement of water in the numerical simulation model of the field was expected to occur in the Middle Unit. A stochastic approach was used to calculate the effective vertical permeability of the Middle aeolian Unit (layers 3, 4, 5, 6 and 7). For the other layers the influence of vertical flow was negligible.

This Middle dune Unit can have very high $k_v/k_h$ ratios. However, the occasional presence of clay drapes on top of the wadi sands may cause disturbance in the vertical flow paths. The severity of the disturbance depends on the type and number of flow barriers and on their horizontal extent, influencing flow path tortuosity.

*Shales*

Previous studies (Elewaut and Huurdeman, 1986) revealed that most shales within the reservoir are very difficult to correlate. The Rotliegend deposits contain three types of shales that can be seen as transmissibility barriers to vertical flow:

- deterministic or correlatable shale layers UF and LF4;
- shales that have a possible degree of correlation;
- shales that are uncorrelatable.

The deterministic shales UF and LF4 were entered as layers 2 and 8 in the model as a result of the facies correlation.

Comparison with other Rotliegend reservoir studies indicated that the lower part of the Middle Unit was likely to harbour major shale barriers consisting of shales that have a possible degree of correlation. To account for the larger effect of these shale barriers, the transmissibility to flow between layers 7 and 8 and layers 8 and 9 was reduced to zero for, respectively, 40% and 30% of the total field area. These areas were randomly distributed over the field.

The uncorrelatable shale barriers have been treated stochastically as described below.

*Flow-path tortuosity*

Tortuosity is defined as the length of the flow path between two points divided by the shortest distance between these two points. The actual length of the flow paths consists of the shortest vertical distance, increased by a number of horizontal displacements or step-out distances ($d_{so}$) resulting from the presence of flow barriers (Fig. 5). The definition of the tortuosity $T$ becomes

$$T=\frac{h_{\min}+\Sigma d_{so}}{h_{\min}} \quad (2)$$

*Method*

Haldorsen and Lake (1984) assume that the vertical effective permeability is related to the horizontal permeability according to

$$k_{v,\text{eff}}=k_h\,(1-F_S)\,\frac{1}{T^2} \quad (3)$$

where $F_S$ is the volume fraction of shale. Their method of dealing with the shale intercalations includes generating a pattern of deterministically and stochastically distributed shales. From this pattern, the tortuosity of the flow paths is calculated for each grid block.

We simplified this part of the method by not assigning any locations to the shales in the reservoir volume. We assume the shales to be present somewhere within this volume. We allow a fluid particle that travels through the volume to collide with the shales without pointing to the location of these shales. For each grid block the stochastically averaged length of the flow path of a fluid particle passing through the uncorrelatable shales by travelling through this volume is determined and expressed as flow-path tortuosity. In view of the high permeability values of the dune sands, the vertical permeability of the shales is here considered to be zero.

The total number of clay layers present in the vertical sequences of layers 3, 4, 5, 6 and 7 was assessed in each well, and expressed as a fraction of the total reservoir thickness. All the shales were assumed to have a uniform thickness of 0.5 m. This enabled an iso-shale density map (Fig. 6) to be constructed. To calculate the average tortuosity of a flow path through a grid block, one has to know the number of shales in the block. The iso-shale density map enabled the number of shales to be determined for each grid-block volume.

The discrete occurrence of a phenomenon can be described through the Poisson probability distribution (Candall and Stuart, 1969). The introduction of the Poisson distribution accounts for two effects:

- the uncertainties in the interpolation of the iso-shale density map in the interwell area;
- the probability of a travelling fluid particle colliding with a shale.

For each grid block, the number of shales ($N_S$) determined from the iso-shale density map was used to ascertain the probability that $N_S$ is representative and that the average length of the flow path is influenced by each of the $N_S$ shales. This enabled the statistically averaged number of step-outs ($N_{SO}$) to be determined.

Using information presented in the literature (Weber, 1982; Tunbridge, 1984) we constructed a cumulative frequency plot for the widths and lengths of the shale lenses within the reservoir. The length of each step-out distance was established by generating a random number. Using this number for each shale, the cumulative shale–width frequency plot was entered, to find the actual length of the shale to be encountered. We assumed that on average a fluid particle would always seek to cover the shortest possible flow path, thus the maximum step-out would never be more than 30 m. The expected value for the step-out ($d_{so}$) is 1/4 of the shale width.

Using the tortuosity formula (3), the statistically averaged tortuosity can be calculated for each grid block. From the known $k_h$, the calculated tortuosity and the volumetrically summed shale fraction it is possible to calculate the vertical effective permeability by simply entering these numbers in formula (4) for each grid block.

*Procedure*

The procedure is illustrated in Fig. 7.

(1) Construct an iso-shale density map, related to the isopach map, to define the number of shale barriers per grid block.
(2) Sample the number of shale barriers ($N_S$) per grid block.
(3) Construct a cumulative frequency plot to characterize both the widths and lengths of the shale lenses.
(4) For each grid block establish a Poisson probability distribution expressed as

$$P(k)=\frac{e^{-N_S}\cdot N_S^{\,k}}{k!} \quad (4)$$

(5) Generate two random numbers, from a uniform distribution.
(6) Use the first random number and enter the cumulative probability plot to calculate stochastically the

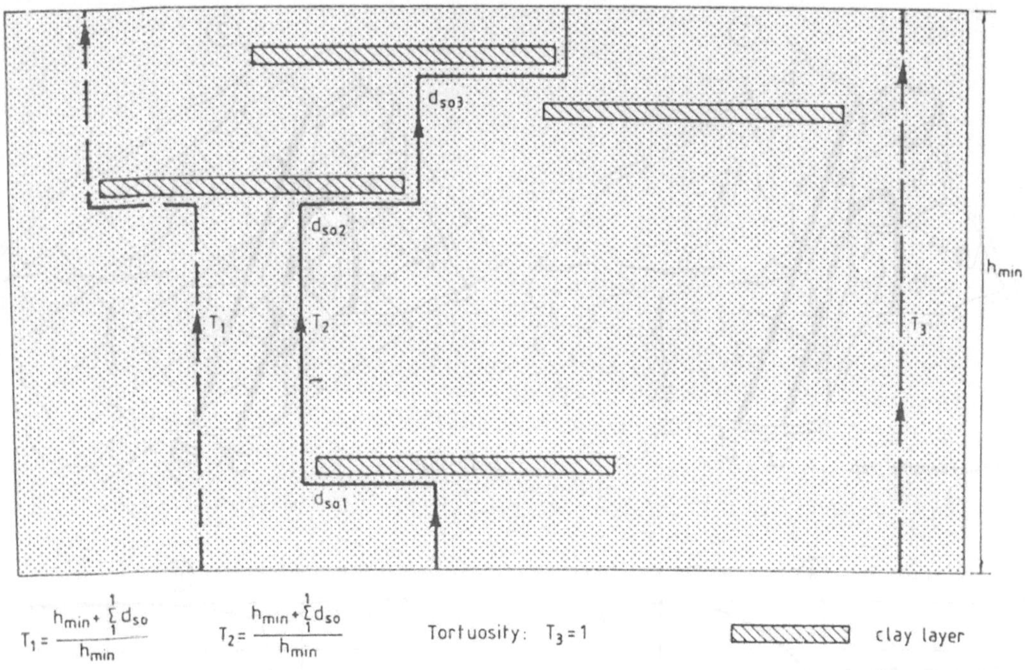


**Fig. 5.** The effect of tortuosity on a flow path.

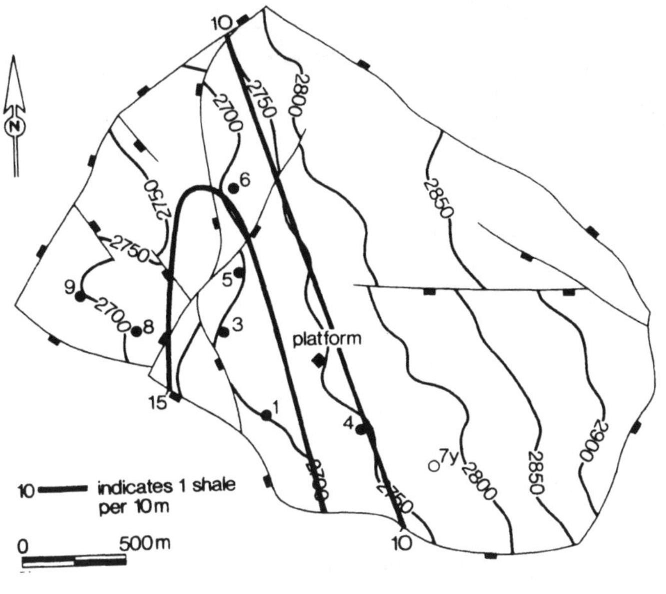


**Fig. 6.** Iso-shale density map.

number of shales with which an averaged fluid particle will collide, i.e. the number of step-outs ($N_{so}$) per grid block.

(7) Use the second random number and enter the cumulative frequency plot to establish the shale width. Divide this value by four (expected value). Repeat this step for $N_{so}$ shales.

(8) Calculate the tortuosity using formula (2) and Fs. The volume fraction of shale is determined by summing the volumes of each of the shales per grid block column and dividing this value by the total volume of the grid block column.

(9) Calculate the vertical effective permeability for each grid block using the formula (3).

The procedure described above was programmed on a mainframe computer and written according to a specific simulator format.

### *Data*

The calculated $k_{v,\text{eff}}$ for the Middle aeolian Unit varies between 0.2 and 0.7 times $k_h$. Figure 8 illustrates the effect of the introduction of the Poisson distribution. The distribution with an average value of $k_{v,\text{eff}}/k_h = 0.35$ was used in the simulation model.

The calculated pattern for the vertical effective permeability is thus preconditioned by the distribution of the $k_h$ and statistically distributed shale barriers, each having a characteristic shape. An earlier study (Elewaut and Huurdeman, 1986) indicated that the $k_{v,\text{eff}}/k_h$ ratios of layers 1 and 2 (the Upper Unit) and 8 and 9 (the Lower Unit) vary between 0.5 and 0.8.

When the method described above is used, the direction and location of the shale bodies is not important. This method of dealing with shales allows for a simple 3D approach and eliminates many disadvantages of models dealing with shales (Kool, 1988), e.g.,

- the discontinuity problem resulting from a deterministic approach;
- ambiguity in the determination of shale widths from outcrop studies.

### *Validation*

During the geological modelling phase, new core and log data came available from a new well (well 8) drilled updip structure. The data were checked to see whether they were consistent with the geological model described above and were found to fit the geological model reasonably well. On the whole, the new data fit into the geological model.

However, the correlation of the calculated porosities with the measured core porosities was poor. A new relation was found, using cross-plots of core porosities with a travel time that was considerably lower. A possible explanation for this discrepancy is that the well was drilled with an oil-based mud, whereas a water-based mud was used in the other wells.

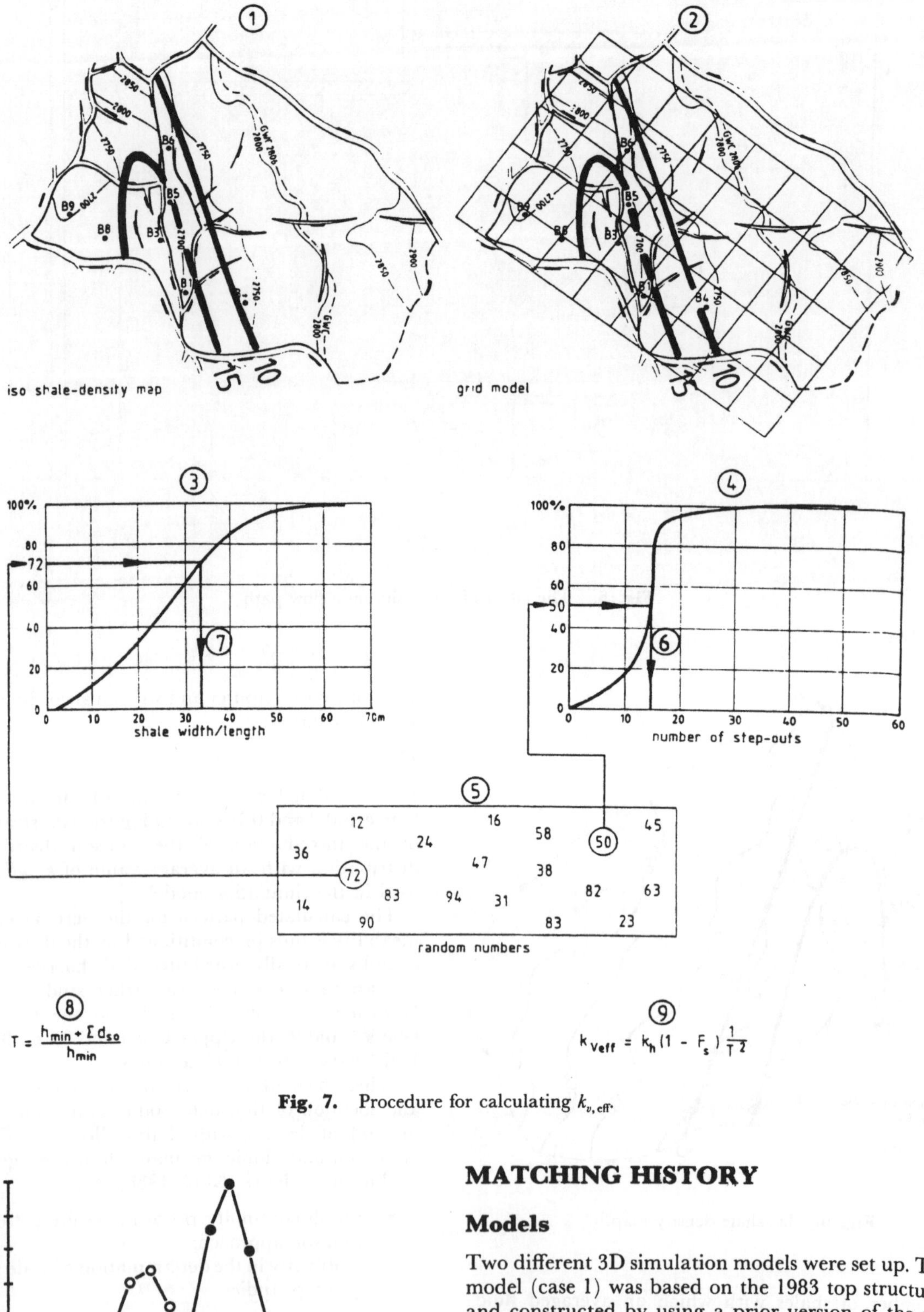


**Fig. 7.** Procedure for calculating $k_{v,\mathrm{eff}}$.

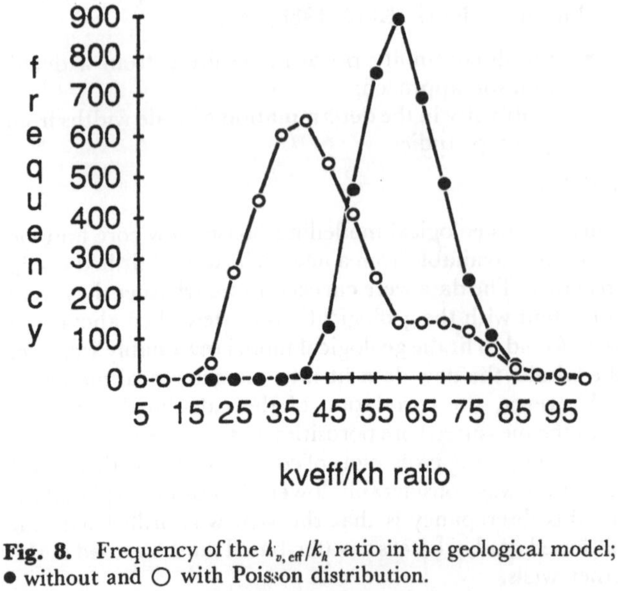


**Fig. 8.** Frequency of the $k_{v,\mathrm{eff}}/k_h$ ratio in the geological model; ● without and ○ with Poisson distribution.

## MATCHING HISTORY

### Models

Two different 3D simulation models were set up. The first model (case 1) was based on the 1983 top structure map and constructed by using a prior version of the method described above (Fig. 9). This model was acceptably matched (Haak and Elewaut, 1988).

At the end of 1987, an interpretation of a 3D survey of seismic data resulted in an adjusted top structure map (Fig. 9). On the basis of this structure map, a second geological and numerical simulation model (case 2) was constructed using the hybrid modelling technique described above. This model was also matched satisfactorily.

*Criteria*

Both models were matched against:

the observered average field pressures;
the flowing bottomhole pressures;

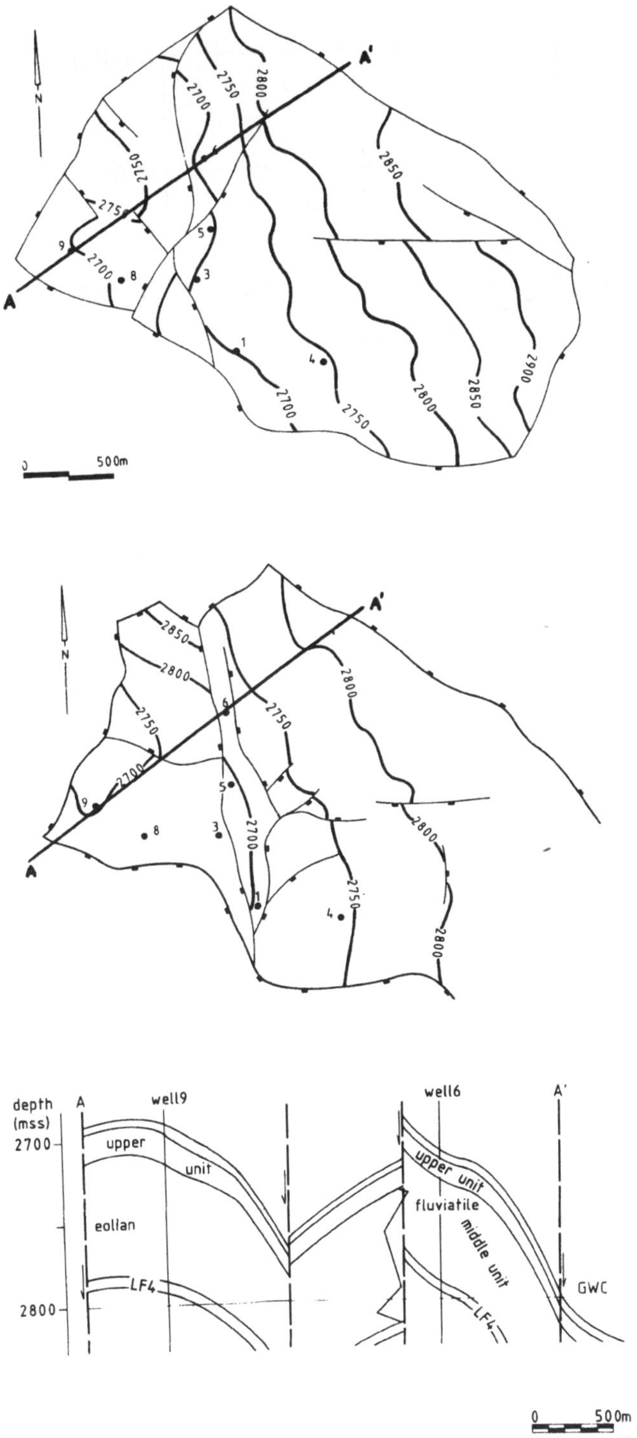


**Fig. 9.** Schematic structure maps of the gas field.

the saturation distributions in the field;
the water breakthrough in well 4.

The matching period extended from January 1983 up to March 1987, a 4-year production history, which for case 2 was extended until March 1988.

Saturation distributions in the field were matched in two ways. First, we used the rise of the GWC from the PNL logs of wells 1, 3, and 9, and in case 2 also of well 8. From the PNL logs of well 4 no significant rise in the GWC could be seen until water breakthrough occurred. Second, water breakthrough that had occurred in well 4 was used to match the water movement in the field.

### Matching parameters

The detailed reservoir description allowed a pressure and water match to be achieved with the adjustment of only a few variables. The pressure could be matched by adjusting the aquifer influx parameters in the model. The saturation distribution was matched by adjusting the values of the endpoint mobility data of the relative permeability curves for water.

No data were available on the $k_v/k_h$ ratio of the dune sands. During the matching phase an overall reduction of the transmissibility using a factor $x$ was introduced to reduce vertical flow of water in the aeolian parts of the Middle Unit. Since no exact data were available, values for $x$ ranging from 0.1 to 0.6 for the grid blocks of the layers 5, 6, 7 and 8 were tested, to incorporate the combined effect of the possible degree of correlation of the shale barriers in the lower part of the Middle Unit, and the ratio of $k_v/k_h$ for the aeolian sediments. A value of 0.3 for the overall transmissibility was found to be appropriate.

## RESULTS

### Pressure match

The pressure matches of both the average field pressure decline as well as the flowing bottomhole pressures were considered to be accurate in both cases (Fig. 10).

### Water match

*Water breakthrough well 4*

In both cases the water matches correctly modelled the water breakthrough of well 4 (Fig. 11). The modelled breakthrough of well 4 (March 1987) resulted from production from the lower parts of the perforated interval, indicating that the GWC had reached the lower perforation of well 4. In the field the actual breakthrough occured at the upper part of the perforation where a highly permeable layer is located. A production log indicated that the lower perforations did not contribute to the gas or water production from this well. It was attempted to model the water overruning the gas in a detailed 2D simulation study. Some of the models for this also included semi-continuous shales, though without success. It is not clear whether the preferential water flow around well 4 should be seen as a superimposed effect (e.g. of coning) upon the rise of the GWC. After plugging-off the perforations that produced water and lowering the production rate below the critical coning rate of well 4 in the field, the water production decreased temporarily. Soon afterwards the perforations of well 4 had to be abandoned in the Middle Unit because of excess water production, indicating that the GWC was near the perforation interval.

*Rise of the GWC*

The results of the water matches are summarized in Table II. The modelled locations of the GWC for case 1 in wells 1, 3, and 9 differ from PNL observations by −5, +5, and 0 m respectively. After water breakthrough at well 4 occurred in the model, the water production from well 4 increased faster than observed. This is the result of the modelled production strategy. The model could be validated against the GWC encountered in the newly drilled well 8. The GWC was found at 2782 m ss, about 4 m deeper than was modelled in case 1.

Similar results were achieved with the second model (case 2). The 1987 location of the GWC in wells 1, 3, 8 and 9 differ from PNL observations by +3, −5, −4 and 0 m

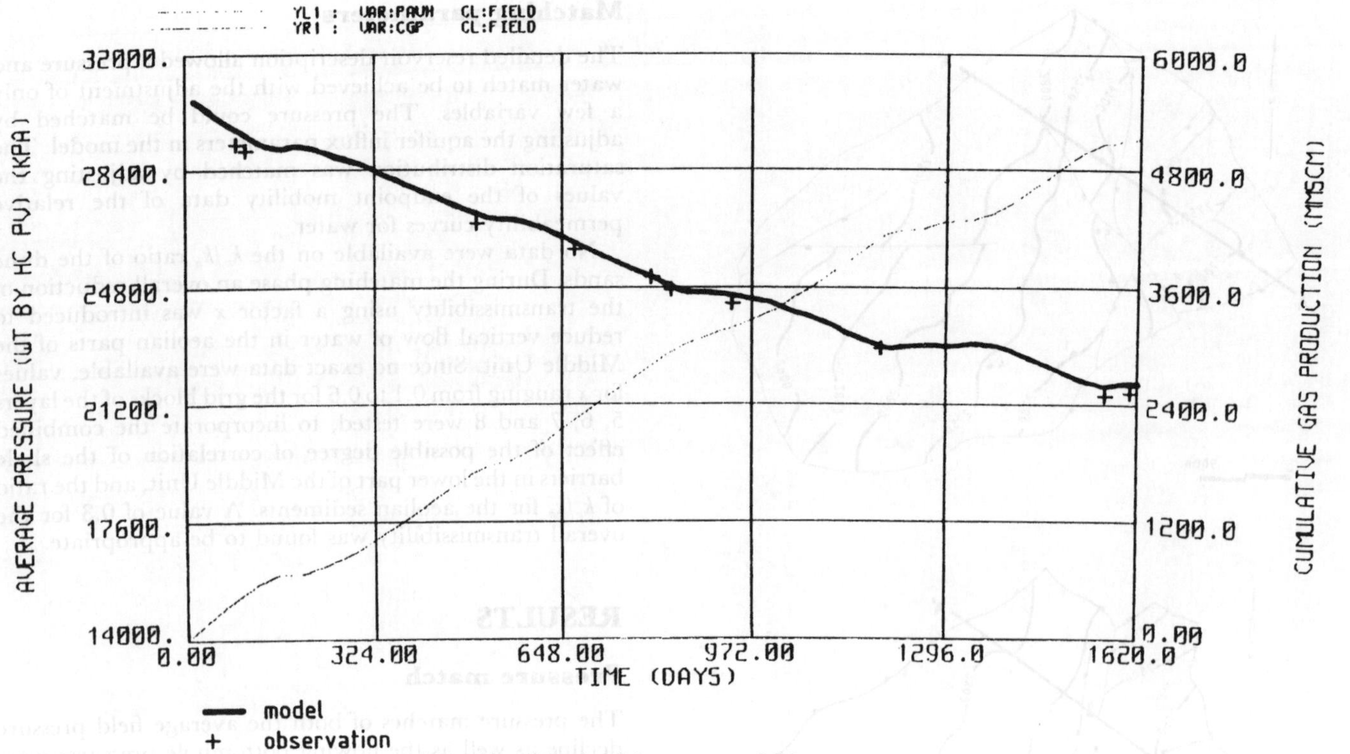


**Fig. 10.** Match of the average field pressure.

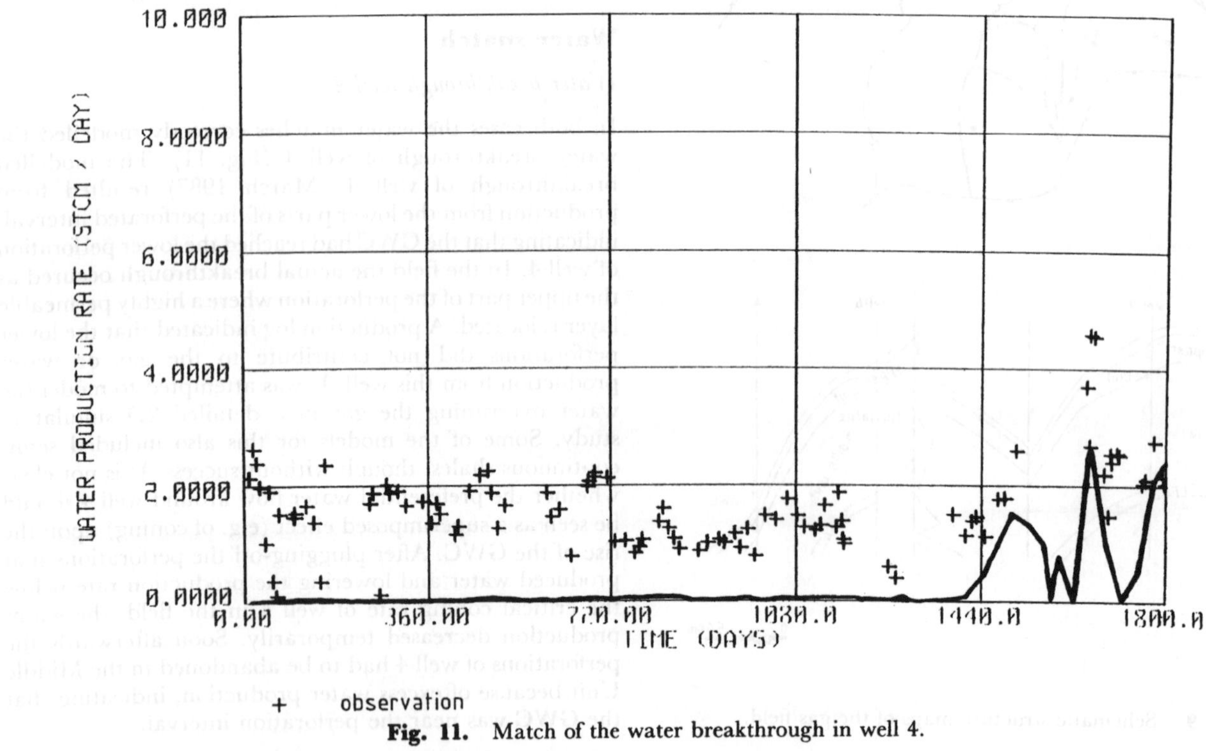


**Fig. 11.** Match of the water breakthrough in well 4.

**Table II.** Simulation results on the rise of the GWC (m ss)

| *Well* | *Log 87* | *Model 1/87* | *+/−* | *Model 2/87* | *+/−* |
|---|---|---|---|---|---|
| 1 | 2788 | 2793 | −5 | 2785 | +3 |
| 3 | 2781 | 2776 | +5 | 2786 | −5 |
| 4 | Water breakthrough occurred | | | | |
| 8 | 2782 | 2778 | +4 | 2785 | −3 |
| 9 | 2791 | 2791 | 0 | 2791 | 0 |
| *Well* | *Log 88* | *Model 1/88* | *+/−* | *Model 2/88* | *+/−* |
| 1 | 2778 | Not modelled | | 2781 | −3 |
| 3 | 2777 | Not modelled | | 2783 | −6 |
| 8 | 2779 | Not modelled | | 2783 | −4 |
| 9 | 2791 | Not modelled | | 2791 | 0 |

respectively. The 1988 locations of the GWC in well 1, 3, 8, and 9 differ by −3, −6, −4 and 0 m respectively.

It should be mentioned here that the good match for well 9 in both models is caused by the presence of the shaly LF4 member beneath the perforations at this depth, which prohibits further rise of the GWC both in the model and in the well.

## DISCUSSION

The approach initially used to model this reservoir, which was based on an assumption of homogeneous facies, failed to model both the moment of water breakthrough and the saturation distribution accurately. Applying the hybrid modelling technique enables the rise of the GWC to be modelled satisfactorily. It is not possible to calculate the location of the GWC from the modelled saturation distribution with greater accuracy than $\pm 5$ m

The rise of the GWC varied in the different field regions (Fig. 12). Further investigation of the logs indicated that the observed stepwise rise of the GWC was influenced by the presence of continuous and discontinuous shale breaks and thus by heterogeneity. Therefore, we concluded that the results of the PNL measurements do not necessarily represent the actual field situation at macroscopic scale. In case 1 the modelled rise of the GWC for wells 3 and 8 was approximately 5.7 m/year. The average calculated rise of the GWC in case 2 (wells 1, 3, and 8) is 4.3 m/year. The observed rise of the GWC in these wells is approximately 5.0 m/year.

The slower rise of the GWC in case 2 indicates a larger reduction of the $k_{v,\mathrm{eff}}$ in the model. Both models represent the actual observed rise of the GWC and the water breakthrough of well 4 satisfactorily.

In the matching phase of case 2 it was calculated that well 6 would start producing water in December 1988. This did not happen in the field. The model was used to study the sensitivity of parameters (e.g. porosity distribution, directional permeability, end point mobilities) that could have caused the early breakthrough in the model. This analysis clearly showed that the 1988 structure map was not correct in certain areas of the reservoir. These areas were reinterpreted, the model was adjusted accordingly and the problem disappeared.

Neither model was sufficiently detailed to consider the effect of water coning. The most recent model only allows field-scale interpretation of the saturation distribution.

## CONCLUSIONS

From this and previous phases of the reservoir study the following conclusions can be drawn.

(1) A proper reservoir description can eliminate the uncertainty about modelling parameters. It reveals the models's sensitivity to different matching parameters.

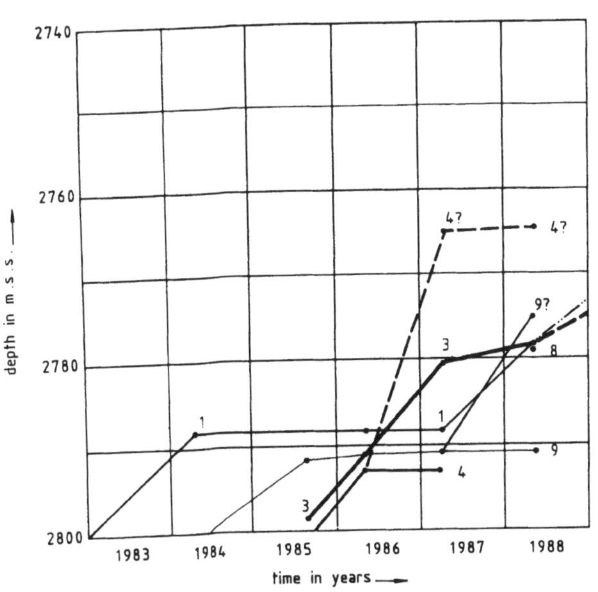


**Fig. 12.** Observed rise of the GWC.

(2) Where reservoir data are sparse, the hybrid approach described here of combining here of combining deterministic and stochastic geological modelling can help provide the engineer with adequate reservoir data.

(3) The method applied to cope with shales can be used to obtain an approximation of the average flow path of a fluid particle in a grid block volume of the reservoir. The actual locations and orientations of the shale barriers are of no importance if the barriers are randomly distributed, are small in relation to grid block size, and do not show marked preferred orientation.

(4) The most recent reservoir model is suitable for interpreting the effects of large-scale saturation distribution on present and future field performance.

(5) The pressure match was easy to accomplish because only the parameters of the aquifer influx functions had to be considered;

(6) Under the specified conditions, the presence of continuous and discontinuous shale breaks in the reservoir influences displacement processes.

## NOMENCLATURE

| | |
|---|---|
| $d_{so}$ | Step-out distance resulting from the presence of the flow barriers (m) |
| $dt_{\log}$ | Measured travel time sonic log |
| $dt_{\mathrm{ma}}$ | Matrix travel time ($\mu$s/m) |
| $dt_{\mathrm{fl}}$ | Fluid travel time ($\mu$s/m) |
| $\phi$ | Average porosity (%) |
| $\phi_{\mathrm{ac}}$ | Porosity from acoustic log (%) |
| $F_S$ | Dimensionless shale fraction |
| GIIP | Gas-initially-in-place (Sm$^3$) |
| GWC | Gas–water-contact (m) |
| $h_{\min}$ | Thickness of a reservoir block volume |
| $k_h$ | Average horizontal permeability (m$^2$) |
| $k_{v,\mathrm{eff}}$ | Vertical effective permeability (m$^2$) |
| MD | Measured depth (m) |
| m ss | Metres subsealevel (m) |
| $N_{\mathrm{SO}}$ | Total number of shales in a grid block |
| $P(k)$ | Possibility of occurrence of $k$ |
| PLT | Production logging tool |
| PNL | Pulse neutron log |
| Sm$^3$ | Standard cubic metre (1 bar, 0°C) |
| $T$ | Tortuosity |
| TVD | True vertical depth (m) |

## REFERENCES


Begg, S.H. and King, P.R. (1985). Modelling the effects of shales on reservoir performance: calculation of effective vertical permeability. SPE paper no. 13539.

Begg, S.H., Chang, D.M. and Haldorsen, H.H. (1985). A simple statistical method for calculating the effective vertical permeability of a reservoir containing discontinuous shales. SPE paper no. 14271.

Bifani, R.G.T. and Lever, A. (1987). Geological and reservoir characteristics of the Rotliegend sandstones in the Argyll Field. In: Brooks, J. and Glennie, K.W. (eds.) *Petroleum Geology of North West Europe*, *pp.* 509–522. Graham and Trotman, London.

Candall, M.G. and Stuart, A. (1969). *The Advanced Theory of Statistics*, Vol. 1. Griffin, London.

Elewaut, E. and Huurdeman, A.J.M. (1986). Reservoir study of a North Sea gas field, part 2: vertical displacement study. TNO Institute of Applied Geoscience report no. HC 86-10, unpublished.

Haak, A.M. and Elewaut, E. (1988). Modelling heterogeneity to improve 3-D simulation results: A case study on a North Sea gas field. *SPE Gas Technology Symposium, Dallas, June* 1988, paper no. SPE 17763.

Haldorsen, H.H. and Lake, L.W. (1984). A new approach to shale management in field-scale models. *Soc. Petroleum Eng. J.* (August), 447–457.

Kool, H. (1988). The influence of shales on fluid flow in reservoirs: A critical review of shale models. Paper in preparation. TNO Institute of Applied Geoscience.

Rust, B.R. (1981). Sedimentation in an arid-zone anastomosing fluvial system: Cooper's Creek, Central Australia *J. Sediment. Petrol.* **51** (3), 745–755.

Tunbridge, I.P. (1984). Facies model for a sandy ephemeral stream and clay playa complex; The Middle Devonian Trentishoe Formation in North Devon, U.K. *Sedimentology* **31**, 697–715.

Weber, K.J. (1982). Influence of common sedimentary structures on fluid flow in reservoir models. *J. Petroleum Technol.* (March).

# 28

# A Stochastic Reservoir Model and its Use in Evaluations of Uncertainties in the Results of Recovery Processes

*J.O. Aasen*[a], *J.K. Silseth*[a], *L. Holden*[b], *H. Omre*[b], *K.B. Halvorsen*[b] *and J. Høiberg*[b]

[a]Continental Shelf and Petroleum Technology Research Institute A/S
[b]Norwegian Computing Centre, Oslo, Norway

A stochastic reservoir description model is defined and a simulation procedure is implemented. The model is based on a discrete facies model with a continuous model for petrophysical variables superimposed. The facies model characterizes the reservoir as discrete facies with user-defined sizes and interactions. The petrophysical variables are defined as porosity, permeabilities and initial saturations. Their characteristics are dependent on the facies in which they occur and they may be correlated. The petrophysical variables constitute the input to the reservoir production simulators. On the basis of the stochastic model and with conditioning on available well information, several realizations of the reservoir description can be provided through simulation.

The reservoir description model is used to study the effects of varying heterogeneities on gas injection on a vertical cross-section. For a given set of model parameters, and observed values in two wells, a number of cross-section descriptions are simulated, with varying net to gross ratios, porosities and permeabilities. Gas injection with a fixed set of production parameters is simulated on each cross-section, and empirical distributions of breakthrough times and recoveries are calculated.

The results show that the variability inherent in the description of the cross-section gives rise to a sizable variability in production, so that no single realization or average is capable of giving representative results. Hence the distributions of the production results become important. Stochastic simulation appears as a simple, though laborious, way of calculating these distributions. The representability of the calculated, empirical distributions as indicators of the possible outcomes of a production scheme depends on the number of realizations chosen as well as the ability of the stochastic model to represent the physical variability of the porous medium.

## INTRODUCTION

Reservoir production simulation* is used to calculate reservoir performance and estimate recovery for a given production scheme, to evaluate the effects on recovery of altered operating conditions, and to compare economics of different recovery methods. To give reliable results, the simulations must be based on accurate numerical models of the fluids involved, their dynamical behaviour, and the porous medium in which they flow. Today's simulations are relatively advanced in their ability to calculate fluid properties, and much effort has been spent in designing numerical models of fluid flow. On the other hand, the reservoir description models used in the simulations have often been overly simplified, e.g. by assigning a single set of rock parameters to large regions of a reservoir, without taking the heterogeneities into consideration. Studies of enhanced oil-recovery methods have, however, led to an increasing awareness of the significance of an accurate reservoir description for the predictability of reservoir potential. Improved reservoir description methods embody several features, viz.

- improved techniques for measuring reservoir parameters;
- development of refined methods for inferring reservoir properties from available measurements;

*The term *simulation* has different meanings when used for flow forecasts by the petroleum engineer and for reservoir description by the geostatistician. To distinguish them we use production simulation in the first meaning and description simulation in the latter.

*North Sea Oil and Gas Reservoirs—II*

- a greater knowledge of the effects of various types of rock variability on displacement processes;
- the ability to represent and utilize the improved reservoir descriptions in production simulators.

Our knowledge about a reservoir under study will always remain incomplete. Hence the objective should be to establish a model characterizing the reservoir parameters with associated uncertainties. Simulations based on this model, assuming consistency with observed data, provide possible realizations of the reservoir parameters. In this contribution a stochastic approach is used taking both facies dependence and spatial variability in the petrophysical parameters into account.

Stochastic reservoir modelling allows a quantification of uncertainties in reservoir parameters by defining them as stochastic variables. Hence, the results of a production simulation—pressures, saturations, breakthrough times, gas or water cuts, recovery—also become stochastic variables, and a study of their statistical properties can lead to a quantification of uncertainties in predicted production results as functions of uncertainties in the reservoir description. To our knowledge, only a few papers in the petroleum literature have addressed this problem (Brown and Smith, 1984; Augedal *et al.*, 1986; Hewett and Behrens, 1988). Analytical results are scarce, based on linearizations, and applicable to one-phase flow only. In some situations, stochastic description simulations can be used. Reservoir production simulations are performed on a set of realizations, and empirical distributions for the production results are obtained. We present an example of this kind. The example concerns gas injection on a cross-section. We calculate empirical distributions of production results for varying facies distributions, porosities, and permeabilities.

In the following section the stochastic model for reservoir description is defined. The underlying assumptions, conditioning and program systems are presented. The next section contains an outline on the assumptions on which the production simulator is based. The results from the example are then presented and discussed. The final section contains some concluding remarks.

## A MODEL FOR RESERVOIR DESCRIPTION

In the computer program system presented here, a discrete process, which may be associated with facies, is overlaid a continuous process having properties conditioned on the actual facies. Conditioning on actual observations in wells can also be performed. Various realizations based on this model concept can be generated and used as input parameters to the reservoir production simulator in order to evaluate its sensitivity to uncertainty in the reservoir description.

In the presentation, it is convenient to divide the variables into two categories.

- Primary variables represent the reservoir characteristics that have direct influence on the flow through the reservoir. Hence, they are input parameters to the reservoir production simulators. These variables are modelled as continuous variables in the three-dimensional (3D) space with model characteristics dependent on the secondary variables present. Relevant primary variables are porosity, initial water saturation and absolute permeability.
- Secondary variables represent reservoir characteristics that influence the value of the primary variables, and hence indirectly the flow rate. The secondary variables may represent a facies model. This entails modelling of reservoir units having reasonably homogeneous characteristics. These facies are modelled as discrete processes in 3D space.

### Model assumptions

In the presentation, the reservoir volume is denoted $V$; $x$ is a location in the reservoir; and $h$ is a shift vector. It seems natural to present the secondary variables first:

*Secondary variables*

The following discrete stochastic model is used for the facies variables

$$\{L(x);\ x \in V\}$$

with

$$L(x) \in \{l^0, l^1, l^2\}:$$

Each location in the reservoir can be assigned one, and only one, facies. The facies model distinguishes three types of facies (see Fig. 1):

- *Matrix facies*, $l^0$, which constitutes the reservoir matrix. It can be considered the major sedimentation process in the creation of the reservoir.
- *Eroding facies*, $l^1$, which are bodies distributed in the reservoir. They can be considered as eroded into the matrix facies. Each facies body has a given shape and a size according to a specified probability distribution for size, $F_v(v)$. Each body also has a reference location in $V$. These reference locations are distributed in $V$ according to a point process with a pairwise interaction function $r^1(h)$. This makes it possible to simulate repulsion between the bodies. The total number of eroding facies bodies is determined by specifying, $ng^1$, the net-to-gross ratio.
- *Barrier facies*, $l^2$, which are horizontal planes in the reservoir. They can be associated with barriers in the reservoir caused by either shales or carbonate cementation. Each barrier has a given shape and an area

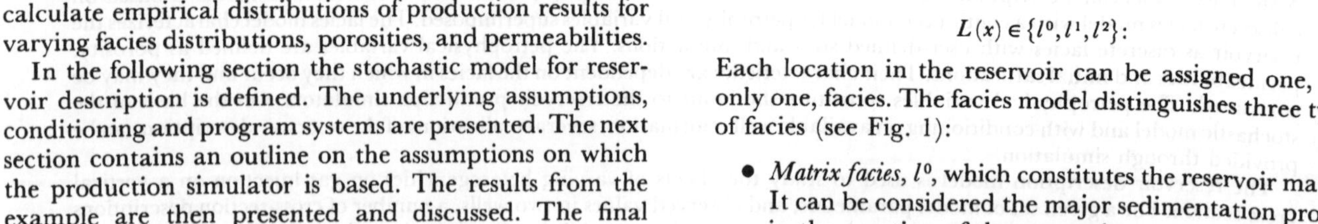


**Fig. 1.** Facies types generated by the facies model.

according to a specified probability distribution for area, $F_A(a)$. Each plane has a reference location in $V$ and pairwise interaction is defined through $r^2(h)$. This allows for repulsion between barriers. The total number of barriers, $n^2$, must be specified.

In the present version of the program, only one eroding facies can be used, an extension to allow for more is possible. Similarly, there is no interaction between eroding and barrier facies, but this can be included relatively simply.

*Primary variables*

The following continuous stochastic model is used for a primary variable.

$$\{P(x);\ x \in V\}$$

The primary variable has an expectation that is related to the secondary variable present:

$$E\{P(x) \mid L(x)\} = \mu^P_{L(x)}; \qquad x \in V$$

This means that, for example, expected porosity is dependent on the actual facies present.

The spatial covariance is

$$\mathrm{Cov}\{P(x'), P(x'') \mid L(x'), L(x'')\} = \sigma^P_{L(x')}\,\sigma^P_{L(x'')}\,\rho^P(x'-x'');\ x', x'' \in V$$

with the spatial correlation function $\rho^P(\cdot)$ dependent on distance $x'-x''$ only. Note that anisotropy is allowed.

A consequence of the model is

$$\mathrm{Var}\{P(x) \mid L(x)\} = (\sigma^P_{L(x)})^2$$

Hence, the primary variable has a variability that is related to the secondary variable present, while the spatial correlation is constant throughout the reservoir. This means that the variability in, for example, porosity in a given location is dependent on the actual facies present.

In this formation only one primary variable is defined. In Berteig *et al.* (1987) a model with several intercorrelated variables was defined and simulated.

At this stage the stochastic model is specified up to second order, i.e. the expectation and variance, and simulations based on a Gaussian assumption can be performed. Unvariate transformations may be performed in order to reproduce skew distributions.

## Conditioning

Well observations in the reservoir can be taken into consideration. These observations should contain information about both primary and secondary variables. For each well one has to specify a well location $(x^1, x^2)$ and observed values

$$\{L(x^3),\ P(x^3);\ x^3 \in [0, x^3_{\max}]\}$$

where $x^1, x^2, x^3$ are the coordinates of the reference variable $x$ in $V$. This entails that facies bodies, barrier planes and petrophysical parameters may be specified in locations down the well. It is assumed that the wells are so far apart that no facies bodies or barrier planes are penetrated by more than one well.

The simulation of the reservoir has to be conditioned on these observations, i.e. they have to be reproduced in all realizations, because they are values considered as known for the reservoir.

## Simulation procedure

The following procedure is used.

- *Secondary variables.* The secondary variables are simulated as marked point processes by an algorithm known as the Ripley–Kelly algorithm (see Stoyan *et al.*, 1987; Ripley, 1981). A death/birth algorithm is used to allow for spatial intersection. Conditioning on the facies observed in the wells is ensured.
- *Primary variables.* The primary variables are continuous and are simulated according to a Gaussian random field as described in Journel and Huijbregts (1978), with the extensions defined in Omre *et al.* (1987). Conditioning on values observed in the wells is ensured.

## Program system

The program stores the reservoir repesentations in a grid format in order to communicate easily with the reservoir production simulator. Hence,

$$\{x \in V\} \text{ is discretized to } \{x_{ijk};\ (i,j,k) \in D_v\}$$

with $D_v$ being a set of indices on a grid covering $V$.

For the actual application of evaluating the flow in reservoirs, the primary variables should be defined by $P(x_{ijk})$ where represents the value of $P$ in a grid block with centre $x_{ijk}$. For the secondary variables the following are required.

- The eroding facies must coincide with the grid intersections and a set of grid blocks.
- The barrier planes must coincide with the grid intersections and a set of grid squares.

The parameters required for running the computer program are specified in Halvorsen *et al.* (1989).

# A MODEL FOR FLUID FLOW

The reservoir production simulator is used to study variability in production results due to variations in reservoir parameters. The fluid is described by standard black-oil equations, in standard notation (Aziz and Settari, 1979)

$$\partial_t(\phi S_w/B_w) + \nabla\cdot(\mathbf{u}_w/B_w) + q_w = 0$$

$$\partial_t(\phi S_o/B_o) + \nabla\cdot(\mathbf{u}_o/B_o) + q_o = 0$$

$$\partial_t(\phi S_g/B_g + \phi S_o R_s/B_o) + \nabla\cdot(\mathbf{u}_g/B_g + R_s\mathbf{u}_o/B_o) + q_g = 0$$

$$\mathbf{u}_j = -\mathbf{K}(k_{rj}/\mu_j)(\nabla p_j - \rho_j g \nabla z), \qquad j = o,\ g,\ w$$

$$p_j = p_{coj} + p_o, \qquad j = w,\ g$$

In this study the fluid parameters ($B_o$, $B_g$, etc.) and the rock-fluid parameters ($k_{ro}$, $p_{cog}$, etc.) are held fixed, while porosity and absolute permeabilities are allowed to vary. In the production simulations we have used the simulator ECLIPSE.

# EXAMPLE

The example is a study of gas injection on a vertical cross-section. The cross-section has length 3600 m, height 45 m and a dip of 7°, and contains a gas cap, an oil zone and an aquifer. Details of the cross-section are shown in Fig. 2.

To construct the porosity and permeability fields we divide the cross-section into 45 layers, each containing 36 blocks of length 100 m. We assume that the cross-section consists of two facies:

- a matrix facies $l^0$, corresponding to a massive sandstone;
- an eroding facies $l^1$, corresponding to a layered sandstone.

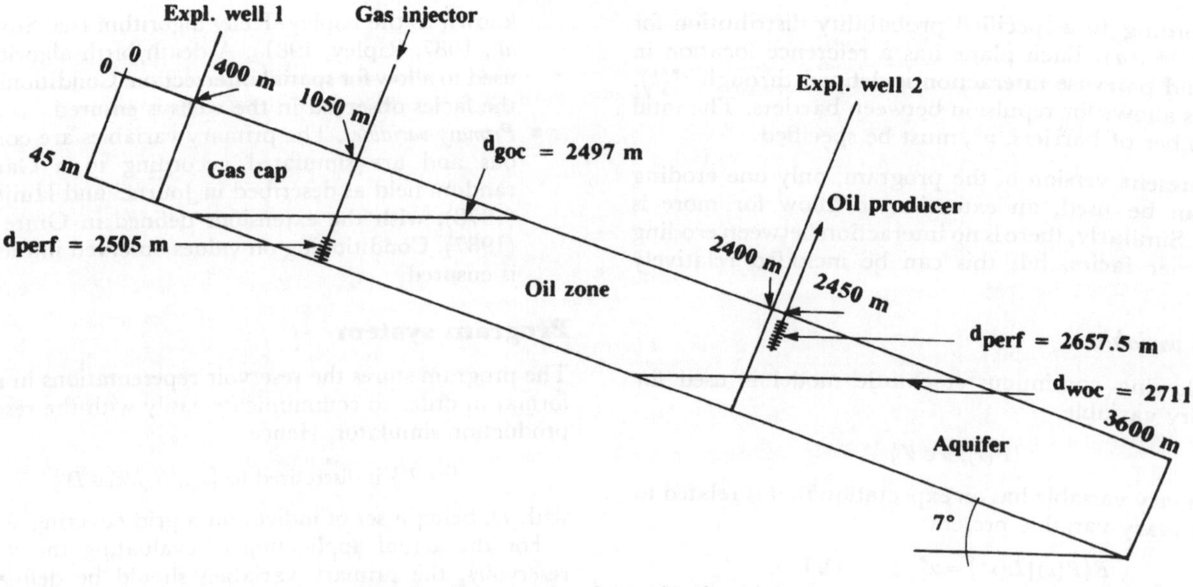


**Fig. 2.** Cross-section for study of gas injection.

The layered sandstone is specified by a net-to-gross ratio (defined as the ratio between the area of layered sandstone and the area of the cross-section), and probability distributions for length and thickness. Length and thickness are Gaussian with expectation 1000 m and 5 m, and standard deviations 500 m and 2 m respectively. The repulsion function is

$$r^1(h)=\begin{cases}0, & h<1\\ (h-1)/19, & 1<h<20\\ 1, & h>20\end{cases}$$

This means that the distance between sand body centres cannot be smaller that 1 m and that bodies with centres further away than 20 m are independent.

The net-to-gross ratio is allowed to vary. A histogram for calculated net-to-gross values is shown in Fig. 3. The primary variable is defined as $P=\log K_H$ where $K_H$ is horizontal permeability. The variable $P$ has the following characteristics.

Expectations and standard deviations are

$$\mu^P_{l0}=7.14, \qquad \sigma^P_{l0}=0.7$$
$$\mu^P_{l0}=5.67, \qquad \sigma^P_{l0}=0.9$$

and the correlation function is

$$\rho^P(x,z)=1-\tfrac{3}{2}h-\tfrac{1}{2}h^3, \qquad h=[(x/200)^2+(z/2)^2]^{1/2}$$

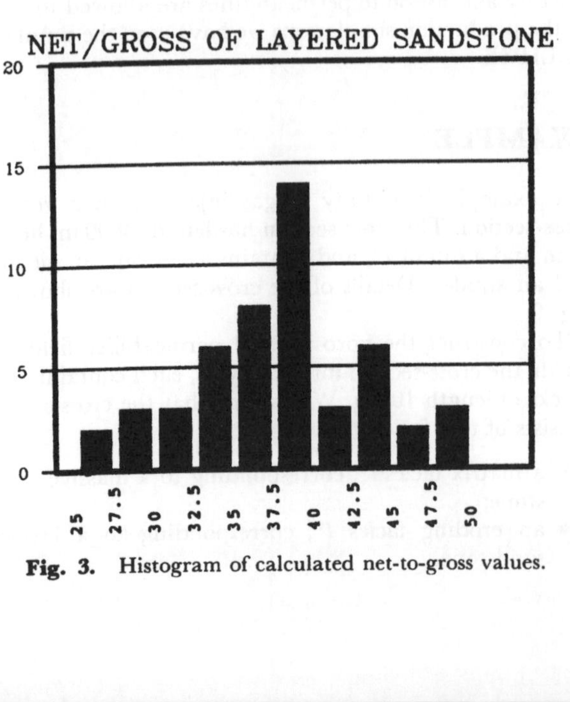


**Fig. 3.** Histogram of calculated net-to-gross values.

After generating $K_H$ by simulating $P$, we calculate vertical permeability $K_v$ by

$$K_V=\begin{cases}0.3K_H & \text{in massive sand}\\ 0.1K_H & \text{in layered sand}\end{cases}$$

and porosity $\phi$ through

$$\phi=0.15+0.01\log K_H+\varepsilon$$

where $\varepsilon$ is a noise term. A porosity/permeability plot for one of the realizations is shown in Fig. 4.

We assume that observations of facies type and horizontal permeabilities are available in two exploration wells, as shown in Fig. 2. The observations are given in Fig. 5. The observed net-to-gross ratios are 82% in well 1 and 18% in well 2.

Generation of realizations of the description model, conditioned on the well observations, uses a 36 × 45 grid. In the production simulations the number of layers was reduced to 15. Consequently, a homogenization of porosites and permeabilities on three vertically adjacent grid blocks had to be performed. Arithmetic averaging was used for porosity and horizontal permeability, and harmonic averaging was used for vertical permeability. In this way, 50 realizations of a cross-section containing 36 × 15 grid blocks were generated. Figures 6 to 8 show the resulting facies distributions, porosities, and permeabilities of three of the realizations. The legend for the raster plots is shown in Fig. 9.

Arithmetic averages of porosity and permeability were calculated for each realization. The calculated average porosities varied very little among the realizations. The lowest value was 21.30% and the highest 21.86%. With fluid parameters as shown in Table I oil in place varied between $3.32 \times 10^6$ Sm$^3$ and $3.40 \times 10^6$ Sm$^3$ for the realizations. The average horizontal permeabilities varied between 924 mD and 1354 mD, and the average vertical permeabilities between 205 mD and 355 mD. Histograms of the distributions show bimodality only for vertical permeability. For porosity and horizontal permeability, the variability inside a facies, the relative closeness of the expectations, and the averaging have made the resulting distributions unimodal.

We have simulated oil production by gas injection on each of the 50 realizations. A gas injector was completed in the lowest five layers of column 11 and an oil producer was completed in the highest five layers of column 25. The

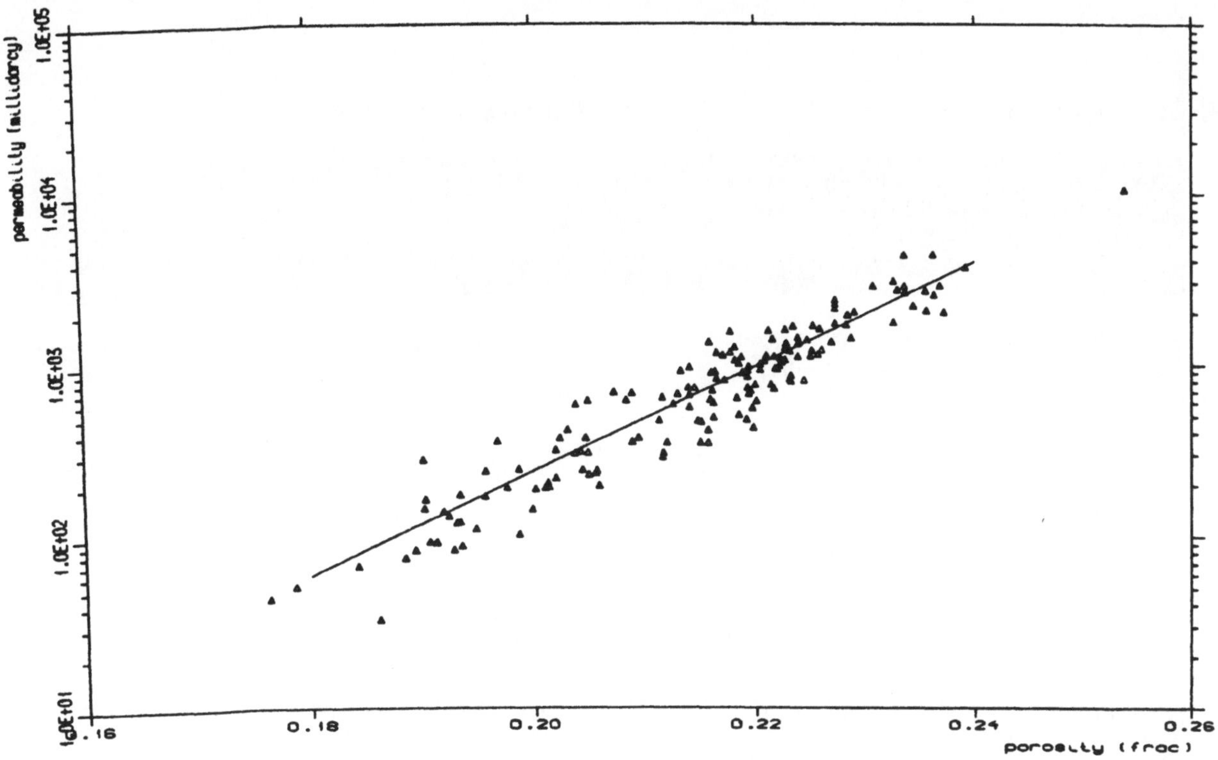


**Fig. 4.** Porosity – permeability plot for a typical realization.

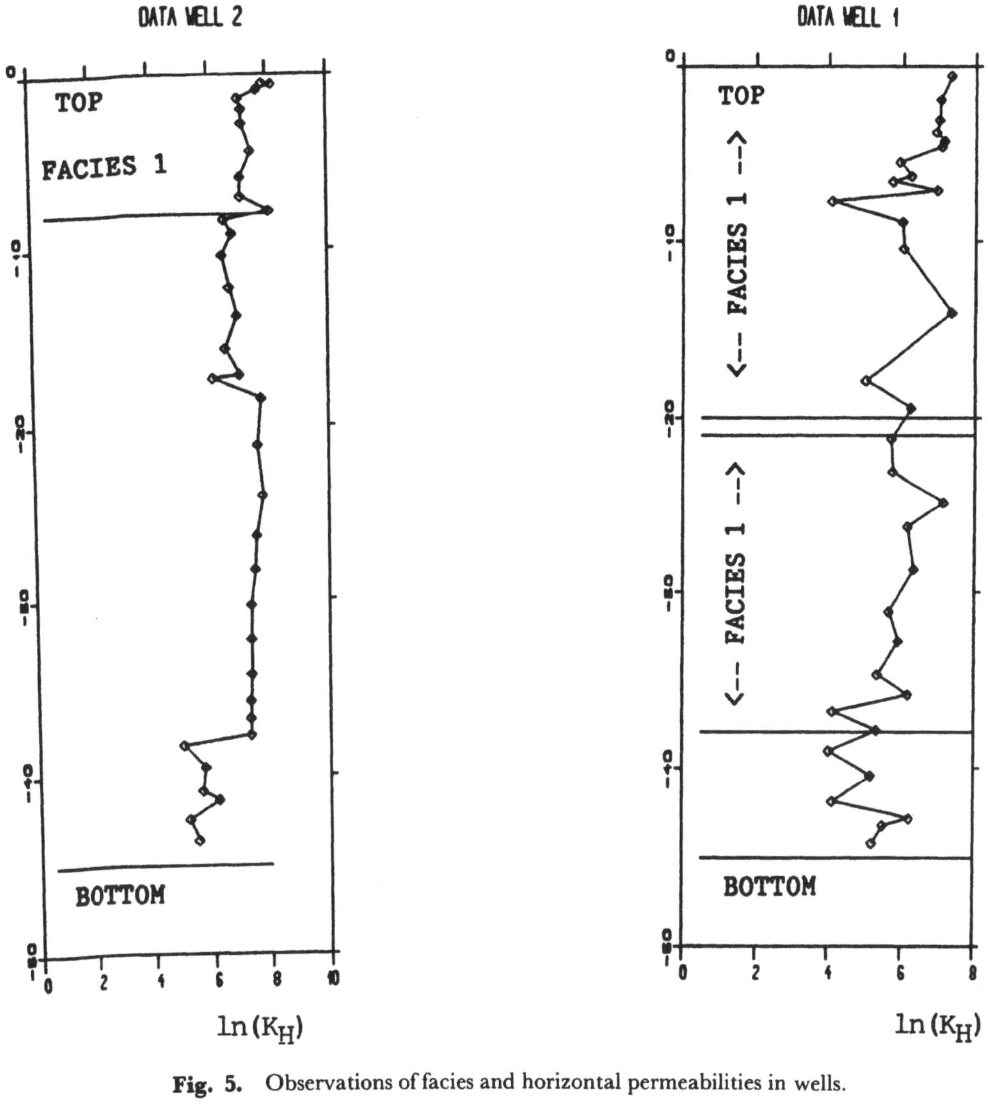


**Fig. 5.** Observations of facies and horizontal permeabilities in wells.

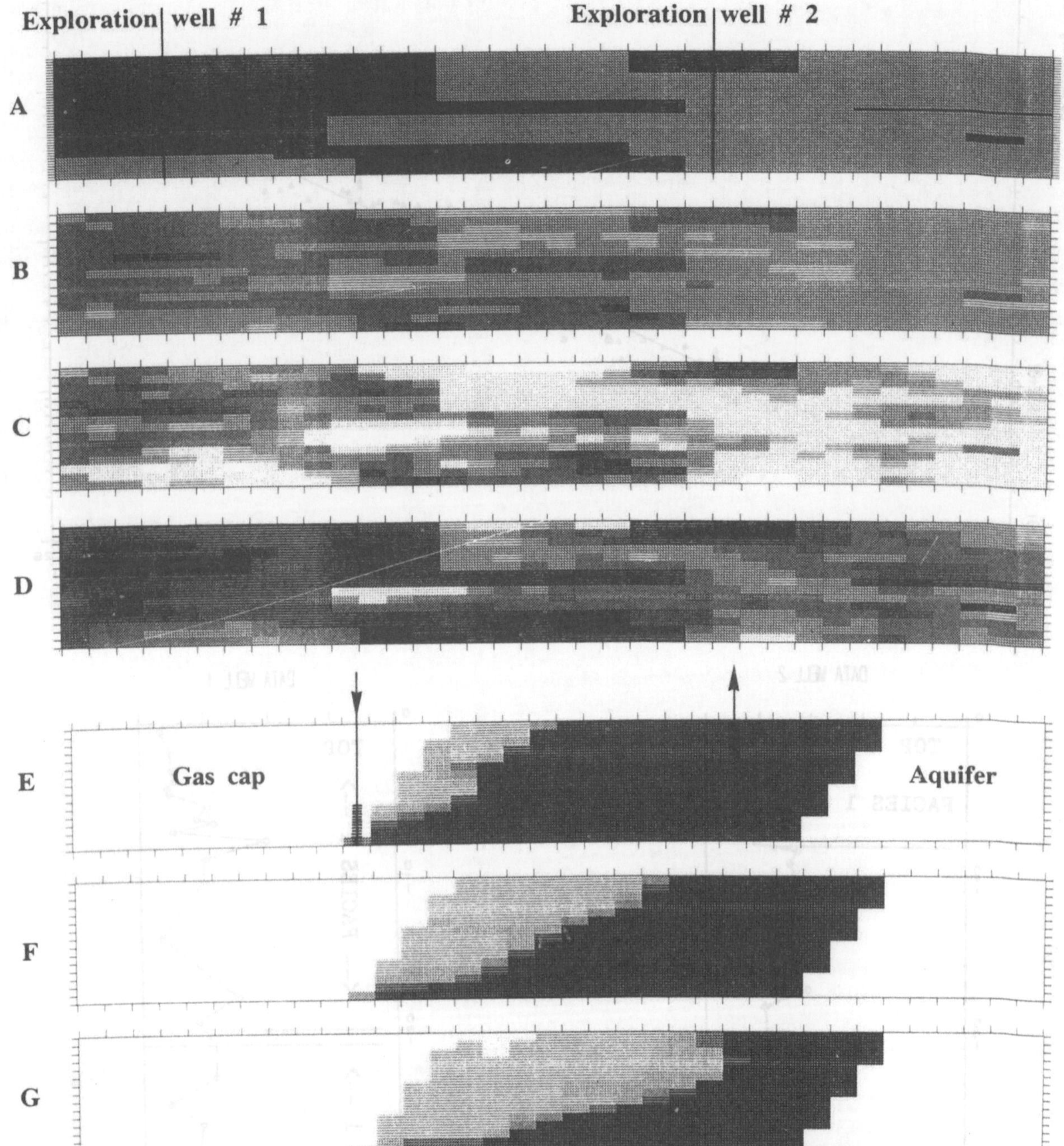


**Fig. 6.** A, facies distribution; B, porosity; C, horizontal permeability; D, vertical permeability; E–I, gas saturation at 2, 4, 6, 8, 10 years.

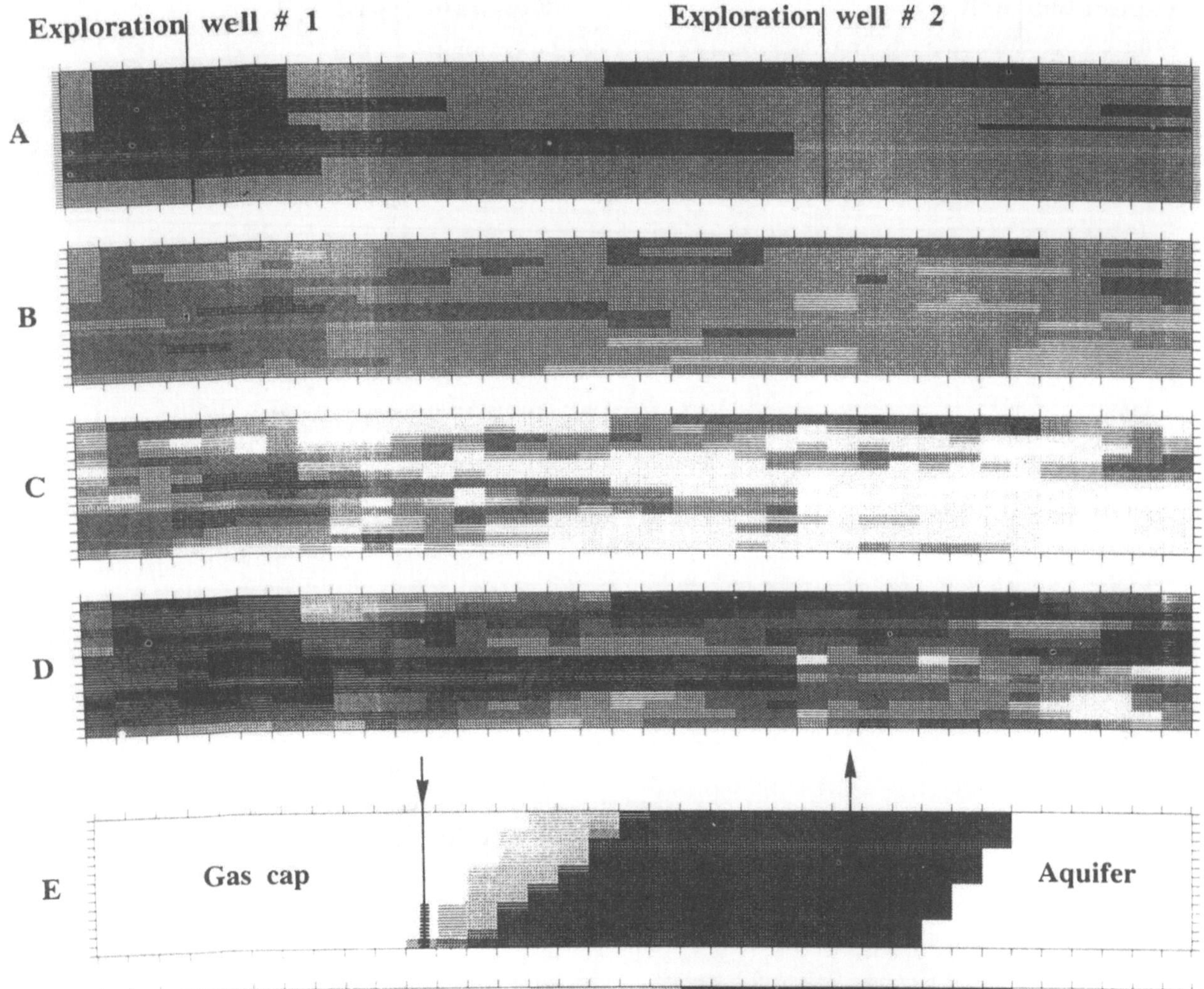


**Fig. 7.** A, facies distribution; B, porosity; C, horizontal permeability; D, vertical permeability; E–I, gas saturation at 2, 4, 6, 8, 10 years.

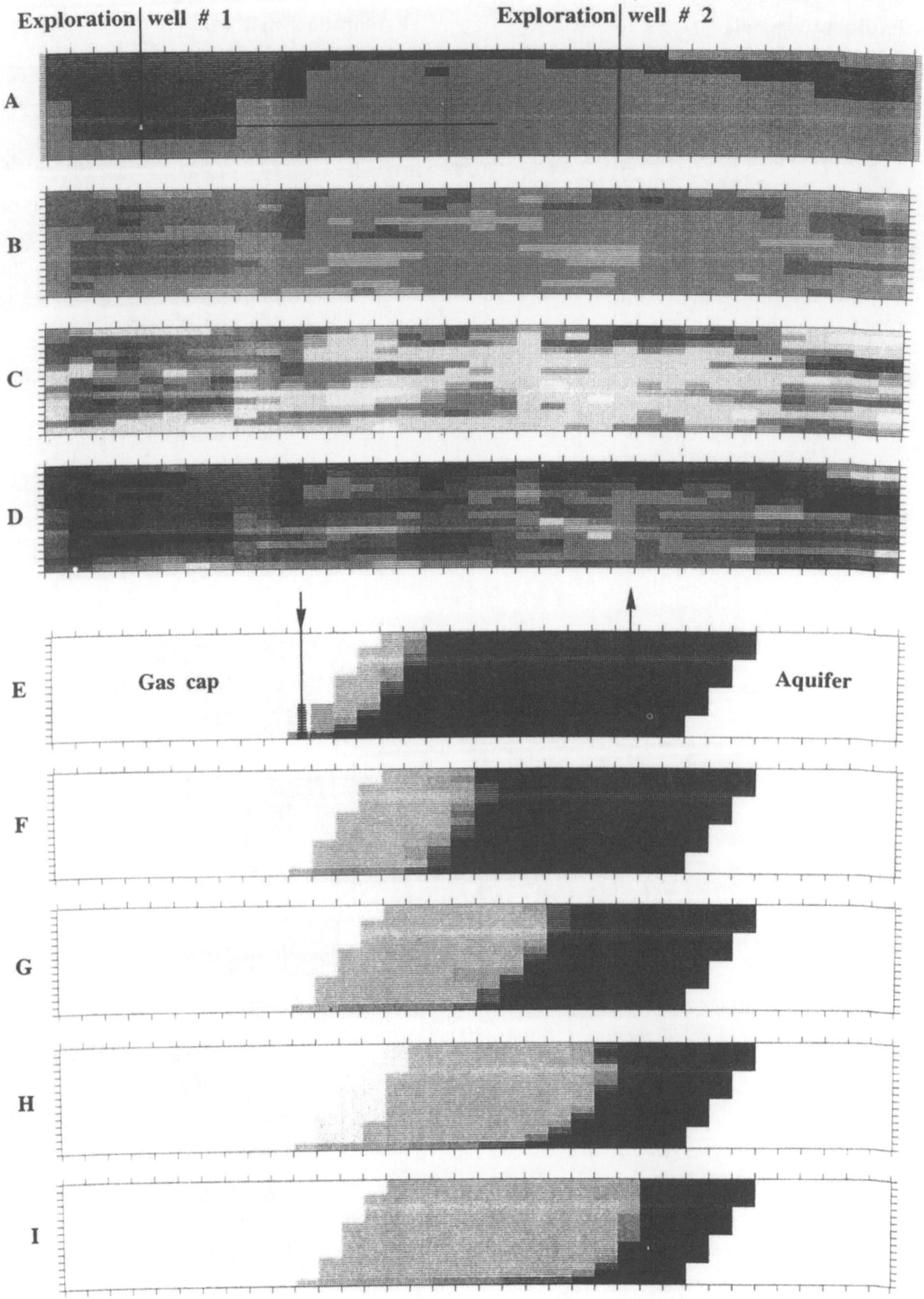


**Fig. 8.** A, facies distribution; B, porosity; C, horizontal permeability; D, vertical permeability; E–I, gas saturation at 2, 4, 6, 8, 10 years.

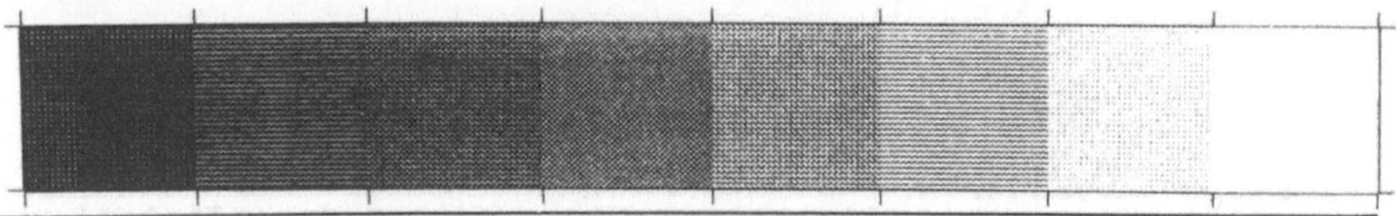

| | | | | | | | |
|---|---|---|---|---|---|---|---|
| **Porosity (fraction)** | **0.15** | **0.17** | **0.19** | **0.21** | **0.23** | **0.25** | **0.27** |
| **Permeability (mD)** | **20** | **80** | **180** | **400** | **750** | **1000** | **3000** |
| **Saturation (fraction)** | **0.01** | **0.10** | **0.20** | **0.30** | **0.40** | **0.50** | **0.65** |

**Fig. 9.** Interpretation of raster plots in Figs 6–8.

production period was 10 years. The injection rate was held fixed at 300 000 $Sm^3/d$ for the whole period, with no restrictions on the well pressure. The oil production rate was held fixed at 400 $Sm^3/d$ until the bottomhole pressure had reached 275 bars. Thereafter, the bottomhole pressure was held fixed at this value. Initial conditions, including saturations, were the same for all simulations. The production parameters used in the simulations can be found in Table I.

Figures 10 to 12 show gas–oil ratios, production rates and recoveries for the simulations. We have defined breakthrough time as the time at which the gas–oil ratio reaches the value 315 $Sm^3/Sm^3$. Histograms of breakthrough times and recoveries after 10 years are shown in Figs. 13 and 14. We have also simulated the gas injection scheme on a cross-section with porosities and permeabilities calculated as arithmetic averages of the 50 realisations. The results are included in Figs. 10 to 12.

The breakthrough times vary between 6.26 years and 8.64 years, with an average of 7.88 years and a standard deviation of 0.50 year. The recoveries after 10 years, as percentages of original oil in place, vary between 36.9% and 43.0% with an average of 40.6% and a standard deviation of 1.5%. The breakthrough time and recovery for the averaged cross-section are 8.49 years and 41.9%.

The distributions of breakthrough time and recovery are both skew, with tails toward shorter times and smaller recoveries. The results from the averaged cross-section are on the high side both for breakthrough time and recovery. However, these results do not vary significantly from the arithmetic average of the results of the 50 realizations.

The average porosities for the realizations vary very

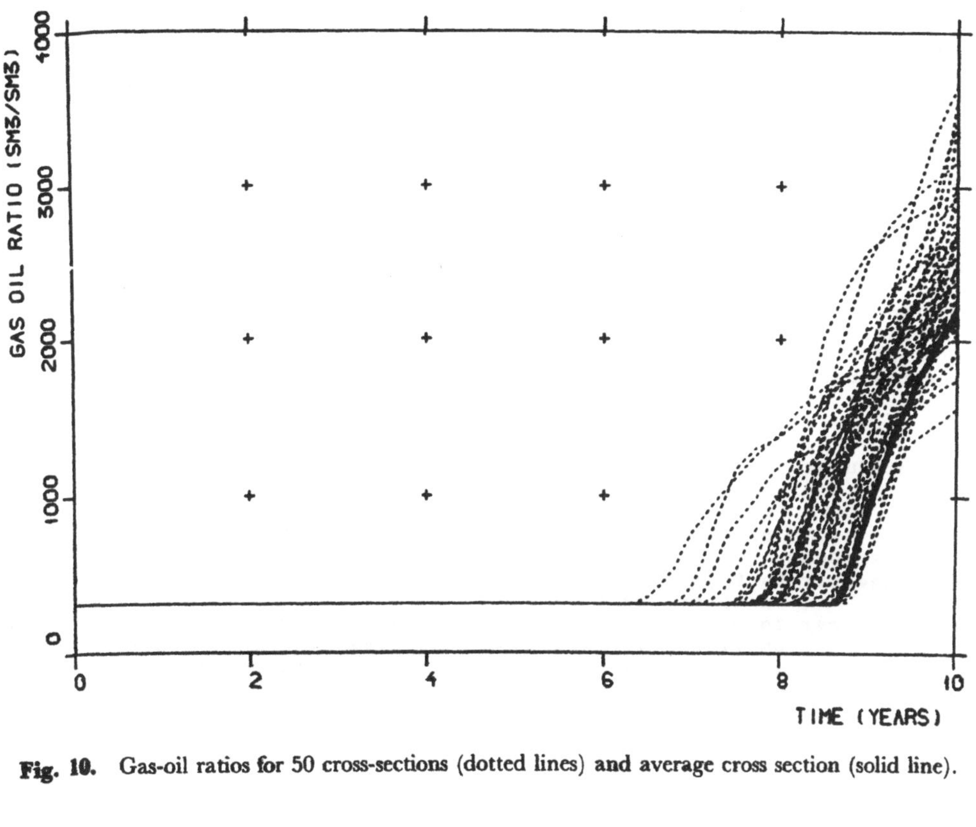


**Fig. 10.** Gas-oil ratios for 50 cross-sections (dotted lines) and average cross section (solid line).

**Table I.** Production data for gas injection.

**GRID** discretization, blocks in x-dir,y-dir,z-dir: 36,1,15.
regular grid with size 100m in x-dir 450m in y-dir 3m in z-dir
depth to top center of block (1,1,1) 2345.8m 7 deg down-dip in x-dir

**ROCK**
compressibility pref= 248 bars crock= 0.0006 1/bar

**RELATIVE PERMEABILITIES**

| water sat SWAT frac | w rel perm KRW frac | PCOW | gas sat SGAS frac | g rel perm KRG frac | PCOG | oil sat SOIL frac | wat-oil relperm KROW frac | gas-oil (1) relperm KROG frac |
|---|---|---|---|---|---|---|---|---|
| 0.2 | 0. | 0. | 0. | 0. | 0. | 0. | 0. | 0. |
| 0.25 | 0.01 | 0. | 0.110 | 0 | 0. | 0.284 | 0.00005 | 0. |
| 0.30 | 0.03 | 0. | 0.234 | 0 | 0. | 0.309 | 0.00018 | 0. |
| 0.40 | 0.07 | 0. | 0.317 | 0.005 | 0. | 0.322 | 0.00025 | 0.0008 |
| 0.50 | 0.13 | 0. | 0.369 | 0.025 | 0. | 0.341 | 0.00054 | 0.0014 |
| 0.547 | 0.163 | 0. | 0.435 | 0.075 | 0. | 0.365 | 0.00200 | 0.0022 |
| 0.570 | 0.174 | 0. | 0.459 | 0.125 | 0. | 0.431 | 0.00800 | 0.0082 |
| 0.582 | 0.179 | 0. | 0.478 | 0.19 | 0. | 0.493 | 0.03450 | 0.0206 |
| 0.614 | 0.202 | 0. | 0.491 | 0.41 | 0. | 0.565 | 0.07000 | 0.0514 |
| 0.645 | 0.226 | 0. | 0.8 | 1.0 | 0. | 0.690 | 0.22000 | 0.3660 |
| 0.658 | 0.242 | 0. | | | | 0.770 | 0.47000 | 0.7600 |
| 0.670 | 0.260 | 0. | | | | 0.800 | 1.00000 | 1.0000 |
| 0.681 | 0.275 | 0 | | | | | | |
| 0.696 | 0.296 | 0. | | | | | | |
| 0.701 | 0.305 | 0. | | | | | | |
| 0.706 | 0.316 | 0. | | | | | | |
| 0.731 | 0.360 | 0. | | | | | | |

(1) measured in presence of a irreducible water saturation (siw)
kro = (sg*krog+(sw-siw)*krow)/(sg+sw-siw)

**FLUID PROPERTIES**
WATER at 248 bars: fvf= 1.054 compr.= 4.75e-5 1/bars visc.= 0.33 cp
oil density at stc ............ ogrv = 765.1 kg/m3
water density at stc .......... wgrw = 1030. kg/m3
gas density at stc ............ ggrv = 1.05 kg/m3

Dry gas properties

| PGAS bars | BGAS rm3/sm3 | VISGAS cp |
|---|---|---|
| 1.0 | 0.98631 | 0.0096 |
| 13.8 | 0.06975 | 0.0118 |
| 55.2 | 0.01650 | 0.0134 |
| 82.7 | 0.01072 | 0.0144 |
| 110.3 | 0.00788 | 0.0157 |
| 137.9 | 0.00622 | 0.0173 |
| 165.5 | 0.00516 | 0.0193 |
| 193.1 | 0.00444 | 0.0217 |
| 220.6 | 0.00394 | 0.0243 |
| 238.2 | 0.00370 | 0.0262 |
| 248.2 | 0.00358 | 0.0273 |
| 262.0 | 0.00344 | 0.0288 |
| 275.8 | 0.00331 | 0.0304 |
| 310.3 | 0.00308 | 0.0340 |
| 314.4 | 0.00306 | 0.0353 |

Saturated oil properties

| | RS sm3/sm3 | POIL bars | FVFO rm3/sm3 | VISO cp |
|---|---|---|---|---|
| | 0. | 1. | 1.0357 | 1.6655 |
| | 27.89 | 13.8 | 1.1603 | 1.1395 |
| | 60.96 | 55.2 | 1.2644 | 0.8164 |
| | 79.38 | 82.7 | 1.3155 | 0.6977 |
| | 98.59 | 110.3 | 1.3663 | 0.6012 |
| | 118.99 | 137.9 | 1.4210 | 0.5213 |
| | 140.93 | 165.5 | 1.4789 | 0.4544 |
| | 164.80 | 193.1 | 1.5413 | 0.3978 |
| | 191.01 | 220.6 | 1.6111 | 0.3496 |
| | 209.44 | 238.2 | 1.6601 | 0.3222 |
| | 220.62 | 248.2 | 1.6900 | 0.3076 |
| | 236.96 | 262.0 | 1.7339 | 0.2887 |
| | 254.54 | 275.8 | 1.7815 | 0.2709 |
| | 304.85 | 310.3 | 1.9184 | 0.2313 |
| (2) | 311.77 | 314.4 | 1.9373 | 0.2270 |
| | 414.52 | 375.0 | 2.1200 | 0.1600 |

(2) bubble point pressure is initially 314.4 bars everywhere
undersaturated oil fvf slope ............... -5.60E-3 1/bars
oil viscosity slope ......... 4.60E-3 1/bars

**INITIAL CONDITION**
pressure at the gas oil contact = bubblepoint pressure = 314.4 bars
depth to the gas oil contact ...................... dgoc = 2497 m
depth to the oil water contact .................... dwoc = 2711 m
no variation of rs with depth

**WELL SPECIFICATION**
gas injector — structurally high - stratigraphically low
in block i=11,j=1 perforated in layer 11,12,13,14,15
well diam 0.21 m depth mid perf 2505 m
prod index calc from permeability thickness product
oil producer — structurally low - stratigraphically high
in block i=25,j=1 perforated in layer 1,2,3,4,5
well diam 0.21 m depth mid perf 2657.5 m
prod index cal from permeability thickness product

**PRODUCTION SCEDULE**
gas injector — to inject 300000 sm3/day for 10 years with no restriction on upper pressure limit
oil producer — to produce at a target plateau rate of 400 sm3/day until a lower limit on well pressure of 275 bars is reached - then convert to constant bottomhole pressure.

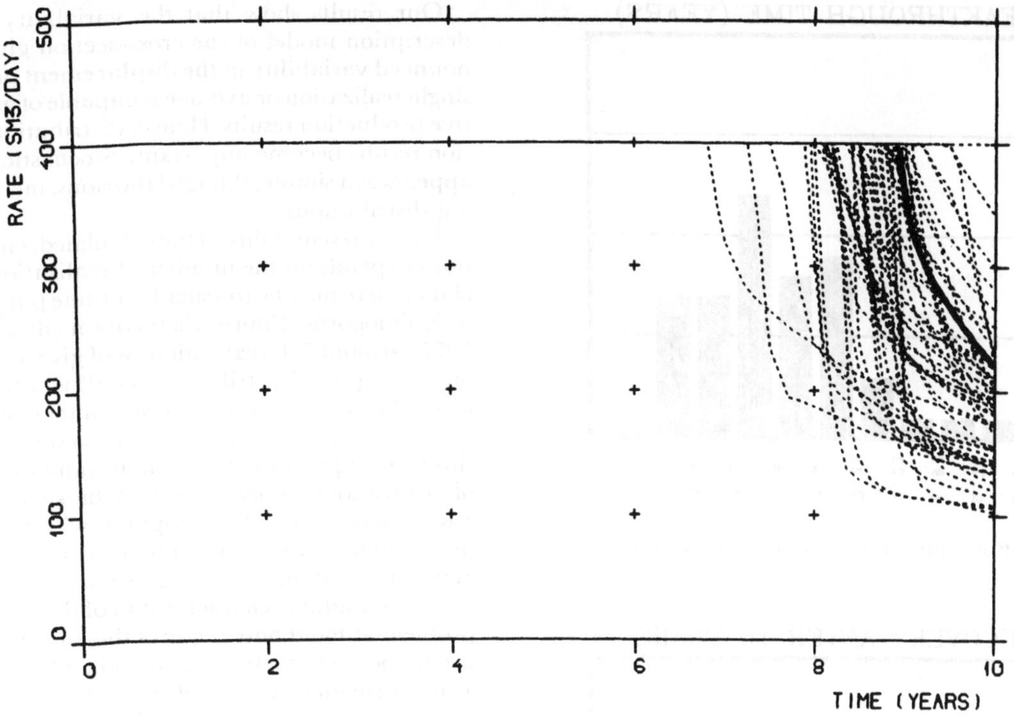


**Fig. 11.** Production rates for 50 cross sections (dotted lines) and average cross section (solid line).

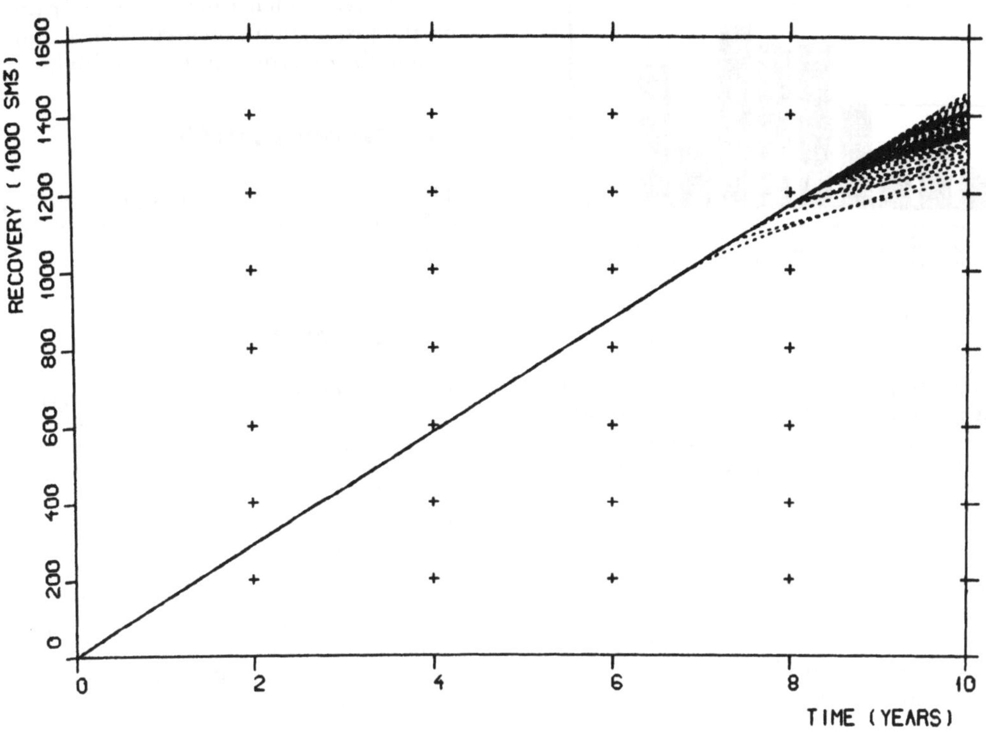


**Fig. 12.** Cumulative oil production for 50 cross-sections.

little, hence the distribution for breakthrough time will have the same form if times are expressed in terms of pore volumes injected. A plot of breakthrough times against net-to-gross ratios of the region between the wells shows that these quantities are uncorrelated, indicating that the effect of facies distribution has been masked by the variability of permeabilities inside each facies and the averaging included in the layer reductions. Estimates of critical velocities based on Dietz's formula and typical fluid and rock data indicate that the displacement may be unstable in some of the cross-sections. A study of the gas saturations confirms this view. Figure 6 shows gas saturations at 2, 4, 6 and 10 years of the realization that gave both the shortest breakthrough time and the lowest recovery. Figures 7 and 8 show gas saturations at the same times for the realizations with the highest recovery and largest

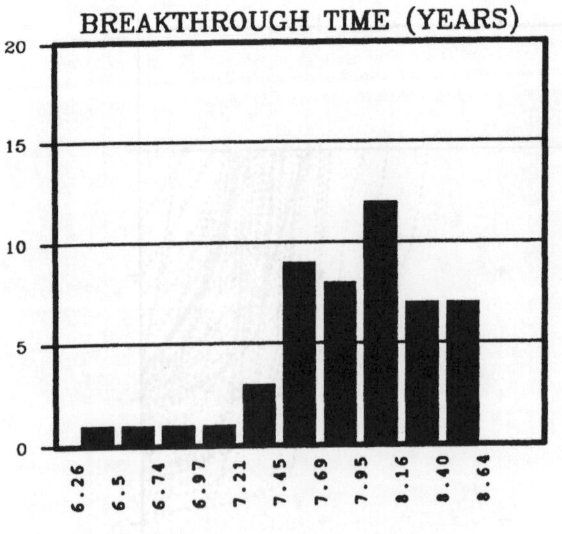


**Fig. 13.** Breakthrough times from 50 cross-sections.

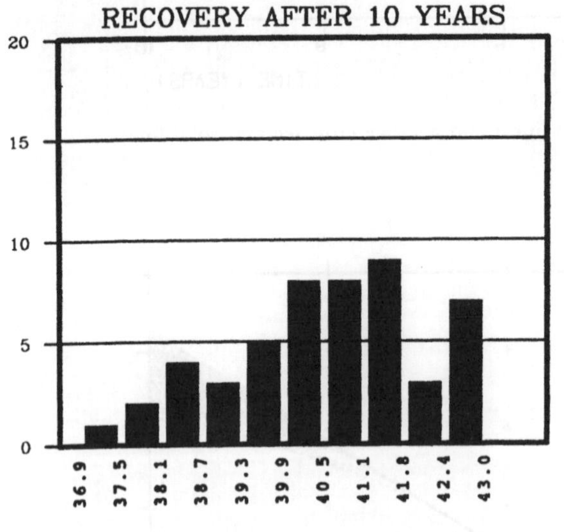


**Fig. 14.** Recoveries from 50 cross-sections.

breakthrough time, respectively. The gas fronts appear to move stably, at a constant angle, in the last two figures, while a gravity tongue seems to have formed in Fig. 6. These results suggest a relation between breakthrough times and viscous-to-gravity ratios. We have tried this, with the square root of the product of average horizontal and vertical permeabilities from the region between the wells as a typical permeability in the viscous-to-gravity ratio. We found no correlation between the two quantities, and conclude that a single aggregate measure characterizing each cross-section is insufficient to explain the variations in the production results.

## CONCLUDING REMARKS

We have presented a stochastic description model with both a discrete process, and a continuous process having properties conditioned on the discrete process. The model also allows conditioning on observations in wells. The model has been used to generate data for simulations of gas injection on a cross-section. Through production simulations using the realizations, we have generated empirical distributions for breakthrough times and recovery as examples of variables characterizing the displacement process.

Our results show that the variability inherent in the description model of the cross-section gives rise to a pronounced variability in the displacement process, so that no single realization or average is capable of giving representative production results. Hence, distributions of the production results become important. Stochastic simulation then appears as a simple, though laborious, method for calculating distributions.

The representability of the calculated, empirical distributions depends on the number of realizations. The number chosen here may be too small for some purposes: according to Kolmogorov–Smirnov's theorem (Bickel and Doksum, 1977) around 740 realizations will give a confidence band for the empirical distribution of half-width 0.05 at the 95% level. To do simulations on a number of realizations as large as this is out of the question, so we will never achieve this level of precision by random simulation. The number of realizations necessary would be smaller if one could select a set of reservoir descriptions with associated weights in an intelligent manner to represent the variability. The actual realizations and weights may be determined from easily computable characteristics of the descriptions. If the problem at hand only requires the knowledge of expected production characteristics associated with a variance, a much smaller number of realizations is required.

The present study was based on a schematic description of a reservoir, although some of the parameter values were selected from a North Sea reservoir. In order to evaluate a given reservoir, a much more realistic description must be provided. The present system can be used to evaluate the relative influence of various description characteristics on the production potential. The authors plan to use the system for determining this sensitivity.

## ACKNOWLEDGEMENT

This paper is based on work performed under the SPOR-GAS programme, directed by the Norwegian Petroleum Directorate.

## REFERENCES

Augedal, H.O., Stanley, K.O. and Omre, H. (1986). SISABOSA, a program for stochastic modelling and evaluation of reservoir geology. In: Nitteberg, J. (ed.) *Proceedings from Conference on Reservoir Description and Simulation with Emphasis on EOR, Oslo, September 1986.* Institute for Energy Technology.

Aziz, K. and Settari, A. (1985). *Petroleum Reservoir Simulation.* Elsevier, Amsterdam.

Berteig, V., Holden, L. and Omre, H. (1987). Examples of simulation of reservoir parameters. *NCC-Note SAND/16/87.*

Bickel, P.J. and Doksum, K.A. (1977). *Mathematical Statistics.* Holden-Day, San Francisco.

Brown, E.C. and Smith, P.J. (1984) The evaluation of uncertainty in surfactant EOR performance prediction. *59th SPE Annual Technical Conference, Houston 1984*, paper no. SPE 13237.

Halvorsen, K., Høiberg, J., Omre, H. (1989). Programsystem for stokastisk reservoarbeskrivelsessimulering, *Norsk Regnesentral*, Oslo 1989 (in Norwegian).

Hewett, T.A. and Behrens, R.A. (1988). Conditional simulation of reservoir heterogeneity with fractals. *63rd SPE Annual Technical Conference, Houston*, 1988, paper no. SPE 18326.

Journel, A.G. and Huijbregts, C.H. (1978). *Mining Geostatistics.* Academic Press, London.

Omre, H., Halvorsen, K.B. and Berteig, V. (1987). Prediction of hydrocarbon pore volume in petroleum reservoirs. *NCC-Note SAND/11/87.*

Ripley, B.D. (1987). *Stochastic Simulation.* Wiley, New York.

Stoyan, D., Kendall, W.S. and Mecke, J. (1987). *Stochastic Geometry and Its Applications.* Wiley, New York.

# 29

# The Movement of Oil Initially Bypassed behind Stochastic Shale Barriers

*J.M.D. Thomas**

Petroleum Reservoir Technology Division, Winfrith Atomic Energy Establishment, Dorchester, UK

This work considers the escape of oil that has been initially bypassed behind stochastic shale barriers by an advancing waterfront. This problem was considered by Richardson *et al.* (1978) for a gas–oil reservoir, where it was found that the principal factor controlling oil escape was the density difference between oil and gas. The contribution of viscous forces was ignored in Richardson *et al.*'s paper.

It is shown that viscous forces are often of fundamental importance in controlling the rate of oil release from behind shale barriers in an undersaturated oil reservoir subject to waterflood.

Fine-scale computer simulations have been used to investigate the oil-trapping potential of various individual shale barrier configurations up to 1000 ft in length. In virtually all cases the combination of viscous and gravity forces releases oil initially trapped behind the shale barrier on a timescale considerably less than the reservoir lifetime.

A fine-scale cross-sectional model has also been set up of an idealized oil–water reservoir containing a high density of explicitly represented stochastic shale barriers, which was subject to a vertical drive waterflood. The shale barriers lead to severe distortion of the oil–water contact, which was reflected in the shape of pseudorelative permeabilities derived from the model. However, the ultimate oil recovery at watercuts of 40% or above from this model was essentially identical to that predicted for an equivalent homogeneous model that did not contain any shale barriers. All of the calculations use realistic flooding rates for North Sea conditions.

It is concluded from this study that stochastic shale barriers less than 1000 ft in length are unlikely to cause significant long-term bypassing of oil under the favourable mobility ratio, gravity-stable displacement waterflooding conditions typically experienced in the North Sea.

## INTRODUCTION

Increasing interest is being paid to the improved treatment of oil reservoir heterogeneities in reservoir simulation models (e.g. Lake and Carroll, 1985; Kortekaas, 1985). Particular attention is being given to heterogeneities that are too small to be represented in a full-field simulation model—because the length scale of the heterogeneity is smaller than the size of individual grid blocks in the full-field model—but that may have significant impact on the ultimate oil recovery from the reservoir. A typical example of this class of heterogeneity is a stochastic shale that may be considered as a horizontal impermeable barrier to flow. The generally accepted definition of stochastic shale barriers is that they are uncorrelatable (between neighbouring wells, for example and are randomly distributed in space. This is in contrast to deterministic shales, which are either correlatable and/or whose position can be fairly accurately predicted. Many sedimentary structures (e.g. submarine fan deposits) contain large numbers of stochastic shale barriers that are much smaller in area than a typical full-field model grid block.

Several attempts have been made to represent the effects of stochastic shales on oil recovery in full-field simulation models. Richardson *et al.* (1978) considered stochastic shales in a generic reservoir model containing oil and gas and produced by gas injection into the overlying gas cap. The advancing gas cap left volumes of bypassed oil lying on the top of each horizontal shale. These escaped slowly owing to the horizontal pressure difference between the top of the oil column at the shale centre and the corresponding point at the shale edge, arising from the fluid density difference between the oil and the gas. Richardson *et al.* did not consider any movement of initially bypassed oil due to viscous pressure differences, however.

Haldorsen and Lake (1984) and Begg and co-workers

*Presently at Phillips Petroleum Company UK Ltd, Woking, Surrey, UK.

*North Sea Oil and Gas Reservoirs—II*

(1985a,b, 1987) have made a very detailed study of the reduction in effective reservoir vertical permeability resulting from the presence of a random shale distribution, and have produced straightforward analytical algorithms for calculating the effective vertical permeability in the presence of stochastic shales that have been verified by numerical simulation. These workers (Davies and Haldorsen, 1987) have also studied the effect of shales on two-phase flow relative-permeability functions.

Very little work has been carried out to assess the degree of long-term trapping of oil below shale barriers in oil–water systems, although it is sometimes the practice to increase the effective residual oil saturation in full-field simulation models to reflect such trapping. There has also been little detailed study of the importance of viscous forces in the escape of oil trapped behind shale barriers.

I present here a study of oil bypassed behind stochastic shale barriers in an oil reservoir subjected to a favourable mobility ratio, vertical waterflood. It is shown that viscous forces are normally at least as effective as gravity forces in enabling oil to escape from below shale barriers and that it is surprisingly difficult to keep oil trapped for an extended period of time compared to reservoir life. An example generic reservoir, containing a high density of randomly positioned stochastic shales, has been taken from the literature and it is shown that the oil recovery at even moderate watercuts (above 40%) is negligibly different from the recovery obtained from the equivalent homogeneous (i.e. shale-free) reservoir.

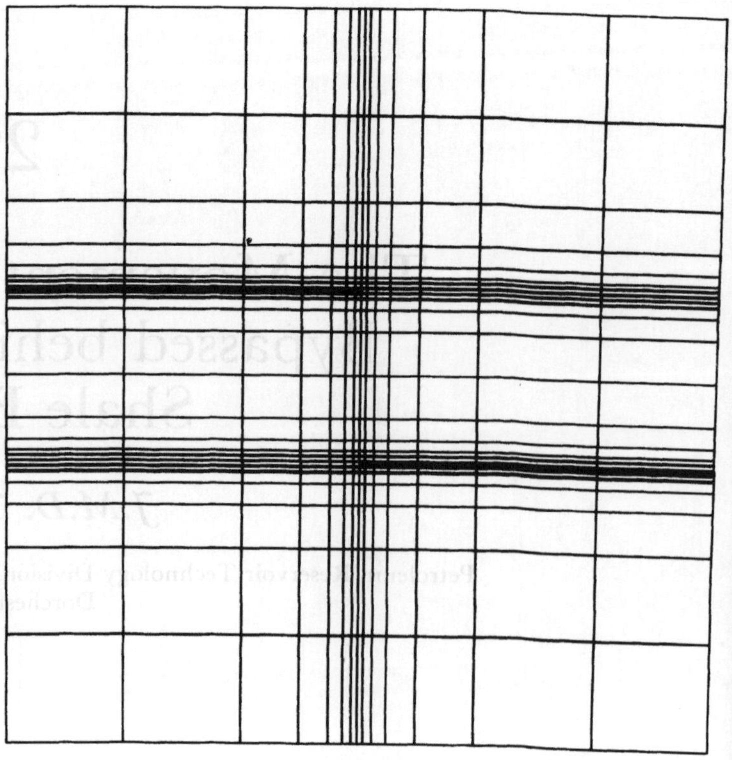

**Fig. 1.** Geometry for Richardson *et al.* model.

## STUDY OF OIL TRAPPING BY INDIVIDUAL SHALE BARRIERS: THE RICHARDSON *ET AL.* MODEL

### Model geometry

The model described in this section was initially proposed by Richardson *et al.* (1978) and is illustrated in Fig. 1. By symmetry, the model represents a dis'ribution of repeated shale barriers of identical size and constant offset. The grid geometry is refined near to the shale barriers, where fluid saturations are changing most rapidly. The model height varied between 34 ft and 136 ft (24 layers) and the width (14 columns) varied between 60 and 240 ft depending on shale size. In the smallest model, the grid block size varied from 10 × 5 ft at the edge of the edge of the model to 0.5 ft × 0.25 ft adjacent to the shale barriers. The model was produced at a constant rate from one of the faces (top face for an oil–water system and bottom face for a gas–oil system) and voidage replacement of gas or water occurred along the opposite face of the model. All simulation calculations were performed using the PORES black-oil reservoir simulator (Cheshire *et al.*, 1980).

### Gas–oil problem

The first case analysed replicated Richardson *et al*'s original study in that gas was used to displace oil with the possibility of oil being trapped above the shale barriers.

These workers provided an approximate analytical treatment of oil drainage from the top of shales in an oil–gas reservoir. Their treatment considers the lateral escape of oil under the driving force resulting from the gradient in the gravity head towards the edge of the shale and provides an estimate for the drainage time of oil initially trapped by an advancing gas flood (Fig. 2).

The results of our gas–oil model were in close agreement with the predictions of Richardson *et al.*, indicating that the release of trapped oil in this case is almost entirely due to the effects of gravity drainage. Note that there is a rate dependency introduced into the oil recovery by the oil trapping above the shale barrier. The drainage time for the trapped oil depends on gravity forces alone and thus has negligible dependence on the gas frontal advance rate in the model (which is controlled by viscous forces). Thus, a low frontal advance rate would lead to higher recovery at a given well gas–oil ratio (GOR) from the model, as more time would be allowed for gravity to drain the oil column lying on top of the shale barrier.

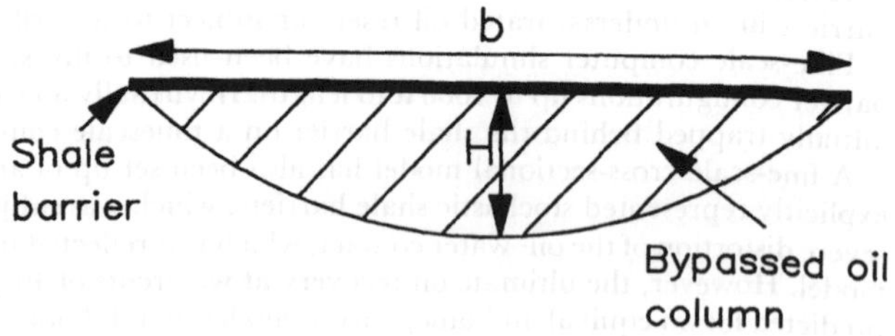


$$Q_{oil} = 1.56 \times 10^{-5} \frac{kH^2\Delta\rho}{\mu_o b}$$

where $Q_{oil}$ = Oil flow rate from centre of shale to shale edge (rbpd/ft)
$k$ = Horizontal permeability (mD)
$H$ = Height of oil column below shale centre (ft)
$\Delta\rho$ = Oil–Water density difference (lb/ft³)
$\mu_o$ = Oil viscosity
$b$ = Shale length (ft)

**Fig. 2.** Gravity drainage of oil from behind a shale barrier.

The effect of shale dip angle was also examined. As expected, the oil trapping time decreased with increasing shale dip angle. For a 600 ft shale, a 2.5° dip angle halved the time for 50% of the initial oil to escape, and a 5° dip angle reduced the trapping time by two-thirds. The oil trapping time can be shown to decrease approximately linearly with increasing dip angle. Note that the trapping time may still be substantial in the case of a dipping shale if the shale is large enough.

### Water–oil problem

Richardson's work was extended to assess trapping effects in an undersaturated oil reservoir subject to water drive (designated as an oil–water reservoir for the purposes of this paper). The same model geometry was adopted as for the gas–oil case (Fig. 1) with a constant fluid offtake rate at the top face of the model and voidage replacement of produced fluids at the bottom face of the model.

The fluid and rock properties assumed in this model are described in Table I and represent typical North Sea values. Relative permeability curves were chosen to give a favourable mobility ratio displacement with a very small Buckley–Leverett tail behind the waterflood front. Unfavourable mobility ratio displacements have not been considered for the oil–water system. The relative permeability endpoints were set at 0 and 1 for simplicity and a vertical frontal advance rate of approximately 20 ft/year was assumed. The effects of shales of sizes between 60 ft and 240 ft were investigated.

**Table I.** Reservoir properties for the single-shale model

| | |
|---|---|
| Horizontal permeability | 600 mD |
| Vertical permeability | 60 mD |
| Porosity | 0.2 |
| Surface density of water | 66 lb/ft$^3$ |
| Surface density of oil | 52 lb/ft$^3$ |
| Water compressibility | $3.0 \times 10^{-6}$/psi |
| Oil compressibility | $9.5 \times 10^{-6}$/psi |
| Water viscosity | 0.36 cP |
| Oil viscosity | 0.7 cP |
| Water formation vol. factor | 1.03 |
| Oil formation vol. factor | 1.24 |
| Rock compressibility | $4.5 \times 10^{-6}$/psi |
| Rock formation vol. factor | 1.00 |

Capillary effects were also ignored in this model and in all other models discussed in our paper. This assumption is considered justified because we are primarily interested in shale trapping effects and capillary effects will reduce oil trapping in a water-wet reservoir (as North Sea fields predominately are) since water will tend to imbibe into the trapped oil lens.

Figure 3 illustrates the predicted saturation distribution for a model containing 60 ft shale barriers at 4 months (after approximately 0.17 pore volumes (PV) fluid throughput), 8 months (0.33 PV), 12 months (0.5 PV) and 24 months (1 PV). It can be seen that the recovery of oil initially trapped behind the shale barriers is excellent and that only a negligible amount of oil remains trapped after 1 PV sweep.

According to the Richardson *et al.* gravity-drainage formula, the time taken for 90% of the trapped oil to escape from underneath the shale barrier would be approximately 670 days. This should be compared with a value of only 250 days estimated fom the saturation plots in Fig. 3.

It is therefore evident that gravity forces alone are unable to explain the speed at which oil is released from below the shale barriers in this model and that viscous forces must be dominant. This conclusion was verified by re-running the case with the oil and water density set equal at reservoir conditions, thus removing the gravity contribution to oil escape from beneath the shale. The results were virtually indistinguishable from the case in which gravity was included.

Similar conclusions were reached for the case of a 240 ft shale barrier. The oil column thickness directly below the centre of the shale barrier decreased by 80% in less than 2 years after the waterfront had passed the shale edge, whereas Richardson *et al.*'s gravity-drainage formula predicted that a time delay of over 13 years would be required to drain 80% of the oil.

### Conclusions from Richardson *et al.* model

The principal conclusion to be drawn from the calculations with the Richardson *et al.* model is that gravity forces appear to be dominant in controlling the release of oil initialy trapped by shale barriers in an oil–gas reservoir, whereas viscous forces appear to be dominant in controlling the trapping time in oil–water systems. This conclusion was verified by simple analytical calculations which considered the pressure gradients that develop at the interface between the trapped oil layer and the displacing phase (water or gas) due to both gravity and viscous forces.

These conclusions may be understood on a straightforward physical basis as follows. Consider first the oil–gas reservoir example. The gas front passes the shale barriers from above and leaves a column of trapped oil above the shale. The low gas viscosity results in gas behind the floodfront continuing to sweep past the trapped oil with very little viscous drag at the boundary of the oil lens. The oil lens is thus unaffected by the viscous forces in the surrounding gas and can only escape by gravity-induced pressure differences.

In the case of an undersaturated oil reservoir undergoing waterflood, oil is initially trapped below shale barriers as the flood front rises above the shale edge. The relatively low water mobility results in water behind the floodfront continuing to sweep past the trapped oil with significant viscous drag at the boundary of the oil lens. This viscous drag will tend to push the trapped oil away from the shale barrier and allow it to escape. The effect of gravity forces will be much less than in a gas–oil system because of the relatively small density difference between oil and water.

We therefore conclude that trapping times for oil initially bypassed behind stochastic shale barriers in a favourable mobility ratio waterdrive are much less than those that would be calculated by the formula of Richardson *et al.*, which ignores viscous forces.

## A RESERVOIR-SCALE EXAMPLE

### Introduction

The studies described above have shown that viscous effects can be far more important than gravity effects in determining the release time for oil trapped by stochastic shale barriers in undersaturated oil reservoirs subject to waterflood under typical North Sea conditions. These studies have looked in detail at the trapping near to isolated barriers. We have also considered a generic example reservoir containing a large number of stochastic shale barriers and examined the effect of these barriers on

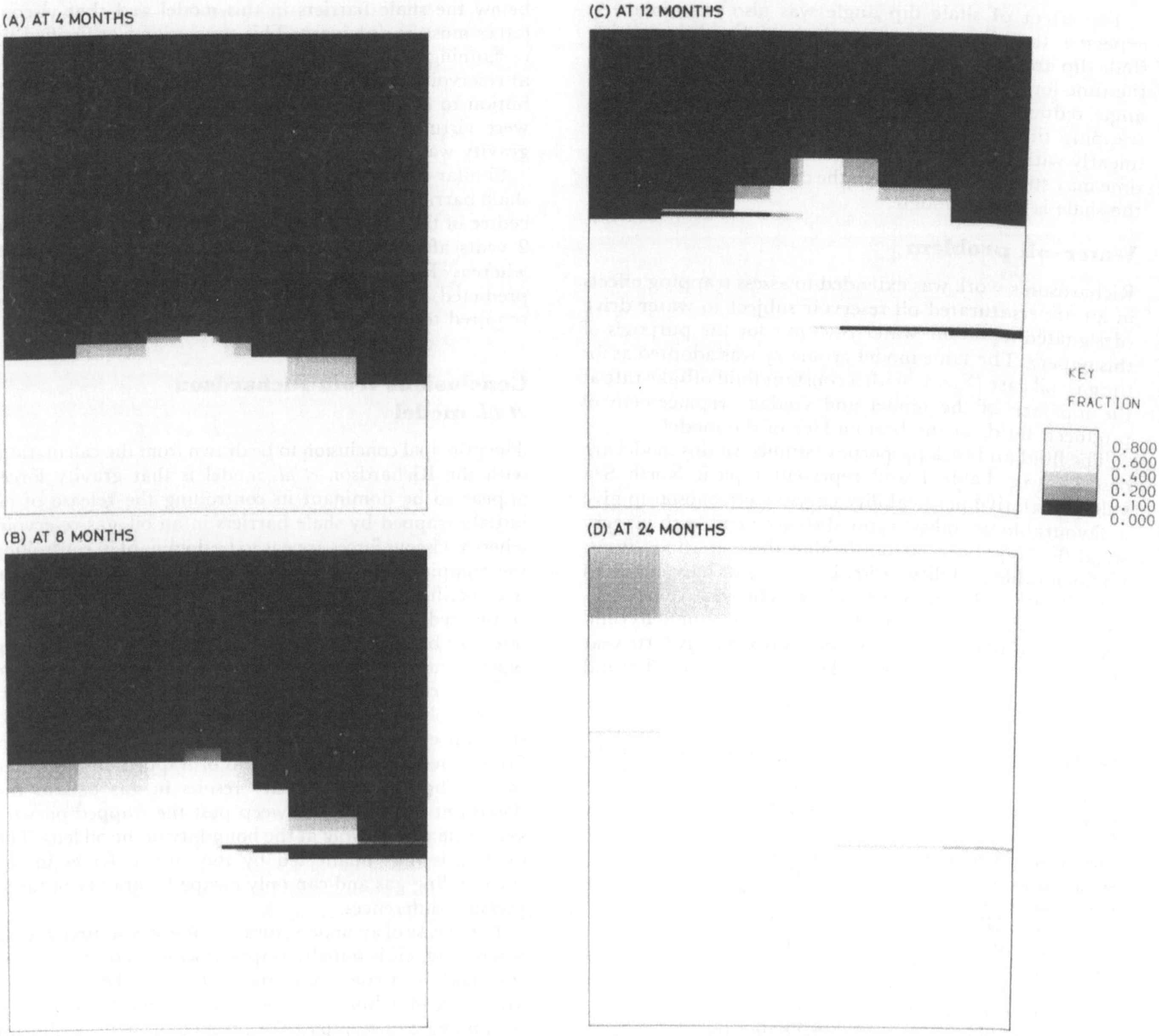


**Fig. 3.** Results of Richardson *et al.* model for the oil–water case.

two of the parameters of most interest to the reservoir engineer; the ultimate economic oil recovery from the reservoir and the watercut history.

## The model

The idealized reservoir model chosen is based on one recently described in the literature (Davies and Haldorsen, 1987). It consists of a vertical section through a constant-thickness non-dipping reservoir, with one water-injection and one producing well (Fig. 4). The reservoir was discretized into 24 layers of 100 blocks each. Each cell in the model was 15 ft thick by 300 ft wide (total model dimensions 3000 ft by 360 ft). Davies and Haldorsen did not specify the rock and fluid properties assumed in their model and we have again used the values in Table I. Capillary effects were ignored as they would lessen the effects of oil trapping behind shales if they were included (assuming a water-wet system). Davies and Haldorsen also ignored capillary effects.

The model contained 150 stochastically generated zero-permeability shale barriers of varying length, introduced by setting the transmissibility between blocks on either side of the barrier to zero. The shales were generated according to the shale length distribution given by Davies and Haldorsen, which is illustrated in Fig. 4. It should be noted that the shale length distribution was discretized so that the length of any generated shale was an integer number of grid block lengths (i.e. shale length $= \mathcal{N} \times 30$ ft where $\mathcal{N}$ is an integer).

The length of the individual shales varied from 30 ft to 900 ft. The random shales were then emplaced in the model using a program that randomly positions shales in a reservoir simulation model according to a given size distribution and ensures that neighbouring shales do not overlap or directly touch (which would alter the shale size distribution). The final shale distribution adopted in our model is illustrated in Fig. 5.

The model production rate was adjusted to give a vertical flood front advance rate of approximately 20 ft per year. The injector provided voidage replacement of produced fluids into the aquifer zone. The reservoir was

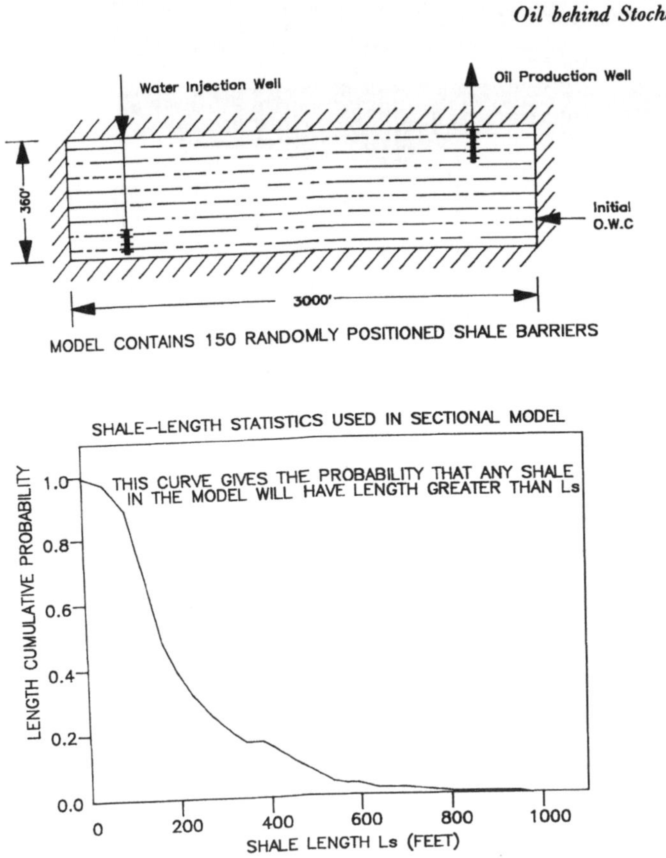


**Fig. 4.** Grid geometry and shale distribution for multishale idealized reservoir model.

initially assumed to contain oil and connate water with values of 0.38 and 0.27 for the connate water saturation and residual oil saturation respectively.

After a preliminary calculation, our model was redefined to increase the vertical resolution from 24 to 96 blocks and to double the horizontal resolution, without altering the shale distribution previously generated. This was done because we were primarily interested in the movement of oil from behind shale barriers and this increase in vertical resolution enabled us to represent the region under each shale by four layers rather than by only one (as in Davies and Hadorsen's model).

## The results

### *Saturation distribution*

The results of this model are illustrated in the form of water saturation versus time in Fig. 5. These plots contain many interesting features.

The most obvious feature is that the presence of the shale barriers leads to significant disturbance of the oil–water contact, with much evidence of tongues of water moving through the gaps between the shale barriers. In later field life this behaviour manifests itself as coning towards the production well. The disturbance of the oil–water contact is reflected in the shape of pseudorelative permeability curves derived from the model.

Initial trapping of oil behind shale barriers is also very evident. The saturation distribution at 2192 days (after 0.33 PV throughput) illustrated in Fig. 5 shows how oil escaping from under one shale barrier may drain and subsequently be trapped behind a shale barrier higher in the formation. This process is known as "recharging" and may extend the time for release of trapped oil. It is evident from these plots that oil may escape from behind the shale barriers over a considerable timescale. The escape paths take the form of narrow vertical "streamers" of relatively high oil saturation adjacent to the shale barriers.

The most important feature of these plots is that, despite the major effect of the shales on the shape of the developing oil–water contact and the evidence of oil trapping, very little oil remains trapped behind the flood front after 7305 days (i.e. after 1 PV injected water sweep).

### *Ultimate oil recovery*

It is evident from the results described above that the presence of discontinuous stochastic shale barriers leads to early water breakthrough in the idealized reservoir model. A more important consideration is how the shales affect the ultimate recovery from the reservoir. Figure 6 compares the water cut and cumulative recovery profiles for the case where shales are explicity represented and for an equivalent homogenous and anisotropic reservoir where the effect of the shales is merely represented by an adjustment to absolute vertical permeability (from 60 to 3.75 mD). The early water breakthrough in the explicit shale model is very evident. Note, however, that the watercut in the equivalent homogenous reservoir eventually overtakes the value in the explicit shale case. This may be explained by the baffling effect of the shales on the rising oil–water contact. Water advances more rapidly through the gaps between the shales, but is held up below the shales themselves. This effect can lead to a reduction of coning at wells surrounded by shale barriers, as reported in the literature. Figure 7 illustrates the oil recovery from both models as a function of water cut and is perhaps the most interesting result of this study. It will be seen that the recovery at water breakthrough is far worse in the model where shales are explicitly represented and is also worse for water cuts of less than about 40%. However the recovery at water cuts in excess of 40% is essentially identical for the two models and is in fact slightly better for the model where shales are explicitly represented for water cuts between 40% and 80%. This may be ascribed to the effect of the shales acting as "baffles" that channel water into areas of possible poor sweep.

We thus conclude that although the shale distribution explicitly represented in this idealized reservoir example leads to significant distortion of the rising oil–water contact and to subsequent early water breakthrough at the producing well, it does not affect the ultimate oil recovery at economically realistic watercut limits (say 50–80%).

## Sensitivity studies

A number of sensitivity cases were run to support the conclusions of the idealized reservoir model.

### *Effect of offtake rate*

The fine-grid explicit shale model was re-run with the offtake rate doubled to provide one reservoir PV sweep in 10 years (equivalent to a vertical rate of flood front rise of 40 ft/year). The results of this case were very similar to the base case at the same pore volume sweep.

### *Effect of grid resolution*

The results predicted by the fine-grid 200 × 96 model (in which the minimum vertical separation between vertical shales is 4 grid blocks) were compared to those predicted by the original 100 × 20 model of Davies and Haldorsen (in which the minimum vertical separation between vertical shales is only 1 grid block). The results of the two cases were broadly similar excepting that the finer-scale simulation

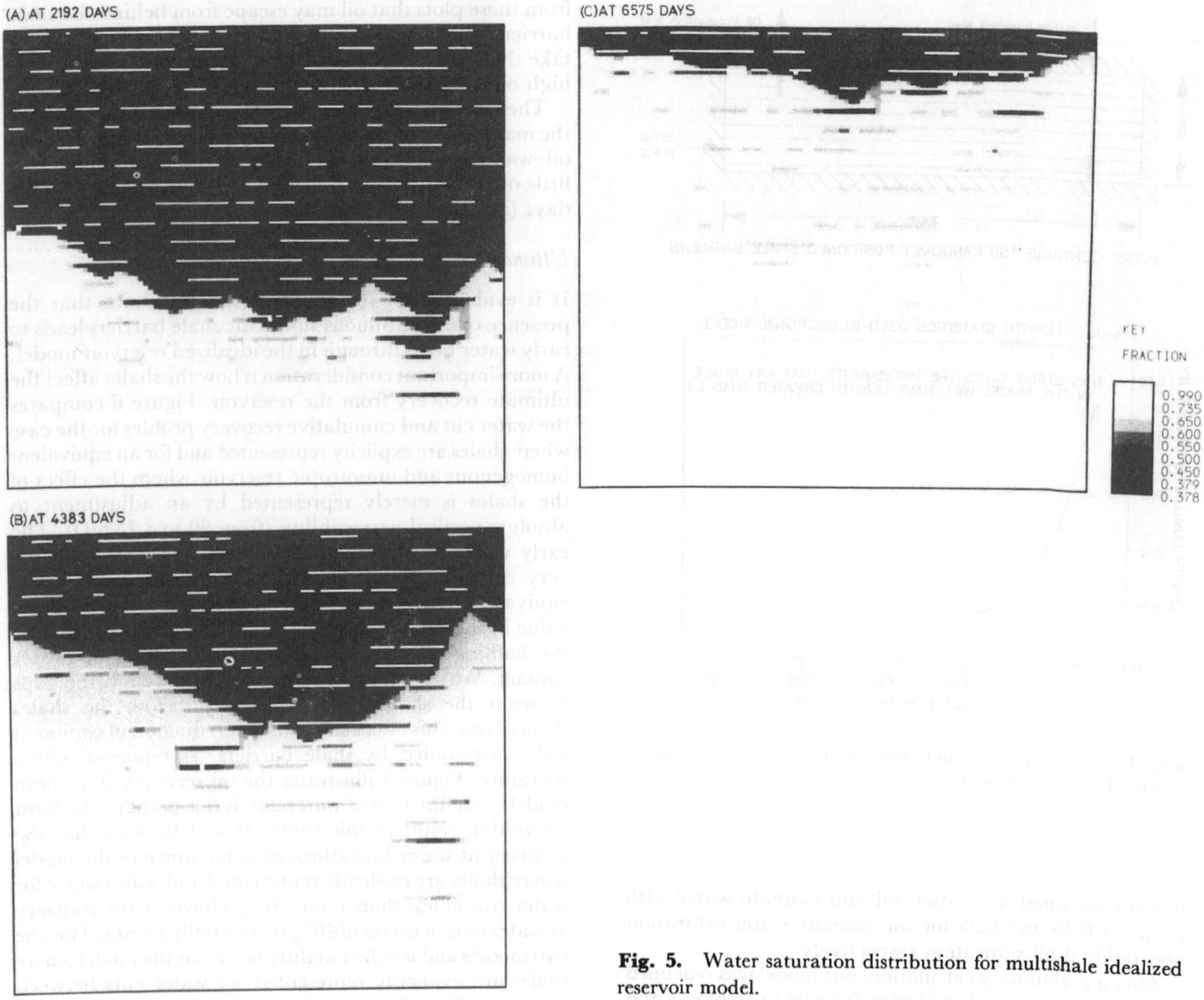


**Fig. 5.** Water saturation distribution for multishale idealized reservoir model.

predicted later water breadthrough (by about 1 year). This may be attributed to a more realistic representation of the water movement through the gaps between the shales in the fine-grid model. The behaviour after water breakthrough was very similar for the two models.

*Effect of low permeability adjacent to shale barriers*

The studies discussed in this report have assumed that the stochastic shale distribution exists within a uniform-permeability sandstone matrix. This is the simplest assumption to make and is the one generally assumed in the literature when analysing the effect of stochastic shale distributions.

In some reservoirs it is more realistic geologically to reduce the permeability in the region of sand below each shale barrier. Shales are evidence of quiet water deposition conditions in a geological sense and if the transition to quiet water conditions was a gradual one it might be expected that the rock layers deposited just before the shale layer are of lower than average permeability.

As a sensitivity case, the fine-grid model was rerun with the permeability in the cells directly underneath each shale barrier reduced to 1/4 of the average value and the permeability in the cells directly below these being reduced to 1/2 the average value. All other cells in the model retained their normal permeability values.

The reduction in permeability beneath the shale barriers leads to an increase in the trapping time of oil beneath the shale barriers. However, very little oil remains trapped behind the shale barriers at the time when the oil–water contact has reached the top of the reservoir, and the recovery from this model still exceeds that from the equivalent homogeneous model (i.e. a model in which the effect of the shales is merely represented by a reduction in the effective vertical permeability).

# DISCUSSION AND CONCLUSIONS

This paper has analysed the phenomenon of oil trapping behind discontinuous stochastic shale barriers in undersaturated reservoirs subject to gravity-stable waterflood.

Oil initially trapped behind a shale barrier (or other low-permeability obstacle) does not remain stationary, but can escape by two principal mechanisms. The first mechanism arises from the density difference between the trapped oil and the displacing water phase and was considered by Richardson *et al.* The leakage rate of trapped oil due to gravity effects is independent of the flow rate in the water flood and this can lead to a rate-dependent ultimate recovery. A slower reservoir offtake rate would allow more time for gravity-induced oil leakage to occur and thus lead to an increased oil recovery.

The rate of trapped oil escape due to gravity is initially

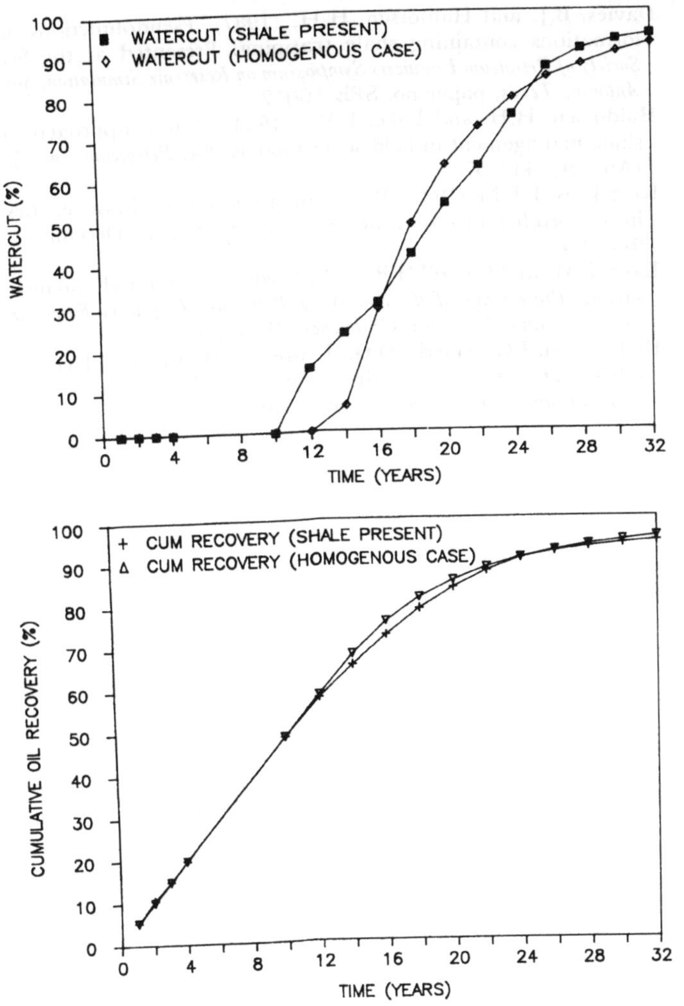


**Fig. 6.** Water cut and cumulative recovery for multishale idealized reservoir model.

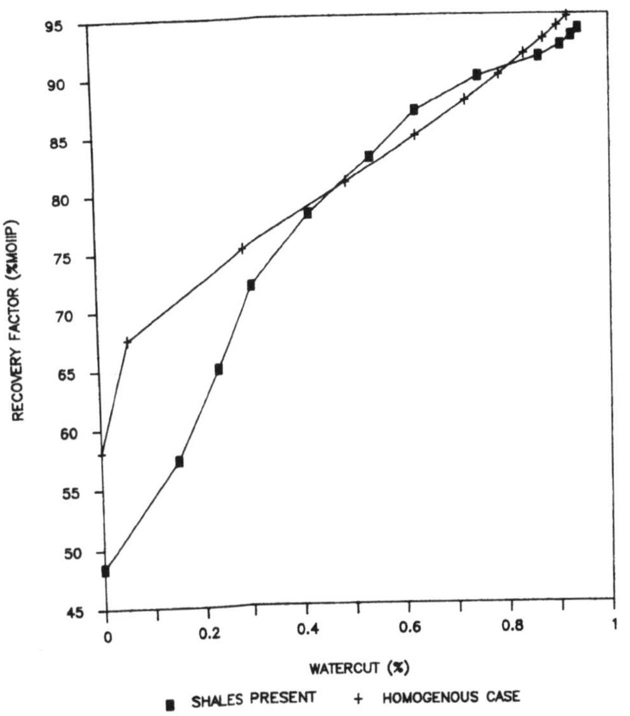


**Fig. 7.** Cumulative oil recovery versus watercut for multishale idealized reservoir model.

quite fast, but diminishes rapidly as the thickness of the bypassed oil column beneath the shale barrier is reduced. The escape rate also depends on the square of the shale length and thus reduces quickly as the shale size increases.

The other important escape mechanism for bypassed oil is caused by viscous forces. These forces arise from the viscous drag of the displacing phase (water) on the lens of bypassed oil. The effectiveness of viscous forces in releasing trapped oil can be gauged by streamline analysis, which reveals that flow close to the shale boundary is tangential to the shale and this is the flowpath of the escaping oil. There is a stagnant flow point for viscous flow directly underneath the centre of the shale, but this is also the point at which the oil flow due to gravity forces is greatest. Thus, there are no points of completely stagnant flow if both viscous and gravity forces are taken into account.

The relative importance of viscous and gravity forces in controlling the escape of bypassed oil can be assessed by considering the flow gradients that develop at the boundary between the lens of bypassed oil and the displacing phase. In simple physical terms it may be said that gravity forces are likely to be dominant in an oil reservoir subject to a gas flood, or in a waterflooded oil reservoir where the mobility ratio is unfavourable. In these cases the displacing phase has a tendency to "slide" past the bypassed oil lens with very little viscous drag and the oil may be considered as an isolated lens draining purely by gravity forces. In a reservoir where the mobility ratio is favourable, the viscous drag of the displacing phase on the oil lens will be significant and will tend to move the oil from behind the shale barrier. In this case, viscous forces may be orders of magnitude more effective than gravity forces in releasing trapped oil. The effective trapping time of oil behind isolated shale barriers is then small compared to typical reservoir lifetimes and the effect of the shales on ultimate oil recovery will be small.

Calculations have been performed on an idealized reservoir cross-section containing a large number of stochastic shale barriers. Oil trapping behind the shale barriers in the model was evident, but the amount of oil left behind the shale barriers on the time of the reservoir lifetime was very small. In fact it was found that the shales had the beneficial effect of inhibiting coning towards the production well, and the ultimate recovery at water cuts above about 40% was better than in a comparable model that did not contain shale barriers.

We conclude that stochastic shale barriers less than about 1000 ft in length are unlikely to cause long-term trapping of oil under the favourable waterflood conditions generally encountered in the North Sea, with viscous forces being effective in sweeping bypassed oil from under the shale barriers. However, it should be noted that this work has only considered the effect of stochastic shales that are considered to be small on the scale of a full-field model grid block and are separated from each other and predominately horizontal. We do not discount the trapping of oil behind larger-scale shale barriers, which may form stratigraphic traps for bypassed oil.

## ACKNOWLEDGEMENTS

This work was funded by the United Kingdom Department of Energy.

## REFERENCES

Begg, S.H. and King, P.R. (1985a). Modelling the effects of shales on reservoir performance: Calculations of effective vertical permeability. Presented at the *8th Society of Petroleum Engineers Symposium on Reservoir Simulation, Dallas*, paper no. SPE 13529.

Begg, S.H. Chang, D.M., and Haldorsen, H.H. (1985b). A simple statistical method for calculating the effective vertical permeability of a reservoir containing discontinuous shales. Presented at the *20th Annual Technical Conference of the Society of Petroleum Engineers, Los Angeles*, Paper no. SPE 14271.

Begg, S.H., Carter, R.R. and Dranfield, P. (1987). Assigning effective values to simulator grid-block parameters for heterogeneous reservoirs. Presented at the *62nd Annual Technical Conference of the Society of Petroleum Engineers, Dallas*, paper no. SPE 16754.

Cheshire, I.M., Appleyard, A.M., Banks, D., Crozier, R.J. and Holme, J.A. (1980). An efficient fully implicit simulator *European Offshore Petroleum Conference and Exhibition Proceedings*, Vol. 1 pp. 325–336, Paper no. EUR 179.

Davies, B.J. and Haldorsen, H.H. (1987). Pseudofunctions in formations containing shale trapping. Presented at the *9th Society of Petroleum Engineers Symposium on Reservoir Simulation, San Antonio, Texas*, paper no. SPE 16012.

Haldorsen, H.H. and Lake, L.W. (1984). A new approach to shale management in field-scale models. *Soc. Petroleum Eng. J.* (August), 447–57.

Kortekaas, T.F.M. (1985). Water–oil displacement characteristics in crossbedded reservoir zones. *Soc. Petroleum Eng. J.* (December), 917–927.

Lake, L.W. and Carroll H.B. (1985) (eds.) Reservoir characterisation. *Proceedings of the Society of Petroleum Engineers Reservoir Characterisation Technical Conference, Dallas, Texas.*

Richardson, J.G., Harris, D.G., Rossen, R.H. and Von Hee, G. (1978). The effect of small discontinuous shales on oil recovery. *J. Petroleum Technol.* (November), 1531.

# 30

# Practical Application of Supercomputing in Black-Oil Reservoir Simulation

*A. Henríquez, T. Kårstad and T. Steihaug*

Statoil AS, Stavanger, Norway

The optimization of hydrocarbon recovery by numerical reservoir simulation requires very high computing performance. Statoil has vectorized the iterative linear solver of the reservoir simulator currently in use in our IBM 3090-200 with satisfying speed-up factors. The vectorization of a direct linear solver has been investigated. As in scalar versions, a direct solver is faster than an iterative solver only for a few hundred cells. The increasing requirements of more sophisticated models require the power of parallel computing. We review the practical implementations of novel methods of using these supercomputing techniques in a reservoir simulator.

Our approach to parallelization has been gradual, beginning with uncoupled structures and continuing with weakly coupled systems; at present we are investigating the possibilities for strongly coupled systems. This approach, based on the physical processes involved, is reflected in the level of difficulty of finding and implementing new methods. A self-imposed constraint is that an acceptable parallelization method should reduce not only the elapsed time, but also the execution time. Our first step took advantage of the fact that some hydrocarbon reservoirs consist of isolated geological structures that are coupled through the production control at the level of several wells. In a new advance, it has been made possible to simulate reservoirs in which only oil and water (or some other two phases) are present in some structures while all three phrases are present in others. The speed-up factors achieved are shown to be as expected.

New possibilities of parallelization are being investigated for weakly coupled systems. These systems appear in simulation studies where the flux between parts of a reservoir is insignificant or when local phenomena are studied in full-scale reservoir studies. Possible advances for strongly coupled systems are under study, with the parallelization of nested factorization for few processors in a mainframe environment or for many processors in a local memory multiprocessor machine.

The vectorized and parallelized code has been in production for some time. Results are shown for realistic cases, showing the significant reductions achieved in elapsed and actual CPU times. As a consequence, larger simulation models reflecting more faithfully the geological description of the reservoir are economically feasible.

## INTRODUCTION

The objective of optimizing hydrocarbon recovery by reservoir simulation requires a realistic geological description and an accurate reservoir model. In several cases, a faithful translation of the geological data to a simulation grid requires a large number of blocks. Grid sizes for full-field simulations may be several tens of thousands of cells, and large computer resources are used in pursuing optimal recovery strategies.

In a tight economic environment, the need to study different production scenarios increases, and this need has to be met within engineering and CPU time limitations. Statoil has made a sustained effort in recent years to utilize and expand new techniques in the area of supercomputing. Black-oil reservoir simulation, which places a large load on the computer resources, has been one of the targets of this effort. Since most of the calculations are performed to solve the large systems of linear equations arising at each time step, it is natural to focus efforts at vectorization and parallelization on this part of the code.

The vectorization of the linear solver and other smaller parts of the black-oil simulator, which will be described briefly in the next sections, was achieved with satisfying speed-up factors (Henriquez, 1986).

Further reductions in execution time have been needed in order to obtain realistic turnaround times for some of the largest full-field simulations. These reductions may only be possible through the use of parallelization. This area of supercomputing is just taking its first steps in applications to real problems in the industry. Therefore, our work was begun systematically with physically decoupled systems and advanced through weakly coupled systems into the investigation of strongly coupled systems.

As the first stage of our parallelization work, we have

*North Sea Oil and Gas Reservoirs—II*

taken advantage of the decoupling that is possible for isolated geological structures (Kårstad *et al.*, 1987). The parallelization of the code at this level led naturally to the investigation of the case in which some of the reservoirs have two phases present, while others have all three phases (Jarosch *et al.* 1988).

The successful effort with isolated geological structures the parallelization of weakly coupled systems appear promising. Such systems arise when there are near-well effects, such as gas or water conning, or condensate liquid dropout. The timescale and volumetric scale of these effects are much smaller than those of field phenomena, and they have to be studied in a finer grid and with shorter time steps than most of the reservoir. In order to solve this problem, a local grid-refinement capability with small time steps, immersed in the global grid with global time steps, has been implemented in our reservoir simulator in such a way that it is parallelizable (Cheshire and Henriquez, 1989).

Weak coupling also appears when parts of the reservoir have small fluxes to the rest of the reservoir. Several physical situations may give rise to this phenomenon. One case would be the presence of a thin oil zone in a region of the reservoir, which can be recovered, while another part of the reservoir is mainly a gas accumulation. Another case is when the timing of the platforms or the wells essential isolates drainage areas, which for history-matching or prediction purposes can be simulated on its own with a higher degree of detail. It would be a definite advantage to simulate the show reservoir and store the fluxes at the limiting surfaces of these weakly isolated areas, using the fluxes as boundary conditions in simulations of the areas. This capability has recently been coded in our reservoir simulator, and in the light of our experience in using it in practical problems, the possibility of parallelizing these weakly coupled structures will be considered.

Finally, the case of non-decoupled structures may be parallelized at the level of the algorithms used in solving the linear systems of equations with the few processors available in large mainframes (Van der Vorst, 1986) or by making use of the new massively parallelized machines recently becoming available to the industry such as the Hypercube.

In the following sections, we will give a short description of the algorithm used for the solution of the linear system of equations, as needed for the understanding of the methods developed. We will then give some details of the implementations in our machine and the mechanisms used for vectorization and multitasking. Results and conclusions will be given in the last two sections.

## A brief description of the linear equations

In black-oil reservoir simulation, the solution of a large number of coupled non-linear equations must be found. The use of Newton's method involves the repeated solution of sets of linear equations.

A seven-point discretization scheme in three dimensions leads to a block seven banded matrix, represented in Eq. (1), below by the submatrix $A$. The size of the blocks depends on the number of phases.

The modelling of multicell wells introduces extra degrees of coupling between cells, represented by submatrices $B$ and $C$. The interaction between wells produced by the interdependence of the well flow rate targets introduces coupling between well elements of the matrix, represented by submatrix $D$. The derivatives of the flow rates with respect to all other wells should be included in the Jacobian matrix equation.

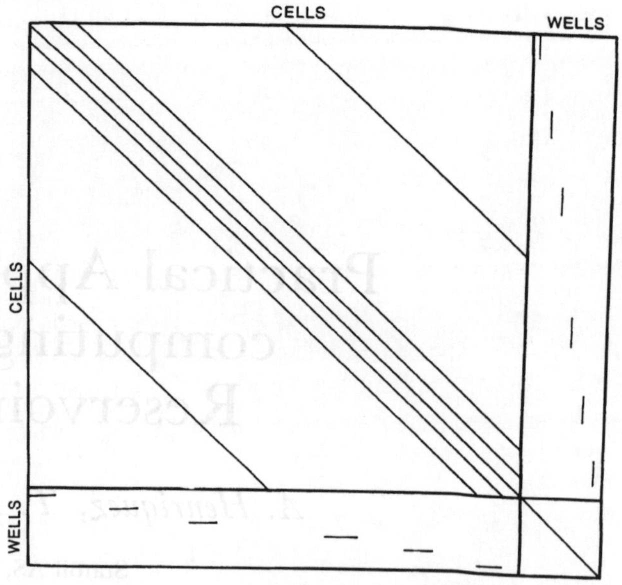


**Fig. 1.** Symbolic representation of the Jacobian matrix composed of cell–cell, well–cell and well–well interactions.

Including all the derivatives of the well fow rates in the submatrix $D$ would make it very dense and difficult to invert. In our simulator, only the diagonal elements are included. As we will see later, this makes it possible the full matrix into one matrix for each independent structure in the reservoir. The resulting matrix is reprensented symbolically in Fig. 1. In the following we describe briefly the solution procedure for this problem. For details, see Holmes (1983).

Let Ncell be the total number of active grid cells and Nwell be the number of wells. Further, if every element in the matrix is a $p \cdot p$ block matrix where $p$ is the number of phases, then

- $A$ is a banded Ncell × Ncell matrix;
- $B$ is a sparse Ncell × Nwell matrix;
- $C$ is a sparse Nwell × Ncell matrix;
- $D$ is a diagonal Nwell × Nwell matrix;
- $b$ is a vector of length Ncell;
- $b_w$ is a vector of length Nwell,

Then the matrix equation, shown graphically in Fig. 1, can be written as

$$\begin{bmatrix} A & B \\ C & D \end{bmatrix} \times \begin{bmatrix} X \\ X_w \end{bmatrix} = \begin{bmatrix} b \\ b_w \end{bmatrix} \qquad (1)$$

## VECTORIZATION

Since the solution of the linear equations in (1) are the most CPU time intensive calculations in most reservoir simulation computations, the first objective of a vectorization project of a reservoir simulator is the linear solver. The linear equations in (1) may be solved by using an iterative technique or with a direct solver.

### Direct solver

The vectorization of a direct solver has been investigated for our simulator. The main reason for this investigation are that there exist simulation studies with a small number of cells in which an iterative solver, usually faster than a direct solver, converges very slowly. This may be the case for conning studies or well test analysis by history matching, where the time steps may be very small and the number of cells a few hundred. Another reason is that a direct solver could be used in parallel with an iterative linear solver in the solution of problems where grid refinement is used, as will be discussed later.

The solution of sparse linear systems of equations by direct methods has been an area of intensive research. Most of the effort has been directed towards a combination of Gaussian elimination with some reordering of the equations and unknowns. The goal is to obtain a permuted system that can be solved more easily. The solver that was used in this work is very fast if the pivot sequence (the symbolic factorization) is known (Houbak, 1985). Unfortunately, this is not true in our simulator. At each non-linear iteration the solution vector for each cell may change, owing to changes in the reservoir conditions. Then the structure of the linear equation may change, and the solver has to perform a new factorization.

Several runs indicate that for a few hundred grid cells this solver is more effective than iterative methods, especially for problems where there are large changes in the solution in a short timescale. Four hundred cells was the number determined for a specific case of gas injection in an initially undersaturated reservoir as the point at which the iterative solver discussed in the next section is faster, but this number depends of course on the problem being studied.

This result is valid where the symbolic factorization does not change, and the cross-over point occurs at a lower number of cells when there are changes. As changes are to be expected in problems where a direct solver may be efficient, we believe that further work must be done in the area of vectorized direct direct solvers, probably with D-4 ordering of the cells.

## Iterative solver

It is generally more effective to use an iterative method with preconditioning. Simplifying the notation used in Eq. (1), we assume that, $A$ is the Jacobian matrix and $B1$ an approximation. Then we have to solve

$$A \cdot x = b. \tag{2}$$

For a given approximation $X_m$, the Orthomin method of order $k$ (Vinsome, 1976) calculates a correction $S_m$ as follows. First solve

$$B1 \cdot q_m = r_m \tag{3}$$

where

$$r_m = b - A \cdot X_m \tag{4}$$

Then find $S_m$ as a linear combination of $q_m$ and the $k$ last corrections,

$$S_m = a_m q_m - \sum_{i=m-1}^{m-k} a_i s_i \tag{5}$$

where the coefficients are chosen such that $As_m$ is orthogonal to $As_i$, $i = m-1, \cdots, m-k$, and the length of $r_m - As_m$ is minimized.

The differences between most linear solvers are due to the methods used for finding the approximation $B1$ to $A$. The simulator we work with, Eclipse, implements nested factorization. This incomplete block factorization of the Jacobian matrix is described by Appleyard and Cheshire (1983). Essentially, it consists of approximating the hepta-diagonal banded matrix $A$ arising from a 3D simulation.

$$A = D1 + L_1 + U_1 + L_2 + U_2 + L_3 + U_3 \tag{6}$$

where $D1$ is the diagonal, $L_1$, $L_2$ and $L_3$ are lower bands and $U_1$, $U_2$ and $U_3$ are the corresponding upper bands. Nested factorization exploits the structure of $A$ by defining the preconditioning matrix in the sequence

$$B1 = (P + L_3) \cdot P^{-1} \cdot (P + U_3) \tag{7}$$

$$P = (T + L_2) \cdot T^{-1} \cdot (T + U_2) \tag{8}$$

$$T = (M + L_1) \cdot M^{-1} \cdot (M + U_1) \tag{9}$$

where, finally, $M$ is a diagonal matrix. If $M$ is defined by

$$M = D1 - L_1 \cdot M^{-1} \cdot U_1 - colsum(L_2 \cdot T^{-1} \cdot U_2) - colsum(L_3 \cdot P^{-1} \cdot U_3) \tag{10}$$

then $colsum(A) = colsum B1$. This definition guarantees that the material balance error is zero (Appleyard and Cheshire, 1983).

Before starting the linear iterations it is necessary to compute $M^{-1}$ and $L_1 \cdot M^{-1}$. In this calculation, only $colsum(L_3 \cdot P^{-1} \cdot U_3)$ is worth vectorizing, as the other calculations have short vectors.

The solution after initialization is hierarchical in nature. The standard nested factorization technique for solving the equation, $B1 \cdot x = r$ is to work forwards one plane at a time solving $y_{i+1} = r_i - L_3 \cdot P^{-1} y_i$, where $y = (P + U_3) \cdot x$ and then to work backwards one plane at a time solving $x_{i-1} = P^{-1} \cdot (y_{i-1} - U_3 \cdot x_i)$. This technique is possible because $L_3 \cdot P^{-1} y_i$ and $U_3 \cdot x_i$ involves the solution of the previously calculated plane.

For the forward solution, calculations of $P^{-1} y_i$ are performed by solving, working forwards and backwards one line at a time. For the backward solution, the equation $Px_{i-1} = v$(where $v$ is defined as $y_{i-1} - U_3 \cdot x_i$) is solved working forwards and backwards one line at a time.

Finally, within a line, a similar equation for $T$ is solved working forwards and then backwards one cell at a time. This last step suffers from short vector lengths, and the usual trick of cyclic reduction would not help much.

The total number of floating-point multiplications for solving $B1 \cdot x = r$ are $18p^2$ per cell (Appleyard and Cheshire, 1983). To this must be added the work $7p^2$ required for a matrix vector multiplication, to update the residual. The conjugate gradient algorithm requires $5p$ multiplications per cell. This adds up to $25p^2 + 5p$ multiplications per cell for a 3D problem.

An estimate of the vectorizability of the linear solver based on these considerations gives up to 35% for one iteration, and up to 65% for four linear iterations. This loose estimate matches quite well with the vector-to-scalar speed ratio of the IBM 3090, which is between 3 and 5, depending on the floating-point arithmetic content of the vectorizable part of the code. In some levels of the hierarchy, with the definition of intermediate variables, it has been possible to write loops that are totally vectorizable, with stride 1, and with vector lenghts of the size of the number of cells in the reservoir. For a vectorization fraction of 50%, Amdahl's law would predict a relative job performance of approximately 1.6 between vectorized and scalar code, as discussed in Henriquez (1986).

## Vectorization procedure

The now classical procedure for code vectorization is followed for our simulator. The first step is to choose some representative benchmark data sets.

Several automatic vectorizers are now available, most of them are specially designed to take advantage of special features of the machine which they usually accompany. However, it is widely recognized that no automatic vectorizer is able to recognize all the data dependencies in the program or have the knowledge of the program to be vectorized that is needed to detect opportunities of vectorizing.

Hand-tailoring of crucial subroutines is essential. This

work is time consuming, and it requires expert knowledge of the machine architecture, the peculiarities of the FORTRAN implementation in this architecture and the code that is being vectorized. Since in most large production codes more than 80% of the CPU time is spent in less than 10% of the lines of code, the effort of hand vectorization is focused on those subroutines that perform the largest part of the work. A "hot-spot" analyser is the tool used in this step. For several data sets, execution times are collected during simulations, with user-defined sampling times. The reports on the program performance must be at the level of detail required by the user, and this depends on the weight of the calculations supported by the routines that are being studied. We typically used sampling times of 2 ms and initial reporting at the subroutine level. As expected, typically two-thirds of the CPU time was spent in the linear solver.

The next step is the vectorization of the subroutines and lines of code detected in the previous step. In this stage of the vectorization process, access to the assembler code is essential. Even a superficial knowledge of vector assembler language will reveal the need for qualified assistance in difficult parts of the code. At this level, the use of microcoded hardware instructions and the library subroutines provided for vector operations must be considered.

A major inhibitor for vectorization is the short length of the vectors. For one of the key subroutines in the solver it was possible to increase the length from a typical line length to a length of the total number of cells.

Several loops in the linear solver subroutines were modified to increase the number of compound operations of the type multiply–add: this is very effective in the 3090 because of chaining.

Optimization of the use of vector registers played a limited role in our example. As most vectors are accessible to the calculations with stride 1, owing to the extended use of indirect addressing in the simulator, there was no need for cache optimization.

Approximately one-half of the subroutines that were vectorized by the compiler were used without any further changes, as they could not be optimized further.

Figure 2 shows the performance gain as a function of the time used in vectorizing the code. It should be noticed that a large fraction of time was spent verifying that the results of the vectorized code were the same as those of the scalar code.

### Performance of the vectorized simulator

As the simulator is multipurpose, the performance measurements will depend on the options used. The optimized code was therefore tested on a variety of data sets to confirm our results, before it was released for production runs.

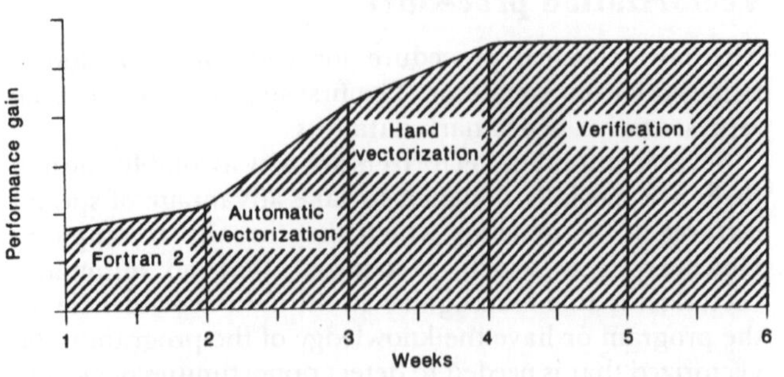


**Fig. 2.** Migration timescale for a reservoir simulator on an IBM 3090-200.

The speed-up between scalar and vectorized versions in the 3090-200 was 1.6. This is an average taken over several data sets. The data sets were chosen to give a representative sample of the simulations performed at Statoil, with both two- and three-dimensional examples. The figure 1.6 is as expected, according to the discussion in the previous section. In some subroutines of the linear solver the speed up factors were in the range of 2–6.

The greatest performance improvements were achieved for subroutines handling the longest vectors. One or two subroutines, handling very short vectors, had a speed-up factor smaller than 1.0 for some data sets. This is an effect caused by the machine cycles used in starting vector calculations. However, these subroutines had adequate speed-up on other data sets, so we kept them vectorized.

For several supercomputers the reduction in execution time due to vectorization may be offset by large elapsed times due to slow access to disk or virtual storage. This may be caused by the large memory requirement of the program being tested or because of competition with some other programs sharing the CPU. We did two tests to investigate how use of virtual storage influences the performance of our simulator. In the first, we ran a 15 000-cell simulation; in the second, we ran two 15 000-cell simulations. Both tests were performed with few or no other activities running concurrently. The machine was able to handle this situation with no significant degradation of the elapsed time for the jobs, in spite of the heavy load on the memory systems. However, a detailed study of mean paging rates between expanded and main storage shows that a very high CPU utilization, with other concurrent tasks in addition to two large model simulations, would require an increase of expanded storage.

An important issue is the time spent in obtaining the performance gains we experienced. As commented earlier, an expert knowledge of the supercomputer architecture, the peculiarities of the FORTRAN implementation, and the code to be vectorized are needed. If this knowledge is available, the vectorization process does not require a long time to bear fruit. In Fig. 2 we show the speed-up factors obtained in the different stages of the process. As noted in previous sections, automatic vectorization must be complemented with programmer intervention in order to obtain higher performance. In a later section, we comment on the relatively large fraction of time spent in the verification process.

## PARALLELIZATION

The 3090 family of machines has made available the possibility of executing separate, independent subroutines on different processors of the same machine. This is done by the Multitasking Facility (hereafter called the MTF), which is part of the standard FORTRAN provided by IBM. The MTF calls runtime library routines which use standard operating-system facilities to schedule the specified subroutines that can be executed independently on the available CPUs. Synchronization is done after the subroutines have been executed. The subroutines can be vectorized or scalar.

The MTF is suitable for coarse granularity of the independent tasks. The overhead involved in setting up the parallelization is approximately 75 μs for an empty subroutine. Several thousand instructions are therefore required for each independent task to experience execution time reduction through parallelization.

We have chosen a systematic approach to the new field of parallelization in reservoir simulation. We focus on the degree of coupling of the physical processes being simulated, recognizing that this will be reflected in the level of difficulty of the coding and the algorithms to be employed. Therefore, we have advanced from uncoupled systems to weakly coupled systems and are considering strongly coupled systems.

A self-imposed constraint that a parallelization method has to satisfy in order to be implemented in the production code is that the execution time should be reduced, and not only the elapsed time. The consequence of this constraint is that the simulator would also perform better in a sequential machine. The reason for the constraint is that most accounting systems are based on CPU time actually used and not on elapsed time. The constraint is strong, because parallelizing methods usually increase the number of calculations to be performed.

## Uncoupled systems

The first step has been based on the recognition that in the North Sea many reservoirs consist of geological structures totally isolated from each other, as exemplified in Fig. 3. They are coupled only through the offshore platforms or through clusters of wells, and through the production control that has to be provided at this level for complying with sales contracts or optimal exploitation. With the data structure implemented in our simulator, it is possible to decouple the Jacobian matrix in non-interacting submatrices, consisting of the cells in each structure and their interaction with the wells, such that the linear solver now can be applied independently to each structure.

Parallelization at this level has several benefits. Implementation in an already existing code is straightforward, and the constraint imposed by the present MTR of a very coarse granularity of the independent tasks is fulfilled. In

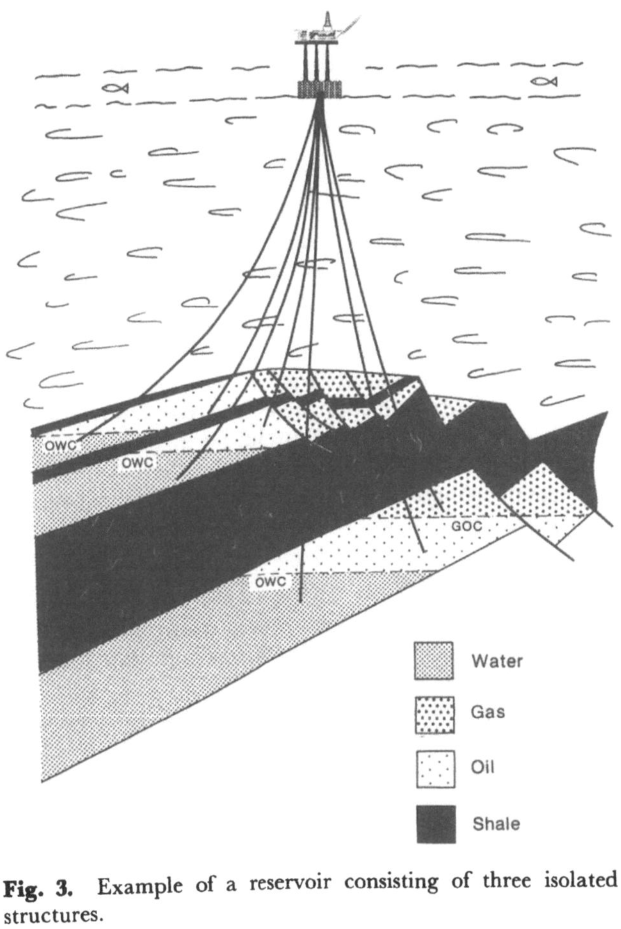


**Fig. 3.** Example of a reservoir consisting of three isolated structures.

reservoir simulation, different orderings of the grid cells have been used. These methods usually treat the reservoir as just one box of cells. Parallelization at the level of isolated structures is achieved by numbering the cells and the wells consecutively in the first structures and continuing with the next structures in the same way. This method of numbering the grid cells can be used on serial and parallel machines.

Figure 4(a) shows schematically how the linear equations are solved on a serial machine and Fig. 4(b) shows how they are solved on a parallel machine.

Figure 5 shows symbolically the matrix partitioning for the reservoir shown in Fig. 3, which consists of three isolated structures. Here it is essential that the $D$ submatrix mentioned previously is diagonal. On each of these independent submatrices we can use the solution method described previously.

We have used natural ordering of the cells within each structure, but other types of ordering schemes could be used. The only information the program needs is given in the input data file, and consists of the number of independent structures (three in this example) and a description of the boundaries of the structures.

All black-oil simulators so far must have the same number of phases for all structures in the reservoir. Isolated geological structures may have a different number of phases. As the number of equations to be solved is proportional to the number of phases, this implies that structures with just two phases waste a considerable amount of CPU time, calculating more than twice as many equations as really needed.

Access to parallelization makes it possible to expand the work of solving each structure separately, to take into account the number of phases in each of these structures. In a project in collaboration with IBM Scientific Center Bergen (Jarosch, *et al.*, 1988) we have done this; the results are presented later. The changes in the code were substantial, since both input and output of tables of fluid properties for both two and three phases had to be implemented.

The iterative linear solver part of the simulator consists of several FORTRAN subroutines, including the routine with the main iteration loop controlling the convergence. Multitasking is achieved by despatching all the linear solver routines and the table look-up subroutines for two and three phases for each of the independent reservoirs. Synchronization is done after the linear equations for all reservoirs are solved for both two and three phases. In our case, the independent tasks execute several million instructions and the overhead involved in using MTF is therefore negligible.

An important consequence of this parallelization method is that, since the amount of work for solving a matrix equation is proportional to a power of the number of elements, the decomposition of the matrices results in a smaller number of operations for the total job. It should be noticed that this is also the case if the code is run in a serial mode.

## Weakly coupled systems

Previous work has been constrained to uncoupled systems. A conceptual search for weakly coupled systems led to the recognition that in most reservoirs, the largest changes in pressure and saturations occur in the neighborhood of the wells. Near-well effects have a decisive influence on the behaviour of some reservoirs. Two examples are liquid dropout in condensate fields when the pressure falls under the dewpoint in the vicinity of the well, and gas or water coning in thin oil zone reservoirs. Furthermore, the time

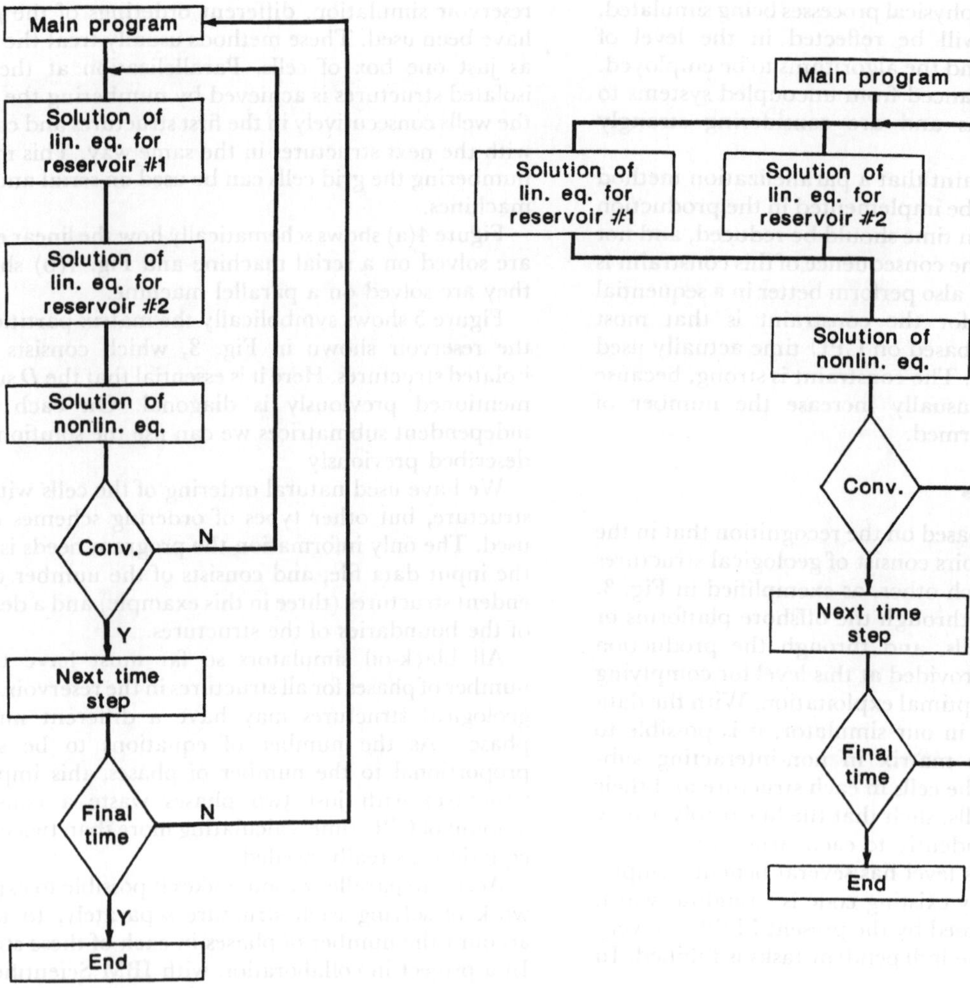


**Fig. 4.** (a) Serial method of solution of the linear equations. (b) Parallel method of solution of the linear equations.

offset in the drilling of the wells or opening and closing of wells during the usual production schedules causes time transients. All these effects must be resolved in reservoir simulation with small time steps and small cells. The need for a larger resolution is propagated to the whole simulation grid, which otherwise could support much larger cells and larger time steps.

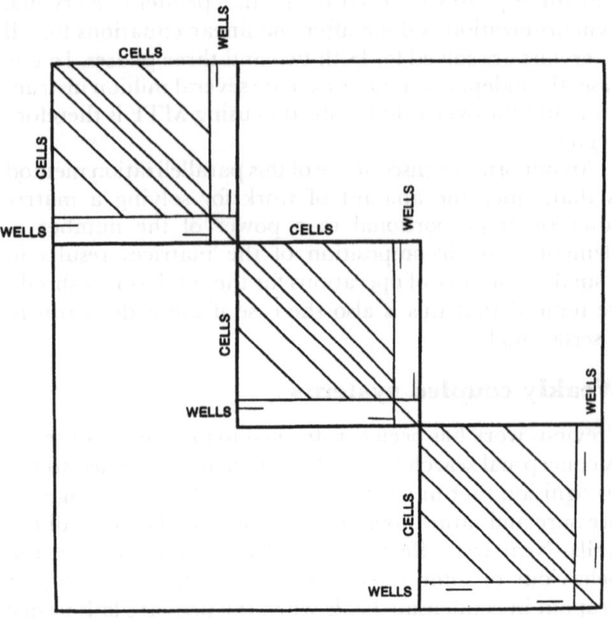


**Fig. 5.** Symbolic representation of the Jacobian matrix parallelized for three isolated structures.

In order to solve this problem, a local grid refinement with the possibility of taking small time steps in the refined area has been implemented in Eclipse (Cheshire and Henriquez, 1989). Since each of these refined areas is solved separately within the larger time step of the global grid, the data structure has been chosen so that their solution is parallelizable. Since direct solvers take less time than iterative solvers for the few hundred cells that are typical of grid-refined areas, they should be implemented for this task. The parallelization will depend on the practical use made of this feature by the reservoir engineers. Use of few refinements within larger grids would not benefit considerably from parallelizing, whereas the use of many small refinements would certainly make parallelization attractive.

## Coupled systems

The experiences gained in the previous stages have led to the investigation of a parallelization of the nested factorization algorithm itself.

A new kind of architecture for parallel processing is represented by multi-processor machines with local memory. In a joint project between Christian Michelsens Institutt, Bergen (CMI), Exploration Consultants Limited (ECL), and Statoil, the possibility of running Eclipse on a Hypercube has been investigated. The linear solver takes most of the execution time, so again the work has focused on that part of the program.

One way of parallelizing the linear solver on a local memory machine is to split the preconditioning part of the solver. This is done by assigning a group of solver planes to each processor. Flow between plane groups is not included in the resultant preconditioning. Material balance is, however, preserved by adding the outer band terms connecting the plane groups onto the diagonal. For the preconditioning, this leads to the same decoupling of the matrix as shown in Fig. 5. However, this decoupling does not reflect any physical process or structure of the reservoir. Probably as a result, the number of iterations needed to get convergence increases.

A simulation of this method has been implemented, and was tested using between 1 and 16 plane groups. Figure 6 shows the total number of iterations divided by the number of processors, which is approximately proportional to the elapsed time. This method will be implemented on CMI's Intel iPSC/2 Hypercube.

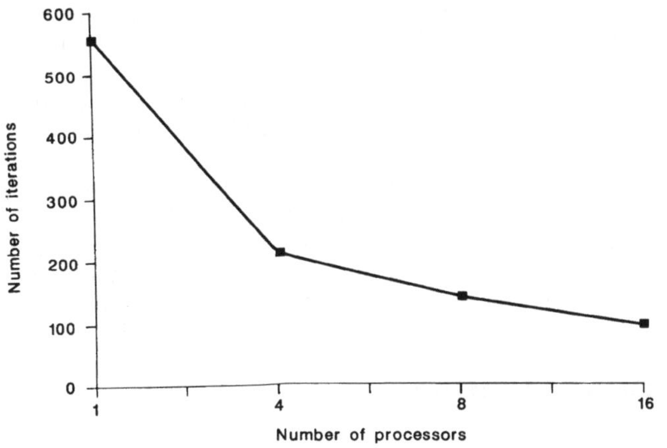


**Fig. 6.** Number of iterations in the linear solver per processor, as parallelized in the Hypercube.

Preliminary studies show that the method proposed by Van der Vorst (1986), with small modifications, can be applied to the reservoir simulator we use. For two processors, the algorithm starts with one processor beginning at the first plane working forwards until half of all the planes have been processed. At the same time, processor 2 starts at the last plane and works backwards until the second half have been processed. Then the first processor starts working backwards one plane at a time and the second processor starts working forwards one plane at a time.

The main advantage of this method is that no restrictions to special reservoir geometries or weak coupling of the physical processes will be present. An additional advantage is that this method can be still be applied in conjunction with these special cases.

The preliminary studies performed indicate that the method may be extended to 16 processors.

## Performance of the parallelized simulator

The CPU time used in solving the linear equations with nested factorization depends on the number of cells raised to a power close to 2. To test the efficiency of the parallelization at the level of isolated geological structures, four cases were run with an increasing number of cells in three structures. The test results are shown in Fig. 7. The execution and elapsed times increase nearly linearly with the number of grid cells for this range of number of grid

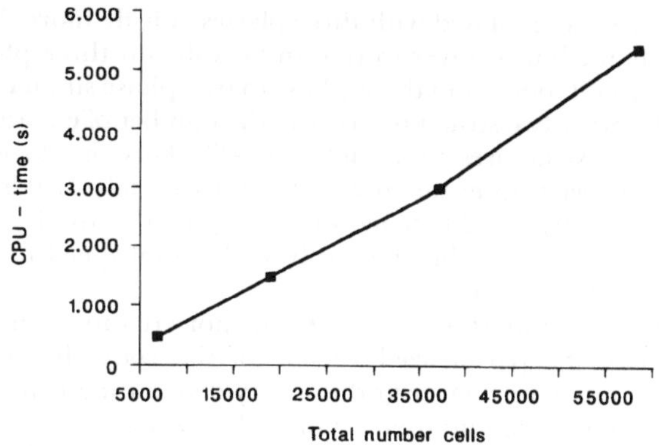


**Fig. 7.** Execution time for three isolated structures with varying total number of cells.

cells, owing to the parallelizzation of the code. This behaviour depends on the particular data case.

Some specific results for a large oilfield that Statoil operates, consisting of three isolated structures, are given in the following paragraphs. The number of cells used in the numerical simulation study is 8700 cells for structure 1, 6960 for structure 2, and 12 180 for structure 3.

Tasks 1 and 2 used 41% of the CPU time spent in the linear solver and 59% was used by task 3. Since the linear solver, which has been parallelized, takes 46% of the total execution time in this example, the savings in turnaround time are significant. It is more difficult to measure, in this realistic example, the gains achieved by the decomposition of the matrices into smaller matrices and by the use of just the amount of linear iterations needed in each reservoir. An indication is obtained by the fact that for this simulation the old vectorized code often needed 25 linear iterations to converge. By decoupling the matrix into three smaller matrices, one for each reservoir, reservoir 3 still needed 25 linear iterations but reservoirs 1 and 2 converged after 2 or 3 linear iterations.

One measure of the total efficiency may also be given by the fact that the total average CPU time used by the previous vectorized version was 1.6 times more than the present parallelized and vectorized version. We also note that in several of the runs the elapsed time has been significantly less than the total CPU time used.

We emphasize that our approach is not limited to two or three isolated structures, but in this particular case the number of processors available, the relative size of the reservoirs and the amount of work needed to solve the numerical problem was such that this was a convenient division.

The performance gains of the parallelized version, where takes advantage of the presence of three phases in only one of three structures, is shown in Table I. Here C333 denotes the case in which all reservoirs must be handled as a three-phase case, and C233 the case where two of the reservoirs are handled as two-phase structures.

The speed-up in the linear solver is 1.34. This factor may be explained in the following way. When all of the

**Table I.** CPU time in seconds for three reservoirs with three phases (C333) and only one reservoir with three phases (C223)

| | *C333* | *C223* | *Diff.* | *C333/C223* |
|---|---|---|---|---|
| CPU time | 628 | 542 | 86 | 1.16 |
| Solver | 327 | 244 | 83 | 1.34 |
| Solver% | 52 | 45 | — | — |

reservoir is simulated with three phases, a little more than half of the linear solver work is in the difficult three-phase reservoir. Going from three-phase to two-phase simulation in the other two structures reduces the number of equations in those structures by a factor of 4/9. This should give a total speed-up in the solver of approximately $1/(0.55+(0.45\cdot(4/9)))=1.33$. Since only 52% of the total CPU time was spent in the linear solver, the total speed-up for the whole run was 1.16.

Another important issue is the memory requirement for running the parallelized version of the code. Running a simulation on a large grid is a very demanding task, not only from a computational point of view but also on the memory system of the computer. Some of the larger simulations require more than 60–70 Mbytes for the main working arrays that need to stay in physical memory during the whole simulation. The total size of the program is such that this is the amount of real memory that is needed in order to fully utilize the CPU (100% utilization) without being slowed down by the virtual memory system.

Running the parallel version of the simulator on a full-field simulation consisting of approximately 20 000 cells required an amount of memory that must be available in real memory of approximately 9000 pages (36 Mbytes). The total size of the program was much bigger. However, because of the virtual memory system on the 3090 family of machines, parts of the program are moved to secondary storage as soon as they have not been used during a system-defined period, in order to release main memory. The 36 Mbytes are the minimum that are needed to let the program run efficiently, and consists mainly of the key working arrays used by the linear solver.

The parallel version of the simulator is running together with other batch programs and interactive users on the 3090-200/VF and the system runs at almost 100% utilization for long periods of time. However, we have no indications of any negative effect on the system or on the other users when running our parallelized code using the vector units simultaneously. The main reason for this is that the tasks that have been despatched for parallel execution are treated as any other task in the system asking for CPU resources and therefore under control of the priority system in MVS/XA.

To summarize the performance gain of the application of supercomputing to reservoir simulation, Table II show the efficiency ratio of the modified codes to the scalar version. The vectorized program was on average 1.6 times faster than the scalar, the parallelized code at the level of isolated structures was in turn 1.6 times faster than the vectorized code, and the possibility of simulating separately two- and three-phase structures gives a speed-up factor of 1.16 with respect to the "normally" parallelized version.

**Table II.** Speed-up factors due to vectorization and parallelization of Eclipse on the IBM-3090 at Statoil

| | *Scalar* | *Vector* | *Parallel* | *2–3 phases* |
|---|---|---|---|---|
| Ratio | 1.0 | 1.6 | 2.6 | 3.0 |

## VERIFICATION AND DEBUGGING

Verification of the results is an important part of the migration to a supercomputing environment. In an industrial environment, we have to be absolutely sure that the software and the hardware are completely reliable and that no errors have been introduced in the customizing process. In our case, we spent an appreciable amount of time on this step because the FORTRAN functions provided with the VS FORTRAN 2 compilers are more accurate than in previous compilers. Because the simulator is driven forwards in time by convergence criteria, small differences in intermediate results may vary the total number of time steps; the number of linear iterations is not the same for vector or parallel and scalar runs, and the numerical results are not identical. The times for closing and opening wells can also vary somewhat, since they are being controlled by numerical limits. The results must then be studied very carefully, to ensure that no bugs creep in during the hand-tailoring of the subroutines.

Another problem in comparing the results arises from different accumulation sequences in the scalar and vector versions, as computer arithmetic operations are not commutative.

In Fig. 2, we show that the time spent in verifying the results of the program against the scalar calculations took one-third of the total time used before it could be released to the users.

Debugging a parallelized program is a non-trivial exercise. The updating of the variables and its protection depends on the undetermined time when they are calculated by the different processors. However, the relatively coarse way in which this is done in the MTF, where all the variables are passed on to the subroutines through a COMMON statement, makes this part of the parallelization somewhat easier than in microtasking environments.

## CONCLUSIONS

We summarize our work in the following point.

(1) The performance obtained is, for all practical purposes, as close as possible to the highest attainable without a major change in the data structure and in the linear solver. The total speed-up of the linear solver is significant; the CPU time was reduced by a factor of 3 and the use of multitasking also reduced the total elapsed time.

(2) A new linear solver with higher vectorizability would not have a much better performance unless the vector/scalar ratio of the machine were improved substantially (an effect of Amdahl's law). Any change we can foresee in the linear solver will not change the relative amounts of short and long vectors in our simulator. Therfore, a dynamical reconfiguration of the vector registers, in the sense that the optimal length of each vector register is accommodated to the actual data at execution time for each loop, would be very desirable.

(3) A vectorized direct solver does not lend itself to the usual reservoir simulation problems, but it might be useful in extremely difficult single-well studies. Further parallelization work in the case of weakly coupled structures would benefit from using a direct solver in the locally refined areas.

(4) Running large simulations using both the IBM 3090 vector units and multitasking together produced no negative effects on the other concurrent users of the system.

(5) Multiprocessor machines with local memory considerably reduce the elapsed time, but the actual execution time increases.

(6) In reservoir simulator codes, parallelization and vectorization techniques may complement each other.

Since most of the vectorizing effort was concentrated in getting vectors with a length equal to the total number of active cells, the decomposition resulted in vector lengths that still made successful use of the vector facility. It is interesting to note that, once physical processes have been identified, inhibitors of parallelization are weaker than for vectorization.

(7) The method of independent structures can be used on other parts of the simulator to increase the parallelization and therefore reduce the elapsed time even further.

(8) The verification and debugging phases involved in supercomputing are very time consuming. Debugging tasks that run in parallel requires a conceptually different approach from that commonly used for serial codes.

## ACKNOWLEDGEMENTS

Dr Ian Cheshire, at ECL and Knut Korsell, Oddvar Gjerde and Hans Jarosch, at IBM have given us their support in this work, which we acknowledge.

Terje Kårstad acknowledges the collaboration of C. Davidson and D. Warhurst at CMI in the work of parallelizing Eclipse on CMI's Intel iPSC/Hypercube, and the collaboration of N. Houbak and S. Nørsett at NTH in the work of vectorizing a direct linear solver.

Comments from anonymous referees have contributed to clarifications of some points.

The permission of Statoil to publish this work is gratefully acknowledged.

## NOMENCLATURE

- $A$ Matrix of coefficients for the linear equations
- $B$ Matrix of coefficients for the linear equations showing interactions between the wells and the cells
- $B1$ Preconditioning matrix
- $C$ Matrix of coefficients for the linear equations showing interactions between the wells and the cells
- $D$ Matrix of the wells interactions
- $x$ Vector of solutions of the linear equations
- $x_w$ Vector of solutions of the well linear equations
- $b$ Vector of residuals of the cells
- $b_w$ Vector of residuals of the wells
- $D1$ Main diagonal of $A$
- $L_i$ $i$th lower diagonal band of $A$
- $U_i$ $i$th upper diagonal band of $A$
- $s_m$ Search direction vector, $m$th iteration
- $r_m$ Residual vector, $m$th iteration
- $p$ Number of simultaneous solutions variables in each cell, one per phase
- $a_i$ Scalar used in the conjugate-gradient method

## REFERENCES

Appleyard, J.R. and Cheshire, I.M. (1983). Nested factorization. Presented at the *Seventh Reservoir Simulation Symposium, San Francisco, November, 1983*, paper no. SPE 12264.

Cheshire, I. and Henríquez, A. (1989). Local grid refinement. *International Conference on Mathematical Methods in Supercomputing, Cambridge*, June 1989.

Henríquez, A. (1986). Vectorization and parallelization of a reservoir simulator. *Proceedings of the Second Seminar on Reservoir Description and Simulation with Emphasis on EOR, Oslo, September 1986.*

Holmes, J.A. (1983). Enhancements to the strongly coupled, fully implicit well model: wellbore crossflow modelling and collective well control. Presented at the *Seventh Reservoir Simulation Symposium, San Francisco, November*, 1983, paper no. SPE 12259.

Houbak, N. (1985). SESYS—a sparse matrix linear equation solver. Users guide. Report 12 in the Series: Olie- og Gasreservoirmodeller, Forsøkgsanlæg Risø, Denmark.

Kårstad, T., Henríquez, A. and Korsell, K. (1987). Parallelization of a reservoir simulator. *Proceedings of the International Confernce on Supercomputing, Athens, June 1987.*

Jarosch, H., Gjerde, O. and Kårstad, T. (1988). Improvements to the black oil simulator Eclipse 100. Presented at the *Second International Conference on Vector and Parallel Computing, Tromsø, June 1988.*

Van der Vorst, H. (1986). Large tridiagonal and cell tridiagonal linear systems on vector and parallel computers. Report 86–25, Delft University of Technology, Delft.

Vinsome, P.K.W. (1976). Orthomin, an iterative method for solving sparse banded sets of simultaneous linear equations. paper SPE 5729 Presented at the *Fourth Symposium on Numerical Simulation of Reservoir Performance, Los Angeles, 1976*, paper no. SPE 5729.

MIX
Papier aus verantwortungsvollen Quellen
Paper from responsible sources
FSC® C105338

If you have any concerns about our products,
you can contact us on
**ProductSafety@springernature.com**

In case Publisher is established outside the EU,
the EU authorized representative is:
**Springer Nature Customer Service Center GmbH**
**Europaplatz 3, 69115 Heidelberg, Germany**

Printed by Libri Plureos GmbH
in Hamburg, Germany